Formulas from Geometry

Formulas for Area (A), Perimeter (P), Circumference (C), and Volume (V):

Square
$A = s^2$

$P = 4s$

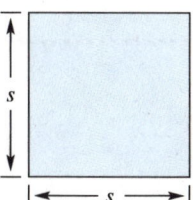

Rectangle
$A = lw$

$P = 2l + 2w$

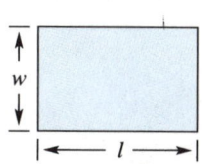

Circle
$A = \pi r^2$

$C = 2\pi r$

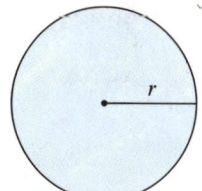

Triangle
$A = \frac{1}{2}bh$

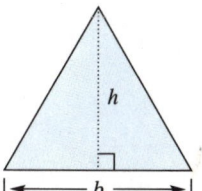

Trapezoid
$A = \frac{1}{2}h(b_1 + b_2)$

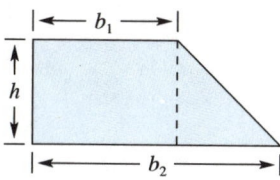

Parallelogram
$A = bh$

$P = 2a + 2b$

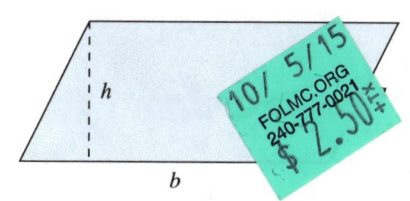

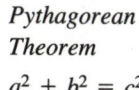

Pythagorean Theorem
$a^2 + b^2 = c^2$

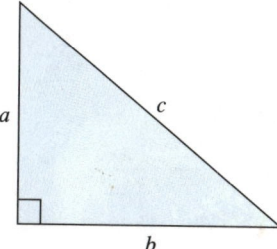

Cube
$V = s^3$

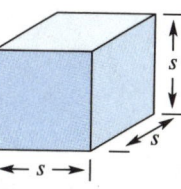

Rectangular Solid
$V = lwh$

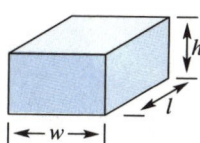

Circular Cylinder
$V = \pi r^2 h$

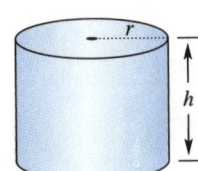

Sphere
$V = \frac{4}{3}\pi r^3$

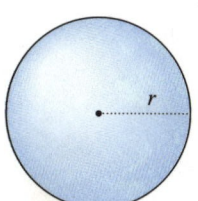

Elementary and Intermediate Algebra

A Combined Course

Second Edition

Roland E. Larson

Robert P. Hostetler

The Pennsylvania State University
The Behrend College

with the assistance of
David E. Heyd
The Pennsylvania State University
The Behrend College

D. C. Heath and Company
Lexington, Massachusetts Toronto

Address editorial correspondence to:
D. C. Heath and Company
125 Spring Street
Lexington, MA 02173

Acquisitions Editors: Ann Marie Jones, Charles Hartford
Managing Editor: Catherine B. Cantin
Development Editor: Emily Keaton
Production Editor (art): Rachel D'Angelo Wimberly
Marketing Manager: Christine Hoag
Designer: Henry Rachlin
Interior Photo Researchers: Derek Wing and Billie Porter
Production Coordinator: Lisa Merrill
Composition: Meridian Creative Group
Art: Folium, Inc.; Meridian Creative Group; Patrice Rossi; Illustrious, Inc.
Cover Photo Researcher: Martha L. Shethar

Trademark Acknowledgments: TI is a registered trademark of Texas Instruments, Inc. Casio is a registered trademark of Casio, Inc. Sharp is a registered trademark of Sharp Electronics Corp.

Published simultaneously in Canada.

Printed in the United States of America.

International Standard Book Number: 0-669-41764-5

Library of Congress Catalog Number: 95-81733

3456789-DOC-00 99 98 97

Preface

The primary goals of *Elementary and Intermediate Algebra: A Combined Course,* Second Edition, are to encourage students to develop their proficiency in algebra and to show how algebra is a modern modeling language for real-life problems.

Coverage This text was designed to be flexible with respect to the order of coverage of core algebra topics, adapting easily to a wide variety of course syllabi and teaching styles. It begins with Prerequisites, a review chapter. All or part of this material may be covered or it can be omitted. Graphing is first introduced in Chapter 4. The use of graphs encourages visualization to offer an opportunity for more conceptual understanding, strengthens graph-reading skills, and supports a smoother transition from concrete visual ideas to more abstract mathematics. Throughout the text, attention is given to geometry, collecting and interpreting data and statistics, and creating models, as well as to the NCTM Standards and Addenda and the AMATYC Guidelines.

Problem Solving A general problem-solving process for applied problems is stressed throughout the text: form a verbal model, label terms, create a mathematical model, solve, and check the answer in the original statement of the problem (see page 169). This problem-solving process helps students understand the problem, organize their work, and develop facility with verbal, analytical, graphical, and numerical approaches to problem solving. Students are also reminded of specific problem-solving strategies (see page 80) that are reinforced throughout the text in the exercises (see Exercises 81–82 on page 120 and Exercise 125 on page 843).

Exercises The comprehensive exercise sets offer students ample opportunity to practice algebraic techniques (see pages 160–161 and 579–581) and develop their conceptual and critical-thinking skills (see Exercise 45 on page 281, Exercises 75–82 on page 352, Exercise 93 on page 363, Exercises 89 and 90 on page 486, Exercise 41 on page 653, and Exercise 104 on page 765). The broad range of computational, conceptual, and applied problems in each exercise set is carefully graded to provide a smooth transition from routine to more challenging problems. Section and review exercises—as well as mid-chapter quizzes (see page 244) and chapter tests (see page 446)—consistently encourage student mastery of algebraic skills and concepts through practice and self-assessment.

Group Activities Each section ends with a Group Activity. This exercise reinforces students' understanding by exploring mathematical concepts in a variety of ways: You Be the Instructor, Extending the Concept, Problem Solving, Exploring with Technology, and Communicating Mathematically. Some Group Activities encourage interpretation or discovery of mathematical concepts and results (see pages 159, 213, and 760); some provide opportunities for problem posing and error analysis (see pages 22, 324, 594, and 772); and others reinforce methods of interpreting and constructing mathematical models, tables, and graphs (see pages 240, 634, 651, and 794). Designed to be completed in class or as homework assignments, the

Group Activities give students the opportunity to work cooperatively as they think, talk, and write about mathematics.

Technology   Recognizing that graphing technology is becoming increasingly available, the text offers the opportunity to use graphing utilities throughout, but without requiring their use. This is achieved through a combination of features, including—at point of use—discovery opportunities that require scientific or graphing calculators (see pages 306 and 770), graphing utility instructions (see pages 412 and 575), and clearly labeled exercises that require the use of a graphing utility (see Exercise 48 on page 405 and Exercises 67–78 on pages 697 and 698).

Data Analysis/Modeling   Throughout the text, students are offered opportunities to collect and interpret data, make conjectures, and construct mathematical models. Students are exposed to combining mathematical models to make related models (see Exercise 39 on page 455); encouraged to use mathematical models to make predictions and estimates from real data (see Exercise 78 on page 234, Exercise 43 on page 504, and Exercise 95 on page 638); invited to compare two or more models or compare actual data with a model (see Exercise 103 on page 765); and asked to use curve-fitting techniques to write their own models from data (see Exercise 98 on page 407, Exercises 37–39 on page 732, and Exercise 41 on page 745). Students are encouraged to use charts, tables, scatter plots, and graphs to summarize and interpret data.

Applications   To emphasize for students the connection between mathematical concepts and real-world situations, up-to-date, real-life applications are integrated throughout the text. Appearing as examples (see pages 203, 220, 693, and 794), exercises (see Exercise 56 on page 223, Exercise 98 on page 440, Exercise 41 on page 707, and Exercise 116 on page 806), group activities (see page 717), and projects (see page 748), these applications help students validate the material they are learning and offer them frequent opportunities to use and review their problem-solving skills. A wide range of disciplines is represented by the applications—including such areas as physics, chemistry, electronics, the social sciences, biology, and business—as well as the career interviews, covering areas such as insurance, real estate, architecture, engineering, graphic arts, business, education, scuba diving, biochemistry, and economics.

Connections   In addition to highlighting the connections between algebra and areas outside mathematics through real-world applications, this text also emphasizes the connections between algebra and other branches of mathematics, such as probability (see page 465), geometry (see page 134), logic (see Appendix A), and statistics (see Appendix B). Too, many examples and exercises throughout the text reinforce the connections among graphical, numerical, and algebraic representations of important algebraic concepts (see Exercises 27 and 28 on page 606).

There are many other new features of *Elementary and Intermediate Algebra: A Combined Course,* Second Edition, as well, including Discovery, Chapter Opening Applications, Study Tips, Historical Notes, Mid-Chapter Quizzes, Chapter Summaries, Career Interviews, and Chapter Projects. These and other features of the Second Edition are described in greater detail on the following pages.

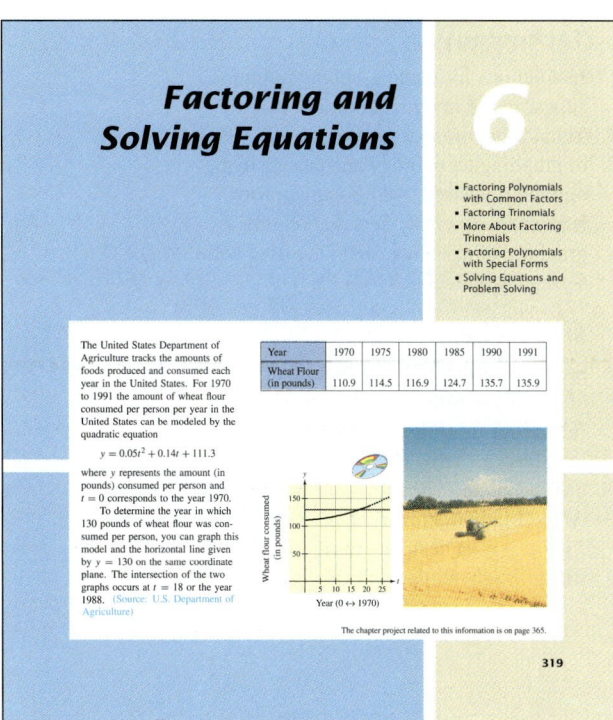

Chapter Opener

Each chapter opens with a look at a real-life application that is explored in depth in the Chapter Project at the end of the chapter. Real data is shown using graphical, numerical, and algebraic techniques. In addition, a list of the section titles shows students how the topics fit into the overall development of algebra.

Section Outline

Each section begins with a list of the major topics covered in that section. These topics are also the subsection titles and can be used for easy reference and review by students.

Historical Notes

To help students understand that algebra has a past, historical notes featuring mathematical artifacts or mathematicians and their work are included in each chapter.

Notes

Notes anticipate students' needs by offering additional insight, pointing out common errors, and describing generalizations.

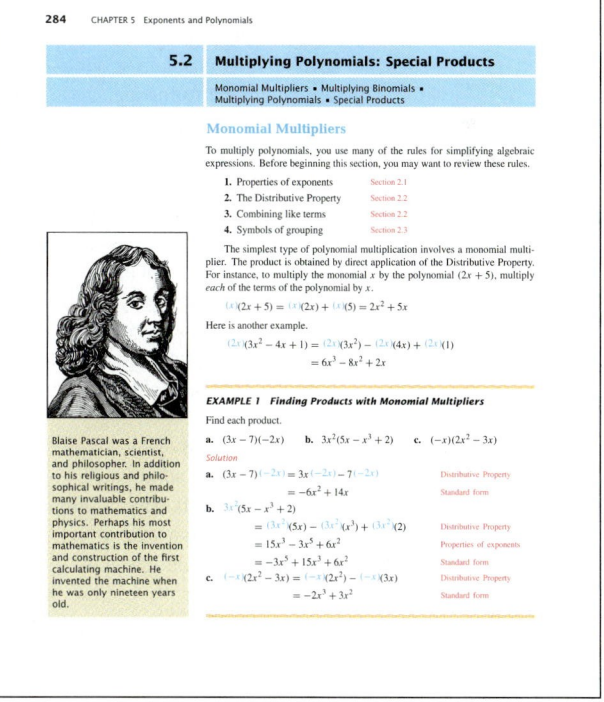

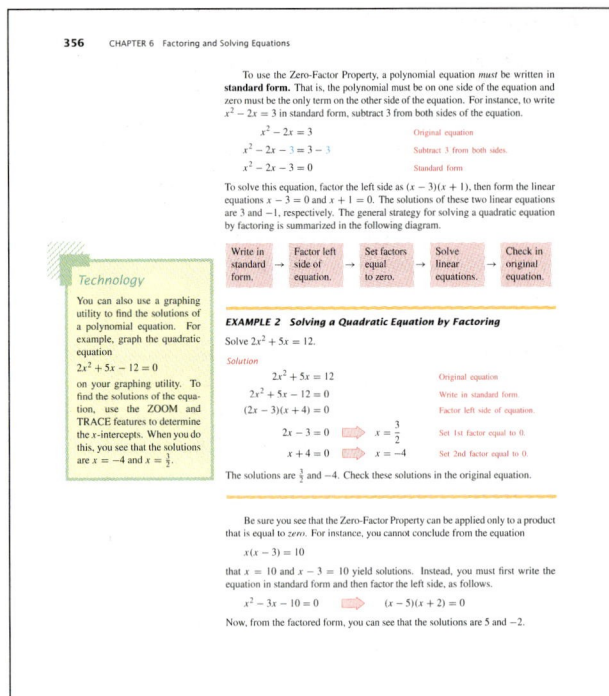

Technology

Instructions for using graphing utilities appear in the margin at point of use. They offer convenient reference for users of graphing technology and can easily be omitted if desired. Additionally, problems in the Exercise Sets that require a graphing utility have been identified with a graphing calculator icon.

Study Tips

Study Tips appear in the margin at point of use. They offer students specific, helpful, and insightful suggestions for studying algebra. "How to Study Algebra" on page xxvii and "Reading and Writing About Mathematics" on page xxx outline a general plan designed to improve student study skills.

Problem Solving

The text provides ample opportunity for students to develop their problem-solving skills. They are taught the following approach to solving applied problems: (1) Construct a verbal model; (2) Label variable and constant terms; (3) Construct an algebraic model; (4) Using the model, solve the problem; and (5) Check the answer in the original statement of the problem. This process has wide applicability, and it is used with verbal, analytical, graphical, and numerical approaches to problem solving. Identifying units of measure and checking solutions is emphasized, and many solutions have explanations and additional help in the form of comments adjacent to the computation. Color is also used to emphasize and clarify the solution steps.

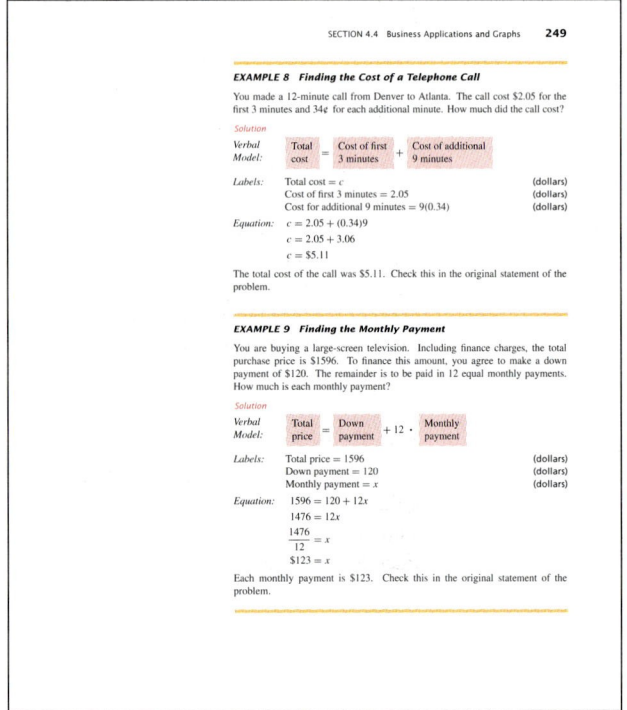

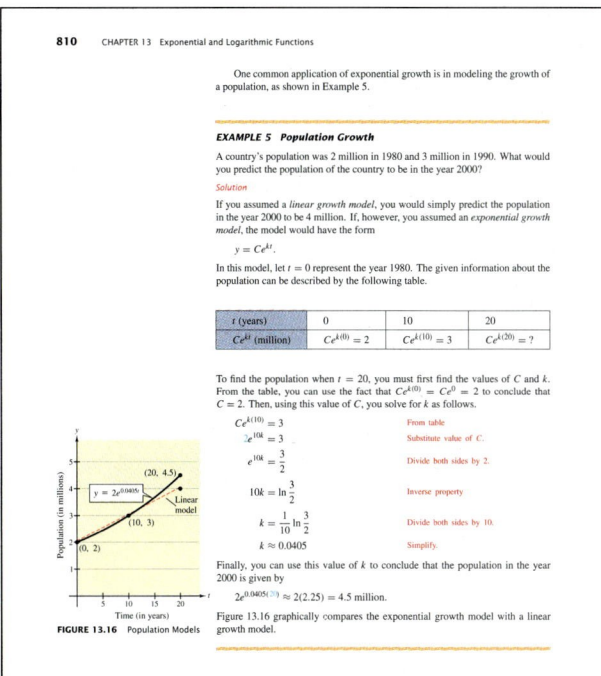

Applications

Real-life applications are integrated throughout the text in examples and exercises. These applications offer students constant review of problem-solving skills and emphasize the relevance of the mathematics. Many of the applications use recent, real data, and all are titled for easy reference. Photographs with captions throughout the text also encourage students to see the link between mathematics and real life.

Examples

Each of the text examples was carefully chosen to illustrate a particular mathematical concept, problem-solving approach, or computational technique, and to enhance students' understanding. The examples in the text cover a wide variety of problem types, including computational, real-life applications (many with real data), and those requiring the use of graphing technology. Each example is titled for easy reference, and real-life applications are labeled. Many examples include side comments in color, which clarify the key steps of the solution process.

Discovery

Throughout the text, Discovery notes encourage active participation by students, often taking advantage of the power of technology (graphing calculators and scientific calculators) to explore mathematical concepts and discover mathematical patterns. Using a variety of approaches, including visualization, verification, pattern recognition, and modeling, students develop an intuitive understanding of algebraic topics.

Definitions and Rules

All of the important rules, formulas, guidelines, properties, definitions, and summaries are highlighted for emphasis. Each is also titled for easy reference.

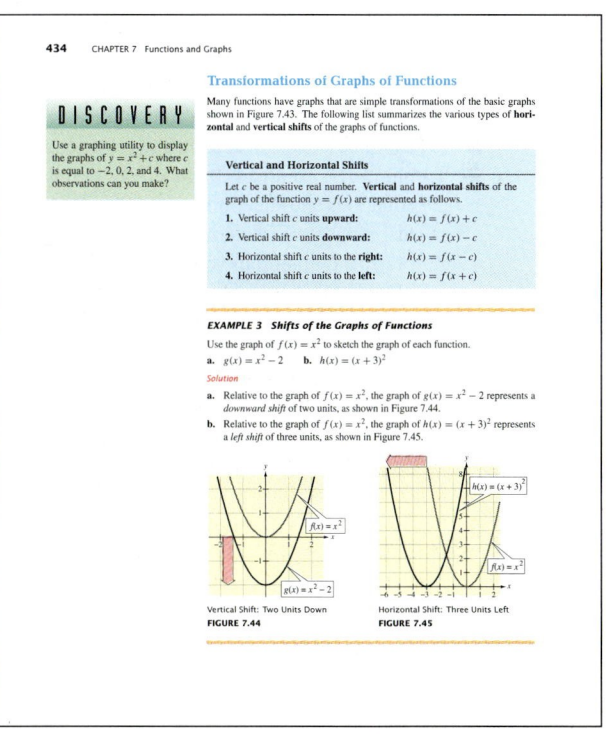

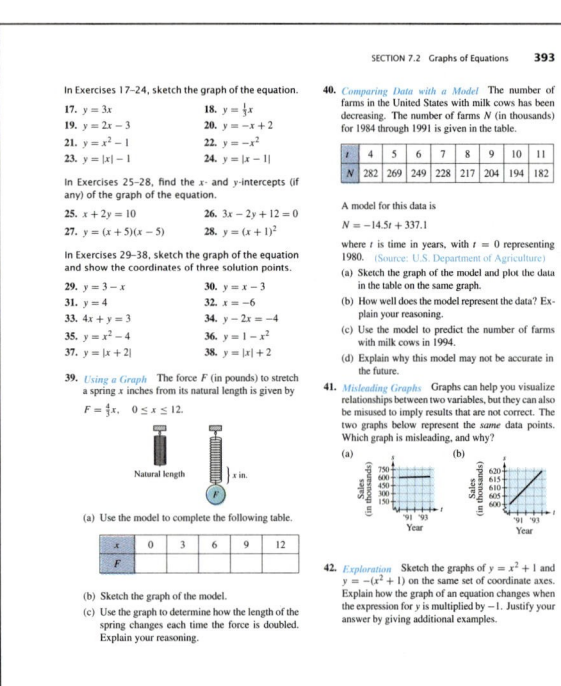

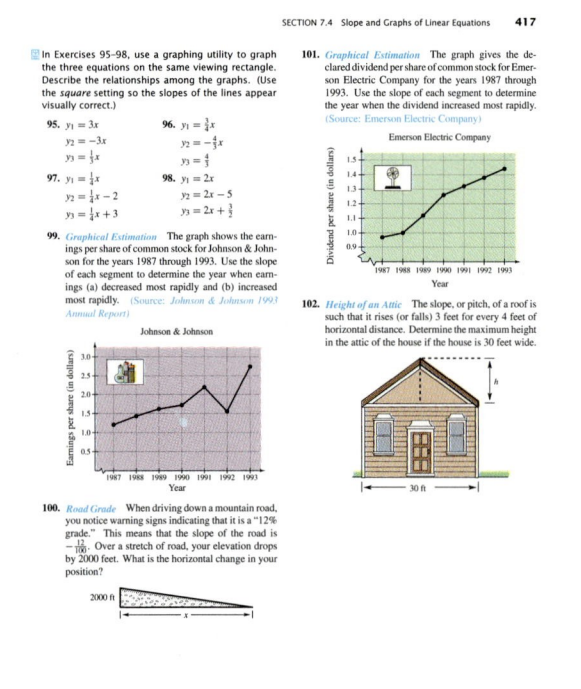

Graphics

Visualization is a critical problem-solving skill. To encourage the development of this ability, the text has numerous figures in examples, exercises, and answers to odd-numbered exercises. Included are graphs of equations and functions, geometric figures, displays of statistical information, scatter plots, and numerous screen outputs from graphing technology. All graphs of equations and functions, computer- or calculator-generated for accuracy, are designed to resemble students' actual screen outputs as closely as possible. Graphics are also used to emphasize graphical interpretation, comparison, and estimation.

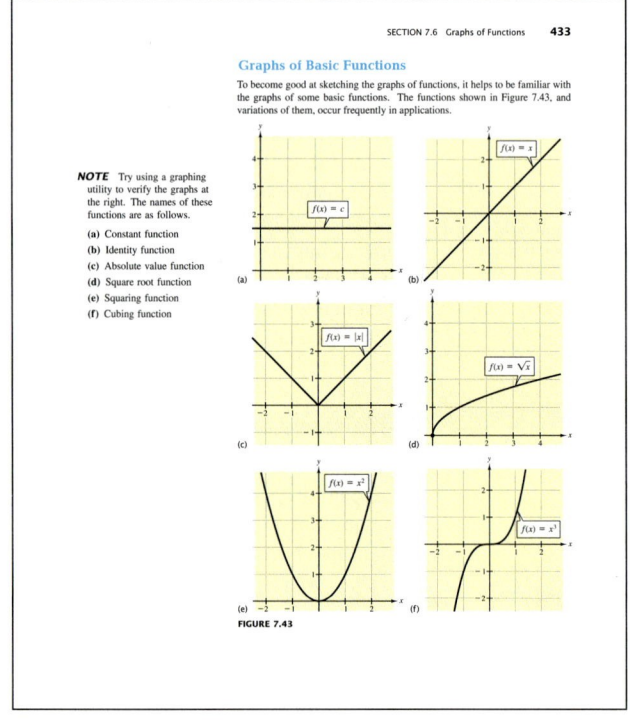

Group Activities   Extending the Concept

A Mathematical Riddle   What is the largest number that can be written using the three digits 2, 3, and 4? The number 432 seems to be the obvious answer. However, if you allow the digits to be exponents, then you can obtain numbers that are much larger than 432. For instance, consider the numbers

$$(32)^4 = 1,048,576$$

and

$$3^{24} \approx 282,430,000,000.$$

In your group, see who can create the largest number using the three digits 2, 3, and 4.

Group Activities   Extending the Concept

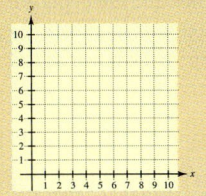

Using Inequalities   Try the following activity. One person picks a point with whole number coordinates on a grid like the one at left without revealing the coordinates. A second person writes the equation of a line passing through the grid region. The first person graphs the line on the grid and indicates whether the secret point lies above, below, or on the line. Continue writing and graphing lines until the second person is able to guess the coordinates of the secret point. Switch roles and try again. What is the fewest number of turns your team required to guess the point?

Group Activities

The Group Activities that appear at the end of sections reinforce students' understanding by approaching mathematical concepts in a variety of ways: Communicating Mathematically, You Be the Instructor, Extending the Concept, Problem Solving, and Exploring with Technology. Designed to be completed as group projects in class or as homework assignments, the Group Activities give students opportunities for interactive learning and to think, talk, and write about mathematics.

Group Activities   Problem Solving

Fitting a Quadratic Model   The data in the table represents the United States government's annual net receipts y (in billions of dollars) from individual income taxes for the year x from 1990 through 1992, where $x = 0$ corresponds to 1990.   (Source: U.S. Department of the Treasury)

x	0	1	2
y	467	468	476

Use a system of three linear equations to find a quadratic model that fits the data. According to your model, what were the annual net receipts from individual income taxes in 1993? The actual annual net receipts for 1993 were $510 billion. How does the value obtained from your quadratic model compare? Suppose you had been involved in planning the 1993 federal budget and had used this model to estimate how much federal income could be expected from 1993 individual income taxes. When you review the actual 1993 tax receipts and see that the model wasn't completely accurate, how do you evaluate the model's prediction performance? Are you satisfied with it? Why or why not?

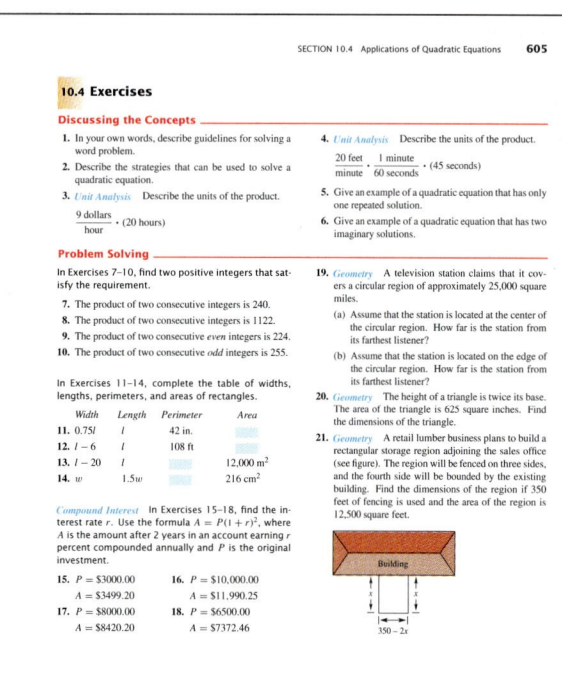

Exercises

Exercise are grouped into four categories: Discussing the Concepts, Problem Solving, Reviewing the Major Concepts, and Additional Problem Solving. The numerous computational, conceptual, and applied problems include multi-part, exploration and discovery, writing, estimation, numeracy, geometry, and challenging exercises, as well as real-life applications, mathematical modeling, graphical comparisons, data interpretation and analysis, fitting a line to data, and exercises that require graphing technology. Designed to build competence, skill, and understanding, each part of the exercise set is graded in difficulty to allow students to gain confidence as they progress. Detailed solutions to all odd-numbered exercises are given in the Student Solutions Guide, and answers to all odd-numbered exercises appear in the back of the text.

Geometry

Geometric formulas and concepts are reviewed throughout the text. For reference, common formulas are listed inside the back cover of this text.

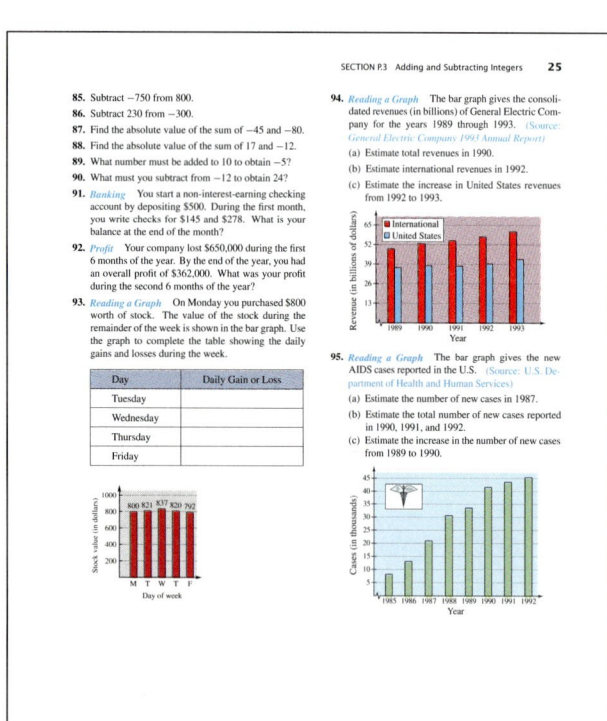

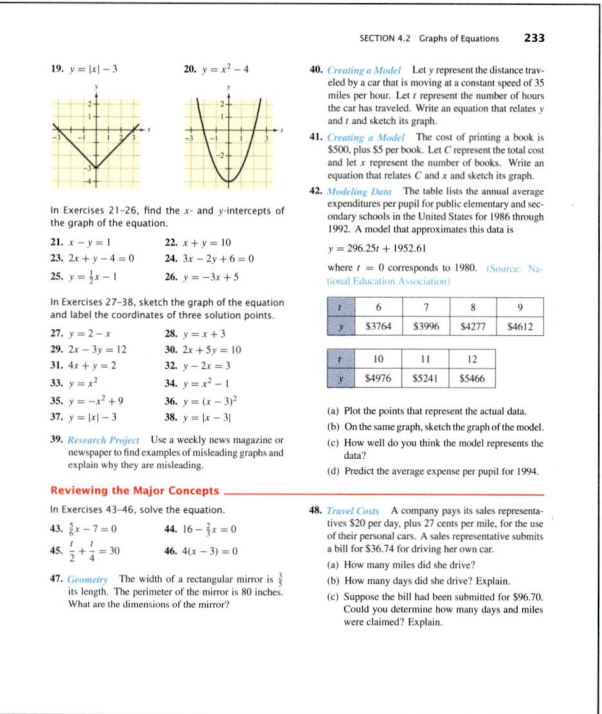

CAREER INTERVIEW

Lisa M. Deitemeyer

Civil Engineer

Johnson-Brittain & Associates, Inc.

Tucson, AZ 85701

Johnson-Brittain does highway design work, primarily for the Arizona Department of Transportation. I am responsible for drainage design of roadways and intersections. It is important that water properly drain off the road surface to avoid flooding problems. One strategy for removing excess water is to use a pipe drainage system that empties into a retention pond. When designing a pipe system and choosing pipe size, I use the equation $V = Q/A$ to find the velocity V of water moving at flow rate Q (volume per unit time) through a given pipe of cross-sectional area A. Finding the water velocity is very important. If it is too fast, erosion can occur in the retention pond. If it is too slow, sedimentation can clog the pipe. As you can see, algebra is very important to my work. I am always solving for different variables that are needed for drainage design.

Career Interviews

Appearing in each chapter, Career Interviews with people who use algebra in their jobs help students understand that algebra is a modern, problem-solving language.

Math Matters

Each chapter contains a Math Matters feature that engages student interest by discussing a historical note or mathematical problem. For those features that pose a question, the answers appear in the back of the text.

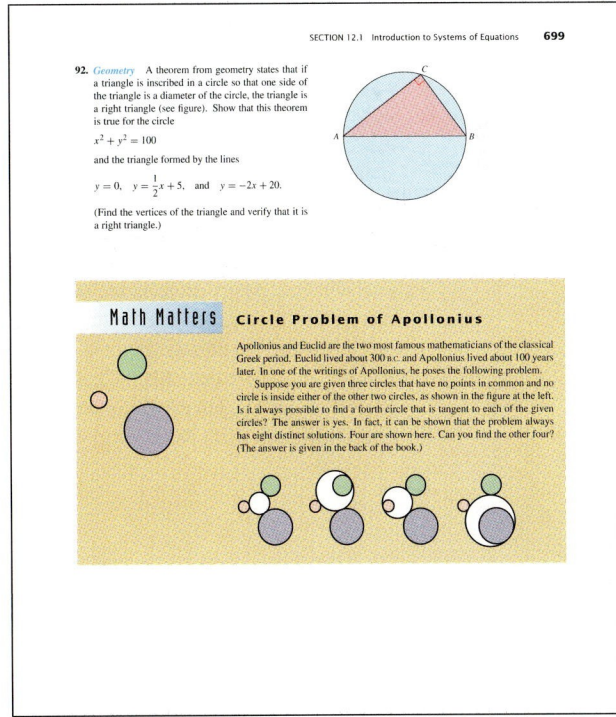

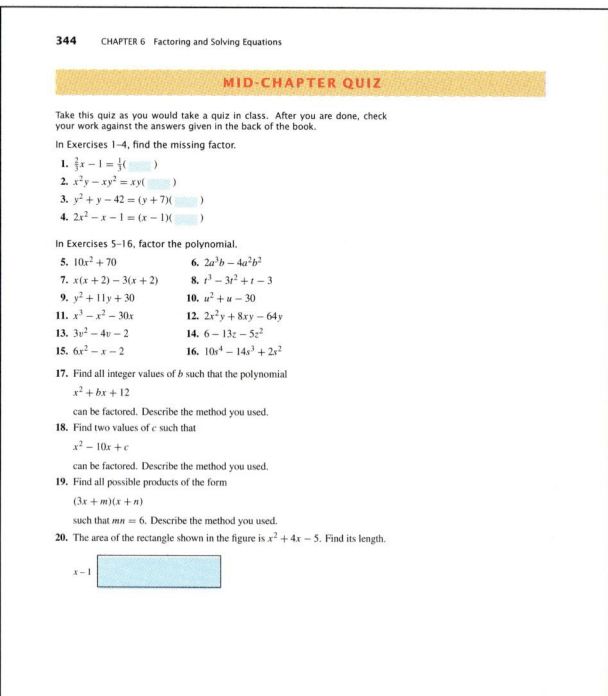

Mid-Chapter Quizzes

Each chapter contains a Mid-Chapter Quiz with answers in the back of the text. This feature allows the student to perform a self-assessment midway through the chapter.

Chapter Project

Chapter Projects, referenced in the chapter opener, are engaging applications that use real data, graphs, and modeling to enhance students' understanding of mathematical concepts. Designed as individual or group projects, they offer additional opportunities to think, discuss, and write about mathematics. Many projects include research assignments that give students the opportunity to collect, analyze, and interpret their own data. Each Chapter Project is also available in an interactive, multimedia, CD-ROM format.

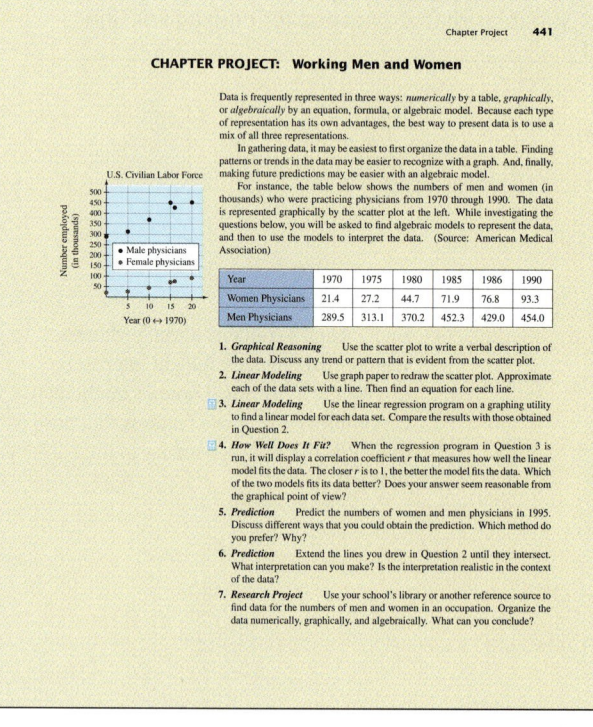

Chapter Summary

The Chapter Summary reviews the skills covered in the chapter. Section references for the major topics make this an effective study tool, and correlation to the review exercises offers guided practice.

Review Exercises

The Review Exercises at the end of each chapter offer the student an opportunity for additional practice. Each set of review exercises includes both computational and applied problems covering a wide range of topics.

Chapter Test

Chapter Tests allow students to assess their own level of success.

Cumulative Tests

The Cumulative Tests that appear after Chapters 3, 6, 10, and 13 help students judge their mastery of previously covered material, as well as reinforce the knowledge students have been accumulating throughout the text—preparing them for other exams and for future courses.

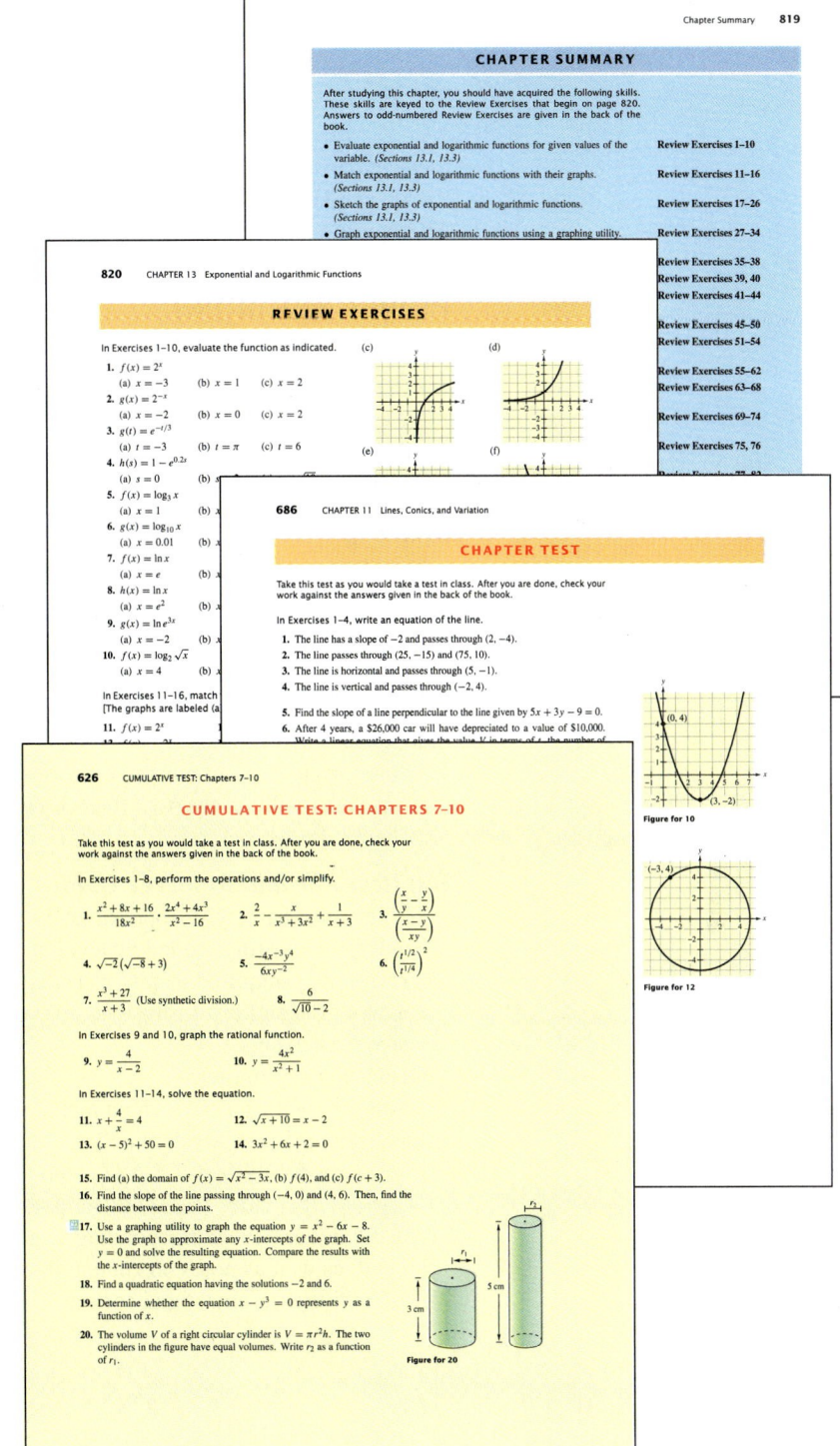

Supplements

Elementary and Intermediate Algebra: A Combined Course, Second Edition, by Larson and Hostetler, is accompanied by a comprehensive supplements package. All items are keyed to the text.

Printed Resources

Student Solutions Guide
- Detailed, step-by-step solutions to all odd-numbered section exercises (except Discussing the Concept) and review exercises
- Detailed, step-by-step solutions to all Mid-Chapter Quiz, Chapter Test, and Cumulative Test questions

Graphing Technology Keystroke Guide: Algebra by Benjamin N. Levy
- Keystroke instructions for Texas Instruments, Sharp, Casio, and Hewlett-Packard graphing calculators
- Examples with step-by-step solutions
- Extensive graphics screen output
- Technology tips

Instructor's Guide
- Solutions to even-numbered exercises
- Answers to all Group Activities, Technology Boxes, Discovery Boxes, and Chapter Projects

Test Item File and Resource Guide
- Printed test bank with approximately 4000 test items (multiple-choice, open-ended, and writing) coded by level of difficulty
- Technology-required test items coded for easy reference
- Bank of chapter test forms with answer keys
- Two final exams
- Transparency masters
- Notes to the Instructor, which includes information on standardized tests such as the Texas Academic Skills Program (TASP), Florida College Level Academic Skills Test (CLAST), and the California State University Entry Level Mathematics (ELM) Examination and provides a list of skills covered by the test and the corresponding section(s) in the text where the topic can be found, as well as notes on contemporary instructional strategies such as alternative assessment and cooperative learning

Media Resources

Tutor (IBM, Macintosh)
- Extensive additional practice

Videotapes by Dana Mosely
- Comprehensive coverage keyed to the text by section
- Detailed explanation of important concepts
- Numerous examples and applications, often illustrated via computer-generated animations
- Discussion of study skills
- For media resource centers; also available for student purchase

D. C. Heath Interactive Math Series CD-ROM Projects
- Real-life applications in an interactive, multimedia CD-ROM format
- IBM PC for Windows; Macintosh
- See page xvi for a description.

Computerized Testing
- Test-generating software for both IBM and Macintosh computers
- Approximately 4500 test items
- Also available as a printed test bank

CD-ROM Projects
for Elemtentary and Intermediate Algebra:
A Combined Course, Second Edition

To accommodate a variety of teaching and learning styles, a series of real-life applications is available in a multimedia, interactive CD-ROM format. Suitable for individual or group assignments, these projects reinforce a variety of mathematical concepts. For each text chapter project is a CD-ROM project, allowing students to explore interactively questions that expand upon the topic and goals of the text project. Students have the opportunity to discover the nature of data sets through exploration, using a combination of graphical, numerical, and algebraic approaches in a guided learning environment. Throughout the text, you will notice a CD-ROM icon 💿 that reminds you of the availability of this multimedia software in conjunction with the chapter projects.

These multimedia projects broaden the scope of the text by offering additional opportunities for finding patterns and drawing conclusions, covering related topics and concepts, and providing practice with interpreting graphs, charts, and tables. The multimedia format provides access to extensive real data sets and facilitates hands-on data manipulation for practicing data analysis and modeling techniques. In addition, the projects include animations, color photographs, and audio enhancements.

Each multimedia project is presented in four parts: Introduction, Data, Exploration, and Exercises. The Introduction explains the goals of the project and the background of the project topic. The Data section presents all of the data that may be manipulated in the context of the project in a format that is appropriate to the placement in the text; additional history or pertinent facts may often be found in this section. The Exploration section enables students to manipulate data and discover certain facts about or patterns within the data. For example, the Transportation project allows students to use graphs to find patterns and interactively experiment with placing a line on a scatter plot of actual data to approximate a best fitting line. The Exercises section is a set of questions designed to guide the student to the types of discoveries that may be made from exploration of the data. For example, with the Transportation project students are asked to interpret slopes and y-intercepts, consider predictions, and compare various models.

The CD-ROM Projects for *Elementary and Intermediate Algebra: A Combined Course*, Second Edition, are available for use with multimedia Macintosh or IBM with Windows computers. They cover the following topics:

Chapter P	Population Growth Patterns	**Chapter 8**	Air Resistance and Parachutes
Chapter 1	Musical Sound	**Chapter 9**	Fractals
Chapter 2	Temperature	**Chapter 10**	Gravitation
Chapter 3	Solar Eclipses	**Chapter 11**	Transportation
Chapter 4	Sporting Goods Sales	**Chapter 12**	Retail Sales of Companies
Chapter 5	Chemistry and Color	**Chapter 13**	Half-Life and Radioactivity
Chapter 6	Food Consumption	**Chapter 14**	Banking and Personal Finance
Chapter 7	Working Men and Women		

Acknowledgments

We would like to thank the many people who have helped us at various stages of this project to prepare the text and supplements package. Their encouragement, criticisms, and suggestions have been invaluable to us.

Advisory Panel: Mary Jean Brod, University of Montana; Terry Fung, Kean College of New Jersey; Kenneth Johnston, Hinds Community College; Ruth Meyering, Grand Valley State University; Beverly Michael, University of Pittsburgh; Sally Sestini; Cerritos College.

Reviewers: Delaine Cochran, Indiana University—Southeast; Laura Dyer, Bellville Area College; Cynthia Fleck, Wright State University; Lisa Grenier, Pima College—Downtown; Bruce Hoelter, Raritan Valley Community College; Kian Kaviani, Los Angelos City College; Brenda Lackey, University of Tennessee at Martin; Nancy Long, Trinity Valley Community College; Wanda Long, St. Charles County Community College; Russell Lundstrom; Judith Marwick, Prairie State College; Marilyn Moss, Collin County Community College; Derek Mpinga, North Lake College; Jon Odell, Richland Community College; John Squires, Cleveland State Community College; Debbie Singleton, Lexington Community College; Pat Stanley, Ball State University; Peg Williamson, Milwaukee Area Technical College.

Second Edition Survey Respondents: Over 300 professors took time to respond to an Algebra Survey. We appreciate your comments.

Career Interviews: Our thanks to Laura Balaoro, Amy L. Bick, Jacquelyn Bick, Dean R. Brookie, Mary Kay Brown, Alfred A. Campos, Melanie Cansler, Sidney W. Carter, Alice L. Chi, Lisa M. Deitemeyer, Amy L. Dines, Candy J. Dwyer, Catie Hanley, Bernie Khoo, Kimberly R. Lee, Peggy Murray, Lt. Robert Orr, Rob Prester, Jeremy Reiner, Timothy J. Renning, Richard S. Schroeder, Alex J. Swaneck, and John A. Yirga for their help in creating the career interviews. We appreciate their time and effort.

Thanks to all of the people at D. C. Heath and Company who worked with us in the development and production of the text, especially Charles Hartford and Ann Marie Jones, Mathematics Acquisitions Editors; Cathy Cantin, Managing Editor; Emily Keaton, Developmental Editor; Carolyn Johnson, Editorial Associate; Karen Carter, Andrea Cava, and Rachel Wimberly, Production Editors; Henry Rachlin, Designer; Gary Crespo, Art Editor; Lisa Merrill, Production Coordinator; and Billie L. Porter, Photo Researcher.

We would also like to thank the staff at Larson Texts, Inc., who assisted with proofreading the manuscript; preparing and proofreading the art package; and checking and typesetting the supplements.

On a personal level, we are grateful to our wives, Deanna Gilbert Larson and Eloise Hostetler, for their love, patience, and support. Also, a special thanks goes to R. Scott O'Neil.

If you have suggestions for improving the text, please feel free to write to us. Over the past two decades, we have received many useful comments from both instructors and students, and we value these very much.

Roland E. Larson
Robert P. Hostetler

Contents

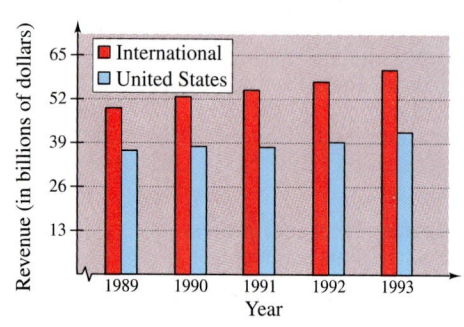
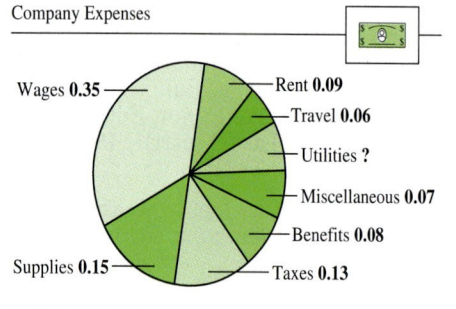

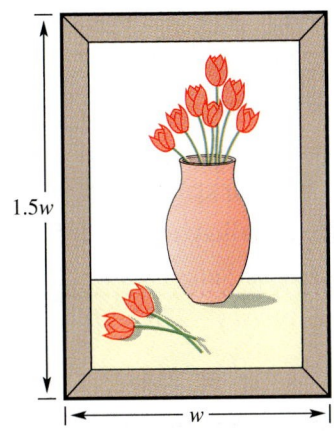

Chapter 2 Fundamentals of Algebra 91

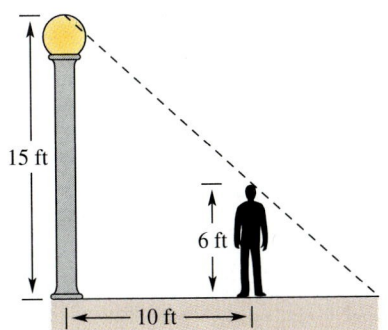

Chapter 3 Linear Equations and Problem Solving 151

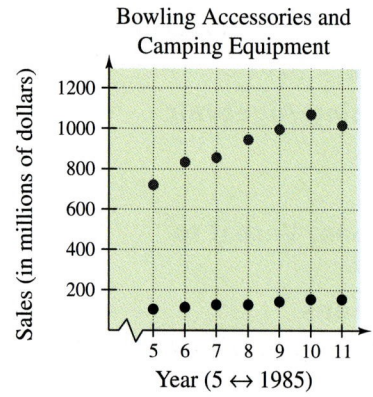

Bowling Accessories and Camping Equipment

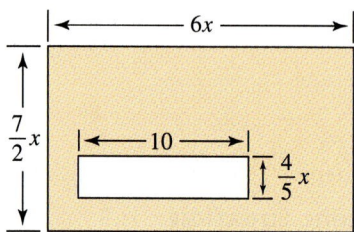

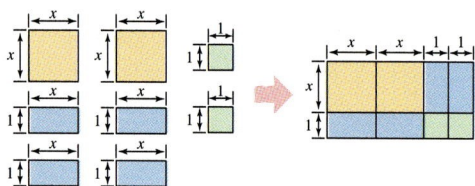

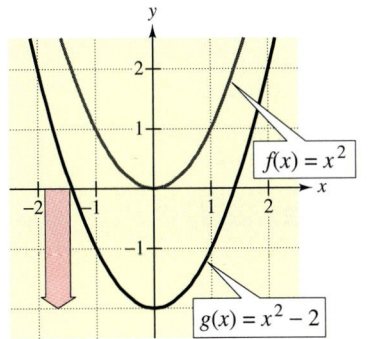
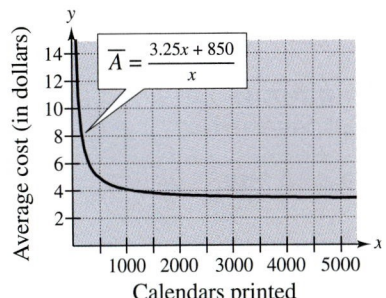

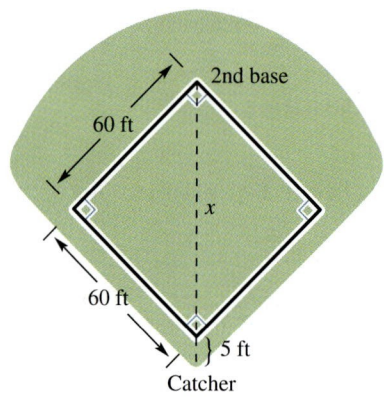

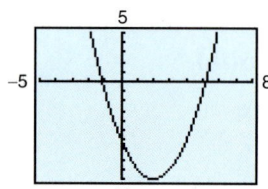

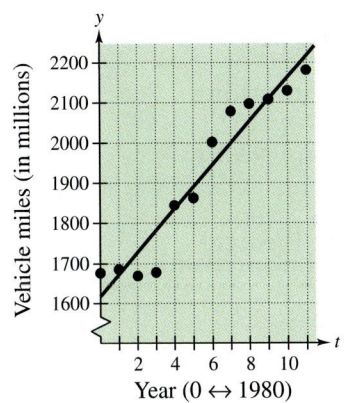

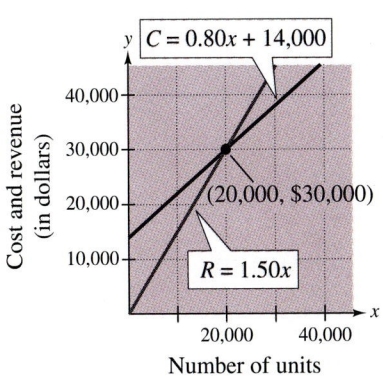

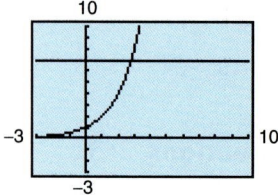

Appendices

To the Student

- *How to Study Algebra*
- *Reading and Writing About Mathematics*
- *What Is Algebra?*

How to Study Algebra

After years of teaching and guiding students through algebra courses, we have compiled the following list of suggestions for studying algebra. These study tips may take some time and effort—but they work!

Making a Plan   Make your own course plan right now! Determine the number of hours you need to spend on algebra each week. Write your plans on your calendar or some other schedule planner, and then *stick to your plan*.

Preparing for Class   Before attending class, read the portion of the text that is to be covered. This takes a lot of self-discipline, but it pays off. By going to class prepared, you will be able to benefit much more from your instructor's presentation. Algebra, like most other technical subjects, is easier to understand the second or third time you hear it.

Attending Class   Attend every class. Arrive on time with your text, a pen or pencil, paper for notes, and your calendar.

Participating in Class   As you are reading the text before class, write down any questions that you have about the material. Then, ask your instructor during class.

Taking Notes   Take notes in class, especially on definitions, examples, concepts, and rules. Then, as soon after class as possible, read through your notes, adding any explanations that are necessary to make your notes understandable *to you*.

Doing the Homework   Learning algebra is like learning to play the piano or learning to play basketball. You cannot become skilled by just watching someone else do it. You must also do it yourself. A general guideline is to spend two to four hours of study outside of class for each hour in class. When working exercises, your ultimate goal is to be able to solve the problems accurately and quickly. When you start a new exercise set, however, understanding is much more important than speed.

Finding a Study Partner   When you get stuck on a problem, it may help to try to work with someone else. Even if you feel you are giving more help than you are getting, you will find that an excellent way to learn is by teaching others.

Working in a Group Agree on what you have to do and make a plan. Listen to each other's ideas, and try to build on them. Ask for help when you need it; give help when asked. Finish the project together. Discuss what you did well together and what you could do differently next time.

Building a Math Library Start building a library of books that can help you with this and future math courses. You might consider using the *Study and Solutions Guide* that accompanies the text. Also, since you will probably be taking other math courses after you finish this course, we suggest that you keep the text. It will be a valuable reference book. Tutorial software and videos available with this text will also be valuable additions to your mathematics library.

Keeping Up with the Work Don't let yourself fall behind in the course. If you think that you are having trouble, seek help immediately. Ask your instructor, attend your school's tutoring services, talk with your study partner, use additional study aids such as videos or software tutorials—but do something. If you are having trouble with the material in one chapter of your algebra text, there is a good chance that you will also have trouble in later chapters.

Getting Stuck *Everyone* who has ever taken a math course has had this experience: You are working on a problem and cannot see how to solve it, or you have solved it but your answer does not agree with the answer given in the back of the book. People have different approaches to this sort of problem. You might ask for help, take a break to clear your thoughts, sleep on it, rework the problem, or reread the section in the text. The point is, try not to get frustrated or spend too much time on a single problem.

Assessing Your Progress In the middle of each chapter is a *Mid-Chapter Quiz.* Take the quiz as you would if you were in class, then check your answers in the back of the text.

Checking Your Work One of the nice things about algebra is that you don't have to wonder whether your solution is correct. You can tell whether it is correct by checking it in the original statement of the problem. If, in addition to your "solving skills," you work on your "checking skills," you should find your test scores improving.

Preparing for Exams Cramming for algebra exams seldom works. If you have kept up with the work and followed the suggestions given here, you should be almost ready for the exam. At the end of each chapter, we have included three features that should help as a final preparation. Read the *Chapter Summary,* work the *Review Exercises,* and set aside an hour to take the sample *Chapter Test.*

Taking Exams   Most instructors suggest that you do *not* study right up to the minute you are taking a test. This tends to make people anxious. The best cure for anxiousness during tests is to prepare well before taking the test. Once the test has begun, read the directions carefully, and try to work at a reasonable pace. (You might want to read the entire test first, then work the problems in the order with which you feel most comfortable.) Hurrying tends to cause people to make careless errors. If you finish early, take a few moments to clear your thoughts and then take time to go over your work.

Learning from Mistakes   When you get an exam back, be sure to go over any errors that you might have made. Don't be too quick to pass off an error as just a "dumb mistake." Take advantage of any mistakes by hunting for ways to continually improve your test-taking abilities.

Reading and Writing About Mathematics

Reading a Mathematics Textbook

The following suggestions can help you read your textbook most effectively.

- Before each class, read the portion of the text that will be covered during class. Read the material carefully, and keep a pen or pencil and your calculator nearby and ready to use. Work the examples *before* reading the solution, and try the Discovery activities.

- It may help to take notes as you read, paying particular attention to terms, special symbols, and new ideas. Don't expect to read a mathematics textbook as quickly as you would a novel or magazine. A mathematics textbook takes a little more time and your full attention.

- See how the text is organized. Notice that the major parts of a section are listed at the beginning of each section. These are the key concepts and objectives for that section. Notice that important terms are highlighted in boldface type. Make sure you understand this vocabulary. Study the side comments next to solution steps that show how to proceed from one step to the next. Note the Study Tips in the margin.

- Especially when reading definitions, consider every sentence, word, and symbol carefully. It is helpful to read these more than once—first to get the general idea of the statement and then a second time for details, such as under what conditions the statement is true.

- Ask questions in class based on what you have read.

- After class, reread the text. Take your time. Have a pen or pencil and calculator in hand. Make sure that you now understand any material that was unclear during your first reading.

- Be patient. The ability to read mathematics (or any technical material) is a skill that will be useful to you in this course, in other mathematics courses, and for most jobs.

Writing About Mathematics

Mathematics is a language—a way of communicating ideas using symbols. As a mathematics student, you will use the mathematics language in many different ways. You will show how a problem can be solved in a logical series of steps. You may also write about mathematical ideas and what they mean in the real world. Being able to think about a problem in a logical way is a useful skill. Here are some suggestions to help you write about mathematics.

- When asked to discuss or explain a mathematical finding, begin by making a short list of important points.
- Put the list of points in order, and see if you have left out anything. Imagine that you are explaining your answer to a friend. Would he or she understand it?
- Use complete statements and explanations when writing mathematics for others to read. (This is just like using complete sentences in English.)
- If you have difficulty writing a solution, look at the solutions in the text to see if you can use one as a model.
- You may find it easier to begin the solution to a problem by describing the problem in your own words.
- It may help to give an example or counterexample as part of your explanation.
- List any assumptions and define all variables used in your writing.
- When writing a solution to a problem, try to explain or justify your reasoning for *all* steps.
- Use a graph or a table of data if you think it will make your answer easier to understand.
- Ask yourself if you have covered all the possible outcomes of the situation.
- Check your written answers to make sure that they are organized logically, have detailed steps, and are convincing.
- Be neat.
- If research is involved, list your sources fully and accurately.
- The time you spend learning to write about mathematics is worth it. The ability to write a logical mathematical argument or explanation is a skill that will be useful in your algebra course, in other mathematics courses, and in the workplace.

What Is Algebra?

To some, algebra is manipulating symbols or performing mathematical operations with letters instead of numbers. To others, it is factoring, solving equations, or solving word problems. And to still others, algebra is a mathematical language that can be used to model real-world problems. In fact, algebra is all of these!

As you study this text, it is helpful to view algebra from the "big picture"—to see how the various rules, operations, and strategies fit together.

The rules of arithmetic form the foundation of algebra. These rules are generalized through the use of symbols and letters to form the basic rules of algebra, which are used to *rewrite* algebraic expressions and equations in new, more useful forms. The ability to rewrite algebraic expressions and equations is the common skill involved in the three major components of algebra—*simplifying* algebraic expressions, *solving* algebraic equations, and *graphing* algebraic functions. The following chart shows how this college algebra text fits into the "big picture" of algebra.

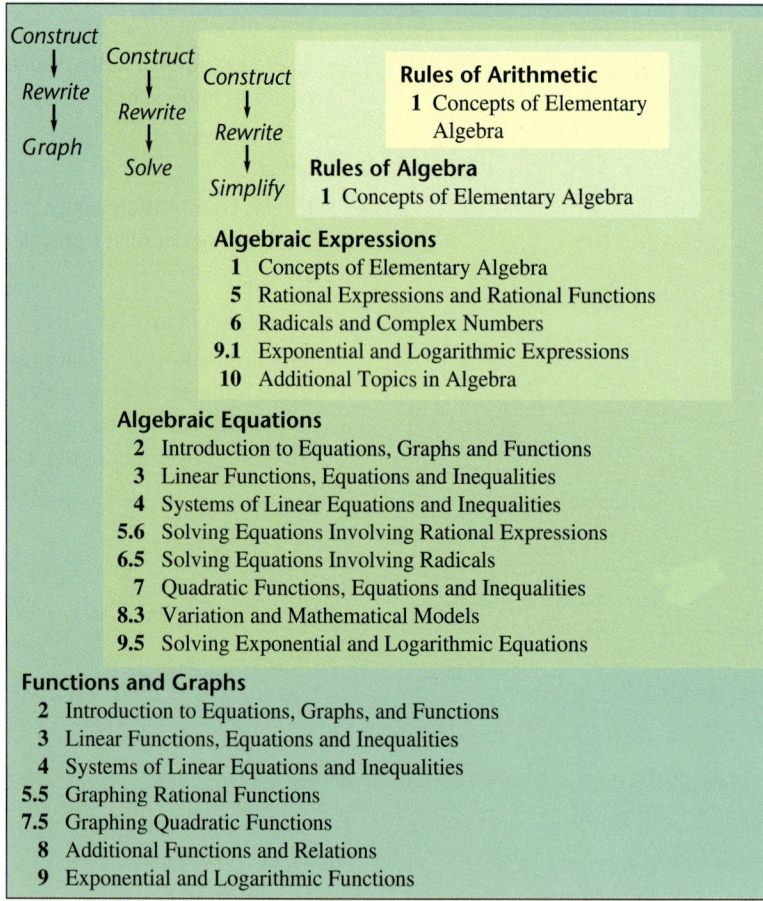

Construct
↓
Rewrite
↓
Graph

Construct
↓
Rewrite
↓
Solve

Construct
↓
Rewrite
↓
Simplify

Rules of Arithmetic
  1 Concepts of Elementary Algebra

Rules of Algebra
  1 Concepts of Elementary Algebra

Algebraic Expressions
  1 Concepts of Elementary Algebra
  5 Rational Expressions and Rational Functions
  6 Radicals and Complex Numbers
  9.1 Exponential and Logarithmic Expressions
  10 Additional Topics in Algebra

Algebraic Equations
  2 Introduction to Equations, Graphs and Functions
  3 Linear Functions, Equations and Inequalities
  4 Systems of Linear Equations and Inequalities
  5.6 Solving Equations Involving Rational Expressions
  6.5 Solving Equations Involving Radicals
  7 Quadratic Functions, Equations and Inequalities
  8.3 Variation and Mathematical Models
  9.5 Solving Exponential and Logarithmic Equations

Functions and Graphs
  2 Introduction to Equations, Graphs, and Functions
  3 Linear Functions, Equations and Inequalities
  4 Systems of Linear Equations and Inequalities
  5.5 Graphing Rational Functions
  7.5 Graphing Quadratic Functions
  8 Additional Functions and Relations
  9 Exponential and Logarithmic Functions

Prerequisites: Arithmetic Review

P

- Real Numbers: Order and Absolute Value
- Integers and Prime Factorization
- Adding and Subtracting Integers
- Multiplying and Dividing Integers

Year	1950	1960	1970	1980	1990
Population	10,586	15,717	19,971	23,668	29,760

Newspapers and magazines organize data for their readers using tables and graphs. The entries in a table enable the reader to interpret (and manipulate) data in an efficient manner. Data from a table can then be organized into a graph, which provides a visual interpretation of the data.

The table at the right shows the population (in thousands) of the state of California for five different years. From the numbers in the table you can see that the population of California increased each decade.

The numbers in the table can be used to create a bar graph that represents the population. From the graph you can see that the largest increase in population occurred between 1980 and 1990. (Source: U.S. Bureau of the Census)

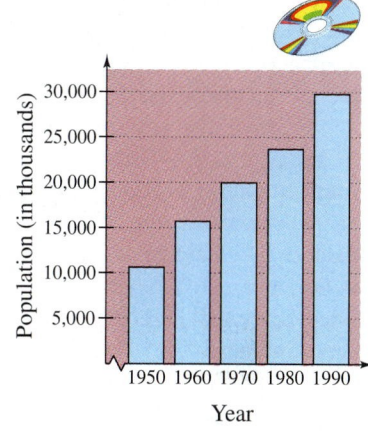

The chapter project related to this data is on page 37.

1

P.1	**Real Numbers: Order and Absolute Value**
	Sets and Real Numbers ▪ The Real Number Line ▪ Ordering Real Numbers ▪ Absolute Value

Sets and Real Numbers

The ability to communicate precisely is an essential part of a modern society, and it is the primary goal of this text. Specifically, this text concerns the language that is used to communicate numerical concepts.

The formal term that is used in mathematics to talk about a collection of objects is the word **set.** For instance, the set {1, 2, 3} contains the three numbers 1, 2, and 3. Note that a pair of braces { } is used to list the members of the set. Parentheses ( ) and brackets [ ] are used to represent other ideas.

The set of numbers that is used in arithmetic is called the set of **real numbers.** The term *real* distinguishes real numbers from *imaginary* numbers—a type of number that is used in some mathematics courses. You will not study imaginary numbers in this course.

If each member of a set *A* is a member of set *B*, then *A* is called a **subset** of *B*. The set of real numbers has many important subsets, each with a special name. For instance, the set

$$\{1,\ 2,\ 3,\ 4,\ \ldots\} \qquad \text{A subset of the set of real numbers}$$

is the set of **natural numbers** or **positive integers.** Note that the three dots indicate that the pattern continues. For instance, the set also contains the numbers 5, 6, 7, and so on. Every positive integer is a real number, but there are many real numbers that are not positive integers. For example, the numbers -2, 0, and $\frac{1}{2}$ are real numbers, but they are not positive integers.

Positive integers can be used to describe many things that you encounter in everyday life. For instance, you might be taking four classes this term, or you might be paying $180 a month for rent. But even in everyday life, positive integers cannot describe some concepts accurately. For instance, you could have a zero balance in your checking account, or the temperature could be $-10°$ (ten degrees below zero). To describe such quantities you need to expand the set of positive integers to include **zero** and the **negative integers.** The expanded set is called the set of **integers.**

$$\underbrace{\{\ldots,\ -3,\ -2,\ -1,}_{\text{Negative integers}}\ \overset{\text{Zero}}{0},\ \underbrace{1,\ 2,\ 3,\ \ldots\}}_{\text{Positive integers}} \qquad \text{Set of integers}$$

The set of integers is also a subset of the set of real numbers.

Even with the set of integers, there are still many quantities in everyday life that you cannot describe accurately. The costs of many items are not in whole-dollar amounts, but in parts of dollars, such as $1.19 or $39.98. You might work $8\frac{1}{2}$ hours, or you might miss the first half of a movie. To describe such quantities, you can expand the set of integers to include **fractions.** The expanded set is called the set of **rational numbers.** In the formal language of mathematics, a real number is **rational** if it can be written as a ratio of two integers. Thus, $\frac{3}{4}$ is a rational number; so is 0.5 (it can be written as $\frac{1}{2}$); and so is every integer. Each of the sets of numbers mentioned—natural numbers, integers, and rational numbers—is a subset of the set of real numbers, as shown in Figure P.1.

NOTE   A real number that is not rational is called **irrational** and cannot be written as the ratio of two integers. One example of an irrational number is $\sqrt{2}$ (read as the positive square root of 2). You will study this type of real number later in the text.

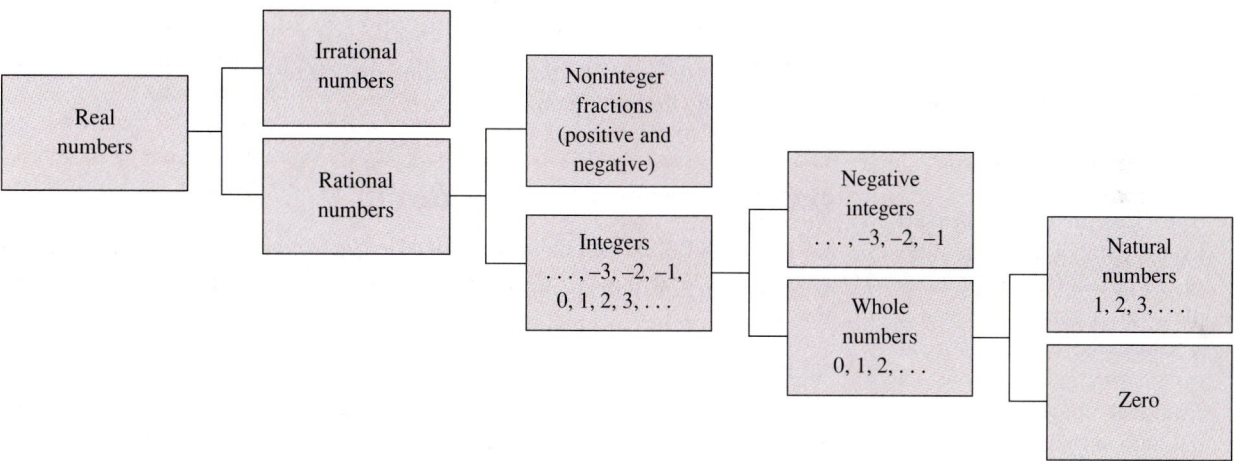

FIGURE P.1   Subsets of Real Numbers

EXAMPLE 1   *Classifying Real Numbers*

Determine which numbers in the following set are (a) natural numbers, (b) integers, and (c) rational numbers.

$$\left\{ \tfrac{1}{2}, \ -1, \ 0, \ 4, \ -\tfrac{5}{8}, \ \tfrac{4}{2}, \ -\tfrac{3}{1}, \ 0.86 \right\}$$

Solution

a.   Natural numbers: $\left\{ 4, \ \tfrac{4}{2} = 2 \right\}$

b.   Integers: $\left\{ -1, \ 0, \ 4, \ \tfrac{4}{2} = 2, \ -\tfrac{3}{1} = -3 \right\}$

c.   Rational numbers: $\left\{ \tfrac{1}{2}, \ -1, \ 0, \ 4, \ -\tfrac{5}{8}, \ \tfrac{4}{2}, \ -\tfrac{3}{1}, \ 0.86 \right\}$

The Real Number Line

The picture that is used to represent the real numbers is called the **real number line.** It consists of a horizontal line with a point (the **origin**) labeled 0. Numbers to the left of 0 are **negative** and numbers to the right of 0 are **positive,** as shown in Figure P.2. The real number zero is neither positive nor negative. Thus, the term **nonnegative** implies that a number may be positive *or* zero.

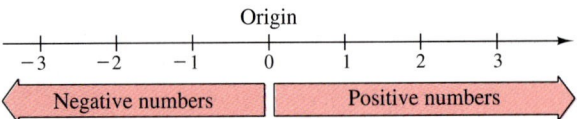

FIGURE P.2   The Real Number Line

Drawing the point on the number line that corresponds to a real number is called **plotting** the real number.

EXAMPLE 2   Plotting Real Numbers

a.   In Figure P.3(a), the point corresponds to the real number $-\frac{1}{2}$.

b.   In Figure P.3(b), the point corresponds to the real number 2.

c.   In Figure P.3(c), the point corresponds to the real number $-\frac{3}{2}$.

d.   In Figure P.3(d), the point corresponds to the real number 1.

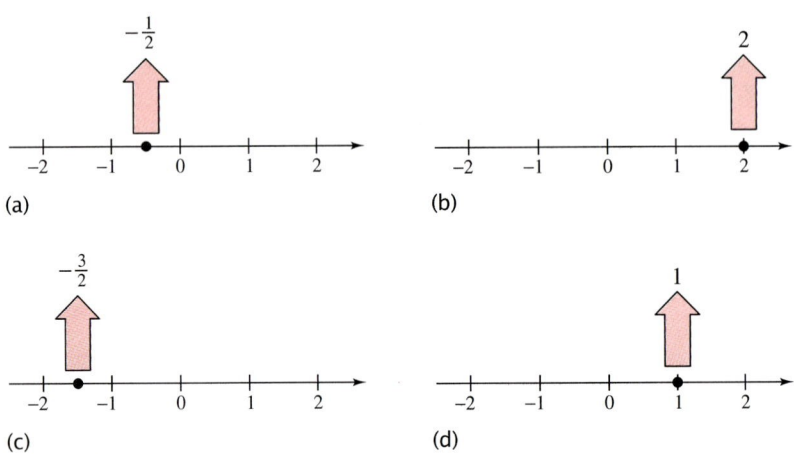

FIGURE P.3

Example 2 illustrates the following principle. *Each point on the real number line corresponds to exactly one real number, and each real number corresponds to exactly one point on the real number line.*

Ordering Real Numbers

If you choose any two numbers on the real number line, one of the numbers must be to the left of the other number. The number to the left is **less than** the number to the right, and the number to the right is **greater than** the number to the left. For example, from Figure P.4 you can see that -3 is less than 2 because -3 lies to the left of 2 on the number line. A "less than" comparison is denoted by the **inequality symbol** $<$. For instance, "-3 is less than 2" is denoted by $-3 < 2$.

Similarly, the inequality symbol $>$ is used to denote a "greater than" comparison. For instance, "2 is greater than -3" is denoted by $2 > -3$. The inequality symbol $\leq$ means **less than or equal to,** and the inequality symbol $\geq$ means **greater than or equal to.**

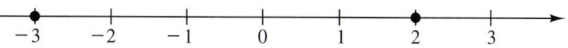

FIGURE P.4 -3 lies to the left of 2.

When you are asked to **order** two numbers, you are simply being asked to say which of the two numbers is greater.

EXAMPLE 3   *Ordering Integers*

Place the correct inequality symbol ($<$ or $>$) between the two numbers.

a. 3 ⬚ 5   **b.** -3 ⬚ -5

c. 4 ⬚ 0   **d.** -2 ⬚ 2

Solution

See Figure P.5.

a. $3 < 5$, because 3 lies to the *left* of 5.

b. $-3 > -5$, because -3 lies to the *right* of -5.

c. $4 > 0$, because 4 lies to the *right* of 0.

d. $-2 < 2$, because -2 lies to the *left* of 2.

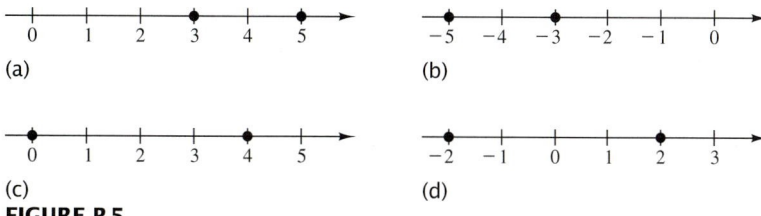

(a)

(b)

(c)

(d)

FIGURE P.5

EXAMPLE 4   Ordering Decimals

Place the correct inequality symbol ($<$ or $>$) between the two numbers.

a. -3.1 �_____ 2.8 **b.** -1.09 �_____ -1.90

Solution

See Figure P.6.

a. $-3.1 < 2.8$, because -3.1 lies to the *left* of 2.8.

b. $-1.09 > -1.90$, because -1.09 lies to the *right* of -1.90.

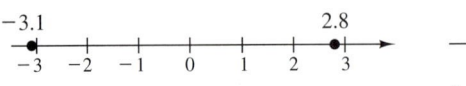
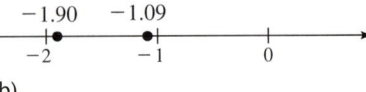

(a)   (b)

FIGURE P.6

There are two ways to order fractions: you can write both fractions with the same denominator, or you can rewrite both fractions in decimal form. Here are two examples.

$$\frac{1}{3} = \frac{4}{12} \quad \text{and} \quad \frac{1}{4} = \frac{3}{12} \qquad \Longrightarrow \qquad \frac{1}{3} > \frac{1}{4}$$

$$\frac{11}{131} \approx 0.084 \quad \text{and} \quad \frac{19}{209} \approx 0.091 \qquad \Longrightarrow \qquad \frac{11}{131} < \frac{19}{209}$$

EXAMPLE 5   Ordering Fractions

Place the correct inequality symbol ($<$ or $>$) between the two numbers.

a. $\dfrac{1}{3}$ �_____ $\dfrac{1}{5}$ **b.** $-\dfrac{3}{2}$ �_____ $\dfrac{1}{2}$

Solution

See Figure P.7.

a. $\frac{1}{3} > \frac{1}{5}$, because $\frac{1}{3} = \frac{5}{15}$ lies to the *right* of $\frac{1}{5} = \frac{3}{15}$.

b. $-\frac{3}{2} < \frac{1}{2}$, because $-\frac{3}{2}$ lies to the *left* of $\frac{1}{2}$.

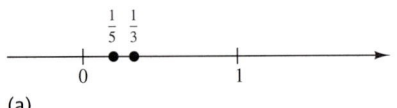
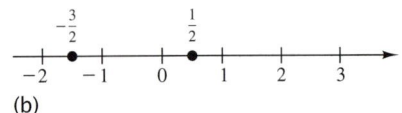

(a)   (b)

FIGURE P.7

Absolute Value

Two real numbers are **opposites** of each other if they lie the same distance from, but on opposite sides of, zero. For example, -2 is the opposite of 2, and 4 is the opposite of -4, as shown in Figure P.8.

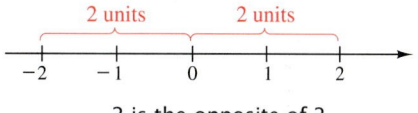

2 is the opposite of 2.

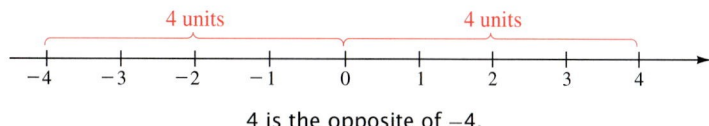

4 is the opposite of -4.

FIGURE P.8

Parentheses are useful for denoting the opposite of a negative number. For example, $-(-3)$ means the opposite of -3, which you know to be 3. That is,

$$-(-3) = 3.$$ The opposite of -3 is 3.

For any real number, its distance from zero (on the real number line) is its **absolute value.** A pair of vertical bars, $|\ \ |$, is used to denote absolute value. Here are two examples.

$$|5| = \text{``distance between 5 and 0''} = 5$$
$$|-8| = \text{``distance between } -8 \text{ and 0''} = 8$$

Because opposite numbers lie the same distance from 0 on the real number line, they have the same absolute value. Thus,

$$|5| = 5 \qquad \text{and} \qquad |-5| = 5.$$

You can write this more simply as $|5| = |-5| = 5$.

NOTE   The absolute value of a real number is either positive or zero (never negative). Moreover, zero is the only real number whose absolute value is 0. That is, $|0| = 0$.

The word **expression** means a collection of numbers and symbols such as $3 + 5$ or $|-4|$. When you are asked to **evaluate** an expression, you are being asked to find the *number* that is equal to the expression.

EXAMPLE 6   Evaluating Absolute Value

Evaluate the following expressions.

a. $|-10|$ **b.** $\left|\dfrac{3}{4}\right|$ **c.** $|-3.2|$ **d.** $-|-6|$

Solution

a. $|-10| = 10$, because the distance between -10 and 0 is 10.

b. $\left|\frac{3}{4}\right| = \frac{3}{4}$, because the distance between $\frac{3}{4}$ and 0 is $\frac{3}{4}$.

c. $|-3.2| = 3.2$, because the distance between -3.2 and 0 is 3.2.

d. $-|-6| = -(6) = -6$.

Note in Example 6(d) that $-|-6| = -6$ does not contradict the fact that the absolute value of a real number cannot be negative.

EXAMPLE 7   Comparing Absolute Values

Place the correct symbol ($<$, $>$, or $=$) between the two numbers.

a. $|-9|$ _____ $|9|$ **b.** 0 _____ $|-5|$

c. -4 _____ $-|-4|$ **d.** $|12|$ _____ $|-15|$

Solution

a. $|-9| = |9|$, because both are equal to 9.

b. $0 < |-5|$, because $|-5| = 5$ and 0 is less than 5.

c. $-4 = -|-4|$, because both numbers are equal to -4.

d. $|12| < |-15|$, because $|12| = 12$ and $|-15| = 15$, and 12 is less than 15.

Group Activities   Communicating Mathematically

Interpreting Inequalities   Is the statement "$5 \geq 5$" true? Does everyone in your group give the same answer? Together, write an explanation to support the agreed-upon answer.

P.1 Exercises

Discussing the Concepts

1. Explain why $\frac{8}{4}$ is a natural number, but $\frac{7}{4}$ is not.

2. On the real number line, how many numbers are three units from 0? Explain your reasoning.

3. On the real number line, which lies farther from 0?

(a) -25 (b) 10

Explain your reasoning.

4. On the real number line, which lies farther from -7?

(a) 3   (b) -10

Explain your reasoning.

5. Explain how to determine the smaller of two real numbers.

6. Which is smaller: $\frac{3}{8}$ or 0.35?

Problem Solving

7. Which numbers in the following set are (a) natural numbers, (b) integers, and (c) rational numbers?

$$\left\{-3, 2, -\tfrac{3}{2}, \tfrac{9}{3}, 4.5\right\}$$

8. Plot each number in the set in Exercise 7 on the real number line.

In Exercises 9–12, write the real numbers shown by the points on the real number line and place the correct inequality symbol ($<$ or $>$) between the two numbers.

9.

![Number line with points at 2 and 5, marked 0 to 6]

10.

![Number line with points at -5 and -3, marked -5 to 0]

11.

![Number line with points at -9/2 and -3, marked -5 to 0]

12.

![Number line with points at 50.5 and 53.5, marked 49 to 54]

In Exercises 13–16, show each real number as a point on the real number line and place the correct inequality symbol between the real numbers.

13. 4 ___ $-\frac{7}{2}$

14. $-\frac{7}{3}$ ___ $-\frac{7}{2}$

15. -4.6 ___ 1.5

16. $-\frac{3}{8}$ ___ $-\frac{5}{8}$

In Exercises 17 and 18, on the real number line, what is the distance between a and zero?

17. $a = -4$

18. $a = 5$

In Exercises 19 and 20, find the opposite number.

19. -3

20. 2

In Exercises 21–24, evaluate the expression.

21. $|-3.4|$

22. $|76.3|$

23. $-|-23.6|$

24. $-|91|$

In Exercises 25–28, place the correct symbol ($<$, $>$, or $=$) between the two real numbers.

25. $|-4|$ ___ $|3|$

26. $|525|$ ___ $|-525|$

27. $-|-48.5|$ ___ $|-48.5|$

28. $\left|-\frac{7}{8}\right|$ ___ $\left|\frac{4}{3}\right|$

In Exercises 29 and 30, plot the numbers.

29. $\frac{5}{2}$, π, -2, $-|-3|$

30. 3.7, $\frac{16}{3}$, $|-1.9|$, $-\frac{1}{2}$

In Exercises 31 and 32, find all real numbers whose distance from a is d.

31. $a = 8$, $d = 12.5$

32. $a = 21.3$, $d = 6$

Additional Problem Solving

In Exercises 33 and 34, which numbers in the set are (a) natural numbers, (b) integers, and (c) rational numbers? Plot the numbers on the real number line.

33. $\left\{-\frac{5}{2}, 6.5, -4.5, \frac{8}{4}, \frac{3}{4}\right\}$

34. $\left\{8, -1, \frac{4}{3}, -3.25, -\frac{10}{2}\right\}$

In Exercises 35–38, write the real numbers shown by the points on the real number line and place the correct inequality symbol ($<$ or $>$) between the two numbers.

35.

36.

37.

38.

In Exercises 39–46, plot each real number on the real number line and place the correct inequality symbol between the real numbers.

39. $\frac{1}{3}$ ___ 4 **40.** 6 ___ -2

41. -2π ___ -10 **42.** 2 ___ π

43. $\frac{7}{16}$ ___ $\frac{5}{8}$ **44.** 28.60 ___ -3.75

45. 0 ___ $-\frac{7}{16}$ **46.** 2π ___ π

In Exercises 47 and 48, on the real number line, what is the distance between a and zero?

47. $a = -3$ **48.** $a = -\frac{1}{2}$

In Exercises 49–52, find the opposite of the number.

49. 5 **50.** -3.6

51. $-\frac{5}{2}$ **52.** $\frac{3}{4}$

In Exercises 53–60, evaluate the expression.

53. $|7|$ **54.** $|-16.2|$

55. $\left|-\frac{7}{2}\right|$ **56.** $\left|-\frac{9}{16}\right|$

57. $-|4.09|$ **58.** $-|-43.8|$

59. $-|-3.2|$ **60.** $|0|$

In Exercises 61–68, place the correct symbol ($<$, $>$, or $=$) between the two real numbers.

61. $|-15|$ ___ $|15|$ **62.** $|16|$ ___ $|-25|$

63. $|32|$ ___ $|-50|$ **64.** $|1026|$ ___ $|800|$

65. $\left|\frac{3}{16}\right|$ ___ $\left|\frac{3}{2}\right|$ **66.** $-|-64|$ ___ $|-50|$

67. $|-\pi|$ ___ $-|-2\pi|$ **68.** $|-4.9|$ ___ $|-10.2|$

In Exercises 69 and 70, show the numbers on the real number line.

69. $\frac{3}{2}, -2\pi, 3.2, |-4|$

70. $3.5, 2, -|-1|, -\frac{1}{2}$

In Exercises 71 and 72, find all real numbers whose distance from a is given by d.

71. $a = -2$, $d = 3.5$

72. $a = 42.5$, $d = 7$

True or False?   In Exercises 73–78, decide whether the statement is true or false. Explain your reasoning.

73. The absolute value of any real number is positive.

74. The absolute value of a number is equal to the absolute value of its opposite.

75. The absolute value of a rational number is a rational number.

76. A given real number corresponds to exactly one point on the real number line.

77. The opposite of a positive number is a negative number.

78. Every rational number is an integer.

P.2 Integers and Prime Factorization

Factors ▪ Prime and Composite Numbers ▪
Common Factors and Multiples

In the 5th century B.C., the followers of the Greek mathematician, Pythagoras, believed that numbers revealed the basic structure of the universe. They devoted their lives to the study, discovery, and proof of number patterns. Their works laid the foundation for the field of mathematics now called *number theory*.

Factors

The set of positive integers,

$$\{1, \ 2, \ 3, \ \ldots\}$$

is one subset of the real numbers that has intrigued mathematicians for many centuries.

Historically, an important number concept has been *factors* of positive integers. From experience, you know that in a multiplication problem such as $3 \cdot 7 = 21$, the numbers 3 and 7 are called *factors* of 21.

$$\underbrace{3 \cdot 7}_{\text{Factors}} = \underbrace{21}_{\text{Product}}$$

It is also correct to call the numbers 3 and 7 *divisors* of 21, because 3 and 7 each divide evenly into 21.

Factor (or Divisor)

If a and b are positive integers, then a is a **factor** (or **divisor**) of b if and only if there is a positive integer c such that $a \cdot c = b$.

EXAMPLE 1 *Factors and Divisors*

a. 3 is a factor of 12 because

$$3 \cdot 4 = 12.$$

b. 6 is a divisor of 12 because

$$6 \cdot 2 = 12.$$

c. 5 is not a divisor of 12 because there is no positive integer c such that $5 \cdot c = 12$.

d. The set of divisors of 12 is

$$\{1, \ 2, \ 3, \ 4, \ 6, \ 12\}. \qquad \text{Divisors of 12}$$

Prime and Composite Numbers

The concept of factors allows you to classify positive integers into three groups: *prime* numbers, *composite* numbers, and the number 1.

Prime and Composite Numbers

---

1. A positive integer greater than 1 with no factors other than itself and 1 is called a **prime number,** or simply a **prime.**

2. A positive integer greater than 1 with more than two factors is called a **composite number,** or simply a **composite.**

The numbers 2, 3, 5, 7, and 11 are primes because they have only themselves and 1 as factors. The numbers 4, 6, 8, 9, and 10 are composites because each has more than two factors. The number 1 is neither prime nor composite because 1 is its only factor.

Every composite number can be expressed as a unique product of prime factors. Here are some examples.

$$6 = 2 \cdot 3, \quad 15 = 3 \cdot 5, \quad 18 = 2 \cdot 3 \cdot 3, \quad 42 = 2 \cdot 3 \cdot 7, \quad 124 = 2 \cdot 2 \cdot 31$$

One strategy for factoring a composite number into prime numbers is to begin by finding an easily recognized factor. Dividing this factor into the number yields a *companion* factor. For instance, 5 is a readily recognized divisor of 45 and its companion factor is $9 = 45 \div 5$. Continue hunting for factors and companion factors until each factor is prime. As shown in Figure P.9, a *tree diagram* is a nice way to record your work. From the tree diagram, you can see that the prime factorization of 45 is $45 = 3 \cdot 3 \cdot 5$.

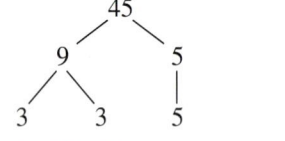

FIGURE P.9   Tree Diagram

EXAMPLE 2   *Prime Factorization*

Find the prime factorization of each of the following.

a. 84   **b.** 78   **c.** 133   **d.** 43

Solution

a. 4 is a recognized divisor of 84. Thus, $84 = 4 \cdot 21 = 2 \cdot 2 \cdot 3 \cdot 7$.

b. 2 is a recognized divisor of 78. Thus, $78 = 2 \cdot 39 = 2 \cdot 3 \cdot 13$.

c. If you do not recognize a divisor of 133, you can get started by dividing any of the prime numbers 2, 3, 5, 7, 11, 13, etc., into 133. You will find 7 to be the first prime to divide 133. Thus, $133 = 7 \cdot 19$ (19 is prime).

d. In this case, none of the primes less than 43 divides 43. Thus, 43 is prime.

Prime Factor Test

To find prime factors of a number n, you need only to search among prime numbers less than or equal to the square root of n.

Other aids to prime factoring a composite number include the following divisibility tests.

Divisibility Tests

	Example
1. A number is divisible by 2 if it is *even*.	364 is divisible by 2 because it is even.
2. A number is divisible by 3 if the sum of its digits is divisible by 3.	261 is divisible by 3 because $2 + 6 + 1 = 9$.
3. A number is divisible by 9 if the sum of its digits is divisible by 9.	738 is divisible by 9 because $7 + 3 + 8 = 18$.
4. A number is divisible by 5 if its units digit is 0 or 5.	325 is divisible by 5 because its units digit is 5.
5. A number is divisible by 10 if its units digit is 0.	120 is divisible by 10 because its units digit is 0.

EXAMPLE 3   *Using Divisibility Tests*

Determine whether the following numbers are prime or composite.

a. 247   **b.** 401   **c.** 1701

Solution

a. The divisibility tests show that 247 is not divisible by 2, 3, 5, 9, or 10. By testing the remaining primes less than or equal to $\sqrt{247} \approx 16$, you will find that $247 = 13 \cdot 19$. Thus, 247 is composite.

b. The divisibility tests yield no factors of 401. By testing the remaining primes less than or equal to $\sqrt{401} \approx 20$, you can conclude that 401 is a prime number.

c. The divisibility tests show that 1701 is divisible by 9, so 1701 is composite. Its prime factorization is $1701 = 3 \cdot 3 \cdot 3 \cdot 3 \cdot 3 \cdot 7$.

Common Factors and Multiples

To simplify fractions in arithmetic and algebra, it is useful to find *common* factors for two numbers or algebraic expressions. Among all common factors, the *greatest* common factor is often the most useful.

Greatest Common Factor

The **greatest common factor** (**GCF**) of two (or more) positive integers a and b is the largest positive integer that is a factor of both a and b.

EXAMPLE 4   *Finding the Greatest Common Factor*

Find the greatest common factor of the following.

a. 24 and 60   **b.** 35 and 42

c. 32, 80, and 96   **d.** 15 and 68

Solution

a. Prime factorization yields $24 = 2 \cdot 2 \cdot 2 \cdot 3$ and $60 = 2 \cdot 2 \cdot 3 \cdot 5$. Because two 2's and one 3 are common to both numbers, the greatest common factor is $2 \cdot 2 \cdot 3 = 12$.

b. Prime factorization yields $35 = 5 \cdot 7$ and $42 = 2 \cdot 3 \cdot 7$. Hence, the greatest common factor is 7.

c. Prime factorization yields $32 = 2 \cdot 2 \cdot 2 \cdot 2 \cdot 2$, $80 = 2 \cdot 2 \cdot 2 \cdot 2 \cdot 5$, and $96 = 2 \cdot 2 \cdot 2 \cdot 2 \cdot 2 \cdot 3$. Because four 2's are common to all three numbers, the greatest common factor is $2 \cdot 2 \cdot 2 \cdot 2 = 16$.

NOTE   Two numbers, such as 15 and 68, that have no common prime factors are called **relatively prime.**

d. Prime factorization yields $15 = 3 \cdot 5$ and $68 = 2 \cdot 2 \cdot 17$. Because there are no common prime factors, the greatest common factor is 1.

Common *multiples* of two or more numbers are also useful in working with fractions.

EXAMPLE 5   *Listing Multiples of a Number*

a. The multiples of 6 are 6, 12, 18, 24, 30, ...

b. The multiples of 15 are 15, 30, 45, 60, 75, ...

c. The multiples of 21 are 21, 42, 63, 84, 105, ...

From parts (a) and (b) of Example 5, you can see that 6 and 15 have a *common multiple* of 30. By extending the lists you can find additional common multiples of 6 and 15. To add (or subtract) fractions, it is useful to be able to find the *least common multiple* of two or more numbers.

Least Common Multiple

The **least common multiple** (**LCM**) of two (or more) positive integers a and b is the smallest positive integer that is a multiple of both a and b.

Prime factorization can be used to find least common multiples. Remember that the least common multiple must contain each prime factor, repeated the maximum number of times it occurs in any one of the factorizations.

EXAMPLE 6 *Finding the Least Common Multiple*

Find the least common multiple of the following.

a. 9 and 12   **b.** 7 and 15   **c.** 8, 14, and 24

Solution

a. By prime factorization, $9 = 3 \cdot 3$ and $12 = 2 \cdot 2 \cdot 3$. To be a multiple of 9, the LCM must contain two 3's, and to be a multiple of 12 it must also contain two 2's. Thus, the least common multiple is $2 \cdot 2 \cdot 3 \cdot 3 = 36$.

b. By prime factorization, $7 = 1 \cdot 7$ and $15 = 3 \cdot 5$. Hence, the least common multiple is $3 \cdot 5 \cdot 7 = 105$.

c. Because $8 = 2 \cdot 2 \cdot 2$, $14 = 2 \cdot 7$, and $24 = 2 \cdot 2 \cdot 2 \cdot 3$, the LCM must contain three 2's, one 7, and one 3. Therefore, the least common multiple is $2 \cdot 2 \cdot 2 \cdot 3 \cdot 7 = 168$.

Group Activities   Extending the Concept

Proper Factors   The **proper factors** of a number are all its factors less than the number itself. A number is **perfect** if the sum of its proper factors is equal to the number. A number is **abundant** if the sum of its proper factors is greater than the number. Which numbers less than 25 are perfect? Which are abundant? Compare your answers with those of your group and resolve any differences. Try to find the first perfect number greater than 25.

P.2 Exercises

Discussing the Concepts

1. Explain why 14 is not a prime number.
2. Is the number 1543 prime? Explain your reasoning.
3. Explain how the word *factor* can be used as a noun or as a verb.
4. Explain how you can check the factorization of a real number.

5. Let L be the least common multiple of two positive integers a and b. Is $L \leq a$ or is $L \geq a$? Explain your reasoning.
6. Let G be the greatest common factor of two positive integers a and b. Is $G \leq a$ or is $G \geq a$? Explain your reasoning.

Problem Solving

In Exercises 7–10, find the product.

7. $3 \cdot 5 \cdot 11$ **8.** $7 \cdot 13 \cdot 23$
9. $2 \cdot 2 \cdot 5 \cdot 5$ **10.** $3 \cdot 7 \cdot 7 \cdot 11$

In Exercises 11–14, is the number prime or composite?

11. 2400   **12.** 257
13. 3911   **14.** 12,801

In Exercises 15–18, write the prime factorization.

15. 210   **16.** 561
17. 525   **18.** 264

In Exercises 19–22, find the greatest common factor.

19. 20, 45   **20.** 45, 90
21. 18, 84, 90   **22.** 240, 300, 360

In Exercises 23 and 24, list the first five positive integer multiples of the number.

23. 12   **24.** 25

In Exercises 25–28, find the least common multiple.

25. 10, 18   **26.** 20, 25
27. 6, 9, 14   **28.** 18, 27, 45

Explanation In Exercises 29 and 30, verify that the numbers are relatively prime.

29. 63, 1375   **30.** 21, 20, 143

31. *Writing* What is the only even prime number? Explain why there are no other even prime numbers.
32. *Investigation* Twin primes are prime numbers that differ by 2. For instance, 3 and 5 are twin primes. How many other twin primes are less than 100?

Additional Problem Solving

In Exercises 33–38, find the product.

33. $5 \cdot 17 \cdot 23$ **34.** $2 \cdot 29 \cdot 37$
35. $5 \cdot 5 \cdot 13 \cdot 17$ **36.** $3 \cdot 11 \cdot 11 \cdot 37$
37. $2 \cdot 2 \cdot 2 \cdot 7 \cdot 7 \cdot 7$ **38.** $13 \cdot 13 \cdot 37 \cdot 37$

In Exercises 39–44, is the number prime or composite?

39. 643   **40.** 533
41. 8324   **42.** 3555
43. 1321   **44.** 1323

In Exercises 45–50, write the prime factorization.

45. 120   **46.** 52
47. 192   **48.** 245
49. 2535   **50.** 1521

In Exercises 51–56, find the greatest common factor.

51. 28, 52   **52.** 48, 64
53. 84, 98, 192   **54.** 117, 195, 507
55. 134, 225, 315, 945   **56.** 80, 144, 214, 504

In Exercises 57 and 58, list the first five positive integer multiples of the number.

57. 14   **58.** 32

In Exercises 59–64, find the least common multiple.

59. 15, 10   **60.** 36, 54

61. 12, 20, 25   **62.** 8, 28, 56

63. 4, 14, 28, 49   **64.** 18, 20, 30, 36

Relatively Prime Numbers In Exercises 65–70, decide whether the numbers are relatively prime.

65. 495, 784   **66.** 621, 1496

67. 403, 899   **68.** 24, 65, 161

69. 51, 85, 119   **70.** 84, 289, 325

71. *Investigation* The numbers 14, 15, and 16 are an example of three consecutive composite numbers. Is it possible to find ten consecutive composite numbers? If so, list an example. If not, explain why.

72. *Think About It* The number 1997 is not divisible by any prime number that is less than 45. Explain why this implies that 1997 is a prime number.

73. *The Sieve of Eratosthenes* Write the integers from 1 through 100 in ten lines of ten numbers each.

(a) Cross out the number 1. Cross out all multiples of 2 other than 2 itself. Do the same for 3, 5, and 7.

(b) Of what type are the remaining numbers? Explain why this is the only type of number left.

Math Matters   History of Numbers

Today, we take numbers for granted. We use numbers to count, to measure, to order things, and to calculate.

Early people knew little about numbers. They could see that an antelope had four legs, but if they saw a group of several antelope, they wouldn't have been able to tell others how many were in the group. We know that primitive people were aware of numbers because in some prehistoric caves there are pictures of animals with lines or dots scratched beside the pictures—perhaps indicating the number of animals killed in a hunt.

The fact that early people couldn't count might not have been important to tribes that relied primarily on hunting. But, as people began to farm, counting became more and more important. Farmers had to keep track of the number of animals they owned. To do this, people invented a way of writing numbers. By 2800 B.C. both the Sumerians and the Egyptians had devised ways of writing numbers. Moreover, as people traveled from one civilization to another, they borrowed writing techniques. For instance, the Romans borrowed their

Lascaux Cave, France

method of writing numbers from the Greeks.

Our present system of writing numbers was derived from many ancient civilizations, primarily Babylonian, Egyptian, Arabian, Greek, and Indian.

MID-CHAPTER QUIZ

Take this quiz as you would take a quiz in class. After you are done, check your work against the answers given in the back of the book.

In Exercises 1–4, show each real number as a point on the real line and place the correct inequality symbol (< or >) between the real numbers.

1. -2.5 ▢ -4 **2.** $\frac{3}{16}$ ▢ $\frac{3}{8}$

3. -3.1 ▢ 2.7 **4.** 2π ▢ 6

In Exercises 5 and 6, evaluate the expression.

5. $-|-0.75|$ **6.** $|25.2|$

In Exercises 7 and 8, place the correct symbol (<, >, or =) between the real numbers.

7. $\left|\frac{7}{2}\right|$ ▢ $|-3.5|$ **8.** $\left|\frac{3}{4}\right|$ ▢ $-|0.75|$

In Exercises 9 and 10, copy the number line, write the opposites of a and b, and plot the opposites on the number line.

9.

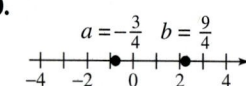

$a = -\frac{3}{2}$ $b = \frac{5}{2}$

10.

$a = -\frac{3}{4}$ $b = \frac{9}{4}$

In Exercises 11 and 12, find the product.

11. $7 \cdot 11 \cdot 11 \cdot 13$ **12.** $5 \cdot 23 \cdot 29$

In Exercises 13 and 14, determine whether the number is prime or composite, and explain your reasoning.

13. 457 **14.** 1341

In Exercises 15 and 16, write the prime factorization.

15. 354 **16.** 945

In Exercises 17 and 18, find the greatest common factor.

17. $50, 60$ **18.** $276, 1035$

In Exercises 19 and 20, find the least common multiple.

19. $12, 28$ **20.** $6, 10, 15, 45$

P.3	**Adding and Subtracting Integers**

Adding Integers ▪ Subtracting Integers

Adding Integers

In this section, you will study two of the four operations of arithmetic on the set of integers: **addition** and **subtraction.** There are many examples of integer addition in real life. For instance, suppose that your business had a gain of $550 during one week and a loss of $600 the next week. Over the two-week period, your business would have had a combined profit of

$$550 + (-600) = -50,$$

which means you had a loss of $50.

The number line is a good visual model for demonstrating addition of integers. To add a positive integer, move to the right. To add a negative integer, move to the left.

STUDY TIP

As you continue through this chapter, try to capture the overall picture of a *mathematical system,* and note the particular features discussed in each section. If you haven't read it already, go back and read the material entitled "What Is Algebra?" on page xxxi.

EXAMPLE 1   *Adding Integers with a Number Line*

Addition	*Visual Model*	*Sum*
a. $5 + 2$		$5 + 2 = 7$
b. $5 + (-2)$		$5 + (-2) = 3$
c. $-5 + 2$		$-5 + 2 = -3$
d. $-5 + (-2)$		$-5 + (-2) = -7$
e. $5 + (-5)$		$5 + (-5) = 0$

Example 1 illustrates a *graphical approach* to adding integers. It is more common to use an *analytic approach*, as summarized by the following rules. The result of an addition problem is called its **sum.**

Addition of Integers

Example

1. To add two integers *with like signs,* add their absolute values and attach the common sign to the result.

$$-3 + (-7) = -(|-3| + |-7|)$$
$$= -(3 + 7)$$
$$= -10$$

NOTE When applying the rules at the right, most people perform the absolute value steps mentally.

2. To add two integers with *different* signs, subtract the smaller absolute value from the larger absolute value and attach the sign of the integer with the larger absolute value.

$$3 + (-7) = -(|-7| - |3|)$$
$$= -(7 - 3)$$
$$= -4$$

EXAMPLE 2   Adding Integers

a.   Different signs: $22 + (-17) = 22 - 17 = 5$

b.   Different signs: $-84 + 14 = -(84 - 14) = -70$

c.   Like signs: $-138 + (-62) = -(138 + 62) = -200$

There are different ways to add three or more integers. You can use the **carrying algorithm** with a vertical format with nonnegative integers, as shown in Figure P.10, or you can add them two at a time, as illustrated in Example 3.

$$
\begin{array}{r}
1\ 1 \\
1\ 4\ 8 \\
6\ 2 \\
+\ 5\ 3\ 6 \\
\hline
7\ 4\ 6
\end{array}
$$

FIGURE P.10   Carrying Algorithm

EXAMPLE 3   Adding Three or More Integers

a. $27 + (-52) + 13 = [27 + (-52)] + 13$ Group first two terms.

$$= -25 + 13$$ Add first two terms.

$$= -12$$ Add remaining terms.

b. $-18 + 42 + (-29) + 5 = (-18 + 42) + (-29 + 5)$

$$= 24 + (-24)$$

$$= 0$$

Subtracting Integers

Subtraction can be thought of as "taking away." For instance, $8 - 5$ can be thought of as "8 take away 5," which leaves 3. On the number line, you first move 8 units to the right, then 5 units to the left, as shown in Figure P.11(a). This same result is accomplished by "adding the opposite," as shown in Figure P.11(b).

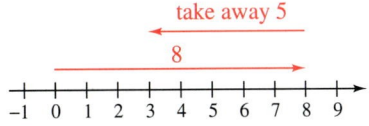

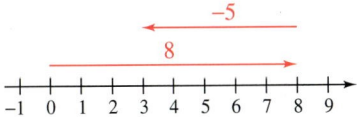

(a) "Taking 5 Away from 8"   (b) "Adding the Opposite of 5 to 8"
FIGURE P.11

Subtraction of Integers

To **subtract** one integer from another, add the opposite of the integer being subtracted to the other integer. The result is called the **difference** of the two integers.

The **opposite** of an integer is also called its **additive inverse.** For instance, the additive inverse of 5 is -5. The name *additive inverse* comes from the fact that the sum of an integer and its additive inverse is 0. For instance, $5 + (-5) = 0$.

EXAMPLE 4   Subtracting Integers

a. $3 - 8 = 3 + (-8) = -5$

b. $10 - (-13) = 10 + 13 = 23$

c. $-5 - 12 = -5 + (-12) = -17$

d. $-4 - (-17) - 23 = -4 + 17 + (-23) = -10$

$$\begin{array}{r} 3\ 10\ 15 \\ \cancel{4}\ \cancel{1}\ \cancel{5} \\ -\ 2\ 7\ 6 \\ \hline 1\ 3\ 9 \end{array}$$

FIGURE P.12   Borrowing Algorithm

Be sure that you understand that the terminology involving subtraction is not the same as that used for negative numbers. For instance, -5 is read as "negative 5," but $8 - 5$ is read as "8 minus 5."

For subtraction problems involving only two nonnegative integers, you can use the **borrowing algorithm** shown in Figure P.12.

EXAMPLE 5   Subtracting with Symbols of Grouping

$$3 - (11 - 5) = 3 - [11 + (-5)] \qquad \text{Add the opposite.}$$
$$= 3 - 6 \qquad\qquad\quad\ \text{Add.}$$
$$= 3 + (-6) \qquad\qquad \text{Add the opposite.}$$
$$= -3 \qquad\qquad\qquad \text{Add.}$$

Technology

The keys $\boxed{+/-}$ and $\boxed{(-)}$ change a number to its opposite and $\boxed{-}$ is the subtraction key. For instance, the keystrokes $\boxed{-}\,4\,\boxed{-}\,5\,\boxed{\text{ENTER}}$ will not produce the result given in Example 6.

This text includes several examples and exercises that use a calculator. As each new calculator application is encountered, you will be given general instructions for using a calculator. These instructions, however, may not agree precisely with the steps required by *your* calculator, so be sure you are familiar with the use of the keys on your own calculator.

NOTE   For each of the calculator examples in the text, we will give two possible keystroke sequences: one for a standard *scientific* calculator, and one for a *graphing* calculator.

EXAMPLE 6   Evaluating Expressions with a Calculator

To evaluate the expression $-4 - 5$, use the following keystrokes.

Keystrokes	*Display*	
4 $\boxed{+/-}$ $\boxed{-}$ 5 $\boxed{=}$	-9	Scientific
$\boxed{(-)}$ 4 $\boxed{-}$ 5 $\boxed{\text{ENTER}}$	-9	Graphing

Group Activities   You Be the Instructor

$$9 - (3 - 1) = 9 - 3 - 1$$
$$= 9 - 4$$
$$= 5$$

Error Analysis   Imagine that you are teaching algebra. One of your students wrote the steps shown at the left. What is wrong with the procedure? Decide as a group what you could say to help your student avoid this type of error.

P.3 Exercises

Discussing the Concepts

1. Explain why the sum of two negative numbers is a negative number.

2. When you are adding two numbers with opposite signs, how do you determine the sign of the sum?

In Exercises 3 and 4, an addition problem is shown visually on the real number line. (a) Write the addition problem and find the sum. (b) State the rule for the addition of integers demonstrated. (c) Suppose the numbers represent the yards gained in two consecutive downs of a football game. How would the sportscasters announce the plays?

3.

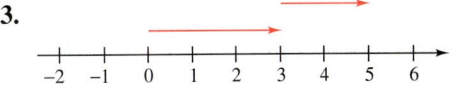

4.

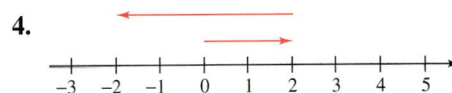

In Exercises 5 and 6, rewrite the subtraction problem as an addition problem and evaluate the result.

5. $8 - (-6)$ **6.** $-25 - 13$

Problem Solving

In Exercises 7–10, evaluate the sum. Then sketch the addition on the real number line.

7. $2 + 7$ **8.** $10 + (-3)$

9. $-6 + 4$ **10.** $(-8) + (-3)$

In Exercises 11–18, find the sum.

11. $-23 + 4$

12. $10 + (-10)$

13. $-10 + 6 + 34$

14. $-15 + (-3) + 8$

15. $32 + (-32) + (-16)$

16. $-312 + (-564) + (-100)$

17. $49 + (-|-17|)$

18. $|-10| + |35|$

In Exercises 19–22, write the subtraction problem as an addition problem and evaluate the result.

19. $12 - 9$ **20.** $4 - (-1)$

21. $-4 - (-4)$ **22.** $9 - (-6)$

In Exercises 23–30, evaluate the expression.

23. $55 - 20$ **24.** $39 - 13$

25. $1000 - (-500)$ **26.** $2500 - (-600)$

27. $-210 - 400$ **28.** $-110 - (-30)$

29. $23 - (15 - 8)$

30. $-32 - [25 - (10 + 8) + 3]$

31. Find the sum of 250 and -300.

32. Find the sum of -40 and -60.

33. Subtract -120 from 380.

34. Find the absolute value of the sum of -35 and 15.

35. *Temperature Change*   The temperature at 6 A.M. was $-10°$F. By noon, the temperature had increased by $22°$F. What was the temperature at noon?

36. *Balance in an Account*   At the beginning of a month, your balance was $2750. During the month you withdrew $350 and $500, deposited $450, and earned an interest of $6.42. What is your balance at the end of the month?

37. *Flying Altitude*   An airliner flying at an altitude of 31,000 feet is instructed to descend to an altitude of 24,000 feet (see figure). How many feet must the aircraft descend?

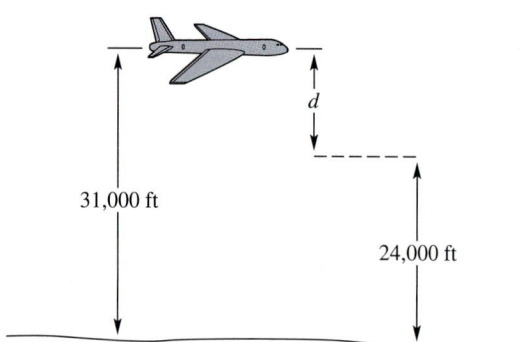

38. *Reading a Graph*   The total student enrollment at a college is shown in the accompanying bar graph. Determine the annual enrollment gain or loss from the previous year.

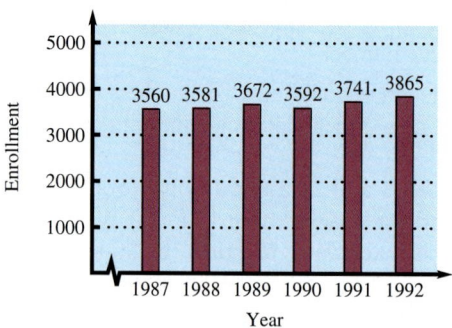

Additional Problem Solving

In Exercises 39–54, find the sum.

39. $14 + (-14)$

40. $-45 + 45$

41. $5 + |-3|$

42. $-|-12| + |-16|$

43. $-18 + (-12)$

44. $-340 + (-160)$

45. $-32 + 16$

46. $-75 + 100$

47. $-82 + (-36) + 82$

48. $150 + (-75) + (-75)$

49. $1200 + 1300 + (-275)$

50. $104 + 203 + 613 + (-214)$

51. $1875 + (-3143) + 5826$

52. $4365 + (-2145) + (-1873) + 40,084$

53. $|-890| + (-|-82|) + 90$

54. $-770 + |492| + (-|-383|)$

In Exercises 55–68, perform the subtraction.

55. $453 - 354$

56. $278 - 574$

57. $-714 - 320$

58. $-84 - 106$

59. $-10 - (-4)$

60. $-804 - (-1408)$

61. $-942 - (-942)$

62. $1043 - (-4831)$

63. $|15| - |-7|$

64. $|-100| - |25|$

65. $53 - (25 - 9)$

66. $-125 - (64 - 100)$

67. $-32 - [18 - (25 + 8)]$

68. $515 - [175 + (330 - 160)]$

In Exercises 69–80, perform the operation.

69. $0 - (-12)$

70. $-36 + 0$

71. $-130 + 130$

72. $60 - (-60)$

73. $72 - 85$

74. $400 - 525$

75. $-12 - 2 + |-3|$

76. $8 - |-7 + 11| + (-4)$

77. $550 + (-1625) + (-4060) + 7132$

78. $-730 + 1820 + 3150 + (-10,000)$

79. $34 - [54 - (-16 + 4) + 6]$

80. $-120 - (-20 + 13) - (25 - 4)$

81. Find the sum of 72 and -37.

82. Find the sum of -25 and -15.

83. Subtract 1500 from 2500.

84. Subtract 600 from 250.

85. Subtract −750 from 800.

86. Subtract 230 from −300.

87. Find the absolute value of the sum of −45 and −80.

88. Find the absolute value of the sum of 17 and −12.

89. What number must be added to 10 to obtain −5?

90. What must you subtract from −12 to obtain 24?

91. *Banking*   You start a non-interest-earning checking account by depositing $500. During the first month, you write checks for $145 and $278. What is your balance at the end of the month?

92. *Profit*   Your company lost $650,000 during the first 6 months of the year. By the end of the year, you had an overall profit of $362,000. What was your profit during the second 6 months of the year?

93. *Reading a Graph*   On Monday you purchased $800 worth of stock. The value of the stock during the remainder of the week is shown in the bar graph. Use the graph to complete the table showing the daily gains and losses during the week.

Day	Daily Gain or Loss
Tuesday	
Wednesday	
Thursday	
Friday	

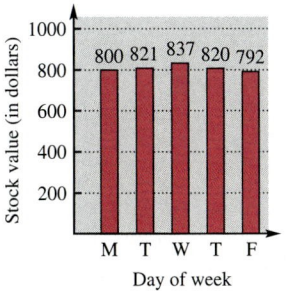

94. *Reading a Graph*   The bar graph gives the consolidated revenues (in billions) of General Electric Company for the years 1989 through 1993. *(Source: General Electric Company 1993 Annual Report)*

(a) Estimate total revenues in 1990.

(b) Estimate international revenues in 1992.

(c) Estimate the increase in United States revenues from 1992 to 1993.

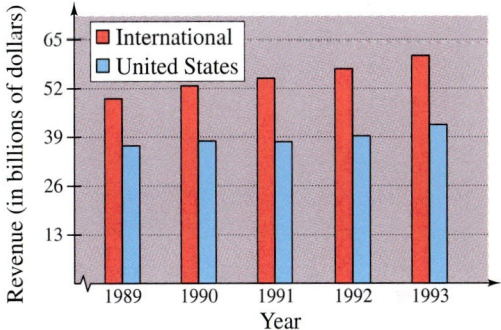

95. *Reading a Graph*   The bar graph gives the new AIDS cases reported in the U.S. *(Source: U.S. Department of Health and Human Services)*

(a) Estimate the number of new cases in 1987.

(b) Estimate the total number of new cases reported in 1990, 1991, and 1992.

(c) Estimate the increase in the number of new cases from 1989 to 1990.

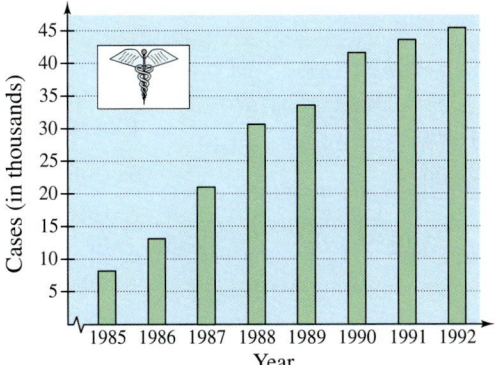

P.4	**Multiplying and Dividing Integers**
	Multiplying Integers ▪ Dividing Integers ▪ Summary of Definitions, Rules, and Properties

Multiplying Integers

Multiplication of two integers can be described as repeated addition or subtraction. Here are two examples.

Multiplication   *Repeated Addition*

$3 \times 5 = 15$ $\underbrace{5 + 5 + 5}_{\text{Add 5 three times.}} = 15$

$4 \times (-2) = -8$ $\underbrace{(-2) + (-2) + (-2) + (-2)}_{\text{Add } -2 \text{ four times.}} = -8$

Multiplication is denoted in a variety of ways. For instance,

$$7 \times 3, \quad 7 \cdot 3, \quad 7(3), \quad (7)3 \quad \text{and} \quad (7)(3)$$

all denote the product of "7 times 3," which is 21.

D I S C O V E R Y

When finding the product of more than two integers, you can determine whether the product is positive or negative by applying a simple rule. Can you discover the rule?

Use your calculator to evaluate the following:

$(-1)(-1)(-1)$,

$(-1)(-1)(-1)(-1)$, and

$(-1)(-1)(-1)(-1)(-1)$.

Describe the pattern. Write a rule for multiplying more than two integers. Without using your calculator, evaluate $(-1)(-1)(-1)$ $(-1)(-1)(-1)(-1)$.

Rules for Multiplying Integers

1. The product of an integer and zero is 0.

2. The product of two integers with *like* signs is *positive*.

3. The product of two integers with *different* signs is *negative*.

EXAMPLE 1   Multiplying Integers

a. $-6 \cdot 9 = -54$ **b.** $(-5)(-7) = 35$

c. $3(-12) = -36$ **d.** $-12 \cdot 0 = 0$

e. To find the product of more than two numbers, first find the product of their absolute values. If there is an even number of negative factors, then the product is positive. If there is an odd number of negative factors, then the product is negative. For instance,

$$5(-3)(-4)(7) = 420.$$

Be careful to properly distinguish between expressions such as $3(-5)$ and $3 - 5$ or $-3(-5)$ and $-3 - 5$. The first of each pair is a multiplication problem, whereas the second is a subtraction problem.

Multiplication	*Subtraction*
$3(-5) = -15$	$3 - 5 = -2$
$-3(-5) = 15$	$-3 - 5 = -8$

To multiply two integers having two or more digits, we suggest the **vertical multiplication algorithm** demonstrated in the next example. The sign of the product is determined by the usual multiplication rule.

EXAMPLE 2   *Using the Vertical Multiplication Algorithm*

Multiply -34 and 78 using the vertical multiplication algorithm.

Solution

Using the vertical algorithm with the absolute values of the factors, you can write the following.

$$
\begin{array}{r}
78 \\
\times \quad 34 \\
\hline
312 \\
234 \quad\;\; \\
\hline
2652 \\
\end{array}
$$

⟵ Multiply 4 times 78.

⟵ Multiply 3 times 78.

⟵ Add columns.

Now, because the factors have unlike signs, it follows that

$$-34 \times 78 = -2652.$$

Dividing Integers

Just as subtraction can be expressed in terms of addition, you can express division in terms of multiplication. Here are some examples.

Division		*Related Multiplication*
$12 \div 4 = 3$	because	$12 = 3 \cdot 4$
$15 \div 3 = 5$	because	$15 = 5 \cdot 3$
$15 \div (-3) = -5$	because	$15 = (-5) \cdot (-3)$
$-15 \div (-3) = 5$	because	$-15 = 5 \cdot (-3)$

The result of dividing one integer by another is called the **quotient** of the integers. Division is denoted by the symbol ÷, or by /, or by a horizontal line. For example,

$$30 \div 6, \quad 30/6, \quad \text{and} \quad \frac{30}{6}$$

each denotes the quotient of 30 and 6, which is 5. Using the form $30 \div 6$, 30 is called the **dividend** and 6 is the **divisor.** In the forms 30/6 and $\frac{30}{6}$, 30 is the **numerator** and 6 is the **denominator.**

It is important to know how to use 0 in a division problem. Zero divided by a nonzero integer is always 0. For instance,

$$\frac{0}{13} = 0 \quad \text{because} \quad 0 = 0 \cdot 13.$$

On the other hand, division by zero is *undefined.*

Because division can be described in terms of multiplication, the rules for dividing two integers with like or unlike signs are the same as those for multiplying such integers.

DISCOVERY

Does $\frac{1}{0} = 0$? $\frac{2}{0} = 0$? Write the division above in terms of multiplication. What does this tell you about division by zero? What does your calculator display when you perform the division?

Rules for Dividing Integers

1. Zero divided by a nonzero integer is 0, whereas a nonzero integer divided by zero is *undefined.*

2. The quotient of two nonzero integers with *like* signs is *positive.*

3. The quotient of two nonzero integers with *different* signs is *negative.*

EXAMPLE 3   Dividing Integers

a. $\dfrac{-42}{-6} = 7$ because $-42 = 7(-6)$.

b. $36 \div (-9) = -4$ because $(-4)(-9) = 36$.

c. $\dfrac{0}{-13} = 0$ because $(0)(-13) = 0$.

d. $-105 \div 7 = -15$ because $(-15)(7) = -105$.

When dividing large numbers, the **long division** algorithm can be used. For instance, the long division algorithm shown in Figure P.13 implies that

$$\frac{351}{13} = 27.$$

```
      27
  13)351
      26
      ──
      91
      91
      ──
```

FIGURE P.13 Long Division Algorithm

All four operations on integers (addition, subtraction, multiplication, and division) are used in the following real-life example.

EXAMPLE 4   An Application: Stock Purchase

On Monday you bought $500 worth of stock in a company. During the rest of that week, you recorded the following gains and losses in your stock's values.

Tuesday	Wednesday	Thursday	Friday
Gained $15	Lost $18	Lost $23	Gained $10

a. What was the value of the stock at the close of Tuesday?

b. What was the value of the stock at the close of Wednesday?

c. What was the value of the stock at the end of the week?

d. What would the total loss have been if Thursday's loss had occurred each of the four days?

e. What was the average daily gain (or loss) for the four days recorded?

Solution

a. Because the original value of the stock was $500, and the stock gained $15 by the close of Tuesday, its value at the close of Tuesday was

$$500 + 15 = \$515.$$

b. Using the result of part (a), the value at the close of Wednesday was

$$515 - 18 = \$497.$$

c. The value of the stock at the end of the week was

$$500 + 15 - 18 - 23 + 10 = \$484.$$

d. The loss on Thursday was $23. If this loss had occurred each day, the total loss would have been

$$4(23) = \$92.$$

e. To find the average of the four gains and losses, we add and divide by 4. Thus, the average is

$$\text{Average} = \frac{15 + (-18) + (-23) + 10}{4} = \frac{-16}{4} = -4.$$

This means that during the four days, the stock had an average loss of $4 per day.

Summary of Definitions, Rules, and Properties

So far in this chapter, we have described rules and procedures more with words than with symbols. For instance, subtraction is verbally defined as "adding the opposite of the number being subtracted." As you move to higher and higher levels of mathematics, it becomes more and more convenient to use symbols to describe rules and procedures. For instance, subtraction is symbolically defined as $a - b = a + (-b)$.

At its simplest level, algebra is a symbolic form of arithmetic. This arithmetic–algebra connection can be illustrated in the following way.

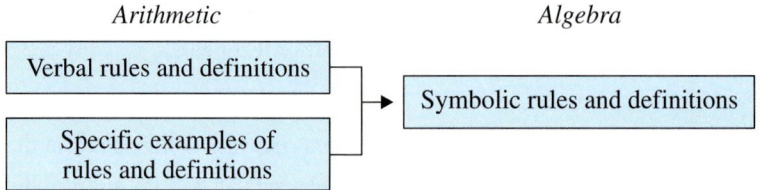

An illustration of this connection is the **Commutative Property of Addition**.

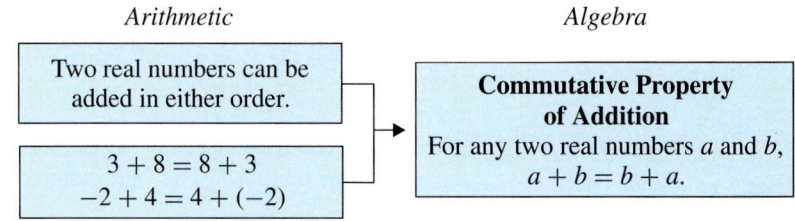

EXAMPLE 5 *Writing a Rule of Arithmetic in Symbolic Form*

Write an example and an algebraic description of the arithmetic rule: *The product of two integers with unlike signs is negative.*

Solution

Example	*Algebraic Description*
For integers 3 and 7,	If a and b are positive integers, then

$$(-3) \cdot 7 = 3 \cdot (-7)$$
$$= -(3 \cdot 7)$$
$$= -21.$$

$$(-a) \cdot b = a \cdot (-b) = -(a \cdot b).$$

Unlike signs   Unlike signs   Negative product

The following list summarizes the algebraic versions of important definitions and rules of arithmetic and properties of real numbers. In each case a specific example is included for clarification.

Arithmetic Summary

Definitions: Let a, b, and c be integers.

| *Definition* | *Example* |

1. Subtraction:

$$a - b = a + (-b)$$

$$5 - 7 = 5 + (-7)$$

2. Multiplication: (a is a positive integer)

$$a \cdot b = \underbrace{b + b + \cdots + b}_{a \text{ terms}}$$

$$3 \cdot 5 = 5 + 5 + 5$$

3. Division: ($b \neq 0$)

$$a \div b = c, \text{ if and only if } a = c \cdot b.$$

$$12 \div 4 = 3 \text{ because } 12 = 3 \cdot 4$$

4. Less than:

$a < b$ if there is a positive real number c such that $a + c = b$.

$-2 < 1$ because $-2 + 3 = 1$

5. Absolute value: $|a| = \begin{cases} a, & \text{if } a \geq 0 \\ -a, & \text{if } a < 0 \end{cases}$

$$|-3| = -(-3) = 3$$

6. Divisor:

a is a divisor of b if and only if there is an integer c such that $a \cdot c = b$.

7 is a divisor of 21 because $7 \cdot 3 = 21$

Rules: Let a and b be integers.

| *Rule* | *Example* |

1. Addition:

(a) If a and b have *like* signs, evaluate $|a| + |b|$ and attach the common sign to the result.

$$3 + 7 = |3| + |7| = 10$$

(b) If a and b have *different* signs, evaluate which difference, $|a| - |b|$ or $|b| - |a|$, is positive and attach the sign of the integer with the larger absolute value.

$$-5 + 8 = |8| - |-5|$$
$$= 8 - 5$$
$$= 3$$

2. Multiplication:

(a) $a \cdot 0 = 0 = 0 \cdot a$

$$3 \cdot 0 = 0 = 0 \cdot 3$$

(b) Like signs: $a \cdot b > 0$

$$(-2)(-5) = 10$$

(c) Different signs: $a \cdot b < 0$

$$(2)(-5) = -10$$

3. Division:

Division by zero is undefined.

Properties of Real Numbers: Let a, b, and c be integers.

Property   Example

1. Commutative Property of Addition:
   Two real numbers can be added in either order.

   $$a + b = b + a$$ $3 + 5 = 5 + 3$

2. Commutative Property of Multiplication:
   Two real numbers can be multiplied in either order.

   $$ab = ba$$ $4 \cdot (-7) = -7 \cdot 4$

3. Associative Property of Addition:
   When three real numbers are added, it makes no difference which two are added first.

   $$(a + b) + c = a + (b + c)$$ $(2 + 6) + 5 = 2 + (6 + 5)$

4. Associative Property of Multiplication:
   When three real numbers are multiplied, it makes no difference which two are multiplied first.

   $$(ab)c = a(bc)$$ $(3 \cdot 5) \cdot 2 = 3 \cdot (5 \cdot 2)$

5. Distributive Property:
   Multiplication distributes over addition.

   $$a(b + c) = ab + ac$$ $3(8 + 5) = 3 \cdot 8 + 3 \cdot 5$
   $$(b + c)a = ba + ca$$ $(8 + 5)3 = 8 \cdot 3 + 5 \cdot 3$

6. Additive Identity Property:
   The sum of zero and a real number equals the number itself.

   $$a + 0 = 0 + a = a$$ $3 + 0 = 0 + 3 = 3$

7. Multiplicative Identity Property:
   The product of one and a real number equals the number itself.

   $$a \cdot 1 = 1 \cdot a = a$$ $4 \cdot 1 = 1 \cdot 4 = 4$

8. Additive Inverse Property:
   The sum of a real number and its opposite is zero.

   $$a + (-a) = 0$$ $3 + (-3) = 0$

9. Multiplicative Inverse Property:
   The product of a nonzero real number and its reciprocal is one.

   $$a \cdot \frac{1}{a} = 1, a \neq 0$$ $8 \cdot \dfrac{1}{8} = 1$

EXAMPLE 6   *Identifying Rules and Properties*

a. Use the definition of multiplication to complete the following.

$$6 + 6 + 6 + 6 = \boxed{}$$

b. Use the rule for adding integers with unlike signs to complete the following.

$$-7 + 3 = \boxed{}$$

c. Name the property that justifies the statement $3(-5) = (-5)(3)$.

d. Name the property that justifies the statement $5(7 + 2) = 5(7) + 5(2)$.

e. Name the property that justifies the statement $9 + (-9) = 0$.

Solution

a. $6 + 6 + 6 + 6 = 4 \cdot 6$

b. $-7 + 3 = -(|-7| - |3|) = -4$

c. Commutative Property of Multiplication

d. Distributive Property

e. Additive Inverse Property

Group Activities   Exploring with Technology

Finding a Pattern   Complete the patterns below. As a group, decide which rules the patterns demonstrate. Use a calculator to confirm your answers.

$3 \cdot (3) = 9$	$-3 \cdot (3) = -9$
$3 \cdot (2) = 6$	$-3 \cdot (2) = -6$
$3 \cdot (1) = 3$	$-3 \cdot (1) = -3$
$3 \cdot (0) = 0$	$-3 \cdot (0) = 0$
$3 \cdot (-1) = \boxed{?}$	$-3 \cdot (-1) = \boxed{?}$
$3 \cdot (-2) = \boxed{?}$	$-3 \cdot (-2) = \boxed{?}$
$3 \cdot (-3) = \boxed{?}$	$-3 \cdot (-3) = \boxed{?}$

P.4 Exercises

Discussing the Concepts

1. Give a verbal description of "$3(-5)$."

2. In your own words, state the rules for determining the sign of the product or quotient of two numbers.

3. Explain why the product of an even integer and any other integer is even. What can you conclude about the product of two odd integers?

4. Explain how to check the result of a division problem.

5. An integer n is divided by 2 and the quotient is an even integer. What does this tell you about n? Give some examples.

6. Which of the following is (are) undefined: $\frac{1}{1}$, $\frac{0}{1}$, $\frac{1}{0}$, $\frac{0}{0}$?

Problem Solving

In Exercises 7 and 8, write each multiplication as repeated addition and find the product.

7. $3 \cdot 2$

8. $5 \times (-2)$

In Exercises 9–16, find the product.

9. 7×30

10. $0 \cdot 20$

11. $4(-8)$

12. $(-20)(-8)$

13. $5(-3)(-6)$

14. $7(3)(-1)$

15. $|-3|(-3)(4)$

16. $|8(-9)|$

In Exercises 17 and 18, use the vertical multiplication algorithm to find the product.

17. $75(-632)$

18. $(-136)(-2012)$

In Exercises 19–26, perform the indicated division, if possible. If not possible, state the reason.

19. $27 \div 9$

20. $-35 \div -5$

21. $72 \div (-12)$

22. $0 \div 8$

23. $\frac{8}{0}$

24. $\frac{-125}{-25}$

25. $\frac{-180}{-45}$

26. $\left| \frac{52}{-4} \right|$

In Exercises 27–30, use the long division algorithm to find the quotient.

27. $1440 \div -45$

28. $-1312 \div -16$

29. $\frac{2209}{47}$

30. $\left| \frac{-8700}{116} \right|$

In Exercises 31–34, evaluate the expression.

31. $\dfrac{5 - 3 + 16}{10 - 4}$

32. $\dfrac{40 - (12 - 36)}{8}$

33. $\dfrac{8(-45)}{18}$

34. $\dfrac{|14(-9)|}{-21}$

35. *Temperature Change*   The temperature measured by a weather balloon is decreasing approximately $3°$ for each 1000-foot increase in altitude. The balloon rises 8000 feet. Describe its temperature change.

36. *Stock Prices*   The Dow Jones average loses 11 points on each of 4 consecutive days. What is the cumulative loss during the 4 days?

37. *Exam Scores*   A student has a total of 328 points after four 100-point exams.

(a) What is the average number of points scored per exam?

(b) The scores on the four exams are 87, 73, 77, and 91. Plot each of the scores and the average score on the real number line.

(c) Find the difference between each score and the average score. Find the sum of these differences and give a possible explanation of the result.

38. *Travel Time*   A commuter train travels a distance of 195 miles between two cities in 3 hours. What is the average speed of the train?

Additional Problem Solving

In Exercises 39–42, write each multiplication as repeated addition and find the product.

39. 4×5 **40.** $2(6)$

41. $5 \times (-3)$ **42.** $6(-2)$

In Exercises 43–54, find the product.

43. $(-6)(-12)$ **44.** $(8)(8)$

45. $(-6)(8)$ **46.** $10(-25)$

47. $(310)(-32)$ **48.** $(-500)(-6)$

49. $(-2)(-3)(-5)$ **50.** $(-10)(-4)(-2)$

51. $|3(-5)(6)|$ **52.** $|6(20)(4)|$

53. $(-3)(4)(0)(6)(8)$ **54.** $(-2)(-1)(-3)(-5)$

In Exercises 55–58, use the vertical multiplication algorithm to find the product.

55. 26×130 **56.** $(-14) \times 2423$

57. $(-72)(866)$ **58.** $(-145)(-585)$

In Exercises 59–70, perform the indicated division, if possible. If not possible, state the reason.

59. $54 \div 9$ **60.** $(-28) \div 4$

61. $\dfrac{-81}{-3}$ **62.** $\dfrac{72}{12}$

63. $\dfrac{-58}{2}$ **64.** $\dfrac{48}{4}$

65. $\dfrac{0}{81}$ **66.** $\dfrac{32}{0}$

67. $\dfrac{6}{-1}$ **68.** $\dfrac{-12}{1}$

69. $144 \div 0$ **70.** $(-27) \div (-27)$

In Exercises 71–76, use the long division algorithm to find the quotient.

71. $-1248 \div 48$ **72.** $-1312 \div (-16)$

73. $\dfrac{2209}{47}$ **74.** $\left|\dfrac{-8700}{116}\right|$

75. $2750 \div 25$ **76.** $22{,}010 \div 71$

In Exercises 77–82, evaluate the expression.

77. $\dfrac{25 - 10}{-2 - 3}$ **78.** $\dfrac{|7 - 15|}{2}$

79. $\dfrac{|-200|}{10(-4)}$ **80.** $\dfrac{72}{2 \cdot 4 \cdot 3}$

81. $\dfrac{8 + 3 \cdot 4}{5}$ **82.** $\dfrac{|-3| \cdot |8| - (-4)}{4}$

In Exercises 83–88, use a calculator to evaluate the expression.

83. $5(1650) - 3710$ **84.** $332(516) + 467(-125)$

85. $\dfrac{44{,}290}{515}$ **86.** $\dfrac{33{,}511}{47}$

87. $\dfrac{169{,}290}{162}$ **88.** $\dfrac{1{,}027{,}500}{250}$

Mental Math In Exercises 89 and 90, find the product mentally. Explain your strategy.

89. $(-2)(532)(500)$ **90.** $72(8)(25)$

91. *Savings Plan* After you decide to save $50 per month for 10 years, what is the total amount you will have deposited?

92. *Total Cost* Computer printer ribbons cost $8 per ribbon. Each box has ten ribbons. How much will six boxes cost?

93. *Loss Leaders* To attract customers, a grocery store runs a sale on bananas. The bananas are *loss leaders*, which means the store loses money on the bananas but hopes to make it up on other items. The store sells 800 pounds at a loss of 26 cents per pound. What is the total loss?

94. *Geometry* Find the area of a football field.

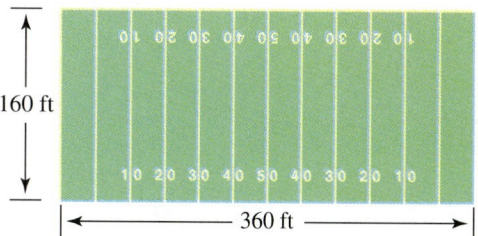

95. *Geometry*   Find the area of the floor of the building.

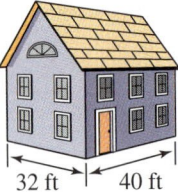

32 ft   40 ft

Geometry   In Exercises 96 and 97, find the volume of the prism. The volume of a rectangular prism is found by multiplying the length, width, and height of the prism.

96.

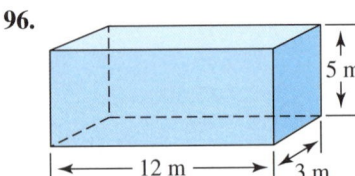

5 m

12 m   3 m

97.

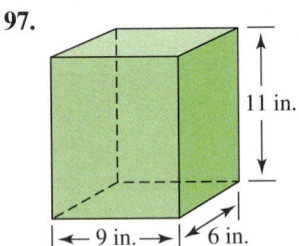

11 in.

9 in.   6 in.

98. *Geometry*   The rectangular prism shown below has a volume of 4095 cubic feet. What is the height of the prism?

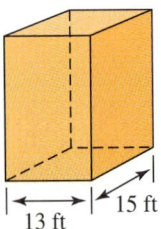

13 ft   15 ft

99. *Temperature*   The temperature fell 28° in 4 hours. Find the average rate of change per hour.

100. *Average Speed*   A family on vacation traveled 504 miles in 9 hours. What was their average speed?

101. A nonzero product has 25 factors, 17 of which are negative. What is the sign of the product?

102. A nonzero product has 25 factors, 16 of which are negative. What is the sign of the product?

103. A number is divisible by 11 if the difference of the sum of the alternate digits and the sum of the remaining digits is divisible by 11. For example, 909,556,857 is divisible by 11 because

$$(9 + 9 + 5 + 8 + 7) - (0 + 5 + 6 + 5) = 22$$

is divisible by 11. Use a calculator to verify this result.

104. Use the algorithm in Exercise 103 to find a six-digit number of the form 205,?92 that is divisible by 11.

CHAPTER PROJECT: Population Growth Patterns

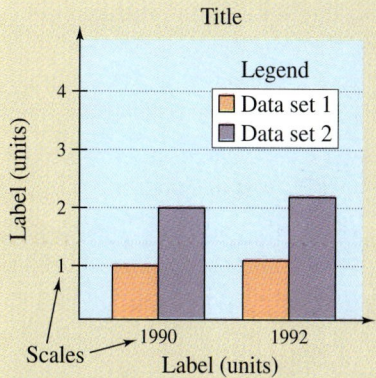

FIGURE A

Graphs are a form of communication. For the reader to interpret the data represented in a graph, the graph must contain all the information needed to make it understandable.

A standard bar graph, such as that in Figure A, consists of a labeled vertical axis with a scale, a labeled horizontal axis with a scale, and a title. When more than one set of data is being represented on a single graph, a legend is needed to distinguish between the different sets.

The double bar graph in Figure B compares the populations of California and New York. Notice that each state is represented by a different color.

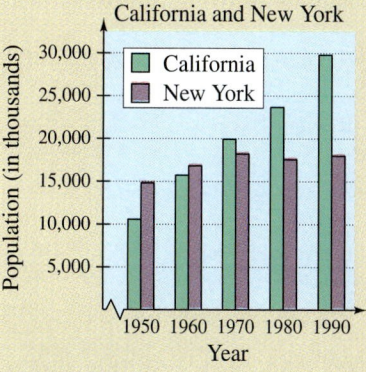

FIGURE B

Populations (in thousands)

State	1950	1960	1970	1980	1990
California	10,586	15,717	19,971	23,668	29,760
New York	14,830	16,782	18,241	17,558	17,990
Texas	7,711	9,580	11,199	14,229	16,987
Florida	2,771	4,952	6,791	9,746	12,938
Pennsylvania	10,498	11,319	11,801	11,864	11,882

(Source: U.S. Bureau of the Census)

Use the table and graphs to investigate the following questions.

1. Look at Figure B. Between which two years did the population of California become greater than the population of New York? Explain how you can determine this from the graph.

2. Your partner on a project has started constructing the bar graph shown in Figure C. You must complete the graph. You know only that the graph is to compare the populations of two of the five states from the table. Which two states is the graph comparing? Explain how to determine which states and describe what needs to be done to complete the graph.

3. Construct a bar graph that compares the populations of all five states in the table.

4. *Research Project* Find a table in a newspaper or magazine that compares two or more sets of data. Construct a bar graph to represent the data. Keep a log describing the steps and the decisions you made in creating the bar graph.

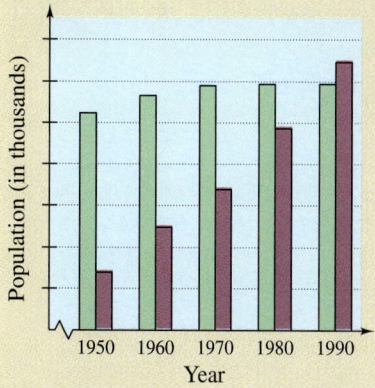

FIGURE C

CHAPTER SUMMARY

After studying this chapter, you should have acquired the following skills. These skills are keyed to the Review Exercises that begin on page 39. Answers to odd-numbered Review Exercises are given in the back of the book.

- Compare two real numbers and plot them on the real number line. *(Section P.1)*   **Review Exercises 1–4**

```
  +----+----+----+----+----+----+----+----+--->
 -4   -3   -2   -1    0    1    2    3    4
```

- Find the opposite of a number. *(Section P.1)*   **Review Exercises 5–8**

- Evaluate expressions with absolute value. *(Section P.1)*   **Review Exercises 9–12**

- Compare two real numbers with absolute value. *(Section P.1)*   **Review Exercises 13–16**

- Decide whether a number is prime or composite. *(Section P.2)*   **Review Exercises 17–20**

- Write the prime factorization of a number. *(Section P.2)*   **Review Exercises 21–24**

- Find the greatest common factor. *(Section P.2)*   **Review Exercises 25–28**

- Find the least common multiple. *(Section P.2)*   **Review Exercises 29–32**

- Evaluate sums, differences, products, and quotients. *(Sections P.3, P.4)*   **Review Exercises 33–60**

- Use the long division algorithm. *(Section P.4)*   **Review Exercises 61, 62**

- Use a calculator to evaluate expressions. *(Sections P.3, P.4)*   **Review Exercises 63–66**

- Translate verbal sentences into arithmetic expressions and solve. *(Sections P.3, P.4)*   **Review Exercises 67–70, 73–78**

- Order real-life data. *(Section P.1)*   **Review Exercises 71, 72**

Study Tips If you haven't yet read the study tips entitled "How to Study Algebra" on pages xxvi–xxviii, we suggest you do so now. The suggestions on those pages have proven to be helpful to many students in beginning algebra courses.

REVIEW EXERCISES

In Exercises 1–4, plot each real number as a point on the real number line and place the correct symbol ($<$, $>$, or $=$) between the real numbers.

1. $-\frac{1}{10}$ ___ 4 **2.** $\frac{25}{3}$ ___ $\frac{5}{3}$

3. -3 ___ -7 **4.** 10.6 ___ -3.5

In Exercises 5–8, find the opposite of the number.

5. 152 **6.** -10.4

7. $-\frac{7}{3}$ **8.** $\frac{2}{3}$

In Exercises 9–12, evaluate the expression.

9. $|-8.5|$ **10.** $|3.4|$

11. $-|-8.5|$ **12.** $|-9.6|$

In Exercises 13–16, place the correct symbol ($<$, $>$, or $=$) between the two real numbers.

13. $|-84|$ ___ $|84|$ **14.** $|-10|$ ___ $|4|$

15. $\left|\frac{3}{10}\right|$ ___ $-\left|\frac{4}{5}\right|$ **16.** $|2.3|$ ___ $-|2.3|$

In Exercises 17–20, decide whether the number is prime or composite.

17. 839 **18.** 909

19. 1764 **20.** 1847

In Exercises 21–24, write the prime factorization of the number.

21. 378 **22.** 858

23. 1612 **24.** 1787

In Exercises 25–28, find the greatest common factor.

25. $54, 90$ **26.** $154, 220$

27. $63, 84, 441$ **28.** $99, 132, 253$

In Exercises 29–32, find the least common multiple.

29. $16, 18$ **30.** $15, 45$

31. $14, 20, 24$ **32.** $10, 16, 40$

In Exercises 33–60, evaluate the expression, if possible. If it is not possible, state the reason.

33. $32 + 68$ **34.** $14 + 54$

35. $16 + (-5)$ **36.** $-125 + 30$

37. $350 - 125 + 15$ **38.** $35 - 25 - 10$

39. $-114 + 76 - 230$ **40.** $-448 - 322 + 100$

41. $|-86| - |124|$ **42.** $67 + |-53|$

43. $122 - [45 - (32 + 8) - 23]$

44. $-58 - (48 - 12) - (-30 - 4)$

45. 15×3 **46.** -22×4

47. $-300(-5)$ **48.** $18(3200)$

49. $131(-6)(3)$ **50.** $(-46)(-5)(-2)$

51. $\dfrac{-162}{9}$ **52.** $\dfrac{-52}{-4}$

53. $815 \div 0$ **54.** $-48 \div 6$

55. $\dfrac{78 - |-78|}{5}$ **56.** $\dfrac{144}{2 \cdot 3 \cdot 3}$

57. $\dfrac{54 - 4 \cdot 3}{6}$ **58.** $\dfrac{3 \cdot 5 + 125}{10}$

59. $\dfrac{6 \cdot 4 - 36}{4}$ **60.** $\dfrac{300}{15 - |-15|}$

In Exercises 61 and 62, use the long division algorithm to find the quotient.

61. $33,768 \div (-72)$ **62.** $-144,512 \div (-32)$

In Exercises 63–66, use a calculator to perform the operations.

63. $7(5207) - 52,318$

64. $783(1995) + 75(-832)$

65. $\dfrac{345,582}{438}$

66. $\dfrac{1,111,521}{89}$

67. Subtract -549 from 613.

68. Find the absolute value of the sum of 693 and -420.

69. What must you add to 75 to obtain -27?

70. What must you subtract from -83 to obtain 43?

71. *Reading a Table* The ten NFL coaches with the most wins in regular season games through 1990 are listed in the table. Order the coaches in decreasing order according to the number of wins.

Coach	Wins	Coach	Wins
Paul Brown	166	Tom Landry	250
Bud Grant	158	Chuck Noll	177
George Halas	319	Steve Owen	151
Chuck Knox	155	Don Shula	269
Earl Lambeau	226	Hank Stram	131

72. *Reading a Table* The lowest temperatures ever recorded in various cities are listed below. Order the cities in increasing order by their lowest temperatures.

City	Low	City	Low
Baltimore	$-7°$	Honolulu	$53°$
Cincinnati	$-25°$	Indianapolis	$-22°$
Columbia	$-1°$	Jackson	$2°$
Great Falls	$-43°$	Phoenix	$17°$
Hartford	$-26°$	Seattle	$0°$

73. *Unreadable Receipt* A manufacturing company purchased components from a supplier at a cost of $9 each. The total on the receipt read $19,?02. (The digit represented by ? was unreadable.)

(a) Determine the missing digit and give a reason for your answer.

(b) How many components were purchased?

74. *Total Cost* You have purchased a television set. You make a down payment of $75.00, plus nine monthly payments of $25.00 each. What is the total amount you will pay for the product?

75. *Reading a Table* The costs of adult and student tickets for a concert are $25 and $10, respectively. The following table gives the number of tickets sold for the first four days of sales.

Day	1	2	3	4
Adult	162	98	148	186
Student	98	64	81	105

(a) Find the revenue from ticket sales for each day.

(b) Find the revenue from ticket sales for each type of ticket.

(c) Find the total revenue from ticket sales. Explain how you can obtain this result from parts (a) and (b). How could you use this to check your work?

76. *Reading a Table* The table records the number of miles driven by two delivery vans for one 5-day work week. The fuel efficiencies of Van #1 and Van #2 are 12 miles per gallon and 16 miles per gallon, respectively.

Day	1	2	3	4	5
Van #1	212	153	84	186	48
Van #2	178	310	145	162	235

(a) Determine the total number of miles each van was driven during the week.

(b) Calculate the number of gallons of gasoline used by each van during the week.

(c) Find the cost of the fuel for each van if the price of fuel is $1.189.

77. *Think About It* You rotate the tires on your car, including the spare, so that all five tires are used equally. After 40,000 miles, how many miles has each tire been used?

78. *Think About It* A small business began in 1989 and the revenue from sales for that year was $175,000. For each of the years through 1994, revenues increased by approximately $5000. Find the total revenue for the years 1989 through 1994.

79. *Think About It*   Which is smaller: $\frac{2}{3}$ or 0.6?

80. *Think About It*   An integer n is divisible by 3 and the quotient is also divisible by 3. What does this tell you about n? Give some examples.

81. *True or False?*   The sum of two integers, one negative and one positive, is negative. Explain.

82. *True or False?*   The product of two integers, one negative and one positive, is negative. Explain.

83. *Reading a Graph*   The bar graph shows the popular votes (in millions) cast for president in the presidential elections from 1972 through 1992. (Source: U.S. Bureau of the Census)

   (a) Estimate the total popular votes cast for the candidates in 1980.

   (b) Estimate the difference between the votes cast for the winning candidate and the independent candidate in 1972.

   (c) Estimate the total popular votes cast for each of the elections. Did the number of votes cast increase or decrease with time?

   (d) Describe how you can use the graph to determine whether the winning candidate received more than one-half of the popular vote. Did this always occur for the elections shown on the graph?

In the six presidential elections from 1972 to 1992, the Democratic party won twice and the Republican party won four times.

84. *Reading a Graph*   The bar graph shows the per capita food consumption (in pounds) of selected meats for the years 1970, 1980, and 1991. (Source: U.S. Bureau of the Census)

   (a) Estimate the change in consumption of beef from 1970 through 1991.

   (b) Estimate the change in consumption of chicken from 1970 through 1991.

   (c) Estimate the total per capita consumption of the four kinds of meat in 1991.

   (d) Estimate the total number of pounds of chicken consumed in 1980 if the population of the United States was approximately 226 million.

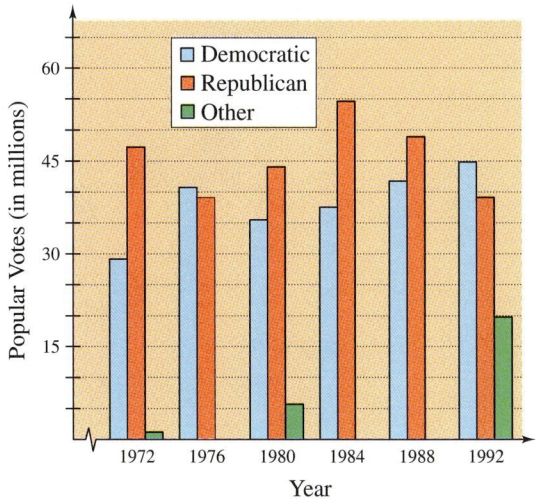

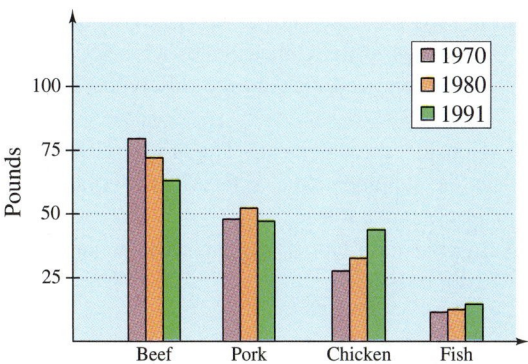

CHAPTER TEST

Take this test as you would take a test in class. After you are done, check your work against the answers given in the back of the book.

1. Which of the following are (a) natural numbers? (b) integers? (c) rational numbers?

$$-10, \ 8, \ \frac{3}{4}, \ \frac{12}{4}, \ 6.5$$

2. Place the correct inequality symbol ($<$ or $>$) between the real numbers:

$-\frac{3}{5}$ ⬚ $-|-2|$.

3. Find all real numbers whose distance from 10 is 18.

4. Write the prime factorization of 234.

5. Find the greatest common factor of 90 and 150.

6. Find the least common multiple of 56 and 84.

In Exercises 7–14, evaluate the expression.

7. $16 + (-20)$ **8.** $-50 - (-60)$ **9.** $7 + |-3|$ **10.** $64 - (25 - 8)$

11. $-5(32)$ **12.** $\dfrac{-72}{-9}$ **13.** $\dfrac{12 + 9}{7}$ **14.** $-\dfrac{(-2)(5)}{10}$

15. A company's quarterly profits are shown in the bar graph at the right. What is the company's total profit for the year?

16. A cord of wood is a pile 8 feet long, 4 feet wide, and 4 feet high. The volume of a rectangular solid is its length times its width times its height. Find the number of cubic feet in a cord of wood.

17. It is necessary to cut a 90-foot rope into six pieces of equal length. What is the length of each piece?

18. In 1989 the expenditures for electricity and natural gas in commercial buildings in the United States were $55.9 million and $9.2 million, respectively. Calculate the difference in cost for these two energy sources.

19. Consider the statement, "The sum of two negative integers is positive." Is the statement true or false? If it is false, suggest any change that would make it true.

20. In your own words, explain why the sum of two odd integers is an even integer.

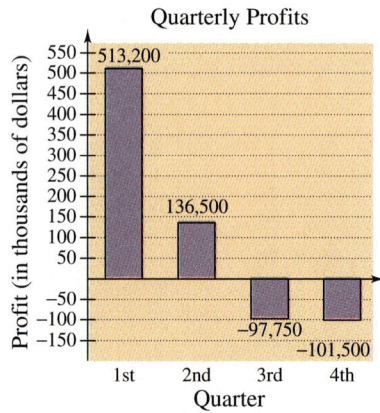

Figure for 15

The Real Number System

- Adding and Subtracting Fractions
- Multiplying and Dividing Fractions
- Exponents and Order of Operations
- Algebra and Problem Solving

Whether your favorite music is rock, country, or classical, you enjoy listening to it because it "sounds" good. But why does it sound good? What is music's appeal?

Note	A	B	C	D	E	F	G
Frequency	220	246.96	261.63	293.66	329.63	349.23	392

All sound is produced by vibrations. For instance, when the strings of a piano vibrate, they produce sound waves, which you interpret as sound. Sound waves can be described by their frequencies. To produce identical musical sounds, each instrument and voice must emit sound waves that have the same frequency.

The partial piano keyboard at the right shows several notes contained in one octave. The frequency of each note is listed in the table. Notice that as you move to the right on the keyboard, the frequency of the notes increases.

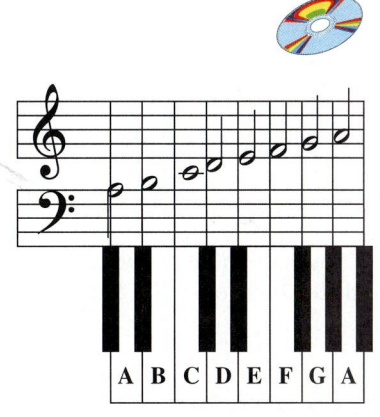

The chapter project related to this information is on page 86.

43

1.1	## Adding and Subtracting Fractions
	Rewriting Fractions ▪ Adding Fractions ▪ Subtracting Fractions

Rewriting Fractions

A **fraction** is a number that is written as a quotient, with a *numerator* and *denominator*. Some examples are

$$\frac{3}{8}, \frac{-12}{5}, \frac{1}{7}, \text{ and } \frac{3}{1}.$$

NOTE The words *fraction* and *rational number* are related, but are not exactly the same. The word *fraction* refers to a number's form, whereas the word *rational number* refers to a number's essence. For instance, the number 2 is a rational number regardless of how it is written. When this number is written as $\frac{2}{1}$, it is called a fraction.

Rule of Signs for Fractions

1. If the numerator and denominator of a fraction have *like* signs, the value of the fraction is *positive*.

2. If the numerator and denominator of a fraction have *unlike* signs, the value of the fraction is *negative*.

EXAMPLE 1 *Positive and Negative Fractions*

a. All of the following fractions are positive and are equivalent to $\frac{2}{3}$.

$$\frac{2}{3}, \frac{-2}{-3}, -\frac{-2}{3}, -\frac{2}{-3}$$

b. All of the following fractions are negative and are equivalent to $-\frac{2}{3}$.

$$-\frac{2}{3}, \frac{-2}{3}, \frac{2}{-3}, -\frac{-2}{-3}$$

In both arithmetic and algebra, it is often beneficial to write a fraction in **reduced form** or **lowest terms,** which means that the numerator and denominator have no common factors (other than 1).

Writing a Fraction in Reduced Form

To reduce a fraction to lowest terms, divide both the numerator and denominator by their greatest common factor (GCF).

You can obtain an **equivalent fraction** by multiplying the numerator and denominator by the same nonzero number or by dividing the numerator and denominator by the same nonzero number. Here are some examples.

Fraction	Equivalent Fraction	Operation
$\dfrac{9}{12} = \dfrac{\cancel{3} \cdot 3}{\cancel{3} \cdot 4}$	$\dfrac{3}{4}$	Divide numerator and denominator by 3.
$\dfrac{6}{5} = \dfrac{6 \cdot 2}{5 \cdot 2}$	$\dfrac{12}{10}$	Multiply numerator and denominator by 2.
$\dfrac{-8}{12} = -\dfrac{\cancel{2} \cdot \cancel{2} \cdot 2}{\cancel{2} \cdot \cancel{2} \cdot 3}$	$-\dfrac{2}{3}$	Divide numerator and denominator by 4.

EXAMPLE 2   Writing Fractions in Reduced Form

Write each fraction in reduced form.

a. $\dfrac{18}{24}$ **b.** $\dfrac{35}{21}$ **c.** $\dfrac{24}{72}$

Solution

a. $\dfrac{18}{24} = \dfrac{\cancel{2} \cdot \cancel{3} \cdot 3}{2 \cdot 2 \cdot \cancel{2} \cdot \cancel{3}} = \dfrac{3}{4}$ Divide out GCF of 6.

b. $\dfrac{35}{21} = \dfrac{5 \cdot \cancel{7}}{3 \cdot \cancel{7}} = \dfrac{5}{3}$ Divide out GCF of 7.

c. $\dfrac{24}{72} = \dfrac{\cancel{2} \cdot \cancel{2} \cdot \cancel{2} \cdot \cancel{3}}{\cancel{2} \cdot \cancel{2} \cdot \cancel{2} \cdot \cancel{3} \cdot 3} = \dfrac{1}{3}$ Divide out GCF of 24.

EXAMPLE 3   Writing Equivalent Fractions

Write an equivalent fraction with the indicated denominator.

a. $\dfrac{2}{3} = \dfrac{}{15}$ **b.** $\dfrac{4}{7} = \dfrac{}{42}$

Solution

a. $\dfrac{2}{3} = \dfrac{2 \cdot 5}{3 \cdot 5} = \dfrac{10}{15}$ Multiply numerator and denominator by 5.

b. $\dfrac{4}{7} = \dfrac{4 \cdot 6}{7 \cdot 6} = \dfrac{24}{42}$ Multiply numerator and denominator by 6.

Adding Fractions

Suppose a pizza is cut into 12 equal-sized pieces, as shown in Figure 1.1. If you eat 3 pieces and your friend eats 4 pieces, what portion of the pizza has been eaten? Because 7 of 12 pieces were eaten, you can reason that $\frac{7}{12}$ of the pizza was eaten. In other words, to find the portion that was eaten, you simply add the fractions $\frac{3}{12}$ and $\frac{4}{12}$ to get

$$\frac{3}{12} + \frac{4}{12} = \frac{3+4}{12} = \frac{7}{12}.$$

Now suppose that you ate $\frac{1}{4}$ of the pizza and your friend ate $\frac{1}{3}$ of it. In this case, what portion of the pizza was eaten? This seems to be a more difficult problem. But why? Isn't it because in the first instance, the fractions $\frac{3}{12}$ and $\frac{4}{12}$ have *like* denominators, whereas in the second instance, the fractions $\frac{1}{4}$ and $\frac{1}{3}$ have *unlike* denominators?

To add two fractions with unlike denominators, first rewrite the two fractions so that they have a common denominator. Once this is done, you can add the fractions by simply adding their numerators and writing the sum over the common denominator.

Let's see how this works with $\frac{1}{4}$ and $\frac{1}{3}$. First, find the least common multiple of both 3 and 4. In this case, the least common multiple is 12. Then, use this number as the common denominator for rewriting the given fractions. To write the fraction $\frac{1}{3}$ so that it has a denominator of 12, multiply both the numerator and the denominator by 4.

$$\frac{1}{3} = \frac{1(4)}{3(4)} = \frac{4}{12}$$

Similarly, to write the fraction $\frac{1}{4}$ so that it has a denominator of 12, multiply its numerator and denominator by 3.

$$\frac{1}{4} = \frac{1(3)}{4(3)} = \frac{3}{12}$$

Because the two fractions now have like denominators, you can add them as follows.

$$\frac{1}{4} + \frac{1}{3} = \frac{3}{12} + \frac{4}{12} = \frac{3+4}{12} = \frac{7}{12}$$

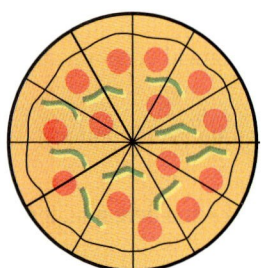

FIGURE 1.1

NOTE Adding fractions with unlike denominators is an example of a basic problem-solving strategy that is used in mathematics—rewriting a given problem in a simpler or more familiar form.

Addition of Fractions

1. To add two fractions *with like denominators,* add their numerators and write the sum over the like (or common) denominator.

2. To add two fractions *with unlike denominators,* rewrite both fractions so that they have like denominators. Then use the rule for adding fractions with like denominators.

EXAMPLE 4   Adding Fractions with Like Denominators

Find the following sum.

$$\frac{3}{7} + \frac{5}{7}$$

Solution

These two fractions have like denominators. Therefore, you can add them by adding their numerators and writing the sum over the like denominator.

$$\frac{3}{7} + \frac{5}{7} = \frac{8}{7}$$

To add fractions with unlike denominators, a common denominator is needed. Section P.2 in the prerequisites chapter shows how prime factorization can be used to find a common multiple of two numbers. The same procedure can be used to find a common denominator for two fractions.

EXAMPLE 5   Adding Fractions with Unlike Denominators

Find the following sum.

$$\frac{3}{8} + \frac{-5}{12}$$

Solution

These two fractions have unlike denominators. In order to add them, you should first find a common denominator. In this case, 24 is a common multiple of 8 and 12. Thus, you add the two fractions as follows.

$$\frac{3}{8} + \frac{-5}{12} = \frac{3(3)}{8(3)} + \frac{(-5)(2)}{12(2)} \qquad \text{Find common denominator.}$$

$$= \frac{9}{24} + \frac{-10}{24} \qquad \text{Like fractions}$$

$$= \frac{9 + (-10)}{24} \qquad \text{Add numerators.}$$

$$= \frac{-1}{24} \qquad \text{Simplify.}$$

$$= -\frac{1}{24}$$

You can add *two* fractions, without first finding a common denominator, by using the following rule.

Alternative Rule for Adding Two Fractions

If a, b, c, and d are integers with $b \neq 0$ and $d \neq 0$, then

$$\frac{a}{b} + \frac{c}{d} = \frac{ad + bc}{bd}.$$

Note how the result of Example 5 can be found by this alternative rule for adding fractions.

$$\frac{3}{8} + \frac{-5}{12} = \frac{3(12) + 8(-5)}{8(12)}$$ Apply alternative rule.

$$= \frac{36 - 40}{96}$$ Simplify.

$$= \frac{-4}{96}$$ Simplify.

$$= -\frac{1}{24}$$ Reduce to lowest terms.

EXAMPLE 6   Adding a Fraction and an Integer

Find the following sum.

$$\frac{1}{4} + 2$$

Solution

Considering 2 to have a denominator of 1, you use the rule for adding fractions with unlike denominators.

$$\frac{1}{4} + \frac{2}{1} = \frac{1}{4} + \frac{2(4)}{1(4)}$$ Find common denominator.

$$= \frac{1}{4} + \frac{8}{4}$$ Like fractions

$$= \frac{1 + 8}{4}$$ Add numerators.

$$= \frac{9}{4}$$ Simplify.

EXAMPLE 7   Adding a Mixed Number and a Fraction

Find the following sum.

$$1\tfrac{4}{5} + \frac{11}{15}$$

Solution

To begin, rewrite the **mixed number** $1\tfrac{4}{5}$ as a fraction.

$$1\tfrac{4}{5} = 1 + \frac{4}{5} = \frac{5}{5} + \frac{4}{5} = \frac{9}{5}$$

Then add the two fractions as follows.

$$\begin{aligned}
1\tfrac{4}{5} + \frac{11}{15} &= \frac{9}{5} + \frac{11}{15} \\
&= \frac{9(3)}{5(3)} + \frac{11}{15} \\
&= \frac{27}{15} + \frac{11}{15} \\
&= \frac{38}{15}
\end{aligned}$$

NOTE   In Example 7, a common shortcut for writing $1\tfrac{4}{5}$ as $\tfrac{9}{5}$ is to multiply 1 by 5, add the result to 4, and then divide by 5, as follows.

$$1\tfrac{4}{5} = \frac{1(5) + 4}{5} = \frac{9}{5}$$

When you obtain a fraction as an answer to a problem, it is a good idea to check if the fraction can be reduced. Although it is not a hard and fast rule, the reduced fraction $\tfrac{2}{3}$ is normally considered to be a better answer than the fraction $\tfrac{8}{12}$.

Technology

You can use a calculator to add or subtract fractions. When you do this, however, remember that the calculator may introduce roundoff error. For instance, to add the fractions $\tfrac{1}{3}$ and $\tfrac{1}{4}$ with a calculator, you can use the following keystrokes.

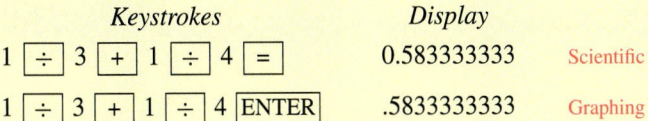

	Keystrokes		*Display*	
	1 ÷ 3 + 1 ÷ 4 =		0.583333333	Scientific
	1 ÷ 3 + 1 ÷ 4 ENTER		.5833333333	Graphing

No matter how many decimal places you write with the answer, it does not have the same accuracy as the exact answer, which is $\tfrac{7}{12}$.

Subtracting Fractions

To subtract fractions, use the strategy used for integers. Add the opposite of the fraction. Note how this is done in Examples 8 and 9.

EXAMPLE 8 *Subtracting Fractions*

Perform the following subtractions.

a. $\dfrac{3}{7} - \dfrac{5}{7}$ **b.** $\dfrac{7}{9} - \dfrac{11}{12}$

Solution

a.
$$\frac{3}{7} - \frac{5}{7} = \frac{3}{7} + \frac{-5}{7} \qquad \text{To subtract, add the opposite.}$$
$$= \frac{3 + (-5)}{7} \qquad \text{Add numerators.}$$
$$= \frac{-2}{7} \qquad \text{Simplify.}$$
$$= -\frac{2}{7}$$

b.
$$\frac{7}{9} - \frac{11}{12} = \frac{7(4)}{9(4)} + \frac{-11(3)}{12(3)} \qquad \text{Find common denominator, 36.}$$
$$= \frac{28}{36} + \frac{-33}{36} \qquad \text{Like fractions}$$
$$= \frac{28 + (-33)}{36} \qquad \text{Add numerators.}$$
$$= \frac{-5}{36} \qquad \text{Simplify.}$$
$$= -\frac{5}{36}$$

EXAMPLE 9 *Subtracting Fractions*

$$\frac{5}{16} - \left(-\frac{7}{30}\right) = \frac{5}{16} + \frac{7}{30} \qquad \text{Add the opposite.}$$
$$= \frac{5(30) + 16(7)}{16(30)} = \frac{150 + 112}{480} \qquad \text{Apply alternative rule.}$$
$$= \frac{262}{480} \qquad \text{Simplify.}$$
$$= \frac{131}{240} \qquad \text{Reduce to lowest terms.}$$

EXAMPLE 10 *Combining Three or More Fractions*

Evaluate the following.

$$\frac{5}{6} - \frac{7}{15} + \frac{3}{10} - 1$$

Solution

The least common denominator of 6, 15, and 10 is 30. Therefore, you can rewrite the given expression as follows.

$$\frac{5}{6} - \frac{7}{15} + \frac{3}{10} - 1 = \frac{5(5)}{6(5)} + \frac{(-7)(2)}{15(2)} + \frac{3(3)}{10(3)} + \frac{(-1)(30)}{30}$$

$$= \frac{25}{30} + \frac{-14}{30} + \frac{9}{30} + \frac{-30}{30}$$

$$= \frac{25 - 14 + 9 - 30}{30}$$

$$= \frac{-10}{30}$$

$$= -\frac{1(\cancel{10})}{3(\cancel{10})} \qquad \text{Reduce fraction.}$$

$$= -\frac{1}{3}$$

Group Activities   You Be the Instructor

Error Analysis   Is the following solution correct or incorrect? If it is correct, explain why. If it is incorrect, identify the error, and explain how you could help someone avoid this type of error.

$$\frac{3}{5} - \frac{2}{3} - \frac{1}{2} = \frac{18}{30} - \frac{20}{30} - \frac{15}{30}$$

$$= \frac{18}{30} - \frac{5}{30}$$

$$= \frac{13}{30}$$

In your group, list some common errors that people might make when adding or subtracting fractions.

1.1 Exercises

Discussing the Concepts

1. A pie was served at dinner and $\frac{5}{6}$ of it was eaten. In the context of this statement, interpret the meaning of the numerator and denominator.

2. Explain why $\frac{12}{15} = \frac{4}{5}$.

3. Explain how to find the least common denominator of $\frac{3}{8}$ and $\frac{7}{20}$. How does your method compare with the procedure for finding the least common multiple of 8 and 20?

4. In your own words, describe how to find the difference of two fractions with unlike denominators.

5. Is it true that the sum of two fractions of like signs is positive? If not, give an example that shows the statement is false.

6. Is the following statement valid? Explain.

$$\frac{2}{3} + \frac{3}{2} = \frac{2+3}{3+2} = 1$$

Problem Solving

In Exercises 7–10, write the fraction in reduced form.

7. $\dfrac{2}{8}$

8. $\dfrac{21}{28}$

9. $\dfrac{60}{192}$

10. $\dfrac{45}{225}$

Geometry In Exercises 11–14, each figure is divided into regions of equal area. Find the sum of the two fractions indicated by the shaded regions.

11. 12.

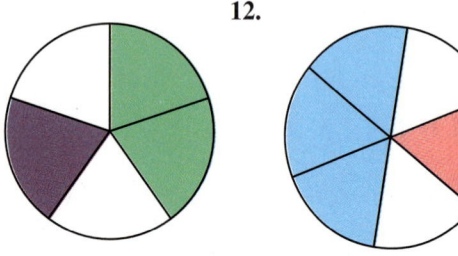

13.

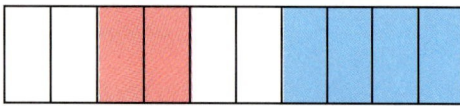

14.

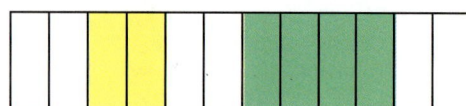

In Exercises 15–18, add or subtract. Then simplify.

15. $\dfrac{7}{15} + \dfrac{2}{15}$

16. $\dfrac{13}{35} + \dfrac{5}{35}$

17. $\dfrac{3}{4} - \dfrac{5}{4}$

18. $\dfrac{3}{8} - \left| -\dfrac{5}{8} \right|$

In Exercises 19 and 20, write an equivalent fraction with the indicated denominator.

19. $\dfrac{3}{8} = \dfrac{}{16}$

20. $\dfrac{4}{5} = \dfrac{}{15}$

In Exercises 21–28, add or subtract. Then simplify.

21. $\dfrac{1}{2} + \dfrac{1}{3}$

22. $\dfrac{3}{5} + \dfrac{1}{2}$

23. $\left| -\dfrac{1}{8} \right| - \dfrac{1}{6}$

24. $\dfrac{13}{8} - \dfrac{3}{4}$

25. $4 - \dfrac{8}{3}$

26. $\dfrac{17}{25} + 2$

27. $1 + \dfrac{2}{3} - \dfrac{5}{6}$

28. $2 - \dfrac{15}{16} - \dfrac{7}{8} + |-5|$

In Exercises 29–32, write the mixed number as a fraction.

29. $4\frac{3}{5}$

30. $7\frac{2}{3}$

31. $3\frac{7}{10}$

32. $-1\frac{3}{4}$

In Exercises 33–36, add or subtract. Then simplify.

33. $3\frac{1}{2} + 5\frac{2}{3}$ **34.** $5\frac{3}{4} + 8\frac{1}{10}$

35. $15\frac{5}{6} - 20\frac{1}{4}$ **36.** $6 - 3\frac{5}{8}$

In Exercises 37 and 38, determine the unknown fractional part of the pie graph.

37.

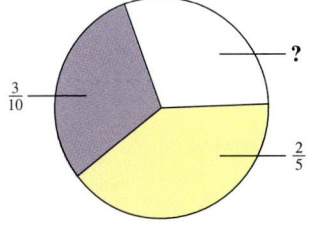

38.

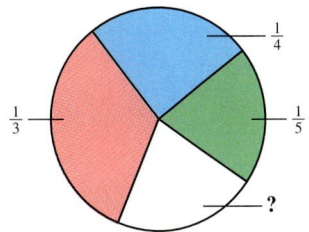

39. *Stock Price*   On Monday, a stock closed at $\$52\frac{5}{8}$ per share. On Tuesday, it closed at $\$54\frac{1}{4}$ per share. Determine the increase in the price.

40. *Sewing*   A pattern requires $3\frac{1}{6}$ yards of material to make a skirt and an additional $2\frac{3}{4}$ yards to make a matching jacket. Find the total amount of material required.

Reviewing the Major Concepts

41. Find the greatest common factor of the numbers 28, 126, and 294.

42. Find the least common multiple of the numbers 9, 12, and 30.

43. What must you subtract from -15 to obtain 8?

44. State the rules for determining the sign of the sum when adding integers.

45. *Profit and Loss*   A company had a first-quarter loss of $312,500, a second-quarter profit of $275,500, a third-quarter profit of $297,750, and a fourth-quarter profit of $71,300. What was the profit for the year?

46. *Average Speed*   You drive 371 miles in 7 hours. What is your average speed?

Additional Problem Solving

In Exercises 47–50, write the fraction in reduced form.

47. $\dfrac{12}{18}$ **48.** $\dfrac{16}{56}$

49. $\dfrac{28}{350}$ **50.** $\dfrac{88}{154}$

In Exercises 51–58, add or subtract. Then simplify.

51. $\dfrac{9}{11} + \dfrac{5}{11}$ **52.** $\dfrac{5}{6} + \dfrac{13}{6}$

53. $\dfrac{9}{16} - \dfrac{3}{16}$ **54.** $\dfrac{15}{32} - \dfrac{7}{32}$

55. $-\dfrac{23}{11} + \dfrac{12}{11}$ **56.** $\dfrac{46}{13} - \dfrac{20}{13}$

57. $\dfrac{13}{15} + \left|-\dfrac{11}{15}\right| - \dfrac{4}{15}$ **58.** $\dfrac{5}{8} - \left(-\dfrac{13}{8}\right) + \dfrac{3}{8}$

In Exercises 59–72, add or subtract. Then simplify.

59. $\dfrac{1}{4} - \dfrac{1}{3}$ **60.** $\dfrac{2}{3} + \dfrac{1}{6}$

61. $\dfrac{3}{16} + \dfrac{3}{8}$ **62.** $\dfrac{2}{3} + \dfrac{4}{9}$

63. $-\dfrac{5}{6} - \left(-\dfrac{3}{4}\right)$

64. $\dfrac{3}{4} - \dfrac{2}{5}$

65. $-\dfrac{7}{8} - \dfrac{5}{6}$

66. $-\dfrac{5}{12} - \dfrac{1}{9}$

67. $\dfrac{5}{12} - \dfrac{3}{8} + \dfrac{5}{16}$

68. $-\dfrac{3}{7} + \dfrac{5}{14} + \dfrac{3}{4}$

69. $\dfrac{7}{4} - 3$

70. $20 - \dfrac{4}{5}$

71. $2 - \dfrac{25 - 15}{6} + \dfrac{3}{4}$

72. $3 + \left|\dfrac{12 - 8}{3}\right| + \dfrac{1}{9}$

In Exercises 73–76, write the mixed number as a fraction.

73. $8\frac{2}{3}$

74. $2\frac{5}{8}$

75. $-10\frac{5}{11}$

76. $3\frac{1}{100}$

In Exercises 77–80, add or subtract. Then simplify.

77. $1\frac{3}{16} - 2\frac{1}{4}$

78. $5\frac{7}{8} - 2\frac{1}{2}$

79. $-5\frac{2}{3} - 4\frac{5}{12}$

80. $-2\frac{3}{4} - 3\frac{1}{5}$

81. *Animal Feed*   During the months of January, February, and March, an animal shelter bought $8\frac{3}{4}$ tons, $7\frac{1}{5}$ tons, and $9\frac{3}{8}$ tons of feed. Find the total amount of feed purchased during the first quarter of the year.

82. *Recipe*   You are making a batch of chocolate chip cookies. You have placed 2 cups of flour, $\frac{1}{3}$ cup shortening, $\frac{1}{3}$ cup butter, $\frac{1}{2}$ cup brown sugar, and $\frac{1}{3}$ cup granulated sugar in the mixing bowl. How many cups of ingredients are in the mixing bowl?

83. *Volume*   The fuel gauge on a gasoline tank indicates that the tank is $\frac{3}{8}$ full. What fraction of the tank is empty?

84. *Work Progress*   The state highway workers have a sign by a construction project indicating what fraction of the work has been completed. At the beginnings of May and June, the fractions of work completed were $\frac{5}{16}$ and $\frac{2}{3}$, respectively. What fraction of the work was completed during the month of May?

Estimation   In Exercises 85 and 86, mentally estimate the sum or difference to the nearest integer.

85. $\dfrac{3}{11} + \dfrac{7}{10}$

86. $\dfrac{5}{8} + \dfrac{9}{7}$

87. *Think About It*   Determine the placement of the digits 3, 4, 5, and 6 in the following addition problem so that you obtain the specified sum.

$$\boxed{}\ \ + \ \ \boxed{}\ \ = \dfrac{13}{10}$$

CAREER INTERVIEW

Rob Prester
Musician
New York, NY 10128

I am both a jazz musician and a composer—my primary instrument is piano/keyboards. I play in jazz clubs, teach jazz and classical music, have recorded my own CD of original work, and also try to spend much of my time composing new pieces.

Music is very mathematical; its rhythm is based on subdividing units of time. In a musical piece, a *measure* is the basic unit of time. In $\frac{4}{4}$ time, each measure has four beats, each of which can be divided into subbeats. In jazz music it is very common to repeat several times a musical figure or phrase that may not completely fill a whole measure. Suppose I use a figure consisting of seven notes, each worth one-half beat, for a total of three and one-half beats for the figure. If I repeat this figure four times, then I need to add up the fractional rhythm values for each figure $(3\frac{1}{2} + 3\frac{1}{2} + 3\frac{1}{2} + 3\frac{1}{2})$ and subtract from the value of four full measures (16) to find how many beats are left over at the end of the last measure. In this case I would need two more beats to fill out the last measure.

1.2	**Multiplying and Dividing Fractions**

Multiplying Fractions ▪ Dividing Fractions ▪
Operations with Decimals ▪ Application

Multiplying Fractions

The procedure for multiplying fractions is simpler than those for adding and subtracting fractions. Regardless of whether the fractions have like or unlike denominators, you can find the product of two fractions by multiplying the numerators and multiplying the denominators.

Multiplication of Fractions

To multiply two fractions, multiply the two numerators to form the numerator of the product, and multiply the two denominators to form the denominator of the product.

EXAMPLE 1   *Multiplying Fractions*

a. $\dfrac{5}{8} \cdot \dfrac{3}{2} = \dfrac{5(3)}{8(2)}$ Multiply numerators and denominators.

$= \dfrac{15}{16}$ Simplify.

b. $\left(-\dfrac{7}{9}\right)\left(-\dfrac{5}{21}\right) = \dfrac{7}{9} \cdot \dfrac{5}{21}$ Product of two negatives is positive.

$= \dfrac{7(5)}{9(21)}$ Multiply numerators and denominators.

$= \dfrac{7(5)}{9(3)(7)}$ Factor and reduce.

$= \dfrac{5}{27}$ Simplify.

c. $\dfrac{8}{15}\left(-\dfrac{6}{10}\right) = -\dfrac{8(6)}{(15)(10)}$ Product of negative and positive is negative.

$= -\dfrac{(8)(2)(3)}{(3)(5)(2)(5)}$ Factor and reduce.

$= -\dfrac{8}{25}$ Simplify.

EXAMPLE 2   Multiplying Fractions

a. $\left(3\dfrac{1}{5}\right)\left(-\dfrac{7}{6}\right)\left(\dfrac{5}{3}\right) = \left(\dfrac{16}{5}\right)\left(-\dfrac{7}{6}\right)\left(\dfrac{5}{3}\right)$ Rewrite mixed number as a fraction.

$$= -\frac{(8)(2)(7)(5)}{(5)(3)(2)(3)}$$ Factor and reduce.

$$= -\frac{56}{9}$$ Simplify.

b. $(-3)\left(\dfrac{-6}{14}\right) = \dfrac{-3}{1} \cdot \dfrac{-6}{14}$ Rewrite -3 as a fraction.

$$= \frac{(-3)(-6)}{(1)(14)}$$ Multiply numerators and denominators.

$$= \frac{18}{14}$$ Simplify.

$$= \frac{2 \cdot 9}{2 \cdot 7}$$ Factor and reduce.

$$= \frac{9}{7}$$ Simplify.

c. $\left(\dfrac{8}{12}\right)\left(\dfrac{3}{2}\right)(5) = \left(\dfrac{8}{12}\right)\left(\dfrac{3}{2}\right)\left(\dfrac{5}{1}\right)$ Rewrite 5 as a fraction.

$$= \frac{(8)(3)(5)}{(12)(2)(1)}$$ Multiply numerators and denominators.

$$= \frac{4 \cdot 2 \cdot 3 \cdot 5}{4 \cdot 3 \cdot 2 \cdot 1}$$ Factor and reduce.

$$= 5$$ Simplify.

Technology

Try verifying some of the products shown in Example 2 with your calculator. For instance, you can verify that $\frac{8}{12}\left(\frac{3}{2}\right)(5) = 5$ as follows.

$\boxed{(}\ \boxed{8}\ \boxed{\div}\ \boxed{12}\ \boxed{)}\ \boxed{\times}\ \boxed{(}\ \boxed{3}\ \boxed{\div}\ \boxed{2}\ \boxed{)}\ \boxed{\times}\ \boxed{5}\ \boxed{\text{ENTER}}$

Similarly, you can verify that $\left(3\frac{1}{5}\right)\left(-\frac{7}{6}\right)\left(\frac{5}{3}\right) = -\frac{56}{9}$ as follows.

$\boxed{(}\ \boxed{3}\ \boxed{+}\ \boxed{1}\ \boxed{\div}\ \boxed{5}\ \boxed{)}\ \boxed{\times}\ \boxed{(}\ \boxed{(\text{-})}\ \boxed{7}\ \boxed{\div}\ \boxed{6}\ \boxed{)}\ \boxed{\times}$

$\boxed{(}\ \boxed{5}\ \boxed{\div}\ \boxed{3}\ \boxed{)}\ \boxed{\text{ENTER}}$

Dividing Fractions

The **reciprocal** of a number is the number by which it must be multiplied to obtain 1. For instance, the reciprocal of 3 is $\frac{1}{3}$ because $3\left(\frac{1}{3}\right) = 1$. Similarly, the reciprocal of $-\frac{2}{3}$ is $-\frac{3}{2}$ because

$$\left(-\frac{2}{3}\right)\left(-\frac{3}{2}\right) = 1.$$

To divide two fractions, multiply the first fraction by the reciprocal of the second fraction. Another way of saying this is "invert the divisor and multiply."

Division of Fractions

The quotient of $\dfrac{a}{b}$ and $\dfrac{c}{d}$ is

$$\frac{a}{b} \div \frac{c}{d} = \frac{a}{b} \cdot \frac{d}{c}.$$

EXAMPLE 3   *Dividing Fractions*

Perform the following divisions and write the answers in reduced form.

a. $\dfrac{5}{8} \div \dfrac{20}{12}$ **b.** $\dfrac{6}{13} \div \left(-\dfrac{9}{26}\right)$

Solution

a.
$$\frac{5}{8} \div \frac{20}{12} = \frac{5}{8} \cdot \frac{12}{20}$$ Invert divisor and multiply.

$$= \frac{(5)(12)}{(8)(20)}$$ Multiply numerators and denominators.

$$= \frac{(5)(3)(4)}{(8)(4)(5)}$$ Factor and reduce.

$$= \frac{3}{8}$$ Simplify.

b.
$$\frac{6}{13} \div \left(-\frac{9}{26}\right) = \frac{6}{13} \cdot \left(-\frac{26}{9}\right)$$ Invert divisor and multiply.

$$= -\frac{(6)(26)}{(13)(9)}$$ Multiply numerators and denominators.

$$= -\frac{(2)(3)(2)(13)}{(13)(3)(3)}$$ Factor and reduce.

$$= -\frac{4}{3}$$ Simplify.

NOTE   In Example 3(a), note that the expression $\frac{5}{8} \div \frac{20}{12}$ can also be written in the **complex fraction** form

$$\frac{5/8}{20/12}.$$

The same rule applies—invert the divisor (the bottom fraction) and multiply to obtain $\frac{5}{8} \cdot \frac{12}{20}$.

EXAMPLE 4   Dividing Fractions

Perform the following divisions and write the answers in reduced form.

a. $-\dfrac{8}{3} \div 6$ **b.** $3\dfrac{9}{13} \div 1\dfrac{5}{8}$

Solution

a. $-\dfrac{8}{3} \div 6 = -\dfrac{8}{3} \div \dfrac{6}{1}$ Write 6 as a fraction.

$= -\dfrac{8}{3} \cdot \dfrac{1}{6}$ Invert divisor and multiply.

$= -\dfrac{(8)(1)}{(3)(6)}$ Multiply numerators and denominators.

$= -\dfrac{(2)(4)}{(3)(2)(3)}$ Factor and reduce.

$= -\dfrac{4}{9}$ Simplify.

b. $3\dfrac{9}{13} \div 1\dfrac{5}{8} = \dfrac{48}{13} \div \dfrac{13}{8}$ Write mixed numbers as fractions.

$= \dfrac{48}{13} \cdot \dfrac{8}{13}$ Invert divisor and multiply.

$= \dfrac{(48)(8)}{(13)(13)}$ Multiply numerators and denominators.

$= \dfrac{384}{169}$ Simplify.

Technology

When dividing fractions on a calculator, be sure that you understand the order of operations that your calculator uses. For instance, which of the following keystroke sequences would you enter to divide $\frac{4}{3}$ by $\frac{2}{3}$?

a.   4 $\boxed{\div}$ 3 $\boxed{\div}$ 2 $\boxed{\div}$ 3 $\boxed{=}$

b. $\boxed{(}$ 4 $\boxed{\div}$ 3 $\boxed{)}$ $\boxed{\div}$ $\boxed{(}$ 2 $\boxed{\div}$ 3 $\boxed{)}$ $\boxed{=}$

If your calculator follows standard conventions, the first set of keystrokes would not produce the correct result. In the second set of keystrokes, note the use of parentheses to separate the dividend from the divisor.

Operations with Decimals

Rational numbers can be represented as **terminating** or **repeating decimals.** Here are some examples.

Terminating Decimals	*Repeating Decimals*
$\dfrac{1}{4} = 0.25$	$\dfrac{1}{6} = 0.1666\ldots$ or $0.1\overline{6}$
$\dfrac{3}{8} = 0.375$	$\dfrac{1}{3} = 0.3333\ldots$ or $0.\overline{3}$
$\dfrac{2}{10} = 0.2$	$\dfrac{1}{12} = 0.0833\ldots$ or $0.08\overline{3}$
$\dfrac{5}{16} = 0.3125$	$\dfrac{8}{33} = 0.2424\ldots$ or $0.\overline{24}$

Note that the *bar* notation is used to indicate the *repeated* digit (or digits) in the decimal notation. You can obtain the decimal representation of any fraction by long division. For instance, the decimal representation for $\frac{5}{12}$ is $0.41\overline{6}$, as can be seen from the following long division algorithm.

$$0.4166\ldots = 0.41\overline{6}$$

$$
\begin{array}{r}
12\overline{)\,5.000} \\
\underline{4\,8} \\
20 \\
\underline{12} \\
80 \\
\underline{72} \\
80
\end{array}
$$

For calculations involving decimals such as $0.41666\ldots$, you must **round the decimal.** For instance, rounded to two decimal places, the number $0.41666\ldots$ is 0.42. Similarly, rounded to three decimal places, the number $0.41666\ldots$ is 0.417.

Technology

You can use a calculator to round decimals. For instance, to round 0.9375 to two decimal places on a scientific calculator, enter

$\boxed{\text{FIX}}\;\boxed{2}\;.9375\;\boxed{=}\,.$

On a TI-82 graphing calculator, enter

round(.9375, 2) $\boxed{\text{ENTER}}$.

Without using a calculator, round -0.88247 to three decimal places. Verify your answer with a calculator. Name the rounding and decision digits.

Rounding a Decimal

1. Determine the number of digits of accuracy you wish to keep. The digit in the last position you keep is called the **rounding digit,** and the digit in the first position you discard is called the **decision digit.**

2. If the decision digit is 5 or greater, round up by adding 1 to the rounding digit.

3. If the decision digit is 4 or less, round down by leaving the rounding digit unchanged.

EXAMPLE 5 *Operations with Decimals*

a. To add decimals, align the decimal points and proceed as in integer addition.

$$
\begin{array}{r}
1\ 1 \\
0.583 \\
1.06 \\
+\ 2.9104 \\
\hline
4.5534
\end{array}
$$

b. To multiply decimals, use integer multiplication and then place the decimal point (in the product) so that the number of decimal places equals the sum of the decimal places in the two factors.

$$
\begin{array}{rl}
-3.57 & \text{Two decimal places} \\
\times\quad 0.032 & \text{Three decimal places} \\
\hline
714 & \\
1071 & \\
\hline
-0.11424 & \text{Five decimal places}
\end{array}
$$

EXAMPLE 6 *Dividing Decimal Fractions*

To divide 1.483 by 0.56, convert the divisor to an integer by moving its decimal point to the right. Move the decimal point in the dividend an equal number of places to the right. Place the decimal point in the quotient directly above the new decimal point in the dividend and then divide as with integers.

$$
\begin{array}{r}
2.648 \\
56\,)\overline{148.3} \\
\underline{112} \\
36\ 3 \\
\underline{33\ 6} \\
2\ 70 \\
\underline{2\ 24} \\
460 \\
\underline{448}
\end{array}
$$

Rounded to two decimal places, the answer is 2.65. This answer can be written as

$$
\frac{1.483}{0.56} \approx 2.65
$$

where the symbol $\approx$ means **approximately equal to.**

Application

The following example is similar to the stock investment example on page 29. The difference is that this time the gains and losses are given in fractional form.

EXAMPLE 7   An Application: Stock Purchase

On Monday you bought 50 shares of stock at $\$48\frac{1}{2}$ per share. During the week the stocks rose and fell, as shown in the table.

Tuesday	Wednesday	Thursday	Friday
Up $\frac{3}{8}$	Up $1\frac{3}{4}$	Down $\frac{1}{2}$	Up $2\frac{7}{8}$

a. What was the value of the stock at the close on Tuesday?

b. What was the value of the stock at the close on Wednesday?

c. What was the value of the stock at the end of the week?

d. What would the total gain have been if Friday's gain had occurred each of the four days?

e. What was the average daily gain (loss) for the four days recorded?

Solution

a. Because the original value of the stock was $50\left(48\frac{1}{2}\right) = \$2425$, and each of the 50 shares gained $\$\frac{3}{8}$ by the close of Tuesday, the total value of your stock at the close of Tuesday was
$$2425 + 50\left(\tfrac{3}{8}\right) = 2425 + 18.75 = \$2443.75.$$

b. Using the result of part (a), the value at the close of Wednesday was
$$2443.75 + 50\left(1\tfrac{3}{4}\right) = 2443.75 + 87.5 = \$2531.25.$$

c. The value of the stock at the end of the week was
$$2425 + 50\left(\tfrac{3}{8}\right) + 50\left(1\tfrac{3}{4}\right) + 50\left(-\tfrac{1}{2}\right) + 50\left(2\tfrac{7}{8}\right) = \$2650.00.$$

d. The gain on Friday was $2\frac{7}{8}$ per share. If this gain had occurred each day, the total gain would have been
$$4(50)\left(2\tfrac{7}{8}\right) = \$575.00.$$

e. The total gain for the four days was $2650 - 2425 = \$225$. Thus, the average daily gain for four days was Average $= \frac{225}{4} = \$56.25$.

Summary of Rules for Fractions

Let a, b, c, and d be real numbers.

Rule   *Example*

1. Addition of fractions:

$$\frac{a}{b} + \frac{c}{d} = \frac{ad + bc}{bd}, \quad b \neq 0, d \neq 0$$

$$\frac{1}{3} + \frac{2}{7} = \frac{1 \cdot 7 + 3 \cdot 2}{3 \cdot 7} = \frac{13}{21}$$

2. Subtraction of fractions:

$$\frac{a}{b} - \frac{c}{d} = \frac{ad - bc}{bd}, \quad b \neq 0, d \neq 0$$

$$\frac{1}{3} - \frac{2}{7} = \frac{1 \cdot 7 - 3 \cdot 2}{3 \cdot 7} = \frac{1}{21}$$

3. Multiplication of fractions:

$$\frac{a}{b} \cdot \frac{c}{d} = \frac{a \cdot c}{b \cdot d}, \quad b \neq 0, d \neq 0$$

$$\frac{1}{3} \cdot \frac{2}{7} = \frac{1(2)}{3(7)} = \frac{2}{21}$$

4. Division of fractions:

$$\frac{a}{b} \div \frac{c}{d} = \frac{a}{b} \cdot \frac{d}{c}, \quad b \neq 0, \, d \neq 0, \, c \neq 0$$

$$\frac{1}{3} \div \frac{2}{7} = \frac{1}{3} \cdot \frac{7}{2} = \frac{7}{6}$$

5. Rule of signs for fractions:

$$\frac{a}{b} = \frac{-a}{-b} \text{ and } \frac{-a}{b} = \frac{a}{-b} = -\frac{a}{b}$$

$$\frac{12}{4} = \frac{-12}{-4}, \, \frac{-12}{4} = \frac{12}{-4} = -\frac{12}{4}$$

6. Equivalent fractions:

$$\frac{a}{b} = \frac{c}{d}, \text{ if and only if } ad = bc; \, b \neq 0, \, d \neq 0$$

$$\frac{1}{4} = \frac{3}{12} \text{ because } 1 \cdot 12 = 3 \cdot 4$$

Group Activities   Exploring with Technology

To Round or Not to Round?   When using a calculator to perform operations with decimals, you should try to get in the habit of rounding your answers *only* after all the calculations are done. If you round the answer at a preliminary stage, you can introduce unnecessary roundoff error. Suppose $l = 5.24$, $w = 3.03$, and $h = 2.749$ are the dimensions of a box. Have part of your group find the volume, $l \cdot w \cdot h$, by multiplying the given numbers and then rounding the answer to one decimal place. Have the others in your group use a second method, first rounding each dimension to one decimal place and then multiplying them. Compare your answers, and discuss which of these techniques produces the more accurate answer.

1.2 Exercises

Discussing the Concepts

1. In your own words, describe the rule for determining the sign of the product of two fractions.

2. Two-thirds of a pizza was eaten at dinner.

    (a) How much of the pizza was left?

    (b) For a midnight snack you ate $\frac{1}{2}$ of the pizza that was left. What fraction of the whole pizza did you eat as a midnight snack?

    (c) Make a sketch of the pizza and show how it could be cut to obtain the results of parts (a) and (b). Explain how you could obtain the result of part (a) by multiplying fractions.

3. Consider two fractions that are between 0 and 1. Explain why the product of the fractions is less than either factor.

4. Is it true that $\frac{2}{3} = 0.67$? Explain your reasoning.

5. Use the figure to determine how many one-fourths are in 3? Explain how to obtain the same result by division.

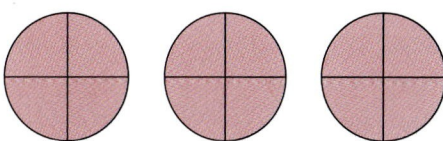

6. Use the figure to determine how many one-sixths are in $\frac{2}{3}$. Explain how to obtain the same result by division.

Problem Solving

In Exercises 7–14, evaluate the expression. Write your result in reduced form.

7. $\dfrac{1}{2} \times \dfrac{3}{4}$

8. $-\dfrac{2}{3} \times \dfrac{5}{7}$

9. $\dfrac{2}{3} \times \left(-\dfrac{9}{16}\right)$

10. $\left(-\dfrac{3}{4}\right)\left(-\dfrac{4}{9}\right)$

11. $\left(-\dfrac{3}{2}\right)\left(-\dfrac{15}{16}\right)\left(\dfrac{12}{25}\right)$

12. $\left(\dfrac{1}{2}\right)\left(-\dfrac{4}{15}\right)\left(-\dfrac{5}{24}\right)$

13. $2\dfrac{3}{4} \times 3\dfrac{2}{3}$

14. $-5\dfrac{2}{3} \times 4\dfrac{1}{2}$

In Exercises 15–20, find the reciprocal of the number. Show that the product of the number and its reciprocal is 1.

15. 7

16. 14

17. $\dfrac{4}{7}$

18. $-\dfrac{5}{9}$

19. $2\dfrac{1}{2}$

20. $-3\dfrac{2}{3}$

In Exercises 21–28, evaluate the expression. If it is not possible, explain why. If it is possible, write the result in reduced form.

21. $\dfrac{3}{8} \div \dfrac{3}{4}$

22. $\dfrac{5}{16} \div \dfrac{25}{8}$

23. $-\dfrac{5}{12} \div \dfrac{45}{32}$

24. $\left(-\dfrac{16}{21}\right) \div \left(-\dfrac{12}{27}\right)$

25. $\dfrac{-7/15}{-14/25}$

26. $\dfrac{-5/9}{0}$

27. $3\dfrac{3}{4} \div 2\dfrac{5}{8}$

28. $1\dfrac{5}{6} \div 2\dfrac{1}{3}$

In Exercises 29–32, write the fraction in decimal form. Use the bar notation for repeating digits.

29. $\dfrac{3}{4}$

30. $\dfrac{5}{8}$

31. $\dfrac{5}{11}$

32. $\dfrac{8}{15}$

In Exercises 33–40, evaluate the expression. Round the answer to two decimal places.

33. $1.21 + 4.06 - 3$

34. $-3.4 + 1.062 - 5.13$

35. $(-6.3)(9.05)$

36. $(-0.05)(-85.95)$

37. $4.69 \div 0.12$

38. $1.062 \div (-2.1)$

39. $\dfrac{(-15.1)(-6.02)}{-9.6}$

40. $\dfrac{5.42 - 3(1.2)}{4.02}$

41. *Estimation*   Each day for a week, you practiced the saxophone for $\frac{2}{3}$ hour.

(a) Explain how to use mental math to estimate the number of hours of practice in a week.

(b) Determine the actual number of hours you practiced during the week. Write the result in decimal form, rounding to one decimal place.

42. *Estimation*   You are buying 200 shares of stock at $\$23\frac{5}{8}$ per share and 300 shares at $\$96\frac{1}{4}$ per share.

(a) Mentally estimate the total cost of the stock.

(b) Use a calculator to find the total cost of the stock.

43. *Walking Time*   Your apartment is $\frac{3}{4}$ mile from the subway. If you walk at the rate of $3\frac{1}{4}$ miles per hour, how long will it take you to walk from your apartment to the subway?

44. *Interpreting a Pie Graph*   The percentages of the U.S. population with different blood types are shown in the accompanying pie graph. What percentage of the population has blood type O^+?   (Source: U.S. Department of Health and Human Services)

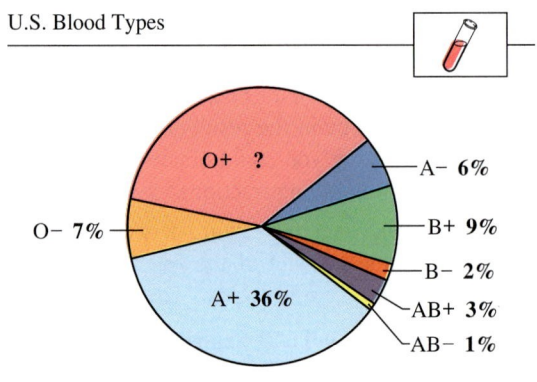

U.S. Blood Types

O+ ?
A− 6%
B+ 9%
B− 2%
AB+ 3%
AB− 1%
A+ 36%
O− 7%

Reviewing the Major Concepts

In Exercises 45–50, evaluate the expression.

45. $-13 - 7$

46. $13 - 16$

47. $\dfrac{5}{16} - \dfrac{3}{10}$

48. $\left|-\dfrac{5}{9}\right| + 2$

49. $-7\frac{3}{5} - 3\frac{1}{2}$

50. $\dfrac{9}{16} + 2\frac{3}{12}$

51. *Relay Race*   In a $\frac{3}{4}$-mile relay race, the last change of runners occurs at the $\frac{2}{3}$-mile marker. How far does the last person run?

52. *Weight Limit*   Three trucks, weighing $10\frac{1}{3}$ tons, $7\frac{3}{5}$ tons, and $12\frac{5}{6}$ tons, are on a bridge at the same time. The weight limit for the bridge is 30 tons. Is there a problem? Explain.

Additional Problem Solving

In Exercises 53–74, evaluate the expression.

53. $\dfrac{2}{3} \times \dfrac{1}{2}$

54. $\dfrac{1}{8} \times \left(-\dfrac{4}{5}\right)$

55. $\left(-\dfrac{7}{16}\right)\left(-\dfrac{12}{5}\right)$

56. $\left(\dfrac{5}{3}\right)\left(-\dfrac{3}{5}\right)$

57. $\left(\dfrac{11}{12}\right)\left(-\dfrac{9}{44}\right)$

58. $\left(\dfrac{5}{18}\right)\left(\dfrac{3}{4}\right)$

59. $9\left(\dfrac{4}{15}\right)$

60. $24\left(-\dfrac{7}{18}\right)$

61. $\left(-\dfrac{3}{11}\right)\left(-\dfrac{11}{3}\right)$

62. $35\left(\dfrac{3}{5}\right)\left(\dfrac{5}{3}\right)$

63. $2\frac{4}{5} \times 6\frac{2}{3}$

64. $-8\frac{1}{2} \times 3\frac{2}{5}$

65. $\dfrac{3}{5} \div 0$

66. $\dfrac{3}{5} \div \dfrac{7}{5}$

67. $\dfrac{15/8}{3/8}$

68. $\dfrac{11/12}{-5/6}$

69. $\dfrac{-5}{15/16}$ **70.** $\dfrac{-35/12}{-14}$

71. $-10 \div \dfrac{1}{9}$ **72.** $0 \div -33$

73. $3\frac{3}{4} \div 1\frac{1}{2}$ **74.** $2\frac{4}{9} \div 5\frac{1}{3}$

In Exercises 75–82, write the fraction in decimal form. Use the bar notation for repeating digits.

75. $\dfrac{7}{10}$ **76.** $\dfrac{9}{4}$

77. $\dfrac{9}{16}$ **78.** $\dfrac{7}{20}$

79. $\dfrac{2}{3}$ **80.** $\dfrac{5}{6}$

81. $\dfrac{7}{12}$ **82.** $\dfrac{5}{21}$

In Exercises 83–90, evaluate the expression. Round the result to two decimal places.

83. $132.1 + (-25.45)$ **84.** $-0.0005 - 2.01 + 0.111$

85. $(-0.09)(-0.45)$ **86.** $3.7(-14.8)$

87. $\dfrac{-10.5}{0.75}$ **88.** $\dfrac{135.85}{2.05}$

89. $\dfrac{(-0.2)(0.05)}{-1.5}$ **90.** $\dfrac{12(7.58)}{3.25}$

91. *Profit or Loss?* Your company had a loss of $5519.80 in July, a profit of $2337.06 in August, and a profit of $3615.11 in September. Did your company have a profit or a loss for the third quarter? Explain.

92. *What Time Is It?* You leave California at 7:00 A.M. and fly to Florida. The flight includes $6\frac{3}{4}$ hours of air time and two $1\frac{1}{2}$-hour stops. Find the local time when you arrive in Florida. (Eastern Standard Time is 3 hours later than Pacific Time.)

93. *Grocery Purchase* At a convenience store, you buy two gallons of milk at $2.23 per gallon and three loaves of bread at $1.23 per loaf. You give the clerk a twenty-dollar bill. How much change should you receive? (Assume there is no sales tax.)

94. *Telephone Charge* A telephone company charges $1.16 for the first minute and $0.85 for each additional minute. Find the cost of a 7-minute phone call.

95. *Making Breadsticks* You make 60 ounces of dough for breadsticks. If each breadstick requires $\frac{5}{4}$ ounces of dough, how many breadsticks can you make?

96. *Gasoline Price* The price per gallon of regular unleaded gasoline at three service stations is $1.159, $1.269, and $1.179, respectively. Find the average price per gallon.

97. *Annual Fuel Cost* The sticker on a new car gives the fuel efficiency as 22.3 miles per gallon. The average cost of fuel is $1.159 per gallon. Estimate the annual fuel cost for a car that will be driven approximately 12,000 miles per year.

98. *Unit Price* A $2\frac{1}{2}$-pound can of food costs $4.95. What is the cost per pound?

99. *Estimation* Use mental math to decide whether $\left(5\frac{3}{4}\right) \times \left(4\frac{1}{8}\right)$ is less than 20. Explain your reasoning.

100. *Geometry* Sketch a rectangle. Draw a line horizontally through the rectangle so that it divides the rectangle into two equal parts. Then draw two vertical lines so that these lines divide the original rectangle into three equal parts.

(a) This process has divided the original rectangle into how many regions of equal size?

(b) What mathematical operation have you shown geometrically?

True or False? In Exercises 101 and 102, determine whether the statement is true or false.

101. The reciprocal of every nonzero integer is an integer.

102. The reciprocal of every nonzero rational number is a rational number.

103. *Think About It* Think of different ways that you could correctly complete the following sentence. "If the product of two real numbers is 15, then . . . ."

MID-CHAPTER QUIZ

Take this quiz as you would take a quiz in class. After you are done, check your work against the answers given in the back of the book.

1. Write the fraction $\frac{105}{126}$ in reduced form.

2. Write the fraction $\frac{3}{7}$ in decimal form rounded to three decimal places.

3. Is it true that $\frac{1}{3} = 0.3$? Explain your answer.

4. Write the mixed number $-8\frac{4}{9}$ as a fraction.

In Exercises 5–16, evaluate the expression. Write fractions in reduced form.

5. $\dfrac{9}{11} - \dfrac{3}{11}$

6. $\dfrac{5}{8} + \dfrac{3}{4}$

7. $3 + \dfrac{5}{12}$

8. $2 - \dfrac{3}{4} + 2\frac{4}{5}$

9. $\dfrac{2}{5} \times \dfrac{5}{8}$

10. $\dfrac{9}{16} \times \dfrac{28}{54}$

11. $6\left(\dfrac{5}{21}\right)$

12. $12\left(-\dfrac{9}{4}\right)\left(\dfrac{25}{81}\right)$

13. $-\dfrac{5}{8} \div \dfrac{3}{16}$

14. $\dfrac{114}{215} \div 2$

15. $\dfrac{-9/25}{-21/10}$

16. $4\frac{3}{14} \div 3\frac{1}{2}$

In Exercises 17 and 18, evaluate the expression. Round the answer to two decimal places.

17. $(3.8)(8.25)$

18. $25.63 \div 2.7$

19. A $1\frac{1}{2}$-pound can of food costs $3.29. What is the cost per ounce?

20. One hundred shares of stock were bought for $35\frac{3}{8}$ per share and sold for $47\frac{1}{2}$ per share. Determine the profit from this investment.

1.3	**Exponents and Order of Operations**
	Positive Integer Exponents ▪ Order of Operations ▪ Calculators

Positive Integer Exponents

In Section P.4, you learned that multiplication by a positive integer can be described as repeated addition.

Repeated Addition   *Multiplication*

$\underbrace{7 + 7 + 7 + 7}_{\text{4 terms of 7}}$ 4×7

In a similar way, repeated multiplication can be written in **exponential form.**

Repeated Multiplication   *Exponential Form*

$\underbrace{7 \cdot 7 \cdot 7 \cdot 7}_{\text{4 factors of 7}}$ 7^4

DISCOVERY

When raising a negative number to a power, the use of parentheses is very important. To discover why, use a calculator to evaluate $(-5)^4$ and -5^4. Write a statement explaining the results. Then use a calculator to evaluate $(-5)^3$ and -5^3. If necessary, write a new statement explaining your discoveries.

In the exponential form 7^4, 7 is the **base** and it specifies the repeated factor. The number 4 is the **exponent** and it indicates how many times the base occurs as a factor.

When you write the exponential form 7^4, you can say that you are raising 7 to the fourth **power.** When a number is raised to the first power, you usually do not write the exponent 1. For instance, we usually write 5^1 simply as 5. Here are some examples of how exponential expressions are read.

Exponential Expression	*Verbal Statement*
7^2	"seven to the second power" or "seven squared"
4^3	"four to the third power" or "four cubed"
$(-2)^4$	"negative two to the fourth power"
-2^4	"the opposite of two to the fourth power"

It is important to recognize how exponential forms such as $(-2)^4$ and -2^4 differ.

$$(-2)^4 = (-2)(-2)(-2)(-2) \qquad \text{The negative sign is part of the base.}$$
$$= 16 \qquad \text{The value of the expression is positive.}$$
$$-2^4 = -(2 \cdot 2 \cdot 2 \cdot 2) \qquad \text{The negative sign is not part of the base.}$$
$$= -16 \qquad \text{The value of the expression is negative.}$$

Keep in mind that an exponent applies only to the factor (number) directly preceding it. Parentheses are needed to include a negative sign or other factors as part of the base.

EXAMPLE 1   *Evaluating Exponential Expressions*

a. $2^5 = 2 \cdot 2 \cdot 2 \cdot 2 \cdot 2$ Rewrite expression as a product.

$\quad\quad = 32$ Simplify.

b. $5^2 = 5 \cdot 5$ Rewrite expression as a product.

$\quad\quad = 25$ Simplify.

c. $(-3)^3 = (-3)(-3)(-3)$ Rewrite expression as a product.

$\quad\quad\quad = -27$ Simplify.

d. $(-3)^4 = (-3)(-3)(-3)(-3)$ Rewrite expression as a product.

$\quad\quad\quad = 81$ Simplify.

e. $-3^4 = -(3 \cdot 3 \cdot 3 \cdot 3)$ Rewrite expression as a product.

$\quad\quad\quad = -81$ Simplify.

NOTE   In parts (c) and (d) of the above example, note that when a negative number is raised to an odd power, the result is negative, and when a negative number is raised to an even power, the result is positive.

EXAMPLE 2   *Evaluating Exponential Expressions*

a. $1^6 = 1 \cdot 1 \cdot 1 \cdot 1 \cdot 1 \cdot 1 = 1$ 1 to any positive power is 1.

b. $-1^6 = -(1^6) = -1$

c. $0^5 = 0 \cdot 0 \cdot 0 \cdot 0 \cdot 0 = 0$ 0 to any positive power is 0.

d. $\left(\dfrac{2}{3}\right)^4 = \dfrac{2}{3} \cdot \dfrac{2}{3} \cdot \dfrac{2}{3} \cdot \dfrac{2}{3}$ Rewrite expression as a product.

$\quad\quad\quad = \dfrac{2 \cdot 2 \cdot 2 \cdot 2}{3 \cdot 3 \cdot 3 \cdot 3}$ Multiply fractions.

$\quad\quad\quad = \dfrac{16}{81}$ Simplify.

e. $\left(-\dfrac{1}{5}\right)^3 = \left(-\dfrac{1}{5}\right)\left(-\dfrac{1}{5}\right)\left(-\dfrac{1}{5}\right)$ Rewrite expression as a product.

$\quad\quad\quad = \dfrac{(-1)(-1)(-1)}{(5)(5)(5)}$ Multiply fractions.

$\quad\quad\quad = -\dfrac{1}{125}$ Simplify.

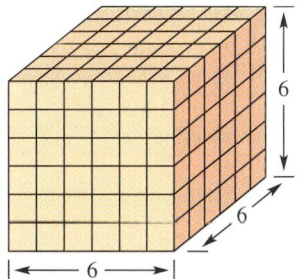

FIGURE 1.2

EXAMPLE 3   An Application: Transporting Capacity

A truck can transport a load of motor oil that is six cases high, six cases wide, and six cases long. Each case contains six quarts of motor oil. How many quarts can the truck transport?

Solution

A sketch can help you solve this problem. From Figure 1.2, you can see that 6 occurs as a factor four times. That is, there are $6 \cdot 6 \cdot 6$ cases of motor oil and each case contains six quarts, which implies that the total number of quarts is

$$6^4 = 1296.$$

Thus, the truck can transport 1296 quarts of oil.

Order of Operations

Up to this point in the text, you have studied five operations of arithmetic—addition, subtraction, multiplication, division, and exponentiation (repeated multiplication). When you use more than one operation in a given problem, you face the question of which operation to do first. For example, without further guidelines, you could evaluate $4 + 3 \cdot 5$ in two ways.

Add First	*Multiply First*
$4 + 3 \cdot 5 \overset{?}{=} (4 + 3) \cdot 5$	$4 + 3 \cdot 5 \overset{?}{=} 4 + (3 \cdot 5)$
$= 7 \cdot 5$	$= 4 + 15$
$= 35$	$= 19$

According to the established **order of operations,** the second evaluation is correct. The reason for this is that multiplication has a higher priority than addition. The accepted priorities for order of operations are summarized below.

Order of Operations

1. Perform operations inside *symbols of grouping* or *absolute value symbols,* starting with the innermost symbol.

2. Evaluate all *exponential* expressions.

3. Perform all *multiplications* and *divisions* from left to right.

4. Perform all *additions* and *subtractions* from left to right.

The symbols used to represent operations were introduced over time. The minus and plus signs were first used in Germany in 1489. The plus sign evolved from the Latin word *et*, meaning "and." The equal sign, the square root notation, and variables were introduced in the 1500s in Europe. In the 1630s, × was introduced to indicate multiplication. Gottfried Wilhelm von Leibniz thought it was easily confused with the variable *x* and proposed the dot symbol (·) in 1698.

In the priorities for order of operations, note that the highest priority is given to **symbols of grouping** such as parentheses or brackets. This means that when you want to be sure that you are communicating an expression correctly, you can insert symbols of grouping to specify which operations you intend to be performed first. For instance, if you want to make sure that $4 + 3 \cdot 5$ will be evaluated correctly, you can write it as $4 + (3 \cdot 5)$.

EXAMPLE 4   Order of Operations

a.
$$7 - [(5 \cdot 3) + 2^3] = 7 - [15 + 2^3]$$ Multiply inside the parentheses.
$$= 7 - [15 + 8]$$ Evaluate exponential expression.
$$= 7 - 23$$ Add inside the brackets.
$$= -16$$ Subtract.

b.
$$[36 \div (3^2 \cdot 2)] - 6 = [36 \div (9 \cdot 2)] - 6$$ Evaluate exponential expression.
$$= [36 \div 18] - 6$$ Multiply inside the parentheses.
$$= 2 - 6$$ Divide inside the brackets.
$$= -4$$ Subtract.

When you are using symbols of grouping in an expression, we suggest that you alternate between parentheses and brackets. For instance, the expression
$$10 - (3 - [4 - (5 + 7)])$$
is easier to understand than $10 - (3 - (4 - (5 + 7)))$.

EXAMPLE 5   Order of Operations

a.
$$36 - [3^2 \cdot (2 \div 6)] = 36 - \left[3^2 \cdot \frac{1}{3}\right]$$ Divide inside the parentheses.
$$= 36 - \left[9 \cdot \frac{1}{3}\right]$$ Evaluate exponential expression.
$$= 36 - 3$$ Multiply inside the brackets.
$$= 33$$ Subtract.

b.
$$10 - 2(8 + |5 - 7|) = 10 - 2(8 + |-2|)$$ Subtract inside absolute value.
$$= 10 - 2(10)$$ Add inside the parentheses.
$$= 10 - 20$$ Multiply.
$$= -10$$ Subtract.

STUDY TIP

Often in mathematics, there is no "best way" to solve a problem. For instance, here is a different way to evaluate the expression in Example 6(b). Which way do you prefer?

$$\frac{8}{3}\left(\frac{1}{6}+\frac{1}{4}\right)$$

$$= \frac{8}{3} \cdot \frac{1}{6} + \frac{8}{3} \cdot \frac{1}{4}$$

$$= \frac{8}{18} + \frac{8}{12}$$

$$= \frac{16}{36} + \frac{24}{36}$$

$$= \frac{40}{36}$$

$$= \frac{10}{9}$$

EXAMPLE 6 Order of Operations

a. $\dfrac{3}{7} \div \dfrac{8}{7} + \left(-\dfrac{3}{5}\right)\left(\dfrac{1}{3}\right) = \dfrac{3}{7} \cdot \dfrac{7}{8} + \left(-\dfrac{3}{5}\right)\left(\dfrac{1}{3}\right)$ *Invert divisor and multiply.*

$$= \frac{3}{8} + \left(-\frac{1}{5}\right) \qquad \text{\textit{Multiply fractions.}}$$

$$= \frac{15}{40} + \frac{-8}{40} \qquad \text{\textit{Find common denominator.}}$$

$$= \frac{7}{40} \qquad \text{\textit{Add fractions.}}$$

b. $\dfrac{8}{3}\left(\dfrac{1}{6}+\dfrac{1}{4}\right) = \dfrac{8}{3}\left(\dfrac{2}{12}+\dfrac{3}{12}\right)$ *Find common denominator.*

$$= \frac{8}{3}\left(\frac{5}{12}\right) \qquad \text{\textit{Add inside the parentheses.}}$$

$$= \frac{40}{36} \qquad \text{\textit{Multiply fractions.}}$$

$$= \frac{10}{9} \qquad \text{\textit{Simplify.}}$$

EXAMPLE 7 Order of Operations

Evaluate the expression $6 + \dfrac{15}{3^2 - 4} - (-5)$.

Solution

Using the established order of operations, you can evaluate the expression as follows.

$$6 + \frac{15}{3^2 - 4} - (-5) = 6 + \frac{15}{9 - 4} - (-5) \qquad \text{\textit{Evaluate exponential expression.}}$$

$$= 6 + \frac{15}{5} - (-5) \qquad \text{\textit{Subtract in denominator.}}$$

$$= 6 + 3 - (-5) \qquad \text{\textit{Divide.}}$$

$$= 9 + 5 \qquad \text{\textit{Add.}}$$

$$= 14 \qquad \text{\textit{Add.}}$$

NOTE In Example 7, note that a fraction bar acts as a symbol of grouping. For instance, the expressions

$$\frac{15}{3^2 - 4} \quad \text{and} \quad 15 \div 3^2 - 4$$

do not represent the same quantity. Can you see why?

Calculators

Try the following evaluations on your own calculator to see whether the indicated keystrokes give the same display.

EXAMPLE 8 *Evaluating Expressions on a Calculator*

a. To evaluate the expression $-3^2 + 4$, use the following keystrokes.

Keystrokes	*Display*	
3 $\boxed{x^2}$ $\boxed{+/-}$ $\boxed{+}$ 4 $\boxed{=}$	-5	Scientific
$\boxed{(-)}$ 3 $\boxed{x^2}$ $\boxed{+}$ 4 $\boxed{\text{ENTER}}$	-5	Graphing

b. To evaluate the expression $5/(4 + 3 \cdot 2)$, use the following keystrokes.

Keystrokes	*Display*	
5 $\boxed{\div}$ $\boxed{(}$ 4 $\boxed{+}$ 3 $\boxed{\times}$ 2 $\boxed{)}$ $\boxed{=}$	0.5	Scientific
5 $\boxed{\div}$ $\boxed{(}$ 4 $\boxed{+}$ 3 $\boxed{\times}$ 2 $\boxed{)}$ $\boxed{\text{ENTER}}$	0.5	Graphing

c. To evaluate $-5^3 + (-2)^4$, use the following keystrokes.

Keystrokes	*Display*	
5 $\boxed{+/-}$ $\boxed{y^x}$ 3 $+$ $\boxed{(}$ 2 $\boxed{+/-}$ $\boxed{)}$ $\boxed{y^x}$ 4 $\boxed{=}$	-109	Scientific
$\boxed{(-)}$ 5 $\boxed{\wedge}$ 3 $+$ $\boxed{(}$ $\boxed{(-)}$ 2 $\boxed{)}$ $\boxed{\wedge}$ 4 $\boxed{\text{ENTER}}$	-109	Graphing

---

Group Activities   Problem Solving

Compound Interest   The formula

$$A = 1000 \left(1 + \frac{0.04}{N}\right)^{10N}$$

gives the balance A in an account in which $1000 is deposited for 10 years. The annual percentage rate is 4%, and the interest is compounded N times per year. Each person in your group should choose a different value of N from $N = 1, 4, 12,$ or 365. Use your calculator to find the balance A. List and compare your results. Describe a relationship between the amount A and the value of N.

1.3 Exercises

Discussing the Concepts

1. Consider the expression 3^5.

   (a) What is the number 3 called?

   (b) What is the number 5 called?

   (c) Rewrite the expression as a product.

2. Are -6^2 and $(-6)^2$ equal? Explain.

3. Are $2 \cdot 5^2$ and 10^2 equal? Explain.

4. In your own words, describe the priorities for the established order of operations.

5. In the expression $12 + 48 \div 6 - 5$, where would you insert symbols of grouping to help someone understand that the value is 15?

6. Is the following solution correct? If not, correct it.

$$-9 + \frac{9+20}{3(5)} - (-3) = -9 + \frac{9}{3} + \frac{20}{5} - (-3)$$
$$= -9 + 3 + 4 - (-3)$$
$$= 1$$

Problem Solving

In Exercises 7 and 8, rewrite in exponential form.

7. $2 \cdot 2 \cdot 2 \cdot 2 \cdot 2$

8. $\left(-\frac{1}{4}\right)\left(-\frac{1}{4}\right)\left(-\frac{1}{4}\right)$

In Exercises 9 and 10, rewrite as a product.

9. $(-3)^6$

10. $\left(\frac{3}{8}\right)^5$

True or False? In Exercises 11 and 12, decide whether the statement is true or false. If it is false, state the reason.

11. -2^4 is positive.

12. $(-2)^3$ is negative.

In Exercises 13–26, evaluate the expression. If it is not possible, state the reason.

13. 3^2

14. $-(-3)^2$

15. $\dfrac{1}{4^3}$

16. $\left(\dfrac{4}{5}\right)^3$

17. $(-1.2)^3$

18. $(1.5)^4$

19. $\dfrac{1-3^2}{-2}$

20. $\dfrac{3^2+4^2}{5}$

21. 2.1×10^2

22. 4.85×10^4

23. $\dfrac{8.4}{10^3}$

24. $\dfrac{6.23}{10^2}$

25. $\dfrac{3^2+1}{0}$

26. $\dfrac{0}{5^2+1}$

In Exercises 27–38, evaluate the expression. Write fractional answers in reduced form.

27. $4 - [3(4-9) - |10-3|]$

28. $12(7+2) - 3|5-8|$

29. $16 + 3 \cdot 4$

30. $25 - 32 \div 4$

31. $(16-5) \div (3-5)$

32. $(10-16) \cdot (20-26)$

33. $(-4)^2 - 3 \cdot 2^4$

34. $3 \cdot 4^2 - 32$

35. $4\left(-\dfrac{2}{3} + \dfrac{4}{3}\right)$

36. $18\left(\dfrac{1}{2} + \dfrac{2}{3}\right)$

37. $\dfrac{3 \cdot 6 - 4 \cdot 6}{5+1}$

38. $\dfrac{3 + [15 \div (-3)]}{16}$

In Exercises 39 and 40, use a calculator to evaluate the expression. Round your result to two decimal places.

39. $3.4^2 - 6(1.2)^3$

40. $1000 \div \left(1 + \dfrac{0.09}{4}\right)^8$

In Exercises 41 and 42, explain why the statement is true. (The symbol $\neq$ means *is not equal to*.)

41. $4 \cdot 6^2 \neq 24^2$

42. $4 - (6-2) \neq 4 - 6 - 2$

43. *Geometry* The land area of the earth is approximately 5.75×10^7 square miles. Evaluate this quantity.

44. *School Enrollment* The projected number of elementary and secondary teachers for the year 2000 is 3.24×10^6. Evaluate this quantity. (Source: U.S. National Center for Education Statistics)

Geometry In Exercises 45 and 46, find the area.

45.

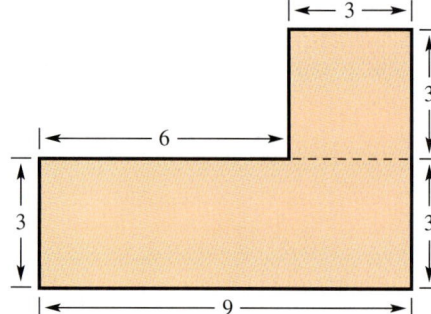

46.

47. *Geometry* The volume of a cube with each edge of length 7 inches is given by $V = 7 \cdot 7 \cdot 7$. Write the volume using exponential notation. What is unit measure for the volume V?

48. *Balance in an Account* $3000 is deposited in an account earning 8% compounded yearly. The balance at the end of 4 years is $A = 3000(1.08)^4$. Use a calculator to find the balance.

Reviewing the Major Concepts

In Exercises 49–52, evaluate the expression. Write the result in reduced form.

49. $-\dfrac{0}{32}$

50. $-2 + \dfrac{3}{2}$

51. $\left(-\dfrac{4}{3}\right)\left(-\dfrac{9}{16}\right)$

52. $\dfrac{\frac{1}{3} + \frac{5}{6}}{\frac{5}{12}}$

53. *Sewing* A pattern requires $2\frac{2}{3}$ yards of material to make a skirt and an additional $1\frac{1}{8}$ yards to make a vest. Find the total amount of material required.

54. *Annual Fuel Cost* A new car has a fuel efficiency of 24.8 miles per gallon. The average cost of fuel is $1.109 per gallon. Estimate the annual fuel cost if the car is driven 15,000 miles per year.

Additional Problem Solving

In Exercises 55–58, rewrite in exponential form.

55. $(-5)(-5)(-5)(-5)$

56. $\left(\dfrac{7}{3}\right)\left(\dfrac{7}{3}\right)\left(\dfrac{7}{3}\right)\left(\dfrac{7}{3}\right)\left(\dfrac{7}{3}\right)\left(\dfrac{7}{3}\right)\left(\dfrac{7}{3}\right)$

57. $\left(\dfrac{5}{8}\right)\left(\dfrac{5}{8}\right)\left(\dfrac{5}{8}\right)\left(\dfrac{5}{8}\right)\left(\dfrac{5}{8}\right)$

58. $(1.6)(1.6)(1.6)(1.6)(1.6)$

In Exercises 59–62, rewrite as a product.

59. $(9.8)^3$

60. $(0.01)^8$

61. $\left(-\dfrac{1}{2}\right)^5$

62. $\left(\dfrac{3}{11}\right)^4$

In Exercises 63–66, is the value positive or negative?

63. -2^2

64. $(-2)^4$

65. -5^3

66. $-(-5)^3$

In Exercises 67–82, evaluate the expression. If it is not possible, state the reason.

67. 2^6

68. 4^3

69. $(-5)^3$

70. 3^4

71. -5^3

72. $-(-3)^4$

73. $\left(\dfrac{2}{3}\right)^4$

74. $\left(\dfrac{1}{4}\right)^3$

75. $\dfrac{3^2 - 4^2}{0}$

76. $\dfrac{0}{3^2 - 4^2}$

77. $\dfrac{5^2 + 12^2}{13}$

78. $\dfrac{4^2 - 2^3}{4}$

79. 5.84×10^3

80. 3.28×10^5

81. $\dfrac{732}{10^2}$

82. $\dfrac{8235}{10^4}$

In Exercises 83–98, evaluate the expression. Write fractional answers in reduced form.

83. $4 - 6 + 10$

84. $5 - (8 - 15)$

85. $-|2 - (6 + 5)|$

86. $125 - |10 - (25 - 3)|$

87. $5 + (2^2 \cdot 3)$

88. $181 - (13 \cdot 3^2)$

89. $(-6)^2 - (5^2 \cdot 4)$

90. $(-3)^3 + (12 \div 2^2)$

91. $(45 \div 10) \cdot 2$

92. $[360 - (8 + 12)] \div 10$

93. $\left(3 \cdot \dfrac{5}{9}\right) + 1 - \dfrac{1}{3}$

94. $\dfrac{2}{3}\left(\dfrac{3}{4}\right)^2 + 2$

95. $\dfrac{3}{2}\left(\dfrac{2}{3} + \dfrac{1}{6}\right)$

96. $\dfrac{7}{25}\left(\dfrac{7}{16} - \dfrac{1}{8}\right)$

97. $\dfrac{\frac{7}{3}\left(\frac{2}{3}\right)}{\frac{28}{15}}$

98. $\dfrac{3}{8}\left(\dfrac{1}{5}\right) \div \dfrac{25}{32}$

In Exercises 99 and 100, use a calculator to evaluate the expression. Round your result to two decimal places.

99. $300\left(1 + \dfrac{0.1}{12}\right)^{24}$

100. $\dfrac{1.32 + 4(3.68)}{1.5}$

In Exercises 101 and 102, is the statement true? Explain.

101. $-3^2 \neq (-3)(-3)$

102. $\dfrac{8 - 6}{2} \neq 4 - 6$

103. *Interpreting a Pie Graph*   The portions of total expenses for a company are shown in the pie graph. What portion of the total expenses is spent on utilities? If the total expenses are $450,000, how much is spent on utilities?

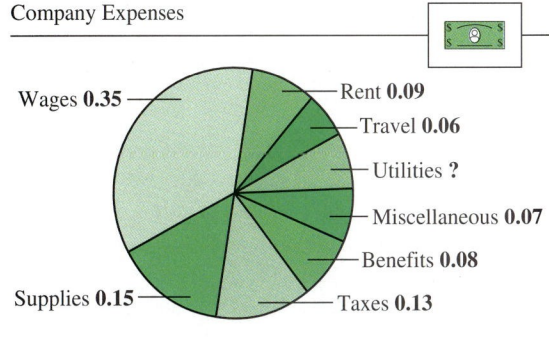

104. *Interpreting a Pie Graph*   The portions of payroll deductions for an employee are shown in the pie graph. What portion of the gross amount is represented by the net amount (the take-home pay)? If the gross amount is $1800, what is the net amount?

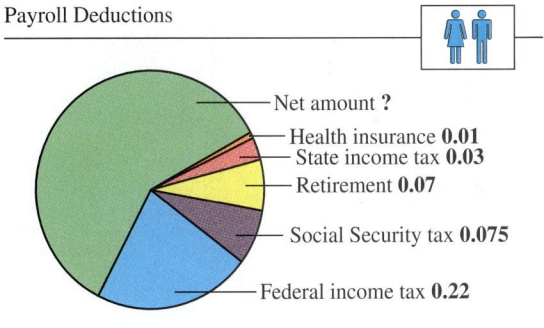

105. *Total Cost*   A car is purchased for $750 down and 48 monthly payments of $215 each. What is the total amount paid for the car?

106. *Think About It*   Your daughter suggests the following plan for an allowance during a month with 30 days. The first day of the month she will receive 1 cent, the second day 2 cents, the third day 4 cents, and so on. If the amount continues to double each day, what will her allowance be on day 30?

Math Matters Finding the Hidden Picture

In the diagram at the right, do you see three complete cubes or five complete cubes? The illusion you see is created by the repeating pattern of the blocks. Your perception of what you see changes, but the pattern remains the same.

With the capabilities of computers, complex repeating patterns can be created. Although the picture below consists of columns of roses, you may also see the three-dimensional illusion that is present.

One way to view the illusion is to hold the page up to your nose. Relax your eyes and try to look through the page. Slowly move the page away from your eyes until the illusion appears.

Another way to see the illusion is to cross your eyes while holding the page at a comfortable distance. Relax your eyes and the illusion should appear.

What Is Algebra?

Algebra is a problem-solving language that is used to solve real-life problems. It has four basic components, which tend to nest within each other, as indicated in Figure 1.3.

1. Symbolic representations and applications of the rules of arithmetic

2. Rewriting (reducing, simplifying, factoring) algebraic expressions into equivalent forms

3. Creating and solving equations

4. Studying relationships among variables by the use of functions and graphs

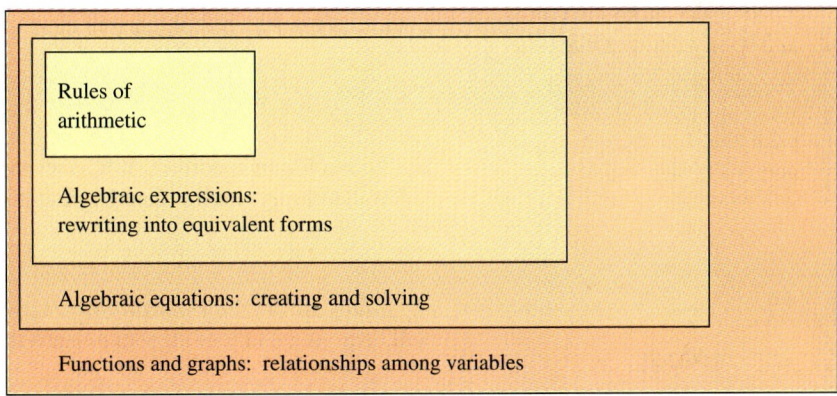

FIGURE 1.3

Notice that one of the components deals with expressions and another deals with equations. As you study algebra, it is important to understand the difference between simplifying or rewriting an algebraic *expression,* and solving an algebraic *equation.* In general, remember that a mathematical expression *has no equal sign,* whereas a mathematical equation *must have an equal sign.*

When you use an equal sign to *rewrite* an expression, you are merely indicating the *equivalence* of the new expression and the previous one.

Original Expression   Equivalent Expression

$$a + b = b + a$$
$$a - b = a + (-b)$$
$$(a + b)c = ac + bc$$

Properties of Real Numbers

Another thing that you need to remember during your study of algebra is the rules of arithmetic, as summarized on pages 31 and 32. Whether you apply these rules to numbers, variables, or expressions, it is important to remember that algebra is a logical system that depends on the adherence to rules.

EXAMPLE 1   Error Analysis

A friend of yours has written the following solutions. Are they correct? If not, what is wrong?

a. $\dfrac{1}{2} + \dfrac{1}{3} = \dfrac{1+1}{2+3} = \dfrac{2}{5}$ **b.** $\dfrac{1}{2} \cdot \dfrac{1}{3} = \dfrac{1 \cdot 1}{2 \cdot 3} = \dfrac{1}{6}$

Solution

a. This solution is *not* correct. It is an example of one of the "made-up" rules that students sometimes use in algebra. You cannot add fractions by adding their numerators and denominators. The correct way to solve this problem is to use the *rule for adding fractions*, as follows.

$$\frac{1}{2} + \frac{1}{3} = \frac{1 \cdot 3}{2 \cdot 3} + \frac{1 \cdot 2}{3 \cdot 2} = \frac{3}{6} + \frac{2}{6} = \frac{3+2}{6} = \frac{5}{6}$$

b. This solution is correct. It follows the rule for multiplying fractions. That is, you multiply two fractions by multiplying their numerators and denominators.

Many of the rules of arithmetic have names, as illustrated in Example 2. It will help your study of algebra to learn these names.

EXAMPLE 2   Identifying Properties of Real Numbers

Name the property of real numbers that justifies the given statement.

a. $3(a + 2) = 3 \cdot a + 3 \cdot 2$ **b.** $5 \cdot \dfrac{1}{5} = 1$

c. $7 + (5 + b) = (7 + 5) + b$ **d.** $(b + 3) + 0 = b + 3$

Solution

a. This statement is justified by the Distributive Property.

b. This statement is justified by the Multiplicative Inverse Property.

c. This statement is justified by the Associative Property of Addition.

d. This statement is justified by the Additive Identity Property.

STUDY TIP

To help avoid the type of error shown in Example 1(a), you should consider checking a problem using a different solution method. For instance, if you want to check whether the sum of $\frac{1}{2}$ and $\frac{1}{3}$ is $\frac{2}{5}$, you can rewrite each number in decimal form. In that form, it is clear that the sum of 0.5 and 0.333 is not 0.4.

Technology

The Multiplicative Inverse Property is another rule of arithmetic that can be added to the list on page 32. It states that *The product of a nonzero real number and its reciprocal is one.*

$$a \cdot \frac{1}{a} = 1, a \neq 0$$

This property can be illustrated on your calculator by using the reciprocal key $\boxed{1/x}$ or $\boxed{x^{-1}}$. Try doing this with $a = \frac{2}{3}$.

EXAMPLE 3   Using the Properties of Real Numbers

Complete each statement using the specified property of real numbers.

a. Multiplicative Identity Property:

$(3b)1 = $ ___

b. Associative Property of Addition:

$(c + 2) + 7 = $ ___

c. Additive Inverse Property:

$0 = 3a + $ ___

d. Distributive Property:

$3 \cdot a + 3 \cdot 4 = $ ___

Solution

a. By the Multiplicative Identity Property, you can write

$(3b)1 = 3b.$

b. By the Associative Property of Addition, you can write

$(c + 2) + 7 = c + (2 + 7).$

c. By the Additive Inverse Property, you can write

$0 = 3a + (-3a).$

d. By the Distributive Property, you can write

$3 \cdot a + 3 \cdot 4 = 3(a + 4).$

One of the distinctive things about algebra is that its rules make sense. You don't have to accept them on "blind faith"—instead, you can learn the reasons that the rules work. For instance, the next example looks at some basic differences among the operations of addition, multiplication, subtraction, and division.

EXAMPLE 4   Properties of Real Numbers

In the summary of properties of real numbers on page 32, why are all the properties listed in terms of addition and multiplication and not subtraction and division?

Solution

The reason for this is that subtraction and division fail to possess many of the properties listed in the summary. For instance, subtraction and division are not commutative. To see this, consider the following.

$$7 - 5 \neq 5 - 7 \quad \text{and} \quad 12 \div 4 \neq 4 \div 12$$

Similarly, subtraction and division are not associative.

$$9 - (5 - 3) \neq (9 - 5) - 3 \quad \text{and} \quad 12 \div (4 \div 2) \neq (12 \div 4) \div 2$$

Problem-Solving Strategies

We already mentioned that algebra is a problem-solving language that can be used to solve real-life problems. This use of algebra involves writing verbal models that can be translated into algebraic models, as discussed in Section 2.4. In addition to mathematical modeling, there are also many other problem-solving strategies that can help you succeed in this course.

Summary of Problem-Solving Strategies

1. **Guess, Check, and Revise** Guess a reasonable solution based on the given data. Check the guess, and revise it, if necessary. Continue guessing, checking, and revising until a correct solution is found.

2. **Make a Table/Look for a Pattern** Make a table using the data in the problem. Look for a number pattern. Then use the pattern to complete the table or find a solution.

3. **Draw a Diagram** Draw a diagram that shows the facts from the problem. Use the diagram to visualize the action of the problem. Use algebra to find a solution. Then check the solution against the facts.

4. **Solve a Simpler Problem** Construct a simpler problem that is similar to the given problem. Solve the simpler problem. Then use the same procedure to solve the given problem.

EXAMPLE 5   *Guess, Check, and Revise*

You deposit \$500 in an account that earns 6% interest compounded quarterly. The balance in the account after t years is

$$A = 500 \left(1 + \frac{0.06}{4} \right)^{4t}.$$

How long will it take for your investment to double?

Solution

You can solve this problem using a *Guess, Check, and Revise* strategy. For instance, you might guess that it takes 5 years to double. The balance in 5 years is

$$A = 500 \left(1 + \frac{0.06}{4} \right)^{4(5)} \approx \$673.43.$$

Because the amount has not yet doubled, you increase your guess. After trying several, you can determine that your balance doubles in about 11.7 years.

EXAMPLE 6   *Make a Table/Look for a Pattern*

Find the following products. Then describe the pattern and use your description to find the product of 14 and 16.

$$1 \cdot 3, \quad 2 \cdot 4, \quad 3 \cdot 5, \quad 4 \cdot 6, \quad 5 \cdot 7, \quad 6 \cdot 8, \quad 7 \cdot 9$$

Solution

One way to help find a pattern is to organize the results in a table.

Numbers	$1 \cdot 3$	$2 \cdot 4$	$3 \cdot 5$	$4 \cdot 6$	$5 \cdot 7$	$6 \cdot 8$	$7 \cdot 9$
Product	3	8	15	24	35	48	63

From the table, you can see that each of the products is 1 less than a perfect square. For instance, 3 is 1 less than 2^2 or 4, 8 is 1 less than 3^2 or 9, 15 is 1 less than 4^2 or 16, and so on.

If this pattern continues for other numbers, you can hypothesize that the product of 14 and 16 is 1 less that 15^2 or 225. That is,

$$14 \cdot 16 = 15^2 - 1 = 224.$$

You can confirm this result by actually multiplying 14 and 16.

EXAMPLE 7   *Draw a Diagram*

The outer dimensions of a rectangular apartment are 25 feet by 40 feet. The combination living-room, dining-room, and kitchen areas occupy two-fifths of the apartment's area. Find the area of the remaining rooms.

Solution

For this problem, it helps to draw a diagram, as shown in Figure 1.4. From the figure, you can see that the total area in the apartment is

$$
\begin{aligned}
\text{Area} &= (\text{length})(\text{width}) \\
&= (40)(25) \\
&= 1000 \text{ square feet.}
\end{aligned}
$$

The area occupied by the living room, dining room, and kitchen is

$$\frac{2}{5}(1000) = 400 \text{ square feet.}$$

This implies that the remaining rooms must have a total area of 600 square feet.

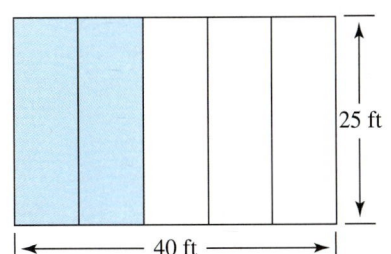

FIGURE 1.4

EXAMPLE 8 Solve a Simpler Problem

You are driving on an interstate highway and are averaging 60 miles per hour. How far will you travel in $12\frac{1}{2}$ hours?

Solution

One way to solve the problem is to use the formula that relates distance, rate, and time. Suppose, however, that you have forgotten the formula. To help you remember, you could solve some simpler problems.

- If you travel 60 miles per hour for 1 hour, you will travel 60 miles.
- If you travel 60 miles per hour for 2 hours, you will travel 120 miles.
- If you travel 60 miles per hour for 3 hours, you will travel 180 miles.

From these examples, it appears that you can find the total miles traveled by multiplying the rate times the time. Thus, if you travel 60 miles per hour for $12\frac{1}{2}$ hours, you will travel a distance of

$$(60)(12.5) = 750 \text{ miles.}$$

Group Activities Communicating Mathematically

Order of Operations   Using the established order of operations, the value of $7 \cdot 8 + 12$ is

$$7 \cdot 8 + 12 = 56 + 12 = 68.$$

By inserting parentheses into the expression, you can obtain a value of

$$7 \cdot (8 + 12) = 7(20) = 140.$$

Using the established order of operations, which of the following expressions has a value of 72? For those that don't, decide whether you can insert parentheses into the expression so that its value is 72.

a. $4 + 2^3 - 7$ **b.** $4 + 8 \cdot 6$

c. $93 - 25 - 4$ **d.** $70 + 10 \div 5$

e. $60 + 20 \div 2 + 32$ **f.** $35 \cdot 2 + 2$

1.4 Exercises

Discussing the Concepts

1. In your own words, state the Commutative Properties of Addition and Multiplication. Give an example of each.

2. In your own words, state the Associative Properties of Addition and Multiplication. Give an example of each.

3. Consider the operation of addition.

   (a) In your own words, describe the Additive Identity Property. Give an example.

   (b) In your own words, describe the Additive Inverse Property. Give an example.

4. Consider the operation of multiplication.

   (a) In your own words, describe the Multiplicative Identity Property. Why do you think it has this name? Give an example of the property's use.

   (b) In your own words, describe the Multiplicative Inverse Property. Why do you think it has this name? Give an example of the property's use.

5. Explain why $5(7 + 12) \neq 5 \cdot 7 + 12$.

6. Consider the rectangle shown in the figure.

   (a) Find the area of the large rectangle by adding the areas of Regions I and II.

   (b) Find the area of the large rectangle by multiplying its length by its width.

   (c) Explain how the results of parts (a) and (b) relate to the Distributive Property.

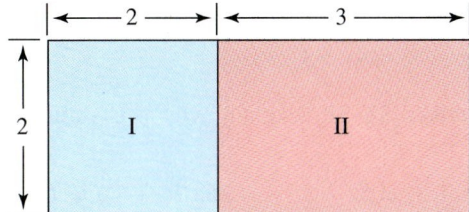

Problem Solving

In Exercises 7–16, state the property of real numbers that justifies the statement.

7. $6(-3) = -3(6)$

8. $16 + 10 = 10 + 16$

9. $0 + 15 = 15$

10. $1 \cdot 4 = 4$

11. $(2 \cdot 3)4 = 2(3 \cdot 4)$

12. $6(3 + x) = 6 \cdot 3 + 6x$

13. $7\left(\dfrac{1}{7}\right) = 1$

14. $14 + (-14) = 0$

15. $(10 + 3) + 2 = 10 + (3 + 2)$

16. $(14 - 2)3 = 14 \cdot 3 - 2 \cdot 3$

In Exercises 17 and 18, use the Commutative Property of Addition or Multiplication to rewrite the expression.

17. $5(u + v) =$

18. $10(-3) =$

In Exercises 19 and 20, use the Distributive Property to rewrite the expression.

19. $6(x + 2) =$

20. $5(u + v) =$

In Exercises 21–24, find (a) the additive inverse and (b) the multiplicative inverse of the quantity.

21. 50

22. 12

23. -1

24. $-\dfrac{1}{2}$

In Exercises 25–28, rewrite the expression using the Associative Property of Addition or Multiplication.

25. $10 + (8 + 2)$

26. $(z + 5) + 15$

27. $(2 \cdot 3) \cdot 4$

28. $10 \cdot (6x)$

In Exercises 29–32, simplify the expression.

29. $3(6 + 10)$

30. $4(8 - 3)$

31. $\frac{2}{3}(9z + 24)$

32. $\frac{1}{2}(4 - 2u)$

In Exercises 33 and 34, identify the property of real numbers used to justify each rewritten step.

33. $3 + 10(x + 1) = 3 + 10x + 10$
$$= 3 + 10 + 10x$$
$$= (3 + 10) + 10x$$
$$= 13 + 10x$$

34. $4(2 + x) = 4(x + 2)$
$$= 4x + 8$$

In Exercises 35–38, explain why the statement is true.

35. $5(x + 3) \neq 5x + 3$

36. $7(x - 2) \neq 7x - 2$

37. $\frac{8}{0} \neq 0$

38. $5\left(\frac{1}{5}\right) \neq 0$

39. *Sales Tax*   You purchase an item for x dollars. There is a 6% sales tax, which implies that the total amount you must pay is $x + 0.06x$.

(a) Use the Distributive Property to rewrite the expression.

(b) How much must you pay if the item costs $25.95?

40. *Geometry*   You measure the width of a movie screen and find that it is 30 feet. You don't have a ladder to measure its height, but you are told that it is 8 feet less than the width. Write an expression for the area of the movie screen. Use the Distributive Property to rewrite the expression.

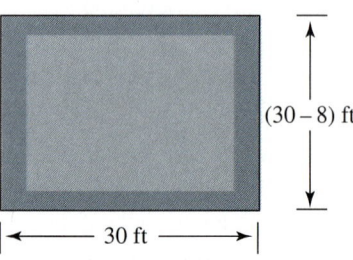

$(30 - 8)$ ft

$\longleftarrow$ 30 ft $\longrightarrow$

Reviewing the Major Concepts

In Exercises 41–44, evaluate the expression.

41. $\dfrac{4 + 5 \cdot 2 - 14}{7}$

42. $\dfrac{7}{8} \div \dfrac{3}{16}$

43. $(-4)^2 - (30 \div 5)$

44. $(8 \cdot 9) + (-4)^3$

45. *Work Progress*   At the beginning of May, a construction project sign stated that $\frac{1}{6}$ of the project was complete. At the beginning of June, $\frac{2}{5}$ of the project was complete. What fraction of the work was completed during the month of May?

46. *Balance* $2500 is deposited in an account earning 6% compounded yearly. After 5 years, the balance is $A = 2500(1.06)^5$. Find the balance.

Additional Problem Solving

In Exercises 47–58, state the property of real numbers that justifies the statement.

47. $x + 10 = 10 + x$

48. $8x = x(8)$

49. $-16 + 16 = 0$

50. $25 + (-25) = 0$

51. $4(3 \cdot 10) = (4 \cdot 3)10$

52. $\dfrac{1}{a}(3 + y) = \dfrac{1}{a}(3) + \dfrac{1}{a}(y)$

53. $10(6 - y) = 10 \cdot 6 - 10 \cdot y$

54. $10(6 - y) = (6 - y)10$

55. $(4 + x)(2 - x) = 4(2 - x) + x(2 - x)$

56. $(32 + 8) + 5 = 32 + (8 + 5)$

57. $x + (y + 3) = (x + y) + 3$

58. $[(x + y)u]v = (x + y)(uv)$

In Exercises 59 and 60, use the Commutative Property of Addition or Multiplication to rewrite the expression.

59. $(3 + x)7 =$ ▢   **60.** $y + 5 =$ ▢

In Exercises 61 and 62, use the Associative Property of Addition or Multiplication to rewrite the expression.

61. $3x + (2y + 5) =$ ▢   **62.** $12(3 \cdot 4) =$ ▢

In Exercises 63 and 64, use the Distributive Property to rewrite the expression.

63. $(4 + y)25 =$ ▢   **64.** $x(4 - y) =$ ▢

In Exercises 65–68, find (a) the additive inverse and (b) the multiplicative inverse of the expression.

65. $2x$ **66.** $5y$

67. ab **68.** uv

In Exercises 69–72, rewrite the expression using the Associative Property of Addition or Multiplication.

69. $(x + 3) + 2$ **70.** $16 + (4 + 3)$

71. $2 \cdot (3y)$ **72.** $3 \cdot (5 \cdot 10)$

In Exercises 73–76, rewrite the expression using the Distributive Property, and simplify the answer.

73. $3(2x - 4)$ **74.** $10(15 - 3t)$

75. $\dfrac{3}{5}(10y - 45)$ **76.** $x(3 + x)$

In Exercises 77 and 78, identify the property of real numbers used to justify each step in rewriting the expression.

77. $7x + 9 + 2x = 7x + 2x + 9$
$$= (7x + 2x) + 9$$
$$= (7 + 2)x + 9$$
$$= 9x + 9$$
$$= 9(x + 1)$$

78. $2(x + 3) + x = 2x + 2 \cdot 3 + x$
$$= 2x + x + 6$$
$$= 3x + 6$$
$$= 3(x + 2)$$

79. *Geometry*   Write an expression for the perimeter of the triangle shown in the figure. Use the properties of real numbers to simplify the expression.

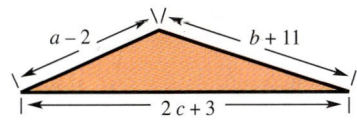

80. *Geometry*   Find the area of the blue rectangle in two different ways. Explain how the results are related to the Distributive Property.

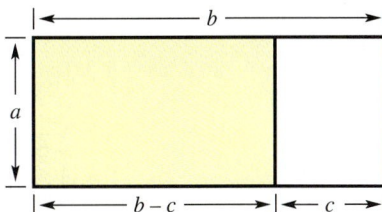

Think About It   In Exercises 81 and 82, determine whether the order in which the two activities are performed is "commutative." That is, do you obtain the same results regardless of which activity is performed first?

81. (a) "Drain the used oil from the engine."

(b) "Fill the crankcase with 5 quarts of new oil."

82. (a) "Weed the flower beds."

(b) "Mow the lawn."

CHAPTER PROJECT: Musical Sound

Musical sound (like any sound) is what you hear when a sound wave traveling through the air strikes your ears. A graph of a sound wave is shown at the left.

Every sound wave has a frequency. Frequency is the number of complete cycles of the wave that occur in one second. Frequency is measured in cycles per second.

$$\text{Frequency} = \frac{\text{number of cycles}}{1 \text{ second}}$$

There are only 88 different sounds that are considered to be musical sounds. The only musical instrument that can produce each musical sound is the piano. That is why a piano has 88 keys. Other musical instruments, such as a violin or trombone, can play only a portion of these sounds.

In the figure below, 44 piano keys along with their notes and frequencies are shown. You will be asked to complete the piano keyboard along with other investigations in the following questions.

Sound Wave

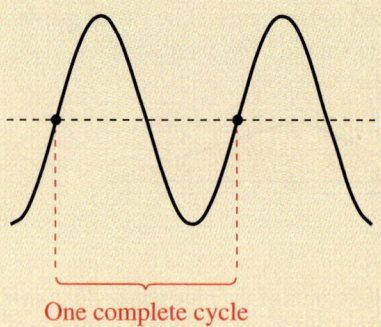

One complete cycle

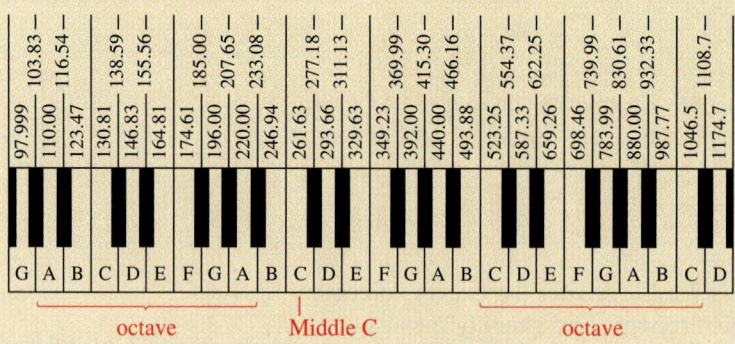

octave   Middle C   octave

1. Find the frequency. Then list the corresponding note.

   (a) 882 cycles in $4\frac{1}{2}$ seconds

   (b) $261\frac{5}{8}$ cycles in $\frac{1}{2}$ second

2. The period of a sound wave is defined as the reciprocal of the sound wave's frequency. Write this definition mathematically. What units are used to measure the period of a sound wave?

3. Consider two musical sounds. One has a period of about 0.00455 seconds per cycle and the other has a period of about 0.00681 seconds per cycle. Which of the two has a higher frequency? What are the names of these two musical sounds (notes)?

4. In music, an octave consists of any 13 consecutive notes. (See figure.) Note that the beginning and ending notes in an octave have the same name. How do the frequencies of such notes compare?

5. The lowest and highest musical sounds that can be produced by a piano have frequencies of 27.5 cycles per second and 4186 cycles per second. Using the results of Question 4, find the names of these musical sounds.

6. Using the results of Questions 4 and 5, construct a diagram of a complete piano keyboard. Label the notes and the frequencies that correspond to each key.

CHAPTER SUMMARY

After studying this chapter, you should have acquired the following skills. These skills are keyed to the Review Exercises that begin on page 88. Answers to odd-numbered Review Exercises are given in the back of the book.

- Write an equivalent fraction with the indicated denominator. *(Section 1.1)*

  Review Exercises 1–4

- Evaluate sums and differences of fractions, integers, and mixed numbers with like and unlike denominators. *(Section 1.1)*

  Review Exercises 5–16

- Use a table to solve real-life problems. *(Section 1.2)*

  Review Exercise 17

- Translate verbal sentences into arithmetic expressions and solve. *(Sections 1.1, 1.2, 1.3)*

  Review Exercises 18, 39–42, 55

- Find the reciprocal of a number. *(Section 1.2)*

  Review Exercises 19–22

- Evaluate products and quotients of fractions, integers, and decimals. *(Section 1.2)*

  Review Exercises 23–34

- Write fractions in decimal form. *(Section 1.2)*

  Review Exercises 35–38

- Evaluate expressions with exponents. *(Section 1.3)*

  Review Exercises 43–46

- Compare two real numbers with exponents. *(Section 1.3)*

  Review Exercises 47–50

- Use a calculator to evaluate expressions. *(Section 1.3)*

  Review Exercises 51–54

- Solve problems involving geometry. *(Sections 1.2, 1.3, 1.4)*

  Review Exercises 56, 87

- Use a calculator to experiment with exponents. *(Section 1.3)*

  Review Exercises 57, 58

- Evaluate sums, differences, products, and quotients. *(Sections 1.1, 1.2, 1.3)*

  Review Exercises 59–76

- Name the property of real numbers. *(Section 1.4)*

  Review Exercises 77–86

REVIEW EXERCISES

In Exercises 1–4, complete the statement.

1. $\dfrac{2}{3} = \dfrac{}{15}$

2. $\dfrac{3}{7} = \dfrac{}{28}$

3. $\dfrac{6}{10} = \dfrac{}{25}$

4. $\dfrac{9}{12} = \dfrac{}{16}$

In Exercises 5–16, evaluate the expression. Write the result in reduced form.

5. $\dfrac{3}{25} + \dfrac{7}{25}$

6. $\dfrac{9}{64} + \dfrac{7}{64}$

7. $\dfrac{27}{16} - \dfrac{15}{16}$

8. $-\dfrac{5}{12} + \dfrac{1}{12}$

9. $-\dfrac{5}{9} + \dfrac{2}{3}$

10. $\dfrac{7}{15} - \dfrac{2}{25}$

11. $\dfrac{25}{32} + \dfrac{7}{24}$

12. $-\dfrac{7}{8} - \dfrac{11}{12}$

13. $5 - \dfrac{15}{4}$

14. $\dfrac{12}{5} - 3$

15. $5\tfrac{3}{4} - 3\tfrac{5}{8}$

16. $-3\tfrac{7}{10} + 1\tfrac{1}{20}$

17. *Reading a Table* Initially, a share of stock cost $35\tfrac{1}{4}$. The daily changes in closing values during the week are shown in the table. Determine the closing price of a share on Friday.

Day	Mon	Tue	Wed	Thu	Fri
Change	$-\tfrac{3}{8}$	$-\tfrac{1}{2}$	$-\tfrac{1}{8}$	$+1\tfrac{1}{4}$	$+\tfrac{1}{2}$

18. *Fuel Consumption* The morning and evening readings of the fuel gauge on a car were $\tfrac{7}{8}$ and $\tfrac{1}{3}$. What fraction of a tank of fuel did the car use that day?

In Exercises 19–22, find the reciprocal of the number.

19. 9

20. 15

21. $-\dfrac{5}{3}$

22. $\dfrac{4}{15}$

In Exercises 23–34, evaluate the expression. If it is not possible, explain why.

23. $\dfrac{5}{8} \cdot \dfrac{-2}{15}$

24. $\dfrac{3}{32} \cdot \dfrac{32}{3}$

25. $35 \left(\dfrac{1}{35} \right)$

26. $-\dfrac{5}{12} \left(-\dfrac{4}{25} \right)$

27. $\dfrac{5}{14} \div \dfrac{15}{28}$

28. $-\dfrac{7}{10} \div \dfrac{4}{15}$

29. $\dfrac{-\tfrac{3}{4}}{-\tfrac{7}{8}}$

30. $\dfrac{\tfrac{15}{32}}{-5}$

31. $\dfrac{\tfrac{5}{9}}{0}$

32. $\dfrac{0}{12}$

33. $\dfrac{5.25}{0.25}$

34. $(5.2)(16.8)$

In Exercises 35–38, write the fraction in decimal form. (Use the bar notation for repeating digits.)

35. $\dfrac{15}{8}$

36. $\dfrac{11}{25}$

37. $\dfrac{5}{12}$

38. $\dfrac{5}{11}$

39. *Telephone Charge* A telephone call costs $0.64 for the first minute plus $0.72 for each additional minute. Find the cost of a 5-minute call.

40. *Total Charge* To buy a television, you make a down payment of $75, plus nine monthly payments of $25 each. What is the total amount you will pay?

41. *Unit Cost* A container of food weighing 22 ounces is purchased for $1.43. Find the cost per ounce.

42. *Snowfall Rate* During an 8-hour period, $6\tfrac{3}{4}$ inches of snow fell. What was the average rate of snowfall per hour?

In Exercises 43–46, evaluate the expression.

43. 7^3

44. $(-5)^2$

45. $(-7)^3$

46. $-(-2)^4$

In Exercises 47–50, insert the correct symbol ($<$, $>$, or $=$) between the numbers.

47. 2^2 ▭ 2^4

48. $(-3)^2$ ▭ $(-3)^3$

49. $\dfrac{3}{4}$ ▭ $\left(\dfrac{3}{4}\right)^2$

50. $\left(\dfrac{2}{3}\right)^3$ ▭ $\left(\dfrac{2}{3}\right)^2$

In Exercises 51–54, use a calculator to evaluate the expression. Round your answer to two decimal places.

51. $(5.8)^4 - (3.2)^5$

52. $\dfrac{(15.8)^3}{(2.3)^8}$

53. $\dfrac{3000}{(1.05)^{10}}$

54. $500\left(1 + \dfrac{0.07}{4}\right)^{40}$

55. *Depreciation*   After 3 years, the value of a $16,000 car is given by $16{,}000\left(\frac{3}{4}\right)^3$.

(a) What is its value after 3 years?

(b) How much has the car depreciated during the 3 years?

56. *Geometry*   The volume of water in a hot tub is given by $V = 6^2 \cdot 3$. How many cubic feet of water will the hot tub hold? Find the total weight of the water in the tub. (Use the fact that 1 cubic foot of water weighs 62.4 pounds.)

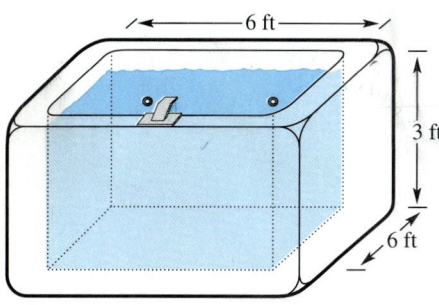

57. *Calculator Experiment*   Enter any number between 0 and 1 in a calculator. Square the number, and then square the result. Continue this process. What number does the calculator display seem to be approaching? Explain.

58. *Calculator Experiment*   Use a calculator to calculate 15^4 in two ways. First, use the exponential key y^x. Second, enter 15 and press the square key x^2 twice. Why do these two methods give the same result?

In Exercises 59–76, evaluate the expression. Write fractional results in reduced form.

59. $3.2 - 1.5 + 11.4$

60. $-25 + 16.5 + 3.75$

61. $\left(\dfrac{3}{5}\right)^4$

62. $\dfrac{2}{6^3}$

63. $240 - (4^2 \cdot 5)$

64. $5^2 - (625 \cdot 5^2)$

65. $3^2(10 - 2^2)$

66. $-5(16 - 5^2)$

67. $\left(\dfrac{3}{4}\right)\left(\dfrac{5}{6}\right) + 4$

68. $75 - 24 \div 2^3$

69. $\dfrac{3}{5}\left(\dfrac{3}{4} - \dfrac{2}{3}\right)$

70. $1 + \left(-\dfrac{2}{3}\right)\left(\dfrac{1}{8}\right) + \dfrac{5}{4}$

71. $\dfrac{\frac{3}{2} - \frac{1}{4}}{\frac{3}{8}}$

72. $\dfrac{-\frac{3}{5} - \frac{1}{2}}{\frac{9}{10}}$

73. $\left(\dfrac{16}{25} - \dfrac{3}{5}\right) \div \dfrac{3}{10}$

74. $\left(\dfrac{11}{12} - \dfrac{3}{4}\right) \div \dfrac{5}{6}$

75. $\dfrac{\frac{3}{8} + \frac{7}{8}}{\frac{5}{16}}$

76. $\dfrac{\frac{3}{5}}{\frac{1}{2} - \frac{1}{5}}$

In Exercises 77–86, state the property of real numbers that justifies the statement.

77. $123 - 123 = 0$

78. $9 \cdot \dfrac{1}{9} = 1$

79. $14(3) = 3(14)$

80. $5(3x) = (5 \cdot 3)x$

81. $17 \cdot 1 = 17$

82. $10 + 6 = 6 + 10$

83. $r + (2s + 3) = (r + 2s) + 3$

84. $8(7 + 5) = 8 \cdot 7 + 8 \cdot 5$

85. $-2(7 + x) = -2 \cdot 7 + (-2)x$

86. $4 + (3 + x) = (4 + 3) + x$

87. *Geometry*   Sketch a diagram of a triangle. Label its base as $2x + 6$ and its height as 7. Write an expression for the area of the triangle. (The area of a triangle is one-half its base times its height.) Use the Distributive Property to rewrite the expression.

CHAPTER TEST

Take this test as you would take a test in class. After you are done, check your work against the answers given in the back of the book.

1. Write the fraction $\frac{30}{72}$ in reduced form.

2. Write the fraction $\frac{5}{9}$ in decimal form rounded to three decimal places.

3. Explain why the value of -3^4 is not equal to $(-3)^4$.

4. State the order of operations for the expression $32 - 3 \cdot 2^3$.

In Exercises 5–14, evaluate the expression. Write fractions in reduced form.

5. $\dfrac{5}{16} + \dfrac{3}{16}$

6. $\dfrac{5}{6} - \dfrac{1}{8}$

7. $\left(-\dfrac{3}{4}\right)\left(-\dfrac{6}{15}\right)$

8. $-27\left(\dfrac{5}{6}\right)$

9. $\dfrac{7}{16} \div \dfrac{21}{28}$

10. $\dfrac{-8.1}{0.3}$

11. $(-4)^3$

12. $-\left(\dfrac{2}{3}\right)^2$

13. $35 - (50 \div 5^2)$

14. $\dfrac{1}{4}\left(\dfrac{3}{5} - \dfrac{1}{10}\right)$

In Exercises 15–18, state the property of real numbers that justifies the statement.

15. $3(4 + 6) = 3 \cdot 4 + 3 \cdot 6$ *Distributive*

16. $5 \cdot \frac{1}{5} = 1$ *— Multiplicative inverse*

17. $3 + (4 + 8) = (3 + 4) + 8$ *Associative*

18. $3(x + 2) = (x + 2)3$ *Commutative*

19. You purchase three boxes of cereal at $2.79 per box and two cans of pineapple at $0.59 per can. If you give the cashier $20, how much change should you receive? (Assume there is no sales tax.)

20. Copy the figure shown below. Then shade two-thirds of the figure. Write two different fractions that are represented by the shaded region. Which of these is in reduced form?

Fundamentals of Algebra

2

- Algebraic Expressions and Exponents
- Basic Rules of Algebra
- Rewriting and Evaluating Algebraic Expressions
- Translating Expressions: Verbal to Algebraic
- Introduction to Equations

In 1593, the first known thermometer was invented by the Italian astronomer, Galileo. Galileo's thermometer, called a thermoscope, was not always accurate. As a result, an accurate thermometer containing alcohol was developed in 1641. In 1714, German physicist Gabriel D. Fahrenheit built a mercury thermometer similar to that used today. He also gave the temperature scale, Fahrenheit, its name. In 1742, Swedish astronomer Anders Celsius devised the Celsius temperature scale. The table at the right shows common Fahrenheit and Celsius equivalents.

The most common thermometers used today are called liquid-in-glass. When measuring outdoor temperatures, meteorologists use liquid-in-glass thermometers.

Property	°C	°F
Freezing Point of Water	0	32
Normal Human Body Temperature	37	98.6
Boiling Point of Water	100	212

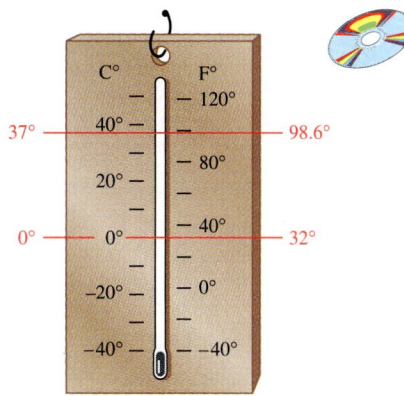

The chapter project related to this data is on page 145.

2.1	**Algebraic Expressions and Exponents**

Variables and Algebraic Expressions ▪ Positive Integer Exponents ▪ Properties of Exponents

Variables and Algebraic Expressions

One of the distinguishing characteristics of algebra is its use of symbols to represent quantities whose numerical values are unknown. Here is a simple example.

EXAMPLE 1 Writing an Algebraic Expression

Suppose you accept a part-time job for $6 an hour. The job offer states that you will be expected to work between 15 and 30 hours a week. Because you don't know how many hours you will work during a given week, your total income for a week is unknown. Moreover, your income would likely *vary* from week to week. By representing the variable quantity (the number of hours worked) by the letter x, you can represent the weekly income by the following *algebraic expression*.

$6 per hour   Number of hours worked

$$6x$$

In the product $6x$, the number 6 is a *constant* and the letter x is a *variable*.

Algebraic Expression

A collection of letters (called **variables**) and real numbers (called **constants**) combined by using addition, subtraction, multiplication, or division is called an **algebraic expression.**

Some examples of algebraic expressions are

$$3x + y, \quad -5a^3, \quad 2W - 7, \quad \frac{x}{y+3}, \quad \text{and} \quad x^2 - 4x + 5.$$

The **terms** of an algebraic expression are those parts that are separated by *addition*. For example, the expression $x^2 - 4x + 5$ has three terms: x^2, $-4x$, and 5. Note that $-4x$, rather than $4x$, is a term of $x^2 - 4x + 5$ because

$$x^2 - 4x + 5 = x^2 + (-4x) + 5. \qquad \text{To subtract, add the opposite.}$$

For variable terms such as x^2 and $-4x$, the numerical factor is called the **coefficient** of the term. Here, the coefficient of x^2 is 1 and the coefficient of $-4x$ is -4.

EXAMPLE 2   Identifying the Terms of an Algebraic Expression

Identify the terms of each algebraic expression.

a. $3x - \dfrac{1}{2}$ **b.** $2y - 5x - 7$

c. $5(x - 3) + 3x - 4$ **d.** $4 - 6x + \dfrac{x + 9}{3}$

Solution

Algebraic Expression	Terms
a. $3x - \dfrac{1}{2}$	$3x, \ -\dfrac{1}{2}$
b. $2y - 5x - 7$	$2y, \ -5x, \ -7$
c. $5(x - 3) + 3x - 4$	$5(x - 3), \ 3x, \ -4$
d. $4 - 6x + \dfrac{x + 9}{3}$	$4, \ -6x, \ \dfrac{x + 9}{3}$

The terms of an algebraic expression depend on the way the expression is written. Rewriting the expression can (and in fact, usually does) change its terms. For instance, the expression $2 + 4 - x$ has three terms, but the equivalent expression $6 - x$ only has two terms.

EXAMPLE 3   Identifying Coefficients

Identify the coefficient of each of the following terms.

a. $-5x^2$ **b.** x^3 **c.** $\dfrac{2x}{3}$ **d.** $3(2 - x)$ **e.** $-x^3$

Solution

Term	Coefficient	Comment
a. $-5x^2$	-5	Note that $-5x^2 = (-5)x^2$.
b. x^3	1	Note that $x^3 = 1 \cdot x^3$.
c. $\dfrac{2x}{3}$	$\dfrac{2}{3}$	Note that $\dfrac{2x}{3} = \dfrac{2}{3}(x)$.
d. $3(2 - x)$	3	Note that $3(2 - x) = 6 + (-3)x$.
e. $-x^3$	-1	Note that $-x^3 = (-1)x^3$.

Positive Integer Exponents

You know from Section 1.3 that exponents can be used to denote repeated multiplication. For example, 7^4 represents the product obtained by multiplying 7 by itself four times.

$$\overset{\text{Exponent}}{7^4} = \underbrace{7 \cdot 7 \cdot 7 \cdot 7}_{\text{4 factors}}$$

In general, for any positive integer n and any real number a, you have

$$a^n = \underbrace{a \cdot a \cdot a \cdots a}_{n \text{ factors}}.$$

This rule applies to factors that are variables as well as to factors that are algebraic expressions.

Definition of Exponential Form

Let n be a positive integer and let a be a real number, a variable, or an algebraic expression.

$$a^n = \underbrace{a \cdot a \cdot a \cdots a}_{n \text{ factors}}$$

In this definition remember that the letter a can be a number, a variable, or an algebraic expression. It may be helpful to think of a as a box into which you can place any algebraic expression.

EXAMPLE 4 *Interpreting Exponential Expressions*

Write each expression as a product.

a. 3^4 **b.** x^4 **c.** $(-2x)^4$ **d.** $(y+2)^4$ **e.** $(5m^2)^4$

Solution

a. $3^4 = 3 \cdot 3 \cdot 3 \cdot 3$

b. $x^4 = x \cdot x \cdot x \cdot x$

c. $(-2x)^4 = (-2x)(-2x)(-2x)(-2x)$

d. $(y+2)^4 = (y+2)(y+2)(y+2)(y+2)$

e. $(5m^2)^4 = (5m^2)(5m^2)(5m^2)(5m^2)$

EXAMPLE 5   Writing Products in Exponential Form

Write each product in exponential form.

a. $5 \cdot 5 \cdot 5 \cdot 5$ **b.** $3 \cdot x \cdot x \cdot x$ **c.** $3x \cdot 3x \cdot 3x$

d. $xxyyy$ **e.** $2 \cdot 2 \cdot 2 \cdot aab$ **f.** $xzzyzy$

Solution

a. $\underbrace{5 \cdot 5 \cdot 5 \cdot 5}_{\text{4 factors}} = 5^4$

b. $3 \cdot \underbrace{x \cdot x \cdot x}_{\text{3 factors}} = 3 \cdot x^3 = 3x^3$ 3 *is not* a factor in the base.

c. $\underbrace{3x \cdot 3x \cdot 3x}_{\text{3 factors}} = (3x)^3$ 3 *is* a factor in the base.

d. $xxyyy = x^2y^3$

e. $2 \cdot 2 \cdot 2 \cdot aab = 2^3a^2b$

f. $xzzyzy = xyyzzz = xy^2z^3$ Commutative Property of Multiplication

Be sure you understand the priorities for order of operations involving exponents. Here are two examples that tend to cause problems.

Expression	Correct Evaluation	Incorrect Evaluation
-3^2	$-(3 \cdot 3) = -9$	$\cancel{(-3)(-3) = 9}$
$3x^2$	$3 \cdot x \cdot x$	$\cancel{(3x)(3x)}$

EXAMPLE 6   Writing Exponential Forms as Products

Write each expression as a product of factors.

a. $(2)^3 \cdot (-3)^2$ **b.** $(5x)^2y^3$ **c.** $5x^2y^3$

d. $-x^4z$ **e.** $(-x)^4z$

Solution

a. $(2)^3 \cdot (-3)^2 = 2 \cdot 2 \cdot 2 \cdot (-3)(-3)$

b. $(5x)^2y^3 = (5x)(5x)yyy = 5 \cdot 5xxyyy$

c. $5x^2y^3 = 5xxyyy$

d. $-x^4z = -(x^4)z = -xxxxz$

e. $(-x)^4z = (-x)(-x)(-x)(-x)z = xxxxz$

Use the definition of exponential form to write each of the following expressions as a single power of 2. From the results, can you find a general rule for simplifying expressions of the form $a^m \cdot a^n$?

a. $2^1 \cdot 2^2$

b. $2^1 \cdot 2^4$

c. $2^2 \cdot 2^2$

d. $2^3 \cdot 2^5$

e. $2^4 \cdot 2^2$

Properties of Exponents

When multiplying two exponential expressions that have the *same* base, you add exponents. To see why this is true, consider the product

$$a^3 \cdot a^2.$$

Because the first expression represents $a \cdot a \cdot a$ and the second represents $a \cdot a$, it follows that the product of the two expressions represents $a \cdot a \cdot a \cdot a \cdot a$, as follows.

$$a^3 \cdot a^2 = \underbrace{(a \cdot a \cdot a)}_{\substack{\text{Three} \\ \text{factors}}} \cdot \underbrace{(a \cdot a)}_{\substack{\text{Two} \\ \text{factors}}} = \underbrace{(a \cdot a \cdot a \cdot a \cdot a)}_{\substack{\text{Five} \\ \text{factors}}} = a^5 = a^{3+2}$$

This property of exponents is summarized as follows.

Multiplying Exponential Forms with the Same Base

Let m and n be positive integers, and let a be a real number, a variable, or an algebraic expression.

$$a^m \cdot a^n = a^{m+n}$$

This rule extends to three or more factors of a raised to positive integer powers. For example,

$$a^m \cdot a^n \cdot a^k = a^{m+n+k}.$$

In the next example you can see how this rule can be used to simplify products involving exponential forms.

EXAMPLE 7 *Simplifying Products Involving Exponential Forms*

Simplify each expression.

a. $5^2 \cdot 5^6 \cdot 5$ b. $b^4 b^2 b$

c. $3^2 x^3 \cdot x$ d. $(-9x^2)(-3x^5)$

Solution

a. $5^2 \cdot 5^6 \cdot 5 = 5^{2+6+1} = 5^9$

b. $b^4 b^2 b = b^{4+2+1} = b^7$

c. $3^2 x^3 \cdot x = (3^2)(x^{3+1}) = 9x^4$

d. $(-9x^2)(-3x^5) = (-9)(-3)(x^2 \cdot x^5) = 27(x^{2+5}) = 27x^7$

EXAMPLE 8 *Simplifying Products Involving Exponential Forms*

Simplify each expression.

a. $(-3x^2y)(5xy)(2y^2)$ **b.** $(-2x^4)^3$ **c.** $2xy^3(3x^2y)^2$

Solution

a. $(-3x^2y)(5xy)(2y^2) = (-3)(5)(2)(x^2 \cdot x)(y \cdot y \cdot y^2)$
$$= -30x^3y^4$$

b. $(-2x^4)^3 = (-2x^4)(-2x^4)(-2x^4)$
$$= (-2)(-2)(-2)(x^4)(x^4)(x^4)$$
$$= -8x^{12}$$

c. $2xy^3(3x^2y)^2 = 2xy^3(3x^2y)(3x^2y)$
$$= (2 \cdot 3 \cdot 3)(x \cdot x^2 \cdot x^2)(y^3 \cdot y \cdot y)$$
$$= 18x^5y^5$$

Be sure you see the difference between the expressions

$$x^3 \cdot x^4 \quad \text{and} \quad x^3 + x^4.$$

The first is a product of exponential forms, whereas the second is a sum of exponential forms. The rule for multiplying exponential forms having the same base can be applied to the first expression, but *not* to the second expression.

EXAMPLE 9 *Simplifying Expressions with More than One Term*

a. $(3x)^2 - 5x^3 = (3x)(3x) - 5x^3$
$$= (3 \cdot 3)(x \cdot x) - 5x^3$$
$$= 9x^2 - 5x^3$$

b. $(-x^2y)^3 + 2(x^2y)^2$
$$= (-x^2y)(-x^2y)(-x^2y) + 2(x^2y)(x^2y)$$
$$= (-1)(x^2y)(-1)(x^2y)(-1)(x^2y) + 2(x^2y)(x^2y)$$
$$= (-1)(-1)(-1) \cdot x^2 \cdot x^2 \cdot x^2 \cdot y \cdot y \cdot y + 2 \cdot x^2 \cdot x^2 \cdot y \cdot y$$
$$= (-1)x^6y^3 + 2x^4y^2$$
$$= -x^6y^3 + 2x^4y^2$$

There are two other rules of exponents that you need to know. The first rule deals with an exponential expression that is itself raised to a power. For instance, how would you evaluate the expression $(a^m)^n$? As an example, try writing out the repeated multiplication for $(a^2)^3$.

$$(a^2)^3 = (a^2)(a^2)(a^2) = a^{2+2+2} = a^6 = a^{2 \cdot 3}$$

From this result, it appears that the rule is

$$(a^m)^n = a^{mn}.$$

The second rule deals with a product that is raised to a power. For instance, how would you evaluate $(ab)^m$? Consider a simple example, say $(ab)^3$. By writing the repeated multiplication and applying the Commutative Property of Multiplication you have

$$(ab)^3 = (ab)(ab)(ab) = (a \cdot a \cdot a)(b \cdot b \cdot b) = a^3 b^3.$$

Thus, it appears that the general rule is

$$(ab)^m = a^m b^m.$$

Rules of Exponents

Let m and n be positive integers, and let a and b be real numbers, variables, or algebraic expressions. Then the following properties are true.

1. $a^m \cdot a^n = a^{m+n}$ **2.** $(a^m)^n = a^{m \cdot n}$ **3.** $(ab)^m = a^m b^m$

EXAMPLE 10 Applying the Rules of Exponents

Use the rules of exponents to simplify each of the following.

a. $(2^3)^4$ **b.** $(y^2)^3$ **c.** $[(x+2)^3]^3$

d. $(3x)^3$ **e.** $(-x)^4$ **f.** $(2x^2)^3$

Solution

a. $(2^3)^4 = 2^{3 \cdot 4} = 2^{12} = 4096$

b. $(y^2)^3 = y^{2 \cdot 3} = y^6$

c. $[(x+2)^3]^3 = (x+2)^{3 \cdot 3} = (x+2)^9$

d. $(3x)^3 = 3^3 \cdot x^3 = 27x^3$

e. $(-x)^4 = (-1)^4 x^4 = x^4$

f. $(2x^2)^3 = 2^3 (x^2)^3 = 2^3 x^{2 \cdot 3} = 8x^6$

EXAMPLE 11   *Applying the Rules of Exponents*

Use the rules of exponents to simplify each of the following.

a. $(5x)^2(5x)^4$

b. $(x^2)^3(2x)^2$

c. $(-3x)^3(2x^2)^2 + 3x^3$

d. $(xy^2)^2(x^2y)^3$

Solution

a. $(5x)^2(5x)^4 = (5x)^{2+4}$

$$= (5x)^6$$

b. $(x^2)^3(2x)^2 = x^6 \cdot 2^2 \cdot x^2$

$$= 4x^8$$

c. $(-3x)^3(2x^2)^2 + 3x^3 = (-3)^3 \cdot x^3 \cdot 2^2 \cdot (x^2)^2 + 3x^3$

$$= -27 \cdot 4 \cdot x^3 \cdot x^4 + 3x^3$$

$$= -108x^7 + 3x^3$$

d. $(xy^2)^2(x^2y)^3 = x^2 \cdot (y^2)^2 \cdot (x^2)^3 \cdot y^3$

$$= x^2 \cdot y^4 \cdot x^6 \cdot y^3$$

$$= x^8y^7$$

Group Activities   Extending the Concept

A Mathematical Riddle   What is the largest number that can be written using the three digits 2, 3, and 4? The number 432 seems to be the obvious answer. However, if you allow the digits to be exponents, then you can obtain numbers that are much larger than 432. For instance, consider the numbers

$$(32)^4 = 1,048,576$$

and

$$3^{24} \approx 282,430,000,000.$$

In your group, see who can create the largest number using the three digits 2, 3, and 4.

2.1 Exercises

Discussing the Concepts

1. Discuss the difference between terms and factors.

2. Is $3x$ a term of $4 - 3x$? Explain.

3. In the expression $(10x)^3$, what is $10x$ called? What is 3 called?

4. Discuss the difference between $(6x)^4$ and $6x^4$.

5. The expressions $4x$ and x^4 each represent repeated operations. What are the operations? Write the expressions showing the repeated operations.

6. Which of the following are equivalent? Explain.
   (a) $12x^8$ (b) $12(x^3)^5$ (c) $12x^3x^5$
   (d) $3 \cdot 2^2(x^2)^4$ (e) $3 \cdot 5x^8$

Problem Solving

In Exercises 7–10, identify the variables and constants in the expression.

7. $x + 3$

8. $y + 1$

9. π

10. $3^2 + z$

In Exercises 11 and 12, identify the terms of the expression.

11. $3x^2 + 5$

12. $5 - 3t^2 - 4t^3$

In Exercises 13 and 14, identify the coefficient of the term.

13. $-6x^2$

14. $\dfrac{3x}{4}$

In Exercises 15–20, rewrite the expression in exponential form.

15. $2 \cdot u \cdot u \cdot u \cdot u$

16. $\dfrac{1}{3}x \cdot x \cdot x \cdot x \cdot x$

17. $2u \cdot 2u \cdot 2u \cdot 2u$

18. $\dfrac{x}{y^2} \cdot \dfrac{x}{y^2} \cdot \dfrac{x}{y^2}$

19. $3(a - b) \cdot 3(a - b) \cdot 3(a - b) \cdot 3(a - b) \cdot 3(a - b)$

20. $2(r + s)^2 \cdot (r + s)^2 \cdot 2 \cdot 2$

In Exercises 21–26, expand the expression as a product of factors.

21. 2^2x^4

22. $3uv^4$

23. $(a^2)^3$

24. $(z^3)^3$

25. $[3(r + s)^2][3(r + s)]^2$

26. $[2(a - b)^3][2(a - b)](a - b)^2$

In Exercises 27–42, simplify the expression.

27. $u^2 \cdot u^4$

28. $4y^3 \cdot y^5$

29. $(-5z^3)z^2$

30. $(-6x^2)x^4$

31. $(t^2)^4$

32. $(v^3)^2$

33. $5(uv)^5$

34. $3(pq)^4$

35. $(-2s)^3$

36. $(-3z)^2$

37. $(x - 2y)(x - 2y)^3$

38. $10(x - 3)^2(x - 3)^5$

39. $(-2x)^3(3x^2)^2 + 5x^2$

40. $(2y)^3 - 3y$

41. $(u^2v^3)(uv^2)^4$

42. $\dfrac{8a^2}{5b} \cdot \dfrac{8a}{5^2b^3}$

In Exercises 43–46, decide whether the expressions are equal. Explain your reasoning.

43. $x^5 \cdot x^3 \overset{?}{=} x^{15}$

44. $(-2x)^4 \overset{?}{=} -2x^4$

45. $-3x^3 \overset{?}{=} -27x^3$

46. $(xy)^2 \overset{?}{=} xy^2$

In Exercises 47 and 48, write the number as a power of 10.

47. $10,000$

48. $1,000,000$

49. *Balance in an Account*   The balance in an account that has an annual interest rate of r, compounded quarterly for 1 year, is given by

$$P\left(1 + \frac{r}{4}\right)\left(1 + \frac{r}{4}\right)\left(1 + \frac{r}{4}\right)\left(1 + \frac{r}{4}\right).$$

Simplify this expression.

50. *Moment of Inertia*   The moment of inertia of a solid is given by

$$\frac{1}{2}m(2a)^2(2L).$$

Simplify this expression.

51. *Think About It*   Find a positive integer n such that $n^3 + 1$ is a prime number.

52. *Geometry*   The square and cube shown below have edges of length x. Use exponential notation to write an expression for the area of the square and the volume of the cube.

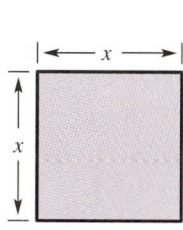 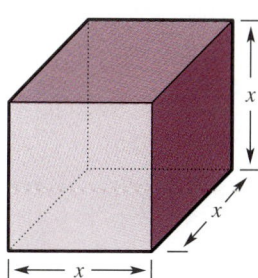

Reviewing the Major Concepts

In Exercises 53 and 54, evaluate the expression.

53. $36 \div 3^2 + 4^2$

54. $-2[3 - (12 - 3)]$

In Exercises 55 and 56, state the property of real numbers that justifies the statement.

55. $-5(3 + 6) = -5 \cdot 3 + (-5)6$

56. $4 + (13 - 5) = (4 + 13) - 5$

57. *Total Cost*   You buy a pickup truck for $1800 down and 36 monthly payments of $625 each.

(a) What is the total amount you will pay?

(b) The cost of the pickup is $19,999. How much extra did you pay in finance charges and other fees?

58. *Ounces and Pounds*   How many $\frac{3}{16}$-ounce pieces of chocolate are in 6 pounds of chocolate?

Additional Problem Solving

In Exercises 59–66, identify the terms of the expression.

59. $\frac{5}{3} - 3y^3$

60. $6x - \frac{2}{3}$

61. $2x - 3y + 1$

62. $x^2 + 18xy + y^2$

63. $3(x + 5) + 10$

64. $16 - (x + 1)$

65. $\frac{3}{x + 2} - 3x + 4$

66. $x^2 + \frac{3x + 1}{x - 1} + 4$

In Exercises 67–74, identify the coefficient of the term.

67. $-\frac{1}{3}y$

68. $\frac{1}{8}n$

69. $-\frac{3x}{2}$

70. $\frac{y}{2}$

71. $-120x^2$

72. $25y^5$

73. $-4.7u$

74. $5.32b$

In Exercises 75–82, write the expression in exponential form.

75. $a \cdot a \cdot a \cdot b \cdot b$

76. $y \cdot y \cdot z \cdot z \cdot z \cdot z$

77. $4 \cdot x \cdot x \cdot y \cdot x \cdot y$

78. $u \cdot 7 \cdot v \cdot v \cdot 7 \cdot u$

79. $3 \cdot (x - y) \cdot (x - y) \cdot 3 \cdot 3$

80. $(u - v) \cdot (u - v) \cdot 8 \cdot 8 \cdot 8 \cdot (u - v)$

81. $\left(\dfrac{x^2}{2}\right)\left(\dfrac{x^2}{2}\right)\left(\dfrac{x^2}{2}\right)$

82. $\dfrac{r - s}{5} \cdot \dfrac{r - s}{5} \cdot \dfrac{r - s}{5} \cdot \dfrac{r - s}{5}$

In Exercises 83–94, expand the expression as a product of factors.

83. $4y^2z^3$

84. 5^3x^2

85. $(-2y)^3$

86. $(v^3)^2$

87. $5x^3 \cdot x^4$

88. $a^2y^2 \cdot y^3$

89. $(ab)^3$

90. $2(xz)^4$

91. $(x + y)^2$

92. $(s - t)^5$

93. $\left(\dfrac{a}{3s}\right)^4$

94. $\left(\dfrac{2}{x + 1}\right)^3$

In Exercises 95–106, simplify the expression.

95. $5x \cdot (x^6)$

96. $5z^3 \cdot z$

97. $(-2x^2)(4x)$

98. $(-xz)(-2y^2z)$

99. $2b^4(-ab)(3b^2)$

100. $(4xy)(-3x^2)(-2y^3)$

101. $-2(3x)^2(3x) + 5x^2$

102. $(x^3)^2(x^2)^4 + 7x^3$

103. $(a^2b)^3(ab^2)^4$

104. $(-2st^3)^5(s^2t)^4$

105. $\dfrac{7x}{9y^2} \cdot \dfrac{7x^3}{9^2y^3}$

106. $\dfrac{(x + y)^2}{3x} \cdot \dfrac{(x + y)^4}{3x^2}$

107. What power of 3 is 81?

108. What power of 10 is 1,000,000,000?

In Exercises 109 and 110, evaluate the expression.

109. $8 \cdot 10^3 + 3 \cdot 10^2 + 9 \cdot 10^1$

110. $3 \cdot 10^6 + 5 \cdot 10^4 + 7 \cdot 10^2$

111. *Reasonable Wages*   You are offered a job for 25 days with the following pay schedule. On the first day the wages are 10¢, the second day 20¢, the third day 40¢, and so on. Complete the following table which gives the wages for selected days in the 25-day assignment. Would you accept these wages?

t	1	5	10	20	25
$10(2^{t-1})$	10¢	?	?	?	?

112. *Volume of a Safe*   A fireproof safe has a cubical shape. Use the formula for the volume of a cube from Exercise 52 to find the volume of the storage space in the safe.

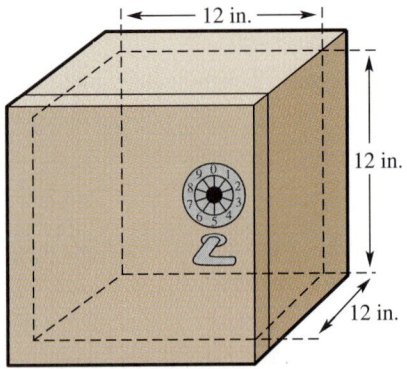

113. *Moment of Inertia*   The moment of inertia of a circular cylinder is given by

$$k\pi a^2 L\left(\dfrac{a^2}{2}\right).$$

Simplify this expression.

114. *Balance in an Account*   Suppose you deposit $1000 into a savings account that has an annual percentage rate of 6%, compounded annually. The balance in the account after 3 years is given by

$$1000(1.06)(1.06)(1.06).$$

(a) Rewrite the product in exponential form.

(b) Use a calculator to find the balance in the account.

In Exercises 115 and 116, write the expression that corresponds to the calculator steps. Then evaluate the expression.

115. 3 $\boxed{\times}$ 4 $\boxed{x^y}$ 3 $\boxed{=}$ *Scientific*

3 $\boxed{\times}$ 4 $\boxed{\wedge}$ 3 $\boxed{\text{ENTER}}$ *Graphing*

116. 2 $\boxed{x^y}$ 5 $\boxed{=}$ $\boxed{+/-}$ *Scientific*

$\boxed{(-)}$ 2 $\boxed{\wedge}$ 5 $\boxed{\text{ENTER}}$ *Graphing*

2.2	**Basic Rules of Algebra**
	Basic Rules of Algebra ▪ Combining Like Terms

Basic Rules of Algebra

The properties of real numbers listed on page 32 are often used to rewrite algebraic expressions. An expanded version of these properties is listed below. Note that the examples now involve algebraic expressions.

Basic Rules of Algebra

Let a, b, and c represent real numbers, variables, or algebraic expressions.

Property	*Example*
Commutative Property of Addition:	
$a + b = b + a$	$3x + x^2 = x^2 + 3x$
Commutative Property of Multiplication:	
$ab = ba$	$(5 + x)x^3 = x^3(5 + x)$
Associative Property of Addition:	
$(a + b) + c = a + (b + c)$	$(2x + 7) + x^2 = 2x + (7 + x^2)$
Associative Property of Multiplication:	
$(ab)c = a(bc)$	$(2x \cdot 5y) \cdot 7 = 2x \cdot (5y \cdot 7)$
Distributive Property:	
$a(b + c) = ab + ac$	$4x(7 + 3x) = 4x \cdot 7 + 4x \cdot 3x$
$(a + b)c = ac + bc$	$(2y + 5)y = 2y \cdot y + 5 \cdot y$
Additive Identity Property:	
$a + 0 = a$	$3y^2 + 0 = 3y^2$
Multiplicative Identity Property:	
$a \cdot 1 = 1 \cdot a = a$	$(-2x^3) \cdot 1 = 1 \cdot (-2x^3) = -2x^3$
Additive Inverse Property:	
$a + (-a) = 0$	$3y^2 + (-3y^2) = 0$
Multiplicative Inverse Property:	
$a \cdot \dfrac{1}{a} = 1, \quad a \neq 0$	$(x^2 + 2) \cdot \dfrac{1}{x^2 + 2} = 1$

NOTE   Because subtraction is defined as "adding the opposite," the Distributive Property is also true for subtraction. That is,

$$a(b - c) = ab - ac$$

and

$$(a - b)c = ac - bc.$$

EXAMPLE 1 *Identifying the Basic Rules of Algebra*

Identify the rule of algebra illustrated in each of the following.

a. $(3x^2)7 = 7(3x^2)$

b. $8x \cdot \dfrac{1}{8x} = 1, \quad x \neq 0$

c. $(4x^2 + x) - (4x^2 + x) = 0$

d. $(2 + x^2) + 2x^2 = 2 + (x^2 + 2x^2)$

e. $(y - 5)2 + (y - 5)y = (y - 5)(2 + y)$

Solution

a. This equation illustrates the Commutative Property of Multiplication. You obtain the same result whether you multiply $3x^2$ by 7 or multiply 7 by $3x^2$.

b. This equation illustrates the Multiplicative Inverse Property. It is important that x is nonzero because the reciprocal of zero is undefined.

c. This equation illustrates the Additive Inverse Property. Informally, this property states that when any expression is subtracted from itself the result is zero.

d. This equation illustrates the Associative Property of Addition. To form the sum $2 + x^2 + 2x^2$, it doesn't matter whether 2 and x^2 are added first or x^2 and $2x^2$ are added first.

e. This equation illustrates the Distributive Property.

$$ab + ac = a(b + c) \qquad \text{Distributive Property}$$

$$(y - 5)2 + (y - 5)y = (y - 5)(2 + y)$$

Note in this case that $a = y - 5$, $b = 2$, and $c = y$.

EXAMPLE 2 *Using the Basic Rules of Algebra*

a. What is the additive inverse of $3x$? What is the result when this additive inverse is added to $3x$?

b. What is the multiplicative inverse of $3x$? What is the result when $3x$ is multiplied by this multiplicative inverse?

Solution

a. The additive inverse of $3x$ is $-3x$. The sum is $3x + (-3x) = 0$.

b. If $x = 0$, then $3x$ has no multiplicative inverse. If $x \neq 0$, then the multiplicative inverse of $3x$ is $1/(3x)$. The product is

$$3x \left(\frac{1}{3x} \right) = 1.$$

EXAMPLE 3   Applying the Basic Rules of Algebra

Use the indicated rule to complete the statement.

a. Additive Identity Property: $(x - 2) + \boxed{} = x - 2$

b. Commutative Property of Multiplication: $5(y + 6) = \boxed{}$

c. Commutative Property of Addition: $5(y + 6) = \boxed{}$

d. Distributive Property: $5(y + 6) = \boxed{}$

e. Associative Property of Addition: $(x^2 + 3) + 7 = \boxed{}$

Solution

a. $(x - 2) + 0 = x - 2$

b. $5(y + 6) = (y + 6)5$

c. $5(y + 6) = 5(6 + y)$

d. $5(y + 6) = 5y + 5(6)$

e. $(x^2 + 3) + 7 = x^2 + (3 + 7)$

Example 4 illustrates some common uses of the Distributive Property. Study this example carefully. Such uses of the Distributive Property are very important in algebra.

NOTE Applying the Distributive Property as illustrated in Example 4 is called **expanding** an algebraic expression. For instance, you can expand $3(x + y)$ by writing the expression as $3x + 3y$.

EXAMPLE 4   Using the Distributive Property

Use the Distributive Property to expand each expression.

a. $2(7 - x)$ **b.** $(10 - 2y)3$ **c.** $2x(x + 4)$ **d.** $-3(1 - 2y + x)$

Solution

a. $\begin{aligned} 2(7 - x) &= 2 \cdot 7 - 2 \cdot x \\ &= 14 - 2x \end{aligned}$

b. $\begin{aligned} (10 - 2y)3 &= 10(3) - 2y(3) \\ &= 30 - 6y \end{aligned}$

c. $\begin{aligned} 2x(x + 4) &= 2x(x) + 2x(4) \\ &= 2x^2 + 8x \end{aligned}$

d. $\begin{aligned} -3(1 - 2y + x) &= (-3)(1) - (-3)(2y) + (-3)(x) \\ &= -3 + 6y - 3x \end{aligned}$

Combining Like Terms

Two or more terms of an algebraic expression can be combined only if they are *like terms*.

NOTE Factors such as x in $5x$ or ab in $6ab$ are called **variable factors**.

Definition of Like Terms
In an algebraic expression, two terms are said to be **like terms** if they are both constant terms or if they have the same variable factor(s).

The terms $5x$ and $-3x$ are like terms because they have the same variable factor. Similarly, $3x^2y$, $-x^2y$, and $\frac{1}{3}(x^2y)$ are like terms.

EXAMPLE 5 Identifying Like Terms in Expressions

Expression	Like Terms
a. $5xy + 1 - xy$	$5xy$ and $-xy$
b. $12 - x^2 + 3x - 5$	12 and -5
c. $7x - 3 - 2x + 5$	$7x$ and $-2x$, -3 and 5

To combine like terms in an algebraic expression, you can simply add their respective coefficients and attach the common variable factor. This is actually an application of the Distributive Property, as shown in Example 6.

EXAMPLE 6 Using the Distributive Law to Combine Like Terms

Simplify each expression by combining like terms.

a. $5x + 2x - 4$ **b.** $-5 + 8 + 7y - 5y$ **c.** $2y - 3x - 4x$

Solution

a. $5x + 2x - 4 = (5 + 2)x - 4$　　　Distributive Property

$\qquad\qquad\qquad = 7x - 4$　　　Simplest form

b. $-5 + 8 + 7y - 5y = (-5 + 8) + (7 - 5)y$　　　Distributive Property

$\qquad\qquad\qquad\qquad = 3 + 2y$　　　Simplest form

c. $2y - 3x - 4x = 2y - (3 + 4)x$　　　Distributive Property

$\qquad\qquad\qquad = 2y - 7x$　　　Simplest form

Often, you need to use other rules of algebra before you can apply the Distributive Property to combine like terms. This is illustrated in the next example.

EXAMPLE 7   Using Rules of Algebra to Combine Like Terms

Simplify each expression by combining like terms.

a. $7x + 3y - 4x$ **b.** $12a - 5 - 3a + 7$ **c.** $y - 4x - 7y + 9y$

Solution

a. $\begin{aligned} 7x + 3y - 4x &= 3y + 7x - 4x & &\text{Commutative Property} \\ &= 3y + (7x - 4x) & &\text{Associative Property} \\ &= 3y + (7 - 4)x & &\text{Distributive Property} \\ &= 3y + 3x & &\text{Simplest form} \end{aligned}$

b. $\begin{aligned} 12a - 5 - 3a + 7 &= 12a - 3a - 5 + 7 & &\text{Commutative Property} \\ &= (12a - 3a) + (-5 + 7) & &\text{Associative Property} \\ &= (12 - 3)a + (-5 + 7) & &\text{Distributive Property} \\ &= 9a + 2 & &\text{Simplest form} \end{aligned}$

c. $\begin{aligned} y - 4x - 7y + 9y &= -4x + (y - 7y + 9y) & &\text{Collect like terms.} \\ &= -4x + (1 - 7 + 9)y & &\text{Distributive Property} \\ &= -4x + 3y & &\text{Simplest form} \end{aligned}$

STUDY TIP

As you gain experience with the rules of algebra, you may want to combine some of the steps in your work. For instance, you might feel comfortable listing only the following steps to solve part (b) of Example 7.

$12a - 5 - 3a + 7$
$= (12a - 3a) + (-5 + 7)$
$= 9a + 2$

Group Activities   Communicating Mathematically

Expressing Algebraic Rules in Words   This section lists ten basic rules of algebra.

- 2 commutative properties
- 2 associative properties
- 2 distributive properties
- 2 identity properties
- 2 inverse properties

Write an example of a real-life nonmathematical operation that is not commutative. Share your example with your group to see if it makes sense to them. Together, try to make up a nonmathematical example of the distributive property. Can you write similar examples for the other properties?

2.2 Exercises

Discussing the Concepts

1. Explain why $3(x + 9) \neq 3x + 9$.

2. What is the multiplicative inverse of $2x$? How did you find this inverse? The product of an expression and its multiplicative inverse will always equal what value?

3. Use the Associative Property of Addition to rewrite the expression $3 + (4 + x)$. Explain how the Associative Property helps to simplify the expression.

4. In your own words, state the definition of like terms. Give an example of like terms and an example of unlike terms.

5. Describe how to combine like terms. What operations are used? Give an example of an expression that can be simplified by combining like terms.

6. Explain how to use the Distributive Property to rewrite $-(x - 4)$ as an expression with two terms.

Problem Solving

In Exercises 7–14, identify the rule (or rules) of algebra illustrated by the equation.

7. $x + 2y = 2y + x$

8. $-10(xy^2) = (-10x)y^2$

9. $(3x + 2y) + z = 3x + (2y + z)$

10. $2zy = 2yz$

11. $16xy \cdot \dfrac{1}{16xy} = 1, \; xy \neq 0$

12. $(5m + 3) - (5m + 3) = 0$

13. $x(y + z) = xy + xz$

14. $3y + (z^3 - z^3) = 3y$

In Exercises 15–20, complete the statement. State the rule of algebra that you used.

15. $(x + 1) - \boxed{} = 0$

16. $(-5r)s = -5(\boxed{})$

17. $v(2) = \boxed{}$

18. $(4x - 3y) + \boxed{} = 4x - 3y$

19. $(t + 5)(t - 2) = t(\boxed{}) + 5(\boxed{})$

20. $(s - 5) \cdot (\boxed{}) = s - 5$

In Exercises 21–28, use the Distributive Property to expand the expression.

21. $-5(2x - y)$

22. $\frac{1}{8}(16 + 8z)$

23. $(x + 2)(3)$

24. $(4 - t)(-6)$

25. $x(x + xy + y^2)$

26. $r(r - t + s)$

27. $z[z^2 - 2(z - 1)]$

28. $-(4 - 2x - x^2)$

In Exercises 29 and 30, identify the terms of the expression and the coefficient of each term.

29. $6x^2 - 3xy + y^2$

30. $-4xy + 2xz - yz$

In Exercises 31 and 32, identify the like terms.

31. $16t^3 + 4 - 5 + 3t^3$

32. $a^2 + 5ab^2 - 3b^2 + 7a^2b - ab^2 + a^2$

In Exercises 33–38, simplify the expression by combining like terms.

33. $3y - 5y$

34. $7s + 3 - 3s - 4$

35. $x^2 - 2xy + 4 + xy$

36. $5z - 5 + 10z + 2z + 16$

37. $3\left(\dfrac{1}{x}\right) - \dfrac{1}{x} + 8$

38. $z^3 + 2z^2 + z + z^2 + 2z + 1$

In Exercises 39 and 40, state why the two expressions are not like terms.

39. $\frac{1}{2}x^2y, \; -\frac{5}{2}xy^2$

40. $-16x^2y^3, \; 7x^2y$

Mental Math   In Exercises 41 and 42, use the Distributive Property to perform the required arithmetic *mentally.* For example, suppose you work in an industry where the wage is $14 per hour and time and one-half for overtime. Thus, your hourly wage for overtime is

$14(1.5) = 14\left(1 + \frac{1}{2}\right) = 14 + 7 = \$21.$

41. $8(52) = 8(50 + 2)$

42. $12(11.95) = 12(12 - 0.05)$

43. *Geometry*   Write an expression for the perimeter of the triangle shown in the figure. Use the Properties of Real Numbers to simplify the expression.

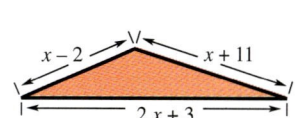

Figure for 43

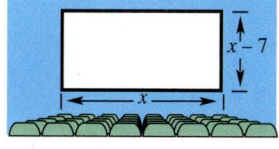

Figure for 44

44. *Geometry*   Write an expression for the area of the movie screen shown in the figure. Use the Distributive Property to rewrite the expression.

Reviewing the Major Concepts

In Exercises 45–48, simplify the expression.

45. $(-2x)^2 x^4$

46. $(u^3)^2$

47. $5z^3(z^2)^2$

48. $(a + 3)^2(a + 3)^5$

49. *Balance in an Account*   The balance in an account that has an annual percentage rate of r, compounded annually for 4 years, is given by

$P(1 + r)(1 + r)(1 + r)(1 + r).$

Simplify this expression.

50. *Reading a Graph*   The bar graph at the right shows the revenues (in millions of dollars) of the Walt Disney Company for 1989 through 1993. (Source: *The Walt Disney Company 1993 Annual Report*)

(a) Approximate the largest and smallest annual increases in revenue.

(b) Approximate the average annual increase in revenues.

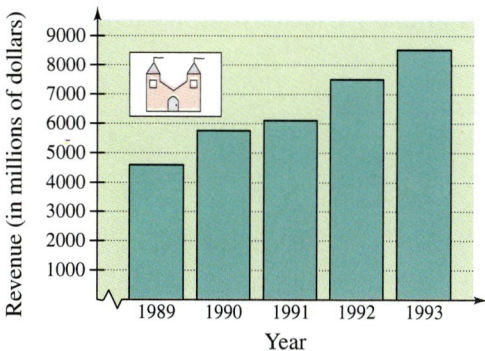

Figure for 50

Additional Problem Solving

In Exercises 51–58, identify the rule (or rules) of algebra illustrated by the equation.

51. $(9x)y = 9(xy)$

52. $rt + 0 = rt$

53. $(x^2 + y^2) \cdot 1 = x^2 + y^2$

54. $(x^2 + 2x) - 7 = x^2 + (2x - 7)$

55. $(x + 2)y = y(x + 2)$

56. $(x + y) \cdot \dfrac{1}{x + y} = 1$

57. $x^2 + (y^2 - y^2) = x^2$

58. $(x + 2)(x + y) = x(x + y) + 2(x + y)$

In Exercises 59–64, complete the statement. State the rule of algebra that you used.

59. $5x \cdot \boxed{} = 1$

60. $x + 2y + \boxed{} = x + 2y$

61. $(2z - 3) - \boxed{} = 0$

62. $(u + 3) \boxed{} = u + 3$

63. $12 + (8 - x) = \boxed{} - x$

64. $(2x - y)(-3) = -3 \boxed{}$

In Exercises 65–76, use the Distributive Property to expand the expression.

65. $3(x^2y + x)$

66. $4(2y^2 - y)$

67. $(-1)(u - v)$

68. $(-1)(x^2 - 2xy + y^2)$

69. $x(8 + x)$

70. $(x - 14)x$

71. $2x(x + 9)$

72. $(12 - 5y)y$

73. $-4y(3y - 4)$

74. $-z(5 - 2z)$

75. $a(a^2 + ab + b^2)$

76. $-(5x - 7)$

In Exercises 77 and 78, identify the like terms.

77. $6x^2y + 2xy - 4x^2y$

78. $-\frac{1}{4}x^2 - 3x + \frac{3}{4}x^2 + x$

In Exercises 79–86, simplify the expression.

79. $x + 5y - 3x - y$

80. $-16x + 25x$

81. $2x^2 + 9x^2 + 4$

82. $10x - 4 - 5x$

83. $3x^2 - x^2y + 4xy^2 + 3x^2y - xy^2 + y^2$

84. $3x^2y + xy - xy^2 - 6xy$

85. $1.2 \left(\dfrac{1}{x} \right) + 3.8 \left(\dfrac{1}{x} \right) - 4x$

86. $16 \left(\dfrac{a}{b} \right) - 6 \left(\dfrac{a}{b} \right) + \dfrac{3}{2} - \dfrac{1}{2}$

True or False? In Exercises 87–90, decide whether the statement is true or false.

87. $3(x - 4) \overset{?}{=} 3x - 4$

88. $-3(x - 4) \overset{?}{=} -3x - 12$

89. $6x - 4x \overset{?}{=} 2x$

90. $12y^2 + 3y^2 \overset{?}{=} 36y^2$

Mental Math In Exercises 91 and 92, use mental math to evaluate the expression.

91. $5(7.98) = 5(8 - 0.02)$

92. $6(29) = 6(30 - 1)$

93. *Geometric Model* The figure shows two adjoining rectangles. Demonstrate the Distributive Property by filling in the blanks to express the combined area of the two rectangles in two ways.

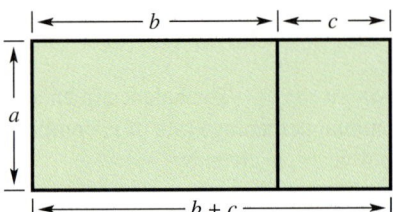

94. *Geometric Model* The figure shows two adjoining rectangles. Demonstrate the subtraction version of the Distributive Property by filling in the blanks to express the area on the left in two ways.

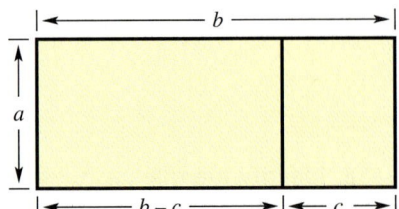

95. *Exploration* You have defined a new mathematical operation using the symbol $\odot$. This operation is defined as $a \odot b = 2 \cdot a + b$. Is this operation commutative? Is it associative? Explain your reasoning.

Math Matters

Finding the Day of the Week

Many people are interested in finding the day of the week on which they were born. There is a way to do this (without looking at old calendars) by using the following number values.

Month	Jan	Feb	Mar	Apr	May	Jun	Jul	Aug	Sep	Oct	Nov	Dec
Value	0	3	3	6	1	4	6	2	5	0	3	5

Day	Sun	Mon	Tues	Wed	Thurs	Fri	Sat
Value	0	1	2	3	4	5	6

The constellation Leo represents the Zodiac sign for those born between July 23 and August 22.

To use these values to find the day of the week for any date between January 1, 1901, and December 31, 1999, use the following steps. (This example uses August 16, 1967.)

Step		*Example*	
1. Divide the year of the date by 4, and ignore the remainder.* (This process determines the number of leap years between January 1, 1901 and the date.)	16	$67 \div 4$
2. Add this to the day of the month,	16	
3. the value of the month (from the Month table), and	2	August
4. the year of the century.	$+\ \underline{67}$	1967
	101	Total
5. Divide total by 7; keep remainder.	$101 \div 7 = 14$	Remainder = 3
6. Use the remainder to determine the day of the week according to the Day table.	3	Wednesday

Thus, August 16, 1967, occurred on a Wednesday.

---

*Do not count the leap year if the date is earlier than March 1 on a leap year. For instance, there were only four leap years between January 1, 1901 and February 28, 1920.

2.3	**Rewriting and Evaluating Algebraic Expressions**
	Simplifying Algebraic Expressions ▪ Symbols of Grouping ▪ Evaluating Algebraic Expressions

Simplifying Algebraic Expressions

Simplifying an algebraic expression by rewriting it into a more usable form is one of the three most frequently used skills in algebra. You will study the other two—solving an equation and sketching the graph of an equation—later in this text.

To "simplify an algebraic expression" generally means to remove symbols of grouping and combine like terms. For instance, the expression $x + (3 + x)$ can be simplified as $2x + 3$.

EXAMPLE 1   Simplifying Algebraic Expressions

Simplify each expression.

a. $-3(-5x)$ **b.** $7(-x)$

Solution

a. $-3(-5x) = (-3)(-5)x$ Associative Property

$\qquad\qquad\;\; = 15x$ Simplest form

b. $7(-x) = 7(-1)(x)$ Coefficient of $-x$ is -1.

$\qquad\quad\; = -7x$ Simplest form

EXAMPLE 2   Simplifying Algebraic Expressions

a. $\dfrac{5x}{3} \cdot \dfrac{3}{5} = \left(\dfrac{5}{3} \cdot x\right) \cdot \dfrac{3}{5}$ Coefficient of $\dfrac{5x}{3}$ is $\dfrac{5}{3}$.

$\qquad\quad\; = \left(\dfrac{5}{3} \cdot \dfrac{3}{5}\right) \cdot x$ Commutative and Associative Properties

$\qquad\quad\; = 1 \cdot x$ Multiplicative Inverse

$\qquad\quad\; = x$ Multiplicative Identity

b. $x^2(-2x^3) = (-2)(x^2 \cdot x^3)$ Commutative and Associative Properties

$\qquad\qquad\;\; = -2x^{2+3}$ Rule of Exponents

$\qquad\qquad\;\; = -2x^5$

Symbols of Grouping

The main tool for removing symbols of grouping is the Distributive Property, as illustrated in Example 3.

EXAMPLE 3 Removing Symbols of Grouping

Simplify each expression.

a. $-(2y - 7)$ **b.** $5x + (x - 7)2$

c. $-2(4x - 1) + 3x$ **d.** $3(y - 5) - (2y - 7)$

Solution

a. $-(2y - 7) = -2y + 7$ Distributive Property

b. $5x + (x - 7)2 = 5x + 2x - 14$ Distributive Property

            $= 7x - 14$ Combine like terms.

c. $-2(4x - 1) + 3x = -8x + 2 + 3x$ Distributive Property

            $= -8x + 3x + 2$ Commutative Property

            $= -5x + 2$ Combine like terms.

d. $3(y - 5) - (2y - 7) = 3y - 15 - (2y - 7)$ Distributive Property

            $= 3y - 15 - 2y + 7$ Distributive Property

            $= (3y - 2y) + (-15 + 7)$ Group like terms.

            $= y - 8$ Combine like terms.

EXAMPLE 4 Removing Nested Symbols of Grouping

a. $5x - 2[4x + 3(x - 1)]$

       $= 5x - 2[4x + 3x - 3]$ Distributive Property

       $= 5x - 2[7x - 3]$ Combine like terms.

       $= 5x - 14x + 6$ Distributive Property

       $= -9x + 6$ Combine like terms.

b. $-7y + 3[2y - (3 - 2y)] - 5y + 4$

       $= -7y + 3[2y - 3 + 2y] - 5y + 4$ Distributive Property

       $= -7y + 3[4y - 3] - 5y + 4$ Combine like terms.

       $= -7y + 12y - 9 - 5y + 4$ Distributive Property

       $= (-7y + 12y - 5y) + (-9 + 4)$ Group like terms.

       $= -5$ Combine like terms.

EXAMPLE 5 *Simplifying Algebraic Expressions*

a. $(-3x)(5x^4) + 7x^5 = (-3)(5)x \cdot x^4 + 7x^5$ Commutative and Associative Properties

$$= -15x^5 + 7x^5$$ Rule of Exponents

$$= -8x^5$$ Combine like terms.

b. $2x(x + 3y) + 4(5 - xy) = 2x^2 + 6xy + 20 - 4xy$ Distributive Property

$$= 2x^2 + 6xy - 4xy + 20$$ Commutative Property

$$= 2x^2 + 2xy + 20$$ Combine like terms.

The next example illustrates the use of the Distributive Property with fractional expressions.

EXAMPLE 6 *Simplifying Fractional Expressions*

a. $\dfrac{3x}{5} - \dfrac{x}{5} = \dfrac{3}{5}x - \dfrac{1}{5}x$ Write with fractional coefficients.

$$= \left(\dfrac{3}{5} - \dfrac{1}{5}\right)x$$ Distributive Property

$$= \dfrac{2}{5}x$$ Subtract fractions.

b. $\dfrac{x}{4} + \dfrac{2x}{7} = \dfrac{1}{4}x + \dfrac{2}{7}x$ Write with fractional coefficients.

$$= \left(\dfrac{1}{4} + \dfrac{2}{7}\right)x$$ Distributive Property

$$= \left(\dfrac{1(7)}{4(7)} + \dfrac{2(4)}{7(4)}\right)x$$ Common denominator

$$= \dfrac{15}{28}x$$ Simplest form.

c. $\dfrac{3x}{10} - \dfrac{4x}{15} = \dfrac{3}{10}x - \dfrac{4}{15}x$ Write with fractional coefficients.

$$= \left(\dfrac{3}{10} - \dfrac{4}{15}\right)x$$ Distributive Property

$$= \left(\dfrac{3(3)}{10(3)} - \dfrac{4(2)}{15(2)}\right)x$$ Common denominator

$$= \dfrac{1}{30}x$$ Simplest form.

Evaluating Algebraic Expressions

In applications of algebra, you are often required to **evaluate** an algebraic expression. This means you are to find the *value* of an expression when its variables are replaced by real numbers. For instance, when $x = 2$, the value of the expression $2x + 3$ is as follows.

Expression	*Replace x by 2.*	*Value of Expression*
$2x + 3$	$2(2) + 3$	7

When finding the value of an algebraic expression, be sure to replace every occurrence of the specified variable with the appropriate real number. For instance, when $x = -2$, the value of $x^2 - x + 3$ is

$$(-2)^2 - (-2) + 3 = 4 + 2 + 3 = 9.$$

EXAMPLE 7 *Evaluating Algebraic Expressions*

Evaluate each expression when $x = -3$ and $y = 5$.

a. $-x$ **b.** $3x + 2y$ **c.** $y - 2(x + y)$ **d.** $y^2 - 3y - 6$

Solution

a. When $x = -3$, the value of $-x$ is

$$-x = -(-3) \qquad\qquad \text{Substitute } -3 \text{ for } x.$$
$$= 3. \qquad\qquad\qquad \text{Simplify.}$$

b. When $x = -3$ and $y = 5$, the value of $3x + 2y$ is

$$3x + 2y = 3(-3) + 2(5) \qquad \text{Substitute } -3 \text{ for } x \text{ and } 5 \text{ for } y.$$
$$= -9 + 10 \qquad\qquad\quad \text{Simplify.}$$
$$= 1. \qquad\qquad\qquad\quad \text{Simplify.}$$

c. When $x = -3$ and $y = 5$, the value of $y - 2(x + y)$ is

$$y - 2(x + y) = 5 - 2[(-3) + 5] \qquad \text{Substitute } -3 \text{ for } x \text{ and } 5 \text{ for } y.$$
$$= 5 - 2(2) \qquad\qquad\quad \text{Simplify.}$$
$$= 1. \qquad\qquad\qquad\quad \text{Simplify.}$$

d. When $y = 5$, the value of $y^2 - 3y - 6$ is

$$y^2 - 3y - 6 = (5)^2 - 3(5) - 6 \qquad \text{Substitute } 5 \text{ for } y.$$
$$= 25 - 15 - 6 \qquad\qquad \text{Simplify.}$$
$$= 4. \qquad\qquad\qquad\quad \text{Simplify.}$$

NOTE   Notice in parts (a) and (c) of Example 7 that it is a good idea to use parentheses when substituting a negative number for a variable.

EXAMPLE 8 Evaluating Algebraic Expressions

Evaluate each expression when $x = 4$ and $y = -6$.

a. y^2 **b.** $-y^2$ **c.** $y - x$ **d.** $|y - x|$ **e.** $|x - y|$

Solution

a. When $y = -6$, the value of the expression y^2 is

$$y^2 = (-6)^2 = 36.$$

b. When $y = -6$, the value of the expression $-y^2$ is

$$-y^2 = -(y^2) = -(-6)^2 = -36.$$

c. When $x = 4$ and $y = -6$, the value of the expression $y - x$ is

$$y - x = (-6) - 4 = -6 - 4 = -10.$$

d. When $x = 4$ and $y = -6$, the value of the expression $|y - x|$ is

$$|y - x| = |-6 - 4| = |-10| = 10.$$

e. When $x = 4$ and $y = -6$, the value of the expression $|x - y|$ is

$$|x - y| = |4 - (-6)| = |4 + 6| = |10| = 10.$$

EXAMPLE 9 Evaluating Algebraic Expressions

Evaluate each expression when $x = -5$, $y = -2$, and $z = 3$.

a. $\dfrac{y + 2z}{5y - xz}$ **b.** $(y + 2z)(z - 3y)$

Solution

a. When $x = -5$, $y = -2$, and $z = 3$, the value of the expression is

$$\frac{y + 2z}{5y - xz} = \frac{-2 + 2(3)}{5(-2) - (-5)(3)} \qquad \text{Substitute for } x, y, \text{ and } z.$$

$$= \frac{-2 + 6}{-10 + 15} \qquad \text{Simplify.}$$

$$= \frac{4}{5}. \qquad \text{Simplify.}$$

b. When, $y = -2$ and $z = 3$, the value of the expression is

$$(y + 2z)(z - 3y) = [(-2) + 2(3)][3 - 3(-2)]$$

$$= (-2 + 6)(3 + 6)$$

$$= 4(9)$$

$$= 36.$$

On occasion you may need to evaluate an algebraic expression for *several* values of x. In such cases, you should simplify the expression as much as possible *before* substituting for x.

Technology

If you have a graphing calculator, try using it to store and evaluate the expression given in Example 10. For instance, here are the steps that will evaluate $-9x + 6$ when $x = 2$ on a *TI-82*.

- Use Y= key to store expression as Y_1.

- 2nd QUIT

- Store 2 in X.

  2 STO▷ X,T,θ ENTER

- Display Y_1.

  Y-vars ENTER ENTER

  and then press

  ENTER again.

EXAMPLE 10 *Repeated Evaluation of an Expression*

Evaluate the expression

$$5x - 2[4x + 3(x - 1)]$$

when $x = 0$, $x = 2$, $x = -1$, and $x = -5$.

Solution

In Example 4(a), this expression was simplified as

$$5x - 2[4x + 3(x - 1)] = -9x + 6.$$

Clearly, it is easier to substitute values of x into the simplified expression than to substitute them into the original expression.

When $x = 0$: $-9x + 6 = -9(0) + 6 = 6$

When $x = 2$: $-9x + 6 = -9(2) + 6 = -12$

When $x = -1$: $-9x + 6 = -9(-1) + 6 = 15$

When $x = -5$: $-9x + 6 = -9(-5) + 6 = 51$

Try evaluating the original expression for the same four values of x to see how much longer it takes.

Group Activities

You Be the Instructor

Error Analysis   Suppose you are teaching an algebra class and one of your students hands in the following problem. What is the error in this work?

$$5[9 + 4(x - 2)] = 5[9 + 4x - 2]$$
$$= 5[7 + 4x]$$
$$= 35 + 20x$$

What are some possible related errors? Discuss ways of helping students avoid these types of errors.

2.3 Exercises

Discussing the Concepts

1. In your own words, describe the procedure for removing nested symbols of grouping.

2. What basic rule of algebra is the main tool for removing symbols of grouping?

3. In your own words, describe the priorities for order of operations.

4. Does the expression $[x - (3 \cdot 4)] \div 5$ change if the parentheses are removed? Does it change if the brackets are removed? Explain each of your answers.

5. *Error Analysis*   Describe the error in the equation.

$$\frac{x}{3} + \frac{4x}{3} = \frac{5x}{6}$$

What is the correct sum?

6. Is it possible to evaluate the expression

$$\frac{x + 2}{y - 3}$$

when $x = 5$ and $y = 3$? Explain.

Problem Solving

In Exercises 7–10, simplify the expression.

7. $-2(6x)$

8. $(-32t)(-0.5t^2)$

9. $\dfrac{5x}{8} \cdot \dfrac{16}{5}$

10. $(7r^2s^3)(3rs)$

In Exercises 11–20, simplify the expression by removing symbols of grouping and combining like terms.

11. $x - (x + 4)$

12. $5y - (y - 3)$

13. $3n + (n - 3)$

14. $5m - 2(m - 1)$

15. $2(x - 2y) + y$

16. $6(2s - t) + s + 4t$

17. $3 - 2[6 + (4 - x)]$

18. $10x + 5[6 - (2x + 3)]$

19. $-3t(4 - t) + t(t + 1)$

20. $4y[5 - (y + 1)] + 3y(y + 1)$

In Exercises 21–26, use the Distributive Property to simplify the expression.

21. $\dfrac{x}{2} + \dfrac{x}{3}$

22. $\dfrac{3n}{4} - \dfrac{2n}{5}$

23. $\dfrac{2x}{3} - \dfrac{x}{3}$

24. $\dfrac{7y}{8} - \dfrac{3y}{8}$

25. $\dfrac{3x}{10} - \dfrac{x}{10} + \dfrac{4x}{5}$

26. $\dfrac{3z}{4} - \dfrac{z}{2} - \dfrac{z}{3}$

In Exercises 27–34, evaluate the algebraic expression for the given values of the variables. If it is not possible, state the reason.

Expression	Values
27. $2x - 1$	(a) $x = \frac{1}{2}$
	(b) $x = 4$
28. $-|x|$	(a) $x = \frac{4}{3}$
	(b) $x = -\frac{4}{3}$
29. $x - 3(x - y)$	(a) $x = 3, y = 3$
	(b) $x = 4, y = -4$
30. $-3x + 2(x + y)$	(a) $x = -2, y = 2$
	(b) $x = 0, y = 5$
31. $\dfrac{5x}{y - 3}$	(a) $x = 2, y = 4$
	(b) $x = 2, y = 3$
32. $\dfrac{2x - y}{y^2 + 1}$	(a) $x = 1, y = 2$
	(b) $x = 1, y = 3$

33. *Area of a Triangle*
$\frac{1}{2}bh$

    (a) $b = 3, h = 5$

    (b) $b = 2, h = 10$

34. *Volume of a Rectangular Prism*
lwh

    (a) $l = 4, w = 2, h = 9$

    (b) $l = 10, w = 5, h = 20$

35. *Finding a Pattern*

(a) Complete the table by evaluating the expression $3x - 2$.

x	-1	0	1	2	3	4
$3x - 2$						

(b) Use the table to find the increase in the value of the expression for each one-unit increase in x.

(c) From the pattern in parts (a) and (b), predict the increase in the expression $\frac{2}{3}x + 4$ for each one-unit increase in x. Then verify your prediction.

36. *Finding a Pattern*

(a) Complete the table by evaluating the expression $4.5x + 1.2y - 9.5$ for the given values of x and y.

y \ x	12	14	16	18	20
12					
14					
16					
18					
20					

(b) Determine the increase in the value of the algebraic expression for each two-unit increase in x. Does the answer depend on the value of y?

Balance in an Account   In Exercises 37 and 38, the balance in an account with an initial deposit of P dollars, at an annual interest rate of r for t years, is $P(1 + r)^t$. Find the balance for the given values of P, r, and t.

37. $P = 10,000,$ $r = 0.08,$ $t = 10$

38. $P = 10,000,$ $r = 0.04,$ $t = 10$

Geometry   In Exercises 39–42, find an expression for the area of the figure. Then evaluate the expression for the given value(s) of the variable(s).

39. $n = 80$

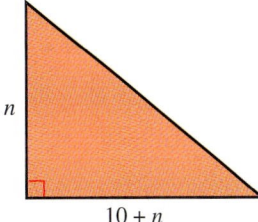

$10 + n$

40. $a = 5, b = 4$

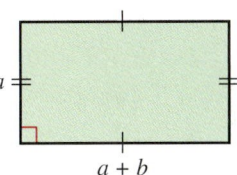

$a + b$

41. $x = 10, y = 3$

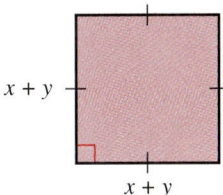

$x + y$

$x + y$

42. $x = 9$

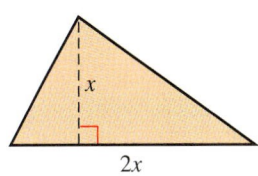

$2x$

Reviewing the Major Concepts

In Exercises 43 and 44, use the Distributive Property to expand the expression.

43. $4(2x - 5)$

44. $-z(xz - 2y^2)$

In Exercises 45 and 46, simplify the expression.

45. $\dfrac{5x}{3} - \dfrac{2x}{3} - 4$

46. $2x^2 - 4 + 5 - 3x^2$

47. *Profit*   A company had a loss of $1,530,000 during the first 6 months of the year. At the end of the year, the company ended up with an overall profit of $832,000. What was the profit during the last 2 quarters of the year?

48. *Average Speed*   You traveled 676 miles in 13 hours. What was your average speed?

Additional Problem Solving

In Exercises 49–56, simplify the expression.

49. $-7(5a)$

50. $-4(-3y)$

51. $(-2x)(-3x)$

52. $(-5z)(2z^2)$

53. $\dfrac{18a}{5} \cdot \dfrac{15}{6}$

54. $\left(\dfrac{4x}{3}\right)\left(\dfrac{3x}{2}\right)$

55. $(-1.5x^2)(4x^3)$

56. $(12xy^2)(-2x^3y^2)$

In Exercises 57–68, simplify the expression by re-moving symbols of grouping and combining like terms.

57. $-3(x + 1) - 2$

58. $(2x - 1)(2) + x$

59. $\frac{2}{3}(12x + 15) + 16$

60. $5l - 6(3l - 5l)$

61. $\frac{3}{8}(4 - y) - \frac{5}{2} + 10$

62. $m - 3(m - 5) + 25$

63. $7x(2 - x) - 4x$

64. $-6x(x - 1) + x^2$

65. $4x^2 + x(5 - x)$

66. $-z(z - 2) + 3z^2 + 5$

67. $3(r - 2s) - 5(3r - 5s)$

68. $-6(1 - 2x) + 10(5 - x)$

In Exercises 69–72, use the Distributive Property to simplify the expression.

69. $\dfrac{4z}{5} + \dfrac{3z}{5}$

70. $\dfrac{5t}{12} + \dfrac{7t}{12}$

71. $\dfrac{x}{3} - \dfrac{5x}{4}$

72. $\dfrac{5x}{7} + \dfrac{2x}{3}$

In Exercises 73–80, evaluate the algebraic expression for the given values of the variables. If it is not possible, state the reason.

	Expression	*Values*
73.	$2x^2 + 4x - 5$	(a) $x = -2$
		(b) $x = 3$
74.	$64 - 16t^2$	(a) $t = 2$
		(b) $t = 3$
75.	$3x - 2y$	(a) $x = 4, y = 3$
		(b) $x = \frac{2}{3}, y = 1$
76.	$10u - 3v$	(a) $u = 3, v = 10$
		(b) $u = -2, v = -7$

	Expression	*Values*
77.	$b^2 - 4ac$	(a) $a = 2, b = -3, c = -1$
		(b) $a = -4, b = 6, c = -2$
78.	$(2a - b) - (-a)$	(a) $a = \frac{3}{2}, b = 3$
		(b) $a = -2, b = 4$
79.	$\dfrac{x - 2y}{x + 2y}$	(a) $x = 4, y = 2$
		(b) $x = 4, y = -2$
80.	$\dfrac{-y}{x^2 + y^2}$	(a) $x = 0, y = 5$
		(b) $x = 1, y = -3$

81. *Finding a Pattern*

(a) Complete the table by evaluating $3 - 2x$.

x	-1	0	1	2	3	4
$3 - 2x$						

(b) Use the table to find the decrease in the value of the expression for each one-unit increase in x.

(c) From the pattern in parts (a) and (b), predict the decrease in the expression $4 - \frac{3}{2}x$ for each one-unit increase in x. Then verify your prediction.

82. *Finding a Pattern*

(a) Complete the table by evaluating the expression $2x + 5y$ for the given values of x and y.

y \ x	-1	0	1	2	3
-1					
0					
1					
2					
3					

(b) Determine the increase in the value of the algebraic expression for each one-unit increase in x. Does the answer depend on the value of y?

83. *Geometry* The basic floor plan for a one-story house is shown in the figure.

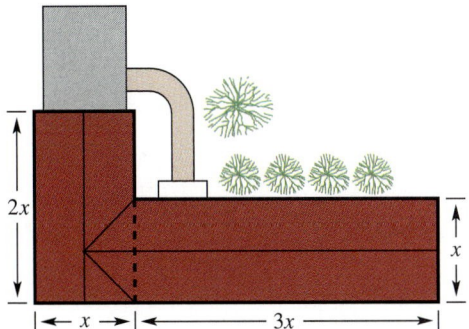

(a) Write a simplified algebraic expression for the amount of floor space in the house.

(b) Determine the amount of floor space if $x = 20$ feet. What units of measure are used for the amount of floor space?

84. *Geometry* The area of the trapezoid with parallel bases of length b_1 and b_2 and height h (see figure) is given by $\frac{1}{2}h(b_1 + b_2)$.

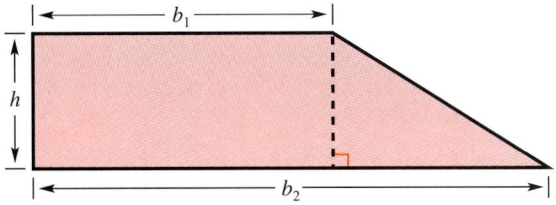

(a) Show that the area can also be expressed as $b_1h + \frac{1}{2}(b_2 - b_1)h$, and give a geometric explanation for the area represented by each term in this expression.

(b) Find the area of a trapezoid with $b_1 = 7$, $b_2 = 12$, and $h = 3$.

Geometry In Exercises 85 and 86, use the formula for the area of a trapezoid, $\frac{1}{2}h(b_1 + b_2)$, to find the area of the trapezoidal house lot and tile.

85.

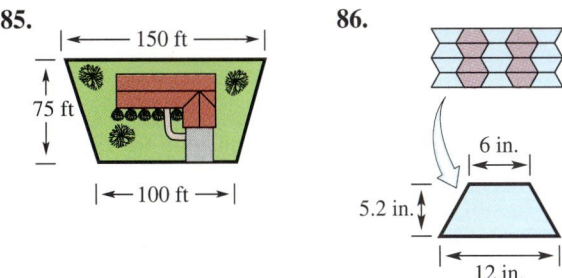

86.

CAREER INTERVIEW

Timothy J. Reuning

Senior Programmer/ Project Leader

Wisconsin Packing Company, Incorporated

Butler, WI 53007

Wisconsin Packing produces hamburger patties and other specialty food items. We have our own computer system to monitor production on the plant floor, and I am responsible for maintaining the programs used by this system and assisting with user problems.

Algebra is essential to programming. One subroutine we often use converts a date entered by a plant worker in the form *MM/DD/YY* to a *Julian date*—the number of days since January 1, 1900. To convert a date occurring after March 31 in a year prior to 2000, the program evaluates a series of algebraic expressions:

Y is $YY \cdot 365.25$ (ignore remainder),

M is $(MM + 1) \cdot 30.6001$ (ignore remainder),

and finally, the Julian date is given as

$J = Y + M + DD$.

Suppose the date 07/03/98 is entered in the program. Then $Y = 98 \cdot 365.25 = 35{,}794$, $M = (7 + 1) \cdot 30.6001 = 244$, and the Julian date is $J = 35{,}794 + 244 + 3 = 36{,}041$ days since the start of the 20th century.

MID-CHAPTER QUIZ

Take this quiz as you would take a quiz in class. After you are done, check your work against the answers given in the back of the book.

1. Identify the coefficients of the terms (a) $-5xy^2$ and (b) $\dfrac{5z}{16}$.

2. Rewrite the expression in exponential form.

   (a) $3y \cdot 3y \cdot 3y \cdot 3y$ (b) $2 \cdot (x-3) \cdot (x-3) \cdot 2 \cdot 2$

In Exercises 3–8, simplify the expression.

3. $x^4 \cdot x^3$

4. $(v^2)^5$

5. $(-3y)^2 y^3$

6. $8(x-4)^2(x-4)^4$

7. $\dfrac{2z^2}{3y} \cdot \dfrac{5z}{7y^3}$

8. $\left(\dfrac{x}{y}\right)^2 \left(\dfrac{x}{y}\right)^5$

In Exercises 9–12, identify the rule of algebra illustrated by the equation.

9. $-3(2y) = (-3 \cdot 2)y$

10. $(x+2)y = xy + 2y$

11. $3y \cdot \dfrac{1}{3y} = 1, \quad y \neq 0$ *Multiplicative Inverse*

12. $x - x^2 + 2 = -x^2 + x + 2$ *commutative*

In Exercises 13 and 14, simplify the expression by combining like terms.

13. $y^2 - 3xy + y + 7xy$

14. $10\left(\dfrac{1}{u}\right) - 7\left(\dfrac{1}{u}\right) + 3u$

In Exercises 15 and 16, simplify the expression by removing symbols of grouping and combining like terms.

15. $5(a-2b) + 3(a+b)$

16. $4x + 3[2 - 4(x+6)]$

In Exercises 17 and 18, evaluate the algebraic expression for the specified values of the variables. If it is not possible, state the reason.

17. $x^2 - 3x$ (a) $x = 3$ (b) $x = -2$ (c) $x = 0$

18. $\dfrac{x}{y-3}$ (a) $x = 2, \ y = 4$ (b) $x = 0, \ y = -1$ (c) $x = 5, \ y = 3$

19. Simplify the following expression for the moment of inertia of a cone of height h and radius r.

   $\left(\tfrac{1}{3}\pi r^2 h\right)\left(\tfrac{3}{10}r^2\right)$

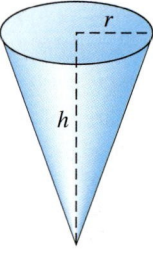

Figure for 19

20. Evaluate the expression $4 \cdot 10^4 + 5 \cdot 10^3 + 7 \cdot 10^2$.

2.4   **Translating Expressions: Verbal to Algebraic**

Introduction ▪ Translating Phrases ▪
Hidden Operations

Introduction

In the first three sections of this chapter, you studied techniques for rewriting and simplifying algebraic expressions. In this section you will study ways to *construct* algebraic expressions.

Let's take another look at Example 1 (page 92). In that example you were paid $6 an hour and your weekly pay was represented by

$$\boxed{\text{Pay per hour}} \quad \cdot \quad \boxed{\text{Number of hours}} = \boxed{6 \text{ dollars}} \quad \cdot \quad \boxed{x \text{ hours}} = 6x.$$

Note the hidden operation of multiplication in this expression. Nowhere in the verbal problem does it say you are to multiply 6 times x. It is *implied* in the problem—you just use common sense to come up with the product. This is often the case when using algebra to solve real-life problems.

Your students might be interested to know that, in 1990, 1 million tons of aluminum containers were recycled. This accounted for over 38% of all aluminum containers produced.   (Source: Franklin Associates, Ltd.)

EXAMPLE 1   Constructing an Algebraic Expression

You are paid 5¢ for each aluminum soda can, and 3¢ for each glass soda bottle you collect. Write an algebraic expression that represents the total weekly income for this recycling activity.

Solution

Before writing an algebraic expression for the weekly income, it is helpful to construct an informal verbal model. For instance, the following verbal model could be used.

$$\boxed{\text{Pay per can}} \quad \cdot \quad \boxed{\text{Number of cans}} \quad + \quad \boxed{\text{Pay per bottle}} \quad \cdot \quad \boxed{\text{Number of bottles}}$$

Note that the word *and* in the problem indicates addition. Because both the number of cans and the number of bottles can vary from week to week, you can use the two variables x and y, respectively, to write the following algebraic expression.

$$\boxed{5 \text{ cents}} \quad \cdot \quad \boxed{x \text{ cans}} \quad + \quad \boxed{3 \text{ cents}} \quad \cdot \quad \boxed{y \text{ bottles}} = 5x + 3y$$

Translating Phrases

When you are translating verbal sentences and phrases into algebraic expressions, it is sometimes helpful to watch for key words and phrases that indicate the four different operations of arithmetic. The following list gives several examples.

Translating Phrases into Algebraic Expressions

Key Words and Phrases	Verbal Description	Algebraic Expression
Addition: Sum, plus, greater, increased by, more than, exceeds, total of	The sum of 6 and x	$6 + x$
	Eight more than y	$y + 8$
Subtraction: Difference, minus, less, decreased by, subtracted from, reduced by, the remainder	Five decreased by a	$5 - a$
	Four less than z	$z - 4$
Multiplication: Product, multiplied by, twice, times, percent of	Five times x	$5x$
Division: Quotient, divided by, ratio, per	The ratio of x to 3	$\dfrac{x}{3}$

EXAMPLE 2   Translating Phrases Having Specified Variables

Translate each of the following into an algebraic expression.

a. Three less than the product of five and m

b. y decreased by the sum of ten and x^2

Solution

a. Three less than the product of five and m

$$5m - 3 \qquad \text{Think: 3 subtracted from what?}$$

b. y decreased by the sum of ten and x^2

$$y - (10 + x^2) \qquad \text{Think: What is subtracted from } y?$$

EXAMPLE 3   Translating Phrases Having Specified Variables

Translate each of the following into an algebraic expression.

a.   Six times the sum of x and seven

b.   The product of four and x, all divided by three

Solution

a.   Six times the sum of x and seven

$$6(x + 7)$$ Think: 6 multiplied by what?

b.   The product of four and x, all divided by three

$$\frac{4x}{3}$$ Think: What is divided by 3?

In most applications of algebra, the variables are not specified and it is your task to assign variables to the *appropriate* quantities.   Though similar to the translations in Examples 2 and 3, the translations in the next example may seem more difficult because variables have not been assigned to the unknown quantities.

EXAMPLE 4   Translating Phrases Having No Specified Variable

Translate each of the following into a variable expression.

a.   The sum of three and a number

b.   Five decreased by the product of three and a number

c.   The difference of a number and three, divided by twelve

Solution

In each case, let x be the unspecified number.

a.   The sum of three and a number

$$3 + x$$ Think: 3 added to what?

b.   Five decreased by the product of three and a number

$$5 - 3x$$ Think: What is subtracted from 5?

c.   The difference of a number and three, divided by twelve

$$\frac{x - 3}{12}$$ Think: What is divided by 12?

A good way to learn algebra is to do it *forward* and *backward*. In the next example, algebraic expressions are translated into verbal form. Keep in mind that other key words could be used to describe the operations in each expression. Your goal is to use key words or phrases that keep the verbal expressions clear and concise.

EXAMPLE 5   *Translating Algebraic Expressions into Verbal Form*

Without using a variable, write a verbal description for each of the following.

a. $7x - 12$ **b.** $7(x - 12)$ **c.** $5 + \dfrac{x}{2}$ **d.** $\dfrac{5 + x}{2}$

Solution

a. *Algebraic expression:* $7x - 12$
   Primary operation: Subtraction
   Terms: $7x$ and 12
   Verbal description: Twelve less than the product of seven and a number

b. *Algebraic expression:* $7(x - 12)$
   Primary operation: Multiplication
   Factors: 7 and $(x - 12)$
   Verbal description: Twelve is subtracted from a number and the result is multiplied by seven.

c. *Algebraic expression:* $5 + \dfrac{x}{2}$
   Primary operation: Addition
   Terms: 5 and $\dfrac{x}{2}$
   Verbal description: Five added to the quotient of a number and two

d. *Algebraic expression:* $\dfrac{5 + x}{2}$
   Primary operation: Division
   Numerator, denominator: Numerator is $5 + x$; denominator is 2
   Verbal description: The sum of five and a number, all divided by two

NOTE   Translating algebraic expressions into verbal phrases is more difficult than it may appear. It is easy to write a phrase that is ambiguous. For instance, what does the phrase "the sum of five and a number times two" mean? Without further information, this phrase could mean

$$5 + 2x \quad \text{or} \quad 2(5 + x).$$

Hidden Operations

Most real-life problems do not contain verbal expressions that clearly identify the arithmetic operations involved. You need to rely on common sense, past experience, and the physical nature of the problem in order to identify the operations hidden in the problem statement. Watch for *hidden products* in the next two examples.

EXAMPLE 6   *Discovering Hidden Products*

a. A cash register contains x quarters. Write an expression for this amount of money in dollars.

b. A cash register contains n nickels and d dimes. Write an expression for this amount of money in cents.

c. Write an expression showing how far a person can ride a bicycle in t hours if the person travels at a constant rate of 15 miles per hour.

Solution

a. The amount of money is a product.

Verbal Model: | Value of coin | · | Number of coins |

Labels: Value of coin $= 0.25$ (dollars)
             Number of coins $= x$
Expression: $0.25x$ (dollars)

b. The amount of money is a sum of products.

Verbal Model: | Value of nickel | · | Number of nickels | + | Value of dime | · | Number of dimes |

Labels: Value of nickel $= 5$ (cents)
             Number of nickels $= n$
             Value of dime $= 10$ (cents)
             Number of dimes $= d$
Expression: $5n + 10d$ (cents)

c. The distance traveled is a product.

Verbal Model: | Rate of travel | · | Time traveled |

Labels: Rate of travel $= 15$ (miles per hour)
             Time traveled $= t$ (hours)
Expression: $15t$ (miles)

STUDY TIP

In Example 6(c), the final answer is listed in terms of miles. This makes sense in the following way.

$$15 \, \frac{\text{miles}}{\text{hour}} \cdot t \, \text{hours}$$

Note that the hours "cancel," leaving the answer in terms of miles. This technique, called *unit analysis*, can be very helpful in determining the final unit of measure.

EXAMPLE 7 *Discovering Hidden Operations*

a. A person added k liters of a fluid containing 45% antifreeze to a car radiator. Write an expression that indicates how much antifreeze was added.

b. A consumer received a 30% discount on an item priced at d dollars. Write an expression, in dollars, for the discount.

c. A person paid x dollars plus a 6% sales tax for an automobile. Write an expression for the total cost of the automobile.

Solution

a. The amount of antifreeze is a product.

Verbal Model: $\boxed{\text{Percent of antifreeze}} \cdot \boxed{\text{Number of liters}}$

Labels: Percent of antifreeze $= 0.45$ (decimal form)
Number of liters $= k$

Expression: $0.45k$ (liters)

b. The amount of the discount is a product.

Verbal Model: $\boxed{\text{Percent of discount}} \cdot \boxed{\text{Original price}}$

Labels: Percent of discount $= 0.3$ (decimal form)
Original price $= d$ (dollars)

Expression: $0.3d$ (dollars)

c. The total cost is a sum.

Verbal Model: $\boxed{\text{Cost of automobile}} + \boxed{\text{Percent of sales tax}} \cdot \boxed{\text{Cost of automobile}}$

Labels: Percent of sales tax $= 0.06$ (decimal form)
Cost of automobile $= x$ (dollars)

Expression: $x + 0.06x = (1 + 0.06)x = 1.06x$

Hidden operations are often involved when variable names (labels) are assigned to two unknown quantities. A good strategy is to use a *specific* case to help you write a model for the *general* case. For instance, a specific case of

$$3, 3 + 1, \text{ and } 3 + 2$$

may help you write a general case for n, $n + 1$, and $n + 2$. This strategy is illustrated in Examples 8 and 9.

EXAMPLE 8   *Using a Specific Case to Find a General Case*

In each of the following, use the given variable to label the unknown quantity.

a. A person's weekly salary is d dollars. What is the annual salary?

b. A person's annual salary is y dollars. What is the monthly salary?

Solution

a. There are 52 weeks in a year.

Specific case: If the weekly salary is $200, then the annual salary (in dollars) is $52 \cdot 200$.

General case: If the weekly salary is d dollars, then the annual salary (in dollars) is $52 \cdot d$ or $52d$.

b. There are 12 months in a year.

Specific case: If the annual salary is $24,000, then the monthly salary (in dollars) is $24,000 \div 12$.

General case: If the annual salary is y, then the monthly salary (in dollars) is $y \div 12$ or $y/12$.

EXAMPLE 9   *Using a Specific Case to Find a General Case*

In each of the following, use the given variable to label the unknown quantity.

a. One person is k inches shorter than another person. The first person is 60 inches tall. How tall is the second person?

b. A consumer buys g gallons of gasoline for a total of d dollars. What is the price per gallon?

Solution

a. The first person is k inches shorter than the second person.

Specific case: If the first person is 10 inches shorter than the second person, then the second person is $60 + 10$ inches tall.

General case: If the first person is k inches shorter than the second person, then the second person is $60 + k$ inches tall.

b. To obtain the price per gallon, divide the price by the number of gallons.

Specific case: If the total price is $11.50 and the total number of gallons is 10, then the price per gallon is $11.50 \div 10$ dollars per gallon.

General case: If the total price is d dollars and the total number of gallons is g, then the price per gallon is $d \div g$ or d/g dollars per gallon.

In mathematics it is useful to know how to represent certain types of integers algebraically. For instance, consider the set $\{2, 4, 6, 8, \ldots\}$ of even integers. What algebraic symbol could we use to denote an even integer? Because every even integer has 2 as a factor,

$$2 = 2 \cdot 1, \quad 4 = 2 \cdot 2, \quad 6 = 2 \cdot 3, \quad 8 = 2 \cdot 4, \ldots,$$

it follows that any integer n multiplied by 2 must be an *even* integer. Moreover, if $2n$ is even, then $2n - 1$ and $2n + 1$ must be *odd* integers. For example, choose $n = 5$. Then $2n = 2 \cdot 5 = 10$ is even, whereas

$$2n - 1 = 10 - 1 = 9 \quad \text{and} \quad 2n + 1 = 10 + 1 = 11$$

are both odd.

Two integers are called **consecutive integers** if they differ by 1. Hence, for any integer n, its next two larger consecutive integers are $n + 1$ and $(n + 1) + 1$ or $n + 2$. Thus, you can denote three consecutive integers by n, $n + 1$, and $n + 2$.

Expressions for Special Types of Integers

Let n be an integer. Then the following expressions can be used to denote even integers, odd integers, and consecutive integers.

1. $2n$ denotes an *even* integer.

2. $2n - 1$ and $2n + 1$ denote *odd* integers.

3. The set $\{n, n + 1, n + 2\}$ denotes three *consecutive* integers.

Group Activities   Problem Solving

Enough Information?   Most of the verbal problems you encounter in a mathematics text have precisely the right amount of information necessary to solve the problem. In real life, however, you may need to collect additional information. In your group, decide what additional information would be needed to solve the following problem.

During a given week, a person worked 48 hours for the same employer. The hourly rate for overtime is $12. Write an expression for the person's gross pay for the week, including any pay received for overtime.

2.4 Exercises

Discussing the Concepts

1. The word *difference* indicates what operation?

2. The word *ratio* indicates what operation?

3. Which of the following are equivalent to the expression $n + 4$?

(a) 4 more than n

(b) the sum of n and 4

(c) n less than 4

(d) the ratio of n to 4

(e) the total of 4 and n

4. Is order important when translating the following phrases into algebraic expressions? Explain.

(a) x is increased by 10

(b) 10 is decreased by x

(c) the product of x and 10

(d) the quotient of x and 10

5. Give two interpretations of "the quotient of 5 and a number times 3."

6. Let n be an integer. Explain why the expressions $2n$ and $2n + 2$ both represent even integers.

Problem Solving

In Exercises 7–16, translate the phrase into an algebraic expression. (Let x represent the number.)

7. A number increased by 5

8. A number decreased by 7

9. Six less than a number

10. Ten more than a number

11. Twice a number

12. A number divided by 100

13. The ratio of a number to 50

14. The product of 30 and a number

15. A number is tripled and the result is increased by 5

16. A number is increased by 5 and the result is tripled

In Exercises 17–22, write a verbal description of the algebraic expression. Use words only—do not use the variable. (There is more than one correct answer.)

17. $3x + \frac{1}{3}$

18. $4 - 7x$

19. $\dfrac{t + 1}{2}$

20. $\dfrac{1}{2} - \dfrac{t}{5}$

21. $\frac{1}{2}(x + 1)$

22. $-(2x + 4)$

In Exercises 23–26, translate the sentence into an algebraic expression. Simplify the expression.

23. The sum of x and 3 is multiplied by x.

24. The sum of 25 and x is added to x.

25. The difference of 9 and x is multiplied by 3.

26. The sum of 4 and x is added to the sum of x and -8.

27. *Total Amount of Money*   A cash register contains n dimes. Write an algebraic expression that represents the total amount of money (in dollars).

28. *Income Tax*   The state income tax on a gross income of I dollars is 2.2%. Write an algebraic expression that represents the total amount of income tax. (To find 2.2% of a quantity, multiply the quantity by 0.022.)

29. *Weekly Paycheck*   You worked n hours in a week. Your hourly wage is $8.25 per hour. Write an algebraic expression that represents your gross pay for the week.

30. *Revenue*   You sell x units of a product. The price per unit is $18.75. Write an algebraic expression that represents your revenue.

31. *Travel Time*   A truck travels 100 miles at an average speed of r miles per hour, as shown in the figure. Write an algebraic expression that represents the total travel time.

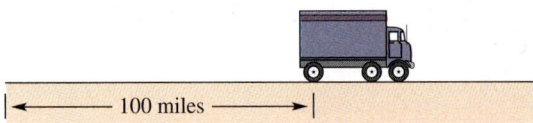

|← 100 miles →|

32. *Hourly Wage*   The hourly wage for an employee is $6.50 per hour plus 75 cents for each q units produced during the hour. Write an algebraic expression that represents the total hourly earnings for the employee.

33. *Computer Screen*   A computer screen has sides that are s inches long, as shown in the figure. Write an algebraic expression that represents the area of the screen. What are the units of measure for the area of the screen?

|← s →|

34. *Geometry*   A rectangle has sides of length $3w$ and w. Write an algebraic expression that represents the perimeter of the rectangle.

35. *Sum*   Write an expression that represents the sum of three consecutive integers, the first of which is n.

36. *Sum*   Write an expression that represents the sum of two consecutive even integers, the first of which is $2n$.

37. *Finding a Pattern*

(a) Complete the table. The third row in the table consists of the difference between consecutive entries of the second row.

n	0	1	2	3	4	5
$2n - 1$						
Differences						

(b) Describe the pattern of the third row. What would the pattern be if the algebraic expression were $an + b$ rather than $2n - 1$?

38. *Fibonacci Sequence*   Each term in the Fibonacci Sequence is the sum of the two previous terms.

(a) Let n and m represent two consecutive terms in the sequence. Write an algebraic expression for the next term.

(b) The first three terms are 1, 1, and 2. Write the next five terms.

Reviewing the Major Concepts ───────────────────

In Exercises 39 and 40, simplify the algebraic expression.

39. $-y^2(y^2 + 4) + 6y^2$ **40.** $5t(2 - t) + t^2$

In Exercises 41 and 42, evaluate the expression.

41. $x^2 - y^2$ (a) $x = 4, y = 3$ (b) $x = -5, y = 3$

42. $z^2 + x^2$ (a) $x = 1, z = 1$ (b) $x = 2, z = 2$

43. *Parcel Service*   You are shipping 133 pounds of a product. The parcel service you are using will not accept packages that weigh more than 30 pounds. What is the fewest number of packages you can use to ship the product?

44. *Geometry*   The length of a rectangle is one and one-half its width. Its width is 8 meters. Find its perimeter.

Additional Problem Solving

In Exercises 45–50, match the verbal phrase with the corresponding algebraic expression.

(a) $11 + \frac{1}{3}x$ (b) $3x - 12$ (c) $3(x - 12)$

(d) $12 - 3x$ (e) $11x + \frac{1}{3}$ (f) $12x + 3$

45. Twelve decreased by 3 times a number

46. Eleven more than $\frac{1}{3}$ of a number

47. Eleven times a number plus $\frac{1}{3}$

48. Three increased by 12 times a number

49. The difference between 3 times a number and 12

50. Three times the difference of a number and 12

In Exercises 51–64, translate the phrase into an algebraic expression. (Let x represent the number.)

51. 25 more than a number

52. A number decreased by 25

53. One-fourth of a number

54. One-fifth of a number

55. A number divided by 3

56. Three-tenths of a number

57. Five times a number, plus 8

58. The ratio of a number to 5, plus 15

59. Ten times the sum of a number and 4

60. Seven more than 5 times a number

61. The absolute value of the sum of a number and 4

62. The absolute value of 4 less than twice a number

63. The square of a number increased by 1

64. Twice the cube of a number, increased by 4

In Exercises 65–72, write a verbal description of the algebraic expression. Use words only; do not use the variable. (There is more than one correct answer.)

65. $x - 10$ **66.** $x + 9$

67. $7x + 4$ **68.** $9 - \frac{1}{4}x$

69. $3(2 - x)$ **70.** $-10(t - 6)$

71. $x^2 + 5$ **72.** $x^3 - 1$

In Exercises 73–76, translate the sentence into an algebraic expression. Simplify the expression.

73. The sum of 6 and n is multiplied by 5.

74. The square of x is added to the product of x and $x + 1$.

75. The product of eight times the sum of x and 24 is divided by 2.

76. Fifteen is subtracted from x and the difference is multiplied by 4.

77. *Sales Tax*   The sales tax on a purchase of L dollars is 6%. Write an algebraic expression that represents the total amount of sales tax. (To find 6% of a quantity, multiply the quantity by 0.06.)

78. *Total Amount of Money*   A cash register contains m dimes and n quarters. Write an algebraic expression that represents the total amount of money (in dollars).

79. *Camping Fee*   A campground charges $15 for adults and $2 for children. Write an algebraic expression that represents the total camping fee for m adults and n children.

80. *Distance Traveled*   A plane travels at the rate of r miles per hour for 3 hours. Write an algebraic expression that represents the total distance traveled by the plane.

81. *Geometry*   Write an algebraic expression that represents the perimeter of the picture frame.

$1.5w$

w

82. *Geometry* A square has sides of length s. Write an algebraic expression that represents the perimeter of the square.

Geometry In Exercises 83–86, write an algebraic expression that represents the area of the region. Then use the rules of algebra to simplify the expression.

83.

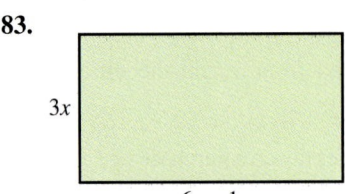

$3x$

$6x - 1$

84.

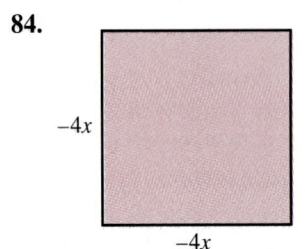

$-4x$

$-4x$

85.

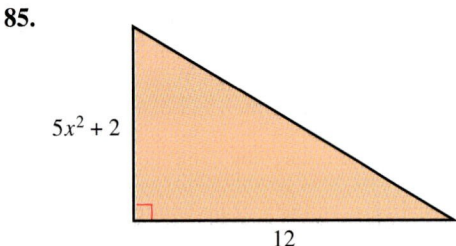

$5x^2 + 2$

12

86.

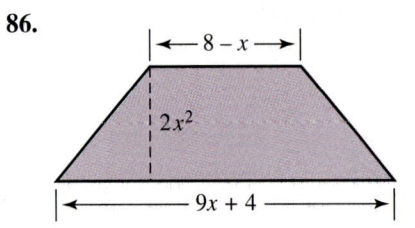

$\longleftarrow 8 - x \longrightarrow$

$2x^2$

$\longleftarrow \quad 9x + 4 \quad \longrightarrow$

87. *Sum* Write an algebraic expression that represents the sum of two consecutive odd integers, the first of which is $2n + 1$.

88. *Sum* Write an algebraic expression that represents the sum of the squares of two consecutive even integers, the first of which is $2n$.

89. *Finding a Pattern*

(a) Complete the table. The third row in the table consists of the difference between consecutive entries of the second row.

n	0	1	2	3	4	5
$2n + 5$						
Differences						

(b) Describe the pattern of the third row. What would the pattern be if the algebraic expression were $an + b$, rather than $2n + 5$?

Exploration In Exercises 90 and 91, find a and b so that the expression $an + b$ would yield the table.

90.

n	0	1	2	3	4	5
$an + b$	4	9	14	19	24	29

91.

n	0	1	2	3	4	5
$an + b$	1	5	9	13	17	21

2.5 | **Introduction to Equations**

Equations ▪ Forming Equivalent Equations ▪
Constructing Equations

Equations

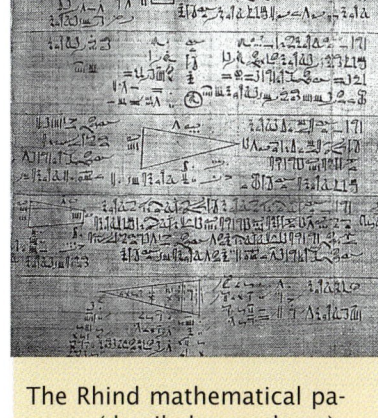

The Rhind mathematical papyrus (detail shown above), discovered in 1858, contains one of the earliest examples of mathematical writing in existence. The papyrus itself dates back to around 1600 B.C., but it is actually a copy of writings from two centuries earlier. The algebraic equations on the papyrus were written in words.

An **equation** is a statement in which two mathematical expressions are equal. Here are some examples:

$$x = 3, \quad 5x - 2 = 8, \quad 3x - 12 = 3(x - 4), \quad \text{and} \quad x^2 - 9 = 0.$$

To **solve** an equation involving x means to find all values of x for which the equation is true. Such values are called **solutions,** and we say that the solutions **satisfy** the equation. For instance, 3 is a solution of $x = 3$ because $3 = 3$ is a true statement.

The **solution set** of an equation is the set of all real numbers that are solutions of the equation. Sometimes an equation will have the set of all real numbers as its solution set. Such an equation is called an **identity.** For instance, the equation

$$3x - 12 = 3(x - 4) \qquad \qquad \text{Identity}$$

is an identity because the equation is true for all real values of x. Try values such as 0, 1, -2, and 5 in this equation to see that each one is a solution.

An equation whose solution set is not the entire set of real numbers is called a **conditional equation.** For instance, the equation

$$x^2 - 9 = 0 \qquad \qquad \text{Conditional equation}$$

is a conditional equation because it has only two solutions, 3 and -3. Examples 1 and 2 show how to **check** whether a given value of x is a solution of an equation.

NOTE When checking a solution, we suggest that you write a question mark over the equal sign to indicate that you are not sure of the validity of the equation.

EXAMPLE 1   *Checking a Solution of an Equation*

Determine whether -2 is a solution of $x^2 - 5 = 4x + 7$.

Solution

$$x^2 - 5 = 4x + 7 \qquad \qquad \text{Original equation}$$

$$(-2)^2 - 5 \overset{?}{=} 4(-2) + 7 \qquad \qquad \text{Substitute } -2 \text{ for } x.$$

$$4 - 5 \overset{?}{=} -8 + 7 \qquad \qquad \text{Simplify.}$$

$$-1 = -1 \qquad \qquad \text{Solution checks.} \quad ✓$$

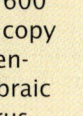

Because both sides of the equation turn out to be the same number, you can conclude that -2 is a solution of the original equation.

Just because you have found one solution of an equation, you should not conclude that you have found all of the solutions. For instance, you can check that 6 is also a solution of the equation in Example 1 as follows.

$$x^2 - 5 = 4x + 7 \qquad \text{Original equation}$$

$$(6)^2 - 5 \overset{?}{=} 4(6) + 7 \qquad \text{Substitute 6 for } x.$$

$$36 - 5 \overset{?}{=} 24 + 7 \qquad \text{Simplify.}$$

$$31 = 31 \qquad \text{Solution checks.} ✔$$

EXAMPLE 2 A Trial Solution That Does Not Check

Determine whether 2 is a solution of $x^2 - 5 = 4x + 7$.

Solution

$$x^2 - 5 = 4x + 7 \qquad \text{Original equation}$$

$$(2)^2 - 5 \overset{?}{=} 4(2) + 7 \qquad \text{Substitute 2 for } x.$$

$$4 - 5 \overset{?}{=} 8 + 7 \qquad \text{Simplify.}$$

$$-1 \neq 15 \qquad \text{2 is not a solution.} ✗$$

Because the two sides of the equation turn out to be different, you can conclude that 2 is not a solution of the original equation.

Be sure that you understand the following distinction between an algebraic expression and an algebraic equation.

1. An algebraic *expression* contains no equal sign. You can rewrite or simplify an expression, but you cannot solve it. When you rewrite an expression, you use an equal sign to show that the rewritten expression is equivalent to the original expression. For instance, you can rewrite the expression $4(x - 1)$ as $4x - 4$ and indicate the equivalence of the two expressions as

    $$4(x - 1) = 4x - 4.$$

    This equation is an identity. That is, the expression on the left yields the same value as the expression on the right when x is replaced by any real number.

2. An algebraic *equation* has an equal sign. An equation may be either an identity or a conditional equation. You do not try to solve an equation that is an identity because (by definition) the solution set consists of all real numbers. You can, however, try to solve conditional equations. In doing so you try to find all real numbers that are solutions of the conditional equation.

Forming Equivalent Equations

It is helpful to think of an equation as having two sides that are in balance. Consequently, when you try to solve an equation, you must be careful to maintain that balance by performing the same operation on both sides.

Two equations that have the same set of solutions are called **equivalent.** For instance, the equations $x = 3$ and $x - 3 = 0$ are equivalent because both have only one solution—the number 3. When any one of the operations in the following list is applied to an equation, the resulting equation is equivalent to the original equation.

Forming Equivalent Equations: Properties of Equality

An equation can be transformed into an *equivalent equation* using one or more of the following procedures.

	Original Equation	*Equivalent Equation*
1. *Simplify either side:* Remove symbols of grouping, combine like terms, or reduce fractions on one or both sides of the equation.	$3x - x = 8$	$2x = 8$
2. *Apply the Addition Property of Equality:* Add (or subtract) the same quantity to (from) *both* sides of the equation.	$x - 2 = 5$	$x = 7$
3. *Apply the Multiplication Property of Equality:* Multiply (or divide) *both* sides of the equation by the same *nonzero* quantity.	$3x = 9$	$x = 3$
4. *Interchange the sides of the equation.*	$7 = x$	$x = 7$

The second and third operations in this list can be used to eliminate terms or factors in an equation. For example, to solve the equation $x - 5 = 1$, we need to eliminate the term -5 on the left side. This is accomplished by adding its opposite, 5, to both sides.

$$x - 5 = 1 \qquad \text{Original equation}$$
$$x - 5 + 5 = 1 + 5 \qquad \text{Add 5 to both sides.}$$
$$x + 0 = 6 \qquad \text{Combine like terms.}$$
$$x = 6 \qquad \text{Solution}$$

All four of the equations listed above are equivalent, and we call them the **steps** of the solution.

The next example shows how the properties of equality can be used to solve equations. You will get many more opportunities to practice these skills in the next chapter. For now, your goal should be to understand why each step in the solution is valid. For instance, the second step in part (a) is valid because the Addition Property of Equality states that you can add the same quantity to both sides of an equation.

EXAMPLE 3 Operations Used to Solve Equations

a. To eliminate the term -2 on the left side of the following equation, add 2 to both sides of the equation.

$$x - 2 = 5 \qquad \text{Original equation}$$
$$x - 2 + 2 = 5 + 2 \qquad \text{Add 2 to both sides.}$$
$$x = 7 \qquad \text{Solution}$$

b. To eliminate the term $2x$ on the right side of the following equation, subtract $2x$ from both sides of the equation.

$$3x = 2x + 4 \qquad \text{Original equation}$$
$$3x - 2x = 2x - 2x + 4 \qquad \text{Subtract } 2x \text{ from both sides.}$$
$$x = 4 \qquad \text{Solution}$$

c. To eliminate the 5 in the denominator of the left side of the equation, multiply both sides of the equation by 5.

$$\frac{x}{5} = -2 \qquad \text{Original equation}$$
$$\frac{x}{5}(5) = -2(5) \qquad \text{Multiply both sides by 5.}$$
$$x = -10 \qquad \text{Solution}$$

d. To eliminate the factor 4 on the left side of the equation, divide both sides of the equation by 4.

$$4x = 9 \qquad \text{Original equation}$$
$$\frac{4x}{4} = \frac{9}{4} \qquad \text{Divide both sides by 4.}$$
$$x = \frac{9}{4} \qquad \text{Solution}$$

NOTE In Example 3(b), $2x$ was *subtracted* from both sides of the equation to eliminate $2x$ on the right side. You could just as easily have *added* $-2x$ to both sides. Both techniques are legitimate—the one you choose to use is a matter of personal preference.

Constructing Equations

Phase 1

Verbal Description of Problem

↓

Verbal Model

Phase 2

Assign Labels

↓

Algebraic Equation

It is helpful to use two phases in constructing equations that model real life. In the first phase, you translate the verbal description into a *verbal model.* In the second phase, you assign labels and translate the verbal model into a *mathematical model* or *algebraic equation.* Here are two examples of verbal models.

1. The sale price of a basketball is $18. The sale price is $7 less than the original price. What is the original price?

Verbal Model: Sale price = Original price − Discount

$18 = Original price − $7

2. The original price of a basketball is $25. The original price is discounted by $7. What is the sale price?

Verbal Model: Sale price = Original price − Discount

Sale price = $25 − $7

EXAMPLE 4   Using Verbal Models to Construct Equations

Write an algebraic equation for the problem.

The total income that an employee received in 1992 was $21,550. Of that, $750 represented a bonus given at the end of the year. How much was the employee paid each week? Assume that each weekly paycheck contained the same amount, and that the year consisted of 52 weeks.

Solution

Verbal Model: Income for year = 52 · Weekly pay + Bonus

Labels: Income for year = 21,550   (dollars)
         Weekly pay = x (dollars)
         Bonus = 750   (dollars)

Algebraic Model: $21,550 = 52x + 750$

EXAMPLE 5 Using Verbal Models to Construct Equations

Write an algebraic equation for the problem.

Tickets for a concert were $15 for each floor seat and $10 for each stadium seat. There were 800 seats on the main floor, and these were sold out. If the total revenue from ticket sales was $52,000, how many stadium seats were sold?

Solution

Verbal Model:

Total revenue	=	Revenue from floor seats	+	Revenue from stadium seats

Labels:

Total revenue = 52,000	(dollars)
Price per floor seat = 15	(dollars per seat)
Number of floor seats = 800	(seats)
Price per stadium seat = 10	(dollars per seat)
Number of stadium seats = x	(seats)

Algebraic Model:

$$52,000 = 15(800) + 10x$$

NOTE In Example 5, you can use the following *unit analysis* to check that both sides of the equation are measured in dollars.

$$52,000 \text{ dollars} = \left(\frac{15 \text{ dollars}}{\text{seat}} \right) (800 \text{ seats}) + \left(\frac{10 \text{ dollars}}{\text{seat}} \right) (x \text{ seats})$$

In the next chapter, you will study techniques for solving the equations constructed in Examples 4 and 5.

Group Activities   P r o b l e m   S o l v i n g

Red Herring   When constructing an equation to represent a word problem, you are occasionally given too much information. The unnecessary information in a word problem is sometimes called a "red herring." In your group, find the red herring in the following problem.

Returning to college after spring break, a student travels 3 hours and stops for lunch. If it takes 45 minutes to complete the last 36 miles of the 180 mile trip, find the average speed during the first 3 hours of the trip.

Together, decide what question to ask that uses all the given information.

2.5 Exercises

Discussing the Concepts

1. In your own words, explain the difference between a conditional equation and an identity.

2. Explain how to decide whether a real number is a solution of an equation. Give an example of an equation with a solution that checks and one that does not check.

3. Explain the difference between simplifying an expression and solving an equation. Give an example of each.

4. In your own words, explain what is meant by the term *equivalent equations*.

5. Describe, from memory, the steps that are used to transform an equation into an equivalent equation.

6. Describe a real-life problem that uses the following verbal model.

$$
\boxed{\begin{array}{c} \text{Revenue} \\ \text{of \$840} \end{array}} = \boxed{\begin{array}{c} \text{\$35 per} \\ \text{case} \end{array}} \cdot \boxed{\begin{array}{c} \text{Number} \\ \text{of cases} \end{array}}
$$

Problem Solving

In Exercises 7–14, decide whether the value of x is a solution of the equation.

Equation	*Values*	
7. $2x - 6 = 0$	(a) $x = 3$	(b) $x = 1$
8. $5x - 25 = 0$	(a) $x = 10$	(b) $x = 5$
9. $x + 5 = 2x$	(a) $x = -1$	(b) $x = 5$
10. $2x - 3 = 5x$	(a) $x = 0$	(b) $x = -1$
11. $x^2 - 4 = x + 2$	(a) $x = 3$	(b) $x = -2$
12. $x^2 = 2(4 - x)$	(a) $x = 2$	(b) $x = -4$
13. $\dfrac{2}{x} - \dfrac{1}{x} = 1$	(a) $x = 3$	(b) $x = \dfrac{1}{3}$
14. $\dfrac{5}{x - 1} + \dfrac{1}{x} = 5$	(a) $x = 3$	(b) $x = \dfrac{1}{6}$

In Exercises 15–18, use a calculator to decide whether the value of x is a solution of the equation.

Equation	*Values*	
15. $2x^2 - x = 10$	(a) $x = 2.5$	(b) $x = -1.09$
16. $x = \dfrac{3}{4x + 1}$	(a) $x = 0.75$	(b) $x = -0.25$
17. $x^3 - 1.728 = 0$	(a) $x = \frac{6}{5}$	(b) $x = -\frac{6}{5}$
18. $20x - 560 = 0$	(a) $x = 27.5$	(b) $x = -27.5$

In Exercises 19–22, justify each step of the solution.

19. $4x = -28$

$$\frac{4x}{4} = \frac{-28}{4}$$

$$x = -7$$

20. $x + 12 = 22$

$$x + 12 - 12 = 22 - 12$$

$$x = 10$$

21. $2x - 2 = x + 3$

$$-x + 2x - 2 = -x + x + 3$$

$$x - 2 = 3$$

$$x - 2 + 2 = 3 + 2$$

$$x = 5$$

22. $x + 6 = -24 + 6x$

$$-x + x + 6 = -x - 24 + 6x$$

$$6 = 5x - 24$$

$$6 + 24 = 5x - 24 + 24$$

$$30 = 5x$$

$$\frac{30}{5} = \frac{5x}{5}$$

$$6 = x$$

Mental Math In Exercises 23–26, solve the equation mentally and check your solution.

23. $x + 4 = 6$ **24.** $x - 10 = 5$

25. $3x = 30$ **26.** $\dfrac{x}{4} = 12$

Modeling In Exercises 27–34, construct an equation for the word problem. Do not solve the equation.

27. *Test Score* After your instructor added 6 points to each student's test score, your score is 94. What was your original score?

28. *List Price* The sale price of a coat is $225.98. The discount is $64. What is the list price?

29. Four times the sum of a number and 6 is 100. What is the number?

30. The sum of a number and 8, divided by 4, is 32. What is the number?

31. *Fund Raising* A student group is selling boxes of greeting cards at a profit of $1.75 each. The group needs $2000 to have enough money for a trip to Washington, D.C. How many boxes does the group need to sell to earn $2000?

32. *Thunderstorm* You hear thunder 3 seconds after you see lightning. The speed of sound is 1100 feet per second. How far away is the lightning?

33. *Average Speed* After traveling for 3 hours, your family is still 25 miles from completing a 160-mile trip (see figure). What was your average speed during the first 3 hours?

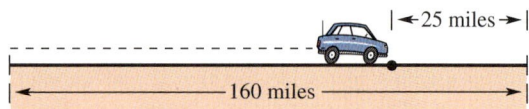

34. *Height of a Box* Find the height of a rectangular box if its base is 4 feet by 6 feet and its volume is 72 cubic feet (see figure).

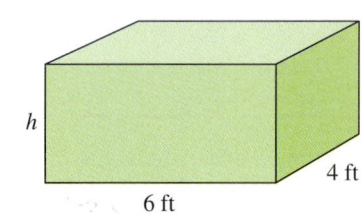

In Exercises 35–38, write a verbal description of the equation. Use words only; do not use the variable. (There is more than one correct answer.)

35. $x + 8 = 25$ **36.** $3x = 52$

37. $2(x - 5) = 12$ **38.** $\dfrac{x + 1}{3} = 8$

Reviewing the Major Concepts

In Exercises 39 and 40, simplify the algebraic expression.

39. $3x - 2(x - 5)$ **40.** $(3a^2)(4ab)$

In Exercises 41 and 42, translate the phrase into an algebraic expression. (Let x represent the number.)

41. Four more than twice a number

42. A number is increased by 25 and the sum is divided by 2.

43. *Volume* At the beginning of the day, a gasoline tank was full. The tank holds 20 gallons. At the end of the day the fuel gauge indicates that the tank is $\frac{5}{8}$ full. How many gallons of gasoline were used?

44. If a negative number is used as a factor 8 times, what is the sign of the product? Give a reason for your answer.

Geometry In Exercises 45 and 46, write expressions for the perimeter and area of the figure.

45.

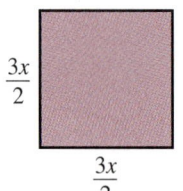

46.

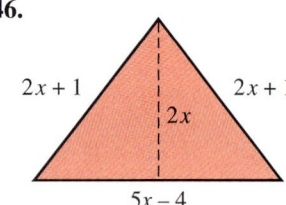

Additional Problem Solving

In Exercises 47–54, decide whether the value of x is a solution of the equation.

Equation	Values
47. $2x + 4 = 2$	(a) $x = 0$ (b) $x = -1$
48. $3x + 10 = 4$	(a) $x = -2$ (b) $x = 2$
49. $x + 3 = 2(x - 4)$	(a) $x = 11$ (b) $x = -5$
50. $5x - 1 = 3(x + 5)$	(a) $x = 8$ (b) $x = -2$
51. $2x + 10 = 7(x + 1)$	(a) $x = \frac{3}{5}$ (b) $x = \frac{2}{3}$
52. $9x + 6 = 9 - x$	(a) $x = -\frac{3}{4}$ (b) $x = \frac{3}{10}$
53. $\dfrac{4}{x} + \dfrac{2}{x} = 1$	(a) $x = 0$ (b) $x = 6$
54. $\dfrac{3}{x - 2} = x$	(a) $x = -1$ (b) $x = 3$

In Exercises 55–58, use a calculator to determine whether the value of x is a solution of the equation.

Equation	Values
55. $x + 3 = 3.5x$	(a) $x = 1.2$ (b) $x = 4.8$
56. $3 - x = 3(x + 10)$	(a) $x = \frac{21}{4}$ (b) $x = -\frac{27}{4}$
57. $x(22 - 5x) = 17$	(a) $x = 1$ (b) $x = 3.4$
58. $\dfrac{1}{x} - \dfrac{9}{x - 4} = 1$	(a) $x = 0$ (b) $x = -2$

In Exercises 59–62, justify each step of the solution.

59.
$$\tfrac{2}{3}x = 12$$
$$\tfrac{3}{2}\left(\tfrac{2}{3}x\right) = \tfrac{3}{2}(12)$$
$$x = 18$$

60.
$$14 - 3x = 5$$
$$14 - 3x - 14 = 5 - 14$$
$$14 - 14 - 3x = -9$$
$$-3x = -9$$
$$\frac{-3x}{-3} = \frac{-9}{-3}$$
$$x = 3$$

61.
$$x = -2(x + 3)$$
$$x = -2x - 6$$
$$2x + x = 2x - 2x - 6$$
$$3x = 0 - 6$$
$$3x = -6$$
$$\frac{3x}{3} = \frac{-6}{3}$$
$$x = -2$$

62.
$$\frac{x}{3} = x + 1$$
$$3\left(\frac{x}{3}\right) = 3(x + 1)$$
$$x = 3x + 3$$
$$-3x + x = -3x + 3x + 3$$
$$-2x = 0 + 3$$
$$-2x = 3$$
$$\frac{-2x}{-2} = \frac{3}{-2}$$
$$x = -\frac{3}{2}$$

Modeling   In Exercises 63–80, construct an equation for the word problem. Do not solve the equation.

63. *Computer Purchase*   You have $3650 saved for the purchase of a new computer that will cost $4532. How much more must you save?

64. *Rainfall*   With the 1.2-inch rainfall today, the total for the month is 4.5 inches. How much has been recorded for the month before today's rainfall?

65. *Travel Costs*   A company pays its sales representatives 25 cents per mile for the use of their personal cars. A sales representative submitted a bill to be reimbursed for $105.75 for driving. How many miles did the sales representative drive?

66. *Total Amount*   A cash register has n quarters and seven $1 bills that total $8.75. How many quarters are in the cash register?

67. The sum of a number and 12 is 45. What is the number?

68. The sum of three times a number and 4 is 16. What is the number?

69. Find a number such that two times the number decreased by 14 equals the number divided by 3.

70. Find a number such that 6 times the number subtracted from 120 is 96.

71. The sum of three consecutive even integers is 18. Find the first number.

72. The sum of two consecutive odd integers is 100. Find the first number.

73. *Dimensions of a Mirror*   The width of a rectangular mirror is one-third its length, as shown in the figure. The perimeter of the mirror is 96 inches. What are the dimensions of the mirror?

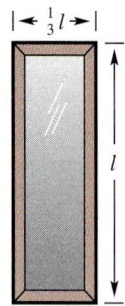

74. *Average Speed*   A group of students plans to take two cars to a soccer game. The first car leaves on time, travels at an average speed of 45 miles per hour, and arrives at the game in 3 hours. What average speed must be obtained by the students in the second car if they leave one-half hour after the first car and arrive at the game at the same time as the students in the first car?

75. *Price of a Product*   The price of a product has increased by $45 over the past year. It is now selling for $375. What was the price one year ago?

76. *Dow Jones Average*   The Dow Jones Average fell 38 points during a week and was 3695 at the close of the market on Friday. What was the average at the close of the market on the previous Friday?

77. *Annual Depreciation*   A corporation buys equipment with an initial purchase price of $750,000. It is estimated that its useful life will be 3 years and at that time its value will be $75,000. The total depreciation is divided equally among the three years. Determine the amount of depreciation declared each year.

78. *Car Payments*   Suppose you make 48 monthly payments of $158 each to buy a used car. The total amount financed is $6000. Find the amount of interest that you paid.

79. *Average Speed*   After traveling for 4 hours, you are still 24 miles from completing a 200-mile trip. It will take one-half hour to travel the last 24 miles. Find the average speed during the first 4 hours of the trip.

80. *Price of a Product*   The price of a product increased by $1432 during the past year. The price of the product was $9850 two years ago and $10,120 one year ago. What is its current price?

In Exercises 81–84, write a verbal description of the equation. Use words only; do not use the variable. (There is more than one correct answer.)

81. $9 - x^2 = 0$

82. $x + 9 = 52$

83. $10(x - 3) = 8x$

84. $\dfrac{x - 2}{10} = 6$

Unit Analysis   In Exercises 85–90, simplify the expression. State the units of the simplified value.

85. $\dfrac{3 \text{ dollars}}{\text{unit}} \cdot (5 \text{ units})$

86. $\dfrac{25 \text{ miles}}{\text{gallon}} \cdot (15 \text{ gallons})$

87. $\dfrac{3 \text{ dollars}}{\text{pound}} \cdot (5 \text{ pounds})$

88. $\dfrac{12 \text{ dollars}}{\text{hour}} \cdot \dfrac{1 \text{ hour}}{60 \text{ minutes}} \cdot (45 \text{ minutes})$

89. $\dfrac{5 \text{ feet}}{\text{second}} \cdot \dfrac{60 \text{ seconds}}{\text{minute}} \cdot (20 \text{ minutes})$

90. $\dfrac{100 \text{ centimeters}}{\text{meter}} \cdot (2.4 \text{ meters})$

CHAPTER PROJECT: Average Daily Temperatures

Meteorologists use liquid-in-glass thermometers to measure outside temperatures. Given a temperature in degrees Fahrenheit (F), you can find its equivalence in degrees Celsius (C), and vice versa. These equivalences can be modeled by $C = \frac{5}{9}(F - 32)$ and $F = \frac{9}{5}C + 32$. These equations allow you to convert from one temperature scale to the other.

In the United States, temperature is usually measured in degrees Fahrenheit. The bar graph below shows the average daily temperatures (in degrees Fahrenheit) for selected cities.   (Source: *PC USA*)

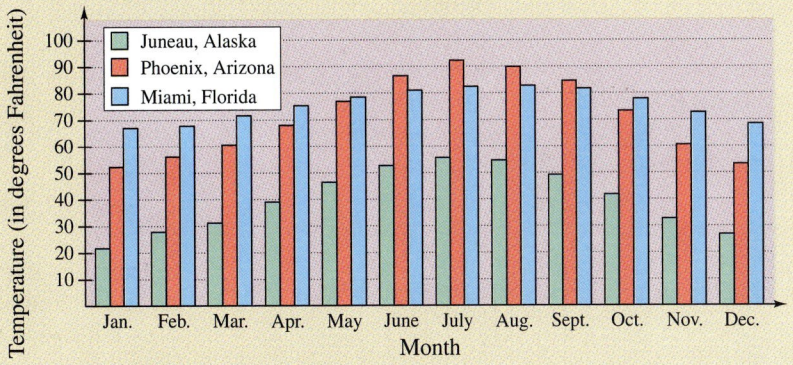

Use this information on temperatures to investigate the following questions.

1. Write each of the temperature conversion equations as a verbal statement. Use the form: "To convert from Fahrenheit to Celsius, ..."

2. Copy and complete the table at the left by using the temperature conversion equations.

3. Which city in the bar graph has the greatest annual temperature range? Which has the least?

4. Which city in the graph has the greatest decrease in temperature for any two consecutive months? Which has the greatest increase?

5. Use the bar graph to estimate the average annual temperature for Juneau, Alaska.

6. *Research Project*   Use your school's library or some other reference source to find the record high temperatures for ten cities in the United States. Write each temperature in degrees Celsius. Construct a bar graph of the data. Find the latitude of each city. Is there a relationship between the latitude and the record high temperature? Explain.

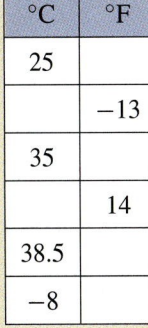

°C	°F
25	
	−13
35	
	14
38.5	
−8	

CHAPTER SUMMARY

After studying this chapter, you should have acquired the following skills. These skills are keyed to the Review Exercises that begin on page 147. Answers to odd-numbered Review Exercises are given in the back of the book.

- Identify the terms and coefficients of algebraic expressions. *(Section 2.1)*
  Review Exercises 1–4

- Rewrite products in exponential form. *(Section 2.1)*
  Review Exercises 5–8

- Simplify expressions with exponents. *(Section 2.1)*
  Review Exercises 9–16, 82

- Simplify expressions with exponents that represent real-life situations and geometry. *(Section 2.1)*
  Review Exercises 17, 18

- Identify the rule of algebra illustrated by the equation. *(Section 2.2)*
  Review Exercises 19–24

- Use the Distributive Property to expand expressions. *(Section 2.2)*
  Review Exercises 25–32

- Simplify expressions by combining like terms. *(Section 2.2)*
  Review Exercises 33–44

- Simplify expressions by removing symbols of grouping and combining like terms. *(Section 2.3)*
  Review Exercises 45–56

- Evaluate expressions for values of the variables. *(Section 2.3)*
  Review Exercises 57–60

- Translate phrases into algebraic expressions. *(Section 2.4)*
  Review Exercises 61–70, 79

- Write verbal descriptions of algebraic expressions. *(Section 2.4)*
  Review Exercises 71–74

- Translate real-life sentences into algebraic expressions. *(Section 2.4)*
  Review Exercises 75, 76, 78, 80

- Write algebraic expressions involving geometry. *(Section 2.4)*
  Review Exercises 77, 81

- Complete the table and describe the pattern of its entries. *(Section 2.4)*
  Review Exercises 83, 84

- Determine whether values of x are solutions of the equation. *(Section 2.5)*
  Review Exercises 85–94

- Translate statements into algebraic equations. *(Section 2.5)*
  Review Exercises 95, 96

- Write algebraic equations involving geometry. *(Section 2.5)*
  Review Exercises 97, 98

REVIEW EXERCISES

In Exercises 1–4, identify the terms and coefficients.

1. $4 - \frac{1}{2}x^3$

2. $5x^2 - 3x + 10$

3. $y^2 - 10yz + \frac{2}{3}z^2$

4. $\dfrac{x + 2y}{3} - \dfrac{4x}{y}$

In Exercises 5–8, rewrite the product in exponential form.

5. $5z \cdot 5z \cdot 5z$

6. $\frac{3}{8}y \cdot \frac{3}{8}y \cdot \frac{3}{8}y \cdot \frac{3}{8}y$

7. $a(b - c) \cdot a(b - c)$

8. $3 \cdot (y - x) \cdot (y - x) \cdot 3 \cdot 3$

In Exercises 9–16, simplify the expression.

9. $(x^3)^2$

10. $y^2 \cdot y^3 \cdot y$

11. $t^4(-2t^2)$

12. $(-3u^2)^2(3u^2)$

13. $(xy)(-5x^2y^3)$

14. $5v^2(3uv)(-2uv^2)$

15. $(-2y^2)^3(8y)$

16. $2(x - y)^4(x - y)^2$

17. *Depreciation* You pay P dollars for new equipment. Its value after 5 years is given by

$$P\left(\frac{9}{10}\right)\left(\frac{9}{10}\right)\left(\frac{9}{10}\right)\left(\frac{9}{10}\right)\left(\frac{9}{10}\right).$$

Simplify this expression.

18. *Geometry* The height of a triangle is $1\frac{1}{2}$ times its base. Its area is given by $\frac{1}{2}b\left(\frac{3}{2}b\right)$. Simplify this expression.

In Exercises 19–24, identify the rule of algebra illustrated by the equation.

19. $xy \cdot \dfrac{1}{xy} = 1$

20. $u(vw) = (uv)w$

21. $(x - y)(2) = 2(x - y)$

22. $(a + b) + 0 = a + b$

23. $2x + (3y - z) = (2x + 3y) - z$

24. $x(y + z) = xy + xz$

In Exercises 25–32, use the Distributive Property to expand the expression.

25. $4(x + 3y)$

26. $\frac{3}{4}(8s - 12t)$

27. $-5(2u - 3v)$

28. $-3(-2x - 8y)$

29. $x(8x + 5y)$

30. $-u(3u - 10v)$

31. $-(-a + 3b)$

32. $(7 - 2j)(-6)$

In Exercises 33–44, simplify the expression.

33. $3a - 5a$

34. $6c - 2c$

35. $3p - 4q + q + 8p$

36. $10x - 4y - 25x + 6y$

37. $\frac{1}{4}s - 6t + \frac{7}{2}s + t$

38. $\frac{2}{3}a + \frac{3}{5}a - \frac{1}{2}b + \frac{2}{3}b$

39. $x^2 + 3xy - xy + 4$

40. $uv^2 + 10 - 2uv^2 + 2$

41. $5x - 5y + 3xy - 2x + 2y$

42. $y^3 + 2y^2 + 2y^3 - 3y^2 + 1$

43. $5\left(1 + \dfrac{r}{n}\right)^2 - 2\left(1 + \dfrac{r}{n}\right)^2$

44. $-7\left(\dfrac{1}{u}\right) + 4\left(\dfrac{1}{u^2}\right) + 3\left(\dfrac{1}{u}\right)$

In Exercises 45–56, simplify the expression.

45. $5(u - 4) + 10$

46. $16 - 3(v + 2)$

47. $3s - (r - 2s)$

48. $50x - (30x + 100)$

49. $10 - [8(5 - x) + 2]$

50. $3[2(4x - 5) + 4] - 3$

51. $2[x + 2(y - x)]$

52. $2t[4 - (3 - t)] + 5t$

53. $-3(1 - 10z) + 2(1 - 10z)$

54. $8(15 - 3y) - 5(15 - 3y)$

55. $\frac{1}{3}(42 - 18z) - 2(8 - 4z)$

56. $\frac{1}{4}(100 + 36s) - (15 - 4s)$

In Exercises 57–60, evaluate the expression for the values of the variables.

Expression	Values	
57. $x^2 - 2x + 5$	(a) $x = 0$	(b) $x = 2$
58. $x^3 - 8$	(a) $x = 2$	(b) $x = 4$
59. $y^2 - y(y + 1)$	(a) $y = -1$	(b) $y = 2$
60. $\dfrac{y + 5}{y}$	(a) $y = 3$	(b) $y = -1$

In Exercises 61–70, translate the phrase into an algebraic expression. (Let x represent the number.)

61. Two thirds of a number, plus 5

62. One hundred, decreased by 5 times a number

63. Ten less than twice a number

64. The ratio of a number to 10

65. Fifty, increased by the product of 7 and a number

66. Ten, decreased by the quotient of a number and 2

67. The sum of a number and 10 divided by 8

68. The product of 15 and a number, decreased by 2

69. The sum of the square of a number and 64

70. The absolute value of the sum of a number and -10

In Exercises 71–74, write a verbal description of the expression without using the variable. (There is more than one correct answer.)

71. $x + 3$ 　　　　　**72.** $3x - 2$

73. $\dfrac{y - 2}{3}$ 　　　　**74.** $4(x + 5)$

75. *Income Tax* The income tax rate on a taxable income of I dollars is 28%. Write an algebraic expression that represents the total amount of income tax. (To find 28% of a quantity, multiply the quantity by 0.28.)

76. *Total Amount of Money* A person has n nickels and q quarters. Write an algebraic expression that represents the total amount of money.

77. *Geometry* The width of a refrigerator is w feet. Its height is 3 feet greater than its width (see figure). Write an algebraic expression that represents the area of the front of the refrigerator.

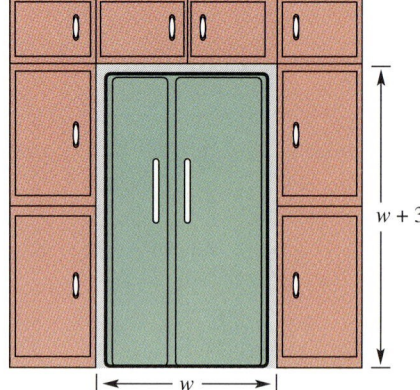

78. *Distance Traveled* A car travels for 10 hours at an average speed of s miles per hour. Write an algebraic expression that represents the total distance traveled.

79. *Sum* Write an algebraic expression that represents the sum of three consecutive odd integers, the first of which is $2n - 1$.

80. *Rental Income* Write an expression that represents the rent for n months if the monthly rent is $625.

81. *Geometry* The face of a tape deck has the dimensions shown in the figure. Find an algebraic expression that represents the area of the face of the tape deck. (*Hint:* The area is given by the difference of the areas of the two rectangles.)

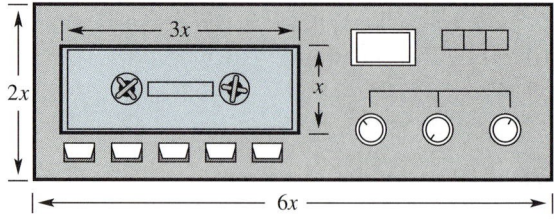

82. Perform the indicated operations and simplify.

$$7 \cdot 10^4 + 2 \cdot 10^3 + 8 \cdot 10^1$$

83. *Finding a Pattern*

(a) Complete the table. The third row is the difference of consecutive entries of the second row. The fourth row is the difference of consecutive entries of the third row.

n	0	1	2	3	4	5
$n^2 + 3n + 2$						
Differences					
Differences				

(b) Describe the patterns of the third and fourth rows.

84. *Finding a Pattern* Find values for a and b such that the expression $an + b$ agrees with the values given in the table.

n	0	1	2	3	4	5
$an + b$	5	8	11	14	17	20

In Exercises 85–94, decide whether the value of x is a solution of the equation.

Equation		*Values*	
85. $5x + 6 = 36$	(a) $x = 3$	(b) $x = 6$
86. $17 - 3x = 8$	(a) $x = 3$	(b) $x = -3$
87. $3x - 12 = x$	(a) $x = -1$	(b) $x = 6$
88. $8x + 24 = 2x$	(a) $x = 0$	(b) $x = -4$
89. $4(2 - x) = 3(2 + x)$	(a) $x = \frac{2}{7}$	(b) $x = -\frac{2}{3}$
90. $5x + 2 = 3(x + 10)$	(a) $x = 14$	(b) $x = -10$
91. $\dfrac{4}{x} - \dfrac{2}{x} = 5$	(a) $x = -1$	(b) $x = \frac{2}{5}$
92. $\dfrac{x}{3} + \dfrac{x}{6} = 1$	(a) $x = \frac{2}{9}$	(b) $x = -\frac{2}{9}$
93. $x(x - 7) = -12$	(a) $x = 3$	(b) $x = 4$
94. $x(x + 1) = 2$	(a) $x = 1$	(b) $x = -2$

In Exercises 95–98, write an equation that represents the statement. (Identify the letters you choose as labels.)

95. *Sum* The sum of a number and its reciprocal is $\frac{37}{6}$.

96. *Distance* An automobile travels 135 miles in t hours with an average speed of 45 miles per hour (see figure).

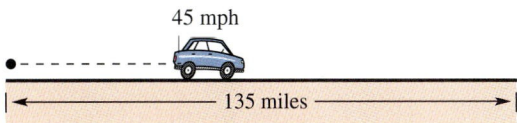

45 mph

135 miles

97. *Geometry* The area of the shaded region in the figure is 24 square inches.

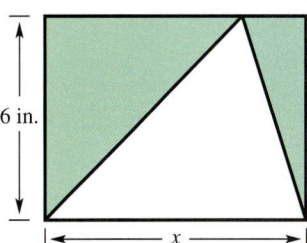

6 in.

x

98. *Geometry* The perimeter of the face of the rectangular traffic light is 72 inches (see figure).

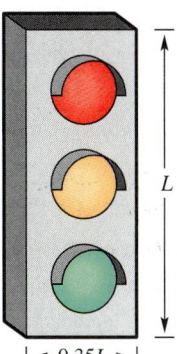

L

$0.35L$

CHAPTER TEST

Take this test as you would take a test in class. After you are done, check your work against the answers given in the back of the book.

1. Identify the terms and coefficients of the expression.

   $2x^2 - 7xy + 3y^3$

2. Rewrite the following product in exponential form.

   $x \cdot (x + y) \cdot x \cdot (x + y) \cdot x$

In Exercises 3–6, identify the rule of algebra demonstrated.

3. $(5x)y = 5(xy)$

4. $2 + (x - y) = (x - y) + 2$

5. $7xy - 7xy = 0$

6. $1 \cdot (x + 5) = (x + 5)$

Fiye of this

In Exercises 7 and 8, use the Distributive Property to expand the expression.

7. $3(x + 8)$

8. $-y(3 - 2y)$

In Exercises 9–14, simplify the expression.

9. $(c^2)^4$

10. $-5uv(2u^3)$

11. $3b - 2a + a - 10b$

12. $15(u - v) - 7(u - v)$

13. $3z - (4 - z)$

14. $2[10 - (t + 1)]$

Six of this

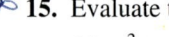 15. Evaluate the expression when $x = 3$ and $y = -12$. *one of this*

   (a) $x^3 - 2$ (b) $x^2 + 4(y + 2)$

16. Explain why it is not possible to evaluate $\dfrac{a + 2b}{3a - b}$ when $a = 2$ and $b = 6$.

17. Translate the phrase, "one-fifth of a number, increased by two," into an algebraic expression. Let n represent the number.

18. (a) Write expressions for the perimeter and area of the rectangle at the right.

    (b) Simplify the expressions.

    (c) Identify the unit of measure for each expression.

    (d) Evaluate each expression when $w = 12$ feet.

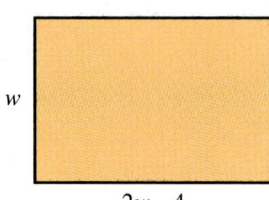

w

$2w - 4$

Figure for 18

19. Write an algebraic expression for the income from a concert if the prices of the tickets for adults and children are \$3 and \$2, respectively. Let n represent the number of adults in attendance and let m represent the number of children.

 20. Determine whether the values of x are solutions of $6(3 - x) - 5(2x - 1) = 7$. *only one*

   (a) $x = -2$ (b) $x = 1$

Linear Equations and Problem Solving

3

- Solving Linear Equations
- Percents and the Percent Equation
- More About Solving Linear Equations
- Ratio and Proportion
- Linear Inequalities and Applications

Have you ever seen an eclipse? An eclipse is the darkening of one celestial body by another. In our solar system, eclipses occur in two ways: solar (the sun is darkened) and lunar (a moon is darkened.) An example of a total solar eclipse is shown at the right. This type of eclipse occurs when a moon's image exactly covers the sun's image. When this occurs, the sun's corona (the outermost part of the sun's atmosphere) becomes visible.

Because earth and its moon move in predictable orbits in the solar system, astronomers are able to determine when eclipses on earth will occur, as shown in the table. The next total solar eclipse in the United States will occur August 21, 2017. It will be seen in a 70-mile-wide path from Salem, Oregon to Charleston, South Carolina.

Date	Location
October 24, 1995	Southeast Asia, Indonesia
March 9, 1997	Mongolia, Eastern Siberia
February 26, 1998	Colombia, Venezuela
August 11, 1999	Europe, Middle East, Southern Asia

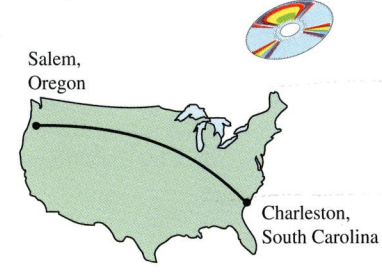

Salem, Oregon

Charleston, South Carolina

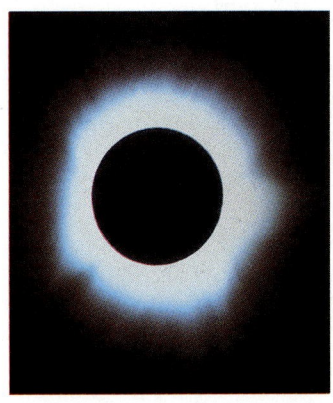

The chapter project related to this data is on page 207.

151

3.1	**Solving Linear Equations**
	Linear Equations ▪ Solving a Linear Equation in Standard Form ▪ Solving a Linear Equation in Nonstandard Form ▪ Applications

Linear Equations

Diophantus, a Greek of Alexandria who lived around 250 A.D., is often called the "Father of Algebra." He was the first to use abbreviated word forms in equations. Diophantus introduced this symbolism in the *Arithmetica*, a collection of problems comprising 13 books.

This is an important step in your study of algebra. In the first three chapters, you were introduced to the rules of algebra, and you learned to use these rules to rewrite and simplify algebraic expressions. In Sections 2.4 and 2.5, you gained experience in translating verbal expressions and problems into algebraic forms. You are now ready to use these skills and experiences to *solve equations*.

In this section, you will learn how the rules of algebra and the properties of equality can be used to solve the most common type of equation—a linear equation in one variable.

Definition of Linear Equation

A **linear equation** in one variable x is an equation that can be written in the standard form

$$ax + b = c$$

where a, b, and c are real numbers with $a \neq 0$.

A linear equation in one variable is also called a **first-degree equation** because its variable has an (implied) exponent of one. Some examples of linear equations in standard form are

$$2x = 3, \quad x - 7 = 5, \quad 4x + 6 = 0, \quad \text{and} \quad \frac{x}{2} - 1 = \frac{5}{3}.$$

Remember that to *solve* an equation involving x means that you are to find all values of x that satisfy the equation. For the linear equation $ax + b = c$, the goal is to *isolate* x by rewriting the equation in the form

$$x = \boxed{\text{a number}}.$$ Isolate the variable x.

To obtain this form, you are to use the techniques discussed in Section 2.5. That is, beginning with the original equation, you write a sequence of equivalent equations, each having the same solution as the original equation. For instance, to solve the linear equation $x - 2 = 0$, you can add 2 to both sides of the equation to obtain $x = 2$. As mentioned in the previous section, each equivalent equation is called a **step** of the solution.

EXAMPLE 1   *Solving a Linear Equation*

Solve $3x - 2 = 10$.

Solution

$3x - 2 = 10$	Original equation
$3x - 2 + 2 = 10 + 2$	Add 2 to both sides.
$3x = 12$	Combine like terms.
$\dfrac{3x}{3} = \dfrac{12}{3}$	Divide both sides by 3.
$x = 4$	Simplify.

It appears that the solution is 4. Here is the check.

Check

$3x - 2 = 10$	Original equation
$3(4) - 2 \overset{?}{=} 10$	Substitute 4 for x.
$12 - 2 \overset{?}{=} 10$	Simplify.
$10 = 10$	Solution checks. ✔

In Example 1, notice that the equation has only one solution. This is true of all linear equations. To see why, consider an arbitrary linear equation of the form $ax + b = c$, where $a \neq 0$.

$ax + b = c$	Original equation
$ax = c - b$	Subtract b from both sides.
$x = \dfrac{c - b}{a}$	Divide both sides by a.

Thus, the *linear* equation has exactly one solution: $x = (c - b)/a$. Remember that other types of equations may have two or more solutions. For instance, the *nonlinear* equation $x^2 = 4$ has two solutions: 2 and -2.

Solution of a Linear Equation

A linear equation of the form

$$ax + b = c, \text{ where } a \neq 0$$

has exactly one solution.

Solving a Linear Equation in Standard Form

A common question in algebra is

"How do I know which step to do *first* to isolate x?"

The answer is that you need practice. By solving many linear equations, you will find that your skill will improve. The key thing to remember is that you can "get rid of" terms and factors by using inverse operations. Here are some guidelines and examples.

Guideline	*Equation*	*Operation*
1. Subtract to remove a sum.	$x + 3 = 4$	Subtract 3 from both sides.
2. Add to remove a difference.	$x - 5 = 7$	Add 5 to both sides.
3. Divide to remove a product.	$4x = 20$	Divide both sides by 4.
4. Multiply to remove a quotient.	$\dfrac{x}{8} = 2$	Multiply both sides by 8.

For additional examples, review Example 3 on page 138. In each case of that example, note how inverse operations are used to isolate the variable.

EXAMPLE 2 *Solving a Linear Equation in Standard Form*

Solve $2x + 7 = 3$.

Solution

$$2x + 7 = 3 \qquad \text{Original equation}$$
$$2x + 7 - 7 = 3 - 7 \qquad \text{Subtract 7 from both sides.}$$
$$2x = -4 \qquad \text{Combine like terms.}$$
$$\frac{2x}{2} = -\frac{4}{2} \qquad \text{Divide both sides by 2.}$$
$$x = -2 \qquad \text{Simplify.}$$

Check

$$2x + 7 = 3 \qquad \text{Original equation}$$
$$2(-2) + 7 \stackrel{?}{=} 3 \qquad \text{Substitute } -2 \text{ for } x.$$
$$-4 + 7 \stackrel{?}{=} 3 \qquad \text{Simplify.}$$
$$3 = 3 \qquad \text{Solution checks.} \checkmark$$

Thus, the solution is -2.

EXAMPLE 3   *Solving a Linear Equation in Standard Form*

Solve $5x = 9$.

Solution

$$5x = 9 \qquad \text{Original equation}$$

$$\frac{5x}{5} = \frac{9}{5} \qquad \text{Divide both sides by 5.}$$

$$x = \frac{9}{5} \qquad \text{Simplify.}$$

The solution is $\frac{9}{5}$. Check this in the original equation.

STUDY TIP

To eliminate a fractional coefficient, it is easier to multiply both sides by the *reciprocal* of the fraction than to divide by the fraction itself. Here is an example.

$$-\frac{2}{3}x = 4$$

$$\left(-\frac{3}{2}\right)\left(-\frac{2}{3}\right)x = \left(-\frac{3}{2}\right)4$$

$$x = -\frac{12}{2}$$

$$x = -6$$

EXAMPLE 4   *Solving a Linear Equation in Standard Form*

Solve $\dfrac{x}{3} - 2 = -4$.

Solution

$$\frac{x}{3} - 2 = -4 \qquad \text{Original equation}$$

$$\frac{x}{3} - 2 + 2 = -4 + 2 \qquad \text{Add 2 to both sides.}$$

$$\frac{x}{3} = -2 \qquad \text{Combine like terms.}$$

$$3\left(\frac{x}{3}\right) = 3(-2) \qquad \text{Multiply both sides by 3.}$$

$$x = -6 \qquad \text{Simplify.}$$

The solution is -6. Check this in the original equation.

As you gain experience in solving linear equations, you will probably find that you can perform some of the solution steps in your head. For instance, you might solve the equation given in Example 4 by writing only the following steps.

$$\frac{x}{3} - 2 = -4 \qquad \text{Original equation}$$

$$\frac{x}{3} = -2 \qquad \text{Add 2 to both sides.}$$

$$x = -6 \qquad \text{Multiply both sides by 3.}$$

Remember, however, that you should not skip the final step—checking your solution. You may find your calculator to be useful for checking solutions.

Solving a Linear Equation in Nonstandard Form

The definition of a linear equation contains the phrase "that can be written" in the standard form

$$ax + b = c.$$

This suggests that some linear equations may come in nonstandard or disguised form.

A common nonstandard form of linear equations is one in which the variable terms are not combined into one term. Some examples are

$$3x + 8 - 5x = 4, \quad 7x = 4x + 9, \quad \text{and} \quad x + 2 = 2x - 6.$$

In such cases, you can begin the solution by rewriting the equation in standard form. Note how this is done in the next two examples.

EXAMPLE 5 Solving a Linear Equation in Nonstandard Form

Solve $3y + 8 - 5y = 4$.

Solution

$3y + 8 - 5y = 4$	Original equation
$3y - 5y + 8 = 4$	Collect like terms.
$-2y + 8 = 4$	Standard form
$-2y + 8 - 8 = 4 - 8$	Subtract 8 from both sides.
$-2y = -4$	Combine like terms.
$\dfrac{-2y}{-2} = \dfrac{-4}{-2}$	Divide both sides by -2.
$y = 2$	Simplify.

Check

$3y + 8 - 5y = 4$	Original equation
$3(2) + 8 - 5(2) \stackrel{?}{=} 4$	Substitute 2 for y.
$6 + 8 - 10 \stackrel{?}{=} 4$	Simplify.
$4 = 4$	Solution checks. ✓

Thus, the solution is 2.

NOTE In Example 5, note that the variable in the equation doesn't always have to be x. Any letter can be used for the variable.

In Example 5, note that the solution began by collecting like terms. You can use any of the rules of algebra to help attain your goal of "isolating the variable." For instance, the next example shows how to solve a linear equation using the Distributive Property.

EXAMPLE 6   *Using the Distributive Property*

Solve $x + 2 = 2(x - 3)$.

Solution

$$\begin{array}{ll}
x + 2 = 2(x - 3) & \text{Original equation} \\
x + 2 = 2x - 6 & \text{Apply Distributive Property.} \\
x - 2x + 2 = 2x - 2x - 6 & \text{Subtract } 2x \text{ from both sides.} \\
-x + 2 = -6 & \text{Combine like terms.} \\
-x + 2 - 2 = -6 - 2 & \text{Subtract 2 from both sides.} \\
-x = -8 & \text{Combine like terms.} \\
(-1)(-x) = (-1)(-8) & \text{Multiply both sides by } -1. \\
x = 8 & \text{Simplify.}
\end{array}$$

The solution is 8. Check this in the original equation.

At the beginning of this section, you saw that every linear equation in *standard form* has exactly one solution. When an equation is written in *nonstandard form,* you cannot be sure that it has exactly one solution. For instance, the equation

$$2x + 3 = 2(x + 4)$$

has no solution. To see this, try to write the equation in standard form, as follows.

$$\begin{array}{l}
2x + 3 = 2(x + 4) \\
2x + 3 = 2x + 8 \\
2x - 2x + 3 = 2x - 2x + 8 \\
3 = 8
\end{array}$$

Because there are no values of x that make the last equation true, you can conclude that the original equation has no solution.

It is also possible that a linear equation in nonstandard form has infinitely many solutions. For instance, every real number is a solution of the equation

$$2(x + 3) = 2x + 6.$$

Watch out for these types of equations in the exercise set.

Applications

The next example reexamines a problem that was introduced in Section 2.5.

EXAMPLE 7 An Application: Ticket Sales

Tickets for a concert were $15 for each floor seat and $10 for each stadium seat. There were 800 seats on the main floor, and these were sold out. The total revenue from ticket sales was $52,000. How many stadium seats were sold?

Solution

Verbal Model:

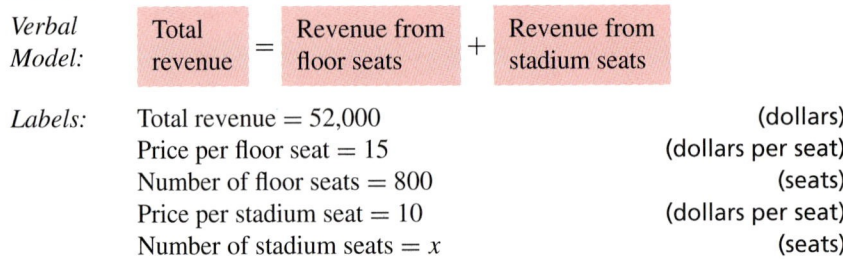

$$\boxed{\text{Total revenue}} = \boxed{\text{Revenue from floor seats}} + \boxed{\text{Revenue from stadium seats}}$$

Labels:

Total revenue $= 52{,}000$	(dollars)
Price per floor seat $= 15$	(dollars per seat)
Number of floor seats $= 800$	(seats)
Price per stadium seat $= 10$	(dollars per seat)
Number of stadium seats $= x$	(seats)

Algebraic Model:

$$52{,}000 = 15(800) + 10x$$

Now that you have written an algebraic equation to represent the problem, you can solve the equation as follows.

$52{,}000 = 15(800) + 10x$	Original equation
$52{,}000 = 12{,}000 + 10x$	Simplify.
$52{,}000 - 12{,}000 = 12{,}000 - 12{,}000 + 10x$	Subtract 12,000 from both sides.
$40{,}000 = 10x$	Combine like terms.
$\dfrac{40{,}000}{10} = \dfrac{10x}{10}$	Divide both sides by 10.
$4000 = x$	Simplify.

There were 4000 stadium seats sold. To check this solution, you should go back to the original statement of the problem. In this case there were 4000 stadium seats sold and 800 seats sold on the main floor. Thus, the total revenue is

Stadium seats   Floor seats

$$10(4000) + 15(800) = 40{,}000 + 12{,}000$$
$$= \$52{,}000$$

which agrees with the original statement.

EXAMPLE 8   An Application: Consecutive Integers

Find three consecutive integers whose sum is 48.

Solution

Verbal
Model: $\boxed{\text{First integer}} + \boxed{\text{Second integer}} + \boxed{\text{Third integer}} = 48$

Labels: First integer $= n$
Second integer $= n + 1$
Third integer $= n + 2$

Algebraic
Model: $n + (n + 1) + (n + 2) = 48$

You can solve this equation as follows.

$$n + (n + 1) + (n + 2) = 48 \qquad \text{Original equation}$$
$$3n + 3 = 48 \qquad \text{Combine like terms.}$$
$$3n + 3 - 3 = 48 - 3 \qquad \text{Subtract 3 from both sides.}$$
$$3n = 45 \qquad \text{Combine like terms.}$$
$$\frac{3n}{3} = \frac{45}{3} \qquad \text{Divide both sides by 3.}$$
$$n = 15 \qquad \text{Simplify.}$$

The solution is $n = 15$. This implies that the three consecutive integers are 15, 16, and 17. Check this in the original statement of the problem.

Group Activities   Problem Solving

When solving a word problem, be sure to ask yourself whether your solution makes sense. In your group, decide why the following answers don't make sense.

a. A problem asks you to find the volume of an oil drum. The answer you obtain is 20 square feet.

b. A problem asks you to find the price per bar of a packet of candy bars. The answer you obtain is 0.42¢.

c. A problem asks you to find the net weight of a carton of oranges. The answer you obtain is 12.5 liters.

3.1 Exercises

Discussing the Concepts

1. Give two examples of linear equations and two examples of nonlinear equations.

2. The scale below is balanced. Each blue box weighs 1 ounce. How much does the red box weigh? If you removed three blue boxes from each side, would the scale still balance? What property of equality does this illustrate?

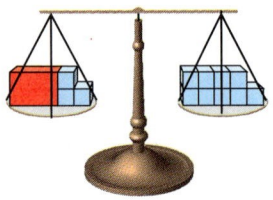

3. In your own words, describe the steps that can be used to transform an equation into an equivalent equation.

4. Explain how to solve the equation

$$x + 5 = 32.$$

What basic rule of algebra are you using?

5. Explain how to solve the equation

$$3x = 5.$$

What basic rule of algebra are you using?

6. When solving a word problem to determine the average speed of a moving van on a trip from Pittsburgh to Chicago, you obtained an answer of 94.5 kilometers per hour. Can this answer be correct? Explain.

Problem Solving

In Exercises 7–10, solve the equation mentally.

7. $x + 6 = 14$ 8. $u - 3 = 8$

9. $4s = 12$ 10. $6z = 18$

In Exercises 11 and 12, justify each step of the solution.

11. $$5x + 15 = 0$$
$$5x + 15 - 15 = 0 - 15$$
$$5x = -15$$
$$\frac{5x}{5} = \frac{-15}{5}$$
$$x = -3$$

12. $$22 = 10 + 3x$$
$$22 - 10 = 10 + 3x - 10$$
$$12 = 3x$$
$$\frac{12}{3} = \frac{3x}{3}$$
$$4 = x$$

In Exercises 13–32, solve the equation and check your solution. (Some equations have no solution.)

13. $8x - 2 = 20$ 14. $-7x + 24 = 3$

15. $10 - 4x = -6$ 16. $6x + 1 = -11$

17. $-5x = 30$ 18. $12x = 18$

19. $6x - 4 = 0$ 20. $8z + 10 = 0$

21. $4 - 7x = 5x$ 22. $2s - 13 = 28s$

23. $15x - 3 = 15 - 3x$ 24. $2x - 5 = 7x + 10$

25. $-6t = 0$ 26. $4z = 10$

27. $t - \frac{1}{3} = \frac{1}{2}$ 28. $z + \frac{2}{5} = -\frac{3}{10}$

29. $2s + \frac{3}{2} = 2s + 2$ 30. $14 - 5s = -2 + 5s$

31. $2y - 18 = -5y - 4$ 32. $6 - 21x - 12 = -21x$

In Exercises 33–36, solve the equation. Round the solution to two decimal places.

33. $0.234x + 1 = 2.805$ 34. $275x - 3130 = 512$

35. $\dfrac{x}{3.155} = 2.850$ 36. $2x + \dfrac{1}{3.7} = \dfrac{3}{4}$

37. *Geometry* The length of a rectangular tennis court is 6 feet more than twice the width (see figure). The length is 78 feet. What is the width?

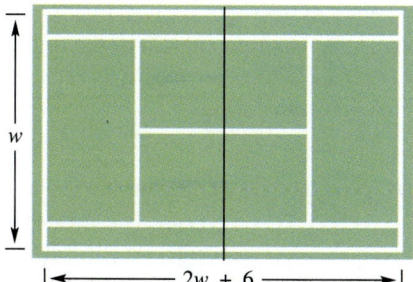

w

$\longleftarrow\ 2w + 6\ \longrightarrow$

38. *Car Repair* A portion of the bill (including parts and labor) for the repair of your car is shown below. Some of the bill is unreadable. From what is given, can you determine how many hours were spent on labor? Explain.

| Parts .... $285.00 |
| Labor ($32 per hour) .... $ |
| Total .... **$357.00** |

39. Find a number such that the sum of that number and 45 is 75.

40. The sum of three consecutive even integers is 192. Find the three integers.

Reviewing the Major Concepts

In Exercises 41 and 42, evaluate the expression.

41. $5 + 4 \cdot 3$

42. $[6 - 3(-5)] \div 7$

In Exercises 43 and 44, simplify the expression.

43. $-3(3x - 2y) + 5y$

44. $3v - (4 - 5v)$

In Exercises 45 and 46, translate the sentence into an algebraic expression. (Let x represent the number.)

45. A number is decreased by 10 and the difference is doubled.

46. A number is decreased by one-third.

Additional Problem Solving

In Exercises 47–50, solve the equation mentally.

47. $x - 9 = 4$

48. $a + 5 = 11$

49. $7y = 28$

50. $4z = -36$

In Exercises 51 and 52, justify each step of the solution.

51. $-2x = 8$

$$\frac{-2x}{-2} = \frac{8}{-2}$$

$$x = -4$$

52. $7x - 14 = 0$

$$7x - 14 + 14 = 0 + 14$$

$$7x = 14$$

$$\frac{7x}{7} = \frac{14}{7}$$

$$x = 2$$

In Exercises 53–76, solve the equation and check your solution. (Some equations have no solution.)

53. $-14x = 42$

54. $9x = -21$

55. $25x - 4 = 46$

56. $15x - 18 = 12$

57. $3y - 2 = 2y$

58. $24 - 5x = x$

59. $4 - 5t = 16 + t$

60. $3x + 4 = x + 10$

61. $-3t = 0$

62. $4z + 2 = 4z$

63. $-3t + 5 = -3t$

64. $4z - 8 = 2$

65. $2x + 4 = -3x + 6$

66. $4y + 4 = -y + 5$

67. $2x = -3x$

68. $2x = 3x - 3$

69. $2x - 5 + 10x = 3$

70. $-4x + 10 + 10x = 4$

71. $\dfrac{x}{3} = 10$

72. $-\dfrac{x}{2} = 3$

73. $x - \frac{1}{3} = \frac{4}{3}$

74. $2x + \frac{5}{2} = \frac{9}{2}$

75. $3x + \frac{1}{4} = \frac{3}{4}$

76. $2x - \frac{3}{8} = \frac{5}{8}$

In Exercises 77–80, solve the equation. Round the solution to two decimal places.

77. $0.02x - 0.96 = 1.50$

78. $135x + 1450 = 6340$

79. $\dfrac{x}{3.25} + 1 = 2.08$

80. $\dfrac{3x}{4.5} = \dfrac{1}{8}$

81. The sum of two consecutive odd integers is 72. Find the two integers.

82. Five times the sum of a number and 16 is 100. Find the number.

83. *Geometry* The perimeter of a rectangle is 240 inches. Find the dimensions of the rectangle if the length is twice the width.

84. *Geometry* The sign shown below has the shape of an equilateral triangle. The perimeter of the sign is 225 centimeters. Find the length of the sides of the sign. (An equilateral triangle is one whose sides have the same length.)

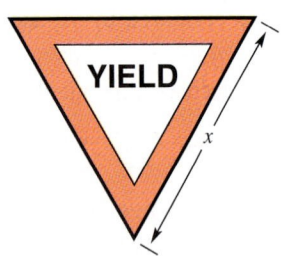

85. *Construction* You are asked to cut a 12-foot board into three pieces. Two pieces are to have the same length and the third is to be twice as long as the others. How long are the pieces?

86. *Summer Jobs* You have two summer jobs. In the first job, you work 40 hours a week and earn $6.25 an hour. In the second job, you work as many hours as you want and earn $5.50 an hour. If you want to earn $316 a week, how many hours must you work at the second job?

87. *Car Repair* The bill for the repair of your car was $415. The cost for parts was $265. The cost for labor was $25 per hour. How many hours did the repair work take?

88. *Height of a Fountain* Water is forced out of a fountain with an initial velocity of 28 feet per second, as shown in the figure. The velocity of the water stream is given by $v = -32t + 28$, where t is time in seconds. The height of the water is given by $h = -16t^2 + 28t$. What is the maximum height of the fountain? (*Hint:* Find the time when the velocity is zero, and then find the height for that time.)

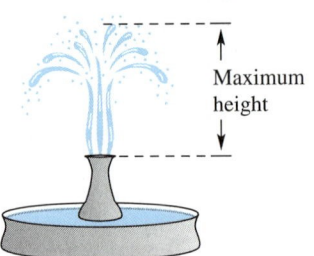

89. *Finding a Pattern* The length of a rectangle is t times its width, as shown in the figure. The rectangle has a perimeter of 1200 meters, which implies that

$$2w + 2(tw) = 1200$$

where w is the width of the rectangle.

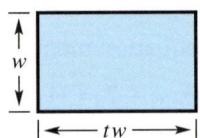

(a) Complete the table.

t	1	1.5	2	3	4	5
Width						
Length						
Area						

(b) Use the completed table to draw a conclusion concerning the area of a rectangle of given perimeter as its length increases relative to its width.

3.2	**Percents and the Percent Equation**
	Percents ▪ The Percent Equation ▪ Applications

Percents

Numbers that describe rates, increases, decreases, and discounts are often given as percents, denoted by the symbol %. **Percent** means *per hundred,* or *parts of 100.* (The Latin word for 100 is *centum.*) For example, 30% means 30 parts of 100, which is equivalent to the fraction $\frac{30}{100}$ or $\frac{3}{10}$.

In applications involving percents, you must usually convert the percent to decimal (or fraction) form before performing any arithmetic operations. Consequently, you need to be able to convert from percents to decimals (or fractions) and conversely. The following verbal model can be used to perform the conversions.

$$\boxed{\text{Decimal or fraction}} \cdot \boxed{100\%} = \boxed{\text{Percent}}$$

For example, the decimal 0.38 corresponds to 38 percent. That is,

$$0.38(100\%) = 38\%.$$

DISCOVERY

Some scientific calculators can convert percents to decimal and fraction forms. For example, to convert 39% to decimal form and then fraction form, use the following keystrokes.

Keystroke	*Display*
39 %	0.39
F⊃D	39/100

Repeat this procedure using other percents. Write a verbal model that represents the conversions. How does your verbal model compare to the one shown at the right? Use your calculator to show that 60% is equal to $\frac{3}{5}$ in Example 1.

EXAMPLE 1   *Converting Decimals and Fractions to Percents*

Convert each number to a percent.

a. $\dfrac{3}{5}$ **b.** 1.20

Solution

a. *Verbal Model:* $\boxed{\text{Fraction}} \cdot \boxed{100\%} = \boxed{\text{Percent}}$

Equation: $\dfrac{3}{5}(100\%) = \dfrac{300}{5}\%$

$= 60\%$

Thus, the fraction $\frac{3}{5}$ corresponds to 60%.

b. *Verbal Model:* $\boxed{\text{Decimal}} \cdot \boxed{100\%} = \boxed{\text{Percent}}$

Equation: $(1.20)(100\%) = 120\%$

Thus, the decimal 1.20 corresponds to 120%.

NOTE   Note in Example 1(b) that it is possible to have percents that are larger than 100%.

EXAMPLE 2 Converting Percents to Decimals and Fractions

a. Convert 3.5% to a decimal.

b. Convert 55% to a fraction.

Solution

a.
Verbal Model: $\boxed{\text{Decimal}} \cdot \boxed{100\%} = \boxed{\text{Percent}}$

Label: $x = $ decimal

Equation: $x(100\%) = 3.5\%$

$$x = \frac{3.5\%}{100\%}$$

$$x = 0.035$$

Thus, 3.5% corresponds to the decimal 0.035.

b.
Verbal Model: $\boxed{\text{Fraction}} \cdot \boxed{100\%} = \boxed{\text{Percent}}$

Label: $x = $ fraction

Equation: $x(100\%) = 55\%$

$$x = \frac{55\%}{100\%}$$

$$x = \frac{11}{20}$$

Thus, 55% corresponds to the fraction $\frac{11}{20}$.

Some percents occur so commonly that it is helpful to memorize their conversions. For instance, 100% corresponds to 1 and 200% corresponds to 2. The table below shows the decimal and fraction conversions for several percents.

Percent	10%	$12\frac{1}{2}\%$	20%	25%	$33\frac{1}{3}\%$	50%	$66\frac{2}{3}\%$	75%
Decimal	0.1	0.125	0.2	0.25	$0.\overline{3}$	0.5	$0.\overline{6}$	0.75
Fraction	$\frac{1}{10}$	$\frac{1}{8}$	$\frac{1}{5}$	$\frac{1}{4}$	$\frac{1}{3}$	$\frac{1}{2}$	$\frac{2}{3}$	$\frac{3}{4}$

In applications involving percent, many people like to state percent in terms of a portion. For instance, the statement "20% of the population lives in apartments" is often stated as "1 out of every 5 people lives in an apartment."

The Percent Equation

The primary use of percents is to compare two numbers. For example, 2 is 50% of 4, and 5 is 25% of 20. The following model is helpful.

Verbal Model: $\boxed{a} = \boxed{p \text{ percent of } b}$

Labels: b = base number
p = percent (in decimal form)
a = number being compared to b

Equation: $a = p \cdot b$

EXAMPLE 3 *Solving Percent Equations*

a. What is 30% of 70?　　**b.** Fourteen is 25% of what?

c. One hundred thirty-five is what percent of 27?

Solution

a. *Verbal Model:* $\boxed{\text{What number}} = \boxed{30\% \text{ of } 70}$

Label: a = unknown number

Equation: $a = (0.3)(70) = 21$

Therefore, 21 is 30% of 70.

b. *Verbal Model:* $\boxed{14} = \boxed{25\% \text{ of what number}}$

Label: b = unknown number

Equation: $14 = 0.25b$

$$\frac{14}{0.25} = b$$

$$56 = b$$

Therefore, 14 is 25% of 56.

c. *Verbal Model:* $\boxed{135} = \boxed{\text{What percent of } 27}$

Label: p = unknown percent (in decimal form)

Equation: $135 = p(27)$

$$\frac{135}{27} = p$$

$$5 = p$$

Therefore, 135 is 500% of 27.

EXAMPLE 4   Solving Percent Equations

a. $761.25 is $14\frac{1}{2}\%$ of what?

b. 19 is what percent of 95?

Solution

a.   *Verbal Model:* $761.25\ =\ 14\frac{1}{2}\%$ of what number

    Labels:

Base number $= b$	(dollars)
Percent $= 0.145$	(in decimal form)
Number being compared to $b = 761.25$	(dollars)

    Equation: $761.25 = 0.145b$

$$\frac{761.25}{0.145} = b$$

$$5250 = b$$

Therefore, $761.25 is $14\frac{1}{2}\%$ of $5250. Check this by multiplying 0.145 by 5250 to obtain 761.25.

b.   *Verbal Model:* $19\ =\ $ What percent of 95

    Label: $p = $ unknown percent (in decimal form)

    Equation: $19 = p(95)$

$$\frac{19}{95} = p$$

$$0.2 = p$$

Therefore, 19 is 20% of 95.

From Examples 3 and 4, you can see that there are three basic types of percent problems. Each can be solved by substituting the two given quantities into the percent equation and solving for the third quantity.

Question	*Given*	*Percent Equation*
a is what percent of b?	a and b	Solve for p.
What is p percent of b?	p and b	Solve for a.
a is p percent of what?	a and p	Solve for b.

For instance, part (a) of Example 4 fits the form "a is p percent of what?" What forms do the other parts of Examples 3 and 4 fit?

Applications

In most real-life applications, the base number b and the number a are much more disguised than they are in Examples 3 and 4. It sometimes helps to think of a as a "new" amount and b as the "original" amount.

EXAMPLE 5   Real Estate Commission

A real estate agency receives a commission of $5167.50 for the sale of a $79,500 house. What percent commission is this?

Solution

Verbal Model:
$$\boxed{\text{Commission}} = \boxed{\begin{array}{c}\text{Percent of}\\\text{sale price}\end{array}}$$

Labels:
Commission $= 5167.50$	(dollars)
Percent $= p$	(in decimal form)
Sale price $= 79{,}500$	(dollars)

Equation:
$$5167.50 = p \cdot (79{,}500)$$
$$\frac{5167.50}{79{,}500} = p$$
$$0.065 = p$$

Therefore, the real estate agency receives a commission of 6.5%. Check this solution in the original statement of the problem.

EXAMPLE 6   Cost-of-Living Raise

A union negotiates for a cost-of-living raise of 7%. What is the raise for a union member whose salary is $17,240? What is this person's new salary?

Solution

Verbal Model:
$$\boxed{\text{Raise}} = \boxed{\begin{array}{c}\text{Percent of}\\\text{salary}\end{array}}$$

Labels:
Raise $= a$	(dollars)
Percent $= 0.07$	(in decimal form)
Salary $= 17{,}240$	(dollars)

Equation: $a = 0.07(17{,}240) = 1206.80$

Therefore, the raise is $1206.80 and the new salary is $17{,}240.00 + 1206.80$ or $18,446.80. Check this solution in the original statement of the problem.

EXAMPLE 7   Course Grade

Suppose you missed an A in your chemistry course by only three points. Your point total for the course is 402. How many points were possible in the course? (Assume that you needed 90% of the course total for an A.)

Solution

Verbal Model:
$$\boxed{\text{Your points}} + \boxed{\text{3 points}} = \boxed{\text{90\% of total points}}$$

Labels:
Your points $= 402$ (points)
Percent $= 0.9$ (in decimal form)
Total points for course $= b$ (points)

Equation: $405 = 0.9b$

$$\frac{405}{0.9} = b$$

$$450 = b$$

Therefore, there were 450 total points for the course. Check this solution in the original statement of the problem.

EXAMPLE 8   Seating Capacity

The seating capacity of a university football stadium was increased from 68,000 to 78,500. What percent increase in seating capacity does this represent?

Solution

Verbal Model:
$$\boxed{\text{Increase}} = \boxed{\text{Percent of original capacity}}$$

Labels:
Increase $= 78{,}500 - 68{,}000 = 10{,}500$ (seats)
Percent $= p$ (in decimal form)
Original capacity $= 68{,}000$ (seats)

Equation: $10{,}500 = p \cdot 68{,}000$

$$\frac{10{,}500}{68{,}000} = p$$

$$0.154 \approx p$$

Therefore, the increase in seating capacity is approximately 15.4%. Check this solution in the original statement of the problem.

The following guidelines summarize the problem-solving strategy that we recommend for word problems.

Guidelines for Solving Word Problems

1. Write a *verbal model* that describes the problem.

2. Assign *labels* to fixed quantities and variable quantities.

3. Rewrite the verbal model as an *algebraic equation* using the assigned labels.

4. *Solve* the algebraic equation.

5. *Check* to see that your solution satisfies the word problem as stated.

Group Activities

Problem Solving

Comparing Growth Patterns In 1996, your annual salary is $28,000. You are given two options for an 8-year contract. In the first option, you will be given a $1500 raise each year. In the second option, you will be given a 5% raise each year. Use the following spreadsheet to decide which option you would choose. Discuss the advantages and disadvantages of each option.

Year	Salary for Option 1	Raise for Option 1	Salary for Option 2	Raise for Option 2
1996	$28,000.00	$1500.00	$28,000.00	$1400.00
1997	$29,500.00	$1500.00	$29,400.00	$1470.00
1998	$31,000.00	$1500.00	$30,870.00	$1543.50
1999	$32,500.00	$1500.00	$32,413.50	$1620.68
2000	$34,000.00	$1500.00	$34,034.18	$1701.71
2001	$35,500.00	$1500.00	$35,735.89	$1786.79
2002	$37,000.00	$1500.00	$37,522.68	$1876.13
2003	$38,500.00	$1500.00	$39,398.81	$1969.94

Which option would you choose if it were a three-year contract? a four-year contract? Explain your reasoning to your group.

3.2 Exercises

Discussing the Concepts

1. Explain the meaning of the word "percent."

2. In your own words, explain how to change a percent to a fraction. Give an example.

3. In your own words, explain how to change a decimal to a percent. Give an example.

4. In your own words, explain how to change a fraction to a percent. Give an example.

5. Can any positive decimal be written as a percent? Explain.

6. Is it true that $\frac{1}{2}\% = 50\%$? Explain.

Problem Solving

In Exercises 7–10, complete the table showing the equivalent forms of a percent.

7.

Percent	Parts out of 100	Decimal	Fraction
40%			

8.

Percent	Parts out of 100	Decimal	Fraction
	10.5		

9.

Percent	Parts out of 100	Decimal	Fraction
		0.155	

10.

Percent	Parts out of 100	Decimal	Fraction
			$\frac{3}{20}$

In Exercises 11–14, change the percent to a decimal.

11. 12.5% **12.** 95%

13. 250% **14.** 0.3%

In Exercises 15-18, change the decimal to a percent.

15. 0.075 **16.** 0.57

17. 0.62 **18.** 1.75

In Exercises 19-22, change the fraction to a percent.

19. $\frac{4}{5}$ **20.** $\frac{5}{4}$

21. $\frac{7}{20}$ **22.** $\frac{2}{3}$

In Exercises 23 and 24, what percent of the figure is shaded?

23. 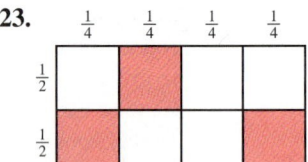 **24.**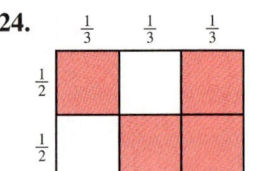

25. What is 30% of 150?

26. What is 62% of 1200?

27. What is $\frac{3}{4}\%$ of 56?

28. What is 0.2% of 100,000?

29. $12\frac{1}{2}\%$ of what number is 275?

30. 250% of what number is 210?

31. 1000 is what percent of 200?

32. 148.8 is what percent of 960?

33. *Rent Payment* You spend 17% of your monthly income of $2500 for rent. What is your monthly payment?

34. *Enrollment*   Thirty-five percent of the students enrolled in a college are freshmen. The enrollment of the college is 2800. How many of the students are freshmen?

35. *Snowfall*   During the winter, there were 120 inches of snow. Of that, 86 inches fell in December. What percent of the snow fell in December?

36. *Layoff*   Because of slumping sales, a small company laid off 30 of its 153 employees.

(a) What percent of the work force was laid off?

(b) Complete the statement: "About 1 out of every _____ workers was laid off."

37. *Defective Parts*   A quality control engineer tested several parts and found two to be defective. The engineer reported that 2.5% were defective. How many were tested?

38. *Price*   The price of a new van is approximately 110% of what it was 3 years ago. The current price is $22,850. What was the approximate price 3 years ago?

39. *Membership Drive*   Because of a membership drive for a public television station, the current membership is 125% of what it was a year ago. The current number is 7815. How many members did the station have last year?

40. *Course Grade*   You were 6 points shy of a B in your mathematics course. Your point total for the course is 394. How many points were possible in the course? (Assume that you needed 80% of the course total for a B.)

41. *Graphical Estimation*   Every year, approximately 8000 Americans suffer spinal cord injuries. The graph classifies the major causes of these injuries. Estimate the number of Americans who enter each of these classifications annually. (Source: *U.S. News & World Report*, January 24, 1994)

Causes of U.S. Spinal Cord Injuries

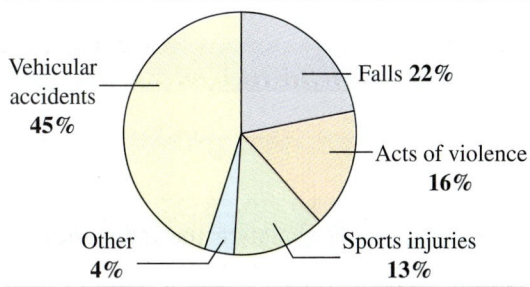

42. *Graphical Estimation*   The graph shows the number (in thousands) of criminal cases commenced in United States District Courts from 1984 through 1992. Estimate the percent increase in cases over the period from 1984 through 1992. (Source: Administrative Office of the U.S. Courts)

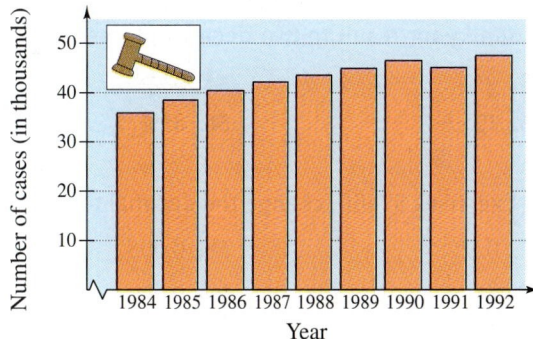

Reviewing the Major Concepts

In Exercises 43–46, solve the linear equation. Write a justification for each step of your solution. Then check your solution.

43. $2x - 5 = x + 9$

44. $6x + 8 = 8 - 2x$

45. $2x + \frac{3}{2} = \frac{3}{2}$

46. $-10x + \frac{2}{3} = \frac{7}{3} - 5x$

47. *Telephone Charge*   A telephone company charges $1.37 for the first minute and $0.95 for each additional minute. Find the cost of a 15-minute call.

48. *Distance Traveled*   A train travels at the rate of r miles per hour for 5 hours. Write an algebraic expression that represents the total distance traveled.

Additional Problem Solving

In Exercises 49–52, complete the table showing the equivalent forms of a percent.

49.

Percent	Parts out of 100	Decimal	Fraction
	63		

50.

Percent	Parts out of 100	Decimal	Fraction
150%			

51.

Percent	Parts out of 100	Decimal	Fraction
			$\frac{3}{5}$

52.

Percent	Parts out of 100	Decimal	Fraction
		0.80	

In Exercises 53–56, change the percent to a decimal. (Round your result to two decimal places.)

53. $\frac{3}{4}\%$ **54.** $33\frac{1}{3}\%$

55. 125%   **56.** 85%

In Exercises 57–60, change the decimal to a percent.

57. 0.20   **58.** 0.005

59. 2.5   **60.** 0.38

In Exercises 61–64, change the fraction to a percent.

61. $\frac{1}{4}$ **62.** $\frac{3}{2}$

63. $\frac{5}{6}$ **64.** $\frac{6}{5}$

65. What is 9.5% of 816?   **66.** What is $33\frac{1}{3}\%$ of 516?

67. What is 200% of 88?   **68.** What is 325% of 450?

69. 43% of what number is 903?

70. 85% of what number is 425?

71. 450% of what number is 594?

72. $66\frac{2}{3}\%$ of what number is 814?

73. 0.6% of what number is 2.16?

74. 0.08% of what number is 51.2?

75. 576 is what percent of 800?

76. 1950 is what percent of 5000?

77. 45 is what percent of 360?

78. 38 is what percent of 5700?

In Exercises 79 and 80, what percent of the figure is shaded?

79. 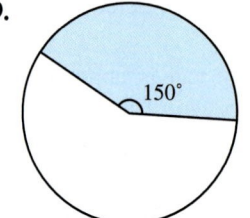   **80.**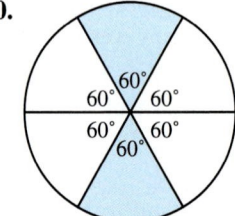

81. *Cost of Housing*   You budget 30% of your annual after-tax income for housing. Your after-tax income is $32,500. What amount can you spend on housing?

82. *Retirement Plan*   You budget $7\frac{1}{2}\%$ of your gross income for an individual retirement plan. Your annual gross income is $29,800. How much will you put in your retirement plan each year?

83. *Original Price*   A coat sells for $250 during a 20% storewide clearance sale. What was the original price of the coat?

84. *Decision Making*   A new car you want to buy costs $17,800. If you wait another month to buy the car, the price will increase by 6%. However, to buy it now, you will have to pay an interest penalty of $450 for the early withdrawal of a certificate of deposit. Should you buy the car now or wait a month? Explain.

85. *Eligible Voters*   The news media reported that 6432 votes were cast in the last election and this represented 63% of the eligible voters of a district. How many eligible voters are in the district?

86. *Target Size*   A circular target is attached to a rectangular board, as shown in the figure. The radius of the circle is $4\frac{1}{2}$ inches, and the dimensions of the board are 12 inches by 15 inches. What percent of the board is covered by the target? (The area of a circle is $A = \pi r^2$, where r is the radius of the circle.)

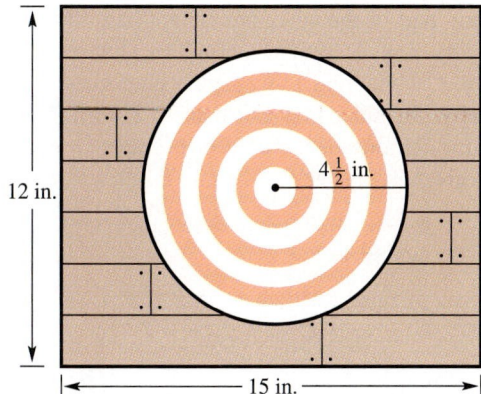

12 in.

$4\frac{1}{2}$ in.

15 in.

87. *Interpreting a Table*   The table shows the number of women scientists in the United States as a percentage of the total number of scientists in each field in 1983 and 1992.   (Source: U.S. Bureau of Labor and Statistics)

(a) Find the total number of mathematical and computer scientists (men and women) in 1992.

(b) Find the total number of chemists (men and women) in 1983.

(c) Explain how the number of women in biology can increase while the percent of women in biology decreases.

Women Scientists in the United States

Field	1983 Number	1983 %	1992 Number	1992 %
Math/Computer	137,000	30.6%	313,200	33.5%
Chemistry	22,800	23.3%	36,100	30.1%
Geology	11,700	18.0%	6,100	11.8%
Biology	22,400	40.8%	32,100	33.8%

88. *Rearranging Furniture*   Three hundred people were surveyed and reported that they rearranged furniture in their homes for several different reasons (see figure). How many people in the survey said that they rearranged furniture for each of the given reasons?   (Source: Southwestern Bell)

Why We Rearrange Furniture

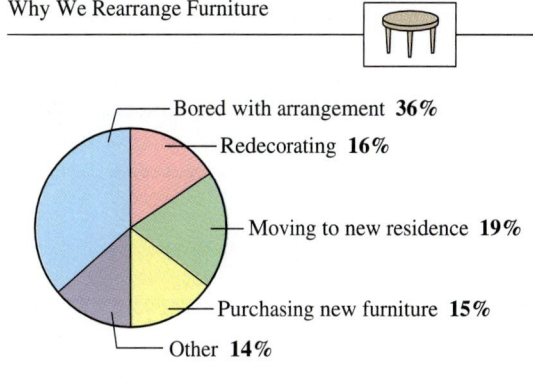

Bored with arrangement **36%**

Redecorating **16%**

Moving to new residence **19%**

Purchasing new furniture **15%**

Other **14%**

Estimation   In Exercises 89 and 90, figure (a) was put into a photocopier and reduced or enlarged to produce figure (b). Estimate what percent the copy is of the original.

89. (a)   (b)

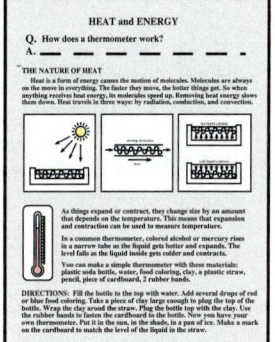   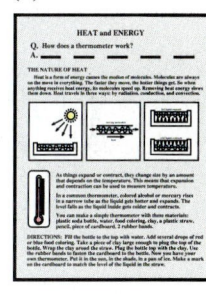

90. (a)   (b)

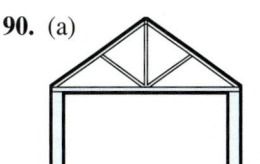   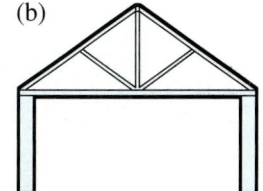

Math Matters   Finding Your Way Through the Maze

The following maze was written by Robert Abbott. This is the first of 20 mazes that appear in his book *Mad Mazes*.* Robert Abbott's mazes have appeared in *Scientific American*, *Games*, and *Discover* magazines.

This maze is called the "Arctic Caves of Norway" and concerns a skier who was trapped in a cave by an avalanche. Some parts of the cave have showers of icy waters (shown in blue) and other parts are filled with steam from hot springs (shown in red). To get to the other side of the mountain, the skier can travel through any part of the cave, but cannot pass through two icy spots in succession or through two hot spots in succession. Can you find a way for the skier to escape? (Incidentally, the skier may not make a U-turn when halfway through one of the cold or hot areas.) (The answer is given in the back of the text.)

---

*Abbott, R. *Mad Mazes.* Holbrook, MA: Bob Adams, Inc. Copyright © 1990. Used by permission.

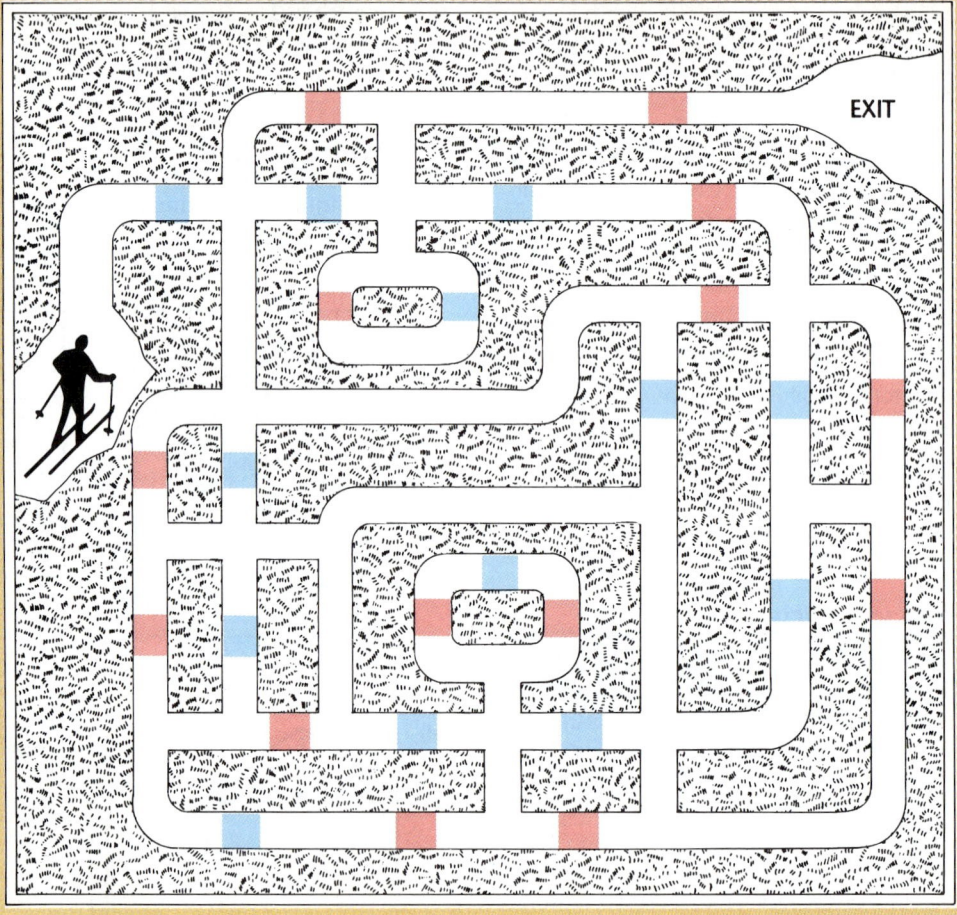

3.3	**More About Solving Linear Equations**

Equations Containing Symbols of Grouping   ▪
Equations Involving Fractions or Decimals

Equations Containing Symbols of Grouping

In Section 3.1 you studied techniques for solving linear equations in standard form. You also looked at techniques for solving some simpler linear equations in nonstandard form such as $3x + 8 - 5x = 4$ and $x + 2 = 2x - 6$.

In this section you will continue your study of linear equations by looking at more complicated nonstandard forms. To solve a linear equation that contains symbols of grouping, first remove the symbols of grouping from each side. Then combine like terms and proceed to solve the resulting linear equation the usual way using properties of equality.

Study the next two examples carefully. Pay attention to the mental steps that can usually be performed without being written down.

EXAMPLE 1   *Solving a Linear Equation Involving Parentheses*

Solve $4(x - 3) = 8$.

Solution

$4(x - 3) = 8$	Original equation
$4 \cdot x - 4 \cdot 3 = 8$	Distributive Property (mental step)
$4x - 12 = 8$	Simplify.
$4x - 12 + 12 = 8 + 12$	Add 12 to both sides. (mental step)
$4x = 20$	Combine like terms.
$\dfrac{4x}{4} = \dfrac{20}{4}$	Divide both sides by 4. (mental step)
$x = 5$	Simplify.

Check

$4(x - 3) = 8$	Original equation
$4(5 - 3) \stackrel{?}{=} 8$	Substitute 5 for x.
$4(2) \stackrel{?}{=} 8$	Simplify.
$8 = 8$	Solution checks. ✔

The solution is 5.

EXAMPLE 2 *Solving a Linear Equation Involving Parentheses*

Solve $3(2x - 1) + x = 11$.

Solution

$3(2x - 1) + x = 11$	Original equation
$3 \cdot 2x - 3 \cdot 1 + x = 11$	Distributive Property (mental step)
$6x - 3 + x = 11$	Simplify.
$6x + x - 3 = 11$	Collect like terms. (mental step)
$7x - 3 = 11$	Combine like terms.
$7x - 3 + 3 = 11 + 3$	Add 3 to both sides. (mental step)
$7x = 14$	Combine like terms.
$\dfrac{7x}{7} = \dfrac{14}{7}$	Divide both sides by 7. (mental step)
$x = 2$	Simplify.

The solution is 2. Check this in the original equation.

In the examples that follow, we will not write some of the mental steps illustrated in Examples 1 and 2. Some people call this "skipping steps." We don't like that description, because it sounds as if part of the solution is being omitted. Instead, we like to say that we are "performing the steps mentally." Thus, we aren't skipping the steps—we just aren't writing them down. When studying the next few examples, look for the steps that were performed mentally.

EXAMPLE 3 *Solving a Linear Equation Involving Parentheses*

Solve $5(x + 2) = 2(x - 1)$.

Solution

$5(x + 2) = 2(x - 1)$	Original equation
$5x + 10 = 2x - 2$	Distributive Property
$5x - 2x + 10 = -2$	Subtract $2x$ from both sides.
$3x + 10 = -2$	Combine like terms.
$3x = -2 - 10$	Subtract 10 from both sides.
$3x = -12$	Combine like terms.
$x = -4$	Divide both sides by 3.

The solution is -4. Check this in the original equation.

EXAMPLE 4 Solving a Linear Equation Involving Parentheses

Solve $2(x - 7) - 3(x + 4) = 4 - (5x - 2)$.

Solution

$$2(x - 7) - 3(x + 4) = 4 - (5x - 2)$$ Original equation

$$2x - 14 - 3x - 12 = 4 - 5x + 2$$ Distributive Property

$$-x - 26 = -5x + 6$$ Combine like terms.

$$-x + 5x - 26 = 6$$ Add $5x$ to both sides.

$$4x - 26 = 6$$ Combine like terms.

$$4x = 32$$ Add 26 to both sides.

$$x = 8$$ Divide both sides by 4.

The solution is 8. Check this in the original equation.

The linear equation in the next example involves both brackets and parentheses. Watch out for nested symbols of grouping such as these. We suggest that the innermost symbols of grouping be removed first.

EXAMPLE 5 An Equation Involving Nested Symbols of Grouping

Solve $5x - 2[4x + 3(x - 1)] = 8 - 3x$.

Solution

$$5x - 2[4x + 3(x - 1)] = 8 - 3x$$ Original equation

$$5x - 2[4x + 3x - 3] = 8 - 3x$$ Distributive Property

$$5x - 2[7x - 3] = 8 - 3x$$ Combine like terms inside brackets.

$$5x - 14x + 6 = 8 - 3x$$ Distributive Property

$$-9x + 6 = 8 - 3x$$ Combine like terms.

$$-9x + 3x + 6 = 8$$ Add $3x$ to both sides.

$$-6x = 2$$ Combine like terms and subtract 6 from both sides.

$$x = \frac{2}{-6}$$ Divide both sides by -6.

$$x = -\frac{1}{3}$$ Simplify.

The solution is $-\frac{1}{3}$. Check this in the original equation.

Technology

Try using your calculator to check the solution found in Example 5. This will give you practice working with nested parentheses on a calculator.

Equations Involving Fractions or Decimals

You can solve a linear equation such as $2x - \frac{3}{4} = 1$ by adding $\frac{3}{4}$ to both sides and then dividing by 2 as follows.

$$2x - \frac{3}{4} + \frac{3}{4} = 1 + \frac{3}{4}$$

$$2x = \frac{7}{4}$$

$$x = \frac{7}{8}$$

For linear equations that contain more than one fraction, however, it is usually better to *clear the equation of fractions* by multiplying both sides of the equation by the least common multiple of the denominators of all the fractions. For example, the equation

$$\frac{3x}{2} - \frac{1}{3} = \frac{x}{4} + 2$$

can be cleared of fractions by multiplying both sides by 12, the least common multiple of 2, 3, and 4. Notice how this is done in the next example. Section P.2 of the Prerequisites Chapter contains more details about finding the least common multiple of two or more integers.

EXAMPLE 6 *Solving a Linear Equation Involving Fractions*

Solve $\dfrac{3x}{2} - \dfrac{1}{3} = \dfrac{x}{4} + 2$.

Solution

$$\frac{3x}{2} - \frac{1}{3} = \frac{x}{4} + 2 \qquad \text{Original equation}$$

$$12\left(\frac{3x}{2} - \frac{1}{3}\right) = 12\left(\frac{x}{4} + 2\right) \qquad \text{Multiply both sides by 12.}$$

$$12 \cdot \frac{3x}{2} - 12 \cdot \frac{1}{3} = 12 \cdot \frac{x}{4} + 12 \cdot 2 \qquad \text{Distributive Property}$$

$$18x - 4 = 3x + 24 \qquad \text{Clear fractions.}$$

$$15x - 4 = 24 \qquad \text{Subtract } 3x \text{ from both sides.}$$

$$15x = 28 \qquad \text{Add 4 to both sides.}$$

$$x = \frac{28}{15} \qquad \text{Divide both sides by 15.}$$

The solution is $\frac{28}{15}$. Check this in the original equation.

EXAMPLE 7   *Solving a Linear Equation Involving Fractions*

Solve $\dfrac{x}{5} + \dfrac{3x}{4} = 19$.

Solution

$$\dfrac{x}{5} + \dfrac{3x}{4} = 19 \qquad\qquad \text{Original equation}$$

$$20\left(\dfrac{x}{5}\right) + 20\left(\dfrac{3x}{4}\right) = 20(19) \qquad\qquad \text{Multiply both sides by 20.}$$

$$4x + 15x = 380 \qquad\qquad \text{Simplify.}$$

$$19x = 380 \qquad\qquad \text{Combine like terms.}$$

$$x = 20 \qquad\qquad \text{Divide both sides by 19.}$$

Check

$$\dfrac{x}{5} + \dfrac{3x}{4} = 19 \qquad\qquad \text{Original equation}$$

$$\dfrac{20}{5} + \dfrac{3(20)}{4} \overset{?}{=} 19 \qquad\qquad \text{Substitute 20 for } x.$$

$$4 + 15 \overset{?}{=} 19 \qquad\qquad \text{Simplify.}$$

$$19 = 19 \qquad\qquad \text{Solution checks.} ✔$$

The solution is 20.

EXAMPLE 8   *Solving a Linear Equation Involving Fractions*

Solve $\dfrac{x-2}{4} + \dfrac{2x+1}{6} = \dfrac{17}{12}$.

Solution

$$\dfrac{x-2}{4} + \dfrac{2x+1}{6} = \dfrac{17}{12} \qquad\qquad \text{Original equation}$$

$$12 \cdot \dfrac{x-2}{4} + 12 \cdot \dfrac{2x+1}{6} = 12 \cdot \dfrac{17}{12} \qquad\qquad \text{Multiply both sides by 12.}$$

$$3(x-2) + 2(2x+1) = 17 \qquad\qquad \text{Clear fractions.}$$

$$3x - 6 + 4x + 2 = 17 \qquad\qquad \text{Distributive Property}$$

$$7x - 4 = 17 \qquad\qquad \text{Combine like terms.}$$

$$7x = 21 \qquad\qquad \text{Add 4 to both sides.}$$

$$x = 3 \qquad\qquad \text{Divide both sides by 7.}$$

The solution is 3. Check this in the original equation.

A common type of linear equation is one that equates two fractions. To solve such equations, we consider the fractions to be **equivalent** and use **cross-multiplication.** That is, if

$$\frac{a}{b} = \frac{c}{d}, \quad \text{then} \quad a \cdot d = b \cdot c.$$

Note how cross-multiplication is used in the next example.

EXAMPLE 9   *Using Cross-Multiplication*

Use cross-multiplication to solve $\dfrac{x+2}{3} = \dfrac{8}{5}$.

Solution

$\dfrac{x+2}{3} = \dfrac{8}{5}$	Original equation
$5(x+2) = 3(8)$	Cross-multiply.
$5x + 10 = 24$	Distributive Property
$5x = 14$	Subtract 10 from both sides.
$x = \dfrac{14}{5}$	Divide both sides by 5.

Check

$\dfrac{x+2}{3} = \dfrac{8}{5}$	Original equation
$\dfrac{\left(\frac{14}{5} + 2\right)}{3} \overset{?}{=} \dfrac{8}{5}$	Substitute $\frac{14}{5}$ for x.
$\dfrac{\left(\frac{14}{5} + \frac{10}{5}\right)}{3} \overset{?}{=} \dfrac{8}{5}$	Write 2 as $\frac{10}{5}$.
$\dfrac{\frac{24}{5}}{3} \overset{?}{=} \dfrac{8}{5}$	Simplify.
$\dfrac{24}{5}\left(\dfrac{1}{3}\right) \overset{?}{=} \dfrac{8}{5}$	Invert and multiply.
$\dfrac{8}{5} = \dfrac{8}{5}$	Solution checks. ✔

The solution is $\frac{14}{5}$.

More extensive applications of cross-multiplication will be discussed when you study ratios and proportions in the next section.

Many real-life applications of linear equations involve decimal coefficients. To solve such an equation, you can clear it of decimals in much the same way you clear an equation of fractions. Multiply both sides by a power of 10 that converts all decimal coefficients to integers, as shown in the next example.

EXAMPLE 10   *Solving a Linear Equation Involving Decimals*

Solve $0.3x + 0.2(10 - x) = 0.15(30)$.

Solution

NOTE   There are several ways to solve the decimal equation in Example 10. You could immediately clear the equation of decimals by multiplying both sides by 100. Or, you could keep the decimals and use a calculator to find the solution. Which method do you prefer?

$0.3x + 0.2(10 - x) = 0.15(30)$	Original equation
$0.3x + 2 - 0.2x = 4.5$	Distributive Property
$0.1x + 2 = 4.5$	Combine like terms.
$10(0.1x + 2) = 10(4.5)$	Multiply both sides by 10.
$x + 20 = 45$	Clear decimals.
$x = 25$	Subtract 20 from both sides.

The solution is 25. Check this in the original equation.

Group Activities

You Be the Instructor

Error Analysis   Suppose you are teaching an algebra class and one of your students hands in the following problem. In your group, decide if the answer the student obtained is correct. If not, write an explanation for the student.

$4(x + 2) - 8 = 3x$	Given equation
$4x + 8 - 8 = 3x$	Distributive Property
$4x = 3x$	Additive inverse
$4 = 3$	Divide both sides by x.

No solution because 4 is not equal to 3.

Explain what happens when you divide both sides of an equation by a variable factor.

3.3 Exercises

Discussing the Concepts

1. In your own words, describe the procedure for removing symbols of grouping. Give some examples.

2. Describe the error in the following.

$$-2(x - 5) = 8$$
$$-2x - 5 = 8$$

3. You could solve $3(x - 7) = 15$ by applying the Distributive Property as a first step. However, there is another way to begin. What is it?

4. What is meant by the least common denominator of two or more fractions? Discuss the method for finding the least common denominator of fractions.

5. When solving an equation that contains fractions, what is accomplished by multiplying both sides of an equation by the least common denominator of the fractions?

6. When simplifying an algebraic *expression* involving fractions, why can't you simplify the expression by multiplying by the least common denominator?

Problem Solving

In Exercises 7–10, solve the equation mentally.

7. $\dfrac{x}{10} = \dfrac{1}{5}$

8. $\dfrac{t}{6} = \dfrac{2}{3}$

9. $\dfrac{z + 2}{3} = 4$

10. $\dfrac{u - 4}{4} = 8$

In Exercises 11–30, solve the equation and check your result. (Some equations have no solution.)

11. $2(x - 3) = 4$

12. $4(x + 1) = 24$

13. $3 - (2x - 4) = 3$

14. $16 - (3x - 10) = 5$

15. $8(t - 3) = 0$

16. $4(u + 5) = 4(u - 1) + 1$

17. $7 = 3(x + 2) - 3(x - 5)$

18. $0.24 = 0.12(z + 1) - 0.03(z - 2)$

19. $0.6(x + 4) = 2(x + 4)$

20. $-8(x - 6) = 3(x - 6)$

21. $2(3x + 5) - 7 = 3(5x - 2)$

22. $6[x - (2x + 3)] = 8 - 5x$

23. $\dfrac{x}{2} = \dfrac{3}{2}$

24. $\dfrac{t}{4} = \dfrac{3}{8}$

25. $\dfrac{6x}{25} = \dfrac{3}{5}$

26. $-\dfrac{8x}{9} = \dfrac{2}{3}$

27. $\dfrac{5x}{4} + \dfrac{1}{2} = 0$

28. $\dfrac{y}{4} - \dfrac{5}{8} = 2$

29. $\dfrac{100 - 4u}{3} = \dfrac{5u + 6}{4} + 6$

30. $\dfrac{8 - 3x}{2} - 4 = \dfrac{x}{6}$

In Exercises 31–34, solve the equation by first cross-multiplying.

31. $\dfrac{x - 2}{5} = \dfrac{2}{3}$

32. $\dfrac{2x + 1}{3} = \dfrac{5}{2}$

33. $\dfrac{x}{4} = \dfrac{1 - 2x}{3}$

34. $\dfrac{x + 1}{6} = \dfrac{3x}{10}$

35. *Time to Complete a Task*   Two people can complete 80% of a task in t hours, where t must satisfy the equation

$$\dfrac{t}{10} + \dfrac{t}{15} = 0.8.$$

Solve this equation for t.

36. *Course Grade*   To get an A in a course you must have an average of at least 90 points for four tests of 100 points each. For the first three tests, your scores are 87, 92, and 84. What must you score on the fourth test to earn a 90% average for the course?

In Exercises 37 and 38, use $W_1x = W_2(a - x)$.

37. *Balancing a Seesaw* Find the position of the fulcrum so the seesaw shown in the figure will balance. ($W_1 = 90$, $W_2 = 60$, and $a = 10$.)

38. *Raising a Weight* The fulcrum of a 6-foot-long lever is 6 inches from a weight, as shown in the figure. Find the maximum weight that a 190-pound person can lift using this lever. ($W_1 = 190$, $x = 5\frac{1}{2}$, and $a = 6$.)

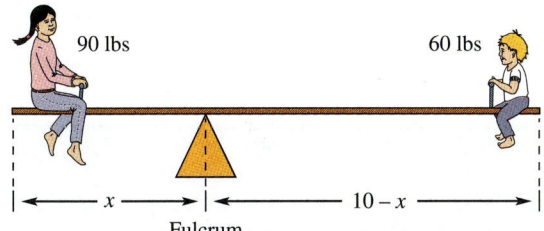

Fulcrum

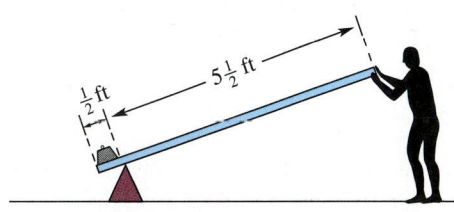

Fulcrum

Reviewing the Major Concepts

In Exercises 39–42, solve the equation.

39. $3x - 42 = 0$

40. $64 - 16x = 0$

41. $2 - 3x = 14 + x$

42. $7 + 5x = 7x - 1$

43. What is $\frac{1}{2}$% of 6000?

44. A car rents for \$37.50 plus 27¢ per mile. Your rental budget is limited to \$75. How many miles can you drive the rental car and stay within your budget?

Additional Problem Solving

In Exercises 45–48, solve the equations mentally.

45. $\frac{3}{2}x = 9$

46. $\frac{a + 3}{10} = 2$

47. $2(y - 4) = 12$

48. $-3(x + 1) = 18$

In Exercises 49–72, solve the equation and check your result. (Some equations have no solution.)

49. $7(x + 5) = 49$

50. $25(z - 2) = 60$

51. $4 - (z + 6) = 8$

52. $25 - (y + 3) = 15$

53. $-3(t + 5) = 0$

54. $4(z - 2) = 0$

55. $-3(t + 5) = 6$

56. $4(z - 2) = 32$

57. $7x - 2(x - 2) = 12$

58. $15(x + 1) - 8x = 29$

59. $6 = 3(y + 1) - 4(1 - y)$

60. $100 = 4(y - 6) - (y - 1)$

61. $7(2x - 1) = 4(1 - 5x) + 6$

62. $-3(5x + 2) + 5(1 + 3x) = 0$

63. $\frac{y}{5} = \frac{3}{5}$

64. $\frac{z}{3} = -\frac{5}{3}$

65. $\frac{y}{5} = -\frac{3}{10}$

66. $\frac{v}{4} = \frac{4}{3}$

67. $\frac{t + 4}{6} = \frac{2}{3}$

68. $\frac{x - 6}{10} = \frac{3}{5}$

69. $0.2x - 0.5x = 1$

70. $0.04x + 0.03x = 0.014$

71. $0.24(z + 5) - 0.03(z + 24) = 0$

72. $1.5x + 0.25(x - 2) = 10$

In Exercises 73–76, solve the equation by first cross-multiplying.

73. $\frac{5x - 4}{4} = \frac{2}{3}$

74. $\frac{10x + 3}{6} = \frac{1}{2}$

75. $\frac{10 - x}{2} = \frac{x + 4}{5}$

76. $\frac{2x + 3}{5} = \frac{3 - 4x}{8}$

77. *Fireplace Construction* A fireplace is 93 inches wide. Each brick in the fireplace has a length of 8 inches and there is $\frac{1}{2}$ inch of mortar between adjoining bricks. Let n be the number of bricks per row.

(a) Explain why the number of bricks per row is the solution of the equation $8n + \frac{1}{2}(n-1) = 93$.

(b) Find the number of bricks per row in the fireplace.

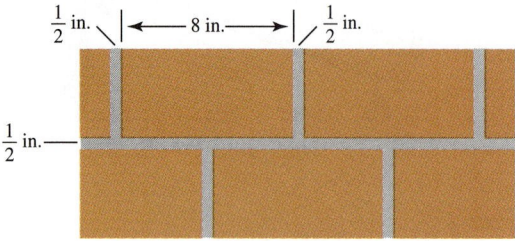

78. *Time to Complete a Task* The time to complete a task is given by the solution of the equation

$$\frac{t}{10} + \frac{t}{15} = 1.$$

Find the required time t.

In Exercises 79–82, solve the following equation for x.

$$p_1 x + p_2(a - x) = p_3 a$$

79. *Mixture Problem* Determine the number of quarts of a 10% solution that must be mixed with a 30% solution to obtain 100 quarts of a 25% solution. ($p_1 = 0.1$, $p_2 = 0.3$, $p_3 = 0.25$, and $a = 100$.)

80. *Mixture Problem* Determine the number of gallons of a 25% solution that must be mixed with a 50% solution to obtain 5 gallons of a 30% solution. ($p_1 = 0.25$, $p_2 = 0.5$, $p_3 = 0.3$ and $a = 5$.)

81. *Mixture Problem* An 8-quart automobile cooling system is filled with coolant that is 40% antifreeze. Determine the amount that must be withdrawn and replaced with pure antifreeze so that the 8 quarts of coolant will be 50% antifreeze. ($p_1 = 1$, $p_2 = 0.4$, $p_3 = 0.5$, and $a = 8$.)

82. *Mixture Problem* A grocer mixes two kinds of nuts priced at \$2.49 per pound and \$3.89 per pound to make 100 pounds of a mixture to be priced at \$3.19 per pound. How many pounds of the \$2.49 per pound nuts must be put into the mixture? ($p_1 = 2.49$, $p_2 = 3.89$, $p_3 = 3.19$, and $a = 100$.)

83. *Exploration* Review Exercises 79–82, and describe what p_1, p_2, p_3, and a represent.

CAREER INTERVIEW

Amy L. Bick

Advancement Director

The Greater Cincinnati Foundation

Cincinnati, OH 45202

GCF is our area's community foundation. We are a nonprofit foundation that serves donors by providing a flexible vehicle for charitable giving in perpetuity. With income from funds contributed by individuals, families, and businesses, GCF works to improve the quality of life in Greater Cincinnati by making grants to local nonprofit organizations. My job is to seek donors interested in charitable giving.

We often use algebra at GCF for planning purposes. Recently I was asked to estimate the value of gifts needed each year to reach our goal of \$240 million in fund assets by January 2000. In January 1994, GCF had fund assets equal to \$120 million. I estimated that amount to earn \$50.22 million in simple interest by January 2000, ignoring any interest that might be earned by new gifts. Finding the necessary value of annual gifts uses the same process as solving the equation: Goal = principal + interest + (number of years)x, where x = annual gifts. Inserting the known values gives \$240,000,000 = \$120,000,000 + \$50,220,000 + 6x. I found that no more than \$11.63 million in annual gifts is needed each year through 1999 to reach our goal of \$240 million in assets.

MID-CHAPTER QUIZ

Take this quiz as you would take a quiz in class. After you are done, check your work against the answers given in the back of the book.

In Exercises 1–8, solve the equation.

1. $120 - 3y = 0$

2. $10(y - 8) = 0$

3. $3x + 1 = x + 20$

4. $6x + 8 = 8 - 2x$

5. $-10x + \dfrac{2}{3} = \dfrac{7}{3} - 5x$

6. $\dfrac{x}{5} + \dfrac{x}{8} = 1$

7. $\dfrac{9 + x}{3} = 15$

8. $4 - 0.3(1 - x) = 7$

In Exercises 9 and 10, solve the equation. Round the solution to two decimal places. In your own words, explain how to check the solution.

9. $32.86 - 10.5x = 11.25$

10. $\dfrac{x}{5.45} + 3.2 = 12.6$

11. What is 62% of 25?

12. What is $\frac{1}{2}$% of 8400?

13. 300 is what percent of 150?

14. 145.6 is 32% of what number?

15. The perimeter of a rectangle is 60 meters. Find the dimensions of the rectangle if the length is one and one-half times the width.

16. You have two jobs. In the first job, you work 40 hours a week and earn $7.50 per hour. In the second job you earn $6.00 per hour and can work as many hours as you want. If you want to earn $360 a week, how many hours must you work at the second job?

17. A region has an area of 42 square meters. It must be divided into three subregions so that the second has twice the area of the first, and the third has twice the area of the second. Determine the area of each subregion.

18. To get an A in a course you must have an average of at least 90 points for three tests of 100 points each. For the first two tests, your scores are 84 and 93. What must you score on the third test to earn a 90% average for the course?

19. The price of a television set is approximately 108% of what it was two years ago. The current price is $535. What was the approximate price two years ago?

20. The figure at the right shows the economic costs per year for Alzheimer's. What percent of the total cost is the value of time of unpaid caregivers? (Source: *American Journal of Public Health*)

Alzheimer's Economic Costs

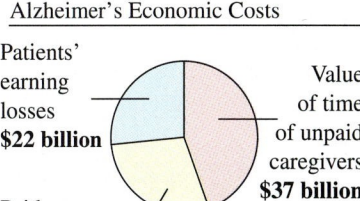

Patients' earning losses
$22 billion

Value of time of unpaid caregivers
$37 billion

Paid care
$24 billion

Figure for 20

3.4	**Ratio and Proportion**
	Setting Up Ratios ▪ Unit Prices ▪ Solving Proportions ▪ The Consumer Price Index

Setting Up Ratios

A **ratio** is a comparison of one number to another by division. For example, in a class of 29 students made up of 16 women and 13 men, the ratio of women to men is 16 to 13 or $\frac{16}{13}$. Some other ratios for this class are as follows.

Men to women: $\frac{13}{16}$ Men to students: $\frac{13}{29}$ Students to women: $\frac{29}{16}$

Note the order implied by a ratio. The ratio of a to b means a/b, whereas the ratio of b to a means b/a.

Definition of Ratio

The **ratio** of the real number a to the real number b is given by

$$\frac{a}{b}.$$

The ratio of a to b is sometimes written as $a : b$.

EXAMPLE 1   Finding Ratios

a. The ratio of 7 to 5 is given by

$$\frac{7}{5}.$$

b. The ratio of 12 to 8 is given by

$$\frac{12}{8} = \frac{3}{2}.$$

Note that the fraction $\frac{12}{8}$ can be written in reduced form as $\frac{3}{2}$.

c. The ratio of $3\frac{1}{2}$ to $5\frac{1}{4}$ is given by

$$\frac{3\frac{1}{2}}{5\frac{1}{4}} = \frac{\frac{7}{2}}{\frac{21}{4}} = \frac{7}{2} \cdot \frac{4}{21} = \frac{2}{3}.$$

Conversion Factors
Length:
1 foot = 12 inches
1 yard = 3 feet
1 mile = 5280 feet
1 meter = 100 centimeters
1 kilometer = 1000 meters
Weight:
1 pound = 16 ounces
1 ton = 2000 pounds
1 kilogram = 1000 grams
Volume:
1 pint = 16 fluid ounces
1 quart = 2 pints
1 gallon = 4 quarts
1 liter = 1000 milliliters
Time:
1 minute = 60 seconds
1 hour = 60 minutes

There are many real-life applications of ratios. For instance, they are used to describe opinion surveys (for/against), populations (male/female, unemployed/employed), and mixtures (oil/gasoline, water/alcohol).

When comparing two *measurements* by a ratio, you should use the same unit of measurement in both the numerator and the denominator. For example, to find the ratio of 4 feet to 8 inches, you could convert 4 feet to 48 inches (by multiplying by 12) to obtain

$$\frac{4 \text{ feet}}{8 \text{ inches}} = \frac{48 \text{ inches}}{8 \text{ inches}} = \frac{48}{8} = \frac{6}{1}.$$

Or you could convert 8 inches to $\frac{8}{12}$ feet (by dividing by 12) to obtain

$$\frac{4 \text{ feet}}{8 \text{ inches}} = \frac{4 \text{ feet}}{\frac{8}{12} \text{ feet}} = 4 \cdot \frac{12}{8} = \frac{6}{1}.$$

If you use different units of measurement in the numerator and denominator, then you *must* include the units. If you use the same units of measurement in the numerator and denominator, then it is not necessary to write the units. A list of common conversion factors is given at the left.

EXAMPLE 2 Comparing Measurements

Find a ratio to compare the relative sizes of the following.

a. 5 gallons to 7 gallons   **b.** 3 meters to 40 centimeters

c. 200 cents to 3 dollars   **d.** 30 months to $1\frac{1}{2}$ years

Solution

a. Because the units of measurement are the same, the ratio is $\frac{5}{7}$.

b. Because the units of measurement are different, begin by converting meters to centimeters *or* centimeters to meters. Here, it is easier to convert meters to centimeters by multiplying by 100.

$$\frac{3 \text{ meters}}{40 \text{ centimeters}} = \frac{3(100) \text{ centimeters}}{40 \text{ centimeters}} = \frac{300}{40} = \frac{15}{2}$$

c. Because 200 cents is the same as 2 dollars, the ratio is

$$\frac{200 \text{ cents}}{3 \text{ dollars}} = \frac{2 \text{ dollars}}{3 \text{ dollars}} = \frac{2}{3}.$$

d. Because $1\frac{1}{2}$ years = 18 months, the ratio is

$$\frac{30 \text{ months}}{1\frac{1}{2} \text{ years}} = \frac{30 \text{ months}}{18 \text{ months}} = \frac{30}{18} = \frac{5}{3}.$$

Unit Prices

As a consumer, you must be able to determine the unit prices of items you buy in order to make the best use of your money. The **unit price** of an item is given by the ratio of the total price to the total units.

$$\text{Unit price} = \frac{\text{Total price}}{\text{Total units}}$$

To state unit prices, we usually use the word *per*. For instance, the unit price for a particular brand of coffee might be 4.69 dollars *per* pound, or $4.69 per pound.

EXAMPLE 3 Finding a Unit Price

Find the unit price (in dollars per ounce) for a 5-pound, 4-ounce box of detergent that sells for $4.62.

Solution

Begin by writing the weight in ounces. That is, 5 pounds and 4 ounces is equal to

$$5 \text{ lb} + 4 \text{ oz} = 84 \text{ oz}.$$

Next, determine the unit price as follows.

Verbal Model: $\text{Unit price} = \dfrac{\text{Total price}}{\text{Total units}}$

Unit Price: $\dfrac{\$4.62}{84 \text{ ounces}} = \$0.055 \text{ per ounce}$

EXAMPLE 4 Comparing Unit Prices

Which has the lower unit price: a 12-ounce box of breakfast cereal for $2.69 or a 16-ounce box of the same cereal for $3.49?

Solution

The unit price for the smaller box is

$$\text{Unit price} = \frac{\text{total price}}{\text{total units}} = \frac{\$2.69}{12 \text{ ounces}} \approx \$0.224 \text{ per ounce.}$$

The unit price for the larger box is

$$\text{Unit price} = \frac{\text{total price}}{\text{total units}} = \frac{\$3.49}{16 \text{ ounces}} \approx \$0.218 \text{ per ounce.}$$

Thus, the larger box has a slightly lower unit price.

Solving Proportions

A **proportion** is a statement that equates two ratios. For example, if the ratio of a to b is the same as the ratio of c to d, we can write the proportion as

$$\frac{a}{b} = \frac{c}{d}.$$

In typical applications, you know the values for three of the letters (quantities) and are required to find the value of the fourth. To solve such a fractional equation, you can use the *cross-multiplication* procedure introduced in Section 3.3.

Solving a Proportion

If $\dfrac{a}{b} = \dfrac{c}{d}$, then $ad = bc$. The quantities a and d are called the **extremes** of the proportion, whereas b and c are called the **means** of the proportion.

EXAMPLE 5   *Solving Proportions*

Solve the following proportions for x.

a. $\dfrac{50}{x} = \dfrac{2}{28}$ **b.** $\dfrac{x}{3} = \dfrac{10}{6}$

Solution

a.

$\dfrac{50}{x} = \dfrac{2}{28}$ Proportion

$50(28) = 2x$ Cross-multiply.

$\dfrac{1400}{2} = x$ Divide both sides by 2.

$700 = x$ Simplify.

Thus, the ratio of 50 to 700 is the same as the ratio of 2 to 28.

b.

$\dfrac{x}{3} = \dfrac{10}{6}$ Proportion

$x = \dfrac{30}{6}$ Multiply both sides by 3.

$x = 5$ Simplify.

Thus, the ratio of 5 to 3 is the same as the ratio of 10 to 6.

EXAMPLE 6   Gasoline Cost

You are driving from New York to Arizona, a trip of 2750 miles. You begin the trip with a full tank of gas. After traveling 416 miles, you refill the tank for $24. How much should you plan to spend on gasoline for the entire trip?

Solution

Verbal Model: $$\boxed{\dfrac{\text{Cost of long trip}}{\text{Cost of short trip}}} = \boxed{\dfrac{\text{Miles for long trip}}{\text{Miles for short trip}}}$$

Labels: Cost of long trip $= x$ (dollars)
Cost of short trip $= 24$ (dollars)
Miles for long trip $= 2750$ (miles)
Miles for short trip $= 416$ (miles)

Proportion: $$\frac{x}{24} = \frac{2750}{416}$$

$$x = (24)\left(\frac{2750}{416}\right)$$

$$x \approx \$158.65$$

You should plan to spend approximately $160.00.

EXAMPLE 7   *Using a Proportion to Calculate Taxes*

You have just moved into a new house valued at $110,000. Your next-door neighbor pays $1150 in real estate taxes each year on a house valued at $89,000. How much a year should you expect to pay in real estate taxes?

Solution

Verbal Model: $$\boxed{\dfrac{\text{Your taxes}}{\text{Neighbor's taxes}}} = \boxed{\dfrac{\text{Your house value}}{\text{Neighbor's house value}}}$$

Labels: Your real estate taxes $= x$ (dollars)
Neighbor's real estate taxes $= 1150$ (dollars)
Your house value $= 110,000$ (dollars)
Neighbor's house value $= 89,000$ (dollars)

Proportion: $$\frac{x}{1150} = \frac{110,000}{89,000}$$

$$x = 1150 \cdot \frac{110,000}{89,000}$$

$$x \approx \$1421$$

You should expect your real estate taxes to be about $1421.

The Consumer Price Index

The rate of inflation is important to all of us. Simply stated, *inflation* is an economic condition in which the price of a fixed amount of goods or services increases. Thus, a fixed amount of money buys less in a given year than in previous years.

The most widely used measurement of inflation in the United States is the *Consumer Price Index* (CPI), often called the *Cost-of-Living Index*. The table below shows the "All Items" or general index for the years 1950 to 1992. (Source: U.S. Bureau of Labor and Statistics)

Year	CPI	Year	CPI	Year	CPI	Year	CPI
1950	24.1	1961	29.9	1972	41.8	1983	99.6
1951	26.0	1962	30.2	1973	44.4	1984	103.9
1952	26.5	1963	30.6	1974	49.3	1985	107.6
1953	26.7	1964	31.0	1975	53.8	1986	109.6
1954	26.9	1965	31.5	1976	56.9	1987	113.6
1955	26.8	1966	32.4	1977	60.6	1988	118.3
1956	27.2	1967	33.4	1978	65.2	1989	124.0
1957	28.1	1968	34.8	1979	72.6	1990	130.7
1958	28.9	1969	36.7	1980	82.4	1991	136.2
1959	29.1	1970	38.8	1981	90.9	1992	140.3
1960	29.6	1971	40.5	1982	96.5		

To determine (from the CPI) the change in the buying power of a dollar from one year to another, use the following proportion.

$$\frac{\text{Price year } n}{\text{Price year } m} = \frac{\text{index year } n}{\text{index year } m}$$

For instance, if you paid $15,000 for a house in 1950, then the amount you could expect to pay for the same house in 1990 is given by the following proportion.

$$\frac{\text{Price in 1990}}{\text{Price in 1950}} = \frac{\text{index in 1990}}{\text{index in 1950}}$$

$$\frac{x}{15,000} = \frac{130.7}{24.1} \quad \Longrightarrow \quad x \approx \$81,350$$

EXAMPLE 8 *Using the Consumer Price Index*

You purchased a piece of jewelry for $750 in 1980. What would you expect the replacement value of the jewelry to have been in 1988?

Solution

To answer this question, you can use the Consumer Price Index, as follows.

Verbal Model:
$$\frac{\text{Price in 1988}}{\text{Price in 1980}} = \frac{\text{Index in 1988}}{\text{Index in 1980}}$$

Labels:

Price in 1988 = x	(dollars)
Price in 1980 = 750	(dollars)
Index in 1988 = 118.3	
Index in 1980 = 82.4	

Proportion:
$$\frac{x}{750} = \frac{118.3}{82.4}$$
$$x = 750 \cdot \frac{118.3}{82.4}$$
$$x \approx \$1077$$

You should expect the replacement value of the jewelry to have been approximately $1077 in 1988. Check this solution in the original statement of the problem.

Group Activities   Exploring with Technology

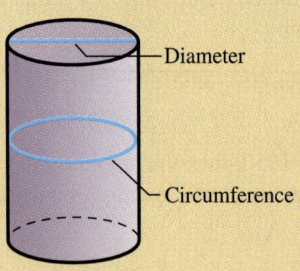

Diameter

Circumference

The Value of Pi   One of the best known ratios in mathematics is denoted by the Greek letter π, pronounced "pie." This number represents the ratio of the circumference of *any* circle to its diameter. To estimate the value of π, try the following experiment in your group. Measure the diameter of a circular cylinder (such as a soda can), and then measure the circumference of the cylinder, as shown in the figure at the left. Then use your calculator to approximate the value of π as follows.

$$\pi = \frac{\text{Circumference}}{\text{Diameter}}$$

Try to find this ratio for several different sizes of circular objects. Does the ratio depend on the size of the circumference of the circle?

3.4 Exercises

Discussing the Concepts

1. In your own words, describe the term *ratio*.

2. You are told that the ratio of men to women in a class is 2 to 1. Does this information tell you the total number of people in the class? Explain.

3. Explain the following statement. "When setting up a ratio, be sure you are comparing apples to apples and not apples to oranges."

4. In your own words, describe the term *proportion*.

Problem Solving

In Exercises 5–8, write the ratio as a fraction in reduced form.

5. 36 to 9

6. 24 to 32

7. 14 : 21

8. 60 : 45

In Exercises 9–16, express the ratio as a fraction in reduced form. (Use the same units for both quantities.)

9. Thirty-six inches to 24 inches

10. Twenty-four pounds to 30 pounds

11. One quart to 1 gallon

12. Three inches to 2 feet

13. Seventy-five centimeters to 2 meters

14. Sixty milliliters to 1 liter

15. Ninety minutes to 2 hours

16. Three thousand pounds to 5 tons

In Exercises 17–20, express the statement as a ratio in reduced form. (Use the same units for both quantities.)

17. *Study Hours* You study 6 hours per day and are in class 3 hours per day. Find the ratio of the number of study hours to class hours.

18. *Income Tax* You have $10 of state tax withheld from your paycheck each week. Your weekly gross pay is $500. Find the ratio of state tax to gross pay.

19. *Price-Earnings Ratio* The ratio of the price of a stock to its earnings is called the *price-earnings ratio*. A certain stock sells for $78 per share and earns $6.50 per share. What is the price-earnings ratio?

20. *Compression Ratio* The *compression ratio* of an engine is the ratio of the expanded volume of gas in one of its cylinders to the compressed volume of gas in the cylinder (see figure). A cylinder in an engine has an expanded volume of 345 cubic centimeters and a compressed volume of 17.25 cubic centimeters. What is the compression ratio of this engine?

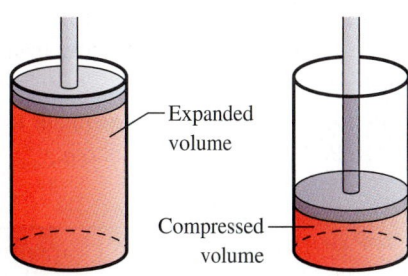

Expanded volume

Compressed volume

In Exercises 21–24, find the unit price (in $/oz).

21. A 20-ounce can of pineapple for 79¢

22. An 18-ounce box of cereal for $3.19

23. A 1-pound, 4-ounce loaf of bread for $1.29

24. A 1-pound package of cheese for $2.89

In Exercises 25 and 26, which product has the smaller unit price?

25. A $27\frac{3}{4}$-ounce can of spaghetti sauce for $1.19, or a 32-ounce jar for $1.45

26. A 16-ounce package of margarine quarters for $1.29, or a 3-pound tub for $3.29

In Exercises 27–32, solve the proportion.

27. $\dfrac{3}{5} = \dfrac{y}{20}$ **28.** $\dfrac{x}{9} = \dfrac{5}{18}$

29. $\dfrac{8}{3} = \dfrac{t}{6}$ **30.** $\dfrac{7}{12} = \dfrac{x}{6}$

31. $\dfrac{x+6}{3} = \dfrac{x-5}{2}$ **32.** $\dfrac{x-2}{4} = \dfrac{x+10}{10}$

33. *Gasoline Cost*   A car uses 20 gallons of gasoline for a trip of 360 miles. How many gallons would be used on a trip of 400 miles?

34. *Force on a Spring*   A force of 50 pounds stretches a spring 4 inches. How many pounds of force are required to stretch the spring 6 inches?

35. *Amount of Gasoline*   The gasoline-to-oil ratio for a two-cycle engine is 40 to 1. How much gasoline is required to produce a mixture that contains one-half pint of oil?

36. *Enlarging a Recipe*   Two cups of flour are required to make one batch of cookies. How many cups are required for $2\frac{1}{2}$ batches?

37. *Estimation*   Use the map to estimate the distance between Philadelphia and Pittsburgh.

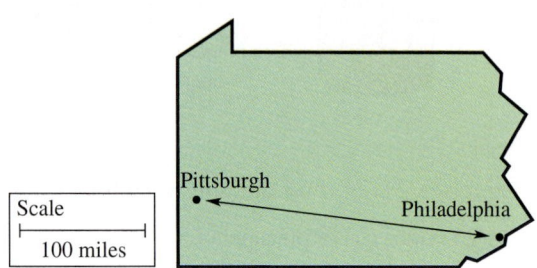

38. *Estimation*   On a map, $1\frac{1}{2}$ inches represents 40 miles. Estimate the distance between two cities that are 4 inches apart on the map.

Geometry   In Exercises 39 and 40, use the fact that if two triangles are similar, their corresponding sides are proportional.

39. *Similar Triangles*   The two triangles are similar. Find the length x.

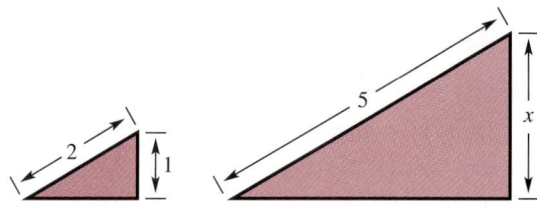

40. *Shadow Length*   A man who is 6 feet tall walks directly toward the tip of the shadow of a tree. When the man is 100 feet from the tree, he starts forming his own shadow beyond the shadow of the tree. The length of the shadow of the tree beyond this point is 8 feet. Find the height of the tree.

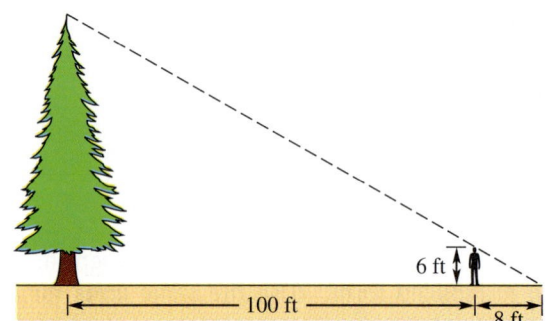

Reviewing the Major Concepts

In Exercises 41–44, solve the linear equation and check your solution.

41. $50 - z = 15$ **42.** $x - 6 = 3x + 10$

43. $\dfrac{x}{6} + \dfrac{x}{3} = 1$ **44.** $\dfrac{x}{5} + \dfrac{1}{5} = \dfrac{7}{10}$

45. *Geometry*   The perimeter of an isosceles triangle is 120 inches. The third side is $\frac{1}{2}$ the length of the two sides of equal length. Find the lengths of the sides of the triangle.

46. *List Price*   The price of a lawn tractor has been discounted 20%. The sale price is $2495. Find the original price of the tractor.

Additional Problem Solving

In Exercises 47–50, write the ratio as a fraction in reduced form.

47. 27 to 54

48. 50 to 15

49. 144:16

50. 12:30

In Exercises 51–60, express the ratio as a fraction in reduced form. (Use the same units for both quantities.)

51. Fifteen feet to 12 feet

52. Forty dollars to $60

53. Seven nickels to 3 quarters

54. Twenty-four ounces to 3 pounds

55. Three hours to 90 minutes

56. Twenty-one feet to 35 yards

57. Two meters to 75 centimeters

58. Fifty cubic centimeters to 1 liter

59. Five and one-half pints to 2 quarts

60. Twelve thousand pounds to 2 tons

In Exercises 61–64, express the statement as a ratio in reduced form. (Use the same units for both quantities.)

61. *Student-Teacher Ratio*   There are 2921 students and 127 faculty members at your school. Find the ratio of the number of students to the number of faculty.

62. *Turn Ratio*   The *turn ratio* of a transformer is the ratio of the number of turns on the secondary winding to the number of turns on the primary winding. A transformer has a primary winding with 250 turns and a secondary winding with 750 turns. What is its turn ratio?

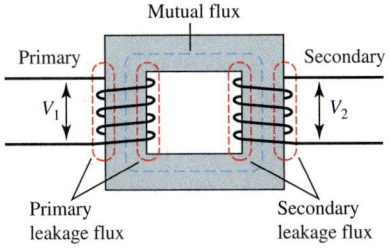

63. *Gear Ratio*   The *gear ratio* of two gears is the ratio of the number of teeth in one gear to the number of teeth in the other gear. Find the gear ratio of the larger gear to the smaller gear for the following gears.

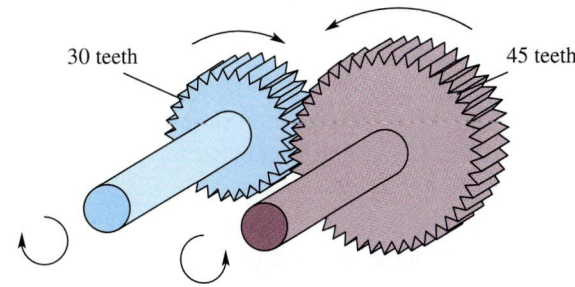

30 teeth   45 teeth

64. *Reading a Table*   On a 5-speed bicycle, the ratio of the pedal gear to the axle gear depends on which axle gear is engaged. Use the following table to find the gear ratios for the five different gears. For which gear is it easiest to pedal? Why?

Gear	1st	2nd	3rd	4th	5th
Teeth on Pedal Gear	52	52	52	52	52
Teeth on Axle Gear	28	24	20	17	14

65. *Geometry*   Find the ratio of the area of the larger pizza to the area of the smaller pizza in the figure. (*Note:* The area of a circle is $A = \pi r^2$.)

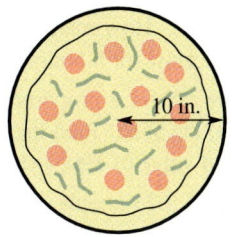 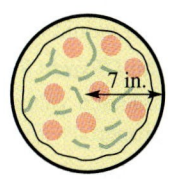

10 in.   7 in.

66. *Specific Gravity* The *specific gravity* of a substance is the ratio of its weight to the weight of the same volume of water. Kerosene weighs 0.82 grams per cubic centimeter and water weighs 1 gram per cubic centimeter. What is the specific gravity of kerosene?

In Exercises 67–70, which product has the smaller unit price?

67. A 10-ounce package of frozen green beans for 59¢, or a 16-ounce package for 89¢

68. An 18-ounce jar of peanut butter for $1.39, or a 28-ounce jar for $2.19

69. A 2-liter bottle (67.6 ounces) of soft drink for $1.09, or six 12-ounce cans for $1.69

70. A 1-quart container of oil for $1.29, or a 2.5-gallon container for $11.20

In Exercises 71–78, solve the proportion.

71. $\dfrac{t}{4} = \dfrac{25}{2}$

72. $\dfrac{y}{25} = \dfrac{12}{10}$

73. $\dfrac{x}{5} = \dfrac{2}{3}$

74. $\dfrac{z}{35} = \dfrac{5}{14}$

75. $\dfrac{x+1}{5} = \dfrac{3}{10}$

76. $\dfrac{z-3}{8} = \dfrac{3}{16}$

77. $\dfrac{x+2}{8} = \dfrac{x-1}{3}$

78. $\dfrac{x-4}{5} = \dfrac{x}{6}$

79. *Building Material* One hundred cement blocks are needed to build a 16-foot wall. How many blocks are needed to build a 40-foot wall?

80. *Amount of Fuel* A tractor uses 4 gallons of diesel fuel to plow for 90 minutes. How many gallons of fuel would be required to plow for 8 hours?

81. *Real Estate Taxes* The tax on a property with an assessed value of $65,000 is $825. Find the tax on a property with an assessed value of $90,000.

82. *Polling Results* In a poll, 624 people out of 1100 indicated they would vote for a specific candidate. How many votes can the candidate expect to receive from 40,000 votes cast?

83. *Pounds of Sand* The ratio of cement to sand in an 80-pound bag of dry mix is 1 to 4. Find the number of pounds of sand in the bag. (*Note:* Dry mix is composed of only cement and sand.)

Geometry In Exercises 84 and 85, use the fact that if two triangles are similar, their corresponding sides are proportional.

84. *Similar Triangles* Use the fact that the following triangles are similar to find the length x.

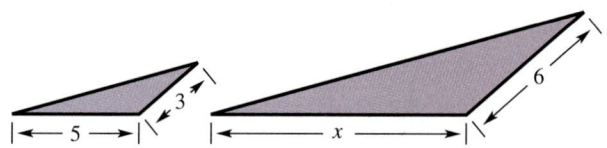

85. *Shadow Length* In the figure below, how long is the man's shadow? (*Hint:* Use similar triangles to create a proportion.)

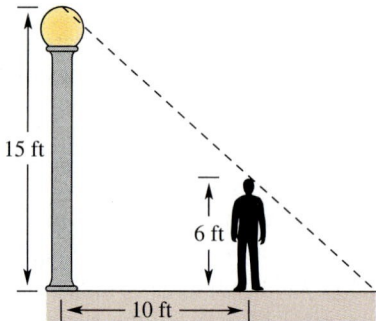

Estimation In Exercises 86–89, use the Consumer Price Index to estimate the price of the item in the indicated year.

86. 1992 price of a tractor that cost $2875 in 1978

87. 1992 price of a watch that cost $58 in 1973

88. 1960 price of a gallon of milk that cost $2.07 in 1992

89. 1950 price of a coat that cost $225 in 1990

3.5   Linear Inequalities and Applications

Inequalities and Their Graphs ▪ Properties of Inequalities ▪
Solving Inequalities ▪ Solving a Double Inequality ▪ Applications

Inequalities and Their Graphs

In this section you will study **algebraic inequalities,** which are inequalities that contain one or more variable terms. Here are some examples.

$$x \le 3, \quad x \ge -2, \quad x - 5 < 2, \quad \text{and} \quad 5x - 7 < 3x + 9$$

Each of these inequalities is a **linear inequality** in the variable x because the (implied) exponent of x is 1.

As with an equation, you can **solve an inequality** in the variable x by finding all values of x for which the inequality is true. Such values are **solutions** and are said to **satisfy** the inequality. The **solution set** of an inequality is the set of all real numbers that are solutions of the inequality.

Often, the solution set of an inequality will consist of infinitely many real numbers. To get a visual image of the solution set, it is helpful to sketch its **graph** on the real number line. For instance, the graph of the solution set of $x < 2$ consists of all points on the real number line that are to the left of 2.

EXAMPLE 1   *Graphs of Inequalities*

Inequality	*Graph of Solution Set*	*Verbal Description*
a. $x < 2$	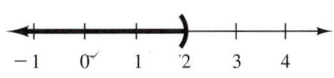	x is less than 2.
b. $x \ge -2$		x is greater than or equal to -2.
c. $-1 \le x \le 2$		x is greater than or equal to -1 *and* less than or equal to 2.
d. $2 \le x < 5$	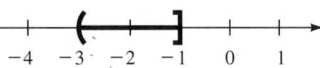	x is greater than or equal to 2 *and* less than 5.
e. $-3 < x \le -1$		x is greater than -3 *and* less than or equal to -1.

NOTE   When the solution set of an inequality is graphed, a parenthesis is used to show that a point *is not* included in the solution set when the inequality symbol is $<$ or $>$, and a square bracket is used to show that a point *is* included in the solution set when the inequality symbol is $\le$ or $\ge$.

Properties of Inequalities

The procedures for solving linear inequalities in one variable are much like those for solving linear equations. To isolate the variable, you can use the **properties of inequalities.** These properties are similar to the properties of equality, but there are two important exceptions. When both sides of an inequality are multiplied or divided by a negative number, the direction of the inequality symbol must be reversed. Here is an example.

$$-2 < 5 \qquad \text{\color{red} Original inequality}$$
$$(-3)(-2) > (-3)(5) \qquad \text{\color{red} Multiply both sides by } -3 \text{ and reverse inequality.}$$
$$6 > -15 \qquad \text{\color{red} Simplify.}$$

Two inequalities that have the same solution set are called **equivalent.** The following list describes operations that can be used to create equivalent inequalities.

Graphing Utility

Linear inequalities can be graphed using a graphing utility. For instance, to graph $x > -2$, use the following keystrokes and range setting.

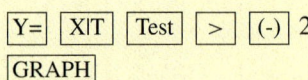

$$\begin{aligned}
&\text{Xmin} = -10 \\
&\text{Xmax} = 10 \\
&\text{Xscl} = 1 \\
&\text{Ymin} = -5 \\
&\text{Ymax} = 5 \\
&\text{Yscl} = 1
\end{aligned}$$

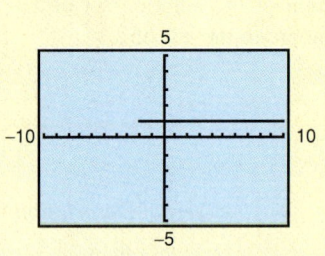

Properties of Inequalities

Let a, b, and c be real numbers, variables, or algebraic expressions.

Property	*Verbal and Algebraic Descriptions*
Addition:	Add the same quantity to both sides.
	If $a < b$, then $a + c < b + c$.
Subtraction:	Subtract the same quantity from both sides.
	If $a < b$, then $a - c < b - c$.
Multiplication:	Multiply both sides by a *positive* quantity.
	If $a < b$ and c is positive, then $ac < bc$.
	Multiply both sides by a *negative* quantity and reverse the inequality symbol.
	If $a < b$ and c is negative, then $ac > bc$.
Division:	Divide both sides by a *positive* quantity.
	If $a < b$ and c is positive, then $\dfrac{a}{c} < \dfrac{b}{c}$.
	Divide both sides by a *negative* quantity and reverse the inequality symbol.
	If $a < b$ and c is negative, then $\dfrac{a}{c} > \dfrac{b}{c}$.
Transitive:	If $a < b$ and $b < c$, then $a < c$.

NOTE Each of the properties to the right is true if the symbol $<$ is replaced by $\leq$ and the symbol $>$ is replaced by $\geq$. Moreover, the letters a, b, and c can be real numbers, variables, or algebraic expressions. Note that we cannot multiply or divide both sides of an inequality by zero.

Solving Inequalities

STUDY TIP

Checking the solution set of an inequality is not as simple as checking the solution set of an equation. (There are usually too many x-values to substitute back into the original inequality.) You can, however, get an indication of the validity of a solution set by substituting a few convenient values of x. For instance, in Example 2 the solution of $x + 5 < 8$ was found to be $x < 3$. Try checking that $x = 0$ satisfies the original inequality, whereas $x = 4$ does not.

In the next four examples, pay special attention to the steps in which the inequality symbol is reversed. Remember that when you multiply or divide an inequality by a negative number, you must reverse the inequality symbol.

EXAMPLE 2   Solving a Linear Inequality

Solve and graph the inequality $x + 5 < 8$.

Solution

$$x + 5 < 8 \qquad\qquad \text{Original inequality}$$
$$x + 5 - 5 < 8 - 5 \qquad\qquad \text{Subtract 5 from both sides.}$$
$$x < 3 \qquad\qquad \text{Solution set}$$

The graph of the solution set is shown in Figure 3.1.

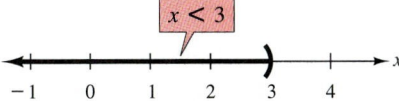

FIGURE 3.1   All real numbers that are less than 3

EXAMPLE 3   Solving a Linear Inequality

Solve and graph the inequality $3y - 1 \leq -7$.

Solution

$$3y - 1 \leq -7 \qquad\qquad \text{Original inequality}$$
$$3y - 1 + 1 \leq -7 + 1 \qquad\qquad \text{Add 1 to both sides.}$$
$$3y \leq -6 \qquad\qquad \text{Combine like terms.}$$
$$\frac{3y}{3} \leq \frac{-6}{3} \qquad\qquad \text{Divide both sides by (positive) 3.}$$
$$y \leq -2 \qquad\qquad \text{Solution set}$$

The graph of the solution set is shown in Figure 3.2.

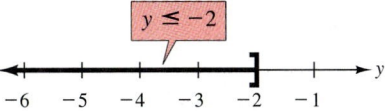

FIGURE 3.2   All real numbers that are less than or equal to -2

EXAMPLE 4   Solving a Linear Inequality

Solve and graph the inequality $12 - 2x > 10$.

Solution

$$
\begin{array}{ll}
12 - 2x > 10 & \text{Original inequality} \\
12 - 12 - 2x > 10 - 12 & \text{Subtract 12 from both sides.} \\
-2x > -2 & \text{Combine like terms.} \\
\dfrac{-2x}{-2} < \dfrac{-2}{-2} & \text{Divide both sides by } -2 \text{ and reverse inequality.} \\
x < 1 & \text{Solution set}
\end{array}
$$

The graph of the solution set is shown in Figure 3.3.

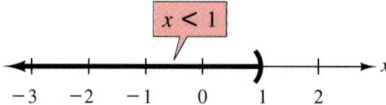

FIGURE 3.3   All real numbers that are less than 1.

EXAMPLE 5   Solving a Linear Inequality

Solve and graph the inequality $1 - \dfrac{3x}{2} \geq x - 4$.

Solution

$$
\begin{array}{ll}
1 - \dfrac{3x}{2} \geq x - 4 & \text{Original inequality} \\
2 - 3x \geq 2x - 8 & \text{Multiply both sides by 2.} \\
-3x \geq 2x - 10 & \text{Subtract 2 from both sides.} \\
-5x \geq -10 & \text{Subtract } 2x \text{ from both sides.} \\
x \leq 2 & \text{Divide both sides by } -5 \text{ and reverse inequality.}
\end{array}
$$

The graph of the solution set is shown in Figure 3.4.

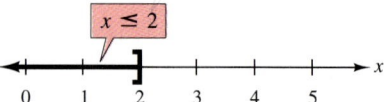

FIGURE 3.4   All real numbers that are less than or equal to 2.

Solving a Double Inequality

Sometimes it is convenient to write two inequalities as a **double inequality.** For instance, you can write the two inequalities

$$-3 \le 6x - 1 \text{ and } 6x - 1 < 3$$

more simply as

$$-3 \le 6x - 1 < 3.$$

This form allows us to solve the two given inequalities together, as demonstrated in Example 6.

EXAMPLE 6   Solving a Double Inequality

Solve the inequality $-3 \le 6x - 1 < 3$.

Solution

$-3 \le 6x - 1 < 3$	Original inequality
$-3 + 1 \le 6x - 1 + 1 < 3 + 1$	Add 1 to all three parts.
$-2 \le 6x < 4$	Combine like terms.
$\dfrac{-2}{6} \le \dfrac{6x}{6} < \dfrac{4}{6}$	Divide each part by 6.
$-\dfrac{1}{3} \le x < \dfrac{2}{3}$	Solution set

The graph of the solution set is shown in Figure 3.5.

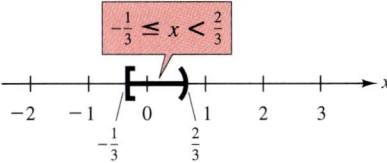

FIGURE 3.5 All real numbers that are greater than or equal to $-\frac{1}{3}$ and less than $\frac{2}{3}$.

The double inequality in Example 6 could have been solved in two parts as follows.

$$
\begin{array}{lcl}
-3 \le 6x - 1 & \text{and} & 6x - 1 < 3 \\
-2 \le 6x & & 6x < 4 \\
-\dfrac{1}{3} \le x & & x < \dfrac{2}{3}
\end{array}
$$

The solution set consists of all real numbers that satisfy *both* inequalities. In other words, the solution set is the set of all values of x for which $-\frac{1}{3} \le x < \frac{2}{3}$.

Applications

Before looking at applications, we give some examples of translation of verbal statements into inequalities.

EXAMPLE 7 Translating Verbal Statements

Verbal Statement	Inequality
a. x is at most 2.	$x \leq 2$
b. x is no more than 2.	$x \leq 2$
c. x is at least 2.	$x \geq 2$
d. x is more than 2.	$x > 2$
e. x is less than 2.	$x < 2$

NOTE When translating inequalities, remember that "at most" means "less than or equal to," and "at least" means "greater than or equal to." Also, be sure to distinguish between the *sum* "2 more than a number" $(x + 2)$ and the *inequality* "2 is more than a number" $(2 > x)$.

EXAMPLE 8 Car Rental

A subcompact car can be rented from Company A for $190 per week with no extra charge for mileage. A similar car can be rented from Company B for $100 per week, plus 20¢ for each mile driven. How many miles must you drive in a week to make the rental fee for Company A less than that for Company B?

Solution

Verbal Model: [Weekly cost for A] $<$ [Weekly cost for B]

Labels:
Number of miles driven in one week $= m$ (miles)
Weekly cost for A $= 190$ (dollars)
Weekly cost for B $= 100 + 0.2m$ (dollars)

Inequality:
$$190 < 100 + 0.2m$$
$$90 < 0.2m$$
$$450 < m$$

Note that the inequality $450 < m$ is equivalent to writing $m > 450$. Thus, the car from Company A is cheaper if you plan to drive more than 450 miles in a week.

EXAMPLE 9   Course Grade

Suppose you are taking a college course in which your grade is based on six 100-point exams. To earn an A in the course, you must have a total of at least 90% of the points. On the first five exams, your scores were 85, 92, 88, 96, and 87. How many points do you have to obtain on the sixth test in order to earn an A in the course?

Solution

*Verbal
Model:* $\boxed{\text{Total points}} \geq \boxed{90\% \text{ of } 600}$

Labels: Score for sixth exam $= x$ (points)
           Total points $= (85 + 92 + 88 + 96 + 87) + x$ (points)

Inequality: $(85 + 92 + 88 + 96 + 87) + x \geq 0.9(600)$

$$448 + x \geq 540$$
$$x \geq 540 - 448$$
$$x \geq 92$$

You must get at least 92 points on the sixth exam to earn an A in the course. Check this solution in the original statement of the problem.

Group Activities   Extending the Concept

Misuse of a Double Inequality   Suppose you are to find the solution set to an inequality that says: "$x - 2$ is greater than 3 and less then -3." In your group, determine if it is appropriate to write the inequality in the form

$$3 < x - 2 < -3.$$

If not, explain why. Then find the solution set using appropriate steps.

3.5 Exercises

Discussing the Concepts

1. Give a verbal description of each of the symbols $<$, $\leq$, $>$, $\geq$, and $=$.

2. Is adding -5 to both sides of an inequality the same as subtracting 5 from both sides? Explain.

3. Is dividing both sides of an inequality by 5 the same as multiplying both sides by $\frac{1}{5}$? Explain.

4. How many numbers are in a solution set of a linear inequality? Give an example.

5. Explain the effect on an inequality when both sides are multiplied or divided by a negative number. Give examples demonstrating your explanation.

6. Compare solving equations to solving inequalities.

Problem Solving

In Exercises 7 and 8, describe the inequality verbally and sketch its graph.

7. $x \geq 3$ 8. $-3 < x < 4$

In Exercises 9–12, write the inequality symbolically.

9.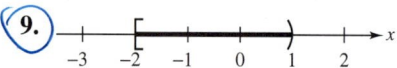

10.

11. x greater than 0 and less than or equal to 6
12. x greater than or equal to -4 and less than 3

In Exercises 13–16, determine whether the value of x is a solution of the inequality.

Inequality	Values

13. $5x - 12 > 0$ (a) $x = 3$ (b) $x = -3$
                     (c) $x = \frac{5}{2}$ (d) $x = \frac{3}{2}$

14. $x + 1 < \frac{2}{3}x$ (a) $x = 0$ (b) $x = 4$
                     (c) $x = -4$ (d) $x = -3$

15. $0 < \frac{1}{4}(x - 2) < 2$ (a) $x = 4$ (b) $x = 10$
                     (c) $x = 0$ (d) $x = \frac{7}{2}$

16. $-1 < \frac{1}{2}(3 - x) \leq 1$ (a) $x = 0$ (b) $x = 3$
                     (c) $x = 1$ (d) $x = 5$

In Exercises 17–22, match the inequality with its graph. [The graphs are labeled (a), (b), (c), (d), (e), and (f).]

(a)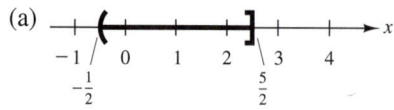

(b)

(c)

(d)

(e)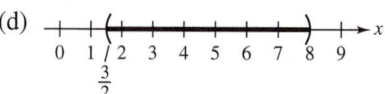

(f)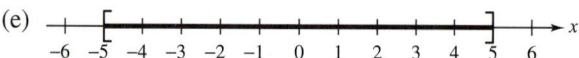

17. $x < 4$ 18. $x \geq 6$ 19. $-3 \leq x < 2$
20. $-\frac{1}{2} < x \leq \frac{5}{2}$ 21. $8 > x > \frac{3}{2}$ 22. $5 \geq x \geq -5$

In Exercises 23–26, use inequality notation to denote the statement.

23. x is nonnegative.   24. P is no more than 2.
25. z is at least 3.   26. y is more than -6.

In Exercises 27–40, solve and graph the inequality.

27. $t - 3 \geq 2$

28. $t + 1 < 6$

29. $4x < 12$

30. $2x > 3$

31. $-\frac{3}{4}x > -3$

32. $-\frac{1}{6}x < -2$

33. $4 - 2x < 3$

34. $14 - 3x > 5$

35. $6 < 3(y + 1) - 4(1 - y)$

36. $6[x - (2x + 3)] < 8 - 5x$

37. $\frac{x}{5} - \frac{x}{2} \leq 1$

38. $\frac{x}{3} + \frac{x}{4} \geq 1$

39. $-4 < \frac{2x - 3}{3} < 4$

40. $0 \leq \frac{x + 3}{2} < 5$

41. *Annual Operating Budget*   A utility company has a fleet of vans. The annual operating cost C (in dollars) per van is

$$C = 0.32m + 2300$$

where m is the number of miles traveled by a van in a year. What number of miles will yield an annual operating cost that is less than \$10,000?

42. *Profit*   The revenue for selling x units of a product is $R = 115.95x$. The cost of producing x units is

$$C = 95x + 750.$$

To obtain a profit, the revenue must be greater than the cost. For what values of x will this product produce a profit?

43. *Planet Distances*   Mars is farther from the sun than Venus, and Venus is farther from the sun than Mercury, as shown in the figure. What can be said about the relationship between the distances from the sun to Mars and Mercury?

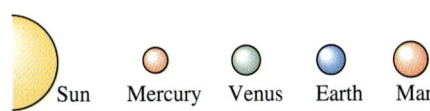

Sun   Mercury   Venus   Earth   Mars

44. *Geometry*   The lengths of the sides of the triangle in the figure are a, b, and c. Find the inequality that relates $a + b$ and c.

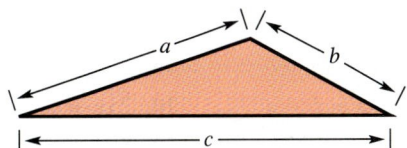

Reviewing the Major Concepts

In Exercises 45 and 46, place the correct inequality symbol between the two real numbers.

45. (a) $-\frac{1}{2}$ ▭ -7 (b) $-\frac{1}{3}$ ▭ $-\frac{1}{6}$

46. (a) $-\pi$ ▭ -3 (b) -6 ▭ $-\frac{13}{2}$

In Exercises 47–50, solve the equation and check your solution.

47. $\frac{9 + x}{3} = 15$

48. $16 + 2l = 64$

49. $4 - 3(1 - x) = 7$

50. $6(t - 6) = 0$

Geometry   In Exercises 51 and 52, write expressions for the perimeter and area of the triangle. Then simplify the expressions.

51.

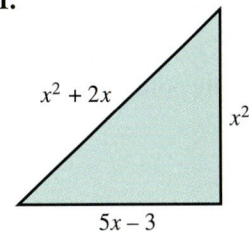

$x^2 + 2x$

x^2

$5x - 3$

52.

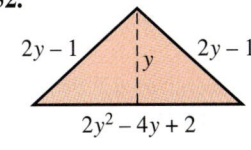

$2y - 1$ y $2y - 1$

$2y^2 - 4y + 2$

Additional Problem Solving

In Exercises 53–56, determine whether the value of x is a solution of the inequality.

	Inequality	*Values*

53. $3 - \frac{1}{2}x \geq 0$ (a) $x = 10$ (b) $x = 6$
   (c) $x = -\frac{3}{4}$ (d) $x = 0$

54. $2(x + 3) \leq 3(x + 4)$ (a) $x = -5$ (b) $x = -1$
   (c) $x = -7$ (d) $x = \frac{2}{3}$

55. $-2 < \frac{1}{3}(x - 1) \leq 5$ (a) $x = 0$ (b) $x = 25$
   (c) $x = 16$ (d) $x = -5$

56. $0 \leq 2(x - 4) \leq 12$ (a) $x = 0$ (b) $x = 15$
   (c) $x = 5$ (d) $x = 10$

In Exercises 57–82, solve and graph the inequality.

57. $x + 4 \leq 6$ **58.** $z - 2 > 0$

59. $-10x < 40$ **60.** $-6x > 18$

61. $-3n > -9$ **62.** $-7n < -21$

63. $\frac{2}{3}x \leq 12$ **64.** $\frac{5}{8}x \geq 10$

65. $2x - 5 > 7$ **66.** $3x + 2 \leq 14$

67. $5 - x \leq 1$ **68.** $3 - y \geq -3$

69. $2x - 5 > 6 - x$ **70.** $25x + 4 \leq 10x + 19$

71. $10(1 - y) < 3 - 2y$ **72.** $8(t - 3) < 4(t - 3)$

73. $6(3 - z) \geq 5(3 + z)$ **74.** $-2(z + 1) \geq 3(z + 1)$

75. $\frac{5x}{4} + \frac{1}{2} > 0$ **76.** $\frac{y}{4} - \frac{5}{8} < 2$

77. $1 < 2x + 3 < 9$ **78.** $-9 \leq -3(x - 2) < 12$

79. $6 > \frac{x - 2}{-3} > -2$ **80.** $-2 < \frac{x - 4}{-2} \leq 3$

81. $\frac{3}{4} > x + 1 > \frac{1}{4}$ **82.** $-1 < -\frac{x}{3} < 1$

In Exercises 83–86, use inequality notation to denote the statement.

83. x is at least 4.

84. t is less than 8.

85. y is no more than 25.

86. x is a real number greater than or equal to -2 and less than 5.

87. *Budgets* Department A's budget is less than Department B's budget, and Department B's budget is less than Department C's budget. What can you say about the relationship between the budgets of Department A and Department C? Identify the property of inequalities that is demonstrated.

88. *Budget for a Trip* You have $2500 budgeted for a trip. The transportation for the trip will cost $900. To stay within your budget, all other costs must be no more than what amount?

89. *Telephone Cost* The cost for a long-distance telephone call is $0.46 for the first minute and $0.31 for each additional minute. The total cost of the call cannot exceed $4. Find the interval of time that is available for the call.

90. *Cargo Weight* The weight of a truck is 4350 pounds. The legal gross weight of the loaded truck is 6000 pounds. Find an interval for the number of crates that the truck can haul if each crate weighs 48 pounds.

91. *Distance* The minimum and maximum speeds on a highway are 45 miles per hour and 65 miles per hour. You travel nonstop for 4 hours on this highway. Assuming that you stay within the speed limits, give an interval for the distance you travel.

92. *Comparing Distances* You live 3 miles from college and 2 miles from the business where you work, as shown in the figure. Let d represent the distance between your work and the college. Write an inequality involving d.

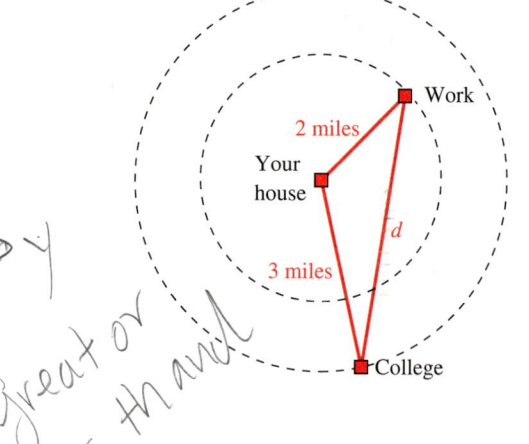

CHAPTER PROJECT: Solar Eclipses

There are three types of solar eclipses: total, partial, and annular. In this project, you will study total and annular eclipses. An annular eclipse, shown at the left, occurs when a moon's image is centered on, but never completely covers, the sun's image. When this occurs, the image of the sun is seen as an annulus or ring. A spectacular annular eclipse occurred on May 10, 1994 in the United States.

Our solar system has many moons. Each moon travels around its planet and each planet travels around the sun. The diagram at lower left shows the positions of the sun, a moon, and a planet when an eclipse occurs. The ratios, R/r and D/d can be used to determine the type of eclipse (total or annular) that occurs.

The table below lists several planets in our solar system and one moon of each planet (except Venus), along with the values of r, R, d, and D (in miles).

Planet	R	D	Moon	r	d
Venus	432,000	67,000,000	None		
Earth	432,000	93,000,000	Moon	1080	232,490
Mars	432,000	142,000,000	Phobos	8	5,800
Jupiter	432,000	484,000,000	Metis	12	79,500
Uranus	432,000	1,784,000,000	Cordelia	12	30,900

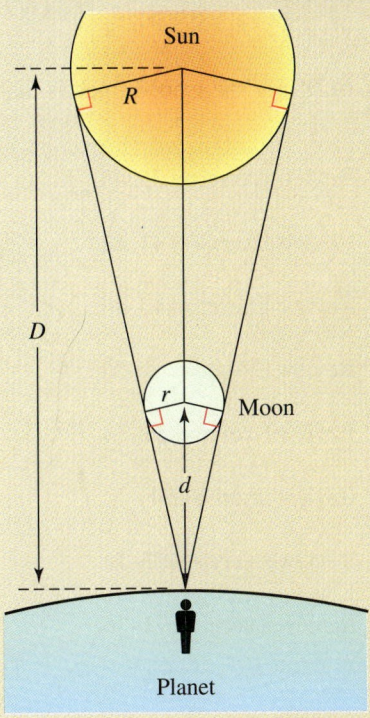

D is the distance between the viewer and the sun's center. d is the distance between the viewer and the moon's center. r is the radius of the moon. R is the radius of the sun.

Use this information to investigate the following questions.

1. Find the ratios R/r and D/d for each planet listed in the table. Then copy the table and include two additional columns for these ratios.

2. A total eclipse occurs when $R/r \leq D/d$. What is the relationship between R/r and D/d for annular eclipses?

3. Classify each eclipse in your table as total or annular.

4. *Research Project*   Find D, d, and r for the four planets in our solar system not given in the table. Repeat Questions 1 and 3 for these four planets.

5. The sun's corona (outer atmosphere) is visible only during an eclipse in which $R/r = D/d$. At which locations in the solar system could an astronomer study the sun's corona? How could you verify your answer?

6. On earth, why are some solar eclipses annular and some total?

CHAPTER SUMMARY

After studying this chapter, you should have acquired the following skills. These skills are keyed to the Review Exercises that begin on page 209. Answers to odd-numbered Review Exercises are given in the back of the book.

- Use mental math to solve linear equations. *(Sections 3.1, 3.3)* — **Review Exercises 1–4**

- Justify each step of the solution of a linear equation. *(Sections 3.1, 3.3)* — **Review Exercises 5, 6**

- Solve linear equations and check the solutions. *(Sections 3.1, 3.3)* — **Review Exercises 7–26**

- Use a calculator to solve linear equations. *(Sections 3.1, 3.3)* — **Review Exercises 27–30**

- Translate real-life sentences into linear equations and solve. *(Sections 3.1, 3.3)* — **Review Exercises 31, 32**

- Complete a table showing the equivalent forms of a percent. *(Section 3.2)* — **Review Exercises 33, 34**

- Solve percent equations. *(Section 3.2)* — **Review Exercises 35–40**

- Solve real-life problems involving percents. *(Section 3.2)* — **Review Exercises 41, 42**

- Express ratios as fractions in reduced form. *(Section 3.4)* — **Review Exercises 43–46**

- Solve the proportion. *(Section 3.4)* — **Review Exercises 47–50**

- Use proportions to solve real-life problems. *(Section 3.4)* — **Review Exercises 51–53**

- Use proportions to solve problems involving geometry. *(Section 3.4)* — **Review Exercise 54**

- Solve and graph inequalities. *(Section 3.5)* — **Review Exercises 55–70**

- Translate verbal sentences into inequalities. *(Section 3.5)* — **Review Exercises 71–76**

REVIEW EXERCISES

In Exercises 1–4, solve the equation mentally.

1. $y - 25 = 10$

2. $z + 5 = 12$

3. $\dfrac{x}{4} = 7$

4. $6u = 30$

In Exercises 5 and 6, justify each step of the solution.

5.
$$10x - 12 = 18$$
$$10x - 12 + 12 = 18 + 12$$
$$10x = 30$$
$$\dfrac{10x}{10} = \dfrac{30}{10}$$
$$x = 3$$

6.
$$\dfrac{t}{4} + \dfrac{t}{3} = 1$$
$$3t + 4t = 12$$
$$7t = 12$$
$$\dfrac{7t}{7} = \dfrac{12}{7}$$
$$t = \dfrac{12}{7}$$

In Exercises 7–26, solve the equation and check your solution.

7. $10x = 50$

8. $-3x = 21$

9. $8x + 7 = 39$

10. $12x - 5 = 43$

11. $24 - 7x = 3$

12. $13 + 6x = 61$

13. $15x - 4 = 16$

14. $3x - 8 = 2$

15. $3x - 2(x + 5) = 10$

16. $4x + 2(7 - x) = 5$

17. $2x + 3 = 2x - 2$

18. $8(x - 2) = 4(2x - 4)$

19. $\dfrac{x}{5} = 4$

20. $-\dfrac{x}{14} = \dfrac{1}{2}$

21. $\dfrac{2}{3}x - \dfrac{1}{6} = \dfrac{9}{2}$

22. $\dfrac{1}{8}x + \dfrac{3}{4} = \dfrac{5}{2}$

23. $\dfrac{x}{3} = \dfrac{1}{9}$

24. $\dfrac{4 - x}{8} = 7$

25. $\dfrac{u}{10} + \dfrac{u}{5} = 6$

26. $\dfrac{x}{3} + \dfrac{x}{5} = 1$

In Exercises 27–30, solve the equation. Round your result to two decimal places.

27. $516x - 875 = 3250$

28. $2.825x + 3.125 = 12.5$

29. $\dfrac{x}{4.625} = 48.5$

30. $5x + \dfrac{1}{4.5} = 18.125$

31. *Driving Distances*   On a 1200-mile trip, you drive about $1\frac{1}{2}$ times as much as your friend. Approximate the number of miles each of you drives.

32. *Hourly Wage*   Your hourly wage is $4.30 per hour plus 60 cents for each unit you produce. How many units must you produce in an hour so your hourly wage is $11.50?

In Exercises 33 and 34, complete the table.

33.

Percent	Parts out of 100	Decimal	Fraction
35%			

34.

Percent	Parts out of 100	Decimal	Fraction
			$\frac{4}{5}$

35. What is 125% of 16?

36. What is 0.8% of 3250?

37. $37\frac{1}{2}$% of what number is 150?

38. 95% of what number is 323?

39. 150 is what percent of 250?

40. 130.6 is what percent of 3265?

41. *Revenue*   The revenues for a corporation (in millions of dollars) in 1993 and 1994 were $4521.4 and $4679.0, respectively. Find the percent increase in revenue from 1993 to 1994.

42. *Price Increase*   The manufacturer's suggested retail price for a certain model car is $18,459. Estimate the price of a comparably equipped car for the next model year. Assume that car prices will increase by $4\frac{1}{2}\%$.

In Exercises 43–46, express the ratio as a fraction in reduced form.   (Use the same units for both quantities.)

43. Eighteen inches to 4 yards

44. One pint to 2 gallons

45. Two hours to 90 minutes

46. Four meters to 150 centimeters

In Exercises 47–50, solve the proportion.

47. $\dfrac{7}{16} = \dfrac{z}{8}$

48. $\dfrac{x}{12} = \dfrac{5}{4}$

49. $\dfrac{x+2}{4} = \dfrac{x-1}{3}$

50. $\dfrac{x-4}{1} = \dfrac{9}{4}$

In Exercises 51–54, use a proportion to solve the problem.

51. *Real Estate Taxes*   The tax on property with an assessed value of $75,000 is $1150. Find the tax on property with an assessed value of $110,000.

52. *Recipe Proportions*   One and one-half cups of milk are needed to make one batch of pudding. How much is required to make 3 batches?

53. *Map Distance*   The scale represents 100 miles on the map. Use the map to approximate the distance between St. Petersburg and Tallahassee.

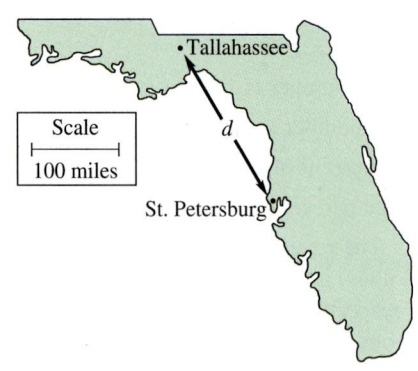

54. *Geometry*   Solve for the length x. Assume the two triangles are similar, and use the fact that corresponding sides of similar triangles are proportional.

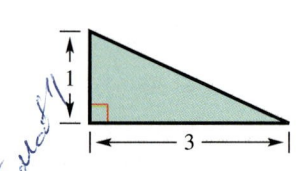 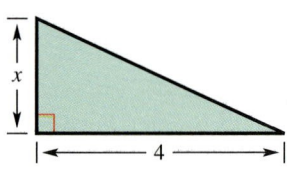

In Exercises 55–70, solve and graph the inequality.

55. $x + 5 \geq 7$

56. $x - 2 \leq 1$

57. $3x - 8 < 1$

58. $4x + 3 > 15$

59. $-11x \leq -22$

60. $-7x \geq 21$

61. $\frac{4}{5}x > 8$

62. $\frac{2}{3}n < -4$

63. $14 - \frac{1}{2}t < 12$

64. $32 + \frac{7}{8}k > 11$

65. $3(1 - y) \geq 2(4 + y)$

66. $4 - 3y \leq 8(10 - y)$

67. $-2 < \dfrac{x}{3} \leq 2$

68. $-5 \leq 3 - 4x < 5$

69. $3 > \dfrac{x+1}{-2} > 0$

70. $5 \geq \dfrac{x-3}{3} > 2$

In Exercises 71–76, write an inequality that represents the given statement.

71. z is at least 10.

72. x is nonnegative.

73. y is more than 8 but less than 12.

74. The area A is no more than 100 square feet.

75. The volume V is less than 12 cubic feet.

76. The perimeter P is at least 24 inches.

CHAPTER TEST

Take this test as you would take a test in class. After you are done, check your work against the answers given in the back of the book.

In Exercises 1–4, solve the equation and check your solution.

1. $4x - 3 = 18$

2. $10 - (2 - x) = 2x + 1$

3. $\dfrac{5x}{4} = \dfrac{5}{2} + x$

4. $\dfrac{t + 2}{3} = \dfrac{2t}{5}$

5. Solve $4.08(x + 10) = 9.50(x - 2)$. Round the result to two decimal places.

6. When the sum of a number and 6 is divided by 8, the result is 7. Find the number.

7. The bill (including parts and labor) for the repair of a home appliance is $134. The cost for parts is $62. How many hours were spent repairing the appliance if the cost of labor was $18 per hour?

8. Express the fraction $\frac{3}{8}$ as a percent and a decimal.

9. 324 is 27% of what number?

10. 90 is what percent of 250?

11. The price of a product increased by 20% during the past year. It is now selling for $240. What was the price 1 year ago?

12. Express the ratio of 40 inches to 2 yards as a fraction in reduced form. Use the same units for both quantities, and explain how you made this conversion.

In Exercises 13 and 14, solve the proportion.

13. $\dfrac{5}{8} = \dfrac{x}{12}$

14. $\dfrac{2x}{3} = \dfrac{x + 14}{5}$

15. On the map at the right, 1 centimeter represents 55 miles. Approximate the distance between the two indicated cities.

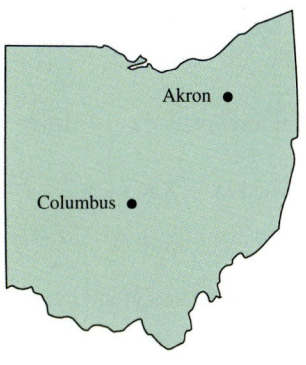

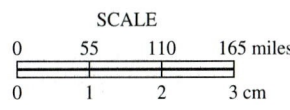

SCALE

| 0 | 55 | 110 | 165 miles |

| 0 | 1 | 2 | 3 cm |

Figure for 15

In Exercises 16–19, solve and graph the inequality.

16. $x + 3 \le 7$

17. $-\dfrac{2x}{3} > 4$

18. $-3 < 2x - 1 \le 3$

19. $2 \ge \dfrac{3 - x}{2} > -1$

20. Use inequality notation to express each phrase mathematically.

(a) y is no more than 10.

(b) t is greater than or equal to 10.

CUMULATIVE TEST: CHAPTERS P–3

Take this test as you would take a test in class. After you are done, check your work against the answers given in the back of the book.

1. Place the correct symbol ($<$ or $>$) between the numbers: $-\frac{3}{4}$ [   ] $\left|-\frac{7}{8}\right|$.

In Exercises 2–5, evaluate the expression.

2. $(-200)(2)(-3)$

3. $\frac{3}{8} - \frac{5}{6}$

4. $-\frac{2}{9} \div \frac{8}{75}$

5. $-(-2)^3$

6. Use exponential form to write the product $3 \cdot (x + y) \cdot (x + y) \cdot 3 \cdot 3$.

7. Use the Distributive Property to expand $-2x(x - 3)$.

8. Identify the rule of algebra illustrated by $2 + (3 + x) = (2 + 3) + x$.

In Exercises 9–11, simplify the expression.

9. $(3x^3)(5x^4)$

10. $(a^3b^2)(ab)^5$

11. $2x^2 - 3x + 5x^2 - (2 + 3x)$

In Exercises 12–14, solve the equation and check your solution.

12. $12x - 3 = 7x + 27$

13. $2x - \dfrac{5x}{4} = 13$

14. $2(x - 3) + 3 = 12 - x$

15. Solve and graph the inequality
$$-1 \le \frac{x + 3}{2} < 2.$$

16. The sticker on a new car gives the fuel efficiency as 28.3 miles per gallon. In your own words, explain how to estimate the annual fuel cost for the buyer if the car will be driven approximately 15,000 miles per year and the fuel cost is $1.179 per gallon.

17. Express the ratio "24 ounces to 2 pounds" as a fraction in reduced form.

18. The sum of two consecutive even integers is 494. Find the two numbers.

19. The suggested retail price of a camcorder is $1150. The camcorder is on sale for "20% off" the list price. Find the sale price.

20. The figure at the right shows two pieces of property. The assessed values of the properties are proportional to their areas. The value of the larger piece is $95,000. What is the value of the smaller piece?

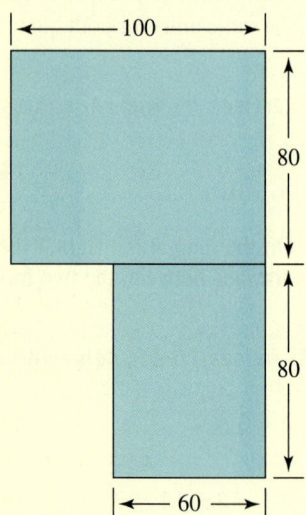

Figure for 20

Graphs and Linear Applications

4

- Ordered Pairs and Graphs
- Graphs of Equations
- Graphs and Graphing Utilities
- Business Applications and Graphs
- Formulas and Scientific Applications

An important factor in managing a successful business is being able to accurately predict future sales of a product. One way to do this is to consider how much money consumers spent on the product in previous years.

The table at the right shows the sales (in millions of dollars) of athletic and sporting equipment for the years 1987 through 1992. From the numbers in the table, you can see that the sales of athletic and sporting equipment increased each year.

Using the numbers in the table, a scatter plot can be created which represents the sales of athletic and sporting equipment. From the scatter plot, you can see that the smallest increase in sales occurred in 1991. (Source: National Sporting Goods Association)

Year	1987	1988	1989	1990	1991	1992
Sales	9,900	10,705	11,503	11,965	11,982	12,416

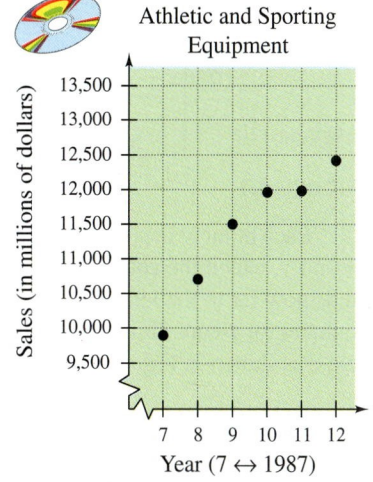

The chapter project related to this data is on page 268.

4.1 Ordered Pairs and Graphs

The Rectangular Coordinate System ▪
Ordered Pairs as Solutions of Equations ▪ Application

The Rectangular Coordinate System

Just as you can represent real numbers by points on the real number line, you can represent ordered pairs of real numbers by points in a plane. This plane is called a **rectangular coordinate system** or the **Cartesian plane,** after the French mathematician René Descartes (1596–1650).

A rectangular coordinate system is formed by two real lines intersecting at right angles, as shown in Figure 4.1. The horizontal number line is usually called the *x*-**axis** and the vertical number line is usually called the **y-axis.** (The plural of axis is *axes*.) The point of intersection of the two axes is the **origin,** and the axes separate the plane into four regions called **quadrants.**

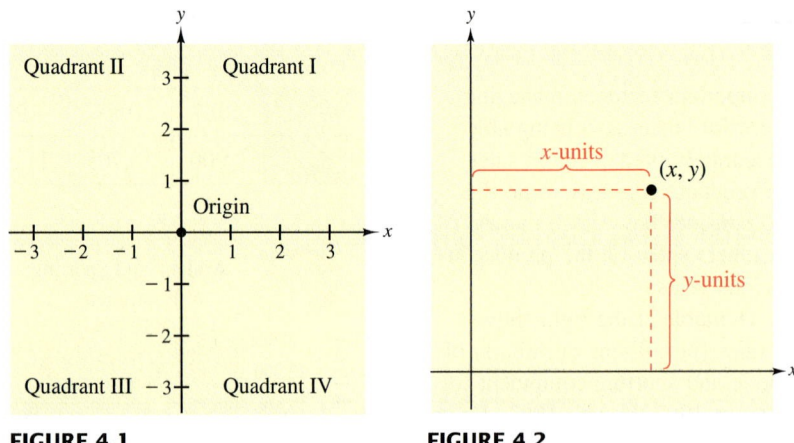

FIGURE 4.1   **FIGURE 4.2**

Each point in the plane corresponds to an **ordered pair** (x, y) of real numbers x and y, called the **coordinates** of the point. The first number (or *x*-**coordinate**) tells how far to the left or right the point is from the vertical axis, and the second number (or **y-coordinate**) tells how far up or down the point is from the horizontal axis, as shown in Figure 4.2.

> **NOTE** A positive x-coordinate implies that the point lies to the *right* of the vertical axis; a negative x-coordinate implies that the point lies to the *left* of the vertical axis; and an x-coordinate of zero implies that the point lies *on* the vertical axis. Similar statements can be made about y-coordinates. For instance, a positive y-coordinate implies that the point lies *above* the horizontal axis.

Locating a point in a plane is called **plotting** the point. This procedure is demonstrated in Example 1.

EXAMPLE 1   Plotting Points on a Rectangular Coordinate System

Plot the points $(-1, 2)$, $(3, 0)$, $(2, -1)$, $(3, 4)$, $(0, 0)$, and $(-2, -3)$ on a rectangular coordinate system.

Solution

The point $(-1, 2)$ is one unit to the *left* of the vertical axis and two units *above* the horizontal axis.

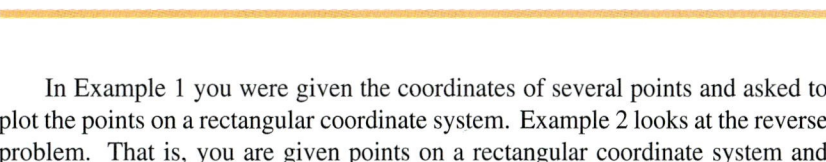

Similarly, the point $(3, 0)$ is three units to the right of the vertical axis and *on* the horizontal axis. (It is on the horizontal axis because the y-coordinate is zero.) The other four points can be plotted in a similar way, as shown in Figure 4.3.

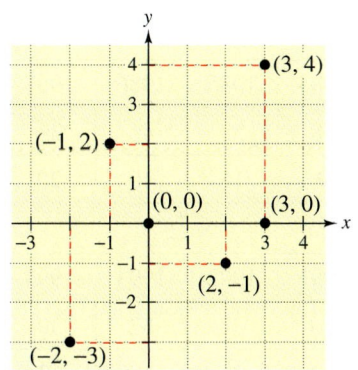

FIGURE 4.3

In Example 1 you were given the coordinates of several points and asked to plot the points on a rectangular coordinate system. Example 2 looks at the reverse problem. That is, you are given points on a rectangular coordinate system and asked to determine their coordinates.

EXAMPLE 2   Finding Coordinates of Points

Determine the coordinates for each of the points shown in Figure 4.4.

Solution

Point A lies 3 units to the *left* of the vertical axis, and 2 units *above* the horizontal axis. Therefore, point A must be given by the ordered pair $(-3, 2)$. The coordinates of the other four points can be determined in a similar way, and we summarize the results as follows.

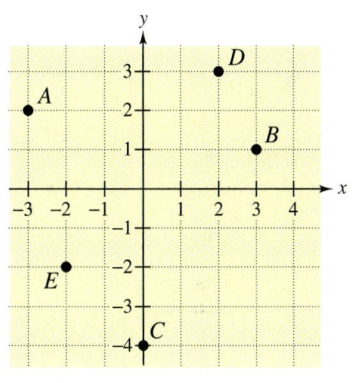

FIGURE 4.4

Point	Position	Coordinates
A	3 units *left*, 2 units *up*	$(-3, 2)$
B	3 units *right*, 1 unit *up*	$(3, 1)$
C	0 units *left*(or *right*), 4 units *down*	$(0, -4)$
D	2 units *right*, 3 units *up*	$(2, 3)$
E	2 units *left*, 2 units *down*	$(-2, -2)$

Each year since 1967, the winners of the American Football Conference and National Football Conference have played in the Super Bowl. The first Super Bowl was played between the Green Bay Packers and the Kansas City Chiefs.

EXAMPLE 3 *Super Bowl Scores*

The scores of the winning and losing football teams for the Super Bowl games from 1981 through 1994 are given in the table below. Plot these points on a rectangular coordinate system. (Source: National Football League)

Year	1981	1982	1983	1984	1985	1986	1987
Winning Score	27	26	27	38	38	46	39
Losing Score	10	21	17	9	16	10	20

Year	1988	1989	1990	1991	1992	1993	1994
Winning Score	42	20	55	20	37	52	30
Losing Score	10	16	10	19	24	17	13

Solution

Plot the years on the x-axis and the winning and losing scores on the y-axis. In Figure 4.5, the winning scores are shown as black dots, and the losing scores are shown as blue dots. Note that the break in the x-axis indicates that the numbers between 0 and 1981 have been omitted.

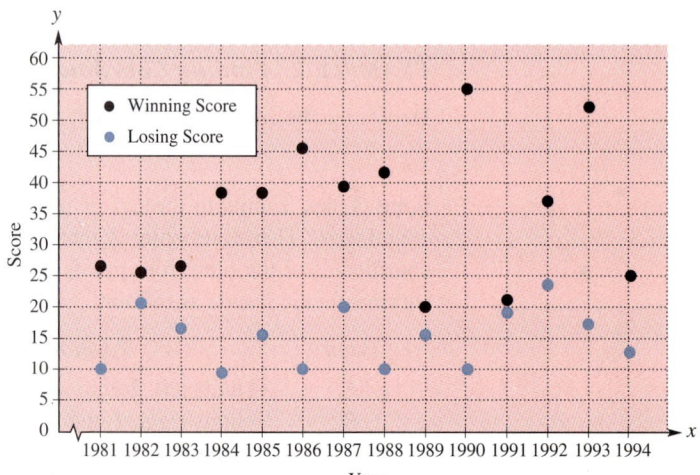

FIGURE 4.5

Ordered Pairs as Solutions of Equations

In Example 3, the relationship between the year and the Super Bowl scores was given by a **table of values.** In mathematics, the relationship between the variables x and y is often given by an equation. From the equation, you must then construct your own table of values. For instance, consider the equation

$$y = 2x + 1.$$

To construct a table of values for this equation, choose several x-values and then calculate the corresponding y-values. For example, if you choose $x = 1$, the corresponding y-value is

$$y = 2(1) + 1 = 3.$$

The corresponding ordered pair $(x, y) = (1, 3)$ is a **solution point** (or simply a **solution**) of the equation. The table below is a table of values (and the corresponding solution points) using x-values of $-3, -2, -1, 0, 1, 2,$ and 3.

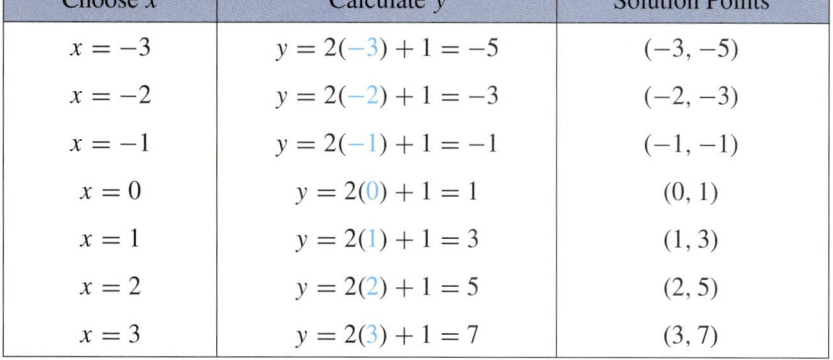

Choose x	Calculate y	Solution Points
$x = -3$	$y = 2(-3) + 1 = -5$	$(-3, -5)$
$x = -2$	$y = 2(-2) + 1 = -3$	$(-2, -3)$
$x = -1$	$y = 2(-1) + 1 = -1$	$(-1, -1)$
$x = 0$	$y = 2(0) + 1 = 1$	$(0, 1)$
$x = 1$	$y = 2(1) + 1 = 3$	$(1, 3)$
$x = 2$	$y = 2(2) + 1 = 5$	$(2, 5)$
$x = 3$	$y = 2(3) + 1 = 7$	$(3, 7)$

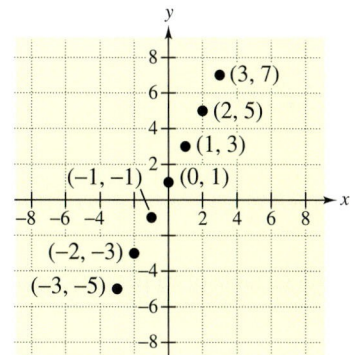

FIGURE 4.6

Once you have constructed a table of values, you can get a visual idea of the relationship between the variables x and y by plotting the solution points on a rectangular coordinate system. For instance, the solution points shown in the table are plotted in Figure 4.6.

NOTE   In many places throughout this course, you will see that approaching a problem in different ways can help you understand the problem better. For instance, the above discussion looks at solutions of an equation in three ways.

1. **Algebraic Approach** Use algebra to find several solutions.

2. **Numerical Approach** Construct a table that shows several solutions.

3. **Graphical Approach** Draw a graph that shows several solutions.

When making up a table of values for an equation, it is helpful to first solve the equation for y. For instance, the equation $4x + 2y = 7$ can be solved for y as follows.

$4x + 2y = 7$	Original equation
$4x - 4x + 2y = 7 - 4x$	Subtract $4x$ from both sides.
$2y = 7 - 4x$	Combine like terms.
$\dfrac{2y}{2} = \dfrac{7 - 4x}{2}$	Divide both sides by 2.
$y = \dfrac{7}{2} - 2x$	Simplify.

This procedure is further demonstrated in Example 4.

EXAMPLE 4 *Making up a Table of Values*

Make up a table of values showing five solution points for the equation

$$6x - 2y = 4.$$

Then plot the solution points on a rectangular coordinate system. (Choose x-values of $-2, -1, 0, 1,$ and 2.)

Solution

$6x - 2y = 4$	Original equation
$6x - 6x - 2y = 4 - 6x$	Subtract $6x$ from both sides.
$-2y = -6x + 4$	Combine like terms.
$\dfrac{-2y}{-2} = \dfrac{-6x + 4}{-2}$	Divide both sides by -2.
$y = 3x - 2$	Simplify.

Now, using the equation $y = 3x - 2$, you can construct a table of values, as shown below.

x	-2	-1	0	1	2
$y = 3x - 2$	-8	-5	-2	1	4
Solution Points	$(-2, -8)$	$(-1, -5)$	$(0, -2)$	$(1, 1)$	$(2, 4)$

Finally, from the table you can plot the five solution points on a rectangular coordinate system, as shown in Figure 4.7.

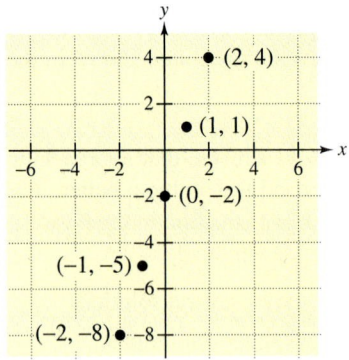

FIGURE 4.7

To determine whether an ordered pair is a solution of an equation, use the following steps.

1. Substitute the values of x and y into the equation.
2. Simplify both sides of the equation.
3. If both sides contain the same number, the ordered pair is a solution. If both sides contain different numbers, the ordered pair is not a solution.

EXAMPLE 5   *Verifying Solutions of an Equation*

Determine which of the following ordered pairs is a solution of $x + 3y = 6$.

a. $(1, 2)$ **b.** $\left(-2, \frac{8}{3}\right)$ **c.** $(-6, 0)$ **d.** $(0, 2)$ **e.** $\left(\frac{8}{3}, \frac{10}{9}\right)$

Solution

a. For the ordered pair $(x, y) = (1, 2)$, substitute $x = 1$ and $y = 2$ into the original equation.

$$x + 3y = 6 \qquad \text{Original equation}$$
$$1 + 3(2) \stackrel{?}{=} 6 \qquad \text{Substitute } x = 1 \text{ and } y = 2.$$
$$7 \neq 6 \qquad \text{Simplify.}$$

Because the substitution did not satisfy the original equation, the ordered pair $(1, 2)$ *is not* a solution of the equation.

b. For the ordered pair $(x, y) = \left(-2, \frac{8}{3}\right)$, substitute $x = -2$ and $y = \frac{8}{3}$ into the original equation.

$$x + 3y = 6 \qquad \text{Original equation}$$
$$(-2) + 3\left(\frac{8}{3}\right) \stackrel{?}{=} 6 \qquad \text{Substitute } x = -2 \text{ and } y = \frac{8}{3}.$$
$$-2 + 8 = 6 \qquad \text{Simplify.}$$

Because the substitution satisfies the original equation, the ordered pair $\left(-2, \frac{8}{3}\right)$ *is* a solution of the equation.

c. The ordered pair $(-6, 0)$ *is not* a solution of the original equation because

$$-6 + 3(0) = -6 \neq 6.$$

d. The ordered pair $(0, 2)$ *is* a solution of the original equation because

$$0 + 3(2) = 6.$$

e. The ordered pair $\left(\frac{8}{3}, \frac{10}{9}\right)$ *is* a solution of the original equation because

$$\frac{8}{3} + 3\left(\frac{10}{9}\right) = \frac{8}{3} + \frac{10}{3} = \frac{18}{3} = 6.$$

Application

EXAMPLE 6 A Business Application

You are setting up a small business to assemble computer keyboards. Your initial investment is $120,000, and your unit cost to assemble each keyboard is $40. Write an equation that relates your total cost to the number of keyboards produced. Then plot the total costs of producing 1000, 2000, 3000, 4000, and 5000 keyboards.

Solution

Let x represent the number of keyboards and C represent the total cost of producing x keyboards. The total cost equation must represent both the initial cost and the unit cost. For instance, to produce 1 or 2 keyboards, your total cost is as follows.

$$C = 40(1) + 120,000 = \$120,040$$
$$C = 40(2) + 120,000 = \$120,080$$

From this pattern, you can see that the total cost of producing x keyboards is given by the equation

$$C = 40x + 120,000.$$

Using this equation, construct a table of values. Then, plot the ordered pairs given in your table, as shown in Figure 4.8.

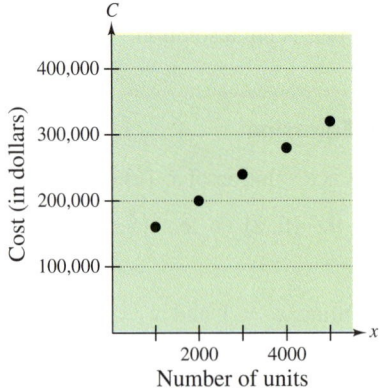

FIGURE 4.8

Group Activities   Extending the Concept

Misleading Graphs   Although graphs can help us visualize relationships between two variables, they can also be misleading. The graphs shown below represent the same data points. Which graph is misleading? Why?

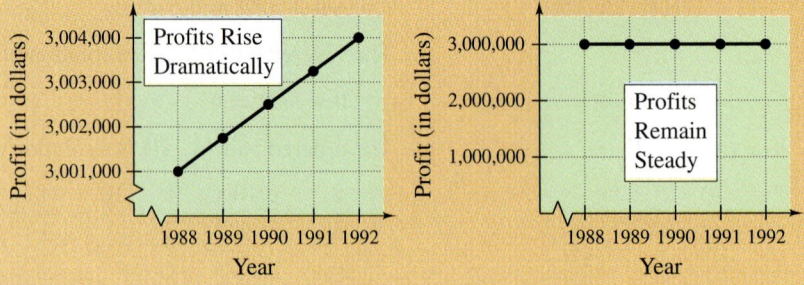

4.1 Exercises

Discussing the Concepts

1. Discuss the significance of the word "ordered" when referring to an ordered pair (x, y).

2. When the point (x, y) is plotted, what does the x-coordinate measure? What does the y-coordinate measure?

3. What is the x-coordinate of any point on the y-axis? What is the y-coordinate of any point on the x-axis?

4. Describe the signs of the x- and y-coordinates for points that lie in the first and second quadrants.

5. Describe the signs of the x- and y-coordinates for points that lie in the third and fourth quadrants.

6. In a rectangular coordinate system, must the scales on the x-axis and y-axis be the same? If not, give an example in which the scales differ.

Problem Solving

In Exercises 7–10, plot the points on a rectangular coordinate system.

7. $(3, 2), (-4, 2), (2, -4)$

8. $(-1, 6), (-1, -6), (4, 6)$

9. $\left(\frac{3}{2}, -1\right), \left(-3, \frac{3}{4}\right), \left(\frac{1}{2}, -\frac{1}{2}\right)$

10. $\left(-\frac{2}{3}, 4\right), \left(\frac{1}{2}, -\frac{5}{2}\right), \left(-4, -\frac{5}{4}\right)$

Graphical Estimation In Exercises 11 and 12, estimate the coordinates of the points.

11.
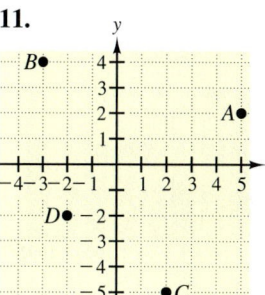

12.
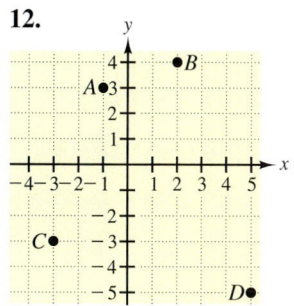

Geometry In Exercises 13–16, plot the points and connect them with line segments to form the figure.

13. Triangle: $(-1, 1), (2, -1), (3, 4)$

14. Rectangle: $(2, 1), (4, 2), (-1, 7), (1, 8)$

15. Parallelogram: $(5, 2), (7, 0), (1, -2), (-1, 0)$

16. Rhombus: $(0, 0), (1, 2), (2, 1), (3, 3)$

In Exercises 17–20, determine the quadrant in which the point is located.

17. $\left(-3, \frac{1}{8}\right)$

18. $(4.2, -3.05)$

19. $(-100, -365.6)$

20. $\left(\frac{3}{11}, \frac{7}{8}\right)$

In Exercises 21 and 22, determine the quadrant or quadrants in which the point must be located.

21. $(-5, y)$, y is a real number

22. $(x, 3)$, x is a real number

23. *Organizing Data* The table gives the normal temperature y (Fahrenheit) in Anchorage, Alaska for each month of the year. The months are numbered 1 through 12, with 1 corresponding to January. (Source: National Oceanic and Atmospheric Administration)

x	1	2	3	4	5	6
y	13	18	24	35	46	54

x	7	8	9	10	11	12
y	58	56	48	35	22	14

(a) Plot the data given in the table.

(b) Did you use the same scale on both axes? Explain.

(c) Using the graph, find the three consecutive months when the normal temperature changes the least.

24. *Organizing Data* The table gives the speed of a car x (in kilometers per hour) and the approximate stopping distance y in meters.

x	50	70	90	110	130
y	20	35	60	95	148

(a) Plot the data given in the table.

(b) The x-coordinates increase in equal increments of 20 kilometers per hour. Describe the pattern for the y-coordinates. What are the implications for the driver?

In Exercises 25–28, decide whether the ordered pairs are solutions of the equation.

25. $y = 2x + 4$ (a) $(3, 10)$ (b) $(-1, 3)$
                   (c) $(0, 0)$ (d) $(-2, 0)$

26. $x - 8y = -100$ (a) $(10, -10)$ (b) $(-60, 5)$
                      (c) $(100, 0)$ (d) $(20, -10)$

27. $y = \frac{2}{3}x$ (a) $(6, 6)$ (b) $(-9, -6)$
            (c) $(0, 0)$ (d) $\left(-1, \frac{2}{3}\right)$

28. $y = \frac{5}{8}x - 2$ (a) $(8, 3)$ (b) $(0, 0)$
                (c) $(-16, -12)$ (d) $(32, 18)$

In Exercises 29–32, complete the table. Plot the resulting data on a rectangular coordinate system.

29.

x	-2	0	2	4	6
$y = 3x - 4$					

30.

x	-2	0	2	4	6
$y = \frac{1}{4}x + 1$					

31.

x	-5	$\frac{3}{2}$	5	10	20
$y = -\frac{3}{2}x + 5$					

32.

x	-2	-1	0	$\frac{1}{2}$	$\frac{5}{4}$
$y = -\frac{1}{2}x + \frac{5}{4}$					

33. *Think About It* Review the tables in Exercises 29–32 and observe that in some cases the y-coordinates of the solution points increase and in others the y-coordinates decrease. What factor in the equation causes this? Explain.

34. *Hourly Wages* When an employee produces x units per hour, the hourly wage y is $y = 0.50x + 10$. Complete the table to determine the hourly wages for producing the specified number of units. Plot your results on a rectangular coordinate system.

x	2	5	8	10	20
$y = 0.50x + 10$					

Graphical Estimation In Exercises 35–38, use the bar graph showing new housing starts (in thousands) in the United States from 1980 through 1992. (Source: U.S. Bureau of the Census)

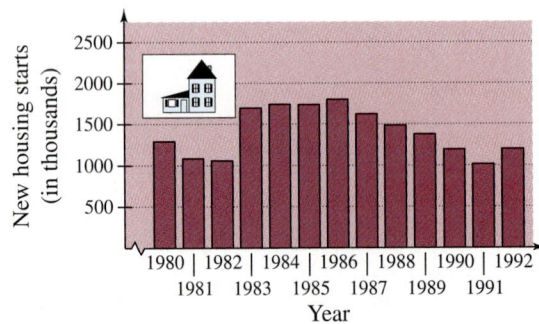

35. Estimate the number of new housing starts in 1986.

36. Estimate the number of new housing starts in 1990.

37. Estimate the increase and the percent increase in housing starts from 1982 to 1983.

38. Estimate the decrease and the percent decrease in housing starts from 1990 to 1991.

Reviewing the Major Concepts

In Exercises 39–42, solve the equation.

39. $125(r - 1) = 625$ **40.** $2(3 - y) = 7y + 5$

41. $20 - \frac{1}{9}x = 4$ **42.** $0.35x = 70$

43. *Housing Costs*   The total cost of a lot and house is $154,000. The cost of constructing the house is seven times the cost of the lot. What is the cost of the lot?

44. *Summer Jobs*   You have two summer jobs. In the first job, you work 40 hours a week and earn $7.50 an hour. In the second job, you work as many hours as you want and earn $6.25 an hour. If you plan to earn $375 a week, how many hours a week should you work at the second job?

Additional Problem Solving

In Exercises 45–48, plot the points on a rectangular coordinate system.

45. $(-10, -4), (4, -4), (4, 3)$

46. $(-6, 4), (0, 0), (3, -2)$

47. $(3, -4), (5, 0), (0, 3)$

48. $\left(\frac{5}{2}, 2\right), \left(-3, \frac{4}{3}\right), \left(\frac{3}{4}, \frac{9}{4}\right)$

Graphical Estimation   In Exercises 49 and 50, estimate the coordinates of the points.

49.

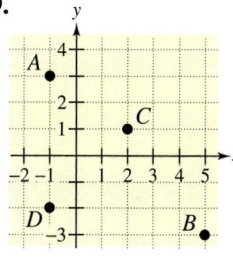

50.

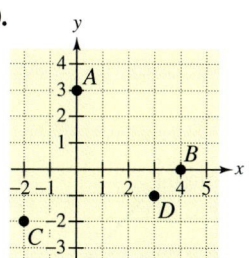

Geometry   In Exercises 51–54, plot the points and connect them with line segments to form the figure.

51. Square: $(2, 4), (5, 1), (2, -2), (-1, 1)$

52. Triangle: $(0, 3), (-1, -2), (4, 8)$

53. Rhombus: $(0, 0), (3, 2), (2, 3), (5, 5)$

54. Parallelogram: $(-1, 1), (0, 4), (4, -2), (5, 1)$

55. *Graphical Interpretation*   The table gives the hours x that a student studied for five different algebra exams and the resulting scores y.

x	3.5	1	8	4.5	0.5
y	72	67	95	81	53

(a) Plot the data given in the table.

(b) Use the graph to describe the relationship between the number of hours studied and the resulting exam score.

56. *Graphical Interpretation*   The table gives the net income y per share of common stock of the H. J. Heinz Company for 1984 through 1993. The year is represented by t.   (Source: H. J. Heinz Company)

t	1984	1985	1986	1987	1988
y	$0.85	$0.96	$1.10	$1.24	$1.45

t	1989	1990	1991	1992	1993
y	$1.67	$1.90	$2.13	$2.40	$1.53

(a) Plot the data given in the table.

(b) Use the graph to find the year that had the greatest increase and the year that had the greatest decrease in the income per share.

In Exercises 57–60, decide whether the ordered pairs are solutions of the equation.

57. $y = 3 - 4x$ (a) $(-1, 7)$ (b) $(2, 5)$
                         (c) $(0, 0)$ (d) $(2, -5)$

58. $y = 5x - 2$ (a) $(2, 0)$ (b) $(-2, -12)$
                         (c) $(6, 28)$ (d) $(1, 1)$

59. $2y - 3x + 1 = 0$ (a) $(1, 1)$ (b) $(5, 7)$
                         (c) $(-3, -1)$ (d) $\left(-2, -\frac{7}{2}\right)$

60. $y = \frac{3}{2}x + 1$ (a) $\left(0, \frac{3}{2}\right)$ (b) $(4, 7)$
                         (c) $\left(\frac{2}{3}, 2\right)$ (d) $(-2, -2)$

In Exercises 61–64, complete the table. Plot the results on a rectangular coordinate system.

61.

x	-2	-1	0	1	2
$y = 2x - 1$					

62.

x	-2	-1	0	1	2
$y = |x|$					

63.

x	-2	-1	0	1	2
$y = 4 - x^2$					

64.

x	-2	0	$\frac{1}{2}$	2	4
$y = -\frac{7}{2}x + 3$					

65. *Organizing Data*   The cost y of producing x units of a product is given by $y = 35x + 5000$. Complete the table to determine the cost for producing the specified number of units. Plot the results on a rectangular coordinate system.

x	100	150	200	250	300
$y = 35x + 5000$					

66. *Organizing Data*   Your company buys a new copier for \$9500. Its value y after x years is given by $y = -800x + 9500$. Complete the table to determine the value of the copier at the specified times. Plot the results on a rectangular coordinate system.

x	0	2	4	6	8
$y = -800x + 9500$					

Graphical Comparisons   In Exercises 67 and 68, the graphs of two types of depreciation are shown. In one type, called *straight-line depreciation*, the value depreciates by the same amount each year. In the other type, called *declining balances*, the value depreciates by the same percent each year. Which is which?

67.

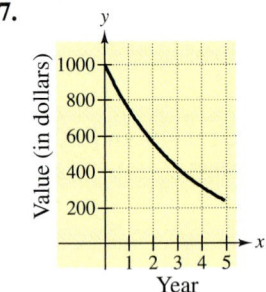

68.

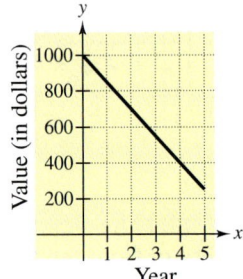

69. *Graphical Interpretation*   In Exercises 67 and 68, what is the original cost of the equipment that is being depreciated?

70. *Writing*   Compare the benefits and disadvantages of the two types of depreciation shown in Exercises 67 and 68.

Graphical Estimation   In Exercises 71 and 72, use the bar graph showing the per capita personal income in the United States from 1984 through 1992. (Source: U.S. Bureau of Economic Analysis)

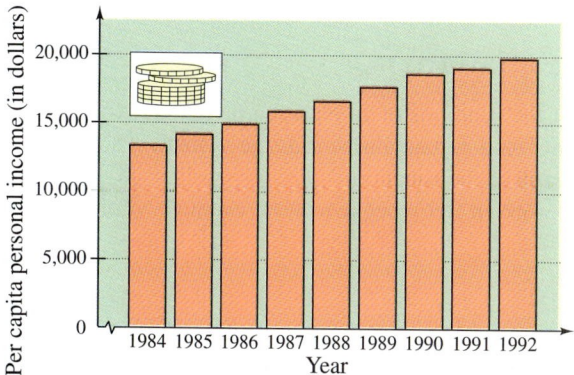

71. Estimate the increase in per capita personal income from 1984 to 1989.

72. Estimate the percent increase in per capita personal income from 1991 to 1992.

Graphical Estimation   In Exercises 73 and 74, use the bar graph, which compares the percents of gross domestic product spent on health care in selected countries in 1991. (Source: Organization for Economic Cooperation and Development)

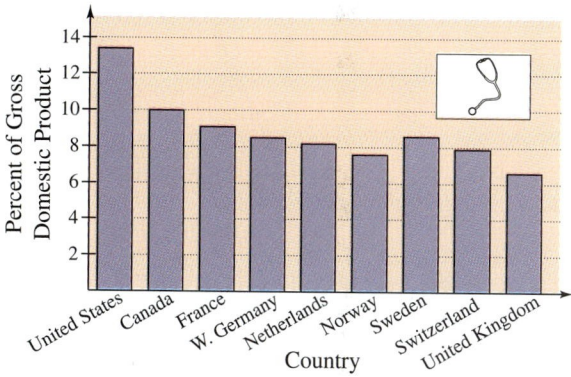

73. Estimate the percent of gross domestic product spent on health care in Sweden.

74. Estimate the percent of gross domestic product spent on health care in the United States.

75. (a) Plot the points $(3, 2)$, $(-5, 4)$, and $(6, -4)$ on a rectangular coordinate system.

 (b) Change the sign of the x-coordinate of each point and plot the three new points on the same axes.

 (c) What can you infer about the location of a point when the sign of the x-coordinate is changed?

76. (a) Plot the points $(3, 2)$, $(-5, 4)$, and $(6, -4)$ on a rectangular coordinate system.

 (b) Change the sign of the y-coordinate of each point and plot the three new points on the same axes.

 (c) What can you infer about the location of a point when the sign of the y-coordinate is changed?

CAREER INTERVIEW

Jeremy Reiner

Chief Meteorologist

KAAL—TV6

Austin, MN 55912

At KAAL, I am responsible for formulating weather forecasts (for up to five days in advance) and delivering them in three news broadcasts each day. In my job I act as scientist while formulating the forecast, as artist while creating weather maps and graphics for the broadcast, and as actor while delivering the forecast on the air.

I routinely use algebra as a tool while formulating my forecasts, from computing equivalent rainfall amounts for given snowfall depths to predicting how long a storm system will take to reach our viewing area. I also use my knowledge of algebra in finding *wind chill equivalent temperatures*. To do this, I use a coordinate system chart showing temperature on the y-axis and wind speed on the x-axis. To find the wind chill equivalent temperature for, say, $-10°$F and a 30-mph wind speed, all I have to do is locate the coordinate point $(30, -10)$ on the chart and read off the wind chill factor listed at that point, which in this case is $-70°$F.

4.2   Graphs of Equations

The Graph of an Equation ▪ Intercepts: An Aid to Sketching Graphs ▪ Application

The Graph of an Equation

You have already seen that the solutions of an equation involving two variables can be represented by points on a rectangular coordinate system. The set of all such points is called the **graph** of the equation.

To see how to sketch a graph, let's begin with an example. For instance, consider the equation

$$y = 2x - 1.$$

To begin sketching the graph of this equation, construct a table of values, as shown at the left. Next, plot the solution points on a rectangular coordinate system, as shown in Figure 4.9(a). Finally, find a pattern for the plotted points and use the pattern to connect the points with a smooth curve or line, as shown in Figure 4.9(b).

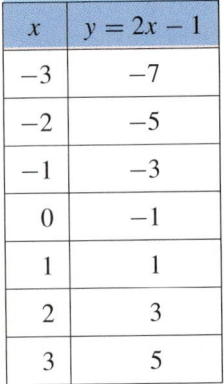

x	$y = 2x - 1$
-3	-7
-2	-5
-1	-3
0	-1
1	1
2	3
3	5

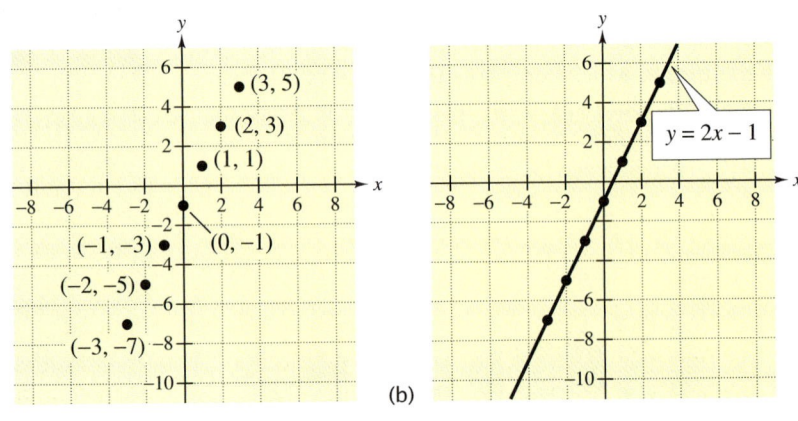

FIGURE 4.9   (a)   (b)

The Point-Plotting Method of Sketching a Graph

1. If possible, rewrite the equation by isolating one of the variables.

2. Make up a table of values showing several solution points.

3. Plot these points on a rectangular coordinate system.

4. Connect the points with a smooth curve or line.

EXAMPLE 1 *Sketching the Graph of an Equation*

Sketch the graph of $3x + y = 5$.

Solution

To begin, rewrite the equation so that y is isolated on the left.

$3x + y = 5$ Original equation

$3x - 3x + y = -3x + 5$ Subtract $3x$ from both sides.

$y = -3x + 5$ Simplify.

Next, create a table of values, as shown below.

x	-2	-1	0	1	2	3
$y = -3x + 5$	11	8	5	2	-1	-4
Solution	$(-2, 11)$	$(-1, 8)$	$(0, 5)$	$(1, 2)$	$(2, -1)$	$(3, -4)$

Plot the six solution points, as shown in Figure 4.10(a). It appears that all six points lie on a line, so you can complete the sketch by drawing a line through the six points, as shown in Figure 4.10(b).

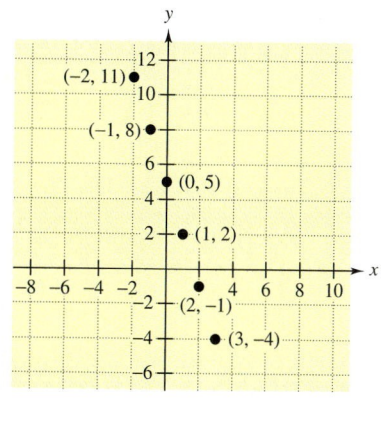

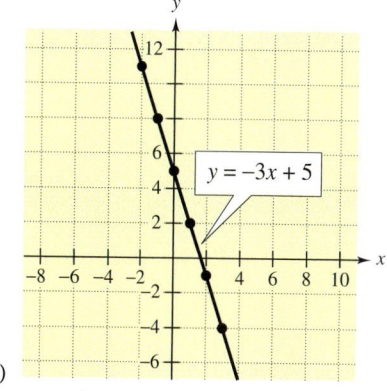

FIGURE 4.10   (a)   (b)

> **STUDY TIP**
>
> When you are creating a table of values, you are generally free to choose any x-values. When doing this, however, remember that the more x-values you choose, the easier it will be to recognize a pattern.

The equation in Example 1 is an example of a **linear equation** in two variables—the variables are raised to the first power and its graph is a straight line. As shown in the next two examples, graphs of nonlinear equations are not straight lines.

EXAMPLE 2   Sketching the Graph of a Nonlinear Equation

Sketch a graph of $x^2 + y = 4$.

Solution

To begin, rewrite the equation so that y is isolated on the left.

$$x^2 + y = 4 \qquad \text{Original equation}$$
$$x^2 - x^2 + y = -x^2 + 4 \qquad \text{Subtract } x^2 \text{ from both sides.}$$
$$y = -x^2 + 4 \qquad \text{Simplify.}$$

Next, create a table of values, as shown below. Be careful with the signs of the numbers when creating a table. For instance, when $x = -3$, the value of y is

$$y = -(-3)^2 + 4 = -9 + 4 = -5.$$

x	-3	-2	-1	0	1	2	3
$y = -x^2 + 4$	-5	0	3	4	3	0	-5
Solution	$(-3, -5)$	$(-2, 0)$	$(-1, 3)$	$(0, 4)$	$(1, 3)$	$(2, 0)$	$(3, -5)$

Plot the seven solution points, as shown in Figure 4.11(a). Finally, connect the points with a smooth curve, as shown in Figure 4.11(b).

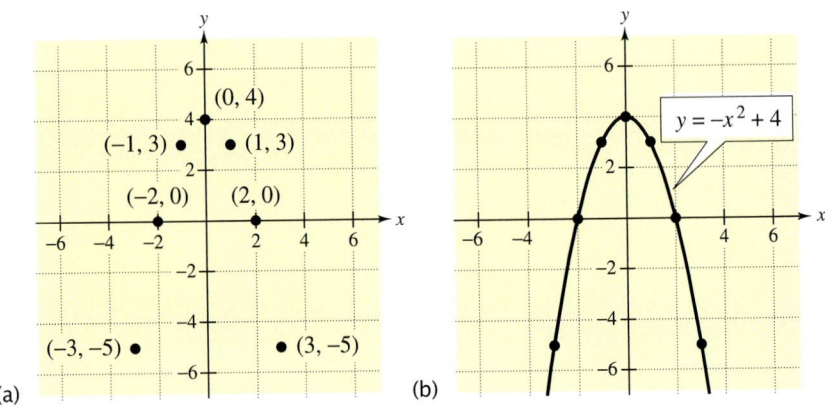

(a)   (b)

FIGURE 4.11

The graph of the equation in Example 2 is called a **parabola.** You will study this type of graph in detail in Chapter 11.

Example 3 examines the graph of an equation that involves an absolute value. Remember that to find the absolute value of a number, you disregard the sign of the number. For instance, $|-5| = 5$, $|2| = 2$, and $|0| = 0$.

EXAMPLE 3   The Graph of an Equation Having an Absolute Value

Sketch the graph of $y = |x - 1|$.

Solution

This equation is already written in a form with y isolated on the left. You can begin by creating a table of values, as shown below. Be sure to check the values in this table to make sure that you understand how the absolute value is working. For instance, when $x = -2$, the value of y is

$$y = |-2 - 1| = |-3| = 3.$$

Similarly, when $x = 2$, the value of y is $|2 - 1|$ or 1.

x	-2	-1	0	1	2	3	4
$y = |x - 1|$	3	2	1	0	1	2	3
Solution	$(-2, 3)$	$(-1, 2)$	$(0, 1)$	$(1, 0)$	$(2, 1)$	$(3, 2)$	$(4, 3)$

Plot the seven solution points, as shown in Figure 4.12(a). It appears that the points lie in a "V-shaped" pattern, with the point $(1, 0)$ lying at the bottom of the "V." Following this pattern, you can connect the points to form the graph shown in Figure 4.12(b).

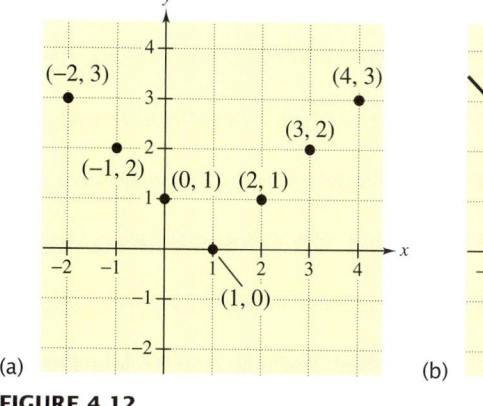

(a)

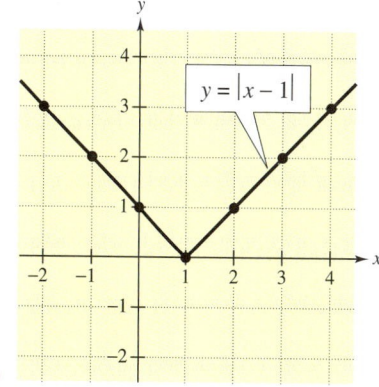

(b)

FIGURE 4.12

Intercepts: An Aid to Sketching Graphs

Two types of solution points that are especially useful are those having zero as either the x- or y-coordinate. Such points are called **intercepts** because they are the points at which the graph intersects the x- or y-axis.

Definition of Intercepts

The point $(a, 0)$ is called an **x-intercept** of the graph of an equation if it is a solution point of the equation. To find the x-intercept(s), let y be zero and solve the equation for x.

The point $(0, b)$ is called a **y-intercept** of the graph of an equation if it is a solution point of the equation. To find the y-intercept(s), let x be zero and solve the equation for y.

EXAMPLE 4   Finding the Intercepts of a Graph

Find the intercepts and sketch the graph of $y = 2x - 5$.

Solution

To find any x-intercepts, let $y = 0$ and solve the resulting equation for x.

$$y = 2x - 5 \qquad \text{Original equation}$$

$$0 = 2x - 5 \qquad \text{Let } y = 0.$$

$$\frac{5}{2} = x \qquad \text{Solve equation for } x.$$

Therefore, the graph has one x-intercept, which occurs at the point $\left(\frac{5}{2}, 0\right)$. To find any y-intercepts, let $x = 0$ and solve the resulting equation for y.

$$y = 2x - 5 \qquad \text{Original equation}$$

$$y = 2(0) - 5 \qquad \text{Let } x = 0.$$

$$y = -5 \qquad \text{Solve equation for } y.$$

Therefore, the graph has one y-intercept, which occurs at the point $(0, -5)$. To sketch the graph of the equation, create a table of values. Then plot the points and connect the points with a line, as shown in Figure 4.13.

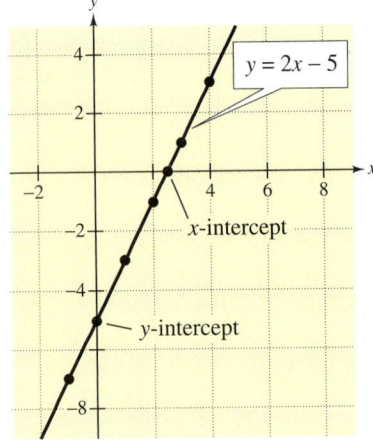

FIGURE 4.13

NOTE   When you create a table of values, we suggest that you include any intercepts you have found. We also suggest that you include points to the left and to the right of the intercepts.

Application

EXAMPLE 5   Depreciation

You are depreciating a $25,500 van over 10 years. At the end of the 10 years, the salvage value is expected to be $1500. Find an equation that relates the value of the van to the number of years. Then sketch the graph of the equation. (The depreciation is the same each year.)

Solution

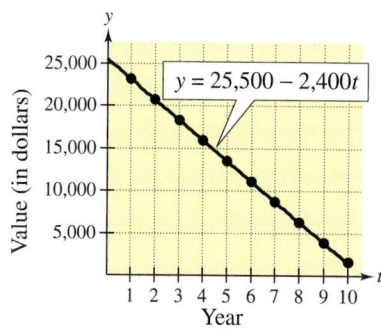

FIGURE 4.14

The total depreciation over the 10 years is $25,500 - 1500 = \$24,000$. Because the same amount is depreciated each year, it follows that the annual depreciation is $24,000/10 = \$2400$. Thus, after 1 year, the value of the van is

Value after 1 year $= 25,500 - 2400 = \$23,100.$

By similar reasoning, you can see that the value after 2 years is $25,500 - 2400(2)$, or $20,700. By letting y represent the value of the van after t years, you can use the pattern given by the first two years and conclude that the equation is

$$y = 25,500 - 2400t.$$

A sketch of the graph of this equation is shown in Figure 4.14.

Group Activities   Extending the Concept

Interpreting a Graph   The graph below shows the speed of a delivery van on a 22-minute trip. Write a short paragraph describing the trip.

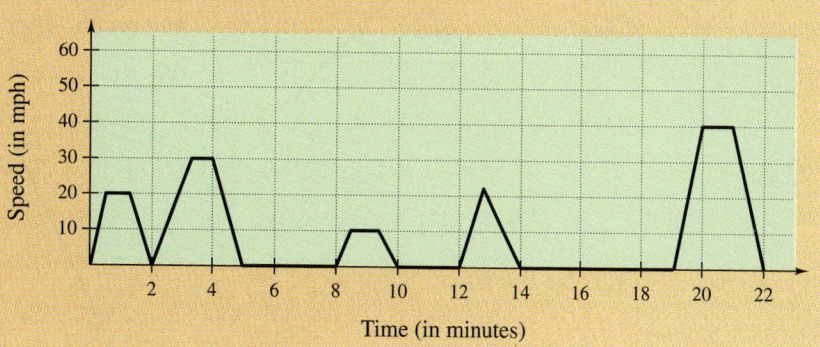

4.2 Exercises

Discussing the Concepts

1. In your own words, define what is meant by the *graph* of an equation.

2. How many solution points can an equation have? How many points do you need to plot to be able to find the general shape of the graph?

3. In your own words, describe the point-plotting method of sketching the graph of an equation.

4. In your own words, describe how you can check that an ordered pair (x, y) is a solution of an equation.

5. Explain how to find the x- and y-intercepts of a graph.

6. You are walking toward an object. Let x represent the time (in seconds) and let y represent the distance between you and the object (in feet). Sketch a graph that shows how x and y are related.

Problem Solving

In Exercises 7–10, match the equation with its graph. [The graphs are labeled (a), (b), (c), and (d).]

(a)
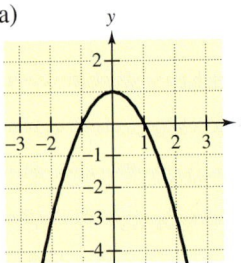

(b)
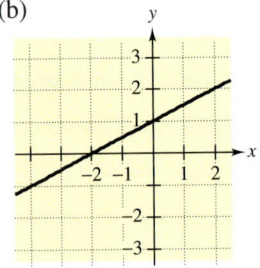

(c)
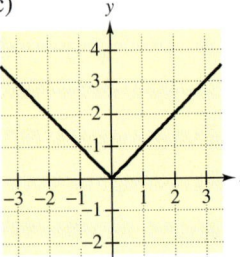

(d)
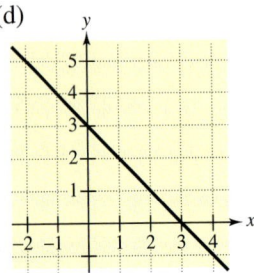

7. $y = 3 - x$

8. $y = \frac{1}{2}x + 1$

9. $y = -x^2 + 1$

10. $y = |x|$

In Exercises 11–14, solve the equation for y.

11. $3x + 4y = 12$

12. $2x + 3y = 6$

13. $x - 2y = 8$

14. $-x - 3y = 9$

In Exercises 15 and 16, complete the table and use the results to sketch the graph of the equation.

15. $y = 9 - x$

x	-2	-1	0	1	2
y					

16. $3x - 2y = 6$

x	-2	0	2	4	6
y					

Graphical Solution/Algebraic Check In Exercises 17–20, graphically estimate the x- and y-intercepts of the graph. Then check your results algebraically.

17. $4x - 2y = -8$

18. $5y - 2x = 10$

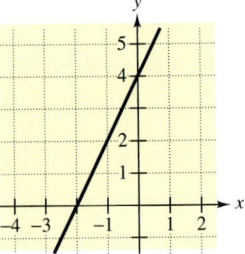

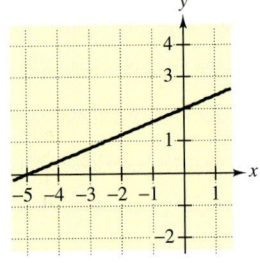

19. $y = |x| - 3$

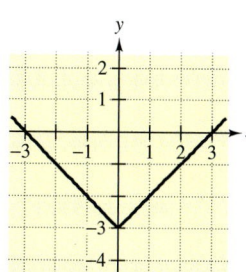

20. $y = x^2 - 4$

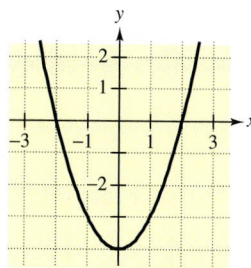

In Exercises 21–26, find the x- and y-intercepts of the graph of the equation.

21. $x - y = 1$

22. $x + y = 10$

23. $2x + y - 4 = 0$

24. $3x - 2y + 6 = 0$

25. $y = \frac{1}{2}x - 1$

26. $y = -3x + 5$

In Exercises 27–38, sketch the graph of the equation and label the coordinates of three solution points.

27. $y = 2 - x$

28. $y = x + 3$

29. $2x - 3y = 12$

30. $2x + 5y = 10$

31. $4x + y = 2$

32. $y - 2x = 3$

33. $y = x^2$

34. $y = x^2 - 1$

35. $y = -x^2 + 9$

36. $y = (x - 3)^2$

37. $y = |x| - 3$

38. $y = |x - 3|$

39. *Research Project*   Use a weekly news magazine or newspaper to find examples of misleading graphs and explain why they are misleading.

40. *Creating a Model*   Let y represent the distance traveled by a car that is moving at a constant speed of 35 miles per hour. Let t represent the number of hours the car has traveled. Write an equation that relates y and t and sketch its graph.

41. *Creating a Model*   The cost of printing a book is $500, plus $5 per book. Let C represent the total cost and let x represent the number of books. Write an equation that relates C and x and sketch its graph.

42. *Modeling Data*   The table lists the annual average expenditures per pupil for public elementary and secondary schools in the United States for 1986 through 1992. A model that approximates this data is

$$y = 296.25t + 1952.61$$

where $t = 0$ corresponds to 1980.   (Source: National Education Association)

t	6	7	8	9
y	$3764	$3996	$4277	$4612

t	10	11	12
y	$4976	$5241	$5466

(a) Plot the points that represent the actual data.

(b) On the same graph, sketch the graph of the model.

(c) How well do you think the model represents the data?

(d) Predict the average expense per pupil for 1994.

Reviewing the Major Concepts

In Exercises 43–46, solve the equation.

43. $\frac{5}{6}x - 7 = 0$

44. $16 - \frac{2}{3}x = 0$

45. $\frac{t}{2} + \frac{t}{4} = 30$

46. $4(x - 3) = 0$

47. *Geometry*   The width of a rectangular mirror is $\frac{3}{5}$ its length. The perimeter of the mirror is 80 inches. What are the dimensions of the mirror?

48. *Travel Costs*   A company pays its sales representatives $20 per day, plus 27 cents per mile, for the use of their personal cars. A sales representative submits a bill for $36.74 for driving her own car.

(a) How many miles did she drive?

(b) How many days did she drive? Explain.

(c) Suppose the bill had been submitted for $96.70. Could you determine how many days and miles were claimed? Explain.

Additional Problem Solving

In Exercises 49–52, complete the table and use the results to sketch the graph of the equation.

49. $y = 4x - 2$

x	-2	-1	0	1	2
y					

50. $y = x - 1$

x	-2	-1	0	1	2
y					

51. $y = |x + 1|$

x	-3	-2	-1	0	1
y					

52. $y = (x - 1)^2$

x	-1	0	1	2	3
y					

Graphical Solution/Algebraic Check In Exercises 53 and 54, graphically estimate the x- and y-intercepts of the graph. Then check your results algebraically.

53. $y = 4 - |x|$

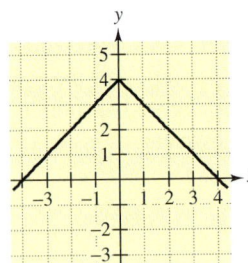

54. $y = x^2 - 16$

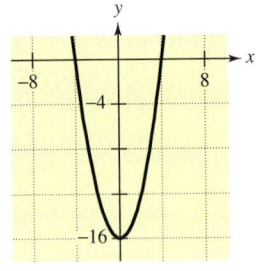

In Exercises 55–60, find the x- and y-intercepts of the graph of the equation.

55. $y = 6x + 2$
56. $y = -\frac{1}{2}x + 3$
57. $2x + 6y = 9$
58. $\frac{1}{2}x + \frac{2}{3}y = 1$
59. $\frac{3}{4}x - \frac{1}{2}y = 3$
60. $2x - 5y + 50 = 0$

In Exercises 61–76, sketch the graph of the equation and label the coordinates of three solution points.

61. $y = x - 1$
62. $y = 5 - x$
63. $4x + 5y = 20$
64. $x - 7y = 14$
65. $y = \frac{3}{8}x + 15$
66. $y = 14 - \frac{2}{3}x$
67. $y = \frac{1}{3}(2x - 5)$
68. $y = \frac{3}{4}(2x + 3)$
69. $y = 3x$
70. $y = -2x$
71. $y = -x^2$
72. $y = (x - 2)^2$
73. $y = x^2 + 3$
74. $y = x(x - 4)$
75. $y = |x - 5|$
76. $y = 5 - |x|$

77. *Interpreting Intercepts* The model $5F - 9C = 160$ relates the temperature in degrees Celsius C and degrees Fahrenheit F.

(a) Graph the equation where F is measured on the horizontal axis.

(b) Explain what the intercepts represent.

78. *Modeling Data* The table gives the life expectancy y (in years) in the United States for a child at birth. A model for the life expectancy during this period is $y = 0.3t + 65.6$, with $t = 0$ corresponding to 1950. (Source: U. S. Bureau of Census)

t	-20	-10	0	10	20	30	40
y	59.7	62.9	68.2	69.7	70.8	73.7	75.6

(a) Graph the actual data and the model.

(b) Predict the life expectancy in the year 2000.

4.3   Graphs and Graphing Utilities

Introduction ▪ Using a Graphing Utility ▪
Using the Special Features of a Graphing Utility

Introduction

When Charles Babbage strove to make his mechanical computer a reality in the 1830's and 1840's, his assistant was Augusta Ada Byron, Countess of Lovelace. In 1842, a paper describing Babbage's ideas for his machine was published in French. Countess Lovelace translated the paper and added a description of a rudimentary computer program. Some consider Countess Lovelace to be the world's first programmer. In honor of her contribution to computer science, a programming language developed for the United States Department of Defense in the 1980's was named Ada. Today Ada is used by many organizations in different applications.

In Section 4.2 you studied the point-plotting method for sketching the graph of an equation. One of the disadvantages of the point-plotting method is that to get a good idea about the shape of a graph you need to plot *many* points. By plotting only a few points, you can badly misrepresent the graph.

For instance, consider the equation $y = x^3$. To graph this equation, suppose you calculated only the following three points.

x	-1	0	1
$y = x^3$	-1	0	1

By plotting these three points, as shown in Figure 4.15(a), you might assume that the graph of the equation is a straight line. This, however, is not correct. By plotting several more points, as shown in Figure 4.15(b), you can see that the actual graph is not straight at all!

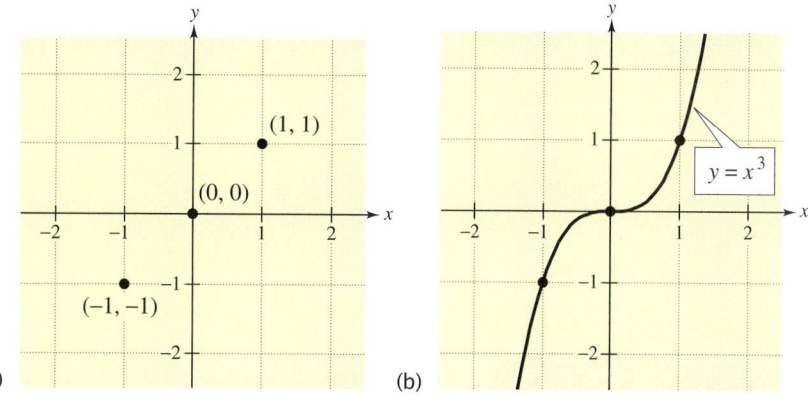

(a)   (b)

FIGURE 4.15

Thus, the point-plotting method leaves you with a dilemma. On the one hand, the method can be very inaccurate if only a few points are plotted. But, on the other hand, it is very time consuming to plot a dozen (or more) points. Technology can help you solve this dilemma. Plotting several (even hundreds of points) on a rectangular coordinate system is something that a computer or graphing calculator can do easily.

Using a Graphing Utility

There are many different graphing utilities: some are graphing packages for computers and some are hand-held graphing calculators. In this section we describe the steps used to graph an equation with a graphing utility. We will often give keystroke sequences for illustration; however, these may not agree precisely with the steps required by *your* calculator.*

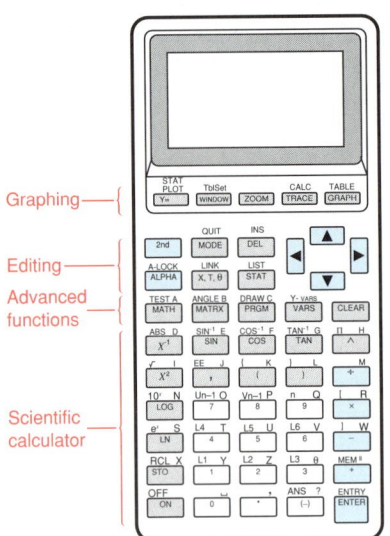

Graphing
Editing
Advanced functions
Scientific calculator

FIGURE 4.16 Keypad of a TI-82 Graphics Calculator

Graphing an Equation with a TI-82 Graphing Calculator

Before performing the following steps, set your calculator so that all of the standard defaults are active. For instance, all of the options at the left of the MODE screen should be highlighted.

1. Set the viewing window for the graph. (See Example 3.) To set the standard viewing window, press ZOOM 6 ENTER .

2. Rewrite the equation so that y is isolated on the left side of the equation.

3. Press the Y= key. Then enter the right side of the equation on the first line of the display. (The first line is labeled $Y_1=$.)

4. Press the GRAPH key.

EXAMPLE 1   Graphing a Linear Equation

Sketch the graph of $2y + x = 4$.

Solution

To begin, solve the given equation for y in terms of x.

$2y + x = 4$	Original equation
$2y = -x + 4$	Subtract x from both sides.
$y = -\dfrac{1}{2}x + 2$	Divide both sides by 2.

Press the Y= key, and enter the following keystrokes.

(-)   X,T,θ   ÷   2   +   2

The top row of the display should now be as follows.

$$Y_1 = -X/2 + 2$$

Press the GRAPH key, and the screen should look like that shown in Figure 4.17.

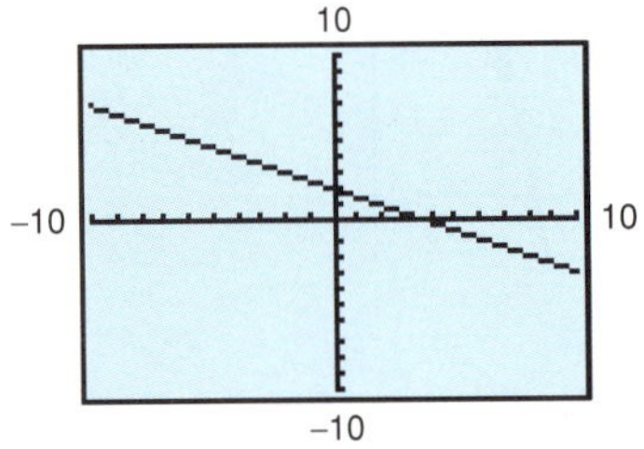

FIGURE 4.17

---

*The graphing calculator keystrokes given in this section correspond to the TI-82 graphing utility by Texas Instruments. For other graphing utilities, the keystrokes may differ.

In Figure 4.17, notice that the calculator screen does not label the tick marks on the x-axis or the y-axis. To see what the tick marks represent, you can press WINDOW . If you set your calculator to the standard graphing defaults before working Example 1, the screen should show the following values.

Xmin= -10   The minimum x-value is -10.
Xmax=10   The maximum x-value is 10.
Xscl=1   The x-scale is one unit per tick mark.
Ymin= -10   The minimum y-value is -10.
Ymax=10   The maximum y-value is 10.
Yscl=1   The y-scale is one unit per tick mark.

These settings are summarized visually in Figure 4.18.

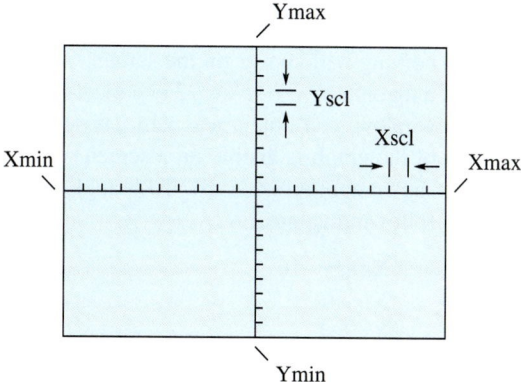

FIGURE 4.18

EXAMPLE 2   *Graphing an Equation Involving Absolute Value*

Sketch the graph of $y = |x - 3|$.

Solution

This equation is already written so that y is isolated on the left side of the equation. Press the Y= key, and enter the following keystrokes.

ABS (X,T,θ $-$ 3)

The top row of the display should now be as follows.

$Y_1 = \text{abs}\ (X - 3)$

Press the GRAPH key, and the screen should look like that shown in Figure 4.19.

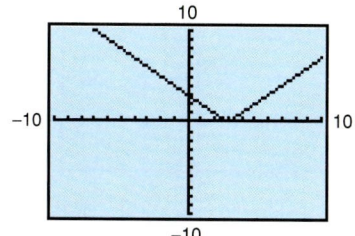

FIGURE 4.19

DISCOVERY

Another special feature of a graphing utility is the TRACE feature. This feature is used to find solution points of an equation. For example, you can approximate the x- and y-intercepts of $y = 3x + 6$ by first graphing the equation, then using the TRACE feature, and finally pressing the $\boxed{\triangleleft}$ $\boxed{\triangleright}$ keys. To get a better approximation of a solution point, you can use the following keystrokes repeatedly.

$\boxed{\text{ZOOM}}$ 2 $\boxed{\text{ENTER}}$

Check to see that you get an x-intercept of $(-2, 0)$ and a y-intercept of $(0, 6)$. Use the TRACE feature to find the x- and y-intercepts of $y = \frac{1}{2}x - 4$.

Using the Special Features of a Graphing Utility

To use your graphing utility to its best advantage, you must learn how to set the viewing window, as illustrated in the next example.

EXAMPLE 3 Setting the Viewing Window

Sketch the graph of $y = x^2 + 12$.

Solution

Press $\boxed{\text{Y=}}$ and enter $x^2 + 12$ on the first line.

$\boxed{\text{X,T,}\theta}$ $\boxed{x^2}$ $\boxed{+}$ 12

Press the $\boxed{\text{GRAPH}}$ key. If your calculator is set to the standard viewing window, nothing will appear on the screen. The reason for this is that the lowest point on the graph of $y = x^2 + 12$ occurs at the point $(0, 12)$. Using the standard viewing window, you obtain a screen whose largest y-value is 10. In other words, none of the graph is visible on a screen whose y-values vary between -10 and 10, as shown in Figure 4.20(a). To change these settings, press $\boxed{\text{WINDOW}}$ and enter the following values.

Xmin= -10	The minimum x-value is -10.
Xmax=10	The maximum x-value is 10.
Xscl=1	The x-scale is 1 unit per tick mark.
Ymin= -10	The minimum y-value is -10.
Ymax=30	The maximum y-value is 30.
Yscl=5	The y-scale is 5 units per tick mark.

Press $\boxed{\text{GRAPH}}$ and you will obtain the graph shown in Figure 4.20(b). On this graph, note that each tick mark on the y-axis represents 5 units because you changed the y-scale to 5. Also note that the highest point on the y-axis is now 30 because you changed the maximum value of y to 30.

NOTE If you changed the y-maximum and y-scale on your utility as indicated in Example 3, you should return to the standard settings before working Example 4. To do this, press $\boxed{\text{ZOOM}}$ $\boxed{6}$ $\boxed{\text{ENTER}}$.

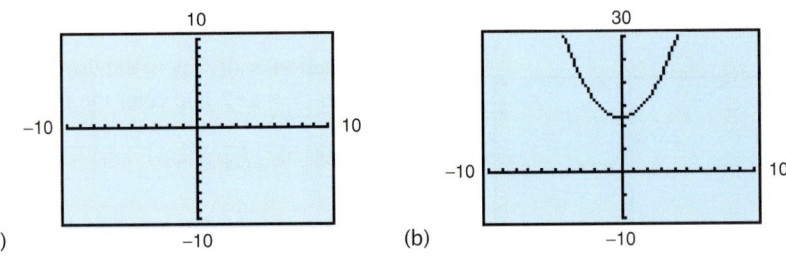

(a)   (b)

FIGURE 4.20

EXAMPLE 4 Using a Square Setting

Sketch the graph of $y = x$. The graph of this equation is a straight line that makes a 45° angle with the x-axis and with the y-axis. From the graph on your utility, does the angle appear to be 45°?

Solution

Press $\boxed{Y=}$ and enter x on the first line.

$\quad Y_1 = X$

Press the $\boxed{GRAPH}$ key and you will obtain the graph shown in Figure 4.21(a). Note that the angle the line makes with the x-axis doesn't appear to be 45°. The reason for this is that the screen is wider than it is tall. This makes the tick marks on the x-axis appear farther apart than the tick marks on the y-axis. To obtain the same distance between tick marks on both axes, you can change the graphing settings from "standard" to "square." To do this, press the following keys.

$\boxed{ZOOM}\quad\boxed{5}\quad\boxed{ENTER}$ Square setting

The screen should look like that shown in Figure 4.21(b). Note in this figure that the square setting has changed the viewing window so that the x-values vary between -15 and 15.

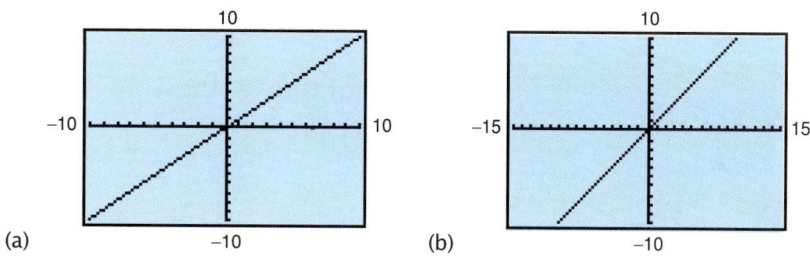

(a)   (b)

FIGURE 4.21

NOTE There are many possible square settings on a graphing utility. To create a square setting, you need the following ratio to be $\frac{2}{3}$.

$$\frac{\text{Ymax} - \text{Ymin}}{\text{Xmax} - \text{Xmin}}$$

For instance, the setting in Example 4 is square because $(\text{Ymax} - \text{Ymin}) = 20$ and $(\text{Xmax} - \text{Xmin}) = 30$.

EXAMPLE 5 *Sketching More than One Graph on the Same Screen*

Sketch the graphs of the following equations on the same screen.

$$y = -x + 4, \quad y = -x, \quad \text{and} \quad y = -x - 4$$

Solution

To begin, press $\boxed{Y=}$ and enter all three equations on the first three lines. The display should now be as follows.

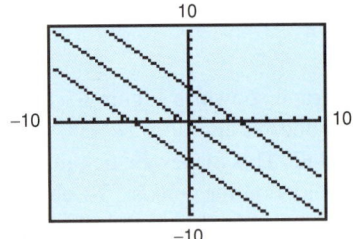
10
−10   10
−10

FIGURE 4.22

$Y_1 = -X + 4$	$\boxed{(-)}$ $\boxed{X,T,\theta}$ $\boxed{+}$ 4
$Y_2 = -X$	$\boxed{(-)}$ $\boxed{X,T,\theta}$
$Y_3 = -X - 4$	$\boxed{(-)}$ $\boxed{X,T,\theta}$ $\boxed{-}$ 4

Press the $\boxed{GRAPH}$ key and you will obtain the graph shown in Figure 4.22. Note that the graph of each equation is a straight line, and that the lines are parallel to each other.

Group Activities   Exploring with Technology

Determining a Viewing Rectangle Set all calculators in your group to a standard setting and sketch graphs of $y = 10x - x^2$. The graph will look similar to the one shown below. To find a better window setting, have part of your group use the TRACE feature and others use the ZOOM feature to find a window setting that shows all the features of the graph. Which method seems to give the better view of the graph?

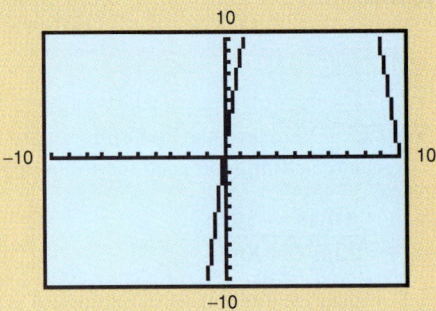
10
−10   10
−10

4.3 Exercises

Discussing the Concepts

1. What should you do first to graph the equation $2x + 4y = 10$ on a graphing utility?

2. In your own words, describe the meanings of Xmin, Xmax, Xscl, Ymin, Ymax, and Yscl.

3. You have correctly entered an equation on a graphing utility. However, no graph appears on the graphing window when you press $\boxed{\text{GRAPH}}$. Explain how you might remedy this problem.

4. You and a friend have identical calculators and you use them to graph the same equation. However, the graphs do not look the same. How is this possible?

5. Describe the change on the display if Xscl is increased from 1 to 2.

6. A crucial part of a graph is cut off at the top of a display. Explain how to change the calculator settings to obtain the missing part.

Problem Solving

In Exercises 7–14, use a graphing utility to graph the equation. (Use a standard setting.)

7. $y = x + 1$

8. $y = -2x$

9. $y = -\frac{1}{3}x$

10. $y = \frac{1}{2}x$

11. $y = -2x^2 + 5$

12. $y = x^2 + 0.5x - 7.5$

13. $y = |x + 1| - 2$

14. $y = 4 - |x - 2|$

In Exercises 15–18, use a graphing utility to graph the equation.

15. $y = 25 - 3x^2$

| Xmin = -4 |
| Xmax = 4 |
| Xscl = 1 |
| Ymin = -5 |
| Ymax = 25 |
| Yscl = 5 |

16. $y = 3x^2 + 2x - 3$

| Xmin = -4 |
| Xmax = 2 |
| Xscl = 1 |
| Ymin = -5 |
| Ymax = 20 |
| Yscl = 2 |

17. $y = 3x^2 - 9x$

| Xmin = -2 |
| Xmax = 4 |
| Xscl = 1 |
| Ymin = -8 |
| Ymax = 8 |
| Yscl = 2 |

18. $y = x^2 - 4x + 16$

| Xmin = -3 |
| Xmax = 7 |
| Xscl = 1 |
| Ymin = -5 |
| Ymax = 30 |
| Yscl = 5 |

In Exercises 19–24, use a graphing utility to find a viewing window that matches the one shown.

19. $y = \frac{1}{2}x + 2$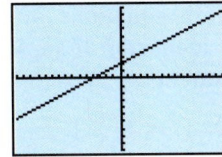

20. $y = 2x - 1$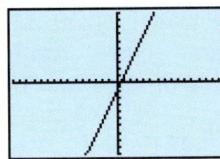

21. $y = \frac{1}{4}x^2 - 4x + 12$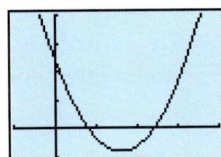

22. $y = 16 - 4x - x^2$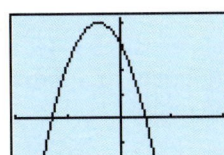

23. $y = |2x - 1|$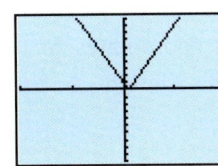

24. $y = |x^2 + 5x + 9|$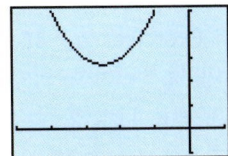

In Exercises 25–28, match the equation with its graph. [The graphs are labeled (a), (b), (c), and (d).]

25. $y = x^2$

26. $y = 2x^2$

27. $y = \frac{1}{4}x^2$

28. $y = -\frac{1}{4}x^2$

(a)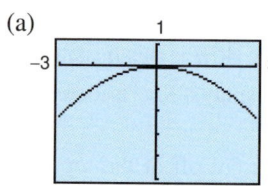

(b)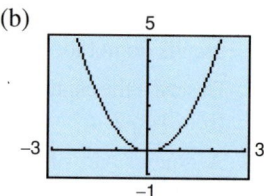

(c)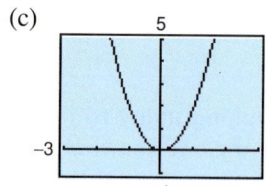

(d)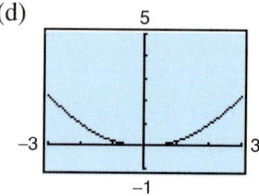

In Exercises 29–32, graph both equations on the same screen. Are the graphs identical? If so, what Rule of Algebra is being illustrated?

29. $y_1 = \frac{1}{3}x - 1$
   $y_2 = -1 + \frac{1}{3}x$

30. $y_1 = 3\left(\frac{1}{4}x\right)$
   $y_2 = \left(3 \cdot \frac{1}{4}\right)x$

31. $y_1 = x(x - 2)$
   $y_2 = x^2 - 2x$

32. $y_1 = 2 + (x + 4)$
   $y_2 = (2 + x) + 4$

In Exercises 33–36, use the TRACE feature of a graphing utility to approximate the x- and y-intercepts of the graph.

33. $y = 2x - 5$

34. $y = 4 - |x|$

35. $y = x^2 + 1.5x - 1$

36. $y = x^3 - 4x$

Modeling Data   In Exercises 37 and 38, use the following models, which give the revenues from stamps and postal cards and from postage paid under permit and meter.

Stamps and postal cards

$$y = 0.06x^2 + 1.59x + 18.08, \quad 4 \le x \le 11$$

Permit and metered mail

$$y = 0.03x^2 + 1.19x + 10.91, \quad 4 \le x \le 11$$

In these models, y is the revenue (in billions of dollars) and x is the year, with $x = 0$ corresponding to 1980.   (Source: U. S. Postal Service)

37. Use the following setting to graph both models on the same display of a graphing utility.

Xmin $= 4$
Xmax $= 11$
Xscl $= 1$
Ymin $= 10$
Ymax $= 45$
Yscl $= 5$

38. (a) Were revenues increasing or decreasing over time?

   (b) Is the distance between the graphs increasing or decreasing over time? What does this mean to the U. S. Postal Service?

In Exercises 39–42, find a viewing window that shows the important characteristics of the graph.

39. $y = 15 + |x - 12|$

40. $y = 15 + (x - 12)^2$

41. $y = -15 + |x + 12|$

42. $y = -15 + (x + 12)^2$

Reviewing the Major Concepts

In Exercises 43–46, write the coordinates of three points that are solutions of the equation.

43. $y = \frac{4}{5}x + 2$

44. $y = 3 - \frac{5}{6}x$

45. $y = 8 - 0.75x$

46. $y = 2 + 0.6x$

In Exercises 47–50, simplify the expression.

47. $v^2 \cdot v^3$

48. $-y^2(-2y)^3$

49. $x(x - 3) - (x^2 + 6x)$

50. $3[x - (2x + 5)]$

Additional Problem Solving

In Exercises 51–62, use a graphing utility to graph the equation. (Use the standard setting.)

51. $y = -3x$

52. $y = x - 4$

53. $y = \frac{3}{4}x - 6$

54. $y = -3x + 2$

55. $y = \frac{1}{2}x^2$

56. $y = -\frac{2}{3}x^2$

57. $y = x^2 - 4x + 2$

58. $y = -0.5x^2 - 2x + 2$

59. $y = |x - 3|$

60. $y = |x + 4|$

61. $y = |x^2 - 4|$

62. $y = |x - 2| - 5$

In Exercises 63–66, use a graphing utility to graph the equation.

63. $y = 27x + 100$

```
Xmin = 0
Xmax = 5
Xscl = .5
Ymin = 75
Ymax = 250
Yscl = 25
```

64. $y = 50,000 - 6000x$

```
Xmin = 0
Xmax = 7
Xscl = .5
Ymin = 0
Ymax = 50000
Yscl = 5000
```

65. $y = 0.001x^2 + 0.5x$

```
Xmin = -500
Xmax = 200
Xscl = 50
Ymin = -100
Ymax = 100
Yscl = 20
```

66. $y = 100 - 0.5|x|$

```
Xmin = -300
Xmax = 300
Xscl = 60
Ymin = -100
Ymax = 100
Yscl = 20
```

In Exercises 67–70, graph both equations on the same screen. Are the graphs identical? If so, what Rule of Algebra is being illustrated?

67. $y_1 = 2x + (x + 1)$
    $y_2 = (2x + x) + 1$

68. $y_1 = \frac{1}{2}(3 - 2x)$
    $y_2 = \frac{3}{2} - x$

69. $y_1 = 2\left(\frac{1}{2}\right)$
    $y_2 = 1$

70. $y_1 = x(0.5x)$
    $y_2 = (0.5x)x$

In Exercises 71–74, use the TRACE feature of a graphing utility to approximate the x- and y-intercepts of the graph.

71. $y = 9 - x^2$

72. $y = 3x^2 - 2x - 5$

73. $y = 6 - |x + 2|$

74. $y = |x - 2|^2 - 3$

Geometry In Exercises 75–78, graph the equations on the same display. Using a "square setting," determine the geometrical shape bounded by the graphs.

75. $y = -4, \ y = -|x|$

76. $y = |x|, \ y = 5$

77. $y = |x| - 8, \ y = -|x| + 8$

78. $y = -\frac{1}{2}x + 7, \ y = \frac{8}{3}(x + 5), \ y = \frac{2}{7}(3x - 4)$

Modeling Data In Exercises 79 and 80, use the following models, which give the number of births and the number of deaths in the United States from 1980 through 1991.

$y = 0.004x^2 + 0.010x + 3.602, \ 0 \le x \le 11$ Births

$y = -0.001x^2 + 0.027x + 1.959, \ 0 \le x \le 11$ Deaths

In these models, y is the number of births and deaths (in millions), and x is the year, with $x = 0$ corresponding to 1980. (Source: National Center for Health Statistics)

79. Graph both models on the same screen.

```
Xmin = 0
Xmax = 11
Xscl = .5
Ymin = 1.5
Ymax = 5
Yscl = .5
```

80. What conclusions can you make about the change in population of the United States during the years 1980–1991? (The models do not reflect any changes in population due to immigration.)

MID-CHAPTER QUIZ

Take this quiz as you would take a quiz in class. After you are done, check your work against the answers given in the back of the book.

1. Plot the points $(4, -2)$ and $\left(-1, -\frac{5}{2}\right)$ on a rectangular coordinate system.

2. Determine the quadrants in which the points $(x, 5)$ must be located. (x is a real number.)

3. Decide whether the ordered pairs are solutions of the equation $y = 9 - |x|$.
   (a) $(2, 7)$ (b) $(-3, 12)$ (c) $(-9, 0)$ (d) $(0, -9)$

4. Complete the table for $y = -x + 3$. Plot the resulting data.

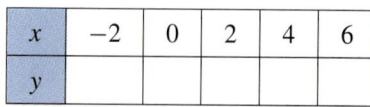

x	-2	0	2	4	6
y					

Table for 4

5. The bar graph shows the average number (in millions) of shares traded per day on the New York Stock Exchange for the years 1988 through 1992. Estimate the percent increase in the average number of shares traded per day from 1990 to 1992.   (Source: The New York Stock Exchange)

6. What is the y-coordinate of any point on the x-axis?

In Exercises 7 and 8, find the intercepts of the graph of the equation.

7. $x - 3y = 12$ 8. $y = 6 - 4x$

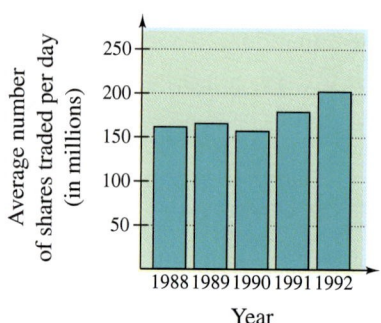

Figure for 5

In Exercises 9–14, sketch the graph of the equation and label the intercepts.

9. $y = x - 1$ 10. $y = 5 - 2x$ 11. $y = 4 - x^2$
12. $y = (x + 2)^2$ 13. $y = x^3$ 14. $y = 1 - |x|$

In Exercises 15–16, use a graphing utility to graph the equation. Find a viewing window that shows the important characteristics of the graph. In your own words, describe the graph's important characteristics.

15. $y = \frac{1}{2}x^2 - 12$ 16. $y = x^3 - x$

17. Use the TRACE feature of a graphing utility to approximate (to two decimal places) the x- and y-intercepts of the graph of $y = x^2 - 6x - 7$.

18. Use a graphing utility to graph the equations $y_1 = x(x + 3)$ and $y_2 = x^2 + 3x$ on the same screen. Are the graphs identical? If so, what Rule of Algebra is illustrated?

19. Use a graphing utility to graph the equations $y_1 = \sqrt{25 - x^2}$ and $y_2 = -\sqrt{25 - x^2}$. The resulting graphs should form a circle. What change to the calculator settings is required if the graph does not appear circular?

20. The cost of a new computer system is approximately \$3000 and depreciates at the rate of \$500 per year for 4 years. Let y represent the value of the system after x years. Write an equation that relates y and x and sketch its graph.

4.4 Business Applications and Graphs

Markup and Discounts ■ Bills, Charges, and Payments ■ Wages, Salaries, and Commissions ■ Business Graphs

Markup and Discounts

You may have had the experience of buying an item at one store and later finding that you could have paid less for the same item at another store. The basic reason for this price difference is **markup,** which is the difference between the **cost** (the amount a retailer pays for the item) and the **price** (the amount at which the retailer sells the item to the consumer). A verbal model for this problem is as follows.

$$\boxed{\text{Selling price}} = \boxed{\text{Cost}} + \boxed{\text{Markup}}$$

In such a problem, the markup may be known or it may be expressed as a percent of the cost. This percent is called the **markup rate.**

$$\boxed{\text{Markup}} = \boxed{\text{Markup rate}} \cdot \boxed{\text{Cost}}$$

EXAMPLE 1 *Finding the Selling Price*

A sporting goods store uses a markup rate of 55% on all items. The cost of a golf bag is $35. What is the selling price of the bag?

Solution

Verbal Model: $\boxed{\dfrac{\text{Selling}}{\text{price}}} = \boxed{\text{Cost}} + \boxed{\text{Markup}}$

Labels: Selling price $= x$ (dollars)
        Cost $= 35$ (dollars)
        Markup rate $= 0.55$ (percent in decimal form)
        Markup $= (0.55)(35)$ (dollars)

Equation: $x = 35 + (0.55)(35)$
           $= 35 + 19.25$
           $= \$54.25$

The selling price is $54.25. Check this in the original statement of the problem.

In Example 1, you are given the cost and are asked to find the selling price. Example 2 illustrates the reverse problem. That is, in Example 2 you are given the selling price and asked to find the cost.

EXAMPLE 2   *Finding the Cost of an Item*

The selling price of a pair of ski boots is $98. The markup rate is 55%. What is the cost of the boots?

Solution

Verbal Model: | Selling price | = | Cost | + | Markup |

Labels: Selling price = 98   (dollars)
            Cost = x (dollars)
            Markup rate = 0.55   (percent in decimal form)
            Markup = 0.55x (dollars)

Equation: $98 = x + 0.55x$

                   $98 = 1.55x$

                   $\dfrac{98}{1.55} = x$

                   $\$63.23 \approx x$

The cost is $63.23. Check this in the original statement of the problem.

EXAMPLE 3   *Finding the Markup Rate*

A pair of shoes sells for $60. The cost of the shoes is $24. What is the markup rate?

Solution

Verbal Model: | Selling price | = | Cost | + | Markup |

Labels: Selling price = 60   (dollars)
            Cost = 24   (dollars)
            Markup rate = p (percent in decimal form)
            Markup = $p(24)$ (dollars)

Equation: $60 = 24 + p(24)$

               $36 = 24p$

               $\dfrac{36}{24} = p$

               $1.5 = p$

Because $p = 1.5$, it follows that the markup rate is 150%. Check this in the original statement of the problem.

The mathematics of a discount is similar to that of a markup. The model for this situation is

$$\boxed{\text{Sale price}} \; = \; \boxed{\text{List price}} \; - \; \boxed{\text{Discount}}$$

where the **discount** is given in dollars, and the **discount rate** is given as a percent of the list price.

$$\boxed{\text{Discount}} \; = \; \boxed{\text{Discount rate}} \; \cdot \; \boxed{\text{List price}}$$

EXAMPLE 4 *Finding the Discount Rate*

During a mid-summer sale, a lawn mower listed at $199.95 is on sale for $139.95. What is the discount rate?

Solution

Verbal Model: $\boxed{\text{Discount}} \; = \; \boxed{\begin{array}{c}\text{Discount}\\\text{rate}\end{array}} \; \cdot \; \boxed{\begin{array}{c}\text{List}\\\text{price}\end{array}}$

Labels: Discount $= 199.95 - 139.95 = 60$ (dollars)
           List price $= 199.95$ (dollars)
           Discount rate $= p$ (percent in decimal form)

Equation: $60 = p(199.95)$

Try solving this equation. You should obtain a discount rate of 30%.

EXAMPLE 5 *Finding the Sale Price*

A drug store advertises 40% off the prices of all summer tanning products. A bottle of suntan oil lists for $3.49. What is the sale price?

Solution

Verbal Model: $\boxed{\begin{array}{c}\text{Sale}\\\text{price}\end{array}} \; = \; \boxed{\begin{array}{c}\text{List}\\\text{price}\end{array}} \; - \; \boxed{\text{Discount}}$

Labels: List price $= 3.49$ (dollars)
           Discount rate $= 0.4$ (percent in decimal form)
           Discount $= 0.4(3.49)$ (dollars)
           Sale price $= x$ (dollars)

Equation: $x = 3.49 - (0.4)(3.49)$
              $\approx \$2.09$

The sale price is $2.09. Check this in the original statement of the problem.

Bills, Charges, and Payments

EXAMPLE 6 Finding the Hours of Labor

An auto repair bill of $250 lists $110 for parts and the rest for labor. The cost of labor is $28 per hour. How many hours of labor did it take to repair the auto?

Solution

Verbal Model:

| Total bill | = | Price of parts | + | Price of labor |

Labels:
Total bill = 250	(dollars)
Price of parts = 110	(dollars)
Number of hours of labor = x	(hours)
Hourly rate for labor = 28	(dollars per hour)
Price of labor = $28x$	(dollars)

Equation:
$$250 = 110 + 28x$$
$$140 = 28x$$
$$\frac{140}{28} = x$$
$$5 = x$$

It took 5 hours of labor. Check this in the original statement of the problem.

EXAMPLE 7 Finding the Tip Rate

For dinner at a restaurant, a customer left $18.50 for a meal that cost $15.95. What percent is the tip?

Solution

Verbal Model:

| Tip | = | Percent | · | Price of meal |

Labels:
Tip = 18.50 − 15.95 = 2.55	(dollars)
Percent = p	(percent in decimal form)
Price of meal = 15.95	(dollars)

Equation:
$$2.55 = p(15.95)$$
$$\frac{2.55}{15.95} = p$$
$$0.16 \approx p$$

The customer left a tip that amounted to approximately 16% of the price of the meal. Check this in the original statement of the problem.

EXAMPLE 8   Finding the Cost of a Telephone Call

You made a 12-minute call from Denver to Atlanta. The call cost $2.05 for the first 3 minutes and 34¢ for each additional minute. How much did the call cost?

Solution

Verbal Model:

$$\boxed{\text{Total cost}} = \boxed{\text{Cost of first 3 minutes}} + \boxed{\text{Cost of additional 9 minutes}}$$

Labels: Total cost $= c$ (dollars)
   Cost of first 3 minutes $= 2.05$ (dollars)
   Cost for additional 9 minutes $= 9(0.34)$ (dollars)

Equation: $c = 2.05 + (0.34)9$

   $c = 2.05 + 3.06$

   $c = \$5.11$

The total cost of the call was $5.11. Check this in the original statement of the problem.

EXAMPLE 9   Finding the Monthly Payment

You are buying a large-screen television. Including finance charges, the total purchase price is $1596. To finance this amount, you agree to make a down payment of $120. The remainder is to be paid in 12 equal monthly payments. How much is each monthly payment?

Solution

Verbal Model:

$$\boxed{\text{Total price}} = \boxed{\text{Down payment}} + 12 \cdot \boxed{\text{Monthly payment}}$$

Labels: Total price $= 1596$ (dollars)
   Down payment $= 120$ (dollars)
   Monthly payment $= x$ (dollars)

Equation: $1596 = 120 + 12x$

   $1476 = 12x$

   $\dfrac{1476}{12} = x$

   $\$123 = x$

Each monthly payment is $123. Check this in the original statement of the problem.

Wages, Salaries, and Commissions

EXAMPLE 10 *Finding Hours of Overtime*

An employee's regular hourly rate is $9.50, and the overtime hourly rate is $14.25 (for each hour over 40 hours in a week). During a given week the employee earned $551. How many hours of overtime did the employee work?

Solution

Verbal Model: $\boxed{\text{Total pay}} = \boxed{\text{Regular pay}} + \boxed{\text{Overtime pay}}$

Labels:
Total pay = 551	(dollars)
Regular pay = (9.50)(40)	(dollars)
Number of hours of overtime = x	(hours)
Overtime pay = 14.25x	(dollars)

Equation: $551 = (9.50)(40) + 14.25x$

Try solving this equation for x. You should find that the employee worked 12 hours of overtime during the week. Check this in the original statement of the problem.

EXAMPLE 11 *Finding Commission Rate*

A sales representative receives a weekly salary of $250 plus a commission on the total weekly sales. For a given week, the sales representative's total income was $530, and total sales were $8000. Find the commission rate paid to the sales representative.

Solution

Verbal Model: $\boxed{\text{Total income}} = \boxed{\text{Salary}} + \boxed{\text{Commission}}$

Labels:
Total income = 530	(dollars)
Salary = 250	(dollars)
Commission rate = p	(percent in decimal form)
Commission = $p(8000)$	(dollars)

Equation:
$$530 = 250 + p(8000)$$
$$280 = 8000p$$
$$0.035 = p$$

The commission rate is 3.5%. Check this in the original statement of the problem.

Business Graphs

Many of the examples in this section can be represented graphically. For instance, the next example shows how to graphically represent the cost of a telephone call.

EXAMPLE 12 *Graphing the Cost of a Telephone Call*

A telephone call costs $2.05 for the first 3 minutes and $0.34 for each additional minute. Graphically display the cost of such a call.

Solution

One way to graphically represent the cost is to use a bar graph. You can begin by creating a table of values, as follows.

Minutes	1	2	3	4	5	6	7
Cost	$2.05	$2.05	$2.05	$2.39	$2.73	$3.07	$3.41

Minutes	8	9	10	11	12	13	14
Cost	$3.75	$4.09	$4.43	$4.77	$5.11	$5.45	$5.79

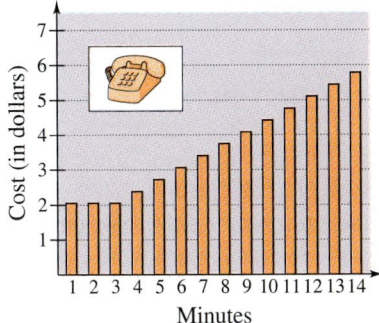

FIGURE 4.23

Using the data in this table, you can create the graph shown in Figure 4.23.

Group Activities   Extending the Concept

Comparing Markup Rates   You buy a camera that has a retail price of $120. The markup on the item is $60.

- The seller claims that the markup rate is only 50% (this is called the markup rate on sales price).
- You claim that the markup rate is 100% (this is called the markup rate on cost).

In your group, determine how each of these two types of markup rates is calculated. Which markup rate seems more suitable to the *consumer*? Which markup rate seems more suitable to the *seller*? Ask a local merchant how prices are set—by markup on cost or by competitive pricing.

4.4 Exercises

Discussing the Concepts

1. In your own words, describe the difference between the *cost* of an item and the *price* of an item.

2. In your own words, explain the difference between the markup rate and the markup.

3. Describe how to find the sale price of an item that is on sale for *r* percent off the list price. Give an example.

4. What factors would affect your decision on whether to take a regular weekly salary or take a smaller weekly salary plus commissions on your sales?

5. Explain how to change a decimal to a percent. Give examples.

6. Explain how to change a percent to a decimal. Give examples.

Problem Solving

In Exercises 7–10, find the missing quantities. The markup rate is a percent of the cost.

	Cost	Selling Price	Markup	Markup Rate
7.	$26.97	$49.95		
8.		$224.87	$75.08	
9.		$74.38		81.5%
10.	$680.00			$33\frac{1}{3}\%$

In Exercises 11–14, find the missing quantities. The discount rate is a percent of the list price.

	List Price	Sale Price	Discount	Discount Rate
11.	$39.95	$29.95		
12.	$18.95		$8.00	
13.		$18.95		20%
14.		$259.97	$135.00	

15. *Labor Charge* An appliance repair shop charges $30 for a service call and $\frac{1}{2}$ hour of service. For each additional half hour of labor there is a charge of $16. Find the length of a service call if the bill is $78.

16. *Tip Rate* A customer left $65 for a meal that cost $56.52. Determine the tip rate.

17. *Telephone Charge* The weekday rate for a telephone call is $0.55 for the first minute plus $0.40 for each additional minute, as shown in the graph. Determine the length of a call that costs $2.95. What would a call of the same length have cost if it had been made during the weekend when there is a 60% discount?

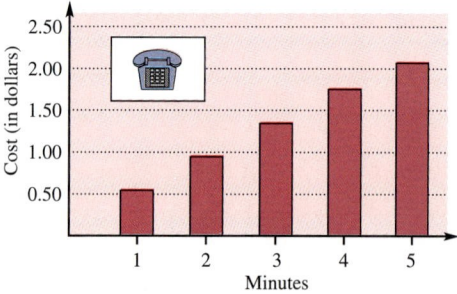

18. *Insurance Premium* The annual insurance premium for a policyholder is normally $925. Find the annual premium if the policyholder must pay a 75% surcharge because of a traffic violation.

19. *Amount Financed* You buy a motor bike that costs $1450, plus 6% sales tax.

  (a) Find the amount of the sales tax and the total bill.

  (b) How much of the total bill must you finance if you make a down payment of $500?

20. *Weekly Pay*   The weekly salary of an employee is $400 plus a 6% commission on the employee's total sales. Find the pay for a week when sales amounted to $6000.

21. *Commission Rate*   Determine the commission rate for an employee who earned $500 in commissions on sales of $4000.

22. *Overtime Rate*   You worked 48 hours during a given week (40 hours at the regular hourly rate and 8 hours at the overtime rate). Your gross pay was $672. Your regular hourly rate is $12. How much is your overtime rate?

23. *Comparing Prices*   A mail-order catalog lists automotive shock absorbers for $48.99 a pair, plus a shipping charge of $4.69. A local store has a special sale with 25% off a list price of $63.99. Which is the better bargain?

24. *Price per Pound*   The produce manager of a supermarket pays $27 for a 100-pound box of bananas. From past experience, the manager estimates that 10% of the bananas will spoil before they are sold. At what price per pound should the bananas be sold to give the supermarket an average markup rate of 30% on cost?

25. *Graphical Estimation*   As a sales representative, you receive a weekly salary of $300 plus a 3% commission on all sales.

(a) Write a linear equation giving your weekly salary y in terms of your sales x.

(b) Use a graphing utility to graph the equation in part (a). Use the following settings.

Xmin = 0	Ymin = 200
Xmax = 10000	Ymax = 600
Xscl = 1000	Yscl = 50

(c) Use the graph to estimate your weekly salary if your sales are $5500.

26. *Graphical Estimation*   You live in a state where the state income tax is 2.8% of your taxable income.

(a) Write a linear equation giving your state income tax y in terms of your taxable income x.

(b) Use a graphing utility to graph the equation in part (a). Use the following settings.

Xmin = 0	Ymin = 0
Xmax = 100000	Ymax = 3000
Xscl = 10000	Yscl = 200

(c) Use the graph to estimate your tax liability if your taxable income is $40,000.

Reviewing the Major Concepts

In Exercises 27–30, solve the equation.

27. $\dfrac{x}{3} = \dfrac{4}{9}$

28. $\dfrac{t}{16} = \dfrac{1}{4}$

29. $0.3x + 4.5 = 20.7$

30. $2(x - 3) = 5x - 24$

31. *Modeling*   Write an algebraic expression that represents the distance traveled by a train traveling at 65 miles per hour for t hours.

32. Sketch the graph of $y = -2x + 4$.

Additional Problem Solving

In Exercises 33–38, find the missing quantities. The markup rate is a percent of the cost.

	Cost	Selling Price	Markup	Markup Rate
33.		$125.98	$56.69	
34.	$71.97	$119.95		
35.		$15,900.00	$2650.00	

	Cost	Selling Price	Markup	Markup Rate
36.		$350.00	$80.77	
37.	$107.97			85.2%
38.		$69.99		55.5%

In Exercises 39–44, find the missing quantities. The discount rate is a percent of the list price.

	List Price	Sale Price	Discount	Discount Rate
39.	$189.99		$30.00	
40.	$50.99	$45.99		
41.	$119.96			50%
42.	$84.95			65%
43.		$695.00	$300.00	
44.		$189.00		40%

45. *Labor Charge* A computer company charges $50 for a service call and $\frac{1}{2}$ hour of service. Each additional half hour of labor costs $19. Find the length of a service call if the bill is $145.

46. *Labor Charge* An auto repair bill of $450 lists $178 for parts and the rest for labor. The labor costs $32 per hour. How long did it take to repair the car?

47. *Tip Rate* A customer left $20 for a meal that cost $16.95. Determine the tip rate.

48. *Tip Rate* A customer gave a barber $15 for a service that cost $12.50. Determine the tip rate.

49. *Telephone Charge* The weekday rate for a telephone call is $0.85 for the first minute plus $0.70 for each additional minute. How long is a call that costs $16.25? What would such a call cost during the weekend when there is a 60% discount?

50. *Telephone Charge* The weekday rate for a telephone call is $0.72 for the first minute plus $0.55 for each additional minute. Determine the length of a call that costs $5.12. What would such a call cost during the evening when there is a 35% discount?

51. *Insurance Premium* The annual insurance premium for a policyholder is normally $739. However, after having an automobile accident, the policyholder was charged an additional 30%. What is the new annual premium?

52. *Monthly Payment* Including finance charges, the total purchase price of a diamond ring is $1050. Of this amount, $450 is given as a down payment and the remainder is paid in nine monthly payments. How much is each payment?

53. *Amount of Sales* The monthly salary of an employee is $1000 plus a 7% commission on the total sales. How much must the employee sell in order to obtain a monthly salary of $3500?

54. *Commission Rate* Determine the commission rate for an employee who earned $812.50 in commissions on sales of $12,500.

55. *Hours of Overtime* An employee is paid $11.25 per hour for the first 40 hours and $16 for each additional hour (see figure). During the first week on the job, the employee's gross pay was $622. How many hours of overtime did the employee work?

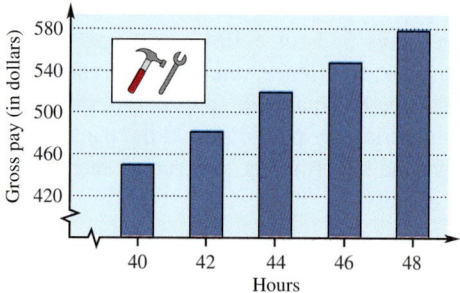

56. *Hours of Overtime* An employee worked 6 hours overtime during a given week. The wage for overtime is $18 per hour. The employee's gross pay for the week was $588. Determine the hourly wage for the 40 hours of the regular work week.

57. *Wholesale Cost* The list price of an automobile tire is $88. During a promotional sale, the fourth tire is free with the purchase of three at the list price. Counting the free tire, the markup rate on cost is 10%. Find the cost of each tire.

58. *Comparing Prices* A department store is offering a discount of 20% on a sewing machine with a list price of $239.95. A mail-order catalog has the same machine for $188.95 plus $4.32 for shipping. Which is the better bargain?

4.5 Formulas and Scientific Applications

Using Formulas ▪ Solving Mixture Problems ▪
Solving Work-Rate Problems

Using Formulas

Some formulas occur so frequently in problem solving that it is to your benefit to memorize them. For instance, the following formulas for area, perimeter, and volume are often used to create verbal models for word problems.

NOTE In the geometry formulas at the right, A represents area, P represents perimeter, C represents circumference, and V represents volume.

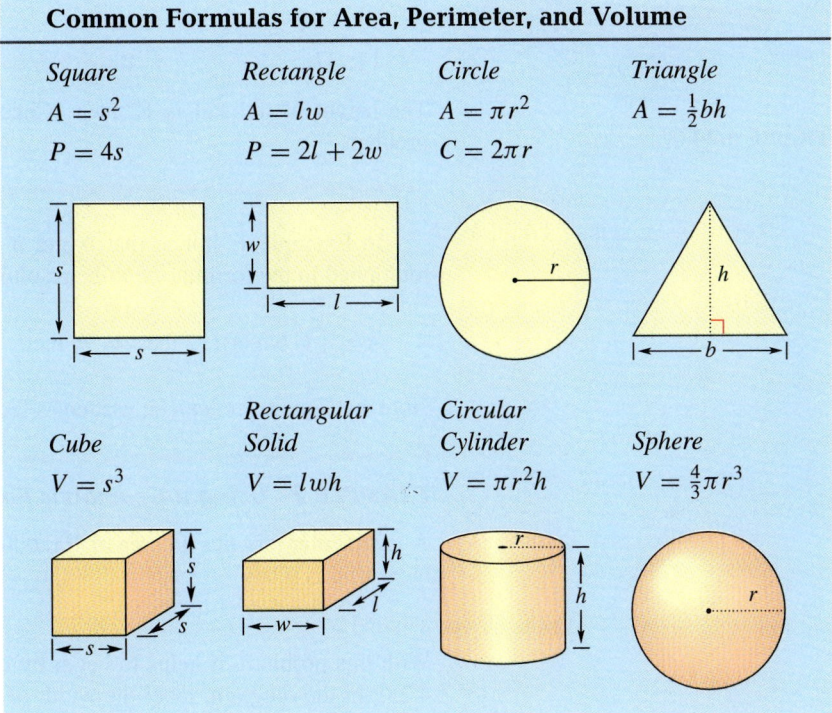

Common Formulas for Area, Perimeter, and Volume

Square
$A = s^2$
$P = 4s$

Rectangle
$A = lw$
$P = 2l + 2w$

Circle
$A = \pi r^2$
$C = 2\pi r$

Triangle
$A = \frac{1}{2}bh$

Cube
$V = s^3$

Rectangular Solid
$V = lwh$

Circular Cylinder
$V = \pi r^2 h$

Sphere
$V = \frac{4}{3}\pi r^3$

When you are solving problems involving perimeter, area, or volume, be sure you list the units of measure for your answers.

- Perimeter is always measured in linear units, such as inches, feet, miles, centimeters, meters, and kilometers.
- Area is always measured in square units, such as square inches, square feet, square centimeters, and square meters.
- Volume is always measured in cubic units, such as cubic inches.

EXAMPLE 1   Using a Geometric Formula

A sailboat has a triangular sail with an area of 96 square feet and a base that is 16 feet long, as shown in Figure 4.24. What is the height of the sail?

Solution

Because the sail is triangular, and you are given its area, you should begin with the formula for the area of a triangle.

$$A = \frac{1}{2}bh \qquad \text{Area of a triangle}$$

$$96 = \frac{1}{2}(16)h \qquad \text{Substitute 96 for } A \text{ and 16 for } b.$$

$$96 = 8h \qquad \text{Simplify.}$$

$$12 = h \qquad \text{Divide both sides by 8.}$$

The height of the sail is 12 feet. Check this in the original statement of the problem.

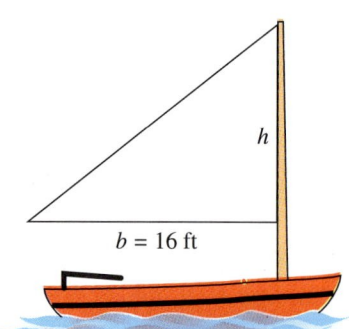

FIGURE 4.24

In Example 1, notice that b and h are measured in feet. When they are multiplied in the formula $\frac{1}{2}bh$, the resulting area is measured in *square* feet.

$$A = \frac{1}{2}(16 \text{ feet})(12 \text{ feet}) = 96 \text{ feet}^2$$

Note that square feet can be written as feet^2.

EXAMPLE 2   Using a Geometric Formula

A rectangular plot has an area of 100,000 square feet. The plot is 200 feet wide. How long is it?

Solution

With this problem, it helps to begin by drawing a diagram, as shown in Figure 4.25. In the diagram, label the width of the rectangle as $w = 200$ feet and the unknown length as l.

$$A = lw \qquad \text{Area of a rectangle}$$

$$100,000 = l(200) \qquad \text{Substitute 100,000 for } A \text{ and 200 for } w.$$

$$100,000 = 200l \qquad \text{Simplify.}$$

$$500 = l \qquad \text{Divide both sides by 200.}$$

The length of the rectangular plot is 500 feet. Check this in the original statement of the problem.

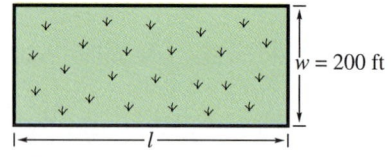

FIGURE 4.25

Miscellaneous Common Formulas

Temperature: F = degrees Fahrenheit, C = degrees Celsius

$$F = \frac{9}{5}C + 32$$

Simple interest: I = interest, P = principal, r = interest rate, t = time

$$I = Prt$$

Distance: d = distance traveled, r = rate, t = time

$$d = rt$$

In some applications, it helps to rewrite a common formula by solving for a different variable. For instance, you can obtain a formula for C (degrees Celsius) in terms of F (degrees Fahrenheit) as follows.

$$F = \frac{9}{5}C + 32 \qquad \text{Temperature formula}$$

$$F - 32 = \frac{9}{5}C \qquad \text{Subtract 32 from both sides.}$$

$$\frac{5}{9}(F - 32) = C \qquad \text{Multiply both sides by } \tfrac{5}{9}.$$

$$C = \frac{5}{9}(F - 32) \qquad \text{Formula}$$

EXAMPLE 3   Simple Interest

An amount of $5000 was deposited in an account paying simple interest. After 6 months, the account earned $162.50 in interest. What is the annual interest rate for this account?

Solution

$$I = Prt \qquad \text{Simple interest formula}$$

$$162.50 = 5000(r)\left(\tfrac{1}{2}\right) \qquad \text{Substitute for } I,\ P,\ \text{and } t.$$

$$162.50 = 2500r \qquad \text{Simplify.}$$

$$\frac{162.50}{2500} = r \qquad \text{Divide both sides by 2500.}$$

$$0.065 = r \qquad \text{Simplify.}$$

The annual interest rate is $r = 0.065$ (or 6.5%). Check this solution in the original statement of the problem.

One of the most familiar and often used formulas in real life is the formula that relates distance, rate (or speed), and time: $d = rt$. For instance, if you are traveling at a constant (or average) rate of 50 miles per hour for 45 minutes, the total distance traveled is given by

$$\left(50\ \frac{\text{miles}}{\text{hour}}\right) \cdot \left(\frac{45}{60}\ \text{hour}\right) = 37.5\ \text{miles}.$$

As with all problems involving applications, be sure to check that the units in the model make sense. For instance, in this problem the rate is given in *miles per hour.* Therefore, in order for the solution to be given in *miles,* we must convert the time (from minutes) to *hours.* In the model, you can think of canceling the two "hours," as follows.

$$\left(50\ \frac{\text{miles}}{\cancel{\text{hour}}}\right) \cdot \left(\frac{45}{60}\ \cancel{\text{hour}}\right) = 37.5\ \text{miles}$$

EXAMPLE 4   A Distance-Rate-Time Problem

You can jog at an average rate of 8 kilometers per hour. How long will it take you to jog 14 kilometers?

Solution

Verbal Model:

$$\boxed{\text{Distance}} = \boxed{\text{Rate}} \cdot \boxed{\text{Time}}$$

Labels:
Distance = 14   (kilometers)
Rate = 8   (kilometers per hour)
Time = t (hours)

Equation:
$$14 = 8(t)$$
$$\frac{14}{8} = t$$
$$1.75 = t$$

It will take you 1.75 hours (or 1 hour and 45 minutes). Check this in the original statement of the problem.

If you are having trouble solving a distance-rate-time problem, consider making a table such as that shown below.

Time (hours)	0.25	0.50	0.75	1.00	1.25	1.50	1.75	2.00
Distance (kilometers)	2	4	6	8	10	12	14	16

Solving Mixture Problems

Many real-world problems involve combinations of two or more quantities that make up a new or different quantity. Such problems are called **mixture problems.** They are usually composed of the sum of two or more "hidden products" that fit the following verbal model.

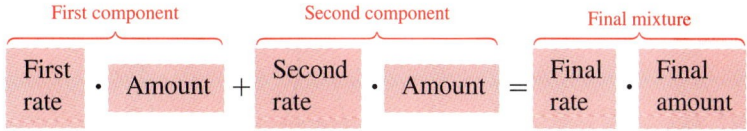

EXAMPLE 5   A Nut Mixture Problem

A grocer wants to mix cashew nuts worth $7 per pound with 15 pounds of peanuts worth $2.50 per pound. To obtain a nut mixture worth $4 per pound, how many pounds of cashews are needed? How many pounds of mixed nuts will be produced for the grocer to sell?

Solution

Verbal Model:

$$\boxed{\begin{array}{c}\text{Total cost}\\\text{of cashews}\end{array}} + \boxed{\begin{array}{c}\text{Total cost}\\\text{of peanuts}\end{array}} = \boxed{\begin{array}{c}\text{Total cost of}\\\text{mixed nuts}\end{array}}$$

Labels:
Cost of cashews $= 7$ (dollars per pound)
Cost of peanuts $= 2.5$ (dollars per pound)
Cost of mixed nuts $= 4$ (dollars per pound)
Amount of cashews $= x$ (pounds)
Amount of peanuts $= 15$ (pounds)
Amount of mixed nuts $= x + 15$ (pounds)

Equation:
$$7(x) + 2.5(15) = 4(x + 15)$$
$$7x + 37.5 = 4x + 60$$
$$3x = 22.5$$
$$x = \frac{22.5}{3}$$
$$x = 7.5$$

The grocer needs 7.5 pounds of cashews. This will result in $7.5 + 15$ or 22.5 pounds of mixed nuts. You can check these results as follows.

$$\overbrace{(\$2.50/\text{lb})(15 \text{ lb})}^{\text{Peanuts}} + \overbrace{(\$7.00/\text{lb})(7.5 \text{ lb})}^{\text{Cashews}} = \overbrace{(\$4.00/\text{lb})(22.5 \text{ lb})}^{\text{Mixed Nuts}}$$
$$\$37.50 + \$52.50 = \$90.00$$

EXAMPLE 6   A Solution Mixture Problem

A pharmacist needs to strengthen a 15% alcohol solution so that it contains 32% alcohol. How much pure alcohol should be added to 100 milliliters of the 15% solution? (See Figure 4.26.)

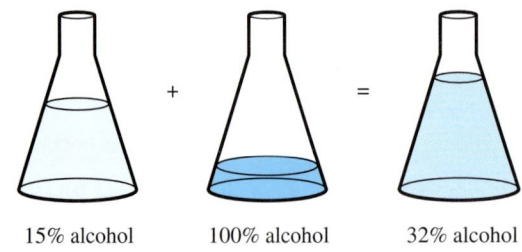

15% alcohol   100% alcohol   32% alcohol

FIGURE 4.26

Solution

Verbal Model:

| Amount of alcohol in original solution | + | Amount of alcohol in pure solution | = | Amount of alcohol in final solution |

Labels:

Alcohol percent of original solution $= 0.15$	(% in decimal form)
Alcohol amount of original solution $= 100$	(milliliters)
Alcohol percent of pure solution $= 1.00$	(% in decimal form)
Alcohol amount of pure solution $= x$	(milliliters)
Alcohol percent of final solution $= 0.32$	(% in decimal form)
Alcohol amount of final solution $= x + 100$	(milliliters)

Equation:
$$0.15(100) + 1.00(x) = 0.32(100 + x)$$
$$15 + x = 32 + 0.32x$$
$$0.68x = 17$$
$$x = \frac{17}{0.68} = 25 \text{ ml}$$

The pharmacist should add 25 milliliters of pure alcohol to the original solution. You can check this in the original statement of the problem as follows.

$$\overbrace{0.15(100)}^{\text{Original}} + \overbrace{1.00(25)}^{\text{Pure}} = \overbrace{0.32(125)}^{\text{Final}}$$
$$15 + 25 = 40$$

Problems such as those shown in Examples 5 and 6 can be difficult to solve. One good way to get started is to use a *Guess, Check, and Revise* approach.

Solving Work-Rate Problems

In **work-rate problems,** the work rate is the *reciprocal* of the time needed to do the entire job. For instance, if it takes 7 hours to complete a job, the per-hour work rate is

$$\frac{1}{7} \text{ job per hour.}$$

Similarly, if it takes $4\frac{1}{2}$ minutes to complete a job, the per-minute rate is

$$\frac{1}{4\frac{1}{2}} = \frac{1}{\frac{9}{2}} = \frac{2}{9} \text{ job per minute.}$$

EXAMPLE 7   A Work-Rate Problem

Consider two machines in a paper manufacturing plant. Machine 1 can produce 2000 pounds of paper in 3 hours. Machine 2 is newer and can produce 2000 pounds of paper in $2\frac{1}{2}$ hours. How long will it take the two machines working together to produce 2000 pounds of paper?

Solution

Verbal Model:

Work done	=	Portion done by Machine 1	+	Portion done by Machine 2

Labels:

Work done $= 1$	(job)
Rate (Machine 1) $= \frac{1}{3}$	(job per hour)
Time (Machine 1) $= t$	(hours)
Rate (Machine 2) $= \frac{2}{5}$	(job per hour)
Time (Machine 2) $= t$	(hours)

Equation:

$$1 = \left(\frac{1}{3}\right)(t) + \left(\frac{2}{5}\right)(t)$$

$$1 = \left(\frac{1}{3} + \frac{2}{5}\right)(t)$$

$$1 = \left(\frac{11}{15}\right)(t)$$

$$\frac{15}{11} = t$$

It would take $\frac{15}{11}$ hours (or about 1.36 hours) for both machines to complete the job. Check this solution in the original statement of the problem.

NOTE   Note in Example 7 that the "2000 pounds" of paper was unnecessary information. We simply represented the 2000 pounds as "one complete job." This unnecessary information was a red herring.

EXAMPLE 8 A Fluid-Rate Problem

An above-ground swimming pool has a capacity of 15,600 gallons, as shown in Figure 4.27. A drain pipe can empty the pool in $6\frac{1}{2}$ hours. At what rate (in gallons per minute) does the water flow through the drain pipe?

Solution

To begin, change the time from hours to minutes by multiplying by 60. That is, $6\frac{1}{2}$ hours is equal to $(6.5)(60)$ or 390 minutes.

Verbal Model: | Volume of pool | = | Rate | · | Time |

Labels:
$$\text{Volume} = 15,600 \qquad \text{(gallons)}$$
$$\text{Rate} = r \qquad \text{(gallons per minute)}$$
$$\text{Time} = 390 \qquad \text{(minutes)}$$

Equation:
$$15,600 = r(390)$$
$$\frac{15,600}{390} = r$$
$$40 = r$$

The water is flowing through the drain pipe at the rate of 40 gallons per minute. You can check this as follows.

$$\left(\frac{40 \text{ gallons}}{\text{minute}}\right)(390 \text{ minutes}) = 15,600 \text{ gallons}.$$

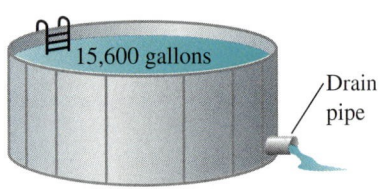

15,600 gallons

Drain pipe

FIGURE 4.27

Group Activities   Extending the Concept

Creating a Formula   You are to design a package from a rectangular 9-inch by 12-inch piece of material by cutting out a square from each corner and folding up the sides, as shown in the figure. In your group, calculate the volume of the different boxes formed by cutting out 1-inch, 2-inch, and 3-inch squares. Together, create a formula for the volume of a box made in this manner.

|← 12 in. →|

9 in.

4.5 Exercises

Discussing the Concepts

1. In your own words, describe the units of measure used for perimeter, area, and volume. Give some examples of each.

2. If the height of a triangle doubled, would the area of the triangle double? Explain.

3. If the radius of a circle doubled, would its circumference double? Would its area double? Explain.

4. It takes you 4 hours to drive 180 miles. Explain how to use mental math to find your average speed. Then explain how your method is related to the formula $d = rt$.

5. It takes you 5 hours to complete a job. What portion do you complete in each hour?

6. Give an example of a mixture problem.

Problem Solving

7. *Geometry* Each room in the floor plan of a house is square (see figure). The perimeter of the bathroom is 32 feet. The perimeter of the kitchen is 80 feet. Find the area of the living room.

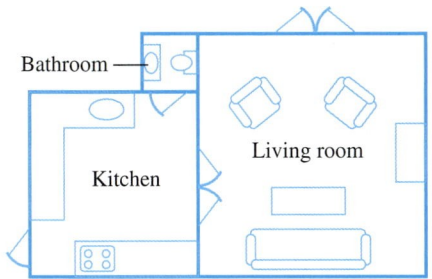

8. *Geometry* The circumference of the wheel in the figure is 30π inches. Find the diameter of the wheel.

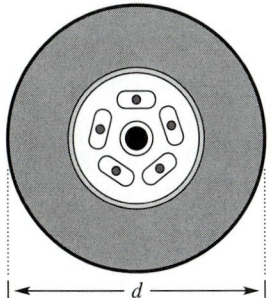

9. *Geometry* A circle has a circumference of 15 meters. What is the radius of the circle? Round your result to two decimal places.

10. *Geometry* A rectangle has a perimeter of 10 feet and a width of 2 feet. Find the length of the rectangle.

Simple Interest In Exercises 11–14, use the formula for simple interest.

11. Find the interest on a $1000 bond paying an annual rate of 9% for 6 years.

12. Find the annual rate on a certificate of deposit that accumulated $128.98 interest in 1 year on a principal of $1500.

13. You borrow $15,000 for $\frac{1}{2}$ year. You promise to pay back the principal and the interest in one lump sum. The annual interest rate is 13%. What is your payment?

14. Six thousand dollars is divided between two investments earning 7% and 9% simple interest. The total interest for 1 year is $500. How much is in each investment?

Geometry In Exercises 15 and 16, evaluate the formula. List the units of measure for your result.

15. The volume of a right circular cylinder is $V = \pi r^2 h$. Find the volume of a right circular cylinder that has a radius of 2 meters and a height of 3 meters.

16. Find the volume of a rectangular solid that has a width of 2 inches, a height of 4 inches, and a length of 5 inches.

In Exercises 17–20, solve each formula for the specified variable.

17. Solve for h: $A = \frac{1}{2}bh$

18. Solve for l: $P = 2l + 2w$

19. Solve for C: $S = C + RC$

20. Solve for b: $A = \frac{1}{2}(a + b)h$

In Exercises 21–26, find the missing distance, rate, or time.

	Distance, d	Rate, r	Time, t
21.		55 mi/hr	3 hr
22.		32 ft/sec	10 sec
23.	500 km	90 km/hr	
24.	128 ft	16 ft/sec	
25.	5280 ft		$\frac{5}{2}$ sec
26.	432 mi		9 hr

27. *Space Shuttle Time*   The speed of a space shuttle is 17,000 miles per hour (see figure). How long will it take the shuttle to travel a distance of 3000 miles?

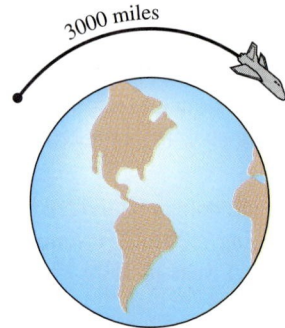

3000 miles

28. *Time*   Two cars start at the same point and travel in the same direction at average speeds of 40 miles per hour and 55 miles per hour. How much time must elapse before the two cars are 5 miles apart?

29. *Speed*   Determine the average speed of an Olympic runner who completes the 10,000-meter race in 27 minutes and 45 seconds.

30. *Distance*   Two planes leave an airport at approximately the same time and fly in opposite directions (see figure). Their speeds are 510 miles per hour and 600 miles per hour. How far apart will the planes be after $1\frac{1}{2}$ hours?

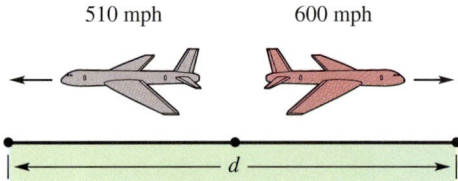

510 mph   600 mph

d

31. *Interpreting a Table*   An agricultural corporation places an order for 100 tons of cattle feed. The feed is to be a mixture of soybeans, which cost $200 per ton, and corn, which costs $125 per ton. Complete the table, where x is the number of tons of corn in the mixture.

Corn Weight x	Soybean Weight $100 - x$	Price per Ton of Mixture
0		
20		
40		
60		
80		
100		

(a) How does the increase in the corn weight affect the soybean weight?

(b) How does the increase in the number of tons of corn affect the price per ton of the mixture?

(c) If there were an equal number of tons of corn and soybeans in the mixture, how would the resulting price of the mixture relate to the price of each component?

32. *Number of Stamps* You have 100 stamps that have a total value of $18. Some of the stamps are worth 15¢ each and others are worth 30¢ each. How many stamps of each type do you have?

33. *Number of Coins* A person has 50 coins in dimes and quarters with a combined value of $7.70. Determine the number of coins of each type.

34. *Nut Mixture* You buy a mixture of 3 pounds of peanuts at $3 per pound and 4 pounds of cashews at $5 per pound. What is the price per pound for the mixture?

35. *Flower Order* A floral shop receives an order for flowers that totals $384. The price per dozen for the roses and carnations are $18 and $12, respectively. The order contains twice as many roses as carnations. How many of each type of flower are in the order?

36. *Work Rate* Suppose you can mow a lawn in 2 hours using a riding mower, and in 3 hours using a push mower. Using both machines together, how long will it take you and a friend to mow the lawn?

Reviewing the Major Concepts

In Exercises 37–42, evaluate the expression.

37. $-50 - 4(3 - 8)$

38. $(-5)^2 + 3$

39. $\dfrac{-|7 + 3^2|}{4}$

40. $\left(-\dfrac{7}{12}\right)\left(\dfrac{3}{28}\right)$

41. $|-3 + 5(6 - 8)|$

42. $|-2 - 9| - |-7 + 14|$

43. *Sales Tax* You buy a computer for $2750 plus 6% sales tax. What is your total bill?

44. *Comparing Prices* A mail-order catalog lists an area rug for $109.95, plus a shipping charge of $14.25. A local store has a special sale on the same rug with 20% off a list price of $139.99. Which is the better bargain?

Additional Problem Solving

45. *Geometry* A triangle has an area of 48 square meters and a height of 12 meters. Find the length of the base.

46. *Geometry* A circular cylinder with radius 6 inches has a volume of 144π cubic inches. Find the height of the cylinder.

47. *Geometry* The perimeter of a square is 48 feet. Find its area.

48. *Geometry* A circle has a circumference of 25 meters. Find the radius and area of the circle. Round your results to two decimal places.

In Exercises 49–52, use the closed rectangular box shown in the figure to answer the question.

49. Find the area of the base.

50. Find the perimeter of the base.

51. Find the volume of the box.

52. Find the surface area of the box. (*Note:* This is the combined areas of the six surfaces.)

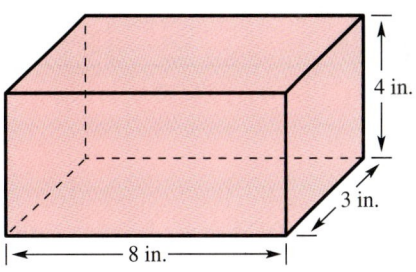

Figure for 49–52

Simple Interest In Exercises 53–56, use the formula for simple interest.

53. Find the annual interest rate on a savings account that earns $110 interest in 1 year on a principal of $1000.

54. How long must $1000 be invested at an annual interest rate of $7\frac{1}{2}\%$ to earn $225 interest?

55. Find the principal required to earn $408 interest in 4 years, if the annual interest rate is $8\frac{1}{2}\%$.

56. An inheritance of $25,000 is divided into two investments earning 8.5% and 10% simple interest, respectively. How much is in each investment if the total interest for 1 year is $2350?

In Exercises 57–64, solve for the specified variable.

57. Solve for R: $E = IR$.

58. Solve for r: $C = 2\pi r$.

59. Solve for l: $V = lwh$.

60. Solve for h: $V = \pi r^2 h$.

61. Solve for r: $A = P + Prt$.

62. Solve for L: $S = L - RL$.

63. Solve for r: $V = \frac{1}{3}\pi h^2(3r - h)$.

64. Solve for b: $V = \frac{4}{3}\pi a^2 b$.

65. *Distance*   Two cars start at a given point and travel in the same direction at average speeds of 45 miles per hour and 52 miles per hour (see figure). How far apart will they be in 4 hours?

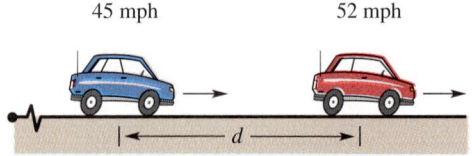

45 mph   52 mph

66. *Speed of Light*   The speed of light is 670,616,625.6 miles per hour, and the distance between earth and the sun is 93,000,000 miles. How long does it take light from the sun to reach earth?

67. *Speed*   Determine the average speed of an experimental plane that can travel 3000 miles in 2.6 hours.

68. *Time*   Suppose on the first part of a 225-mile automobile trip you averaged 55 miles per hour. On the remainder of the trip you averaged 48 miles per hour because of increased traffic congestion. The total trip took 4 hours and 15 minutes. Find the amount of time at each speed.

In Exercises 69–72, determine the numbers of units of alcohol solutions 1 and 2 needed to obtain the desired amount and the alcohol concentration of the final solution.

	Concentration Solution 1	Concentration Solution 2	Concentration Final Solution	Amount of Final Solution
69.	10%	30%	25%	100 gal
70.	25%	50%	30%	5 liters
71.	15%	45%	30%	10 qt
72.	70%	90%	75%	25 gal

73. *Antifreeze*   The cooling system in a truck contains 4 gallons of coolant that is 30% antifreeze. How much must be withdrawn and replaced with 100% antifreeze to bring the coolant in the system to 50% antifreeze?

74. *Interpreting a Table*   A metallurgist is making 5 ounces of an alloy out of metal A, which costs $52 per ounce, and metal B, which costs $16 per ounce. Complete the following table where x is the number of ounces of metal A in the alloy.

Metal A x	Metal B $5 - x$	Price/Ounce of the Alloy
0		
1		
2		
3		
4		
5		

(a) How does the increase in the number of ounces of metal A in the alloy affect the number of ounces of metal B in the alloy?

(b) How does the increase in the number of ounces of metal A affect the price of the alloy?

(c) If there were equal amounts of metal A and metal B in the alloy, how would the price of the alloy relate to the price of each of the components?

75. *Nut Mixture* A grocer mixes two kinds of nuts that cost $2.49 and $3.89 per pound to make 100 pounds of a mixture that costs $3.47 per pound. How many pounds of each kind of nut were put into the mixture?

76. *Ticket Sales* Ticket sales for a play total $1700. The number of tickets sold to adults is three times the number sold to children. The prices of the tickets for adults and children are $5 and $2, respectively. How many of each type were sold?

77. *Number of Coins* A person has 20 coins in nickels and dimes with a combined value of $1.60. Determine the number of coins of each type.

78. *Number of Stamps* You have 100 stamps that have a total value of $25.20. Some of the stamps are worth 15¢ each and others are worth 30¢ each. How many stamps of each type do you have?

79. *Poll Results* One thousand people were surveyed in an opinion poll. Candidates A and B received approximately the same number of votes. Candidate C received twice as many votes as each of the other two candidates. How many votes did each candidate receive?

80. *Poll Results* One thousand people were surveyed in an opinion poll. The numbers of votes for candidates A, B, and C were in the ratio of 5 to 3 to 2, respectively. How many voted for each candidate?

81. *Work Rate* One person can complete a typing project in 6 hours, and another can complete the same project in 8 hours. If they both work on the project, in how many hours can it be completed?

82. *Work Rate* One worker can complete a task in h hours while a second can complete the task in $3h$ hours. Show that by working together they can complete the task in $t = \frac{3}{4}h$ hours.

83. *Ages* A mother was 30 years old when her son was born. How old will the son be when his age is $\frac{1}{3}$ his mother's age?

84. *Ages* The difference in age between a father and daughter is 32 years. Determine the age of the father when his age is twice that of his daughter.

Math Matters

Composition of the Earth's Atmosphere

The atmosphere of the earth is a gaseous envelope of air that encloses our planet. It consists of colorless, odorless, and tasteless gases, and particles of dust. The composition of the air that we breathe is relatively constant throughout the world, although there are some slight variations depending on the zone and altitude. The eight basic gases that make up our atmosphere are as follows.

Gas	Percent	Gas	Percent
Nitrogen	78.08%	Neon	0.0018%
Oxygen	20.95%	Helium	0.0005%
Argon	0.93%	Krypton	0.0001%
Carbon dioxide	0.03%	Xenon	0.00001%

If you add these percentages, you will see that the total is not 100%. The remaining 0.00759% consists of small amounts of hydrocarbons, hydrogen, peroxide, sulfur, water vapor, and dust.

CHAPTER PROJECT: Sporting Goods Sales

Standard Scatter Plot

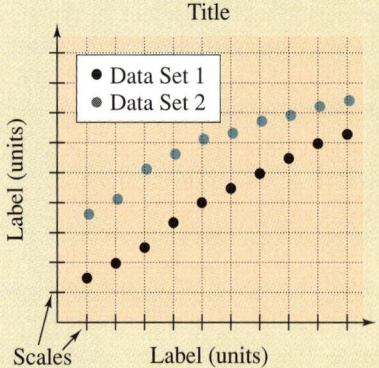

Figure A

Scatter plots can be used to recognize patterns in data. A standard scatter plot, such as that shown in Figure A, consists of a labeled vertical axis with a scale, a labeled horizontal axis with a scale, and usually a title. To analyze more than one set of data in a scatter plot, a legend is needed to distinguish between the data.

If the data in a scatter plot appears to be linear, then it is possible to find a linear model of the form $y = mx + b$ that best fits the data. After determining the linear model, you can use it to make predictions about future data, such as predicting the future sales of a product.

The scatter plots shown in Figures B and C represent sales of golfing equipment, and of bowling accessories and camping equipment, for several years (see table below). In Figure B, the best-fitting linear model is also shown.

Sales of Athletic and Sporting Equipment (in millions of dollars)

Equipment	1985	1986	1987	1988	1989	1990	1991
Bowling Accessories	106	114	129	129	143	155	155
Camping Equipment	724	833	858	945	996	1072	1016
Golfing Equipment	730	828	946	1111	1167	1219	1149
Water Skis	125	132	148	160	96	88	63

(Source: National Sporting Goods Association)

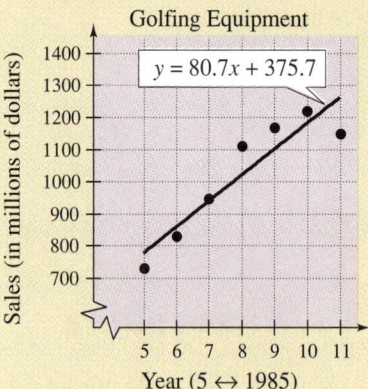

Figure B

Use the scatter plots and table to investigate the following questions.

1. Use the linear model given for the sales of golfing equipment in Figure 4B to predict the sales of golfing equipment in 1995.

2. Is the linear model in Figure B an appropriate model? Explain.

3. In Figure C, which color denotes sales of bowling accessories?

4. The best-fitting linear models for bowling accessories and camping equipment are listed below. Which linear model corresponds to each type of equipment?
   (a) $y = 53.3x + 494.3$ (b) $y = 8.7x + 63.6$

5. Would a linear model accurately predict the sales of water skis (see table) in 1996? Explain using a scatter plot.

6. *Research Project*   Use your school's library or some other reference source to find data for the sales of some commodity. Organize the data graphically. Then describe how you could use the data to predict future sales of the commodity.

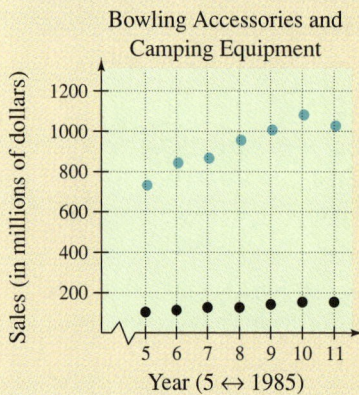

Figure C

CHAPTER SUMMARY

After studying this chapter, you should have acquired the following skills. These skills are keyed to the Review Exercises that begin on page 270. Answers to odd-numbered Review Exercises are given in the back of the book.

- Plot points on a rectangular coordinate system. *(Section 4.1)*   **Review Exercises 1, 2**

- Determine the quadrants in which points are located. *(Section 4.1)*   **Review Exercises 3–6**

- Decide whether ordered pairs are solutions of equations. *(Section 4.1)*   **Review Exercises 7, 8**

- Interpreting real-life data represented by graphs. *(Section 4.1)*   **Review Exercise 9**

- Graph and interpret real-life data given in tables. *(Section 4.1)*   **Review Exercise 10**

- Sketch graphs of equations and determine any intercepts. *(Section 4.2)*   **Review Exercises 11–18**

- Translate real-life statements into algebraic equations and graphs. *(Section 4.2)*   **Review Exercise 27**

- Use a graphing utility to graph equations using a standard setting. *(Section 4.3)*   **Review Exercises 19–22**

- Use a graphing utility to graph equations with a given range. *(Section 4.3)*   **Review Exercises 23, 24**

- Use the TRACE feature of a graphing utility to approximate the x- and y-intercepts of graphs. *(Section 4.3)*   **Review Exercises 25, 26**

- Graph and interpret real-life data modeled by algebraic equations. *(Section 4.3)*   **Review Exercise 28**

- Translate real-life statements into algebraic equations and solve. *(Section 4.4)*   **Review Exercises 29–34**

- Solve problems using geometric formulas. *(Section 4.5)*   **Review Exercises 37, 38**

- Use formulas to solve real-life problems. *(Section 4.5)*   **Review Exercises 35, 36, 39, 40, 43, 44**

- Solve formulas for specified variables. *(Section 4.5)*   **Review Exercises 41, 42**

REVIEW EXERCISES

In Exercises 1 and 2, plot the points on a rectangular coordinate system.

1. $(-2, 0)$, $\left(\frac{3}{2}, 4\right)$, $(-1, -3)$

2. $\left(3, -\frac{5}{2}\right)$, $\left(-5, 2\frac{3}{4}\right)$, $(4, 6)$

In Exercises 3–6, determine the quadrant or quadrants in which the points must be located.

3. $(-5, 3)$ **4.** $(4, -6)$

5. $(x, 5)$, $x < 0$ **6.** (x, y), $y > 0$

In Exercises 7 and 8, decide whether each ordered pair is a solution of the equation.

7. $y = \frac{2}{3}x + 3$ (a) $(3, 5)$ (b) $(-2, 0)$

8. $y = 4 - |x + 2|$ (a) $(-4, 6)$ (b) $(6, 0)$

9. *Graphical Interpretation*   The line graph shows the receipts (in billions of dollars) for businesses associated with lodging, food, and recreation in the travel industry in the United States for the years 1981 through 1990.   (Source: U. S. Travel Data Center)

(a) Approximate the receipts for eating places in the travel industry in 1985.

(b) In what year was approximately 40 billion dollars spent on recreation in the travel industry?

(c) Approximate the percent increase in receipts of lodging establishments from 1988 to 1989.

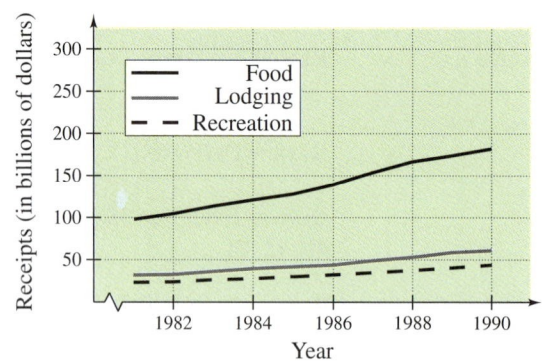

10. *Organizing Data*   The data from a study measuring the relationship between the wattage x of a standard light bulb and the energy rate y (in lumens) is given in the table.   (Source: Standard Handbook for Mechanical Engineers)

x	25	40	60	100	150	200
y	266	470	840	1750	2700	4000

(a) Plot the data given in the table.

(b) Describe the relationship between x and y.

(c) Estimate the value of y when $x = 125$. Is it easier to use the table or the graph to estimate this value? Explain your reasoning.

In Exercises 11–18, sketch the graph of the equation and determine any intercepts of the graph.

11. $y = 4 - \frac{1}{2}x$ **12.** $y = \frac{3}{2}x - 1$

13. $y - 2x - 3 = 0$ **14.** $3x + 2y + 6 = 0$

15. $y = 3 - |x|$ **16.** $y = |x - 2|$

17. $y = (x - 4)^2$ **18.** $y = x(4 - x)$

In Exercises 19–22, use a graphing utility to graph the equation. (Use a standard setting.)

19. $y = \frac{7}{8}x + 1$ **20.** $y = 5 - 2x$

21. $y = -\frac{1}{4}x^2 + x$ **22.** $y = x(x^2 - 4)$

In Exercises 23 and 24, graph the equation using a graphing utility.

23. $y = 250 - 50x$ **24.** $y = 800 + 9x$

| Xmin = -5 |
| Xmax = 5 |
| Xscl = 1 |
| Ymin = 0 |
| Ymax = 500 |
| Yscl = 25 |

| Xmin = 0 |
| Xmax = 100 |
| Xscl = 10 |
| Ymin = 500 |
| Ymax = 2000 |
| Yscl = 200 |

In Exercises 25 and 26, use the TRACE feature of a graphing utility to approximate the x- and y-intercepts of the graph.

25. $y = 5.2 - |x|$ **26.** $y = x^2 - 4x + 3$

27. *Business Expense*   A company reimburses its sales representatives $125 per day for lodging and meals plus 27¢ per mile driven. Write an equation giving the daily cost y to the company in terms of x, the number of miles driven. Graph the equation.

28. *Graphical Estimation*   A person who weighs 180 pounds begins a diet of 1500 calories per day. The person's weight, y, after x weeks of dieting is approximated by

$$y = 0.014x^2 - 1.218x + 180, \quad 0 \le x \le 26.$$

(a) Use a graphing utility to graph the equation.

(b) Approximate the person's weight after 26 weeks.

(c) Approximate the time required to lose 5 pounds.

29. *Price*   A food processor that costs a retailer $85 is marked up 35%. Find the price for the consumer.

30. *Markup Rate*   A retailer pays $155 for a set of tools that is resold for $199.98. Find the markup rate.

31. *Sale Price*   While shopping for a new suit you find one you like with a list price of $279. The list price has been reduced by 35% to form the sale price. Find the sale price of the suit.

32. *Tip Rate*   A customer left $40 for a meal that cost $34.25. Determine the tip rate.

33. *Comparing Two Prices*   A mail-order catalog lists the price of luggage at $89 plus $4 for shipping and handling. A local department store has the same luggage for $112.98. The department store has a special 20%-off sale. Which is the better price?

34. *Sales Commission*   The weekly salary of an employee is $250 plus a 6% commission on the employee's total sales. How much must the employee sell in order to obtain a weekly salary of $600?

35. *Time*   A train's average speed is 60 miles per hour. How long will it take the train to travel 562 miles?

36. *Speed*   For the first hour of a 350-mile trip, your average speed is 40 miles per hour. Determine the average speed that must be maintained for the remainder of the trip if you want the average speed for the entire trip to be 50 miles per hour.

37. *Dimensions of a Swimming Pool*   The width of a rectangular swimming pool is 4 feet less than its length. The perimeter of the pool is 112 feet. Find the dimensions of the pool.

38. *Dimensions of a Triangle*   The perimeter of an isosceles triangle is 65 centimeters. Find the length of the two equal sides if each is 10 centimeters longer than the third side. (An isosceles triangle has two equal sides, and its perimeter is the sum of the lengths of its three sides.)

39. *Simple Interest*   Find the principal required to have an annual interest income of $25,000 if the annual interest rate on the principal is 8.75%.

40. *Simple Interest*   You invest $2500 in a certificate of deposit that has an annual interest rate of 7%. After 6 months, the interest is computed and added to the principal. During the second 6 months, the interest is computed using the original investment plus the interest earned during the first 6 months. What is the total interest earned during the first year of the investment?

In Exercises 41 and 42, solve the formula for the indicated variable.

41. Solve for θ: $A = \dfrac{r^2 \theta}{2}$.

42. Solve for n: $S = a + (n-1)d$.

43. *Work Rate*   Find the time for two people working together to complete a task that, if they work individually, takes them 5 hours and 6 hours, respectively.

44. *Work Rate*   Suppose the person in Exercise 43 who can complete the task in 5 hours has already worked 1 hour when the second person starts. How long will they work together to complete the task?

CHAPTER TEST

Take this test as you would take a test in class. After you are done, check your work against the answers given in the back of the book.

1. Plot the points $(-1, 2)$, $(1, 4)$, and $(2, -1)$ on a rectangular coordinate system. Connect the points with line segments to form a right triangle.

2. When an employee produces x units per hour, the hourly wage is $y = 0.75x + 4$. Complete the table at the right to determine the hourly wages for producing the specified numbers of units. Plot the results.

3. Which ordered pairs are solutions of $y = |x| + |x - 2|$?

   (a) $(0, -2)$ (b) $(0, 2)$ (c) $(-4, 10)$ (d) $(-2, 2)$

4. Find the x- and y-intercepts of the graph of $3x - 4y + 12 = 0$.

In Exercises 5–8, use a graphing utility to sketch the graph of the equation.

5. $x - 2y = 6$ 6. $y = |x + 2|$

7. $y = 0.6(x - 2)$ 8. $y = 9 - (x - 3)^2$

9. A calculator is marked up from $80 to $112. What is the markup rate?

10. A person has 20 coins in dimes and quarters with a combined value of $3.80. Determine the number of coins of each type.

11. Roses cost $18 per dozen and carnations cost $9 per dozen. How many roses are in a $12 arrangement of one dozen roses and carnations?

12. You traveled 264 miles in $5\frac{1}{2}$ hours. What was your average speed?

13. You can paint a building in 9 hours. Your friend would take 12 hours. Working together, how long will it take the two of you to paint the building?

14. Find three consecutive integers whose sum is 93.

15. Solve for R in the following formula: $S = C + RC$.

16. Find the total interest for a 6-month $1000 bond paying an annual interest rate of 8%. (Use the formula for simple interest.)

17. How much must you deposit to earn $500 per year at 8% simple interest?

In Exercises 18–20, use the figure at the right.

18. Find the area of the blue region.

19. Find the perimeter of the green region. Describe the method you used.

20. Find the area of the green region. Describe the method you used.

x	$y = 0.75x + 4$
2	
4	
6	
8	
10	
12	

Table for 2

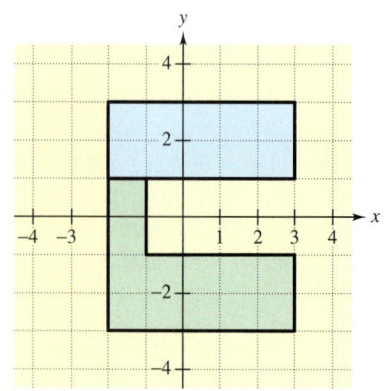

Figure for 18–20

Exponents and Polynomials

5

- Adding and Subtracting Polynomials
- Multiplying Polynomials: Special Products
- Dividing Polynomials
- Negative Exponents and Scientific Notation

Sir Isaac Newton, a 17th century English scientist, astronomer, and mathematician, discovered that white light passed through a prism produces bands of colors. This is called the continuous spectrum. The portion of this that can be seen by the human eye is called the visible spectrum, as shown at the right.

Newton's discovery has enabled scientists to determine the chemical compositions of substances with a spectroscope. When an element is heated to a high temperature, it gives off light. This light forms a special pattern called a line spectrum. For example, when the element nitrogen is heated, it forms the line spectrum shown at the upper right. Each element's spectrum is unique. For instance, oxygen's spectrum is different from nitrogen's spectrum.

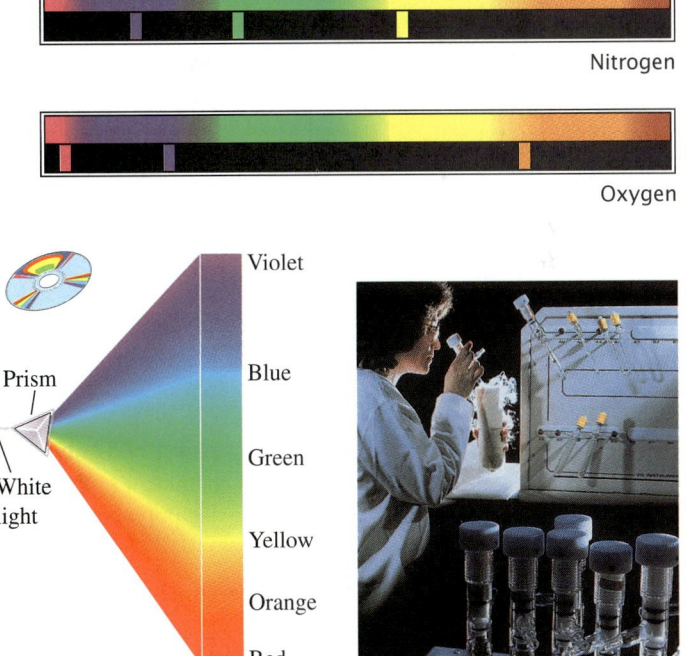

Nitrogen

Oxygen

Violet

Blue

Green

Yellow

Orange

Red

Prism

White light

The chapter project related to this data is on page 313.

5.1   Adding and Subtracting Polynomials

Basic Definitions ▪ Adding Polynomials ▪ Subtracting Polynomials

Basic Definitions

To work with polynomials, you need to know the following rules for exponents, which were discussed in Section 2.1.

1. $a^m \cdot a^n = a^{m+n}$ Multiply powers having the same base.

2. $(a^m)^n = a^{m \cdot n}$ Raise a power to a power.

3. $(ab)^m = a^m b^m$ Raise a product to a power.

Additional rules for exponents will be introduced in this chapter.

Remember that the *terms* of an algebraic expression are those parts separated by addition. An algebraic expression whose terms are all of the form ax^k, where a is any real number and k is a nonnegative integer, is called a **polynomial in one variable,** or simply a **polynomial.** Here are some examples of polynomials in one variable.

$$2x + 5, \quad x^2 - 3x + 7, \quad \text{and} \quad x^3 + 8$$

In the term ax^k, a is the **coefficient** of the term and k is the **degree** of the term. Because a polynomial is an algebraic sum, the coefficients take on the signs between the terms. For instance,

$$x^4 + 2x^3 - 5x^2 + 7 = (1)x^4 + 2x^3 + (-5)x^2 + (0)x + 7$$

has coefficients 1, 2, −5, 0, and 7. For this polynomial, the last term, 7, is the **constant term.** Polynomials are usually written in the order of descending powers of the variable. This is called **standard form.** Here are two examples.

Nonstandard Form	Standard Form
$4 + x$	$x + 4$
$3x^2 - 5 - x^3 + 2x$	$-x^3 + 3x^2 + 2x - 5$

The **degree of a polynomial** is the degree of the term with the highest power, and the coefficient of this term is the **leading coefficient** of the polynomial. For instance, the polynomial

Leading coefficient

$$-3x^4 + 4x^2 + x + 7$$

is of fourth degree, and its leading coefficient is −3. The reasons why the degree of a polynomial is important will become clear as you study factoring and problem solving in Chapter 6.

Definition of a Polynomial in x

Let a_n, $a_{n-1} \ldots$, a_2, a_1, a_0 be real numbers and let n be a *nonnegative integer.* A **polynomial in x** is an expression of the form

$$a_n x^n + a_{n-1} x^{n-1} + \cdots + a_2 x^2 + a_1 x + a_0$$

where $a_n \neq 0$. The polynomial is of **degree** n, and the number a_n is the **leading coefficient.** The number a_0 is the **constant term.**

NOTE   The following are *not* polynomials for the reasons stated.

$$2x^{-1} + 5 \qquad\quad \text{Exponent in } 2x^{-1} \text{ is } not \text{ nonnegative.}$$
$$x^3 + 3x^{1/2} \qquad\quad \text{Exponent in } 3x^{1/2} \text{ is } not \text{ an integer.}$$

EXAMPLE 1   *Identifying Degrees and Leading Coefficients*

	Polynomial	Standard Form	Degree	Leading Coefficient
a.	$4x^2 - 5x^7 - 2 + 3x$	$-5x^7 + 4x^2 + 3x - 2$	7	-5
b.	$4 - 9x^2$	$-9x^2 + 4$	2	-9
c.	8	8	0	8
d.	$2 + x^3 - 5x^2$	$x^3 - 5x^2 + 2$	3	1

In part (c) note that a polynomial with *only* a constant term has a degree of zero.

A polynomial with only one term is called a **monomial.** Polynomials with two *unlike* terms are called **binomials,** and those with three *unlike* terms are called **trinomials.**

EXAMPLE 2   *Classifying Polynomials by Number of Terms*

Classify the following polynomials.

a. $3x^2$ **b.** $-3x + 1$ **c.** $4x^2 - 5x + 6$

Solution

a. The polynomial $3x^2$ is a *monomial* because it has only one term.

b. The polynomial $-3x + 1$ is a *binomial* because it has two terms.

c. The polynomial $4x^2 - 5x + 6$ is a *trinomial* because it has three terms.

NOTE   The prefix *poly* means many. For instance, a polygon is a many-sided figure. Similarly, the prefix *mono* means one, the prefix *bi* means two, and the prefix *tri* means three.

Adding Polynomials

As with algebraic expressions, the key to adding two polynomials is to recognize *like* terms—those having the *same* degree. By the Distributive Property, you can then combine the like terms using either a horizontal or a vertical arrangement of terms. For instance, the polynomials $2x^2 + 3x + 1$ and $x^2 - 2x + 2$ can be added horizontally to obtain

$$(2x^2 + 3x + 1) + (x^2 - 2x + 2) = (2x^2 + x^2) + (3x - 2x) + (1 + 2)$$
$$= (2 + 1)x^2 + (3 - 2)x + (1 + 2)$$
$$= 3x^2 + x + 3$$

or they can be added vertically to obtain the same result.

$$\begin{array}{r} 2x^2 + 3x + 1 \\ x^2 - 2x + 2 \\ \hline 3x^2 + x + 3 \end{array}$$

STUDY TIP

When you use this vertical arrangement to add polynomials, be sure that you line up the *like terms*.

Technology

You can use a graphing utility to check the results of adding or subtracting polynomials. For instance, try graphing

$$y = (2x + 1) + (-3x - 4)$$

and

$$y = -x - 3$$

on the same screen, as shown below. Because both graphs are the same, you can reason that

$$(2x+1)+(-3x-4)=-x-3.$$

This graphing technique is called "graph the left side and graph the right side."

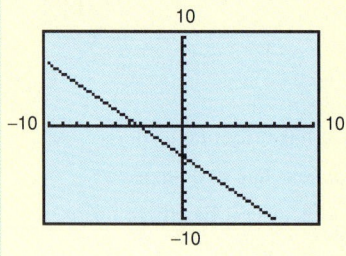

EXAMPLE 3   Adding Polynomials Horizontally

Use a horizontal arrangement to find the sum.

a. $(x^3 + 2x^2 + 4) + (3x^2 - x + 5)$ Original polynomials

$\qquad = (x^3) + (2x^2 + 3x^2) + (-x) + (4 + 5)$ Group like terms.

$\qquad = x^3 + 5x^2 - x + 9$ Standard form

b. $(2x^2 - x + 3) + (4x^2 - 7x + 2) + (-x^2 + x - 2)$

$\qquad = (2x^2 + 4x^2 - x^2) + (-x - 7x + x) + (3 + 2 - 2)$

$\qquad = 5x^2 - 7x + 3$

EXAMPLE 4   Adding Polynomials Vertically

Use a vertical arrangement to find the sum.

a. $(-4x^3 - 2x^2 + x - 5) + (2x^3 + 3x + 4)$

b. $(5x^3 + 2x^2 - x + 7) + (3x^2 - 4x + 7) + (-x^3 + 4x^2 - 2x - 8)$

Solution

a. $\begin{array}{r} -4x^3 - 2x^2 + x - 5 \\ 2x^3 + 3x + 4 \\ \hline -2x^3 - 2x^2 + 4x - 1 \end{array}$

b. $\begin{array}{r} 5x^3 + 2x^2 - x + 7 \\ 3x^2 - 4x + 7 \\ -x^3 + 4x^2 - 2x - 8 \\ \hline 4x^3 + 9x^2 - 7x + 6 \end{array}$

Subtracting Polynomials

To subtract one polynomial from another, you *add the opposite* by changing the sign of each term of the polynomial that is being subtracted and then adding the resulting like terms. For instance, you can subtract the polynomial $(x^2 - 1)$ from the polynomial $(2x^2 - 4)$ as follows.

$$(2x^2 - 4) - (x^2 - 1) = 2x^2 - 4 - x^2 + 1 \qquad \text{Distributive Property}$$
$$= (2x^2 - x^2) + (-4 + 1) \qquad \text{Group like terms.}$$
$$= x^2 - 3 \qquad \text{Combine like terms.}$$

Recall from the Distributive Property that

$$-(x^2 - 1) = (-1)(x^2 - 1)$$
$$= -x^2 + 1.$$

EXAMPLE 5   Subtracting Polynomials Horizontally

Perform the following operations.

a. $(2x^2 + 3) - (3x^2 - 4)$

b. $(3x^3 - 4x^2 + 3) - (x^3 + 3x^2 - x - 4)$

c. $(x^2 - 2x + 1) - [(x^2 + x - 3) + (-2x^2 - 4x)]$

Solution

a. $(2x^2 + 3) - (3x^2 - 4) = 2x^2 + 3 - 3x^2 + 4 \qquad \text{Distributive Property}$
$$= (2x^2 - 3x^2) + (3 + 4) \qquad \text{Group like terms.}$$
$$= -x^2 + 7 \qquad \text{Combine like terms.}$$

b. $(3x^3 - 4x^2 + 3) - (x^3 + 3x^2 - x - 4) \qquad \text{Original polynomials}$
$$= 3x^3 - 4x^2 + 3 - x^3 - 3x^2 + x + 4 \qquad \text{Distributive Property}$$
$$= (3x^3 - x^3) + (-4x^2 - 3x^2) + (x) + (3 + 4) \qquad \text{Group like terms.}$$
$$= 2x^3 - 7x^2 + x + 7 \qquad \text{Combine like terms.}$$

c. $(x^2 - 2x + 1) - [(x^2 + x - 3) + (-2x^2 - 4x)] \qquad \text{Original polynomials}$
$$= (x^2 - 2x + 1) - [(x^2 - 2x^2) + (x - 4x) + (-3)] \qquad \text{Group like terms.}$$
$$= (x^2 - 2x + 1) - [-x^2 - 3x - 3] \qquad \text{Combine like terms.}$$
$$= x^2 - 2x + 1 + x^2 + 3x + 3 \qquad \text{Distributive Property}$$
$$= (x^2 + x^2) + (-2x + 3x) + (1 + 3) \qquad \text{Group like terms.}$$
$$= 2x^2 + x + 4 \qquad \text{Combine like terms.}$$

Be especially careful to use the correct signs when subtracting one polynomial from another. One of the most common mistakes in algebra is to forget to change signs correctly when subtracting one expression from another. Here is an example.

Wrong sign
↓

$$(x^2 + 3) - (x^2 + 2x - 2) \neq x^2 + 3 - x^2 + 2x - 2 \qquad \text{Common Error}$$

↑
Wrong sign

Note that the error is forgetting to change two of the signs in the polynomial that is being subtracted. Here is the correct way to perform the subtraction.

Correct sign
↓

$$(x^2 + 3) - (x^2 + 2x - 2) = x^2 + 3 - x^2 - 2x + 2 \qquad \text{Correct}$$

↑
Correct sign

Just as you did for addition, you can use a vertical arrangement to subtract one polynomial from another. (The vertical arrangement doesn't work well with subtractions involving three or more polynomials.) When using a vertical arrangement, write the polynomial being subtracted underneath the one it is being subtracted from. Be sure to align like terms in vertical columns.

EXAMPLE 6   Subtracting Polynomials Vertically

Use a vertical arrangement to perform the following operations.

a. $(3x^2 + 7x - 6) - (3x^2 + 7x)$

b. $(4x^4 - 2x^3 + 5x^2 - x + 8) - (3x^4 - 2x^3 + 3x - 4)$

Solution

a.
$$\begin{array}{l} (3x^2 + 7x - 6) \\ -(3x^2 + 7x \qquad) \end{array} \quad \Longrightarrow \quad \begin{array}{l} 3x^2 + 7x - 6 \\ -3x^2 - 7x \\ \hline \qquad\qquad - 6 \end{array}$$

Change signs and add.

Combine like terms.

b.
$$\begin{array}{l} (4x^4 - 2x^3 + 5x^2 - x + 8) \\ -(3x^4 - 2x^3 \qquad + 3x - 4) \end{array} \quad \Longrightarrow \quad \begin{array}{l} 4x^4 - 2x^3 + 5x^2 - x + 8 \\ -3x^4 + 2x^3 \qquad\quad - 3x + 4 \\ \hline x^4 \qquad\quad + 5x^2 - 4x + 12 \end{array}$$

NOTE   In Example 6, try using a horizontal arrangement to perform the subtractions. Which arrangement do you prefer?

EXAMPLE 7 Combining Polynomials

Perform the indicated operations.

a. $(3x^2 - 7x + 2) - (4x^2 + 6x - 1) + (-x^2 + 4x + 5)$

b. $(-2x^2 + 4x - 3) - [(4x^2 - 5x + 8) - (-x^2 + x + 3)]$

c. $3(x^2 - 2x + 1) - 2(x^2 + x - 3)$

Solution

a. $(3x^2 - 7x + 2) - (4x^2 + 6x - 1) + (-x^2 + 4x + 5)$

$$= 3x^2 - 7x + 2 - 4x^2 - 6x + 1 - x^2 + 4x + 5$$

$$= (3x^2 - 4x^2 - x^2) + (-7x - 6x + 4x) + (2 + 1 + 5)$$

$$= -2x^2 - 9x + 8$$

b. $(-2x^2 + 4x - 3) - [(4x^2 - 5x + 8) - (-x^2 + x + 3)]$

$$= (-2x^2 + 4x - 3) - [4x^2 - 5x + 8 + x^2 - x - 3]$$

$$= (-2x^2 + 4x - 3) - [(4x^2 + x^2) + (-5x - x) + (8 - 3)]$$

$$= (-2x^2 + 4x - 3) - [5x^2 - 6x + 5]$$

$$= -2x^2 + 4x - 3 - 5x^2 + 6x - 5$$

$$= (-2x^2 - 5x^2) + (4x + 6x) + (-3 - 5)$$

$$= -7x^2 + 10x - 8$$

c. $3(x^2 - 2x + 1) - 2(x^2 + x - 3) = 3x^2 - 6x + 3 - 2x^2 - 2x + 6$

$$= (3x^2 - 2x^2) + (-6x - 2x) + (3 + 6)$$

$$= x^2 - 8x + 9$$

Group Activities   Extending the Concept

Adding Polynomials   Write a paragraph that explains how the adage "You can't add apples and oranges" might relate to adding two polynomials. Include several examples to illustrate the applicability of this statement. Share your paragraph and examples with your group to see if they make sense to others.

5.1 Exercises

Discussing the Concepts

1. Explain the difference between the degree of a term of a polynomial and the degree of a polynomial.

2. Determine which of the two statements is always true. Is the statement not selected always false? Explain.

(a) "A polynomial is a trinomial."

(b) "A trinomial is a polynomial."

3. In your own words, define "like terms." What is the only factor of like terms that can differ?

4. Describe how to combine like terms. What operations are used?

5. Is a polynomial an algebraic expression? Explain.

6. In your own words, explain how to subtract polynomials. Give an example.

Problem Solving

In Exercises 7–10, write the polynomial in standard form. Then find its degree and leading coefficient.

7. $5 - 32x$

8. $5x^3 - 3x^2 + 10$

9. $8x + 2x^5 - x^2 - 1$

10. -32

In Exercises 11–14, determine whether the polynomial is a monomial, binomial, or trinomial.

11. $x^3 - 4$

12. $u^2 - 3u + 5$

13. 5

14. $16 - z^2$

In Exercises 15–18, determine whether the expression is a polynomial. If it is not, explain why.

15. $\dfrac{6}{x}$

16. $t^2 - 4$

17. $9 - z$

18. $9 - z^{1/2}$

In Exercises 19–22, give an example of a polynomial that fits the description. (*Note:* There are many correct answers.)

19. A binomial in one variable of degree 3

20. A trinomial in one variable of degree 4

21. A monomial in one variable of degree 2

22. A binomial in one variable of degree 5

In Exercises 23–26, use a vertical arrangement to perform the polynomial addition.

23. $\begin{array}{r} -x^3 \qquad\ + 3 \\ \underline{3x^3 + 2x^2 + 5} \end{array}$

24. $\begin{array}{r} 2z^3 \qquad + 3z - 2 \\ \underline{z^2 - 2z} \end{array}$

25. $(2 - 3y) + (y^4 + 3y + 2)$

26. $(a^2 + 3a - 2) + (5a - a^2 - 6a) + (a^2 + 2)$

In Exercises 27–30, use a horizontal arrangement to perform the polynomial addition.

27. $(3z^2 - z + 2) + (z^2 - 4)$

28. $(6x^4 + 8x) + (4x - 6)$

29. $(2a - 3) + (a^2 - 2a) + (4 - a^2)$

30. $(uv - 3) + (4uv + 1)$

31. *Comparing Two Formats* Add the two polynomials $6x^2 + 5$ and $3 - 2x^2$.

(a) Use a horizontal arrangement.

(b) Use a vertical arrangement. Which format do you prefer? Explain.

32. *Comparing Two Formats* Add the two polynomials $2z - 8z^2 - 3$ and $z^2 + 5z$.

(a) Use a horizontal arrangement.

(b) Use a vertical arrangement. Which format do you prefer? Explain.

In Exercises 33–38, use a vertical arrangement to perform the polynomial subtraction.

33. $2x^2 - x + 2$
       $- (3x^2 + x - 1)$

34. $y^4 - 2$
       $- (y^4 + 2)$

35. $(4t^3 - 3t + 5) - (3t^2 - 3t - 10)$

36. $(-s^2 - 3) - (2s^2 + 10s)$

37. Subtract $7x^3 - 4x + 5$ from $10x^3 + 15$.

38. Subtract $y^5 - y^4$ from $y^2 + 3y^4$.

In Exercises 39 and 40, use a horizontal arrangement to perform the polynomial subtraction.

39. $(4 - 2x - x^3) - (3 - 2x + 2x^3)$

40. $(t^4 - 2t^2) - (3t^2 - t^4 - 5)$

In Exercises 41–44, perform the operations.

41. $2(x^4 + 2x) + (5x + 2)$

42. $(z^4 - 2z^2) + 3(z^4 + 4)$

43. $5z - [3z - (10z + 8)]$

44. $(y^3 + 1) - [(y^2 + 1) + (3y - 7)]$

45. *Comparing Models*   From 1970 through 1991, the average number of pounds of beef, B, and poultry, P, consumed by Americans can be modeled by

$B = -0.04t^2 - 0.79t + 76.86,\quad -10 \le t \le 11$

$P = 0.06t^2 + 1.14t + 39.04,\quad -10 \le t \le 11$

where $t = 0$ represents 1980.   (Source: U.S. Department of Agriculture)

(a) Add the polynomials to find a model for the total, T, amount of beef and poultry consumed.

(b) Use a graphing utility to graph the models B, P, and T.

(c) Use the graphs of part (b) to determine whether Americans are increasing or decreasing their consumption of each type of meat. Is the model for the sum increasing or decreasing?

46. *Geometry*   Find the polynomial representing the sum of the areas of the regions.

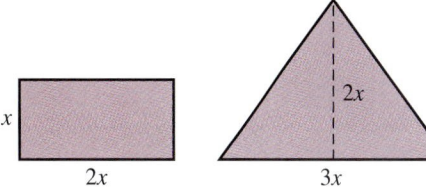

Geometry   In Exercises 47 and 48, find the perimeter of the figure.

47.   **48.**

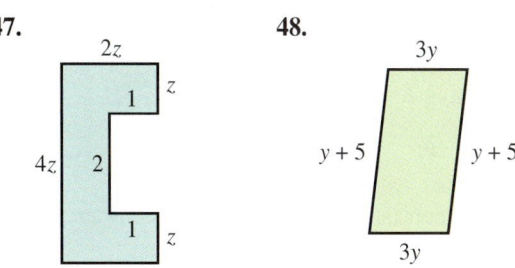

Geometry   In Exercises 49 and 50, write a polynomial that represents the area of the shaded portion of the figure.

49.

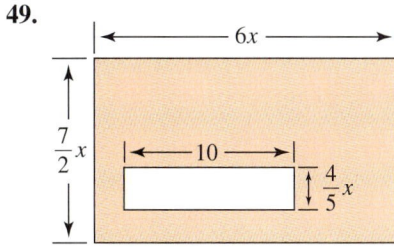

50.

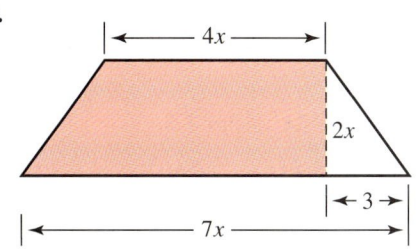

Reviewing the Major Concepts

In Exercises 51–54, use the exponential form to rewrite the expression. Then state the degree of the expression.

51. $3z \cdot 3z \cdot 3z \cdot 3z$

52. $8 \cdot y \cdot y \cdot 8 \cdot 8 \cdot y \cdot y$

53. $4 \cdot 4 \cdot m \cdot m \cdot m^2$

54. $(6 \cdot 6 \cdot b \cdot b)(6b^2)^3$

In Exercises 55 and 56, expand the expression as a product of factors.

55. $3^2 x^4 y^3$ **56.** $5(uv)^4$

In Exercises 57 and 58, use the Distributive Property to expand the expression.

57. $-2(3t - 4)$ **58.** $-(-15x - 12z)$

Additional Problem Solving

In Exercises 59–64, write the polynomial in standard form. Then find its degree and leading coefficient.

59. $x^3 - 4x^2 + 9$ **60.** $9 - 2y^4$

61. 10 **62.** $3x$

63. $3r + \pi r^2$ **64.** $v_0 t - 16t^2$
    (π is a constant.)   (v_0 is a constant.)

In Exercises 65–68, determine whether the expression is a polynomial. If it is not, explain why.

65. $t^3 - 3t + 4$ **66.** $1 - \dfrac{4}{x}$

67. $z^{-1} + z^2 - 2$ **68.** $9 - |y|$

In Exercises 69–76, use a vertical arrangement to perform the polynomial addition.

69. $3x^4 - 2x^3 - 4x^2 + 2x - 5$
$$\underline{ x^2 - 7x + 5}$$

70. $x^5 - 4x^3 + x + 9$
$$\underline{ 2x^4 + 3x^3 - 3}$$

71. $(n^2 + 1) + (2n^2 - 3)$

72. $(3m + m^2 + 1) + (m^2 + 3m + 3)$

73. $(x^2 - 4) + (2x^2 + 6)$

74. $(x^3 + 2x - 3) + (4x + 5)$

75. $(x^2 - 2x + 2) + (x^2 + 4x) + 2x^2$

76. $(5y + 10) + (y^2 - 3y - 2)$

In Exercises 77–80, use a horizontal arrangement to perform the polynomial addition.

77. $b^2 + (b^3 - 2b^2 + 3) + (b^3 - 3)$

78. $(3 + 6x + 8x^2 + 9x^3) + (3 - 2x + 4x^2 - 5x^3)$

79. $\left(\frac{2}{3}y^2 - \frac{3}{4}\right) + \left(\frac{5}{6}y^2 + 2\right)$

80. $(0.1t^3 - 3.4t^2) + (1.5t^3 - 7.3)$

In Exercises 81–86, use a vertical arrangement to perform the polynomial subtraction.

81. $ - 3x^3 - 4x^2 + 2x - 5$
$$\underline{-(2x^4 + 2x^3 - 4x + 5)}$$

82. $ 12x^3 + 25x^2 - 15$
$$\underline{-(-2x^3 + 18x^2 - 3x)}$$

83. $(2 - x^3) - (2 + x^3)$

84. $(4z^3 - 6) - (-z^3 + z - 2)$

85. $(6x^3 - 3x^2 + x) - [(x^3 + 3x^2 + 3) + (x - 3)]$

86. $(y^2 - y) - [(2y^2 + y) - (4y^2 - y + 2)]$

In Exercises 87–90, use a horizontal arrangement to perform the polynomial subtraction.

87. $(x^2 - x) - (x - 2)$

88. $(x^2 - 4) - (x^2 - 4)$

89. $10 - (u^2 + 5)$

90. $(z^3 + z^2 + 1) - z^2$

In Exercises 91–100, perform the operations.

91. $(6x - 5) - (8x + 15)$

92. $(2x^2 + 1) + (x^2 - 2x + 1)$

93. $-(x^3 - 2) + (4x^3 - 2x)$

94. $-(5x^2 - 1) - (-3x^2 + 5)$

95. $(15x^2 - 6) - (-8x^3 - 14x^2 - 17)$

96. $(15x^4 - 18x - 19) - (-13x^4 - 5x + 15)$

97. $2(t^2 + 5) - 3(t^2 + 5) + 5(t^2 + 5)$

98. $-10(u + 1) + 8(u - 1) - 3(u + 6)$

99. $8v - 6(3v - v^2) + 10(10v + 3)$

100. $3(x^2 - 2x + 3) - 4(4x + 1) - (3x^2 - 2x)$

Geometry In Exercises 101–104, write a polynomial that represents the area of the shaded portion of the figure.

101.

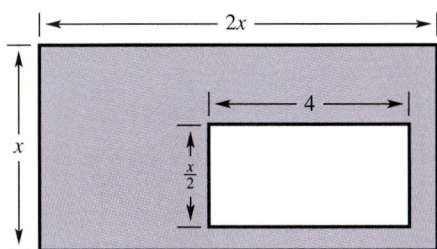

102.

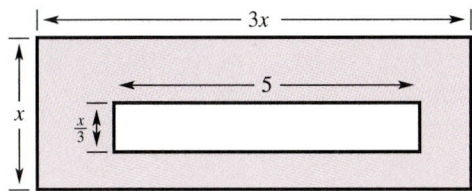

103.

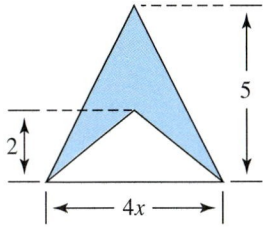

104.

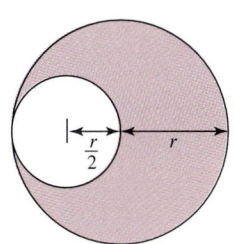

105. *Comparing Business Models* The cost of producing x units is $C = 100 + 30x$. The revenue for selling x units is $R = 90x - x^2$, where $0 \le x \le 40$. The profit is given by the revenue minus the cost.

(a) Perform the subtraction required to find the polynomial representing profit.

(b) Use a graphing utility to graph the polynomial representing profit.

(c) Determine the profit when $x = 30$ units are produced and sold. Use the graph of part (b) to predict the change in profit if x is some value other than 30.

From 1980 to 1990 over 25% of all college freshmen were majoring in business.

5.2 | Multiplying Polynomials: Special Products

Monomial Multipliers ▪ Multiplying Binomials ▪
Multiplying Polynomials ▪ Special Products

Monomial Multipliers

To multiply polynomials, you use many of the rules for simplifying algebraic expressions. Before beginning this section, you may want to review these rules.

1. Properties of exponents   Section 2.1
2. The Distributive Property   Section 2.2
3. Combining like terms   Section 2.2
4. Symbols of grouping   Section 2.3

The simplest type of polynomial multiplication involves a monomial multiplier. The product is obtained by direct application of the Distributive Property. For instance, to multiply the monomial x by the polynomial $(2x + 5)$, multiply *each* of the terms of the polynomial by x.

$$(x)(2x + 5) = (x)(2x) + (x)(5) = 2x^2 + 5x$$

Here is another example.

$$(2x)(3x^2 - 4x + 1) = (2x)(3x^2) - (2x)(4x) + (2x)(1)$$
$$= 6x^3 - 8x^2 + 2x$$

EXAMPLE 1   *Finding Products with Monomial Multipliers*

Find each product.

  a. $(3x - 7)(-2x)$ **b.** $3x^2(5x - x^3 + 2)$ **c.** $(-x)(2x^2 - 3x)$

Solution

a. $(3x - 7)(-2x) = 3x(-2x) - 7(-2x)$ Distributive Property
$$= -6x^2 + 14x$$ Standard form

b. $3x^2(5x - x^3 + 2)$
$$= (3x^2)(5x) - (3x^2)(x^3) + (3x^2)(2)$$ Distributive Property
$$= 15x^3 - 3x^5 + 6x^2$$ Properties of exponents
$$= -3x^5 + 15x^3 + 6x^2$$ Standard form

c. $(-x)(2x^2 - 3x) = (-x)(2x^2) - (-x)(3x)$ Distributive Property
$$= -2x^3 + 3x^2$$ Standard form

Blaise Pascal was a French mathematician, scientist, and philosopher. In addition to his religious and philosophical writings, he made many invaluable contributions to mathematics and physics. Perhaps his most important contribution to mathematics is the invention and construction of the first calculating machine. He invented the machine when he was only nineteen years old.

Multiplying Binomials

To multiply two binomials, you can use both (left and right) forms of the Distributive Property. For example, if you treat the binomial $(5x + 7)$ as a single quantity, you can multiply $(3x - 2)$ by $(5x + 7)$ as follows.

$$(3x - 2)(5x + 7) = 3x(5x + 7) - 2(5x + 7)$$
$$= (3x)(5x) + (3x)(7) - (2)(5x) - 2(7)$$
$$= 15x^2 + 21x - 10x - 14$$

Product of **First terms**	Product of **Outer terms**	Product of **Inner terms**	Product of **Last terms**

$$= 15x^2 + 11x - 14$$

With practice you should be able to multiply two binomials without writing out all of the above steps. In fact, the four products in the boxes above suggest that you can write the product of two binomials in just one step. This is referred to as the **FOIL Method.** Note that the words *first, outer, inner,* and *last* refer to the positions of the terms in the original product.

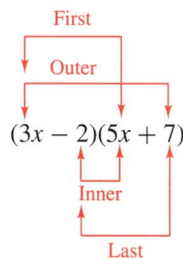

First
Outer
$(3x - 2)(5x + 7)$
Inner
Last

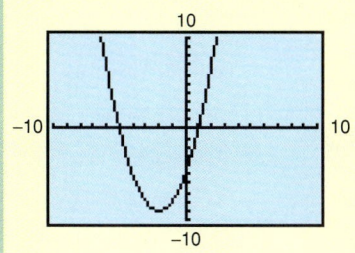

EXAMPLE 2   *Multiplying with the Distributive Property*

Use the Distributive Property to find the product.

$$(x - 1)(x + 5)$$

Solution

$$(x - 1)(x + 5) = x(x + 5) - (1)(x + 5) \qquad \text{Right Distributive Property}$$
$$= x^2 + 5x - x - 5 \qquad \text{Left Distributive Property}$$
$$= x^2 + (5x - x) - 5 \qquad \text{Group like terms.}$$
$$= x^2 + 4x - 5 \qquad \text{Combine like terms.}$$

EXAMPLE 3 Multiplying Binomials Using the FOIL Method

Use the FOIL Method to find the product.

a. $(3x + 5)(2x + 1)$ **b.** $(x - 4)(x + 4)$

Solution   F   O   I   L

a. $(3x + 5)(2x + 1) = 6x^2 + 3x + 10x + 5$

$$= 6x^2 + 13x + 5 \qquad \text{Combine like terms.}$$

NOTE In Example 3(b), note that the outer and inner products add up to zero.

                              F   O   I   L

b. $(x - 4)(x + 4) = x^2 + 4x - 4x - 16$

$$= x^2 - 16 \qquad \text{Combine like terms.}$$

EXAMPLE 4 Simplifying Polynomial Expressions

Simplify the expression and write the result in standard form.

$$5x(x^2 + 2x - 5) - (x^3 + 10x)$$

Solution

$$5x(x^2 + 2x - 5) - (x^3 + 10x)$$

$$= 5x^3 + 10x^2 - 25x - x^3 - 10x \qquad \text{Distributive Property}$$

$$= 4x^3 + 10x^2 - 35x \qquad \text{Combine like terms.}$$

EXAMPLE 5 Simplifying Polynomial Expressions

Simplify each expression and write the result in standard form.

a. $(3x^2 - 2)(4x + 7) - (4x)^2$ **b.** $(4x + 5)^2$

Solution

a. $(3x^2 - 2)(4x + 7) - (4x)^2$

$$= 12x^3 + 21x^2 - 8x - 14 - (4x)^2 \qquad \text{Multiply binomials.}$$

$$= 12x^3 + 21x^2 - 8x - 14 - 16x^2 \qquad \text{Square monomial.}$$

$$= 12x^3 + 5x^2 - 8x - 14 \qquad \text{Combine like terms.}$$

b. $(4x + 5)^2 = (4x + 5)(4x + 5) \qquad \text{Repeated multiplication}$

$$= 16x^2 + 20x + 20x + 25 \qquad \text{Multiply binomials.}$$

$$= 16x^2 + 40x + 25 \qquad \text{Combine like terms.}$$

Multiplying Polynomials

The FOIL Method for multiplying two binomials is simply a device for guaranteeing that *each term of one binomial is multiplied by each term of the other binomial.*

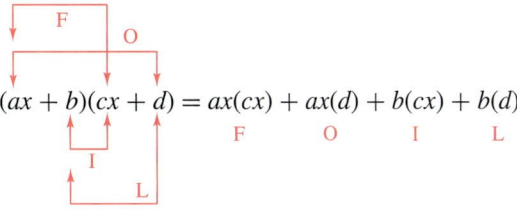

$$(ax + b)(cx + d) = ax(cx) + ax(d) + b(cx) + b(d)$$
$$\qquad\qquad\qquad\qquad\quad \text{F} \qquad \text{O} \qquad \text{I} \qquad \text{L}$$

This same rule applies to the product of two polynomials: *each term of one polynomial must be multiplied by each term of the other polynomial.* This can be accomplished using either a horizontal or vertical format.

EXAMPLE 6   Multiplying Polynomials (Horizontal Format)

Use a horizontal format to find the product: $(2x^2 - 7x + 1)(4x + 3)$.

Solution

$$(2x^2 - 7x + 1)(4x + 3)$$
$$= (2x^2 - 7x + 1)(4x) + (2x^2 - 7x + 1)(3) \qquad \text{Distributive Property}$$
$$= 8x^3 - 28x^2 + 4x + 6x^2 - 21x + 3 \qquad\qquad \text{Distributive Property}$$
$$= 8x^3 - 22x^2 - 17x + 3 \qquad\qquad\qquad\qquad \text{Combine like terms.}$$

EXAMPLE 7   Multiplying Polynomials (Vertical Format)

Use a vertical format to find the product: $(3x^2 + x - 5)(2x - 1)$.

Solution

With a vertical format, line up like terms in the same vertical columns much like you align digits in whole number multiplication.

$$
\begin{array}{r}
3x^2 + x - 5 \\
\times \qquad 2x - 1 \\
\hline
-3x^2 - x + 5 \\
6x^3 + 2x^2 - 10x \phantom{{}+5} \\
\hline
6x^3 - x^2 - 11x + 5
\end{array}
$$

Place polynomial with most terms on top.

$-1(3x^2 + x - 5)$

$2x(3x^2 + x - 5)$

Combine like terms in columns.

When multiplying two polynomials, it is best to write each in standard form before using either the horizontal or vertical format. This is illustrated in the next example.

EXAMPLE 8 *Multiplying Polynomials*

Multiply the polynomials.

$$(x + 3x^2 - 4)(5 + 3x - x^2)$$

Solution

$$
\begin{array}{r}
3x^2 + x - 4 \\
\times \quad -x^2 + 3x + 5 \\
\hline
15x^2 + 5x - 20 \\
9x^3 + 3x^2 - 12x \\
-3x^4 - x^3 + 4x^2 \\
\hline
-3x^4 + 8x^3 + 22x^2 - 7x - 20
\end{array}
$$

Standard form
Standard form

$5(3x^2 + x - 4)$
$3x(3x^2 + x - 4)$
$-x^2(3x^2 + x - 4)$

EXAMPLE 9 *Multiplying Polynomials*

Find the product.

$$(x - 3)^3$$

Solution

To raise $(x - 3)$ to the third power, you can use two steps. First, because $(x - 3)^3 = (x - 3)^2(x - 3)$, find the square of $(x - 3)$.

$$(x - 3)(x - 3) = x^2 - 3x - 3x + 9 \qquad \text{Find } (x-3)^2.$$
$$= x^2 - 6x + 9 \qquad \text{Combine like terms.}$$

Now, using a vertical arrangement, find $(x - 3)^3$.

$$
\begin{array}{r}
x^2 - 6x + 9 \\
\times \qquad x - 3 \\
\hline
- 3x^2 + 18x - 27 \\
x^3 - 6x^2 + 9x \\
\hline
x^3 - 9x^2 + 27x - 27
\end{array}
$$

Thus, $(x - 3)^3 = x^3 - 9x^2 + 27x - 27$.

Special Products

Some binomial products such as those in Example 3(b) and Example 5(b) have special forms that occur frequently in algebra. Let's look at those products again. The product $(x + 4)(x - 4)$ is called a **product of the sum and difference of two terms.** With such products, the two middle terms cancel, as follows.

$$(x + 4)(x - 4) = x^2 - 4x + 4x - 16 \qquad \text{Sum and difference of two terms}$$
$$= x^2 - 16 \qquad \text{Product has no middle term.}$$

Another common type of product is the **square of a binomial.** With this type of product, the middle term is always twice the product of the terms in the binomial.

$$(4x + 5)^2 = (4x + 5)(4x + 5) \qquad \text{Square of a binomial}$$
$$= 16x^2 + 20x + 20x + 25$$
$$= 16x^2 + 40x + 25 \qquad \text{Middle term is twice the product of the terms of the binomial.}$$

You should learn to recognize the patterns of these two special products. We give the general form of these special products in the following statements. The FOIL Method can be used to verify each rule.

Special Products

Let a and b be real numbers, variables, or algebraic expressions.

Special Product   *Example*

Sum and Difference of Two Terms:

$$(a + b)(a - b) = a^2 - b^2 \qquad (2x - 5)(2x + 5) = 4x^2 - 25$$

Square of a Binomial:

$$(a + b)^2 = a^2 + 2ab + b^2 \qquad (3x + 4)^2 = 9x^2 + 2(3x)(4) + 16$$
$$= 9x^2 + 24x + 16$$

$$(a - b)^2 = a^2 - 2ab + b^2 \qquad (x - 7)^2 = x^2 - 2(x)(7) + 49$$
$$= x^2 - 14x + 49$$

NOTE   When a binomial is squared, the resulting middle term is always *twice* the product of the two terms.

$$\underbrace{(a + b)^2}_{\text{Terms}} = a^2 + \underbrace{2(ab)}_{\substack{\text{Twice the} \\ \text{product of} \\ \text{the terms}}} + b^2$$

Be sure to include the middle term. For instance, $(a + b)^2$ is *not* equal to $a^2 + b^2$.

EXAMPLE 10 Finding Sum and Difference Products

Find each product.

a. $(5x - 6)(5x + 6)$ **b.** $(2 + 3x)(2 - 3x)$

Solution

a.
$$(5x - 6)(5x + 6) = (5x)^2 - (6)^2$$
$$= 25x^2 - 36$$

b.
$$(2 + 3x)(2 - 3x) = (2)^2 - (3x)^2$$
$$= 4 - 9x^2$$

EXAMPLE 11 Squaring a Binomial

Find the product: $(4x - 9)^2$.

Solution

$$(4x - 9)^2 = (4x)^2 - 2(4x)(9) + (9)^2$$
$$= 16x^2 - 72x + 81$$

Group Activities   Extending the Concept

Pascal's Triangle   The following triangular pattern of numbers shows the first seven rows of **Pascal's Triangle,** named after the French mathematician Blaise Pascal (1623–1662). In your group, try to discover the pattern formed by the numbers in the triangle. Then use the pattern to write out the expansion of $(x + 1)^7$.

$$
\begin{array}{ccccccccccccc}
 & & & & & & 1 & & & & & & \\
 & & & & & 1 & & 1 & & & & & \\
 & & & & 1 & & 2 & & 1 & & & & \\
 & & & 1 & & 3 & & 3 & & 1 & & & \\
 & & 1 & & 4 & & 6 & & 4 & & 1 & & \\
 & 1 & & 5 & & 10 & & 10 & & 5 & & 1 & \\
1 & & 6 & & 15 & & 20 & & 15 & & 6 & & 1 \\
\end{array}
$$

$$(x + 1)^0 = 1$$
$$(x + 1)^1 = x + 1$$
$$(x + 1)^2 = x^2 + 2x + 1$$
$$(x + 1)^3 = x^3 + 3x^2 + 3x + 1$$
$$(x + 1)^4 = x^4 + 4x^3 + 6x^2 + 4x + 1$$
$$(x + 1)^5 = x^5 + 5x^4 + 10x^3 + 10x^2 + 5x + 1$$
$$(x + 1)^6 = x^6 + 6x^5 + 15x^4 + 20x^3 + 15x^2 + 6x + 1$$

Discuss how to determine the number of terms in the expansion of $(x + 1)^{14}$.

5.2 Exercises

Discussing the Concepts

1. Explain why an understanding of the Distributive Property is essential in multiplying polynomials. Illustrate your explanation with an example.

2. Describe the properties of exponents that are used to multiply polynomials. Give examples.

3. Discuss the difference between the expressions $(3x)^2$ and $3x^2$.

4. Explain the meaning of each letter of FOIL as it relates to multiplying two binomials.

5. What is the degree of the product of two polynomials of degrees m and n? Explain.

6. *True or False?*   Because the product of two monomials is a monomial, it follows that the product of two binomials is a binomial.

Problem Solving

In Exercises 7–20, multiply and simplify.

7. $x(-2x)$

8. $9x\left(\dfrac{x}{12}\right)$

9. $(-2b^2)(-3b)$

10. $(-4m)(3m^2)$

11. $2x(3x)^2$

12. $(4z)^3\left(\dfrac{z}{2}\right)^2$

13. $(x+3)(x+4)$

14. $(x-5)(x+10)$

15. $(5x+7)(2x-3)$

16. $(3x-1)(4x-5)$

17. $-4x(3+3x^2-6x^3)$

18. $5v(5-4v+5v^2)$

19. $(s-2t)(s+t)-(s-2t)(s-t)$

20. $2u(u-v)+3v(u+v)$

In Exercises 21–24, multiply using a horizontal format.

21. $(x^3-2x+1)(x-5)$

22. $(x+1)(x^2-x+1)$

23. $(x-2)(x^2+2x+4)$

24. $(x^2+9)(x^2-x-4)$

In Exercises 25–30, multiply using a vertical format.

25. $\begin{array}{r} x^2-3x+9 \\ \times \qquad x+3 \\ \hline \end{array}$

26. $\begin{array}{r} 4x^4-6x^2+9 \\ \times \qquad 2x^2+3 \\ \hline \end{array}$

27. $(x^2-x+2)(x^2+x-2)$

28. $(x^2+2x+5)(2x^2-x-1)$

29. $(x^3+x+3)(x^2+5x-4)$

30. $(x-1)(x^2+x+1)$

In Exercises 31–38, use a special product pattern to find the product.

31. $(x+2)(x-2)$

32. $(3z+4)(3z-4)$

33. $(x+6)^2$

34. $(2x-8)^2$

35. $(2x-5y)^2$

36. $(4s+3t)^2$

37. $[u-(v-3)]^2$

38. $[2u+(v+1)]^2$

39. *Geometry*   The base of a triangular sail is $2x$ feet and its height is $x+10$ feet (see figure). Find the area A of the sail.

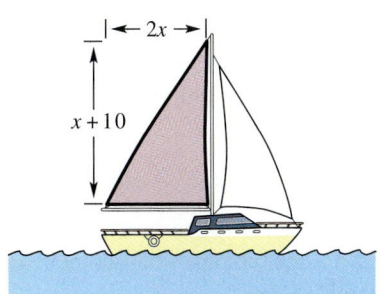

40. *Geometry* Add the areas of the four rectangular regions shown in the figure. What special product does the geometric model represent?

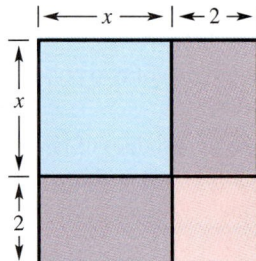

41. *Using Mathematical Models* From 1980 through 1992, each American's share, S, of the debt of the federal government can be modeled by

$$S = 32.26t^2 + 594.49t + 3798.18, \quad 0 \le t \le 12$$

where $t = 0$ represents 1980. The population, P (in millions), during the same time can be modeled by

$$P = 2.26t + 227.42, \quad 0 \le t \le 12.$$

(Source: U.S. Bureau of the Census)

(a) Use a graphing utility to graph the model of the per capita debt S.

(b) Multiply the polynomials representing the population P and the per capita debt S.

(c) Use the product of part (b) to estimate the total federal debt for 1990. (*Note:* The answer will be in millions of dollars.)

Reviewing the Major Concepts

In Exercises 45–48, simplify the expression.

45. $2(x - 4) + 5x$ **46.** $4(3 - y) + 2(y + 1)$

47. $-3(z - 2) - (z - 6)$ **48.** $(u - 2) - 3(2u + 1)$

49. *Amount of Sales* You earn a sales commission rate of 5.5%. Your commission is $1600. How much did you sell?

42. *Finding a Pattern* Perform each multiplication.

(a) $(x - 1)(x + 1)$

(b) $(x - 1)(x^2 + x + 1)$

(c) $(x - 1)(x^3 + x^2 + x + 1)$

(d) Use the pattern formed in the first three products to find the pattern for

$$(x - 1)(x^4 + x^3 + x^2 + x + 1).$$

Verify your pattern by multiplying.

Geometry In Exercises 43 and 44, what polynomial product is represented? Explain.

43.

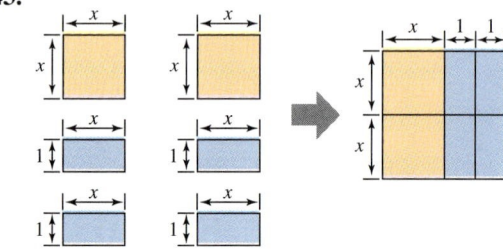

44.

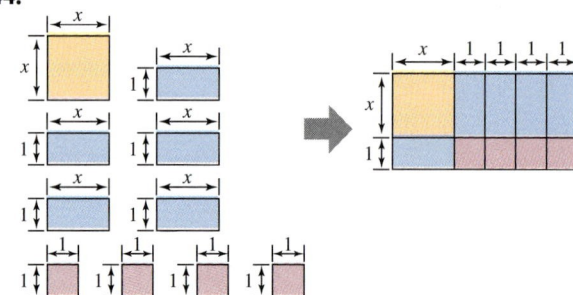

50. *Drawing a Diagram* A jogger leaves a location on a fitness trail running at a rate of 4 miles per hour. Fifteen minutes later, a second jogger leaves from the same location running at 5 miles per hour. How long will it take the second jogger to overtake the first jogger and how far will each have run at that point? Use a diagram to help answer the question.

Additional Problem Solving

In Exercises 51–80, multiply and simplify.

51. $\left(\dfrac{x}{4}\right)(10x)$

52. $(-5y)(-3y)$

53. $t^2(4t)$

54. $3u(u^4)$

55. $(-3t)^3$

56. $(-4z)^2$

57. $y(3-y)$

58. $z(z-3)$

59. $-x(x^2-4)$

60. $-t(10-3t)$

61. $3t(2t-5)$

62. $-5u(u^2+4)$

63. $3x(x^2-2x+1)$

64. $y^2(4y^2+2y-3)$

65. $2x(x^2-2x+8)$

66. $-3x^2(x-3)$

67. $-2x(-3x)(5x+2)$

68. $4x(-2x)(x^2-1)$

69. $(x-7)(x-9)$

70. $(x-8)(x+2)$

71. $(3x-5)(2x+1)$

72. $(7x-2)(4x-3)$

73. $(x+y)(x+2y)$

74. $(2x-y)(x-2y)$

75. $2x(6x^4)-3x^2(2x^2)$

76. $-8y(-5y^4)-2y^2(5y^3)$

77. $5x(x+1)-3x(2x-4)$

78. $(2x-3)(x+3)+3(2x-1)$

79. $(u-1)(2u+3)(2u+1)$

80. $(2x+5)(x-2)(5x-3)$

In Exercises 81–94, use a special product pattern to find the product.

81. $(x+5)(x-5)$

82. $(n-4)(n+4)$

83. $(y+9)(y-9)$

84. $(2u+3)(2u-3)$

85. $(2x+3y)(2x-3y)$

86. $(5u+12v)(5u-12v)$

87. $(a-2)^2$

88. $(x+10)^2$

89. $(3x+2)^2$

90. $(t-3)^2$

91. $(8-3z)^2$

92. $(1-5t)^2$

93. $[(x+1)+y]^2$

94. $[(x-3)-y]^2$

In Exercises 95 and 96, perform the multiplication. Then simplify.

95. $(x+2)^2-(x-2)^2$

96. $(u+5)^2+(u-5)^2$

Think About It   In Exercises 97 and 98, is the equation an identity? Explain your reasoning.

97. $(x+y)^3 = x^3+3x^2y+3xy^2+y^3$

98. $(x-y)^3 = x^3-3x^2y+3xy^2-y^3$

In Exercises 99 and 100, use the results of Exercises 97 and 98 to find the product.

99. $(x+2)^3$

100. $(x+1)^3$

101. *Geometry*   The height of a rectangular sign is twice its width w (see figure). Find (a) the perimeter and (b) the area of the rectangle.

102. *Geometry*   The inner and outer radii of a washer are x and $x+2$ centimeters, respectively (see figure). Find the area of one side of the washer.

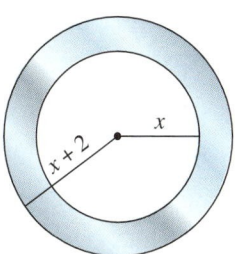

Geometry In Exercises 103 and 104, find a polynomial product that represents the area of the region. Then simplify the product.

103.

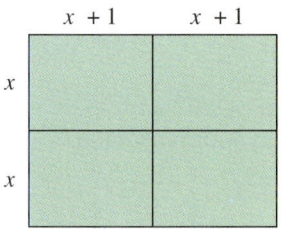

104.

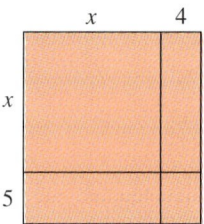

Geometry In Exercises 105 and 106, find two different expressions that represent the area of the shaded portion of the figure.

105.

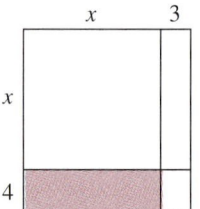

106.

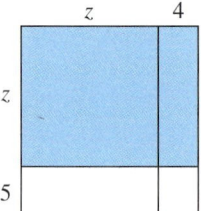

107. *Interpreting Graphs* When x units are sold, the revenue R is given by $R = x(900 - 0.5x)$.

(a) Use a graphing utility to graph the expression.

(b) Multiply the expression for revenue and use a graphing utility to graph the product. Verify that the graph is the same as in part (a).

(c) Find the revenue when $x = 500$. Use the graph to determine whether the revenue would increase or decrease if more units were sold.

108. *Compound Interest* After 2 years, a $500 investment, compounded annually at interest rate r, will yield the amount $500(1 + r)^2$. Find this product.

CAREER INTERVIEW

Kimberly R. Lee

Marketing Associate, Shoebox Greetings

Hallmark Cards, Inc.

Kansas City, MO 64141

As a marketing associate, I plan strategies for promoting, pricing, and distributing Hallmark greeting card products. I need to understand the greeting card market and our customers' sending needs.

One numerical measure that is useful when determining whether a card line has adequate balance to meet the needs of consumers who prefer elaborate, heavily processed cards, as well as those who are looking for cards that are editorially strong but simpler in design, is the *average price offered*. To begin, I find how many different card prices are projected for the line given the different processes used and then calculate the percent of the line offered at each price. The average price offered is then the sum of (price × % of line) for each projected price, giving me an idea of the mix of premium and value-priced cards in the line.

I have found that algebra is also an invaluable tool in answering "What if . . . ?" questions—such as "If we offer more lower-priced cards, how is the number of units sold affected?"—when formulating marketing plans.

MID-CHAPTER QUIZ

Take this quiz as you would take a quiz in class. After you are done, check your work against the answers given in the back of the book.

1. Explain why $x^2 + 2x - \dfrac{3}{x}$ is not a polynomial.

2. Determine the degree and the leading coefficient of the polynomial $-3x^4 + 2x^2 - x$.

3. Give an example of a trinomial in one variable of degree 5.

4. *True or False?* The product of two binomials is a binomial. If false, give an example to show it is false.

In Exercises 5–14, perform the indicated operation and simplify.

5. $y + (4 + 3y)$

6. $(3v^2 - 5) - (v^3 + 2v^2 - 6v)$

7. $9s - [6 - (s - 5) + 7s]$

8. $-3(4 - x) + 4(x^2 + 2) - (x^2 - 2x)$

9. $2r(5r)^2$

10. $m(-2m)^3$

11. $6a\left(\dfrac{2a}{3}\right)^3$

12. $(2y - 3)(y + 5)$

13. $(4 - 3x)^2$

14. $(2u - 3)(2u + 3)$

In Exercises 15–18, perform the indicated operation using a vertical format.

15.
$$
\begin{array}{r}
5x^4 \quad + 2x^2 + \ x - 3 \\
+ \quad 3x^3 - 2x^2 - 3x + 5 \\
\hline
\end{array}
$$

16.
$$
\begin{array}{r}
2x^3 + \ x^2 \qquad - 8 \\
- \quad (\ 5x^2 - 3x - 9) \\
\hline
\end{array}
$$

17.
$$
\begin{array}{r}
3x^2 + 7x + 1 \\
\times \qquad 2x - 5 \\
\hline
\end{array}
$$

18.
$$
\begin{array}{r}
5x^3 - 6x^2 + \ 3 \\
\times \qquad x^2 - 3x \\
\hline
\end{array}
$$

19. Find the perimeter of the figure.

20. Find the area of the figure.

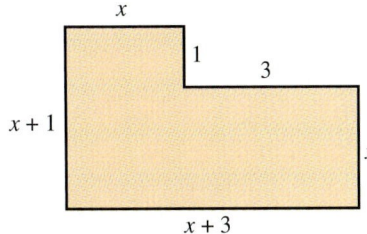

| **5.3** | **Dividing Polynomials** |

Dividing a Monomial by a Monomial ▪ Dividing a Polynomial by a Monomial ▪ Dividing a Polynomial by a Binomial

Dividing a Monomial by a Monomial

In this section, you will learn how to divide a polynomial by a monomial and a binomial.

To begin, let's consider division problems in which both the numerator *and* the denominator are monomials. To divide a monomial by a monomial, you make use of the *subtraction* property of exponents illustrated in the following examples. (In each of the following, assume that the variable is *not zero*.)

By Reducing

$$\frac{x^4}{x^2} = \frac{x \cdot x \cdot \cancel{x} \cdot \cancel{x}}{\cancel{x} \cdot \cancel{x}} = x^2$$

$$\frac{y^3}{y^3} = \frac{\cancel{y} \cdot \cancel{y} \cdot \cancel{y}}{\cancel{y} \cdot \cancel{y} \cdot \cancel{y}} = 1$$

$$\frac{5y^7}{2y^5} = \frac{5 \cdot y \cdot y \cdot \cancel{y} \cdot \cancel{y} \cdot \cancel{y} \cdot \cancel{y} \cdot \cancel{y}}{2 \cdot \cancel{y} \cdot \cancel{y} \cdot \cancel{y} \cdot \cancel{y} \cdot \cancel{y}} = \frac{5y^2}{2}$$

$$\frac{2x^2}{x^5} = \frac{2 \cdot \cancel{x} \cdot \cancel{x}}{x \cdot x \cdot x \cdot \cancel{x} \cdot \cancel{x}} = \frac{2}{x^3}$$

By Subtracting Exponents

$$\frac{x^4}{x^2} = x^{4-2} = x^2$$

$$\frac{y^3}{y^3} = y^{3-3} = y^0 = 1$$

$$\frac{5y^7}{2y^5} = \frac{5y^{7-5}}{2} = \frac{5y^2}{2}$$

$$\frac{2x^2}{x^5} = \frac{2}{x^{5-2}} = \frac{2}{x^3}$$

The above examples show that you can divide one monomial by another by subtracting the smaller exponent from the larger exponent.

Properties of Exponents

Let m and n be positive integers and let a represent a real number, a variable, or an algebraic expression.

1. $\dfrac{a^m}{a^n} = a^{m-n}$, if $m > n$

2. $\dfrac{a^m}{a^n} = \dfrac{1}{a^{n-m}}$, if $n > m$

3. $\dfrac{a^n}{a^n} = 1 = a^0$

Note the special definition for raising a *nonzero* quantity to the zero power. That is, if $a \neq 0$, then $a^0 = 1$.

EXAMPLE 1   Dividing a Monomial by a Monomial

Perform each division. (In each case, assume that $x \neq 0$ and $y \neq 0$.)

a. $16x^4 \div 4x^2$

b. $16x^4 \div 3x$

c. $32y^3 \div 8y^5$

d. $8x^3 \div \frac{1}{2}x^3$

e. $12x^3 \div 4x^4$

Solution

a. $\dfrac{16x^4}{4x^2} = \dfrac{16}{4} \cdot \dfrac{x^4}{x^2} = \dfrac{16}{4} \cdot x^{4-2} = 4x^2$

b. $\dfrac{16x^4}{3x} = \dfrac{16}{3} \cdot \dfrac{x^4}{x} = \dfrac{16}{3} \cdot x^{4-1} = \dfrac{16}{3}x^3$

c. $\dfrac{32y^3}{8y^5} = \dfrac{32}{8} \cdot \dfrac{y^3}{y^5} = \dfrac{32}{8} \cdot \dfrac{1}{y^{5-3}} = 4 \cdot \dfrac{1}{y^2} = \dfrac{4}{y^2}$

d. $\dfrac{8x^3}{\frac{1}{2}x^3} = \dfrac{8}{\frac{1}{2}} \cdot \dfrac{x^3}{x^3} = 8\left(\dfrac{2}{1}\right)(1) = 16$

e. $\dfrac{12x^3}{4x^4} = \dfrac{12}{4} \cdot \dfrac{x^3}{x^4} = 3 \cdot \dfrac{1}{x^{4-3}} = \dfrac{3}{x}$

NOTE   The subtraction property of exponents works only for division of monomials with the *same variable* for a base. For instance, the subtraction property does not apply to x^5/y^3 because no cancellations can occur. Thus, for the fraction

$$\dfrac{x^5}{y^3} = \dfrac{x \cdot x \cdot x \cdot x \cdot x}{y \cdot y \cdot y}$$

no simplifying is possible.

Although the division problems in Example 1 are straightforward, you should study the example carefully. Be sure you can justify each step in the example. Also remember that there are often several ways to solve a given problem in algebra. As you gain practice and confidence, you will discover that you like some techniques better than others. For instance, which one of the following techniques seems best to you?

1. $\dfrac{6x^3}{2x} = \dfrac{3 \cdot 2 \cdot x \cdot x \cdot \cancel{x}}{2 \cdot \cancel{x}} = 3x^2, \quad x \neq 0$

2. $\dfrac{6x^3}{2x} = \left(\dfrac{6}{2}\right)\left(\dfrac{x^3}{x}\right) = (3)(x^{3-1}) = 3x^2, \quad x \neq 0$

3. $\dfrac{6x^3}{2x} = \dfrac{3x^3}{x} = 3x^{3-1} = 3x^2, \quad x \neq 0$

When two different people (even math instructors) are writing out the steps of a solution, rarely will the steps be the same, so don't worry if your steps don't look exactly like someone else's. If you feel comfortable with writing more steps, then you should write more steps. Just be sure that each step can be justified by the rules of algebra.

Dividing a Polynomial by a Monomial

The preceding examples show how to divide a *monomial* by a monomial. To divide a *polynomial* by a monomial, use the reverse form of the rule for adding two fractions with a common denominator. In Section 1.1, you added two fractions with like denominators using the rule

$$\frac{a}{c} + \frac{b}{c} = \frac{a+b}{c}. \qquad \text{Adding fractions}$$

Here you can use the rule in the *reverse* order and divide a polynomial by a monomial by dividing each term of the polynomial by the monomial. That is,

$$\frac{a+b}{c} = \frac{a}{c} + \frac{b}{c}. \qquad \text{Dividing by a monomial}$$

Here is an example.

$$\frac{x^3 - 5x^2}{x^2} = \frac{x^3}{x^2} - \frac{5x^2}{x^2} = x - 5, \quad x \neq 0$$

Note that the essence of this problem is to separate the original division problem into *two* division problems, each involving the division of a monomial by a monomial.

EXAMPLE 2   Dividing a Polynomial by a Monomial

a. $\dfrac{6x - 5}{3} = \dfrac{6x}{3} - \dfrac{5}{3} = 2x - \dfrac{5}{3}$

b. $\dfrac{4x^2 - 3x}{3x} = \dfrac{4x^2}{3x} - \dfrac{3x}{3x} = \dfrac{4x}{3} - 1, \quad x \neq 0$

c. $\dfrac{8x^3 - 6x^2 + 10x}{2x} = \dfrac{8x^3}{2x} - \dfrac{6x^2}{2x} + \dfrac{10x}{2x} = 4x^2 - 3x + 5, \quad x \neq 0$

EXAMPLE 3   Dividing a Polynomial by a Monomial

Perform the division. (Assume $x \neq 0$.)

$$(5x^3 - 4x^2 - x + 6) \div 2x$$

Solution

$$\frac{5x^3 - 4x^2 - x + 6}{2x} = \frac{5x^3}{2x} - \frac{4x^2}{2x} - \frac{x}{2x} + \frac{6}{2x} \qquad \text{Divide each term separately.}$$

$$= \frac{5x^2}{2} - 2x - \frac{1}{2} + \frac{3}{x} \qquad \text{Use properties for dividing monomials.}$$

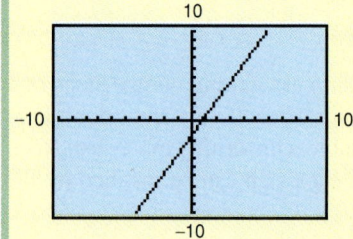

Dividing a Polynomial by a Binomial

To divide a polynomial by a *binomial*, follow the *long division* pattern used for dividing whole numbers. Recall that you divide 6982 by 27 as follows.

Think $\frac{69}{27} \approx 2$

Think $\frac{158}{27} \approx 5$

Think $\frac{232}{27} \approx 8$

$$
\begin{array}{r}
258 \\
27{\overline{\smash{\big)}\,6982}} \\
\underline{54} \\
158 \\
\underline{135} \\
232 \\
\underline{216} \\
16
\end{array}
$$

Multiply 2 by 27.
Subtract and bring down 8.
Multiply 5 by 27.
Subtract and bring down 2.
Multiply 8 by 27.
Remainder

You can express the result as $\frac{6982}{27} = 258\frac{16}{27}$ or $258 + \frac{16}{27}$.

EXAMPLE 4   *Dividing a Polynomial by a Binomial*

Divide $(x^2 + 3x + 5)$ by $(x + 1)$.

Solution

Think $\frac{x^2}{x} = x$

Think $\frac{2x}{x} = 2$

$$
\begin{array}{r}
x + 2 \\
x + 1{\overline{\smash{\big)}\,x^2 + 3x + 5}} \\
\underline{x^2 + x} \\
2x + 5 \\
\underline{2x + 2} \\
3
\end{array}
$$

Multiply $x(x + 1)$.
Subtract and bring down 5.
Multiply $2(x + 1)$.
Subtract.

Considering the remainder as a fractional part of the divisor, you can write

$$
\underbrace{\frac{\overbrace{x^2 + 3x + 5}^{\text{Dividend}}}{\underbrace{x + 1}_{\text{Divisor}}}} = \overbrace{x + 2}^{\text{Quotient}} + \frac{\overset{\text{Remainder}}{3}}{\underbrace{x + 1}_{\text{Divisor}}}.
$$

EXAMPLE 5 A Binomial Divisor

Divide $(6x^3 - 19x^2 + 16x - 4)$ by $(x - 2)$.

Solution

Think $\frac{6x^3}{x} = 6x^2$

Think $-\frac{7x^2}{x} = -7x$

Think $\frac{2x}{x} = 2$

$$
\begin{array}{r}
6x^2 - 7x + 2 \\
x - 2 \overline{)\ 6x^3 - 19x^2 + 16x - 4} \\
\underline{6x^3 - 12x^2} \\
-\ 7x^2 + 16x \\
\underline{-\ 7x^2 + 14x} \\
2x - 4 \\
\underline{2x - 4} \\
0
\end{array}
$$

Multiply $6x^2(x - 2)$.
Subtract and bring down $16x$.
Multiply $-7x(x - 2)$.
Subtract and bring down -4.
Multiply $2(x - 2)$.
Remainder

NOTE In Example 5, the remainder is zero. In such cases, the denominator (or divisor) is said to **divide evenly** into the numerator (or dividend).

Thus, you have

$$\frac{6x^3 - 19x^2 + 16x - 4}{x - 2} = 6x^2 - 7x + 2, \quad x \neq 2.$$

EXAMPLE 6 A Binomial Divisor

Divide $(-13x^3 + 10x^4 + 8x - 7x^2 + 4)$ by $(3 - 2x)$.

Solution

First write the divisor and dividend in standard polynomial form.

$$
\begin{array}{r}
-5x^3 - x^2 + 2x - 1 \\
-2x + 3 \overline{)\ 10x^4 - 13x^3 - 7x^2 + 8x + 4} \\
\underline{10x^4 - 15x^3} \\
2x^3 - 7x^2 \\
\underline{2x^3 - 3x^2} \\
-4x^2 + 8x \\
\underline{-4x^2 + 6x} \\
2x + 4 \\
\underline{2x - 3} \\
7
\end{array}
$$

Using the fractional form of the remainder, you can write

$$\frac{10x^4 - 13x^3 - 7x^2 + 8x + 4}{-2x + 3} = -5x^3 - x^2 + 2x - 1 + \frac{7}{-2x + 3}.$$

EXAMPLE 7   A Binomial Divisor

Use the long division algorithm to simplify.

$$\frac{x^3 - 1}{x - 1}.$$

Solution

Because there are no x^2- or x-terms in the dividend, you can line up the subtractions by using *zero* coefficients (or by leaving spaces) for the missing terms.

$$
\begin{array}{r}
x^2 + x + 1 \\
x - 1 \overline{) x^3 + 0x^2 + 0x - 1} \\
\underline{x^3 - x^2} \\
x^2 \\
\underline{x^2 - x} \\
x - 1 \\
\underline{x - 1} \\
0
\end{array}
$$

Thus, $x - 1$ divides evenly into $x^3 - 1$ and you can write

$$\frac{x^3 - 1}{x - 1} = x^2 + x + 1, \quad x \neq 1.$$

Group Activities   You Be the Instructor

Creating Practice Problems   You are tutoring a friend in algebra and you want to create some division problems for practice. You want to find several division problems that involve third-degree polynomials that are evenly divisible by first-degree polynomials. For instance, in Example 7, $x^3 - 1$ is evenly divisible by $x - 1$ because

$$\frac{x^3 - 1}{x - 1} = x^2 + x + 1.$$

In your group, develop a method for finding a third-degree polynomial that is evenly divisible by a first-degree polynomial. Demonstrate your method by finding a third-degree polynomial that is evenly divisible by $2x - 1$.

5.3 Exercises

Discussing the Concepts

1. Match each part of the equation with its name.

$$\frac{x^2+2}{x-3} = x+3+\frac{11}{x-3}$$

    (a) Dividend   (b) Divisor

    (c) Quotient   (d) Remainder

2. Explain how you can check the result of a division problem algebraically *and* graphically.

3. Give an example of using the subtraction property of exponents to divide a monomial by a monomial.

4. Describe the method of dividing a polynomial by a monomial.

5. What does it mean when a divisor divides *evenly* into the dividend?

6. If the degree of the dividend is 5 and the degree of the divisor is 3, what is the degree of the quotient?

Problem Solving

In Exercises 7–10, perform the indicated division by cancellation *and* by subtracting exponents. (Assume the denominator is not zero.)

7. $\dfrac{x^5}{x^2}$

8. $\dfrac{3^5 x^7}{3^3 x^4}$

9. $\dfrac{4^5 x^3}{4x^5}$

10. $\dfrac{6z^5}{6z^5}$

In Exercises 11–18, simplify the expression. (Assume the denominator is not zero.)

11. $\dfrac{-3x^2}{6x}$

12. $\dfrac{-4a^6}{-2a^2}$

13. $\dfrac{-18s^4}{-12r^2 s}$

14. $\dfrac{54x^2}{-24x^4}$

15. $\dfrac{(-3z)^2}{18z^3}$

16. $\dfrac{(4ab)^3}{(-8a)^2 b^4}$

17. $\dfrac{24(u^2 v)^4}{18u^2 v^6}$

18. $\dfrac{15x^3 y^0}{27x^3}$

In Exercises 19–30, perform the division and simplify. (Assume the denominator is not zero.)

19. $\dfrac{3z+3}{3}$

20. $\dfrac{8u-24}{8}$

21. $\dfrac{8z^3+3z^2-2z}{z}$

22. $\dfrac{14y^4+21y^3}{-7y^3}$

23. $\dfrac{x^2-x-2}{x+1}$

24. $\dfrac{x^2-5x+6}{x-2}$

25. $\dfrac{3y^2+4y-4}{3y-2}$

26. $\dfrac{7t^2-10t-8}{7t+4}$

27. Divide $4x^2+3x+1$ by $x+1$.

28. Divide $7x^2+4x+3$ by $x+2$.

29. $(x^3-4x^2+9x-7) \div (x-2)$

30. $(2x^3-2x^2+3x+9) \div (x+1)$

In Exercises 31 and 32, simplify the expression. (Assume the denominator is not zero.)

31. $\dfrac{4x^3}{x^2}-2x$

32. $\dfrac{x^2+2x+1}{x+1}-(3x-4)$

In Exercises 33–36, determine whether the cancellation is valid.

33. $\dfrac{3+4}{3} = \dfrac{\cancel{3}+4}{\cancel{3}} = 4$

34. $\dfrac{4+7}{4+11} = \dfrac{\cancel{4}+7}{\cancel{4}+11} = \dfrac{7}{11}$

35. $\dfrac{7 \cdot 12}{19 \cdot 7} = \dfrac{\cancel{7} \cdot 12}{19 \cdot \cancel{7}} = \dfrac{12}{19}$

36. $\dfrac{24}{43} = \dfrac{24}{\cancel{43}} = \dfrac{2}{3}$

Reviewing the Major Concepts

In Exercises 37–40, solve the percent problem.

37. What is 62% of 25?

38. What percent of 900 is 600?

39. 32% of what number is 145.6?

40. 0.8% of what number is 2?

41. Write an algebraic expression that represents the product of two consecutive odd integers, the first of which is $2n + 1$.

42. *Average Speed*   After traveling 3 hours, you are still 24 miles from completing a 180-mile trip. It takes you one-half hour to travel the last 24 miles. Find the average speed during the trip.

Additional Problem Solving

In Exercises 43–50, perform the division by cancellation *and* by subtracting exponents. (Assume the denominator is not zero.)

43. $\dfrac{2^3 y^4}{2^2 y^2}$

44. $\dfrac{z^6}{5z^4}$

45. $\dfrac{z^4}{z^7}$

46. $\dfrac{y^8}{y^3}$

47. $\dfrac{x^5}{x^5}$

48. $\dfrac{5^3 x}{5^3 x^2}$

49. $\dfrac{3^4 (ab)^2}{3(ab)^3}$

50. $\dfrac{8^2 u^4 v^5}{8^3 u^4 v^2}$

In Exercises 51–58, simplify the expression. (Assume the denominator is not zero.)

51. $\dfrac{-12z^3}{-3z}$

52. $\dfrac{16y^5}{8y^3}$

53. $\dfrac{32b^4}{12b^3}$

54. $\dfrac{-7c^2}{8c^5}$

55. $\dfrac{-22y^2}{4y}$

56. $-\dfrac{16}{v^2}$

57. $\dfrac{(-3xy)^3}{-9x^6 y^2}$

58. $\dfrac{10(x^2 y)^2}{5^2 (xy^3)^4}$

In Exercises 59–92, perform the division and simplify. (Assume the denominator is not zero.)

59. $\dfrac{4z - 12}{4}$

60. $\dfrac{7x + 7}{7}$

61. $(5x^2 - 2x) \div x$

62. $(16a^2 + 5a) \div a$

63. $\dfrac{25z^3 + 10z^2}{-5z}$

64. $\dfrac{12c^4 - 36c}{-6c}$

65. $-\dfrac{4x^2 - 3x}{x}$

66. $\dfrac{3x^4 + 5x^3 - 4x^2}{x^2}$

67. $\dfrac{m^3 + 3m - 4}{m}$

68. $\dfrac{l^2 - 4l + 8}{l}$

69. $\dfrac{x^2 + 9x + 20}{x + 4}$

70. $\dfrac{x^2 - 7x - 30}{x - 10}$

71. Divide $9x^2 - 1$ by $3x + 1$.

72. Divide $25y^2 - 4$ by $5y - 2$.

73. Divide $2z^2 + 5z - 3$ by $z + 3$.

74. Divide $4u^2 - 18u - 10$ by $u - 5$.

75. $(18t^2 - 21t - 4) \div (3t - 4)$

76. $(20t^2 + 32t - 16) \div (2t + 4)$

77. $\dfrac{x^3 - 8}{x - 2}$

78. $\dfrac{x^3 + 27}{x + 3}$

79. $\dfrac{x^3 - 7x + 6}{x - 2}$

80. $\dfrac{x^3 - 28x - 48}{x + 4}$

81. $(7x + 3) \div (x + 2)$

82. $(8x - 5) \div (2x + 1)$

83. $\dfrac{x^2 + 9}{x + 3}$

84. $\dfrac{y^2 + 3}{y + 3}$

85. $\dfrac{2x^2 + 7x + 8}{2x + 3}$

86. $\dfrac{6x^2 + x - 9}{3x - 4}$

87. $\dfrac{4z^2 + 3z}{2z - 1}$

88. $\dfrac{10y^2 - 3y}{y + 3}$

89. $\dfrac{3t^3 + 7t^2 + 3t - 2}{t + 2}$

90. $\dfrac{5(x + 3)^2 + 3(x + 3)}{x + 3}$

91. Divide $x^4 - 1$ by $x - 1$.

92. Divide x^4 by $x - 1$.

In Exercises 93 and 94, simplify the expression. (Assume the denominator is not zero.)

93. $\dfrac{8u^2v}{2u} + \dfrac{(uv)^2}{uv}$

94. $\dfrac{25x^2y}{10x} + \dfrac{3xy^2}{2y}$

95. *Comparing Ages*   You have two children: one is 18 years old and the other is 8 years old. In t years, their ages will be $t + 18$ and $t + 8$.

(a) Use long division to rewrite the ratio of your older child's age to your younger child's age.

(b) Complete the table.

t	0	10	20	30	40	50	60
$\dfrac{t+18}{t+8}$							

(c) What happens to the values of the ratio as t increases? Use the result of part (a) to explain your conclusion.

96. *Exploration*   Consider the equation
$$2x^3 - 5x^2 + 2x - 5 = (2x - 5)(x^2 + 1).$$

(a) Use a graphing utility to verify the equation by graphing both the left side and right side of the equation. Are the graphs the same?

(b) Verify the equation by multiplying the polynomials on the right side of the equation.

(c) Verify the equation by performing the long division
$$\frac{2x^3 - 5x^2 + 2x - 5}{2x - 5}.$$

97. *Exploration*   Use a graphing utility to compare the graphs of the following equations. On most graphing utilities, the graphs appear the same. There is, however, a slight difference in the graphs. What is it?
$$y = \frac{x^2 - 3x + 2}{x - 1} \quad \text{and} \quad y = x - 2$$

In Exercises 98–100, match the graph with *two* of the expressions.

(a) $x - 1$

(b) $1 - x$

(c) $\dfrac{x^2 - 1}{x + 1}$

(d) $\dfrac{x^2 + 2x + 1}{x + 1}$

(e) $\dfrac{1 - 2x + x^2}{1 - x}$

(f) $\dfrac{x^2 - 1}{x - 1}$

98.

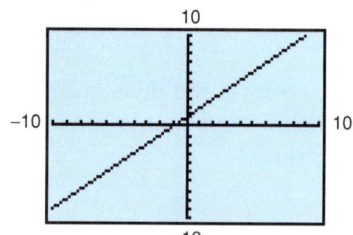

99.

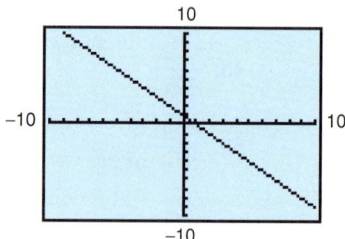

100.

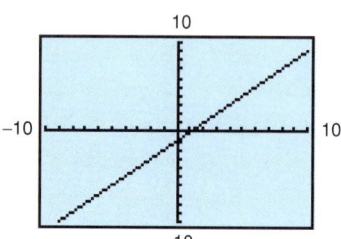

| **5.4** | **Negative Exponents and Scientific Notation** |

Negative Exponents ▪ Rules of Exponents ▪
Scientific Notation

Negative Exponents

This section extends the properties of exponents to include **negative exponents.**
Consider the property

$$a^m \cdot a^n = a^{m+n}, \quad a \neq 0.$$

If this property is to hold for negative exponents, then the statement

$$a^2 \cdot a^{-2} = a^{2+(-2)} = a^0 = 1$$

implies that a^{-2} is the *reciprocal* of a^2. In other words, it must be true that

$$a^{-2} = \frac{1}{a^2}.$$

Informally, you can think of this property as allowing you to "move" powers from
the numerator to the denominator by changing the sign of the exponent.

Negative Exponent Rule

Let n be an integer and let a be a real number, variable, or algebraic
expression such that $a \neq 0$.

$$a^{-n} = \frac{1}{a^n}$$

STUDY TIP

Be sure you see that the
negative exponent rule allows you to move only
factors in a numerator (or
denominator), *not terms.*
For example,

$$\frac{x^{-2} \cdot y}{4} = \frac{y}{4x^2}$$

whereas the following is not
true.

$$\frac{x^{-2} + y}{4} \neq \frac{y}{4x^2}$$

EXAMPLE 1   *Monomials Involving Negative Exponents*

a. $x^{-7} = \dfrac{1}{x^7}$ Move x^{-7} to denominator and change the sign of the exponent.

b. $5x^{-4} = \dfrac{5}{x^4}$ Move x^{-4} to denominator and change the sign of the exponent.

c. $\dfrac{1}{2x^{-3}} = \dfrac{x^3}{2}$ Move x^{-3} to numerator and change the sign of the exponent.

d. $x^{-2}y^3 = \dfrac{y^3}{x^2}$ Move x^{-2} to denominator and change the sign of the exponent.

Rules of Exponents

All of the rules of exponents apply to negative exponents. For convenience, these rules are summarized below. Remember that these rules apply to real numbers, variables, or algebraic expressions.

Rules of Exponents

Let m and n be integers, and let a and b be real numbers, variables, or algebraic expressions such that $a \neq 0$ and $b \neq 0$.

Property	*Example*
1. $a^m a^n = a^{m+n}$	$y^2 \cdot y^4 = y^{2+4} = y^6$
2. $\dfrac{a^m}{a^n} = a^{m-n}$	$\dfrac{x^7}{x^4} = x^{7-4} = x^3$
3. $a^{-n} = \dfrac{1}{a^n}$	$y^{-4} = \dfrac{1}{y^4}$
4. $a^0 = 1$	$(x^2 + 1)^0 = 1$
5. $(ab)^m = a^m b^m$	$(5x)^4 = 5^4 x^4$
6. $\left(\dfrac{a}{b}\right)^m = \dfrac{a^m}{b^m}$	$\left(\dfrac{2}{x}\right)^3 = \dfrac{2^3}{x^3}$
7. $(a^m)^n = a^{mn}$	$(y^3)^{-4} = y^{3(-4)} = y^{-12}$

NOTE There is more than one way to solve problems such as those in Example 2. For instance, you might prefer to write Example 2(b) as

$$\frac{y^{-2}}{3y^{-5}} = \frac{y^5}{3y^2}$$

$$= \frac{y^{5-2}}{3}$$

$$= \frac{y^3}{3}.$$

EXAMPLE 2 Using Rules of Exponents

a.
$$(-3ab^4)(4ab^{-3}) = (-3)(4)(a)(a)(b^4)(b^{-3}) \qquad \text{Regroup factors.}$$

$$= (-12)(a^{1+1})(b^{4-3}) \qquad \text{Apply rules of exponents.}$$

$$= -12a^2 b \qquad \text{Simplify.}$$

b.
$$\frac{y^{-2}}{3y^{-5}} = \frac{y^{-2}y^5}{3} \qquad \text{Apply rules of exponents.}$$

$$= \frac{y^{-2+5}}{3} \qquad \text{Multiply powers with like bases.}$$

$$= \frac{y^3}{3} \qquad \text{Simplify.}$$

EXAMPLE 3 Using Rules of Exponents

Use rules of exponents to write each expression without negative exponents. (Assume that no variable is equal to zero.)

a. $3x^{-1}(-4x^2y)^0$ **b.** $\left(\dfrac{5x^3}{y^{-1}}\right)^2$ **c.** $\left(\dfrac{a^2}{3}\right)^{-2}$

Solution

a. This problem is a little tricky. Note that the factor $(-4x^2y)$ is raised to the *zero* power. Because any number (other than zero) raised to the zero power is 1, you can write

$$3x^{-1}(-4x^2y)^0 = 3x^{-1}(1) = \frac{3}{x}.$$

b. This problem can also be tricky. The important thing to realize is that the *entire fraction* $(5x^3/y^{-1})$ is raised to the second power. This means that you must apply the exponent of 2 to each factor of the numerator and denominator, as follows.

$$\left(\frac{5x^3}{y^{-1}}\right)^2 = \frac{(5x^3)^2}{(y^{-1})^2} = \frac{5^2(x^3)^2}{y^{-2}} = \frac{25x^6}{y^{-2}} = 25x^6y^2$$

c. In this problem, note that the *entire expression* $(a^2/3)$ is raised to the -2 power.

$$\left(\frac{a^2}{3}\right)^{-2} = \frac{a^{-4}}{3^{-2}} = \frac{3^2}{a^4} = \frac{9}{a^4}$$

Technology

Calculators and Negative Exponents

The keystrokes used to evaluate expressions with negative exponents vary. For instance, to evaluate 13^{-2} on a calculator, you can try one of the following keystroke sequences.

Keystrokes

13 $\boxed{x^y}$ 2 $\boxed{+/-}$ $\boxed{=}$ Scientific

13 $\boxed{\wedge}$ $\boxed{(-)}$ 2 $\boxed{\text{ENTER}}$ Graphing

With either of these sequences, your calculator should display .00591716. If it doesn't, consult the user's manual for your calculator to find the correct keystrokes.

Scientific Notation

Exponents provide an efficient way of writing and computing with the very large (or very small) numbers used in science. For instance, a drop of water contains more than 33 billion billion molecules. That is 33 followed by 18 zeros.

$$33,000,000,000,000,000,000$$

It is convenient to write such numbers in **scientific notation.** This notation has the form $c \times 10^n$, where $1 \le c < 10$ and n is an integer. Thus, the number of molecules in a drop of water can be written in scientific notation as follows.

$$3.3 \times 10^{19} = 33,000,000,000,000,000,000$$

19 places

The *positive* exponent 19 indicates that the number is large (10 or more) and that the decimal point has been moved 19 places.

A *negative* exponent in scientific notation indicates that the number is *small* (less than 1). For instance, the mass (in grams) of one electron is approximately as follows.

$$9.0 \times 10^{-28} = 0.00000000000000000000000000009$$

28 places

EXAMPLE 4 Converting from Decimal to Scientific Notation

a. $0.0000782 = 7.82 \times 10^{-5}$ Small number yields negative exponent.

Five places

b. $836,100,000.0 = 8.361 \times 10^{8}$ Large number yields positive exponent.

Eight places

EXAMPLE 5 Converting from Scientific to Decimal Notation

a. $1.345 \times 10^{2} = 134.5$ Positive exponent yields large number.

Two places

b. $9.36 \times 10^{-6} = 0.00000936$ Negative exponent yields small number.

Six places

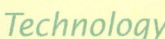

Calculators and Scientific Notation

Most scientific calculators automatically switch to scientific notation when they are displaying large (or small) numbers that exceed the display range. Try multiplying $98{,}900{,}000 \times 5000$. If your calculator follows standard conventions, its display should show

| 4.945 11 | or | 4.945 E 11 |.

This means that $c = 4.945$ and the exponent of 10 is $n = 11$, which implies that the number is 4.945×10^{11}. For *entering* numbers in scientific notation, your calculator should have an exponential entry key labeled EE or EXP.

DISCOVERY

Use a calculator to evaluate $(7.8)(2.4) \times (10^4)(10^9)$. How does the result relate to Example 6? This is an illustration of what property of multiplication?

EXAMPLE 6 Using Scientific Notation

Use a calculator to evaluate $78{,}000 \times 2{,}400{,}000{,}000$.

Solution

Because $78{,}000 = 7.8 \times 10^4$ and $2{,}400{,}000{,}000 = 2.4 \times 10^9$, you can evaluate the product as follows.

7.8 EXP 4 × 2.4 EXP 9 =   *Scientific*

7.8 EE 4 × 2.4 EE 9 ENTER   *Graphing*

After these keystrokes have been entered, the calculator display should show 1.872 14. Therefore, the product of the two numbers is

$$(7.8 \times 10^4)(2.4 \times 10^9) = 1.872 \times 10^{14} = 187{,}200{,}000{,}000{,}000.$$

Group Activities   E x p l o r i n g   w i t h   T e c h n o l o g y

Exponential Expressions   As a team, find as many equivalent pairs as possible among the following exponential expressions:

$$\frac{2}{x^{-3}}, \frac{1}{2x^3}, 2x^{-3}, \frac{1}{(2x)^{-3}}, 2x^3, \frac{x^3}{8}, \frac{1}{8x^3}, \frac{x^{-3}}{2}, 8x^3, (2x)^{-3}.$$

Next, use a calculator, with $x = 3$, to illustrate the equivalence of each pair. Organize your work into a table with four columns—an expression, the equivalent expression, and each expression evaluated at $x = 3$.

5.4 Exercises

Discussing the Concepts

True or False? In Exercises 1–3, state whether the equation is valid. If it is invalid, find values of x and y for which the equation is not true.

1. $x^3y^3 = xy^3$

2. $x^{-1}y^{-1} = \dfrac{1}{xy}$

3. $x^{-1} + y^{-1} = \dfrac{1}{x + y}$

4. Without looking back at page 306, how many of the seven rules of exponents can you state?

5. Give examples of large and small numbers written in scientific notation.

6. Justify each step.
$$
\begin{aligned}
(3 \times 10^5)(4 \times 10^6) &= (3 \times 10^5)(10^6 \times 4) \\
&= 3(10^5 \times 10^6)(4) \\
&= 3(10^{5+6})(4) \\
&= (3 \cdot 4)10^{11} \\
&= 12 \times 10^{11} \\
&= 1.2 \times 10^{12}
\end{aligned}
$$

Problem Solving

In Exercises 7–10, rewrite with positive exponents.

7. $5x^{-4}$

8. $6x^{-2}y^{-3}$

9. $\dfrac{1}{2z^{-4}}$

10. $\dfrac{7x^2}{y^{-3}}$

In Exercises 11–14, rewrite with positive exponents. Then evaluate the expression.

11. 3^{-2}

12. $(-6)^{-2}$

13. $\dfrac{2^{-4}}{3^{-2}}$

14. $\left(\dfrac{3}{4}\right)^{-3}$

In Exercises 15–18, use a calculator to evaluate the expression.

15. 3.8^{-4}

16. $8(6.2)^{-3}$

17. $\dfrac{100}{1.06^{-15}}$

18. $500(1.08)^{-20}$

In Exercises 19–30, rewrite with positive exponents. (Assume that no variable is equal to zero.)

19. $\dfrac{x^2}{x^{-3}}$

20. $\dfrac{x^{-3}}{x^2}$

21. $(2x^2)^{-2}$

22. $(2x^5)^0$

23. $(4a^{-2}b^3)^{-3}$

24. $(-2s^{-1}t^{-2})^{-1}$

25. $(-2x^2)^3(4x^3)^{-1}$

26. $(4y^{-2})(3y^4)$

27. $\left(\dfrac{x}{10}\right)^{-1}$

28. $\left(\dfrac{x^{-3}y^4}{5}\right)^{-3}$

29. $\dfrac{3}{2} \cdot \left(\dfrac{-2}{3}\right)^{-3}$

30. $\dfrac{(-x)^3}{-x^{-2}}$

In Exercises 31–34, write in decimal form.

31. 1.09×10^6

32. 2.345×10^8

33. 8.52×10^{-3}

34. 7.021×10^{-5}

In Exercises 35–38, write in scientific notation.

35. $1,637,000,000$

36. 67.8

37. 0.000435

38. 0.008367

In Exercises 39–42, use a calculator to evaluate the expression.

39. $8,000,000 \times 623,000$

40. $(4.85 \times 10^5)(2.04 \times 10^8)$

41. $\dfrac{67,000,000}{0.0052}$

42. $\dfrac{8.6 \times 10^4}{3.9 \times 10^7}$

43. *Time for Light to Travel*   Light travels from the sun to the earth in approximately

$$\frac{9.3 \times 10^7}{1.1 \times 10^7}$$

minutes. Write this time in decimal form.

44. *Distance to a Star*   The star Beta Andromeda is approximately 76 light years from the earth (see figure). (A light year is the distance light can travel in 1 year.) Estimate the distance to this star if a light year is approximately 5.8746×10^{12} miles.

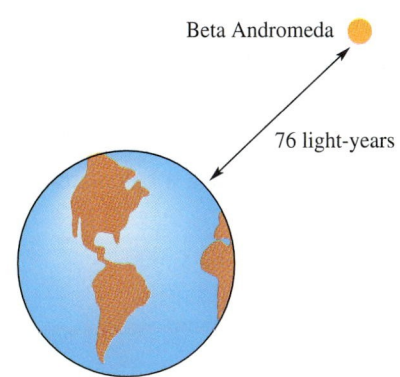

Beta Andromeda

76 light-years

Figure for 44

Reviewing the Major Concepts

In Exercises 45 and 46, evaluate the expression.

45. $25 - 3^2 \cdot 2$

46. $(12 - 9)^3 \cdot 4 + 36$

In Exercises 47 and 48, graph the equation using a graphing utility.

47. $y = x^2 - 2x + 1$

48. $y = |2x + 1|$

In Exercises 49 and 50, evaluate the expression for the specified values. (If it is not possible, state the reason.)

Expression	*Values*
49. $10 + 3x - 2x^2$	(a) $x = -2$ (b) $x = 0$
50. $\dfrac{5x}{x - 10}$	(a) $x = 20$ (b) $x = 10$

Additional Problem Solving

In Exercises 51–58, rewrite with positive exponents. Then evaluate the result.

51. $(-4)^{-3}$

52. 5^{-3}

53. $\dfrac{1}{16^{-1}}$

54. $\dfrac{4}{3^{-2}}$

55. $\dfrac{4^{-3}}{2}$

56. $\dfrac{4^{-2}}{3^{-4}}$

57. $\left(\frac{2}{3}\right)^{-2}$

58. $\left(\frac{5}{4}\right)^{-3}$

In Exercises 59–78, rewrite with positive exponents and simplify. (Assume that no variable equals zero.)

59. $\dfrac{y^{-5}}{y}$

60. $\dfrac{z^4}{z^{-2}}$

61. $\dfrac{x^{-4}}{x^{-2}}$

62. $\dfrac{t^{-5}}{t^{-1}}$

63. $(y^{-3})^2$

64. $(z^{-2})^3$

65. $(s^2)^{-1}$

66. $(a^3)^{-3}$

67. $\dfrac{b^2 \cdot b^{-3}}{b^4}$

68. $\dfrac{c^{-3} \cdot c^4}{c^{-1}}$

69. $(3x^2y)^{-2}$

70. $(4x^{-3}y^2)^{-3}$

71. $(5x^2y^4)^3(5x^2y^4)^{-3}$

72. $(x^2 + 3)^4(x^2 + 3)^{-4}$

73. $\left(\dfrac{3z^2}{x}\right)^{-2}$

74. $\left(\dfrac{4}{z}\right)^{-2}$

75. $(2x^2y)^0$

76. $\dfrac{(3z)^{-2}}{(3z)^{-2}}$

77. $\dfrac{3}{8} \cdot \left(\dfrac{-5}{2}\right)^3$

78. $\dfrac{4}{5}\left(\dfrac{2x^{-3}}{(5x)^{-1}}\right)$

In Exercises 79–84, write in decimal form.

79. 6.21×10^0

80. 9.4675×10^4

81. 8.67×10^{-2}

82. 4.73×10^0

83. $(8 \times 10^3) + (3 \times 10^0) + (5 \times 10^{-2})$

84. $(6 \times 10^4) + (9 \times 10^3) + (4 \times 10^{-1})$

In Exercises 85–88, write in scientific notation.

85. 93,000,000

86. 900,000,000

87. 0.004392

88. 0.00000045

In Exercises 89–94, use a calculator to evaluate the expression.

89. $0.000345 \times 8,980,000,000$

90. $93,200,000 \times 1,657,000$

91. $3,200,000^5$

92. $(7.5 \times 10^4)^5$

93. $(3.28 \times 10^{-6})^4$

94. $\dfrac{8.48 \times 10^8}{1.62 \times 10^6}$

95. *Boltzmann's Constant* The study of the kinetic energy of an ideal gas uses Boltzmann's constant. This constant k is given by

$$k = \frac{8.31 \times 10^7}{6.02 \times 10^{23}}.$$

Perform the division, leaving your result in scientific notation.

96. *Hydraulic Compression* A hydraulic cylinder in a large press contains 2 gallons of oil. When the cylinder is under full pressure, the actual volume of oil will decrease by

$$2(150)(20 \times 10^{-6})$$

gallons. Write this volume in decimal form.

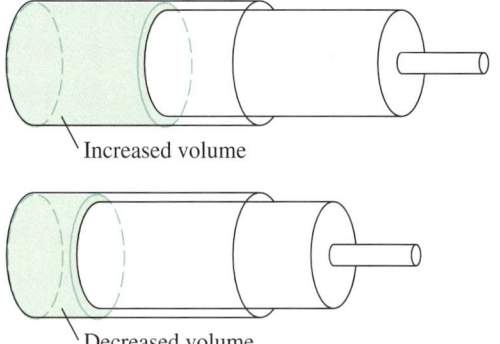

Increased volume

Decreased volume

97. *Numerical and Graphical Approach* A new car is purchased for $24,000. Its value after t years is

$$V = 24,000(1.2)^{-t}.$$

(a) Use the model to complete the table.

t	0	2	4	6	8
$24{,}000(1.2)^{-t}$					

(b) Represent the data in the table graphically.

(c) *Guess, Check, and Revise* When will the car be valued at less than $100?

98. *Numerical and Graphical Approach*

(a) Complete the table by evaluating the indicated powers of 2.

x	-1	-2	-3	-4	-5
2^x					

(b) Graphically represent the data in the table.

(c) Use the table or the graph to describe the value of 2^{-n} when n is very large. Will the value of 2^{-n} ever be negative?

99. *An Infinite Sum* Successively remove from the unit square in the figure the fractional parts given in the table in Exercise 98. If this process were continued, what do you think would happen? Use this to estimate the sum

$$\frac{1}{2} + \frac{1}{4} + \frac{1}{8} + \frac{1}{16} + \cdots.$$

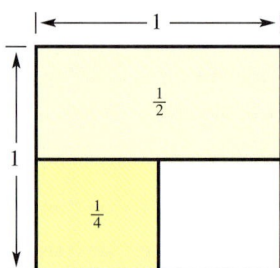

CHAPTER PROJECT: Chemistry and Color

When an element is heated to a high temperature, it emits bands of light called its line spectrum. Normal "white light" consists of many colors, each of which is characterized by a wavelength, measured in nanometers (1 nanometer $= 1 \times 10^{-9}$ meters). The colors of light in the visible spectrum correspond to the range of wavelengths from about 400 to about 700 nanometers, as shown on page 317.

Every element has its own line spectrum. This allows chemists to identify the chemical composition of a substance. The table below shows several elements and the three prominent wavelengths of their line spectra.

Name	Symbol	Line Spectrum Values		
Hydrogen	H	434	486	656
Carbon	C	427	515	589
Nitrogen	N	501	568	594
Oxygen	O	419	465	616
Sodium	Na	569	589	590
Sulfur	S	493	545	561
Chlorine	Cl	480	522	542
Calcium	Ca	502	531	646

The line spectrum of a chemical compound consists of the line spectra of the elements that make up the compound. For instance, the line spectrum of water is made up of the line spectra of hydrogen and oxygen.

Investigate the following questions using the line spectrum values for the given elements.

1. Write the line spectrum values for H, C, N, and O in meters. Use decimal *and* scientific notation.

2. Describe the compositions of the following compounds.

Compound	Line Spectrum Values					
Carbon dioxide	419	427	465	515	589	616
Salt	480	522	542	569	589	590
Methane	427	434	486	515	589	656
Calcium sulfide	493	502	531	545	561	646
Ammonia	434	486	501	568	594	656

3. Explain how a scientist can determine the chemical compositions of distant objects such as comets and stars. Could this same method be used to analyze a planet? Explain your reasoning.

4. Identify the colors present in the line spectra of the elements and compounds below by comparing the line spectrum values with the wavelength values in the Math Matters table on page 317.

Hydrogen	Calcium	Carbon
Carbon dioxide	Nitrogen	Salt
Oxygen	Methane	Sodium
Calcium sulfide	Sulfur	Ammonia
Chlorine		

CHAPTER SUMMARY

After studying this chapter, you should have acquired the following skills. These skills are keyed to the Review Exercises that begin on page 315. Answers to odd-numbered Review Exercises are given in the back of the book.

- Write polynomials in standard form and find their degrees and leading coefficients. *(Section 5.1)*

  Review Exercises 1–4

- Give examples of polynomials that satisfy given conditions. *(Section 5.1)*

  Review Exercises 5, 6

- Determine whether expressions are equal. *(Sections 5.1, 5.2, 5.3, 5.4)*

  Review Exercises 7–18

- Add, subtract, multiply, and divide polynomials. *(Sections 5.1, 5.2, 5.3)*

  Review Exercises 19–48

- Use one of the special product patterns shown below to expand the expression. *(Section 5.2)*

  Review Exercises 49, 50, 51, 52, 56, 58

  1. $(a + b)^2 = a^2 + 2ab + b^2$ Square of a binomial

  2. $(a - b)^2 = a^2 - 2ab + b^2$ Square of a binomial

- Use the special product pattern shown below to expand the expression. *(Section 5.2)*

  Review Exercises 53, 54, 55, 57

  $(a + b)(a - b) = a^2 - b^2$ Sum and Difference of Two Terms

- Evaluate expressions involving negative exponents and scientific notation. *(Section 5.4)*

  Review Exercises 59–70

- Rewrite expressions having negative exponents and simplify. *(Section 5.4)*

  Review Exercises 71–78

- Solve problems involving geometry. *(Sections 5.1, 5.2)*

  Review Exercises 79–83

- Solve real-life problems involving polynomials and scientific notation. *(Sections 5.1, 5.2, 5.4)*

  Review Exercises 84–87

REVIEW EXERCISES

In Exercises 1–4, write the polynomial in standard form. Then find its degree and leading coefficient.

1. $10x - 4 - 5x^3$

2. $2x^2 + 9$

3. $2(x - 5) + 10$

4. $100 - 3(x^2 + 25)$

In Exercises 5 and 6, give an example of a polynomial that satisfies the given conditions. (*Note:* There are many correct answers.)

5. A trinomial of degree 4

6. A monomial of degree 2

In Exercises 7–18, determine whether the expressions are equal. If not, rewrite the right side so that it is a simplification of the left side.

7. $\dfrac{3}{8} + \dfrac{1}{8} \overset{?}{=} \dfrac{4}{16}$

8. $\dfrac{1}{3}(2x) \overset{?}{=} \dfrac{2}{3} \cdot \dfrac{x}{3}$

9. $(2x)^4 \overset{?}{=} 2x^4$

10. $(-x)^4 \overset{?}{=} -x^4$

11. $3^{-2} \overset{?}{=} -9$

12. $\dfrac{y^2}{y^{-2}} \overset{?}{=} y^{2-2}$

13. $\dfrac{7-x}{7} \overset{?}{=} 1 - x$

14. $(x - 3)^0 \overset{?}{=} 0$

15. $-3(x - 2) \overset{?}{=} -3x - 6$

16. $3x - (x + 5) \overset{?}{=} 2x + 5$

17. $(x + 2)^2 \overset{?}{=} x^2 + 4$

18. $(x + 3)(x - 3) \overset{?}{=} x^2 + 6x - 9$

In Exercises 19–48, perform the operations and simplify. (Assume that no denominator equals zero.)

19. $(2x + 3) + (x - 4)$

20. $\left(\frac{1}{2}x + \frac{2}{3}\right) + \left(4x + \frac{1}{3}\right)$

21. $(t - 5) - (t^2 - t - 5)$

22. $(y^2 + 3) - (y^2 - 9)$

23. $(4 - x^2) + 2(x - 2)$

24. $(3u + 4u^2) + 5(u + 1) + 3u^2$

25. $(-x^3 - 3x) - 2(2x^3 + x + 1)$

26. $(z^2 + 6z) - 3(z^2 + 2z)$

27. $4y^2 - [y - 3(y^2 + 2)]$

28. $(6a^3 + 3a) - 2[a + (a^3 - 2)]$

29. $2x(x + 4)$

30. $3y(-4y)(y + 1)$

31. $(x + 3)(2x - 4)$

32. $(3y + 2)(4y - 3)$

33. $(x^2 + 5x + 2)(2x + 3)$

34. $(s^3 + 4s - 3)(s - 3)$

35. $2u(u - 5) - (u + 1)(u - 5)$

36. $(3v - 2)(-2v) + 2v(3v - 2)$

37. $(5x^2 + 15x) \div 5x$

38. $(8u^3 + 4u^2) \div 2u$

39. $7(x + 4)^3 \div 3(x + 4)^2$

40. $(x^2 + x) \div x$

41. $\dfrac{x^2 - x - 6}{x - 3}$

42. $\dfrac{x^2 + x - 20}{x + 5}$

43. $\dfrac{24x^2 - x - 8}{3x - 2}$

44. $\dfrac{4x + 7}{3x - 2}$

45. $\dfrac{2x^3 + 2x^2 - x + 2}{x - 1}$

46. $\dfrac{6x^4 - 4x^3 - 27x^2 + 18x}{3x - 2}$

47. $\dfrac{x^4 - 3x^2 + 2}{x^2 - 1}$

48. $\dfrac{3x^4}{x^2 - 1}$

In Exercises 49–58, use a special product pattern to expand the expression.

49. $(x + 3)^2$

50. $(x - 5)^2$

51. $(4x - 7)^2$

52. $(9 - 2x)^2$

53. $(u - 6)(u + 6)$

54. $(3a + 8)(3a - 8)$

55. $(3t - 1)(3t + 1)$

56. $(4 + 3b)^2$

57. $[(a - 1) + b][(a - 1) - b]$

58. $[(a - 1) + b]^2$

In Exercises 59–70, evaluate the expression.

59. 4^{-2}

60. 3^{-4}

61. $6^{-4}6^2$

62. $(2^2 \cdot 3^2)^{-1}$

63. $\left(\dfrac{3}{5}\right)^{-3}$

64. $\left(\dfrac{2^{-2}}{3}\right)^2$

65. $\left(-\dfrac{2}{5}\right)^3 \left(\dfrac{5}{2}\right)^2$

66. $\dfrac{2^2 \cdot 3^{-2}}{2^{-2} \cdot 3^{-1}}$

67. $(3 \times 10^3)^2$

68. $(4 \times 10^{-3})(5 \times 10^7)$

69. $\dfrac{1.85 \times 10^9}{5 \times 10^4}$

70. $\dfrac{1}{(4 \times 10^{-2})^3}$

In Exercises 71–78, rewrite with positive exponents and simplify. (Assume that no variable equals zero.)

71. $(-3a^2)^{-2}$

72. $(3u^{-3})(9u^6)$

73. $(x^2y^{-3})^2$

74. $5(x+3)^0$

75. $\dfrac{t^{-4}}{t^{-1}}$

76. $\dfrac{a^3 \cdot a^{-2}}{a^{-1}}$

77. $\left(\dfrac{y}{5}\right)^{-2}$

78. $(3x^2y^4)^3(3x^2y^4)^{-3}$

Geometry In Exercises 79–82, find a polynomial that represents the area of the shaded portion of the figure.

79.

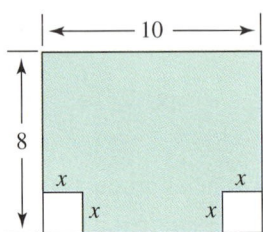

80.

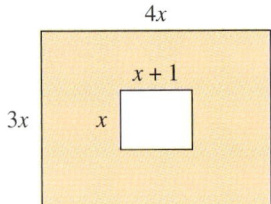

81.

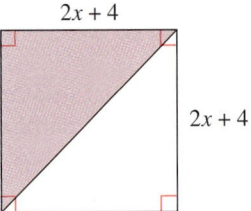

82.

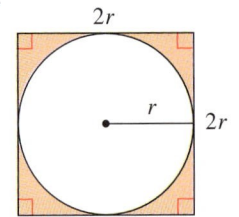

83. *Geometry* The length of a rectangular wall is x units, and its height is $x - 3$ units (see figure). Find (a) the perimeter and (b) the area of the rectangle.

84. *Graphical Interpretation* The cost of producing x units is

$$C = 15 + 26x.$$

The revenue for selling x units is

$$R = 40x - \tfrac{1}{2}x^2, \quad 0 \le x \le 20.$$

The profit is the difference between revenue and cost.

(a) Perform the subtraction required to find the polynomial representing profit.

(b) Use a graphing utility to graph the polynomial representing profit.

(c) Determine the profit when $x = 14$ units. Use the graph of part (b) to describe the profit when x is less than or greater than 14.

85. *Probability* The probability of three successes in five trials of an experiment is

$$10p^3(1 - p)^2.$$

Find this product.

86. *Metal Expansion* When the temperature of a 150-foot iron steam pipe is increased $100°C$, the length of the pipe will increase by $100(150)(10 \times 10^{-6})$ feet, as shown in the figure. Write the amount of increase in decimal form.

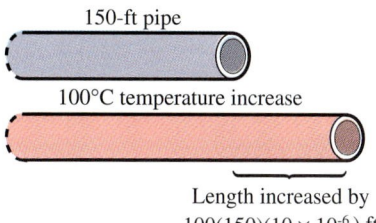

150-ft pipe

100°C temperature increase

Length increased by
$100(150)(10 \times 10^{-6})$ ft

87. *Special Product* What special product does the figure illustrate? Explain your reasoning.

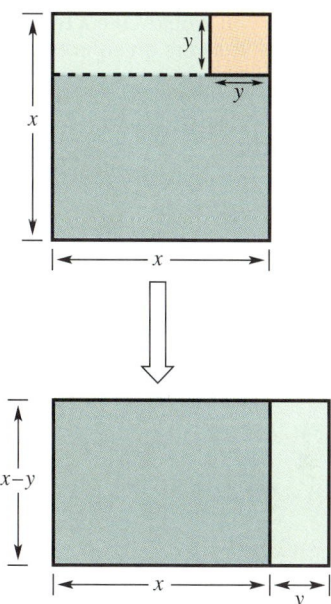

Math Matters

The Spectrum

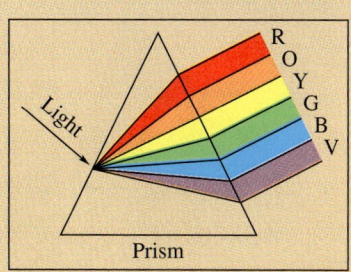

Prism

The spectrum is the visible result produced when light is resolved into its various wavelengths or frequencies. For instance, when light is passed through a prism, as shown in the accompanying figure, the light is separated into a rainbow of colors ranging from red to violet.

The frequencies and wavelengths of the colors of the rainbow are as follows.

Color of Light	Frequency (10^{14} hertz)	Wavelength (10^{-7} meters)
Red	4.284–4.634	6.470–7.000
Orange	4.634–5.125	5.850–6.470
Yellow	5.125–5.215	5.750–5.850
Green	5.215–6.104	4.912–5.750
Blue	6.104–7.115	4.240–4.912
Violet	7.115–7.495	4.000–4.240

CHAPTER TEST

Take this test as you would take a test in class. After you are done, check your work against the answers given in the back of the book.

1. Explain how to determine the degree and identify the leading coefficient of $-3x^4 - 5x^2 + 2x - 10$.

2. Give an example of a trinomial in one variable of degree 4.

In Exercises 3–14, perform the indicated operation and simplify. (Assume that no variable or denominator equals zero.)

3. $(3z^2 - 3z + 7) + (8 - z^2)$

4. $(8u^3 + 3u^2 - 2u - 1) - (u^3 + 3u^2 - 2u)$

5. $6y - [2y - (3 + 4y - y^2)]$

6. $-5(x^2 - 1) + 3(4x + 7) - (x^2 + 26)$

7. $(5b + 3)(2b - 1)$

8. $4x\left(\dfrac{3x}{2}\right)^2$

9. $(z + 2)(2z^2 - 3z + 5)$

10. $(x - 5)^2$

11. $(2x - 3)(2x + 3)$

12. $\dfrac{15x + 25}{5}$

13. $\dfrac{x^3 - x - 6}{x - 2}$

14. $\dfrac{4x^3 + 10x^2 - 2x - 5}{2x + 1}$

In Exercises 15 and 16, simplify the expression. (Assume that no variable equals zero.)

15. $\dfrac{-6a^2b}{-9ab}$

16. $(3x^{-2}y^3)^{-2}$

17. Evaluate the expression *without* using a calculator. Show your work.

    (a) $\dfrac{2^{-3}}{3^{-1}}$ (b) $(1.5 \times 10^5)^2$

18. Write an expression that represents the area of the triangle (see figure). Explain your reasoning.

19. The mean distance from earth to the moon is 3.84×10^8 meters. Write this distance in decimal form.

20. The standard atmospheric pressure is 101,300 newtons per square meter. Write this pressure in scientific notation.

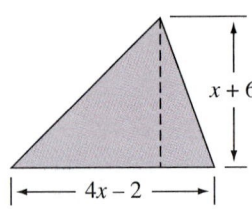

Figure for 18

Factoring and Solving Equations

6

- Factoring Polynomials with Common Factors
- Factoring Trinomials
- More About Factoring Trinomials
- Factoring Polynomials with Special Forms
- Solving Equations and Problem Solving

The United States Department of Agriculture tracks the amounts of foods produced and consumed each year in the United States. For 1970 to 1991 the amount of wheat flour consumed per person per year in the United States can be modeled by the quadratic equation

$$y = 0.05t^2 + 0.14t + 111.3$$

where y represents the amount (in pounds) consumed per person and $t = 0$ corresponds to the year 1970.

To determine the year in which 130 pounds of wheat flour was consumed per person, you can graph this model and the horizontal line given by $y = 130$ on the same coordinate plane. The intersection of the two graphs occurs at $t = 18$ or the year 1988. (Source: U.S. Department of Agriculture)

Year	1970	1975	1980	1985	1990	1991
Wheat Flour (in pounds)	110.9	114.5	116.9	124.7	135.7	135.9

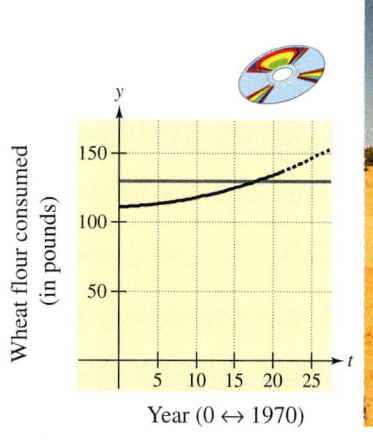

Wheat flour consumed (in pounds)

Year (0 ↔ 1970)

The chapter project related to this information is on page 365.

319

6.1	**Factoring Polynomials with Common Factors**
	Greatest Common Factor ■ Common Monomial Factors ■ Factoring by Grouping

Greatest Common Factor

In Chapter 5, you used the Distributive Property to multiply polynomials. In this chapter, you will study the reverse process, which is **factoring**.

Multiplying Polynomials   *Factoring Polynomials*

$2x(7 - 3x)$ ⟹ $14x - 6x^2$ $14x - 6x^2$ ⟹ $2x(7 - 3x)$

To factor an expression efficiently, you need to understand the concept of the *greatest common factor* of two (or more) integers or terms. In Section P.2, you learned that the **greatest common factor** of two or more integers is the greatest integer that is a factor of each integer. For example, the greatest common factor of 12 and 30 is 6.

EXAMPLE 1   *Finding the Greatest Common Factor*

Find the greatest common factor of $5x^2y^2$ and $30x^3y$.

Solution

From the factorizations

$$5x^2y^2 = 5 \cdot x \cdot x \cdot y \cdot y = (5x^2y)(y)$$
$$30x^3y = 2 \cdot 3 \cdot 5 \cdot x \cdot x \cdot x \cdot y = (5x^2y)(6x)$$

you can conclude that the greatest common factor is $5x^2y$.

EXAMPLE 2   *Finding the Greatest Common Factor*

Find the greatest common factor of $8x^5$, $20x^3$, and $16x^4$.

Solution

From the factorizations

$$8x^5 = 2 \cdot 2 \cdot 2 \cdot x \cdot x \cdot x \cdot x \cdot x = (4x^3)(2x^2)$$
$$20x^3 = 2 \cdot 2 \cdot 5 \cdot x \cdot x \cdot x = (4x^3)(5)$$
$$16x^4 = 2 \cdot 2 \cdot 2 \cdot 2 \cdot x \cdot x \cdot x \cdot x = (4x^3)(4x)$$

you can conclude that the greatest common factor is $4x^3$.

Common Monomial Factors

Consider the three terms listed in Example 2 as terms of the polynomial

$$8x^5 + 16x^4 + 20x^3.$$

The greatest common factor, $4x^3$, of these terms is the **greatest common monomial factor** of the polynomial. When you use the Distributive Property to remove this factor from each term of the polynomial, you are **factoring out** the common monomial factor.

$$8x^5 + 16x^4 + 20x^3 = 4x^3(2x^2) + 4x^3(4x) + 4x^3(5) \qquad \text{Factor each term.}$$

$$= 4x^3(2x^2 + 4x + 5) \qquad \begin{array}{l}\text{Factor out common}\\ \text{monomial factor.}\end{array}$$

EXAMPLE 3 Greatest Common Monomial Factor

Factor out the greatest common monomial factor from $6x - 18$.

Solution

The greatest common factor of $6x$ and 18 is 6.

$$6x - 18 = 6(x) - 6(3) \qquad \text{Greatest common monomial factor is 6.}$$

$$= 6(x - 3) \qquad \text{Factor 6 out of each term.}$$

EXAMPLE 4 Greatest Common Monomial Factor

Factor out the greatest common monomial factor from $10y^3 - 25y^2$.

Solution

The greatest common factor of $10y^3$ and $25y^2$ is $5y^2$.

$$10y^3 - 25y^2 = (5y^2)(2y) - (5y^2)(5) \qquad \text{Greatest common factor is } 5y^2.$$

$$= 5y^2(2y - 5) \qquad \text{Factor } 5y^2 \text{ out of each term.}$$

NOTE From Examples 3, 4, and 5, you can make the following generalizations about a polynomial with integer coefficients.

1. The coefficient a of the greatest common monomial factor is the greatest integer that *divides* each coefficient of the polynomial.

2. The variable factor x^n of the greatest common monomial factor has the *least power* of x of all terms of the polynomial.

EXAMPLE 5 Greatest Common Monomial Factor

Factor out the greatest common monomial factor from $45x^3 - 15x^2 - 15$.

Solution

The greatest common factor of $45x^3$, $15x^2$, and 15 is 15.

$$45x^3 - 15x^2 - 15 = 15(3x^3) - 15(x^2) - 15(1)$$

$$= 15(3x^3 - x^2 - 1)$$

EXAMPLE 6 Common Monomial Factors

Factor each polynomial.

a. $35y^3 - 7y^2 - 14y$ **b.** $6y^5 + 3y^3 - 2y^2$

Solution

a. $35y^3 - 7y^2 - 14y = 7y(5y^2) - 7y(y) - 7y(2)$ $7y$ is common factor.

$= 7y(5y^2 - y - 2)$ Factor $7y$ out of each term.

b. $6y^5 + 3y^3 - 2y^2 = y^2(6y^3) + y^2(3y) - y^2(2)$ y^2 is common factor.

$= y^2(6y^3 + 3y - 2)$ Factor y^2 out of each term.

The greatest common monomial factor of the terms of a polynomial is usually considered to have a positive coefficient. However, sometimes it is convenient to factor a negative number out of a polynomial.

EXAMPLE 7 A Negative Common Monomial Factor

Factor the polynomial $-2x^2 + 8x - 12$ in two ways.

a. Factor out a common monomial factor of 2.

b. Factor out a common monomial factor of -2.

Solution

a. To factor out the common monomial factor of 2, write the following.

$$-2x^2 + 8x - 12 = 2(-x^2) + 2(4x) - 2(6)$$
$$= 2(-x^2 + 4x - 6)$$

b. To factor -2 out of the polynomial, write the following.

$$-2x^2 + 8x - 12 = -2(x^2) + (-2)(-4x) + (-2)(6)$$
$$= -2(x^2 - 4x + 6)$$

Check this result by multiplying $(x^2 - 4x + 6)$ by -2. When you do, you will obtain the original polynomial.

NOTE   With experience, you should be able to omit writing the first step shown in Examples 6 and 7. For instance, to factor -2 out of $-2x^2 + 8x - 12$, you could simply write

$$-2x^2 + 8x - 12 = -2(x^2 - 4x + 6).$$

Factoring by Grouping

There are occasions when the common factor of a polynomial is not simply a monomial. For instance, the polynomial

$$x^2(x-2) + 3(x-2)$$

has the common *binomial* factor $(x-2)$. Factoring out this common factor produces

$$x^2(x-2) + 3(x-2) = (x-2)(x^2+3).$$

EXAMPLE 8   *Common Binomial Factors*

Factor each polynomial.

a. $5x^2(7x-1) - 3(7x-1)$ **b.** $2x(3x-4) + (3x-4)$

Solution

a. Each of the terms of this polynomial has a binomial factor of $(7x-1)$.

$$5x^2(7x-1) - 3(7x-1) = (7x-1)(5x^2-3)$$

b. Each of the terms of this polynomial has a binomial factor of $(3x-4)$.

$$2x(3x-4) + (3x-4) = (3x-4)(2x+1)$$

Be sure you see that when $(3x-4)$ is factored out of itself, you are left with 1. This follows from the fact that $(3x-4)(1) = (3x-4)$.

In Example 8, the polynomials were already grouped so that it was easy to determine the common binomial factors. In practice, you will have to do the grouping as well as the factoring. To see how this works, consider the expression

$$x^3 + 2x^2 + 3x + 6$$

and try to *factor* it. Note first that there is no common monomial factor to take out of all four terms. But suppose you *group* the first two terms together and the last two terms together.

$$x^3 + 2x^2 + 3x + 6 = (x^3 + 2x^2) + (3x+6) \qquad \text{Group terms.}$$
$$= x^2(x+2) + 3(x+2) \qquad \text{Distributive Property}$$
$$= (x+2)(x^2+3) \qquad \text{Distributive Property}$$

This process is called **factoring by grouping.** Try grouping the polynomial as

$$(x^3 + 3x) + (2x^2 + 6).$$

How can you use this grouping to factor the polynomial?

EXAMPLE 9   Factoring by Grouping

Factor $x^3 - 2x^2 + x - 2$.

Solution

$$x^3 - 2x^2 + x - 2 = (x^3 - 2x^2) + (x - 2) \qquad \text{Group terms.}$$
$$= x^2(x - 2) + (x - 2) \qquad \text{Distributive Property}$$
$$= (x - 2)(x^2 + 1) \qquad \text{Distributive Property}$$

EXAMPLE 10   Factoring by Grouping

Factor $3x^2 + 12x - 5x - 20$.

Solution

$$3x^2 + 12x - 5x - 20 = (3x^2 + 12x) - (5x + 20)$$
$$= 3x(x + 4) - 5(x + 4)$$
$$= (x + 4)(3x - 5)$$

NOTE   You can always check to see that you have factored an expression correctly by multiplying and comparing the result with the original expression. Try using multiplication to check the results of Examples 9 and 10.

Group Activities   You Be the Instructor

Factoring by Grouping   Suppose you are tutoring someone in algebra and you want to create several polynomials for your student to factor. In your group, develop a procedure for creating polynomials that contain a common factor or that can be factored by the grouping method of this section. Create a list of practice problems and have another member of your group factor them.

6.1 Exercises

Discussing the Concepts

1. Give an example of a polynomial that is written in factored form.

2. Give an example of a trinomial whose greatest common monomial factor is $3x$.

3. In your own words, describe a method for finding the greatest common factor of a polynomial.

4. How can you check your result when factoring a polynomial?

5. Explain how the word *factor* can be used as a noun and as a verb.

6. Give several examples of the use of the Distributive Property in factoring.

Problem Solving

In Exercises 7–14, find the greatest common factor.

7. $24, 90$

8. $60, 80, 90$

9. $2x^2, 12x$

10. $36x^4, 18x^3$

11. $9yz^2, -12y^2z^3$

12. $-15x^6y^3, 45xy^3$

13. $14x^2, 1, 7x^4$

14. $5x^0y^4, 10x^2y^2, 15xy$

In Exercises 15–30, factor the polynomial.

15. $3x + 3$

16. $5y + 5$

17. $8t - 16$

18. $3u + 12$

19. $-25x - 10$

20. $7z^3 + 21$

21. $12x^2 - 2x$

22. $12u + 9u^2$

23. $10ab + 10a^2b$

24. $21x^2z - 35xz$

25. $100 + 75z - 50z^2$

26. $42t^3 - 21t^2 + 7$

27. $x(x - 3) + 5(x - 3)$

28. $x(x + 6) + 3(x + 6)$

29. $z^2(z + 5) + (z + 5)$

30. $(y - 8)^2 + y^2(y - 8)^2$

In Exercises 31–34, factor the polynomial by grouping.

31. $x^2 - 5x + xy - 5y$

32. $kx^2 + 10kx + x + 10$

33. $t^3 - 3t^2 + 2t - 6$

34. $3s^3 + 6s^2 + 2s + 4$

In Exercises 35–38, factor a negative real number from the polynomial. Then use the Commutative Property to write the polynomial factor with a positive leading coefficient.

35. $5 - 10x$

36. $3 - x$

37. $4 + 2x - x^2$

38. $18 - 12x - 6x^2$

In Exercises 39 and 40, complete the factorization.

39. $\frac{1}{2}x + \frac{3}{4} = \frac{1}{4}\left(\right)$

40. $\frac{2}{3}x - \frac{1}{6} = \frac{1}{6}\left(\right)$

41. *Geometry*   The front of a microwave oven has an area of $44h - h^2$. Factor this expression to determine the length of the microwave in terms of h.

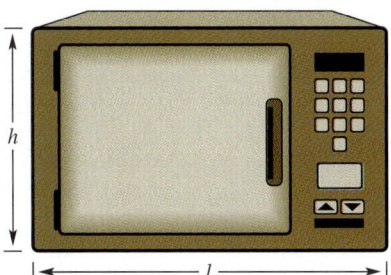

42. *Chemical Reaction*   The rate of change in a chemical reaction is $kQx - kx^2$, where Q is the original amount, x is the new amount, and k is a constant of proportionality. Factor the expression.

Reviewing the Major Concepts

In Exercises 43–46, use a graphing utility to graph the equation. Then identify the coordinates of at least three solution points.

43. $y = 8 - 4x$

44. $y = 3x - 6$

45. $y = -\frac{1}{2}x^2$

46. $y = |x + 2|$

47. *Commission Rate* Determine the commission rate for an employee who earned $1620 in commissions on sales of $54,000.

48. *Work Rate* One person can complete a typing project in 10 hours. Another person can complete the project in 6 hours. Working together, how long will they take to complete the project?

Additional Problem Solving

In Exercises 49–56, find the greatest common factor.

49. 18, 150, 100

50. 20, 45

51. $z^2, -z^6$

52. t^4, t^7

53. u^2v, u^3v^2

54. $r^6s^4, -rs$

55. $28a^4b^2, 14a^3b^3, 42a^2b^5$

56. $16x^2y, 12xy^2, 36x^2y^2$

In Exercises 57–84, factor the polynomial. (*Note:* Some of the polynomials have no common factor.)

57. $6z - 6$

58. $3x - 3$

59. $24y^2 - 18$

60. $-14y - 7$

61. $x^2 + x$

62. $-s^3 - s$

63. $25u^2 - 14u$

64. $36t^4 + 24t^2$

65. $2x^4 + 6x^3$

66. $9z^6 + 27z^4$

67. $7s^2 + 9t^2$

68. $12x^2 - 5y^3$

69. $-10r^3 - 35r$

70. $-144a^2 + 24a^2$

71. $16a^3b^3 + 24a^4b^3$

72. $6x^4y + 12x^2y$

73. $12x^2 + 16x - 8$

74. $9 - 3y - 15y^2$

75. $9x^4 + 6x^3 + 18x^2$

76. $32a^5 - 2a^3 + 6a$

77. $5u^2 + 5u^2 + 5u$

78. $11y^3 - 22y^2 + 11y^2$

79. $t(s + 10) - 8(s + 10)$

80. $y(q - 5) - 10(q - 5)$

81. $a^2(b + 2) - a(b + 2)$

82. $x^3(x + 4) + x^2(x + 4)$

83. $(a + b)(c + 7) + (a + b)(c + 7)^2$

84. $(x + y)(x - y) - x(x - y)$

In Exercises 85–92, factor by grouping.

85. $a^2 - 4a + ab - 4b$

86. $x^2 + 25x + xy + 25y$

87. $xy^2 - 4xy + 2y - 8$

88. $ay^2 + 3ay + 3y + 9$

89. $x^3 + 2x^2 + x + 2$

90. $x^3 - 5x^2 + x - 5$

91. $z^3 + 3z^2 - 2z - 6$

92. $4u^3 - 2u^2 - 6u + 3$

In Exercises 93–96, factor a negative real number from the polynomial. Then use the Commutative Property to write the polynomial factor with a positive leading coefficient.

93. $3000 - 3x$

94. $9 - 2x^2$

95. $4 + 12x - 2x^2$

96. $x - 2x^2 - x^4$

In Exercises 97–100, complete the factorization.

97. $2y - \frac{1}{5} = \frac{1}{5} \left(\right)$

98. $3z + \frac{3}{4} = \frac{1}{4} \left(\right)$

99. $\frac{7}{8}x + \frac{5}{16}y = \frac{1}{16} \left(\right)$

100. $\frac{5}{12}u - \frac{5}{8}v = \frac{1}{24} \left(\right)$

In Exercises 101–104, use a graphing utility to graph both equations on the same screen. Use the graphs to verify the factorization.

101. $y_1 = 9 - 3x$
$y_2 = -3(x - 3)$

102. $y_1 = x^2 - 4x$
$y_2 = x(x - 4)$

103. $y_1 = 6x - x^2$
$y_2 = x(6 - x)$

104. $y_1 = x(x + 2) - 3(x + 2)$
$y_2 = (x + 2)(x - 3)$

Geometry   In Exercises 105 and 106, find the length of the rectangle.

105. Area $= 2x^2 + 2x$

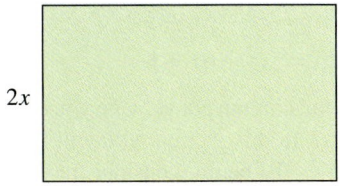

$2x$

106. Area $= x^2 + 2x + 10x + 20$

$x + 2$

107. *Geometry*   The surface area of a right circular cylinder is

$$S = 2\pi r^2 + 2\pi rh.$$

Factor the expression for the surface area.

108. *Simple Interest*   The amount after t years when a principal of P dollars is invested at $r\%$ simple interest is given by $P + Prt$. Factor the expression for simple interest.

109. *Unit Price*   The revenue R for selling x units of a product at a price of p dollars per unit is given by $R = xp$. For a particular commodity the revenue is

$$900x - 0.1x^2.$$

Factor the revenue model and determine an expression that represents the price p in terms of x.

CAREER INTERVIEW

Sidney W. Carter

Professional Engineer
Self-employed in environmental engineering, specializing in soil and groundwater reclamation and remediation
Lakeland, FL 33813

Florida has long been mined for phosphates for use in fertilizers. Mining companies are required to reclaim the areas they have mined, and one of my jobs is to design the appearance of the reclaimed areas.

Because the amount of fill dirt that may be available on the site of a reclamation project is limited, it is important to know before planning a design how much earth is already available at a mined site. I routinely obtain a viable estimate by working with a scale aerial photo of the mined site and the *end-area method* for computing volume. For each mound of earth in the photo, I identify the cross-sectional shapes of the two ends of the mound. I then find the areas, A_1 and A_2, of each end using measurements from the photo and formulas for standard geometric shapes such as a trapezoid or a triangle. The last step in the end-area method is to average the areas of the two ends and multiply by the length L of the mound: Volume $= [(A_1 + A_2)/2]L$.

6.2	**Factoring Trinomials**

Factoring Trinomials of the Form $x^2 + bx + c$ ▪
Factoring Trinomials in Two Variables ▪ Factoring Completely

Factoring Trinomials of the Form $x^2 + bx + c$

From Section 5.2, you know that the product of two binomials is often a trinomial. Here are some examples.

Factored Form	F O I L	*Trinomial Form*
$(x - 1)(x + 5)$	$= x^2 + 5x - x - 5$	$= x^2 + 4x - 5$
$(x - 3)(x - 3)$	$= x^2 - 3x - 3x + 9$	$= x^2 - 6x + 9$
$(x + 5)(x + 1)$	$= x^2 + x + 5x + 5$	$= x^2 + 6x + 5$
$(x - 2)(x - 4)$	$= x^2 - 4x - 2x + 8$	$= x^2 - 6x + 8$

Try covering the factored forms in the left-hand column above. Can you determine the factored forms from the trinomial forms? In this section, you will learn how to factor trinomials of the form $x^2 + bx + c$. To begin, consider the following factorization.

$$(x + m)(x + n) = x^2 + nx + mx + mn$$
$$= x^2 + (n + m)x + mn$$

Sum of terms ↓   Product of terms ↓

$$= x^2 + \boxed{b}\, x + \boxed{c}$$

Thus, to *factor* a trinomial $x^2 + bx + c$ into a product of two binomials, you must find two numbers m and n whose product is c and whose sum is b.

There are many different techniques that people use to factor trinomials. The most common is to use *Guess, Check, and Revise* with mental math.

EXAMPLE 1   Factoring a Trinomial

Factor $x^2 + 5x + 6$.

Solution

You need to find two numbers whose product is 6 and whose sum is 5. Using mental math, you can determine that the numbers are 2 and 3.

$$x^2 + 5x + 6 = (x + 2)(x + 3)$$

EXAMPLE 2 Factoring Trinomials

Factor each trinomial.

a. $x^2 + 5x - 6$ **b.** $x^2 - x - 6$ **c.** $x^2 - 5x + 6$

Solution

a. You need to find two numbers whose product is -6 and whose sum is 5.

The product of -1 and 6 is -6.

$$x^2 + 5x - 6 = (x - 1)(x + 6)$$

The sum of -1 and 6 is 5.

b. You need to find two numbers whose product is -6 and whose sum is -1.

The product of -3 and 2 is -6.

$$x^2 - x - 6 = (x - 3)(x + 2)$$

The sum of -3 and 2 is -1.

c. You need to find two numbers whose product is 6 and whose sum is -5.

The product of -2 and -3 is 6.

$$x^2 - 5x + 6 = (x - 2)(x - 3)$$

The sum of -2 and -3 is -5.

If you have trouble factoring a trinomial, it helps to make a list of all the distinct pairs of factors and then check each by multiplying. For instance, consider the trinomial

$$x^2 - 5x - 24.$$

For this trinomial, you need to find two numbers whose product is -24 and whose sum is -5.

Factors of -24	*Multiply Binomials*	
$(1)(-24)$	$(x + 1)(x - 24) = x^2 - 23x - 24$	
$(-1)(24)$	$(x - 1)(x + 24) = x^2 + 23x - 24$	
$(2)(-12)$	$(x + 2)(x - 12) = x^2 - 10x - 24$	
$(-2)(12)$	$(x - 2)(x + 12) = x^2 + 10x - 24$	
$(3)(-8)$	$(x + 3)(x - 8) = x^2 - 5x - 24$	Correct choice
$(-3)(8)$	$(x - 3)(x + 8) = x^2 + 5x - 24$	
$(4)(-6)$	$(x + 4)(x - 6) = x^2 - 2x - 24$	
$(-4)(6)$	$(x - 4)(x + 6) = x^2 + 2x - 24$	

With experience, you will be able to *mentally* narrow the list of possible factors to only two or three possibilities whose sums could then be tested to determine the correct factorization. Here are some suggestions for narrowing down the list.

Guidelines for Factoring $x^2 + bx + c$

To factor $x^2 + bx + c$, you need to find two numbers m and n whose product is c and whose sum is b.

$$x^2 + bx + c = (x + m)(x + n)$$

1. If c is *positive*, then m and n have like signs which match the sign of b.

2. If c is *negative*, then m and n have unlike signs.

3. If $|b|$ is small relative to $|c|$, first try those factors of c that are closest to each other in absolute value.

EXAMPLE 3   Factoring Trinomials

Factor the following trinomials.

a. $x^2 - 2x - 15$ **b.** $x^2 + 20x + 36$

Solution

a. You need to find two numbers whose product is -15 and whose sum is -2.

The product of -5 and 3 is -15.

$$x^2 - 2x - 15 = (x - 5)(x + 3)$$

The sum of -5 and 3 is -2.

b. You need to find two numbers whose product is 36 and whose sum is 20.

The product of 2 and 18 is 36.

$$x^2 + 20x + 36 = (x + 2)(x + 18)$$

The sum of 2 and 18 is 20.

Not all trinomials are factorable using integer factors. For instance,

$$x^2 - 2x - 6$$

is not factorable using integer factors. (List the factors and test them.) Such nonfactorable trinomials are called **prime polynomials.**

Factoring Trinomials in Two Variables

The first three examples each involved trinomials of the form

$$x^2 + bx + c. \qquad \text{Trinomial in one variable}$$

The next two examples show how to factor trinomials of the form

$$x^2 + bxy + cy^2. \qquad \text{Trinomial in two variables}$$

Note that this trinomial has two variables, x and y. However, from the factorization

$$x^2 + bxy + cy^2 = (x + my)(x + ny)$$
$$= x^2 + (m + n)xy + mny^2$$

you can see that you still need to find two factors of c whose sum is b.

EXAMPLE 4   Factoring a Trinomial in Two Variables

Factor the trinomial.

$$x^2 - xy - 12y^2$$

Solution

You need to find two numbers whose product is -12 and whose sum is -1.

The product of –4 and 3 is –12.

$$x^2 - xy - 12y^2 = (x - 4y)(x + 3y)$$

The sum of –4 and 3 is –1.

Check this result by multiplying $(x - 4y)$ by $(x + 3y)$.

EXAMPLE 5   Factoring a Trinomial in Two Variables

Factor the trinomial.

$$y^2 - 6xy + 8x^2$$

Solution

You need to find two numbers whose product is 8 and whose sum is -6.

The product of –2 and –4 is 8.

$$y^2 - 6xy + 8x^2 = (y - 2x)(y - 4x)$$

The sum of –2 and –4 is –6.

Check this result by multiplying $(y - 2x)$ by $(y - 4x)$.

Factoring Completely

Some trinomials have a common monomial factor. In such cases you should first factor out the common monomial factor. Then you can try to factor the resulting trinomial by the methods in this section. This "multiple-stage factoring process" is called **factoring completely.** For instance, the trinomial

$$2x^2 - 4x - 6 = 2(x^2 - 2x - 3) \qquad \text{Factor out 2.}$$
$$= 2(x - 3)(x + 1) \qquad \text{Factor trinomial.}$$

is factored completely.

EXAMPLE 6 Factoring Completely

Factor each trinomial completely.

a. $2x^2 - 12x + 10$ **b.** $3x^3 - 27x^2 + 54x$ **c.** $4y^4 + 32y^3 + 28y^2$

Solution

a. $2x^2 - 12x + 10 = 2(x^2 - 6x + 5)$ Factor out 2.

$\qquad\qquad\qquad\quad = 2(x - 5)(x - 1)$ Factor trinomial.

b. $3x^3 - 27x^2 + 54x = 3x(x^2 - 9x + 18)$ Factor out $3x$.

$\qquad\qquad\qquad\qquad = 3x(x - 3)(x - 6)$ Factor trinomial.

c. $4y^4 + 32y^3 + 28y^2 = 4y^2(y^2 + 8y + 7)$ Factor out $4y^2$.

$\qquad\qquad\qquad\qquad = 4y^2(y + 1)(y + 7)$ Factor trinomial.

Check these results by multiplying the factors to see that you obtain the original trinomials.

Group Activities

Extending the Concept

Factoring Polynomials Discuss this question in your group: "Is it possible to factor a polynomial such as $x^3 + 5x^2 - 3x - 15$ by the method used for trinomials in this section?" Try this method on $x^3 + 5x^2 - 3x - 15$ and $x^3 - 7x^2 + 2x - 14$, and then factor these polynomials by grouping. Which method do you prefer? Explain your preference.

6.2 Exercises

Discussing the Concepts

1. In factoring of $x^2 - 4x + 3$, why is it unnecessary to test $(x - 1)(x + 3)$ and $(x + 1)(x - 3)$?

2. Explain why $(x + 1)(2x - 6)$ is not a complete factorization of $2x^2 - 4x - 6$.

3. In your own words, explain how to factor a trinomial. Give examples with your explanation.

4. What is meant by a prime trinomial?

5. Can you completely factor a trinomial into two different sets of prime factors? Explain.

6. In factoring of the trinomial $x^2 + bx + c$, is the process easier if c is a prime number such as 5 or a composite number such as 120? Explain.

Problem Solving

In Exercises 7–10, find the missing factor. Then check your answer by multiplying the factors.

7. $x^2 + 4x + 3 = (x + 3)()$

8. $c^2 + 2c - 3 = (c + 3)()$

9. $y^2 - 2y - 15 = (y + 3)()$

10. $z^2 - 4z + 3 = (z - 3)()$

In Exercises 11–14, find all possible products of the form $(x + m)(x + n)$ where mn is the specified product. (Assume m and n are integers.)

11. $mn = 11$

12. $mn = 10$

13. $mn = 12$

14. $mn = 18$

In Exercises 15–26, factor the trinomial. (*Note:* Some of the trinomials may be prime.)

15. $x^2 + 6x + 8$

16. $x^2 + 13x + 12$

17. $x^2 - x - 6$

18. $s^2 - 7s - 25$

19. $x^2 - 17x + 72$

20. $x^2 - 8x - 240$

21. $y^2 + 5y + 11$

22. $x^2 + x - 6$

23. $u^2 - 22u - 48$

24. $c^2 - 6c + 10$

25. $a^2 + 2ab - 15b^2$

26. $y^2 + 4yz - 60z^2$

Think About It   In Exercises 27 and 28, find all integer values of b such that the trinomial can be factored.

27. $x^2 + bx + 15$

28. $x^2 + bx - 48$

Think About It   In Exercises 29 and 30, find two values of c such that the trinomial can be factored.

29. $x^2 - 6x + c$

30. $x^2 + 12x + c$

In Exercises 31–36, factor the trinomial completely. (*Note:* Some of the trinomials may be prime.)

31. $3x^2 + 21x + 30$

32. $4x^2 - 32x + 60$

33. $3z^2 + 5z + 6$

34. $-5x^2z + 15xz + 50z$

35. $2x^3y + 4x^2y^2 - 6xy^3$

36. $x^4y^2 + 3x^3y^3 + 2x^2y^4$

Geometric Modeling of Factoring   In Exercises 37–40, factor the trinomial and draw a geometric model of the result. [The sample shows a geometric model for factoring $x^2 + 3x + 2$ as $(x + 1)(x + 2)$.]

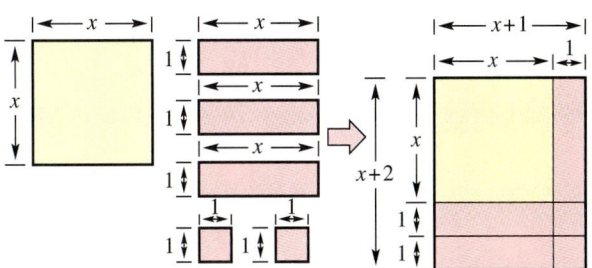

37. $x^2 + 4x + 3$

38. $x^2 + 5x + 4$

39. $x^2 + 5x + 6$

40. $x^2 + 6x + 5$

Reviewing the Major Concepts

In Exercises 41–44, find the product.

41. $-a^2(a-1)$ **42.** $(2x+5)(2x-5)$

43. $(u-8)(u+3)$ **44.** $(v-1)(v-6)$

In Exercises 45 and 46, graph the equation.

45. $y = x^2 - 2x - 8$ **46.** $y = (x+4)(x-3)$

47. *Profit*   A company had a loss of \$2,500,000 during the first 6 months of the year. The company ended the year with an overall profit of \$1,475,000. What was the profit during the second 6 months of the year?

48. *Total Cost*   Computer printer ribbons cost \$11.95 per ribbon. There are 12 ribbons per box. The sales tax is 4%. What is the total cost of 5 boxes of ribbons?

Additional Problem Solving

In Exercises 49–52, find the missing factor. Then check your answer by multiplying the factors.

49. $a^2 + a - 6 = (a+3)()$

50. $x^2 + 5x + 6 = (x+3)()$

51. $z^2 - 5z + 6 = (z-3)()$

52. $y^2 - 4y - 21 = (y+3)()$

In Exercises 53–70, factor the trinomial. (*Note:* Some of the trinomials may be prime.)

53. $x^2 - 13x + 40$ **54.** $x^2 - 9x + 14$

55. $z^2 - 7z + 12$ **56.** $x^2 + 10x + 24$

57. $x^2 + 2x - 15$ **58.** $b^2 - 2b - 15$

59. $x^2 + 3x - 70$ **60.** $x^2 + 21x + 108$

61. $y^2 - 6y + 10$ **62.** $x^2 - x - 36$

63. $x^2 + 19x + 60$ **64.** $r^2 - 30r + 216$

65. $x^2 + xy - 2y^2$ **66.** $x^2 - 5xy + 6y^2$

67. $x^2 + 8xy + 15y^2$ **68.** $u^2 - 4uv - 5v^2$

69. $x^2 - 7xz - 18z^2$ **70.** $x^2 + 15xy + 50y^2$

In Exercises 71–76, factor the trinomial completely. (*Note:* Some of the trinomials may be prime.)

71. $x^3 - 13x^2 + 30x$ **72.** $x^3 + x^2 - 2x$

73. $4y^2 - 8y - 12$ **74.** $7x^2 + 5x + 10$

75. $10x^3 + 50x^2y + 60xy^2$

76. $x^2y - 6xy^2 + y^3$

In Exercises 77–80, use a graphing utility to graph the two equations on the same viewing rectangle. What can you conclude?

77. $y_1 = x^2 - x - 6$

    $y_2 = (x+2)(x-3)$

78. $y_1 = x^2 - 10x + 16$

    $y_2 = (x-2)(x-8)$

79. $y_1 = x^3 + x^2 - 20x$

    $y_2 = x(x-4)(x+5)$

80. $y_1 = 2x - x^2 - x^3$

    $y_2 = x(1-x)(2+x)$

81. *Exploration*   An open box is to be made from a 4-foot by 6-foot sheet of metal by cutting equal squares from each corner and turning up the sides (see figure). The volume of the box can be modeled by

$$V = 4x^3 - 20x^2 + 24x, \quad 0 < x < 2.$$

(a) Factor the trinomial modeling the volume of the box. Use the factored form to explain how the model was found.

(b) Use a graphing utility to graph the trinomial over the specified interval. Use the graph to approximate the size of the squares to be cut from each corner so the volume of the box is greatest.

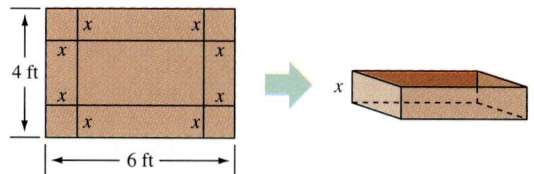

Math Matters

Prime Numbers

A prime number is a positive integer that is greater than 1 and cannot be factored as the product of smaller integers. For instance, the numbers 2, 3, and 5 are prime numbers, whereas the number 4 is not prime (because $4 = 2 \cdot 2$). Prime numbers have fascinated people for hundreds of years, and part of this fascination stems from the fact that no one has ever discovered a simple pattern for prime numbers. There are 168 prime numbers that are less than 1000. Can you determine the next three prime numbers in the list?

2	3	5	7	11	13	17	19	23	29	31	37
41	43	47	53	59	61	67	71	73	79	83	89
97	101	103	107	109	113	127	131	137	139	149	151
157	163	167	173	179	181	191	193	197	199	211	223
227	229	233	239	241	251	257	263	269	271	277	281
283	293	307	311	313	317	331	337	347	349	353	359
367	373	379	383	389	397	401	409	419	421	431	433
439	443	449	457	461	463	467	479	487	491	499	503
509	521	523	541	547	557	563	569	571	577	587	593
599	601	607	613	617	619	631	641	643	647	653	659
661	673	677	683	691	701	709	719	727	733	739	743
751	757	761	769	773	787	797	809	811	821	823	827
829	839	853	857	859	863	877	881	883	887	907	911
919	929	937	941	947	953	967	971	977	983	991	997

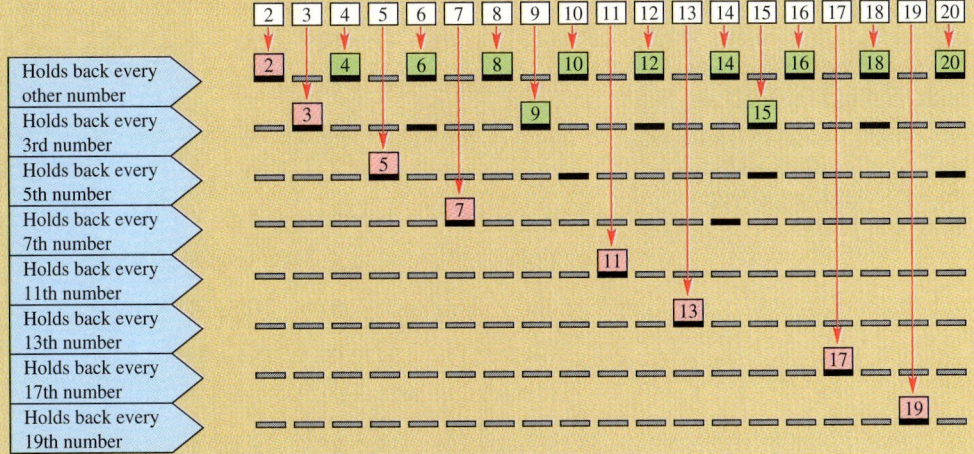

Holds back every other number
Holds back every 3rd number
Holds back every 5th number
Holds back every 7th number
Holds back every 11th number
Holds back every 13th number
Holds back every 17th number
Holds back every 19th number

The first number held back on each level is prime.

6.3	**More About Factoring Trinomials**

Factoring Trinomials of the Form $ax^2 + bx + c$ ■
Factoring Completely ■ Factoring by Grouping

Factoring Trinomials of the Form $ax^2 + bx + c$

In this section you will learn how to factor a trinomial whose leading coefficient is *not* 1. To see how this works, consider the following.

Factors of a

$$ax^2 + bx + c = (\boxed{}x + \boxed{})(\boxed{}x + \boxed{})$$

Factors of c

The goal is to find a combination of factors of a and c so that the outer and inner products add up to the middle term bx.

EXAMPLE 1 Factoring a Trinomial of the Form $ax^2 + bx + c$

Factor $6x^2 + 5x - 4$.

Solution

In this trinomial, $a = 6$ and $c = -4$. There are many combinations of factors of these numbers, as seen in the following list.

STUDY TIP

If the original trinomial has no common monomial factors, then its binomial factors can't have common monomial factors. Thus, in Example 1, you don't have to test factors such as $(6x - 4)$ that have a common monomial factor of 2.

$(x + 1)(6x - 4) = 6x^2 + 2x - 4$	$(6x - 4)$ has common factor 2.
$(x - 1)(6x + 4) = 6x^2 - 2x - 4$	$(6x + 4)$ has common factor 2.
$(x + 4)(6x - 1) = 6x^2 + 23x - 4$	$O + I \neq 5x$
$(x - 4)(6x + 1) = 6x^2 - 23x - 4$	$O + I \neq 5x$
$(x + 2)(6x - 2) = 6x^2 + 10x - 4$	$(6x - 2)$ has common factor 2.
$(x - 2)(6x + 2) = 6x^2 - 10x - 4$	$(6x + 2)$ has common factor 2.
$(2x + 1)(3x - 4) = 6x^2 - 5x - 4$	$O + I$ is opposite of $5x$.
$(2x - 1)(3x + 4) = 6x^2 + 5x - 4$	Correct factorization
$(2x + 4)(3x - 1) = 6x^2 + 10x - 4$	$(2x + 4)$ has common factor 2.
$(2x - 4)(3x + 1) = 6x^2 - 10x - 4$	$(2x - 4)$ has common factor 2.
$(2x + 2)(3x - 2) = 6x^2 + 2x - 4$	$(2x + 2)$ has common factor 2.
$(2x - 2)(3x + 2) = 6x^2 - 2x - 4$	$(2x - 2)$ has common factor 2.

Thus, the correct factorization is $6x^2 + 5x - 4 = (2x - 1)(3x + 4)$.

The following guidelines can help shorten the list of possible factorizations you need to test.

Guidelines for Factoring $ax^2 + bx + c$, $(a > 0)$

1. First, factor out any common monomial factor.

2. Because the resulting trinomial has no common monomial factors, you don't have to test any binomial factors that have a common monomial factor.

3. If the middle-term test (O + I) yields the opposite of b, switch the signs on the factors of c.

Using these guidelines, you can shorten the list in Example 1 to the following.

$$(x + 4)(6x - 1) = 6x^2 + 23x - 4$$
$$(x - 4)(6x + 1) = 6x^2 - 23x - 4$$
$$(2x + 1)(3x - 4) = 6x^2 - 5x - 4$$
$$(2x - 1)(3x + 4) = 6x^2 + 5x - 4 \qquad \text{Correct factorization}$$

Do you see why you can cut the list from 12 possible factorizations to only four?

EXAMPLE 2   Factoring a Trinomial of the Form $ax^2 + bx + c$

Factor $2x^2 + x - 15$.

Solution

In this trinomial, $a = 2$, which factors as $(1)(2)$, and $c = -15$, which factors as $(1)(-15)$, $(-1)(15)$, $(3)(-5)$, and $(-3)(5)$.

$$(2x + 1)(x - 15) = 2x^2 - 29x - 15$$
$$(2x + 15)(x - 1) = 2x^2 + 13x - 15$$
$$(2x + 3)(x - 5) = 2x^2 - 7x - 15$$
$$(2x + 5)(x - 3) = 2x^2 - x - 15 \qquad \text{Middle term has incorrect sign.}$$
$$(2x - 5)(x + 3) = 2x^2 + x - 15 \qquad \text{Correct factorization}$$

Therefore, the correct factorization is

$$2x^2 + x - 15 = (2x - 5)(x + 3).$$

NOTE   In Example 2, do you see what to do when the middle term has the incorrect sign?

Technology

As with other types of factoring, you can use a graphing utility to check your results. For instance, try graphing

$y = 2x^2 + x - 15$ and
$y = (2x - 5)(x + 3)$

on the same screen, as shown below. Because both graphs are the same, you can reason that

$2x^2 + x - 15 = (2x - 5)(x + 3).$

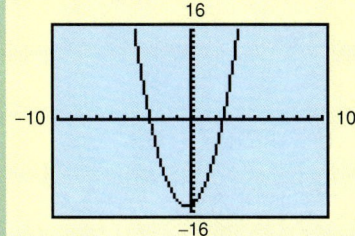

Factoring Completely

Remember that if a trinomial has a common monomial factor, the common monomial factor should be factored out first.

EXAMPLE 3   Factoring Completely

Factor $4x^2 - 30x + 14$.

Solution

Begin by factoring out the common monomial factor.

$$4x^2 - 30x + 14 = 2(2x^2 - 15x + 7)$$

Now, for the trinomial $2x^2 - 15x + 7$, $a = 2$ and $c = 7$. The possible factorizations of this trinomial are listed below.

$$(2x - 7)(x - 1) = 2x^2 - 9x + 7$$
$$(2x - 1)(x - 7) = 2x^2 - 15x + 7 \qquad \text{Correct factorization}$$

Therefore, the complete factorization of the original trinomial is

$$4x^2 - 30x + 14 = 2(2x^2 - 15x + 7) = 2(2x - 1)(x - 7).$$

In factoring a trinomial with a negative leading coefficient, we suggest that you first factor -1 out of the trinomial.

EXAMPLE 4   A Negative Leading Coefficient

Factor $-5x^2 + 7x + 6$.

Solution

Begin by factoring -1 out of the trinomial.

$$-5x^2 + 7x + 6 = (-1)(5x^2 - 7x - 6)$$

Now, for the trinomial $5x^2 - 7x - 6$, $a = 5$ and $c = -6$. After testing the possible factorizations, you can conclude that

$$(x - 2)(5x + 3) = 5x^2 - 7x - 6. \qquad \text{Correct factorization}$$

Thus, a correct factorization is

$$-5x^2 + 7x + 6 = (-1)(x - 2)(5x + 3) = (-x + 2)(5x + 3).$$

Another correct factorization is $(x - 2)(-5x - 3)$.

Factoring by Grouping

The examples in this and the previous section have shown how to use *Guess, Check, and Revise* to factor trinomials. An alternative technique that some people like to use is factoring by grouping. For instance, suppose you rewrite the trinomial $2x^2 + x - 15$ as

$$2x^2 + x - 15 = 2x^2 + 6x - 5x - 15.$$

By grouping the first two terms and the last two terms, you can factor the polynomial as shown.

$$\begin{aligned} 2x^2 + x - 15 &= 2x^2 + (6x - 5x) - 15 && \text{Rewrite middle term.} \\ &= (2x^2 + 6x) - (5x + 15) && \text{Group terms.} \\ &= 2x(x + 3) - 5(x + 3) && \text{Distributive Property} \\ &= (x + 3)(2x - 5) && \text{Distributive Property} \end{aligned}$$

The key to this method of factoring is knowing how to rewrite the middle term. In general, *to factor a trinomial $ax^2 + bx + c$ by grouping, choose factors of the product ac that add up to b and use these factors to rewrite the middle term.*

EXAMPLE 5   *Factoring a Trinomial by Grouping*

Use factoring by grouping to factor the trinomial.

$$2x^2 + 5x - 3$$

Solution

In the trinomial $2x^2 + 5x - 3$, $ac = 2(-3) = -6$, which has factors 6 and -1 that add up to 5. Therefore, rewrite the middle term as $5x = 6x - x$. This produces the following.

$$\begin{aligned} 2x^2 + 5x - 3 &= 2x^2 + 6x - x - 3 && \text{Rewrite middle term.} \\ &= (2x^2 + 6x) - (x + 3) && \text{Group terms.} \\ &= 2x(x + 3) - (x + 3) && \text{Distributive Property} \\ &= (x + 3)(2x - 1) && \text{Distributive Property} \end{aligned}$$

Therefore, the trinomial factors as

$$2x^2 + 5x - 3 = (x + 3)(2x - 1).$$

What do you think of this grouping technique? Many people think it is more efficient than the *Guess, Check, and Revise* strategy, especially when the coefficients a and c have many factors. Try factoring $6x^2 + 5x - 4$ by grouping, and compare your solution with that in Example 1.

Group Activities

Exploring with Technology

Using a Computer Program   Try entering the factoring program below on a computer that uses the BASIC language. Use the program to factor the following trinomials.

$$2x^2 + 5x - 12$$
$$6x^2 - 11x + 5$$
$$5x^2 + 7x - 6$$
$$4x^2 + 4x - 3$$

Note that the coefficients of the trinomial should be substituted for those on line 210. Check the computer's answers by multiplying the factors.

Basic Program:

```
10 READ A,B,C
20 IF C=0 THEN GOTO 160
30 FOR I = −ABS(A) TO ABS(A)
40 IF I=0 THEN GOTO 130
50 A1=I: A2=INT(A/I)
60 IF A1*A2<>A THEN GOTO 130
70 FOR J = −ABS(C) TO ABS(C)
80 IF J=0 THEN GOTO 120
90 C1=J: C2=INT(C/J)
100 IF C1*C2<>C THEN GOTO 120
110 IF A1*C2+A2*C1=B THEN GOTO 170
120 NEXT J
130 NEXT I
140 PRINT ''TRINOMIAL IS NOT FACTORABLE''
150 GOTO 200
160 A1=−A: C1=−B: A2=−1:  C2=0
170 PRINT ''TRINOMIAL FACTORS
    AS (A1X + C1)(A2X + C2) WITH ''
180 PRINT ''A1 = '';−A1;'' C1 = '';−C1
190 PRINT ''A2 = '';−A2;'' C2 = '';−C2
200 END
210 DATA 2,1,0
```

6.3 Exercises

Discussing the Concepts

1. Explain the meaning of each letter of FOIL.

2. Without multiplying, why is $(2x + 3)(x + 5)$ not a factorization of $2x^2 + 7x - 15$?

3. Find the error.

$$9x^2 - 9x - 54 = (3x + 6)(3x - 9)$$
$$= 3(x + 2)(x - 3)$$

4. In factoring of $ax^2 + bx + c$, how many possible factorizations must be tested if a and c are prime? Explain your reasoning.

5. Give an example of a prime trinomial that is of the form $ax^2 + bx + c$.

6. Give an example of a trinomial of the form $ax^3 + bx^2 + cx$ that has a common monomial factor of $2x$.

Problem Solving

In Exercises 7–10, find the missing factor.

7. $5x^2 + 18x + 9 = (x + 3)()$

8. $5c^2 + 11c - 12 = (c + 3)()$

9. $4z^2 - 13z + 3 = (z - 3)()$

10. $3y^2 - y - 30 = (y + 3)()$

In Exercises 11–14, find all possible products of the form $(5x + m)(x + n)$ where mn is the specified product. (Assume m and n are integers.)

11. $mn = 3$

12. $mn = 21$

13. $mn = 12$

14. $mn = 36$

In Exercises 15–22, factor the trinomial. (*Note:* Some of the trinomials may be prime.)

15. $2x^2 + 5x + 3$

16. $3x^2 + 5x - 2$

17. $2y^2 - 3y + 1$

18. $3z^2 - z - 2$

19. $2x^2 + x + 3$

20. $6x^2 - 10x + 5$

21. $16z^2 - 34z + 15$

22. $12x^2 - 41x + 24$

In Exercises 23–26, factor the trinomial. (*Note:* The leading coefficient is negative.)

23. $-2x^2 + x + 3$

24. $-5x^2 + x + 4$

25. $1 - 4x - 60x^2$

26. $2 + 5x - 12x^2$

In Exercises 27–32, factor the polynomial completely. (*Note:* Some of the polynomials may be prime.)

27. $x^2 - 3x$

28. $y^2(y + 1) - y(y + 1)$

29. $v^2 + v - 42$

30. $6x^2 - 6x - 36$

31. $2x^2 + 2x + 1$

32. $3x^3 + 4x^2 + 2x$

Think About It In Exercises 33–36, find all integers b such that the trinomial can be factored.

33. $3x^2 + bx + 10$

34. $4x^2 + bx + 3$

35. $2x^2 + bx - 6$

36. $5x^2 + bx - 6$

In Exercises 37–40, factor the trinomial by grouping.

37. $3x^2 + 7x + 2$

38. $5x^2 - 14x - 3$

39. $15x^2 - 11x + 2$

40. $12x^2 - 13x + 1$

41. Consider the following equations.

$$y_1 = 2x^3 + 3x^2 - 5x$$
$$y_2 = x(2x + 5)(x - 1)$$

(a) Factor the trinomial represented by y_1. What is the relationship between y_1 and y_2?

(b) Demonstrate your answer to part (a) graphically by using a graphing utility to graph y_1 and y_2.

(c) Identify the x- and y-intercepts of the graphs of y_1 and y_2.

42. *Geometry* The sandbox shown in the figure has a height of x and a width of $x + 2$. The volume of the box is $2x^3 + 7x^2 + 6x$. Find the length of the box.

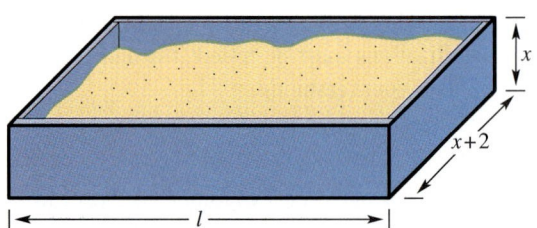

Geometric Model of Factoring In Exercises 43 and 44, factor the trinomial and draw a geometric model of the result. (The sample at the right shows a geometric model for factoring.)

43. $2x^2 + 5x + 2$

44. $3x^2 + 4x + 1$

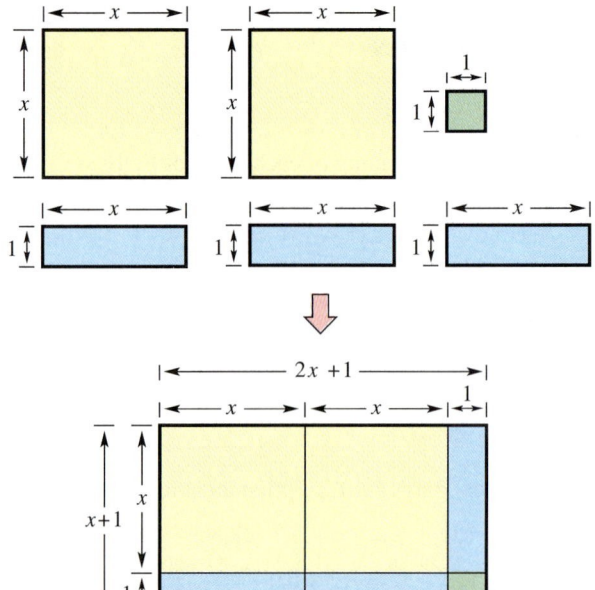

Figure for 43 and 44

This geometric model illustrates the factorization of $2x^2 + 3x + 1$ as $(2x + 1)(x + 1)$.

Reviewing the Major Concepts

In Exercises 45–48, find the product.

45. $-3t(t + 2)$

46. $(2 - x)(2 + x)$

47. $(x + 4)^2$

48. $(4x + 3)(3x - 25)$

49. *Distance* The minimum and maximum speeds on an interstate highway are 40 miles per hour and 55 miles per hour. You travel nonstop for 3 hours on the highway. Give an interval that describes the distance you could have legally traveled.

50. *Earning a Profit* The revenue for selling x units of a product is

$$R = 36.95x.$$

The cost of producing x units is

$$C = 21x + 65.$$

To obtain a profit, the revenue must be greater than the cost. For what values of x will this product earn a profit?

Additional Problem Solving

In Exercises 51–54, find the missing factor.

51. $5a^2 + 12a - 9 = (a + 3)()$

52. $2y^2 - 3y - 27 = (y + 3)()$

53. $5x^2 + 19x + 12 = (x + 3)()$

54. $6z^2 - 23z + 15 = (z - 3)()$

In Exercises 55–70, factor the trinomial. (*Note:* Some of the trinomials may be prime.)

55. $4y^2 + 5y + 1$ **56.** $3x^2 + 7x + 2$

57. $2x^2 - x - 3$ **58.** $3a^2 - 5a + 2$

59. $5x^2 - 2x + 1$ **60.** $4z^2 - 8z + 1$

61. $15a^2 + 14a - 8$ **62.** $12x^2 - 8x - 15$

63. $18u^2 - 9u - 2$ **64.** $6v^2 + v - 2$

65. $10t^2 - 3t - 18$ **66.** $24s^2 + 37s - 5$

67. $15m^2 + 16m - 15$ **68.** $21b^2 - 40b - 21$

69. $5s^2 - 10s + 6$ **70.** $10t^2 + 43t - 9$

In Exercises 71–74, factor the trinomial.

71. $4 - 4x - 3x^2$ **72.** $-4x^2 + 17x + 15$

73. $-6x^2 + 7x + 10$ **74.** $2 + x - 6x^2$

In Exercises 75–86, factor the polynomial completely. (*Note:* Some of the polynomials may be prime.)

75. $15y^2 + 18y$ **76.** $3a^4 - 9a^3$

77. $u(u - 3) + 9(u - 3)$

78. $(x + 1)(x - 8) - 2(x - 8)$

79. $x^2 + 6x - 40$ **80.** $z^2 - 3z - 40$

81. $6x^2 + 8x - 8$ **82.** $-15x^2 - 2x + 8$

83. $15y^2 - 7y^3 - 2y^4$ **84.** $5 + 34x - 7x^2$

85. $9u^2 + 18u - 27$ **86.** $3x^2 - 4x + 2$

In Exercises 87–94, factor the trinomial by grouping.

87. $2x^2 + x - 3$ **88.** $2x^2 + 17x + 21$

89. $6x^2 + 5x - 4$ **90.** $12y^2 + 11y + 2$

91. $3a^2 + 11a + 10$ **92.** $3z^2 - 4z - 15$

93. $16x^2 + 2x - 3$ **94.** $20c^2 + 19c - 1$

Think About It   In Exercises 95 and 96, find two values of c such that the trinomial can be factored.

95. $4x^2 + 3x + c$ **96.** $2x^2 + 5x + c$

In Exercises 97–100, use a graphing utility to graph the equations on the same screen. What can you conclude?

97. $y_1 = 4x^2 - 11x - 45$, $y_2 = (x - 5)(4x + 9)$

98. $y_1 = 6y^2 + 5y - 50$, $y_2 = (2y - 5)(3y + 10)$

99. $y_1 = 2x^3 + 3x^2 - 2x$, $y_2 = x(2x - 1)(x + 2)$

100. $y_1 = -x^2 + 2x + 3$, $y_2 = (3 - x)(1 + x)$

101. *Geometry*   The cake box shown in the figure has a height of x and a width of $x + 1$. The volume of the box is $3x^3 + 4x^2 + x$. Find the length of the box.

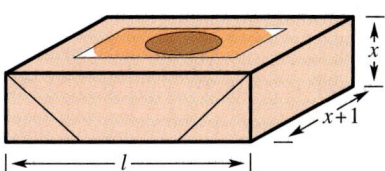

102. *Beam Deflection*   A cantilever beam of length l is fixed at the origin. A load weighing W pounds is attached to the end of the beam (see figure). The deflection y of the beam x units from the origin is

$$y = -\frac{1}{10}x^2 - \frac{1}{120}x^3, \quad 0 \le x \le 3.$$

(a) Factor the expression for the deflection. (Write the binomial factor with positive integer coefficients.)

(b) Use a graphing utility to graph the expression for deflection over the specified interval.

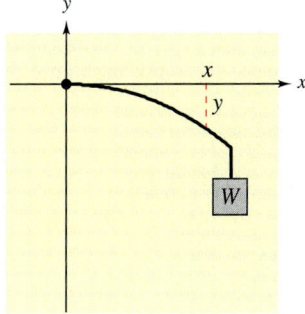

MID-CHAPTER QUIZ

Take this quiz as you would take a quiz in class. After you are done, check your work against the answers given in the back of the book.

In Exercises 1–4, find the missing factor.

1. $\frac{2}{3}x - 1 = \frac{1}{3}(\quad\quad)$
2. $x^2y - xy^2 = xy(\quad\quad)$
3. $y^2 + y - 42 = (y + 7)(\quad\quad)$
4. $2x^2 - x - 1 = (x - 1)(\quad\quad)$

In Exercises 5–16, factor the polynomial.

5. $10x^2 + 70$
6. $2a^3b - 4a^2b^2$
7. $x(x + 2) - 3(x + 2)$
8. $t^3 - 3t^2 + t - 3$
9. $y^2 + 11y + 30$
10. $u^2 + u - 30$
11. $x^3 - x^2 - 30x$
12. $2x^2y + 8xy - 64y$
13. $3v^2 - 4v - 2$
14. $6 - 13z - 5z^2$
15. $6x^2 - x - 2$
16. $10s^4 - 14s^3 + 2s^2$

17. Find all integer values of b such that the polynomial

    $x^2 + bx + 12$

    can be factored. Describe the method you used.

18. Find two values of c such that

    $x^2 - 10x + c$

    can be factored. Describe the method you used.

19. Find all possible products of the form

    $(3x + m)(x + n)$

    such that $mn = 6$. Describe the method you used.

20. The area of the rectangle shown in the figure is $x^2 + 4x - 5$. Find its length.

$x - 1$

6.4   **Factoring Polynomials with Special Forms**

Difference of Two Squares ■ Repeated Factorization ■
Perfect Square Trinomials ■ Sum and Difference of Two Cubes

Difference of Two Squares

One of the easiest special polynomial forms to recognize and to factor is the form
$a^2 - b^2$. It is called a **difference of two squares,** and it factors according to the
following pattern.

DISCOVERY

Use your calculator to verify the
special polynomial form called
the "difference of two squares."
To do so, evaluate the equation
when $a = 16$ and $b = 9$. Try
more values, including negative
values. What can you conclude?

Difference of Two Squares

Let a and b be real numbers, variables, or algebraic expressions.

$$a^2 - b^2 = (a + b)(a - b)$$

Difference   Opposite signs

This pattern can be illustrated geometrically, as shown in Figure 6.1.

NOTE   In Figure 6.1, the area
of the shaded region on the left
is represented by $a^2 - b^2$ (the
area of the larger square minus
the area of the smaller square).
On the right, the *same* area
is represented by a rectangle
whose width is $a + b$ and
whose length is $a - b$.

FIGURE 6.1

To recognize perfect squares, look for coefficients that are squares of integers
and for variables raised to *even* powers. Here are some examples.

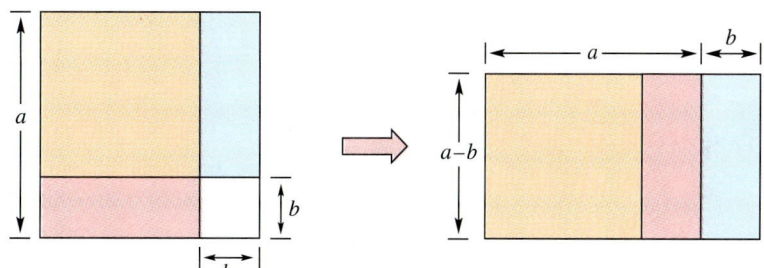

Original Polynomial		*Difference of Squares*		*Factored Form*
$x^2 - 1$	⇨	$(x)^2 - (1)^2$	⇨	$(x + 1)(x - 1)$
$4x^2 - 9$	⇨	$(2x)^2 - (3)^2$	⇨	$(2x + 3)(2x - 3)$
$25 - 64x^4$	⇨	$(5)^2 - (8x^2)^2$	⇨	$(5 + 8x^2)(5 - 8x^2)$

NOTE   When you factor a polynomial, remember that you can check your result by multiplying the factors. For instance, you can check the factorization in Example 1(a) as follows.

$$(x + 6)(x - 6) = x^2 - 36$$

EXAMPLE 1   *Factoring the Difference of Two Squares*

Factor each polynomial.

a. $x^2 - 36$ **b.** $81x^2 - 49$

Solution

a. $x^2 - 36 = x^2 - 6^2$ Write as difference of squares.

$\qquad\qquad = (x + 6)(x - 6)$ Factored form

b. $81x^2 - 49 = (9x)^2 - 7^2$ Write as difference of squares.

$\qquad\qquad\quad = (9x + 7)(9x - 7)$ Factored form

The rule $u^2 - v^2 = (u + v)(u - v)$ applies to polynomials or expressions in which u and v are themselves expressions.

EXAMPLE 2   *Factoring the Difference of Two Squares*

Factor $(x + 1)^2 - 4$.

Solution

$$(x + 1)^2 - 4 = (x + 1)^2 - 2^2 \qquad \text{Write as difference of squares.}$$
$$= [(x + 1) + 2][(x + 1) - 2] \qquad \text{Factored form}$$
$$= (x + 3)(x - 1) \qquad \text{Simplify.}$$

Sometimes the difference of two squares can be hidden by the presence of a common monomial factor. Remember that with all factoring techniques, you should first remove any common monomial factors.

EXAMPLE 3   *Removing a Common Monomial Factor First*

Factor $20x^3 - 5x$.

Solution

$$20x^3 - 5x = 5x(4x^2 - 1) \qquad \text{Factor out } 5x.$$
$$= 5x[(2x)^2 - 1^2] \qquad \text{Write as difference of two squares.}$$
$$= 5x(2x + 1)(2x - 1) \qquad \text{Factored form}$$

Repeated Factorization

To completely factor a polynomial, you should always check to see whether the factors obtained might themselves be factorable. That is, can any of the factors be factored? For instance, after factoring the polynomial $(x^4 - 1)$ once as the difference of two squares

$$x^4 - 1 = (x^2)^2 - 1^2 \qquad \text{Write as difference of squares.}$$
$$= (x^2 + 1)(x^2 - 1) \qquad \text{Factored form}$$

you can see that the second factor is itself the difference of two squares. Thus, to factor the polynomial *completely*, you must continue the factoring process.

$$x^4 - 1 = (x^2 + 1)(x^2 - 1) \qquad \text{Factor as difference of squares.}$$
$$= (x^2 + 1)(x + 1)(x - 1) \qquad \text{Factor completely.}$$

Another example of repeated factoring is shown in the next example.

EXAMPLE 4   Factoring Completely

Factor $x^4 - 16$ completely.

Solution

Recognizing $x^4 - 16$ as a difference of two squares, you can write

$$x^4 - 16 = (x^2)^2 - 4^2 \qquad \text{Write as difference of squares.}$$
$$= (x^2 + 4)(x^2 - 4). \qquad \text{Factored form}$$

Note that the second factor $(x^2 - 4)$ is itself a difference of two squares and therefore

$$x^4 - 16 = (x^2 + 4)(x^2 - 4) \qquad \text{Factor as difference of squares.}$$
$$= (x^2 + 4)(x + 2)(x - 2). \qquad \text{Factor completely.}$$

Note in Example 4 that no attempt was made to factor the *sum of two squares*. The reason for this is that a second-degree polynomial that is the sum of two squares cannot be factored as the product of binomials (using integers as coefficients). For instance, the second-degree polynomials

$$x^2 + 4 \qquad \text{and} \qquad 4x^2 + 9$$

cannot be factored (using integers as coefficients). In general, *the sum of two squares is not factorable.*

Perfect Square Trinomials

A **perfect square trinomial** is the square of a binomial. For instance,

$$x^2 + 4x + 4 = (x + 2)^2$$

is the square of the binomial $(x + 2)$. Perfect square trinomials come in two patterns, one in which the middle term is positive and the other in which the middle term is negative.

Perfect Square Trinomials

Let a and b be numbers, variables, or algebraic expressions.

1. $a^2 + 2ab + b^2 = (a + b)^2$ **2.** $a^2 - 2ab + b^2 = (a - b)^2$

Same sign   Same sign

STUDY TIP

To recognize a perfect square trinomial, remember that the first and last terms must be perfect squares and positive, and the middle term must be twice the product of a and b. (The middle term can be positive or negative.)

EXAMPLE 5   Identifying Perfect Square Trinomials

Which of the following is a perfect square trinomial?

a. $m^2 - 4m + 4$ **b.** $4x^2 - 2x + 1$ **c.** $y^2 + 6y - 9$

Solution

a. This polynomial *is* a perfect square trinomial. It factors as $(m - 2)^2$.

b. This polynomial *is not* a perfect square trinomial because the middle term is not twice the product of $2x$ and 1.

c. This polynomial *is not* a perfect square trinomial because the last term, -9, is not positive.

EXAMPLE 6   Factoring Perfect Square Trinomials

a. $y^2 - 6y + 9 = y^2 - 2(3y) + 3^2$ Recognize the pattern.

$\qquad\qquad\quad = (y - 3)^2$ Write in factored form.

b. $16x^2 + 40x + 25 = (4x)^2 + 2(4x)(5) + 5^2$ Recognize the pattern.

$\qquad\qquad\qquad\quad = (4x + 5)^2$ Write in factored form.

c. $9x^2 - 24xy + 16y^2 = (3x)^2 - 2(3x)(4y) + (4y)^2$ Recognize the pattern.

$\qquad\qquad\qquad\qquad = (3x - 4y)^2$ Write in factored form.

Sum and Difference of Two Cubes

The last type of special factoring that you will study in this section is the sum and difference of two cubes. The patterns for these two special forms are summarized below.

Sum and Difference of Two Cubes

Let a and b be real numbers, variables, or algebraic expressions.

Like signs

1. $a^3 + b^3 = (a + b)(a^2 - ab + b^2)$

Unlike signs

Like signs

2. $a^3 - b^3 = (a - b)(a^2 + ab + b^2)$

Unlike signs

When you are using either of these factoring patterns, pay special attention to the signs, as indicated above. Remembering the "like" and "unlike" patterns for the signs helps.

NOTE It is easy to make arithmetic errors when applying the patterns for factoring the sum or difference of two cubes. When you use these patterns, be sure to check your work by multiplying the two factors. For instance, you can check the factors in Example 7(a) as shown.

$$
\begin{array}{r}
y^2 - 3y + 9 \\
y + 3 \\
\hline
3y^2 - 9y + 27 \\
y^3 - 3y^2 + 9y \\
\hline
y^3 \qquad\qquad + 27
\end{array}
$$

EXAMPLE 7   Factoring Sums and Differences of Two Cubes

Factor the polynomial.

a. $y^3 + 27$ **b.** $64 - x^3$

Solution

a. $y^3 + 27 = y^3 + 3^3$ Write as sum of two cubes.

$\qquad = (y + 3)[y^2 - (y)(3) + 3^2]$ Factored form

$\qquad = (y + 3)(y^2 - 3y + 9)$ Simplify.

b. $64 - x^3 = 4^3 - x^3$ Write as difference of two cubes.

$\qquad = (4 - x)(4^2 + 4x + x^2)$ Factored form

$\qquad = (4 - x)(16 + 4x + x^2)$ Simplify.

Guidelines for Factoring Polynomials

1. Factor out any common factors.

2. Factor according to one of the special polynomial forms: difference of squares, sum or difference of two cubes, or perfect square trinomials.

3. Factor trinomials, $ax^2 + bx + c$, with $a = 1$ or $a \neq 1$.

4. Factor by grouping—for polynomials with four terms.

5. Check to see whether the factors themselves can be factored.

6. Check the results by multiplying the factors.

Group Activities   Extending the Concept

A Three-Dimensional View of a Special Product

The figure below shows two cubes: a large cube whose volume is a^3 and a smaller cube whose volume is b^3. If the smaller cube is removed from the larger, the remaining solid has a volume of $a^3 - b^3$ and is composed of three rectangular boxes, labeled Box 1, Box 2, and Box 3. Find the volume of each box and describe how these results are related to the following special product pattern.

$$a^3 - b^3 = (a - b)(a^2 + ab + b^2)$$
$$= (a - b)a^2 + (a - b)ab + (a - b)b^2$$

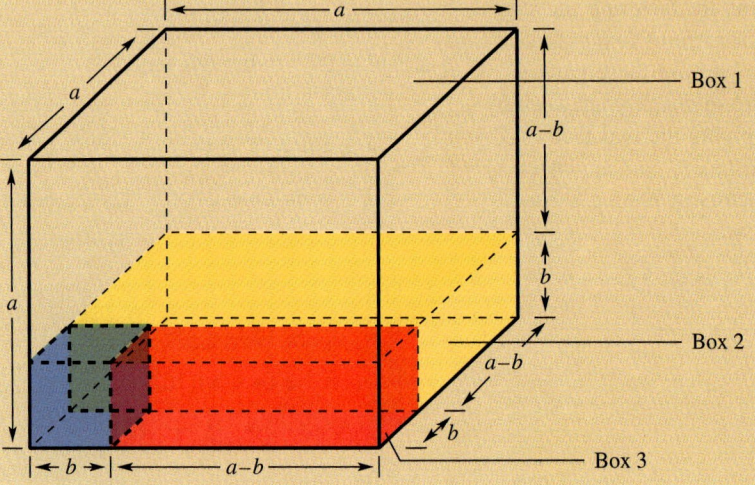

6.4 Exercises

Discussing the Concepts

1. Explain how to identify and factor the difference of two squares.

2. Explain how to identify and factor a perfect square trinomial.

3. Is the expression $x(x + 2) - 2(x + 2)$ in factored form? If not, rewrite it in factored form.

4. Is $x^2 + 4$ equal to $(x + 2)^2$? Explain.

5. *True or False?*   Because the sum of two squares cannot be factored, it follows that the sum of two cubes cannot be factored. Explain your reasoning.

6. In your own words, state the guidelines for factoring polynomials.

Problem Solving

In Exercises 7–10, factor the difference of squares.

7. $x^2 - 36$

8. $9z^2 - 25$

9. $u^2 - \frac{1}{4}$

10. $(x - 1)^2 - 4$

In Exercises 11–14, factor completely.

11. $2x^2 - 72$

12. $27 - 3x^2$

13. $x^4 - 1$

14. $u^4 - 256$

In Exercises 15–22, factor the perfect square trinomial.

15. $x^2 - 4x + 4$

16. $x^2 + 10x + 25$

17. $25y^2 - 10y + 1$

18. $16z^2 + 24z + 9$

19. $b^2 + b + \frac{1}{4}$

20. $4t^2 - \frac{4}{3}t + \frac{1}{9}$

21. $x^2 - 6xy + 9y^2$

22. $9x^2 - 18xy + 9y^2$

Think About It   In Exercises 23–26, find two values of b so that the expression is a perfect square trinomial.

23. $x^2 + bx + 100$

24. $x^2 + bx + \frac{16}{25}$

25. $4x^2 + bx + \frac{1}{4}$

26. $25x^2 + bxy + 4y^2$

In Exercises 27–30, factor the sum or difference of the two cubes.

27. $x^3 - 8$

28. $z^3 + 125$

29. $1 + 8t^3$

30. $64v^3 - 125$

In Exercises 31–42, factor the expression completely. (*Note:* Some of the polynomials may be prime.)

31. $x^3 - 4x^2$

32. $16 + 6x - x^2$

33. $1 - 4x + 4x^2$

34. $13x + 6 + 5x^2$

35. $x(x - 1) + (x - 1)^2$

36. $3x^3 + x^2 + 15x + 5$

37. $9t^2 - 16$

38. $(t - 1)^2 - 49$

39. $2y^2 - 3y - 5$

40. $x^2 + 16$

41. $2t^3 - 16$

42. $81 - y^4$

Geometric Factoring Models   In Exercises 43 and 44, write the factoring problem represented by the geometric factoring model.

43.

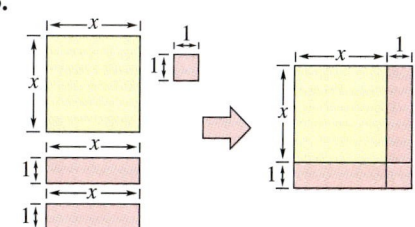

44.

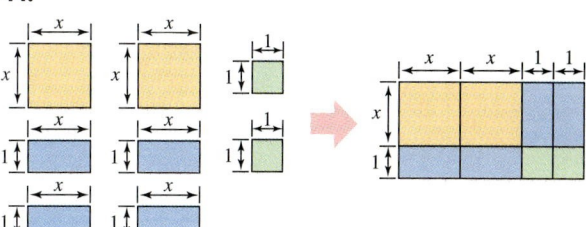

45. *Geometry* An annulus is the region between two concentric circles. The area of the annulus in the figure is $\pi R^2 - \pi r^2$. Give the complete factorization of the expression for the area.

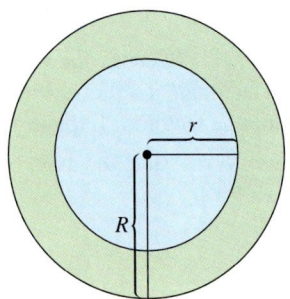

46. *Geometry* The building shown in the figure has a square base and a height of x. The volume of the building is $x^3 - 80x^2 + 1600x$ cubic feet. Find the dimensions of the building.

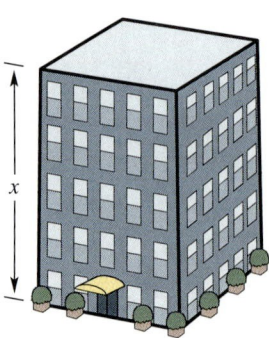

Reviewing the Major Concepts

In Exercises 47–50, solve the equation and check your result.

47. $2(x + 1) = 0$

48. $\dfrac{3x}{4} + \dfrac{1}{2} = 8$

49. $\frac{3}{4}(12x - 8) = 10$

50. $2 - 5(x - 1) = 2[x + 10(x - 1)]$

51. *Membership* The current membership for a public television station is 120% of what it was a year ago. The current number is 8346. How many members did the station have last year?

52. *Budgeting* You budget 26% of your annual after-tax income for housing. Your after-tax income is $46,750. What amount can you spend on housing?

Additional Problem Solving

In Exercises 53–60, factor the difference of squares.

53. $81 - x^2$

54. $y^2 - 49$

55. $t^2 - \frac{1}{16}$

56. $v^2 - \frac{4}{9}$

57. $16y^2 - 9$

58. $100 - 49x^2$

59. $25 - (z + 5)^2$

60. $(a - 2)^2 - 16$

In Exercises 61–64, factor completely.

61. $y^4 - 81$

62. $a^3 - 16a$

63. $8 - 50x^2$

64. $z^4 - 16$

In Exercises 65–74, factor the perfect square trinomial.

65. $z^2 + 6z + 9$

66. $a^2 - 12a + 36$

67. $4t^2 + 4t + 1$

68. $9x^2 - 12x + 4$

69. $4x^2 - x + \frac{1}{16}$

70. $x^2 + \frac{2}{5}x + \frac{1}{25}$

71. $4y^2 + 20yz + 25z^2$

72. $u^2 + 8uv + 16v^2$

73. $9a^2 - 12ab + 4b^2$

74. $\frac{1}{4}m^2 - 2mn + 4n^2$

Think About It In Exercises 75–78, find two values of b so that the expression is a perfect square trinomial.

75. $x^2 + bx + 1$

76. $y^2 + by + \frac{1}{9}$

77. $4x^2 + bx + 9$

78. $9x^2 + bx + 64$

Think About It In Exercises 79–82, find a number c so that the expression is a perfect square trinomial.

79. $x^2 + 6x + c$

80. $x^2 + 10x + c$

81. $y^2 - 4y + c$

82. $z^2 - 14z + c$

In Exercises 83–86, factor the sum or difference of the two cubes.

83. $y^3 + 64$ **84.** $x^3 - 27$

85. $27u^3 + 8$ **86.** $27s^3 + 1$

In Exercises 87–124, factor the polynomials completely. (*Note:* Some of the polynomials may be prime.)

87. $y^4 - 25y^2$ **88.** $6x^2 - 54$

89. $z^4 - \frac{4}{9}z^2$ **90.** $u - \frac{1}{25}u^3$

91. $x^2 - 2x + 1$ **92.** $y^2 + 6y + 9$

93. $4v^2 + 4v + 1$ **94.** $9x^2 - 6x + 1$

95. $2x^2 + 4x - 2x^3$ **96.** $4x^2 + 3x + 1$

97. $9x^2 + 10x + 1$ **98.** $2y^3 - 7y^2 - 15y$

99. $(x - 1)^2 - 2(x - 1)$ **100.** $5(3-4x)^2 - 8(3-4x)$

101. $5 - x + 5x^2 - x^3$ **102.** $3u - 2v + 6 - uv$

103. $x^4 - 4x^3 + x^2 - 4x$ **104.** $y^3 + 3y^2 - 4y - 12$

105. $(y + 2)^2 - 1$ **106.** $(t - 4)^2 - 9$

107. $64 - (z + 8)^2$ **108.** $\frac{1}{4} - (x - 3)^2$

109. $3u^3 - 27u$ **110.** $8a - 2a^3$

111. $8x^2 + 2$ **112.** $8x^2 - 2$

113. $y^3 + \frac{1}{8}$ **114.** $z^4 + z$

115. $16x^3 - 2$ **116.** $24x^3 - 3$

117. $(x - 3)^3 - 1$ **118.** $(y + 2)^3 + 8$

119. $u^3 + 2u^2 + 3u$ **120.** $u^3 + 2u^2 - 3u$

121. $x^4 - 81$ **122.** $2x^4 - 32$

123. $1 - x^4$ **124.** $\frac{1}{16} - y^4$

In Exercises 125–128, use a graphing utility to graph the two equations on the same screen. What can you conclude?

125. $y_1 = x^2 - 36, \quad y_2 = (x + 6)(x - 6)$

126. $y_1 = x^2 - 8x + 16, \quad y_2 = (x - 4)^2$

127. $y_1 = x^3 - 6x^2 + 9x, \quad y_2 = x(x - 3)^2$

128. $y_1 = x^3 + 27, \quad y_2 = (x + 3)(x^2 - 3x + 9)$

Mental Math In Exercises 129–132, evaluate the quantity mentally using the two samples as models.

Samples: $29^2 = (30 - 1)^2$

$$= 30^2 - 2 \cdot 30 \cdot 1 + 1^2$$

$$= 900 - 60 + 1$$

$$= 841$$

$$48 \cdot 52 = (50 - 2)(50 + 2)$$

$$= 50^2 - 2^2$$

$$= 2496$$

129. 21^2 **130.** 49^2

131. $59 \cdot 61$ **132.** $28 \cdot 32$

In Exercises 133 and 134, write the polynomial as the sum of two squares.

133. $x^2 + 6x + 10 = (x^2 + 6x + 9) + 1$

$$= \boxed{}^2 + \boxed{}^2$$

134. $x^2 + 8x + 25 = (x^2 + 8x + 16) + 9$

$$= \boxed{}^2 + \boxed{}^2$$

Geometric Factoring Models In Exercises 135 and 136, write the factoring problem represented by the geometric factoring model.

135.

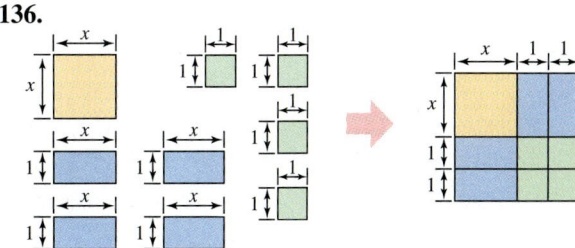

136.

6.5	**Solving Equations and Problem Solving**

The Zero-Factor Property ▪ Solving Quadratic Equations by Factoring ▪ Solving Polynomial Equations by Factoring ▪ Applications

The Zero-Factor Property

You have spent nearly two chapters developing skills for *rewriting* (simplifying and factoring) polynomials. You are now ready to use these skills together with the **Zero-Factor Property** to *solve* equations.

NOTE The Zero-Factor Property is just another way of saying that the only way the product of two or more factors can be zero is if one or more of the factors is zero.

Zero-Factor Property

Let a and b be real numbers, variables, or algebraic expressions. If a and b are factors such that

$$ab = 0$$

then $a = 0$ or $b = 0$. This property also applies to three or more factors.

The Zero-Factor Property is the primary property for solving equations in algebra. For instance, to solve the equation

$$(x - 1)(x + 2) = 0 \qquad \text{Original equation}$$

you can use the Zero-Factor Property to conclude that either $(x - 1)$ or $(x + 2)$ must be zero. Setting the first factor equal to zero implies that $x = 1$ is a solution.

$$x - 1 = 0 \quad \Longrightarrow \quad x = 1 \qquad \text{First solution}$$

Similarly, setting the second factor equal to zero implies that $x = -2$ is a solution.

$$x + 2 = 0 \quad \Longrightarrow \quad x = -2 \qquad \text{Second solution}$$

Thus, the equation $(x - 1)(x + 2) = 0$ has exactly two solutions: 1 and -2. You can check these solutions by substituting them into the original equation.

$$(x - 1)(x + 2) = 0 \qquad \text{Original equation}$$

$$(1 - 1)(1 + 2) \overset{?}{=} 0 \qquad \text{Substitute 1 for } x.$$

$$(0)(3) = 0 \qquad \text{First solution checks.} \checkmark$$

$$(-2 - 1)(-2 + 2) \overset{?}{=} 0 \qquad \text{Substitute } -2 \text{ for } x.$$

$$(-3)(0) = 0 \qquad \text{Second solution checks.} \checkmark$$

Solving Quadratic Equations by Factoring

A **quadratic equation** is an equation of the form $ax^2 + bx + c = 0$. Here are some examples.

$$x^2 - 2x - 3 = 0, \quad 2x^2 + x - 1 = 0, \quad \text{and} \quad x^2 - 5x = 0$$

In the next four examples, note how you can combine your factoring skills with the Zero-Factor Property to solve quadratic equations.

EXAMPLE 1 Using Factoring to Solve a Quadratic Equation

Solve $x^2 - x - 6 = 0$.

Solution

First, check to see that the right side of the equation is zero. Next, factor the left side of the equation. Finally, apply the Zero-Factor Property to find the solutions.

$x^2 - x - 6 = 0$	Original equation
$(x + 2)(x - 3) = 0$	Factor left side of equation.
$x + 2 = 0 \quad \Longrightarrow \quad x = -2$	Set 1st factor equal to 0.
$x - 3 = 0 \quad \Longrightarrow \quad x = 3$	Set 2nd factor equal to 0.

The equation has two solutions: -2 and 3.

$x^2 - x - 6 = 0$	Original equation
$(-2)^2 - (-2) - 6 \stackrel{?}{=} 0$	Substitute -2 for x.
$4 + 2 - 6 \stackrel{?}{=} 0$	Simplify.
$0 = 0 \quad \checkmark$	Solution checks.
$x^2 - x - 6 = 0$	Original equation
$(3)^2 - 3 - 6 \stackrel{?}{=} 0$	Substitute 3 for x.
$9 - 3 - 6 \stackrel{?}{=} 0$	Simplify.
$0 = 0 \quad \checkmark$	Solution checks.

STUDY TIP

In Section 3.1, you learned that the general strategy for solving a linear equation is to *isolate the variable*. Notice in Example 1 that the general strategy for solving a quadratic equation is more complicated.

Factoring and the Zero-Factor Property allow you to solve a quadratic equation by converting it into two *linear* equations, which you already know how to solve. This is a common strategy of algebra—to break down a given problem into simpler parts, each solved by previously learned methods.

To use the Zero-Factor Property, a polynomial equation *must* be written in **standard form.** That is, the polynomial must be on one side of the equation and zero must be the only term on the other side of the equation. For instance, to write $x^2 - 2x = 3$ in standard form, subtract 3 from both sides of the equation.

$$x^2 - 2x = 3 \qquad\qquad \text{Original equation}$$

$$x^2 - 2x - 3 = 3 - 3 \qquad\qquad \text{Subtract 3 from both sides.}$$

$$x^2 - 2x - 3 = 0 \qquad\qquad \text{Standard form}$$

To solve this equation, factor the left side as $(x - 3)(x + 1)$, then form the linear equations $x - 3 = 0$ and $x + 1 = 0$. The solutions of these two linear equations are 3 and -1, respectively. The general strategy for solving a quadratic equation by factoring is summarized in the following diagram.

| Write in standard form. | → | Factor left side of equation. | → | Set factors equal to zero. | → | Solve linear equations. | → | Check in original equation. |

Technology

You can also use a graphing utility to find the solutions of a polynomial equation. For example, graph the quadratic equation

$$2x^2 + 5x - 12 = 0$$

on your graphing utility. To find the solutions of the equation, use the ZOOM and TRACE features to determine the x-intercepts. When you do this, you see that the solutions are $x = -4$ and $x = \frac{3}{2}$.

EXAMPLE 2 Solving a Quadratic Equation by Factoring

Solve $2x^2 + 5x = 12$.

Solution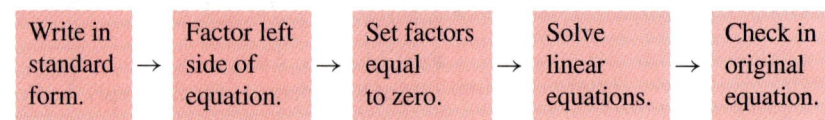

$$2x^2 + 5x = 12 \qquad\qquad \text{Original equation}$$

$$2x^2 + 5x - 12 = 0 \qquad\qquad \text{Write in standard form.}$$

$$(2x - 3)(x + 4) = 0 \qquad\qquad \text{Factor left side of equation.}$$

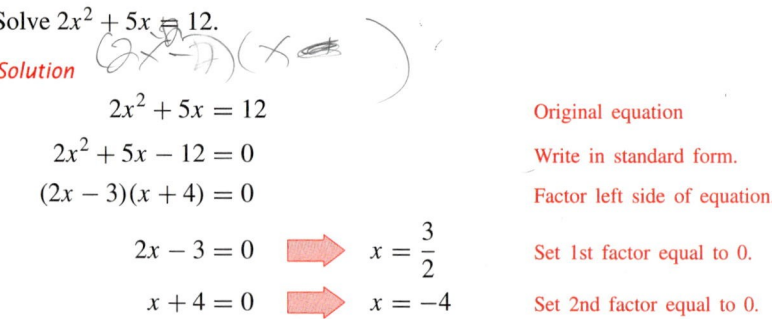

$$2x - 3 = 0 \quad\Longrightarrow\quad x = \frac{3}{2} \qquad \text{Set 1st factor equal to 0.}$$

$$x + 4 = 0 \quad\Longrightarrow\quad x = -4 \qquad \text{Set 2nd factor equal to 0.}$$

The solutions are $\frac{3}{2}$ and -4. Check these solutions in the original equation.

Be sure you see that the Zero-Factor Property can be applied only to a product that is equal to *zero*. For instance, you cannot conclude from the equation

$$x(x - 3) = 10$$

that $x = 10$ and $x - 3 = 10$ yield solutions. Instead, you must first write the equation in standard form and then factor the left side, as follows.

$$x^2 - 3x - 10 = 0 \quad\Longrightarrow\quad (x - 5)(x + 2) = 0$$

Now, from the factored form, you can see that the solutions are 5 and -2.

In Examples 1 and 2, the original equations each involved a second-degree (quadratic) polynomial and each had *two different* solutions. You will sometimes encounter second-degree polynomial equations that have only one (repeated) solution. This occurs when the left side of the equation is a perfect square trinomial, as shown in Example 3.

EXAMPLE 3 A Quadratic Equation with a Repeated Solution

Solve $x^2 - 8x + 20 = 4$.

Solution

$x^2 - 8x + 20 = 4$	Original equation
$x^2 - 8x + 16 = 0$	Write in standard form.
$(x - 4)^2 = 0$	Factor.
$x - 4 = 0$ ⟹ $x = 4$	Set factor equal to 0.

Note that even though the left side of this equation has two factors, the factors are the same. Thus, the only solution of the equation is 4.

$x^2 - 8x + 20 = 4$	Original equation
$(4)^2 - 8(4) + 20 \overset{?}{=} 4$	Substitute 4 for x.
$16 - 32 + 20 \overset{?}{=} 4$	Simplify.
$4 = 4$	Solution checks. ✔

EXAMPLE 4 Solving a Polynomial Equation in Nonstandard Form

Solve $(x + 3)(x + 6) = 4$.

Solution

Begin by multiplying the factors on the left side.

$(x + 3)(x + 6) = 4$	Original equation
$x^2 + 9x + 18 = 4$	Multiply factors.
$x^2 + 9x + 14 = 0$	Standard form
$(x + 2)(x + 7) = 0$	Factor left side of equation.
$x + 2 = 0$ ⟹ $x = -2$	Set 1st factor equal to 0.
$x + 7 = 0$ ⟹ $x = -7$	Set 2nd factor equal to 0.

The equation has two solutions: -2 and -7. Check these in the original equation.

DISCOVERY

In Example 3, graph the equations

$y = x^2 - 8x + 20$

and

$y = 4$

on the same screen. From the graph, determine the number of solutions of the equation. Explain how to use a graphing utility to solve

$x^3 - 2x^2 - 4x + 3 = 0.$

How many solutions does the equation have? How does the number of solutions relate to the degree of the equation?

A common error is setting $x + 3 = 4$ or $x + 6 = 4$. Emphasize the necessity of having a product equal to *zero* before applying the Zero-Factor Property.

Solving Polynomial Equations by Factoring

In the exploration of algebra, general solutions were found for second-, third-, and fourth-degree polynomial equations. (You will study the Quadratic Formula for second-degree equations in Chapter 11.) Attempts to find an algebraic solution for a fifth-degree polynomial equation met with failure.

In 1824, Neils Henrik Abel published a proof that showed that for any degree greater than four, the general polynomial equation could not be solved algebraically. His work included the concept of a group. Subsequent research in group theory gave mathematicians new ways to explore and describe algebraic structures. This marked the beginning of the modern theory of equations.

EXAMPLE 5   Solving a Polynomial Equation with Three Factors

Solve $3x^3 = 12x^2 + 15x$.

Solution

$$3x^3 = 12x^2 + 15x \qquad \text{Original equation}$$

$$3x^3 - 12x^2 - 15x = 0 \qquad \text{Standard form}$$

$$3x(x^2 - 4x - 5) = 0 \qquad \text{Factor out } 3x.$$

$$3x(x - 5)(x + 1) = 0 \qquad \text{Factor completely.}$$

$$3x = 0 \quad \Longrightarrow \quad x = 0 \qquad \text{Set 1st factor equal to 0.}$$

$$x - 5 = 0 \quad \Longrightarrow \quad x = 5 \qquad \text{Set 2nd factor equal to 0.}$$

$$x + 1 = 0 \quad \Longrightarrow \quad x = -1 \qquad \text{Set 3rd factor equal to 0.}$$

There are three solutions: 0, 5, and −1. Check these in the original equation.

Notice that the equation in Example 5 is a third-degree equation and has three solutions. This is not a coincidence. In general, a polynomial equation can have *at most* as many solutions as its degree. For instance, a second-degree equation can have zero, one, or two solutions, but it cannot have three or more solutions.

EXAMPLE 6   Solving a Polynomial Equation with Four Factors

Solve $x^4 + x^3 - 4x^2 - 4x = 0$.

Solution

$$x^4 + x^3 - 4x^2 - 4x = 0 \qquad \text{Original equation}$$

$$x(x^3 + x^2 - 4x - 4) = 0 \qquad \text{Factor out } x.$$

$$x[x^2(x + 1) - 4(x + 1)] = 0 \qquad \text{Distributive Property}$$

$$x[(x + 1)(x^2 - 4)] = 0 \qquad \text{Factor by grouping.}$$

$$x(x + 1)(x + 2)(x - 2) = 0 \qquad \text{Factor completely.}$$

$$x = 0 \quad \Longrightarrow \quad x = 0$$

$$x + 1 = 0 \quad \Longrightarrow \quad x = -1$$

$$x + 2 = 0 \quad \Longrightarrow \quad x = -2$$

$$x - 2 = 0 \quad \Longrightarrow \quad x = 2$$

There are four solutions: 0, −1, −2, and 2. Check these in the original equation.

Applications

EXAMPLE 7   An Application of a Polynomial Equation

The product of two consecutive positive integers is 56. What are the integers?

Solution

Verbal Model:

$$\boxed{\text{First integer}} \cdot \boxed{\text{Second integer}} = \boxed{56}$$

Labels: First integer $= n$
Second integer $= n + 1$

Equation:
$$n(n + 1) = 56$$
$$n^2 + n - 56 = 0$$
$$(n + 8)(n - 7) = 0$$
$$n = -8 \text{ or } 7$$

Because the problem states that the integers be positive, discard -8 as a solution and choose $n = 7$. Thus, the two integers are $n = 7$ and $n + 1 = 8$.

EXAMPLE 8   A Falling Object Model

A rock is dropped from the top of a 256-foot river gorge, as shown in Figure 6.2. The height (in feet) of the rock is modeled by the equation

$$\text{Height} = -16t^2 + 256$$

where t is the time measured in seconds. How long will it take the rock to hit the bottom of the gorge?

Solution

From Figure 6.2, note that the bottom of the gorge corresponds to a height of 0 feet. Thus, substitute a height of 0 into the model and solve for t.

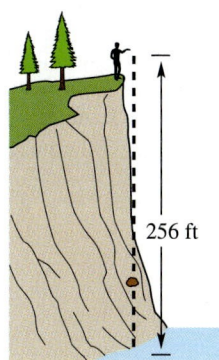

256 ft

FIGURE 6.2

$0 = -16t^2 + 256$	Set height equal to 0.
$16t^2 - 256 = 0$	Standard form
$16(t^2 - 16) = 0$	Factor out 16.
$16(t + 4)(t - 4) = 0$	Factor left side of equation.
$t = -4 \text{ or } 4$	Solutions

Because a time of -4 seconds doesn't make sense in this problem, choose the positive solution and conclude that the rock hits the bottom of the gorge 4 seconds after it is dropped.

EXAMPLE 9   An Application from Geometry

A rectangular family room has an area of 160 feet. The length of the room is 6 feet greater than its width. Find the dimensions of the room.

Solution

To begin, make a sketch of the room, as shown in Figure 6.3. Label the width of the room as x and the length of the room as $x + 6$ because the length is 6 feet greater than the width.

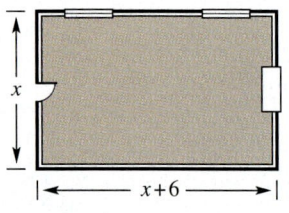

FIGURE 6.3

Verbal Model: $\boxed{\text{Length}} \cdot \boxed{\text{Width}} = \boxed{\text{Area}}$

Labels:		
Width $= x$		(feet)
Length $= x + 6$		(feet)
Area $= 160$		(square feet)

Equation:
$$x(x + 6) = 160$$
$$x^2 + 6x - 160 = 0$$
$$(x + 16)(x - 10) = 0$$
$$x = -16 \ \text{ or } \ 10$$

In this application, the negative solution makes no sense, so discard it and use the positive solution. Thus, the width of the room is 10 feet and the length of the room is 16 feet.

To check this solution, go back to the original statement of the problem. Note that a length of 16 feet is 6 feet greater than a width of 10 feet. Moreover, a rectangular room with dimensions 16 feet by 10 feet has an area of 160 square feet. Thus, the solution checks.

Group Activities   You Be the Instructor

Misleading Factorization   Suppose a student submits the following steps in solving the equation $x^2 + 3x = 10$:

$x(x + 3) = 10$	Factor.
$x = 10$ and $x + 3 = 10$	Set each factor equal to 10.
$x = 7$	

Discuss the error in your group, and write an explanation of why the method does not work. Solve the equation correctly and check your answers.

6.5 Exercises

Discussing the Concepts

1. Use the Zero-Factor Property to complete the statement. If $ab = 0$, then _____.

2. *True or False?*   If $(5x - 1)(x + 3) = 1$, then $5x - 1 = 1$ or $x + 3 = 1$. Explain.

3. In your own words, describe a strategy for solving a quadratic equation by factoring.

4. Explain the difference between a linear equation and a quadratic equation.

5. Is it possible for a quadratic equation to have only one solution? Explain.

6. What is the maximum number of solutions of an nth-degree polynomial equation?

Problem Solving

In Exercises 7–10, use the Zero-Factor Property to solve the equation.

7. $x(x - 5) = 0$

8. $(s - 4)(s - 10) = 0$

9. $y(y - 1)(y + 3) = 0$

10. $(2t - 5)(3t + 1) = 0$

In Exercises 11–26, solve the equation.

11. $x^2 - 16 = 0$

12. $4 - x^2 = 0$

13. $3y^2 - 27 = 0$

14. $(s + 5)^2 - 49 = 0$

15. $6x^2 + 3x = 0$

16. $x^3 + 5x^2 + 6x = 0$

17. $x^2 - 2x - 8 = 0$

18. $x^2 - 8x - 9 = 0$

19. $3 + 5x - 2x^2 = 0$

20. $33 + 5y - 2y^2 = 0$

21. $x(x - 5) = 14$

22. $(x + 1)(x - 2) = 4$

23. $u(u + 2) - 3(u + 2) = 0$

24. $x^3 - 3x^2 - 10x = 0$

25. $x^2(x - 2) - 9(x - 2) = 0$

26. $x^3 - x^2 - 16x + 16 = 0$

27. *Height of an Object*   An object is dropped from a weather balloon 1600 feet above the ground (see figure). Find the time t for the object to reach the ground. The height (above ground) of the object is modeled by

Height $= -16t^2 + 1600$

where the height is measured in feet and the time t is measured in seconds.

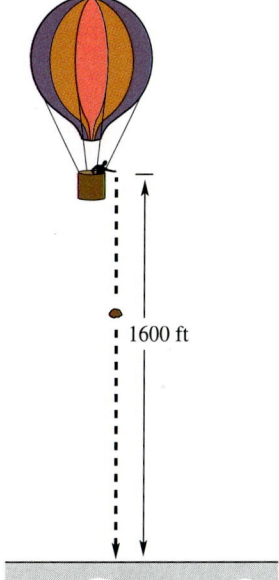

Figure for 27

28. *Geometry*   An open box with a square base is to be constructed from 108 square inches of material. The height of the box is 3 inches.

(a) Sketch the box. Label the sides of the base as x and the height as h.

(b) Find the dimensions of the base of the box. (*Hint:* The surface area is given by $S = x^2 + 4xh$.)

Graphical Estimation In Exercises 29–32, estimate the x-intercepts of the graph visually. Then check your estimates algebraically.

29. $y = x^2 + 2x - 3$

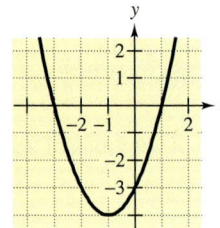

30. $y = 2 + x - x^2$

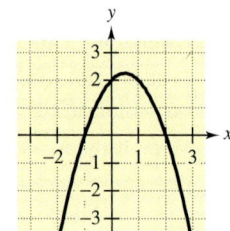

31. $y = x^3 - 6x^2 + 9x$

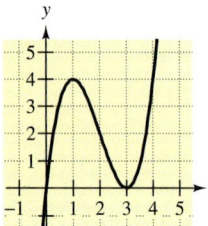

32. $y = 2x^3 + 3x^2 - 5x$

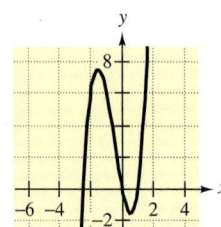

In Exercises 33–36, use a graphing utility to graph the equation. Use the graph to estimate the x-intercepts. Then check your estimates by substituting into the equation.

33. $y = x^2 - 4x$

34. $y = 2x^2 - 5x - 12$

35. $y = x^3 - 4x^2$

36. $y = \frac{1}{4}(x^3 - 2x^2 - x + 2)$

37. *Problem Solving* Find two consecutive positive integers whose product is 72.

38. *Geometry* The length of a rectangle is one and one-half times the width. The area of the rectangle is 600 square inches. Find the dimensions of the rectangle.

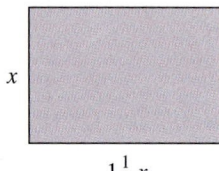

$1\frac{1}{2}x$

Reviewing the Major Concepts

In Exercises 39–42, evaluate the expression.

39. $2(-3) + 9$

40. $\dfrac{|18 - 25|}{6}$

41. $4 - \dfrac{5}{2}$

42. $\dfrac{4}{3} \div \dfrac{5}{6}$

43. *Simple Interest* Find the interest on a $1000 bond paying an annual percentage rate of 7.5% for 10 years.

44. *Speed* A car leaves a town 20 minutes after a truck. The truck's speed is 10 miles per hour less than the car's speed. If the car overtakes the truck in 2 hours, find the speed of each vehicle.

Additional Problem Solving

In Exercises 45–48, use the Zero-Factor Property to solve the equation.

45. $z(z - 3) = 0$

46. $(y - 2)(y - 3) = 0$

47. $x(x - 3)(x + 25) = 0$

48. $x(5x + 3)(x - 8) = 0$

In Exercises 49–78, solve the equation.

49. $(a + 1)(a - 2) = 0$

50. $(t - 3)(t + 8) = 0$

51. $v^2 - 100 = 0$

52. $x^2 - 144 = 0$

53. $4x^2 - 9 = 0$

54. $25z^2 - 100 = 0$

55. $(t - 3)^2 - 25 = 0$

56. $1 - (x + 1)^2 = 0$

57. $x^2 - 2x = 0$

58. $4x^2 - x = 0$

59. $x(x - 8) + 2(x - 8) = 0$

60. $x(x + 2) - 3(x + 2) = 0$

61. $m^2 - 2m + 1 = 0$

62. $a^2 + 6a + 9 = 0$

63. $x^2 + 14x + 49 = 0$

64. $x^2 - 10x + 25 = 0$

65. $4t^2 - 12t + 9 = 0$

66. $16x^2 + 56x + 49 = 0$

67. $6x^2 + 4x - 10 = 0$

68. $12x^2 + 7x + 1 = 0$

69. $z(z + 2) = 15$

70. $x(x - 1) = 6$

71. $y(2y + 1) = 3$

72. $x(5x - 14) = 3$

73. $(x + 1)(x + 4) = 4$

74. $(x - 9)(x + 2) = 12$

75. $2t^3 + 5t^2 - 12t = 0$

76. $3u^3 - 5u^2 - 2u = 0$

77. $y^2(y + 3) - (y + 3) = 0$

78. $a^3 + 2a^2 - 4a - 8 = 0$

Graphical Estimation   In Exercises 79–82, estimate the x-intercepts of the graph visually. Check your estimates algebraically.

79. $y = x^2 - x - 12$

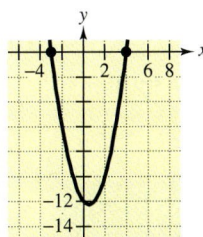

80. $y = 2x^2 + x - 3$

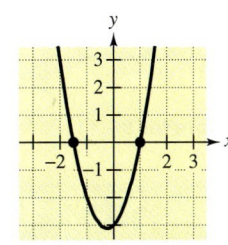

81. $y = 5x - 3x^2 - 2x^3$

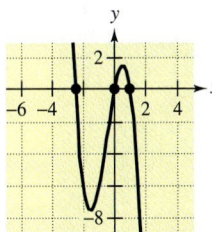

82. $y = x^4 + 2x^3 - 8x^2$

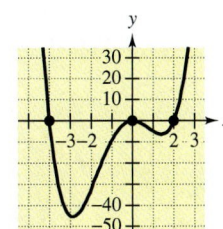

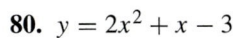 In Exercises 83–86, use a graphing utility to graph the equations. Use the graph to estimate the x-intercepts. Then check your estimates by substituting into the equation.

83. $y = x^2 - 4$

84. $y = 4x^2 + 3x - 10$

85. $y = x^2(x + 2) - 9(x + 2)$

86. $y = \frac{1}{4}(x^3 + 4x^2 - x - 4)$

87. Find two consecutive positive integers whose product is 240.

88. Find two consecutive positive even integers whose product is 440.

89. *Geometry*   The length of a rectangle is 3 inches greater than the width. The area of the rectangle is 108 square inches. Find the dimensions of the rectangle.

90. *Geometry*   The length of a rectangle is 2 times the width. The area of the rectangle is 450 square inches. Find the dimensions of the rectangle.

91. *Profit*   The profit for selling x units is

$$P = -0.4x^2 + 8x - 10.$$

(a) Use a graphing utility to graph the expression for profit.

(b) Use the graph to estimate any values of x that yield a profit of $P = \$20$.

(c) Use factorization to find any values of x that yield a profit of $P = \$20$.

92. *Revenue*   The revenue for selling x units is

$$R = 25x - 0.2x^2.$$

(a) Use a graphing utility to graph the expression for revenue.

(b) Use the graph to estimate the smaller of the two values of x that yield a revenue of $R = \$680$.

(c) Use factorization to find the smaller of the two values of x that yield a revenue of $R = \$680$.

93. *Exploration*   An open box is to be made from a square piece of material by cutting 2-inch squares from each corner and turning up the sides (see figure on next page).

(a) Show that the volume is given by $V = 2x^2$.

(b) Complete the following table.

x	2	4	6	8
V				

(c) Find the dimensions of the original piece of material if $V = 200$ cubic inches.

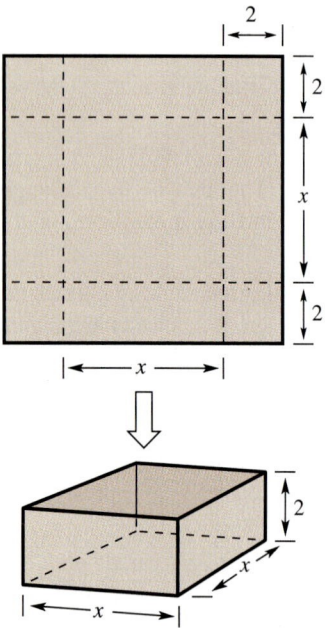

Figure for 93

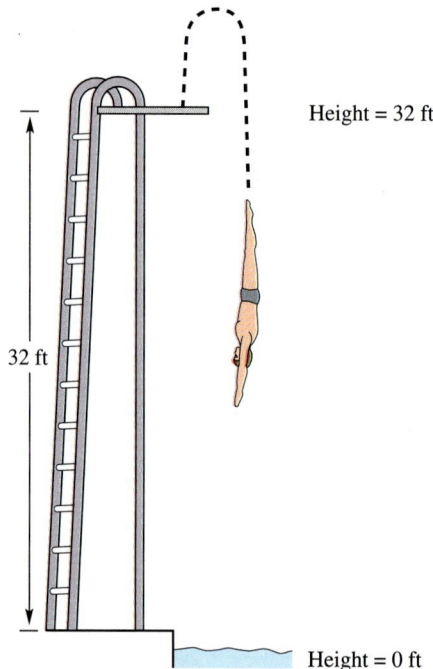

Figure for 96

94. *Exploration* If a and b are nonzero real numbers, show that the equation $ax^2 + bx = 0$ must have two different solutions.

95. *Exploration* If a is a nonzero real number, find two solutions of the equation $ax^2 - ax = 0$.

96. *Height of a Diver* A diver jumps from a diving board 32 feet above the water (see figure). The height of the diver is modeled by

$$\text{Height} = -16t^2 + 16t + 32$$

where the height is measured in feet and the time t is measured in seconds. How many seconds will it take the diver to reach the water?

In national and international competitions, men perform 11 dives: 5 required and 6 optional. Women perform 10 dives: 5 required and 5 optional.

CHAPTER PROJECT: Food Consumption

Factoring is one method used to solve a polynomial equation. Other methods include using the Quadratic Formula and graphing. In this project you will use a graphing utility to graph a quadratic equation and determine its solutions.

Investigate the following questions.

1. Use a graphing utility to graph $y = x^2 - 4x + 3$. What are the x-intercepts of the graph? Substitute each value of x into the equation. Explain why these x-values are solutions to the equation $x^2 - 4x + 3 = 0$.

2. The graph of $y = -4x^2 - 12x - 9$ is shown in the figure at left. Explain the relationship between the number of x-intercepts of the graph and the number of solutions to the equation $-4x^2 - 12x - 9 = 0$.

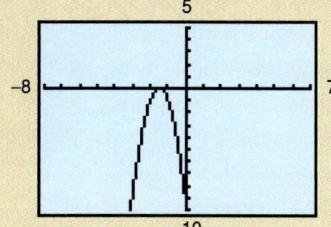

3. Solve the quadratic equation $x^2 + 3x - 15 = 13$ by graphing two different ways. (a) Subtract 13 from both sides, set the quadratic expression equal to y, and graph the result. (b) Set both sides of the equation equal to y. Graph both equations on the same screen of a graphing utility. What do you notice about the x-intercepts of the first method and the x-values of the intersections of the two graphs in the second method? Explain the similarities and differences of the two methods.

In Questions 4–6, use the following quadratic models for food consumption (in pounds) per person in the United States from 1970 to 1991. In each model, $t = 0$ represents 1970.

Food Item	Quadratic Model
Mozzarella Cheese	$y = 0.009t^2 + 0.10t + 1.26$
Flour and Cereal Products	$y = 0.12t^2 - 0.12t + 135.62$
Rye Flour	$y = 0.0015t^2 - 0.061t + 1.216$

Year	Rye Flour Consumption
1970	1.2
1975	1.0
1980	0.7
1985	0.7
1986	0.6
1987	0.6
1988	0.6
1989	0.6
1990	0.6
1991	0.6

4. Use a graphing utility to graphically determine the year in which the consumption of mozzarella cheese was 6 pounds per person. From the graph, is there one or two solutions? Are all the solutions valid? Explain.

5. Use a graphing utility to graphically determine the year in which the consumption of flour and cereal products was 130 pounds per person. Is this solution reasonable? What does the solution tell you about the model before 1970?

6. The table at the left shows the data used to model the equation for rye flour. Evaluate the equation for the same years. Is the model valid? From the table, estimate the consumption per person for the year 2000. Graph the equation using a graphing utility and estimate the consumption per person for the same year. What does this tell you about the model? Would you use this model to estimate future food consumption? Explain.

CHAPTER SUMMARY

After studying this chapter, you should have acquired the following skills. These skills are keyed to the Review Exercises that begin on page 367. Answers to odd-numbered Review Exercises are given in the back of the book.

- Find the greatest common factor of expressions. *(Section 6.1)*

  Review Exercises 1–4

- Factor polynomials completely. *(Sections 6.2, 6.3)*

  Review Exercises 5, 6, 7, 8, 35, 46

- Factor polynomials by grouping. *(Section 6.1)*

  Review Exercises 9, 10, 11, 12, 47, 48, 49, 50

- Factor polynomials having the special form below. *(Section 6.4)*

  Difference of Two Squares
  $$a^2 - b^2 = (a + b)(a - b)$$

  Review Exercises 13, 14, 15, 16, 36

- Factor polynomials having one of the special forms below. *(Section 6.4)*

  Perfect Square Trinomials
  $$a^2 + 2ab + b^2 = (a + b)(a + b)$$
  $$a^2 - 2ab + b^2 = (a - b)(a - b)$$

  Review Exercises 17, 18, 19, 45

- Factor polynomials having one of the special forms below. *(Section 6.4)*

  Sum and Difference of Two Cubes
  $$a^3 + b^3 = (a + b)(a^2 - ab + b^2)$$
  $$a^3 - b^3 = (a - b)(a^2 + ab + b^2)$$

  Review Exercises 43, 44

- Factor polynomials in two variables. *(Section 6.2)*

  Review Exercises 20, 37, 38, 39, 40

- Factor trinomials. *(Sections 6.2, 6.3)*

  Review Exercises 21–24, 31–34, 41, 42

- Find the missing factors of polynomials. *(Sections 6.1, 6.2, 6.3, 6.4)*

  Review Exercises 25–30

- Find all values of the given variable such that the trinomial is factorable. *(Sections 6.2, 6.3, 6.4)*

  Review Exercises 51–58, 59–62

- Graph equations using a graphing utility and draw conclusions. *(Sections 6.1, 6.2, 6.3, 6.4)*

  Review Exercises 63–66

- Solve real-life problems involving geometry. *(Sections 6.1, 6.2, 6.5)*

  Review Exercises 67, 68, 83, 85

- Solve equations using the Zero-Factor Property *(Section 6.5)*

  Review Exercises 69–80, 84

- Solve real-life problems involving quadratic equations. *(Section 6.5)*

  Review Exercises 81, 82

REVIEW EXERCISES

In Exercises 1–4, find the greatest common factor of the expressions.

1. 20, 60, 150

2. $3x^4, 21x^2$

3. $18ab^2, 27a^2b$

4. $14z^2, 1, 21z$

In Exercises 5–24, factor the polynomial.

5. $5x^2 + 10x^3$

6. $7y - 21y^4$

7. $8a - 12a^3$

8. $6u - 9u^2 + 15u^3$

9. $24(x + 1) - 18(x + 1)^2$

10. $(u + 1)(u - 2) + 3(u + 1)$

11. $y^3 + 3y^2 + 2y + 6$

12. $z^3 - 5z^2 - z + 5$

13. $a^2 - 100$

14. $16b^2 - 1$

15. $(u + 1)^2 - 4$

16. $(y - 2)^2 - 9$

17. $x^2 - 8x + 16$

18. $y^2 + 24y + 144$

19. $9s^2 + 12s + 4$

20. $u^2 - 2uv + v^2$

21. $4x^2 + 8x + 3$

22. $8x^2 - 18x + 9$

23. $50 - 5x - x^2$

24. $7 + 5x - 2x^2$

In Exercises 25–30, insert the missing factor.

25. $\frac{1}{3}x + \frac{5}{6} = \frac{1}{6}()$

26. $x^3 - x = x()()$

27. $3x^2 + 14x + 8 = (x + 4)()$

28. $2x^2 + 21x + 10 = (x + 10)()$

29. $x^4 - 2x^2 + 1 = (x + 1)^2()^2$

30. $u^4 - v^4 = (u^2 + v^2)()()$

In Exercises 31–50, factor the polynomial completely. (*Note:* Some of the polynomials may be prime.)

31. $x^2 - 3x - 28$

32. $x^2 - 3x - 40$

33. $6x^2 + 7x + 2$

34. $16x^2 + 13x - 3$

35. $6u^3 + 3u^2 - 30u$

36. $5t - 125t^3$

37. $10x^2 + 9xy + 2y^2$

38. $4u^2 + uv - 5v^2$

39. $s^3t - st^3$

40. $y^3z + 4y^2z^2 + 4yz^3$

41. $2x^2 - 3x + 1$

42. $3x^2 + 8x + 4$

43. $27 - 8t^3$

44. $z^3 - 125$

45. $-16a^3 - 16a^2 - 4a$

46. $8x^3 - 40x^2 + 32x$

47. $x^3 + 2x^2 + x + 2$

48. $x^3 - 5x^2 + 5x - 25$

49. $x^3 - 4x^2 - 4x + 16$

50. $x^3 + 6x^2 - x - 6$

In Exercises 51–58, find all values of b such that the trinomial is factorable.

51. $x^2 + bx + 9$

52. $y^2 + by + 25$

53. $z^2 + bz + 11$

54. $x^2 + bx + 14$

55. $x^2 + bx - 24$

56. $2x^2 + bx - 16$

57. $3x^2 + bx - 20$

58. $3x^2 + bx + 1$

In Exercises 59–62, find two values of c such that the trinomial is factorable.

59. $x^2 + 6x + c$

60. $x^2 - 7x + c$

61. $2x^2 - 4x + c$

62. $5x^2 + 6x + c$

In Exercises 63–66, use a graphing utility to graph the two equations on the same screen. What can you conclude?

63. $y_1 = x^2 - 2x - 3, y_2 = (x + 1)(x - 3)$

64. $y_1 = -x^2 - 6x - 9, y_2 = -(x + 3)^2$

65. $y_1 = x^3 - 4x^2 + 4x, y_2 = x(x - 2)^2$

66. $y_1 = x^3 + 2x^2 - 4x - 8, y_2 = (x + 2)^2(x - 2)$

67. *Geometry*   A rectangular sheet of metal has dimensions 2 feet by 3 feet. An open box is to be made from the metal by cutting equal squares from each corner and turning up the sides. The volume of the box is

$$V = 4x^3 - 10x^2 + 6x, \quad 0 < x < 1.$$

(a) Sketch the rectangular sheet and the open box. Label the height of the box as x.

(b) Factor the expression for the volume and use the result to label the length and width of the box.

(c) Use a graphing utility to graph the volume over the specified interval. Use the graph to approximate the size of the squares to be cut from the corners so that the volume of the box is greatest.

68. *Geometry*   The volume of a cylindrical shell (see figure) is $V = \pi R^2 h - \pi r^2 h$. Completely factor this expression.

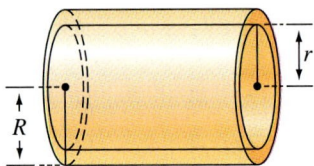

In Exercises 69–80, solve the equation.

69. $x(2x - 3) = 0$ **70.** $3x(5x + 1) = 0$

71. $x^2 - 81 = 0$ **72.** $(x + 1)^2 - 16 = 0$

73. $x^2 - 12x + 36 = 0$ **74.** $2t^2 - 3t - 2 = 0$

75. $4s^2 + s - 3 = 0$ **76.** $y^3 - y^2 - 6y = 0$

77. $x(7 - x) = 12$ **78.** $x(x + 5) = 24$

79. $u^3 + 5u^2 - u = 5$ **80.** $a^3 - 3a^2 - a = -3$

81. *Revenue*   The revenue for selling x units is

$$R = 12x - 0.3x^2.$$

(a) Use a graphing utility to graph the revenue.

(b) Use the graph to estimate the value of x that yields a revenue of $R = \$120$.

(c) Use factorization to find the value of x that yields a revenue of $R = \$120$.

82. *Height of an Object*   A rock is thrown upward from a height of 48 feet with an initial velocity of 32 feet per second. The height of the rock is given by

$$\text{Height} = -16t^2 + 32t + 48$$

where the height is measured in feet and the time t is measured in seconds. Find the time for the rock to reach the water.

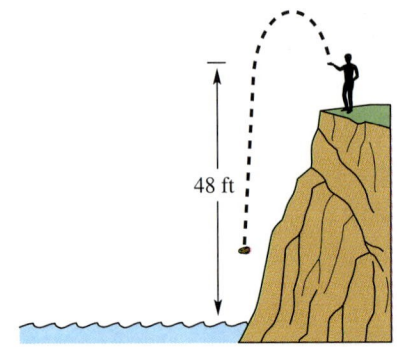

83. *Geometry*   The height of a rectangular window is one and one-half times the width. The area of the window is 2400 square inches. Find the dimensions of the window.

84. The product of two consecutive positive even integers is 168. Find the two integers.

85. *Geometry*   A box with a square base has a surface area of 400 square inches (see figure). The height of the box is 5 inches. Find the dimensions of the box. (*Hint:* The surface area is given by $S = 2x^2 + 4xh$.)

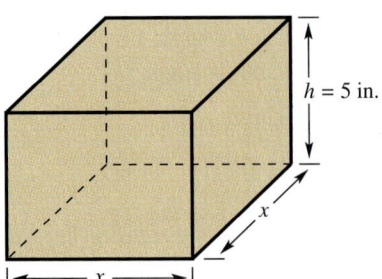

CHAPTER TEST

Take this test as you would take a test in class. After you are done, check your work against the answers given in the back of the book.

In Exercises 1–10, completely factor the polynomial.

1. $7x^2 - 14x^3$

2. $z(z + 7) - 3(z + 7)$

3. $t^2 - 4t - 5$

4. $6x^2 - 11x + 4$

5. $6y^3 + 45y^2 + 75y$

6. $4 - 25v^2$

7. $4x^2 - 20x + 25$

8. $16 - (z + 9)^2$

9. $x^3 + 2x^2 - 9x - 18$

10. $16 - z^4$

11. Find the missing factor: $\frac{2}{3}x - \frac{3}{4} = \frac{1}{12}($ ___ $)$.

12. Find all values of b such that $x^2 + bx + 5$ can be factored.

13. Find a number c such that $x^2 + 12x + c$ is a perfect square trinomial.

14. Explain why $(x + 1)(3x - 6)$ is not a complete factorization of $3x^2 - 3x - 6$.

In Exercises 15–18, solve the equation.

15. $(x + 4)(2x - 3) = 0$

16. $7x^2 - 14x = 0$

17. $3x^2 + 7x - 6 = 0$

18. $y(2y - 1) = 6$

19. The width of a rectangle is 5 inches less than the length. The area of the rectangle is 84 square inches. Find the dimensions of the rectangle.

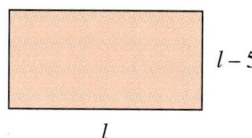

$l - 5$

l

20. An object is dropped from a height of 40 feet. Its height at any time is modeled by

$$\text{Height} = -16t^2 + 40$$

where the height is measured in feet and the time t is measured in seconds. How long will it take the object to fall to a height of 15 feet?

CUMULATIVE TEST: CHAPTERS 4–6

Take this test as you would take a test in class. After you are done, check your work against the answers given in the back of the book.

1. Describe how to identify the quadrants in which the points $(-2, y)$ must be located. (y is a real number.)

2. Determine whether the ordered pairs are solution points of the equation $9x - 4y + 36 = 0$.

   (a) $(-1, -1)$ (b) $(8, 27)$ (c) $(-4, 0)$ (d) $(3, -2)$

In Exercises 3 and 4, sketch the graph of the equation and determine any intercepts of the graph.

3. $y = 2 - |x|$ **4.** $y = \frac{1}{2}x - 2$

In Exercises 5 and 6, use a graphing utility to graph the equation. Use the TRACE feature to approximate the x-intercept(s) of the graph. Check your estimates algebraically.

5. $y = x^2 - 2x - 3$ **6.** $y = x^3 - 9x$

7. Subtract: $(x^3 - 3x^2) - (x^3 + 2x^2 - 5)$ **8.** Multiply: $(6z)(-7z)(z^2)$

9. Multiply: $(3x + 5)(x - 4)$ **10.** Multiply: $(5x - 3)(5x + 3)$

11. Expand: $(5x + 6)^2$ **12.** Divide: $(6x^2 + 72x) \div 6x$

13. Divide: $\dfrac{x^2 - 3x - 2}{x - 4}$ **14.** Evaluate: $(3^2 \cdot 4^{-1})^2$

15. Factor: $2u^2 - 6u$ **16.** Factor and simplify: $(x - 2)^2 - 16$

17. Factor completely: $x^3 + 8x^2 + 16x$ **18.** Factor completely: $x^3 + 2x^2 - 4x - 8$

19. Solve: $u(u - 12) = 0$ **20.** Solve: $5x^2 - 12x - 9 = 0$

21. Rewrite the expression $\left(\dfrac{x}{2}\right)^{-2}$ using positive exponents.

22. An inheritance of \$12,000 is divided between two investments earning 7.5% and 9% simple interest. How much is in each investment if the total interest for 1 year is \$960?

23. A sales representative is reimbursed \$125 per day for lodging and meals, plus \$0.35 per mile driven. Write a linear equation giving the daily cost C to the company in terms of x, the number of miles driven. Explain the reasoning you used to write the model. Find the cost for a day when the representative drives 70 miles.

Functions and Graphs

- The Rectangular Coordinate System
- Graphs of Equations
- Graphs and Graphing Utilities
- Slope and Graphs of Linear Equations
- Relations and Functions
- Graphs of Functions

In the early history of the United States, most people were self-employed in agriculture and men's and women's roles were roughly equivalent. Everyone—men, women, and children—was involved in running the family farm.

As the country became less rural and more industrial, working roles changed. Men's roles became associated with working outside the home, and women's roles became associated with homemaking and caring for children. Now, since World War II, the roles of men and women in the work force are once again becoming more alike.

The table at the right shows the numbers of men and women (in thousands) that made up the civilian labor force in selected years from 1960 to 1992. The graph shows a scatter plot of the data.

Year	1960	1970	1980	1990	1991	1992
Women	21,874	29,688	42,117	53,479	53,284	53,793
Men	43,904	48,990	57,186	64,435	63,593	63,805

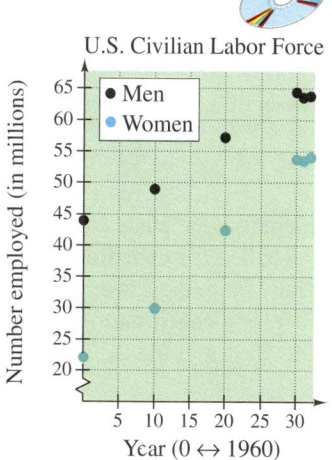

U.S. Civilian Labor Force

The chapter project related to this information is on page 441.

7.1 | The Rectangular Coordinate System

The Rectangular Coordinate System ▪ Ordered Pairs as Solutions ▪
The Distance Formula

The Rectangular Coordinate System

René Descartes (1596–1650) was a French mathematician, philosopher, and scientist. He is sometimes called the father of modern philosophy, and his phrase "I think, therefore I am," has been quoted often. In mathematics, Descartes is known as the father of analytic geometry. Prior to Descartes's time, geometry and algebra were separate mathematical studies—it was Descartes's introduction of the rectangular coordinate system that brought the two studies together.

Just as you can represent real numbers by points on the real number line, you can represent ordered pairs of real numbers by points in a plane. This plane is called a **rectangular coordinate system** or the **Cartesian plane**, after the French mathematician René Descartes.

A rectangular coordinate system is formed by two real lines intersecting at a right angle, as shown in Figure 7.1. The horizontal number line is usually called the **x-axis,** and the vertical number line is usually called the **y-axis.** (The plural of axis is *axes*.) The point of intersection of the two axes is called the **origin,** and the axes separate the plane into four regions called **quadrants.**

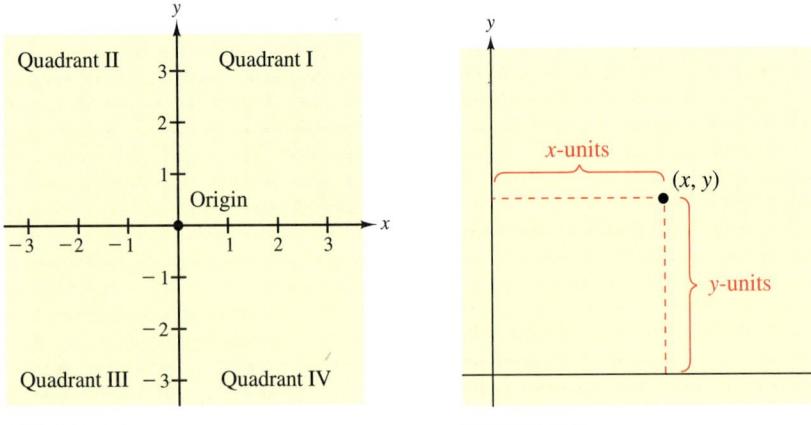

FIGURE 7.1   **FIGURE 7.2**

Each point in the plane corresponds to an **ordered pair** (x, y) of real numbers x and y, called the **coordinates** of the point. The first number (or **x-coordinate**) tells how far to the left or right the point is from the vertical axis, and the second number (or **y-coordinate**) tells how far up or down the point is from the horizontal axis, as shown in Figure 7.2.

NOTE The signs of the coordinates tell you which quadrant the point lies in. For instance, if x and y are positive, the point (x, y) lies in Quadrant I.

Locating a given point in a plane is called **plotting** the point. Example 1 shows how this is done.

EXAMPLE 1   *Plotting Points on a Rectangular Coordinate System*

Plot the points $(-2, 1), (4, 0), (3, -1), (4, 3), (0, 0),$ and $(-1, -3)$ on a rectangular coordinate system.

Solution

The point $(-2, 1)$ is two units to the *left* of the vertical axis and one unit *above* the horizontal axis.

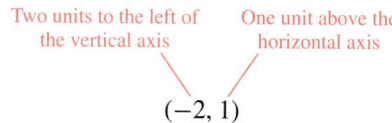

Similarly, the point $(4, 0)$ is four units to the *right* of the vertical axis and *on* the horizontal axis. (It is on the horizontal axis because its y-coordinate is 0.) The other four points can be plotted in a similar way, as shown in Figure 7.3.

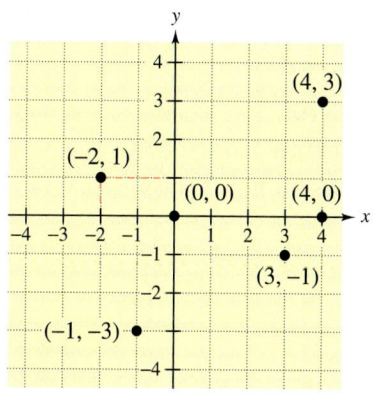

FIGURE 7.3

In Example 1, you were given the coordinates of several points and asked to plot the points on a rectangular coordinate system. Example 2 looks at the reverse problem. That is, you are given points on a rectangular coordinate system and are asked to determine their coordinates.

EXAMPLE 2   *Finding Coordinates of Points*

Determine the coordinates of each of the points shown in Figure 7.4.

Solution

Point A lies two units to the *right* of the vertical axis and one unit *below* the horizontal axis. Therefore, point A must be given by the ordered pair $(2, -1)$. The coordinates of the other four points can be determined in a similar way; the results are summarized as follows.

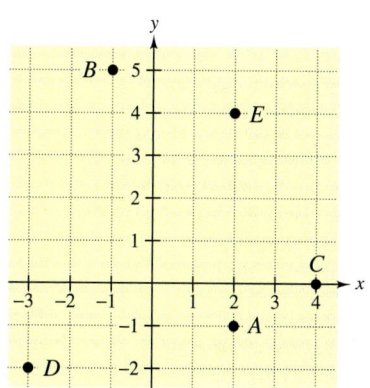

FIGURE 7.4

Point	Position	Coordinates
A	2 units *right*, 1 unit *down*	$(2, -1)$
B	1 unit *left*, 5 units *up*	$(-1, 5)$
C	4 units *right*, 0 units *up*	$(4, 0)$
D	3 units *left*, 2 units *down*	$(-3, -2)$
E	2 units *right*, 4 units *up*	$(2, 4)$

Notice that because point C lies on the x-axis, it has a y-coordinate of 0.

The primary value of a rectangular coordinate system is that it allows you to visualize relationships between two variables. Today, Descartes's ideas are commonly used in virtually every scientific and business-related field.

EXAMPLE 3 *Representing Data Graphically*

The population (in millions) of California from 1975 through 1990 is listed in the table. Plot these points on a rectangular coordinate system.   (Source: U.S. Bureau of Census)

Year	1975	1976	1977	1978	1979	1980	1981	1982
Population	21.2	21.5	21.9	22.3	23.3	23.7	24.3	24.8

Year	1983	1984	1985	1986	1987	1988	1989	1990
Population	25.3	26.4	27.0	27.0	27.7	28.3	28.6	29.8

Solution

Begin by choosing which variable will be plotted on the horizontal axis and which will be plotted on the vertical axis. For these data, it seems natural to plot the years on the horizontal axis (which means that the population must be plotted on the vertical axis). Next, use the data in the table to form ordered pairs. For instance, the first three ordered pairs are (1975, 21.2), (1976, 21.5), and (1977, 21.9). All 16 points are shown in Figure 7.5. Note that the break in the x-axis indicates that the numbers between 0 and 1975 have been omitted. The break in the y-axis indicates that the numbers between 0 and 21,000 have been omitted.

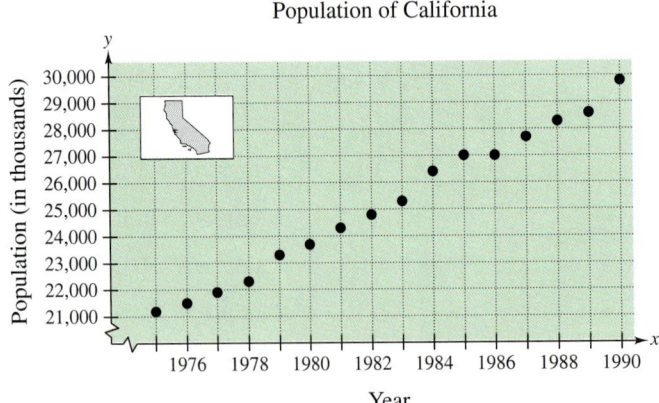

Population of California

FIGURE 7.5

Ordered Pairs as Solutions

In Example 3, the relationship between the year and the population was given by a **table of values.** In mathematics, the relationship between the variables x and y is often given by an equation. From the equation, you can construct your own table of values.

EXAMPLE 4 *Constructing a Table of Values*

Construct a table of values for $y = 3x + 2$. Then plot the solution points on a rectangular coordinate system. Choose x-values of -3, -2, -1, 0, 1, 2, and 3.

Solution

For each x-value, you must calculate the corresponding y-value. For example, if you choose $x = 1$, then the y-value is

$$y = 3(1) + 2 = 5.$$

The ordered pair $(x, y) = (1, 5)$ is a **solution point** (or **solution**) of the equation.

Choose x	Calculate y from $y = 3x + 2$	Solution Point
$x = -3$	$y = 3(-3) + 2 = -7$	$(-3, -7)$
$x = -2$	$y = 3(-2) + 2 = -4$	$(-2, -4)$
$x = -1$	$y = 3(-1) + 2 = -1$	$(-1, -1)$
$x = 0$	$y = 3(0) + 2 = 2$	$(0, 2)$
$x = 1$	$y = 3(1) + 2 = 5$	$(1, 5)$
$x = 2$	$y = 3(2) + 2 = 8$	$(2, 8)$
$x = 3$	$y = 3(3) + 2 = 11$	$(3, 11)$

Once you have constructed a table of values, you can get a visual idea of the relationship between the variables x and y by plotting the solution points on a rectangular coordinate system, as shown in Figure 7.6.

DISCOVERY

In the table of values in Example 4, successive x-values differ by 1. How do successive y-values differ? If successive x-values differed by 1, how would successive y-values differ for the following equations?

a. $y = x + 2$

b. $y = 2x + 2$

c. $y = 4x + 2$

d. $y = -x + 2$

Describe the pattern. *Hint*: If you have a *TI-82*, use it to form the tables, as shown on page 126.

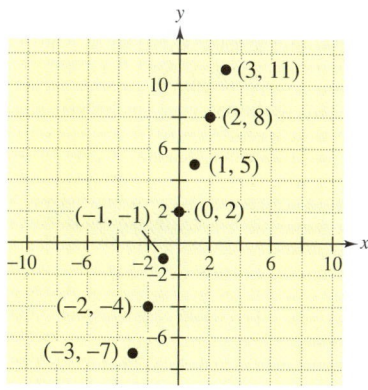

FIGURE 7.6

When making up a table of values for an equation, it is helpful to first solve the equation for y. Here is an example.

$5x + 3y = 4$	Original equation
$3y = -5x + 4$	Subtract $5x$ from both sides.
$y = -\dfrac{5}{3}x + \dfrac{4}{3}$	Divide both sides by 3.

Technology

1. Y=

> Y$_1$=(2/3)X^2 – (5/3)
> Y$_2$=
> Y$_3$=
> Y$_4$=
> Y$_5$=
> Y$_6$=

2. TblSet

> TABLE SETUP
> TblMin=-3
> ΔTbl=1
> Indpnt: Auto Ask
> Depend: Auto Ask

3. TABLE

X	Y$_1$	
-3	4.3333	
-2	1	
-1	-1	
0	-1.667	
1	-1	
2	1	
3	4.3333	

X=-3

Creating a Table with a Graphing Utility

Some graphing utilities, such as the *TI-82*, have built-in programs that can create tables. Suppose, for instance, that you want to create a table of values for the equation $2x^2 - 3y = 5$. To begin, solve the equation for y.

$$2x^2 - 3y = 5 \qquad \text{Original equation}$$

$$-3y = -2x^2 + 5 \qquad \text{Subtract } 2x^2 \text{ from both sides.}$$

$$y = \frac{2}{3}x^2 - \frac{5}{3} \qquad \text{Divide both sides by } -3.$$

Now, using the equation $y = \frac{2}{3}x^2 - \frac{5}{3}$, you can use the following steps.

1. Press the Y= key and enter the equation as Y_1.

2. Use the TblSet feature and enter the values of TblMin and ΔTbl. The value of TblMin is the beginning x-value you want displayed in the table, and the value of ΔTbl is the increment for the x-values in the table. For instance, using TblMin= -3 and ΔTbl=1, the table has x-values of $-3, -2, -1$, and so on.

3. Use the TABLE feature to obtain the table of values. You can use the cursor keys to view x- and y-values that are not shown on the default screen.

You can confirm the values shown on the utility's screen by substituting the appropriate x-values, as shown in the table below. (Note that the utility lists decimal approximations of fractional values. For instance, $-\frac{5}{3}$ is listed as -1.667.)

Choose x	Calculate y from $y = \frac{2}{3}x^2 - \frac{5}{3}$	Solution Point
$x = -3$	$y = \frac{2}{3}(-3)^2 - \frac{5}{3} = \frac{13}{3}$	$\left(-3, \frac{13}{3}\right)$
$x = -2$	$y = \frac{2}{3}(-2)^2 - \frac{5}{3} = 1$	$(-2, 1)$
$x = -1$	$y = \frac{2}{3}(-1)^2 - \frac{5}{3} = -1$	$(-1, -1)$
$x = 0$	$y = \frac{2}{3}(0)^2 - \frac{5}{3} = -\frac{5}{3}$	$\left(0, -\frac{5}{3}\right)$
$x = 1$	$y = \frac{2}{3}(1)^2 - \frac{5}{3} = -1$	$(1, -1)$
$x = 2$	$y = \frac{2}{3}(2)^2 - \frac{5}{3} = 1$	$(2, 1)$
$x = 3$	$y = \frac{2}{3}(3)^2 - \frac{5}{3} = \frac{13}{3}$	$\left(3, \frac{13}{3}\right)$

After creating the table, you can plot the points, as shown at the left.

NOTE   When a substitution of (x, y) produces a true equation, the ordered pair (x, y) is said to **satisfy** the equation.

In the next example, you are given several ordered pairs and are asked to determine whether they are solutions of the given equation. To do this, you need to substitute the values of x and y into the equation. If the substitution produces a true equation, then the ordered pair (x, y) is a solution.

EXAMPLE 5   Verifying Solutions of an Equation

Which of the ordered pairs are solutions of $x^2 - 2y = 6$?

a. $(2, 1)$ **b.** $(0, -3)$ **c.** $(-2, -5)$ **d.** $\left(1, -\frac{5}{2}\right)$

Solution

a. For the ordered pair $(2, 1)$, substitute $x = 2$ and $y = 1$ into the equation.

$$x^2 - 2y = 6 \qquad \text{Original equation}$$
$$(2)^2 - 2(1) \stackrel{?}{=} 6 \qquad \text{Substitute 2 for } x \text{ and 1 for } y.$$
$$2 \neq 6 \qquad \text{Is not a solution} \ \text{✗}$$

Because the substitution does not satisfy the given equation, you can conclude that the ordered pair $(2, 1)$ *is not* a solution of the given equation.

b. For the ordered pair $(0, -3)$, substitute $x = 0$ and $y = -3$ into the equation.

$$x^2 - 2y = 6 \qquad \text{Original equation}$$
$$(0)^2 - 2(-3) \stackrel{?}{=} 6 \qquad \text{Substitute 0 for } x \text{ and } -3 \text{ for } y.$$
$$6 = 6 \qquad \text{Is a solution} \ \text{✓}$$

Because the substitution satisfies the given equation, you can conclude that the ordered pair $(0, -3)$ *is* a solution of the given equation.

c. For the ordered pair $(-2, -5)$, substitute $x = -2$ and $y = -5$ into the equation.

$$x^2 - 2y = 6 \qquad \text{Original equation}$$
$$(-2)^2 - 2(-5) \stackrel{?}{=} 6 \qquad \text{Substitute } -2 \text{ for } x \text{ and } -5 \text{ for } y.$$
$$14 \neq 6 \qquad \text{Is not a solution} \ \text{✗}$$

Because the substitution does not satisfy the given equation, you can conclude that the ordered pair $(-2, -5)$ *is not* a solution of the given equation.

d. For the ordered pair $\left(1, -\frac{5}{2}\right)$, substitute $x = 1$ and $y = -\frac{5}{2}$ into the equation.

$$x^2 - 2y = 6 \qquad \text{Original equation}$$
$$(1)^2 - 2\left(-\frac{5}{2}\right) \stackrel{?}{=} 6 \qquad \text{Substitute 1 for } x \text{ and } -\frac{5}{2} \text{ for } y.$$
$$6 = 6 \qquad \text{Is a solution} \ \text{✓}$$

Because the substitution satisfies the given equation, you can conclude that the ordered pair $\left(1, -\frac{5}{2}\right)$ *is* a solution of the given equation.

The Distance Formula

You know from Section P.1 of the Prerequisites chapter that the distance d between two points a and b on the real number line is simply $d = |b - a|$. The same "absolute value rule" is used to find the distance between two points that lie on the same *vertical or horizontal line*, as shown in Example 6.

EXAMPLE 6   *Finding Horizontal and Vertical Distances*

a. Find the distance between the points $(2, -2)$ and $(2, 4)$.

b. Find the distance between the points $(-3, -2)$ and $(2, -2)$.

Solution

a. Because the x-coordinates are equal, you can visualize a vertical line through the points $(2, -2)$ and $(2, 4)$, as shown in Figure 7.7. The distance between these two points is the absolute value of the difference of their y-coordinates.

$$\text{Vertical distance} = |4 - (-2)| \qquad \text{\color{red}Subtract } y\text{-coordinates.}$$
$$= 6 \qquad \text{\color{red}Evaluate absolute value.}$$

b. Because the y-coordinates are equal, you can visualize a horizontal line through the points $(-3, -2)$ and $(2, -2)$, as shown in Figure 7.7. The distance between these two points is the absolute value of the difference of their x-coordinates.

$$\text{Horizontal distance} = |2 - (-3)| \qquad \text{\color{red}Subtract } x\text{-coordinates.}$$
$$= 5 \qquad \text{\color{red}Evaluate absolute value.}$$

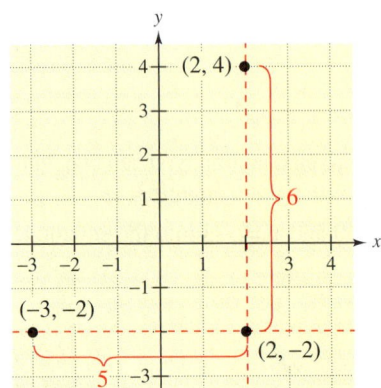

FIGURE 7.7

NOTE   In Figure 7.7, note that the horizontal distance between the points $(-3, -2)$ and $(2, -2)$ is the absolute value of the difference of the x-coordinates, and the vertical distance between the points $(2, -2)$ and $(2, 4)$ is the absolute value of the difference of the y-coordinates.

The technique applied in Example 6 can be used to develop a general formula for finding the distance between two points in the plane. This general formula will work for any two points, even if they do not lie on the same vertical or horizontal line. To develop the formula, you use the **Pythagorean Theorem,** which states that for a right triangle, the hypotenuse c and sides a and b are related by the formula

$$a^2 + b^2 = c^2 \qquad \text{\color{red}Pythagorean Theorem}$$

as shown in Figure 7.8. (The converse is also true. That is, if $a^2 + b^2 = c^2$, then the triangle is a right triangle.)

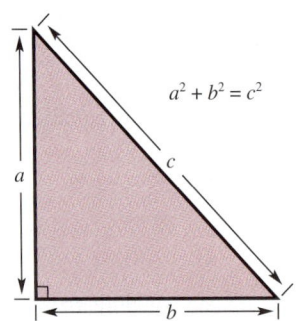

FIGURE 7.8   Pythagorean Theorem

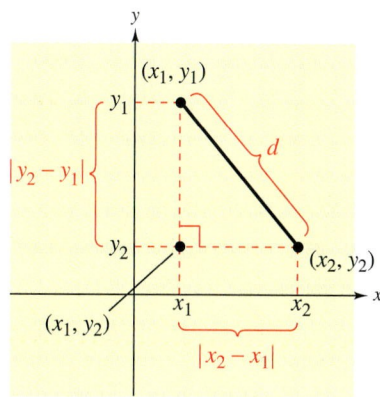

FIGURE 7.9   Distance Between Two Points

To develop a general formula for the distance between two points, let (x_1, y_1) and (x_2, y_2) represent two points in the plane (that do not lie on the same horizontal or vertical line). With these two points, a right triangle can be formed, as shown in Figure 7.9. Note that the third vertex of the triangle is (x_1, y_2). Because (x_1, y_1) and (x_1, y_2) lie on the same vertical line, the length of the vertical side of the triangle is $|y_2 - y_1|$. Similarly, the length of the horizontal side is $|x_2 - x_1|$. By the Pythagorean Theorem, the square of the distance between (x_1, y_1) and (x_2, y_2) is

$$d^2 = |x_2 - x_1|^2 + |y_2 - y_1|^2.$$

Because the distance d must be positive, you can choose the positive square root and write

$$d = \sqrt{|x_2 - x_1|^2 + |y_2 - y_1|^2}.$$

Finally, replacing $|x_2 - x_1|^2$ and $|y_2 - y_1|^2$ by the equivalent expressions $(x_2 - x_1)^2$ and $(y_2 - y_1)^2$ gives you the **Distance Formula.**

DISCOVERY

Plot the points $A(-1, -3)$ and $B(5, 2)$ and sketch the line segment from A to B. How could you verify that point $C(2, -0.5)$ is the midpoint of the segment? Why is it not sufficient to show that the distances from A to C and from C to B are equal? Find a formula for the coordinates of the midpoint of the line segment connecting (x_1, y_1) and (x_2, y_2).

The Distance Formula

The distance d between two points (x_1, y_1) and (x_2, y_2) is

$$d = \sqrt{(x_2 - x_1)^2 + (y_2 - y_1)^2}.$$

Note that for the special case in which the two points lie on the same vertical or horizontal line, the Distance Formula still works. For instance, applying the Distance Formula to the points $(2, -2)$ and $(2, 4)$ produces

$$d = \sqrt{(2 - 2)^2 + [4 - (-2)]^2} = \sqrt{6^2} = 6$$

which is the same result obtained in Example 6.

EXAMPLE 7   *Finding the Distance Between Two Points*

Find the distance between the points $(-1, 2)$ and $(2, 4)$, as shown in Figure 7.10.

Solution

Let $(x_1, y_1) = (-1, 2)$ and $(x_2, y_2) = (2, 4)$, and apply the Distance Formula.

$$d = \sqrt{[2 - (-1)]^2 + (4 - 2)^2} \qquad \text{Substitute coordinates of points.}$$
$$= \sqrt{3^2 + 2^2} \qquad \text{Simplify.}$$
$$= \sqrt{13} \qquad \text{Simplify.}$$
$$\approx 3.61 \qquad \text{Use a calculator.}$$

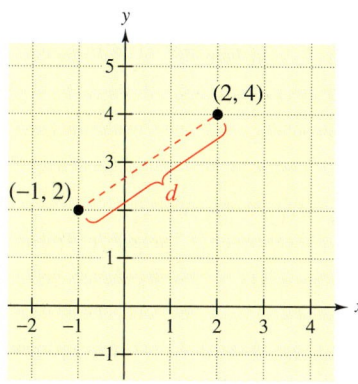

FIGURE 7.10

The Distance Formula has many applications in mathematics. For instance, the next example shows how you can use the Distance Formula and the converse of the Pythagorean Theorem to verify that three points form the vertices of a right triangle.

EXAMPLE 8   An Application of the Distance Formula

Show that the points $(1, 2)$, $(3, 1)$, and $(4, 3)$ are vertices of a right triangle.

Solution

The three points are plotted in Figure 7.11. Using the Distance Formula, you can find the lengths of the three sides of the triangle.

$$d_1 = \sqrt{(3 - 1)^2 + (1 - 2)^2} = \sqrt{4 + 1} = \sqrt{5}$$
$$d_2 = \sqrt{(4 - 3)^2 + (3 - 1)^2} = \sqrt{1 + 4} = \sqrt{5}$$
$$d_3 = \sqrt{(4 - 1)^2 + (3 - 2)^2} = \sqrt{9 + 1} = \sqrt{10}$$

Because

$$d_1{}^2 + d_2{}^2 = 5 + 5$$
$$= 10$$
$$= d_3{}^2$$

you can conclude from the converse of the Pythagorean Theorem that the triangle is a right triangle.

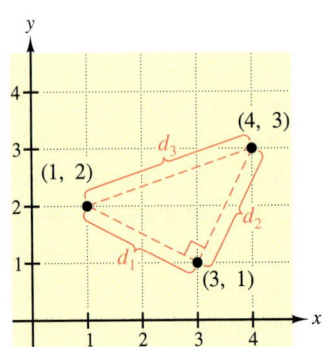

FIGURE 7.11

Group Activities   Extending the Concept

Determining Collinearity   Three or more points are **collinear** if they all lie on the same line. Use the following steps to determine if the set of points $\{A(3, 1), B(5, 4), C(9, 10)\}$ and the set of points $\{A(2, 2), B(4, 3), C(5, 4)\}$ are collinear.

a. For each set of points, use the Distance Formula to find the distances from A to B, from B to C, and from A to C. What relationship exists among these distances for each set of points?

b. Plot each set of points on a rectangular coordinate system. Do all the points of either set appear to lie on the same line?

c. Compare your conclusions from part (a) with the conclusions you made from the graphs in part (b). Make a general statement about how to use the Distance Formula to determine collinearity.

7.1 Exercises

Discussing the Concepts

1. Discuss the significance of the word *order* when referring to an ordered pair (x, y).

2. When plotting the point $(3, -2)$, what does the x-coordinate measure? What does the y-coordinate measure?

3. What is the x-coordinate of any point on the y-axis? What is the y-coordinate of any point on the x-axis?

4. Explain why the ordered pair $(-3, 4)$ is not a solution point of the equation $y = 4x + 15$.

5. When plotting points on the rectangular coordinate system, is it true that the scales on the x- and y-axes must be the same? Explain.

6. State the Pythagorean Theorem and give examples of its use.

Problem Solving

In Exercises 7–10, plot the points on a rectangular coordinate system.

7. $(4, 3)$, $(-5, 3)$, $(3, -5)$

8. $(-2, 5)$, $(-2, -5)$, $(3, 5)$

9. $\left(\frac{5}{2}, -2\right)$, $\left(-2, \frac{1}{4}\right)$, $\left(\frac{3}{2}, -\frac{7}{2}\right)$

10. $\left(-\frac{2}{3}, 3\right)$, $\left(\frac{1}{4}, -\frac{5}{4}\right)$, $\left(-5, -\frac{7}{4}\right)$

In Exercises 11 and 12, approximate the coordinates of the points.

11.

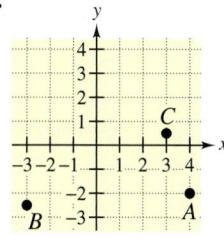

12.

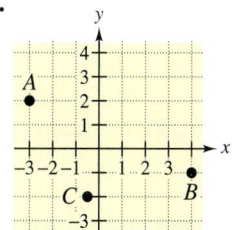

Geometry In Exercises 13–16, plot the points and connect them with line segments to form the figure. (*Note:* A *rhombus* is a parallelogram whose sides are all the same length.)

13. *Triangle:* $(-1, 2)$, $(2, 0)$, $(3, 5)$

14. *Rectangle:* $(7, 0)$, $(9, 1)$, $(4, 6)$, $(6, 7)$

15. *Parallelogram:* $(4, 0)$, $(6, -2)$, $(0, -4)$, $(-2, -2)$

16. *Rhombus:* $(-3, -3)$, $(-2, -1)$, $(-1, -2)$, $(0, 0)$

In Exercises 17–22, determine the quadrant or quadrants in which the point is located.

17. $(-3, -5)$

18. $(4, -2)$

19. $(x, 4)$

20. $(-10, y)$

21. (x, y), $xy < 0$

22. (x, y), $x > 0, y > 0$

In Exercises 23–26, find the coordinates of the point.

23. The point is located five units to the left of the y-axis and two units above the x-axis.

24. The point is located 10 units to the right of the y-axis and four units below the x-axis.

25. The point is on the positive x-axis 10 units from the origin.

26. The point is on the negative y-axis five units from the origin.

In Exercises 27 and 28, plot the points whose coordinates are given in the table.

27. *Exam Score*   The table gives the time x in hours invested in concentrated study for five different algebra exams and the resulting exam score y.

x	5	2	3	6.5	4
y	81	71	88	92	86

28. *Price of Stock* The year-end market prices y per share of common stock in PepsiCo, Inc. for the years 1987 through 1993 are given in the table. The time in years is given by x. (Source: PepsiCo, Inc.)

x	1987	1988	1989	1990	1991	1992	1993
y	$11\frac{1}{4}$	$13\frac{1}{8}$	$21\frac{3}{8}$	$25\frac{3}{4}$	$33\frac{3}{4}$	$42\frac{1}{4}$	$41\frac{7}{8}$

In Exercises 29 and 30, determine whether the ordered pairs are solutions of the equation.

29. $y = 3x + 8$
  (a) $(3, 17)$ (b) $(-1, 10)$
  (c) $(0, 0)$ (d) $(-2, 2)$

30. $5x - 2y + 50 = 0$
  (a) $(-10, 0)$ (b) $(-5, 5)$
  (c) $(0, 25)$ (d) $(20, -2)$

In Exercises 31 and 32, complete the table of values. Then plot the solution points on a rectangular coordinate system.

31.

x	-2	0	2	4	6
$y = 5x - 1$					

32.

x	-2	0	2	4	6
$y = \frac{3}{4}x + 2$					

In Exercises 33 and 34, use a graphing utility to complete the table of values.

33.

x	-2	0	2	4	6
$y = 4x^2 + x - 2$					

34.

x	-2	0	2	4	6
$y = \lvert 3x - 4 \rvert + 1$					

In Exercises 35–38, plot the points and find the distance between them. State whether the points lie on a horizontal or vertical line.

35. $(3, -2), (3, 5)$ **36.** $(-2, 8), (-2, 1)$

37. $\left(\frac{1}{2}, \frac{7}{8}\right), \left(\frac{11}{2}, \frac{7}{8}\right)$ **38.** $\left(\frac{3}{4}, 1\right), \left(\frac{3}{4}, -10\right)$

In Exercises 39–42, find the distance between the points.

39. $(1, 3), (5, 6)$ **40.** $(3, 10), (15, 5)$

41. $(0, 0), (12, -9)$ **42.** $(-5, 0), (3, 15)$

In Exercises 43 and 44, find the coordinates of (x, y), the lengths of the vertical and horizontal sides of the right triangle, and the length of the hypotenuse.

43.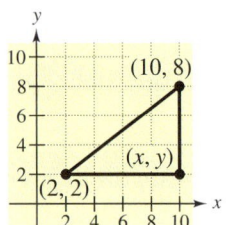

44.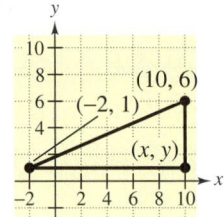

45. *Making a Conjecture* Plot the points $(2, 1), (-3, 5),$ and $(7, -3)$ on a rectangular coordinate system. Then change the sign of the x-coordinate of each point and plot the three new points on the same rectangular coordinate system. What conjecture can you make about the location of a point when the sign of the x-coordinate is changed?

46. *Making a Conjecture* Plot the points $(2, 1), (-3, 5),$ and $(7, -3)$ on a rectangular coordinate system. Then change the sign of the y-coordinate of each point and plot the three new points on the same rectangular coordinate system. What conjecture can you make about the location of a point when the sign of the y-coordinate is changed?

Reviewing the Major Concepts

In Exercises 47–50, solve the inequality.

47. $-4 < 10x + 1 < 6$

48. $-2 \leq 1 - 2x \leq 2$

49. $-3 \leq -\dfrac{x}{2} \leq 3$

50. $-5 < x - 25 < 5$

51. *Price Inflation* A new van costs $32,500, which is about 112% of what it was 3 years ago. What was its price 3 years ago?

52. *Pension Fund* You withhold $3\frac{1}{2}$% of your gross monthly income for your pension. Your gross monthly income is $3100. How much is withheld for pension?

Additional Problem Solving

In Exercises 53–56, plot the points on a rectangular coordinate system.

53. $(-8, -2), (6, -2), (6, 5)$

54. $(0, 4), (0, 0), (3, 0)$

55. $\left(\frac{3}{2}, 1\right), (4, -3), \left(-\frac{4}{3}, \frac{7}{3}\right)$

56. $(-3, -5), \left(\frac{9}{4}, \frac{3}{4}\right), \left(\frac{5}{2}, -2\right)$

In Exercises 57 and 58, approximate the coordinates of the points.

57.

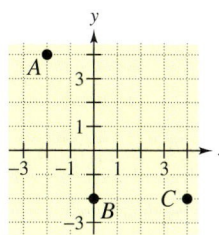

58.

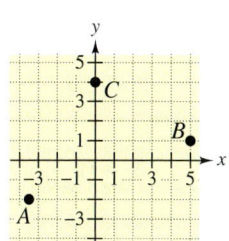

Geometry In Exercises 59–62, plot the points and connect them with line segments to form the figure.

59. *Square:* $(0, 6), (3, 3), (0, 0), (-3, 3)$

60. *Triangle:* $(1, 4), (0, -1), (5, 9)$

61. *Rhombus:* $(0, 0), (3, 2), (2, 3), (5, 5)$

62. *Parallelogram:* $(-1, 1), (0, 4), (4, -2), (5, 1)$

In Exercises 63 and 64, without plotting it, determine the quadrant in which each point is located.

63. (a) $\left(3, -\frac{5}{8}\right)$ (b) $(-6.2, 8.05)$

64. (a) $(200, 1365.6)$ (b) $\left(-\frac{5}{11}, -\frac{3}{8}\right)$

In Exercises 65–68, determine the quadrant or quadrants in which the point is located.

65. $(-3, y)$

66. $(x, 5)$

67. $(x, y),\quad xy > 0$

68. $(x, y),\quad x > 0,\ y < 0$

In Exercises 69–72, find the coordinates of the point.

69. The point is located three units to the right of the y-axis and four units below the x-axis.

70. The point is located two units to the left of the y-axis and five units above the x-axis.

71. The coordinates of the point are equal, and it is located in the third quadrant 10 units to the left of the y-axis.

72. The coordinates of the point are equal in magnitude and opposite in sign, and it is located seven units to the right of the y-axis.

Graphical Interpretation In Exercises 73 and 74, plot the points whose coordinates are given in the table. Then write a paragraph that summarizes the relationship between x and y.

73. *Fuel Efficiency* The table gives the speed x of a car in miles per hour and the approximate fuel efficiency y in miles per gallon.

x	50	55	60	65	70
y	28	26.4	24.8	23.4	22

74. *Average Temperature* The table gives the average temperature y (degrees Fahrenheit) for Duluth, Minnesota for each month of the year, with $x = 1$ representing January. (Source: PC USA)

x	1	2	3	4	5	6
y	6.3	12.0	22.9	38.3	50.3	59.4

x	7	8	9	10	11	12
y	65.3	63.2	54.0	44.2	28.2	13.8

In Exercises 75–78, determine whether the ordered pairs are solutions of the equation.

75. $y = \frac{7}{8}x$ 　　(a) $\left(\frac{8}{7}, 1\right)$ 　　(b) $\left(4, \frac{7}{2}\right)$
　　　　　　　　(c) $(0, 0)$ 　　(d) $(-16, 14)$

76. $y = \frac{5}{8}x - 2$ 　(a) $(0, 0)$ 　　(b) $(2, 2)$
　　　　　　　　(c) $(-4, -7)$ 　(d) $(32, 49)$

77. $4y - 2x = -1$ 　(a) $(0, 0)$ 　　(b) $\left(\frac{1}{2}, 0\right)$
　　　　　　　　(c) $\left(-3, -\frac{7}{4}\right)$ 　(d) $\left(1, -\frac{3}{4}\right)$

78. $y = 10x - 7$ 　(a) $(2, 10)$ 　　(b) $(-2, -27)$
　　　　　　　　(c) $(5, 43)$ 　　(d) $(1, 5)$

In Exercises 79 and 80, complete the table of values. Then plot the solution points on a rectangular coordinate system.

79.

x		-4	$\frac{2}{5}$	4	8	12
$y = -\frac{5}{2}x + 4$						

80.

x		-6	-3	0	$\frac{3}{4}$	10
$y = \frac{4}{3}x - \frac{1}{3}$						

81. *Numerical Interpretation* The cost C of producing x units is given by $C = 28x + 3000$. Use a table to help write a paragraph that describes the relationship between x and C.

82. *Numerical Interpretation* When an employee produces x units per hour, the hourly wage y is given by $y = 0.75x + 8$. Use a table to help write a paragraph that describes the relationship between x and y.

In Exercises 83–86, plot the points and find the distance between them. State whether the points lie on a horizontal or vertical line.

83. $(3, 2), (10, 2)$ 　　　**84.** $(-120, -2), (130, -2)$
85. $\left(-3, \frac{3}{2}\right), \left(-3, \frac{9}{4}\right)$ 　　**86.** $\left(\frac{1}{3}, -4\right), \left(\frac{1}{3}, \frac{5}{2}\right)$

In Exercises 87 and 88, find the coordinates of (x, y), the lengths of the vertical and horizontal sides of the right triangle, and the length of the hypotenuse.

87.

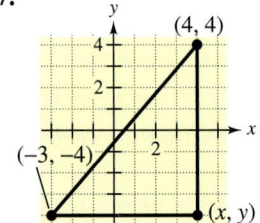

88.

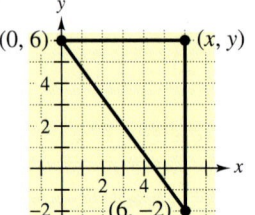

In Exercises 89–92, find the distance between the points.

89. $(-2, -3), (4, 2)$ 　　**90.** $(-5, 4), (10, -3)$
91. $(1, 3), (3, -2)$ 　　　**92.** $\left(\frac{1}{2}, 1\right), \left(\frac{3}{2}, 2\right)$

In Exercises 93–96, use the Distance Formula to determine whether the three points lie on a line.

93. $(2, 3), (2, 6), (6, 3)$
94. $(2, 4), (-1, 6), (-3, 1)$
95. $(8, 3), (5, 2), (2, 1)$
96. $(2, 4), (1, 1), (0, -2)$

Geometry In Exercises 97 and 98, find the perimeter of the triangle having the given vertices.

97. $(-2, 0), (0, 5), (1, 0)$
98. $(-5, -2), (-1, 4), (3, -1)$

In Exercises 99–102, plot the points and the midpoint of the line segment joining the points. The coordinates of the midpoint of the line segment joining the points (x_1, y_1) and (x_2, y_2) are

$$\text{Midpoint} = \left(\frac{x_1 + x_2}{2}, \frac{y_1 + y_2}{2} \right).$$

99. $(-2, 0), (4, 8)$ **100.** $(-3, -2), (7, 2)$

101. $(1, 6), (6, 3)$ **102.** $(2, 7), (9, -1)$

Shifting a Graph In Exercises 103 and 104, the figure is shifted to a new location in the plane. Find the coordinates of the vertices of the figure in its new location.

103.

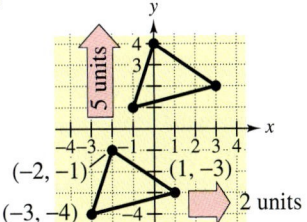

104.

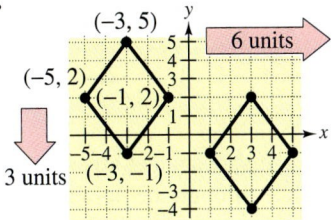

105. *Housing Construction* A house is 30 feet wide and the ridge of the roof is 7 feet above the tops of the walls (see figure). The rafters overhang the edges of the walls by 2 feet. How long are the rafters?

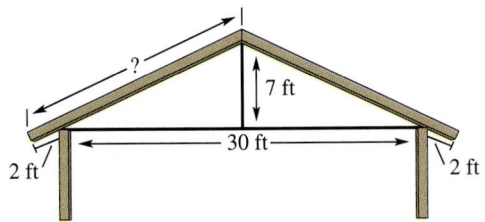

106. *Housing Construction* Determine the length of the handrail over the stairs shown in the figure.

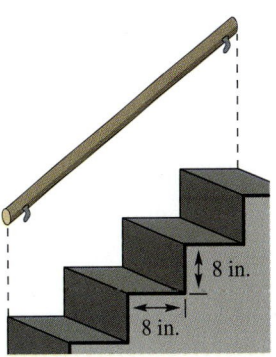

CAREER INTERVIEW

Dean R. Brookie

Architect/Planner

Brookie Architecture and Planning, Inc.

Durango, CO 81301

Mathematics is integral to the work of an architect. I use algebra and geometry in both the analytical and creative components of my work, planning and designing commercial buildings, multifamily housing, and custom-built homes. I routinely develop project budgets within certain parameters, calculate forces on various parts of buildings, and compute areas and volumes of various geometric figures. I have also found that raw geometric shapes are the building blocks for the creative process. Geometry in the form of rhythms and proportions can be used to give a building a harmonious, comforting feel or to create interesting visual tension.

7.2 | Graphs of Equations

The Graph of an Equation ▪ Intercepts: An Aid to Sketching Graphs ▪ Real-Life Application of Graphs

The Graph of an Equation

In Section 7.1, you saw that the solutions of an equation in x and y can be represented by points on a rectangular coordinate system. The set of *all* solutions of an equation is called its **graph.** In this section, you will study a basic technique for sketching the graph of an equation—the **point-plotting method.**

EXAMPLE 1   *Sketching the Graph of an Equation*

Sketch the graph of $3x - y = 2$.

Solution

To begin, solve the equation for y to obtain $y = 3x - 2$. Next, create a table of values. The choice of x-values to use in the table is somewhat arbitrary. However, the more x-values you choose, the easier it will be to recognize a pattern.

NOTE The equation in Example 1 is an example of a **linear equation** in two variables—it is of first degree in both variables and its graph is a straight line. You will study this type of equation in Section 7.4.

x	-2	-1	0	1	2	3
$y = 3x - 2$	-8	-5	-2	1	4	7
Solution	$(-2, -8)$	$(-1, -5)$	$(0, -2)$	$(1, 1)$	$(2, 4)$	$(3, 7)$

Now, plot the points, as shown in Figure 7.12(a). It appears that all six points lie on a line, so complete the sketch by drawing a line through the points, as shown in Figure 7.12(b).

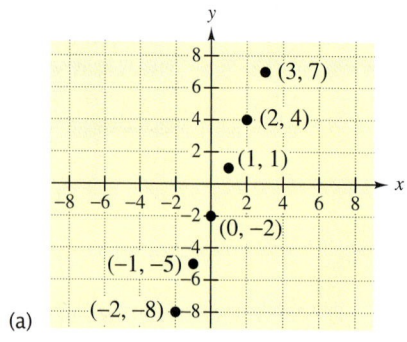

(a)

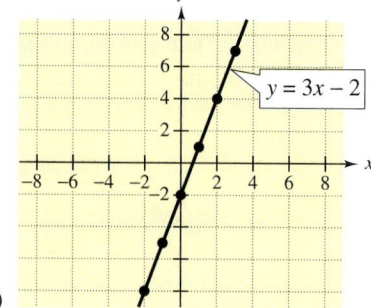

(b)

FIGURE 7.12

NOTE   By drawing a line (curve) through the plotted points, we are implying that every point on this line (curve) is a solution point of the given equation and conversely.

> ### The Point-Plotting Method of Sketching a Graph
>
> 1. If possible, rewrite the equation by isolating one of the variables.
>
> 2. Make up a table of values showing several solution points.
>
> 3. Plot these points on a rectangular coordinate system.
>
> 4. Connect the points with a smooth curve or line.

EXAMPLE 2   *Sketching the Graph of a Nonlinear Equation*

Sketch the graph of $-x^2 + 2x + y = 0$.

Solution

Begin by solving the equation for y to obtain $y = x^2 - 2x$. Next, create a table of values.

x	-2	-1	0	1	2	3	4
$y = x^2 - 2x$	8	3	0	-1	0	3	8
Solution	$(-2, 8)$	$(-1, 3)$	$(0, 0)$	$(1, -1)$	$(2, 0)$	$(3, 3)$	$(4, 8)$

Now, plot the seven solution points, as shown in Figure 7.13(a). Finally, connect the points with a smooth curve, as shown in Figure 7.13(b).

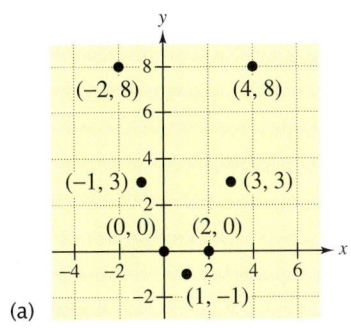

(a)

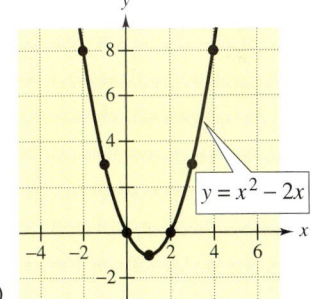

(b)

FIGURE 7.13

NOTE   The graph of the equation given in Example 2 is called a **parabola.** You will study this type of graph in detail in Section 11.3.

Example 3 looks at the graph of an equation that involves an absolute value. Remember that to find the absolute value of a number you disregard the sign of the number. For instance, $|-5| = 5$, $|2| = 2$, and $|0| = 0$.

EXAMPLE 3   The Graph of an Absolute Value Equation

Sketch the graph of $y = |x - 2|$.

Solution

This equation is already written in a form with y isolated on the left. So begin by creating a table of values. Be sure that you understand how the absolute value is evaluated. For instance, when $x = -2$, the value of y is

$$y = |-2 - 2| = |-4| = 4$$

and when $x = 3$, the value of y is

$$y = |3 - 2| = |1| = 1.$$

x	-2	-1	0	1	2	3	4	5
$y = |x - 2|$	4	3	2	1	0	1	2	3
Solution	$(-2, 4)$	$(-1, 3)$	$(0, 2)$	$(1, 1)$	$(2, 0)$	$(3, 1)$	$(4, 2)$	$(5, 3)$

Next, plot the points, as shown in Figure 7.14(a). It appears that the points lie in a "V-shaped" pattern, with the point $(2, 0)$ lying at the bottom of the "V." Following this pattern, connect the points to form the graph shown in Figure 7.14(b).

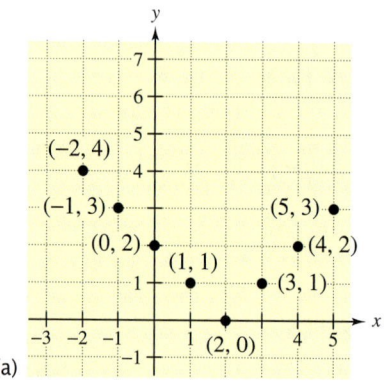
(a)

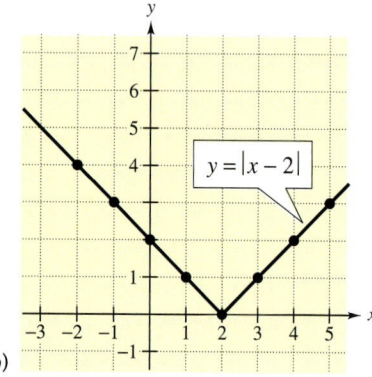
(b)

FIGURE 7.14

Intercepts: An Aid to Sketching Graphs

Two types of solution points that are especially useful are those having zero as the x-coordinate and those having zero as the y-coordinate. Such points are called **intercepts** because they are the points at which the graph intersects, respectively, the y- and x-axes.

Definition of Intercepts

The point $(a, 0)$ is called an ***x*-intercept** of the graph of an equation if it is a solution point of the equation. To find the x-intercepts, let $y = 0$ and solve the equation for x.

The point $(0, b)$ is called a ***y*-intercept** of the graph of an equation if it is a solution point of the equation. To find the y-intercepts, let $x = 0$ and solve the equation for y.

EXAMPLE 4   *Finding the Intercepts of a Graph*

Find the intercepts and sketch the graph of $y = 2x - 3$.

Solution

Find the x-intercept by letting $y = 0$ and solving for x.

$$y = 2x - 3$$
$$0 = 2x - 3$$
$$\tfrac{3}{2} = x$$

Find the y-intercept by letting $x = 0$ and solving for y.

$$y = 2x - 3$$
$$y = 2(0) - 3$$
$$y = -3$$

Therefore, the graph has one x-intercept, which occurs at the point $\left(\tfrac{3}{2}, 0\right)$, and one y-intercept, which occurs at the point $(0, -3)$. To sketch the graph of the equation, create a table of values. (Include the intercepts in the table.) Finally, using the solution points given in the table, sketch the graph of the equation, as shown in Figure 7.15.

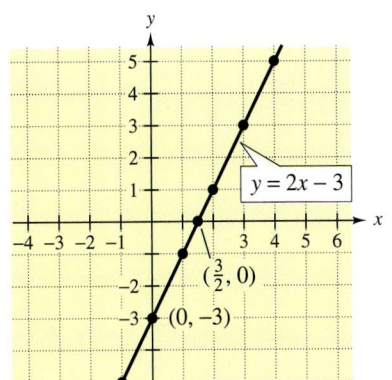

FIGURE 7.15

x	-1	0	1	$\tfrac{3}{2}$	2	3	4
$y = 2x - 3$	-5	-3	-1	0	1	3	5
Solution	$(-1, -5)$	$(0, -3)$	$(1, -1)$	$\left(\tfrac{3}{2}, 0\right)$	$(2, 1)$	$(3, 3)$	$(4, 5)$

It is possible for a graph to have no intercepts or several intercepts. For instance, consider the three graphs in Figure 7.16.

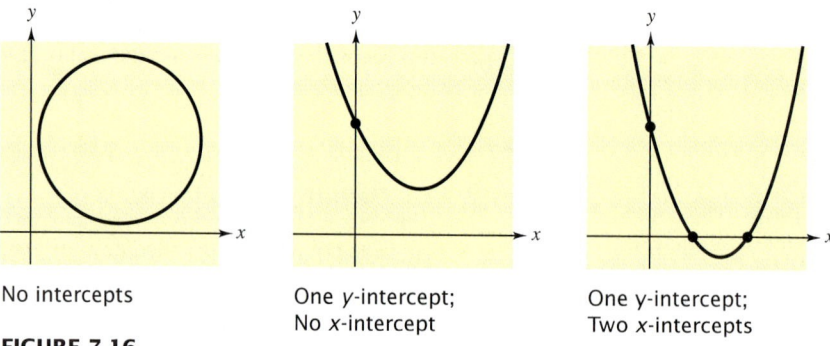

No intercepts

One *y*-intercept;
No *x*-intercept

One y-intercept;
Two *x*-intercepts

FIGURE 7.16

EXAMPLE 5   A Graph That Has Two x-Intercepts

Find the intercepts and sketch the graph of $y = x^2 - 5x + 4$.

Solution

Begin by making a table of values, as shown below. From the table, you can see that the graph has two *x*-intercepts, which occur at $(1, 0)$ and $(4, 0)$, and one *y*-intercept, which occurs at $(0, 4)$.

x	-1	0	1	2	3	4	5
$y = x^2 - 5x + 4$	10	4	0	-2	-2	0	4
Solution	$(-1, 10)$	$(0, 4)$	$(1, 0)$	$(2, -2)$	$(3, -2)$	$(4, 0)$	$(5, 4)$

Next, plot the points given in the table, as shown in Figure 7.17(a). Then connect them with a smooth curve, as shown in Figure 7.17(b).

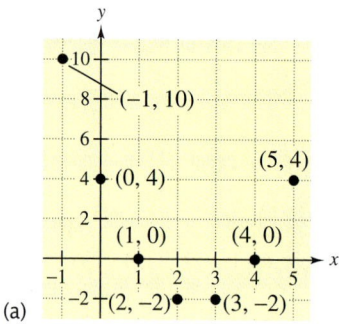

(a)

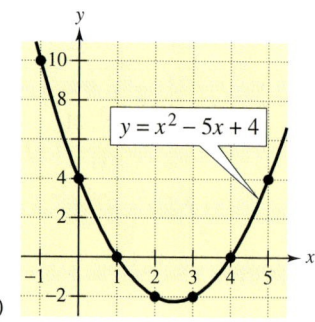

(b)

FIGURE 7.17

Real-Life Application of Graphs

Newspapers and news magazines frequently use graphs to show real-life relationships between variables. Example 6 shows how such a graph can help you visualize the concept of **straight-line depreciation.**

EXAMPLE 6 *Straight-Line Depreciation*

Your small business buys a new printing press for $65,000. For income tax purposes, you decide to depreciate the printing press over a 10-year period. At the end of the 10 years, the salvage value of the printing press is expected to be $5000. Find an equation that relates the depreciated value of the printing press to the number of years. Then sketch the graph of the equation.

Solution

The total depreciation over the 10-year period is $65,000 - 5000 = \$60,000$. Because the same amount is depreciated each year, it follows that the annual depreciation is $60,000/10 = \$6000$. Thus, after 1 year, the value of the printing press is

$$\text{Value after 1 year} = 65,000 - (1)6000 = \$59,000.$$

By similar reasoning, you can see that the value after 2 years is

$$\text{Value after 2 years} = 65,000 - (2)6000 = \$53,000.$$

Let y represent the value of the printing press after t years and follow the pattern determined for the first 2 years to obtain

$$y = 65,000 - 6000t.$$

A sketch of the graph of this equation is shown in Figure 7.18.

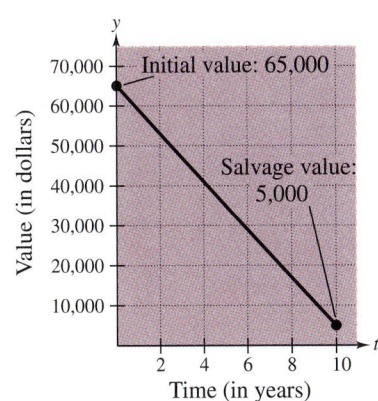

FIGURE 7.18 Straight Line Depreciation

Group Activities   Extending the Concept

Straight-Line Depreciation   In Example 6, suppose that you depreciated the printing press over 8 years instead of 10 years. Write an equation that represents the depreciated value of the printing press during the 8-year period. Then graph both depreciation models on the same rectangular coordinate system and compare the results. What are the advantages and disadvantages of each model?

7.2 Exercises

Discussing the Concepts

1. Define the graph of an equation.

2. How many solution points make up the graph of $y = 2x - 1$? Explain.

3. Explain how to find the intercepts of a graph. Give examples.

4. An equation gives the relationship between profit y and time t. Profit is decreasing at a lower rate than it has in the past. Is it possible to sketch the graph of such an equation? If so, sketch a representative graph.

Problem Solving

In Exercises 5–10, match the equation with its graph. [The graphs are labeled (a), (b), (c), (d), (e), and (f).]

(a)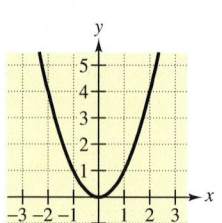

(b)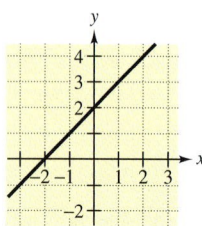

(c)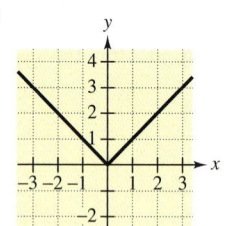

(d)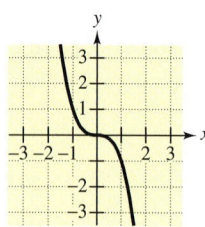

(e)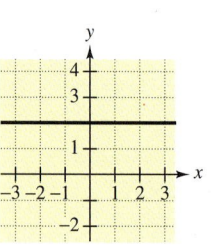

(f)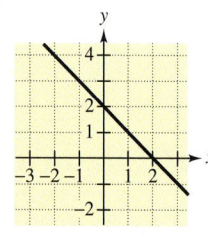

5. $y = 2$

6. $y = 2 + x$

7. $y = 2 - x$

8. $y = x^2$

9. $y = -x^3$

10. $y = |x|$

In Exercises 11–14, complete the table and use the results to sketch the graph of the equation.

11. $2x + y = 3$

x	-4			2	4
y		7	3		

12. $2x - 3y = 6$

x	-3	0		3	
y			$-\frac{2}{3}$		5

13. $y = 4 - x^2$

x		-1		2	
y	0		4		-5

14. $y = \frac{1}{2}x^3 - 4$

x	-1		1		3
y		-4		0	

In Exercises 15 and 16, use a graphing utility to create a table of values. Then sketch the graph.

15. $y = 4 - |x|$

16. $y = 4 - x^2$

In Exercises 17–24, sketch the graph of the equation.

17. $y = 3x$

18. $y = \frac{1}{3}x$

19. $y = 2x - 3$

20. $y = -x + 2$

21. $y = x^2 - 1$

22. $y = -x^2$

23. $y = |x| - 1$

24. $y = |x - 1|$

In Exercises 25–28, find the x- and y-intercepts (if any) of the graph of the equation.

25. $x + 2y = 10$

26. $3x - 2y + 12 = 0$

27. $y = (x + 5)(x - 5)$

28. $y = (x + 1)^2$

In Exercises 29–38, sketch the graph of the equation and show the coordinates of three solution points.

29. $y = 3 - x$

30. $y = x - 3$

31. $y = 4$

32. $x = -6$

33. $4x + y = 3$

34. $y - 2x = -4$

35. $y = x^2 - 4$

36. $y = 1 - x^2$

37. $y = |x + 2|$

38. $y = |x| + 2$

39. *Using a Graph*   The force F (in pounds) to stretch a spring x inches from its natural length is given by

$$F = \frac{4}{3}x, \quad 0 \le x \le 12.$$

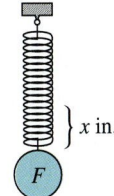

Natural length $\}\, x$ in.

F

(a) Use the model to complete the following table.

x	0	3	6	9	12
F					

(b) Sketch the graph of the model.

(c) Use the graph to determine how the length of the spring changes each time the force is doubled. Explain your reasoning.

40. *Comparing Data with a Model*   The number of farms in the United States with milk cows has been decreasing. The number of farms N (in thousands) for 1984 through 1991 is given in the table.

t	4	5	6	7	8	9	10	11
N	282	269	249	228	217	204	194	182

A model for this data is

$$N = -14.5t + 337.1$$

where t is time in years, with $t = 0$ representing 1980.   (Source: U.S. Department of Agriculture)

(a) Sketch the graph of the model and plot the data in the table on the same graph.

(b) How well does the model represent the data? Explain your reasoning.

(c) Use the model to predict the number of farms with milk cows in 1994.

(d) Explain why this model may not be accurate in the future.

41. *Misleading Graphs*   Graphs can help you visualize relationships between two variables, but they can also be misused to imply results that are not correct. The two graphs below represent the *same* data points. Which graph is misleading, and why?

(a)

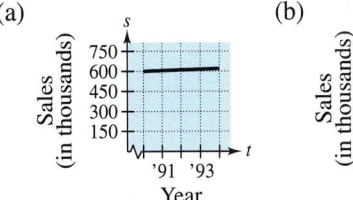

(b)

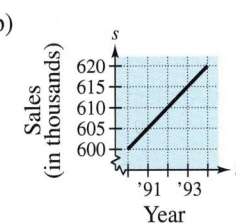

42. *Exploration*   Sketch the graphs of $y = x^2 + 1$ and $y = -(x^2 + 1)$ on the same set of coordinate axes. Explain how the graph of an equation changes when the expression for y is multiplied by -1. Justify your answer by giving additional examples.

Reviewing the Major Concepts

In Exercises 43–46, name the property of real numbers that justifies the statement.

43. $8x \cdot \dfrac{1}{8x} = 1$ **44.** $3x + 0 = 3x$

45. $-4(x + 10) = -4 \cdot x + (-4)(10)$

46. $5 + (-3 + x) = (5 - 3) + x$

47. *Weight of Sand*   The ratio of cement to sand in a 90-pound bag of dry mix is 1 to 4. Find the number of pounds of sand in the bag.

48. *Defective Units*   A quality control engineer for a manufacturer found one defective unit in a sample of 125. At that rate, what is the expected number of defective units in a shipment of 150,000?

Additional Problem Solving

In Exercises 49–54, sketch the graph of the equation.

49. $y = 4 - x$ **50.** $y = \frac{1}{2}x - 2$

51. $y = x^2 - 4$ **52.** $y = x^3$

53. $y = |x|$ **54.** $y = |x + 3|$

In Exercises 55–72, sketch the graph of the equation and show the coordinates of three solution points.

55. $y = 2x - 3$ **56.** $y = -4x + 8$

57. $2x - 3y = 6$ **58.** $3x + 4y = 12$

59. $x + 5y = 10$ **60.** $5x - y = 10$

61. $x - 1 = 0$ **62.** $y + 3 = 0$

63. $y = x^2 - 9$ **64.** $y = 9 - x^2$

65. $y = x(x - 2)$ **66.** $y = -x(x + 4)$

67. $y = x^3 - 1$ **68.** $y = 1 - x^4$

69. $y = |x| - 3$ **70.** $y = |x - 3|$

71. $y = -|x|$ **72.** $y = |x| + |x - 2|$

Graphical and Algebraic Solutions   In Exercises 73–76, graphically estimate the intercepts of the graph. Then check your results algebraically.

73. $y = x^2 + 3$ **74.** $x = 3$

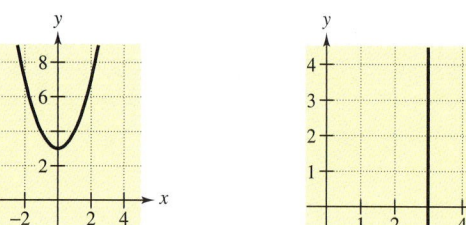

75. $y = |x - 2|$ **76.** $y = (x - 3)(x - 4)$

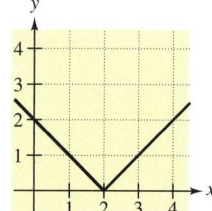   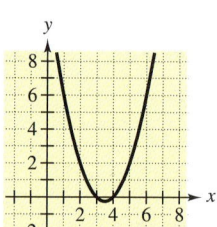

77. *Straight-Line Depreciation*   A manufacturing plant purchases a new molding machine for $225,000. The depreciated value y after t years is given by

$$y = 225{,}000 - 20{,}000t, \quad 0 \le t \le 8.$$

Sketch the graph of this model.

78. *Geometry*   A rectangle of length l and width w has a perimeter of 12 meters.

(a) Show that the width of the rectangle is $w = 6 - l$ and its area is $A = l(6 - l)$.

(b) Sketch the graph of the equation for the area.

(c) From the graph in part (b), which dimensions produce a rectangle of maximum area?

Exploration   In Exercises 79 and 80, graph the equations on the same set of coordinate axes. What conclusions can you make by comparing the graphs?

79. (a) $y = x^2$ **80.** (a) $y = x^2$

    (b) $y = (x - 2)^2$ (b) $y = x^2 - 2$

    (c) $y = (x + 2)^2$ (c) $y = x^2 - 4$

    (d) $y = (x + 4)^2$ (d) $y = x^2 + 4$

7.3   **Graphs and Graphing Utilities**

Introduction ▪ Using a Graphing Utility ▪
Using a Graphing Utility's Special Features

Introduction

In Section 7.2, you studied the point-plotting method of sketching the graph of an equation. One of the disadvantages of the point-plotting method is that to get a good idea about the shape of a graph you need to plot *many* points. By plotting only a few points, you can badly misrepresent the graph.

For instance, consider the equation $y = \frac{1}{30}x(x^4 - 10x^2 + 39)$. To graph this equation, suppose you calculated only the five points in the table below.

x	-3	-1	0	1	3
$y = \frac{1}{30}x(x^4 - 10x^2 + 39)$	-3	-1	0	1	3

By plotting the five points, as shown in Figure 7.19(a), you might assume that the graph of the equation is a straight line. This, however, is not correct. By plotting several more points, as shown in Figure 7.19(b), you can see that the actual graph is not straight at all.

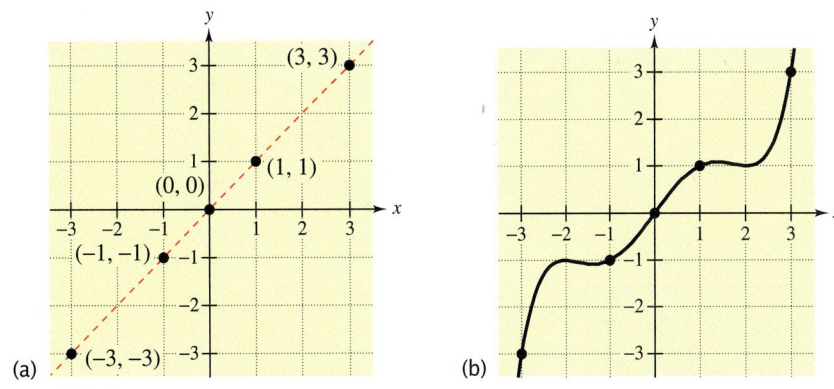

(a)   (b)

FIGURE 7.19

Thus, the point-plotting method leaves you with a dilemma. The method can be very inaccurate if only a few points are plotted, but it is very time consuming to plot a dozen (or more) points. Technology can help you solve this dilemma. Plotting several points (or even hundreds of points) on a rectangular coordinate system is something that a graphing utility can do easily.

Using a Graphing Utility

There are many different graphing utilities: some are graphing packages for computers and some are hand-held graphing calculators. In this section we describe the steps used to graph an equation with a graphing utility. We will often give keystroke sequences for illustration; however, these may not agree precisely with the steps required by *your* graphing utility.*

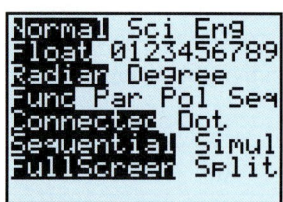

FIGURE 7.20

Graphing an Equation with a *TI-82* Graphing Utility

Before performing the following steps, set your utility so that all the standard defaults are active. For instance, all of the options at the left of the MODE screen should be highlighted (see Figure 7.20).

1. Set the viewing window for the graph. (See Example 3.) To set the standard viewing window, press ZOOM 6.

2. Rewrite the equation so that y is isolated on the left side of the equation.

3. Press the Y= key. Then enter the right side of the equation on the first line of the display. (The first line is labeled $Y_1=$.)

4. Press the GRAPH key.

EXAMPLE 1 Graphing a Linear Equation

Sketch the graph of $3y + x = 5$.

Solution

To begin, solve the given equation for y in terms of x.

$$3y + x = 5 \qquad \text{Original equation}$$
$$3y = -x + 5 \qquad \text{Subtract } x \text{ from both sides.}$$
$$y = -\frac{1}{3}x + \frac{5}{3} \qquad \text{Divide both sides by 3.}$$

Press the Y= key, and enter the following keystrokes.

(-) X,T,θ ÷ 3 + 5 ÷ 3

The top row of the display should now be as follows.

$$Y_1 = -X/3 + 5/3$$

Press the GRAPH key, and the screen should look like that shown in Figure 7.21.

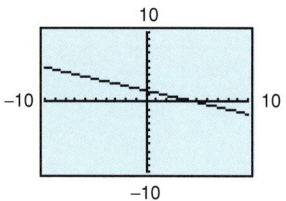

FIGURE 7.21

---

*The keystrokes given in this section correspond to the *TI-82* Graphics Calculator by Texas Instruments. For other graphing utilities, the keystrokes may differ.

In Figure 7.21, notice that the graphing utility screen does not label the tick marks on the x-axis or the y-axis. To see what the tick marks represent, you can press WINDOW. If you set your graphing utility to the standard graphing defaults before working Example 1, the screen should show the following values.

Xmin=-10	The minimum x-value is -10.
Xmax=10	The maximum x-value is 10.
Xscl=1	The x-scale is one unit per tick mark.
Ymin=-10	The minimum y-value is -10.
Ymax=10	The maximum y-value is 10.
Yscl=1	The y-scale is one unit per tick mark.

These settings are summarized visually in Figure 7.22.

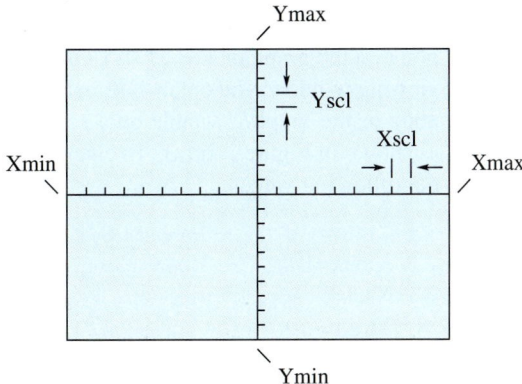

FIGURE 7.22

EXAMPLE 2   Graphing an Equation Involving Absolute Value

Sketch the graph of $y = |x - 4|$.

Solution

This equation is already written so that y is isolated on the left side of the equation. Press the Y= key, and enter the following keystrokes.

ABS (X,T,θ − 4)

The top row of the display should now be as follows.

$Y_1 = \text{abs} (X - 4)$

Press the GRAPH key, and the screen should look like that shown in Figure 7.23.

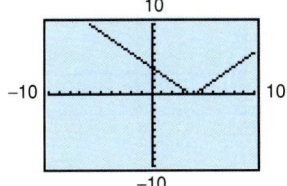

FIGURE 7.23

Using a Graphing Utility's Special Features

To use your graphing utility to its best advantage, you must learn how to set the viewing rectangle, as illustrated in the next example.

EXAMPLE 3   Setting the Viewing Window

Sketch the graph of $y = x^2 + 14$.

Solution

Press [Y=] and enter the following keystrokes.

[X,T,θ] [x^2] [+] 14

Press the [GRAPH] key. If your graphing utility is set to the standard viewing window, nothing will appear on the screen. The reason for this is that the lowest point on the graph of $y = x^2 + 14$ occurs at the point $(0, 14)$. Using the standard viewing window, you obtain a screen whose largest y-value is 10. In other words, none of the graph is visible on a screen whose y-values range from -10 to 10, as shown in Figure 7.24(a). To change these settings, press [WINDOW] and enter the following values.

Xmin=-10	The minimum x-value is -10.
Xmax=10	The maximum x-value is 10.
Xscl=1	The x-scale is one unit per tick mark.
Ymin=-10	The minimum y-value is -10.
Ymax=30	The maximum y-value is 30.
Yscl=5	The y-scale is five units per tick mark.

Press [GRAPH], and you will obtain the graph shown in Figure 7.24(b). On this graph, note that each tick mark on the y-axis represents five units because you changed the y-scale to 5. Also note that the highest point on the y-axis is now 30 because you changed the maximum value of y to 30.

NOTE   If you changed the maximum y-value and the y-scale on your utility as indicated in Example 3, you should return to the standard settings before working Example 4. To do this, press [ZOOM] [6].

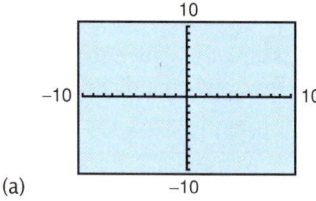

(a)

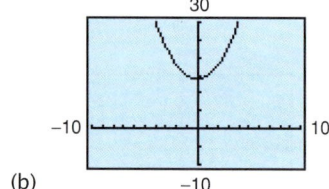

(b)

FIGURE 7.24

EXAMPLE 4 Using a Square Setting

Sketch the graph of $y = x$. The graph of this equation is a straight line that makes a 45° angle with the x-axis and y-axis. From the graph on your graphing utility, does the angle appear to be 45°?

Solution

Press $\boxed{Y=}$ and enter $y = x$ on the first line.

$$Y_1 = X$$

Press the $\boxed{\text{GRAPH}}$ key, and you will obtain the graph shown in Figure 7.25(a). Note that the angle the line makes with the x-axis does not appear to be 45°. The reason for this is that the screen is wider than it is tall. This makes the tick marks on the x-axis appear farther apart than the tick marks on the y-axis. To obtain the same distance between tick marks on both axes, change the graphing settings from "standard" to "square." To do this, press the following keys.

$\boxed{\text{ZOOM}}$ $\boxed{5}$ Square setting

The screen should look like that shown in Figure 7.25(b). Note in this figure that the square setting has changed the viewing window so that the x-values vary between -15 and 15.

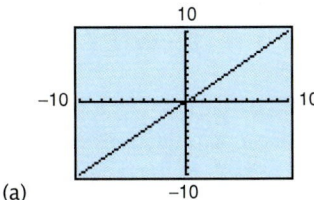 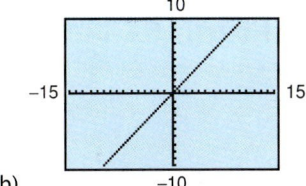

(a)   (b)

FIGURE 7.25

NOTE There are many possible square settings on a graphing utility. To create a square setting, you need the following ratio to be $\frac{2}{3}$.

$$\frac{\text{Ymax} - \text{Ymin}}{\text{Xmax} - \text{Xmin}}$$

For instance, the setting in Figure 7.25(b) is square because $(\text{Ymax} - \text{Ymin}) = 20$ and $(\text{Xmax} - \text{Xmin}) = 30$.

EXAMPLE 5 Sketching More than One Graph on the Same Screen

Sketch the graphs of the following equations on the same screen.

$$y = -x + 5, \quad y = -x, \quad \text{and} \quad y = -x - 5$$

Solution

To begin, press $\boxed{\text{Y=}}$ and enter all three equations on the first three lines. The display should now be as follows.

$Y_1 = \text{-X} + 5$ $\qquad$ $\boxed{(\text{-})}$ $\boxed{\text{X,T,}\theta}$ $\boxed{+}$ 5

$Y_2 = \text{-X}$ $\qquad\qquad$ $\boxed{(\text{-})}$ $\boxed{\text{X,T,}\theta}$

$Y_3 = \text{-X} - 5$ $\qquad$ $\boxed{(\text{-})}$ $\boxed{\text{X,T,}\theta}$ $\boxed{-}$ 5

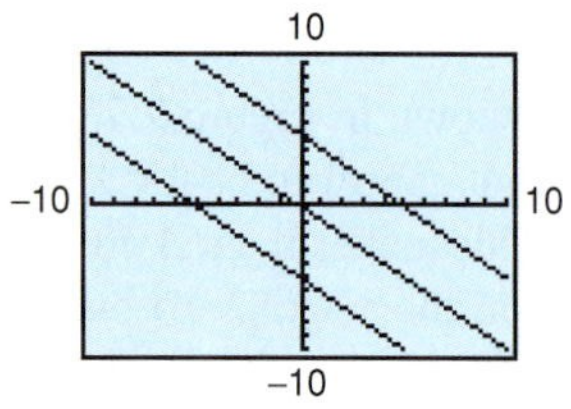

Press the $\boxed{\text{GRAPH}}$ key and you will obtain the graph shown in Figure 7.26. Note that the graph of each equation is a straight line, and that the lines are parallel to each other.

FIGURE 7.26

Group Activities Exploring with Technology

Graphing Equations For each of the following equations, discuss what must first be done to the equation before you can graph it with a graphing utility. If it is possible, graph each equation with a graphing utility and approximate the x-intercept. If it is not possible to graph the equation by entering it into a graphing utility, explain why.

a. $3x + 5y = 15$ $\qquad$ **b.** $2x - 7y = 21$

c. $x = -4$ $\qquad\qquad$ **d.** $-x - 3y - 9 = 0$

Cesar Preza

7.3 Exercises

Discussing the Concepts

1. Suppose you correctly enter an expression for the variable y on a graphing utility. However, no graph appears in the viewing window when you graph the equation. Give a possible explanation and what steps you could take to remedy the problem. Illustrate your explanation with an example.

2. In your own words, explain the change on the display if Ymax is increased from 10 to 20.

3. In your own words, explain the change on the display if Xscl is increased from 1 to 2.

4. You graph $y = x$ and $y = x + 1$ on the same screen with $-100 \le y \le 100$. Why is this inappropriate?

Problem Solving

In Exercises 5–10, use a graphing utility to match the equation with its graph. [The graphs are labeled (a), (b), (c), (d), (e), and (f).]

(a)
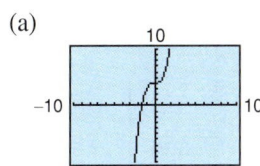

(b)
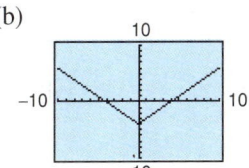

(c)
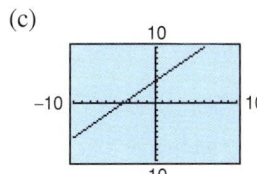

(d)
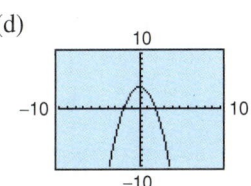

(e)
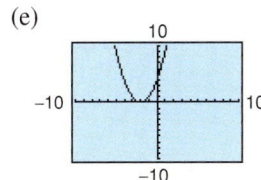

(f)
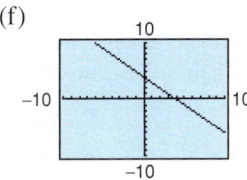

5. $y = 4 + x$

6. $y = 4 - x$

7. $y = 4 - x^2$

8. $y = x^2 + 4x + 4$

9. $y = x^3 + 4$

10. $y = |x| - 4$

In Exercises 11–18, use a graphing utility to graph the equation. Use a standard setting.

11. $y = 2x - 6$

12. $y = x - 2$

13. $y = x^2 - 3$

14. $y = 6 - x^2$

15. $y = \sqrt{x + 4}$

16. $y = 2\sqrt{2x + 3}$

17. $y = |x| - 6$

18. $y = |x - 6|$

In Exercises 19–22, use a graphing utility to graph the equation. Begin by using a standard setting. Then graph the equation a second time using the specified setting. Which setting is better? Explain.

19. $y = \frac{1}{2}x - 10$

| Xmin = -1 |
| Xmax = 20 |
| Xscl = 1 |
| Ymin = -15 |
| Ymax = 5 |
| Yscl = 1 |

20. $y = -2x + 25$

| Xmin = -1 |
| Xmax = 20 |
| Xscl = 1 |
| Ymin = -2 |
| Ymax = 28 |
| Yscl = 1 |

21. $y = 2x^3 - 5x^2$

| Xmin = -5 |
| Xmax = 5 |
| Xscl = 1 |
| Ymin = -5 |
| Ymax = 5 |
| Yscl = 1 |

22. $y = 6\sqrt{x + 4}$

| Xmin = -5 |
| Xmax = 5 |
| Xscl = 1 |
| Ymin = -1 |
| Ymax = 20 |
| Yscl = 2 |

In Exercises 23–26, find a setting on a graphing utility such that the graph of the equation agrees with the given graph.

23. $y = -3x + 15$ **24.** $y = 2x^2 - 16x + 26$

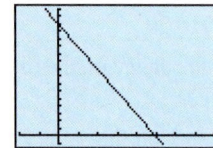   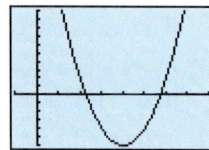

25. $y = x^3 - 3x + 2$ **26.** $y = (x^2 - 4)^2$

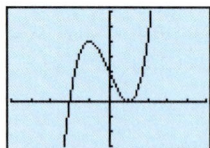   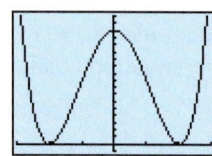

In Exercises 27–30, use the ZOOM feature of a graphing utility to estimate the lowest point on the graph.

27. $y = \frac{1}{4}(x^2 - 1)$ **28.** $y = x^2 + 3x - 2$

29. $y = x^3(3x + 4)$ **30.** $y = x\sqrt{x + 1}$

31. *Think About It*   In Exercises 27–30, you graphically estimated the lowest points on graphs. Explain how you could confirm your estimates algebraically with tables.

In Exercises 32–35, use a graphing utility to approximate the x-intercepts of the graph of the equation. (*Hint:* The ZOOM and TRACE features can help you get a better view of the intercepts.)

32. $y = 14 + 3x - 2x^2$ **33.** $y = 0.4x^2 - 0.2x - 3$

34. $y = x^3 - 3x^2 + 2$ **35.** $y = x^3 + x - 1$

36. *Think About It*   In Exercises 32–35, you graphically estimated the intercepts of graphs. Explain how you could confirm your estimates algebraically.

In Exercises 37–40, explain how to use a graphing utility to verify that $y_1 = y_2$. Then identify the rule of algebra that is illustrated.

37. $y_1 = \frac{1}{2}(x - 4)$ **38.** $y_1 = 3x - x^2$
$\quad\;\; y_2 = \frac{1}{2}x - 2$ $\qquad y_2 = -x^2 + 3x$

39. $y_1 = x + (2x - 1)$ **40.** $y_1 = 2x + 0$
$\quad\;\; y_2 = (x + 2x) - 1$ $\qquad y_2 = 2x$

Stopping Distance   In Exercises 41 and 42, consider the stopping distance y of an automobile (in feet), which is modeled by

$$y = 30.00 + 0.08x^2, \quad 25 \le x \le 75$$

where x is speed in miles per hour.

41. Find a graphing utility viewing rectangle that gives a good view of the graph of this model.

42. Use the graph to approximate the stopping distance when the car's speed is (a) $x = 35$ miles per hour and (b) $x = 55$ miles per hour.

Reviewing the Major Concepts

In Exercises 43–46, solve for y in terms of x.

43. $3x + y = 4$ **44.** $2x + 3y = 2$

45. $x^2 + 3y = 4$ **46.** $x^2 + y - 4 = 0$

In Exercises 47 and 48, use the point-plotting method to sketch the graph of the equation.

47. $2x - 7y + 14 = 0$ **48.** $y = 3 - |x - 2|$

In Exercises 49 and 50, write an algebraic expression that represents the quantity in the verbal statement, and simplify it if possible.

49. The total hourly wage for an employee when the base pay is $9.35 per hour plus 75 cents for each of q units produced per hour

50. The sum of two consecutive odd integers, the first of which is $2n - 1$

Additional Problem Solving

In Exercises 51–64, use a graphing utility to graph the equation. Use a standard setting.

51. $y = \frac{2}{3}x + 1$

52. $y = \frac{1}{4}x - 2$

53. $y = -\frac{3}{4}x + 4$

54. $y = -2x + 5$

55. $y = -x^2 + 4x - 1$

56. $y = \frac{1}{4}x^2 - x$

57. $y = x^3 - 2$

58. $y = 1 - x^3$

59. $y = 2\left(1 - \sqrt{x}\right)$

60. $y = x\sqrt{x + 3}$

61. $y = |x| - 4$

62. $y = 4 - |x|$

63. $y = |x - 4|$

64. $y = \frac{1}{2}|x|$

In Exercises 65–70, use a graphing utility to graph the equation. Begin by using a standard setting. Then graph the equation a second time using the specified setting. Which setting is better? Explain.

65. $y = \frac{1}{6}(2x + 3)$

| Xmin = 0 |
| Xmax = 20 |
| Xscl = 1 |
| Ymin = 0 |
| Ymax = 10 |
| Yscl = 1 |

66. $y = |x| + |x - 5|$

| Xmin = -3 |
| Xmax = 8 |
| Xscl = 1 |
| Ymin = 0 |
| Ymax = 10 |
| Yscl = 1 |

67. $y = 20 - \frac{1}{4}x^2$

| Xmin = -10 |
| Xmax = 10 |
| Xscl = 1 |
| Ymin = -2 |
| Ymax = 22 |
| Yscl = 1 |

68. $y = x^2 - 12x + 36$

| Xmin = -5 |
| Xmax = 15 |
| Xscl = 1 |
| Ymin = -2 |
| Ymax = 20 |
| Yscl = 1 |

69. $y = 0.6x^3 - 2x - 1$

| Xmin = -2.5 |
| Xmax = 2.5 |
| Xscl = 1 |
| Ymin = -3.5 |
| Ymax = 1.5 |
| Yscl = 1 |

70. $y = 0.3x\sqrt{64 - x^2}$

| Xmin = 0 |
| Xmax = 20 |
| Xscl = 1 |
| Ymin = 0 |
| Ymax = 20 |
| Yscl = 1 |

In Exercises 71–74, find a setting on a graphing utility such that the graph of the equation agrees with the given graph.

71. $y = 100 - 3x$

72. $y = \frac{2}{3}x - 8$

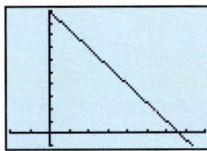
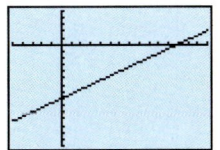

73. $y = -\frac{1}{8}x^2 + x$

74. $y = (x - 5)^2(x - 15)$

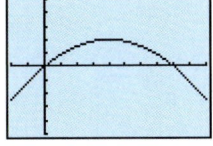
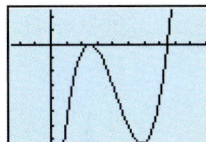

In Exercises 75–78, use a graphing utility to graph both equations on the same screen. Do the graphs intersect? If so, how many times do they intersect?

75. $y = x^2$
$y = 4x - x^2$

76. $y = x + 2$
$y = 2 + 4x - x^2$

77. $y = x$
$y = 3\sqrt{x}$

78. $y = 2x^2$
$y = x^4 - 2x^2$

In Exercises 79–82, use a graphing utility to estimate the x-intercepts of the graph of the equation.

79. $y = x^2 - 4x - 3$

80. $y = -2x^2 - x + 8$

81. $y = -x^2 + 12x - 37$

82. $y = x^2 + 14x + 49$

Exploration In Exercises 83 and 84, use a graphing utility to find a graph with the given intercepts. Explain your strategy.

83. $(2, 0), (0, 2)$

84. $(5, 0), (0, -3)$

Comparing Views   In Exercises 85 and 86, the equation gives the profit y (in thousands of dollars) for selling x units of a product. Which graphing utility viewing rectangle gives a better view of the graph? Explain your reasoning.

85. $y = 0.02x^2 + 0.75x - 10$

Xmin = 0
Xmax = 50
Xscl = 5
Ymin = -10
Ymax = 100
Yscl = 20

Xmin = 0
Xmax = 50
Xscl = 5
Ymin = -10
Ymax = 400
Yscl = 50

86. $y = \dfrac{6(4x - 1)}{x + 1}$

Xmin = 0
Xmax = 10
Xscl = 1
Ymin = -2
Ymax = 25
Yscl = 2

Xmin = 0
Xmax = 10
Xscl = 1
Ymin = -2
Ymax = 100
Yscl = 10

87. *Income*   The net income per share of common stock for McDonald's Corporation from 1984 through 1993 can be approximated by the model

$$y = (0.081x + 0.648)^2, \quad 4 \le x \le 13$$

where y is the net income per share and x represents the year, with $x = 0$ corresponding to 1980. (Source: McDonald's Corporation)

(a) Use a graphing utility to graph the model.

(b) The actual net income per share of stock in 1993 was $2.91. How closely does the model predict this value?

88. *Depreciation*   A manufacturing plant purchases a new numerically controlled machine for $500,000. The depreciated value y after x years is given by

$$y = 500,000 - 60,000x, \quad 0 \le x \le 6.$$

Use the constraints of the model to determine an appropriate viewing rectangle and graph the equation.

Strength of Copper   In Exercises 89 and 90, use the following model, which relates the percent strength of copper y in terms of its Celsius temperature x.

$$y = \sqrt{10,700 - 17.6x}, \quad 100 \le x \le 500$$

The percent strength is relative to 100% at 20°C.

89. Use a graphing utility to sketch the graph of the equation using the following range settings.

Xmin = 0
Xmax = 500
Xscl = 50
Ymin = 30
Ymax = 110
Yscl = 10

90. Use the TRACE key to find the approximate percent strength when the temperature is (a) $x = 150°C$ and (b) $x = 400°C$.

McDonald's restaurant chain owner Ray Kroc opened his first restaurant in Des Plaines, Illinois, shortly after buying business rights in 1955. At that time a regular hamburger cost $0.15 and a regular order of fries cost $0.11.

MID-CHAPTER QUIZ

Take this quiz as you would take a quiz in class. After you are done, check your work against the answers given in the back of the book.

In Exercises 1 and 2, plot the points on a rectangular coordinate system and find the distance between them.

1. $(-1, 5)$, $(3, 2)$ **2.** $(-3, -2)$, $(2, 10)$

3. Determine the quadrants in which the point $(x, 4)$ must be located if x is a real number. Explain your reasoning.

4. Find the coordinates of the point that lies 10 units to the right of the y-axis and three units below the x-axis.

5. Determine whether the following ordered pairs are solution points of the equation $4x - 3y = 10$.

(a) $(2, 1)$ (b) $(1, -2)$ (c) $(2.5, 0)$ (d) $\left(2, -\frac{2}{3}\right)$

6. Find the x- and y-intercepts of the graph of the equation $6x - 8y + 48 = 0$.

In Exercises 7–12, sketch the graph of the equation and show the coordinates of three solution points (including intercepts).

7. $y = 2x - 3$ **8.** $y = 5$

9. $3x + y - 6 = 0$ **10.** $y = x^2 - 4$

11. $y = 6x - x^2$ **12.** $y = |x - 2| - 3$

In Exercises 13–16, use a graphing utility to graph the equation. Use a standard setting.

13. $y = 7 - 2x$ **14.** $y = x^2 + 4x$

15. $y = \sqrt{8 - x}$ **16.** $y = |x - 2| + |x + 2|$

In Exercises 17 and 18, find a setting on a graphing utility such that the graph of the equation is approximately the same as the given graph.

17. $y = -2(x^2 - 8x + 6)$ **18.** $y = x^3 - 5x + 5$

19. Use a graphing utility to approximate the x-intercepts of the graph of $y = x^2 - x - 2.75$.

20. Use a graphing utility to verify that $y_1 = y_2$ if $y_1 = -x + (2x - 2)$ and $y_2 = (-x + 2x) - 2$. Identify the rule of algebra that is illustrated.

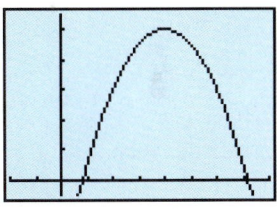

Figure for 17

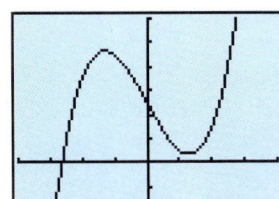

Figure for 18

7.4 Slope and Graphs of Linear Equations

The Slope of a Line ▪ Slope as a Graphing Aid ▪
Parallel and Perpendicular Lines

The Slope of a Line

The **slope** of a nonvertical line is the number of units the line rises or falls vertically for each unit of horizontal change from left to right. For example, the line in Figure 7.27 rises two units for each unit of horizontal change from left to right, and we say that this line has a slope of $m = 2$.

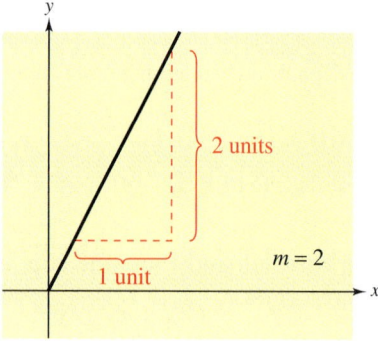

FIGURE 7.27   **FIGURE 7.28**

Definition of the Slope of a Line

The **slope** m of the nonvertical line passing through the points (x_1, y_1) and (x_2, y_2) is

$$m = \frac{y_2 - y_1}{x_2 - x_1} = \frac{\text{change in } y}{\text{change in } x}$$

where $x_1 \neq x_2$ (see Figure 7.28).

When the formula for slope is used, the *order of subtraction* is important. Given two points on a line, you are free to label either of them (x_1, y_1) and the other (x_2, y_2). However, once this is done, you must form the numerator and denominator using the same order of subtraction.

$$m = \frac{y_2 - y_1}{x_2 - x_1} \qquad m = \frac{y_1 - y_2}{x_1 - x_2} \qquad m = \frac{y_2 - y_1}{x_1 - x_2}$$

Correct   Correct   Incorrect

EXAMPLE 1   Finding the Slope of a Line Through Two Points

Find the slope of the line passing through each pair of points.

a. $(1, 2)$ and $(4, 5)$ **b.** $(1, 4)$ and $(3, 4)$ **c.** $(-1, 4)$ and $(2, 1)$

Solution

a. Let $(x_1, y_1) = (1, 2)$ and $(x_2, y_2) = (4, 5)$.

$$m = \frac{y_2 - y_1}{x_2 - x_1}$$

   Difference in y-values
Difference in x-values

$$= \frac{5 - 2}{4 - 1}$$

$$= 1$$

NOTE You can use slope to determine if three points are collinear—i.e., all lie on the same line. Consider any three points A, B, and C. If the slope of the line through points A and B is the same as the slope of the line through points B and C, the three points are collinear.

b. The slope of the line through $(1, 4)$ and $(3, 4)$ is

$$m = \frac{4 - 4}{3 - 1} = \frac{0}{2} = 0.$$

c. The slope of the line through $(-1, 4)$ and $(2, 1)$ is

$$m = \frac{1 - 4}{2 - (-1)} = \frac{-3}{3} = -1.$$

The graphs of the three lines are shown in Figure 7.29.

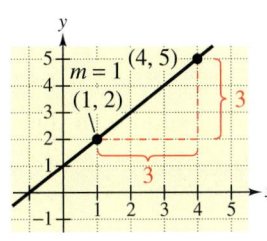
(a) Positive Slope

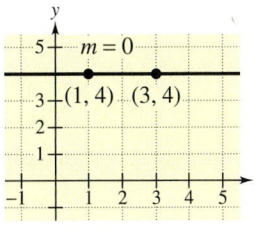
(b) Zero Slope

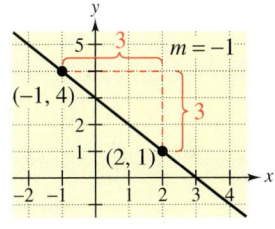
(c) Negative Slope

FIGURE 7.29

The definition of slope does not apply to vertical lines. For instance, consider the points $(3, 1)$ and $(3, 3)$ on the vertical line shown in Figure 7.30. Applying the formula for slope, you have

$$\frac{3 - 1}{3 - 3} = \frac{2}{0}.$$ Undefined

Because division by zero is not defined, the slope of a vertical line is not defined.

FIGURE 7.30   Slope is undefined.

From the slopes of the lines shown in Figures 7.29 and 7.30, you can make the following generalizations about the slope of a line.

1. A line with positive slope ($m > 0$) *rises* from left to right.

2. A line with negative slope ($m < 0$) *falls* from left to right.

3. A line with zero slope ($m = 0$) is *horizontal*.

4. A line with undefined slope is *vertical*.

EXAMPLE 2 Using Slope to Describe Lines

Describe the line through each pair of points.

a. $(2, -1), (2, 3)$ **b.** $(-2, 4), (3, 1)$ **c.** $(1, 3), (4, 3)$ **d.** $(-1, 1), (2, 5)$

Solution

a. Because the slope is undefined, the line is vertical.

$$m = \frac{3 - (-1)}{2 - 2} = \frac{4}{0}$$

Undefined slope (See Figure 7.31a.)

b. Because the slope is negative, the line falls from left to right.

$$m = \frac{1 - 4}{3 - (-2)} = -\frac{3}{5} < 0$$

Negative slope (See Figure 7.31b.)

c. Because the slope is zero, the line is horizontal.

$$m = \frac{3 - 3}{4 - 1} = \frac{0}{3} = 0$$

Zero slope (See Figure 7.31c.)

d. Because the slope is positive, the line rises from left to right.

$$m = \frac{5 - 1}{2 - (-1)} = \frac{4}{3} > 0$$

Positive slope (See Figure 7.31d.)

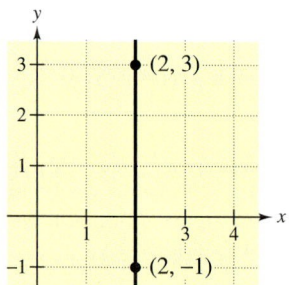

(a) Vertical line
Undefined slope

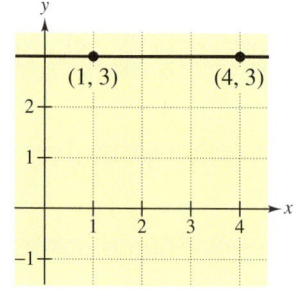

(b) Line falls
Negative slope

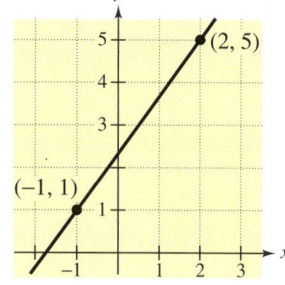

(c) Horizontal line
Zero slope

(d) Line rises
Positive slope

FIGURE 7.31

Any two points on a nonvertical line can be used to calculate its slope. This is demonstrated in the next example.

EXAMPLE 3 *Finding the Slope of a Line*

Sketch the graph of the line given by $2x + 3y = 6$. Then find the slope of the line. (Choose two different pairs of points on the line and show that the same slope is obtained from either pair.)

Solution

Begin by solving the given equation for y.

$$y = -\frac{2}{3}x + 2 \qquad \text{\textcolor{red}{y is a function of x.}}$$

Then construct a table of values, as shown below.

x	-3	0	3	6
$y = -\frac{2}{3}x + 2$	4	2	0	-2
Solution Point	$(-3, 4)$	$(0, 2)$	$(3, 0)$	$(6, -2)$

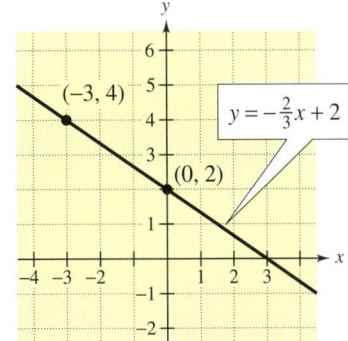

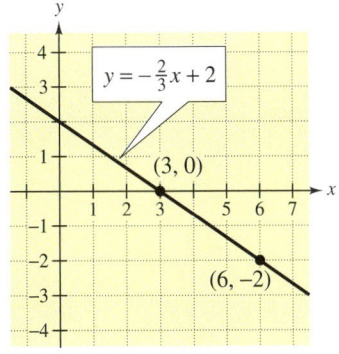

FIGURE 7.32

From the solution points shown in the table, sketch the graph of the line, as shown in Figure 7.32. To calculate the slope of the line using two different sets of points, first use the points $(-3, 4)$ and $(0, 2)$, and obtain a slope of

$$m = \frac{2 - 4}{0 - (-3)} = -\frac{2}{3}.$$

Next, use the points $(3, 0)$ and $(6, -2)$ to obtain a slope of

$$m = \frac{-2 - 0}{6 - 3} = -\frac{2}{3}.$$

Try some other pairs of points on the line to see that you obtain a slope of $m = -\frac{2}{3}$ regardless of which two points you use.

Technology

Setting the viewing window on a graphing utility affects the appearance of its slope. When you are using a graphing utility, remember that you cannot judge whether a slope is steep or shallow *unless* you use a square setting.

Slope as a Graphing Aid

You have seen that, before creating a table of values for an equation, you should first solve the equation for y. When you do this for a linear equation, you obtain some very useful information. Consider the results of Example 3.

$$2x + 3y = 6 \qquad \text{Original equation}$$
$$2x - 2x + 3y = -2x + 6 \qquad \text{Subtract } 2x \text{ from both sides.}$$
$$3y = -2x + 6 \qquad \text{Combine like terms.}$$
$$\frac{3y}{3} = \frac{-2x + 6}{3} \qquad \text{Divide both sides by 3.}$$
$$y = -\frac{2}{3}x + 2 \qquad \text{Simplify.}$$

Observe that the coefficient of x is the slope of the graph of this equation (see Example 3). Moreover, the constant term, 2, gives the y-intercept of the graph.

$$y \uparrow -\frac{2}{3}x + 2$$
$$\text{Slope} \qquad y\text{-intercept } (0, 2)$$

This form is called the **slope-intercept** form of the equation of the line.

NOTE The slope-intercept form of the equation of a line identifies y as a function of x. Hence, the term **linear function** is often used as an alternative description of this form of the equation of a line.

Slope-Intercept Form of the Equation of a Line

The graph of the equation

$$y = mx + b$$

is a line whose slope is m and whose y-intercept is $(0, b)$. (See Figure 7.33.)

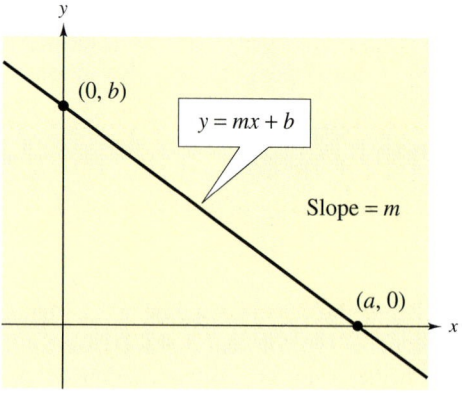

FIGURE 7.33

So far, you have been plotting several points to sketch the equation of a line. However, now that you can recognize equations of lines (linear functions), you don't have to plot as many points—two points are enough. (You might remember from geometry that *two points are all that are necessary to determine a line*.)

EXAMPLE 4   Using the Slope and y-Intercept to Sketch a Line

Use the slope and y-intercept to sketch the graph of

$$y = \frac{2}{3}x + 1.$$

Solution

The equation is already in slope-intercept form.

$$y = mx + b$$

$$y = \frac{2}{3}x + 1 \qquad\qquad \text{Slope-intercept form}$$

Thus, the slope of the line is $m = \frac{2}{3}$ and the y-intercept is $(0, b) = (0, 1)$. Now you can sketch the graph of the line as follows. First, plot the y-intercept. Then, using a slope of $\frac{2}{3}$

$$m = \frac{2}{3} = \frac{\text{change in } y}{\text{change in } x},$$

locate a second point on the line by moving three units to the right and two units up (or two units up and three units to the right), as shown in Figure 7.34(a). Finally, obtain the graph by drawing a line through the two points, as shown in Figure 7.34(b).

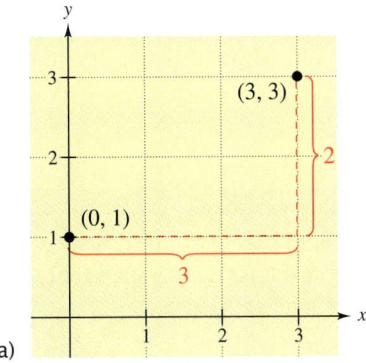

(a)

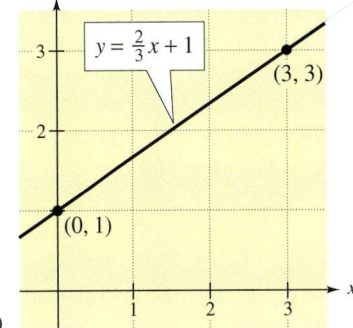

(b)

FIGURE 7.34

Parallel and Perpendicular Lines

You know from geometry that two lines in a plane are *parallel* if they do not intersect. What this means in terms of their slopes is suggested by Example 5.

EXAMPLE 5 Lines That Have the Same Slope

On the same set of axes, sketch the lines given by $y = 2x$ and $y = 2x - 3$.

Solution

For the line given by

$$y = 2x$$

the slope is $m = 2$ and the y-intercept is $(0, 0)$. For the line given by

$$y = 2x - 3$$

the slope is also $m = 2$ and the y-intercept is $(0, -3)$. The graphs of these two lines are shown in Figure 7.35.

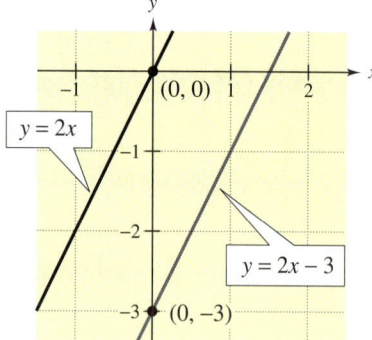

FIGURE 7.35

In Example 5, notice that the two lines have the same slope *and* appear to be parallel. The following rule states that this is always the case. That is, two (nonvertical) lines are parallel *if and only if* they have the same slope.

Parallel Lines

Two distinct nonvertical lines are parallel if and only if they have the same slope.

NOTE The phrase "if and only if" in this rule is used in mathematics as a way to write two statements in one. The first statement says that *if two distinct nonvertical lines have the same slope, they must be parallel.* The second statement says that *if two distinct nonvertical lines are parallel, they must have the same slope.*

Another rule from geometry is that two lines in a plane are *perpendicular* if they intersect at right angles. In terms of their slopes, this means that two nonvertical lines are perpendicular if their slopes are negative reciprocals of each other.

Perpendicular Lines

Consider two nonvertical lines whose slopes are m_1 and m_2. The two lines are perpendicular if and only if their slopes are *negative reciprocals* of each other. That is,

$$m_1 = -\frac{1}{m_2}, \quad \text{or equivalently,} \quad m_1 \cdot m_2 = -1.$$

EXAMPLE 6   *Parallel or Perpendicular?*

Are the following pairs of lines parallel, perpendicular, or neither?

a. $y = -2x + 4$, $y = \frac{1}{2}x + 1$ **b.** $y = \frac{1}{3}x + 2$, $y = \frac{1}{3}x - 3$

Solution

a. The first line has a slope of $m_1 = -2$, and the second line has a slope of $m_2 = \frac{1}{2}$. Because these slopes are negative reciprocals of each other, the two lines must be perpendicular, as shown in Figure 7.36.

b. Each of these two lines has a slope of $m = \frac{1}{3}$. Therefore, the two lines must be parallel, as shown in Figure 7.37.

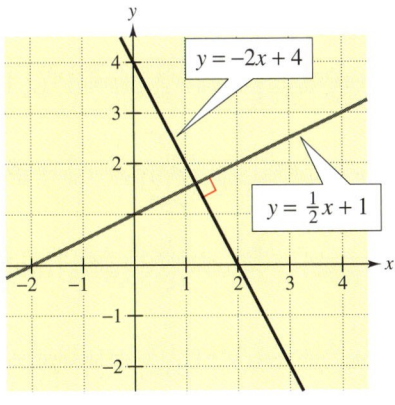

FIGURE 7.36

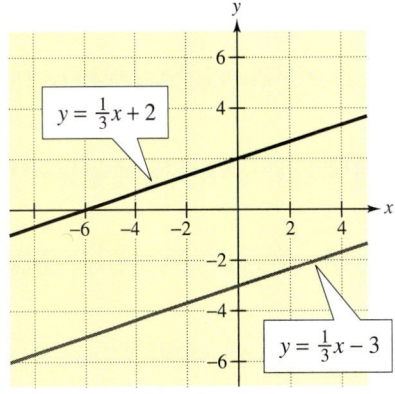

FIGURE 7.37

Group Activities   E x t e n d i n g   t h e   C o n c e p t

Interpreting Slope   Write a function for the given verbal model. Identify the slope and then interpret the slope in the real-life setting.

1.   Total pay per hour $=$ Piecework rate $\cdot$ Number of pieces $+$ Fixed hourly rate

2.   Total cost $=$ Tax rate $\cdot$ List price $+$ List price

7.4 Exercises

Discussing the Concepts

1. Can any pair of points on a line be used to calculate the slope of the line? Explain.

2. In your own words, give interpretations of a negative slope, a zero slope, and a positive slope.

3. The slopes of two lines are -3 and $\frac{3}{2}$. Which is steeper? Explain.

4. In the form $y = mx + b$, what does m represent? What does b represent?

5. What is the relationship between the x-intercept of the line $y = mx + b$ and the solution to the equation $mx + b = 0$? Explain.

6. Is it possible for two lines with positive slopes to be perpendicular to each other? Explain.

Problem Solving

In Exercises 7–12, estimate the slope of the line from its graph.

7.
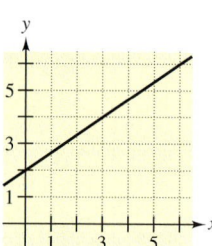

8.
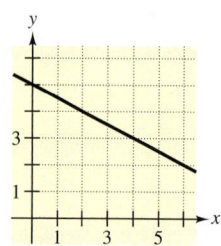

9.
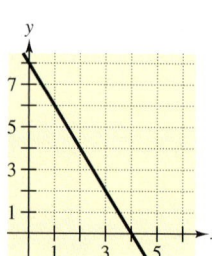

10.
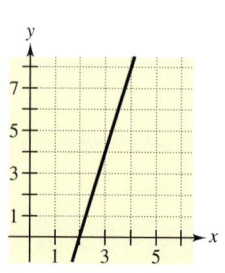

11.
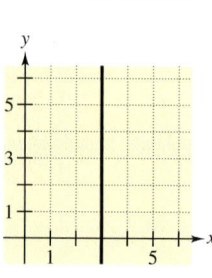

12.
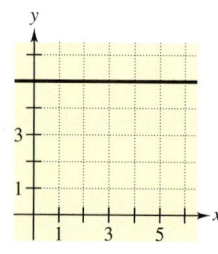

In Exercises 13–20, plot the two points and find the slope (if possible) of the line passing through them. State whether the line is rising, falling, horizontal, or vertical.

13. $(0, 0)$, $(5, -4)$

14. $(0, 0)$, $(-2, -1)$

15. $(-2, -3)$, $(6, 1)$

16. $(3, 6)$, $(5, -2)$

17. $(3, -2)$, $(3, 6)$

18. $(-4, 3)$, $(4, 3)$

19. $\left(\frac{3}{4}, 2\right)$, $\left(\frac{7}{2}, 0\right)$

20. $(-3.5, 1.6)$, $(5.1, 4.3)$

In Exercises 21–24, a point on a line and the slope of the line are given. Find two additional points on the line.

21. $(5, 2)$, $m = 0$

22. $(-4, 3)$, m is undefined.

23. $(3, -4)$, $m = 3$

24. $(-2, 6)$, $m = -3$

In Exercises 25–28, sketch the graph of a line through the point $(3, 2)$ having the given slope.

25. $m = 3$

26. $m = \frac{3}{2}$

27. $m = -\frac{1}{3}$

28. $m = 0$

In Exercises 29 and 30, which line is steeper?

29. $m_1 = 3$, $m_2 = -4$

30. $m_1 = 4$, $m_2 = 5$

In Exercises 31–34, plot the x- and y-intercepts and sketch the graph of the line.

31. $2x - y + 4 = 0$

32. $3x + 5y + 15 = 0$

33. $-5x + 2y - 20 = 0$

34. $3x - 5y - 15 = 0$

In Exercises 35–38, write the equation of the line in slope-intercept form and sketch the line. Use a graphing utility to confirm your sketch.

35. $3x - y - 2 = 0$

36. $x - y - 5 = 0$

37. $3x + 2y - 2 = 0$

38. $y + 4 = 0$

In Exercises 39–42, use a graphing utility to graph the pair of equations on the same viewing rectangle. Are the lines parallel, perpendicular, or neither? (Use the *square* setting so the slopes of the lines appear visually correct.)

39. $L_1: y = \frac{1}{2}x - 2$

  $L_2: y = \frac{1}{2}x + 3$

40. $L_1: y = 3x - 2$

  $L_2: y = 3x + 1$

41. $L_1: y = \frac{3}{4}x - 3$

  $L_2: y = -\frac{4}{3}x + 1$

42. $L_1: y = -\frac{2}{3}x - 5$

  $L_2: y = \frac{3}{2}x + 1$

43. *Geometry* The length and width of a rectangular flower garden are 40 feet and 30 feet, respectively (see figure). A walkway of width x surrounds the garden.

  (a) Write the outside perimeter y of the walkway in terms of x.

(b) Use a graphing utility to graph the equation for the perimeter.

(c) Determine the slope of the graph of part (b). For each additional 1-foot increase in the width of the walkway, determine the increase in its perimeter.

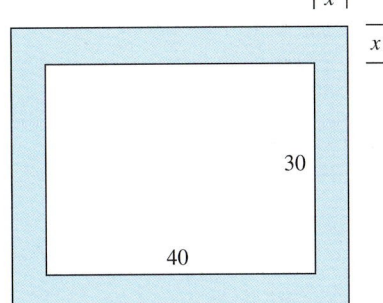

44. *Simple Interest* An inheritance of $8000 is invested in two different mutual funds. One fund pays 6% simple interest and the other pays $7\frac{1}{2}$% simple interest.

  (a) If x dollars is invested in the fund paying 6%, how much is invested in the fund paying $7\frac{1}{2}$%?

  (b) Use the result of part (a) to write the annual interest y in terms of x.

  (c) Use a graphing utility to graph the function in part (b).

  (d) Explain why the slope of the line in part (b) is negative.

Reviewing the Major Concepts

In Exercises 45–48, evaluate the expression.

45. $|-15|$

46. $-|72|$

47. $-|-6|$

48. $|14 - 32|$

49. *Work Rate* Two people can complete a task in t hours where t must satisfy the equation

$$\frac{t}{4} + \frac{t}{6} = 1.$$

Solve this equation and interpret the result.

50. *Maximum Height of an Object* The velocity v of an object projected vertically with an initial velocity of 96 feet per second is given by

$$v = 96 - 32t$$

where t is time in seconds and air resistance is neglected. Find the time when the maximum height (occurs when $v = 0$) of an object is obtained.

Additional Problem Solving

In Exercises 51 and 52, identify the line that has the specified slope m.

51. (a) $m = \frac{3}{4}$ (b) $m = 0$ (c) $m = -3$

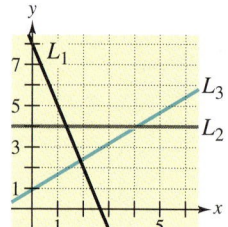

52. (a) $m = -\frac{5}{2}$ (b) m is undefined. (c) $m = 2$

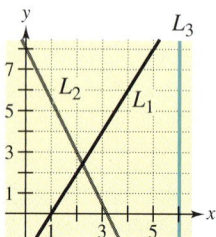

In Exercises 53–66, plot the points and, if possible, find the slope of the line passing through them. State whether the line is rising, falling, horizontal, or vertical.

53. $(0, 0), (7, 5)$

54. $(0, 0), (-3, -4)$

55. $(0, 12), (8, 0)$

56. $(0, -4), (6, 0)$

57. $(-5, -3), (-5, 4)$

58. $(0, -8), (-5, 0)$

59. $(2, -5), (7, -5)$

60. $(-2, 1), (-4, -3)$

61. $\left(\frac{3}{4}, 2\right), \left(5, -\frac{5}{2}\right)$

62. $\left(-\frac{3}{2}, -\frac{1}{2}\right), \left(\frac{5}{8}, \frac{1}{2}\right)$

63. $(4.2, -1), (-4.2, 6)$

64. $(3.4, 0), (3.4, 1)$

65. $(2.5, -2), (4.75, 5.25)$

66. $(0, 4.5), (3, 4.5)$

In Exercises 67 and 68, solve for x so that the line through the points has the given slope.

67. $(4, 5), (x, 7); m = -\frac{2}{3}$

68. $(x, -2), (5, 0); m = \frac{3}{4}$

In Exercises 69 and 70, solve for y so that the line through the points has the given slope.

69. $(-3, y), (9, 3); m = \frac{3}{2}$

70. $(-3, 20), (2, y); m = -6$

In Exercises 71–76, a point on a line and the slope of the line are given. Find two additional points on the line.

71. $(0, 3), m = -1$

72. $(-1, -5), m = 2$

73. $(-5, 0), m = \frac{4}{3}$

74. $(-1, 1), m = -\frac{3}{4}$

75. $(4, 2), m$ is undefined.

76. $(-2, -2), m = 0$

In Exercises 77–82, sketch the graph of a line through the point $(0, 1)$ having the given slope.

77. $m = 2$

78. $m = 0$

79. m is undefined.

80. $m = -1$

81. $m = -\frac{4}{3}$

82. $m = \frac{2}{3}$

In Exercises 83–90, write the equation of the line in slope-intercept form and sketch the line. Use a graphing utility to confirm your sketch.

83. $x + y = 0$

84. $x - y = 0$

85. $x - 4y + 2 = 0$

86. $x - 2y - 2 = 0$

87. $\frac{1}{3}x + \frac{1}{2}y = 1$

88. $8x + 6y - 3 = 0$

89. $y - 2 = 0$

90. $2y + 3 = 0$

In Exercises 91–94, determine whether the lines L_1 and L_2 passing through the given pairs of points are parallel, perpendicular, or neither.

91. L_1: $(1, 3), (2, 1)$; L_2: $(0, 0), (4, 2)$

92. L_1: $(-3, -3), (1, 7)$; L_2: $(0, 4), (5, -2)$

93. L_1: $(-2, 0), (4, 4)$; L_2: $(1, -2), (4, 0)$

94. L_1: $(-5, 3), (3, 0)$; L_2: $(1, 2), \left(3, \frac{22}{3}\right)$

In Exercises 95–98, use a graphing utility to graph the three equations on the same viewing rectangle. Describe the relationships among the graphs. (Use the *square* setting so the slopes of the lines appear visually correct.)

95. $y_1 = 3x$

$y_2 = -3x$

$y_3 = \frac{1}{3}x$

96. $y_1 = \frac{3}{4}x$

$y_2 = -\frac{4}{3}x$

$y_3 = \frac{4}{3}$

97. $y_1 = \frac{1}{4}x$

$y_2 = \frac{1}{4}x - 2$

$y_3 = \frac{1}{4}x + 3$

98. $y_1 = 2x$

$y_2 = 2x - 5$

$y_3 = 2x + \frac{3}{2}$

99. *Graphical Estimation*   The graph shows the earnings per share of common stock for Johnson & Johnson for the years 1987 through 1993. Use the slope of each segment to determine the year when earnings (a) decreased most rapidly and (b) increased most rapidly. *(Source: Johnson & Johnson 1993 Annual Report)*

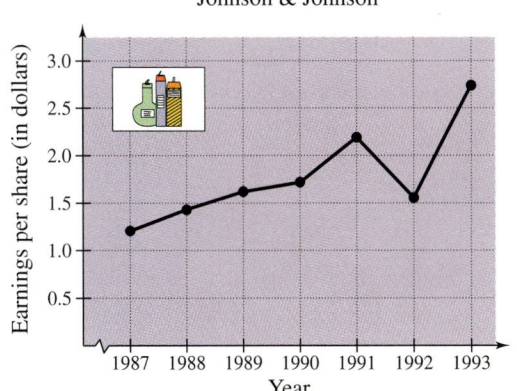

Johnson & Johnson

100. *Road Grade*   When driving down a mountain road, you notice warning signs indicating that it is a "12% grade." This means that the slope of the road is $-\frac{12}{100}$. Over a stretch of road, your elevation drops by 2000 feet. What is the horizontal change in your position?

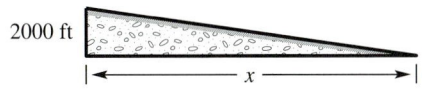

101. *Graphical Estimation*   The graph gives the declared dividend per share of common stock for Emerson Electric Company for the years 1987 through 1993. Use the slope of each segment to determine the year when the dividend increased most rapidly. *(Source: Emerson Electric Company)*

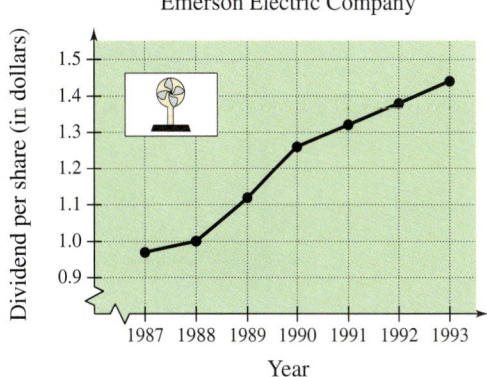

Emerson Electric Company

102. *Height of an Attic*   The slope, or pitch, of a roof is such that it rises (or falls) 3 feet for every 4 feet of horizontal distance. Determine the maximum height in the attic of the house if the house is 30 feet wide.

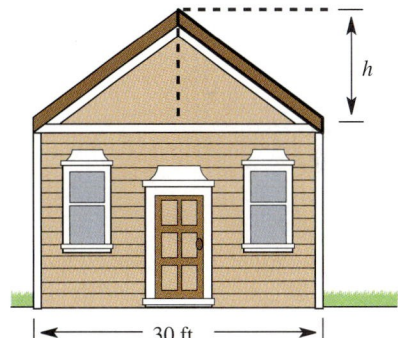

7.5 | Relations and Functions

Relations ▪ Functions ▪ Function Notation ▪
Finding the Domain and Range of a Function ▪ Application

Relations

Many everyday occurrences involve two quantities that are paired or matched with each other by some rule of correspondence. The mathematical term for such a correspondence is a **relation.**

Definition of a Relation

A **relation** is any set of ordered pairs. The set of first components in the ordered pairs is the **domain** of the relation and the set of second components is the **range** of the relation.

EXAMPLE 1   *Analyzing a Relation*

Find the domain and range of {(0, 1), (1, 3), (2, 5), (3, 5), (0, 3)}.

Solution

The domain is the set of all first components of the relation, and the range is the set of all second components.

Domain: {0, 1, 2, 3}

↑   ↑   ↑   ↑   ↑

{(0, 1), (1, 3), (2, 5), (3, 5), (0, 3)}

↓   ↓   ↓   ↓   ↓

Range: {1, 3, 5}

A graphical representation is given in Figure 7.38.

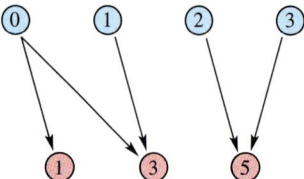

FIGURE 7.38

Functions

In modeling real-life situations, you will work with a special type of relation called a function. A **function** is a relation in which no two ordered pairs have the same first component and different second components. For instance, (2, 3) and (2, 4) could not be ordered pairs of a function.

Definition of a Function

A **function** f from a set A to a set B is a rule of correspondence that assigns to each element x in the set A exactly one element y in the set B.

The set A is called the **domain** (or set of inputs) of the function f, and the set B contains the **range** (or set of outputs) of the function.

The rule of correspondence for a function establishes a set of "input-output" ordered pairs of the form (x, y), where x is an input and y is the corresponding output. In some cases, the rule may generate only a finite set of ordered pairs, whereas in other cases the rule may generate an infinite set of ordered pairs.

EXAMPLE 2   *Input-Output Ordered Pairs for Functions*

a. For the function that pairs the year from 1991 to 1994 with the winner of the Super Bowl, each ordered pair is of the form (year, winner).

$$\{(1991, \text{Giants}), (1992, \text{Redskins}), (1993, \text{Cowboys}), (1994, \text{Cowboys})\}$$

b. For the function given by $y = x - 2$, each ordered pair is of the form (x, y).

$$\{\text{All points on the graph of } y = x - 2\}$$

c. For the function that pairs the positive integers that are less than 7 with their squares, each ordered pair is of the form (n, n^2).

$$\{(1, 1), (2, 4), (3, 9), (4, 16), (5, 25), (6, 36)\}$$

d. For the function that pairs each real number with its square, each ordered pair is of the form (x, x^2).

$$\{\text{All points } (x, x^2), \text{ where } x \text{ is a real number}\}$$

NOTE   In Example 2, the sets in parts (a) and (c) have only finite numbers of ordered pairs, whereas the sets in parts (b) and (d) have infinite numbers of ordered pairs.

Characteristics of a Function

1. Each element in the domain A must be matched with an element in the range, which is contained in the set B.

2. Some elements in the set B may not be matched with any element in the domain A.

3. Two or more elements of the domain may be matched with the same element in the range.

4. No element of the domain is matched with two different elements in the range.

EXAMPLE 3   Test for Functions Represented by Ordered Pairs

Let $A = \{a, b, c\}$ and let $B = \{1, 2, 3, 4, 5\}$. Which represents a function from A to B?

a. $\{(a, 2), (b, 3), (c, 4)\}$ **b.** $\{(a, 4), (b, 5)\}$

c.

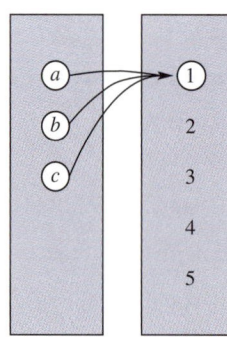

d.

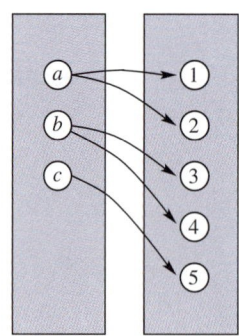

Solution

a. This set of ordered pairs *does* represent a function from A to B. Each element of A is matched with exactly one element of B.

b. This set of ordered pairs *does not* represent a function from A to B. Not all elements of A are matched with an element of B.

c. This diagram *does* represent a function from A to B. It does not matter that each element of A is matched with the same element in B.

d. This diagram *does not* represent a function from A to B. The element a in A is matched with *two* elements, 1 and 2, in B. This is also true of b.

Representing functions by sets of ordered pairs is a common practice in the study of *discrete mathematics*, which deals mainly with finite sets of data or with finite subsets of the set of real numbers. In algebra, however, it is more common to represent functions by equations or formulas involving two variables. For instance, the equation

$$y = x^2 \qquad\qquad \text{Squaring function}$$

represents the variable y as a function of the variable x. The variable x is the **independent variable** and the variable y is the **dependent variable.** In this context, the domain of the function is the set of all *allowable* real values for the independent variable x, and the range of the function is the *resulting* set of all values taken on by the dependent variable y.

EXAMPLE 4   *Testing for Functions Represented by Equations*

Which of the equations represents y as a function of x?

a. $y = x^2 + 1$ **b.** $x - y^2 = 2$ **c.** $-2x + 3y = 4$

Solution

a. From the equation

$$y = x^2 + 1$$

you can see that for each value of x there corresponds just one value of y. For instance, when $x = 1$, the value of y is $1^2 + 1 = 2$. Therefore, y *is* a function of x.

b. By writing the equation $x - y^2 = 2$ in the form

$$y^2 = x - 2$$

you can see that some values of x correspond to *two* values of y. For instance, when $x = 3$, $y^2 = 3 - 2 = 1$ and y can be 1 or -1. Hence, the solution points $(3, 1)$ and $(3, -1)$ show that y *is not* a function of x.

c. By writing the equation $-2x + 3y = 4$ in the form

$$y = \tfrac{2}{3}x + \tfrac{4}{3}$$

you can see that for each value of x there corresponds just one value of y. For instance, when $x = 2$, the value of y is $\tfrac{4}{3} + \tfrac{4}{3} = \tfrac{8}{3}$. Therefore, y *is* a function of x.

NOTE   An equation that defines y as a function of x may or may not also define x as a function of y. For instance, the equation in part (a) does not define x as a function of y, but the equation in part (c) does.

Function Notation

When an equation is used to represent a function, it is convenient to name the function so that it can be easily referenced. For example, the function $y = x^2 + 1$ in Example 4(a) can be given the name "f" and written in **function notation** as

$$f(x) = x^2 + 1.$$

Function Notation

In the notation $f(x)$:

f is the **name** of the function,

x is the **domain** (or input) value, and

$f(x)$ is a **range** (or output) value y for a given x.

The symbol $f(x)$ is read as *the value of f at x* or simply *f of x.*

The process of finding the value of $f(x)$ for a given value of x is called **evaluating a function.** This is accomplished by substituting a given x-value (input) into the equation to obtain the value of $f(x)$ (output). Here is an example.

Function	*x-Value*	*Function Value*
$f(x) = 3 - 4x$	$x = -1$	$f(-1) = 3 - 4(-1) = 3 + 4 = 7$

Although f is often used as a convenient function name and x as the independent variable, you can use other letters. For instance, the equations

$$f(x) = 2x^2 + 5, \quad f(t) = 2t^2 + 5, \quad \text{and} \quad g(s) = 2s^2 + 5$$

all define the same function. In fact, the letters used are simply "placeholders" and this same function is well described by the form

$$f(\quad) = 2(\quad)^2 + 5$$

where the parentheses are used in place of a letter. To evaluate $f(-2)$, simply place -2 in each set of parentheses, as follows.

$$f(-2) = 2(-2)^2 + 5$$
$$= 8 + 5$$
$$= 13$$

When evaluating a function, you are not restricted to substituting only numerical values into the parentheses. For instance, the value of $f(3x)$ is

$$f(3x) = 2(3x)^2 + 5$$
$$= 18x^2 + 5.$$

EXAMPLE 5   *Evaluating a Function*

Let $g(x) = 3x - x^2$ and find the following.

a. $g(1)$ **b.** $g(x + 1)$ **c.** $g(x) + g(1)$

Solution

a. Replacing x by 1 produces $g(1) = 3(1) - (1)^2 = 3 - 1 = 2$.

b. Replacing x with $x + 1$ produces

$$g(x + 1) = 3(x + 1) - (x + 1)^2$$
$$= 3x + 3 - (x^2 + 2x + 1)$$
$$= -x^2 + (3x - 2x) + (3 - 1)$$
$$= -x^2 + x + 2.$$

c. Using the result of part (a), we have

$$g(x) + g(1) = (3x - x^2) + 2 = -x^2 + 3x + 2.$$

Note that $g(x + 1) \neq g(x) + g(1)$. In general, $g(a + b)$ is not equal to $g(a) + g(b)$.

Sometimes a function is defined by more than one equation. An illustration of this is given in Example 6.

EXAMPLE 6   *A Function Defined by Two Equations*

Evaluate the function given by

$$f(x) = \begin{cases} x^2 + 1, & \text{if } x < 0 \\ x - 2, & \text{if } x \geq 0 \end{cases}$$

at (a) $x = -1$, (b) $x = 0$, and (c) $x = 1$.

Solution

a. Because $x = -1 < 0$, we use $f(x) = x^2 + 1$ to obtain

$$f(-1) = (-1)^2 + 1 = 2.$$

b. Because $x = 0 \geq 0$, we use $f(x) = x - 2$ to obtain

$$f(0) = 0 - 2 = -2.$$

c. Because $x = 1 \geq 0$, we use $f(x) = x - 2$ to obtain

$$f(1) = 1 - 2 = -1.$$

Finding the Domain and Range of a Function

The domain of a function may be explicitly described along with the function, or it may be *implied* by the expression used to define the function. The **implied domain** is the set of all real numbers (inputs) that yield real number values for the function. For instance, the function given by

$$f(x) = \frac{1}{x^2 - 9}$$
Domain: all $x \neq \pm 3$

has an implied domain that consists of all real values of x other than $x = \pm 3$. These two values are excluded from the domain because division by zero is undefined. Another common type of implied domain is that used to avoid even roots of negative numbers. For instance, the function given by

$$f(x) = \sqrt{x}$$
Domain: all $x \geq 0$

is defined only for $x \geq 0$. Therefore, its implied domain is the interval $[0, \infty)$. More will be said about the domains of square root functions in Chapter 9.

EXAMPLE 7 Finding the Domain and Range of a Function

Find the domain and range of each function.

a. $f \colon \{(-3, 0), (-1, 2), (0, 4), (2, 4), (4, -1)\}$

b. Area of a circle: $A = \pi r^2$

Solution

a. The domain of f consists of all first coordinates in the set of ordered pairs. The range consists of all second coordinates in the set of ordered pairs. Thus, the domain is

Domain $= \{-3, -1, 0, 2, 4\}$

and the range is

Range $= \{0, 2, 4, -1\}$.

b. For the area of a circle, you must choose nonnegative values for the radius r. Thus, the domain is the set of all real numbers r such that $r \geq 0$. The range is therefore the set of all real numbers A such that $A \geq 0$.

Note in Example 7(b) that the domain of a function can be implied by a physical context. For instance, from the equation $A = \pi r^2$, we would have no reason to restrict r to positive values. However, because we know this function represents the area of a circle, we conclude that the radius must be positive.

Application

To find mathematical models to represent functions, use the same guidelines that you use to model equations.

EXAMPLE 8   *Finding an Equation to Represent a Function*

Is the area of a square a *function* of the length of one of its sides? If so, find an equation that represents this function.

Solution

Figure 7.39 shows a square.

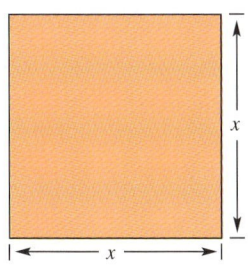

FIGURE 7.39

Verbal Model: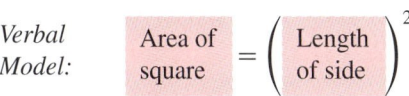

$$\boxed{\text{Area of square}} = \left(\boxed{\text{Length of side}} \right)^2$$

Labels: Area $= A$ (square units)
          Length of side $= x$ (linear units)

Function: $A = x^2$

Thus, the area of a square *is* a function of the length of one of its sides.

Group Activities   **E x t e n d i n g   t h e   C o n c e p t**

Determining Relationships That Are Functions   Compile a list of statements describing relationships in everyday life. For each statement, identify the dependent and independent variables and discuss whether the statement *is* a function or *is not* a function and why. Here are two examples.

a. In the statement, "The number of ceramic tiles required to floor a kitchen is a function of the floor's area," the dependent variable is the required number of ceramic tiles and the independent variable is the area of the floor. This statement *is* a mathematically correct use of the word "function" because for each possible floor area there corresponds exactly one number of tiles needed to do the job.

b. In the statement, "Interest rates are a function of economic conditions," the dependent variable is interest rates and the independent variable is economic conditions. This statement *is not* a mathematically correct use of the word "function" because "economic conditions" is ambiguous; it is difficult to tell if one set of economic conditions would always result in the same interest rates.

7.5 Exercises

Discussing the Concepts

1. Explain the difference between relations and functions.

2. Is every relation a function? Explain.

3. In your own words, explain the meaning of *domain* and *range*.

4. In the equation $|y| = x$, is y a function of x? Explain your reasoning.

5. In the equation $y = |x|$, is y a function of x? Explain your reasoning.

6. Describe an advantage of function notation.

Problem Solving

In Exercises 7–10, state the domain and range of the relation. Then draw a graphic representation.

7. $\{(-2, 0), (0, 1), (1, 4), (0, -1)\}$

8. $\{(3, 10), (4, 5), (6, -2), (8, 3)\}$

9. $\{(0, 0), (4, -3), (2, 8), (5, 5), (6, 5)\}$

10. $\{(-3, 6), (-3, 2), (-3, 5)\}$

In Exercises 11–14, write a set of ordered pairs that represents the rule of correspondence.

11. In a given week, a salesperson travels a distance d in t hours at an average speed of 50 mph. The travel times for each day are 3 hours, 2 hours, 8 hours, 6 hours, and $\frac{1}{2}$ hour.

12. The cubes of all positive integers less than 8

13. The winners of the World Series from 1990 to 1993

14. The men inaugurated president of the United States in 1969, 1974, 1977, 1981, 1989, and 1993.

In Exercises 15–18, decide whether the relation is a function.

15.
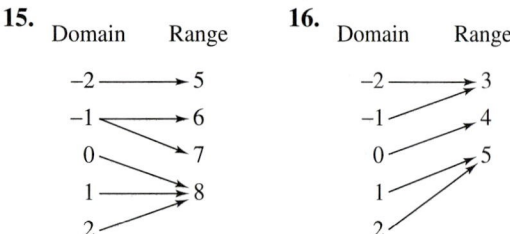

16.

17.
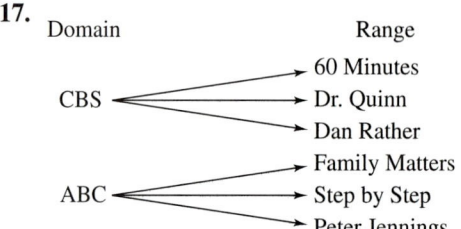

18.
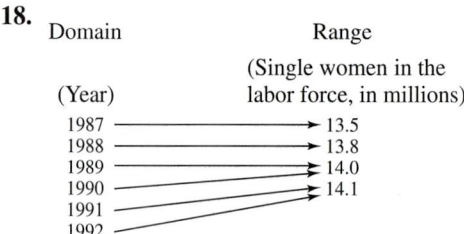

In Exercises 19 and 20, determine which sets of ordered pairs represent functions from A to B.

19. $A = \{0, 1, 2, 3\}$, $B = \{-2, -1, 0, 1, 2\}$

    (a) $\{(0, 1), (1, -2), (2, 0), (3, 2)\}$

    (b) $\{(0, -1), (2, 2), (1, -2), (3, 0), (1, 1)\}$

    (c) $\{(0, 0), (1, 0), (2, 0), (3, 0)\}$

    (d) $\{(0, 2), (3, 0), (1, 1)\}$

20. $A = \{1, 2, 3\}$, $B = \{9, 10, 11, 12\}$

    (a) $\{(1, 10), (3, 11), (3, 12), (2, 12)\}$

    (b) $\{(1, 10), (2, 11), (3, 12)\}$

    (c) $\{(1, 10), (1, 9), (3, 11), (2, 12)\}$

    (d) $\{(3, 9), (2, 9), (1, 12)\}$

In Exercises 21–24, decide whether the equation represents y as a function of x.

21. $y = 10x + 12$

22. $x - 9y + 3 = 0$

23. $|y - 2| = x$

24. $|y| = x + 2$

In Exercises 25 and 26, fill in the blank and simplify.

25. $f(x) = 3x + 5$

   (a) $f(2) = 3(\quad) + 5$

   (b) $f(-2) = 3(\quad) + 5$

   (c) $f(k) = 3(\quad) + 5$

   (d) $f(k + 1) = 3(\quad) + 5$

26. $f(x) = 3 - 2x$

   (a) $f(0) = 3 - 2(\quad)$

   (b) $f(-3) = 3 - 2(\quad)$

   (c) $f(m) = 3 - 2(\quad)$

   (d) $f(t + 2) = 3 - 2(\quad)$

In Exercises 27–32, evaluate the function as indicated, and simplify.

27. $f(x) = 12x - 7$

   (a) $f(3)$ (b) $f\left(\frac{3}{2}\right)$

   (c) $f(a) + f(1)$ (d) $f(a + 1)$

28. $h(x) = x(x - 2)$

   (a) $h(2)$ (b) $h(0)$

   (c) $h(1)$ (d) $h(t + 2)$

29. $f(x) = \begin{cases} x + 8, & \text{if } x < 0 \\ 10 - 2x, & \text{if } x \geq 0 \end{cases}$

   (a) $f(4)$ (b) $f(-10)$

   (c) $f(0)$ (d) $f(6) - f(-2)$

30. $f(x) = \begin{cases} -x, & \text{if } x \leq 0 \\ x^2 - 3x, & \text{if } x > 0 \end{cases}$

   (a) $f(0)$ (b) $f\left(-\frac{3}{2}\right)$

   (c) $f(4)$ (d) $f(-2) + f(25)$

31. $f(x) = 2x + 5$

   (a) $\dfrac{f(x + 2) - f(2)}{x}$ (b) $\dfrac{f(x - 3) - f(3)}{x}$

32. $f(x) = 2 - 3x$

   (a) $f(x + h)$ (b) $\dfrac{f(x + h) - f(x)}{h}$

In Exercises 33–36, find the domain of the function.

33. $f(x) = \dfrac{2x}{x - 3}$ **34.** $h(x) = 4x - 3$

35. $f(x) = \sqrt{2x - 1}$ **36.** $G(x) = \sqrt{x^2 - 25}$

37. *Geometry*   Express the perimeter P of a square as a function of the length x of one of its sides.

38. *Geometry*   Express the surface area S of a cube as a function of the length x of one of its edges.

39. *Geometry*   An open box is made from a square piece of material 24 inches on a side by cutting equal squares from the corners and turning up the sides (see figure). Write the volume V of the box as a function of x.

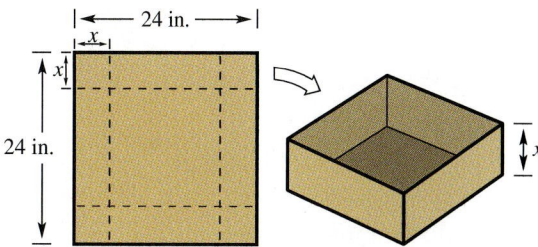

40. *Geometry*   Strips of width x are cut from the four sides of a square that is 32 inches on a side (see figure). Write the area A of the remaining square as a function of x.

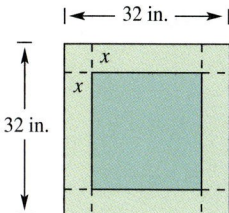

Reviewing the Major Concepts

In Exercises 41–44, simplify the expression.

41. $4 - 3(2x + 1)$

42. $24\left(\dfrac{y}{3} + \dfrac{y}{6}\right)$

43. $5(x + 2) - 4(2x - 3)$

44. $0.12x + 0.05(2000 - 2x)$

45. *Work Rate* You can mow the lawn in 4 hours and your friend can mow it in 5 hours. What fractional part of the lawn can each of you mow in 1 hour? How long will it take both of you to mow the lawn?

46. *Average Speed* A truck traveled at an average speed of 50 miles per hour on a 200-mile trip. On the return trip, the average speed was 42 miles per hour. Find the average speed for the round trip.

Additional Problem Solving

In Exercises 47–54, decide whether the relation is a function.

47.

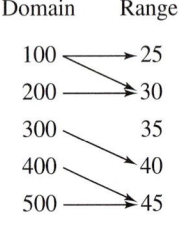

48.

Domain → Range

| 100 → 25 |
| 200 → 30 |
| 300 → 35 |
| 400 → 40 |
| 500 → 45 |

49.

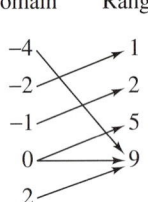

50.

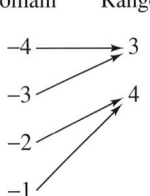

51.

Input Value	0	1	2	3	4
Output Value	0	1	4	9	16

52.

Input Value	0	1	2	1	0
Output Value	1	8	12	15	20

53.

Input Value	4	7	9	7	4
Output Value	2	4	6	8	10

54.

Input Value	0	2	4	6	8
Output Value	5	5	5	5	5

Interpreting a Graph In Exercises 55 and 56, use the graph, which shows numbers of high school and college students in the United States. (Source: U.S. National Center for Education Statistics)

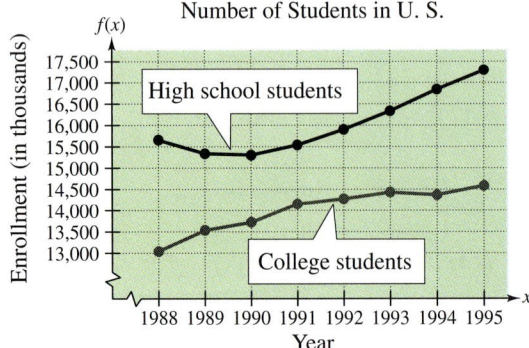

55. Is the high school enrollment a function of the year? Is the college enrollment a function of the year? Explain.

56. Let $f(x)$ represent the number of high school students in year x. Find $f(1993)$.

In Exercises 57–62, decide whether the equation represents y as a function of x.

57. $3x + 7y - 2 = 0$ **58.** $8x + y = 3$

59. $y = x(x - 10)$ **60.** $y = (x + 2)^2 + 3$

61. $|y| = 8 - x$ **62.** $x^2 + y^2 = 25$

In Exercises 63–72, evaluate the function as indicated, and simplify.

63. $g(x) = \frac{1}{2}x^2$

 (a) $g(4)$ (b) $g\left(\frac{2}{3}\right)$

 (c) $g(2y)$ (d) $g(4) + g(6)$

64. $f(x) = 3 - 7x$

 (a) $f(-1)$ (b) $f\left(\frac{1}{2}\right)$

 (c) $f(t) + f(-2)$ (d) $f(w)$

65. $f(x) = \sqrt{x + 5}$

 (a) $f(-1)$ (b) $f(4)$

 (c) $f\left(\frac{16}{3}\right)$ (d) $f(5z)$

66. $g(x) = 8 - |x - 4|$

 (a) $g(0)$ (b) $g(8)$

 (c) $g(16) - g(-1)$ (d) $g(x - 2)$

67. $f(x) = \dfrac{3x}{x - 5}$

 (a) $f(0)$ (b) $f\left(\frac{5}{3}\right)$

 (c) $f(2) - f(-1)$ (d) $f(x + 4)$

68. $g(x) = \dfrac{|x + 1|}{x + 1}$

 (a) $g(2)$ (b) $g\left(-\frac{1}{3}\right)$

 (c) $g(-4)$ (d) $g(3) + g(-5)$

69. $g(x) = 1 - x^2$

 (a) $g(2.2)$ (b) $\dfrac{g(2.2) - g(2)}{0.2}$

70. $f(x) = 3x + 4$

 (a) $\dfrac{f(x + 1) - f(1)}{x}$ (b) $\dfrac{f(x - 5) - f(5)}{x}$

71. $h(x) = \begin{cases} 4 - x^2, & \text{if } x \le 2 \\ x - 2, & \text{if } x > 2 \end{cases}$

 (a) $h(2)$ (b) $h\left(-\frac{3}{2}\right)$

 (c) $h(5)$ (d) $h(-3) + h(7)$

72. $f(x) = \begin{cases} x^2, & \text{if } x < 1 \\ x^2 - 3x + 2, & \text{if } x \ge 1 \end{cases}$

 (a) $f(1)$ (b) $f(-1)$

 (c) $f(2)$ (d) $f(-3) + f(3)$

In Exercises 73–76, find the domain and range of the function.

73. $f: \{(0, 0), (2, 1), (4, 8), (6, 27)\}$

74. $f: \left\{\left(-3, -\frac{17}{2}\right), \left(-1, -\frac{5}{2}\right), (4, 2), (10, 11)\right\}$

75. Circumference of a circle: $C = 2\pi r$

76. Area of a square of side s: $A = s^2$

In Exercises 77–84, find the domain of the function.

77. $h(x) = \dfrac{9}{x^2 + 1}$ **78.** $g(x) = \dfrac{x + 5}{x + 4}$

79. $f(t) = \dfrac{t + 3}{t(t + 2)}$ **80.** $g(s) = \dfrac{s - 2}{(s - 6)(s - 10)}$

81. $g(x) = \sqrt{x + 4}$ **82.** $f(x) = \sqrt{2 - x}$

83. $f(t) = t^2 + 4t - 1$ **84.** $f(x) = |x + 3|$

85. *Geometry* Express the volume V of a cube as a function of the length x of one of its edges.

86. *Distance* A plane is flying at a speed of 230 miles per hour. Express the distance d traveled by the plane as a function of time t in hours.

87. *Cost* The inventor of a new game believes that the variable cost for producing the game is $1.95 per unit and the fixed costs are $8000. Write the total cost C as a function of x, the number of games produced.

88. *Geometry* Strips of width x are cut from two adjacent sides of a square that is 32 inches on a side (see figure). Write the area A of the remaining square as a function of x.

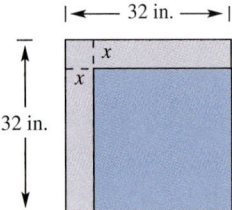

|←—— 32 in. ——→|

32 in.

89. *Safe Load* A solid rectangular beam has a height of 6 inches and a width of 4 inches. The safe load S of the beam with the load at the center is a function of its length L and is approximated by the model

$$S(L) = \frac{128,160}{L}$$

where S is measured in pounds and L is measured in feet. Find (a) $S(12)$ and (b) $S(16)$.

90. *Profit* The marketing department of a business has determined that the profit from selling x units of a product is approximated by the model

$$P(x) = 50\sqrt{x} - 0.5x - 500.$$

Find (a) $P(1600)$ and (b) $P(2500)$.

In Exercises 91 and 92, determine whether the statements use the word *function* in ways that are *mathematically* correct.

91. (a) The sales tax on a purchased item is a function of the selling price.

(b) Your score on the next algebra exam is a function of the number of hours you study the night before the exam.

92. (a) The amount in your savings account is a function of your salary.

(b) The speed at which a freely falling baseball strikes the ground is a function of the height from which it was dropped.

Math Matters   Finding the Shortest Path

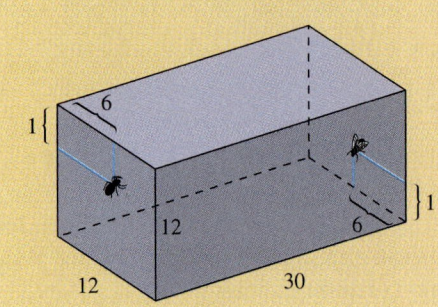

We know that the shortest distance between two points is a straight line. However, when we travel from one place to another, we often cannot travel in a straight line (we must follow the roads or sidewalks). Here is a problem involving the shortest path from one point to another. Suppose a spider and a fly are on opposite walls of a rectangular room, as shown in the figure. The spider wants to visit the fly, and assuming that the spider must travel on the surfaces of the room, what is the shortest path to the fly? (The answer is given in the back of the book.)

7.6   **Graphs of Functions**

The Graph of a Function ▪ The Vertical Line Test ▪
Graphs of Basic Functions ▪ Transformations of Graphs of Functions

The Graph of a Function

Consider a function f whose domain and range are the set of real numbers. The **graph** of f is the set of ordered pairs $(x, f(x))$, where x is in the domain of f.

$x = x$-coordinate of the ordered pair

$f(x) = y$-coordinate of the ordered pair

Figure 7.40 shows a typical graph of such a function.

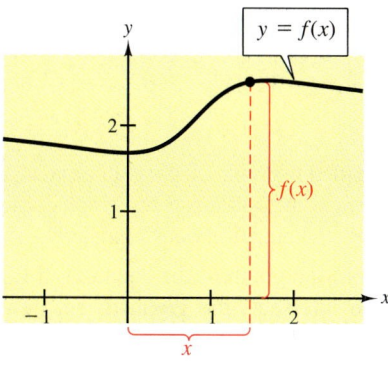

FIGURE 7.40

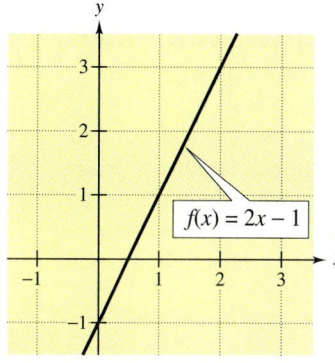

FIGURE 7.41

EXAMPLE 1   *Sketching the Graph of a Function*

Sketch the graph of $f(x) = 2x - 1$.

Solution

Another way to write this function is $y = 2x - 1$. From Section 7.2, you can recognize the graph to be a line, as shown in Figure 7.41.

In Example 1, the (implied) domain of the function is the set of all real numbers. When writing the equation of a function, we sometimes choose to restrict its domain by writing a condition to the right of the equation. For instance, the domain of the function

$$f(x) = 4x + 5, \quad x \geq 0$$

is the set of all nonnegative real numbers (all $x \geq 0$).

The Vertical Line Test

By the definition of a function, at most one y-value corresponds to a given x-value. This implies that any vertical line can intersect the graph of a function at most once.

Vertical Line Test for Functions

A set of points on a rectangular coordinate system is the graph of y as a function of x if and only if no vertical line intersects the graph at more than one point.

EXAMPLE 2   Using the Vertical Line Test

Decide whether the equation represents y as a function of x.

a. $y = x^2 - 3x + \frac{1}{4}$ **b.** $x = y^2 - 1$

Solution

a. From the graph of the equation in Figure 7.42(a), you can see that every vertical line intersects the graph at most once. Therefore, by the Vertical Line Test, the equation does represent y as a function of x.

b. From the graph of the equation in Figure 7.42(b), you can see that a vertical line intersects the graph twice. Therefore, by the Vertical Line Test, the equation does not represent y as a function of x.

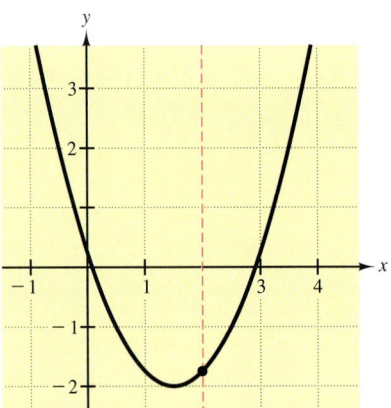

(a) Graph of a function of x.
Vertical line intersects once.

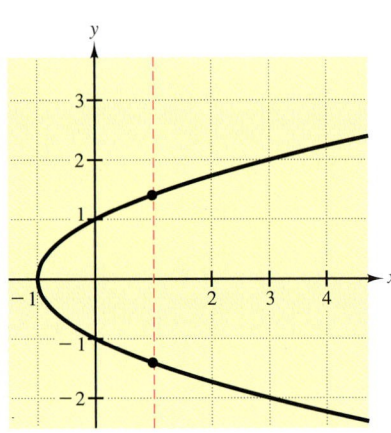

(b) Not a graph of a function of x.
Vertical line intersects twice.

FIGURE 7.42

Graphs of Basic Functions

To become good at sketching the graphs of functions, it helps to be familiar with the graphs of some basic functions. The functions shown in Figure 7.43, and variations of them, occur frequently in applications.

NOTE Try using a graphing utility to verify the graphs at the right. The names of these functions are as follows.

(a) Constant function

(b) Identity function

(c) Absolute value function

(d) Square root function

(e) Squaring function

(f) Cubing function

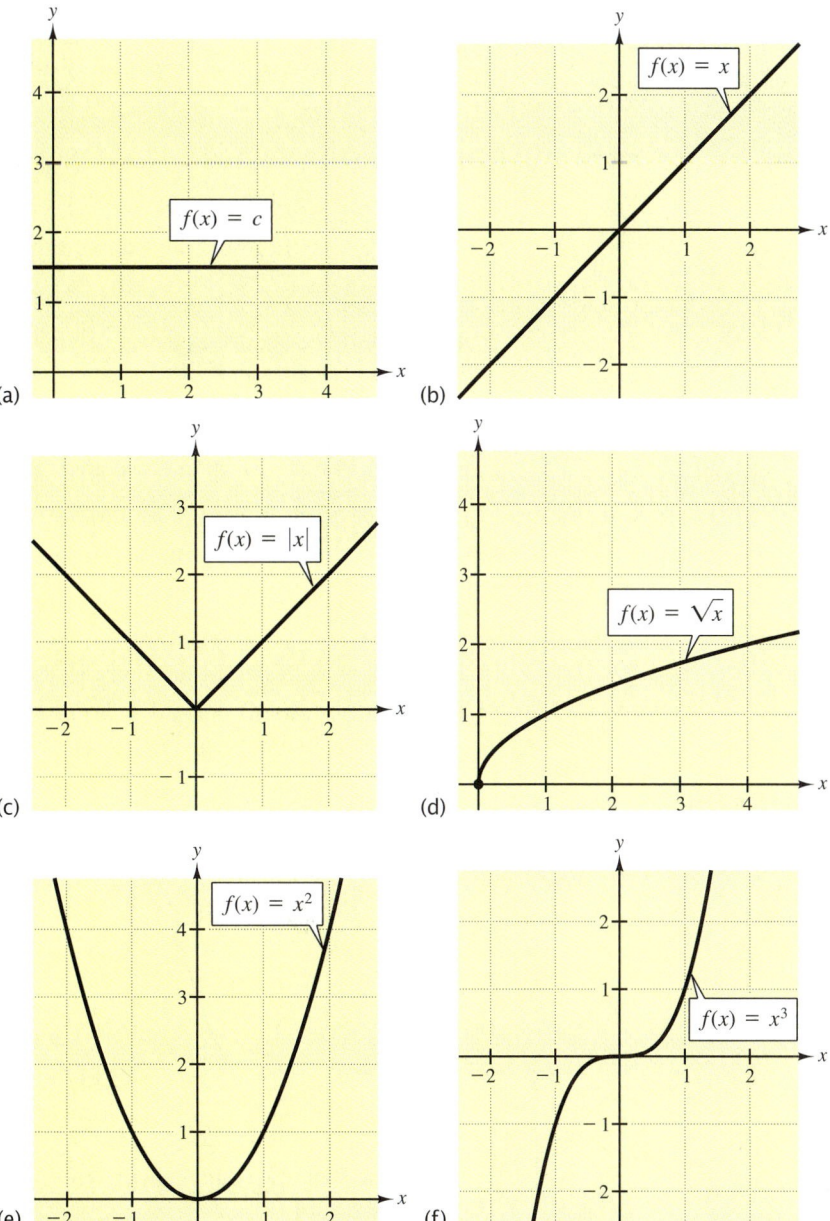

FIGURE 7.43

Transformations of Graphs of Functions

Many functions have graphs that are simple transformations of the basic graphs shown in Figure 7.43. The following list summarizes the various types of **horizontal** and **vertical shifts** of the graphs of functions.

Vertical and Horizontal Shifts

Let c be a positive real number. **Vertical** and **horizontal shifts** of the graph of the function $y = f(x)$ are represented as follows.

1. Vertical shift c units **upward:** $\qquad h(x) = f(x) + c$

2. Vertical shift c units **downward:** $\qquad h(x) = f(x) - c$

3. Horizontal shift c units to the **right:** $\qquad h(x) = f(x - c)$

4. Horizontal shift c units to the **left:** $\qquad h(x) = f(x + c)$

EXAMPLE 3   *Shifts of the Graphs of Functions*

Use the graph of $f(x) = x^2$ to sketch the graph of each function.

a. $g(x) = x^2 - 2$ **b.** $h(x) = (x + 3)^2$

Solution

a. Relative to the graph of $f(x) = x^2$, the graph of $g(x) = x^2 - 2$ represents a *downward shift* of two units, as shown in Figure 7.44.

b. Relative to the graph of $f(x) = x^2$, the graph of $h(x) = (x + 3)^2$ represents a *left shift* of three units, as shown in Figure 7.45.

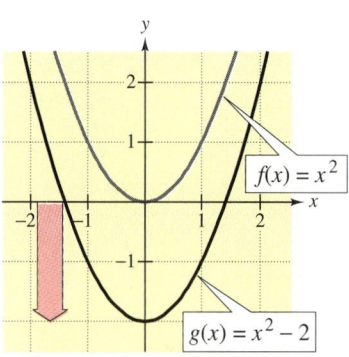

Vertical Shift: Two Units Down
FIGURE 7.44

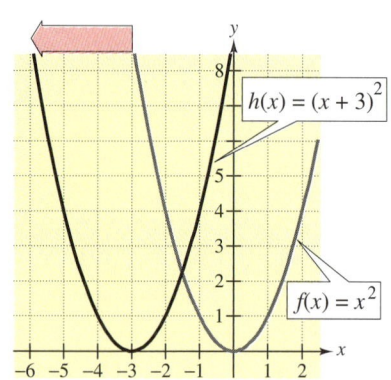

Horizontal Shift: Three Units Left
FIGURE 7.45

Some graphs can be obtained from *combinations* of vertical and horizontal shifts, as shown in part (b) of the next example.

EXAMPLE 4   *Shifts of the Graphs of Functions*

Use the graph of $f(x) = x^3$ to sketch the graph of each function.

a. $g(x) = x^3 + 2$ **b.** $h(x) = (x - 1)^3 + 2$

Solution

a. Relative to the graph of $f(x) = x^3$, the graph of $g(x) = x^3 + 2$ represents an *upward shift* of two units, as shown in Figure 7.46.

b. Relative to the graph of $f(x) = x^3$, the graph of $h(x) = (x - 1)^3 + 2$ represents a *right shift* of one unit, followed by an *upward shift* of two units, as shown in Figure 7.47.

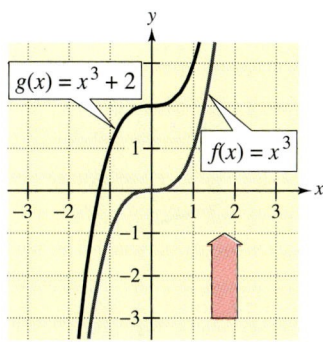

FIGURE 7.46
Vertical Shift: Two Units Up

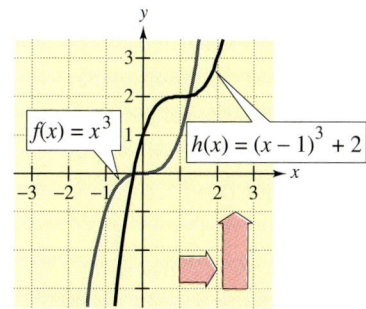

FIGURE 7.47
Horizontal Shift: One Unit Right;
Vertical Shift: Two Units Up

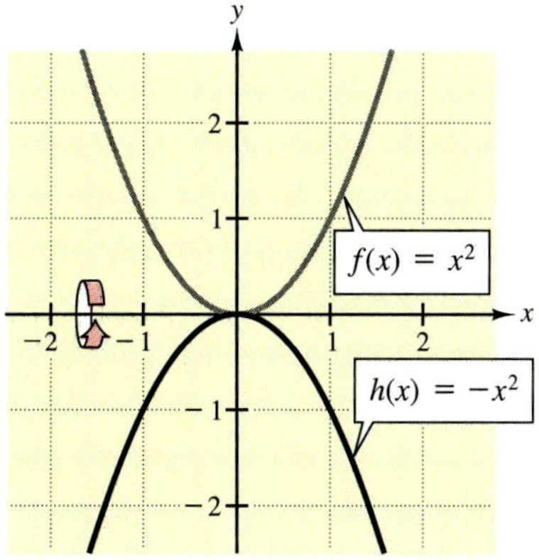

The second basic type of transformation is a **reflection.** For instance, if you imagine that the x-axis represents a mirror, then the graph of $h(x) = -x^2$ is the mirror image (or reflection) of the graph of $f(x) = x^2$, as shown in Figure 7.48.

FIGURE 7.48   Reflection

Reflections in the Coordinate Axes

Reflections of the graph of $y = f(x)$ are represented as follows.

1. Reflection in the x-axis: $h(x) = -f(x)$

2. Reflection in the y-axis: $h(x) = f(-x)$

EXAMPLE 5 *Reflections of the Graphs of Functions*

Use the graph of $f(x) = \sqrt{x}$ to sketch the graph of each function.

a. $g(x) = -\sqrt{x}$ **b.** $h(x) = \sqrt{-x}$

Solution

a. Relative to the graph of $f(x) = \sqrt{x}$, the graph of $g(x) = -\sqrt{x}$ represents a *reflection in the x-axis*, as shown in Figure 7.49.

b. Relative to the graph of $f(x) = \sqrt{x}$, the graph of $h(x) = \sqrt{-x}$ represents a *reflection in the y-axis*, as shown in Figure 7.50.

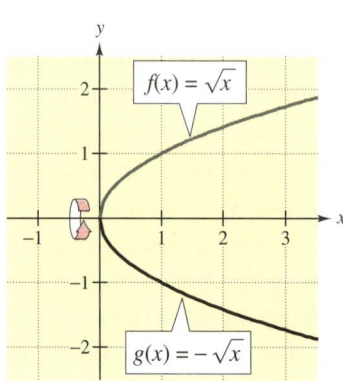

FIGURE 7.49   Reflection in x-Axis

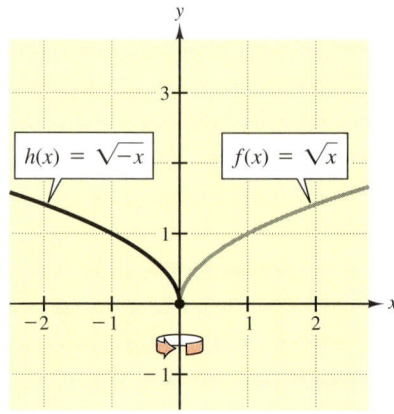

FIGURE 7.50   Reflection in y-Axis

Group Activities   Exploring with Technology

Constructing Transformations   Use a graphing utility to graph $f(x) = x^2 + 2$. Decide how to alter this function to produce each of the following transformation descriptions. Graph each transformation on the same screen with f; confirm that the transformation moved f as described.

a. The graph of f shifted to the left three units.

b. The graph of f shifted downward five units.

c. The graph of f shifted upward one unit.

d. The graph of f shifted to the right two units.

7.6 Exercises

Discussing the Concepts

1. Explain the change in the range of $f(x) = 2x$ if the domain is changed from $[0, 2]$ to $[0, 4]$.

2. In your own words, explain how to use the Vertical Line Test.

3. Describe the four types of shifts of the graph of a function.

4. Describe the relationship between the graphs of $f(x)$ and $g(x) = -f(x)$.

5. Describe the relationship between the graphs of $f(x)$ and $g(x) = f(-x)$.

6. Describe the relationship between the graphs of $f(x)$ and $g(x) = f(x - 2)$.

Problem Solving

In Exercises 7–10, use a graphing utility to graph the function and find its domain and range.

7. $g(x) = 1 - x^2$

8. $f(x) = |x + 1|$

9. $f(x) = \sqrt{x - 2}$

10. $h(t) = \sqrt{4 - t^2}$

In Exercises 11–14, use the Vertical Line Test to determine whether y is a function of x.

11. $y = \frac{1}{3}x^3$

12. $y = x^2 - 2x$

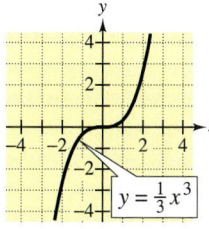

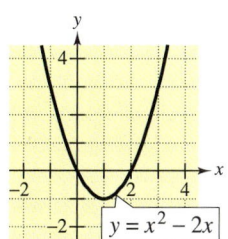

13. $y^2 = x$

14. $y = |x|$

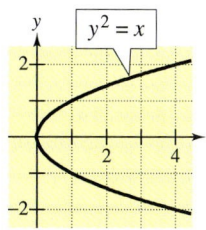

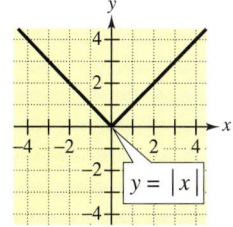

15. *Think About It*   Does the graph in Exercise 11 represent x as a function of y? Explain your reasoning.

16. *Think About It*   Does the graph in Exercise 12 represent x as a function of y? Explain your reasoning.

In Exercises 17–20, match the function with its graph. [The graphs are labeled (a), (b), (c), and (d).]

(a)

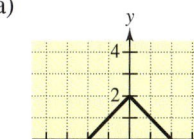

(b)

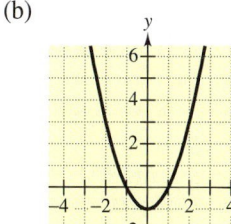

(c)

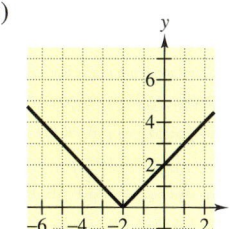

(d)

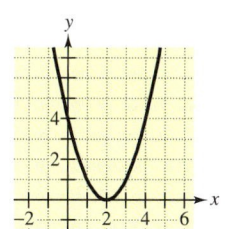

17. $f(x) = x^2 - 1$

18. $f(x) = (x - 2)^2$

19. $f(x) = 2 - |x|$

20. $f(x) = |x + 2|$

In Exercises 21–28, sketch the graph of the function. Then determine its domain and range.

21. $g(x) = \frac{1}{2}x^2$

22. $h(x) = \frac{1}{4}x^2 - 1$

23. $f(x) = -(x - 1)^2$

24. $g(x) = (x + 2)^2 + 3$

25. $K(s) = |s - 4| + 1$

26. $Q(t) = 1 - |t + 1|$

27. $h(x) = \begin{cases} 2x + 3, & \text{if } x < 0 \\ 3 - x, & \text{if } x \geq 0 \end{cases}$

28. $f(t) = \begin{cases} \sqrt{4 + t}, & \text{if } t < 0 \\ \sqrt{4 - t}, & \text{if } t \geq 0 \end{cases}$

In Exercises 29 and 30, select the viewing rectangle that shows the most complete graph of the function.

29. $f(x) = -(x^2 - 20x + 50)$

(a)

Xmin = 0
Xmax = 10
Xscl = 2
Ymin = 0
Ymax = 30
Yscl = 2

(b)

Xmin = 0
Xmax = 20
Xscl = 2
Ymin = -10
Ymax = 60
Yscl = 6

(c)

Xmin = 15
Xmax = 30
Xscl = 2
Ymin = -10
Ymax = 60
Yscl = 5

30. $f(x) = x^4 - 10x^3$

(a)

Xmin = -10
Xmax = 10
Xscl = 1
Ymin = -10
Ymax = 10
Yscl = 1

(b)

Xmin = -5
Xmax = 5
Xscl = 1
Ymin = -10
Ymax = 20
Yscl = 2

(c)

Xmin = -3
Xmax = 12
Xscl = 1
Ymin = -1200
Ymax = 400
Yscl = 100

In Exercises 31–36, identify the transformation of the graph of $f(x) = x^2$ and sketch the graph of h.

31. $h(x) = x^2 + 2$ **32.** $h(x) = x^2 - 4$

33. $h(x) = (x + 2)^2$ **34.** $h(x) = (x - 4)^2$

35. $h(x) = -x^2$ **36.** $h(x) = -x^2 + 4$

In Exercises 37–42, use the graph of $f(x) = \sqrt{x}$ to write a function that represents the graph.

37.

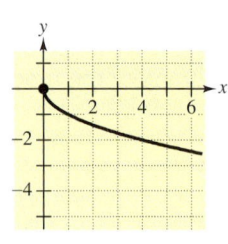

38.

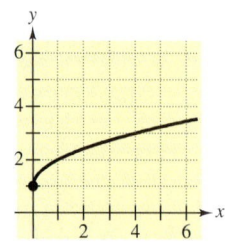

39.

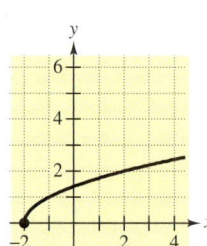

40.

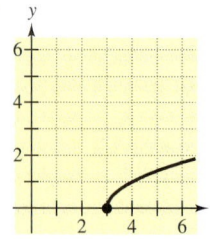

41.

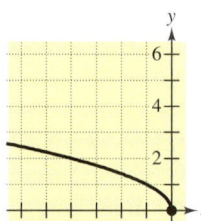

42.

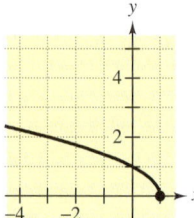

43. *Geometry* The perimeter of a rectangle is 200 meters (see figure).

(a) Show that the area of the rectangle is given by $A = x(100 - x)$, where x is its length.

(b) Use a graphing utility to graph the area function.

(c) Graphically, what value of x yields the largest value of A? Interpret the results.

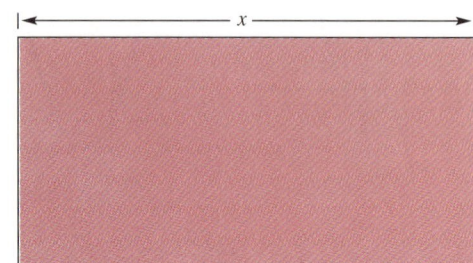

44. *Profit* The profit P when x units of a product are sold is given by $P(x) = 0.47x - 100$ for x in the interval $0 \leq x \leq 1000$.

(a) Use a graphing utility to graph the profit function over the specified domain.

(b) Approximately how many units must be sold for the company to break even $(P = 0)$?

(c) Approximately how many units must be sold for the company to make a profit of $300?

Reviewing the Major Concepts

In Exercises 45–48, solve the equation.

45. $4 - \frac{1}{2}x = 6$ **46.** $500 - 0.75x = 235$

47. $4(x - 3) = 2x$ **48.** $12(3 - x) = 5 - 7x$

49. *Phone Charges*   The cost for a phone call is $1.10 for the first minute and $0.45 for each additional minute. The total cost of a call cannot exceed $11. Find the interval of time that is available for the call.

50. *Operating Cost*   A fuel company has a fleet of trucks. The annual operating cost per truck is

$$C = 0.65m + 4500$$

where m is the number of miles traveled by a truck in a year. What number of miles will yield an annual operating cost that is less than $20,000?

Additional Problem Solving

In Exercises 51–54, use the Vertical Line Test to determine whether y is a function of x.

51. $y = (x + 2)^2$ **52.** $x - 2y^2 = 0$

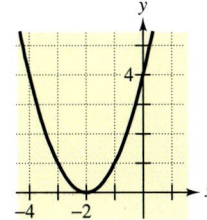
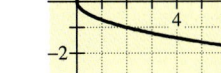

53. $x^2 + y^2 = 16$ **54.** $y = x^3 - 3$

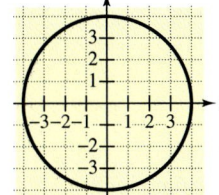
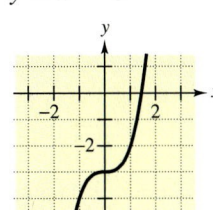

In Exercises 55–58, sketch the graph of the equation. Does the graph represent y as a function of x?

55. $3x - 5y = 15$ **56.** $y = x^2 + 2$

57. $y^2 = x + 1$ **58.** $x = y^4$

In Exercises 59–76, sketch the graph of the function. Then determine its domain and range.

59. $f(x) = 2x - 7$ **60.** $f(x) = 3 - 2x$

61. $h(x) = x^2 - 6x + 8$ **62.** $f(x) = -x^2 - 2x + 1$

63. $f(t) = \sqrt{t - 2}$ **64.** $h(x) = \sqrt{4 - x}$

65. $g(s) = \frac{1}{2}s^3$ **66.** $f(x) = x^3 - 4$

67. $f(x) = |x + 3|$ **68.** $g(x) = 2 - |x - 1|$

69. $f(x) = 6 - 3x, \quad 0 \le x \le 2$

70. $f(x) = \frac{1}{3}x - 2, \quad 6 \le x \le 12$

71. $h(x) = x^3, \quad -2 \le x \le 2$

72. $h(x) = x(6 - x), \quad 0 \le x \le 6$

73. $f(x) = \begin{cases} x + 6, & \text{if } x < 0 \\ 6 - 2x, & \text{if } x \ge 0 \end{cases}$

74. $f(x) = \begin{cases} -x, & \text{if } x \le 0 \\ x^2 - 4x, & \text{if } x > 0 \end{cases}$

75. $h(x) = \begin{cases} 4 - x^2, & \text{if } x \le 2 \\ x - 2, & \text{if } x > 2 \end{cases}$

76. $f(x) = \begin{cases} x^2, & \text{if } x < 1 \\ x^2 - 3x + 2, & \text{if } x \ge 1 \end{cases}$

In Exercises 77–82, identify the transformation of the graph of $f(x) = x^3$ and sketch the graph of h.

77. $h(x) = x^3 + 3$ **78.** $h(x) = x^3 - 5$

79. $h(x) = (x - 3)^3$ **80.** $h(x) = -x^3$

81. $h(x) = 2 - (x - 1)^3$ **82.** $h(x) = (x + 2)^3 - 3$

In Exercises 83–88, identify the transformation of the graph of $f(x) = |x|$ and use a graphing utility to sketch the graph of h.

83. $h(x) = |x - 5|$ **84.** $h(x) = |x + 3|$

85. $h(x) = |x| - 5$ **86.** $h(x) = |-x|$

87. $h(x) = -|x|$ **88.** $h(x) = 5 - |x|$

In Exercises 89–96, use the graph of $f(x) = x^2$ to write an equation that represents the transformation of f given by the graph.

89.

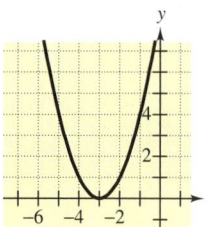

90.

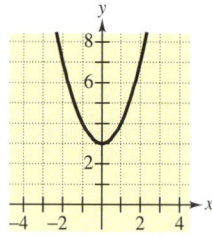

91.

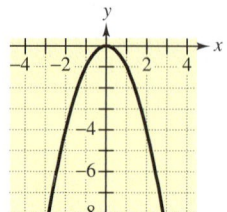

92.

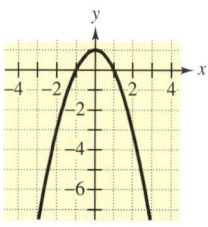

93.

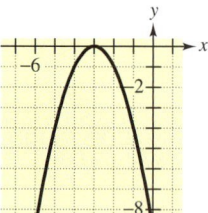

94.

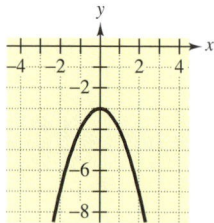

95.

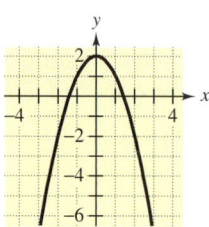

96.

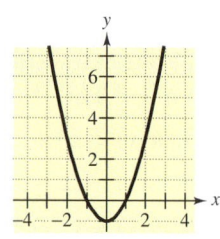

97. *Civilian Population of the United States*   For 1950 through 1990, the civilian population P (in thousands) of the United States can be modeled by

$$P_1(t) = -12t^2 + 2900t + 149,300$$

where $t = 0$ represents 1950.   (Source: U. S. Bureau of Census)

(a) Use a graphing utility to graph the function over the appropriate domain.

(b) In the transformation of the population function

$$P_2(t) = -12(t+20)^2 + 2900(t+20) + 149,300$$

$t = 0$ corresponds to what calendar year? Explain.

(c) Use a graphing utility to graph P_2 over the appropriate domain.

98. *Aircraft Orders*   The following table gives the number N of new civil jet aircraft orders for U.S. aircraft manufacturers for the years 1986 through 1992. (Source: Aerospace Industries Association of America)

Year	1986	1987	1988	1989	1990	1991	1992
N	332	519	956	1015	670	280	231

A model for this data is given by

$$N(t) = \frac{3.13t + 617}{0.20t^2 + 0.58t + 1}$$

where $t = 0$ represents 1990.

(a) What is the domain of the function?

(b) Use a graphing utility to graph the data and the model on the same set of coordinate axes.

(c) For which year does the model most accurately estimate the actual data? During which year is it least accurate?

CHAPTER PROJECT: Working Men and Women

Data is frequently represented in three ways: *numerically* by a table, *graphically*, or *algebraically* by an equation, formula, or algebraic model. Because each type of representation has its own advantages, the best way to present data is to use a mix of all three representations.

In gathering data, it may be easiest to first organize the data in a table. Finding patterns or trends in the data may be easier to recognize with a graph. And, finally, making future predictions may be easier with an algebraic model.

For instance, the table below shows the numbers of men and women (in thousands) who were practicing physicians from 1970 through 1990. The data is represented graphically by the scatter plot at the left. While investigating the questions below, you will be asked to find algebraic models to represent the data, and then to use the models to interpret the data.   (Source: American Medical Association)

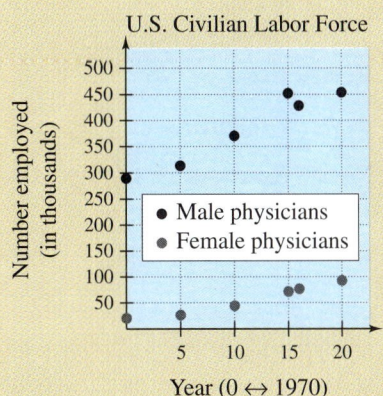

U.S. Civilian Labor Force

● Male physicians
● Female physicians

Year (0 ↔ 1970)

Year	1970	1975	1980	1985	1986	1990
Women Physicians	21.4	27.2	44.7	71.9	76.8	93.3
Men Physicians	289.5	313.1	370.2	452.3	429.0	454.0

1. *Graphical Reasoning*   Use the scatter plot to write a verbal description of the data. Discuss any trend or pattern that is evident from the scatter plot.

2. *Linear Modeling*   Use graph paper to redraw the scatter plot. Approximate each of the data sets with a line. Then find an equation for each line.

3. *Linear Modeling*   Use the linear regression program on a graphing utility to find a linear model for each data set. Compare the results with those obtained in Question 2.

4. *How Well Does It Fit?*   When the regression program in Question 3 is run, it will display a correlation coefficient r that measures how well the linear model fits the data. The closer r is to 1, the better the model fits the data. Which of the two models fits its data better? Does your answer seem reasonable from the graphical point of view?

5. *Prediction*   Predict the numbers of women and men physicians in 1995. Discuss different ways that you could obtain the prediction. Which method do you prefer? Why?

6. *Prediction*   Extend the lines you drew in Question 2 until they intersect. What interpretation can you make? Is the interpretation realistic in the context of the data?

7. *Research Project*   Use your school's library or another reference source to find data for the numbers of men and women in an occupation. Organize the data numerically, graphically, and algebraically. What can you conclude?

CHAPTER SUMMARY

After studying this chapter, you should have acquired the following skills. These skills are keyed to the Review Exercises that begin on page 443. Answers to odd-numbered Review Exercises are given in the back of the book.

- Plot points on a rectangular coordinate system. *(Section 7.1)*   **Review Exercises 1, 2**

- Plot points that form polygons. *(Section 7.1)*   **Review Exercises 3, 4**

- Determine the quadrants in which points are located. *(Section 7.1)*   **Review Exercises 5–8**

- Plot pairs of points and find the distances between them. *(Section 7.1)*   **Review Exercises 9–12**

- Determine whether ordered pairs are solutions of equations. *(Section 7.1)*   **Review Exercises 13, 14**

- Sketch graphs of equations and label the *x*- and *y*-intercepts. *(Section 7.2)*   **Review Exercises 15–20**

- Find the slopes of the lines through pairs of points. *(Section 7.4)*   **Review Exercises 21–26**

- Find missing values so that sets of points are collinear. *(Section 7.4)*   **Review Exercises 27, 28**

- Find additional points on lines given a point on each line and the slope. *(Section 7.4)*   **Review Exercises 29–34**

- Write equations of lines in slope-intercept form and sketch the lines. *(Section 7.4)*   **Review Exercises 35–38**

- Use a graphing utility to graph pairs of equations to determine whether the pairs of lines are parallel, perpendicular, or neither. *(Sections 7.3, 7.4)*   **Review Exercises 39–42**

- Decide whether relations are functions. *(Section 7.5)*   **Review Exercises 43–46**

- Estimate the intercepts of graphs, check them algebraically, and determine whether the graphs represent *y* as a function of *x*. *(Sections 7.2, 7.6)*   **Review Exercises 47–50**

- Evaluate functions and simplify. *(Section 7.5)*   **Review Exercises 51–58**

- Find the domains of functions. *(Section 7.5)*   **Review Exercises 59–62**

- Select viewing rectangles on a graphing utility that show the most complete graphs of functions. *(Sections 7.3, 7.6)*   **Review Exercises 63, 64**

- Sketch graphs of functions and check results using a graphing utility. *(Sections 7.3, 7.6)*   **Review Exercises 65–76**

- Identify and sketch transformations of graphs. *(Section 7.6)*   **Review Exercises 77–80**

- Solve real-life problems modeled by functions. *(Sections 7.5, 7.6)*   **Review Exercises 81–84**

REVIEW EXERCISES

In Exercises 1 and 2, plot the points on a rectangular coordinate system.

1. $(0, -3)$, $\left(\frac{5}{2}, 5\right)$, $(-2, -4)$

2. $\left(1, -\frac{3}{2}\right)$, $\left(-2, 2\frac{3}{4}\right)$, $(5, 10)$

Geometry In Exercises 3 and 4, plot the points and verify that the points form the indicated polygon.

3. *Right Triangle:* $(1, 1)$, $(12, 9)$, $(4, 20)$

4. *Parallelogram:* $(0, 0)$, $(7, 1)$, $(8, 4)$, $(1, 3)$

In Exercises 5–8, determine the quadrant(s) in which the point is located.

5. $(2, -6)$

6. $(-4.8, -2)$

7. $(4, y)$

8. (x, y), $\quad xy > 0$

In Exercises 9–12, plot the points and find the distance between them.

9. $(4, 3)$, $(4, 8)$

10. $(2, -5)$, $(6, -5)$

11. $(-5, -1)$, $(1, 2)$

12. $(-2, 10)$, $(3, -2)$

In Exercises 13 and 14, determine whether the ordered pairs are solutions of the equation.

13. $y = 4 - \frac{1}{2}x$ (a) $(4, 2)$ (b) $(-1, 5)$
                                     (c) $(-4, 0)$ (d) $(8, 0)$

14. $3x - 2y + 18 = 0$ (a) $(3, 10)$ (b) $(0, 9)$
                                     (c) $(-4, 3)$ (d) $(-8, 0)$

In Exercises 15–20, sketch the graph of the equation. Label the x- and y-intercepts.

15. $y = 6 - \frac{1}{3}x$

16. $y = \frac{3}{4}x - 2$

17. $3y - 2x - 3 = 0$

18. $3x + 4y + 12 = 0$

19. $x = |y - 3|$

20. $y = 1 - x^3$

In Exercises 21–26, find the slope of the line through the points.

21. $(-1, 1)$, $(6, 3)$

22. $(-2, 5)$, $(3, -8)$

23. $(-1, 3)$, $(4, 3)$

24. $(7, 2)$, $(7, 8)$

25. $(0, 6)$, $(8, 0)$

26. $(0, 0)$, $\left(\frac{7}{2}, 6\right)$

In Exercises 27 and 28, find t so that the three points are collinear. (*Note:* Collinear means that the points lie on the same straight line.)

27. $(-3, -3)$, $(0, t)$, $(1, 3)$

28. $(2, 1)$, $(1, t)$, $(8, 3)$

In Exercises 29–34, a point on a line and the slope of the line are given. Find two additional points on the line.

29. $(2, -4)$, $m = -3$

30. $\left(-4, \frac{1}{2}\right)$, $m = 2$

31. $(3, 1)$, $m = \frac{5}{4}$

32. $(7, -2)$, $m = 0$

33. $(3, 7)$, m is undefined.

34. $\left(-3, -\frac{3}{2}\right)$, $m = -\frac{1}{3}$

In Exercises 35–38, write the equation of the line in slope-intercept form and sketch the line.

35. $5x - 2y - 4 = 0$

36. $x - 3y - 6 = 0$

37. $x + 2y - 2 = 0$

38. $y - 6 = 0$

In Exercises 39–42, use a graphing utility to graph the equations on the same viewing rectangle. Are the lines parallel, perpendicular, or neither?

39. L_1: $y = \frac{3}{2}x + 1$
       L_2: $y = \frac{2}{3}x - 1$

40. L_1: $y = 2x - 5$
       L_2: $y = 2x + 3$

41. L_1: $y = \frac{3}{2}x - 2$
       L_2: $y = -\frac{2}{3}x + 1$

42. L_1: $y = -0.3x - 2$
       L_2: $y = 0.3x + 1$

In Exercises 43–46, decide whether the relation is a function.

43.
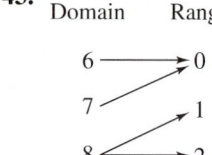

Domain   Range

44.
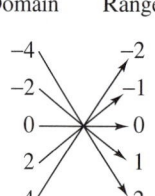

Domain   Range

45.

Domain   Range

46.
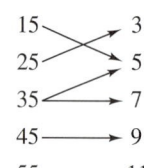

Domain   Range

In Exercises 47–50, graphically estimate the intercepts. Then check your estimate algebraically. Does the graph represent y as a function of x?

47. $9y^2 = 4x^3$

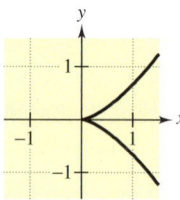

48. $y = 4x^3 - x^4$

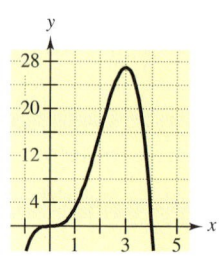

49. $y = x^2(x - 3)$

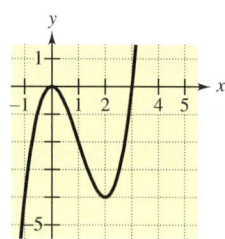

50. $x^3 + y^3 - 6xy = 0$

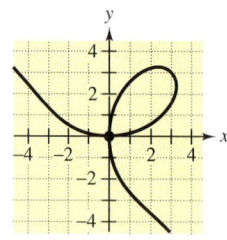

In Exercises 51–58, evaluate the function for the specified values of the independent variable and simplify when possible.

51. $f(x) = 4 - \frac{5}{2}x$

(a) $f(-10)$ (b) $f\left(\frac{2}{5}\right)$

(c) $f(t) + f(-4)$ (d) $f(x + h)$

52. $h(x) = x(x - 8)$

(a) $h(8)$ (b) $h(10)$

(c) $h(-3)$ (d) $h(t + 4)$

53. $f(t) = \sqrt{5 - t}$

(a) $f(-3)$ (b) $f(5)$

(c) $f\left(\frac{11}{3}\right)$ (d) $f(5z)$

54. $g(x) = \dfrac{|x + 4|}{4}$

(a) $g(0)$ (b) $g(-8)$

(c) $g(2) - g(-5)$ (d) $g(x - 2)$

55. $f(x) = \begin{cases} -3x, & \text{if } x \le 0 \\ 1 - x^2, & \text{if } x > 0 \end{cases}$

(a) $f(2)$ (b) $f\left(-\frac{2}{3}\right)$

(c) $f(1)$ (d) $f(4) - f(3)$

56. $h(x) = \begin{cases} x^3, & \text{if } x \le 1 \\ (x - 1)^2 + 1, & \text{if } x > 1 \end{cases}$

(a) $h(2)$ (b) $h\left(-\frac{1}{2}\right)$

(c) $h(0)$ (d) $h(4) - h(3)$

57. $f(x) = 3 - 2x$

(a) $\dfrac{f(x + 2) - f(2)}{x}$ (b) $\dfrac{f(x - 3) - f(3)}{x}$

58. $f(x) = 7x + 10$

(a) $\dfrac{f(x + 1) - f(1)}{x}$ (b) $\dfrac{f(x - 5) - f(5)}{x}$

In Exercises 59–62, find the domain of the function.

59. $h(x) = 4x^2 - 7$

60. $g(s) = \dfrac{s + 1}{(s - 1)(s + 5)}$

61. $f(x) = \sqrt{5 - 2x}$

62. $f(x) = |x - 6| + 10$

In Exercises 63 and 64, select the viewing rectangle on a graphing utility that shows the most complete graph of the function.

63. $f(x) = x^4 - 2x^3$

(a)

Xmin = 0
Xmax = 20
Xscl = 1
Ymin = -10
Ymax = 10
Yscl = 1

(b)

Xmin = -5
Xmax = 5
Xscl = 1
Ymin = -100
Ymax = 100
Yscl = 10

(c)

Xmin = -3
Xmax = 3
Xscl = 1
Ymin = -3
Ymax = 5
Yscl = 1

64. $f(x) = 5x\sqrt{16 - x^2}$

(a)

Xmin = -6
Xmax = 6
Xscl = 1
Ymin = -50
Ymax = 50
Yscl = 10

(b)

Xmin = -20
Xmax = 20
Xscl = 4
Ymin = -600
Ymax = 600
Yscl = 100

(c)

Xmin = -10
Xmax = 10
Xscl = 2
Ymin = -20
Ymax = 20
Yscl = 4

In Exercises 65–76, sketch the graph of the function. Use a graphing utility to confirm your graph.

65. $g(x) = \frac{1}{8}x^2$

66. $y = 4 - (x - 3)^2$

67. $y = (x - 2)^2$

68. $h(x) = 9 - (x - 2)^2$

69. $y = \dfrac{1}{2}x(2 - x)$

70. $f(t) = \sqrt{\dfrac{t}{2}}$

71. $y = 8 - 2|x|$

72. $f(x) = |x + 1| - 2$

73. $g(x) = \frac{1}{4}x^3, \quad -2 \le x \le 2$

74. $h(x) = x(4 - x), \quad 0 \le x \le 4$

75. $f(x) = \begin{cases} 2 - (x - 1)^2, & \text{if } x < 1 \\ 2 + (x - 1)^2, & \text{if } x \ge 1 \end{cases}$

76. $f(x) = \begin{cases} 2x, & \text{if } x \le 0 \\ x^2 + 1, & \text{if } x > 0 \end{cases}$

In Exercises 77–80, identify the transformation of the graph of $f(x) = x^4$ and sketch the graph of h.

77. $h(x) = -x^4$

78. $h(x) = x^4 + 2$

79. $h(x) = (x - 1)^4$

80. $h(x) = 1 - x^4$

81. *Path of a Projectile* The height y (in feet) of a projectile is given by

$$y = -\frac{1}{16}x^2 + 5x$$

where x is the horizontal distance (in feet) from where the projectile was launched.

(a) Sketch the path of the projectile.

(b) How high is the projectile when it is at its maximum height?

(c) How far from the launch point does the projectile strike the ground?

82. *Velocity of a Ball* The velocity of a ball thrown upward from ground level is given by

$$v = -32t + 80$$

where t is time in seconds and v is velocity in feet per second.

(a) Find the velocity when $t = 2$.

(b) Find the time when the ball reaches its maximum height. (*Hint:* Find the time when $v = 0$.)

(c) Find the velocity when $t = 3$.

83. *Power Generation* The power generated by a wind turbine is given by the function

$$P = kw^3$$

where P is the number of kilowatts produced at a wind speed of w miles per hour and k is the constant of proportionality.

(a) Find k if $P = 1000$ when $w = 20$.

(b) Find the output for a wind speed of 25 miles per hour.

84. *Geometry* A wire 100 inches long is to be cut into four pieces to form a rectangle whose shortest side has a length of x. Express the area A of the rectangle as a function of x. Use a graphing utility to graph the function.

CHAPTER TEST

Take this test as you would take a test in class. After you are done, check your work against the answers given in the back of the book.

1. Determine the quadrant in which the point (x, y) lies if $x > 0$ and $y < 0$.

2. Plot the points $(0, 5)$ and $(3, 1)$. Then find the distance between them.

3. Find the x- and y-intercepts of the graph of $y = -3(x + 1)$.

4. Sketch the graph of $y = |x - 2|$.

5. Find the slope of the line through $(-4, 7)$ and $(2, 3)$.

6. Find the slope of the line through $(3, -2)$ and $(3, 6)$.

7. Sketch the line passing through $(0, -6)$ with a slope of $m = \frac{4}{3}$.

8. Plot the x- and y-intercepts of the graph of $2x + 5y = 10$. Use the result to sketch the graph.

9. Write the equation $5x + 3y - 9 = 0$ in slope-intercept form. Find the slope of a line that is perpendicular to this line.

10. The graph of $y^2(4 - x) = x^3$ is shown at the right. Does the graph represent y as a function of x? Explain your reasoning.

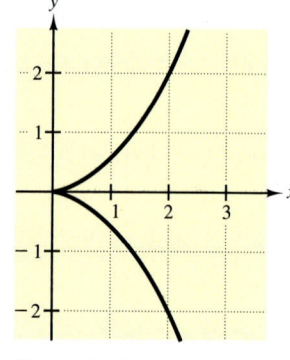

Figure for 10

11. Does $3x^2 - y^2 = 9$ represent y as a function of x? Explain.

12. Does $3x - y = 0$ represent y as a function of x? Explain.

In Exercises 13–15, evaluate $g(x) = x/(x - 3)$ for the indicated value.

13. $g(2)$ 14. $g\left(\frac{7}{2}\right)$ 15. $g(x + 2)$

16. Find the domain of $h(t) = \sqrt{9 - t}$.   17. Find the domain of $f(x) = \dfrac{x + 1}{x - 4}$.

18. Sketch the graph of $g(x) = \sqrt{2 - x}$.

19. Use a graphing utility to graph $y = -0.3x^2 + 2x + 5$. Then use the graph to estimate the intercepts of the graph.

20. Use the graph of $y = |x|$ to write an equation for each graph.

(a)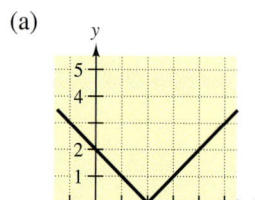

(b)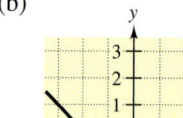

(c)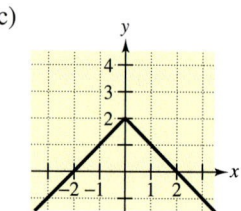

Rational Expressions and Equations

8

- Simplifying Rational Expressions
- Multiplying and Dividing Rational Expressions
- Adding and Subtracting Rational Expressions
- Dividing Polynomials
- Graphing Rational Functions
- Solving Rational Equations

When air resistance is not considered, the velocity v of an object that falls from rest can be modeled by

$$v = -32t \qquad \text{Velocity function}$$

where v is measured in feet per second and t is the time in seconds. With this model, notice that the longer the object falls, the faster it falls. This model works well for compact objects that fall for only a few seconds.

For objects with large surface areas or objects that fall for several seconds, however, air resistance hampers the object's fall so that it reaches a *terminal velocity*. The table and graph at the right compare the velocity of a sky diver with no air resistance (v_1), with air resistance but no parachute (v_2), and with air resistance and a parachute (v_3).

t	0 sec	1 sec	2 sec	3 sec	4 sec
v_1	0 ft/sec	−32 ft/sec	−64 ft/sec	−96 ft/sec	−128 ft/sec
v_2	0 ft/sec	−24 ft/sec	−38 ft/sec	−52 ft/sec	−60 ft/sec
v_3	0 ft/sec	−18 ft/sec	−20 ft/sec	−20 ft/sec	−20 ft/sec

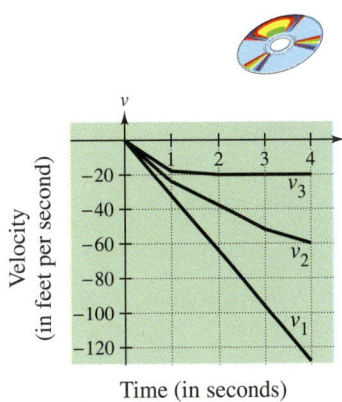

The chapter project related to this information is on page 507.

447

| 8.1 | **Simplifying Rational Expressions** |

The Domain of a Rational Expression ▪
Simplifying Rational Expressions

The Domain of a Rational Expression

Algebra may seem simpler when you realize that it consists primarily of a basic set of operations, including addition, subtraction, multiplication, division, factoring, and simplifying, that are applied to different types of algebraic expressions. In this chapter, you will apply these operations to *rational expressions*.

NOTE   Like polynomials, rational expressions can be used to describe functions. Such functions are called **rational functions.**

Definition of a Rational Expression

Let u and v be polynomials. The algebraic expression

$$\frac{u}{v}$$

is a **rational expression.** The **domain** of this rational expression is the set of all real numbers for which $v \neq 0$.

EXAMPLE 1   *Finding the Domain of a Rational Function*

Find the domains of the following rational functions.

a. $f(x) = \dfrac{4}{x-2}$ **b.** $g(x) = \dfrac{2x+5}{8}$ **c.** $h(x) = 3x^2 + 2x - 5$

Solution

a. The denominator is zero when $x - 2 = 0$ or $x = 2$. Therefore, the domain is all real values of x such that $x \neq 2$. In interval notation, you can write the domain as

$$\text{Domain} = (-\infty, 2) \cup (2, \infty).$$

b. The denominator, 8, is never zero, hence, the domain is the set of *all* real numbers. In interval notation, you can write the domain as

$$\text{Domain} = (-\infty, \infty).$$

c. Note that any polynomial is also a rational expression, because you can consider its denominator to be 1. The domain of this function is the set of all real numbers.

DISCOVERY

Use a graphing utility to graph the equation

$$y = \frac{4}{x-2}.$$

Then use the TRACE feature of the utility to determine the behavior of the graph near $x = 2$. Try graphing equations that correspond to parts (b) and (c) of Example 1. How does each of these graphs differ from the graph of $y = 4/(x - 2)$?

In applications involving rational functions, it is often necessary to further restrict the domain. To indicate such a restriction, write the domain to the right of the fraction. For instance, the domain of the rational function

$$f(x) = \frac{x^2 + 20}{x + 4}, \qquad x > 0$$

is the set of positive real numbers, as indicated by the inequality $x > 0$. (Note that the normal domain of this function would be all real values of x such that $x \neq -4$. However, because "$x > 0$" is listed to the right of the function, the domain is restricted by this inequality.)

EXAMPLE 2 An Application Involving a Restricted Domain

You have started a small manufacturing business. The initial investment for the business is $120,000. The cost of each unit that you manufacture is $15. Thus, your total cost of producing x units is

$$C = 15x + 120,000. \qquad \textcolor{red}{\text{Cost function}}$$

Your average cost per unit depends on the number of units produced. For instance, the average cost per unit $\overline{C}$ for producing 100 units is

$$\overline{C} = \frac{15(100) + 120,000}{100} = \$1215. \qquad \textcolor{red}{\text{Average cost per unit for 100 units}}$$

The average cost per unit decreases as the number of units increases. For instance, the average cost per unit $\overline{C}$ for producing 1000 units is

$$\overline{C} = \frac{15(1000) + 120,000}{1000} = \$135. \qquad \textcolor{red}{\text{Average cost per unit for 1000 units}}$$

In general, the average cost of producing x units is

$$\overline{C} = \frac{15x + 120,000}{x}. \qquad \textcolor{red}{\text{Average cost per unit for } x \text{ units}}$$

What is the domain of this rational function?

Solution

If you were considering this function from only a mathematical point of view, you would say that the domain is all real values of x such that $x \neq 0$. However, because this fraction is a mathematical model representing a real-life situation, you must consider which values of x make sense in real life. For this model, the variable x represents the number of units that you produce. Assuming that you cannot produce a fractional number of units, you conclude that the domain is the set of positive integers. That is,

$$\text{Domain} = \{1, 2, 3, 4, \ldots\}.$$

Simplifying Rational Expressions

As with numerical fractions, a rational expression is said to be **simplified** or **in reduced form** if its numerator and denominator have no factors in common (other than ±1). To reduce fractions, you can apply the following rule.

Cancellation Rule for Fractions

Let u, v, and w represent numbers, variables, or algebraic expressions such that $v \neq 0$ and $w \neq 0$. Then the following Cancellation Rule is valid.

$$\frac{u \not{w}}{v \not{w}} = \frac{u}{v}.$$

In the 19th century, mathematicians were expanding their knowledge of calculus and laying the foundation for complex numbers. At that time, the properties of real numbers had not been finalized. Teaching at the University of Berlin, Karl Weierstrass (1815–1897) recognized the need for a logical foundation for the real number system. His work contributed much to the formal real number system that forms the foundation of modern algebra.

Be sure you see that this Cancellation Rule allows us to cancel only factors, not terms. For instance, consider the following.

$$\frac{\not{2} \cdot 2}{\not{2}(x+5)}$$
      You can cancel common factor 2.

$$\frac{3+x}{3+2x}$$
      You cannot cancel common term 3.

Using the Cancellation Rule to simplify a rational expression requires two steps: (1) completely factor the numerator and denominator and (2) apply the Cancellation Rule to cancel any *factors* that are common to both the numerator and denominator. Thus, your success in simplifying rational expressions actually lies in your ability to *completely factor* the polynomials in both the numerator and denominator.

EXAMPLE 3   *Simplifying a Rational Expression*

Simplify $\dfrac{2x^3 - 6x}{6x^2}$.

Solution

To begin, completely factor both the numerator and denominator.

$$\frac{2x^3 - 6x}{6x^2} = \frac{2x(x^2 - 3)}{2x(3x)}$$
      Factor numerator and denominator.

$$= \frac{\not{2x}(x^2 - 3)}{\not{2x}(3x)}$$
      Cancel common factor $2x$.

$$= \frac{x^2 - 3}{3x}$$
      Simplified form

EXAMPLE 4 *Adjusting the Domain After Simplifying*

Simplify $\dfrac{x^2 + 2x - 15}{3x - 9}$.

Solution

$$\frac{x^2 + 2x - 15}{3x - 9} = \frac{(x + 5)(x - 3)}{3(x - 3)} \qquad \text{Factor numerator and denominator.}$$

$$= \frac{(x + 5)\cancel{(x - 3)}}{3\cancel{(x - 3)}} \qquad \text{Cancel common factor } (x - 3).$$

$$= \frac{x + 5}{3}, \quad x \neq 3 \qquad \text{Simplified form}$$

Canceling common factors from the numerator and denominator of a rational expression can change its domain. For instance, in Example 4 the domain of the original expression is all real values of x such that $x \neq 3$. Thus, the original expression is equal to the simplified expression for all real numbers *except* 3.

EXAMPLE 5 *Simplifying a Rational Expression*

Simplify $\dfrac{x^3 - 16x}{x^2 - 2x - 8}$.

Solution

$$\frac{x^3 - 16x}{x^2 - 2x - 8} = \frac{x(x^2 - 16)}{(x + 2)(x - 4)} \qquad \text{Partially factor.}$$

$$= \frac{x(x + 4)(x - 4)}{(x + 2)(x - 4)} \qquad \text{Factor completely.}$$

$$= \frac{x(x + 4)\cancel{(x - 4)}}{(x + 2)\cancel{(x - 4)}} \qquad \text{Cancel common factor } (x - 4).$$

$$= \frac{x(x + 4)}{x + 2}, \quad x \neq 4 \qquad \text{Simplified form}$$

In this text, when simplifying a rational expression, we follow the convention of listing *by the simplified expression* all values of x that must be specifically excluded from the domain in order to make the domains of the simplified and original expressions agree. For instance, in Example 5 the restriction $x \neq 4$ must be listed with the simplified expression in order to make the two domains agree. (Note that the value of -2 is excluded from both domains, so it is not necessary to list this value.)

STUDY TIP

Be sure to *completely* factor the numerator and denominator of an algebraic fraction before concluding that there is no common factor. This may involve a change in sign to see if further reduction is possible. Note that the Distributive Property allows you to write $(b - a)$ as $-(a - b)$. Watch for this in Example 6.

EXAMPLE 6 Simplification Involving a Change of Sign

Simplify $\dfrac{2x^2 - 9x + 4}{12 + x - x^2}$.

Solution

$$\frac{2x^2 - 9x + 4}{12 + x - x^2} = \frac{(2x - 1)(x - 4)}{(4 - x)(3 + x)} \qquad \text{Factor numerator and denominator.}$$

$$= \frac{(2x - 1)(x - 4)}{-(x - 4)(3 + x)} \qquad (4 - x) = -(x - 4)$$

$$= \frac{(2x - 1)\cancel{(x - 4)}}{-\cancel{(x - 4)}(3 + x)} \qquad \text{Cancel common factor } (x - 4).$$

$$= -\frac{2x - 1}{3 + x}, \quad x \neq 4 \qquad \text{Simplified form}$$

The simplified form is equivalent to the original expression for all values of x except 4. (Note that -3 is excluded from the domains of both the original and simplified expressions.)

In Example 6, be sure you see that when dividing the numerator and denominator by the common factor of $(x - 4)$, you keep the minus sign. In the simplified form of the fraction, we usually like to move the minus sign out in front of the fraction. However, this is a personal preference. All of the following forms are legitimate.

$$-\frac{2x - 1}{3 + x} = \frac{-(2x - 1)}{3 + x} = \frac{2x - 1}{-3 - x} = \frac{2x - 1}{-(3 + x)}$$

In the next two examples, the Cancellation Rule is used to simplify rational expressions that involve more than one variable.

EXAMPLE 7 A Rational Expression Involving Two Variables

Simplify $\dfrac{3xy + y^2}{2y}$.

Solution

$$\frac{3xy + y^2}{2y} = \frac{y(3x + y)}{2y} \qquad \text{Factor numerator and denominator.}$$

$$= \frac{\cancel{y}(3x + y)}{2\cancel{y}} \qquad \text{Cancel common factor } y.$$

$$= \frac{3x + y}{2}, \quad y \neq 0 \qquad \text{Simplified form}$$

EXAMPLE 8 A Rational Expression Involving Two Variables

Simplify $\dfrac{2x^2 + 2xy - 4y^2}{5x^3 - 5xy^2}$.

Solution

$$\frac{2x^2 + 2xy - 4y^2}{5x^3 - 5xy^2} = \frac{2(x - y)(x + 2y)}{5x(x - y)(x + y)} \qquad \text{Factor numerator and denominator.}$$

$$= \frac{2(x - y)(x + 2y)}{5x(x - y)(x + y)} \qquad \text{Cancel common factor } (x - y).$$

$$= \frac{2(x + 2y)}{5x(x + y)}, \quad x \neq y \qquad \text{Simplified form}$$

As you study the examples and work the exercises in this and the following three sections, keep in mind that you are *rewriting expressions in simpler forms.* You are not solving equations. Equal signs are used in the steps of the simplification process only to indicate that the new form of the expression is *equivalent* to the previous one.

Group Activities   You Be the Instructor

Error Analysis   Suppose you are the instructor of an algebra course. One of your students turns in the following incorrect solutions. Find the errors, discuss the student's misconceptions, and construct correct solutions.

a. $\dfrac{3x^2 + 5x - 4}{x} = 3x + 5 - 4 = 3x + 1$

b. $\dfrac{x^2 + 7x}{x + 7} = \dfrac{x^2}{x} + \dfrac{7x}{7} = x + x = 2x$

8.1 Exercises

Discussing the Concepts

1. Define the term *rational expression.*

2. Give an example of a rational function whose domain is the set of all real numbers.

3. How do you determine whether a rational expression is in reduced form?

4. Can you cancel common terms from the numerator and denominator of a rational expression? Explain.

5. *Error Analysis*   Describe the error.

$$\frac{2x^2}{x^2+4} = \frac{2\cancel{x^2}}{\cancel{x^2}+4} = \frac{2}{1+4} = \frac{2}{5}$$

6. Is the following statement true? Explain.

$$\frac{6x-5}{5-6x} = -1$$

Problem Solving

In Exercises 7–12, find the domain of the expression.

7. $\dfrac{5}{x-8}$

8. $\dfrac{9}{x-13}$

9. $\dfrac{x}{x^2+4}$

10. $\dfrac{x}{x^2-4}$

11. $\dfrac{y+5}{y^2-3y}$

12. $\dfrac{3t}{t^2-2t-3}$

In Exercises 13–16, evaluate the function as indicated. If it is not possible, state the reason.

13. $f(x) = \dfrac{4x}{x+3}$

   (a) $f(1)$ (b) $f(-2)$

   (c) $f(-3)$ (d) $f(0)$

14. $g(t) = \dfrac{t-2}{2t-5}$

   (a) $g(2)$ (b) $g\left(\frac{5}{2}\right)$

   (c) $g(-2)$ (d) $g(0)$

15. $h(s) = \dfrac{s^2}{s^2-s-2}$

   (a) $h(10)$ (b) $h(0)$

   (c) $h(-1)$ (d) $h(2)$

16. $f(x) = \dfrac{x^3+1}{x^2-6x+9}$

   (a) $f(-1)$ (b) $f(3)$

   (c) $f(-2)$ (d) $f(2)$

In Exercises 17 and 18, describe the domain.

17. *Inventory Cost*   The inventory cost I when x units of a product are ordered from a supplier is given by $I = (0.25x + 2000)/x.$

18. *Average Cost*   The average cost $\overline{C}$ for a manufacturer to produce x units of a product is given by $\overline{C} = (1.35x + 4570)/x.$

In Exercises 19–22, complete the statement.

19. $\dfrac{5}{6} = \dfrac{5()}{6(x+3)}, \quad x \neq -3$

20. $\dfrac{5x}{12} = \dfrac{25x^2(x-10)}{12()}, \quad x \neq 10$

21. $\dfrac{8x}{x-5} = \dfrac{8x()}{x^2-3x-10}, \quad x \neq -2, \; x \neq 5$

22. $\dfrac{3-z}{z^2} = \dfrac{(3-z)()}{z^3+2z^2}, \quad z \neq -2, \; x \neq 0$

In Exercises 23–34, simplify the expression.

23. $\dfrac{5x}{25}$

24. $\dfrac{32y}{24}$

25. $\dfrac{18x^2y}{15xy^4}$

26. $\dfrac{16y^2z^2}{60y^5z}$

27. $\dfrac{3xy^2}{xy^2+x}$

28. $\dfrac{x+3x^2y}{3xy+1}$

29. $\dfrac{y^2 - 64}{5(3y + 24)}$

30. $\dfrac{x^2 - 25z^2}{x + 5z}$

31. $\dfrac{3 - x}{2x^2 - 3x - 9}$

32. $\dfrac{2y^2 + 13y + 20}{2y^2 + 17y + 30}$

33. $\dfrac{3m^2 - 12n^2}{m^2 + 4mn + 4n^2}$

34. $\dfrac{x^2 + xy - 2y^2}{x^2 + 3xy + 2y^2}$

Using a Table In Exercises 35 and 36, complete the table. What can you conclude?

35.

x	-2	-1	0	1	2	3	4
$\dfrac{x^2 - x - 2}{x - 2}$							
$x + 1$							

36.

x	-2	-1	0	1	2	3	4
$\dfrac{x^2 + 5x}{x}$							
$x + 5$							

Geometry In Exercises 37 and 38, find the ratio of the area of the shaded portion to the total area of the figure.

37.

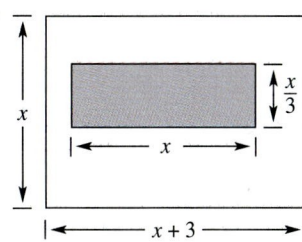

38.

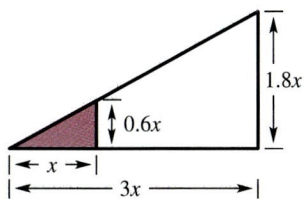

Creating and Using a Model In Exercises 39 and 40, use the following polynomial models, which give the cost of Medicare and the U.S. population aged 65 and older, for the years 1990 to 1995 (see figures).

$C = 107.1 + 12.64t + 0.54t^2$ Cost of Medicare

$P = 31.6 + 0.51t - 0.14t^2$ Population

In these models, C represents the total annual cost of Medicare (in billions of dollars), P represents the U.S. population aged 65 and older, and t represents the year, with $t = 0$ corresponding to 1990. (Source: Congressional Budget Office and U.S. Bureau of Census)

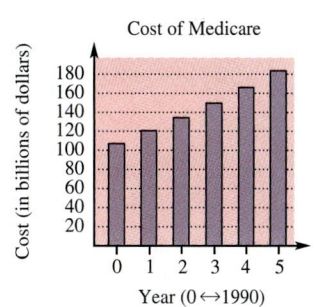

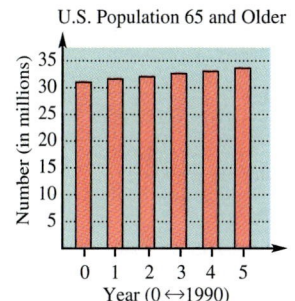

39. Find a rational model that represents the average cost of Medicare *per person aged 65 and older* during the years 1990 to 1995.

40. Use the model found in Exercise 39 to complete the following table, which shows the average cost of Medicare per person aged 65 and older.

Year	Average Cost
1990	
1991	
1992	
1993	
1994	
1995	

Reviewing the Major Concepts

In Exercises 41–44, simplify the expression.

41. $-6x(10 - 7x)$

42. $t(t^2 + 1) - t(t^2 - 1)$

43. $(11 - x)(11 + x)$

44. $(x - 2)(x^2 + 2x + 4)$

In Exercises 45 and 46, write the quantity as a repeated multiplication.

45. $\left(\dfrac{y}{3}\right)^4$

46. $(-2x)^5$

Additional Problem Solving

In Exercises 47–54, find the domain of the expression.

47. $\dfrac{7x}{x + 4}$

48. $x^4 + 2x^2 - 5$

49. $\dfrac{4}{x^2 + 9}$

50. $\dfrac{y^2 - 3}{7}$

51. $\dfrac{5t}{t^2 - 16}$

52. $\dfrac{z + 2}{z(z - 4)}$

53. $\dfrac{u^2}{u^2 - 2u - 5}$

54. $\dfrac{y + 5}{4y^2 - 5y - 6}$

In Exercises 55 and 56, describe the domain.

55. *Geometry* A rectangle of length x inches has an area of 500 square inches. The perimeter of the rectangle is given by

$$P = 2\left(x + \frac{500}{x}\right).$$

56. *Cost* The cost in millions of dollars for the government to seize $p\%$ of a certain illegal drug as it enters the country is given by

$$C = \frac{528p}{100 - p}.$$

In Exercises 57–60, complete the statement.

57. $\dfrac{x}{2} = \dfrac{3x(x + 16)^2}{2()}, \quad x \neq -16$

58. $\dfrac{7}{15} = \dfrac{7()}{45(x - 10)^2}, \quad x \neq 10$

59. $\dfrac{x + 5}{3x} = \dfrac{(x + 5)()}{3x^2(x - 2)}, \quad x \neq 2$

60. $\dfrac{3y - 7}{y + 2} = \dfrac{(3y - 7)()}{y^2 - 4}, \quad y \neq 2$

In Exercises 61–80, simplify the expression.

61. $\dfrac{12y^2}{2y}$

62. $\dfrac{15z^3}{15z^3}$

63. $\dfrac{x^2(x - 8)}{x(x - 8)}$

64. $\dfrac{a^2b(b - 3)}{b^3(b - 3)^2}$

65. $\dfrac{2x - 3}{4x - 6}$

66. $\dfrac{y^2 - 81}{2y - 18}$

67. $\dfrac{5 - x}{3x - 15}$

68. $\dfrac{x^2 - 36}{6 - x}$

69. $\dfrac{a + 3}{a^2 + 6a + 9}$

70. $\dfrac{u^2 - 12u + 36}{u - 6}$

71. $\dfrac{x^2 - 7x}{x^2 - 14x + 49}$

72. $\dfrac{z^2 + 22z + 121}{3z + 33}$

73. $\dfrac{y^3 - 4y}{y^2 + 4y - 12}$

74. $\dfrac{x^2 - 7x}{x^2 - 4x - 21}$

75. $\dfrac{15x^2 + 7x - 4}{15x^2 + x - 2}$

76. $\dfrac{56z^2 - 3z - 20}{49z^2 - 16}$

77. $\dfrac{5xy + 3x^2y^2}{xy^3}$

78. $\dfrac{4u^2v - 12uv^2}{18uv}$

79. $\dfrac{u^2 - 4v^2}{u^2 + uv - 2v^2}$

80. $\dfrac{x^2 + 4xy}{x^2 - 16y^2}$

Think About It In Exercises 81 and 82, explain how you can show that the two expressions are not equivalent.

81. $\dfrac{x - 4}{4} \neq x - 1$

82. $\dfrac{x - 4}{x} \neq -4$

In Exercises 83 and 84, write the rational expression in reduced form. (Assume n is a positive integer.)

83. $\dfrac{x^{2n} - 4}{x^n + 2}$

84. $\dfrac{x^{2n} + x^n - 12}{x^{n+1} + 4x}$

85. *Average Cost*   A machine shop has a setup cost of $2500 for the production of a new product. The cost of labor and material to produce each unit is $9.25.

(a) Write a rational expression that gives the average cost per unit when x units are produced.

(b) Determine the domain of the expression in part (a).

(c) Find the average cost per unit when $x = 100$ units.

86. *Pollution Removal*   The cost in dollars of removing $p\%$ of the air pollutants in the stack emission of a utility company is given by the rational function

$$C = \frac{80,000p}{100 - p}.$$

Determine the domain of the rational function.

87. *Geometry*   One swimming pool is circular and another is rectangular. The rectangular pool's width is three times its depth, and its length is 6 feet more than its width. The circular pool has a diameter that is twice the width of the rectangular pool, and it is 2 feet deeper. Find the ratio of the volume of the circular pool to the volume of the rectangular pool.

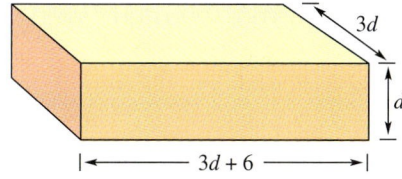

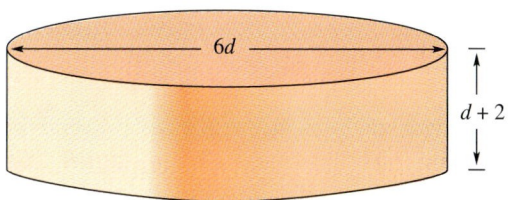

88. *Distance Traveled*   A van starts on a trip and travels at an average speed of 45 miles per hour. Three hours later, a car starts on the same trip and travels at an average speed of 60 miles per hour (see figure).

(a) Find the distance each vehicle has traveled when the car has been on the road for t hours.

(b) Use the result of part (a) to determine the ratio of the distance the car has traveled to the distance the van has traveled.

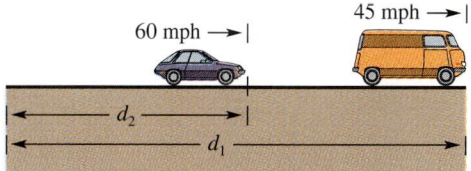

CAREER INTERVIEW

Lisa M. Deitemeyer

Civil Engineer

Johnson-Brittain & Associates, Inc.

Tucson, AZ 85701

Johnson-Brittain does highway design work, primarily for the Arizona Department of Transportation. I am responsible for drainage design of roadways and intersections. It is important that water properly drain off the road surface to avoid flooding problems. One strategy for removing excess water is to use a pipe drainage system that empties into a retention pond. When designing a pipe system and choosing pipe size, I use the equation $V = Q/A$ to find the velocity V of water moving at flow rate Q (volume per unit time) through a given pipe of cross-sectional area A. Finding the water velocity is very important. If it is too fast, erosion can occur in the retention pond. If it is too slow, sedimentation can clog the pipe. As you can see, algebra is very important to my work. I am always solving for different variables that are needed for drainage design.

8.2	**Multiplying and Dividing Rational Expressions**
	Multiplying Rational Expressions ▪ Dividing Rational Expressions ▪ Complex Fractions

Multiplying Rational Expressions

The rule for multiplying rational expressions is the same as the rule for multiplying numerical fractions.

$$\frac{3}{4} \cdot \frac{7}{6} = \frac{21}{24} = \frac{3 \cdot 7}{3 \cdot 8} = \frac{7}{8}$$

That is, you *multiply numerators, multiply denominators, and write the new fraction in reduced form.*

Multiplying Rational Expressions

Let u, v, w, and z be real numbers, variables, or algebraic expressions such that $v \neq 0$ and $z \neq 0$. Then the product of u/v and w/z is given by

$$\frac{u}{v} \cdot \frac{w}{z} = \frac{uw}{vz}.$$

In order to recognize common factors, write the numerators and denominators in factored form, as demonstrated in Example 1.

EXAMPLE 1   *Multiplying Rational Expressions*

NOTE The result in Example 1 can be read as

$$\frac{4x^3y}{3xy^4} \cdot \frac{-6x^2y^2}{10x^4}$$

is equal to

$$-\frac{4}{5y}$$

for all values of x except 0.

Multiply the rational expressions.

$$\frac{4x^3y}{3xy^4} \cdot \frac{-6x^2y^2}{10x^4}$$

Solution

$$\frac{4x^3y}{3xy^4} \cdot \frac{-6x^2y^2}{10x^4} = \frac{(4x^3y) \cdot (-6x^2y^2)}{(3xy^4) \cdot (10x^4)} \quad \text{Multiply numerators and denominators.}$$

$$= \frac{-24x^5y^3}{30x^5y^4} \quad \text{Simplify.}$$

$$= \frac{-4(6)(x^5)(y^3)}{5(6)(x^5)(y^3)(y)} \quad \text{Factor and cancel.}$$

$$= -\frac{4}{5y}, \quad x \neq 0 \quad \text{Simplified form}$$

EXAMPLE 2 *Multiplying Rational Expressions*

Multiply the rational expressions.

$$\frac{x}{5x^2 - 20x} \cdot \frac{x-4}{2x^2 + x - 3}$$

Solution

$$\frac{x}{5x^2 - 20x} \cdot \frac{x-4}{2x^2 + x - 3}$$

$$= \frac{x \cdot (x-4)}{(5x^2 - 20x) \cdot (2x^2 + x - 3)}$$ Multiply numerators and denominators.

$$= \frac{x(x-4)}{5x(x-4)(x-1)(2x+3)}$$ Factor.

$$= \frac{\cancel{x}(\cancel{x-4})}{5\cancel{x}(\cancel{x-4})(x-1)(2x+3)}$$ Cancel common factors.

$$= \frac{1}{5(x-1)(2x+3)}, \quad x \neq 0, \ x \neq 4$$ Simplified form

---

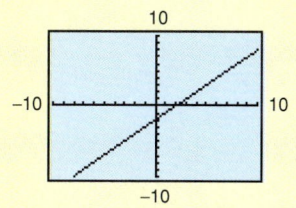

Technology

You can use a graphing utility to check your results when multiplying rational expressions. For instance, in Example 3, try graphing the equations

$$y_1 =$$

$$\frac{4x^2 - 4x}{x^2 + 2x - 3} \cdot \frac{x^2 + x - 6}{4x}$$

and

$$y_2 = x - 2$$

on the same screen. If the two graphs coincide, as shown below, you can conclude that the two functions are equivalent.

EXAMPLE 3 *Multiplying Rational Expressions*

Multiply the rational expressions.

$$\frac{4x^2 - 4x}{x^2 + 2x - 3} \cdot \frac{x^2 + x - 6}{4x}$$

Solution

$$\frac{4x^2 - 4x}{x^2 + 2x - 3} \cdot \frac{x^2 + x - 6}{4x}$$

$$= \frac{4x(x-1)(x+3)(x-2)}{(x-1)(x+3)(4x)}$$ Multiply and factor.

$$= \frac{\cancel{4x}(\cancel{x-1})(\cancel{x+3})(x-2)}{(\cancel{x-1})(\cancel{x+3})(\cancel{4x})}$$ Cancel common factors.

$$= x - 2, \quad x \neq 0, \ x \neq 1, \ x \neq -3$$ Simplified form

---

The rule for multiplying fractions can be extended to cover products involving expressions that are not in fractional form. To do this, rewrite the nonfractional expression as a fraction whose denominator is 1. Here is a simple example.

$$\frac{x+3}{x-2} \cdot (5x) = \frac{x+3}{x-2} \cdot \frac{5x}{1} = \frac{(x+3)(5x)}{x-2} = \frac{5x(x+3)}{x-2}$$

In the next example, note how to divide out a factor that differs only in sign. Note the step in which $(y - x)$ is rewritten as $(-1)(x - y)$.

EXAMPLE 4   Multiplying Rational Expressions

Multiply the rational expressions.

$$\frac{x - y}{y^2 - x^2} \cdot \frac{x^2 - xy - 2y^2}{3x - 6y}$$

Solution

$$
\begin{aligned}
\frac{x - y}{y^2 - x^2} \cdot \frac{x^2 - xy - 2y^2}{3x - 6y} &= \frac{x - y}{(y + x)(y - x)} \cdot \frac{(x - 2y)(x + y)}{3(x - 2y)} \\
&= \frac{x - y}{(y + x)(-1)(x - y)} \cdot \frac{(x - 2y)(x + y)}{3(x - 2y)} \\
&= \frac{(x - y)(x - 2y)(x + y)}{(y + x)(-1)(x - y)(3)(x - 2y)} \\
&= \frac{\cancel{(x - y)}\cancel{(x - 2y)}\cancel{(x + y)}}{\cancel{(x + y)}(-1)\cancel{(x - y)}(3)\cancel{(x - 2y)}} \\
&= -\frac{1}{3}, \quad x \neq y, \ x \neq -y, \ x \neq 2y
\end{aligned}
$$

The rule for multiplying rational expressions can be extended to cover the product of three or more fractions, as shown in Example 5.

EXAMPLE 5   Multiplying Three Rational Expressions

Multiply the rational expressions.

$$\frac{x^2 - 3x + 2}{x + 2} \cdot \frac{3x}{x - 2} \cdot \frac{2x + 4}{x^2 - 5x}$$

Solution

$$
\begin{aligned}
\frac{x^2 - 3x + 2}{x + 2} &\cdot \frac{3x}{x - 2} \cdot \frac{2x + 4}{x^2 - 5x} \\
&= \frac{(x - 1)(x - 2)(3)(x)(2)(x + 2)}{(x + 2)(x - 2)(x)(x - 5)} && \text{Multiply and factor.} \\
&= \frac{(x - 1)\cancel{(x - 2)}(3)\cancel{(x)}(2)\cancel{(x + 2)}}{\cancel{(x + 2)}\cancel{(x - 2)}\cancel{(x)}(x - 5)} && \text{Cancel common factors.} \\
&= \frac{6(x - 1)}{x - 5}, \quad x \neq 0, \ x \neq 2, \ x \neq -2 && \text{Simplified form}
\end{aligned}
$$

Dividing Rational Expressions

To divide two rational expressions, multiply the first fraction by the *reciprocal* of the second. That is, simply *invert the divisor and multiply*. For instance, to perform the following division

$$\frac{x}{x+3} \div \frac{4}{x-1}$$

invert the fraction $4/(x-1)$ and multiply, as follows.

$$\frac{x}{x+3} \div \frac{4}{x-1} = \frac{x}{x+3} \cdot \frac{x-1}{4}$$
$$= \frac{x(x-1)}{(x+3)(4)}$$
$$= \frac{x(x-1)}{4(x+3)}$$

Dividing Rational Expressions

Let u, v, w, and z be real numbers, variables, or algebraic expressions such that $v \neq 0$, $w \neq 0$, and $z \neq 0$. The quotient of u/v and w/z is

$$\frac{u}{v} \div \frac{w}{z} = \frac{u}{v} \cdot \frac{z}{w} = \frac{uz}{vw}.$$

EXAMPLE 6   *Dividing Rational Expressions*

Perform the division.

$$\frac{2x}{3x-12} \div \frac{x^2-2x}{x^2-6x+8}$$

Solution

$$\frac{2x}{3x-12} \div \frac{x^2-2x}{x^2-6x+8}$$

$$= \frac{2x}{3x-12} \cdot \frac{x^2-6x+8}{x^2-2x} \qquad \text{Invert divisor and multiply.}$$

$$= \frac{(2x)(x-2)(x-4)}{(3)(x-4)(x)(x-2)} \qquad \text{Factor.}$$

$$= \frac{(2x)(x-2)(x-4)}{(3)(x-4)(x)(x-2)} \qquad \text{Cancel common factors.}$$

$$= \frac{2}{3}, \quad x \neq 0, \; x \neq 2, \; x \neq 4 \qquad \text{Simplified form}$$

Complex Fractions

Problems involving division of two rational expressions are sometimes written as **complex fractions.** The rules for dividing fractions still apply in such cases.

EXAMPLE 7   Simplifying a Complex Fraction

$$\frac{\left(\dfrac{x^2 + 2x - 3}{x - 3}\right)}{4x + 12} = \frac{\left(\dfrac{x^2 + 2x - 3}{x - 3}\right)}{\left(\dfrac{4x + 12}{1}\right)} \qquad \text{Rewrite denominator.}$$

$$= \frac{x^2 + 2x - 3}{x - 3} \cdot \frac{1}{4x + 12} \qquad \text{Invert divisor and multiply.}$$

$$= \frac{(x - 1)(x + 3)}{(x - 3)(4)(x + 3)} \qquad \text{Factor.}$$

$$= \frac{(x - 1)\cancel{(x + 3)}}{(x - 3)(4)\cancel{(x + 3)}} \qquad \text{Cancel common factors.}$$

$$= \frac{x - 1}{4(x - 3)}, \quad x \neq -3 \qquad \text{Simplified form}$$

Group Activities   Extending the Concept

Using a Table   Complete the following table with the given values of x.

x	60	100	1000	10,000	100,000	1,000,000
$\dfrac{x - 10}{x + 10}$						
$\dfrac{x + 50}{x - 50}$						
$\dfrac{x - 10}{x + 10} \cdot \dfrac{x + 50}{x - 50}$						

What kind of pattern do you see? Try to explain what is going on. Can you see why?

8.2 Exercises

Discussing the Concepts

1. In your own words, explain how to divide rational expressions.

2. Explain how to divide a rational expression by a polynomial. Give an example.

3. Define the term *complex fraction*. Give an example and show how to simplify the fraction.

4. *Error Analysis*   Describe the error.

$$\frac{x^2 - 4}{5x} \div \frac{x + 2}{x - 2} = \frac{5x}{x^2 - 4} \cdot \frac{x + 2}{x - 2}$$

$$= \frac{5x}{(x + 2)(x - 2)} \cdot \frac{x + 2}{x - 2}$$

$$= \frac{5x}{(x - 2)^2}$$

Problem Solving

In Exercises 5 and 6, evaluate the expression for each value of x. If it is not possible, state the reason.

Expression	*Values*	
5. $\dfrac{x - 10}{4x}$	(a) $x = 10$	(b) $x = 0$
	(c) $x = -2$	(d) $x = 12$
6. $\dfrac{x^2 - 4x}{x^2 - 9}$	(a) $x = 0$	(b) $x = 4$
	(c) $x = 3$	(d) $x = -3$

In Exercises 7–10, complete the statement.

7. $\dfrac{7}{3y} = \dfrac{7x^2}{3y()}, \quad x \neq 0$

8. $\dfrac{2x}{x - 3} = \dfrac{14x(x - 3)^2}{(x - 3)()}, \quad x \neq 0$

9. $\dfrac{13x}{x - 2} = \dfrac{13x()}{4 - x^2}, \quad x \neq -2$

10. $\dfrac{x^2}{10 - x} = \dfrac{x^2()}{x^2 - 10x}, \quad x \neq 0$

In Exercises 11–18, multiply and simplify.

11. $\dfrac{7x^2}{3} \cdot \dfrac{9}{14x}$

12. $\dfrac{6}{5a} \cdot (25a)$

13. $\dfrac{8}{3 + 4x} \cdot (9 + 12x)$

14. $\dfrac{1 - 3x}{4} \cdot \dfrac{46}{15 - 45x}$

15. $\dfrac{(2x - 3)(x + 8)}{x^3} \cdot \dfrac{x}{3 - 2x}$

16. $\dfrac{x + 14}{x^3(10 - x)} \cdot \dfrac{x(x - 10)}{5}$

17. $\dfrac{xu - yu + xv - yv}{xu + yu - xv - yv} \cdot \dfrac{xu + yu + xv + yv}{xu - yu - xv + yv}$

18. $\dfrac{t^2 + 4t + 3}{2t^2 - t - 10} \cdot \dfrac{t}{t^2 + 3t + 2} \cdot \dfrac{2t^2 + 4t^3}{t^2 + 3t}$

In Exercises 19–26, divide and simplify.

19. $\dfrac{3x}{4} \div \dfrac{x^2}{2}$

20. $\dfrac{u}{10} \div u^2$

21. $\dfrac{3(a + b)}{4} \div \dfrac{(a + b)^2}{2}$

22. $\dfrac{x^2 + 9}{5(x + 2)} \div \dfrac{x + 3}{5(x^2 - 4)}$

23. $\dfrac{16x^2 + 8x + 1}{3x^2 + 8x - 3} \div \dfrac{4x^2 - 3x - 1}{x^2 + 6x + 9}$

24. $\dfrac{x^2 - 25}{x} \div \dfrac{x^3 - 5x^2}{x^2 + x}$

25. $\dfrac{\left(\dfrac{x^2}{12}\right)}{\left(\dfrac{5x}{18}\right)}$

26. $\dfrac{\left[\dfrac{(3u^2v)^2}{6v^3}\right]}{\left[\dfrac{(uv^3)^2}{3uv}\right]}$

In Exercises 27–30, perform the operations and simplify.

27. $\left[\dfrac{x^2}{9} \cdot \dfrac{3(x+4)}{x^2+2x}\right] \div \dfrac{x}{x+2}$

28. $\left(\dfrac{x^2+6x+9}{x^2} \cdot \dfrac{2x+1}{x^2-9}\right) \div \dfrac{4x^2+4x+1}{x^2-3x}$

29. $\left[\dfrac{xy+y}{4x} \div (3x+3)\right] \div \dfrac{y}{3x}$

30. $\left(\dfrac{3u^2-u-4}{u^2}\right)^2 \div \dfrac{3u^2+12u+4}{u^4-3u^3}$

■ *Graphical Reasoning* In Exercises 31–34, use a graphing utility to graph the two equations in the same viewing rectangle. Use the graphs to verify that the expressions are equivalent.

31. $y_1 = \dfrac{3x+2}{x} \cdot \dfrac{x^2}{9x^2-4}$, $y_2 = \dfrac{x}{3x-2}$

32. $y_1 = \dfrac{x^2-10x+25}{x^2-25} \cdot \dfrac{x+5}{2}$, $y_2 = \dfrac{x-5}{2}$

33. $y_1 = \dfrac{3x+15}{x^4} \div \dfrac{x+5}{x^2}$, $y_2 = \dfrac{3}{x^2}$

34. $y_1 = (x^2+6x+9) \div \dfrac{2x(x+3)}{3}$, $y_2 = \dfrac{3(x+3)}{2x}$

Geometry In Exercises 35 and 36, write an expression for the area of the shaded region. Then simplify the expression.

35.
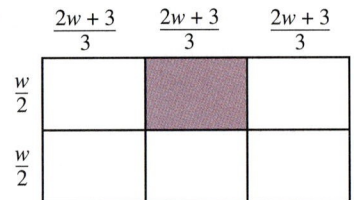

36.
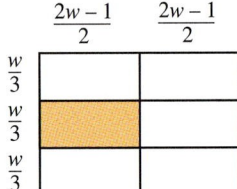

Reviewing the Major Concepts

In Exercises 37–40, factor the expression completely.

37. $3x^2 - 21x$

38. $-x^3 + 3x^2 - x + 3$

39. $4t^2 - 169$

40. $y^3 - 64$

41. *Simple Interest* You borrow $12,000 for 6 months. At simple interest of 12%, how much will you owe?

42. *Geometry* A rectangle is 2 inches longer than it is wide. Write an expression that represents its area.

Additional Problem Solving

In Exercises 43–46, complete the statement.

43. $\dfrac{3x}{x-4} = \dfrac{3x(x+2)^2}{(x-4)()}$, $x \neq -2$

44. $\dfrac{x+1}{x} = \dfrac{(x+1)^3}{x()}$, $x \neq -1$

45. $\dfrac{3u}{7v} = \dfrac{3u()}{7v(u+1)}$, $u \neq -1$

46. $\dfrac{3t+5}{t} = \dfrac{(3t+5)()}{5t^2(3t-5)}$, $t \neq \dfrac{5}{3}$

In Exercises 47–62, multiply and simplify.

47. $\dfrac{8s^3}{9s} \cdot \dfrac{6s^2}{32s}$

48. $\dfrac{3x^4}{7x} \cdot \dfrac{8x^2}{9}$

49. $16u^4 \cdot \dfrac{12}{8u^2}$

50. $\dfrac{25}{8x} \cdot \dfrac{8x}{35}$

51. $\dfrac{8u^2v}{3u+v} \cdot \dfrac{u+v}{12u}$

52. $\dfrac{x+25}{8} \cdot \dfrac{8}{x+25}$

53. $\dfrac{12-r}{3} \cdot \dfrac{3}{r-12}$

54. $\dfrac{8-z}{8+z} \cdot \dfrac{z+8}{z-8}$

55. $\dfrac{6r}{r-2} \cdot \dfrac{r^2-4}{33r^2}$

56. $\dfrac{5y-20}{5y+15} \cdot \dfrac{2y+6}{y-4}$

57. $(u-2v)^2 \cdot \dfrac{u+2v}{u-2v}$

58. $\dfrac{x-2y}{x+2y} \cdot \dfrac{x^2+4y^2}{x^2-4y^2}$

59. $\dfrac{2t^2-t-15}{t+2} \cdot \dfrac{t^2-t-6}{t^2-6t+9}$

60. $\dfrac{y^2-16}{2y^3} \cdot \dfrac{4y}{y^2-6y+8}$

61. $\dfrac{x+5}{x-5} \cdot \dfrac{2x^2-9x-5}{3x^2+x-2} \cdot \dfrac{x^2-1}{x^2+7x+10}$

62. $\dfrac{x^3+3x^2-4x-12}{x^3-3x^2-4x+12} \cdot \dfrac{x^2-9}{x}$

In Exercises 63–70, divide and simplify.

63. $\dfrac{7xy^2}{10u^2v} \div \dfrac{21x^3}{45uv}$

64. $\dfrac{25x^2y}{60x^3y^2} \div \dfrac{5x^4y^3}{16x^2y}$

65. $\dfrac{(x^3y)^2}{(x+2y)^2} \div \dfrac{x^2y}{(x+2y)^3}$

66. $\dfrac{x^2-y^2}{2x^2-8x} \div \dfrac{(x-y)^2}{2xy}$

67. $\dfrac{\left(\dfrac{25x^2}{x-5}\right)}{\left(\dfrac{10x}{5-x}\right)}$

68. $\dfrac{\left(\dfrac{5x}{x+7}\right)}{\left(\dfrac{10}{x^2+8x+7}\right)}$

69. $\dfrac{x(x+3)-2(x+3)}{x^2-4} \div \dfrac{x}{x^2+4x+4}$

70. $\dfrac{t^3+t^2-9t-9}{t^2-5t+6} \div \dfrac{t^2+6t+9}{t-2}$

In Exercises 71 and 72, perform the operations and simplify your answer.

71. $\dfrac{2x^2+5x-25}{3x^2+5x+2} \cdot \dfrac{3x^2+2x}{x+5} \div \left(\dfrac{x}{x+1}\right)^2$

72. $\dfrac{t^2-100}{4t^2} \cdot \dfrac{t^3-5t^2-50t}{t^4+10t^3} \div \dfrac{(t-10)^2}{5t}$

73. *Photocopy Rate*   A photocopier produces copies at a rate of 20 pages per minute.

(a) Determine the time required to copy one page.

(b) Determine the time required to copy x pages.

(c) Determine the time required to copy 35 pages.

74. *Pumping Rate*   The rate for a pump is 15 gallons per minute.

(a) Determine the time required to pump 1 gallon.

(b) Determine the time required to pump x gallons.

(c) Determine the time required to pump 130 gallons.

Probability   In Exercises 75 and 76, consider an experiment in which a marble is tossed into a rectangular box whose dimensions are $3x-2$ by x inches. The probability that the marble will come to rest in the unshaded portion of the box is equal to the ratio of the unshaded area to the total area of the figure. Find the probability.

75.

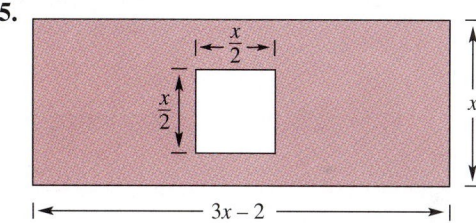

76.

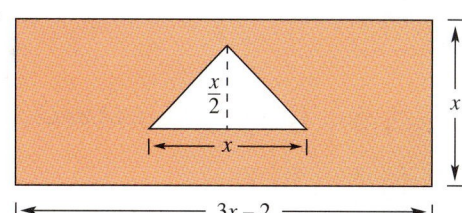

8.3	**Adding and Subtracting Rational Expressions**
	Adding or Subtracting with Like Denominators ▪
	Adding or Subtracting with Unlike Denominators ▪ Complex Fractions

Adding or Subtracting with Like Denominators

As with numerical fractions, the procedure used to add or subtract two rational expressions depends upon whether the expressions have *like* or *unlike* denominators. To add or subtract two rational expressions with *like* denominators, simply combine their numerators and place the result over the common denominator.

Adding or Subtracting with Like Denominators

If u, v, and w are real numbers, variables, or variable expressions, and $w \neq 0$, the following rules are valid.

1. $\dfrac{u}{w} + \dfrac{v}{w} = \dfrac{u + v}{w}$ Add fractions with like denominators.

2. $\dfrac{u}{w} - \dfrac{v}{w} = \dfrac{u - v}{w}$ Subtract fractions with like denominators.

EXAMPLE 1   Adding and Subtracting with Like Denominators

a. $\dfrac{x}{4} + \dfrac{5 - x}{4} = \dfrac{x + (5 - x)}{4} = \dfrac{5}{4}$

b. $\dfrac{7}{2x - 3} - \dfrac{3x}{2x - 3} = \dfrac{7 - 3x}{2x - 3}$

STUDY TIP

After adding or subtracting two (or more) rational expressions, check the resulting fraction to see if it can be simplified, as illustrated in Example 2.

EXAMPLE 2   Subtracting Rational Expressions and Simplifying

$$\frac{x}{x^2 - 2x - 3} - \frac{3}{x^2 - 2x - 3} = \frac{x - 3}{x^2 - 2x - 3} \quad \text{Subtract.}$$

$$= \frac{x - 3}{(x - 3)(x + 1)} \quad \text{Factor.}$$

$$= \frac{(x - 3)(1)}{(x - 3)(x + 1)} \quad \text{Cancel common factor.}$$

$$= \frac{1}{x + 1}, \quad x \neq 3 \quad \text{Simplified form}$$

The rules for adding and subtracting rational expressions with like denominators can be extended to cover sums and differences involving three or more rational expressions, as illustrated in Example 3.

EXAMPLE 3   *Combining Three Rational Expressions*

$$\frac{x^2 - 26}{x - 5} - \frac{2x + 4}{x - 5} + \frac{10 + x}{x - 5} = \frac{(x^2 - 26) - (2x + 4) + (10 + x)}{x - 5}$$

$$= \frac{x^2 - 26 - 2x - 4 + 10 + x}{x - 5}$$

$$= \frac{x^2 - x - 20}{x - 5}$$

$$= \frac{(x - 5)(x + 4)}{x - 5}$$

$$= x + 4, \quad x \neq 5$$

Adding or Subtracting with Unlike Denominators

To add or subtract rational expressions with *unlike* denominators, you must first rewrite each expression using the **least common multiple** of the denominators of the individual expressions. The least common multiple of two (or more) polynomials is the simplest polynomial that is a multiple of each of the original polynomials.

EXAMPLE 4   *Finding Least Common Multiples*

a.   The least common multiple of

$$6x = 2 \cdot 3 \cdot x \quad \text{and} \quad 2x^2 = 2 \cdot x^2$$

is $2 \cdot 3 \cdot x^2 = 6x^2$.

b.   The least common multiple of

$$x^2 - x = x(x - 1) \quad \text{and} \quad 2x - 2 = 2(x - 1)$$

is $2x(x - 1)$.

c.   The least common multiple of

$$3x^2 + 6x = 3x(x + 2) \quad \text{and} \quad x^2 + 4x + 4 = (x + 2)^2$$

is $3x(x + 2)^2$.

To add or subtract rational expressions with *unlike* denominators, you must first rewrite the rational expressions so that they have *like* denominators. The like denominator that you use is the least common multiple of the original denominators and is called the **least common denominator** of the original rational expressions. Once the rational expressions have been written with like denominators, you can simply add or subtract the rational expressions using the rules given at the beginning of this section.

EXAMPLE 5 Adding with Unlike Denominators

Add the rational expressions: $\dfrac{7}{6x} + \dfrac{5}{8x}$.

Solution

The least common denominator of $6x$ and $8x$ is $24x$, so the first step is to rewrite each fraction with this denominator.

$$\frac{7}{6x} + \frac{5}{8x} = \frac{7(4)}{6x(4)} + \frac{5(3)}{8x(3)} \qquad \text{Rewrite fractions using least common denominator.}$$

$$= \frac{28}{24x} + \frac{15}{24x} \qquad \text{Like denominators}$$

$$= \frac{28 + 15}{24x} \qquad \text{Add fractions.}$$

$$= \frac{43}{24x} \qquad \text{Simplified form}$$

Technology

You can use a graphing utility to check your results when adding or subtracting rational expressions. For instance, in Example 5, try graphing the equations

$$y_1 = \frac{7}{6x} + \frac{5}{8x}$$

and

$$y_2 = \frac{43}{24x}$$

on the same screen. If the two graphs coincide, as shown below, you can conclude that the two functions are equivalent.

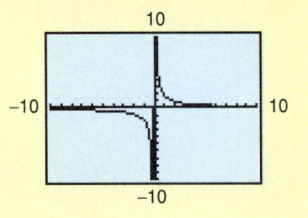

EXAMPLE 6 Subtracting with Unlike Denominators

Subtract the rational expressions: $\dfrac{3}{x-3} - \dfrac{5}{x+2}$.

Solution

The least common denominator is $(x-3)(x+2)$.

$$\frac{3}{x-3} - \frac{5}{x+2} = \frac{3(x+2)}{(x-3)(x+2)} - \frac{5(x-3)}{(x-3)(x+2)}$$

$$= \frac{3x+6}{(x-3)(x+2)} - \frac{5x-15}{(x-3)(x+2)}$$

$$= \frac{(3x+6)-(5x-15)}{(x-3)(x+2)}$$

$$= \frac{3x+6-5x+15}{(x-3)(x+2)}$$

$$= \frac{-2x+21}{(x-3)(x+2)}$$

EXAMPLE 7 *Adding with Unlike Denominators*

NOTE In Example 7, note that the factors in the denominator are $x^2 - 4 = (x + 2)(x - 2)$ and $2 - x$. Because

$$(2 - x) = (-1)(x - 2)$$

the original addition problem can be written as a subtraction problem.

Add the rational expressions: $\dfrac{6x}{x^2 - 4} + \dfrac{3}{2 - x}$.

Solution

$$\frac{6x}{x^2 - 4} + \frac{3}{(-1)(x - 2)} = \frac{6x}{(x + 2)(x - 2)} - \frac{3}{x - 2}$$

$$= \frac{6x}{(x + 2)(x - 2)} - \frac{3(x + 2)}{(x + 2)(x - 2)}$$

$$= \frac{6x}{(x + 2)(x - 2)} - \frac{3x + 6}{(x + 2)(x - 2)}$$

$$= \frac{6x - (3x + 6)}{(x + 2)(x - 2)}$$

$$= \frac{6x - 3x - 6}{(x + 2)(x - 2)}$$

$$= \frac{3x - 6}{(x + 2)(x - 2)}$$

$$= \frac{3(x - 2)}{(x + 2)(x - 2)}$$

$$= \frac{3}{x + 2}, \quad x \neq 2$$

EXAMPLE 8 *Combining Three Rational Expressions*

$$\frac{2x - 5}{6x + 9} - \frac{4}{2x^2 + 3x} + \frac{1}{x} = \frac{(2x - 5)(x)}{3(2x + 3)(x)} - \frac{(4)(3)}{x(2x + 3)(3)} + \frac{3(2x + 3)}{(x)(3)(2x + 3)}$$

$$= \frac{2x^2 - 5x}{3x(2x + 3)} - \frac{12}{3x(2x + 3)} + \frac{6x + 9}{3x(2x + 3)}$$

$$= \frac{2x^2 - 5x - 12 + 6x + 9}{3x(2x + 3)}$$

$$= \frac{2x^2 + x - 3}{3x(2x + 3)}$$

$$= \frac{(x - 1)(2x + 3)}{3x(2x + 3)}$$

$$= \frac{(x - 1)(2x + 3)}{3x(2x + 3)}$$

$$= \frac{x - 1}{3x}, \quad x \neq -\frac{3}{2}$$

Complex Fractions

Complex fractions can have numerators or denominators that are the sums or differences of fractions. To simplify a complex fraction, first combine its numerator and its denominator into single fractions. Then divide by inverting the denominator and multiplying.

EXAMPLE 9 Simplifying a Complex Fraction

$$\dfrac{\left(\dfrac{x}{4}+\dfrac{3}{2}\right)}{\left(2-\dfrac{3}{x}\right)} = \dfrac{\left(\dfrac{x}{4}+\dfrac{6}{4}\right)}{\left(\dfrac{2x}{x}-\dfrac{3}{x}\right)}$$
Find least common denominators.

$$= \dfrac{\left(\dfrac{x+6}{4}\right)}{\left(\dfrac{2x-3}{x}\right)}$$
Add fractions in numerator and denominator.

$$= \dfrac{x+6}{4}\cdot\dfrac{x}{2x-3}$$
Invert divisor and multiply.

$$= \dfrac{x(x+6)}{4(2x-3)}, \qquad x \neq 0$$
Simplified form

Another way to simplify the complex fraction given in Example 9 is to multiply the numerator and denominator by the least common denominator of *every* fraction in the numerator and denominator. For this fraction, notice what happens when we multiply the numerator and denominator by $4x$.

$$\dfrac{\left(\dfrac{x}{4}+\dfrac{3}{2}\right)}{\left(2-\dfrac{3}{x}\right)} = \dfrac{\left(\dfrac{x}{4}+\dfrac{3}{2}\right)}{\left(2-\dfrac{3}{x}\right)}\cdot\dfrac{4x}{4x}$$

$$= \dfrac{\dfrac{x}{4}(4x)+\dfrac{3}{2}(4x)}{2(4x)-\dfrac{3}{x}(4x)}$$

$$= \dfrac{x^2+6x}{8x-12}$$

$$= \dfrac{x(x+6)}{4(2x-3)}, \qquad x \neq 0$$

Which of these two methods do you prefer?

EXAMPLE 10   Complex Fractions

Simplify $\dfrac{\left(\dfrac{2}{x+2}\right)}{\left(\dfrac{1}{x+2}+\dfrac{2}{x}\right)}$.

Solution

$$\frac{\left(\dfrac{2}{x+2}\right)}{\left(\dfrac{1}{x+2}+\dfrac{2}{x}\right)} = \frac{\left(\dfrac{2}{x+2}\right)(x)(x+2)}{\dfrac{1}{x+2}(x)(x+2)+\dfrac{2}{x}(x)(x+2)}$$

$$= \frac{2x}{x+2(x+2)}$$

$$= \frac{2x}{3x+4}, \quad x \neq -2, x \neq 0$$

Group Activities   Extending the Concept

Comparing Two Methods   Each person in your group should evaluate each of the following expressions at the given variable in two different ways: (1) combine and simplify the rational expressions first and then evaluate the simplified expression at the given variable value, and (2) substitute the given value for the variable first and then simplify the resulting expression. Did you get the same result with each method? Discuss which method you prefer and why. List any advantages and/or disadvantages of each method.

a. $\dfrac{1}{m-4} - \dfrac{1}{m+4} + \dfrac{3m}{m^2-16}, m = 2$

b. $\dfrac{x-2}{x^2-9} + \dfrac{3x+2}{x^2-5x+6}, x = 4$

c. $\dfrac{3y^2+16y-8}{y^2+2y-8} - \dfrac{y-1}{y-2} + \dfrac{y}{y+4}, y = 3$

8.3 Exercises

Discussing the Concepts

1. In your own words, describe how to add or subtract rational expressions with like denominators.

2. In your own words, describe how to add or subtract rational expressions with unlike denominators.

3. Is it possible for the least common denominator of two fractions to be the same as one of the fraction's denominators? If so, give an example.

4. *Error Analysis*   Describe the error.

$$\frac{x-1}{x+4} - \frac{4x-11}{x+4} = \frac{x-1-4x-11}{x+4}$$

$$= \frac{-3x-12}{x+4} = \frac{-3(x+4)}{x+4} = -3$$

Problem Solving

In Exercises 5–10, combine and simplify.

5. $\dfrac{5}{8} + \dfrac{7}{8}$

6. $\dfrac{7}{12} - \dfrac{5}{12}$

7. $\dfrac{x}{9} - \dfrac{x+2}{9}$

8. $\dfrac{z^2}{3} + \dfrac{z^2-2}{3}$

9. $\dfrac{5x-1}{x+4} + \dfrac{5-4x}{x+4}$

10. $\dfrac{2x-1}{x(x-3)} + \dfrac{1-x}{x(x-3)}$

In Exercises 11–16, find the least common multiple of the expressions.

11. $5x^2, 20x^3$

12. $14t^2, 42t^5$

13. $15x^2, 3(x+5)$

14. $18y^3, 27y(y-3)^2$

15. $6(x^2-4), 2x(x+2)$

16. $t^3 + 3t^2 + 9t, 2t^2(t^2-9)$

In Exercises 17–20, find the least common denominator of the two fractions and rewrite each fraction using the least common denominator.

17. $\dfrac{n+8}{3n-12}, \dfrac{10}{6n^2}$

18. $\dfrac{8s}{(s+2)^2}, \dfrac{3}{s^3+s^2-2s}$

19. $\dfrac{x-8}{x^2-25}, \dfrac{9x}{x^2-10x+25}$

20. $\dfrac{3y}{y^2-y-12}, \dfrac{y-4}{y^2+3y}$

In Exercises 21–32, perform the operation and simplify.

21. $\dfrac{5}{4x} - \dfrac{3}{5}$

22. $\dfrac{10}{b} + \dfrac{1}{10b^2}$

23. $\dfrac{20}{x-4} + \dfrac{20}{4-x}$

24. $\dfrac{15}{2-t} - \dfrac{7}{t-2}$

25. $25 + \dfrac{10}{x+4}$

26. $\dfrac{100}{x-10} - 8$

27. $\dfrac{x}{x^2-9} + \dfrac{3}{x(x-3)}$

28. $\dfrac{x}{x^2-x-30} - \dfrac{1}{x+5}$

29. $\dfrac{3u}{u^2-2uv+v^2} + \dfrac{2}{u-v}$

30. $\dfrac{1}{x} - \dfrac{3}{y} + \dfrac{3x-y}{xy}$

31. $\dfrac{x+2}{x-1} - \dfrac{2}{x+6} - \dfrac{14}{x^2+5x-6}$

32. $\dfrac{x}{x^2+15x+50} + \dfrac{7}{2(x+10)} - \dfrac{3}{2(x+5)}$

⊞ In Exercises 33 and 34, use a graphing utility to graph the two equations on the same screen. Use the graphs to verify that the expressions are equivalent. Verify the results algebraically.

33. $y_1 = \dfrac{2}{x} + \dfrac{4}{x(x-2)}, y_2 = \dfrac{2x}{x(x-2)}$

34. $y_1 = 3 - \dfrac{1}{x-1}, y_2 = \dfrac{3x-4}{x-1}$

In Exercises 35–40, simplify the complex fraction.

35. $\dfrac{\dfrac{1}{2}}{\left(3+\dfrac{1}{x}\right)}$

36. $\dfrac{\left(\dfrac{1}{t}-1\right)}{\left(\dfrac{1}{t}+1\right)}$

37. $\dfrac{\left(3+\dfrac{9}{x-3}\right)}{\left(4+\dfrac{12}{x-3}\right)}$

38. $\dfrac{\left(x+\dfrac{2}{x-3}\right)}{\left(x+\dfrac{6}{x-3}\right)}$

39. $\dfrac{\left(\dfrac{y}{x}-\dfrac{x}{y}\right)}{\left(\dfrac{x+y}{xy}\right)}$

40. $\dfrac{\left(x-\dfrac{2y^2}{x-y}\right)}{x-2y}$

41. *Parallel Resistance*   When two resistors are connected in parallel (see figure), the total resistance is

$$\dfrac{1}{\left(\dfrac{1}{R_1}+\dfrac{1}{R_2}\right)}.$$

Simplify this complex fraction.

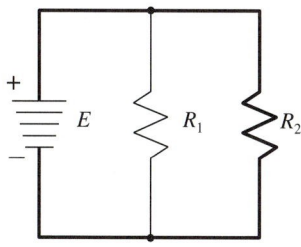

42. *Average of Two Numbers*   Determine the average of the two real numbers $x/3$ and $x/5$. Copy the real number line shown below and plot the average. What can you conclude?

Interpreting a Table   In Exercises 43 and 44, use a graphing utility to complete the table. Comment on the domain and equivalence of the expressions.

43.

x	−3	−2	−1	0	1	2	3
$\dfrac{\left(1-\dfrac{1}{x}\right)}{\left(1-\dfrac{1}{x^2}\right)}$							
$\dfrac{x}{x+1}$							

44.

x	−3	−2	−1	0	1	2	3
$\dfrac{\left(1+\dfrac{4}{x}+\dfrac{4}{x^2}\right)}{\left(1-\dfrac{4}{x^2}\right)}$							
$\dfrac{x+2}{x-2}$							

Reviewing the Major Concepts

In Exercises 45–50, simplify the expression.

45. $-(-3x^2)^3(2x^4)$

46. $(4x^3y^2)(-2xy^3)$

47. $(a^2+b^2)^0$

48. $[(2x^2y)^2]^3$

49. $\left(\dfrac{5}{x^2}\right)^2$

50. $-\dfrac{(2u^2v)^2}{-3uv^2}$

51. *Monthly Wage*   A company offers two wage plans. One plan pays a straight $2500 per month. The second pays $1500 per month plus a commission of 4% on gross sales. Let x represent the gross sales and write an inequality that represents gross sales in which the second plan gives the greater monthly wage. Solve the inequality.

52. Three times a number n must be at least 15 and no more than 60. What interval contains this number?

Additional Problem Solving

In Exercises 53–62, combine and simplify.

53. $\dfrac{4-y}{4} + \dfrac{3y}{4}$

54. $\dfrac{10x^2+1}{3} - \dfrac{10x^2}{3}$

55. $\dfrac{2}{3a} - \dfrac{11}{3a}$

56. $\dfrac{16+z}{5z} - \dfrac{11-z}{5z}$

57. $\dfrac{2x+5}{3} + \dfrac{1-x}{3}$

58. $\dfrac{6x}{13} - \dfrac{7x}{13}$

59. $\dfrac{3y}{3} - \dfrac{3y-3}{3} - \dfrac{7}{3}$

60. $\dfrac{3y-22}{y-6} - \dfrac{2y-16}{y-6}$

61. $\dfrac{-16u}{9} - \dfrac{27-16u}{9} + \dfrac{2}{9}$

62. $\dfrac{7s-5}{2s+5} + \dfrac{3(s+10)}{2s+5}$

In Exercises 63–66, find the least common multiple of the expressions.

63. $9y^3, 12y$

64. $8t(t+2), 14(t^2-4)$

65. $6x^2, 15x(x-1)$

66. $2y^2+y-1, 4y^2-2y$

In Exercises 67–70, find the least common denominator of the two fractions and rewrite each fraction using the least common denominator.

67. $\dfrac{v}{2v^2+2v}, \dfrac{4}{3v^2}$

68. $\dfrac{2}{x^2(x-3)}, \dfrac{5}{x(x+3)}$

69. $\dfrac{4x}{(x+5)^2}, \dfrac{x-2}{x^2-25}$

70. $\dfrac{5t}{2t(t-3)^2}, \dfrac{4}{t(t-3)}$

In Exercises 71–92, perform the operation and simplify.

71. $\dfrac{7}{a} + \dfrac{14}{a^2}$

72. $\dfrac{1}{6u^2} - \dfrac{2}{9u}$

73. $\dfrac{3x}{x-8} - \dfrac{6}{8-x}$

74. $\dfrac{1}{y-6} + \dfrac{y}{6-y}$

75. $\dfrac{3x}{3x-2} + \dfrac{2}{2-3x}$

76. $\dfrac{y}{5y-3} - \dfrac{3}{3-5y}$

77. $-\dfrac{1}{6x} + \dfrac{1}{6(x-3)}$

78. $\dfrac{1}{x} - \dfrac{1}{x+2}$

79. $\dfrac{x}{x+3} - \dfrac{5}{x-2}$

80. $\dfrac{3}{t(t+1)} + \dfrac{4}{t}$

81. $\dfrac{3}{x+1} - \dfrac{2}{x}$

82. $\dfrac{5}{x-4} - \dfrac{3}{x}$

83. $\dfrac{3}{x-5} + \dfrac{2}{x+5}$

84. $\dfrac{7}{2x-3} + \dfrac{3}{2x+3}$

85. $\dfrac{4}{x^2} - \dfrac{4}{x^2+1}$

86. $\dfrac{2}{y} + \dfrac{1}{2y^2}$

87. $\dfrac{4}{x-4} + \dfrac{16}{(x-4)^2}$

88. $\dfrac{3}{x-2} - \dfrac{1}{(x-2)^2}$

89. $\dfrac{y}{x^2+xy} - \dfrac{x}{xy+y^2}$

90. $\dfrac{5}{x+y} + \dfrac{5}{x-y}$

91. $\dfrac{4}{x} - \dfrac{2}{x^2} + \dfrac{4}{x+3}$

92. $\dfrac{5}{2(x+1)} - \dfrac{1}{2x} - \dfrac{3}{2(x+1)^2}$

In Exercises 93–100, simplify the complex fraction.

93. $\dfrac{\left(16x - \dfrac{1}{x}\right)}{\left(\dfrac{1}{x} - 4\right)}$

94. $\dfrac{\left(\dfrac{36}{y} - y\right)}{6+y}$

95. $\dfrac{\left(\dfrac{3}{x^2} + \dfrac{1}{x}\right)}{\left(2 - \dfrac{4}{5x}\right)}$

96. $\dfrac{\left(16 - \dfrac{1}{x^2}\right)}{\left(\dfrac{1}{4x^2} - 4\right)}$

97. $\dfrac{\left(1 - \dfrac{1}{y^2}\right)}{\left(1 - \dfrac{4}{y} + \dfrac{3}{y^2}\right)}$

98. $\dfrac{\left(\dfrac{x+1}{x+2} - \dfrac{1}{x}\right)}{\left(\dfrac{2}{x+2}\right)}$

99. $\dfrac{\left(\dfrac{x}{x-3} - \dfrac{2}{3}\right)}{\left(\dfrac{10}{3x} + \dfrac{x^2}{x-3}\right)}$

100. $\dfrac{\left(\dfrac{1}{2x} - \dfrac{6}{x+5}\right)}{\left(\dfrac{x}{x-5} + \dfrac{1}{x}\right)}$

Difference Quotient In Exercises 101 and 102, use the function to find and simplify the expression for

$$\frac{f(2+h) - f(2)}{h}.$$

This expression is called a difference quotient and is used in calculus.

101. $f(x) = \dfrac{1}{x}$ **102.** $f(x) = \dfrac{x}{x-1}$

103. *Work Rate*   After two workers work together for t hours on a common task, the fractional parts of the job done by the two workers are $t/4$ and $t/6$. What fractional part of the task has been completed?

104. *Work Rate*   After two workers work together for t hours on a common task, the fractional parts of the job done by the two workers are $t/3$ and $t/5$. What fractional part of the task has been completed?

105. *Average of Two Numbers*   Determine the average of the two real numbers $x/4$ and $x/6$.

106. *Average of Three Numbers*   Determine the average of the three real numbers x, $x/2$, and $x/3$.

107. *Equal Parts*   Find three real numbers that divide the real number line between $x/6$ and $x/2$ into four equal parts (see figure).

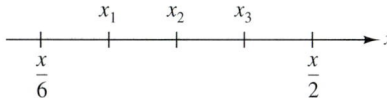

108. *Equal Parts*   Find two real numbers that divide the real number line between $x/5$ and $x/3$ into three equal parts (see figure).

109. *Monthly Payment*   The approximate annual interest rate r of a monthly installment loan is given by

$$r = \frac{\left[\dfrac{24(NM - P)}{N}\right]}{\left(P + \dfrac{MN}{12}\right)}$$

where N is the total number of payments, M is the monthly payment, and P is the amount financed.

(a) Approximate the annual interest rate for a 4-year car loan of $10,000 that has monthly payments of $300.

(b) Simplify the expression for the annual interest rate r, and then rework part (a).

110. *Error Analysis*   Determine whether the following is correct. If it is not, find and correct any errors.

$$\frac{2}{x} - \frac{3}{x+1} + \frac{x+1}{x^2} = \frac{2x(x+1) - 3x^2 + (x+1)^2}{x^2(x+1)}$$

$$= \frac{2x^2 + x - 3x^2 + x^2 + 1}{x^2(x+1)}$$

$$= \frac{x+1}{x^2(x+1)}$$

$$= \frac{1}{x^2}$$

MID-CHAPTER QUIZ

Take this quiz as you would take a quiz in class. After you are done, check your work against the answers given in the back of the book.

1. Determine the domain of $\dfrac{y+2}{y(y-4)}$.

2. Evaluate $h(x) = (x^2 - 9)/(x^2 - x - 2)$ as indicated. If it is not possible, state the reason.

   (a) $h(-3)$ (b) $h(0)$ (c) $h(-1)$ (d) $h(5)$

In Exercises 3–8, write the expression in reduced form.

3. $\dfrac{9y^2}{6y}$

4. $\dfrac{8u^3v^2}{36uv^3}$

5. $\dfrac{4x^2 - 1}{x - 2x^2}$

6. $\dfrac{(z+3)^2}{2z^2 + 5z - 3}$

7. $\dfrac{7ab + 3a^2b^2}{a^2b}$

8. $\dfrac{2mn^2 - n^3}{2m^2 + mn - n^2}$

In Exercises 9–18, perform the operations and simplify your answer.

9. $\dfrac{11t^2}{6} \cdot \dfrac{9}{33t}$

10. $(x^2 + 2x) \cdot \dfrac{5}{x^2 - 4}$

11. $\dfrac{4}{3(x-1)} \cdot \dfrac{12x}{6(x^2 + 2x - 3)}$

12. $\dfrac{5u}{3(u+v)} \cdot \dfrac{2(u^2 - v^2)}{3v} \div \dfrac{25u^2}{18(u-v)}$

13. $\dfrac{\left(\dfrac{9t^2}{3-t}\right)}{\left(\dfrac{6t}{t-3}\right)}$

14. $\dfrac{\left(\dfrac{10}{x^2 + 2x}\right)}{\left(\dfrac{15}{x^2 + 3x + 2}\right)}$

15. $\dfrac{4x}{x+5} - \dfrac{3x}{4}$

16. $4 + \dfrac{x}{x^2 - 4} - \dfrac{2}{x^2}$

17. $\dfrac{\left(1 - \dfrac{2}{x}\right)}{\left(\dfrac{3}{x} - \dfrac{4}{5}\right)}$

18. $\dfrac{\left(\dfrac{3}{x} + \dfrac{x}{3}\right)}{\left(\dfrac{x+3}{6x}\right)}$

19. You start a business with a setup cost of $6000. The cost of material for producing each unit of your product is $10.50.

   (a) Write an algebraic fraction that gives the average cost per unit when x units are produced. Explain your reasoning.

   (b) Find the average cost per unit when $x = 500$ units are produced.

20. Find the ratio of the shaded portion of the figure to the total area of the figure.

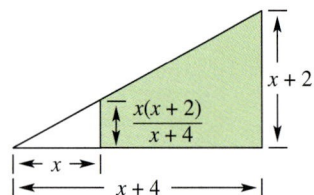

Figure for 20

8.4 Dividing Polynomials

Dividing a Polynomial by a Monomial ▪ Long Division ▪ Synthetic Division ▪ Factoring and Division

Dividing a Polynomial by a Monomial

To divide a polynomial by a monomial, reverse the procedure used to add (or subtract) two rational expressions. Here is an example.

$$2 + \frac{1}{x} = \frac{2x}{x} + \frac{1}{x} = \frac{2x+1}{x}$$ Add fractions.

$$\frac{2x+1}{x} = \frac{2x}{x} + \frac{1}{x} = 2 + \frac{1}{x}$$ Divide by monomial.

Dividing a Polynomial by a Monomial

Let u, v, and w be real numbers, variables, or algebraic expressions such that $w \neq 0$.

1. $\dfrac{u+v}{w} = \dfrac{u}{w} + \dfrac{v}{w}$ **2.** $\dfrac{u-v}{w} = \dfrac{u}{w} - \dfrac{v}{w}$

When dividing a polynomial by a monomial, remember to reduce the resulting expressions to simplest form, as illustrated in Example 1.

EXAMPLE 1   *Dividing a Polynomial by a Monomial*

Perform the division and simplify.

$$\frac{12x^2 - 20x + 8}{4x}$$

Solution

$$\frac{12x^2 - 20x + 8}{4x} = \frac{12x^2}{4x} - \frac{20x}{4x} + \frac{8}{4x}$$ Divide each term by $4x$.

$$= \frac{3(4x)(x)}{4x} - \frac{5(4x)}{4x} + \frac{2(4)}{4x}$$ Cancel common factors.

$$= 3x - 5 + \frac{2}{x}$$ Simplified form

Long Division

In Section 8.2, you learned how to divide one polynomial by another by factoring and canceling common factors. For instance, you can divide $(x^2 - 2x - 3)$ by $(x - 3)$ as follows.

$$(x^2 - 2x - 3) \div (x - 3) = \frac{x^2 - 2x - 3}{x - 3} \qquad \text{Write as fraction.}$$

$$= \frac{(x + 1)(x - 3)}{x - 3} \qquad \text{Factor.}$$

$$= \frac{(x + 1)(x - 3)}{x - 3} \qquad \text{Cancel common factor.}$$

$$= x + 1, \quad x \neq 3 \qquad \text{Simplified form}$$

This procedure works well for polynomials that factor easily. For those that do not, you can use a more general procedure that follows a "long division algorithm" similar to the algorithm used for dividing positive integers. We review that procedure in Example 2.

EXAMPLE 2   Long Division Algorithm for Positive Integers

Use the long division algorithm to divide 6584 by 28.

Solution

Think $\frac{65}{28} \approx 2$.

Think $\frac{98}{28} \approx 3$.

Think $\frac{144}{28} \approx 5$.

$$
\begin{array}{r}
235 \\
28\overline{)6584} \\
\underline{56} \\
98 \\
\underline{84} \\
144 \\
\underline{140} \\
4
\end{array}
$$

Multiply $2 \cdot 28$.
Subtract and bring down 8.
Multiply $3 \cdot 28$.
Subtract and bring down 4.
Multiply $5 \cdot 28$.
Remainder

Thus, you have

$$6584 \div 28 = 235 + \frac{4}{28} = 235 + \frac{1}{7}.$$

In Example 2, the number 6584 is the **dividend,** 28 is the **divisor,** 235 is the **quotient,** and 4 is the **remainder.**

In the next several examples, you will see how the long division algorithm can be extended to cover division of one polynomial by another.

EXAMPLE 3   *Long Division Algorithm for Polynomials*

Use the long division algorithm to perform the division.

$$(x^2 + 2x + 4) \div (x - 1)$$

Solution

$$
\begin{array}{r}
x + 3 \\
x - 1 \overline{\smash{)}\; x^2 + 2x + 4} \\
\underline{x^2 - x} \qquad \text{Multiply } x(x - 1). \\
3x + 4 \qquad \text{Subtract and bring down 4.} \\
\underline{3x - 3} \qquad \text{Multiply } 3(x - 1). \\
7 \qquad \text{Subtract.}
\end{array}
$$

Considering the remainder as a fractional part of the divisor, you can write the result as

$$
\underbrace{\overbrace{\dfrac{x^2 + 2x + 4}{\underbrace{x - 1}_{\text{Divisor}}}}^{\text{Dividend}} = \overbrace{x + 3}^{\text{Quotient}} + \dfrac{\overbrace{7}^{\text{Remainder}}}{\underbrace{x - 1}_{\text{Divisor}}}.}
$$

You can check a long division problem by multiplying. For instance, you can check the results of Example 3 as follows.

$$\frac{x^2 + 2x + 4}{x - 1} \stackrel{?}{=} x + 3 + \frac{7}{x - 1}$$

$$(x - 1)\left(\frac{x^2 + 2x + 4}{x - 1}\right) \stackrel{?}{=} (x - 1)\left(x + 3 + \frac{7}{x - 1}\right)$$

$$x^2 + 2x + 4 \stackrel{?}{=} (x + 3)(x - 1) + 7$$

$$x^2 + 2x + 4 \stackrel{?}{=} (x^2 + 2x - 3) + 7$$

$$x^2 + 2x + 4 = x^2 + 2x + 4 \qquad \text{Result checks.} \; ✔$$

In a long division problem, if the remainder is 0, the divisor is said to **divide evenly** into the dividend. For instance, $x + 2$ divides evenly into $3x^3 + 10x^2 + 6x - 4$:

$$\frac{3x^3 + 10x^2 + 6x - 4}{x + 2} = 3x^2 + 4x - 2.$$

When using the long division algorithm for polynomials, be sure that both the divisor and dividend are written in standard form before beginning the division process.

EXAMPLE 4 Writing in Standard Form Before Dividing

Divide $-13x^3 + 10x^4 + 8x - 7x^2 + 4$ by $3 - 2x$.

Solution

First write the divisor and dividend in standard polynomial form.

$$
\begin{array}{r}
-5x^3 - x^2 + 2x - 1 \\
-2x+3\ \overline{)\ 10x^4 - 13x^3 - 7x^2 + 8x + 4} \\
\underline{10x^4 - 15x^3} \\
2x^3 - 7x^2 \\
\underline{2x^3 - 3x^2} \\
-4x^2 + 8x \\
\underline{-4x^2 + 6x} \\
2x + 4 \\
\underline{2x - 3} \\
7
\end{array}
$$

Multiply $-5x^3(-2x+3)$.

Subtract and bring down $-7x^2$.
Multiply $-x^2(-2x+3)$.

Subtract and bring down $8x$.
Multiply $2x(-2x+3)$.

Subtract and bring down 4.
Multiply $(-1)(-2x+3)$.

This shows that

$$\frac{10x^4 - 13x^3 - 7x^2 + 8x + 4}{-2x + 3} = -5x^3 - x^2 + 2x - 1 + \frac{7}{-2x + 3}.$$

Technology

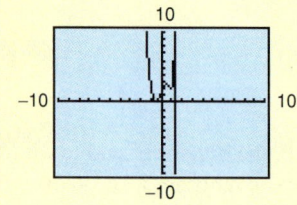

You can check the result of a division problem *algebraically*, as shown at the bottom of page 479. You can check the result *graphically* with a graphing utility by comparing the graphs of the original quotient and the simplified form. For example, the figure at the left shows the graphs of

$$y_1 = \frac{10x^4 - 13x^3 - 7x^2 + 8x + 4}{-2x + 3}$$

and

$$y_2 = -5x^3 - x^2 + 2x - 1 + \frac{7}{-2x + 3}.$$

Because the graphs coincide, it follows that the expressions are equivalent.

When the dividend is missing some powers of x, the long division algorithm requires that you account for the missing powers, as shown in Example 5.

EXAMPLE 5   Accounting for Missing Powers of x

Divide $(x^3 - 2)$ by $(x - 1)$.

Solution

Note how the missing x^2- and x-terms are accounted for.

$$
\begin{array}{r}
x^2 + x + 1 \\
x - 1 \overline{) x^3 + 0x^2 + 0x - 2} \\
\underline{x^3 - x^2} \\
x^2 + 0x \\
\underline{x^2 - x} \\
x - 2 \\
\underline{x - 1} \\
- 1
\end{array}
$$

Multiply $x^2(x - 1)$.

Subtract and bring down $0x$.
Multiply $x(x - 1)$.

Subtract and bring down -2.
Multiply $(1)(x - 1)$.

Subtract.

Thus, you have

$$
\frac{x^3 - 2}{x - 1} = x^2 + x + 1 - \frac{1}{x - 1}.
$$

EXAMPLE 6   A Second-Degree Divisor

Divide $x^4 + 6x^3 + 6x^2 - 10x - 3$ by $x^2 + 2x - 3$.

Solution

NOTE   In each of the long division examples so far, the divisor has been a first-degree polynomial. The long division algorithm works just as well with polynomial divisors of degree 2 or more, as shown in Example 6.

$$
\begin{array}{r}
x^2 + 4x + 1 \\
x^2 + 2x - 3 \overline{) x^4 + 6x^3 + 6x^2 - 10x - 3} \\
\underline{x^4 + 2x^3 - 3x^2} \\
4x^3 + 9x^2 - 10x \\
\underline{4x^3 + 8x^2 - 12x} \\
x^2 + 2x - 3 \\
\underline{x^2 + 2x - 3} \\
0
\end{array}
$$

Multiply $x^2(x^2 + 2x - 3)$.

Subtract and bring down $-10x$.
Multiply $4x(x^2 + 2x - 3)$.

Subtract and bring down -3.
Multiply $(1)(x^2 + 2x - 3)$.

Subtract.

Thus, $x^2 + 2x - 3$ divides evenly into $x^4 + 6x^3 + 6x^2 - 10x - 3$:

$$
\frac{x^4 + 6x^3 + 6x^2 - 10x - 3}{x^2 + 2x - 3} = x^2 + 4x + 1.
$$

Synthetic Division

There is a nice shortcut for division by polynomials of the form $x - k$. It is called **synthetic division,** and is outlined for a third-degree polynomial as follows.

Synthetic Division for a Third-Degree Polynomial

Use synthetic division to divide $ax^3 + bx^2 + cx + d$ by $x - k$ as follows.

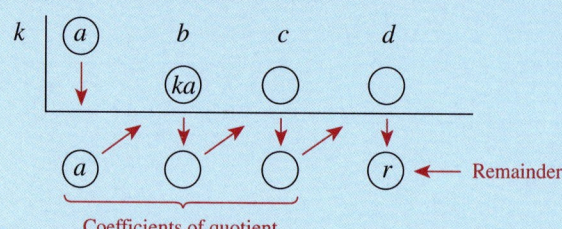

Vertical Pattern: Add terms.
Diagonal Pattern: Multiply by k.

EXAMPLE 7   Using Synthetic Division

Use synthetic division to divide $x^3 + 3x^2 - 4x - 10$ by $x - 2$.

Solution

The coefficients of the dividend form the top row of the synthetic division tableau. Because you are dividing by $(x - 2)$, write 2 at the top left of the tableau. To begin the algorithm, bring down the first coefficient. Then multiply this coefficient by 2, write the result in the second row of the tableau, and add the two numbers in the second column. By continuing this pattern, you obtain the following tableau.

$$
\begin{array}{c|cccc}
2 & 1 & 3 & -4 & -10 \\
  &   & 2 & 10 & 12 \\
\hline
  & 1 & 5 & 6 & ② \leftarrow \text{Remainder}
\end{array}
$$

The bottom row of the tableau shows that the quotient is

$$(1)x^2 + (5)x + (6)$$

and the remainder is 2. Thus, the result of the division problem is

$$\frac{x^3 + 3x^2 - 4x - 10}{x - 2} = x^2 + 5x + 6 + \frac{2}{x - 2}.$$

Factoring and Division

If the remainder in a synthetic division problem turns out to be zero, you can conclude that the divisor divides *evenly* into the dividend.

EXAMPLE 8   Factoring a Polynomial

Completely factor $x^3 - 7x + 6$. Use the fact that $x - 1$ is one of the factors.

Solution

Because you are given one of the factors, divide this factor into the given polynomial to obtain

$$\frac{x^3 - 7x + 6}{x - 1} = x^2 + x - 6.$$

From this result, you can factor the original polynomial as follows.

$$x^3 - 7x + 6 = (x - 1)(x^2 + x - 6)$$
$$= (x - 1)(x + 3)(x - 2)$$

Group Activities   Exploring with Technology

Investigating Polynomials and Their Factors   Use a graphing utility to graph the following polynomials on the same viewing rectangle using the standard setting. Use the TRACE feature to find the x-intercepts. What can you conclude about the polynomials? Verify your conclusion algebraically.

a. $y = (x - 4)(x - 2)(x + 1)$

b. $y = (x^2 - 6x + 8)(x + 1)$

c. $y = x^3 - 5x^2 + 2x + 8$

Now use your graphing utility to graph the function

$$f(x) = \frac{x^3 - 5x^2 + 2x + 8}{x - 2}.$$

Use the TRACE feature to find the x-intercepts. Why does it have only two x-intercepts? To what other function does the graph of $f(x)$ appear to be equivalent? What is the difference between the two graphs? (*Hint:* Zoom in at $x = 2$.)

8.4 Exercises

Discussing the Concepts

1. *Error Analysis*   Describe the error.

$$\frac{6x + 5y}{x} = \frac{6\cancel{x} + 5y}{\cancel{x}} = 6 + 5y$$

2. Create a polynomial division problem and identify the dividend, divisor, quotient, and remainder.

3. What does it mean for a divisor to *divide evenly* into a dividend?

4. Explain how you can check a polynomial division problem. Give an example.

5. *True or False?*   If the divisor divides evenly into the dividend, the divisor and quotient are factors of the dividend. Explain.

6. For synthetic division, what form must the divisor have?

Problem Solving

In Exercises 7–10, perform the division of a polynomial by a monomial and check your result.

7. $\dfrac{50z^3 + 30z}{-5z}$

8. $\dfrac{18c^4 - 24c^2}{-6c}$

9. $(5x^2y - 8xy + 7xy^2) \div 2xy$

10. $(-14s^4t^2 + 7s^2t^2 - 18t) \div 2s^2t$

In Exercises 11–24, perform the division and check your result.

11. $\dfrac{x^2 - 8x + 15}{x - 3}$

12. $\dfrac{t^2 - 18t + 72}{t - 6}$

13. Divide $21 - 4x - x^2$ by $3 - x$.

14. Divide $10t^2 - 7t - 12$ by $2t - 3$.

15. $\dfrac{x^3 - 2x^2 + 4x - 8}{x - 2}$

16. $\dfrac{x^3 - 28x - 48}{x + 4}$

17. $\dfrac{x^2 + 16}{x + 4}$

18. $\dfrac{y^2 + 8}{y + 2}$

19. $\dfrac{6z^2 + 7z}{5z - 1}$

20. $\dfrac{8y^2 - 2y}{3y + 5}$

21. $x^5 \div (x^2 + 1)$

22. $x^4 \div (x - 2)$

23. $(x^3 + 4x^2 + 7x + 6) \div (x^2 + 2x + 3)$

24. $(2x^3 + 2x^2 - 2x - 15) \div (2x^2 + 4x + 5)$

In Exercises 25–28, use synthetic division to perform the division.

25. $\dfrac{x^3 + 3x^2 - 1}{x + 4}$

26. $\dfrac{2x^5 - 3x^3 + x}{x - 3}$

27. $\dfrac{0.1x^2 + 0.8x + 1}{x - 0.2}$

28. $\dfrac{x^3 - 0.8x + 2.4}{x + 1}$

In Exercises 29–32, use synthetic division to perform the division. Use the result to factor the dividend.

29. $\dfrac{x^2 - 15x + 56}{x - 8}$

30. $\dfrac{24 + 13t - 2t^2}{8 - t}$

31. $\dfrac{x^4 - x^3 - 3x^2 + 4x - 1}{x - 1}$

32. $\dfrac{x^4 - 16}{x - 2}$

In Exercises 33 and 34, use a graphing utility to graph the two equations on the same screen. Use the graphs to verify that the expressions are equivalent. Verify the results algebraically.

33. $y_1 = \dfrac{x + 4}{2x}, \quad y_2 = \dfrac{1}{2} + \dfrac{2}{x}$

34. $y_1 = \dfrac{x^2 + 2}{x + 1}, \quad y_2 = x - 1 + \dfrac{3}{x + 1}$

Finding a Pattern   In Exercises 35 and 36, complete the table for the given polynomial. The first row is completed for Exercise 35. What conclusion can you draw as you compare the polynomial values with the remainders? (Use synthetic division to find the remainders.)

x	Polynomial Value	Divisor	Remainder
-2	-8	$x + 2$	-8
-1			
0			
$\frac{1}{2}$			
1			
2			

35. $x^3 - x^2 - 2x$

36. $2x^3 - x^2 - 2x + 1$

Geometry   In Exercises 37 and 38, you are given the expression for the volume of the solid shown. Find an expression for the missing dimension.

37. $V = x^3 + 18x^2 + 80x + 96$

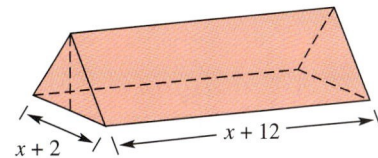

$x + 2$ $x + 12$

38. $V = h^4 + 3h^3 + 2h^2$

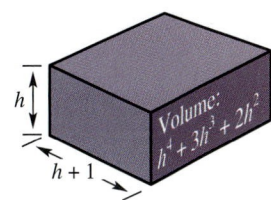

h

Volume: $h^4 + 3h^3 + 2h^2$

$h + 1$

Reviewing the Major Concepts

In Exercises 39–42, multiply.

39. $(x + 1)^2$

40. $(2 - y)(3 + 2y)$

41. $(4 - 5z)(4 + 5z)$

42. $t(t - 4)(2t + 3)$

43. *Geometry*   The base of a triangle is $5x$ and its height is $2x + 9$. Find the area A of the triangle.

44. Given the function $f(x) = 3x - x^2$, find and simplify $f(2 + t) - f(2)$.

Additional Problem Solving

In Exercises 45–54, perform the division of a polynomial by a monomial and check your result.

45. $\dfrac{6z + 10}{2}$

46. $\dfrac{9x + 12}{3}$

47. $\dfrac{10z^2 + 4z - 12}{4}$

48. $\dfrac{4u^2 + 8u - 24}{16}$

49. $(7x^3 - 2x^2) \div x$

50. $(6a^2 + 7a) \div a$

51. $\dfrac{8z^3 + 3z^2 - 2z}{2z}$

52. $\dfrac{6x^4 + 8x^3 - 18x^2}{3x^2}$

53. $\dfrac{m^4 + 2m^2 - 7}{m}$

54. $\dfrac{l^2 - 8}{-l}$

In Exercises 55–76, perform the division and check your result.

55. $\dfrac{4(x + 5)^2 + 8(x + 5)}{x + 5}$

56. $\dfrac{12(r - 9)^3 - 18(r - 9)^2}{4(r - 9)^2}$

57. $(x^2 + 15x + 50) \div (x + 5)$

58. $(y^2 - 6y - 16) \div (y + 2)$

59. Divide $2y^2 + 7y + 3$ by $2y + 1$.

60. Divide $5 + 4x - x^2$ by $1 + x$.

61. $\dfrac{12t^2 - 40t + 25}{2t - 5}$

62. $\dfrac{15 - 14u - 8u^2}{5 + 2u}$

63. $\dfrac{16x^2 - 1}{4x + 1}$

64. $\dfrac{81y^2 - 25}{9y - 5}$

65. $\dfrac{x^3 + 125}{x + 5}$

66. $\dfrac{x^3 - 27}{x - 3}$

67. $\dfrac{2x + 9}{x + 2}$

68. $\dfrac{12x - 5}{2x + 3}$

69. $\dfrac{5x^2 + 2x + 3}{x + 2}$

70. $\dfrac{2x^2 + 5x + 2}{x + 4}$

71. $\dfrac{12x^2 - 17x - 5}{3x + 2}$

72. $\dfrac{8x^2 + 2x + 3}{4x - 1}$

73. $\dfrac{2x^3 - 5x^2 + x - 6}{x - 3}$

74. $\dfrac{5x^3 + 3x^2 + 12x + 20}{x + 1}$

75. $\dfrac{x^6 - 1}{x - 1}$

76. $\dfrac{x^3}{x - 1}$

Think About It   In Exercises 77 and 78, perform the division assuming that n is a positive integer.

77. $\dfrac{x^{3n} + 3x^{2n} + 6x^n + 8}{x^n + 2}$

78. $\dfrac{x^{3n} - x^{2n} + 5x^n - 5}{x^n - 1}$

In Exercises 79–82, use synthetic division to perform the division.

79. $\dfrac{x^4 - 4x^3 + x + 10}{x - 2}$

80. $\dfrac{x^4}{x + 2}$

81. $\dfrac{5x^3 + 12}{x + 5}$

82. $\dfrac{8x + 35}{x - 10}$

In Exercises 83–88, use synthetic division to perform the division. Use the result to factor the dividend.

83. $\dfrac{2a^2 + 14a + 45}{a + 9}$

84. $\dfrac{x^2 + 3x - 154}{x + 14}$

85. $\dfrac{15x^2 - 2x - 8}{x - \frac{4}{5}}$

86. $\dfrac{18x^2 - 9x - 20}{x + \frac{5}{6}}$

87. $\dfrac{2t^3 + 15t^2 + 19t - 30}{t + 5}$

88. $\dfrac{5t^3 - 27t^2 - 14t - 24}{t - 6}$

Think About It   In Exercises 89 and 90, find the constant c so that the denominator will divide evenly into the numerator.

89. $\dfrac{x^3 + 2x^2 - 4x + c}{x - 2}$

90. $\dfrac{x^4 - 3x^2 + c}{x + 6}$

91. *Geometry*   The rectangle's area is $2x^3 + 3x^2 - 6x - 9$. Find its width if its length is $2x + 3$.

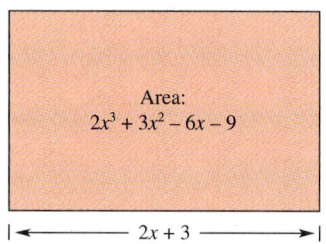

Area:
$2x^3 + 3x^2 - 6x - 9$

|←———— $2x + 3$ ————→|

92. *Geometry*   A rectangular house has a volume of

$$x^3 + 55x^2 + 650x + 2000$$

cubic feet (the space in the attic is not included). The height of the house is $x + 5$ (see figure). Find the number of square feet of floor space *on the first floor* of the house.

$x + 5$

8.5 Graphing Rational Functions

Introduction ▪ Horizontal and Vertical Asymptotes ▪
Graphing Rational Functions ▪ Application

Introduction

Recall that the domain of a rational function consists of all values of x for which the denominator is not zero. For instance, the domain of

$$f(x) = \frac{x+2}{x-1}$$

is all real numbers except $x = 1$. When graphing a rational function, pay special attention to the shape of the graph near x-values that are not in the domain.

EXAMPLE 1 *Sketching the Graph of a Rational Function*

Sketch the graph of $f(x) = \dfrac{x+2}{x-1}$.

Solution

Begin by noticing that the domain is all real numbers except $x = 1$. Next, construct a table of values, including x-values that are close to 1 on the left *and* the right.

x-Values to the Left of 1

x	-3	-2	-1	0	0.5	0.9
$f(x)$	0.25	0	-0.5	-2	-5	-29

x-Values to the Right of 1

x	1.1	1.5	2	3	4	5
$f(x)$	31	7	4	2.5	2	1.75

Plot the points to the left of 1 and connect them with a smooth curve, as shown in Figure 8.1. Do the same for the points to the right of 1. *Do not* connect the two portions of the graph, which are called its **branches.**

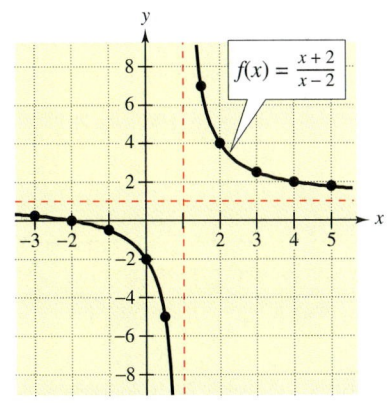

$f(x) = \dfrac{x+2}{x-2}$

FIGURE 8.1

NOTE In Figure 8.1, as x approaches 1 from the left, the values of $f(x)$ approach negative infinity, and as x approaches 1 from the right, the values of $f(x)$ approach positive infinity.

Horizontal and Vertical Asymptotes

An **asymptote** of a graph is a line to which the graph becomes arbitrarily close as $|x|$ or $|y|$ increases without bound. In other words, if a graph has an asymptote, it is possible to move far enough out on the graph so that there is almost no difference between the graph and the asymptote.

The graph in Example 1 has two asymptotes: the line $x = 1$ is a **vertical asymptote,** and the line $y = 1$ is a **horizontal asymptote.** Other examples of asymptotes are shown in Figure 8.2.

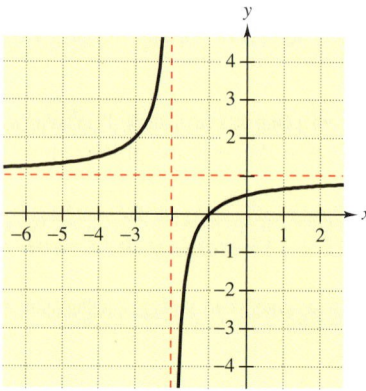

Graph of $y = \dfrac{x+1}{x+2}$

Horizontal asymptote: $y = 1$
Vertical asymptote: $x = -2$

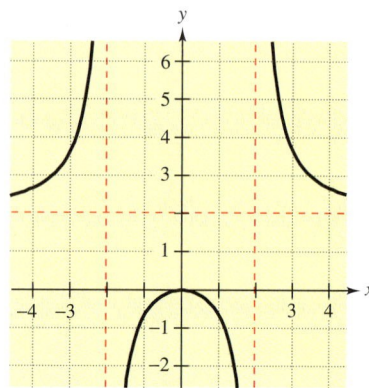

Graph of $y = \dfrac{2x^2}{x^2 - 4}$

Horizontal asymptote: $y = 2$
Vertical asymptotes: $x = \pm 2$

FIGURE 8.2

The graph of a rational function may have no horizontal or vertical asymptotes, or it may have several.

Guidelines for Finding Asymptotes

Let $f(x) = p(x)/q(x)$, where $p(x)$ and $q(x)$ have no common factors.

1. The graph of f has a vertical asymptote at each x-value for which the denominator is zero.

2. The graph of f has at most one horizontal asymptote. If the degree of $p(x)$ is less than the degree of $q(x)$, the line $y = 0$ is a horizontal asymptote. If the degree of $p(x)$ is equal to the degree of $q(x)$, the line $y = a/b$ is a horizontal asymptote, where a is the leading coefficient of $p(x)$ and b is the leading coefficient of $q(x)$. If the degree of $p(x)$ is greater than the degree of $q(x)$, the graph has no horizontal asymptote.

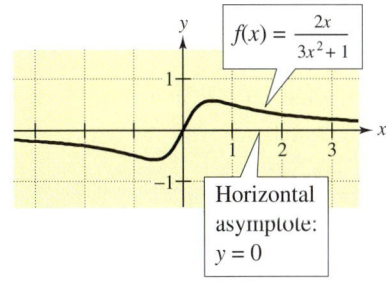

$f(x) = \dfrac{2x}{3x^2 + 1}$

Horizontal asymptote: $y = 0$

FIGURE 8.3

EXAMPLE 2   *Finding Horizontal and Vertical Asymptotes*

Find all horizontal and vertical asymptotes of the graph of

$$f(x) = \frac{2x}{3x^2 + 1}.$$

Solution

For this rational function, the degree of the numerator is less than the degree of the denominator. This implies that the graph has the line

$$y = 0 \hspace{3cm} \textcolor{red}{\text{Horizontal asymptote}}$$

as a horizontal asymptote, as shown in Figure 8.3. To find any vertical asymptotes, set the denominator equal to zero and solve the resulting equation for x.

$$3x^2 + 1 = 0 \hspace{2.5cm} \textcolor{red}{\text{Set denominator equal to zero.}}$$

Because this equation has no real solution, you can conclude that the graph has no vertical asymptote.

Remember that the graph of a rational function can have at most one horizontal asymptote, but it can have several vertical asymptotes. For instance, the graph in Example 3 has two vertical asymptotes.

EXAMPLE 3   *Finding Horizontal and Vertical Asymptotes*

Find all horizontal and vertical asymptotes of the graph of

$$f(x) = \frac{2x^2}{x^2 - 1}.$$

Solution

For this rational function, the degree of the numerator is equal to the degree of the denominator. The leading coefficient of the numerator is 2, and the leading coefficient of the denominator is 1. Thus, the graph has the line

$$y = \frac{2}{1} = 2 \hspace{2.5cm} \textcolor{red}{\text{Horizontal asymptote}}$$

as a horizontal asymptote, as shown in Figure 8.4. To find any vertical asymptotes, set the denominator equal to zero and solve the resulting equation for x.

$$x^2 - 1 = 0 \hspace{2.5cm} \textcolor{red}{\text{Set denominator equal to zero.}}$$

This equation has two real solutions: -1 and 1. Thus, the graph has two vertical asymptotes: the lines $x = -1$ and $x = 1$.

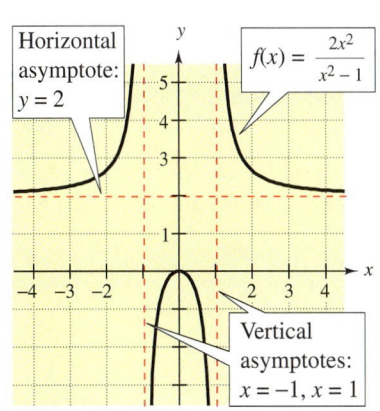

Horizontal asymptote: $y = 2$

$f(x) = \dfrac{2x^2}{x^2 - 1}$

Vertical asymptotes: $x = -1, x = 1$

FIGURE 8.4

Graphing Rational Functions

To sketch the graph of a rational function, we suggest the following guidelines.

Guidelines for Graphing Rational Functions

Let $f(x) = p(x)/q(x)$, where $p(x)$ and $q(x)$ have no common factors.

1. Find and plot the y-intercept (if any) by evaluating $f(0)$.

2. Set the numerator equal to zero and solve the equation for x. The real solutions represent the x-intercepts of the graph. Plot these intercepts.

3. Find and sketch the horizontal and vertical asymptotes of the graph.

4. Plot at least one point both between and beyond each x-intercept and vertical asymptote.

5. Use smooth curves to complete the graph between and beyond the vertical asymptotes.

EXAMPLE 4   Sketching the Graph of a Rational Function

Sketch the graph of $f(x) = \dfrac{2}{x-3}$.

Solution

Begin by noting that the numerator and denominator have no common factors. Following the above guidelines produces the following.

- Because $f(0) = -\frac{2}{3}$, the y-intercept is $\left(0, -\frac{2}{3}\right)$.
- Because the numerator is never zero, there are no x-intercepts.
- Because the denominator is zero when $x = 3$, the line $x = 3$ is a vertical asymptote.
- Because the degree of the numerator is less than the degree of the denominator, the line $y = 0$ is a horizontal asymptote.

Plot the intercepts, asymptotes, and the additional points from the following table. Then complete the graph by drawing two branches, as shown in Figure 8.5. Note that the two branches are not connected.

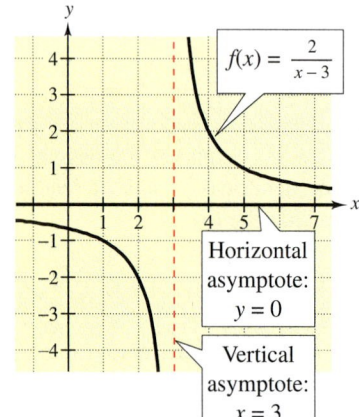

$f(x) = \dfrac{2}{x-3}$

Horizontal asymptote: $y = 0$

Vertical asymptote: $x = 3$

FIGURE 8.5

x	-2	1	2	4	5
$f(x)$	$-\frac{2}{5}$	-1	-2	2	1

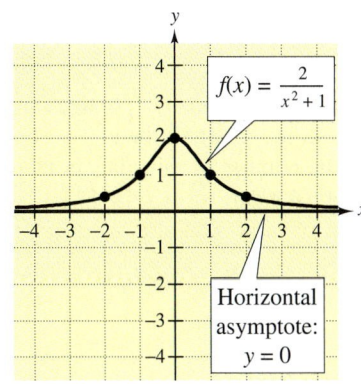

FIGURE 8.6

EXAMPLE 5 *Sketching the Graph of a Rational Function*

Sketch the graph of $f(x) = \dfrac{2}{x^2 + 1}$.

Solution

Begin by noting that the numerator and denominator have no common factors.

- Because $f(0) = 2$, the y-intercept is $(0, 2)$.
- Because the numerator is never zero, there are no x-intercepts.
- Because the denominator is never zero, there are no vertical asymptotes.
- Because the degree of the numerator is less than the degree of the denominator, the line $y = 0$ is a horizontal asymptote.

By plotting the intercepts, asymptotes, and the additional points from the following table, you can obtain the graph shown in Figure 8.6.

x	-2	-1	1	2
$f(x)$	$\frac{2}{5}$	1	1	$\frac{2}{5}$

Technology

Graphing a Rational Function

A graphing utility can help you sketch the graph of a rational function. With most graphing utilities, however, there are problems with graphs of rational functions. If you use the *connected mode*, the graphing utility will try to connect any branches of the graph. If you use the *dot mode*, the graphing utility will draw a dotted (rather than a solid) graph. Both of these options are shown below for the graph of $y = (x - 1)/(x - 3)$.

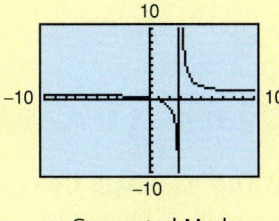

Connected Mode

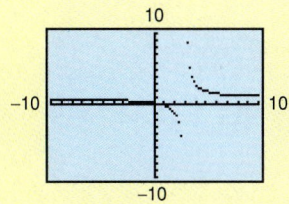

Dot Mode

Application

EXAMPLE 6   Finding the Average Cost

As a fund-raising project, a club is publishing a calendar. The cost of photography and typesetting is \$850. In addition to these "one-time" charges, the unit cost of printing each calendar is \$3.25. Let x represent the number of calendars printed. Write a model that represents the average cost per calendar.

Solution

The total cost C of printing x calendars is

$$C = 3.25x + 850. \qquad \text{Total cost function}$$

The average cost per calendar $\bar{A}$ for printing x calendars is

$$\bar{A} = \frac{3.25x + 850}{x}. \qquad \text{Average cost function}$$

From the graph shown in Figure 8.7, notice that the average cost decreases as the number of calendars increases.

In 1995, about 14,000 individuals were members of the Professional Photographers of America.

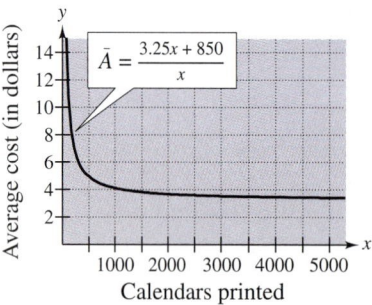

$$\bar{A} = \frac{3.25x + 850}{x}$$

Average cost (in dollars)

Calendars printed

FIGURE 8.7

Group Activities   Extending the Concept

More About the Average Cost   In Example 6, what is the horizontal asymptote of the graph of the average cost function? What is the significance of this asymptote in the problem? Is it possible to sell enough calendars to obtain an average cost of \$3.00 per calendar? Explain your reasoning.

8.5 Exercises

Discussing the Concepts

1. In your own words, describe how to determine the domain of a rational function. Give an example of a rational function whose domain is all real numbers except 2.

2. In your own words, describe what is meant by an *asymptote* of a graph.

3. *True or False?*   If the graph of rational function f has a vertical asymptote at $x = 3$, it is possible to sketch the graph without lifting your pencil from the paper. Explain.

4. Does every rational function have a vertical asymptote? Explain.

Problem Solving

In Exercises 5 and 6, (a) complete each table, (b) determine the vertical and horizontal asymptotes of the graph, and (c) find the domain of the function.

x	0	0.5	0.9	0.99	0.999
y					

x	2	1.5	1.1	1.01	1.001
y					

x	2	5	10	100	1000
y					

5. $f(x) = \dfrac{4}{x-1}$ **6.** $f(x) = \dfrac{2x}{x-1}$

In Exercises 7–12, find the domain of the function and identify any horizontal and vertical asymptotes.

7. $f(x) = \dfrac{5}{x^2}$ **8.** $g(x) = \dfrac{3}{x-5}$

9. $g(t) = \dfrac{3}{t^2+1}$ **10.** $h(s) = \dfrac{2s^2}{s+3}$

11. $y = \dfrac{5x^2}{x^2-1}$ **12.** $y = \dfrac{3x+2}{2x-1}$

In Exercises 13–16, match the function with its graph.

(a)
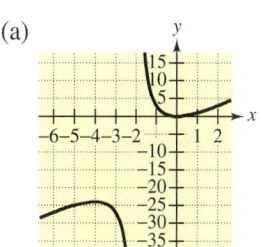

(b)
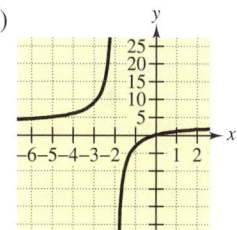

(c)
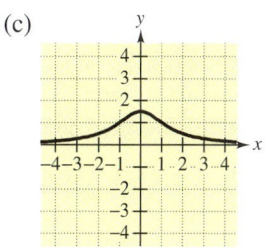

(d)
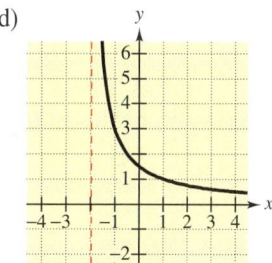

13. $f(x) = \dfrac{3}{x+2}$ **14.** $f(x) = \dfrac{3x}{x+2}$

15. $f(x) = \dfrac{3x^2}{x+2}$ **16.** $f(x) = \dfrac{3}{x^2+2}$

In Exercises 17–24, sketch the graph of the rational function. As sketching aids, check for intercepts, vertical asymptotes, and horizontal asymptotes.

17. $f(x) = \dfrac{1}{x-2}$ **18.** $f(x) = \dfrac{3}{x+1}$

19. $g(x) = \dfrac{1}{2-x}$ **20.** $g(x) = \dfrac{-3}{x+1}$

21. $y = \dfrac{2x+4}{x}$

22. $y = \dfrac{2x}{x+4}$

23. $y = \dfrac{2x^2}{x^2+1}$

24. $y = \dfrac{10}{(x-2)^2}$

In Exercises 25–28, use a graphing utility to graph the function. Give the domain of the function and identify any horizontal or vertical asymptotes.

25. $h(x) = \dfrac{x-3}{x-1}$

26. $h(x) = \dfrac{x^2}{x-2}$

27. $f(t) = \dfrac{6}{t^2+1}$

28. $g(t) = 2 + \dfrac{3}{t+1}$

In Exercises 29–32, use the graph of $f(x) = 1/x$ to sketch the graph of g.

29. $g(x) = -\dfrac{1}{x}$

30. $g(x) = \dfrac{1}{x} + 2$

31. $g(x) = \dfrac{1}{x-2}$

32. $g(x) = \dfrac{1}{x+3}$

33. *Average Cost* The cost of producing x units is $C = 2500 + 0.50x$, $0 < x$.

(a) Write the average cost $\bar{A}$ as a function of x.

(b) Find the average costs of producing $x = 1000$ and $x = 10{,}000$ units.

(c) Use a graphing utility to graph the average cost function. Determine the horizontal asymptote of the graph.

34. *Concentration of a Mixture* A 25-liter container contains 5 liters of a 25% brine solution. You add x liters of a 75% brine solution to the container. The concentration C of the resulting mixture is

$$C = \dfrac{3x+5}{4(x+5)}.$$

(a) Determine the domain of the rational function within the physical constraints of the problem.

(b) Use a graphing utility to graph the function. As the container is filled, what percent does the concentration of the brine appear to approach?

Reviewing the Major Concepts

In Exercises 35–38, solve the inequality and sketch the graph of the solution on the real number line.

35. $2x - 12 \geq 0$

36. $7 - 3x < 4 - x$

37. $|x - 3| < 2$

38. $|x - 5| > 3$

39. Determine all real numbers n such that $\frac{1}{3}n$ must be at least 10 and no more than 50.

40. *Operating Costs* The annual operating cost for a truck is

$$C = 0.45m + 6200$$

where m is the number of miles traveled by the truck in a year. What number of miles will yield an annual operating cost that is less than $15,000?

Additional Problem Solving

In Exercises 41–50, find the domain of the function and identify any horizontal and vertical asymptotes.

41. $f(x) = 2 + \dfrac{1}{x-3}$

42. $f(x) = \dfrac{2}{x-3}$

43. $f(x) = \dfrac{3x}{x^2-9}$

44. $f(x) = \dfrac{5x^2}{x^2-9}$

45. $f(x) = \dfrac{x}{x+8}$

46. $f(u) = \dfrac{u^2}{u-10}$

47. $g(t) = \dfrac{3}{t(t-1)}$

48. $h(x) = 4 - \dfrac{3}{x}$

49. $y = \dfrac{2x^2}{x^2+1}$

50. $y = \dfrac{3-5x}{1-3x}$

In Exercises 51–62, sketch the graph of the rational function. As sketching aids, check for intercepts, vertical asymptotes, and horizontal asymptotes. Use a graphing utility to verify your graph.

51. $g(x) = \dfrac{5}{x}$

52. $g(x) = \dfrac{5}{x - 4}$

53. $f(x) = \dfrac{5}{x^2}$

54. $f(x) = \dfrac{5}{(x - 4)^2}$

55. $g(t) = 3 - \dfrac{2}{t}$

56. $g(v) = \dfrac{2v}{v + 1}$

57. $y = \dfrac{3x}{x + 4}$

58. $y = \dfrac{x - 2}{x}$

59. $y = \dfrac{4}{x^2 + 1}$

60. $y = \dfrac{4x^2}{x^2 + 1}$

61. $y = -\dfrac{x}{x^2 - 4}$

62. $y = \dfrac{3x^2}{x^2 - x - 2}$

In Exercises 63–66, use a graphing utility to graph the function. Give its domain.

63. $y = \dfrac{2(x^2 + 1)}{x^2}$

64. $y = \dfrac{2(x^2 - 1)}{x^2}$

65. $y = \dfrac{3}{x} + \dfrac{1}{x - 2}$

66. $y = \dfrac{x}{2} - \dfrac{2}{x}$

In Exercises 67–70, use the graph of $f(x) = 4/x^2$ to sketch the graph of g.

67. $g(x) = 2 + \dfrac{4}{x^2}$

68. $g(x) = -\dfrac{4}{x^2}$

69. $g(x) = -\dfrac{4}{(x - 2)^2}$

70. $g(x) = 5 - \dfrac{4}{x^2}$

Think About It   In Exercises 71 and 72, use a graphing utility to graph the function. Explain why there is no vertical asymptote when a superficial examination of the function may indicate that there should be one.

71. $g(x) = \dfrac{4 - 2x}{x - 2}$

72. $h(x) = \dfrac{x^2 - 9}{x + 3}$

73. *Medicine*   The concentration of a certain chemical in the bloodstream t hours after injection into the muscle tissue is given by

$$C = \frac{2t}{4t^2 + 25}, \quad 0 \le t.$$

(a) Determine the horizontal asymptote of the function and interpret its meaning in the context of the problem.

(b) Graph the function on a graphing utility. Approximate the time when the concentration is greatest.

74. *Average Cost*   The cost of producing x units is $C = 30{,}000 + 1.25x$, $0 < x$.

(a) Write the average cost $\bar{A}$ as a function of x.

(b) Find the average costs of producing $x = 10{,}000$ and $x = 100{,}000$ units.

(c) Use a graphing utility to graph the average cost function. Determine the horizontal asymptote of the graph.

75. *Geometry*   A rectangular region of length x and width y has an area of 400 square meters.

(a) Verify that the perimeter P is given by

$$P = 2\left(x + \frac{400}{x}\right).$$

(b) Determine the domain of the function within the physical constraints of the problem.

(c) Sketch a graph of the function and approximate the dimensions of the rectangle that has a minimum perimeter.

76. *Sales*   The cumulative number N (in thousands) of units of a product sold over a period of t years on the market is modeled by

$$N = \frac{150t(1 + 4t)}{1 + 0.15t^2}, \quad 0 \le t.$$

(a) Estimate cumulative sales when $t = 1$, $t = 2$, and $t = 4$.

(b) Use a graphing utility to graph the function. Determine the horizontal asymptote of the graph.

(c) Explain the meaning of the horizontal asymptote in the context of the problem.

8.6	**Solving Rational Equations**
	Equations Containing Constant Denominators ▪
	Equations Containing Variable Denominators ▪ Applications

Equations Containing Constant Denominators

In Section 3.1, you studied a strategy for solving equations that contain fractions with *constant* denominators. We review that procedure here because it is the basis for solving more general equations involving fractions. Recall from Section 3.1 that you can "clear an equation of fractions" by multiplying both sides of the equation by the least common denominator of the fractions in the equation. Note how this is done in Example 1.

EXAMPLE 1   An Equation Containing Constant Denominators

Solve $\dfrac{3}{5} = \dfrac{x}{2}$.

Solution

The least common denominator of the two fractions is 10, so begin by multiplying both sides of the equation by 10.

$$\frac{3}{5} = \frac{x}{2} \qquad \text{Original equation}$$

$$10\left(\frac{3}{5}\right) = 10\left(\frac{x}{2}\right) \qquad \text{Multiply both sides by 10.}$$

$$6 = 5x \qquad \text{Simplify.}$$

$$\frac{6}{5} = x \qquad \text{Divide both sides by 5.}$$

The solution is $\frac{6}{5}$. You can check this as follows.

Check

$$\frac{3}{5} = \frac{x}{2} \qquad \text{Original equation}$$

$$\frac{3}{5} \overset{?}{=} \frac{6/5}{2} \qquad \text{Substitute } \tfrac{6}{5} \text{ for } x.$$

$$\frac{3}{5} \overset{?}{=} \frac{6}{5} \cdot \frac{1}{2} \qquad \text{Invert and multiply.}$$

$$\frac{3}{5} = \frac{3}{5} \qquad \text{Solution checks. } \checkmark$$

EXAMPLE 2   An Equation Containing Constant Denominators

Solve $\dfrac{x}{6} = 7 - \dfrac{x}{12}$.

Solution

$$\frac{x}{6} = 7 - \frac{x}{12} \qquad \text{Original equation}$$

$$12\left(\frac{x}{6}\right) = 12\left(7 - \frac{x}{12}\right) \qquad \text{Multiply both sides by 12.}$$

$$2x = 84 - x \qquad \text{Distribute and simplify.}$$

$$3x = 84 \qquad \text{Add } x \text{ to both sides.}$$

$$x = 28 \qquad \text{Divide both sides by 3.}$$

The solution is 28. Check this in the original equation.

EXAMPLE 3   An Equation Containing Constant Denominators

Solve $\dfrac{x+2}{6} - \dfrac{x-4}{8} = \dfrac{2}{3}$.

Solution

$$\frac{x+2}{6} - \frac{x-4}{8} = \frac{2}{3} \qquad \text{Original equation}$$

$$24\left(\frac{x+2}{6} - \frac{x-4}{8}\right) = 24\left(\frac{2}{3}\right) \qquad \text{Multiply both sides by 24.}$$

$$4(x+2) - 3(x-4) = 8(2) \qquad \text{Distribute and simplify.}$$

$$4x + 8 - 3x + 12 = 16 \qquad \text{Distributive Property}$$

$$x + 20 = 16 \qquad \text{Combine like terms.}$$

$$x = -4 \qquad \text{Subtract 20 from both sides.}$$

The solution is -4. You can check this as follows.

Check

$$\frac{x+2}{6} - \frac{x-4}{8} = \frac{2}{3} \qquad \text{Original equation}$$

$$\frac{-4+2}{6} - \frac{-4-4}{8} \overset{?}{=} \frac{2}{3} \qquad \text{Substitute } -4 \text{ for } x.$$

$$-\frac{1}{3} + 1 \overset{?}{=} \frac{2}{3} \qquad \text{Simplify.}$$

$$\frac{2}{3} = \frac{2}{3} \qquad \text{Solution checks. } \checkmark$$

Equations Containing Variable Denominators

Remember that you always *exclude* those values of a variable that make the denominator of a rational expression zero. This is especially critical for solving equations that contain variable denominators. You will see why in the examples that follow.

EXAMPLE 4   *An Equation Containing Variable Denominators*

Solve the equation.

$$\frac{7}{x} - \frac{1}{3x} = \frac{8}{3}$$

Solution

For this equation the least common denominator is $3x$. Therefore, begin by multiplying both sides of the equation by $3x$.

$$\frac{7}{x} - \frac{1}{3x} = \frac{8}{3} \qquad \text{Original equation}$$

$$3x\left(\frac{7}{x} - \frac{1}{3x}\right) = 3x\left(\frac{8}{3}\right) \qquad \text{Multiply both sides by } 3x.$$

$$\frac{21x}{x} - \frac{3x}{3x} = \frac{24x}{3} \qquad \text{Distributive Property}$$

$$21 - 1 = 8x \qquad \text{Simplify.}$$

$$\frac{20}{8} = x \qquad \text{Combine like terms and divide both sides by 8.}$$

$$x = \frac{5}{2} \qquad \text{Simplify.}$$

The solution is $\frac{5}{2}$. You can check this as follows.

Check

$$\frac{7}{x} - \frac{1}{3x} = \frac{8}{3} \qquad \text{Original equation}$$

$$\frac{7}{5/2} - \frac{1}{3(5/2)} \overset{?}{=} \frac{8}{3} \qquad \text{Substitute } \frac{5}{2} \text{ for } x.$$

$$7\left(\frac{2}{5}\right) - \frac{2}{15} \overset{?}{=} \frac{8}{3} \qquad \text{Invert and multiply.}$$

$$\frac{14}{5} - \frac{2}{15} \overset{?}{=} \frac{8}{3} \qquad \text{Simplify.}$$

$$\frac{40}{15} \overset{?}{=} \frac{8}{3} \qquad \text{Combine like terms.}$$

$$\frac{8}{3} = \frac{8}{3} \qquad \text{Solution checks.} \quad$$

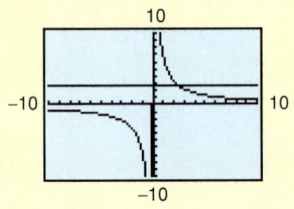

Throughout the text, we have emphasized the importance of checking solutions. Up to this point, the main reason for checking has been to make sure that you did not make errors in the solution process. In the next example you will see that there is another reason for checking solutions in the *original* equation. That is, even with no mistakes in the solution process, it can happen that a "trial solution" does not satisfy the original equation. This type of "solution" is called **extraneous.** An extraneous solution of an equation does not, by definition, satisfy the original equation, and therefore *must not* be listed as an actual solution.

EXAMPLE 5   An Equation with No Solution

Solve the equation.

$$\frac{5x}{x-2} = 7 + \frac{10}{x-2}$$

Solution

The least common denominator for this equation is $x - 2$. Therefore, begin by multiplying both sides of the equation by $x - 2$.

$$\frac{5x}{x-2} = 7 + \frac{10}{x-2} \qquad \text{Original equation}$$

$$(x-2)\left(\frac{5x}{x-2}\right) = (x-2)\left(7 + \frac{10}{x-2}\right) \qquad \text{Multiply both sides by } x - 2.$$

$$5x = 7(x-2) + 10, \quad x \neq 2 \qquad \text{Distribute and simplify.}$$

$$5x = 7x - 14 + 10 \qquad \text{Distributive Property}$$

$$5x = 7x - 4 \qquad \text{Combine like terms.}$$

$$-2x = -4 \qquad \text{Subtract } 7x \text{ from both sides.}$$

$$x = 2 \qquad \text{Divide both sides by } -2.$$

NOTE   In Example 5, can you see why $x = 2$ is extraneous? By looking back at the original equation you can see that 2 is excluded from the domain of two of the fractions that occur in the equation.

At this point, the solution appears to be 2. However, by performing the following check, you will see that this "trial solution" is extraneous.

Check

$$\frac{5x}{x-2} = 7 + \frac{10}{x-2} \qquad \text{Original equation}$$

$$\frac{5(2)}{2-2} \overset{?}{=} 7 + \frac{10}{2-2} \qquad \text{Substitute 2 for } x.$$

$$\frac{10}{0} \overset{?}{=} 7 + \frac{10}{0} \qquad \text{Solution does not check. } \times$$

Because the check results in *division by zero*, 2 is extraneous. Therefore, the original equation has no solution.

EXAMPLE 6 An Equation Containing Variable Denominators

Solve $\dfrac{4}{x-2} + \dfrac{3x}{x+1} = 3$.

Solution

The least common denominator is $(x-2)(x+1)$.

$$\dfrac{4}{x-2} + \dfrac{3x}{x+1} = 3$$

$$(x-2)(x+1)\left(\dfrac{4}{x-2} + \dfrac{3x}{x+1}\right) = 3(x-2)(x+1)$$

$$4(x+1) + 3x(x-2) = 3(x^2 - x - 2), \quad x \neq 2, x \neq -1$$

$$4x + 4 + 3x^2 - 6x = 3x^2 - 3x - 6$$

$$x = -10$$

The solution is -10. Check this in the original equation.

So far in this section, each of the equations has had one solution or no solution. The equation in the next example has two solutions.

DISCOVERY

Use a graphing utility to graph the equation

$$y = \dfrac{3x}{x+1} - \dfrac{12}{x^2-1} - 2.$$

Then use the ZOOM and TRACE features of the utility to determine the x-intercepts. How do the x-intercepts compare with the solutions to Example 7? What can you conclude?

EXAMPLE 7 An Equation That Has Two Solutions

Solve $\dfrac{3x}{x+1} = \dfrac{12}{x^2-1} + 2$.

Solution

The least common denominator is $(x+1)(x-1) = x^2 - 1$.

$$\dfrac{3x}{x+1} = \dfrac{12}{x^2-1} + 2$$

$$(x^2 - 1)\left(\dfrac{3x}{x+1}\right) = (x^2 - 1)\left(\dfrac{12}{x^2-1} + 2\right)$$

$$(x-1)(3x) = 12 + 2(x^2 - 1), \quad x \neq \pm 1$$

$$3x^2 - 3x = 12 + 2x^2 - 2$$

$$x^2 - 3x - 10 = 0$$

$$(x+2)(x-5) = 0$$

$$x + 2 = 0 \quad \Longrightarrow \quad x = -2$$

$$x - 5 = 0 \quad \Longrightarrow \quad x = 5$$

The solutions are -2 and 5. Check these in the original equation.

Applications

EXAMPLE 8   Average Cost

A manufacturing plant can produce x units of a certain item for $26 per unit *plus* an initial investment of $80,000. How many units must be produced to have an average cost of $30 per unit?

Solution

Verbal Model:

$$\boxed{\begin{array}{c}\text{Average cost}\\\text{per unit}\end{array}} = \boxed{\begin{array}{c}\text{Total}\\\text{cost}\end{array}} \div \boxed{\begin{array}{c}\text{Number}\\\text{of units}\end{array}}$$

Labels: Number of units $= x$ (units)
Average cost per unit $= 30$ (dollars per unit)
Total cost $= 26x + 80{,}000$ (dollars)

Equation:
$$30 = \frac{26x + 80{,}000}{x}$$
$$30x = 26x + 80{,}000, \quad x \neq 0$$
$$4x = 80{,}000$$
$$x = 20{,}000$$

The plant should produce 20,000 units. Check this in the original statement of the problem.

EXAMPLE 9   A Work-Rate Problem

With only the cold water valve open, it takes 8 minutes to fill the tub of a washer. With both the hot and cold water valves open, it takes only 5 minutes. How long will it take the tub to fill with only the hot water valve open?

Solution

Verbal Model:

$$\boxed{\begin{array}{c}\text{Rate for}\\\text{cold water}\end{array}} + \boxed{\begin{array}{c}\text{Rate for}\\\text{hot water}\end{array}} = \boxed{\begin{array}{c}\text{Rate for}\\\text{warm water}\end{array}}$$

Labels: Rate for cold water $= \frac{1}{8}$ (tub per minute)
Rate for hot water $= 1/t$ (tub per minute)
Rate for warm water $= \frac{1}{5}$ (tub per minute)

Equation: $\dfrac{1}{8} + \dfrac{1}{t} = \dfrac{1}{5}$

Try solving this equation. You should discover that it takes $13\frac{1}{3}$ minutes to fill the tub with hot water. Check this in the original statement of the problem.

EXAMPLE 10 Batting Average

In this year's playing season, a baseball player has been at bat 140 times and has hit the ball safely 35 times. Thus, the "batting average" for the player is 35/140 = .250. How many consecutive times must the player hit safely to obtain a batting average of .300?

Solution

Verbal Model:

$$\boxed{\text{Batting average}} = \boxed{\text{Total hits}} \div \boxed{\text{Total times at bat}}$$

Labels: Current times at bat = 140
            Current hits = 35
            Additional consecutive hits = x

Equation:

$$.300 = \frac{x + 35}{x + 140}$$

$$.300(x + 140) = x + 35$$

$$.3x + 42 = x + 35$$

$$7 = 0.7x$$

$$10 = x$$

The player must hit safely the next 10 times at bat. After that, the batting average will be 45/150 = .300.

Group Activities   Problem Solving

Interpreting Average Cost   You buy sand in bulk for a construction project. You find that the total cost of your order depends on the weight of the order. Suppose the total cost C in dollars is given by $C = 100 + 50x - 0.2x^2$, $1 \le x \le 50$, where x is the weight in thousands of pounds. Construct a rational function representing the average cost per thousand pounds. Because of cost constraints, you may proceed with the project only if the average cost of the sand is less than $50 per thousand pounds. What is the smallest order you can place and still proceed with the project?

8.6 Exercises

Discussing the Concepts

1. Explain the difference between the following.

$$\frac{5}{x+3} + \frac{5}{3} = 3, \qquad \frac{5}{x+3} + \frac{5}{3} + 3$$

2. Describe how to solve a rational equation.

3. Define the term *extraneous solution*. How do you identify an extraneous solution?

4. Describe the steps that can be used to transform an equation into an equivalent equation.

5. Explain how you can use a graphing utility to estimate the solution of a rational equation.

6. When can you use cross-multiplication to solve a rational equation? Explain.

Problem Solving

In Exercises 7–10, decide whether the values of x are solutions of the equation.

Equation		*Values*

7. $\dfrac{x}{3} - \dfrac{x}{5} = \dfrac{4}{3}$

   (a) $x = 0$ (b) $x = -1$

   (c) $x = \frac{1}{8}$ (d) $x = 10$

8. $x = 4 + \dfrac{21}{x}$

   (a) $x = 0$ (b) $x = -3$

   (c) $x = 7$ (d) $x = -1$

9. $\dfrac{x}{4} + \dfrac{3}{4x} = 1$

   (a) $x = -1$ (b) $x = 1$

   (c) $x = 3$ (d) $x = \frac{1}{2}$

10. $5 - \dfrac{1}{x-3} = 2$

    (a) $x = \frac{10}{3}$ (b) $x = -\frac{1}{3}$

    (c) $x = 0$ (d) $x = 1$

In Exercises 11–28, solve the equation.

11. $\dfrac{x}{4} = \dfrac{3}{8}$

12. $\dfrac{x}{10} = \dfrac{12}{5}$

13. $\dfrac{h}{5} - \dfrac{h+2}{9} = \dfrac{2}{3}$

14. $\dfrac{u}{6} + \dfrac{u+6}{15} = 3$

15. $\dfrac{7}{x} = 21$

16. $\dfrac{9}{t} = -\dfrac{4}{3}$

17. $\dfrac{12}{y+5} + \dfrac{1}{2} = 2$

18. $\dfrac{7}{8} - \dfrac{16}{t-2} = \dfrac{3}{4}$

19. $\dfrac{4}{2x+3} + \dfrac{17}{5(2x-3)} = 3$

20. $\dfrac{2}{6q+5} - \dfrac{3}{4(6q+5)} = \dfrac{1}{28}$

21. $\dfrac{x}{x+4} + \dfrac{4}{x+4} + 2 = 0$

22. $\dfrac{2}{(x-4)(x-2)} = \dfrac{1}{x-4} + \dfrac{2}{x-2}$

23. $\dfrac{32}{t} = 2t$

24. $\dfrac{45}{u} = \dfrac{u}{5}$

25. $\dfrac{1}{x-1} + \dfrac{3}{x+1} = 2$

26. $\dfrac{x+42}{x} = x$

27. $\dfrac{x}{2} = \dfrac{2 - \dfrac{3}{x}}{1 - \dfrac{1}{x}}$

28. $\dfrac{2x}{3} = \dfrac{1 + \dfrac{2}{x}}{1 + \dfrac{1}{x}}$

In Exercises 29–32, (a) use the graph to determine any x-intercepts of the equation, and (b) set $y = 0$ and solve the resulting equation to confirm your result.

29. $y = \dfrac{x+2}{x-2}$

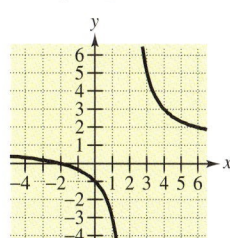

30. $y = \dfrac{2x}{x+4}$

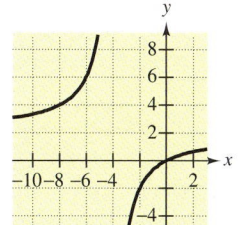

31. $y = x - \dfrac{1}{x}$

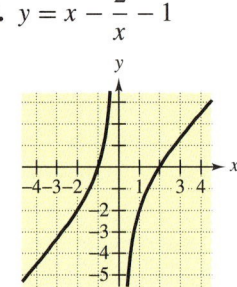

32. $y = x - \dfrac{2}{x} - 1$

$\square$ *Graphical Reasoning* In Exercises 33–36, (a) use a graphing utility to graph the equation and determine any x-intercepts of the equation, and (b) set $y = 0$ and solve the resulting rational equation to confirm the result of part (a).

33. $y = \dfrac{1}{x} + \dfrac{4}{x - 5}$

34. $y = 20 \left(\dfrac{2}{x} - \dfrac{3}{x - 1} \right)$

35. $y = (x + 1) - \dfrac{6}{x}$

36. $y = \dfrac{x^2 - 4}{x}$

37. *Wind Speed* A plane with a speed of 300 miles per hour in still air travels 680 miles with a tail wind in the same time it could travel 520 miles with a head wind of equal speed. Find the speed of the wind.

38. *Average Speed* During the first 4 hours of a 6-hour, 320-mile trip, you travel at an average speed of r miles per hour. During the last part of the trip, you increase your average speed by 10 miles per hour. What were your two average speeds?

39. *Partnership Costs* Some partners buy a piece of property for $78,000 by sharing the cost equally. To ease the financial burden, they look for three additional partners to reduce the cost per person by $1300. How many partners are presently in the group?

40. *Population Model* A biologist introduces 100 insects into a culture. The population P in the culture is approximated by the model

$$P = \dfrac{500(1 + 3t)}{5 + t}$$

where t is the time in hours. Find the time required for the population to increase to 1000 insects.

41. *Swimming Pool* The flow rate of one pipe is $1\frac{1}{4}$ times that of a second pipe. A swimming pool can be filled in 5 hours using both pipes. Find the time required to fill the pool using only the pipe with the lower flow rate.

42. *Swimming Pool* Assume that the pipe with the higher flow rate in Exercise 41 is shut off after 1 hour and that it takes an additional 10 hours to fill the pool. Find the flow rate of each pipe.

Using a Model In Exercises 43 and 44, use the following model, which approximates the total revenue y (in billions of dollars) for the car and truck rental industry in the United States from 1985 to 1991.

$$y = 43.31 - \dfrac{275.25}{t} + \dfrac{654.53}{t^2}, \quad 5 \le t \le 11$$

In this model, $t = 0$ represents 1980. (Source: *Current Business Reports*)

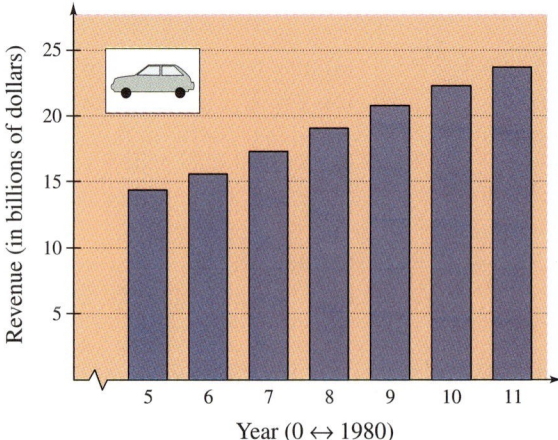

43. (a) Use the bar graph to *graphically* determine the year the total revenue first exceeded $20 billion.

(b) Use the model to *algebraically* confirm your answer to part (a).

(c) What would you estimate the revenue to be in 1996? Explain your reasoning.

$\square$ **44.** Use a graphing utility to graph the model over the specified domain.

Reviewing the Major Concepts

In Exercises 45–48, solve the equation. Show how to use a graphing utility to check your solution.

45. $125 - 50x = 0$

46. $t^2 - 8t = 0$

47. $x^2 + x - 42 = 0$

48. $x(10 - x) = 25$

49. Find two consecutive positive even integers whose product is 624.

50. *Free-Falling Object*   An object is dropped from a construction project 576 feet above the ground. Find the time t for the object to reach the ground by solving the equation $-16t^2 + 576 = 0$.

Additional Problem Solving

In Exercises 51–84, solve the equation.

51. $\dfrac{t}{2} = \dfrac{1}{8}$

52. $\dfrac{y}{5} = \dfrac{3}{2}$

53. $\dfrac{z+2}{3} = \dfrac{z}{12}$

54. $5 + \dfrac{y}{3} = y + 2$

55. $\dfrac{4t}{3} = 15 - \dfrac{t}{6}$

56. $\dfrac{x}{3} + \dfrac{x}{6} = 10$

57. $\dfrac{9}{25 - y} = -\dfrac{1}{4}$

58. $\dfrac{2}{u+4} = \dfrac{5}{8}$

59. $5 - \dfrac{12}{a} = \dfrac{5}{3}$

60. $\dfrac{6}{b} + 22 = 24$

61. $\dfrac{5}{x} = \dfrac{25}{3(x+2)}$

62. $\dfrac{10}{x+4} = \dfrac{15}{4(x+1)}$

63. $\dfrac{8}{3x+5} = \dfrac{1}{x+2}$

64. $\dfrac{500}{3x+5} = \dfrac{50}{x-3}$

65. $\dfrac{3}{x+2} - \dfrac{1}{x} = \dfrac{1}{5x}$

66. $\dfrac{12}{x+5} + \dfrac{5}{x} = \dfrac{20}{x}$

67. $\dfrac{10}{x(x-2)} + \dfrac{4}{x} = \dfrac{5}{x-2}$

68. $\dfrac{x}{x-2} + \dfrac{1}{x-4} = \dfrac{2}{x^2 - 6x + 8}$

69. $\dfrac{10}{x+3} + \dfrac{10}{3} = 6$

70. $3\left(\dfrac{1}{x} + 4\right) = 2 + \dfrac{4}{3x}$

71. $\dfrac{1}{x-5} + \dfrac{1}{x+5} = \dfrac{x+3}{x^2 - 25}$

72. $\dfrac{2}{x-10} - \dfrac{3}{x-2} = \dfrac{6}{x^2 - 12x + 20}$

73. $\dfrac{1}{2} = \dfrac{18}{x^2}$

74. $\dfrac{1}{6} = \dfrac{150}{z^2}$

75. $x + 1 = \dfrac{72}{x}$

76. $t + 4\left(\dfrac{2t+5}{t+4}\right) = 0$

77. $1 = \dfrac{16}{y} - \dfrac{39}{y^2}$

78. $\dfrac{2x}{3x+10} - \dfrac{5}{x} = 0$

79. $\dfrac{2x}{5} = \dfrac{x^2 - 5x}{5x}$

80. $\dfrac{8(x-1)}{x^2 - 4} = \dfrac{4}{x-2}$

81. $\dfrac{2(x+7)}{x+4} - 2 = \dfrac{2x+20}{2x+8}$

82. $\dfrac{10}{x^2 - 2x} + \dfrac{4}{x} = \dfrac{5}{x-2}$

83. $x - \dfrac{24}{x} = 5$

84. $\dfrac{3x}{2} + \dfrac{4}{x} = 5$

Graphical Reasoning   In Exercises 85–88, (a) use a graphing utility to determine any x-intercepts of the equation, and (b) set $y = 0$ and solve the resulting rational equation to confirm the results of part (a).

85. $y = \dfrac{x-4}{x+5}$

86. $y = \dfrac{1}{x} - \dfrac{3}{x+4}$

87. $y = (x-1) - \dfrac{12}{x}$

88. $y = \dfrac{x}{2} - \dfrac{4}{x} - 1$

89. What number can be added to its reciprocal to obtain $\dfrac{65}{8}$?

90. The sum of twice a number and three times its reciprocal is $\dfrac{97}{4}$. What is the number?

91. *Pollution Removal*   The cost C in dollars of removing $p\%$ of the air pollution in the stack emission of a utility company is modeled by

$$C = \frac{120,000p}{100 - p}.$$

(a) Use a graphing utility to graph the model. Use the result to graphically estimate the percent of stack emission that can be removed for $680,000.

(b) Use the model to algebraically determine the percent of stack emission that can be removed for $680,000.

92. *Average Cost*   The average cost for producing x units of a product is given by

$$\text{Average cost} = 1.50 + \frac{4200}{x}.$$

Determine the number of units that must be produced to have an average cost of $2.90.

93. *Comparing Two Speeds*   One person runs 2 miles per hour faster than a second person. The first person runs 5 miles in the same time the second runs 4 miles. Find the speed of each.

94. *Comparing Two Speeds*   The speed of a commuter plane is 150 miles per hour slower than a passenger jet. The commuter plane travels 450 miles in the same time the jet travels 1150 miles. Find the speed of each plane.

95. *Speed*   A boat travels at a speed of 20 miles per hour in still water. It travels 48 miles upstream, and then returns to the starting point, in a total of 5 hours. Find the speed of the current.

96. *Speed*   You travel 72 miles in a certain time. If you had traveled 6 miles per hour faster, the trip would have taken 10 minutes less time. What was your speed?

97. *Partnership Costs*   A group plans to start a new business that will require $240,000 for start-up capital. The partners in the group will share the cost equally. If two additional people join the group, the cost per person will decrease by $4000. How many partners are currently in the group?

98. *Partnership Costs*   A group of people agree to share equally in the cost of a $150,000 endowment to a college. If they could find four more people to join the group, each person's share of the cost would decrease by $6250. How many people are presently in the group?

Using a Table   In Exercises 99 and 100, complete the table by finding the time for two individuals to complete a task. The first two columns in the table give the times required for two individuals to complete the task working alone. (Assume that when they work together their individual rates do not change.)

99.

Person #1	Person #2	Together
6 hours	6 hours	
3 minutes	5 minutes	
5 hours	$2\frac{1}{2}$ hours	

100.

Person #1	Person #2	Together
4 days	4 days	
$5\frac{1}{2}$ hours	3 hours	
a days	b days	

101. *Work Rate*   One landscaper works $1\frac{1}{2}$ times as fast as a second landscaper. Find their individual rates if it takes them 9 hours working together to complete a certain job.

102. *Work Rate*   The slower of the two landscapers in Exercise 101 is given another job after 4 hours. The faster of the two must work an additional 10 hours to complete the task. Find their individual rates.

CHAPTER PROJECT: *Air Resistance*

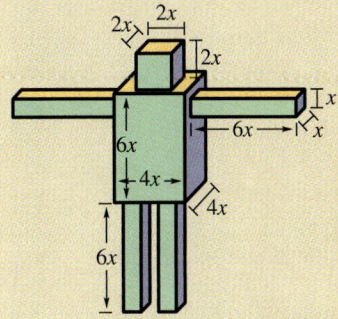

Consider a human, a mouse, and an ant, each of which falls from a height of 30 feet. The mouse and the ant will hit the ground without injury, and simply walk away. The human, on the other hand, is likely to be injured. Why?

The answer is that the mouse and ant encounter such great air resistance that they never fall fast enough to get hurt. The human encounters less air resistance and can fall fast enough to become injured. The amount of air resistance that an object encounters depends on the ratio of its surface area to its volume. A falling object that has a large surface area relative to its volume will encounter great air resistance, but a falling object that has a small surface area relative to its volume will encounter little air resistance.

To model the surface areas and volumes of different animals, consider the geometric model at the left. At first glance, it may appear that the ratio of the geometric model's surface area to its volume is constant. As you investigate the questions below, however, you will discover that the ratio is a function of x. In other words, small animals have different ratios from large animals.

1. Find algebraic models for the total surface area and total volume of the geometric model at the left.

Each Arm and Leg	Head	Trunk
$S = x^2 + 4(6x^2)$	$S = 5(4x^2)$	$S = 2(16x^2) + 4(24x^2) - 4x^2 - 4x^2$
$V = 6x^3$	$V = 8x^3$	$V = 96x^3$

2. Find a rational expression that gives the ratio R of the surface area of the geometric model to its volume.

$$R = \frac{S}{V}$$

Simplify the result.

3. Complete the following table by calculating the ratios of surface area to volume for five animals. What can you conclude?

Animal	Mouse	Squirrel	Cat	Human	Elephant
x	$\frac{1}{24}$	$\frac{1}{12}$	$\frac{1}{6}$	$\frac{2}{5}$	2
Ratio					

4. The geometric model at the left is a better fit for a mouse, cat, human, and elephant than for a squirrel. Why?

5. Approximate the ratio for an ant. Explain your reasoning.

6. Approximate the ratio for a human with a parachute. Explain your reasoning.

CHAPTER SUMMARY

After studying this chapter, you should have acquired the following skills. These skills are keyed to the Review Exercises that begin on page 509. Answers to odd-numbered Review Exercises are given in the back of the book.

- Find the domains of rational expressions. *(Section 8.1)*   **Review Exercises 1–4**

- Simplify rational expressions. *(Section 8.1)*   **Review Exercises 5–12**

- Match functions with their graphs. *(Section 8.5)*   **Review Exercises 13–16**

- Multiply and divide rational expressions. *(Section 8.2)*   **Review Exercises 17–28**

- Add and subtract rational expressions. *(Section 8.3)*   **Review Exercises 29–40**

- Simplify compound fractions. *(Sections 8.2, 8.3)*   **Review Exercises 41–46**

- Divide polynomials. *(Section 8.4)*   **Review Exercises 47–52**

- Divide polynomials using synthetic division. *(Section 8.4)*   **Review Exercises 53–56**

- Graph pairs of equations using a graphing utility, and verify their equivalence. *(Sections 8.2, 8.3, 8.4)*   **Review Exercises 57–60**

- Graph rational functions using a graphing utility. *(Section 8.5)*   **Review Exercises 61–74**

- Solve rational equations. *(Section 8.6)*   **Review Exercises 75–88**

- Graph rational functions using a graphing utility to determine any x-intercepts, and confirm them algebraically. *(Section 8.6)*   **Review Exercises 89, 90**

- Translate real-life situations into rational equations and solve. *(Section 8.6)*   **Review Exercises 91–93**

REVIEW EXERCISES

In Exercises 1–4, find the domain of the expression.

1. $\dfrac{3y}{y-8}$

2. $\dfrac{t+4}{t+12}$

3. $\dfrac{u}{u^2-7u+6}$

4. $\dfrac{x-12}{x(x^2-16)}$

In Exercises 5–12, simplify the expression.

5. $\dfrac{6x^4y^2}{15xy^2}$

6. $\dfrac{2(y^3z)^2}{28(yz^2)^2}$

7. $\dfrac{5b-15}{30b-120}$

8. $\dfrac{4a}{10a^2+26a}$

9. $\dfrac{9x-9y}{y-x}$

10. $\dfrac{x+3}{x^2-x-12}$

11. $\dfrac{x^2-5x}{2x^2-50}$

12. $\dfrac{x^2+3x+9}{x^3-27}$

In Exercises 13–16, match the function with its graph.

(a)

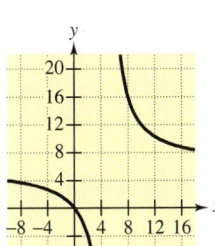

(b)

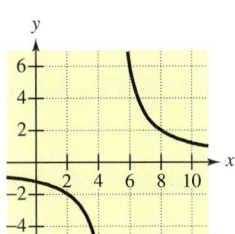

(c)

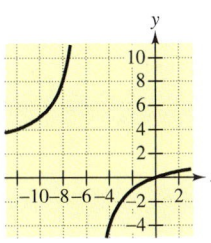

(d)

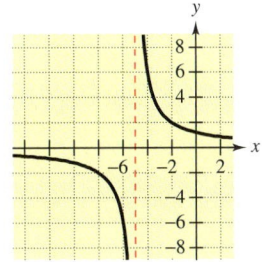

13. $f(x)=\dfrac{5}{x-6}$

14. $f(x)=\dfrac{6}{x+5}$

15. $f(x)=\dfrac{6x}{x-5}$

16. $f(x)=\dfrac{2x}{x+6}$

In Exercises 17–52, perform the operation(s) and simplify your answer.

17. $\dfrac{7}{8}\cdot\dfrac{2x}{y}\cdot\dfrac{y^2}{14x^2}$

18. $\dfrac{15(x^2y)^3}{3y^3}\cdot\dfrac{12y}{x}$

19. $\dfrac{60z}{z+6}\cdot\dfrac{z^2-36}{5}$

20. $\dfrac{1}{6}(x^2-16)\cdot\dfrac{3}{x^2-8x+16}$

21. $\dfrac{u}{u-3}\cdot\dfrac{3u-u^2}{4u^2}$

22. $x^2\cdot\dfrac{x+1}{x^2-x}\cdot\dfrac{(5x-5)^2}{x^2+6x+5}$

23. $\dfrac{6/x}{2/x^3}$

24. $\dfrac{0}{5x^2/2y}$

25. $25y^2\div\dfrac{xy}{5}$

26. $\dfrac{6}{z^2}\div 4z^2$

27. $\dfrac{x^2-7x}{x+1}\div\dfrac{x^2-14x+49}{x^2-1}$

28. $\left(\dfrac{6x}{y^2}\right)^2\div\left(\dfrac{3x}{y}\right)^3$

29. $\dfrac{4}{9}-\dfrac{11}{9}$

30. $\dfrac{2(3y+4)}{2y+1}+\dfrac{3-y}{2y+1}$

31. $\dfrac{15}{16}-\dfrac{5}{24}-1$

32. $-\dfrac{3}{8}+\dfrac{7}{6}-\dfrac{1}{12}$

33. $\dfrac{1}{x+5}+\dfrac{3}{x-12}$

34. $\dfrac{2}{x-10}+\dfrac{3}{4-x}$

35. $5x+\dfrac{2}{x-3}-\dfrac{3}{x+2}$

36. $4-\dfrac{4x}{x+6}+\dfrac{7}{x-5}$

37. $\dfrac{6}{x}-\dfrac{6x-1}{x^2+4}$

38. $\dfrac{5}{x+2}+\dfrac{25-x}{x^2-3x-10}$

39. $\dfrac{5}{x+3}-\dfrac{4x}{(x+3)^2}-\dfrac{1}{x-3}$

40. $\dfrac{8}{y}-\dfrac{3}{y+5}+\dfrac{4}{y-2}$

41. $\dfrac{\left(\dfrac{6x^2}{x^2+2x-35}\right)}{\left(\dfrac{x^3}{x^2-25}\right)}$

42. $\dfrac{\left[\dfrac{24-18x}{(2-x)^2}\right]}{\left(\dfrac{60-45x}{x^2-4x-4}\right)}$

43. $\dfrac{3t}{\left(5 - \dfrac{2}{t}\right)}$

44. $\dfrac{\left(x - 3 + \dfrac{2}{x}\right)}{\left(1 - \dfrac{2}{x}\right)}$

45. $\dfrac{\left(\dfrac{1}{a^2 - 16} - \dfrac{1}{a}\right)}{\left(\dfrac{1}{a^2 + 4a} + 4\right)}$

46. $\dfrac{\left(\dfrac{1}{x^2} - \dfrac{1}{y^2}\right)}{\left(\dfrac{1}{x} + \dfrac{1}{y}\right)}$

47. $(4x^3 - x) \div 2x$

48. $(10x + 15) \div (5x - 2)$

49. $\dfrac{6x^3 + 2x^2 - 4x + 2}{3x - 1}$

50. $\dfrac{4x^4 - x^3 - 7x^2 + 18x}{x - 2}$

51. $\dfrac{x^4 - 3x^2 + 2}{x^2 - 1}$

52. $\dfrac{3x^6}{x^2 - 1}$

In Exercises 53–56, use synthetic division to perform the division.

53. $\dfrac{x^3 + 7x^2 + 3x - 14}{x + 2}$

54. $\dfrac{x^4 - 2x^3 - 15x^2 - 2x + 10}{x - 5}$

55. $(x^4 - 3x^2 - 25) \div (x - 3)$

56. $(2x^3 + 5x - 2) \div \left(x + \frac{1}{2}\right)$

In Exercises 57–60, use a graphing utility to graph the equations on the same screen. Use the graphs to verify that the expressions are equivalent. Verify the results algebraically.

57. $y_1 = \dfrac{x^2 + 6x + 9}{x^2} \cdot \dfrac{x^2 - 3x}{x + 3}, \ y_2 = \dfrac{x^2 - 9}{x}$

58. $y_1 = \dfrac{1}{x} - \dfrac{3}{x(x + 3)}, \ y_2 = \dfrac{1}{x + 3}$

59. $y_1 = \dfrac{\left(\dfrac{1}{x} - \dfrac{1}{2}\right)}{2x}, \ y_2 = \dfrac{2 - x}{4x^2}$

60. $y_1 = \dfrac{x^3 - 2x^2 - 7}{x - 2}, \ y_2 = x^2 - \dfrac{7}{x - 2}$

In Exercises 61–74, use a graphing utility to graph the function.

61. $g(x) = \dfrac{2 + x}{1 - x}$

62. $h(x) = \dfrac{x - 3}{x - 2}$

63. $f(x) = \dfrac{x}{x^2 + 1}$

64. $f(x) = \dfrac{2x}{x^2 + 4}$

65. $P(x) = \dfrac{3x + 6}{x - 2}$

66. $s(x) = \dfrac{2x - 6}{x + 4}$

67. $h(x) = \dfrac{4}{(x - 1)^2}$

68. $g(x) = \dfrac{-2}{(x + 3)^2}$

69. $f(x) = -\dfrac{5}{x^2}$

70. $f(x) = \dfrac{4}{x}$

71. $y = \dfrac{x}{x^2 - 1}$

72. $y = \dfrac{2x}{x^2 - 4}$

73. $y = \dfrac{2x^2}{x^2 - 4}$

74. $y = \dfrac{2}{x + 3}$

In Exercises 75–88, solve the equation.

75. $\dfrac{3x}{8} = -15$

76. $\dfrac{t + 1}{8} = \dfrac{1}{2}$

77. $3\left(8 - \dfrac{12}{t}\right) = 0$

78. $\dfrac{1}{3y - 4} = \dfrac{6}{4(y + 1)}$

79. $\dfrac{2}{y} - \dfrac{1}{3y} = \dfrac{1}{3}$

80. $8\left(\dfrac{6}{x} - \dfrac{1}{x + 5}\right) = 15$

81. $r = 2 + \dfrac{24}{r}$

82. $\dfrac{3}{y + 1} - \dfrac{8}{y} = 1$

83. $\dfrac{2}{x} - \dfrac{x}{6} = \dfrac{2}{3}$

84. $\dfrac{2x}{x - 3} - \dfrac{3}{x} = 0$

85. $\dfrac{12}{x^2 + x - 12} - \dfrac{1}{x - 3} = -1$

86. $\dfrac{3}{x - 1} + \dfrac{6}{x^2 - 3x + 2} = 2$

87. $\dfrac{5}{x^2 - 4} - \dfrac{6}{x - 2} = -5$

88. $\dfrac{3}{x^2 - 9} + \dfrac{4}{x + 3} = 1$

In Exercises 89 and 90, (a) use a graphing utility to determine any x-intercepts of the graph of the equation, and (b) set $y = 0$ and solve the resulting rational equation to confirm the results of part (a).

89. $y = \dfrac{1}{x} - \dfrac{1}{2x + 3}$

90. $y = \dfrac{x}{4} - \dfrac{2}{x} - \dfrac{1}{2}$

91. *Average Speed* You drive 56 miles on a service call for your company. On the return trip, which takes 10 minutes less than the original trip, your average speed is 8 miles an hour faster. What is your average speed on the return trip?

92. *Batting Average* In this year's playing season, a baseball player has been at bat 150 times and has hit the ball safely 45 times. Thus, the batting average for the player is 45/150 = .300. How many consecutive times must the player hit safely to obtain a batting average of .400?

93. *Forming a Partnership* A group of people agree to share equally in the cost of a $60,000 piece of machinery. If they could find two more people to join the group, each person's share of the cost would decrease by $5000. How many people are presently in the group?

Math Matters   Numbers in Other Languages

The numeral systems for several languages are shown below. One numeral system that is not shown is the Roman numeral system. Use a reference book to find how the Roman numeral system works. Then write the standard numerals for the following—each represents a famous date in American history.

(a) MDCCLXXVI   (b) MCDXCII   (c) MDCXX

Modern	1	2	3	4	5	6	7	8	9	10	100
Egyptian											
Babylonian											
Early Roman	I	II	III	IIII	V	VI	VII	VIII	IX	X	C
Chinese											
Hindu											
Mayan											

Take this test as you would take a test in class. After you are done, check your work against the answers given in the back of the book.

1. Find the domain of $\dfrac{3y}{y^2 - 25}$.

2. Find the least common denominator of $\dfrac{3}{x^2}$, $\dfrac{x}{x-3}$, $\dfrac{2x}{x^3(x+3)}$, and $\dfrac{10}{x^2 + 6x + 9}$.

3. Simplify the rational expression. (a) $\dfrac{2-x}{3x-6}$ (b) $\dfrac{2a^2 - 5a - 12}{5a - 20}$

In Exercises 4–15, perform the operation and simplify.

4. $\dfrac{4z^3}{5} \cdot \dfrac{25}{12z^2}$

5. $\dfrac{y^2 + 8y + 16}{2(y-2)} \cdot \dfrac{8y - 16}{(y+4)^3}$

6. $(4x^2 - 9) \cdot \dfrac{2x+3}{2x^2 - x - 3}$

7. $\dfrac{(2xy^2)^3}{15} \div \dfrac{12x^3}{21}$

8. $\dfrac{\left(\dfrac{3x}{x+2}\right)}{\left(\dfrac{12}{x^3 + 2x^2}\right)}$

9. $\dfrac{\left(9x - \dfrac{1}{x}\right)}{\left(\dfrac{1}{x} - 3\right)}$

10. $2x + \dfrac{1 - 4x^2}{x+1}$

11. $\dfrac{5x}{x+2} - \dfrac{2}{x^2 - x - 6}$

12. $\dfrac{3}{x} - \dfrac{5}{x^2} + \dfrac{2x}{x^2 + 2x + 1}$

13. $\dfrac{4}{x+1} + \dfrac{4x}{x+1}$

14. $\dfrac{t^4 + t^2 - 6t}{t^2 - 2}$

15. $\dfrac{2x^4 - 15x^2 - 7}{x - 3}$

16. Sketch the graph of the function. Describe how to find the asymptotes of the function.

   (a) $f(x) = \dfrac{3}{x-3}$ (b) $g(x) = \dfrac{3x}{x-3}$

In Exercises 17–19, solve the equation.

17. $\dfrac{3}{h+2} = \dfrac{1}{8}$

18. $\dfrac{2}{x+5} - \dfrac{3}{x+3} = \dfrac{1}{x}$

19. $\dfrac{1}{x+1} + \dfrac{1}{x-1} = \dfrac{2}{x^2 - 1}$

20. One painter works $1\frac{1}{2}$ times as fast as another. Find their individual rates for painting a room if it takes them 4 hours working together.

Radicals and Complex Numbers

Complex numbers such as -1, i, and $1 + i$ can be plotted in a complex plane, as shown in the graph at the right. The vertical axis represents the imaginary part of the complex number and the horizontal axis represents the real part.

In the graph at the right, the black region is a *fractal* called the *Mandelbrot Set*, after Benoit Mandelbrot. Fractals can be used to model objects in nature, such as coastlines, ferns, or moonscapes. For example, the moonscape at the right is a fractal that was generated by a computer.

A complex number c is in the Mandelbrot Set if the sequence

$$c, \; c^2 + c, \; (c^2 + c)^2 + c, \; \ldots,$$

is bounded. The table at the right indicates that $c = -1$ and $c = i$ are in the Mandelbrot Set, but $c = 1 + i$ is not.

c	$c^2 + c$	$(c^2 + c)^2 + c$	$((c^2 + c)^2 + c)^2 + c$
-1	0	-1	0
i	$-1 + i$	$-i$	$-1 + i$
$1 + i$	$1 + 3i$	$-7 + 7i$	$1 - 97i$

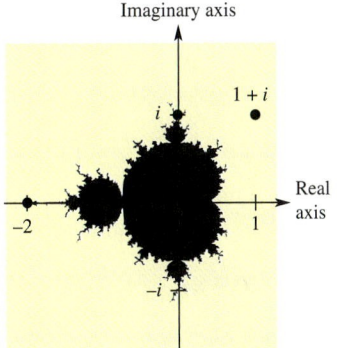

The chapter project related to this information is on page 568.

513

9.1 Integer Exponents and Scientific Notation

Integer Exponents ▪ Scientific Notation

Integer Exponents

So far in the text, all exponents have been positive integers. In this section, the definition of an exponent is extended to include zero and negative integers. If a is a real number such that $a \neq 0$, then a^0 is defined as 1. Moreover, if m is an integer, then a^{-m} is defined as the reciprocal of a^m.

Definitions of Zero Exponents and Negative Exponents

Let a be a real number such that $a \neq 0$, and let m be an integer.

1. $a^0 = 1, \quad a \neq 0$ **2.** $a^{-m} = \dfrac{1}{a^m}, \quad a \neq 0$

These definitions are consistent with the properties of exponents given in Section 5.3. For instance, consider the following.

$$x^0 \cdot x^m = x^{0+m} = x^m = 1 \cdot x^m$$

(x^0 is the same as 1)

EXAMPLE 1 *Zero Exponents and Negative Exponents*

Rewrite each expression without using zero exponents or negative exponents.

a. 3^0 **b.** 0^0 **c.** 2^{-1} **d.** 3^{-2}

Solution

a. $3^0 = 1$

b. 0^0 is undefined. Remember when raising a number to the zero power, you must be sure the number is not zero.

c. $2^{-1} = \dfrac{1}{2^1} = \dfrac{1}{2}$

d. $3^{-2} = \dfrac{1}{3^2} = \dfrac{1}{9}$

The following properties of exponents are valid for all integer exponents, including integer exponents that are zero or negative. The first five properties were listed in Section 5.3.

Properties of Exponents

Let m and n be integers, and let a and b represent real numbers, variables, or algebraic expressions.

Property	Example
1. $a^m \cdot a^n = a^{m+n}$	$x^4(x^3) = x^{4+3} = x^7$
2. $(ab)^m = a^m \cdot b^m$	$(3x)^2 = 3^2(x^2) = 9x^2$
3. $(a^m)^n = a^{mn}$	$(x^3)^3 = x^{3 \cdot 3} = x^9$
4. $\dfrac{a^m}{a^n} = a^{m-n}, \quad a \neq 0$	$\dfrac{x^3}{x} = x^{3-1} = x^2, \quad x \neq 0$
5. $\left(\dfrac{a}{b}\right)^m = \dfrac{a^m}{b^m}, \quad b \neq 0$	$\left(\dfrac{x}{3}\right)^2 = \dfrac{x^2}{3^2} = \dfrac{x^2}{9}$
6. $\left(\dfrac{a}{b}\right)^{-m} = \left(\dfrac{b}{a}\right)^m, \quad a \neq 0, \; b \neq 0$	$\left(\dfrac{x}{3}\right)^{-2} = \left(\dfrac{3}{x}\right)^2 = \dfrac{3^2}{x^2} = \dfrac{9}{x^2}$
7. $a^{-m} = \dfrac{1}{a^m}, \quad a \neq 0$	$x^{-2} = \dfrac{1}{x^2}, \quad x \neq 0$
8. $a^0 = 1, \quad a \neq 0$	$(x^2 + 1)^0 = 1$

STUDY TIP

As you become accustomed to working with negative exponents, you will probably not write as many steps as shown in Example 2. For instance, to rewrite a fraction involving exponents, you might use the following simplified rule. *To move a factor from the numerator to the denominator or vice versa, change the sign of its exponent.* You can apply this rule to the expression in Example 2(c) by "moving" the factor x^{-2} to the numerator and changing the exponent to 2. That is,

$$\frac{3}{x^{-2}} = 3x^2.$$

EXAMPLE 2   Using Properties of Exponents

a. $2x^{-1} = 2(x^{-1}) = 2\left(\dfrac{1}{x}\right) = \dfrac{2}{x}$

b. $(2x)^{-1} = \dfrac{1}{(2x)^1} = \dfrac{1}{2x}$

c. $\dfrac{3}{x^{-2}} = \dfrac{3}{\left(\dfrac{1}{x^2}\right)} = 3\left(\dfrac{x^2}{1}\right) = 3x^2$

d. $\dfrac{1}{(3x)^{-2}} = \dfrac{1}{\left[\dfrac{1}{(3x)^2}\right]} = \dfrac{1}{\left(\dfrac{1}{3^2x^2}\right)} = \dfrac{1}{\left(\dfrac{1}{9x^2}\right)} = (1)\left(\dfrac{9x^2}{1}\right) = 9x^2$

EXAMPLE 3 Using Properties of Exponents

Rewrite each expression using only positive exponents. (For each expression, assume that $x \neq 0$ and $y \neq 0$.)

a. $(-5x^{-3})^2$ **b.** $-\left(\dfrac{7x}{y^2}\right)^{-2}$

Solution

a. $(-5x^{-3})^2 = (-5)^2(x^{-3})^2$ Property 2

$\qquad\qquad\quad = 25x^{-6}$ Property 3

$\qquad\qquad\quad = \dfrac{25}{x^6}$ Property 7

b. $-\left(\dfrac{7x}{y^2}\right)^{-2} = -\left(\dfrac{y^2}{7x}\right)^{2}$ Property 6

$\qquad\qquad\quad = -\dfrac{(y^2)^2}{(7x)^2}$ Property 5

$\qquad\qquad\quad = -\dfrac{y^4}{49x^2}$ Property 3

EXAMPLE 4 Using Properties of Exponents

Rewrite each expression using only positive exponents. (For each expression, assume that $x \neq 0$ and $y \neq 0$.)

a. $\left(\dfrac{8x^{-1}y^4}{4x^3y^2}\right)^{-3}$ **b.** $\dfrac{3xy^0}{x^2(5y)^0}$

Solution

a. $\left(\dfrac{8x^{-1}y^4}{4x^3y^2}\right)^{-3} = \left(\dfrac{2y^2}{x^4}\right)^{-3}$ Simplify.

$\qquad\qquad\qquad = \left(\dfrac{x^4}{2y^2}\right)^{3}$ Property 6

$\qquad\qquad\qquad = \dfrac{x^{12}}{2^3y^6}$ Property 5

$\qquad\qquad\qquad = \dfrac{x^{12}}{8y^6}$ Simplify.

b. $\dfrac{3xy^0}{x^2(5y)^0} = \dfrac{3x(1)}{x^2(1)} = \dfrac{3}{x}$ Property 8

Scientific Notation

Exponents provide an efficient way of writing and computing with very large (or very small) numbers. For instance, a drop of water contains more than 33 billion billion molecules—that is, 33 followed by 18 zeros.

$$33,000,000,000,000,000,000$$

It is convenient to write such numbers in **scientific notation.** This notation has the form $c \times 10^n$, where $1 \leq c < 10$ and n is an integer. Thus, the number of molecules in a drop of water can be written in scientific notation as

$$3.3 \times 10,000,000,000,000,000,000 = 3.3 \times 10^{19}.$$

The *positive* exponent 19 indicates that the number being written in scientific notation is *large* (10 or more) and that the decimal point has been moved 19 places. A *negative* exponent in scientific notation indicates that the number is *small* (less than 1).

EXAMPLE 5   *Writing Scientific Notation*

Write each real number in scientific notation.

a. 0.0000684   **b.** 937,200,000

Solution

a. $0.0000684 = 6.84 \times 10^{-5}$

Five places

b. $937,200,000.0 = 9.372 \times 10^{8}$

Eight places

EXAMPLE 6   *Writing Decimal Notation*

Convert each number from scientific notation to decimal notation.

a. 2.486×10^2 **b.** 1.81×10^{-6}

Solution

a. $2.486 \times 10^{2} = 248.6$

Two places

b. $1.81 \times 10^{-6} = 0.00000181$

Six places

Technology

Using Scientific Notation

Most scientific and graphing calculators automatically switch to scientific notation when they are showing large (or small) numbers that exceed the display range. Try multiplying $86{,}500{,}000 \times 6000$. If your calculator follows standard conventions, its display should be

| 5.19   11 | or | 5.19   E   11 |.

This means that $c = 5.19$ and the exponent of 10 is $n = 11$, which implies that the number is 5.19×10^{11}.

To *enter* numbers in scientific notation, your calculator should have an exponential entry key labeled EE or EXP. If you were to perform the preceding multiplication using scientific notation, you could begin by writing

$$86{,}500{,}000 \times 6000 = (8.65 \times 10^7)(6.0 \times 10^3)$$

and then entering the following.

8.65 EXP 7 × 6 EXP 3 =   *Scientific*

8.65 EE 7 × 6 EE 3 ENTER   *Graphing*

EXAMPLE 7   Using Scientific Notation with a Calculator

Use a scientific or graphing calculator to find the following.

a. $65{,}000 \times 3{,}400{,}000{,}000$ **b.** $0.000000348 \div 870$

Solution

a.   6.5 EXP 4 × 3.4 EXP 9 =   *Scientific*

     6.5 EE 4 × 3.4 EE 9 ENTER   *Graphing*

The calculator display should read | 2.21   14 |, which implies that

$$(6.5 \times 10^4)(3.4 \times 10^9) = 2.21 \times 10^{14} = 221{,}000{,}000{,}000{,}000.$$

b.   3.48 EXP 7 +/− ÷ 8.7 EXP 2 =   *Scientific*

     3.48 EE (−) 7 ÷ 8.7 EE 2 ENTER   *Graphing*

The calculator display should read | 4.0   −10 |, which implies that

$$\frac{3.48 \times 10^{-7}}{8.7 \times 10^2} = 4.0 \times 10^{-10} = 0.0000000004.$$

EXAMPLE 8 *Using Scientific Notation*

Evaluate $\dfrac{(2,400,000,000)(0.00000345)}{(0.00007)(3800)}$.

Solution

Begin by rewriting each number in scientific notation and simplifying.

$$
\begin{aligned}
\frac{(2,400,000,000)(0.00000345)}{(0.00007)(3800)} &= \frac{(2.4 \times 10^9)(3.45 \times 10^{-6})}{(7.0 \times 10^{-5})(3.8 \times 10^3)} \\
&= \frac{(2.4)(3.45)(10^3)}{(7)(3.8)(10^{-2})} \\
&= \frac{(8.28)(10^5)}{26.6} \\
&\approx 0.3112782(10^5) \\
&= 31{,}127.82
\end{aligned}
$$

Group Activities   Communicating Mathematically

Developing a Mathematical Method   Discuss why scientific notation is used and give examples of its usefulness. Develop an easy-to-use saying, description, or method for converting numbers to and from scientific notation. Each member of your group should demonstrate the method to the rest of the group, using one of the following as an example.

a. 0.0000042

b. 293,600,000,000

c. 3.1×10^{-6}

d. 5.12×10^{11}

Discuss how to tell which of two numbers written in scientific notation is larger. Describe a comparison method and test it by deciding which is larger: 7×10^{51} or 8×10^{50}.

9.1 Exercises

Discussing the Concepts

1. In $(3x)^4$, what is $3x$ called? What is 4 called?

2. Discuss any differences between $(-2x)^4$ and $-2x^4$.

3. The expressions $4x$ and x^4 each represent repeated operations. What are the operations? Write the expressions showing the repeated operations.

4. In your own words, describe how you can "move" a factor from the numerator to the denominator or vice versa.

5. Is 32.5×10^5 in scientific notation? Explain.

6. When is scientific notation an efficient way of writing and computing real numbers?

Problem Solving

In Exercises 7–20, evaluate the expression.

7. 5^{-2}

8. 2^{-4}

9. $\dfrac{1}{(-2)^{-5}}$

10. $-\dfrac{1}{6^2}$

11. $\left(\frac{2}{3}\right)^{-1}$

12. $\left(\frac{4}{5}\right)^{-3}$

13. $27 \cdot 3^{-3}$

14. $4^2 \cdot 4^{-3}$

15. $\dfrac{10^3}{10^{-2}}$

16. $\dfrac{10^{-5}}{10^{-6}}$

17. $(4^2 \cdot 4^{-1})^{-2}$

18. $(5^3 \cdot 5^{-4})^{-3}$

19. $(5^0 - 4^{-2})^{-1}$

20. $(32 + 4^{-3})^0$

In Exercises 21–34, rewrite the expression using only positive exponents, and simplify. (Assume any variables in the expression are nonzero.)

21. $y^4 \cdot y^{-2}$

22. $x^{-2} \cdot x^{-5}$

23. $\dfrac{(4t)^0}{t^{-2}}$

24. $\dfrac{(5u)^4}{(5u)^4}$

25. $(-3x^{-3}y^2)(4x^2y^{-5})$

26. $(5s^5t^{-5})\left(\dfrac{3s^{-2}}{50t^{-1}}\right)$

27. $(3x^2y^{-2})^{-2}$

28. $(-4y^{-3}z)^{-3}$

29. $\dfrac{6^2x^3y^{-3}}{12x^{-2}y}$

30. $\dfrac{2^{-4}y^{-1}z^{-3}}{4^{-2}yz^{-3}}$

31. $\left(\dfrac{3u^2v^{-1}}{3^3u^{-1}v^3}\right)^{-2}$

32. $\left(\dfrac{5^2x^3y^{-3}}{125xy}\right)\left(\dfrac{5x}{3y}\right)^{-1}$

33. $\dfrac{a+b}{ba^{-1} - ab^{-1}}$

34. $\dfrac{u^{-1} - v^{-1}}{u^{-1} + v^{-1}}$

In Exercises 35–40, write in scientific notation.

35. *Land Area of Earth:* 57,500,000 square miles

36. *Ocean Area of Earth:* 139,400,000 square miles

37. *Light Year:* 9,461,000,000,000,000 kilometers

38. *Thickness of Soap Bubble:* 0.0000001 meter

39. *Relative Density of Hydrogen:* 0.0000899

40. *One Micron (Millionth of Meter):* 0.00003937 inch

In Exercises 41–46, write in decimal notation.

41. *1993 PepsiCo Beverage Sales:* $\$2.82 \times 10^{10}$

42. *Number of Air Sacs in Lungs:* 3.5×10^8

43. *Temperature of Sun:* 1.3×10^7 degrees Celsius

44. *Width of Air Molecule:* 9.0×10^{-9} meter

45. *Charge of Electron:* 4.8×10^{-10} electrostatic unit

46. *Width of Human Hair:* 9.0×10^{-4} meter

In Exercises 47–50, evaluate without a calculator. Write your answer in scientific notation.

47. $(6.5 \times 10^6)(2 \times 10^4)$

48. $5 \times (10^8)^3$

49. $\dfrac{3.6 \times 10^{12}}{6 \times 10^5}$

50. $\dfrac{72,000,000,000}{0.00012}$

In Exercises 51 and 52, evaluate with a calculator.

51. $\dfrac{(3.82 \times 10^5)^2}{(8.5 \times 10^4)(5.2 \times 10^{-3})}$

52. $\dfrac{(6,200,000)(0.005)^3}{(0.00035)^5}$

53. *Masses of Earth and Sun*   The masses of the earth and sun are approximately 5.975×10^{24} and 1.99×10^{30} kilograms, respectively. The mass of the sun is approximately how many times that of the earth?

54. *Kepler's Third Law*   In 1619, Johannes Kepler, a German astronomer, discovered that the period T (in years) of each planet in our solar system is related to the planet's mean distance R (in astronomical units) from the sun by the equation

$$\frac{T^2}{R^3} = k.$$

Test Kepler's equation for the nine planets in our solar system, using the table at the right. Do you get approximately the same value of k for each planet?

Planet	T	R
Mercury	0.241	0.387
Venus	0.615	0.723
Earth	1.000	1.000
Mars	1.881	1.523
Jupiter	11.861	5.203
Saturn	21.457	9.541
Uranus	84.008	19.190
Neptune	164.784	30.086
Pluto	248.350	39.508

Reviewing the Major Concepts

In Exercises 55–58, factor the expression.

55. $x^2 - 3x + 2$

56. $2x^2 + 5x - 7$

57. $11x^2 + 6x - 5$

58. $4x^2 - 28x + 49$

59. *Comparing Memberships*   The current membership of a public television station is 8415, which is 110% of what it was a year ago. How many members did the station have last year?

60. *Buy Now or Wait?*   A sales representative indicates that if a customer waits another month for a new car that currently costs $23,500, the price will increase by 4%. However, the customer will pay an interest penalty of $725 for the early withdrawal of a certificate of deposit if the car is purchased now. Determine whether the customer should buy now or wait another month.

Additional Problem Solving

In Exercises 61–76, evaluate the expression.

61. -10^{-3}

62. -20^{-2}

63. $(-3)^{-5}$

64. 25^0

65. $\dfrac{1}{4^{-3}}$

66. $\dfrac{1}{-8^{-2}}$

67. $\left(\dfrac{3}{16}\right)^0$

68. $\left(-\dfrac{5}{8}\right)^{-2}$

69. $\dfrac{3^4}{3^{-2}}$

70. $\dfrac{5^{-1}}{5^2}$

71. $(2^{-3})^2$

72. $(-4^{-1})^{-2}$

73. $2^{-3} + 2^{-4}$

74. $4 - 3^{-2}$

75. $\left(\dfrac{3}{4} + \dfrac{5}{8}\right)^{-2}$

76. $\left(\dfrac{1}{2} - \dfrac{2}{3}\right)^{-1}$

In Exercises 77–94, rewrite the expression using only positive exponents, and simplify. (Assume any variables in the expression are nonzero.)

77. $z^5 \cdot z^{-3}$

78. $t^{-1} \cdot t^{-6}$

79. $\dfrac{1}{x^{-6}}$

80. $\dfrac{x^{-3}}{y^{-1}}$

81. $\dfrac{a^{-6}}{a^{-7}}$

82. $\dfrac{6u^{-2}}{15u^{-1}}$

83. $(2x^2)^{-2}$

84. $(4a^{-2}b^3)^{-3}$

85. $\left(\dfrac{x}{10}\right)^{-1}$

86. $\left(\dfrac{4}{z}\right)^{-2}$

87. $\left(\dfrac{a^{-2}}{b^{-2}}\right)\left(\dfrac{b}{a}\right)^3$

88. $\left(\dfrac{a^{-3}}{b^{-3}}\right)\left(\dfrac{b}{a}\right)^3$

89. $\left[(x^{-4}y^{-6})^{-1}\right]^2$

90. $(ab)^{-2}(a^2b^2)^{-1}$

91. $(u + v^{-2})^{-1}$

92. $x^{-2}(x^2 + y^2)$

93. $\left[(2x^{-3}y^{-2})^2\right]^{-2}$

94. $\left[\left(\dfrac{2x^2}{4y}\right)^{-3}\right]^2$

In Exercises 95–98, write in scientific notation.

95. 3,600,000

96. 98,100,000

97. 0.0000000381

98. 0.0007384

In Exercises 99–102, write in decimal notation.

99. 6×10^7

100. 5.05×10^{12}

101. 1.359×10^{-7}

102. 8.6×10^{-9}

In Exercises 103–108, evaluate without a calculator. Write your answer in scientific notation.

103. $(2 \times 10^9)(3.4 \times 10^{-4})$

104. $(5 \times 10^4)^3$

105. $\dfrac{3.6 \times 10^9}{9 \times 10^5}$

106. $\dfrac{2.5 \times 10^{-3}}{5 \times 10^2}$

107. $(4,500,000)(2,000,000,000)$

108. $\dfrac{64,000,000}{0.00004}$

In Exercises 109–112, evaluate with a calculator. Round your answer to two decimal places.

109. $\dfrac{1.357 \times 10^{12}}{(4.2 \times 10^2)(6.87 \times 10^{-3})}$

110. $(8.67 \times 10^4)^7$

111. $\dfrac{(0.0000565)(2,850,000,000,000)}{0.00465}$

112. $\dfrac{(5,000,000)^3(0.000037)^2}{(0.005)^4}$

113. *Distance to the Sun* The distance from the earth to the sun is approximately 93 million miles. Write this distance in scientific notation.

114. *Copper Electrons* A cube of copper with an edge of 1 centimeter has approximately 8×10^{22} free electrons. Write this real number in decimal form.

115. *Light-Year* One light-year (the distance light can travel in 1 year) is approximately 9.45×10^{15} meters. Approximate the time for light to travel from the sun to the earth if that distance is approximately 1.49×10^{11} meters.

116. *Distance to a Star* The star Alpha Andromeda is 90 light years from the earth (see figure). Use the definition of a light year in Exercise 115 to write this distance in meters.

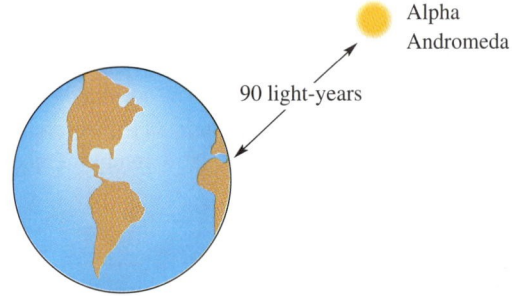

Alpha Andromeda

90 light-years

117. *Federal Debt* In 1993, the population of the United States was 257 million, and the federal debt was 4410 billion dollars. Use these two numbers to determine the amount each person would have had to pay (per capita debt) to remove the debt. (Source: U.S. Bureau of Census)

| **9.2** | **Rational Exponents and Radicals** |

Roots and Radicals ▪ Rational Exponents ▪
Radicals and Calculators ▪ Radical Functions

Roots and Radicals

The **square root** of a number is defined as one of its two equal factors. For example, 5 is a square root of 25 because 5 is one of the two equal factors of 25. In a similar way, a **cube root** of a number is one of its three equal factors.

Number	Equal Factors	Root	Type
$9 = 3^2$	$3 \cdot 3$	3	Square root
$25 = (-5)^2$	$(-5)(-5)$	-5	Square root
$-27 = (-3)^3$	$(-3)(-3)(-3)$	-3	Cube root
$16 = 2^4$	$2 \cdot 2 \cdot 2 \cdot 2$	2	Fourth root

Definition of nth Root of a Number

Let a and b be real numbers and let n be an integer such that $n \geq 2$. If

$$a = b^n$$

then b is an **nth root of a.** If $n = 2$, the root is a **square root,** and if $n = 3$, the root is a **cube root.**

Some numbers have more than one nth root. For example, both 5 and -5 are square roots of 25 because $25 = 5^2$ and $25 = (-5)^2$. To avoid ambiguity about which root of a number you are talking about, the **principal nth root** of a number is defined in terms of a radical symbol $\sqrt[n]{\ }$.

NOTE "Having the same sign" means that the principal nth root of a is positive if a is positive and negative if a is negative. For example, $\sqrt{4} = 2$ and $\sqrt[3]{-8} = -2$.

Principal nth Root of a Number

Let a be a real number that has at least one (real number) nth root. The **principal nth root of a** is the nth root that has the same sign as a, and it is denoted by the **radical symbol**

$$\sqrt[n]{a}. \qquad \text{Principal } n\text{th root}$$

The positive integer n is the **index** of the radical, and the number a is the **radicand.** If $n = 2$, omit the index and write $\sqrt{a}$ rather than $\sqrt[2]{a}$.

EXAMPLE 1 *Finding Roots of a Number*

Find the roots.

a. $\sqrt{36}$ **b.** $-\sqrt{36}$ **c.** $\sqrt{-4}$ **d.** $\sqrt[3]{8}$ **e.** $\sqrt[3]{-8}$

Solution

a. $\sqrt{36} = 6$ because $6 \cdot 6 = 6^2 = 36$.

b. $-\sqrt{36} = -6$ because $6 \cdot 6 = 6^2 = 36$.

c. $\sqrt{-4}$ is not real because there is no real number that when multiplied by itself yields -4.

d. $\sqrt[3]{8} = 2$ because $2 \cdot 2 \cdot 2 = 2^3 = 8$.

e. $\sqrt[3]{-8} = -2$ because $(-2)(-2)(-2) = (-2)^3 = -8$.

Properties of *n*th Roots

1. If a is a positive real number and n is *even,* then a has exactly two (real) *n*th roots, which are denoted by $\sqrt[n]{a}$ and $-\sqrt[n]{a}$.

2. If a is any real number and n is *odd,* then a has only one (real) *n*th root, which is denoted by $\sqrt[n]{a}$.

3. If a is a negative real number and n is *even,* then a has no (real) *n*th root.

Integers such as 1, 4, 9, 16, 49, and 81 are called **perfect squares** because they have integer square roots. Similarly, integers such as 1, 8, 27, 64, and 125 are called **perfect cubes** because they have integer cube roots.

EXAMPLE 2 *Classifying Perfect Roots*

State whether the number is a perfect square root, a perfect cube root, both, or neither.

a. 81   **b.** 64   **c.** 32

Solution

a. 81 is a perfect square root because $9^2 = 81$. It is not a perfect cube root.

b. 64 is a perfect square root because $8^2 = 64$, and it is also a perfect cube root because $4^3 = 64$.

c. 32 is not a perfect square root or a perfect cube root. (It is a perfect 5th root because $2^5 = 32$.)

Raising a number to the nth power and taking the principal nth root of a number can be thought of as *inverse* operations. Here are two examples.

$$\left(\sqrt{4}\right)^2 = 2^2 = 4 \quad \text{and} \quad \sqrt{2^2} = \sqrt{4} = 2$$

$$\left(\sqrt[3]{27}\right)^3 = 3^3 = 27 \quad \text{and} \quad \sqrt[3]{3^3} = \sqrt[3]{27} = 3$$

Inverse Properties of nth Powers and nth Roots

Let a be a real number, and let n be an integer such that $n \geq 2$.

1. If a has a principal nth root, then

$$\left(\sqrt[n]{a}\right)^n = a.$$

2. If n is *odd*, then

$$\sqrt[n]{a^n} = a.$$

If n is *even*, then

$$\sqrt[n]{a^n} = |a|.$$

EXAMPLE 3   *Evaluating Radical Expressions*

Evaluate each radical expression.

a. $\sqrt[3]{5^3}$ **b.** $\sqrt[3]{(-2)^3}$ **c.** $\left(\sqrt{7}\right)^2$ **d.** $\sqrt{(-3)^2}$ **e.** $\sqrt{-3^2}$

Solution

a. Because the index of the radical is odd, you can write

$$\sqrt[3]{5^3} = 5.$$

b. Because the index of the radical is odd, you can write

$$\sqrt[3]{(-2)^3} = -2.$$

c. Using the inverse property of powers and roots, you can write

$$\left(\sqrt{7}\right)^2 = 7.$$

d. Because the index of the radical is even, you must include absolute value signs, and write

$$\sqrt{(-3)^2} = |-3| = 3.$$

e. Because $\sqrt{-3^2} = \sqrt{-9}$ is an even root of a negative number, its value is not a real number.

Rational Exponents

NOTE The numerator of a rational exponent denotes the *power* to which the base is raised, and the denominator denotes the *root* to be taken.

Definition of Rational Exponents

Let a be a real number, and let n be an integer such that $n \geq 2$. If the principal nth root of a exists, we define $a^{1/n}$ to be

$$a^{1/n} = \sqrt[n]{a}.$$

If m is a positive integer that has no common factor with n, then

$$a^{m/n} = (a^{1/n})^m = \left(\sqrt[n]{a}\right)^m \quad \text{and} \quad a^{m/n} = (a^m)^{1/n} = \sqrt[n]{a^m}.$$

It does not matter in which order the two operations are performed, provided the nth root exists. Here is an example.

$$8^{2/3} = \left(\sqrt[3]{8}\right)^2 = 2^2 = 4 \qquad \text{Cube root, then second power}$$
$$8^{2/3} = \sqrt[3]{8^2} = \sqrt[3]{64} = 4 \qquad \text{Second power, then cube root}$$

The properties of exponents that we listed in Section 9.1 also apply to rational exponents (provided the roots indicated by the denominators exist). We relist the first seven of those properties here, with different examples.

DISCOVERY

Use a calculator to evaluate the expressions below.

$$\frac{3.4^{4.6}}{3.4^{3.1}} \text{ and } 3.4^{1.5}$$

How are these two expressions related? Use your calculator to verify some of the other properties of exponents.

Properties of Exponents

Let r and s be rational numbers, and let a and b be real numbers, variables, or algebraic expressions.

Property	Example
1. $a^r \cdot a^s = a^{r+s}$	$4^{1/2}(4^{1/3}) = 4^{5/6}$
2. $(ab)^r = a^r \cdot b^r$	$(2x)^{1/2} = 2^{1/2}(x^{1/2})$
3. $(a^r)^s = a^{rs}$	$(x^3)^{1/3} = x$
4. $\dfrac{a^r}{a^s} = a^{r-s}, \quad a \neq 0$	$\dfrac{x^2}{x^{1/2}} = x^{2-(1/2)} = x^{3/2}$
5. $\left(\dfrac{a}{b}\right)^r = \dfrac{a^r}{b^r}, \quad b \neq 0$	$\left(\dfrac{x}{3}\right)^{1/3} = \dfrac{x^{1/3}}{3^{1/3}}$
6. $\left(\dfrac{a}{b}\right)^{-r} = \left(\dfrac{b}{a}\right)^r, \quad a \neq 0, \quad b \neq 0$	$\left(\dfrac{x}{4}\right)^{-1/2} = \left(\dfrac{4}{x}\right)^{1/2} = \dfrac{2}{x^{1/2}}$
7. $a^{-r} = \dfrac{1}{a^r}, \quad a \neq 0$	$4^{-1/2} = \dfrac{1}{4^{1/2}} = \dfrac{1}{2}$

EXAMPLE 4   Evaluating Expressions with Rational Exponents

a. $8^{4/3} = \left(\sqrt[3]{8}\right)^4 = 2^4 = 16$

b. $(4^2)^{3/2} = \left(\sqrt{4^2}\right)^3 = 4^3 = 64$

c. $25^{-3/2} = \dfrac{1}{25^{3/2}} = \dfrac{1}{\left(\sqrt{25}\right)^3} = \dfrac{1}{5^3} = \dfrac{1}{125}$

d. $\left(\dfrac{64}{125}\right)^{2/3} = \left(\sqrt[3]{\dfrac{64}{125}}\right)^2 = \left(\dfrac{\sqrt[3]{64}}{\sqrt[3]{125}}\right)^2 = \left(\dfrac{4}{5}\right)^2 = \dfrac{16}{25}$

e. $-9^{1/2} = -\sqrt{9} = -3$

f. $(-9)^{1/2} = \sqrt{-9}$ is not a real number.

In parts (e) and (f) of Example 4, be sure that you see the distinction between the expressions $-9^{1/2}$ and $(-9)^{1/2}$.

EXAMPLE 5   Using Properties of Exponents

Rewrite each expression using rational exponents.

a. $x\sqrt[4]{x^3}$ **b.** $\dfrac{\sqrt[3]{x^2}}{\sqrt{x^3}}$

Solution

a. $x\sqrt[4]{x^3} = x(x^{3/4}) = x^{1+(3/4)} = x^{7/4}$

b. $\dfrac{\sqrt[3]{x^2}}{\sqrt{x^3}} = \dfrac{x^{2/3}}{x^{3/2}} = x^{(2/3)-(3/2)} = x^{-5/6} = \dfrac{1}{x^{5/6}}$

EXAMPLE 6   Using Properties of Exponents

Use properties of exponents to rewrite each expression in simpler form.

a. $\sqrt{\sqrt[3]{x}}$ **b.** $\dfrac{(2x-1)^{4/3}}{\sqrt[3]{2x-1}}$

Solution

a. $\sqrt{\sqrt[3]{x}} = \sqrt{x^{1/3}} = (x^{1/3})^{1/2} = x^{1/6}$

b. $\dfrac{(2x-1)^{4/3}}{\sqrt[3]{2x-1}} = \dfrac{(2x-1)^{4/3}}{(2x-1)^{1/3}} = (2x-1)^{(4/3)-(1/3)} = 2x-1$

Radicals and Calculators

There are two methods of evaluating radicals on most calculators. For square roots, you can use the *square root key* $\boxed{\sqrt{\;}}$. For other roots, you should first convert the radical to exponential form and then use the *exponential key* $\boxed{y^x}$ or $\boxed{\wedge}$.

EXAMPLE 7 *Evaluating Roots with a Calculator*

Evaluate the following. Round the result to three decimal places.

a. $\sqrt{5}$ **b.** $\sqrt[5]{25}$ **c.** $\sqrt[3]{-4}$ **d.** $(1.4)^{-2/5}$

Solution

a. 5 $\boxed{\sqrt{\;}}$ Scientific

$\boxed{\sqrt{\;}}$ 5 $\boxed{\text{ENTER}}$ Graphing

The display is 2.236068. Rounded to three decimal places, $\sqrt{5} \approx 2.236$.

b. First rewrite the expression as $\sqrt[5]{25} = 25^{1/5}$. Then use one of the following keystroke sequences.

25 $\boxed{y^x}$ $\boxed{(}$ 1 $\boxed{\div}$ 5 $\boxed{)}$ $\boxed{=}$ Scientific

25 $\boxed{\wedge}$ $\boxed{(}$ 1 $\boxed{\div}$ 5 $\boxed{)}$ $\boxed{\text{ENTER}}$ Graphing

The display is 1.9036539. Rounded to three decimal places, $\sqrt[5]{25} \approx 1.904$.

c. If your calculator does not have a cube root key, use the fact that

$$\sqrt[3]{-4} = \sqrt[3]{(-1)(4)} = \sqrt[3]{-1}\sqrt[3]{4} = -\sqrt[3]{4}$$

and attach the negative sign of the radicand as the last keystroke.

4 $\boxed{y^x}$ $\boxed{(}$ 1 $\boxed{\div}$ 3 $\boxed{)}$ $\boxed{=}$ $\boxed{+/-}$ Scientific

$\boxed{\sqrt[3]{\;}}$ $\boxed{(-)}$ 4 $\boxed{\text{ENTER}}$ Graphing

The display is -1.5874011. Rounded to three decimal places, $\sqrt[3]{-4} \approx -1.587$.

d. 1.4 $\boxed{y^x}$ $\boxed{(}$ 2 $\boxed{\div}$ 5 $\boxed{+/-}$ $\boxed{)}$ $\boxed{=}$ Scientific

1.4 $\boxed{\wedge}$ $\boxed{(}$ $\boxed{(-)}$ 2 $\boxed{\div}$ 5 $\boxed{)}$ $\boxed{\text{ENTER}}$ Graphing

The display is 0.8740752. Rounded to three decimal places, $(1.4)^{-2/5} \approx 0.874$.

NOTE Some calculators have a cube root key or submenu command. If your calculator does, try using it to evaluate the expression in Example 7(c).

Radical Functions

The **domain** of the radical $\sqrt[n]{x}$ is the set of all real numbers such that x has a principal nth root.

Domain of a Radical

Let n be an integer that is greater than or equal to 2.

1. If n is odd, the domain of $\sqrt[n]{x}$ is the set of all real numbers.

2. If n is even, the domain of $\sqrt[n]{x}$ is the set of all nonnegative real numbers.

EXAMPLE 8   *Finding the Domain of a Radical Function*

Describe the domain of each function.

a. $f(x) = \sqrt{x}$ **b.** $f(x) = \sqrt[3]{x}$ **c.** $f(x) = \sqrt{x^2}$ **d.** $f(x) = \sqrt{x^3}$

Solution

a. The domain of $f(x) = \sqrt{x}$ is the set of all nonnegative real numbers. For instance, 2 is in the domain, but -2 is not because $\sqrt{-2}$ is not a real number.

b. The domain of $f(x) = \sqrt[3]{x}$ is the set of all real numbers because for any real number x, the expression $\sqrt[3]{x}$ is a real number.

c. The domain of $f(x) = \sqrt{x^2}$ is the set of all real numbers because for any real number x, the expression x^2 is a nonnegative real number.

d. The domain of $f(x) = \sqrt{x^3}$ is the set of all nonnegative real numbers. For instance, 1 is in the domain, but -1 is not because $\sqrt{(-1)^3} = \sqrt{-1}$ is not a real number.

Group Activities   Exploring with Technology

Describing Domains and Ranges   In your group, discuss the domain and range of each of the following functions. Use a graphing utility to verify your conclusions.

a. $y = x^{3/2}$ **b.** $y = x^2$ **c.** $y = x^{1/3}$ **d.** $y = \left(\sqrt{x}\right)^2$ **e.** $y = x^{-4/5}$

9.2 Exercises

Discussing the Concepts

1. In your own words, define an nth root of a number.

2. Define the *radicand* of a radical and the *index* of a radical.

3. If n is even, what must be true about the radicand for the nth root to be a real number? Explain.

4. Determine the values of x for which $\sqrt{x^2} \neq x$. Explain your answer.

5. Is it true that $\sqrt{2} = 1.414$? Explain.

6. Given a real number x, state the conditions of n for each of the following.

   (a) $\sqrt[n]{x^n} = x$ (b) $\sqrt[n]{x^n} = |x|$

Problem Solving

In Exercises 7–12, complete the statement.

7. Because $7^2 = 49$, _____ is a square root of 49.

8. Because $24.5^2 = 600.25$, 24.5 is a _____.

9. Because $4.2^3 = 74.088$, 4.2 is a _____.

10. Because $6^4 = 1296$, _____ is a fourth root of 1296.

11. Because $45^2 = 2025$, 45 is called the _____ of 2025.

12. Because $12^3 = 1728$, 12 is called the _____ of 1728.

In Exercises 13–26, evaluate the root without a calculator. If it is not possible, state the reason.

13. $\sqrt{81}$

14. $\sqrt{144}$

15. $\sqrt{\frac{9}{16}}$

16. $\sqrt{0.09}$

17. $\sqrt{-64}$

18. $\sqrt{0.36}$

19. $-\sqrt{\frac{4}{9}}$

20. $\sqrt{-\frac{9}{25}}$

21. $\sqrt[3]{125}$

22. $\sqrt[3]{-8}$

23. $\sqrt[4]{81}$

24. $\sqrt[5]{32}$

25. $\sqrt[5]{-0.00243}$

26. $\sqrt[6]{64}$

In Exercises 27–30, evaluate without a calculator.

27. $25^{1/2}$

28. $81^{-3/4}$

29. $32^{-2/5}$

30. $-\left(\frac{1}{125}\right)^{2/3}$

In Exercises 31–36, fill in the missing description.

Radical Form	Rational Exponent Form
31. $\sqrt{16} = 4$	_____
32. $\sqrt[4]{81} = 3$	_____
33. $\sqrt[3]{27^2} = 9$	_____
34. _____	$125^{1/3} = 5$
35. _____	$256^{3/4} = 64$
36. _____	$64^{2/3} = 16$

In Exercises 37–40, use a calculator to evaluate the expression. Round the result to four decimal places.

37. $\dfrac{8 - \sqrt{35}}{2}$

38. $962^{2/3}$

39. $\sqrt[4]{342}$

40. $\sqrt[5]{-35^3}$

In Exercises 41–50, simplify the expression.

41. $\sqrt[3]{t^6}$

42. $\sqrt[4]{z^4}$

43. $3^{1/4} \cdot 3^{3/4}$

44. $(2^{1/2})^{2/3}$

45. $\dfrac{2^{1/5}}{2^{6/5}}$

46. $\dfrac{5^{-3/4}}{5}$

47. $x^{2/3} \cdot x^{7/3}$

48. $z^{3/5} \cdot z^{-2/5}$

49. $\left(\dfrac{x^{1/4}}{x^{1/6}}\right)^3$

50. $\left(\dfrac{3m^{1/6}n^{1/3}}{4n^{-2/3}}\right)^2$

Mathematical Modeling   In Exercises 51 and 52, use the formula for the *declining balances method*

$$r = 1 - \left(\frac{S}{C}\right)^{1/n}$$

to find the depreciation rate r. In the formula, n is the useful life of the item (in years), S is the salvage value (in dollars), and C is the original cost (in dollars).

51. A truck whose original cost is $75,000 is depreciated over an 8-year period, as shown in the graph.

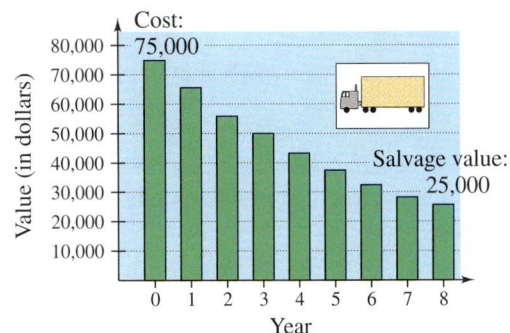

52. A printing press whose original cost is $125,000 is depreciated over a 10-year period, as shown in the graph.

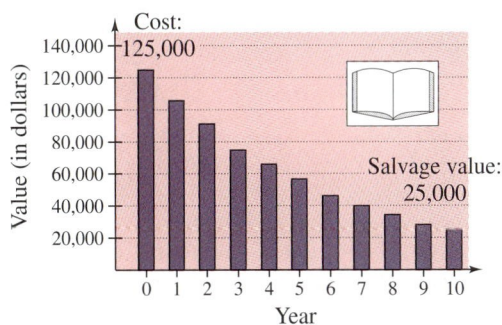

In Exercises 53 and 54, determine the domain of the function.

53. $f(x) = 3\sqrt{x}$ **54.** $g(x) = \dfrac{10}{\sqrt[3]{x}}$

Reviewing the Major Concepts

In Exercises 55–58, evaluate the quantity.

55. $-\dfrac{13}{35} \cdot \dfrac{-25}{104}$ **56.** $\dfrac{5}{12} \cdot \dfrac{9}{75}$

57. $\dfrac{14}{3} \div \dfrac{42}{45}$ **58.** $\dfrac{-8}{105} \div \dfrac{2}{3}$

In Exercises 59 and 60, use the function to find and simplify the expression for

$$\frac{f(2+h) - f(2)}{h}.$$

59. $f(x) = x^2 - 3$ **60.** $f(x) = \dfrac{3}{x + 5}$

Additional Problem Solving

In Exercises 61–74, evaluate without a calculator. If it is not possible, state the reason.

61. $\sqrt{64}$ **62.** $-\sqrt{100}$

63. $\sqrt{-100}$ **64.** $\sqrt{169}$

65. $\sqrt{0.16}$ **66.** $-\sqrt{0.0009}$

67. $\sqrt{49 - 4(2)(-15)}$ **68.** $\sqrt{\dfrac{75}{3}}$

69. $\sqrt[3]{1000}$ **70.** $\sqrt[3]{64}$

71. $\sqrt[3]{-\dfrac{1}{64}}$ **72.** $-\sqrt[3]{0.008}$

73. $-\sqrt[4]{-625}$ **74.** $-\sqrt[4]{\dfrac{1}{625}}$

In Exercises 75–78, determine whether the square root is a rational or irrational number.

75. $\sqrt{6}$ **76.** $\sqrt{\dfrac{9}{16}}$

77. $\sqrt{900}$ **78.** $\sqrt{72}$

In Exercises 79–86, evaluate without a calculator.

79. $49^{1/2}$

80. $-121^{1/2}$

81. $16^{3/4}$

82. $243^{-3/5}$

83. $\left(\dfrac{8}{27}\right)^{2/3}$

84. $\left(\dfrac{256}{625}\right)^{1/4}$

85. $\left(\dfrac{121}{9}\right)^{-1/2}$

86. $\left(\dfrac{27}{1000}\right)^{-4/3}$

In Exercises 87–94, use a calculator to evaluate the expression. (If it is not possible, state the reason.) Round the result to four decimal places.

87. $\sqrt{73}$

88. $\sqrt{-532}$

89. $\dfrac{3-\sqrt{17}}{9}$

90. $\dfrac{-5+\sqrt{3215}}{10}$

91. $1698^{-3/4}$

92. $315^{2/5}$

93. $\sqrt[3]{545^2}$

94. $\sqrt[3]{159}$

In Exercises 95–108, simplify the expression.

95. $\sqrt{t^2}$

96. $\sqrt[3]{z^3}$

97. $\sqrt[3]{y^9}$

98. $\sqrt[4]{a^8}$

99. $\left(\dfrac{2}{3}\right)^{5/3} \cdot \left(\dfrac{2}{3}\right)^{1/3}$

100. $(4^{1/3})^{9/4}$

101. $(3x^{-1/3}y^{3/4})^2$

102. $(-2u^{3/5}v^{-1/5})^3$

103. $\dfrac{18y^{4/3}z^{-1/3}}{24y^{-2/3}z}$

104. $\dfrac{a^{3/4} \cdot a^{1/2}}{a^{5/2}}$

105. $(c^{3/2})^{1/3}$

106. $(k^{-1/3})^{3/2}$

107. $\sqrt{\sqrt[4]{y}}$

108. $\sqrt[3]{\sqrt{2x}}$

In Exercises 109–112, multiply and simplify.

109. $x^{1/2}(2x-3)$

110. $x^{4/3}(3x^2-4x+5)$

111. $y^{-1/3}(y^{1/3}+5y^{4/3})$

112. $(x^{1/2}-3)(x^{1/2}+3)$

In Exercises 113 and 114, determine the domain of the function.

113. $g(x) = \dfrac{2}{\sqrt[4]{x}}$

114. $h(x) = \sqrt[4]{x}$

In Exercises 115–118, use a graphing utility to graph the function. Check the domain of the function algebraically. Did the graphing utility skip part of the domain? If so, complete the graph by hand.

115. $y = \dfrac{5}{\sqrt[4]{x^3}}$

116. $y = 4\sqrt[3]{x}$

117. $g(x) = 2x^{3/5}$

118. $h(x) = 5x^{2/3}$

119. *Perfect Squares* Find all possible "last digits" of perfect squares. (For instance, the last digit of 81 is 1 and the last digit of 64 is 4.) Is it possible that 4,322,788,987 is a perfect square?

120. *Geometry* The usable space in a particular microwave oven is in the form of a cube (see figure). The sales brochure indicates that the interior space of the oven is 2197 cubic inches. Find the inside dimensions of the oven.

121. *Velocity of a Stream* A stream of water moving at the rate of v feet per second can carry particles of size $0.03\sqrt{v}$ inches. Find the particle size that can be carried by a stream flowing at the rate of $\frac{3}{4}$ foot per second.

9.3 | Simplifying and Combining Radicals

Simplifying Radicals ▪ Rationalization Techniques ▪
Adding and Subtracting Radicals ▪ Application of Radicals

Simplifying Radicals

In this section, you will study ways to simplify and combine radicals. For instance, the expression $\sqrt{12}$ can be simplified as

$$\sqrt{12} = \sqrt{4 \cdot 3} = \sqrt{4}\sqrt{3} = 2\sqrt{3}.$$

This rewriting is based on the following rules for multiplying and dividing radicals.

Multiplying and Dividing Radicals

Let u and v be real numbers, variables, or algebraic expressions. If the nth roots of u and v are real, the following properties are true.

1. $\sqrt[n]{u}\,\sqrt[n]{v} = \sqrt[n]{uv}$ Multiplication Property

2. $\dfrac{\sqrt[n]{u}}{\sqrt[n]{v}} = \sqrt[n]{\dfrac{u}{v}}, \quad v \neq 0$ Division Property

---

You can use these properties of radicals to *simplify* radical expressions as follows.

$$\sqrt{48} = \sqrt{16 \cdot 3} = \sqrt{16}\sqrt{3} = 4\sqrt{3}$$

This simplification process is called **removing perfect square factors from the radical.**

EXAMPLE 1   *Removing Constant Factors from Radicals*

Simplify each radical by removing as many factors as possible.

a. $\sqrt{75}$ **b.** $\sqrt{72}$ **c.** $\sqrt{162}$

Solution

a. $\sqrt{75} = \sqrt{25 \cdot 3} = \sqrt{25}\sqrt{3} = 5\sqrt{3}$

b. $\sqrt{72} = \sqrt{36 \cdot 2} = \sqrt{36}\sqrt{2} = 6\sqrt{2}$

c. $\sqrt{162} = \sqrt{81 \cdot 2} = \sqrt{81}\sqrt{2} = 9\sqrt{2}$

When removing *variable* factors from a square root radical, remember that it is not valid to write $\sqrt{x^2} = x$ *unless* you happen to know that x is nonnegative. Without knowing anything about x, the only way you can simplify $\sqrt{x^2}$ is to include absolute value signs when you remove x from the radical.

$$\sqrt{x^2} = |x| \qquad \text{\color{red}Restricted by absolute value signs}$$

When simplifying the expression $\sqrt{x^3}$, it is not necessary to include absolute value signs because the domain of this expression does not include negative numbers.

$$\sqrt{x^3} = \sqrt{x^2(x)} = x\sqrt{x} \qquad \text{\color{red}Restricted by domain of radical}$$

EXAMPLE 2   Removing Variable Factors from Radicals

Simplify the radical expression.

a. $\sqrt{25x^2}$ **b.** $\sqrt{12x^3}, \quad x \geq 0$ **c.** $\sqrt{144x^4}$

Solution

a. $\sqrt{25x^2} = \sqrt{5^2 x^2} = \sqrt{5^2}\sqrt{x^2} = 5|x|$ $\color{red}\sqrt{x^2} = |x|$

b. $\sqrt{12x^3} = \sqrt{2^2 x^2(3x)} = 2x\sqrt{3x}$ $\color{red}\sqrt{2^2}\sqrt{x^2} = 2x, \ x \geq 0$

c. $\sqrt{144x^4} = \sqrt{12^2(x^2)^2} = 12x^2$ $\color{red}\sqrt{12^2}\sqrt{(x^2)^2} = 12|x^2| = 12x^2$

In the same way that perfect squares can be removed from square root radicals, perfect nth powers can be removed from nth root radicals.

EXAMPLE 3   Removing Factors from Radicals

Simplify the radical expressions.

a. $\sqrt[3]{40}$ **b.** $\sqrt[4]{x^5}, \quad x \geq 0$ **c.** $\sqrt[3]{54x^3 y^5}$

Solution

a. $\sqrt[3]{40} = \sqrt[3]{8(5)} = \sqrt[3]{2^3(5)} = 2\sqrt[3]{5}$ $\color{red}\sqrt[3]{2^3} = 2$

b. $\sqrt[4]{x^5} = \sqrt[4]{x^4(x)} = x\sqrt[4]{x}$ $\color{red}\sqrt[4]{x^4} = x, \ x \geq 0$

c. $\sqrt[3]{54x^3 y^5} = \sqrt[3]{27x^3 y^3(2y^2)}$ $\color{red}\sqrt[3]{3^3}\sqrt[3]{x^3}\sqrt[3]{y^3} = 3xy$

$\qquad\qquad = \sqrt[3]{3^3 x^3 y^3(2y^2)}$

$\qquad\qquad = 3xy\sqrt[3]{2y^2}$

Rationalization Techniques

Removing factors from radicals is only one of three techniques that we use to simplify radicals. We summarize all three techniques as follows.

Simplifying Radical Expressions

A radical expression is said to be in simplest form if all three of the following are true.

1. All possible factors have been removed from each radical.

2. No radical contains a fraction.

3. No denominator of a fraction contains a radical.

To meet the last two conditions, you can use a technique called **rationalizing the denominator.** This involves multiplying both the numerator and denominator by a factor that creates a perfect nth power in the denominator.

EXAMPLE 4   *Rationalizing the Denominator*

STUDY TIP

When rationalizing a denominator, remember that for square roots you want a perfect square in the denominator, for cube roots you want a perfect cube, and so on.

a. $\sqrt{\dfrac{3}{5}} = \dfrac{\sqrt{3}}{\sqrt{5}} = \dfrac{\sqrt{3}}{\sqrt{5}} \cdot \dfrac{\sqrt{5}}{\sqrt{5}} = \dfrac{\sqrt{15}}{\sqrt{5^2}} = \dfrac{\sqrt{15}}{5}$ Multiply by $\sqrt{5}/\sqrt{5}$ to create a perfect square in the denominator.

b. $\dfrac{4}{\sqrt[3]{9}} = \dfrac{4}{\sqrt[3]{9}} \cdot \dfrac{\sqrt[3]{3}}{\sqrt[3]{3}} = \dfrac{4\sqrt[3]{3}}{\sqrt[3]{3^3}} = \dfrac{4\sqrt[3]{3}}{3}$ Multiply by $\sqrt[3]{3}/\sqrt[3]{3}$ to create a perfect cube in the denominator.

c. $\dfrac{8}{3\sqrt{18}} = \dfrac{8}{3\sqrt{18}} \cdot \dfrac{\sqrt{2}}{\sqrt{2}} = \dfrac{8\sqrt{2}}{3\sqrt{36}} = \dfrac{8\sqrt{2}}{3(6)} = \dfrac{4\sqrt{2}}{9}$

EXAMPLE 5   *Rationalizing the Denominator*

Simplify the expression.

a. $\sqrt{\dfrac{8x}{12y^5}}$ **b.** $\sqrt[3]{\dfrac{54x^6y^3}{5z^2}}$

Solution

a. $\sqrt{\dfrac{8x}{12y^5}} = \sqrt{\dfrac{2x}{3y^5}} = \dfrac{\sqrt{2x}}{\sqrt{3y^5}} \cdot \dfrac{\sqrt{3y}}{\sqrt{3y}} = \dfrac{\sqrt{6xy}}{\sqrt{9y^6}} = \dfrac{\sqrt{6xy}}{3y^3}$

b. $\sqrt[3]{\dfrac{54x^6y^3}{5z^2}} = \dfrac{\sqrt[3]{(3^3)(2)(x^6)(y^3)}}{\sqrt[3]{5z^2}} \cdot \dfrac{\sqrt[3]{25z}}{\sqrt[3]{25z}} = \dfrac{3x^2y\sqrt[3]{50z}}{\sqrt[3]{5^3z^3}} = \dfrac{3x^2y\sqrt[3]{50z}}{5z}$

Adding and Subtracting Radicals

Two or more radical expressions are *alike* if they have the same radicand and the same index. For instance, $\sqrt{2}$ and $3\sqrt{2}$ are alike, but $\sqrt{3}$ and $\sqrt[3]{3}$ are not alike. Two radical expressions that are alike can be added or subtracted by adding or subtracting their coefficients.

EXAMPLE 6 Combining Radicals

NOTE Notice in Example 6(c) that *before* concluding that two radicals cannot be combined, you should check to see that they are written in simplest form.

a. $\sqrt{7} + 5\sqrt{7} - 2\sqrt{7} = (1 + 5 - 2)\sqrt{7} = 4\sqrt{7}$

b. $6\sqrt{x} - \sqrt[3]{4} - 5\sqrt{x} + 2\sqrt[3]{4} = 6\sqrt{x} - 5\sqrt{x} - \sqrt[3]{4} + 2\sqrt[3]{4}$

$$= (6 - 5)\sqrt{x} + (-1 + 2)\sqrt[3]{4}$$

$$= \sqrt{x} + \sqrt[3]{4}$$

c. $3\sqrt[3]{x} + \sqrt[3]{8x} = 3\sqrt[3]{x} + 2\sqrt[3]{x} = (3 + 2)\sqrt[3]{x} = 5\sqrt[3]{x}$

EXAMPLE 7 Simplifying Radical Expressions

a. $\sqrt{45x} + 3\sqrt{20x} = 3\sqrt{5x} + 6\sqrt{5x} = 9\sqrt{5x}$

b. $5\sqrt{x^3} - x\sqrt{4x} = 5x\sqrt{x} - 2x\sqrt{x} = 3x\sqrt{x}$

c. $\sqrt[3]{54y^5} + 4\sqrt[3]{2y^2} = 3y\sqrt[3]{2y^2} + 4\sqrt[3]{2y^2} = (3y + 4)\sqrt[3]{2y^2}$

In some instances, it may be necessary to rationalize denominators before combining radicals.

EXAMPLE 8 Rationalizing Denominators Before Simplifying

Simplify the expression $\sqrt{7} - \dfrac{5}{\sqrt{7}}$.

Solution

$$\sqrt{7} - \frac{5}{\sqrt{7}} = \sqrt{7} - \left(\frac{5}{\sqrt{7}} \cdot \frac{\sqrt{7}}{\sqrt{7}} \right)$$

$$= \sqrt{7} - \frac{5\sqrt{7}}{7}$$

$$= \left(1 - \frac{5}{7} \right)\sqrt{7}$$

$$= \frac{2}{7}\sqrt{7}$$

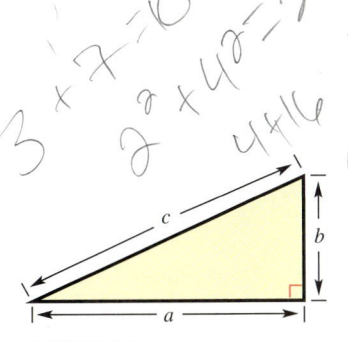

FIGURE 9.1

Application of Radicals

A common use of radicals occurs in applications involving right triangles. Recall that a right triangle is one that contains a right (or 90°) angle, as shown in Figure 9.1. The relationship among the three sides of a right triangle is described by the **Pythagorean Theorem,** which says that if a and b are the lengths of the legs and c is the length of the hypotenuse, then

$$c = \sqrt{a^2 + b^2}.$$ Pythagorean Theorem

For instance, if $a = 6$ and $b = 9$, then

$$c = \sqrt{6^2 + 9^2} = \sqrt{117} = \sqrt{9}\sqrt{13} = 3\sqrt{13}.$$

EXAMPLE 9 An Application of the Pythagorean Theorem

A softball diamond has the shape of a square with 60-foot sides (see Figure 9.2). The catcher is 5 feet behind home plate. How far does the catcher have to throw to reach second base?

Solution

In Figure 9.2, let x be the hypotenuse of a right triangle with 60-foot legs. Thus, by the Pythagorean Theorem, you have the following.

$$x = \sqrt{60^2 + 60^2}$$ Pythagorean Theorem

$$x = \sqrt{7200}$$

$$x \approx 84.9 \text{ feet}$$

Thus, the distance from home plate to second base is approximately 84.9 feet. Because the catcher is 5 feet behind home plate, the catcher must make a throw of

$$x + 5 \approx 84.9 + 5 = 89.9 \text{ feet}.$$

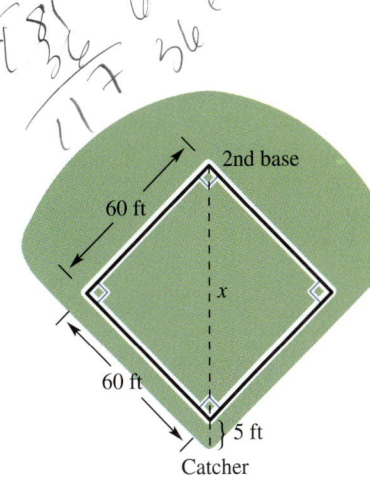

FIGURE 9.2

Group Activities   You Be the Instructor

Error Analysis   Suppose you are an algebra instructor and one of your students hands in the following work. Find and correct the errors, and discuss how you can help your student avoid such errors in the future.

a. $7\sqrt{3} + 4\sqrt{2} = 11\sqrt{5}$ b. $3\sqrt[3]{k} - 6\sqrt{k} = -3\sqrt{k}$

9.3 Exercises

Discussing the Concepts

1. Give an example of multiplying two radicals.

2. Describe the three conditions that characterize a simplified radical expression.

3. Is $\sqrt{2} + \sqrt{18}$ in simplest form? Explain.

4. Describe the steps you would use to simplify $\dfrac{1}{\sqrt{3}}$.

5. For what values of x is $\sqrt{x^2} \neq x$? Explain.

6. Explain what it means for two radical expressions to be alike.

Problem Solving

In Exercises 7–10, write the expression as a single radical.

7. $\sqrt[3]{11} \cdot \sqrt[3]{10}$

8. $\sqrt[4]{35} \cdot \sqrt[4]{3}$

9. $\dfrac{\sqrt{15}}{\sqrt{31}}$

10. $\dfrac{\sqrt[3]{84}}{\sqrt[3]{9}}$

In Exercises 11–14, write the expression as a product or quotient of radicals and simplify.

11. $\sqrt{9 \cdot 35}$

12. $\sqrt[3]{27 \cdot 4}$

13. $\sqrt[3]{\dfrac{1000}{11}}$

14. $\sqrt[5]{\dfrac{243}{2}}$

In Exercises 15–22, simplify the radical.

15. $\sqrt{20}$

16. $\sqrt{50}$

17. $\sqrt[3]{24}$

18. $\sqrt[3]{54}$

19. $\sqrt{\dfrac{15}{4}}$

20. $\sqrt{\dfrac{5}{36}}$

21. $\sqrt[5]{\dfrac{15}{243}}$

22. $\sqrt[3]{\dfrac{1}{1000}}$

In Exercises 23–30, simplify the expression.

23. $\sqrt{9x^5}$

24. $\sqrt{64x^3}$

25. $\sqrt[4]{3x^4 y^2}$

26. $\sqrt[3]{16x^4 y^5}$

27. $\sqrt[5]{\dfrac{32x^2}{y^5}}$

28. $\sqrt[3]{\dfrac{16z^3}{y^6}}$

29. $\sqrt{\dfrac{32a^4}{b^2}}$

30. $\sqrt{\dfrac{18x^2}{z^6}}$

In Exercises 31–38, rationalize the denominator and simplify further, if possible.

31. $\sqrt{\dfrac{1}{3}}$

32. $\sqrt{\dfrac{1}{5}}$

33. $\sqrt[4]{\dfrac{5}{4}}$

34. $\sqrt[3]{\dfrac{9}{25}}$

35. $\dfrac{1}{\sqrt{y}}$

36. $\sqrt{\dfrac{5}{c}}$

37. $\sqrt[3]{\dfrac{2x}{3y}}$

38. $\sqrt[3]{\dfrac{20x^2}{9y^2}}$

In Exercises 39–44, combine the radical expressions, if possible.

39. $3\sqrt{2} - \sqrt{2}$

40. $\dfrac{2}{5}\sqrt{5} - \dfrac{6}{5}\sqrt{5}$

41. $2\sqrt[3]{54} + 12\sqrt[3]{16}$

42. $4\sqrt[4]{48} - \sqrt[4]{243}$

43. $\sqrt{25y} + \sqrt{64y}$

44. $\sqrt[3]{16t^4} - \sqrt[3]{54t^4}$

In Exercises 45 and 46, use a graphing utility to graph the equations on the same screen. Use the graphs to verify that the expressions are equivalent. Verify the results algebraically.

45. $y_1 = \sqrt{\dfrac{3}{x}}$, $y_2 = \dfrac{\sqrt{3x}}{x}$

46. $y_1 = \sqrt[3]{8x^4} - x\sqrt[3]{x}$, $y_2 = x\sqrt[3]{x}$

Geometry In Exercises 47 and 48, find the length of the hypotenuse of the right triangle.

47.

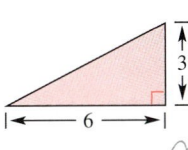

48.

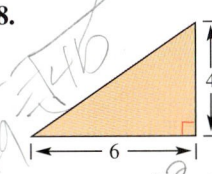

49. *Vibrating String* The frequency f in cycles per second of a vibrating string is given by

$$f = \frac{1}{100}\sqrt{\frac{400 \times 10^6}{5}}.$$

Use a calculator to approximate this number. (Round your answer to two decimal places.)

50. *Geometry* The foundation of a house is 40 feet long and 30 feet wide. The height of the attic is 5 feet (see figure).

(a) Use the Pythagorean Theorem to find the length of the hypotenuse of the right triangle formed by the roof line.

(b) Use the result of part (a) to determine the total area of the roof.

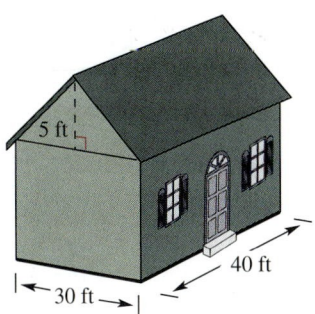

Reviewing the Major Concepts

In Exercises 51–54, simplify the expression.

51. $(x^2 - 3xy)^0$

52. $(-3x^2y^3)^2(4xy^2)$

53. $\dfrac{64r^2s^4}{16rs^2}$

54. $\left(\dfrac{3x}{4y^3}\right)^2$

55. *Geometry* The height of a triangle is 12 inches less than its base. The area is 110 square inches. Find the height and base.

56. *Geometry* An open box with a square base is to be constructed from 825 square inches of material. If the length of an edge of the square base is x inches, the surface area is given by

$$S = x^2 + 4xh.$$

What should be the dimensions of the base if the height of the box is to be 10 inches?

Additional Problem Solving

In Exercises 57–60, write as a single radical.

57. $\sqrt{3} \cdot \sqrt{10}$

58. $\sqrt[5]{9} \cdot \sqrt[5]{19}$

59. $\dfrac{\sqrt[5]{152}}{\sqrt[5]{3}}$

60. $\dfrac{\sqrt[4]{633}}{\sqrt[4]{5}}$

In Exercises 61–64, write the expression as a product or quotient of radicals and simplify.

61. $\sqrt[4]{81 \cdot 11}$

62. $\sqrt[5]{100,000 \cdot 3}$

63. $\sqrt{\dfrac{35}{9}}$

64. $\sqrt[4]{\dfrac{165}{16}}$

In Exercises 65–72, simplify the radical.

65. $\sqrt{27}$

66. $\sqrt{125}$

67. $\sqrt[4]{30,000}$

68. $\sqrt[5]{96}$

69. $\sqrt{\dfrac{15}{49}}$

70. $\sqrt{\dfrac{5}{9}}$

71. $\sqrt[3]{\dfrac{35}{64}}$

72. $\sqrt[4]{\dfrac{5}{16}}$

In Exercises 73–76, simplify the expression.

73. $\sqrt{4 \times 10^{-4}}$ **74.** $\sqrt{8.5 \times 10^2}$

75. $\sqrt[3]{2.4 \times 10^6}$ **76.** $\sqrt[4]{4.4 \times 10^{-4}}$

In Exercises 77–86, simplify the expression.

77. $\sqrt{48y^4}$ **78.** $\sqrt[4]{32x^6}$

79. $\sqrt[3]{x^4y^3}$ **80.** $\sqrt[3]{a^5b^6}$

81. $\sqrt[5]{32x^5y^6}$ **82.** $\sqrt[4]{128u^4v^7}$

83. $\sqrt{\dfrac{13}{25}}$ **84.** $\sqrt{\dfrac{15}{36}}$

85. $\sqrt[3]{\dfrac{54a^4}{b^9}}$ **86.** $\sqrt[4]{\dfrac{3u^2}{16v^8}}$

In Exercises 87–96, rationalize the denominator and simplify further, if possible.

87. $\dfrac{12}{\sqrt{3}}$ **88.** $\dfrac{5}{\sqrt{10}}$

89. $\dfrac{6}{\sqrt[3]{32}}$ **90.** $\dfrac{10}{\sqrt[5]{16}}$

91. $\sqrt{\dfrac{4}{x}}$ **92.** $\dfrac{1}{\sqrt{2x}}$

93. $\dfrac{6}{\sqrt{3b^3}}$ **94.** $\dfrac{1}{\sqrt{xy}}$

95. $\dfrac{a^3}{\sqrt[3]{ab^2}}$ **96.** $\dfrac{3u^2}{\sqrt[4]{8u^3}}$

In Exercises 97–104, combine the radical expressions, if possible.

97. $\sqrt[4]{3} - 5\sqrt[4]{7} - 12\sqrt[4]{3}$

98. $9\sqrt[3]{17} + 7\sqrt[3]{2} - 4\sqrt[3]{17} + \sqrt{2}$

99. $12\sqrt{8} - 3\sqrt[3]{8}$ **100.** $4\sqrt{32} + 7\sqrt{32}$

101. $5\sqrt{9x} - 3\sqrt{x}$ **102.** $3\sqrt{x+1} + 10\sqrt{x+1}$

103. $10\sqrt[3]{z} - \sqrt[3]{z^4}$ **104.** $5\sqrt[3]{24u^2} + 2\sqrt[3]{81u^5}$

In Exercises 105–108, perform the addition or subtraction and simplify your answer.

105. $\sqrt{5} - \dfrac{3}{\sqrt{5}}$ **106.** $\sqrt{10} + \dfrac{5}{\sqrt{10}}$

107. $\sqrt{20} - \sqrt{\dfrac{1}{5}}$ **108.** $\dfrac{x}{\sqrt{3x}} + \sqrt{27x}$

In Exercises 109–112, place the correct symbol ($<$, $>$, or $=$) between the numbers.

109. $\sqrt{7} + \sqrt{18}$ 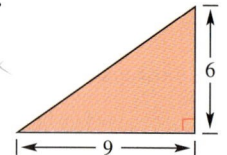 $\sqrt{7+18}$

110. $\sqrt{10} - \sqrt{6}$ 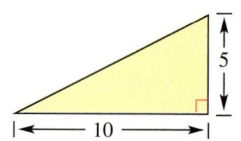 $\sqrt{10-6}$

111. 5 ____ $\sqrt{3^2 + 2^2}$

112. 5 ____ $\sqrt{3^2 + 4^2}$

Geometry In Exercises 113 and 114, find the length of the hypotenuse of the right triangle.

113.

114.

115. *Period of a Pendulum* The period T in seconds of a pendulum (see figure) is given by

$$T = 2\pi\sqrt{\dfrac{L}{32}}$$

where L is the length of the pendulum in feet. Find the period of a pendulum whose length is 4 feet. (Round your answer to two decimal places.)

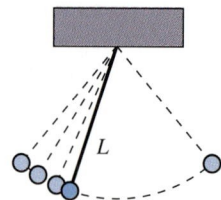

116. *Geometry* All four corners are cut from a 4-foot-by-8-foot sheet of plywood as shown in the figure. Find the perimeter of the remaining piece of plywood.

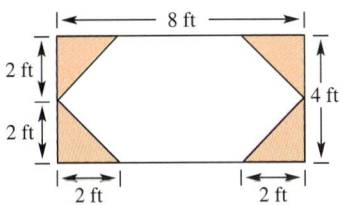

117. *Calculator Experiment* Enter any positive real number in your calculator and find its square root. Then repeatedly take the square root of the result.

$$\sqrt{x}, \sqrt{\sqrt{x}}, \sqrt{\sqrt{\sqrt{x}}}, \ldots$$

What real number does the display appear to be approaching?

118. *Think About It* Square the real number $5/\sqrt{3}$ and note that the radical is eliminated from the denominator. Is this equivalent to rationalizing the denominator? Why or why not?

Math Matters   Standard Metric Prefixes

In the metric system, a *kilo*gram is equal to 1000 grams, a *centi*meter is equal to $\frac{1}{100}$ of a meter, and a *milli*liter is equal to $\frac{1}{1000}$ of a liter. "Kilo," "centi," and "milli" are the most well-known prefixes in the metric system. However, there are several other prefixes that can be used to denote powers of 10 in the metric system.

How many of the prefixes listed have you heard used? Use the list of prefixes to answer the following questions. (The answers are given in the back of the book.)

How many nanoseconds are in a second?
How many millimeters are in a meter?
How many watts are in a kilowatt?

Factor by Which Unit Is Multiplied	Prefix
10^{12}	tera
10^{9}	giga
10^{6}	mega
10^{3}	kilo
10^{2}	hecto
10^{1}	deca
10^{-1}	deci
10^{-2}	centi
10^{-3}	milli
10^{-6}	micro
10^{-9}	nano
10^{-12}	pico
10^{-15}	femto
10^{-18}	atto

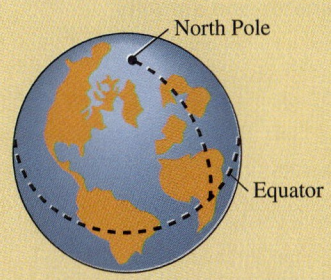

In 1793 the French government adopted the metric system, in which the basic unit of length was defined as one ten-millionth of the distance between the North Pole and the equator. For working standards, a platinum bar was marked with lines a meter apart. Later, when refinements in measuring the earth were discovered, the meter was redefined in terms of the platinum bar that had been constructed.

MID-CHAPTER QUIZ

Take this quiz as you would take a quiz in class. After you are done, check your work against the answers given in the back of the book.

In Exercises 1–4, evaluate the expression.

1. -12^{-2}

2. $\left(\frac{3}{4}\right)^{-3}$

3. $\sqrt{\frac{25}{9}}$

4. $(-64)^{1/3}$

In Exercises 5–8, rewrite the expression using only positive exponents, and simplify. (Assume any variables in the expression are nonzero.)

5. $(t^3)^{-1/2}(3t^3)$

6. $(3x^2y^{-1})(4x^{-2}y)^{-2}$

7. $\dfrac{10u^{-2}}{15u}$

8. $\dfrac{(10x)^0}{(x^2+4)^{-1}}$

9. Write each number in scientific notation: (a) 13,400,000; (b) 0.00075.

10. Evaluate the expression without using a calculator.

(a) $(3 \times 10^3)^4$ (b) $\dfrac{3.2 \times 10^4}{16 \times 10^7}$

In Exercises 11–14, simplify the expression.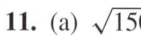

11. (a) $\sqrt{150}$ (b) $\sqrt[3]{54}$ **12.** (a) $\sqrt{27x^2}$ (b) $\sqrt[4]{81x^6}$

13. (a) $\sqrt[4]{\dfrac{5}{16}}$ (b) $\sqrt{\dfrac{24}{49}}$ **14.** (a) $\sqrt{\dfrac{40u^3}{9}}$ (b) $\sqrt[3]{\dfrac{16}{u^6}}$

In Exercises 15 and 16, rationalize the denominator and simplify.

15. (a) $\sqrt{\dfrac{2}{3}}$ (b) $\dfrac{24}{\sqrt{12}}$ **16.** (a) $\dfrac{10}{\sqrt{5x}}$ (b) $\sqrt[3]{\dfrac{3}{2a}}$

In Exercises 17 and 18, combine the radical expressions, if possible.

17. $\sqrt{200y} - 3\sqrt{8y}$ **18.** $6x\sqrt[3]{5x^2} + 2\sqrt[3]{40x^4}$

19. Explain why $\sqrt{5^2 + 12^2} \neq 17$. Determine the correct value of the radical.

20. The four corners are cut from an $8\frac{1}{2}$-inch-by-11-inch sheet of paper as shown in the figure at the right. Find the perimeter of the remaining piece of paper.

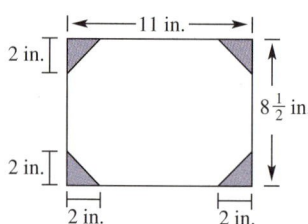

Figure for 20

9.4	**Multiplying and Dividing Radical Expressions**
	Multiplying Radical Expressions ▪ Dividing Radical Expressions

Multiplying Radical Expressions

You can multiply radical expressions by using the Distributive Property or the FOIL Method. In both procedures, you also make use of the Multiplication Property of Radicals. Recall from Section 9.3 that the product of two radicals is given by

$$\sqrt[n]{a}\sqrt[n]{b} = \sqrt[n]{ab}$$

where a and b are real numbers whose nth roots are also real numbers.

EXAMPLE 1   Multiplying Radical Expressions

$$\sqrt{3}\left(2 + \sqrt{5}\right) = 2\sqrt{3} + \sqrt{3}\sqrt{5} \qquad \text{Distributive Property}$$
$$= 2\sqrt{3} + \sqrt{15} \qquad \text{Multiplication Property of Radicals}$$

In Example 1, the product of $\sqrt{3}$ and $\sqrt{5}$ is best left as $\sqrt{15}$. In some cases, however, the product of two radicals can be simplified, as shown in Example 2.

EXAMPLE 2   Multiplying Radical Expressions

Find the products and simplify.

a. $\sqrt{2}\left(4 - \sqrt{8}\right)$ **b.** $\sqrt{6}\left(\sqrt{12} - \sqrt{3}\right)$

Solution

a. $\sqrt{2}\left(4 - \sqrt{8}\right) = 4\sqrt{2} - \sqrt{2}\sqrt{8}$ Distributive Property

$= 4\sqrt{2} - \sqrt{16}$ Multiplication Property of Radicals

$= 4\sqrt{2} - 4$ Simplify.

b. $\sqrt{6}\left(\sqrt{12} - \sqrt{3}\right) = \sqrt{6}\sqrt{12} - \sqrt{6}\sqrt{3}$ Distributive Property

$= \sqrt{72} - \sqrt{18}$ Multiplication Property of Radicals

$= 6\sqrt{2} - 3\sqrt{2}$ Find perfect square factors.

$= 3\sqrt{2}$ Simplify.

In Examples 1 and 2, the Distributive Property was used to multiply radical expressions. In Example 3, note how the FOIL Method can be used to multiply binomial radical expressions.

EXAMPLE 3 Using the FOIL Method

$$
\begin{array}{cc}
& \overset{F}{}\quad\overset{O}{}\quad\overset{I}{}\quad\overset{L}{} \\
\textbf{a.} \quad (2\sqrt{7}-4)(\sqrt{7}+1) = 2(\sqrt{7})^2 + 2\sqrt{7} - 4\sqrt{7} - 4 & \text{FOIL Method} \\
= 2(7) + (2-4)\sqrt{7} - 4 & \text{Combine like radicals.} \\
= 10 - 2\sqrt{7} & \text{Combine like terms.}
\end{array}
$$

$$
\begin{array}{cc}
\textbf{b.} \quad (3-\sqrt{x})(1+\sqrt{x}) = 3 + 3\sqrt{x} - \sqrt{x} - (\sqrt{x})^2 & \\
= 3 + 2\sqrt{x} - x, \quad x \geq 0 & \text{Combine like radicals.}
\end{array}
$$

In addition to the FOIL Method, you can also use special product formulas to multiply binomial radical expressions.

EXAMPLE 4 Using a Special Product Formula

$$
\begin{array}{cc}
\textbf{a.} \quad (2-\sqrt{5})(2+\sqrt{5}) = 2^2 - (\sqrt{5})^2 & \text{Special product formula} \\
= 4 - 5 & \\
= -1 &
\end{array}
$$

$$
\begin{array}{cc}
\textbf{b.} \quad (\sqrt{3}+\sqrt{x})(\sqrt{3}-\sqrt{x}) = (\sqrt{3})^2 - (\sqrt{x})^2 & \text{Special product formula} \\
= 3 - x, \quad x \geq 0 &
\end{array}
$$

In Example 4(a), the expressions $(2-\sqrt{5})$ and $(2+\sqrt{5})$ are called **conjugates** of each other. The product of two conjugates is the difference of two squares, which is given by the special product formula $(a+b)(a-b) = a^2 - b^2$. Here are some other examples.

Expression	*Conjugate*	*Product*
$(1-\sqrt{3})$	$(1+\sqrt{3})$	$(1)^2 - (\sqrt{3})^2 = 1 - 3 = -2$
$(\sqrt{5}+\sqrt{2})$	$(\sqrt{5}-\sqrt{2})$	$(\sqrt{5})^2 - (\sqrt{2})^2 = 5 - 2 = 3$
$(\sqrt{10}-3)$	$(\sqrt{10}+3)$	$(\sqrt{10})^2 - (3)^2 = 10 - 9 = 1$
$\sqrt{x}+2$	$\sqrt{x}-2$	$(\sqrt{x})^2 - (2)^2 = x - 4, \quad x \geq 0$

Dividing Radical Expressions

To simplify a *quotient* involving radicals, we rationalize the denominator. For single-term denominators, you can use the rationalizing process described in Section 9.3. To rationalize a denominator involving two terms, multiply both the numerator and denominator by the *conjugate of the denominator*.

EXAMPLE 5   *Simplifying Quotients Involving Radicals*

a.
$$\frac{\sqrt{3}}{1-\sqrt{5}} = \frac{\sqrt{3}}{1-\sqrt{5}} \cdot \frac{1+\sqrt{5}}{1+\sqrt{5}}$$ Multiply by conjugate of denominator.

$$= \frac{\sqrt{3}\left(1+\sqrt{5}\right)}{1^2 - \left(\sqrt{5}\right)^2}$$ Special product formula

$$= \frac{\sqrt{3}+\sqrt{15}}{1-5}$$ Simplify.

$$= \frac{\sqrt{3}+\sqrt{15}}{-4}$$ Simplify.

b.
$$\frac{4}{2-\sqrt{3}} = \frac{4}{2-\sqrt{3}} \cdot \frac{2+\sqrt{3}}{2+\sqrt{3}}$$ Multiply by conjugate of denominator.

$$= \frac{4\left(2+\sqrt{3}\right)}{2^2 - \left(\sqrt{3}\right)^2}$$ Special product formula

$$= \frac{8+4\sqrt{3}}{4-3}$$ Simplify.

$$= 8+4\sqrt{3}$$ Simplify.

EXAMPLE 6   *Simplifying Quotients Involving Radicals*

$$\frac{5\sqrt{2}}{\sqrt{7}+\sqrt{2}} = \frac{5\sqrt{2}}{\sqrt{7}+\sqrt{2}} \cdot \frac{\sqrt{7}-\sqrt{2}}{\sqrt{7}-\sqrt{2}}$$ Multiply by conjugate of denominator.

$$= \frac{5\left(\sqrt{14}-\sqrt{4}\right)}{\left(\sqrt{7}\right)^2 - \left(\sqrt{2}\right)^2}$$ Special product formula

$$= \frac{5\left(\sqrt{14}-2\right)}{7-2}$$ Simplify.

$$= \frac{5\left(\sqrt{14}-2\right)}{5}$$ Cancel common factor.

$$= \sqrt{14}-2$$ Simplest form

EXAMPLE 7   *Dividing Radical Expressions*

Perform the division and simplify.

a. $6 \div (\sqrt{x} - 2)$ **b.** $(2 - \sqrt{3}) \div (\sqrt{6} + \sqrt{2})$

Solution

a. $\dfrac{6}{\sqrt{x} - 2} = \dfrac{6}{\sqrt{x} - 2} \cdot \dfrac{\sqrt{x} + 2}{\sqrt{x} + 2}$ Multiply by conjugate of denominator.

$= \dfrac{6(\sqrt{x} + 2)}{(\sqrt{x})^2 - 2^2}$ Special product formula

$= \dfrac{6\sqrt{x} + 12}{x - 4}, \quad x \geq 0$ Simplify.

b. $\dfrac{2 - \sqrt{3}}{\sqrt{6} + \sqrt{2}} = \dfrac{2 - \sqrt{3}}{\sqrt{6} + \sqrt{2}} \cdot \dfrac{\sqrt{6} - \sqrt{2}}{\sqrt{6} - \sqrt{2}}$ Multiply by conjugate of denominator.

$= \dfrac{2\sqrt{6} - 2\sqrt{2} - \sqrt{18} + \sqrt{6}}{(\sqrt{6})^2 - (\sqrt{2})^2}$ Special product formula

$= \dfrac{3\sqrt{6} - 2\sqrt{2} - 3\sqrt{2}}{6 - 2}$ Simplify.

$= \dfrac{3\sqrt{6} - 5\sqrt{2}}{4}$ Simplify.

Group Activities

You Be the Instructor

The Golden Section   The ratio of the width of the Temple of Hephaestus to its height is approximately

$$\frac{w}{h} \approx \frac{2}{\sqrt{5} - 1}.$$

This number is called the **golden section.** Early Greeks believed that the most aesthetically pleasing rectangles were those whose sides have this ratio.

a. Rationalize the denominator for this number. Approximate your answer, rounded to two decimal places.

b. Use the Pythagorean Theorem, a straightedge, and a compass to construct a rectangle whose sides have the golden section as their ratio.

9.4 Exercises

Discussing the Concepts

1. Multiply $\sqrt{3}\left(1 - \sqrt{6}\right)$. State an algebraic property to justify each step.

2. Explain the meaning of each letter of the word FOIL in the FOIL Method.

3. Multiply $3 - \sqrt{2}$ by its conjugate. Explain why the result has no radicals.

4. Is the number $3/\left(1 + \sqrt{5}\right)$ in simplest form? If not, explain the steps for writing it in simplest form.

Problem Solving

In Exercises 5–18, multiply and simplify.

5. $\sqrt{3} \cdot \sqrt{6}$

6. $\sqrt{5} \cdot \sqrt{10}$

7. $\sqrt{5}\left(2 - \sqrt{3}\right)$

8. $\sqrt{11}\left(\sqrt{5} - 3\right)$

9. $\left(\sqrt{3} + 2\right)\left(\sqrt{3} - 2\right)$

10. $\left(\sqrt{15} + 3\right)\left(\sqrt{15} - 3\right)$

11. $\left(\sqrt{20} + 2\right)^2$

12. $\left(4 - \sqrt{20}\right)^2$

13. $\sqrt{y}\left(\sqrt{y} + 4\right)$

14. $\left(5 - \sqrt{3v}\right)^2$

15. $\left(9\sqrt{x} + 2\right)\left(5\sqrt{x} - 3\right)$

16. $\left(16\sqrt{u} - 3\right)\left(\sqrt{u} - 1\right)$

17. $\sqrt[3]{4}\left(\sqrt[3]{2} - 7\right)$

18. $\left(\sqrt[3]{9} + 5\right)\left(\sqrt[3]{5} - 5\right)$

In Exercises 19–24, complete the statement.

19. $5\sqrt{3} + 15\sqrt{3} = 5()$

20. $x\sqrt{7} - x^2\sqrt{7} = x()$

21. $4\sqrt{12} - 2x\sqrt{27} = 2\sqrt{3}()$

22. $5\sqrt{50} + 10y\sqrt{8} = 5\sqrt{2}()$

23. $6u^2 + \sqrt{18u^3} = 3u()$

24. $12s^3 - \sqrt{32s^4} = 4s^2\ ()$

In Exercises 25–28, find the conjugate of the number. Then multiply the number by its conjugate.

25. $\sqrt{11} - \sqrt{3}$

26. $\sqrt{10} + \sqrt{7}$

27. $\sqrt{x} - 3$

28. $\sqrt{5a} + \sqrt{2}$

In Exercises 29–36, rationalize the denominator of the expression.

29. $\dfrac{6}{\sqrt{22} - 2}$

30. $\dfrac{3}{2\sqrt{10} - 5}$

31. $\dfrac{8}{\sqrt{7} + 3}$

32. $\dfrac{10}{\sqrt{9} + \sqrt{5}}$

33. $\dfrac{2t^2}{\sqrt{5t} - \sqrt{t}}$

34. $\dfrac{5x}{\sqrt{x} - \sqrt{2}}$

35. $\dfrac{\sqrt{u + v}}{\sqrt{u - v} - \sqrt{u}}$

36. $\dfrac{z}{\sqrt{u + z} - \sqrt{u}}$

In Exercises 37 and 38, evaluate the function.

37. $f(x) = x^2 - 2x - 1$

    (a) $f\left(1 + \sqrt{2}\right)$ (b) $f\left(\sqrt{4}\right)$

38. $g(x) = x^2 - 4x + 1$

    (a) $g\left(1 + \sqrt{5}\right)$ (b) $g\left(2 - \sqrt{3}\right)$

In Exercises 39 and 40, use a graphing utility to graph the two equations on the same screen. Use the graphs to verify that the expressions are equivalent. Verify the results algebraically.

39. $y_1 = \dfrac{10}{\sqrt{x} + 1}$, $y_2 = \dfrac{10\left(\sqrt{x} - 1\right)}{x - 1}$

40. $y_1 = \dfrac{2\sqrt{x}}{2 - \sqrt{x}}$, $y_2 = \dfrac{2\left(2\sqrt{x} + x\right)}{4 - x}$

41. *Strength of a Wooden Beam*   The rectangular cross section of a wooden beam cut from a log of diameter 24 inches (see figure) will have maximum strength if its width w and height h are given by

$$w = 8\sqrt{3} \quad \text{and} \quad h = \sqrt{24^2 - \left(8\sqrt{3}\right)^2}.$$

Find the area of the rectangular cross section and express the area in simplest form.

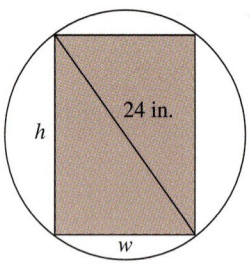

42. *Force*   The force required to slide a 500-pound block across a milling machine is

$$\frac{500k}{\dfrac{1}{\sqrt{k^2+1}} + \dfrac{k^2}{\sqrt{k^2+1}}}$$

where k is the friction constant (see figure). Simplify this expression.

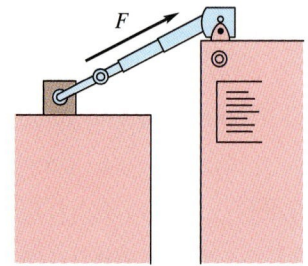

Reviewing the Major Concepts

In Exercises 43–46, solve the equation.

43. $\dfrac{4}{x} - \dfrac{2}{3} = 0$

44. $2x = 3[1 + (4 - x)]$

45. $3x^2 - 13x - 10 = 0$

46. $x(x - 3) = 40$

In Exercises 47–50, use a graphing utility to identify the transformation of the graph of $f(x) = \sqrt{x}$.

47. $h(x) = 4 + \sqrt{x}$

48. $h(x) = \sqrt{4 + x}$

49. $h(x) = -\sqrt{x}$

50. $h(x) = \sqrt{-x}$

Additional Problem Solving

In Exercises 51–68, multiply and simplify.

51. $\sqrt{2} \cdot \sqrt{8}$

52. $\sqrt{6} \cdot \sqrt{18}$

53. $\sqrt{2}\left(\sqrt{20} + 8\right)$

54. $\sqrt{7}\left(\sqrt{14} + 3\right)$

55. $\left(3 - \sqrt{5}\right)\left(3 + \sqrt{5}\right)$

56. $\left(\sqrt{11} + 3\right)\left(\sqrt{11} - 3\right)$

57. $\left(2\sqrt{2} + \sqrt{4}\right)\left(2\sqrt{2} - \sqrt{4}\right)$

58. $\left(4\sqrt{3} + \sqrt{2}\right)\left(4\sqrt{3} - \sqrt{2}\right)$

59. $\left(\sqrt{5} + 3\right)\left(\sqrt{3} - 5\right)$

60. $\left(\sqrt{30} + 6\right)\left(\sqrt{2} + 6\right)$

61. $\left(10 + \sqrt{2x}\right)^2$

62. $\sqrt{x}\left(5 - \sqrt{x}\right)$

63. $\left(3\sqrt{x} - 5\right)\left(3\sqrt{x} + 5\right)$

64. $\left(7 - 3\sqrt{3t}\right)\left(7 + 3\sqrt{3t}\right)$

65. $\left(\sqrt{x} + \sqrt{y}\right)\left(\sqrt{x} - \sqrt{y}\right)$

66. $\left(3\sqrt{u} + \sqrt{3v}\right)\left(3\sqrt{u} - \sqrt{3v}\right)$

67. $\left(\sqrt[3]{2x} + 5\right)^2$

68. $\left(\sqrt[3]{y} + 2\right)\left(\sqrt[3]{y^2} - 5\right)$

In Exercises 69–72, simplify the expression.

69. $\dfrac{4 - 8\sqrt{x}}{12}$

70. $\dfrac{-3 + 27\sqrt{2y}}{18}$

71. $\dfrac{-2y + \sqrt{12y^3}}{8y}$

72. $\dfrac{-t^2 - \sqrt{2t^3}}{3t}$

In Exercises 73–76, find the conjugate of the number. Then multiply the number by its conjugate.

73. $2 + \sqrt{5}$

74. $\sqrt{2} - 9$

75. $\sqrt{2u} - \sqrt{3}$

76. $\sqrt{t} + 7$

In Exercises 77–84, rationalize the denominator of the expression.

77. $\dfrac{2}{6 + \sqrt{2}}$

78. $\dfrac{44\sqrt{5}}{3\sqrt{5} - 1}$

79. $\left(\sqrt{7} + 2\right) \div \left(\sqrt{7} - 2\right)$

80. $\left(5 - 3\sqrt{3}\right) \div \left(3 + \sqrt{3}\right)$

81. $\dfrac{3x}{\sqrt{15} - \sqrt{3}}$

82. $\dfrac{6(y + 1)}{y^2 + \sqrt{y}}$

83. $\left(\sqrt{x} - 5\right) \div \left(2\sqrt{x} - 1\right)$

84. $\left(2\sqrt{t} + 1\right) \div \left(2\sqrt{t} - 1\right)$

In Exercises 85 and 86, evaluate the function as indicated.

85. $f(x) = x^2 - 6x + 1$

    (a) $f\left(2 - \sqrt{3}\right)$ (b) $f\left(3 - 2\sqrt{2}\right)$

86. $g(x) = x^2 + 8x + 11$

    (a) $g\left(-4 + \sqrt{5}\right)$ (b) $g\left(-4\sqrt{2}\right)$

In Exercises 87 and 88, use a graphing utility to graph the two equations on the same screen. Use the graphs to verify that the expressions are equivalent. Verify the results algebraically.

87. $y_1 = \dfrac{4x}{\sqrt{x} + 4}, \quad y_2 = \dfrac{4x\left(\sqrt{x} - 4\right)}{x - 16}$

88. $y_1 = \dfrac{\sqrt{2x} + 6}{\sqrt{2x} - 2}, \quad y_2 = \dfrac{x + 6 + 4\sqrt{2x}}{x - 2}$

89. *Geometry* The areas of the circles in the figure are 15 square centimeters and 20 square centimeters. Find the ratio of the radius of the small circle to the radius of the large circle.

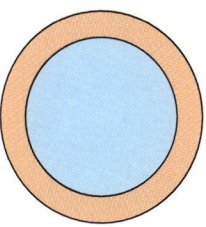

90. *Radical Spiral* Copy the spiral shown below and find the lengths of the sides labeled a, b, c, d, e, and f. Explain your reasoning.

CAREER INTERVIEW

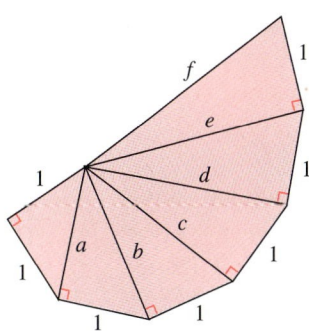

Lt. Robert Orr

Civil Service Lieutenant for Boston Police Department

Boston, MA 02201

I am a member of the Boston Police Dive Team, serving as an underwater investigator for the inner Boston Harbor. Safe execution of a scuba dive depends on making correct computations when preparing for the dive. For instance, it is vital to know how long the compressed air in my tank will last. Pressure under water increases with depth. If I start with a tank containing 70 cubic feet of compressed air at the surface, and dive to a depth at which the pressure is twice that at the surface, the volume of air available for breathing is halved to 35 cubic feet. In general, this relationship is $y = c^{-1}x$, where y is the available underwater air volume, x is the surface air volume, and c is the multiple of surface pressure. Because the body demands the same volume of air while diving as at the surface, the tank of air will last roughly $1/c$ times as long. We use *U.S. Navy Diving Manual* tables for most calculations, but it is useful to understand this relationship to make estimates.

9.5 | Solving Radical Equations

Solving Radical Equations ▪ Applications

Solving Radical Equations

STUDY TIP

You will see in this section that raising both sides of an equation to the nth power often introduces *extraneous* solutions. So when you use this procedure, it is critical that you check each solution in the *original* equation.

Solving equations involving radicals is somewhat like solving equations that contain fractions—you try to get rid of the radicals and obtain a polynomial equation. Then you solve the polynomial equation using the standard procedures. The following property plays a key role.

Raising Both Sides of an Equation to the nth Power

Let u and v be real numbers, variables, or algebraic expressions, and let n be a positive integer. If $u = v$, then it follows that

$$u^n = v^n.$$

This is called **raising both sides of an equation to the nth power.**

To use this property to solve an equation, first try to isolate one of the radicals on one side of the equation.

Technology

To use a graphing utility to check the solution in Example 1, sketch the graph of

$$y = \sqrt{x} - 8$$

as shown below. Notice that the graph crosses the x-axis when $x = 64$, which confirms the solution that was obtained algebraically.

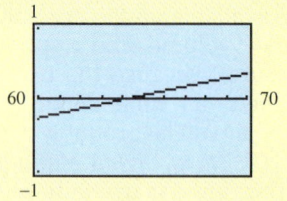

EXAMPLE 1 Solving an Equation Having One Radical

Solve $\sqrt{x} - 8 = 0$.

Solution

$\sqrt{x} - 8 = 0$	Original equation
$\sqrt{x} = 8$	Isolate radical.
$\left(\sqrt{x}\right)^2 = 8^2$	Square both sides.
$x = 64$	Simplify.

Check

$\sqrt{x} - 8 = 0$	Original equation
$\sqrt{64} - 8 \overset{?}{=} 0$	Substitute 64 for x.
$8 - 8 = 0$	Solution checks. ✓

Therefore, the equation has one solution: $x = 64$.

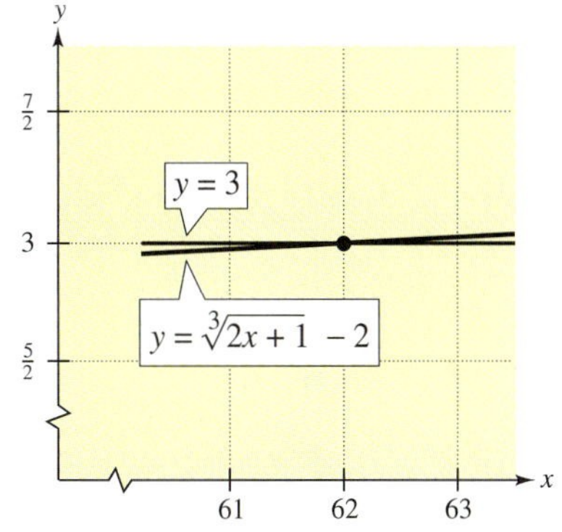

EXAMPLE 2 *Solving an Equation Having One Radical*

Solve $\sqrt[3]{2x+1} - 2 = 3$.

Solution

$\sqrt[3]{2x+1} - 2 = 3$	Original equation
$\sqrt[3]{2x+1} = 5$	Isolate radical.
$\left(\sqrt[3]{2x+1}\right)^3 = 5^3$	Cube both sides.
$2x + 1 = 125$	Simplify.
$2x = 124$	Subtract 1 from both sides.
$x = 62$	Divide both sides by 2.

Check

$\sqrt[3]{2x+1} - 2 = 3$	Original equation
$\sqrt[3]{2(62)+1} - 2 \overset{?}{=} 3$	Substitute 62 for x.
$\sqrt[3]{125} - 2 \overset{?}{=} 3$	Simplify.
$5 - 2 = 3$	Solution checks.

Therefore, the equation has one solution: $x = 62$. You can also check the solution graphically, as shown in Figure 9.3.

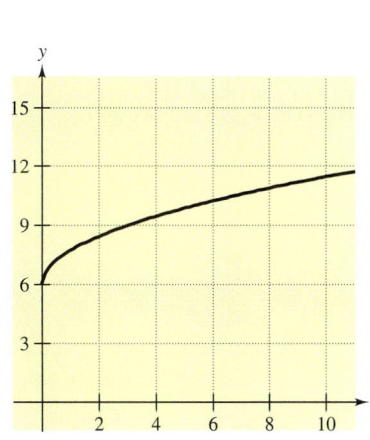

FIGURE 9.3

EXAMPLE 3 *Solving an Equation Having One Radical*

Solve $\sqrt{3x} + 6 = 0$.

Solution

$\sqrt{3x} + 6 = 0$	Original equation
$\sqrt{3x} = -6$	Isolate radical.
$\left(\sqrt{3x}\right)^2 = (-6)^2$	Square both sides.
$3x = 36$	Simplify.
$x = 12$	Divide both sides by 3.

Check

$\sqrt{3x} + 6 = 0$	Original equation
$\sqrt{3(12)} + 6 \overset{?}{=} 0$	Substitute 12 for x.
$6 + 6 \neq 0$	Solution does not check.

Therefore, the equation has no solution. You can also check this graphically, as shown in Figure 9.4.

FIGURE 9.4

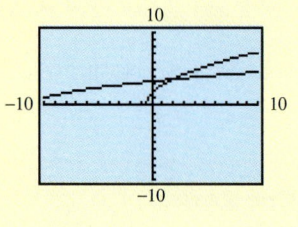

EXAMPLE 4 *Solving an Equation Having Two Radicals*

Solve $\sqrt{5x + 3} = \sqrt{x + 11}$.

Solution

$\sqrt{5x + 3} = \sqrt{x + 11}$	Original equation
$\left(\sqrt{5x + 3}\right)^2 = \left(\sqrt{x + 11}\right)^2$	Square both sides.
$5x + 3 = x + 11$	Simplify.
$5x = x + 8$	Subtract 3 from both sides.
$4x = 8$	Subtract x from both sides.
$x = 2$	Divide both sides by 4.

Check

$\sqrt{5x + 3} = \sqrt{x + 11}$	Original equation
$\sqrt{5(2) + 3} \overset{?}{=} \sqrt{2 + 11}$	Substitute 2 for x.
$\sqrt{13} = \sqrt{13}$	Solution checks. ✔

Therefore, the equation has one solution: $x = 2$.

EXAMPLE 5 *Solving an Equation Having Two Radicals*

Solve $\sqrt[4]{3x} + \sqrt[4]{2x - 5} = 0$.

Solution

$\sqrt[4]{3x} + \sqrt[4]{2x - 5} = 0$	Original equation
$\sqrt[4]{3x} = -\sqrt[4]{2x - 5}$	Isolate radicals.
$\left(\sqrt[4]{3x}\right)^4 = \left(-\sqrt[4]{2x - 5}\right)^4$	Raise both sides to 4th power.
$3x = 2x - 5$	Simplify.
$x = -5$	Subtract $2x$ from both sides.

Check

$\sqrt[4]{3x} + \sqrt[4]{2x - 5} = 0$	Original equation
$\sqrt[4]{3(-5)} + \sqrt[4]{2(-5) - 5} \overset{?}{=} 0$	Substitute -5 for x.
$\sqrt[4]{-15} + \sqrt[4]{-15} \neq 0$	Solution does not check. ✘

The solution does not check because it yields fourth roots of negative radicands. Therefore, this equation has no solution. Try checking this graphically. If you graph both sides of the equation, you will discover that they do not intersect.

EXAMPLE 6   An Equation That Converts to a Quadratic Equation

Solve $\sqrt{x} + 2 = x$.

Solution

$$\sqrt{x} + 2 = x \qquad \text{Original equation}$$

$$\sqrt{x} = x - 2 \qquad \text{Isolate radical.}$$

$$\left(\sqrt{x}\right)^2 = (x - 2)^2 \qquad \text{Square both sides.}$$

$$x = x^2 - 4x + 4 \qquad \text{Simplify.}$$

$$-x^2 + 5x - 4 = 0 \qquad \text{Standard form}$$

$$(-1)(x - 4)(x - 1) = 0 \qquad \text{Factor.}$$

$$x - 4 = 0 \quad \Longrightarrow \quad x = 4 \qquad \text{Set 1st factor equal to 0.}$$

$$x - 1 = 0 \quad \Longrightarrow \quad x = 1 \qquad \text{Set 2nd factor equal to 0.}$$

Try checking each of these solutions. When you do, you will find that $x = 4$ is a valid solution, but that $x = 1$ is extraneous.

When an equation contains two radicals, it may not be possible to isolate both. In such cases, you may have to raise both sides of the equation to a power at *two* different stages in the solution.

EXAMPLE 7   Repeatedly Squaring Both Sides of an Equation

Solve $\sqrt{3t + 1} = 2 - \sqrt{3t}$.

Solution

$$\sqrt{3t + 1} = 2 - \sqrt{3t} \qquad \text{Original equation}$$

$$\left(\sqrt{3t + 1}\right)^2 = \left(2 - \sqrt{3t}\right)^2 \qquad \text{Square both sides (1st time).}$$

$$3t + 1 = 4 - 4\sqrt{3t} + 3t \qquad \text{Simplify.}$$

$$-3 = -4\sqrt{3t} \qquad \text{Isolate radical.}$$

$$(-3)^2 = \left(-4\sqrt{3t}\right)^2 \qquad \text{Square both sides (2nd time).}$$

$$9 = 16(3t) \qquad \text{Simplify.}$$

$$\frac{3}{16} = t \qquad \text{Divide both sides by 48.}$$

The solution is $t = \frac{3}{16}$. Check this in the original equation.

Applications

EXAMPLE 8   An Application Involving Electricity

The amount of power consumed by an electrical appliance is given by

$$I = \sqrt{\frac{P}{R}}$$

where I is the current measured in amps, R is the resistance measured in ohms, and P is the power measured in watts. Find the power used by an electric heater, for which $I = 10$ amps and $R = 16$ ohms.

Solution

$$I = \sqrt{\frac{P}{R}} \qquad \text{Original equation}$$

$$10 = \sqrt{\frac{P}{16}} \qquad \text{Substitute for } I \text{ and } R.$$

$$(10)^2 = \left(\sqrt{\frac{P}{16}}\right)^2 \qquad \text{Square both sides.}$$

$$100 = \frac{P}{16} \qquad \text{Simplify.}$$

$$1600 = P \qquad \text{Multiply both sides by 16.}$$

Therefore, the electric heater uses 1600 watts of power. Check this solution in the original equation.

An alternative way to solve the problem in Example 8 would be to first solve the equation for P.

$$I = \sqrt{\frac{P}{R}} \qquad \text{Original equation}$$

$$I^2 = \left(\sqrt{\frac{P}{R}}\right)^2 \qquad \text{Square both sides.}$$

$$I^2 = \frac{P}{R} \qquad \text{Simplify.}$$

$$I^2 R = P \qquad \text{Multiply both sides by } R.$$

At this stage, you can substitute the known values of I and R to obtain

$$P = (10)^2 16 = 1600.$$

EXAMPLE 9   The Velocity of a Falling Object

The velocity of a free-falling object can be determined from the equation

$$v = \sqrt{2gh}$$

where v is the velocity measured in feet per second, $g = 32$ feet per second per second, and h is the distance (in feet) the object has fallen. Find the height from which a rock has been dropped if it strikes the ground with a velocity of 50 feet per second.

Solution

$v = \sqrt{2gh}$	Original equation
$50 = \sqrt{2(32)h}$	Substitute for v and g.
$(50)^2 = \left(\sqrt{64h}\right)^2$	Square both sides.
$2500 = 64h$	Simplify.
$39 \approx h$	Divide both sides by 64.

Thus, the rock has fallen approximately 39 feet when it hits the ground. Check this solution in the original equation.

Group Activities   Extending the Concept

An Experiment   Without using a stopwatch, you can find the length of time an object has been falling by using the following equation from physics

$$t = \sqrt{\frac{h}{384}}$$

where t is the length of time in seconds and h is the height in inches the object has fallen. How far does an object fall in 0.25 second? in 0.10 second?

Use this equation to test how long it takes members of your group to catch a falling ruler. Hold the ruler vertically while another group member holds his or her hands near the lower end of the ruler ready to catch it. Before releasing the ruler, record the mark on the ruler closest to the top of the catcher's hands. Release the ruler. After it has been caught, again note the mark closest to the top of the catcher's hands. (The difference between these two measurements is h.) Which member of your group reacts most quickly?

9.5 Exercises

Discussing the Concepts

1. In your own words, describe the steps that can be used to solve a radical equation.

2. Does raising both sides of an equation to the nth power always yield an equivalent equation? Explain.

3. One reason for checking a solution in the original equation is to discover errors that were made when solving the equation. Describe another reason.

4. *Error Analysis*   Describe the error.

$$\sqrt{x} + \sqrt{6} = 8$$
$$\left(\sqrt{x}\right)^2 + \left(\sqrt{6}\right)^2 = 8^2$$
$$x + 6 = 64$$
$$x = 58$$

Problem Solving

In Exercises 5–8, determine whether each value of x is a solution of the equation.

Equation		*Values of x*	
5. $\sqrt{x} - 10 = 0$	(a) $x = -4$	(b) $x = -100$	
	(c) $x = \sqrt{10}$	(d) $x = 100$	
6. $\sqrt{3x} - 6 = 0$	(a) $x = \frac{2}{3}$	(b) $x = 2$	
	(c) $x = 12$	(d) $x = -\frac{1}{3}\sqrt{6}$	
7. $\sqrt[3]{x - 4} = 4$	(a) $x = -60$	(b) $x = 68$	
	(c) $x = 20$	(d) $x = 0$	
8. $\sqrt[4]{2x} + 2 = 6$	(a) $x = 128$	(b) $x = 2$	
	(c) $x = -2$	(d) $x = 0$	

In Exercises 9–22, solve the equation. (Some of the equations have no solution.)

9. $\sqrt{x} = 20$

10. $\sqrt{x} = 5$

11. $\sqrt{u} + 13 = 0$

12. $\sqrt{y} + 15 = 0$

13. $\sqrt{3y + 5} - 3 = 4$

14. $\sqrt{5z - 2} + 7 = 10$

15. $\sqrt{x^2 + 5} = x + 3$

16. $\sqrt{x^2 - 4} = x - 2$

17. $\sqrt{3y - 5} - 3\sqrt{y} = 0$

18. $\sqrt{2u + 10} - 2\sqrt{u} = 0$

19. $\sqrt[3]{3x - 4} = \sqrt[3]{x + 10}$

20. $2\sqrt[3]{10 - 3x} = \sqrt[3]{2 - x}$

21. $\sqrt{8x + 1} = x + 2$

22. $\sqrt{3x + 7} = x + 3$

Graphical Reasoning   In Exercises 23–26, use a graphing utility to graph both sides of the equation on the same screen. Then use the graphs to approximate the solution. Check your approximations algebraically.

23. $\sqrt{x} = 2(2 - x)$

24. $\sqrt{2x + 3} = 4x - 3$

25. $\sqrt{x + 3} = 5 - \sqrt{x}$

26. $\sqrt[3]{5x - 8} = 4 - \sqrt[3]{x}$

Geometry   In Exercises 27–30, find the length of the side labeled x. Round to two decimal places.

27.

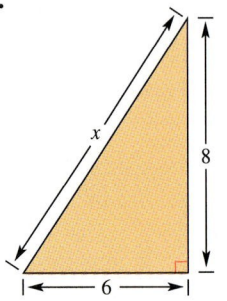

28.

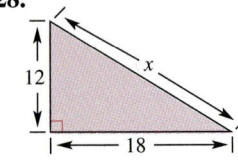

29.

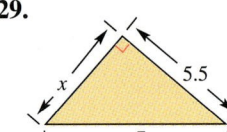

30.

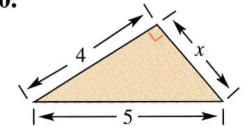

31. *Drawing a Diagram*   A basketball court is 50 feet wide and 94 feet long. Draw a diagram and find the length of the diagonal of the court.

32. *Drawing a Diagram*   For a square "25-inch" television screen, it is the diagonal of the screen that is 25 inches. Draw a diagram that shows a television screen with a 25-inch diagonal. What are its dimensions?

33. *Drawing a Diagram*   A house has a basement floor with dimensions 26 feet by 32 feet. The gas hot water heater and furnace are diagonally across the basement from where the natural gas line enters the house. Draw a diagram showing the gas line and find the length of the gas line across the basement.

34. *Drawing a Diagram*   An 8-foot plank is used to brace a basement wall during the construction of a home. The plank is nailed to the wall 5 feet above the floor. Draw a diagram showing the plank, and find its slope.

Height of an Object   In Exercises 35 and 36, use the following formula, which gives the time t in seconds for a free-falling object to fall d feet.

$$t = \sqrt{\frac{d}{16}}$$

35. A construction worker drops a nail and observes it strike a water puddle after approximately 2 seconds. Estimate the height of the worker.

36. A construction worker drops a nail and observes it strike a water puddle after approximately 3 seconds. Estimate the height of the worker.

Length of a Pendulum   In Exercises 37 and 38, use the following formula, which gives the time t in seconds for a pendulum of length L feet to go through one complete cycle (its period).

$$t = 2\pi\sqrt{\frac{L}{32}}$$

37. How long is the pendulum of a grandfather clock with a period of 1.5 seconds (see figure)?

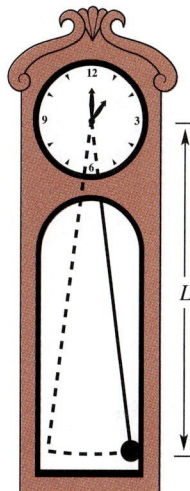

38. How long is the pendulum of a mantle clock with a period of 0.75 seconds?

Reviewing the Major Concepts

In Exercises 39–44, simplify the fraction.

39. $\dfrac{-16x^2}{12x}$

40. $\dfrac{5t^4}{45t^{-2}}$

41. $\dfrac{6u^2v^{-3}}{27uv^3}$

42. $\dfrac{-14r^4s^2}{-98rs^2}$

43. $\left(\dfrac{3x^2}{2y^{-1}}\right)^{-2}$

44. $\left(\dfrac{15a^{-3}b}{21ab^{-2}}\right)^{0}$

45. *Comparing Prices*   A department store is offering a discount of 20% on a sewing machine with a list price of $239.95. A mail-order catalog has the same machine for $188.95 plus $4.32 for shipping. Which is the better bargain?

46. *Insurance Premium*   The annual insurance premium for a policyholder is normally $739. However, after having an automobile accident, the policyholder is charged an additional 30%. What is the new annual premium?

Additional Problem Solving

In Exercises 47–64, solve the equation. (Some of the equations have no solution.)

47. $\sqrt{y} - 7 = 0$

48. $\sqrt{t} - 13 = 0$

49. $\sqrt{a + 100} = 25$

50. $\sqrt{b + 12} = 13$

51. $\sqrt{10x} = 30$

52. $\sqrt{8x} = 6$

53. $5\sqrt{x + 2} = 8$

54. $2\sqrt{x + 4} = 7$

55. $\sqrt{3x + 2} + 5 = 0$

56. $\sqrt{1 - x} + 10 = 4$

57. $\sqrt{x + 3} = \sqrt{2x - 1}$

58. $\sqrt{3t + 1} = \sqrt{t + 15}$

59. $\sqrt[3]{2x + 15} - \sqrt[3]{x} = 0$

60. $\sqrt[4]{2x} + \sqrt[4]{x + 3} = 0$

61. $(x + 4)^{2/3} = 4$

62. $2x^{3/4} = 54$

63. $\sqrt{2x} = x - 4$

64. $\sqrt{x} = x - 6$

Graphical Reasoning In Exercises 65–68, use a graphing utility to graph both sides of the equation on the same screen. Then use the graphs to approximate the solution. Check your approximations algebraically.

65. $\sqrt{x^2 + 1} = 5 - 2x$

66. $\sqrt{8 - 3x} = x$

67. $4\sqrt[3]{x} = 7 - x$

68. $\sqrt[3]{x + 4} = \sqrt{6 - x}$

Geometry In Exercises 69–72, find the length of the side labeled x. Round to two decimal places.

69.

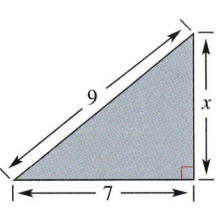

70.

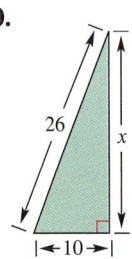

71.

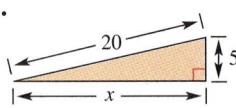

72.

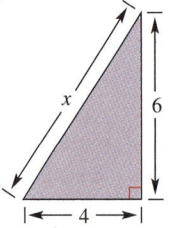

In Exercises 73–78, match the function with its graph. [The graphs are labeled (a), (b), (c), (d), (e), and (f).]

(a)

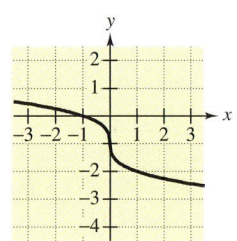

(b)

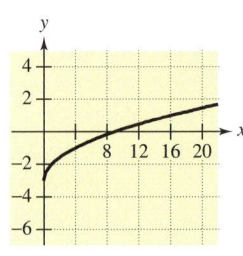

(c)

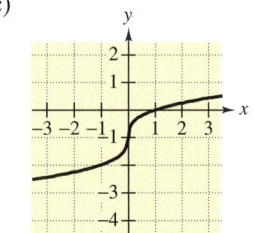

(d)

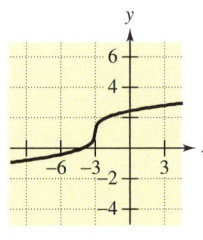

(e)

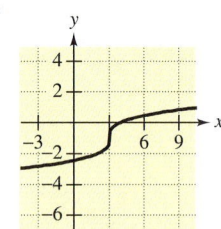

(f)

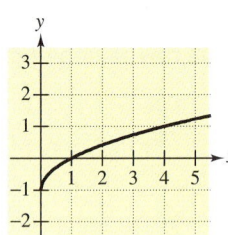

73. $f(x) = \sqrt[3]{x} - 1$

74. $f(x) = \sqrt[3]{x - 3} - 1$

75. $f(x) = \sqrt[3]{x + 3} + 1$

76. $f(x) = -\sqrt[3]{x} - 1$

77. $f(x) = \sqrt{x} - 1$

78. $f(x) = \sqrt{x} - 3$

79. *Length of a Ramp* A ramp is 20 feet long and rests on a porch that is 4 feet high (see figure). Find the distance between the porch and the base of the ramp.

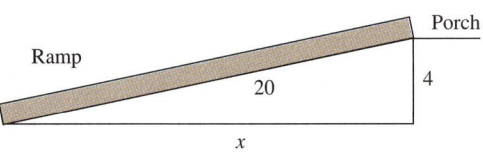

80. *Drawing a Diagram*   A guy wire on a radio tower is attached to the top of a 100-foot tower and to an anchor 50 feet from the base of the tower. Draw a diagram showing the guy wire, and find its length.

Free-Falling Object   In Exercises 81–84, use the equation for the velocity of a free-falling object, $v = \sqrt{2gh}$, as described in Example 9.

81. An object is dropped from a height of 50 feet. Find the velocity of the object when it strikes the ground.

82. An object is dropped from a height of 200 feet. Find the velocity of the object when it strikes the ground.

83. An object that was dropped strikes the ground with a velocity of 60 feet per second. Find the height from which the object was dropped.

84. An object that was dropped strikes the ground with a velocity of 120 feet per second. Find the height from which the object was dropped.

85. *Demand for a Product*   The demand equation for a certain product is

$$p = 50 - \sqrt{0.8(x - 1)}$$

where x is the number of units demanded per day and p is the price per unit. Find the demand if the price is $30.02.

86. *Geometry*   The surface area of a cone is given by

$$S = \pi r \sqrt{r^2 + h^2}$$

as shown in the figure. Solve this equation for h.

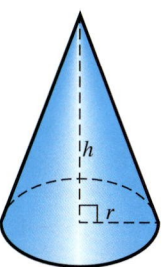

87. *Geometry*   Write a function that gives the radius r of a circle in terms of the circle's area A. Use a graphing utility to graph this function.

88. *Geometry*   Determine the length and width of a rectangle with a perimeter of 68 inches and a diagonal of length 26 inches.

89. *Reading a Graph*   An airline offers daily flights between Chicago and Denver. The total monthly cost of the flights is

$$C = \sqrt{0.2x + 1}, \quad 0 \le x$$

where C is measured in millions of dollars and x is measured in thousands of passengers (see figure). The total cost of the flights for a certain month is 2.5 million dollars. Approximately how many passengers flew that month?

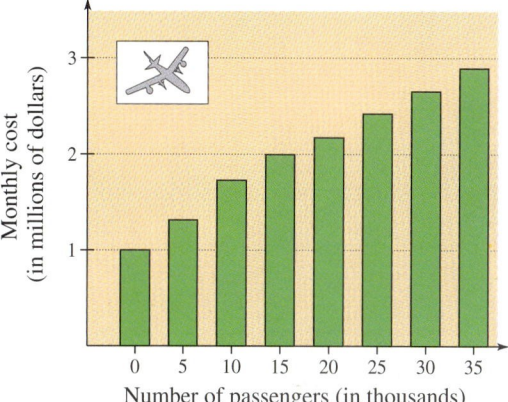

Number of passengers (in thousands)

90. *Congressional Aides*   Each member of the United States Congress has a staff of congressional aides. From 1930 through 1990, the number of aides A assigned to the members of the House of Representatives can be modeled by the function

$$A = 1220\sqrt[3]{t - 42} + 4900$$

where $t = 0$ represents 1930.

(a) Use a graphing utility to graph the function.

(b) In what year were approximately 7340 aides assigned to members of the House of Representatives?

(c) The House of Representatives has 435 members. Write a function that gives the average number of congressional aides per representative. Use a graphing utility to graph this function.

9.6 Complex Numbers

The Imaginary Unit i ▪ Complex Numbers ▪
Operations with Complex Numbers ▪ Complex Conjugates

The Imaginary Unit i

In Section 9.2, you learned that a negative number has no *real* square root. For instance, $\sqrt{-1}$ is not real because there is no real number x such that $x^2 = -1$. Thus, as long as you are dealing only with real numbers, the equation

$$x^2 = -1$$

has no solution. To overcome this deficiency, mathematicians have expanded the set of numbers, using the **imaginary unit i,** defined as

$$i = \sqrt{-1}. \qquad\qquad \text{Imaginary unit}$$

This number has the property that $i^2 = -1$. Thus, the imaginary unit i is a solution of the equation $x^2 = -1$.

The Square Root of a Negative Number

Let c be a positive real number. Then the square root of $-c$ is given by

$$\sqrt{-c} = \sqrt{c(-1)} = \sqrt{c}\sqrt{-1} = \sqrt{c}\,i.$$

When writing $\sqrt{-c}$ in the **i-form,** $\sqrt{c}\,i$, note that i is outside the radical.

DISCOVERY

Use a calculator to evaluate the following radicals. Does one result in an error message? Explain why.

a. $\sqrt{121}$

b. $\sqrt{-121}$

c. $-\sqrt{121}$

EXAMPLE 1 *Writing Numbers in i-Form*

a. $\sqrt{-36} = \sqrt{36(-1)} = \sqrt{36}\sqrt{-1} = 6i$

b. $\sqrt{-\dfrac{16}{25}} = \sqrt{\dfrac{16}{25}(-1)} = \sqrt{\dfrac{16}{25}}\sqrt{-1} = \dfrac{4}{5}i$

c. $\sqrt{-5} = \sqrt{5(-1)} = \sqrt{5}\sqrt{-1} = \sqrt{5}\,i$

d. $\sqrt{-54} = \sqrt{54(-1)} = \sqrt{54}\sqrt{-1} = 3\sqrt{6}\,i$

e. $\dfrac{\sqrt{-48}}{\sqrt{-3}} = \dfrac{\sqrt{48}\sqrt{-1}}{\sqrt{3}\sqrt{-1}} = \dfrac{\sqrt{48}\,i}{\sqrt{3}\,i} = \sqrt{\dfrac{48}{3}} = \sqrt{16} = 4$

f. $\dfrac{\sqrt{-18}}{\sqrt{2}} = \dfrac{\sqrt{18}\sqrt{-1}}{\sqrt{2}} = \dfrac{\sqrt{18}\,i}{\sqrt{2}} = \sqrt{\dfrac{18}{2}}\,i = \sqrt{9}\,i = 3i$

To perform operations with square roots of negative numbers, you must *first* write the numbers in i-form. Once the numbers are written in i-form, you add, subtract, and multiply as follows.

$$ai + bi = (a + b)i \qquad \text{Addition}$$

$$ai - bi = (a - b)i \qquad \text{Subtraction}$$

$$(ai)(bi) = ab(i^2) = ab(-1) = -ab \qquad \text{Multiplication}$$

EXAMPLE 2 *Adding Square Roots of Negative Numbers*

Perform the addition: $\sqrt{-9} + \sqrt{-49}$.

Solution

$$\sqrt{-9} + \sqrt{-49} = \sqrt{9}\sqrt{-1} + \sqrt{49}\sqrt{-1} \qquad \text{Property of radicals}$$

$$= 3i + 7i \qquad \text{Write in } i\text{-form.}$$

$$= 10i \qquad \text{Simplify.}$$

STUDY TIP

When performing operations with numbers in i-form, you sometimes need to be able to evaluate powers of the imaginary unit i. The first several powers of i are as follows.

$$i^1 = i$$

$$i^2 = -1$$

$$i^3 = i(i^2) = i(-1) = -i$$

$$i^4 = (i^2)(i^2) = (-1)(-1) = 1$$

$$i^5 = i(i^4) = i(1) = i$$

$$i^6 = (i^2)(i^4) = (-1)(1) = -1$$

$$i^7 = (i^3)(i^4) = (-i)(1) = -i$$

$$i^8 = (i^4)(i^4) = (1)(1) = 1$$

Note how the pattern of values i, -1, $-i$, and 1 repeats itself for powers greater than 4.

In the next example, notice how you can multiply radicals that involve square roots of negative numbers.

EXAMPLE 3 *Multiplying Square Roots of Negative Numbers*

a. $\sqrt{-15}\sqrt{-15} = (\sqrt{15}\,i)(\sqrt{15}\,i)$ Write in i-form.

$$= (\sqrt{15})^2 i^2 \qquad \text{Multiply.}$$

$$= 15(-1) \qquad \text{Definition of } i$$

$$= -15 \qquad \text{Simplify.}$$

b. $\sqrt{-5}(\sqrt{-45} - \sqrt{-4}) = \sqrt{5}\,i(3\sqrt{5}\,i - 2i)$ Write in i-form.

$$= (\sqrt{5}\,i)(3\sqrt{5}\,i) - (\sqrt{5}\,i)(2i) \qquad \text{Distributive Property}$$

$$= 3(5)(-1) - 2\sqrt{5}(-1) \qquad \text{Multiply.}$$

$$= -15 + 2\sqrt{5} \qquad \text{Simplify.}$$

When multiplying square roots of negative numbers, be sure to write them in i-form *before multiplying*. If you do not you can obtain incorrect answers. For instance, in Example 3(a) be sure you see that

$$\sqrt{-15}\sqrt{-15} \neq \sqrt{(-15)(-15)} = \sqrt{225} = 15.$$

Complex Numbers

A number of the form $a + bi$, where a and b are real numbers, is called a **complex number.**

Definition of a Complex Number

If a and b are real numbers, the number $a + bi$ is a **complex number,** and it is said to be written in **standard form.** If $b = 0$, the number $a + bi = a$ is a real number. If $b \neq 0$, the number $a + bi$ is called an **imaginary number.**

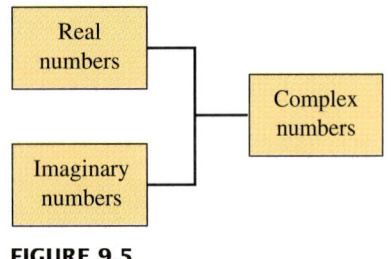

FIGURE 9.5

NOTE Two complex numbers $a + bi$ and $c + di$, in standard form, are equal if and only if $a = c$ and $b = d$.

A number cannot be both real and imaginary. For instance, the numbers -2, 0, 1, $\frac{1}{2}$, and $\sqrt{2}$ are real numbers (but they are *not* imaginary numbers), and the numbers $-3i$, $2 + 4i$, and $-1 + i$ are imaginary numbers (but they are *not* real numbers). The diagram shown in Figure 9.5 further illustrates the relationship among real, complex, and imaginary numbers.

EXAMPLE 4 *Equality of Two Complex Numbers*

a. Are the complex numbers $\sqrt{9} + \sqrt{-48}$ and $3 - 4\sqrt{3}\,i$ equal?

b. Find x and y so that the equation is valid.

$$3x - \sqrt{-25} = -6 + 3yi$$

Solution

a. Begin by writing the first number in standard form.

$$\sqrt{9} + \sqrt{-48} = \sqrt{3^2} + \sqrt{4^2(3)(-1)} = 3 + 4\sqrt{3}\,i$$

From this form, you can see that the two numbers are not equal because they have imaginary parts that differ in sign.

b. Begin by writing the left side of the equation in standard form.

$$3x - \sqrt{-25} = -6 + 3yi \qquad \text{Original equation}$$
$$3x - 5i = -6 + 3yi \qquad \text{Both sides in standard form}$$

For these two numbers to be equal, their real parts must be equal to each other and their imaginary parts must be equal to each other. Thus, $3x = -6$, which implies that $x = -2$, and $-5 = 3y$, which implies that $y = -\frac{5}{3}$.

Operations with Complex Numbers

The real number a is called the **real part** of the complex number $a + bi$, and the real number b is called the **imaginary part** of the complex number. To add or subtract two complex numbers, we add (or subtract) the real and imaginary parts separately. This is similar to combining like terms of a polynomial.

$$(a + bi) + (c + di) = (a + c) + (b + d)i \qquad \text{Addition of complex numbers}$$
$$(a + bi) - (c + di) = (a - c) + (b - d)i \qquad \text{Subtraction of complex numbers}$$

EXAMPLE 5   Adding and Subtracting Complex Numbers

a. $(3 - i) + (-2 + 4i) = (3 - 2) + (-1 + 4)i = 1 + 3i$

b. $3i + (5 - 3i) = 5 + (3 - 3)i = 5$

c. $4 - (-1 + 5i) + (7 + 2i) = [4 - (-1) + 7] + (-5 + 2)i = 12 - 3i$

Note in part (b) that the sum of two complex numbers can be a real number.

The Commutative, Associative, and Distributive Properties of real numbers are also valid for complex numbers.

EXAMPLE 6   Multiplying Complex Numbers

a. $(1 - i)\left(\sqrt{-9}\right) = (1 - i)(3i)$ Write in standard form.

$\qquad\qquad\qquad\quad = (1)(3i) - (i)(3i)$ Distributive Property

$\qquad\qquad\qquad\quad = 3i - 3(i^2)$ Simplify.

$\qquad\qquad\qquad\quad = 3i - 3(-1)$ Definition of i

$\qquad\qquad\qquad\quad = 3 + 3i$ Simplify.

b. $(2 - i)(4 + 3i) = 8 + 6i - 4i - 3i^2$ FOIL Method

$\qquad\qquad\qquad\quad = 8 + 6i - 4i - 3(-1)$ Definition of i

$\qquad\qquad\qquad\quad = 11 + 2i$ Combine like terms.

c. $(3 + 2i)(3 - 2i) = 3^2 - (2i)^2$ Special product formula

$\qquad\qquad\qquad\quad = 9 - 2^2 i^2$ Simplify.

$\qquad\qquad\qquad\quad = 9 - 4(-1)$ Definition of i

$\qquad\qquad\qquad\quad = 9 + 4$ Simplify.

$\qquad\qquad\qquad\quad = 13$ Simplify.

Until very recently it was thought that shapes in nature, such as clouds, coastlines, and mountain ranges, could not be described in mathematical terms. In the 1970s Benoit Mandelbrot discovered that many of these shapes do have a pattern in their irregularity—they are made up of smaller parts that are scaled-down versions of the shape itself. Computers using mathematical terms with complex numbers are able to generate the larger image. Mandelbrot coined the term *fractals* for these shapes and for the geometry used to describe them.

Complex Conjugates

In Example 6(c), note that the product of two complex numbers can be a real number. This occurs with pairs of complex numbers of the form $a + bi$ and $a - bi$, called **complex conjugates.** In general, the product of complex conjugates has the following form.

$$(a + bi)(a - bi) = a^2 - (bi)^2$$
$$= a^2 - b^2 i^2$$
$$= a^2 - b^2(-1)$$
$$= a^2 + b^2$$

Here are some examples.

Complex Number	Complex Conjugate	Product
$4 - 5i$	$4 + 5i$	$4^2 + 5^2 = 41$
$3 + 2i$	$3 - 2i$	$3^2 + 2^2 = 13$
$-2 = -2 + 0i$	$-2 = -2 - 0i$	$(-2)^2 + 0^2 = 4$
$i = 0 + i$	$-i = 0 - i$	$0^2 + 1^2 = 1$

Complex conjugates are used to divide one complex number by another. To do this, multiply the numerator and denominator by the *complex conjugate of the denominator,* as shown in Example 7.

EXAMPLE 7 *Division of Complex Numbers*

Perform the division and write your answer in standard form.

$$5 \div (3 - 2i)$$

Solution

$$\frac{5}{3 - 2i} = \frac{5}{3 - 2i} \cdot \frac{3 + 2i}{3 + 2i} \qquad \text{Multiply by complex conjugate of denominator.}$$

$$= \frac{5(3 + 2i)}{(3 - 2i)(3 + 2i)} \qquad \text{Multiply fractions.}$$

$$= \frac{5(3 + 2i)}{3^2 + 2^2} \qquad \text{Product of complex conjugates}$$

$$= \frac{15 + 10i}{13} \qquad \text{Simplify.}$$

$$= \frac{15}{13} + \frac{10}{13}i \qquad \text{Standard form}$$

EXAMPLE 8   *Division of Complex Numbers*

Divide $2 + 3i$ by $4 - 2i$.

Solution

$$\frac{2 + 3i}{4 - 2i} = \frac{2 + 3i}{4 - 2i} \cdot \frac{4 + 2i}{4 + 2i}$$ Multiply by complex conjugate of denominator.

$$= \frac{8 + 16i + 6i^2}{4^2 + 2^2}$$ Multiply fractions.

$$= \frac{8 + 16i + 6(-1)}{20}$$ Definition of i

$$= \frac{2 + 16i}{20}$$ Combine like terms.

$$= \frac{1}{10} + \frac{4}{5}i$$ Standard form

EXAMPLE 9   *Verifying a Complex Solution of an Equation*

Show that $x = 2 + i$ is a solution of the equation $x^2 - 4x + 5 = 0$.

Solution

$$x^2 - 4x + 5 = 0$$ Original equation

$$(2 + i)^2 - 4(2 + i) + 5 \overset{?}{=} 0$$ Substitute $2 + i$ for x.

$$4 + 4i + i^2 - 8 - 4i + 5 \overset{?}{=} 0$$ Expand.

$$i^2 + 1 \overset{?}{=} 0$$ Combine like terms.

$$(-1) + 1 \overset{?}{=} 0$$ Definition of i

$$0 = 0$$ Solution checks. ✔

Therefore, $2 + i$ is a solution of the original equation.

Group Activities   E x t e n d i n g   t h e   C o n c e p t

Prime Polynomials   The polynomial $x^2 + 1$ is prime *with respect to the integers.* It is not, however, prime *with respect to the complex numbers.* Show how $x^2 + 1$ can be factored using complex numbers.

9.6 Exercises

Discussing the Concepts

1. Define the imaginary unit i.

2. Explain why the equation $x^2 = -1$ does not have real number solutions.

3. *Error Analysis* Describe the error.

$$\sqrt{-3}\sqrt{-3} = \sqrt{(-3)(-3)} = \sqrt{9} = 3$$

4. *True or False?* Some numbers are both real and imaginary. Explain.

5. Find the product of $3 - 2i$ and its complex conjugate.

6. Describe the methods for adding, subtracting, multiplying, and dividing complex numbers.

Problem Solving

In Exercises 7–12, write the number in i-form.

7. $\sqrt{-4}$

8. $\sqrt{-9}$

9. $\sqrt{-27}$

10. $\sqrt{-\frac{8}{25}}$

11. $\sqrt{-7}$

12. $\sqrt{-15}$

In Exercises 13–20, perform the operations and write your answer in standard form.

13. $\sqrt{-16} + 6i$

14. $\sqrt{-\frac{1}{4}} - \frac{3}{2}i$

15. $\sqrt{-50} - \sqrt{-8}$

16. $\sqrt{-500} + \sqrt{-45}$

17. $\sqrt{-8}\sqrt{-2}$

18. $\sqrt{-25}\sqrt{-6}$

19. $\sqrt{-5}\left(\sqrt{-3} - \sqrt{-2}\right)$

20. $\sqrt{-4}\left(\sqrt{-9} + \sqrt{-4}\right)$

In Exercises 21–32, perform the operations and write your answer in standard form.

21. $(4 - 3i) + (6 + 7i)$

22. $22 + (-5 + 8i) + 10i$

23. $13i - (14 - 7i)$

24. $(3 - 2i)^3$

25. $15i - (3 - 25i) + \sqrt{-81}$

26. $(-1 + i) - \sqrt{2} - \sqrt{-2}$

27. $(3i)(12i)$

28. $(8i)^2$

29. $(4 + 3i)(-7 + 4i)$

30. $(7 + i)^2$

31. $\left(-3 - \sqrt{-12}\right)\left(4 - \sqrt{-12}\right)$

32. $(0.05 + 2.50i) - (6.2 + 11.8i)$

In Exercises 33–36, multiply the number by its complex conjugate.

33. $-2 - 8i$

34. $-4 + \sqrt{2}\,i$

35. $1 + \sqrt{-3}$

36. $-3 - \sqrt{-5}$

In Exercises 37–40, perform the division and write your answer in standard form.

37. $\dfrac{-12}{2 + 7i}$

38. $\dfrac{17i}{5 + 3i}$

39. $\dfrac{20}{2i}$

40. $\dfrac{4 - 5i}{4 + 5i}$

Reviewing the Major Concepts

In Exercises 41–44, simplify the expression.

41. $\sqrt{128} + 3\sqrt{50}$

42. $3\sqrt{5}\sqrt{500}$

43. $\dfrac{8}{\sqrt{10}}$

44. $\dfrac{5}{\sqrt{12} - 2}$

45. *Real Estate Taxes* The tax on a property with an assessed value of $145,000 is $2400. Find the tax on a property with an assessed value of $90,000.

46. *Gasoline Cost* A car uses 7 gallons of gas for a trip of 200 miles. How many gallons would be used on a trip of 325 miles?

Additional Problem Solving

In Exercises 47–54, write the number in i-form.

47. $-\sqrt{-144}$ **48.** $\sqrt{-49}$

49. $\sqrt{-\frac{4}{25}}$ **50.** $-\sqrt{-\frac{36}{121}}$

51. $\sqrt{-0.09}$ **52.** $\sqrt{-0.0004}$

53. $\sqrt{-8}$ **54.** $\sqrt{-75}$

In Exercises 55–58, perform the operations and write your answer in standard form.

55. $\sqrt{-18}\sqrt{-3}$ **56.** $\sqrt{-0.16}\sqrt{-1.21}$

57. $\sqrt{-3}\left(\sqrt{-3}+\sqrt{-4}\right)$ **58.** $\sqrt{-12}\left(\sqrt{-3}-\sqrt{-12}\right)$

In Exercises 59–62, determine a and b.

59. $5-4i=(a+3)+(b-1)i$

60. $-10+12i=2a+(5b-3)i$

61. $-4-\sqrt{-8}=a+bi$

62. $\sqrt{-36}-3=a+bi$

In Exercises 63–84, perform the operations and write your answer in standard form.

63. $(-4-7i)+(-10-33i)$

64. $(-10+2i)+(4-7i)$

65. $(15+10i)-(2+10i)$

66. $(-21-50i)+(21-20i)$

67. $(30-i)-(18+6i)+3i^2$

68. $(4+6i)+(15+24i)-(1-i)$

69. $(-2i)(-10i)$ **70.** $(-5i)(4i)$

71. $(-6i)(-i)(6i)$ **72.** $\frac{1}{2}(10i)(12i)(-3i)$

73. $(-3i)^3$ **74.** $(2i)^4$

75. $-5(13+2i)$ **76.** $10(8-6i)$

77. $4i(-3-5i)$ **78.** $-3i(10-15i)$

79. $(-7+7i)(4-2i)$ **80.** $(3+5i)(2+15i)$

81. $(3-4i)^2$ **82.** $(2+i)^3$

83. $\left(-2+\sqrt{-5}\right)\left(-2-\sqrt{-5}\right)$

84. $\sqrt{-9}\left(1+\sqrt{-16}\right)$

In Exercises 85–90, find the conjugate of the number. Then find the product of the number and its conjugate.

85. $2+i$ **86.** $3+2i$

87. $5-\sqrt{6}\,i$ **88.** $10-3i$

89. $10i$ **90.** 20

In Exercises 91–96, perform the division and write your answer in standard form.

91. $\dfrac{4}{1-i}$ **92.** $\dfrac{20}{3+i}$

93. $\dfrac{4i}{1-3i}$ **94.** $\dfrac{15}{2(1-i)}$

95. $\dfrac{2+3i}{1+2i}$ **96.** $\dfrac{1+i}{3i}$

In Exercises 97–100, find the sum or difference and write your answer in standard form.

97. $\dfrac{1}{1-2i}+\dfrac{4}{1+2i}$ **98.** $\dfrac{3i}{1+i}+\dfrac{2}{2+3i}$

99. $\dfrac{i}{4-3i}-\dfrac{5}{2+i}$ **100.** $\dfrac{1+i}{i}-\dfrac{3}{5-2i}$

In Exercises 101–104, decide whether each number is a solution of the equation.

101. $x^2+2x+5=0$

  (a) $x=-1+2i$ (b) $x=-1-2i$

102. $x^2-4x+13=0$

  (a) $x=2-3i$ (b) $x=2+3i$

103. $x^3+4x^2+9x+36=0$

  (a) $x=-4$ (b) $x=-3i$

104. $x^3-8x^2+25x-26=0$

  (a) $x=2$ (b) $x=3-2i$

In Exercises 105–108, perform the operation.

105. $(a+bi)+(a-bi)$ **106.** $(a+bi)(a-bi)$

107. $(a+bi)-(a-bi)$ **108.** $(a+bi)^2+(a-bi)^2$

CHAPTER PROJECT: Fractals

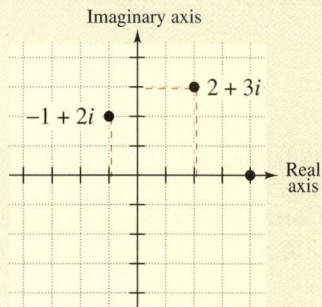

FIGURE A The Complex Plane

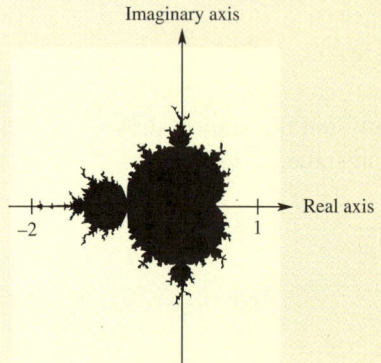

FIGURE B The Mandelbrot Set

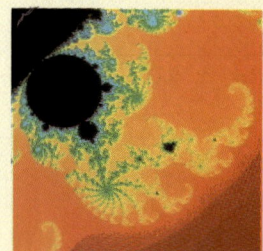

FIGURE C

A complex number $a + bi$ can be represented by a point in the *complex plane,* as shown in Figure A. In the complex plane, the horizontal axis represents the real part of the complex number and the vertical axis represents the imaginary part. For instance, points representing the complex numbers $2 + 3i$ and $-1 + 2i$ are shown in Figure A.

In the hands of a person who understands "fractal geometry," the complex plane can become an easel on which stunning pictures, called *fractals,* can be drawn. The most famous such picture is the *Mandelbrot Set,* named after the Polish-born mathematician Benoit Mandelbrot. To decide whether a complex number is in the Mandelbrot Set, consider the sequence

$$c, \quad c^2 + c, \quad (c^2 + c)^2 + c, \quad ((c^2 + c)^2 + c)^2 + c, \ldots$$

Note that each successive term is obtained by adding c to the square of the previous term. The behavior of this sequence depends on the value of c. Mandelbrot discovered that for some values of c, the sequence remains bounded—the terms of the sequence remain small. It is these values that are members of the Mandelbrot Set, which is represented by the black region in Figure B.

Use this information to investigate the following questions.

1. The table on page 321 shows that the Mandelbrot sequences for $c = -1$ and $c = i$ are bounded and the sequence for $c = 1 + i$ is not bounded. Hence, -1 and i are in the Mandelbrot Set, but $1 + i$ is not. Decide whether the following numbers are in the Mandelbrot Set.

(a) $c = 0$ (b) $c = 2$ (c) $c = \frac{1}{2}i$ (d) $c = -i$ (e) $c = 1$ (f) $c = -2$

2. The absolute value of the complex number $a + bi$ is defined as $|a + bi| = \sqrt{a^2 + b^2}$. If the absolute value of any term in the Mandelbrot sequence exceeds 2, the terms will eventually become unbounded. To add interest to the graph of the Mandelbrot Set, computer scientists discovered that the points that are not in the set can be assigned a variety of colors, as shown in Figure C. A color is assigned to a number c depending on how quickly its Mandelbrot sequence reaches a term whose absolute value is greater than 2. For each item below, find a complex number c such that its Mandelbrot sequence first exceeds an absolute value of 2 in the indicated term.

(a) First term, c

(b) Second term, $c^2 + c$

(c) Third term, $(c^2 + c)^2 + c$

(d) Fourth term, $((c^2 + c)^2 + c)^2 + c$

(e) Fifth term, $(((c^2 + c)^2 + c)^2 + c)^2 + c$

3. ***Research Project*** Use your school's library or some other reference source to find other information about the Mandelbrot Set or about other fractals. Write a paper that describes your findings. Include illustrations in your paper.

CHAPTER SUMMARY

After studying this chapter, you should have acquired the following skills. These skills are keyed to the Review Exercises that begin on page 570. Answers to odd-numbered Review Exercises are given in the back of the book.

- Evaluate numerical expressions involving integer exponents and scientific notation. *(Section 9.1)*   **Review Exercises 1–8**

- Simplify algebraic expressions involving integer exponents. *(Section 9.1)*   **Review Exercises 9–16**

- Evaluate numerical expressions involving integer exponents or radicals. *(Section 9.2)*   **Review Exercises 17–26**

- Translate expressions from radical form to rational exponent form and vice versa. *(Section 9.2)*   **Review Exercises 27–30**

- Evaluate numerical expressions involving rational exponents. *(Section 9.2)*   **Review Exercises 31–36**

- Simplify expressions involving rational exponents or radicals. *(Section 9.2)*   **Review Exercises 37–46**

- Simplify expressions involving radicals by rationalizing the denominator. *(Section 9.3)*   **Review Exercises 47–54**

- Add, subtract, multiply, and divide expressions involving radicals. *(Sections 9.3, 9.4)*   **Review Exercises 55–60**

- Graph functions using a graphing utility. *(Sections 9.2, 9.3, 9.4)*   **Review Exercises 61–64**

- Solve radical equations. *(Section 9.5)*   **Review Exercises 65–74**

- Approximate solutions of equations using a graphing utility. *(Section 9.5)*   **Review Exercises 75, 76**

- Write complex numbers in standard form. *(Section 9.6)*   **Review Exercises 77–82**

- Add, subtract, multiply, and divide complex numbers. *(Section 9.6)*   **Review Exercises 83–92**

- Solve problems using geometry. *(Sections 9.3, 9.5)*   **Review Exercises 93, 94**

- Solve real-life problems modeled by radical equations. *(Section 9.5)*   **Review Exercises 95, 96**

REVIEW EXERCISES

In Exercises 1–8, evaluate the expression.

1. $(2^3 \cdot 3^2)^{-1}$

2. $(2^{-2} \cdot 5^2)^{-2}$

3. $\left(\dfrac{2}{5}\right)^{-3}$

4. $\left(\dfrac{1}{3^{-2}}\right)^2$

5. $(6 \times 10^3)^2$

6. $(3 \times 10^{-3})(8 \times 10^7)$

7. $\dfrac{3.5 \times 10^7}{7 \times 10^4}$

8. $\dfrac{1}{(6 \times 10^{-3})^2}$

In Exercises 9–16, simplify the expression.

9. $\dfrac{4x^2}{2x}$

10. $4(-3x)^3$

11. $(x^3 y^{-4})^2$

12. $5yx^0$

13. $\dfrac{t^{-5}}{t^{-2}}$

14. $\dfrac{a^5 \cdot a^{-3}}{a^{-2}}$

15. $\left(\dfrac{y}{3}\right)^{-3}$

16. $(2x^2 y^4)^4 (2x^2 y^4)^{-4}$

In Exercises 17–22, evaluate the expression.

17. $\sqrt{1.44}$

18. $\sqrt{0.16}$

19. $\sqrt{\dfrac{25}{36}}$

20. $-\sqrt{\dfrac{64}{225}}$

21. $\sqrt{169 - 25}$

22. $\sqrt{16 + 9}$

In Exercises 23–26, evaluate the expression. Round the result to two decimal places.

23. $1800(1 + 0.08)^{24}$

24. $0.0024(7,658,400)$

25. $\sqrt{13^2 - 4(2)(7)}$

26. $\dfrac{-3.7 + \sqrt{15.8}}{2(2.3)}$

In Exercises 27–30, fill in the missing description.

	Radical Form	*Rational Exponent Form*
27.	$\sqrt{49} = 7$	
28.	$\sqrt[3]{0.125} = 0.5$	
29.		$216^{1/3} = 6$
30.		$16^{1/4} = 2$

In Exercises 31–34, evaluate the expression.

31. $27^{4/3}$

32. $16^{3/4}$

33. $25^{3/2}$

34. $243^{-2/5}$

In Exercises 35 and 36, evaluate the expression. Round the result to two decimal places.

35. $75^{-3/4}$

36. $510^{5/3}$

In Exercises 37–46, simplify the expression.

37. $x^{3/4} \cdot x^{-1/6}$

38. $(2y^2)^{3/2} (2y^{-4})^{1/2}$

39. $\dfrac{15x^{1/4} y^{3/5}}{5x^{1/2} y}$

40. $\dfrac{48a^2 b^{5/2}}{14a^{-3} b^{-1/2}}$

41. $\sqrt{360}$

42. $\sqrt{\dfrac{50}{9}}$

43. $\sqrt{0.25x^4 y}$

44. $\sqrt{0.16s^6 t^3}$

45. $\sqrt[3]{48a^3 b^4}$

46. $\sqrt[4]{32u^4 v^5}$

In Exercises 47–54, rationalize the denominator and simplify further when possible.

47. $\sqrt{\dfrac{5}{6}}$

48. $\sqrt{\dfrac{3}{20}}$

49. $\dfrac{3}{\sqrt{12x}}$

50. $\dfrac{4y}{\sqrt{10z}}$

51. $\dfrac{2}{\sqrt[3]{2x}}$

52. $\sqrt[3]{\dfrac{16t}{s^2}}$

53. $\dfrac{6}{7 - \sqrt{7}}$

54. $\dfrac{x}{\sqrt{x} + 1}$

In Exercises 55–60, perform the operations and simplify.

55. $3\sqrt{40} - 10\sqrt{90}$

56. $9\sqrt{50} - 5\sqrt{8} + \sqrt{48}$

57. $\sqrt{25x} + \sqrt{49x} - \sqrt{x}$

58. $(\sqrt{5} + 6)^2$

59. $(3 - \sqrt{x})(3 + \sqrt{x})$

60. $15 \div (\sqrt{x} + 3)$

In Exercises 61–64, use a graphing utility to graph the function.

61. $y = 3\sqrt[3]{2x}$

62. $y = \dfrac{10}{\sqrt[4]{x^2+1}}$

63. $g(x) = 4x^{3/4}$

64. $h(x) = \frac{1}{2}x^{4/3}$

In Exercises 65–74, solve the equation.

65. $\sqrt{y} = 15$

66. $\sqrt{3x} - 9 = 0$

67. $\sqrt{2(a-7)} = 14$

68. $\sqrt{3(2x+3)} = \sqrt{x+15}$

69. $\sqrt{2(x+5)} = x + 5$

70. $\sqrt{5t} = 1 + \sqrt{5(t-1)}$

71. $\sqrt[3]{5x+2} - \sqrt[3]{7x-8} = 0$

72. $\sqrt[4]{9x-2} - \sqrt[4]{8x} = 0$

73. $\sqrt{1+6x} = 2 - \sqrt{6x}$

74. $\sqrt{2+9b} - 1 = 3\sqrt{b}$

In Exercises 75 and 76, use a graphing utility to approximate the solution of the equation.

75. $4\sqrt[3]{x} = 7 - x$

76. $2\sqrt{x} - 1 = \sqrt{10-x}$

In Exercises 77–82, write the complex number in standard form.

77. $\sqrt{-48}$

78. $\sqrt{-0.16}$

79. $10 - 3\sqrt{-27}$

80. $3 + 2\sqrt{-500}$

81. $\frac{3}{4} - 5\sqrt{-\frac{3}{25}}$

82. $-0.5 + 3\sqrt{-1.21}$

In Exercises 83–92, perform the operation and write the answer in standard form.

83. $(-4+5i) - (-12+8i)$

84. $5(3-8i) + (5+12i)$

85. $(-2)(15i)(-3i)$

86. $-10i(4-7i)$

87. $(4-3i)(4+3i)$

88. $(6-5i)^2$

89. $(12-5i)(2+7i)$

90. $\dfrac{4}{5i}$

91. $\dfrac{5i}{2+9i}$

92. $\dfrac{2+i}{1-9i}$

93. *Geometry*   The four corners are cut from an $8\frac{1}{2}$-inch-by-11-inch sheet of paper (see figure). Find the perimeter of the remaining piece of paper.

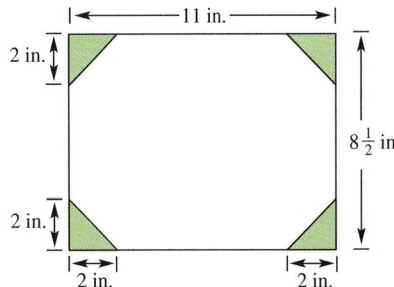

94. *Geometry*   Determine the length and width of a rectangle with a perimeter of 84 inches and a diagonal of 30 inches.

95. *Length of a Pendulum*   The time t in seconds for a pendulum of length L in feet to go through one complete cycle (its period) is given by

$$t = 2\pi\sqrt{\dfrac{L}{32}}.$$

How long is the pendulum of a grandfather clock with a period of 1.3 seconds?

96. *Height of a Bridge*   The time t in seconds for a free-falling object to fall d feet is given by

$$t = \sqrt{\dfrac{d}{16}}.$$

A child drops a pebble from a bridge and observes it strike the water after approximately 4 seconds. Estimate the height of the bridge.

CHAPTER TEST

Take this test as you would take a test in class. After you are done, check your work against the answers given in the back of the book.

In Exercises 1 and 2, evaluate the expressions without using a calculator.

1. (a) $2^{-2} + 2^{-3}$

(b) $\dfrac{6.3 \times 10^{-3}}{2.1 \times 10^2}$

2. (a) $27^{-2/3}$

(b) $\sqrt{2}\sqrt{18}$

3. Write 0.000032 in scientific notation.

4. Write 3.04×10^7 in decimal notation.

In Exercises 5–7, simplify the expressions.

5. (a) $\dfrac{12t^{-2}}{20t^{-1}}$

(b) $(x + y^{-2})^{-1}$

6. (a) $\left(\dfrac{x^{1/2}}{x^{1/3}}\right)^2$

(b) $5^{1/4} \cdot 5^{7/4}$

7. (a) $\sqrt{\dfrac{32}{9}}$

(b) $\sqrt[3]{24}$

8. In your own words, explain the meaning of "rationalize" and demonstrate by rationalizing the denominator: $\dfrac{3}{\sqrt{6}}$

9. Combine: $5\sqrt{3x} - 3\sqrt{75x}$

10. Multiply and simplify: $\sqrt{5}\left(\sqrt{15x} + 3\right)$ **11.** Expand: $\left(4 - \sqrt{2x}\right)^2$

12. Factor: $7\sqrt{27} + 14y\sqrt{12} = 7\sqrt{3}(\ \ \ \ \ \ )$

In Exercises 13 and 14, solve the equation.

13. $\sqrt{x^2 - 1} = x - 2$

14. $\sqrt{x} - x + 6 = 0$

In Exercises 15–18, perform the operation and simplify.

15. $(2 + 3i) - \sqrt{-25}$

16. $(2 - 3i)^2$

17. $\sqrt{-16}\left(1 + \sqrt{-4}\right)$

18. $(3 - 2i)(1 + 5i)$

19. Divide $5 - 2i$ by i. Write the result in standard form.

20. The velocity v (in feet per second) of an object is given by $v = \sqrt{2gh}$, where $g = 32$ feet per second per second and h is the distance (in feet) the object has fallen. Find the height from which a rock has been dropped if it strikes the ground with a velocity of 80 feet per second.

Quadratic Equations and Inequalities

10

- Factoring and Extracting Roots
- Completing the Square
- The Quadratic Formula and the Discriminant
- Applications of Quadratic Equations
- Quadratic and Rational Inequalities

The height s (in feet) of a falling object can be modeled by

$$s = \tfrac{1}{2}gt^2 + v_0 t + s_0 \qquad \text{Position function}$$

where g is the acceleration due to gravity (in feet per second per second), t is the time (in seconds), v_0 is the initial velocity (in feet per second), and s_0 is the initial height (in feet). On earth, $g \approx -32$ feet per second per second. On other planets and moons, the value of g varies significantly. For instance, on earth's moon, $g \approx -5.3$.

The position function for an object that is propelled at a velocity of 32 feet per second straight up from a height of 48 feet is

$$s = -16t^2 + 32t + 48.$$

The height of this object is shown in the table and graph at the right.

t	0.0	0.5	1.0	1.5	2.0	2.5	3.0
s	48	60	64	60	48	28	0

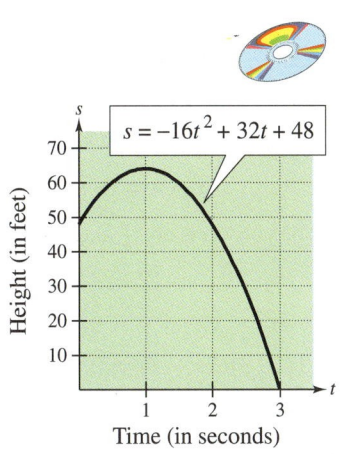

$s = -16t^2 + 32t + 48$

Height (in feet)

Time (in seconds)

The chapter project related to this information is on page 620.

573

10.1 | Factoring and Extracting Roots

Solving Equations by Factoring ▪ Extracting Square Roots ▪ Equations with Complex Solutions ▪ Equations of Quadratic Form

Solving Equations by Factoring

In this chapter, you will study methods for solving quadratic equations and equations of quadratic form. To begin, let's review the method of factoring that you studied in Section 6.5.

Remember that the first step in solving a quadratic equation by factoring is to write the equation in standard form. Next, factor the left side. Finally, set each factor equal to zero and solve for x.

Fermat's Last Theorem states that the equation $x^n + y^n = z^n$ has no solution when x, y, and z are nonzero integers and $n > 2$. In 1637, Pierre de Fermat wrote in the margin of a book that he had discovered a proof of this theorem; however, his proof has never been found. On June 23, 1993, 356 years later, a 200-page proof was presented at a gathering of mathematicians at Cambridge University in England by an American mathematician, Andrew Wiles.

NOTE When the two solutions of a quadratic equation are identical, they are called a **double** or **repeated solution.** This occurs in Example 1(c).

EXAMPLE 1 Solving Quadratic Equations by Factoring

a.

$$x^2 + 5x = 24 \qquad \text{Original equation}$$
$$x^2 + 5x - 24 = 0 \qquad \text{Standard form}$$
$$(x + 8)(x - 3) = 0 \qquad \text{Factor.}$$
$$x + 8 = 0 \quad \Longrightarrow \quad x = -8 \qquad \text{Set 1st factor equal to 0.}$$
$$x - 3 = 0 \quad \Longrightarrow \quad x = 3 \qquad \text{Set 2nd factor equal to 0.}$$

The solutions are -8 and 3. Check these in the original equation.

b.

$$3x^2 = 4 - 11x \qquad \text{Original equation}$$
$$3x^2 + 11x - 4 = 0 \qquad \text{Standard form}$$
$$(3x - 1)(x + 4) = 0 \qquad \text{Factor.}$$
$$3x - 1 = 0 \quad \Longrightarrow \quad x = \tfrac{1}{3} \qquad \text{Set 1st factor equal to 0.}$$
$$x + 4 = 0 \quad \Longrightarrow \quad x = -4 \qquad \text{Set 2nd factor equal to 0.}$$

The solutions are $\tfrac{1}{3}$ and -4. Check these in the original equation.

c.

$$9x^2 + 12 = 3 + 12x + 5x^2 \qquad \text{Original equation}$$
$$4x^2 - 12x + 9 = 0 \qquad \text{Standard form}$$
$$(2x - 3)(2x - 3) = 0 \qquad \text{Factor.}$$
$$2x - 3 = 0 \qquad \text{Set factor equal to 0.}$$
$$x = \frac{3}{2} \qquad \text{Repeated solution}$$

The only solution is $\tfrac{3}{2}$. Check this in the original equation.

Extracting Square Roots

Consider the following equation, where $d > 0$ and u is an algebraic expression.

$$u^2 = d \qquad\qquad\qquad \text{Original equation}$$
$$u^2 - d = 0 \qquad\qquad\qquad \text{Standard form}$$
$$\left(u + \sqrt{d}\right)\left(u - \sqrt{d}\right) = 0 \qquad\qquad\qquad \text{Factor.}$$
$$u + \sqrt{d} = 0 \quad\Longrightarrow\quad u = -\sqrt{d} \qquad \text{Set 1st factor equal to 0.}$$
$$u - \sqrt{d} = 0 \quad\Longrightarrow\quad u = \sqrt{d} \qquad \text{Set 2nd factor equal to 0.}$$

Because the solutions differ only in sign, they can be written together using a "plus or minus sign"

$$u = \pm\sqrt{d}.$$

This form of the solution is read as "u is equal to plus or minus the square root of d." Solving an equation of the form $u^2 = d$ *without* going through the steps of factoring is called **extracting square roots.**

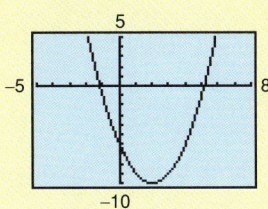

Extracting Square Roots

The equation $u^2 = d$, where $d > 0$, has exactly two solutions:

$$u = \sqrt{d} \qquad \text{and} \qquad u = -\sqrt{d}.$$

These solutions can also be written as $u = \pm\sqrt{d}$.

EXAMPLE 2   *Extracting Square Roots*

a. $3x^2 = 15$ Original equation

$\quad\quad x^2 = 5$ Divide both sides by 3.

$\quad\quad x = \pm\sqrt{5}$ Extract square roots.

The solutions are $\sqrt{5}$ and $-\sqrt{5}$. Check these in the original equation.

b. $(x - 2)^2 = 10$ Original equation

$\quad\quad x - 2 = \pm\sqrt{10}$ Extract square roots.

$\quad\quad\quad x = 2 \pm \sqrt{10}$ Add 2 to both sides.

The solutions are $2 + \sqrt{10} \approx 5.16$ and $2 - \sqrt{10} \approx -1.16$. Check these in the original equation.

Equations with Complex Solutions

Prior to Section 9.6, the only solutions you have been finding have been real numbers. But now that you have studied complex numbers, it makes sense to look for other types of solutions. For instance, although the quadratic equation $x^2 + 1 = 0$ has no solutions that are real numbers, it does have two solutions that are complex numbers: i and $-i$. To check this, substitute i and $-i$ for x.

$$(i)^2 + 1 = -1 + 1 = 0 \qquad \text{Solution checks} \checkmark$$
$$(-i)^2 + 1 = -1 + 1 = 0 \qquad \text{Solution checks} \checkmark$$

One way to find complex solutions of a quadratic equation is to extend the *extraction of square roots* technique to cover the case where d is a negative number.

Extracting Complex Square Roots

The equation $u^2 = d$, where $d < 0$, has exactly two solutions:

$$u = \sqrt{|d|}\, i \qquad \text{and} \qquad u = -\sqrt{|d|}\, i.$$

These solutions can also be written as $u = \pm\sqrt{|d|}\, i$.

Technology

When graphically checking solutions, it is important to realize that only the real solutions appear as x-intercepts—the complex solutions cannot be estimated from the graph. For instance, in Example 3(a), the graph of

$$y = x^2 + 8$$

has no x-intercepts. This agrees with the fact that both of its solutions are complex numbers.

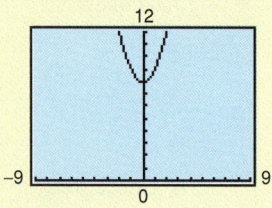

EXAMPLE 3   Extracting Complex Square Roots

a.

$x^2 + 8 = 0$	Original equation
$x^2 = -8$	Subtract 8 from both sides.
$x = \pm\sqrt{8}\, i$	Extract complex square roots.
$x = \pm 2\sqrt{2}\, i$	Simplify.

The solutions are $2\sqrt{2}\, i$ and $-2\sqrt{2}\, i$. Check these in the original equation.

b.

$2(3x - 5)^2 + 32 = 0$	Original equation
$2(3x - 5)^2 = -32$	Subtract 32 from both sides.
$(3x - 5)^2 = -16$	Divide both sides by 2.
$3x - 5 = \pm 4i$	Extract complex square roots.
$3x = 5 \pm 4i$	Add 5 to both sides.
$x = \dfrac{5 \pm 4i}{3}$	Divide both sides by 3.

The solutions are $(5 + 4i)/3$ and $(5 - 4i)/3$. Check these in the original equation.

Equations of Quadratic Form

Both the factoring and extraction of square roots methods can be applied to non-quadratic equations that are of **quadratic form.** An equation is said to be of quadratic form if it has the form

$$au^2 + bu + c = 0$$

where u is an algebraic expression. Here are two examples.

Equation	Written in Quadratic Form
$x^4 + 5x^2 + 4 = 0$	$\left(x^2\right)^2 + 5\left(x^2\right) + 4 = 0$
$x - 5\sqrt{x} + 6 = 0$	$\left(\sqrt{x}\right)^2 - 5\left(\sqrt{x}\right) + 6 = 0$

To solve an equation of quadratic form, it helps to make a substitution and rewrite the equation in terms of u, as demonstrated in Examples 4 and 5.

EXAMPLE 4   Solving an Equation of Quadratic Form

Solve $x^4 - 13x^2 + 36 = 0$.

Solution

Begin by using a graphing utility to graph $y = x^4 - 13x^2 + 36$, as shown in Figure 10.1. The graph indicates that there are four real solutions near $x = \pm 2$ and $x = \pm 3$.

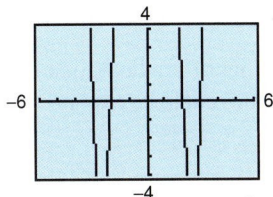

FIGURE 10.1

$$x^4 - 13x^2 + 36 = 0 \qquad \text{Original equation}$$

$$\left(x^2\right)^2 - 13\left(x^2\right) + 36 = 0 \qquad \text{Write in quadratic form.}$$

$$u^2 - 13u + 36 = 0 \qquad \text{Substitute } u \text{ for } x^2.$$

$$(u - 4)(u - 9) = 0 \qquad \text{Factor.}$$

$$u - 4 = 0 \implies u = 4 \qquad \text{Set 1st factor equal to 0.}$$

$$u - 9 = 0 \implies u = 9 \qquad \text{Set 2nd factor equal to 0.}$$

At this point you have found the "u-solutions." To find the "x-solutions," replace u by x^2, as follows.

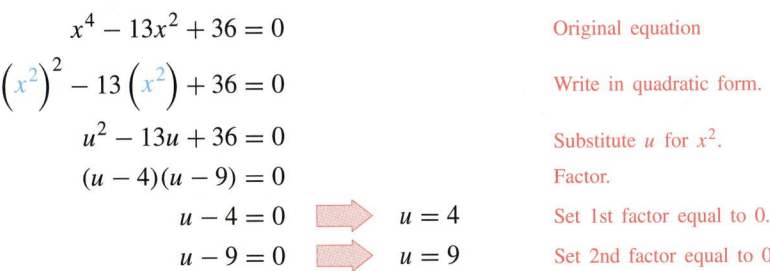

$$u = 4 \implies x^2 = 4 \implies x = \pm 2$$

$$u = 9 \implies x^2 = 9 \implies x = \pm 3$$

The solutions are 2, -2, 3, and -3. Check these in the original equation.

NOTE   Be sure you see in Example 4 that the u-solutions of 4 and 9 represent only a temporary step. They are not solutions of the original equation.

EXAMPLE 5 Solving an Equation of Quadratic Form

Solve $x - 5\sqrt{x} + 6 = 0$ for x.

Solution

This equation is of quadratic form with $u = \sqrt{x}$.

$x - 5\sqrt{x} + 6 = 0$	Original equation
$\left(\sqrt{x}\right)^2 - 5\left(\sqrt{x}\right) + 6 = 0$	Write in quadratic form.
$u^2 - 5u + 6 = 0$	Substitute u for $\sqrt{x}$.
$(u - 2)(u - 3) = 0$	Factor.
$u - 2 = 0 \implies u = 2$	Set 1st factor equal to 0.
$u - 3 = 0 \implies u = 3$	Set 2nd factor equal to 0.

Now, using the u-solutions of 2 and 3, you obtain the following x-solutions.

$$u = 2 \implies \sqrt{x} = 2 \implies x = 4$$

$$u = 3 \implies \sqrt{x} = 3 \implies x = 9$$

The solutions are 4 and 9. Check these in the original equation.

NOTE In Example 5, remember that checking the solutions of a radical equation is especially important because the trial solutions often turn out to be extraneous.

Group Activities

Exploring with Technology

Analyzing Solutions of Quadratic Equations Use a graphing utility to graph each of the following equations. How many times does the graph of each equation cross the x-axis?

a. $y = 2x^2 + x - 15$ **c.** $y = 9x^2 + 24x + 16$

b. $y = -4(x - 2)^2 - 3$ **d.** $y = (3x - 1)^2 + 3$

Now set each equation equal to zero and use the techniques of this section to solve the resulting equations. How many of each type of solution (real or complex) does each equation have? Summarize the relationship between the number of x-intercepts in the graph of a quadratic equation and the number and type of roots found algebraically.

10.1 Exercises

Discussing the Concepts

1. For a quadratic equation $ax^2 + bx + c = 0$ where a, b, and c are real numbers with $a \neq 0$, explain why b and c can equal 0 but a cannot.

2. Explain the Zero-Factor Property and how it can be used to solve a quadratic equation.

3. Is it possible for a quadratic equation to have only one solution? If so, give an example.

4. *True or False?* The only solution of the equation $x^2 = 25$ is $x = 5$. Explain.

5. Describe the steps in solving a quadratic equation by extracting square roots.

6. Describe the procedure for solving an equation of quadratic form. Give an example.

Problem Solving

In Exercises 7–16, solve the equation by factoring.

7. $4x^2 - 12x = 0$

8. $25y^2 - 75y = 0$

9. $x^2 - 12x + 36 = 0$

10. $9x^2 + 24x + 16 = 0$

11. $(y - 4)(y - 3) = 6$

12. $(6 + u)(1 - u) = 10$

13. $3x(x - 6) - 5(x - 6) = 0$

14. $3(4 - x) - 2x(4 - x) = 0$

15. $6x^2 = 54$

16. $\dfrac{x^2}{6} = 24$

In Exercises 17–22, solve the quadratic equation by extracting square roots.

17. $x^2 = 64$

18. $9z^2 = 121$

19. $(x + 4)^2 = 169$

20. $(y - 20)^2 = 625$

21. $(2x + 1)^2 = 50$

22. $(3x - 5)^2 = 48$

In Exercises 23–28, solve the equation by extracting complex square roots.

23. $x^2 + 4 = 0$

24. $x^2 = -9$

25. $9(x + 6)^2 = -121$

26. $4(x - 4)^2 = -169$

27. $(x - 1)^2 = -27$

28. $\left(y - \dfrac{5}{6}\right)^2 = -\dfrac{4}{5}$

In Exercises 29–34, find all real and complex solutions.

29. $2x^2 - 5x = 0$

30. $3x^2 + 8x - 16 = 0$

31. $x^2 - 100 = 0$

32. $(y + 12)^2 + 400 = 0$

33. $(x - 5)^2 + 100 = 0$

34. $(y + 12)^2 - 400 = 0$

In Exercises 35–42, solve the equation. List all real and complex solutions.

35. $x^4 + 7x^2 - 8 = 0$

36. $x^4 - 13x^2 + 36 = 0$

37. $2x - 9\sqrt{x} + 10 = 0$

38. $3x^{2/3} + 8x^{1/3} - 3 = 0$

39. $(x^2 - 2)^2 - 36 = 0$

40. $(u^2 + 4)^2 - 64 = 0$

41. $(x^2 - 5)^2 - 100 = 0$

42. $(y^2 + 12)^2 - 400 = 0$

In Exercises 43–46, use a graphing utility to graph the function. Use the graph to approximate any x-intercepts of the graph. Set $y = 0$ and solve the resulting equation. Compare the result with the x-intercepts of the graph.

43. $y = x^2 - 9$

44. $y = x^2 - 2x - 15$

45. $y = 4 - (x - 3)^2$

46. $y = 4(x + 1)^2 - 9$

In Exercises 47–50, use a graphing utility to graph the function and observe that the graph has no x-intercepts. Set $y = 0$ and solve the resulting equation. What type of roots does the equation have?

47. $y = (x - 1)^2 + 1$

48. $y = (x + 2)^2 + 3$

49. $y = -(x + 3)^2 - 2$

50. $y = -(x - 4)^2 - 4$

Graphical Reasoning   In Exercises 51 and 52, use the model

$$y = (44.17 + 2.82t)^2, \quad 0 \le t \le 22$$

which approximates the amount of fire loss to private property in the United States from 1970 to 1992. In this model, y represents the value of private property lost to fire (in millions of dollars) and t represents the year, with $t = 0$ corresponding to 1970 (see figure).   (Source: Insurance Information Institute)

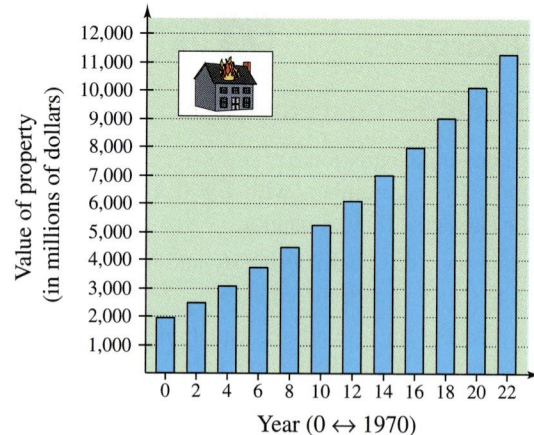

Year (0 ↔ 1970)

51. Graphically estimate the year in which loss of private property due to fire was approximately $4500 million. Verify algebraically.

52. Graphically estimate the year in which loss of private property due to fire was approximately $9500 million. Verify algebraically.

Reviewing the Major Concepts

In Exercises 53–56, factor the expression completely.

53. $16x^2 - 121$

54. $9t^2 - 24t + 16$

55. $x(x - 10) - 4(x - 10)$

56. $4x^3 - 12x^2 + 16x$

57. *Speed*   A boat's still-water speed is 18 miles per hour. It travels 35 miles upstream, and then returns to its starting point, in a total of 4 hours. Find the speed of the current.

58. *Partnership Costs*   A group of people agree to share equally in the cost of a $250,000 endowment to a college. If they could find two more people to join the group, each person's share of the cost would decrease by $6250. How many people are presently in the group?

Additional Problem Solving

In Exercises 59–68, solve the equation by factoring.

59. $u(u - 9) - 12(u - 9) = 0$

60. $16x(x - 8) - 12(x - 8) = 0$

61. $4x^2 - 25 = 0$

62. $16y^2 - 121 = 0$

63. $x^2 + 10x + 600 = 0$

64. $8x^2 - 10x + 3 = 0$

65. $2x(3x + 2) = 5 - 6x^2$

66. $(2z + 1)(2z - 1) = -4z^2 - 5z + 2$

67. $\frac{1}{2}y^2 = 32$

68. $5t^2 = 125$

In Exercises 69–78, solve the quadratic equation by extracting square roots.

69. $25x^2 = 16$

70. $z^2 = 169$

71. $4u^2 - 225 = 0$

72. $16x^2 - 1 = 0$

73. $(x - 3)^2 = 0.25$

74. $(x + 2)^2 = 0.81$

75. $(x - 2)^2 = 7$

76. $(x + 8)^2 = 28$

77. $(4x - 3)^2 - 98 = 0$

78. $(5x + 11)^2 - 300 = 0$

In Exercises 79–88, solve the equation by extracting complex square roots.

79. $u^2 + 17 = 0$

80. $4v^2 + 9 = 0$

81. $(t - 3)^2 = -25$

82. $(x + 5)^2 + 81 = 0$

83. $(2y - 3)^2 + 25 = 0$

84. $(3z + 4)^2 + 144 = 0$

85. $\left(c - \frac{2}{3}\right)^2 + \frac{1}{9} = 0$

86. $\left(u + \frac{5}{8}\right)^2 + \frac{49}{16} = 0$

87. $\left(x + \frac{7}{3}\right)^2 = -\frac{38}{9}$

88. $(2x + 3)^2 = -54$

In Exercises 89–98, find all the real and complex solutions.

89. $x^2 - 900 = 0$

90. $y^2 - 225 = 0$

91. $x^2 + 900 = 0$

92. $y^2 + 225 = 0$

93. $\frac{2}{3}x^2 = 6$

94. $\frac{1}{3}x^2 = 4$

95. $(x - 5)^2 - 100 = 0$

96. $(y + 12)^2 - 400 = 0$

97. $(x - 5)^2 + 100 = 0$

98. $(y + 12)^2 + 400 = 0$

In Exercises 99–114, solve the equation. List all real and complex solutions.

99. $x^4 - 5x^2 + 4 = 0$

100. $4x^4 - 101x^2 + 25 = 0$

101. $x^4 - 5x^2 + 6 = 0$

102. $x^4 - 11x^2 + 30 = 0$

103. $x^4 - 3x^2 - 4 = 0$

104. $x^4 - x^2 - 6 = 0$

105. $(x^2 - 4)^2 + 2(x^2 - 4) - 3 = 0$

106. $(x^2 - 1)^2 + (x^2 - 1) - 6 = 0$

107. $\left(\sqrt{x} - 1\right)^2 + 3\left(\sqrt{x} - 1\right) - 4 = 0$

108. $\left(2 + \sqrt{x}\right)^2 + 4\left(2 + \sqrt{x}\right) - 21 = 0$

109. $\frac{1}{x^2} - \frac{3}{x} + 2 = 0$

110. $\frac{2}{x^2} + \frac{3}{x} - 2 = 0$

111. $3\left(\frac{x}{x+1}\right)^2 + 7\left(\frac{x}{x+1}\right) - 6 = 0$

112. $4\left(\frac{x+1}{x-1}\right)^2 + 19\left(\frac{x+1}{x-1}\right) - 5 = 0$

113. $x^{2/3} - x^{1/3} - 6 = 0$

114. $2x^{2/3} - 7x^{1/3} + 5 = 0$

In Exercises 115 and 116, use a graphing utility to graph the function. Use the graph to approximate any x-intercepts of the graph. Set $y = 0$ and solve the resulting equation. Compare the result with the x-intercepts of the graph.

115. $y = 5x - x^2$

116. $y = 9 - 4(x - 3)^2$

In Exercises 117 and 118, use a graphing utility to graph the function and observe that the graph has no x-intercepts. Set $y = 0$ and solve the resulting equation. What type of roots does the equation have?

117. $y = (x - 2)^2 + 3$

118. $y = (x + 3)^2 + 5$

Think About It In Exercises 119–122, find a quadratic equation having the given solutions.

119. $5, -2$

120. $-2, \frac{1}{3}$

121. $1 + \sqrt{2}, 1 - \sqrt{2}$

122. $1 + \sqrt{2}\,i, 1 - \sqrt{2}\,i$

Free-Falling Object In Exercises 123 and 124, find the time required for an object to reach the ground when it is dropped from a height of s_0 feet. The height h (in feet) is given by

$$h = -16t^2 + s_0$$

where t measures time in seconds from the time the object is released.

123. $s_0 = 256$

124. $s_0 = 48$

125. *Free-Falling Object* The height h (in feet) of an object thrown upward from a tower 144 feet high (see figure) is given by

$$h = 144 + 128t - 16t^2$$

where t measures the time in seconds from the time the object is released. How long does it take for the object to reach the ground?

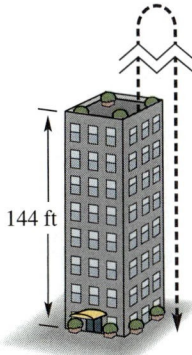

144 ft

126. *Revenue* The revenue R (in dollars) when x units of a product are sold is given by

$$R = x\left(120 - \frac{1}{2}x\right).$$

Determine the number of units that must be sold to produce a revenue of $7000.

Reading a Graph In Exercises 127 and 128, use the following model, which gives the federal funding for health research in the United States from 1985 to 1992.

$$y = (52.9 + 3.9t)^2, \quad 5 \le t \le 12$$

In this model, y represents funding (in millions of dollars) and $t = 0$ represents 1980 (see figure). (Source: U.S. National Science Foundation)

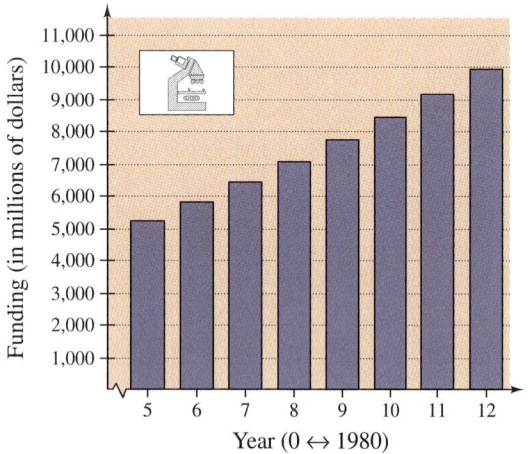

Year (0 ↔ 1980)

127. Use the graph to find the year in which federal funding for health was approximately $6500 million. Verify algebraically.

128. Use the graph to find the year in which federal funding for health was approximately $9200 million. Verify algebraically.

129. *Compound Interest* A principal of $1500 is deposited in an account at an annual interest rate r compounded annually. If the amount after 2 years is $1685.40, the annual interest rate is the solution to the equation $1685.40 = 1500(1 + r)^2$. Find r.

Free-Falling Object In Exercises 130 and 131, find the time required for an object to reach the ground when it is dropped from a height of s_0 feet. The height h (in feet) is given by

$$h = -16t^2 + s_0$$

where t measures time in seconds from the time the object is released.

130. $s_0 = 128$

131. $s_0 = 500$

CAREER INTERVIEW

Peggy Murray

Freelance Artist

HomeArt

Mission Hills, KS 66208

I am self-employed as a freelance artist and work out of my studio in my home on a wide variety of projects including greeting card and announcement designs; pen-and-ink house portraits; finished art, such as flyers and posters; and hand-painted Christmas tree ornaments. Recently I designed and produced a local historical society cookbook. All of my work depends heavily on careful measurement, and I frequently use proportions to change the size of type, pieces of art, or photos to fit a fixed amount of space.

Math is important not only in my creation process but also in my billing process. For most of the work I do, I charge a fixed price for a set size of a particular item, such as a house portrait or a painted ornament. However, when I take on unusual projects or do work for an ad agency, I use a different system of billing that considers my time, supplies, and any extensive driving that was involved. For these cases, I figure the amount I charge as (hourly rate × number of hours worked) + (cost of all supplies) + (rate per mile × number of miles driven).

10.2   Completing the Square

Constructing Perfect Square Trinomials ▪
Solving Equations by Completing the Square

Constructing Perfect Square Trinomials

Consider the quadratic equation

$$(x - 2)^2 = 10. \qquad \text{\color{red}Completed square form}$$

You know from Example 2(b) in the previous section that this equation has two solutions: $2 + \sqrt{10}$ and $2 - \sqrt{10}$. Suppose you had been given the equation in its standard form

$$x^2 - 4x - 6 = 0. \qquad \text{\color{red}Standard form}$$

How would you solve this equation if you were given only the standard form? You could try factoring, but after attempting to do so you would find that the left side of the equation is not factorable (using integer coefficients).

In this section, you will study a technique for rewriting an equation in a completed square form. This technique is called **completing the square.**

Completing the Square

To **complete the square** for the expression $x^2 + bx$, add $(b/2)^2$, which is the square of half the coefficient of x. Consequently,

$$x^2 + bx + \left(\frac{b}{2}\right)^2 = \left(x + \frac{b}{2}\right)^2.$$

EXAMPLE 1   *Creating a Perfect Square Trinomial*

What term should be added to $x^2 - 8x$ so that it becomes a perfect square trinomial?

Solution

For this expression, the coefficient of the x-term is -8. Divide this term by 2, and square the result to obtain $(-4)^2 = 16$. This is the term that should be added to the expression to make it a perfect square trinomial.

$$x^2 - 8x + \textcolor{blue}{16} = x^2 - 8x + (-4)^2 \qquad \text{\color{red}Add 16 to the expression.}$$

$$= (x - 4)^2 \qquad \text{\color{red}Completed square form}$$

Solving Equations by Completing the Square

When completing the square to solve an equation, remember that it is essential to *preserve the equality.* Thus, when you add a constant term to one side of the equation, you must be sure to add the same constant to the other side of the equation.

EXAMPLE 2　Completing the Square: Leading Coefficient Is 1

Solve $x^2 + 12x = 0$.

NOTE　In Example 2, completing the square is used for the sake of illustration. This particular equation would be easier to solve by factoring. Try reworking the problem by factoring to see that you obtain the same two solutions. In Example 3, the equation cannot be solved by factoring (using integer coefficients).

Solution

$$x^2 + 12x = 0 \qquad\qquad \text{Original equation}$$

$$x^2 + 12x + (6)^2 = 36 \qquad\qquad \text{Add } \left(\tfrac{12}{2}\right)^2 = 36 \text{ to both sides.}$$

$$\underset{\text{(half)}^2}{\uparrow}$$

$$(x + 6)^2 = 36 \qquad\qquad \text{Completed square form}$$

$$x + 6 = \pm\sqrt{36} \qquad\qquad \text{Extract square roots.}$$

$$x = -6 \pm 6 \qquad\qquad \text{Subtract 6 from both sides.}$$

$$x = 0 \ \text{ or } \ x = -12 \qquad\qquad \text{Solutions}$$

The solutions are 0 and -12. Check these in the original equation.

EXAMPLE 3　Completing the Square: Leading Coefficient Is 1

Solve $x^2 - 6x + 7 = 0$.

Solution

$$x^2 - 6x + 7 = 0 \qquad\qquad \text{Original equation}$$

$$x^2 - 6x = -7 \qquad\qquad \text{Subtract 7 from both sides.}$$

$$x^2 - 6x + (-3)^2 = -7 + 9 \qquad\qquad \text{Add } (-3)^2 = 9 \text{ to both sides.}$$

$$\underset{\text{(half)}^2}{\uparrow}$$

$$(x - 3)^2 = 2 \qquad\qquad \text{Completed square form}$$

$$x - 3 = \pm\sqrt{2} \qquad\qquad \text{Extract square roots.}$$

$$x = 3 \pm \sqrt{2} \qquad\qquad \text{Solutions}$$

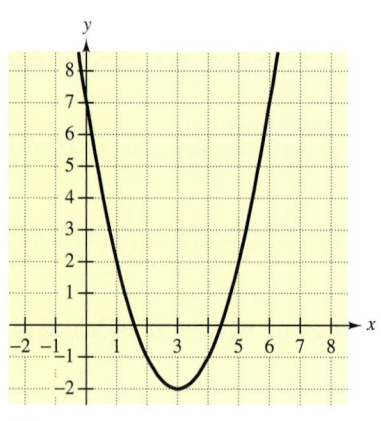

FIGURE 10.2

The solutions are $3 + \sqrt{2}$ and $3 - \sqrt{2}$. Check these in the original equation. Also try graphically checking the solutions, as shown in Figure 10.2.

EXAMPLE 4   A Leading Coefficient That Is Not 1

Solve $3x^2 + 5x = 2$.

Solution

$$3x^2 + 5x = 2 \qquad \text{Original equation}$$

$$x^2 + \frac{5}{3}x = \frac{2}{3} \qquad \text{Divide both sides by 3.}$$

$$x^2 + \frac{5}{3}x + \left(\frac{5}{6}\right)^2 = \frac{2}{3} + \frac{25}{36} \qquad \text{Add } \left(\frac{5}{6}\right)^2 = \frac{25}{36} \text{ to both sides.}$$

$$\left(x + \frac{5}{6}\right)^2 = \frac{49}{36} \qquad \text{Completed square form}$$

$$x + \frac{5}{6} = \pm\frac{7}{6} \qquad \text{Extract square roots.}$$

$$x = -\frac{5}{6} \pm \frac{7}{6} \qquad \text{Subtract } \tfrac{5}{6} \text{ from both sides.}$$

$$x = \frac{1}{3} \quad \text{or} \quad x = -2 \qquad \text{Solutions}$$

The solutions are $\frac{1}{3}$ and -2. Check these in the original equation.

EXAMPLE 5   A Leading Coefficient That Is Not 1

Solve $2x^2 - x - 2 = 0$.

Solution

$$2x^2 - x - 2 = 0 \qquad \text{Original equation}$$

$$2x^2 - x = 2 \qquad \text{Add 2 to both sides.}$$

$$x^2 - \frac{1}{2}x = 1 \qquad \text{Divide both sides by 2.}$$

$$x^2 - \frac{1}{2}x + \left(-\frac{1}{4}\right)^2 = 1 + \frac{1}{16} \qquad \text{Add } \left(-\tfrac{1}{4}\right)^2 = \tfrac{1}{16} \text{ to both sides.}$$

$$\left(x - \frac{1}{4}\right)^2 = \frac{17}{16} \qquad \text{Completed square form}$$

$$x - \frac{1}{4} = \pm\frac{\sqrt{17}}{4} \qquad \text{Extract square roots.}$$

$$x = \frac{1}{4} \pm \frac{\sqrt{17}}{4} \qquad \text{Add } \tfrac{1}{4} \text{ to both sides.}$$

The solutions are $\frac{1}{4}\left(1 \pm \sqrt{17}\right)$. Check these in the original equation.

STUDY TIP

If the leading coefficient of a quadratic expression is not 1, you must divide both sides of the equation by this coefficient *before* completing the square. This process is demonstrated in Examples 4 and 5.

EXAMPLE 6   A Quadratic Equation with Complex Solutions

Solve $x^2 - 4x + 8 = 0$.

Solution

$x^2 - 4x + 8 = 0$	Original equation
$x^2 - 4x = -8$	Subtract 8 from both sides.
$x^2 - 4x + (-2)^2 = -8 + 4$	Add $(-2)^2 = 4$ to both sides.
$(x - 2)^2 = -4$	Completed square form
$x - 2 = \pm 2i$	Extract complex square roots.
$x = 2 \pm 2i$	Add 2 to both sides.

The solutions are $2 + 2i$ and $2 - 2i$. The first of these is checked as follows. Try checking the other.

Check

$x^2 - 4x + 8 = 0$	Original equation
$(2 + 2i)^2 - 4(2 + 2i) + 8 \stackrel{?}{=} 0$	Substitute $2 + 2i$ for x.
$4 + 8i - 4 - 8 - 8i + 8 \stackrel{?}{=} 0$	Simplify.
$0 = 0$	Solution checks. ✔

Group Activities   You Be the Instructor

Error Analysis   Suppose you teach an algebra class and one of your students hands in the following solution. Find and correct the error(s). Discuss how to explain the error(s) to your student.

1. Solve $x^2 + 6x - 13 = 0$ by completing the square.

$$x^2 + 6x = 13$$
$$x^2 + 6x + \left(\frac{6}{2}\right)^2 = 13$$
$$(x + 3)^2 = 13$$
$$x + 3 = \pm\sqrt{13}$$
$$x = -3 \pm \sqrt{13}$$

10.2 Exercises

Discussing the Concepts

1. What is a perfect square trinomial?

2. What term must be added to $x^2 + 5x$ to complete the square? Explain how you found the term.

3. Explain the use of extracting square roots when solving a quadratic equation by the method of completing the square.

4. Is it possible for a quadratic equation to have no real number solution? If so, give an example.

5. When the method of completing the square is used to solve a quadratic equation, what is the first step if the leading coefficient is not 1? Is the resulting equation equivalent to the given equation? Explain.

6. *True or False?*   If you solve a quadratic equation by completing the square and obtain solutions that are rational numbers, you could have solved the equation by factoring. Explain.

Problem Solving

In Exercises 7–10, find the term that must be added to the expression so that it becomes a perfect square trinomial.

7. $x^2 + 8x +$ ▢

8. $y^2 - 2y +$ ▢

9. $t^2 + 5t +$ ▢

10. $a^2 - \frac{1}{3}a +$ ▢

In Exercises 11–14, solve the quadratic equation (a) by completing the square and (b) by factoring.

11. $x^2 - 6x = 0$

12. $t^2 + 9t = 0$

13. $x^2 + 7x + 12 = 0$

14. $y^2 - 8y + 12 = 0$

In Exercises 15–26, solve the quadratic equation by completing the square. Give the solutions in exact form and in decimal form rounded to two decimal places. (The solutions may be complex numbers.)

15. $x^2 - 4x - 3 = 0$

16. $x^2 - 6x + 7 = 0$

17. $x^2 + 2x + 3 = 0$

18. $x^2 - 6x + 12 = 0$

19. $x^2 - \frac{2}{3}x - 3 = 0$

20. $x^2 + \frac{4}{5}x - 1 = 0$

21. $t^2 + 5t + 3 = 0$

22. $u^2 - 9u - 1 = 0$

23. $2x^2 + 8x + 3 = 0$

24. $3x^2 - 24x - 5 = 0$

25. $0.1x^2 + 0.5x + 0.2 = 0$

26. $0.02x^2 + 0.10x - 0.05 = 0$

In Exercises 27–30, find the real solutions.

27. $\dfrac{x}{2} - \dfrac{1}{x} = 1$

28. $\dfrac{x}{2} + \dfrac{5}{x} = 4$

29. $\sqrt{2x + 1} = x - 3$

30. $\sqrt{3x - 2} = x - 2$

In Exercises 31–34, use a graphing utility to approximate any x-intercepts of the graph. Set $y = 0$ and solve the resulting equation. Compare the result with the x-intercepts of the graph.

31. $y = x^2 + 4x - 1$

32. $y = x^2 + 6x - 4$

33. $y = x^2 - 2x - 5$

34. $y = 2x^2 - 6x - 5$

In Exercises 35 and 36, consider a windlass that is used to pull a boat to the dock (see figure).

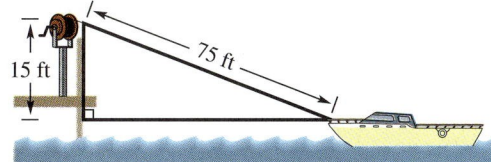

15 ft   75 ft

35. Find the distance from the boat to the dock when the rope is 75 feet long.

36. Find the distance from the boat to the dock when the rope is 50 feet long.

Reviewing the Major Concepts

In Exercises 37–40, simplify the expression.

37. $\sqrt[3]{16x^4y^3}$

38. $\sqrt{72x^2y^3}$

39. $\sqrt{3x^2} - 2\sqrt{12}$

40. $\dfrac{14}{\sqrt{7}}$

41. *Time* A jogger starts running at 6 miles per hour. Five minutes later, another jogger starts on the same trail, running at 8 miles per hour. When will the second jogger overtake the first?

42. Find three consecutive even integers whose sum is 102.

Additional Problem Solving

In Exercises 43–50, find the term that must be added to the expression so that it becomes a perfect square trinomial.

43. $y^2 - 20y +$

44. $x^2 + 12x +$

45. $x^2 - \frac{6}{5}x +$

46. $y^2 + \frac{4}{3}y +$

47. $y^2 - \frac{3}{5}y +$

48. $u^2 + 7u +$

49. $r^2 - 0.4r +$

50. $s^2 + 4.5s +$

In Exercises 51–60, solve the quadratic equation (a) by completing the square and (b) by factoring.

51. $x^2 - 25x = 0$

52. $x^2 + 32x = 0$

53. $t^2 - 8t + 7 = 0$

54. $x^2 + 12x + 27 = 0$

55. $x^2 + 2x - 24 = 0$

56. $z^2 + 3z - 10 = 0$

57. $x^2 - 3x - 18 = 0$

58. $t^2 - 5t - 36 = 0$

59. $2x^2 - 11x + 12 = 0$

60. $3x^2 - 5x - 2 = 0$

In Exercises 61–80, solve the quadratic equation by completing the square. Give the solutions in exact form and in decimal form rounded to two decimal places. (The solutions may be complex numbers.)

61. $x^2 + 4x - 3 = 0$

62. $x^2 + 6x + 7 = 0$

63. $u^2 - 4u + 1 = 0$

64. $a^2 - 10a - 15 = 0$

65. $x^2 - 10x - 2 = 0$

66. $x^2 + 8x - 4 = 0$

67. $y^2 + 20y + 10 = 0$

68. $y^2 + 6y - 24 = 0$

69. $v^2 + 3v - 2 = 0$

70. $z^2 - 7z + 9 = 0$

71. $-x^2 + x - 1 = 0$

72. $1 - x - x^2 = 0$

73. $3x^2 + 9x + 5 = 0$

74. $5x^2 - 15x + 7 = 0$

75. $4y^2 + 4y - 9 = 0$

76. $4z^2 - 3z + 2 = 0$

77. $0.1x^2 + 0.2x + 0.5 = 0$

78. $\frac{1}{2}t^2 + t + 2 = 0$

79. $x(x - 7) = 2$

80. $2x\left(x + \frac{4}{3}\right) = 5$

In Exercises 81–84, use a graphing utility to approximate any x-intercepts of the graph. Set $y = 0$ and solve the resulting equation. Compare the result with the x-intercepts of the graph.

81. $\frac{1}{3}x^2 + 2x - 6 = 0$

82. $\frac{1}{2}x^2 - 3x + 1 = 0$

83. $\dfrac{3}{x} - x - 1 = 0$

84. $\sqrt{x} - x + 2 = 0$

85. *Geometric Modeling*

(a) Find the area of the two adjoining rectangles and the large square in the figure.

(b) Find the area of the small square in the lower right-hand corner of the figure and add it to the area found in part (a).

(c) Find the dimensions and the area of the entire figure after adjoining the small square in the lower right-hand corner. Note that you have shown completing the square geometrically.

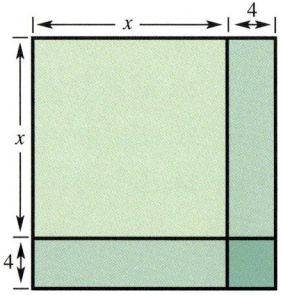

86. *Geometric Modeling*   Use the model in Exercise 85 to geometrically represent completing the square for $x^2 + 6x$.

87. *Geometry*   The area of the rectangle in the figure is 160 square feet. Find the rectangle's dimensions.

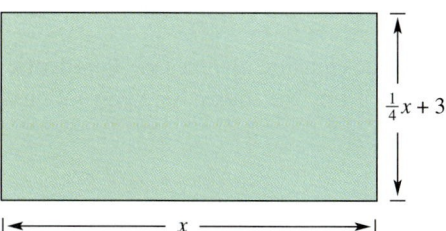

$\frac{1}{4}x + 3$

x

88. *Cutting Across the Lawn*   On the sidewalk, the distance from the dormitory to the cafeteria is 400 meters (see figure). By cutting across the lawn, the walking distance is shortened to 300 meters. How long is each part of the L-shaped sidewalk?

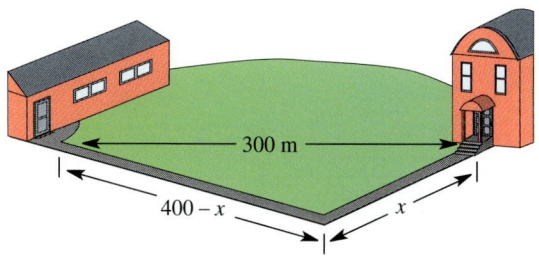

300 m

400 − x   x

89. *Revenue*   The revenue R from selling x units of a certain product is given by

$$R = x\left(50 - \frac{1}{2}x\right).$$

Find the number of units that must be sold to produce a revenue of $1218.

90. *Revenue*   The revenue R from selling x units of a certain product is given by

$$R = x\left(100 - \frac{1}{10}x\right).$$

Find the number of units that must be sold to produce a revenue of $12,000.

91. *Fencing in a Corral*   You have 200 feet of fencing to enclose two adjacent rectangular corrals (see figure). The total area of the enclosed region is 1400 square feet. What are the dimensions of each corral? (The corrals are the same size.)

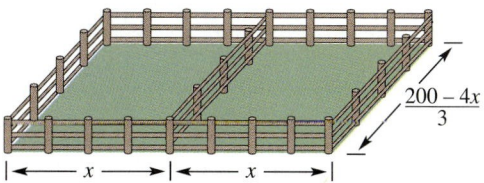

$\frac{200 - 4x}{3}$

x x

92. *Geometry*   An open box with a rectangular base of x inches by $x + 4$ inches has a height of 6 inches (see figure). Find the dimensions of the box if its volume is 840 cubic inches.

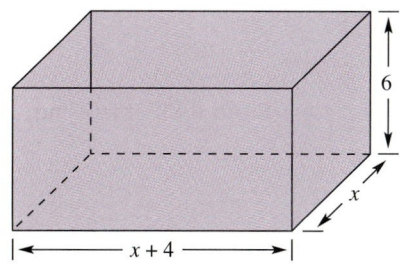

6

x

$x + 4$

93. *Geometry*   A closed box has a square base with edge x feet and a height of 3 feet (see figure). The material for constructing the base costs $1.50 per square foot and the material for the top and sides costs $1 per square foot. Find the dimensions of the box if its volume is 128 cubic feet.

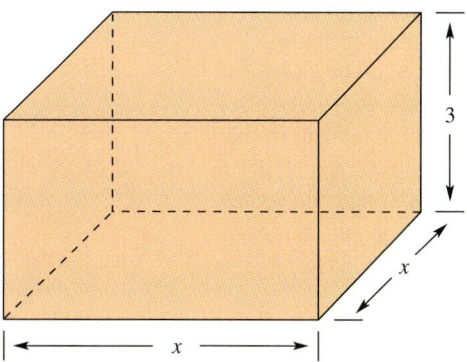

3

x

x

10.3	# The Quadratic Formula and the Discriminant
	The Quadratic Formula ▪ Solving Equations by the Quadratic Formula ▪ The Discriminant

The Quadratic Formula

A fourth technique for solving a quadratic equation involves the **Quadratic Formula.** This formula is obtained by completing the square for a general quadratic equation.

$$ax^2 + bx + c = 0 \qquad \text{Standard form, } a \neq 0$$

$$ax^2 + bx = -c \qquad \text{Subtract } c \text{ from both sides.}$$

$$x^2 + \frac{b}{a}x = -\frac{c}{a} \qquad \text{Divide both sides by } a.$$

$$x^2 + \frac{b}{a}x + \left(\frac{b}{2a}\right)^2 = -\frac{c}{a} + \left(\frac{b}{2a}\right)^2 \qquad \text{Add } \left(\frac{b}{2a}\right)^2 \text{ to both sides.}$$

$$\left(x + \frac{b}{2a}\right)^2 = \frac{b^2 - 4ac}{4a^2} \qquad \text{Simplify.}$$

$$x + \frac{b}{2a} = \pm\sqrt{\frac{b^2 - 4ac}{4a^2}} \qquad \text{Extract square roots.}$$

$$x = -\frac{b}{2a} \pm \frac{\sqrt{b^2 - 4ac}}{2|a|} \qquad \text{Subtract } \frac{b}{2a} \text{ from both sides.}$$

$$x = \frac{-b \pm \sqrt{b^2 - 4ac}}{2a} \qquad \text{Simplify.}$$

STUDY TIP

The Quadratic Formula is one of the most important formulas in algebra, and you should memorize it. We have found that it helps to try to memorize a verbal statement of the rule. For instance, you might try to remember the following verbal statement of the Quadratic Formula: "Minus b, plus or minus the square root of b squared minus $4ac$, all divided by $2a$."

The Quadratic Formula

The solutions of $ax^2 + bx + c = 0$, $a \neq 0$, are given by the **Quadratic Formula**

$$x = \frac{-b \pm \sqrt{b^2 - 4ac}}{2a}.$$

The expression inside the radical, $b^2 - 4ac$, is called the **discriminant.**

1. If $b^2 - 4ac > 0$, the equation has two real solutions.

2. If $b^2 - 4ac = 0$, the equation has one (repeated) real solution.

3. If $b^2 - 4ac < 0$, the equation has no real solutions.

Solving Equations by the Quadratic Formula

When using the Quadratic Formula, remember that *before* the formula can be applied, you must first write the quadratic equation in standard form.

EXAMPLE 1   The Quadratic Formula: Two Distinct Solutions

$x^2 + 6x = 16$	Original equation
$x^2 + 6x - 16 = 0$	Write in standard form.
$x = \dfrac{-b \pm \sqrt{b^2 - 4ac}}{2a}$	Quadratic Formula
$x = \dfrac{-6 \pm \sqrt{6^2 - 4(1)(-16)}}{2(1)}$	Substitute: $a = 1$, $b = 6$, $c = -16$.
$x = \dfrac{-6 \pm \sqrt{100}}{2}$	Simplify.
$x = \dfrac{-6 \pm 10}{2}$	Simplify.
$x = 2 \quad \text{or} \quad x = -8$	Solutions

The solutions are 2 and -8. Check these in the original equation.

NOTE In Example 1, the solutions are rational numbers, which means that the equation could have been solved by factoring. Try solving the equation by factoring.

EXAMPLE 2   The Quadratic Formula: Two Distinct Solutions

$-x^2 - 4x + 8 = 0$	Leading coefficient is negative.
$x^2 + 4x - 8 = 0$	Multiply both sides by -1.
$x = \dfrac{-b \pm \sqrt{b^2 - 4ac}}{2a}$	Quadratic Formula
$x = \dfrac{-4 \pm \sqrt{4^2 - 4(1)(-8)}}{2(1)}$	Substitute: $a = 1$, $b = 4$, $c = -8$.
$x = \dfrac{-4 \pm \sqrt{48}}{2}$	Simplify.
$x = \dfrac{-4 \pm 4\sqrt{3}}{2}$	Simplify.
$x = \dfrac{2(-2 \pm 2\sqrt{3})}{2}$	Cancel common factor.
$x = -2 \pm 2\sqrt{3}$	Solutions

The solutions are $-2 + 2\sqrt{3}$ and $-2 - 2\sqrt{3}$. Check these in the original equation.

STUDY TIP

If the leading coefficient of a quadratic equation is negative, we suggest that you begin by multiplying both sides of the equation by -1, as shown in Example 2. This will produce a positive leading coefficient, which is less cumbersome to work with.

EXAMPLE 3 *The Quadratic Formula: One Repeated Solution*

$18x^2 - 24x + 8 = 0$ Original equation

$9x^2 - 12x + 4 = 0$ Divide both sides by 2.

$$x = \frac{-b \pm \sqrt{b^2 - 4ac}}{2a}$$ Quadratic Formula

$$x = \frac{-(-12) \pm \sqrt{(-12)^2 - 4(9)(4)}}{2(9)}$$

$$x = \frac{12 \pm \sqrt{144 - 144}}{18}$$ Simplify.

$$x = \frac{12 \pm \sqrt{0}}{18}$$ Simplify.

$$x = \frac{2}{3}$$ Solution

The only solution is $\frac{2}{3}$. Check this in the original equation.

Note in the next example how the Quadratic Formula can be used to solve a quadratic equation that has complex solutions.

EXAMPLE 4 *The Quadratic Formula: Complex Solutions*

$2x^2 - 4x + 5 = 0$ Original equation

$$x = \frac{-b \pm \sqrt{b^2 - 4ac}}{2a}$$ Quadratic Formula

$$x = \frac{-(-4) \pm \sqrt{(-4)^2 - 4(2)(5)}}{2(2)}$$

$$x = \frac{4 \pm \sqrt{-24}}{4}$$ Simplify.

$$x = \frac{4 \pm 2\sqrt{6}\,i}{4}$$ Write in i-form.

$$x = \frac{2(2 \pm \sqrt{6}\,i)}{2 \cdot 2}$$ Cancel common factor.

$$x = \frac{2 \pm \sqrt{6}\,i}{2}$$ Solutions

The solutions are $\frac{1}{2}(2 \pm \sqrt{6}\,i)$. Check these in the original equation.

The Discriminant

The radicand in the Quadratic Formula, $b^2 - 4ac$, is called the **discriminant** because it allows you to "discriminate" among different types of solutions.

Using the Discriminant

Let a, b, and c be rational numbers such that $a \neq 0$. The **discriminant** of the quadratic equation $ax^2 + bx + c = 0$ is given by $b^2 - 4ac$, and can be used to classify the solutions of the equation as follows.

Discriminant	*Solution Types*
1. Perfect square	Two distinct rational solutions (Example 1)
2. Positive nonperfect square	Two distinct irrational solutions (Example 2)
3. Zero	One repeated rational solution (Example 3)
4. Negative number	Two distinct imaginary solutions (Example 4)

DISCOVERY

Use a graphing utility to graph the equations below.

a. $y = x^2 - x + 2$

b. $y = 2x^2 - 3x - 2$

c. $y = x^2 - 2x + 1$

d. $y = x^2 - 2x - 10$

Describe the solution type of each equation and check your results with those shown in Example 5. Why do you think the discriminant is used to determine solution types?

EXAMPLE 5   Using the Discriminant

Equation	*Discriminant*	*Solution Types*
a. $x^2 - x + 2 = 0$	$b^2 - 4ac = (-1)^2 - 4(1)(2)$ $= 1 - 8$ $= -7$	Two distinct imaginary solutions
b. $2x^2 - 3x - 2 = 0$	$b^2 - 4ac = (-3)^2 - 4(2)(-2)$ $= 9 + 16$ $= 25$	Two distinct rational solutions
c. $x^2 - 2x + 1 = 0$	$b^2 - 4ac = (-2)^2 - 4(1)(1)$ $= 4 - 4$ $= 0$	One repeated rational solution
d. $x^2 - 2x - 1 = 9$	$b^2 - 4ac = (-2)^2 - 4(1)(-10)$ $= 4 + 40$ $= 44$	Two distinct irrational solutions

You have now studied four ways to solve quadratic equations: (1) factoring, (2) extracting square roots, (3) completing the square, and (4) the Quadratic Formula.

Programming

The *TI-82* program below can be used to solve equations with the Quadratic Formula. (See the Appendix for other calculator models.) Enter a, b, and c when prompted.

```
PROGRAM: QUDRATIC
: Prompt A
: Prompt B
: Prompt C
: B^2 - 4AC → D
: If D < 0
: Then
: Disp ''NO SOLUTION''
: Stop
: End
: (-B + √D̄)/(2A) → S
: Disp S
: (-B - √D̄)/(2A) → S
: Disp S
: Stop
```

EXAMPLE 6 Using a Calculator with the Quadratic Formula

Solve $1.2x^2 - 17.8x + 8.05 = 0$.

Solution

Using the Quadratic Formula, you can write

$$x = \frac{-(-17.8) \pm \sqrt{(-17.8)^2 - 4(1.2)(8.05)}}{2(1.2)}.$$

To evaluate these solutions, begin by calculating the square root.

17.8 $\boxed{+/-}$ $\boxed{x^2}$ $\boxed{-}$ 4 $\boxed{\times}$ 1.2 $\boxed{\times}$ 8.05 $\boxed{=}$ $\boxed{\sqrt{}}$ Scientific

$\boxed{\sqrt{}}$ $\boxed{(}$ $\boxed{(}$ $\boxed{(-)}$ 17.8 $\boxed{)}$ $\boxed{x^2}$ $\boxed{-}$ 4 $\boxed{\times}$ 1.2   Graphing
$\boxed{\times}$ 8.05 $\boxed{)}$ $\boxed{\text{ENTER}}$

The display for either of these keystroke sequences should be 16.67932852. Storing this result and using the recall key, we find the following two solutions.

$$x \approx \frac{17.8 + 16.67932852}{2.4} \approx 14.366$$ Add stored value.

$$x \approx \frac{17.8 - 16.67932852}{2.4} \approx 0.467$$ Subtract stored value.

Group Activities   You Be the Instructor

Problem Posing   Suppose you are writing a quiz that covers quadratic equations. Write four quadratic equations, including one with solutions $x = \frac{5}{3}$ and $x = -2$ and one with solutions $x = 4 \pm \sqrt{3}$, and instruct students to use any of the four solution methods: factoring, extracting square roots, completing the square, and using the Quadratic Formula. Trade quizzes with another member of your group and check one another's work.

10.3 Exercises

Discussing the Concepts

1. State the quadratic formula *in words*.

2. What is the discriminant of $ax^2 + bx + c = 0$? How is the discriminant related to the number of solutions of the equation?

3. Explain how completing the square can be used to develop the Quadratic Formula.

4. Summarize the four methods for solving a quadratic equation.

Problem Solving

In Exercises 5–8, write in standard form.

5. $2x^2 = 7 - 2x$

6. $7x^2 + 15x = 5$

7. $x(10 - x) = 5$

8. $x(3x + 8) = 15$

In Exercises 9–12, solve (a) by the Quadratic Formula and (b) by factoring.

9. $x^2 - 11x + 28 = 0$

10. $x^2 + 9x + 14 = 0$

11. $4x^2 + 12x + 9 = 0$

12. $10x^2 - 11x + 3 = 0$

In Exercises 13–16, use the discriminant to determine the type of solutions of the equation.

13. $2x^2 - 5x - 4 = 0$

14. $10x^2 + 5x + 1 = 0$

15. $3x^2 - x + 2 = 0$

16. $9x^2 - 24x + 16 = 0$

In Exercises 17–26, use the Quadratic Formula to find all real or imaginary solutions.

17. $x^2 - 2x - 4 = 0$

18. $x^2 - 2x - 6 = 0$

19. $t^2 + 4t + 1 = 0$

20. $y^2 + 6y + 4 = 0$

21. $x^2 + 3x + 3 = 0$

22. $2x^2 - x + 1 = 0$

23. $9z^2 + 6z - 4 = 0$

24. $8a^2 - 8a - 1 = 0$

25. $2.5x^2 + x - 0.9 = 0$

26. $0.09x^2 - 0.12x - 0.26 = 0$

In Exercises 27–30, solve by the most convenient method. Find all real or imaginary solutions.

27. $y^2 + 15y = 0$

28. $t^2 = 150$

29. $x^2 + 8x + 25 = 0$

30. $2x^2 + 8x + 4.5 = 0$

Graphical Reasoning   In Exercises 31–34, use a graphing utility to graph the function. Use the graph to approximate any x-intercepts of the graph. Set $y = 0$ and solve the resulting equation. Compare the result with the x-intercepts of the graph.

31. $y = 3x^2 - 6x + 1$

32. $y = x^2 + x + 1$

33. $y = -(4x^2 - 20x + 25)$

34. $y = x^2 - 4x + 3$

In Exercises 35–38, solve the equation.

35. $\dfrac{2x^2}{5} - \dfrac{x}{2} = 1$

36. $\dfrac{x}{3} + \dfrac{1}{x} = 4$

37. $\sqrt{x + 3} = x - 1$

38. $\sqrt{2x - 3} = x - 2$

Exploration   In Exercises 39–42, determine the values of c such that the equation has (a) two real number solutions, (b) one real number solution, and (c) two imaginary number solutions.

39. $x^2 - 6x + c = 0$

40. $x^2 - 12x + c = 0$

41. $x^2 + 8x + c = 0$

42. $x^2 + 2x + c = 0$

43. *Geometry*   A rectangle has a width of x inches, a length of $x + 6.3$ inches, and an area of 58.14 square inches. Find its dimensions.

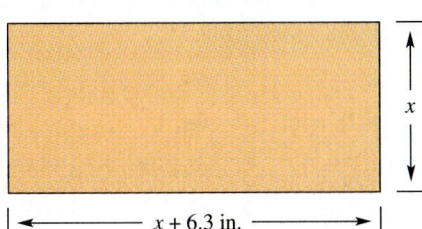

44. *Analyzing Data*   The numbers of cellular phone subscribers (in millions) in the United States for the years 1987 to 1992 are given in the following table. (Source: Cellular Telecommunications Industry Association)

Year	1987	1988	1989	1990	1991	1992
Subscribers	1.23	2.07	3.51	5.28	7.56	11.03

(a) Create a bar graph for the data.

(b) The data can be approximated by the model

$$s = 0.30t^2 + 2.22t + 5.29, \quad -3 \le t \le 2$$

where $t = 0$ corresponds to 1990. Use a graphing utility to graph this model.

(c) Use the model to determine the year in which the cellular phone companies had 2.1 million subscribers.

Reviewing the Major Concepts

In Exercises 45–48, perform the operations and simplify.

45. $\left(\sqrt{x} + 3\right)\left(\sqrt{x} - 3\right)$

46. $\sqrt{u}\left(\sqrt{20} - \sqrt{5}\right)$

47. $\left(2\sqrt{t} + 3\right)^2$

48. $\dfrac{50x}{\sqrt{2}}$

49. *Mixture*   Determine the number of gallons of a 30% solution that must be mixed with a 60% solution to obtain 20 gallons of a 40% solution.

50. *Original Price*   A suit sells for $375 during a 25% store-wide clearance sale. What was the original price of the suit?

Additional Problem Solving

In Exercises 51–56, solve the equation (a) by the Quadratic Formula and (b) by factoring.

51. $x^2 + 6x + 8 = 0$

52. $x^2 - 12x + 27 = 0$

53. $x^2 - \frac{4}{3}x + \frac{4}{9} = 0$

54. $x^2 + x + \frac{1}{4} = 0$

55. $6x^2 - x - 2 = 0$

56. $9x^2 - 30x + 25 = 0$

In Exercises 57–60, use the discriminant to determine the type of solutions of the equation.

57. $x^2 + 7x + 15 = 0$

58. $3x^2 - 2x - 5 = 0$

59. $4x^2 - 12x + 9 = 0$

60. $2x^2 + 10x + 6 = 0$

In Exercises 61–74, use the Quadratic Formula to find all real or imaginary solutions.

61. $x^2 + 6x - 3 = 0$

62. $x^2 + 8x - 4 = 0$

63. $x^2 - 10x + 23 = 0$

64. $u^2 - 12u + 29 = 0$

65. $2v^2 - 2v - 1 = 0$

66. $4x^2 + 6x + 1 = 0$

67. $2x^2 + 4x - 3 = 0$

68. $2x^2 + 3x + 3 = 0$

69. $x^2 - 0.4x - 0.16 = 0$

70. $x^2 + 0.6x - 0.41 = 0$

71. $4x^2 - 6x + 3 = 0$

72. $-5x^2 - 15x + 10 = 0$

73. $9x^2 = 1 + 9x$

74. $x - x^2 = 1 - 6x^2$

In Exercises 75–80, solve by the most convenient method. Find all real or imaginary solutions.

75. $z^2 - 169 = 0$

76. $4u^2 + 49 = 0$

77. $25(x - 3)^2 - 36 = 0$

78. $2y(y - 18) + 3(y - 18) = 0$

79. $x^2 - 24x + 128 = 0$

80. $1.2x^2 - 0.8x - 5.5 = 0$

In Exercises 81–84, use a calculator to solve the equation. Round to three decimal places.

81. $5x^2 - 18x + 6 = 0$

82. $15x^2 + 3x - 105 = 0$

83. $-0.04x^2 + 4x - 0.8 = 0$

84. $3.7x^2 - 10.2x + 3.2 = 0$

85. *Free-Falling Object*   A ball is thrown upward at a velocity of 40 feet per second from a bridge that is 50 feet above the level of the water (see figure). The height h (in feet) of the ball above the water at time t seconds after it is thrown is given by

$$h = -16t^2 + 40t + 50.$$

(a) Find the time when the ball is again at the 50-foot level above the water.

(b) Find the time when the ball strikes the water.

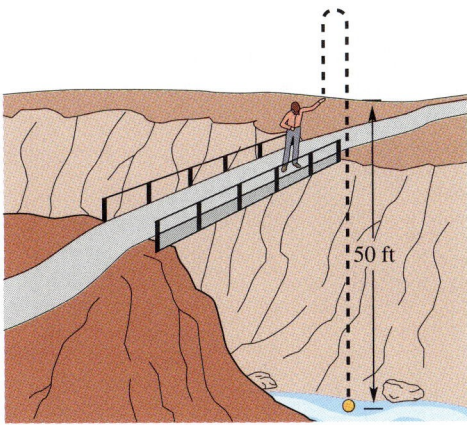

50 ft

86. *Geometry*   Two circular regions are tangent to each other (see figure). The distance between centers is 10 feet.

(a) Find the radius of each circle if their combined area is 52π square feet.

(b) Suppose the distance between the centers remained the same but radii were made equal. Would the combined area of the circles increase or decrease?

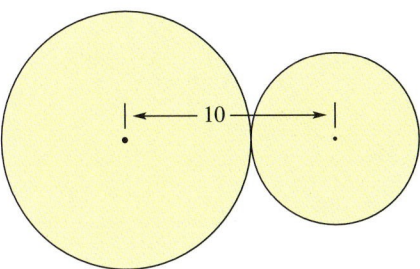

10

Aerospace Employment   In Exercises 87–89, use the following model, which approximates the number of people employed in the aerospace industry in the United States from 1987 to 1992.

$$y = 803.49 - 33.23t - 10.77t^2, \quad -3 \le t \le 2$$

In this model, y represents the number employed in the aerospace industry (in thousands) and t represents the year, with $t = 0$ corresponding to 1990. (Source: U.S. Department of Commerce)

87. Use a graphing utility to graph the model.

88. Use the TRACE feature of a graphing utility to find the year in which there were approximately 750,000 employed in the aerospace industry in the United States.

89. Use the model to estimate the number employed in the aerospace industry in 1993.

In 1991, the aerospace industry's revenue in the United States was about 135 billion dollars.

MID-CHAPTER QUIZ

Take this quiz as you would take a quiz in class. After you are done, check your work against the answers given in the back of the book.

In Exercises 1–8, solve the quadratic equation by the specified method.

1. Factoring:

$2x^2 - 72 = 0$

2. Factoring:

$2x^2 + 3x - 20 = 0$

3. Extracting square roots:

$t^2 = 12$

4. Extracting square roots:

$(u - 3)^2 - 16 = 0$

5. Completing the square:

$s^2 + 10s + 1 = 0$

6. Completing the square:

$2y^2 + 6y - 5 = 0$

7. Quadratic Formula:

$x^2 + 4x - 6 = 0$

8. Quadratic Formula:

$6v^2 - 3v - 4 = 0$

In Exercises 9–16, solve the equation by the most convenient method. (Find all the real *and* imaginary solutions.)

9. $x^2 + 5x + 7 = 0$

10. $36 - (t - 4)^2 = 0$

11. $x(x - 10) + 3(x - 10) = 0$

12. $x(x - 3) = 10$

13. $4b^2 - 12b + 9 = 0$

14. $3m^2 + 10m + 5 = 0$

15. $\dfrac{4}{u} - u = 3$

16. $\sqrt{2x + 5} = x + 1$

In Exercises 17 and 18, use a graphing utility to graph the function. Use the graph to approximate any x-intercepts of the graph. Set $y = 0$ and solve the resulting equation. Write a paragraph comparing the results of your algebraic and graphical solutions.

17. $y = \frac{1}{2}x^2 - 3x - 1$

18. $y = x^2 + 0.45x - 4$

19. The revenue R from selling x units of a certain product is given by

$R = x(20 - 0.2x).$

Find the number of units that must be sold to produce a revenue of $500.

20. The perimeter of a rectangle with sides x and $100 - x$ is 200 meters. Its area A is given by $A = x(100 - x)$. Determine the dimensions of the rectangle if its area is 2275 square meters.

10.4 Applications of Quadratic Equations

Applications of Quadratic Equations

Applications of Quadratic Equations

EXAMPLE 1 An Investment Problem

A car dealer bought a fleet of cars from a car rental agency for a total of $90,000. By the time the dealer had sold all but six of the cars, at an average profit of $2500 each, the original investment of $90,000 had been regained. How many cars did the dealer sell, and what was the average price per car?

Solution

Although this problem is stated in terms of average price and average profit per car, we can use a model that assumes that each car sold for the same price.

Verbal Model:

$$\boxed{\text{Selling price per car}} = \boxed{\text{Cost per car}} + \boxed{\text{Profit per car}}$$

Labels:

Number of cars sold $= x$ (cars)

Number of cars bought $= x + 6$ (cars)

Selling price per car $= \dfrac{90,000}{x}$ (dollars per car)

Cost per car $= \dfrac{90,000}{x + 6}$ (dollars per car)

Profit per car $= 2500$ (dollars per car)

Equation:

$$\frac{90,000}{x} = \frac{90,000}{x + 6} + 2500$$

$$90,000(x + 6) = 90,000x + 2500x(x + 6), \quad x \neq 0, \ x \neq -6$$

$$90,000x + 540,000 = 90,000x + 2500x^2 + 15,000x$$

$$0 = 2500x^2 + 15,000x - 540,000$$

$$0 = x^2 + 6x - 216$$

$$0 = (x - 12)(x + 18)$$

$$x - 12 = 0 \implies x = 12$$

$$x + 18 = 0 \implies x = -18$$

Choosing the positive value, it follows that the dealer sold 12 cars at an average price of $\frac{1}{12}(90,000) = \$7500$ per car. Check this result in the original statement of the problem.

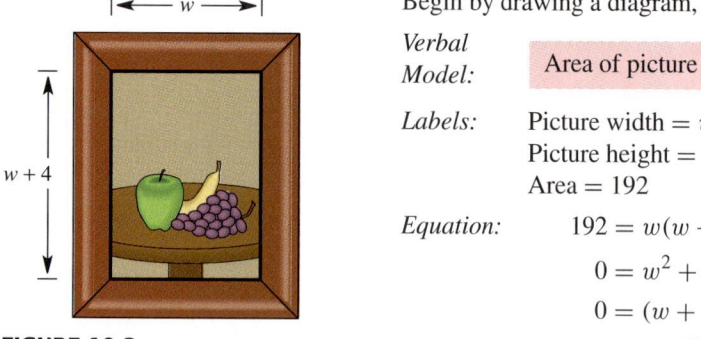

FIGURE 10.3

EXAMPLE 2   *Geometry*

A picture is 4 inches taller than it is wide and has an area of 192 square inches. What are the dimensions of the picture?

Solution

Begin by drawing a diagram, as shown in Figure 10.3.

*Verbal
Model:*   Area of picture   =   Width   ·   Height

Labels: Picture width $= w$ (inches)
Picture height $= w + 4$ (inches)
Area $= 192$ (square inches)

Equation: $192 = w(w + 4)$

$0 = w^2 + 4w - 192$

$0 = (w + 16)(w - 12)$

$w + 16 = 0 \implies w = -16$

$w - 12 = 0 \implies w = 12$

Of the two possible solutions, choose the positive value of w and conclude that the picture is $w = 12$ inches wide and $w + 4$ or 16 inches tall. Check these dimensions in the original statement of the problem.

EXAMPLE 3   *An Interest Problem*

The formula

$$A = P(1 + r)^2$$

represents the amount of money A in an account in an which P dollars is deposited for 2 years at an annual interest rate of r (in decimal form). Find the interest rate if a deposit of $6000 increases to $6933.75 over a 2-year period.

Solution

$A = P(1 + r)^2$ Given formula

$6933.75 = 6000(1 + r)^2$ Substitute for A and P.

$1.155625 = (1 + r)^2$ Divide both sides by 6000.

$\pm 1.075 = 1 + r$ Extract square roots.

$0.075 = r$ Choose positive solution.

The annual interest rate is $r = 0.075 = 7.5\%$. Check this result in the original statement of the problem.

EXAMPLE 4 Reduced Rates

A ski club chartered a bus for a ski trip at a cost of $520. In an attempt to lower the bus fare per skier, the club invited nonmembers to go along. When five nonmembers joined the trip, the fare per skier decreased by $5.20. How many club members are going on the trip?

Solution

Verbal Model:

$$\boxed{\text{Cost per skier}} \cdot \boxed{\text{Number of skiers}} = \boxed{\$520}$$

Labels:

Number of ski club members $= x$ (people)

Number of skiers $= x + 5$ (people)

Original cost per skier $= \dfrac{520}{x}$ (dollars)

New cost per skier $= \dfrac{520}{x} - 5.20$ (dollars)

Equation:

$$\left(\frac{520}{x} - 5.20\right)(x + 5) = 520$$

$$\left(\frac{520 - 5.2x}{x}\right)(x + 5) = 520$$

$$(520 - 5.2x)(x + 5) = 520x, \quad x \neq 0$$

$$520x - 5.2x^2 - 26x + 2600 = 520x$$

$$-5.2x^2 - 26x + 2600 = 0$$

$$x^2 + 5x - 500 = 0$$

$$(x + 25)(x - 20) = 0$$

$$x + 25 = 0 \implies x = -25$$

$$x - 20 = 0 \implies x = 20$$

Choosing the positive value of x implies that there are 20 ski club members. Check this solution in the original equation, as follows.

Check

Number of Skiers	Cost per Skier
20	$\dfrac{520}{20} = \$26.00$
25	$\dfrac{520}{25} = \$20.80$

From these two calculations, you can see that the difference in cost per skier is $\$26.00 - \$20.80 = \$5.20$.

EXAMPLE 5 An Application Involving the Pythagorean Theorem

An L-shaped sidewalk from building A to building B on a college campus is 200 meters long, as shown in Figure 10.4. By cutting diagonally across the grass, students shorten the walking distance to 150 meters. What are the lengths of the two legs of the existing sidewalk?

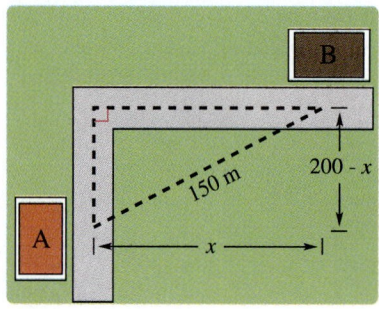

FIGURE 10.4

Solution

Verbal Model:

$$a^2 + b^2 = c^2 \qquad \text{Pythagorean Theorem}$$

Labels:

Length of one leg $= x$	(meters)
Length of other leg $= 200 - x$	(meters)
Length of diagonal $= 150$	(meters)

Equation:

$$x^2 + (200 - x)^2 = (150)^2$$
$$2x^2 - 400x + 40,000 = 22,500$$
$$2x^2 - 400x + 17,500 = 0$$
$$x^2 - 200x + 8750 = 0$$

By the Quadratic Formula, you can find the solutions as follows.

$$x = \frac{200 \pm \sqrt{(-200)^2 - 4(1)(8750)}}{2(1)}$$
$$= \frac{200 \pm \sqrt{5000}}{2}$$
$$= \frac{200 \pm 50\sqrt{2}}{2}$$
$$= 100 \pm 25\sqrt{2}$$

Both solutions are positive, and it does not matter which one you choose. If you let

$$x = 100 + 25\sqrt{2} \approx 135.4 \text{ meters},$$

the length of the other leg is

$$200 - x \approx 200 - 135.4 \approx 64.6 \text{ meters}.$$

NOTE In Example 5, notice that you obtain the same dimensions if you choose the other value of x. That is, if the length of one leg is

$$x = 100 - 25\sqrt{2} \approx 64.6 \text{ meters},$$

the length of the other leg is

$$200 - x \approx 200 - 64.6 \approx 135.4 \text{ meters}.$$

EXAMPLE 6   Work Problem

An office contains two copy machines. Machine B is known to take 12 minutes longer than machine A to copy the company's monthly report. Using both machines together, it takes 8 minutes to reproduce the report. How long would it take each machine alone to reproduce the report?

Solution

Verbal Model:

Work done by machine A	+	Work done by machine B	=	1 complete job

Rate for A	·	Time for both	+	Rate for B	·	Time for both	=	1

Labels:

Time for machine A $= t$	(minutes)
Rate for machine A $= 1/t$	(job per minute)
Time for machine B $= t + 12$	(minutes)
Rate for machine B $= 1/(t + 12)$	(job per minute)
Time for both machines $= 8$	(minutes)
Rate for both machines $= \frac{1}{8}$	(job per minute)

Equation:

$$\frac{1}{t}(8) + \frac{1}{t + 12}(8) = 1$$

$$8\left(\frac{1}{t} + \frac{1}{t + 12}\right) = 1$$

$$8\left[\frac{t + 12 + t}{t(t + 12)}\right] = 1$$

$$8t(t + 12)\left[\frac{2t + 12}{t(t + 12)}\right] = t(t + 12)$$

$$8(2t + 12) = t^2 + 12t$$

$$16t + 96 = t^2 + 12t$$

$$0 = t^2 - 4t - 96$$

$$0 = (t - 12)(t + 8)$$

$$t - 12 = 0 \implies t = 12$$

$$t + 8 = 0 \implies t = -8$$

Choose the positive value for t and find that

Time for machine A $= t = 12$ minutes

Time for machine B $= t + 12 = 24$ minutes.

Check these solutions in the original equation.

EXAMPLE 7   The Height of a Model Rocket

A model rocket is projected straight upward from ground level according to the height equation $h = -16t^2 + 192t$, $t \geq 0$, where h is the height in feet and t is the time in seconds. (a) After how many seconds will the height be 432 feet? (b) When will the rocket hit the ground?

Solution

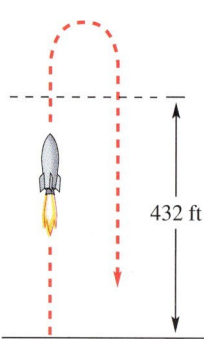

FIGURE 10.5

a.

$h = -16t^2 + 192t$	Original equation
$432 = -16t^2 + 192t$	Substitute 432 for h.
$16t^2 - 192t + 432 = 0$	Standard form
$t^2 - 12t + 27 = 0$	Divide both sides by 16.
$(t - 3)(t - 9) = 0$	Factor.
$t - 3 = 0 \quad\Longrightarrow\quad t = 3$	Set 1st factor equal to 0.
$t - 9 = 0 \quad\Longrightarrow\quad t = 9$	Set 2nd factor equal to 0.

The rocket obtains a height of 432 feet at two different times—once (going up) after 3 seconds, and again (coming down) after 9 seconds. (See Figure 10.5.)

b.   To find the time it takes for the rocket to hit the ground, let the height be 0.

$$0 = -16t^2 + 192t$$
$$0 = t^2 - 12t$$
$$0 = t(t - 12)$$
$$t = 0 \quad \text{or} \quad t = 12$$

The rocket will hit the ground after 12 seconds. (Note that the time of $t = 0$ seconds corresponds to the time of lift-off.)

Group Activities   Exploring with Technology

Analyzing Quadratic Functions   Use a graphing utility to graph $y_1 = 3x^2 + 2x - 1$ and $y_2 = -x^2 + 5x + 4$. For each function, use the ZOOM and TRACE features to find either the maximum or minimum function value. Discuss other methods that you could use to find these values.

10.4 **Exercises**

Discussing the Concepts

1. In your own words, describe guidelines for solving a word problem.

2. Describe the strategies that can be used to solve a quadratic equation.

3. *Unit Analysis* Describe the units of the product.

$$\frac{9 \text{ dollars}}{\text{hour}} \cdot (20 \text{ hours})$$

4. *Unit Analysis* Describe the units of the product.

$$\frac{20 \text{ feet}}{\text{minute}} \cdot \frac{1 \text{ minute}}{60 \text{ seconds}} \cdot (45 \text{ seconds})$$

5. Give an example of a quadratic equation that has only one repeated solution.

6. Give an example of a quadratic equation that has two imaginary solutions.

Problem Solving

In Exercises 7–10, find two positive integers that satisfy the requirement.

7. The product of two consecutive integers is 240.

8. The product of two consecutive integers is 1122.

9. The product of two consecutive *even* integers is 224.

10. The product of two consecutive *odd* integers is 255.

In Exercises 11–14, complete the table of widths, lengths, perimeters, and areas of rectangles.

	Width	Length	Perimeter	Area
11.	$0.75l$	l	42 in.	
12.	$l - 6$	l	108 ft	
13.	$l - 20$	l		$12{,}000 \text{ m}^2$
14.	w	$1.5w$		216 cm^2

Compound Interest In Exercises 15–18, find the interest rate r. Use the formula $A = P(1 + r)^2$, where A is the amount after 2 years in an account earning r percent compounded annually and P is the original investment.

15. $P = \$3000.00$
    $A = \$3499.20$

16. $P = \$10{,}000.00$
    $A = \$11{,}990.25$

17. $P = \$8000.00$
    $A = \$8420.20$

18. $P = \$6500.00$
    $A = \$7372.46$

19. *Geometry* A television station claims that it covers a circular region of approximately 25,000 square miles.

    (a) Assume that the station is located at the center of the circular region. How far is the station from its farthest listener?

    (b) Assume that the station is located on the edge of the circular region. How far is the station from its farthest listener?

20. *Geometry* The height of a triangle is twice its base. The area of the triangle is 625 square inches. Find the dimensions of the triangle.

21. *Geometry* A retail lumber business plans to build a rectangular storage region adjoining the sales office (see figure). The region will be fenced on three sides, and the fourth side will be bounded by the existing building. Find the dimensions of the region if 350 feet of fencing is used and the area of the region is 12,500 square feet.

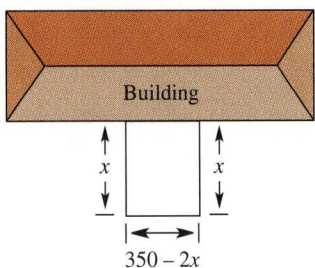

22. *Geometry* Your home is built on a square lot. To add more space to your yard, you purchase an additional 20 feet along the side of the property (see figure). The area of the lot is now 25,500 square feet. What are the dimensions of the new lot?

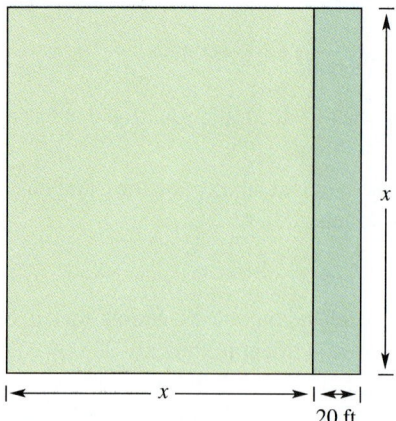

x

x

20 ft

23. *Selling Price* A store owner bought a case of grade A large eggs for $21.60. By the time all but 6 dozen of the eggs had been sold at a profit of $0.30 per dozen, the original investment of $21.60 had been regained. How many dozen eggs did the owner sell, and what was the selling price per dozen?

24. *Selling Price* A manager of a computer store bought several computers of the same model for $27,000. When all but three of the computers had been sold at a profit of $750 per computer, the original investment of $27,000 had been regained. How many computers were sold, and what was the selling price of each computer?

25. *Reduced Ticket Price* A service organization obtains a block of tickets to a ball game for $240. When eight more people decide to go to the game, the price per ticket is decreased by $1. How many people are going to the game?

26. *Reduced Fare* A science club charters a bus to attend a science fair at a cost of $480. In an attempt to lower the bus fare per person, the club invites nonmembers to go along. When two nonmembers join the trip, the fare per person is decreased by $1. How many people are going on the excursion?

27. *Solving Graphically and Algebraically* An adjustable rectangular form has minimum dimensions of 3 meters by 4 meters. The length and width can be expanded by equal amounts x (see figure).

(a) Write the length d of the diagonal as a function of x. Use a graphing utility to graph the function. Use the graph to approximate the value of x when $d = 10$ meters.

(b) Find x algebraically when $d = 10$.

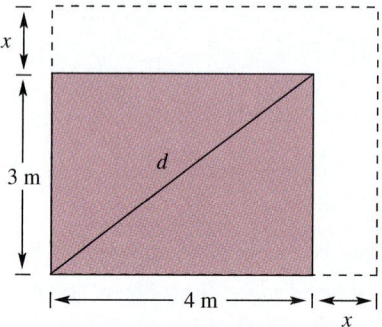

x

3 m

d

4 m

x

28. *Solving Graphically and Numerically* A meteorologist is positioned 100 feet from the point where a weather balloon is launched (see figure).

(a) Write the distance d between the balloon and the meteorologist as a function of the height h of the balloon. Use a graphing utility to graph the function. Use the graph to approximate the value of h when $d = 200$ feet.

(b) Complete the table. Use the table to approximate the value of h when $d = 200$ feet.

h	160	165	170	175	180	185
d						

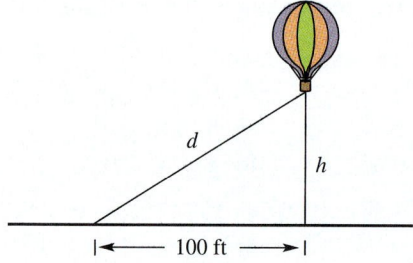

d

h

100 ft

29. *Airspeed*   An airline runs a commuter flight between two cities that are 720 miles apart. If the average speed of the planes could be increased by 40 miles per hour, the travel time would be decreased by 12 minutes. What airspeed is required to obtain this decrease in travel time?

30. *Average Speed*   A truck traveled the first 100 miles of a trip at one speed and the last 135 miles at an average speed of 5 miles per hour less. If the entire trip took 5 hours, what was the average speed for the first part of the trip?

31. *Work Rate*   Working together, two people can complete a task in 5 hours. Working alone, how long would it take each to do the task if one person took 2 hours longer than the other?

32. *Work Rate*   An office contains two printers. Machine B is known to take 3 minutes longer than machine A to produce the company's monthly financial report. Using both machines together, it takes 6 minutes to produce the report. How long would it take each machine to produce the report?

Free-Falling Object   In Exercises 33–36, find the time necessary for an object to fall to ground level from an initial height of h_0 feet if its height h at any time t (in seconds) is given by $h = h_0 - 16t^2$.

33. $h_0 = 144$

34. $h_0 = 625$

35. $h_0 = 1454$ (height of the Sears Tower)

36. $h_0 = 984$ (height of the Eiffel Tower)

Cost, Revenue, and Profit   In Exercises 37 and 38, you are given the cost C of producing x units, the revenue R from selling x units, and the profit P. Find the value of x that will produce the profit P.

37. $C = 100 + 30x$, $R = x(90 - x)$, $P = \$800$

38. $C = 4000 - 40x + 0.02x^2$, $R = x(50 - 0.01x)$, $P = \$63,500$

In Exercises 39 and 40, solve for the specified variable.

39. Solve for r in $A = P(1 + r)^2$.

40. Solve for r in $A = \pi r^2$.

Reviewing the Major Concepts

In Exercises 41–44, find the product.

41. $-2x^5(5x^{-3})$

42. $(3x + 2)(7x - 10)$

43. $(2x - 15)^2$

44. $(x + 3)(x^2 - 3x + 9)$

45. *List Price*   A computer is discounted 15% from its list price to a sale price of $1955. Find the list price.

46. *Investment*   A combined total of $24,000 is invested in two bonds that pay 7.5% and 9% simple interest. The total annual interest is $1935. How much is invested in each bond?

Additional Problem Solving

In Exercises 47–50, find two positive integers that satisfy the requirement.

47. The product of two consecutive *odd* integers is 483.

48. The product of two consecutive *even* integers is 528.

49. The sum of the squares of two consecutive integers is 313.

50. The sum of the squares of two consecutive integers is 421.

In Exercises 51–54, complete the table of widths, lengths, perimeters, and areas of rectangles.

	Width	Length	Perimeter	Area
51.	w	$w + 3$	54 km	
52.	w	$1.5w$	40 m	
53.	$\frac{3}{4}l$	l		2700 in.^{2}
54.	$\frac{1}{3}l$	l		192 in.^{2}

55. *Geometry*   An open-top rectangular conduit for carrying water in a manufacturing process is made by folding up the edges of a sheet of aluminum 48 inches wide (see figure). A cross section of the conduit must have an area of 288 square inches. Find the width and height of the conduit.

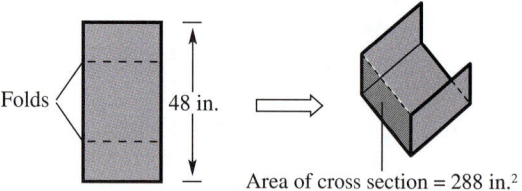

Area of cross section = 288 in.²

56. *Geometry*   A friend built a fence around three sides of his property (see figure). In total, he used 550 feet of fencing. By his calculations, the area of the lot is 1 acre (43,560 square feet). Is this correct? Explain your answer.

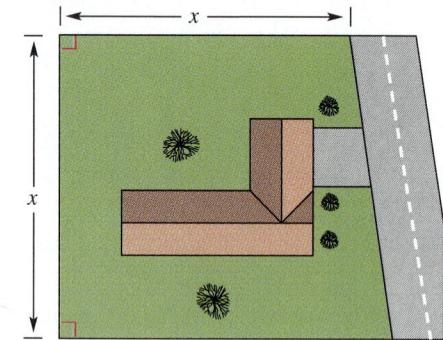

Compound Interest   In Exercises 57–60, find the interest rate r. Use the formula $A = P(1 + r)^2$, where A is the amount after 2 years in an account earning r percent compounded annually and P is the original investment.

57. $P = \$250.00$, $A = \$280.90$

58. $P = \$500.00$, $A = \$572.45$

59. $P = \$10,000.00$, $A = \$11,556.25$

60. $P = \$2000.00$, $A = \$2354.45$

61. *Venture Capital*   Eighty thousand dollars is needed to begin a small business. The cost will be divided equally among investors. Some have made a commitment to invest. If three more investors are found, the amount required from each would decrease by $6000. How many have made a commitment to invest in the business?

62. *Ticket Prices*   A service organization paid $210 for a block of tickets to a ball game. The block contained three more tickets than the organization needed for its members. By inviting three more people to attend (and share in the cost), the organization lowered the price per ticket by $3.50. How many people are going to the game?

63. *Delivery Route*   You are asked to deliver pizza to offices B and C in your city (see figure), and you are required to keep a log of all the mileages between stops. You forget to look at the odometer at stop B, but after getting to stop C you record the total distance traveled from the pizza shop as 18 miles. The return distance from C to A is 16 miles. If the route approximates a right triangle, estimate the distance from A to B.

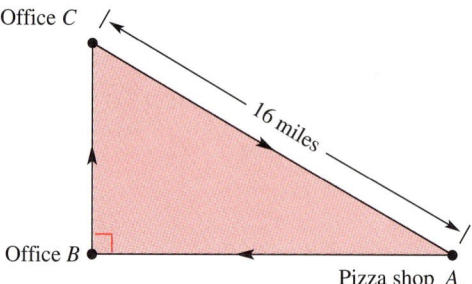

64. *Geometry*   The perimeter of a rectangle is 102 inches and the length of the diagonal is 39 inches. Find the dimensions of the rectangle.

65. *Work Rate*   A builder works with two plumbing companies. Company A is known to take 3 days longer than Company B to do the plumbing in a particular style of house. Using both companies it takes 4 days. How long would it take to do the plumbing using each company individually?

66. *Work Rate*   Working together, two people can complete a task in 6 hours. Working alone, one person takes 2 hours longer than the other. How long would it take each person to do the task alone?

67. *Height of a Baseball*   The height h in feet of a baseball hit 3 feet above the ground is given by

$$h = 3 + 75t - 16t^2$$

where t is time in seconds. Find the time when the ball hits the ground in the outfield.

68. *Hitting Baseballs*   You are hitting baseballs. When you toss the ball into the air, your hand is 5 feet above the ground (see figure). You hit the ball when it falls back to a height of 4 feet. If you toss the ball with an initial velocity of 25 feet per second, the height h of the ball t seconds after leaving your hand is given by

$$h = 5 + 25t - 16t^2.$$

How much time will pass before you hit the ball?

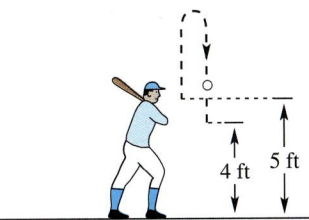

4 ft   5 ft

In Exercises 69–72, solve for the specified variable.

69. Solve for r in $S = 4\pi r^2$.

70. Solve for s in $A = \frac{1}{4}s^2\sqrt{3}$.

71. Solve for b in $I = \frac{1}{12}(M)(a^2 + b^2)$.

72. Solve for h in $S = \pi r\sqrt{r^2 + h^2}$.

73. A small business uses a minivan to make deliveries. The cost per hour for fuel for the van is

$$C = \frac{v^2}{600}$$

where v is the speed in miles per hour. The driver is paid $5 per hour. Find the speed if the cost for wages and fuel for a 110-mile trip is $20.39.

74. *Reading a Graph*   For the years 1983 to 1990, the number of mountain bike owners m (in millions) in the United States can be approximated by the model

$$m = 0.337t^2 - 2.265t + 3.962, \quad 3 \le t \le 10$$

where $t = 3$ represents 1983 (see figure). In which year did 2.5 million people own mountain bikes? (Source: Bicycle Institute of America)

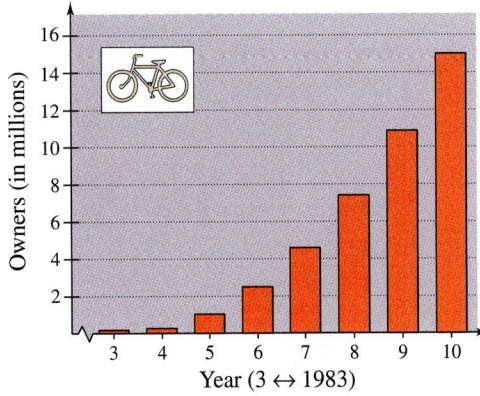

75. *Geometry*   The area A of an ellipse is given by $A = \pi ab$ (see figure). For a certain ellipse it is required that $a + b = 20$.

(a) Show that $A = \pi a(20 - a)$.

(b) Complete the following table.

a	4	7	10	13	16
A					

(c) Find two values of a such that $A = 300$.

(d) Use a graphing utility to graph the area function.

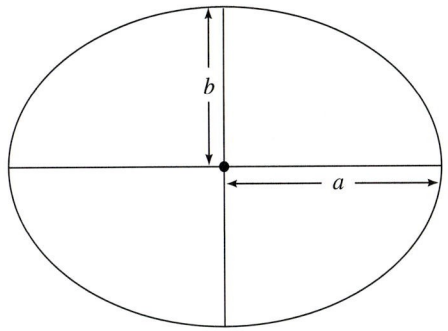

Math Matters   Olympic Games

No.	Year	City	Participants
1	1896	Athens, Greece	13
2	1900	Paris, France	22
3	1904	St. Louis, USA	12
4	1908	London, England	23
5	1912	Stockholm, Sweden	28
6	1916	Berlin, Germany	—
7	1920	Antwerp, Belgium	29
8	1924	Paris, France	44
9	1928	Amsterdam, Holland	46
10	1932	Los Angeles, USA	37
11	1936	Berlin, Germany	49
12	1940	Tokyo, Japan	—
13	1944	London, England	—
14	1948	London, England	59
15	1952	Helsinki, Finland	69
16	1956	Melbourne, Australia	67
17	1960	Rome, Italy	83
18	1964	Tokyo, Japan	93
19	1968	Mexico City, Mexico	112
20	1972	Munich, West Germany	122
21	1976	Montreal, Canada	92
22	1980	Moscow, USSR	81
23	1984	Los Angeles, USA	140
24	1988	Seoul, South Korea	160
25	1992	Barcelona, Spain	172
26	1996	Atlanta, USA	

As this book is being written, the 26th Summer Olympic Games are scheduled to be held in Atlanta, Georgia, USA, in the summer of 1996. The original Olympic Games were held once every four years in ancient Greece. The earliest recorded games took place in 776 B.C., and the games were abandoned in 393 A.D.

In 1896 the games were revived. Since then, they have taken place every four years, except during the two World Wars. Note that the three games that were not held had already been planned, and therefore kept their planned numbers.

The number of participating countries has tended to increase each year. Do you know why the number of participants was down in 1976 and 1980?

Finding Test Intervals ▪ Quadratic Inequalities ▪
Rational Inequalities ▪ Application

Finding Test Intervals

When you are working with polynomial inequalities, it is important to realize that the value of a polynomial can change signs only at its **zeros.** That is, a polynomial can change signs only at the x-values that make the polynomial zero. For instance, the first-degree polynomial $x + 2$ has a zero at -2, and it changes sign at that zero. You can picture this result on the real number line, as shown in Figure 10.6.

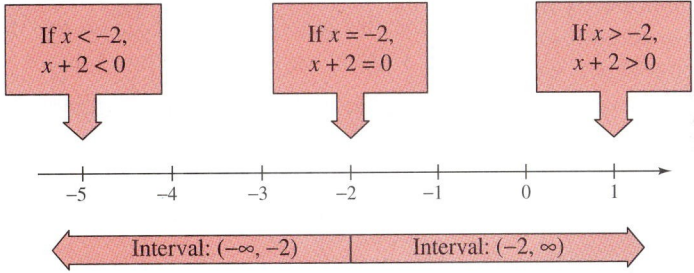

If $x < -2$,
$x + 2 < 0$

If $x = -2$,
$x + 2 = 0$

If $x > -2$,
$x + 2 > 0$

Interval: $(-\infty, -2)$ Interval: $(-2, \infty)$

FIGURE 10.6

Note in Figure 10.6 that the zero of the polynomial partitions the real number line into two **test intervals.** The polynomial is negative for every x-value in the first test interval $(-\infty, -2)$, and it is positive for every x-value in the second test interval $(-2, \infty)$. You can use the same basic approach to determine the test intervals for any polynomial.

Finding Test Intervals for a Polynomial

1. Find all real zeros of the polynomial, and arrange the zeros in increasing order. The zeros of a polynomial are called its **critical numbers.**

2. Use the critical numbers of the polynomial to determine its test intervals.

3. Choose a representative x-value in each test interval and evaluate the polynomial at that value. If the value of the polynomial is negative, the polynomial will have negative values for *every* x-value in the interval. If the value of the polynomial is positive, the polynomial will have positive values for *every* x-value in the interval.

Technology

Most graphing utilities can sketch the graph of a quadratic inequality. For instance, the following steps are used to sketch the graph of $x^2 - 5x < 0$ on a *TI-82*.

 $Y_1 = X^2 - 5X < 0$

GRAPH (Standard setting)

The graph produced by these steps is shown below. Notice that the graph occurs as an interval *above* the x-axis.

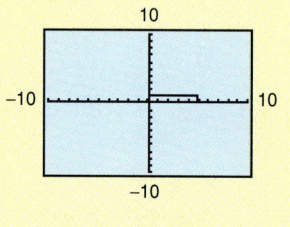

Quadratic Inequalities

The concepts of critical numbers and test intervals can be used to solve nonlinear inequalities, as demonstrated in Examples 1, 2, and 4.

EXAMPLE 1   Solving a Quadratic Inequality

Solve the inequality $x^2 - 5x < 0$.

Solution

Begin by finding the critical numbers of $x^2 - 5x$.

$$x^2 - 5x < 0 \qquad \text{Original inequality}$$
$$x(x - 5) < 0 \qquad \text{Factor.}$$

From this factorization, you can see that the critical numbers of $x^2 - 5x$ are 0 and 5. This implies that the test intervals are

$$(-\infty, 0), \quad (0, 5), \quad \text{and} \quad (5, \infty).$$

To test an interval, choose a convenient number in the interval and compute the sign of $x(x - 5)$. The results are shown in Figure 10.7.

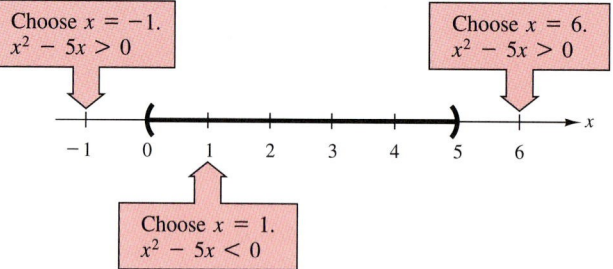

FIGURE 10.7

Because the inequality $x^2 - 5x < 0$ is satisfied only by the middle test interval, you can conclude that the solution set of the inequality is the interval $(0, 5)$.

In Example 1, note that you would have used the same basic procedure if the inequality symbol had been $\leq$, $>$, or $\geq$. For instance, from Figure 10.7, you can see that the solution set of the inequality

$$x^2 - 5x \geq 0$$

consists of the union of the half-open intervals $(-\infty, 0]$ and $[5, \infty)$, which is written as $(-\infty, 0] \cup [5, \infty)$.

Just as with solving quadratic *equations*, the first step in solving a quadratic *inequality* is to write the inequality in **standard form,** with the polynomial on the left and zero on the right, as demonstrated in Example 2.

EXAMPLE 2   Solving a Quadratic Inequality

Solve the inequality $2x^2 + 5x > 12$.

Solution

Begin by writing the inequality in standard form.

$$2x^2 + 5x > 12 \qquad\qquad \text{Original inequality}$$

$$2x^2 + 5x - 12 > 0 \qquad\qquad \text{Write in standard form.}$$

$$(x + 4)(2x - 3) > 0 \qquad\qquad \text{Factor.}$$

From this factorization, you can see that the critical numbers of $2x^2 + 5x - 12$ are -4 and $\frac{3}{2}$. This implies that the test intervals are

$$(-\infty, -4), \quad \left(-4, \tfrac{3}{2}\right), \quad \text{and} \quad \left(\tfrac{3}{2}, \infty\right).$$

To test an interval, choose a convenient number in the interval and compute the sign of $(x + 4)(2x - 3)$. The results are shown in Figure 10.8. From the figure, you can see that the polynomial $2x^2 + 5x - 12$ is positive in the open intervals $(-\infty, -4)$ and $\left(\frac{3}{2}, \infty\right)$. Thus, the solution set is $(-\infty, -4) \cup \left(\frac{3}{2}, \infty\right)$.

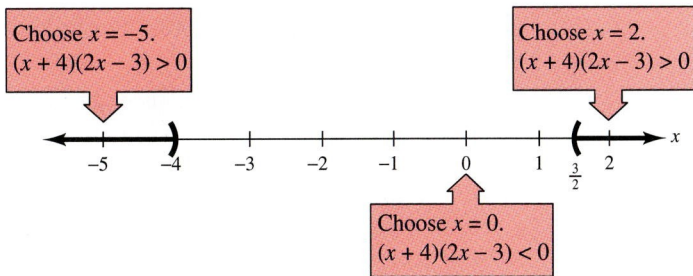

FIGURE 10.8

In Examples 1 and 2, the critical numbers were found by factoring. With quadratic polynomials that do not factor, you can use the Quadratic Formula to find the critical numbers. For instance, to solve the inequality

$$x^2 - 2x - 1 \le 0$$

you can use the Quadratic Formula to determine that the critical numbers are $1 - \sqrt{2} \approx -0.414$ and $1 + \sqrt{2} \approx 2.414$.

The solutions of the quadratic inequalities in Examples 1 and 2 consist, respectively, of a single interval and the union of two intervals. When solving the exercises for this section, you should be on the watch for some unusual solution sets, as illustrated in Example 3.

EXAMPLE 3   *Unusual Solution Sets*

Solve each inequality to verify that the given solution set is correct.

a. The solution set of the quadratic inequality

$$x^2 + 2x + 4 > 0$$

consists of the entire set of real numbers, $(-\infty, \infty)$. In other words, the quadratic $x^2 + 2x + 4$ is positive for every real value of x. (Note that this quadratic inequality has no critical numbers. In such a case, there is only one test interval—the entire real line.)

b. The solution set of the quadratic inequality

$$x^2 + 2x + 1 \leq 0$$

consists of the single real number $\{-1\}$.

c. The solution set of the quadratic inequality

$$x^2 + 3x + 5 < 0$$

is empty. In other words, the quadratic $x^2 + 3x + 5$ is not less than zero for any value of x.

d. The solution set of the quadratic inequality

$$x^2 - 4x + 4 > 0$$

consists of all real numbers *except* the number 2. In interval notation, this solution set can be written as $(-\infty, 2) \cup (2, \infty)$.

Remember that checking the solution set of an inequality is not as straightforward as checking the solutions of an equation, because inequalities tend to have infinitely many solutions. Even so, we suggest that you check several x-values in your solution set to confirm that they satisfy the inequality. Also try checking x-values that are not in the solution set to confirm that they do not satisfy the inequality.

For instance, the solution of $x^2 - 5x < 0$ is $(0, 5)$. Try checking some numbers in this interval to verify that they satisfy the inequality. Then check some numbers outside the interval to verify that they do not satisfy the inequality.

Rational Inequalities

The concepts of critical numbers and test intervals can be extended to inequalities involving rational expressions. To do this, use the fact that the value of a rational expression can change sign only at its *zeros* (the x-values for which its numerator is zero) and its *undefined values* (the x-values for which its denominator is zero). These two types of numbers make up the **critical numbers** of a rational inequality. For instance, the critical numbers of the inequality

$$\frac{x-2}{(x-1)(x+3)} < 0$$

are $x = 2$ (the numerator is zero), and $x = 1$ and $x = -3$ (the denominator is zero). From these three critical numbers you can see that the inequality has *four* test intervals.

$$(-\infty, -3), \quad (-3, 1), \quad (1, 2), \quad \text{and} \quad (2, \infty)$$

EXAMPLE 4   Solving a Rational Inequality

Solve the inequality $\dfrac{x}{x-2} > 0$.

Solution

The numerator is zero when $x = 0$ and the denominator is zero when $x = 2$. Thus, the two critical numbers are 0 and 2, which implies that the test intervals are

$$(-\infty, 0), \quad (0, 2), \quad \text{and} \quad (2, \infty).$$

A test of these intervals, as shown in Figure 10.9, reveals that the rational expression $x/(x-2)$ is positive in the open intervals $(-\infty, 0)$ and $(2, \infty)$. Therefore, the solution set of the inequality is $(-\infty, 0) \cup (2, \infty)$.

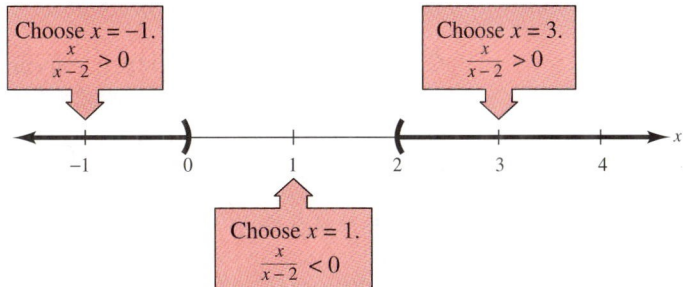

FIGURE 10.9

Application

EXAMPLE 5 The Height of a Projectile

A projectile is fired straight up from ground level with an initial velocity of 256 feet per second, as shown in Figure 10.10, so that its height at any time t is given by

$$h = -16t^2 + 256t$$

where the height h is measured in feet and the time t is measured in seconds. During what interval of time will the height of the projectile exceed 960 feet?

Solution

To solve this problem, find the values of t for which h is greater than 960.

$$-16t^2 + 256t > 960 \qquad \text{Original inequality}$$

$$-16t^2 + 256t - 960 > 0 \qquad \text{Subtract 960 from both sides.}$$

$$t^2 - 16t + 60 < 0 \qquad \text{Divide both sides by } -16 \text{ and reverse the inequality.}$$

$$(t - 6)(t - 10) < 0 \qquad \text{Factor.}$$

Thus, the critical numbers are $t = 6$ and $t = 10$. A test of the intervals $(-\infty, 6)$, $(6, 10)$, and $(10, \infty)$ shows that the solution interval is $(6, 10)$. Therefore, the height of the object will exceed 960 feet for values of t that are greater than 6 seconds and less than 10 seconds.

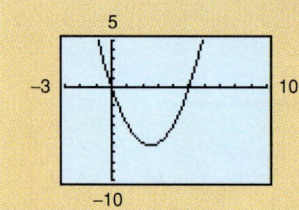

960 ft

Velocity: 256 ft/sec

FIGURE 10.10

Group Activities

Exploring with Technology

Graphing an Inequality You can use a graph on a rectangular coordinate system as an alternative method for solving an inequality. For instance, to solve the inequality in Example 1

$$x^2 - 5x < 0$$

you can sketch the graph of $y = x^2 - 5x$. Using a graphing utility, you can obtain the graph shown at the left. From the graph, you can see that the only part of the curve that lies below the x-axis is the portion for which $0 < x < 5$. Thus, the solution of $x^2 - 5x < 0$ is $0 < x < 5$. Try using this graphing approach to solve some of the other examples in this section.

10.5 Exercises

Discussing the Concepts

1. Explain the change in an inequality that occurs when both sides are multiplied by a negative real number.

2. Give a verbal description of the union of intervals $(-\infty, 5] \cup (10, \infty)$.

3. Define the term *critical number* and explain its use in solving quadratic inequalities.

4. In your own words, describe the procedure for solving quadratic inequalities.

5. Give an example of a quadratic inequality that has no real solution.

6. Explain the distinction between the critical numbers of quadratic inequalities and rational inequalities.

Problem Solving

In Exercises 7–10, find the critical numbers.

7. $x(2x - 5)$

8. $5x(x - 3)$

9. $x^2 - 4x + 3$

10. $3x^2 - 2x - 8$

In Exercises 11–14, determine whether the x-values are solutions of the inequality.

11. $2x^2 - 7x - 4 > 0$

   (a) $x = 0$ (b) $x = 6$

   (c) $x = -\frac{1}{2}$ (d) $x = -\frac{3}{2}$

12. $4x^2 + 3x - 10 \le 0$

   (a) $x = 0$ (b) $x = 3$

   (c) $x = -2$ (d) $x = \frac{1}{2}$

13. $\dfrac{2}{3 - x} \le 0$

   (a) $x = 3$ (b) $x = 4$

   (c) $x = -4$ (d) $x = -\frac{1}{3}$

14. $\dfrac{x - 2}{x + 1} > 0$

   (a) $x = 0$ (b) $x = -1$

   (c) $x = -4$ (d) $x = -\frac{3}{2}$

In Exercises 15–18, determine the intervals for which the polynomial is entirely negative and entirely positive.

15. $x - 4$

16. $7x(3 - x)$

17. $x^2 - 4x - 5$

18. $2x^2 - 4x - 3$

In Exercises 19–30, solve the inequality and graph the solution on the real number line. (Some of the inequalities have no solution.)

19. $2x + 6 \ge 0$

20. $5x - 20 < 0$

21. $3x(2 - x) < 0$

22. $2x(6 - x) > 0$

23. $x^2 + 3x \le 10$

24. $t^2 - 15t + 50 < 0$

25. $x^2 + 4x + 5 < 0$

26. $x^2 + 6x + 10 > 0$

27. $x^2 + 2x + 1 \ge 0$

28. $25 \ge (x - 3)^2$

29. $x(x - 2)(x + 2) > 0$

30. $x^2(x - 2) \le 0$

In Exercises 31–34, determine the critical numbers of the rational expression and plot them on the real number line.

31. $\dfrac{5}{x - 3}$

32. $\dfrac{-6}{x + 2}$

33. $\dfrac{2x}{x + 5}$

34. $\dfrac{x - 2}{x - 10}$

In Exercises 35–38, solve the rational inequality. As part of your solution, include a graph that shows the test intervals.

35. $\dfrac{5}{x - 3} > 0$

36. $\dfrac{x + 5}{3x + 2} \ge 0$

37. $\dfrac{x}{x - 3} \le 2$

38. $\dfrac{x + 4}{x - 5} \ge 10$

In Exercises 39–42, use a graphing utility to solve the inequality. Use the strategies described on page 420 and on page 423.

39. $0.5x^2 + 1.25x - 3 > 0$

40. $\frac{1}{3}x^2 - 3x < 0$

41. $\frac{6x}{x+5} < 2$

42. $x + \frac{1}{x} > 3$

43. *Height of a Projectile*   A projectile is fired straight up from ground level with an initial velocity of 128 feet per second, so that its height at any time t is given by $h = -16t^2 + 128t$, where the height h is measured in feet and the time t is measured in seconds. During what interval of time will the height of the projectile exceed 240 feet?

44. *Annual Interest Rate*   You are investing $1000 in a certificate of deposit for 2 years and you want the interest to exceed $150. The interest is compounded annually. What interest rate should you have?

Reviewing the Major Concepts _____

In Exercises 45–48, solve the inequality.

45. $7 - 3x > 4 - x$ **46.** $2(x + 6) - 20 < 2$

47. $|x - 3| < 2$ **48.** $|x - 5| > 3$

49. *Geometry*   Determine the length and width of a rectangle with a perimeter of 50 inches and a diagonal of length $5\sqrt{13}$ inches.

50. *Demand for a Product*   The demand equation for a certain product is

$$p = 75 - \sqrt{1.2(x - 10)}$$

where x is the number of units demanded per day and p is the price per unit. Find the demand if the price is $59.90.

Additional Problem Solving _____

In Exercises 51–54, find the critical numbers.

51. $4x^2 - 81$ **52.** $y(y - 4) - 3(y - 4)$

53. $4x^2 - 20x + 25$ **54.** $4x^2 - 4x - 3$

In Exercises 55–58, determine the intervals for which the polynomial is entirely negative and entirely positive.

55. $\frac{2}{3}x - 8$ **56.** $3 - x$

57. $2x(x - 4)$ **58.** $x^2 - 9$

In Exercises 59–74, solve the inequality and graph the solution on the real number line. (Some of the inequalities have no solution.)

59. $-\frac{3}{4}x + 6 < 0$ **60.** $3x - 2 \geq 0$

61. $3x(x - 2) < 0$ **62.** $2x(x - 6) > 0$

63. $x^2 - 4 \geq 0$ **64.** $z^2 \leq 9$

65. $-2u^2 + 7u + 4 < 0$ **66.** $-3x^2 - 4x + 4 \leq 0$

67. $x^2 - 4x + 2 > 0$ **68.** $-x^2 + 8x - 11 \leq 0$

69. $(x - 5)^2 < 0$ **70.** $(y + 3)^2 \geq 0$

71. $6 - (x - 5)^2 < 0$ **72.** $(y + 3)^2 - 6 \geq 0$

73. $x^2 - 6x + 9 \geq 0$ **74.** $x^2 + 8x + 16 < 0$

In Exercises 75–84, solve the rational inequality. As part of your solution, include a graph that shows the test intervals.

75. $\frac{4}{x-3} < 0$ **76.** $\frac{3}{4-x} > 0$

77. $\frac{-5}{x-3} > 0$ **78.** $\frac{-3}{4-x} > 0$

79. $\frac{x}{x-4} \leq 0$ **80.** $\frac{z-1}{z+3} < 0$

81. $\frac{y-3}{2y-11} \geq 0$ **82.** $\frac{2(4-t)}{4+t} > 0$

83. $\frac{6}{x-4} > 2$ **84.** $\frac{1}{x+2} > -3$

In Exercises 85–88, use a graphing utility to solve the inequality. Use the strategies described on page 420 and on page 423.

85. $9 - 0.2(x - 2)^2 > 4$ **86.** $8x - x^2 < 10$

87. $\dfrac{3x - 4}{x - 4} < -5$ **88.** $4 - \dfrac{1}{x^2} > 1$

89. *Height of a Projectile*   A projectile is fired straight up from ground level with an initial velocity of 88 feet per second, so that its height at any time t is given by

$$h = -16t^2 + 88t$$

where the height h is measured in feet and the time t is measured in seconds. During what interval of time will the height of the projectile exceed 50 feet?

90. *Company Profits*   The revenue and cost equations for a product are given by

$$R = x(50 - 0.0002x)$$
$$C = 12x + 150,000$$

where R and C are measured in dollars and x represents the number of units sold (see figure). How many units must be sold to obtain a profit of at least $1,650,000?

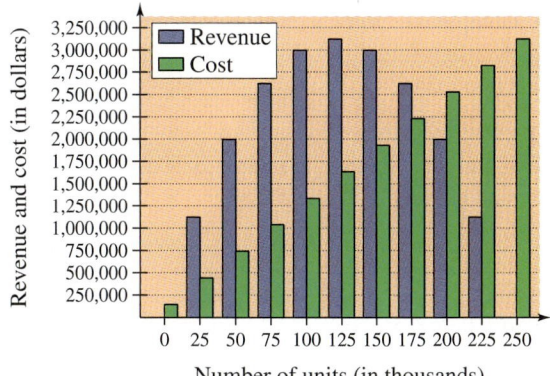

91. *Geometry*   You have 64 feet of fencing to enclose a rectangular region, as shown in the figure. Determine the interval for the length such that the area will exceed 240 square feet.

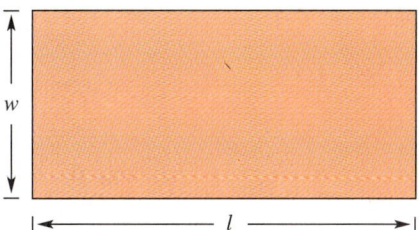

92. *Geometry*   A rectangular playing field with a perimeter of 100 meters is to have an area of at least 500 square meters. Within what bounds must the length of the field lie?

93. *Geometry*   Two circles are tangent to each other (see figure). The distance between their centers is 12 inches.

(a) If x is the radius of one of the circles, express the combined areas of the circles as a function of x. What is the domain of the function?

(b) Use a graphing utility to graph the function found in part (a).

(c) Determine the radii of the circles if the combined areas of the circles must be at least 300 square inches but no more than 400 square inches.

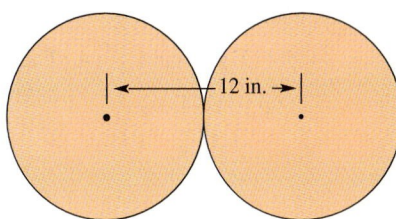

12 in.

CHAPTER PROJECT: Gravitation and Projectile Paths

In this chapter, you studied position functions for falling objects. On earth, position functions take the form

$$s = -16t^2 + v_0 t + s_0.$$
Position function on earth

At other locations in the universe, position functions take the more general form

$$s = -\tfrac{1}{2}gt^2 + v_0 t + s_0.$$
General position function

On the surface of a planet or moon, the value of g (the acceleration due to gravity) depends on both the mass of the planet or moon and its radius. As discovered by Isaac Newton (1642–1727), the value of g is directly proportional to the mass m and inversely proportional to the square of the radius r. That is,

$$g = \frac{-32m}{r^2}$$
Acceleration due to gravity

Planet/Moon	Mass	Radius
Mercury	0.0532	0.3818
Venus	0.8167	0.9500
Earth	1.0000	1.0000
Moon	0.0123	0.2728
Mars	0.1073	0.5305
Jupiter	317.95	10.949
Saturn	95.066	9.1377
Uranus	14.521	3.6837
Neptune	17.177	3.5654
Pluto	0.002	0.1806

where g is measured in feet per second per second, and m and r are the mass and radius *relative to earth's mass and radius*. The table at the left gives the relative masses and radii of earth's moon and the nine planets in our solar system. You are asked to use this data as you investigate the questions below.

1. Find the acceleration due to gravity on earth's moon and on each of the planets. Organize your results in a table.

2. On which planet is the acceleration due to gravity nearest that on earth? Which planet has the smallest value of g? Which has the largest?

3. You throw a rock straight up with an initial velocity of 50 feet per second and an initial height of 6 feet. Compare the motion of the rock on the moon, earth, and Mars. How long is the rock in motion in each location? What is the maximum height of the rock in each location?

4. The circles below represent the moon and the nine planets from smallest to largest. Trace the circles and label them with their values of g. Does the value of g increase as planet size increases? If not, why?

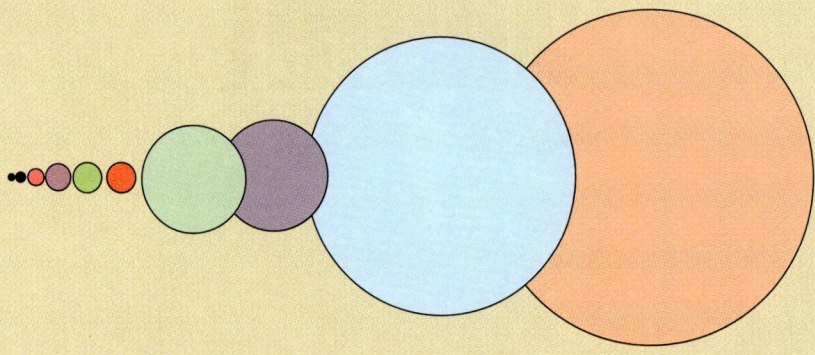

CHAPTER SUMMARY

After studying this chapter, you should have acquired the following skills. These skills are keyed to the Review Exercises that begin on page 622. Answers to odd-numbered Review Exercises are given in the back of the book.

- Solve quadratic equations by factoring. *(Section 10.1)*   **Review Exercises 1–10**

- Solve quadratic equations by extracting square roots. *(Section 10.1)*   **Review Exercises 11–16**

- Solve quadratic equations by completing the square. *(Section 10.2)*   **Review Exercises 17–22**

- Solve quadratic equations using the Quadratic Formula. *(Section 10.3)*   **Review Exercises 23–28**

- Solve quadratic equations by the most convenient method. *(Sections 10.1, 10.2, 10.3)*   **Review Exercises 29–40**

- Solve equations of quadratic form. *(Section 10.1)*   **Review Exercises 41–44**

- Graph functions using a graphing utility. Solve the equations when $y = 0$ and interpret the results. *(Sections 10.1, 10.2, 10.3)*   **Review Exercises 45–50**

- Translate verbal statements into algebraic equations and solve. *(Section 10.4)*   **Review Exercises 51, 52, 57–59, 61–63**

- Solve real-life problems modeled by quadratic equations. *(Sections 10.1, 10.3, 10.4, 10.5)*   **Review Exercises 53, 54, 60, 63, 85**

- Solve problems involving geometry. *(Sections 10.2, 10.3, 10.4)*   **Review Exercises 55, 56**

- Find quadratic equations with the given solutions. *(Section 10.1)*   **Review Exercises 65–68**

- Find quadratic functions that have the given x-intercepts and confirm the results using a graphing utility. *(Sections 10.1, 10.2, 10.3)*   **Review Exercises 69–74**

- Solve inequalities and graph their solutions on the real number line. *(Section 10.5)*   **Review Exercises 75–84**

REVIEW EXERCISES

In Exercises 1–10, solve the equation by factoring.

1. $x^2 + 12x = 0$ **2.** $u^2 - 18u = 0$

3. $3z(z + 10) - 8(z + 10) = 0$

4. $7x(2x - 9) + 4(2x - 9) = 0$

5. $4y^2 - 1 = 0$ **6.** $2z^2 - 72 = 0$

7. $x^2 + \frac{8}{3}x + \frac{16}{9} = 0$ **8.** $4y^2 + 20y + 25 = 0$

9. $2x^2 - 2x = 180$ **10.** $15x^2 - 30x = 45$

In Exercises 11–16, solve the equation by extracting square roots.

11. $x^2 = 10{,}000$ **12.** $x^2 = 98$

13. $y^2 - 2.25 = 0$ **14.** $y^2 - 8 = 0$

15. $(x - 16)^2 = 400$ **16.** $(x + 3)^2 = 0.04$

In Exercises 17–22, solve the equation by completing the square. Find all real or imaginary solutions.

17. $x^2 - 6x - 3 = 0$ **18.** $x^2 + 12x + 6 = 0$

19. $x^2 - 3x + 3 = 0$ **20.** $t^2 + \frac{1}{2}t - 1 = 0$

21. $2y^2 + 10y + 3 = 0$ **22.** $3x^2 - 2x + 2 = 0$

In Exercises 23–28, use the Quadratic Formula to solve the equation. Find all real or imaginary solutions.

23. $y^2 + y - 30 = 0$ **24.** $x^2 - x - 72 = 0$

25. $2y^2 + y - 21 = 0$ **26.** $2x^2 - 3x - 20 = 0$

27. $0.3t^2 - 2t + 5 = 0$ **28.** $-u^2 + 2.5u + 3 = 0$

In Exercises 29–40, solve the equation by the method of your choice. Find all real or imaginary solutions.

29. $(v - 3)^2 = 250$ **30.** $x^2 - 36x = 0$

31. $-x^2 + 5x + 84 = 0$ **32.** $9x^2 + 6x + 1 = 0$

33. $(x - 9)^2 - 121 = 0$ **34.** $60 - (x - 6)^2 = 0$

35. $z^2 - 6z + 10 = 0$ **36.** $z^2 - 14z + 5 = 0$

37. $2y^2 + 3y + 1 = 0$

38. $0.25y^2 + 0.35y - 0.50 = 0$

39. $\dfrac{1}{x} + \dfrac{1}{x + 1} = \dfrac{1}{2}$

40. $x - 5 = \sqrt{x - 2}$

In Exercises 41–44, solve the equation of quadratic form.

41. $(x^2 - 2x)^2 - 4(x^2 - 2x) - 5 = 0$

42. $\left(\sqrt{x} - 2\right)^2 + 2\left(\sqrt{x} - 2\right) - 3 = 0$

43. $6\left(\dfrac{1}{x}\right)^2 + 7\left(\dfrac{1}{x}\right) - 3 = 0$

44. $\left(\dfrac{x}{x - 2}\right)^2 + 4\left(\dfrac{x}{x - 2}\right) + 3 = 0$

In Exercises 45–48, use a graphing utility to graph the function. Use the graph to approximate any x-intercepts of the graph. Set $y = 0$ and solve the resulting equation. Compare the result with the x-intercepts of the graph.

45. $y = x^2 - 7x$ **46.** $y = 12x^2 + 11x - 15$

47. $y = \dfrac{1}{16}x^4 - x^2 + 3$ **48.** $y = \left(\dfrac{1}{x}\right)^2 + 2\left(\dfrac{1}{x}\right) - 3$

Graphical Reasoning In Exercises 49 and 50, use a graphing utility to graph the function and observe that the graph has no x-intercepts. Set $y = 0$ and solve the resulting equation. Identify the type of roots of the equation.

49. $y = x^2 - 8x + 17$ **50.** $y = -(x + 3)^2 - 2$

51. Find two consecutive positive integers such that the sum of their squares is 265.

52. Find two consecutive positive integers whose product is 156.

53. *Falling Time*   The height h in feet of an object above the ground is given by

$$h = 200 - 16t^2, \quad t \geq 0$$

where t is time in seconds.

(a) How high was the object when it was thrown?

(b) Was the object thrown upward or downward? Explain your reasoning.

(c) Find the time when the object strikes the ground.

54. *Falling Time*   The height h in feet of an object above the ground is given by

$$h = -16t^2 + 64t + 192, \quad t \geq 0$$

where t is time in seconds.

(a) How high was the object when it was thrown?

(b) Was the object thrown upward or downward? Explain your reasoning.

(c) Find the time when the object strikes the ground.

55. *Geometry*   The perimeter of a rectangle of length l and width w is 48 feet (see figure).

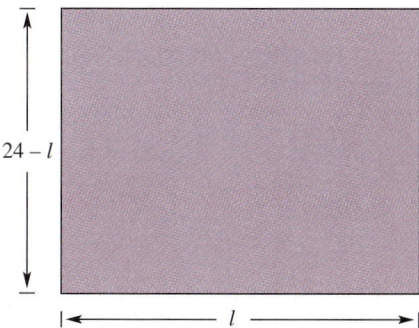

24 − l

l

(a) Show that $w = 24 - l$.

(b) Show that the area is $A = lw = l(24 - l)$.

(c) Use the equation in part (b) to complete the table.

l	2	4	6	8	10	12	14	16	18
A									

56. *Geometry*   Find the dimensions of a triangle if its height is $1\frac{2}{3}$ times its base and its area is 3000 square inches.

57. *Decreased Price*   A Little League baseball team obtains a block of tickets to a ball game for $96. When three more people decide to go to the game, the price per ticket is decreased by $1.60. How many people are going to the game?

58. *Average Speed*   A train travels the first 220 miles of a trip at one speed and the last 180 miles at an average speed of 5 miles per hour faster. If the entire trip takes 7 hours, find the higher speed.

59. *Work Rate*   Working together, two people can complete a task in 6 hours. Working alone, how long would it take each to do the task if one person takes 3 hours longer than the other?

60. *Path of a Projectile*   The path y of a projectile is given by

$$y = -\frac{1}{16}x^2 + 5x$$

where y is the height (in feet) and x is the horizontal distance (in feet) from where the projectile was launched.

(a) Sketch the path of the projectile.

(b) How high is the projectile when it is at its maximum height?

(c) How far from the launch point does the projectile strike the ground?

61. *Reduced Fare*   The college wind ensemble charters a bus to attend a concert at a cost of $360. In an attempt to lower the bus fare per person, the ensemble invites nonmembers to go along. When eight nonmembers join the trip, the fare is decreased by $1.50. How many people are going on the excursion?

62. *Think About It*   When six nonmembers go along on the excursion described in Exercise 61, the fare is decreased by $16. Describe how it is possible to have fewer nonmembers and a greater decrease in the fare.

63. *Work Rate*   Working together, two people can complete a task in 10 hours. Working alone, how long would it take each to do the task if one person takes 2 hours longer than the other?

64. *Solving Graphically, Numerically, and Algebraically*
A launch control team is located 3 miles from the point where a rocket is launched (see figure).

(a) Write the distance d between the rocket and the team as a function of the height h of the rocket. Use a graphing utility to graph the function. Use the graph to approximate the value of h when $d = 4$ miles.

(b) Complete the table. Use the table to approximate the value of h when $d = 4$ miles.

h	1	2	3	4	5	6	7
d							

(c) Find h algebraically when $d = 4$.

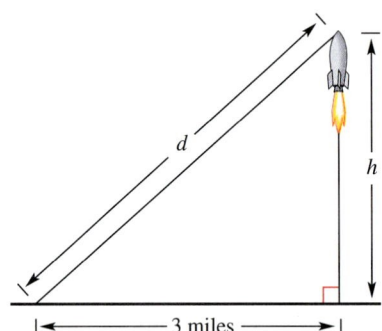

3 miles

In Exercises 65–68, find a quadratic equation with the given solutions.

65. $x = -3, x = 3$

66. $x = -\frac{1}{2}, x = 2$

67. $x = -5, x = -1$

68. $x = -10, x = 8$

In Exercises 69–72, write a quadratic function that has the given x-intercepts. Use a graphing utility to confirm your result.

69. $(1, 0), (3, 0)$

70. $(-2, 0), (2, 0)$

71. $(-4, 0), (-1, 0)$

72. $(2, 0), (5, 0)$

73. *Think About It* Find a *different* quadratic equation that has the same solutions as the equation you found in Exercise 65.

74. *Think About It* Find a *different* quadratic equation that has the same solution as the equation you found in Exercise 66.

In Exercises 75–84, solve the inequality and graph its solution on the real number line. (Some of the inequalities have no solution.)

75. $4x - 12 < 0$

76. $3(x + 2) > 0$

77. $5x(7 - x) > 0$

78. $-2x(x - 10) \leq 0$

79. $16 - (x - 2)^2 \leq 0$

80. $(x - 5)^2 - 36 > 0$

81. $2x^2 + 3x - 20 < 0$

82. $3x^2 - 2x - 8 > 0$

83. $\dfrac{x}{2x - 7} \geq 0$

84. $\dfrac{2x - 9}{x - 1} \leq 0$

85. *College Completion* The percent p of the American population that graduated from college from 1960 to 1990 is approximated by the model

$$p = (2.739 + 0.064t)^2, \quad 0 \leq t$$

where $t = 0$ represents 1960 (see figure). According to this model, when will the percent of college graduates exceed 25% of the population? (Source: U.S. Bureau of Census)

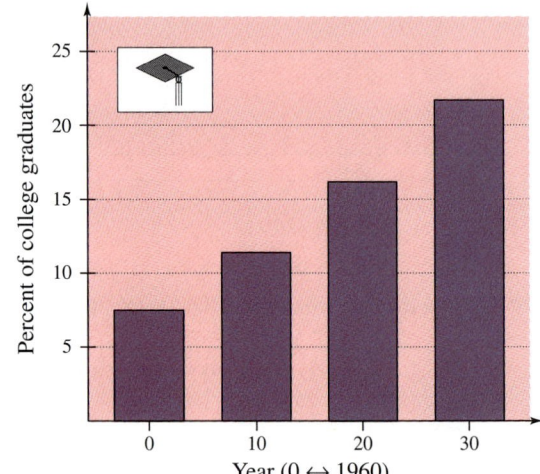

Year (0 ↔ 1960)

CHAPTER TEST

Take this test as you would take a test in class. After you are done, check your work against the answers given in the back of the book.

In Exercises 1 and 2, solve the equation by factoring.

1. $x(x + 5) - 10(x + 5) = 0$ **2.** $8x^2 - 21x - 9 - 0$

In Exercises 3 and 4, solve the equation by extracting square roots.

3. $(x - 2)^2 = 0.09$ **4.** $(x + 3)^2 + 81 = 0$

5. Find the real number c such that $x^2 - 3x + c$ is a perfect square trinomial.

6. Solve by completing the square: $2x^2 - 6x + 3 = 0$

7. Find the discriminant and explain how it may be used to determine the type of solutions of the quadratic equation $5x^2 - 12x + 10 = 0$.

In Exercises 8 and 9, solve by the Quadratic Formula.

8. $3x^2 - 8x + 3 = 0$ **9.** $2y(y - 2) = 7$

10. Solve, and round the result to two decimal places: $\sqrt{3x + 7} - \sqrt{x + 12} = 1$

11. Find a quadratic equation having the solutions -4 and 5.

12. Find a quadratic equation that has $i - 1$ as a solution.

In Exercises 13–16, solve the inequality and sketch the solution.

13. $2x(x - 3) < 0$ **14.** $16 \le (x - 2)^2$

15. $\dfrac{3}{x - 2} > 4$ **16.** $\dfrac{3u + 2}{u - 3} \le 2$

17. Find two consecutive positive integers whose product is 210.

18. The width of a rectangle is 8 feet less than its length. The area of the rectangle is 240 square feet. Find the dimensions of the rectangle.

19. A train traveled the first 125 miles of a trip at one speed and the last 180 miles at an average speed of 5 miles per hour less. If the total time for the trip was $6\frac{1}{2}$ hours, what was the average speed for the first part of the trip?

20. An object is dropped from a height of 75 feet. Its height (in feet) at any time is given by $h = -16t^2 + 75$, where the time t is measured in seconds. Find the time at which the object has fallen to the height of 35 feet.

CUMULATIVE TEST: CHAPTERS 7–10

Take this test as you would take a test in class. After you are done, check your work against the answers given in the back of the book.

In Exercises 1–8, perform the operations and/or simplify.

1. $\dfrac{x^2 + 8x + 16}{18x^2} \cdot \dfrac{2x^4 + 4x^3}{x^2 - 16}$ **2.** $\dfrac{2}{x} - \dfrac{x}{x^3 + 3x^2} + \dfrac{1}{x + 3}$ **3.** $\dfrac{\left(\dfrac{x}{y} - \dfrac{y}{x}\right)}{\left(\dfrac{x - y}{xy}\right)}$

4. $\sqrt{-2}\left(\sqrt{-8} + 3\right)$ **5.** $\dfrac{-4x^{-3}y^4}{6xy^{-2}}$ **6.** $\left(\dfrac{t^{1/2}}{t^{1/4}}\right)^2$

7. $\dfrac{x^3 + 27}{x + 3}$ (Use synthetic division.)   **8.** $\dfrac{6}{\sqrt{10} - 2}$

In Exercises 9 and 10, graph the rational function.

9. $y = \dfrac{4}{x - 2}$ **10.** $y = \dfrac{4x^2}{x^2 + 1}$

In Exercises 11–14, solve the equation.

11. $x + \dfrac{4}{x} = 4$ **12.** $\sqrt{x + 10} = x - 2$

13. $(x - 5)^2 + 50 = 0$ **14.** $3x^2 + 6x + 2 = 0$

15. Find (a) the domain of $f(x) = \sqrt{x^2 - 3x}$, (b) $f(4)$, and (c) $f(c + 3)$.

16. Find the slope of the line passing through $(-4, 0)$ and $(4, 6)$. Then, find the distance between the points.

17. Use a graphing utility to graph the equation $y = x^2 - 6x - 8$. Use the graph to approximate any x-intercepts of the graph. Set $y = 0$ and solve the resulting equation. Compare the results with the x-intercepts of the graph.

18. Find a quadratic equation having the solutions -2 and 6.

19. Determine whether the equation $x - y^3 = 0$ represents y as a function of x.

20. The volume V of a right circular cylinder is $V = \pi r^2 h$. The two cylinders in the figure have equal volumes. Write r_2 as a function of r_1.

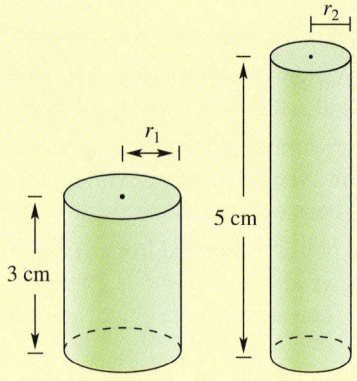

Figure for 20

Lines, Conics, and Variation

11

- Writing Equations of Lines
- Graphs of Linear Inequalities
- Graphs of Quadratic Functions
- Graphs of Second-Degree Equations
- Variation

Mass transportation includes taxis, vans, buses, trains, subways, and airplanes. In 1990, almost four trillion passenger miles were traveled in the United States by these six means of mass transportation. (Passenger miles is the product of the number of vehicle miles and the number of passengers.)

The table at the right shows the number of vehicle miles (in millions) traveled by buses from 1980 through 1991. The data in the table is represented graphically in the scatter plot.

From the scatter plot, you can see that the number of vehicle miles can be approximated by a line. The line that best fits this data is called the *regression line*. It can be used to predict the number of vehicle miles in future years. (Source: Public Transit Association)

t	0	1	2	3	4	5
Miles	1677	1685	1669	1678	1845	1863

t	6	7	8	9	10	11
Miles	2002	2079	2097	2109	2130	2182

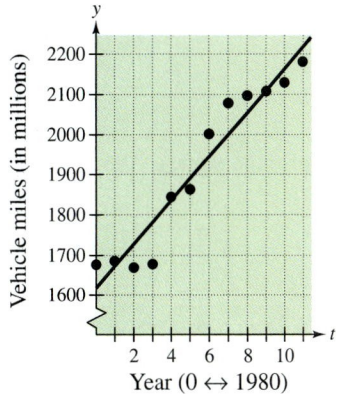

The chapter project related to this information is on page 681.

627

11.1 Writing Equations of Lines

The Point-Slope Equation of a Line ▪
Summary of Equations of Lines ▪ Application

The Point-Slope Equation of a Line

There are two basic types of problems in coordinate geometry.

1. Given an algebraic equation, sketch its graph.

2. Given a description of a graph, write its equation.

The first type of problem can be thought of as moving from algebra to geometry, whereas the second type can be thought of as moving the other way—from geometry to algebra. So far in the text, you have been working primarily with the first type of problem. In this section, you will look at the second problem.

In this section, you will learn that if you know the slope of a line *and* you also know the coordinates of one point on the line, you can find an equation for the line. Before giving a general formula for doing this, let's look at an example.

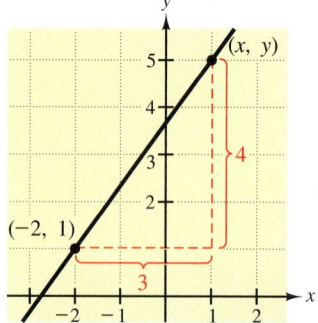

FIGURE 11.1

EXAMPLE 1 Writing an Equation of a Line

A line has a slope of $\frac{4}{3}$ and passes through the point $(-2, 1)$. Find an equation of this line.

Solution

Begin by sketching the line, as shown in Figure 11.1. You know that the slope of a line is the same through any two points on the line. Thus, to find an equation of the line, let (x, y) represent *any* point on the line. Using the representative point (x, y) and the given point $(-2, 1)$, it follows that the slope of the line is

$$m = \frac{y - 1}{x - (-2)}.$$

 Difference in *y*-values
Difference in *x*-values

Because the slope of the line is $m = \frac{4}{3}$, this equation can be rewritten as follows.

$$\frac{4}{3} = \frac{y - 1}{x + 2}$$ Slope formula

$$4(x + 2) = 3(y - 1)$$ Cross-multiply.

$$4x + 8 = 3y - 3$$ Distributive Property

$$4x - 3y = -11$$ Subtract 8 and 3y from both sides.

An equation of the line is $4x - 3y = -11$.

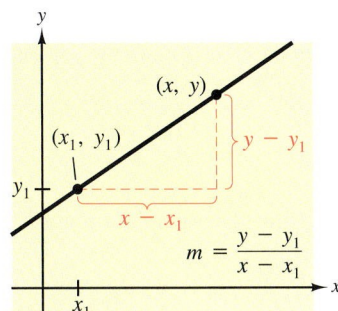

FIGURE 11.2

The procedure in Example 1 can be used to derive a *formula* for the equation of a line, given its slope and a point on the line. In Figure 11.2, let (x_1, y_1) be a given point on the line whose slope is m. If (x, y) is any *other* point on the line, it follows that

$$\frac{y - y_1}{x - x_1} = m.$$

This equation in variables x and y can be rewritten in the form

$$y - y_1 = m(x - x_1)$$

which is called the **point-slope form** of the equation of a line.

Point-Slope Form of the Equation of a Line

The **point-slope form** of the equation of the line that passes through the point (x_1, y_1) and has a slope of m is

$$y - y_1 = m(x - x_1).$$

EXAMPLE 2   *The Point-Slope Form of the Equation of a Line*

Find an equation of the line that passes through the point $(2, -3)$ and has a slope of -2.

Solution

Use the point-slope form with $(x_1, y_1) = (2, -3)$ and $m = -2$.

$$
\begin{array}{ll}
y - y_1 = m(x - x_1) & \text{Point-slope form} \\
y - (-3) = -2(x - 2) & \text{Substitute } y_1 = -3, \ x_1 = 2, \text{ and } m = -2. \\
y + 3 = -2x + 4 & \text{Simplify.} \\
y = -2x + 1 & \text{Subtract 3 from both sides.}
\end{array}
$$

The graph of this line is shown in Figure 11.3.

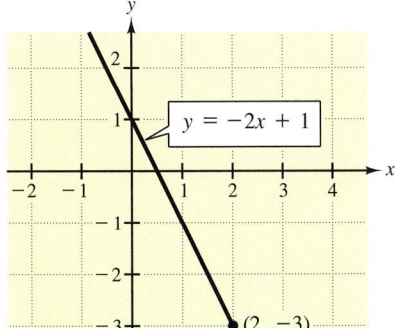

FIGURE 11.3

In Example 2, notice that the final equation is written in slope-intercept form

$$y = mx + b. \qquad \text{Slope-intercept form}$$

You can use this form to check your work. First, observe that the slope is $m = -2$. Then substitute the coordinates of the given point $(2, -3)$ to see that the equation is satisfied.

The point-slope form can be used to find the equation of a line passing through two points (x_1, y_1) and (x_2, y_2). First, use the formula for the slope of the line passing through two points.

$$m = \frac{y_2 - y_1}{x_2 - x_1}$$

Then, once you know the slope, use the point-slope form to obtain the equation

$$y - y_1 = \frac{y_2 - y_1}{x_2 - x_1}(x - x_1). \qquad \text{Two-point form}$$

This is sometimes called the **two-point form** of the equation of a line.

EXAMPLE 3   An Equation of a Line Passing Through Two Points

Find an equation of the line that passes through the points $(4, 2)$ and $(-2, 3)$.

Solution

Let $(x_1, y_1) = (4, 2)$ and $(x_2, y_2) = (-2, 3)$. Then apply the formula for the slope of a line passing through two points as follows.

$$m = \frac{y_2 - y_1}{x_2 - x_1}$$

$$= \frac{3 - 2}{-2 - 4}$$

$$= -\frac{1}{6}$$

Now, using the point-slope form, you can find the equation of the line.

$$y - y_1 = m(x - x_1) \qquad \text{Point-slope form}$$

$$y - 2 = -\frac{1}{6}(x - 4) \qquad \text{Substitute } y_1 = 2, \ x_1 = 4, \text{ and } m = -\tfrac{1}{6}.$$

$$y - 2 = -\frac{1}{6}x + \frac{2}{3} \qquad \text{Simplify.}$$

$$y = -\frac{1}{6}x + \frac{8}{3} \qquad \text{Add 2 to both sides.}$$

The graph of this line is shown in Figure 11.4.

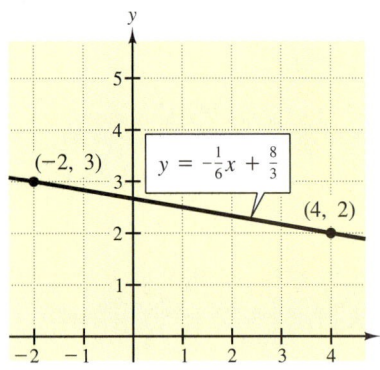

FIGURE 11.4

In Example 3, it does not matter which of the two points is labeled (x_1, y_1) and which is labeled (x_2, y_2). Try switching these labels to

$$(x_1, y_1) = (-2, 3) \quad \text{and} \quad (x_2, y_2) = (4, 2)$$

and reworking the problem to see that you obtain the same equation.

Summary of Equations of Lines

From the slope-intercept form of the equation of a line, you can see that a horizontal line ($m = 0$) has an equation of the form

$$y = (0)x + b \quad \text{or} \quad y = b. \qquad \text{Horizontal line}$$

This is consistent with the fact that each point on a horizontal line through $(0, b)$ has a y-coordinate of b, as shown in Figure 11.5. Similarly, each point on a vertical line through $(a, 0)$ has an x-coordinate of a, as shown in Figure 11.6. Hence, a vertical line has an equation of the form

$$x = a. \qquad \text{Vertical line}$$

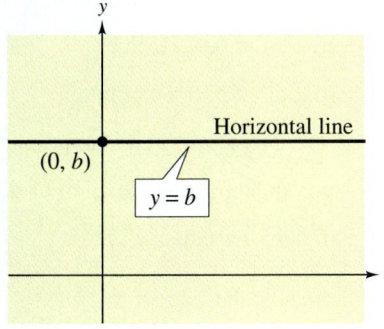

FIGURE 11.5   **FIGURE 11.6**

EXAMPLE 4   *Writing Equations of Horizontal and Vertical Lines*

Write an equation for each line.

a. Vertical line through $(-2, 4)$

b. Line passing through $(-2, 3)$ and $(3, 3)$

c. Line passing through $(-1, 2)$ and $(-1, 3)$

Solution

a. Because the line is vertical and passes through the point $(-2, 4)$, you know that every point on the line has an x-coordinate of -2. Thus, the equation is

$$x = -2.$$

b. The line through $(-2, 3)$ and $(3, 3)$ is horizontal. Thus, its equation is

$$y = 3.$$

c. The line through $(-1, 2)$ and $(-1, 3)$ is vertical. Thus, its equation is

$$x = -1.$$

The equation of a vertical line cannot be written in slope-intercept form because the slope of a vertical line is undefined. However, *every* line has an equation that can be written in the **general form**

$$ax + by + c = 0 \qquad \text{General form}$$

where a and b are not *both* zero.

Summary of Equations of Lines

1. Slope of line through (x_1, y_1) and (x_2, y_2): $m = \dfrac{y_2 - y_1}{x_2 - x_1}$

2. General form of equation of line: $ax + by + c = 0$

3. Equation of vertical line: $x = a$

4. Equation of horizontal line: $y = b$

5. Slope-intercept form of equation of line: $y = mx + b$

6. Point-slope form of equation of line: $y - y_1 = m(x - x_1)$

7. Parallel lines (equal slopes): $m_1 = m_2$

8. Perpendicular lines (negative reciprocal slopes): $m_2 = -\dfrac{1}{m_1}$

EXAMPLE 5   *Parallel and Perpendicular Lines*

Find an equation of the line that passes through the point $(3, -2)$ and is (a) parallel and (b) perpendicular to the line $x - 4y = 6$, as shown in Figure 11.7.

Solution

By writing the given line in slope-intercept form $y = \frac{1}{4}x - \frac{3}{2}$, you can see that it has a slope of $\frac{1}{4}$. Therefore, parallel lines must also have a slope of $\frac{1}{4}$ and perpendicular lines must have a slope of -4.

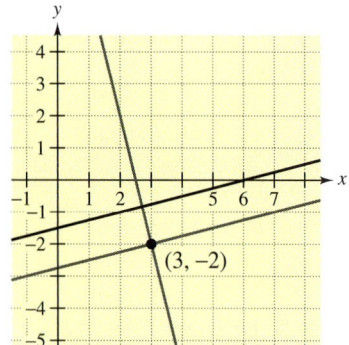

FIGURE 11.7

a. $y - y_1 = m(x - x_1)$ Point-slope form

$y - (-2) = \dfrac{1}{4}(x - 3)$ Substitute $y_1 = -2$, $x_1 = 3$, and $m = \frac{1}{4}$.

$y = \dfrac{1}{4}x - \dfrac{11}{4}$ Equation of parallel line

b. $y - y_1 = m(x - x_1)$ Point-slope form

$y - (-2) = -4(x - 3)$ Substitute $y_1 = -2$, $x_1 = 3$, and $m = -4$.

$y = -4x + 10$ Equation of perpendicular line

Application

EXAMPLE 6   An Application: Total Sales

The total sales of a new computer software company were $500,000 for the second year and $1,000,000 for the fourth year. Using only this information, what would you estimate the total sales to be during the fifth year?

Solution

To solve this problem, use a *linear model*, with y representing the total sales and t representing the year. That is, in Figure 11.8, let (2, 500) and (4, 1000) be two points on the line representing the total sales for the company. The slope of the line passing through these points is

$$m = \frac{1000 - 500}{4 - 2} = 250.$$

Now, using the point-slope form, the equation of the line is

$y - y_1 = m(t - t_1)$	Point-slope form
$y - 500 = 250(t - 2)$	Substitute $y_1 = 500$, $t_1 = 2$, and $m = 250$.
$y = 250t.$	Linear model for sales

Finally, estimate the total sales during the fifth year ($t = 5$) to be

$$y = 250(5) = \$1250 \text{ thousand} = \$1,250,000.$$

FIGURE 11.8

The estimation method illustrated in Example 6 is called **linear extrapolation.** Note in Figure 11.9 that for linear extrapolation, the estimated point lies to the right of the given points. When the estimated point lies *between* two given points, the procedure is called **linear interpolation.**

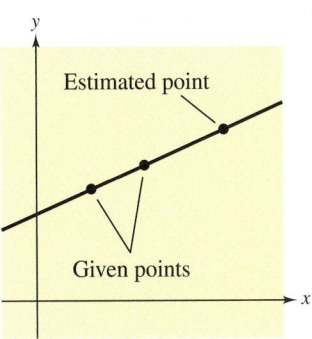

Linear Extrapolation
FIGURE 11.9

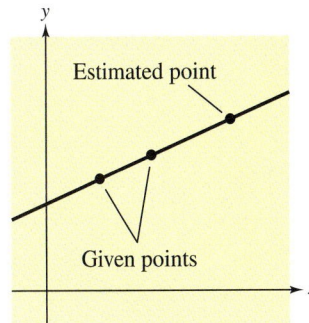

Linear Interpolation

The linear model is the most commonly used type of mathematical model. In the following group activity, you will learn one way to fit a linear model to a set of real data.

Group Activities

Problem Solving

Mathematical Modeling   You can find a linear model that fits a set of real data by first plotting the ordered pairs on a graph known as a **scatter plot.** Using a straightedge, draw a line through the points that seems to have the "best fit." You can then use any of the forms of a linear equation discussed in this section to find the equation of the line.

Suppose your manager asks you to make sense of the following set of data, in which y represents the percent of households with personal computer owners and x represents the year from 1983 to 1994, with $x = 3$ corresponding to 1983.   (Source: Electronic Industries Association Consumer Electronic U.S. Sales, 1982–1994)

x	3	4	5	6	7	8	9	10	11	12	13	14
y	7	13	15	16	20	21	22	27	29	34	35	37

You think that a mathematical model will help you understand the trend in the data and may be useful in predicting what could happen in the future. Construct a scatter plot of the data. Do you think that a linear model would represent the data well? If so, find the equation of the best-fitting line. Interpret the meaning of the slope in the context of the data. Use your model to predict the percent of households with personal computer owners in 1999 (assuming that the trend continues).

11.1 Exercises

Discussing the Concepts

1. Can any pair of points on a line be used to determine an equation of the line? Explain.

2. Write the point-slope form, the slope-intercept form, and the general form of an equation of a line.

3. In the equation $y = 2x + 5$, what does the 2 represent? What does the 5 represent?

4. In the equation of a vertical line, the variable y is missing. Explain why.

5. Is it possible to have two perpendicular lines that rise from left to right? Explain.

6. What is the only type of line for which the concept of slope is not defined?

Problem Solving

In Exercises 7–10, match the equation with its graph. Use a graphing utility to verify your result. [The graphs are labeled (a), (b), (c), and (d).]

(a)
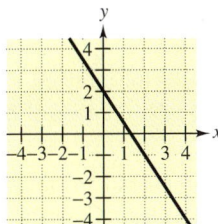

(b)
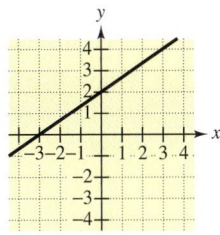

(c)
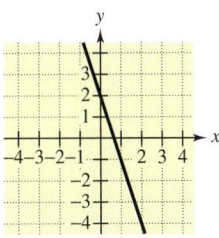

(d)
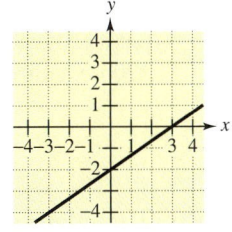

7. $y = \frac{2}{3}x + 2$

8. $y = \frac{2}{3}x - 2$

9. $y = -\frac{3}{2}x + 2$

10. $y = -3x + 2$

In Exercises 11–14, determine the slope of the line and the coordinates of one point through which it passes.

11. $y - 3 = 6(x + 4)$

12. $y = -5x + 12$

13. $3x - 2y = 0$

14. $3x + 4y - 16 = 0$

In Exercises 15–20, find the point-slope form of the equation of the line passing through the given point with the specified slope. Sketch the line.

15. $(0, 0)$, $m = -\frac{1}{2}$

16. $(0, 0)$, $m = \frac{2}{3}$

17. $(0, -4)$, $m = 3$

18. $(0, 6)$, $m = -\frac{3}{4}$

19. $\left(-2, \frac{7}{2}\right)$, $m = -4$

20. $\left(1, -\frac{3}{2}\right)$, $m = \frac{3}{4}$

In Exercises 21–26, find the general form of the equation of the line through the points. Sketch the line.

21. $(0, 0)$, $(2, 3)$

22. $(0, 0)$, $(3, -5)$

23. $(-2, 3)$, $(5, 0)$

24. $(1, -2)$, $(1, 8)$

25. $\left(\frac{3}{2}, 3\right)$, $\left(\frac{9}{2}, -4\right)$

26. $\left(4, \frac{7}{3}\right)$, $\left(-1, \frac{2}{3}\right)$

In Exercises 27–30, write an equation of the line passing through the two points. Use function notation to write y as a function of x. Use a graphing utility to graph the linear function.

27. $(-2, 2)$, $(4, 5)$

28. $(0, 10)$, $(5, 0)$

29. $(-2, 3)$, $(4, 3)$

30. $(-6, -3)$, $(4, 3)$

In Exercises 31–36, write equations of the line through the point that are (a) parallel and (b) perpendicular to the given line.

31. $(2, 1)$, $y = 3x$

32. $(-3, 4)$, $y = -2x$

33. $(-5, 4)$, $x + y = 1$

34. $\left(\frac{5}{8}, \frac{9}{4}\right)$, $5x = 4y$

35. $(-1, 2)$, $y + 5 = 0$

36. $(3, -4)$, $x - 10 = 0$

⊞ In Exercises 37–40, use a graphing utility on a square setting to graph both lines on the same screen. Decide whether the lines are parallel, perpendicular, or neither.

37. $y = -0.6x + 1$
$y = \frac{5}{3}x - 2$

38. $y = \frac{1}{2}(4x - 3)$
$y = \frac{1}{4}(4x + 3)$

39. $y = 0.6x + 1$
$y = \frac{1}{5}(3x + 22)$

40. $y = \frac{3}{2}x - 4$
$y = -\frac{2}{3}x + 1$

41. *Cost* The cost (in dollars) of producing x units of a certain product is given by $C = 20x + 5000$. Use this model to complete the table.

x	0		100		1000
C		6000		15,000	

42. *Temperature Conversion* The relationship between the Fahrenheit and Celsius temperature scales is given by $C = \frac{5}{9}F - \frac{160}{9}$. Use this equation to complete the table.

F	$-20°$		$20°$		$100°$	
C		$-17.8°$		$0°$		$100°$

43. *Rental Occupancy* A real estate office handles an apartment complex with 50 units. When the rent per unit is $450 per month, all 50 units are occupied. However, when the rent is $525 per month, the average number of occupied units drops to 45. Assume the relationship between the monthly rent p and the demand x is linear.

(a) Write the equation of the line giving the demand x in terms of the rent p.

(b) *Linear Extrapolation* Use the equation from part (a) to predict the number of units occupied when the rent is $570.

(c) *Linear Interpolation* Use the equation from part (a) to predict the number of units occupied when the rent is $480.

⊞ **44.** *Graphical Estimation* A sales representative is reimbursed $150 per day for lodging and meals plus $0.34 per mile driven.

(a) Write a linear function giving the daily cost C to the company in terms of x, the number of miles driven.

(b) Use a graphing utility to graphically estimate the daily cost of driving 230 miles. Confirm your estimate algebraically.

(c) Use a graphing utility to graphically estimate the number of miles driven to produce a daily cost of $200. Confirm your estimate algebraically.

Reviewing the Major Concepts

In Exercises 45–48, factor the expression completely.

45. $5x - 20x^2$

46. $64 - (x - 6)^2$

47. $15x^2 - 16x - 15$

48. $8x^3 + 1$

⊞ **49.** *Graphical Reasoning* Consider the equation

$$2x^3 - 3x^2 - 18x + 27 = (2x - 3)(x + 3)(x - 3).$$

(a) Use a graphing utility to verify the equation by graphing both the left side and right side of the equation. Are the graphs the same?

(b) Verify the equation by multiplying the polynomials on the right side of the equation.

⊞ **50.** *Profit* The cost of producing x units is $C = 12 + 8x$. The revenue from selling x units is

$$R = 16x - \frac{1}{4}x^2, \quad 0 \le x \le 20.$$

The profit is given by $P = R - C$.

(a) Perform the subtraction required to find the polynomial representing profit.

(b) Use a graphing utility to graph the polynomial representing profit.

(c) Determine the profit when $x = 16$ units are produced and sold. Use the graph of part (b) to predict the change in profit if x is some value other than 16.

Additional Problem Solving

In Exercises 51–56, find the slope of the line (if possible) and the coordinates of one point through which it passes.

51. $y + \frac{5}{8} = \frac{3}{4}(x + 2)$

52. $y - \frac{3}{4} = \frac{5}{6}\left(x - \frac{9}{10}\right)$

53. $y = \frac{2}{3}x - 2$

54. $y = -2(6x - 1)$

55. $5x - 2y + 24 = 0$

56. $x + 4 = 0$

In Exercises 57–66, find the point-slope form of the equation of the line passing through the given point with the specified slope.

57. $(0, -6), m = \frac{1}{2}$

58. $(0, 9), m = -\frac{1}{3}$

59. $(-2, 8), m = -2$

60. $(0, 0), m = 3$

61. $(5, 0), m = -\frac{2}{3}$

62. $(-3, -7), m = \frac{5}{4}$

63. $(-8, 5), m = 0$

64. $(2, 1), m$ is undefined.

65. $\left(\frac{3}{4}, \frac{5}{2}\right), m = \frac{4}{3}$

66. $\left(-\frac{3}{2}, \frac{1}{2}\right), m = -3$

In Exercises 67–74, find the general form of the equation of the line passing through the points.

67. $(-5, 2), (5, -2)$

68. $(5, 4), (-3, 5)$

69. $\left(10, \frac{1}{2}\right), \left(\frac{3}{2}, \frac{7}{4}\right)$

70. $(-2, 12), (6, 12)$

71. $(7.5, 2), (7.5, 9)$

72. $(2, -8), \left(6, \frac{8}{3}\right)$

73. $(2, 0.6), (8, -4.2)$

74. $(-5, 0.6), (3, -3.4)$

In Exercises 75–78, write equations of the line through the given point that are (a) parallel and (b) perpendicular to the given line.

75. $(3, 7)$

$4x - y - 3 = 0$

76. $(-5, -10)$

$2x + 5y - 12 = 0$

77. $(6, -4)$

$3x + 10y = 24$

78. $\left(\frac{2}{3}, \frac{4}{3}\right)$

$x - 5 = 0$

In Exercises 79–84, use a graphing utility on a square setting to decide whether the lines are parallel, perpendicular, or neither.

79. $y = 3(x - 2)$

$y = 3x + 2$

80. $y = \frac{3}{2}(2x - 3)$

$y = -\frac{1}{3}(x + 3)$

81. $y = 1 - 1.5x$

$y = \frac{2}{3}x - 4$

82. $y = 2.5x - 4$

$y = -2.5x + 3$

83. $x + 2y - 3 = 0$

$-2x - 4y + 1 = 0$

84. $3x - 4y - 1 = 0$

$4x + 3y + 2 = 0$

Exploration In Exercises 85–88, find an equation of the line with intercepts $(a, 0)$ and $(0, b)$ where the equation is given by

$$\frac{x}{a} + \frac{y}{b} = 1, \quad a \neq 0, \quad b \neq 0.$$

85. x-intercept: $(3, 0)$

y-intercept: $(0, 2)$

86. x-intercept: $(-6, 0)$

y-intercept: $(0, 2)$

87. x-intercept: $\left(-\frac{5}{6}, 0\right)$

y-intercept: $\left(0, -\frac{7}{3}\right)$

88. x-intercept: $\left(-\frac{8}{3}, 0\right)$

y-intercept: $(0, -4)$

89. *Sales Commission*   The salary for a sales representative is $1500 per month plus a 3% commission of total monthly sales. Write a linear function giving the salary S in terms of the monthly sales M.

90. *Reimbursed Expenses*   A sales representative is reimbursed $125 per day for lodging and meals plus $0.22 per mile driven. Write a linear function giving the daily cost C to the company in terms of x, the number of miles driven.

91. *Discount Price*   A store is offering a 30% discount on all items in its inventory.

(a) Write a linear function giving the sale price S for an item in terms of its list price L.

(b) Use the function in part (a) to find the sale price of an item that has a list price of $135.

92. *Straight-Line Depreciation*   A small business purchases a photocopier for $7400. After 4 years, its depreciated value will be $1500.

(a) Assuming straight-line depreciation, write a linear function giving the value V of the copier in terms of time t.

(b) Use the function in part (a) to find the value of the copier after 2 years.

93. *College Enrollment* A small college had an enrollment of 1500 students in 1980. During the next 10 years, the enrollment increased by approximately 60 students per year.

(a) Write a linear function giving the enrollment N in terms of the year t. (Let $t = 0$ represent 1980.)

(b) *Linear Extrapolation* Use the function in part (a) to predict the enrollment in the year 2000.

(c) *Linear Interpolation* Use the function in part (a) to estimate the enrollment in 1985.

94. *Soft-Drink Sales* When soft drinks sold for $0.80 per can at football games, approximately 6000 cans were sold. When the price was raised to $1.00 per can, the demand dropped to 4000. Assume the relationship between the price p and demand x is linear.

(a) Write an equation of the line giving the demand x in terms of the price p.

(b) *Linear Extrapolation* Use the equation in part (a) to predict the number of cans of soft drinks sold if the price is $1.10.

(c) *Linear Interpolation* Use the equation in part (a) to predict the number of cans of soft drinks sold if the price is $0.90.

Graphical Interpretation In Exercises 95 and 96, use the graph, which shows the cost C (in billions of dollars) of running the Internal Revenue Service from 1980 to 1990. (Source: Internal Revenue Service)

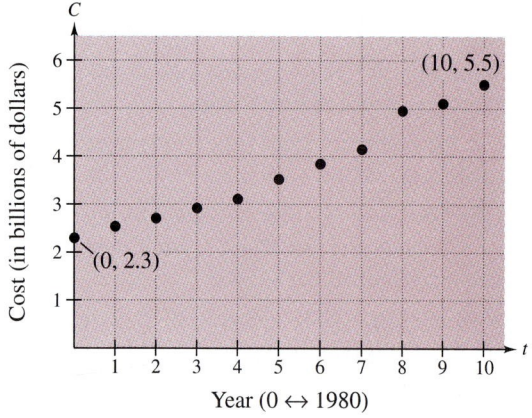

Year (0 ↔ 1980)

95. Using the costs for 1980 and 1990, write a linear model for the average cost, letting $t = 0$ represent 1980. Estimate the cost of running the IRS in 1995.

96. In 1989, the IRS collected about 9.5 trillion dollars. Use the model to approximate the percent of this amount spent to run the IRS.

97. *Best-Fitting Line* An instructor gives 20-point quizzes and 100-point exams in a mathematics course. The average quiz and test scores for six students are given as ordered pairs (x, y) where x is the average quiz score and y is the average test score. The ordered pairs are (18, 87), (10, 55), (19, 96), (16, 79), (13, 76), and (15, 82).

(a) Plot the points.

(b) Use a ruler to sketch the "best-fitting" line through the points.

(c) Find an equation for the line sketched in part (b).

(d) Use the equation of part (c) to estimate the average test score for a person with an average quiz score of 17.

98. *Depth Markers* A swimming pool is 40 feet long, 20 feet wide, 4 feet deep at the shallow end, and 9 feet deep at the deep end. Position the side of the pool on the rectangular coordinate system as shown in the figure and find an equation of the line representing the edge of the inclined bottom of the pool. Use this equation to determine the distances from the deep end at which markers must be placed to indicate each 1-foot change in the depth of the pool.

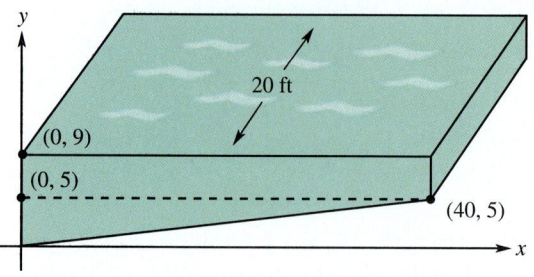

11.2 Graphs of Linear Inequalities

Linear Inequalities in Two Variables ■
The Graph of a Linear Inequality

Linear Inequalities in Two Variables

A **linear inequality** in variables x and y is an inequality that can be written in one of the following forms.

$$ax + by < c, \quad ax + by > c, \quad ax + by \leq c, \quad \text{and} \quad ax + by \geq c$$

Here are some examples.

$$4x - 3y < 7, \quad x - y > -3, \quad x \leq 2, \quad \text{and} \quad y \geq -4$$

An ordered pair (x_1, y_1) is a **solution** of a linear inequality in x and y if the inequality is true when x_1 and y_1 are substituted for x and y, respectively. For instance, the ordered pair $(3, 2)$ is a solution of the inequality $x - y > 0$ because $3 - 2 > 0$ is a true statement.

EXAMPLE 1   *Verifying Solutions of Linear Inequalities*

Decide whether each point is a solution of $2x - 3y \geq -2$.

a. $(0, 0)$ **b.** $(2, 2)$ **c.** $(0, 1)$

Solution

a.
$$2x - 3y \geq -2 \qquad \text{Original inequality}$$
$$2(0) - 3(0) \overset{?}{\geq} -2 \qquad \text{Substitute 0 for } x \text{ and 0 for } y.$$
$$0 \geq -2 \qquad \text{Inequality is satisfied.} ✓$$

Because the inequality is satisfied, the point $(0, 0)$ is a solution.

b.
$$2x - 3y \geq -2 \qquad \text{Original inequality}$$
$$2(2) - 3(2) \overset{?}{\geq} -2 \qquad \text{Substitute 2 for } x \text{ and 2 for } y.$$
$$-2 \geq -2 \qquad \text{Inequality is satisfied.} ✓$$

Because the inequality is satisfied, the point $(2, 2)$ is a solution.

c.
$$2x - 3y \geq -2 \qquad \text{Original inequality}$$
$$2(0) - 3(1) \overset{?}{\geq} -2 \qquad \text{Substitute 0 for } x \text{ and 1 for } y.$$
$$-3 \not\geq -2 \qquad \text{Inequality is not satisfied.} ✗$$

Because the inequality is not satisfied, the point $(0, 1)$ is not a solution.

The Graph of a Linear Inequality

The **graph** of an inequality is the collection of all solution points of the inequality. To sketch the graph of a linear inequality such as

$$4x - 3y < 12 \qquad \text{Original inequality}$$

begin by sketching the graph of the *corresponding linear equation*

$$4x - 3y = 12. \qquad \text{Corresponding equation}$$

Use *dashed* lines for the inequalities $<$ and $>$ and *solid* lines for the inequalities $\le$ and $\ge$. The graph of the equation separates the plane into two **half-planes.** In each half-plane, one of the following must be true.

1. All points in the half-plane are solutions of the inequality.

2. No point in the half-plane is a solution of the inequality.

Thus, you can determine whether the points in an entire half-plane satisfy the inequality by simply testing *one* point in the region.

EXAMPLE 2 *Sketching the Graphs of Linear Inequalities*

Sketch the graph of each linear inequality.

a. $x \ge -3$ **b.** $y < 4$

Solution

a. The graph of the corresponding equation $x = -3$ is a vertical line. The points that satisfy the inequality $x \ge -3$ are those lying on or to the right of this line, as shown in Figure 11.10.

b. The graph of the corresponding equation $y = 4$ is a horizontal line. The points that satisfy the inequality $y < 4$ are those lying below this line, as shown in Figure 11.11.

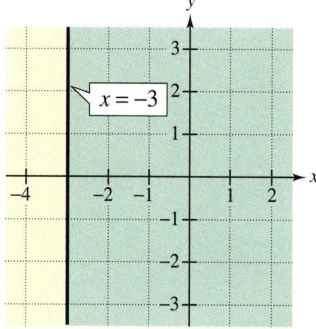

FIGURE 11.10

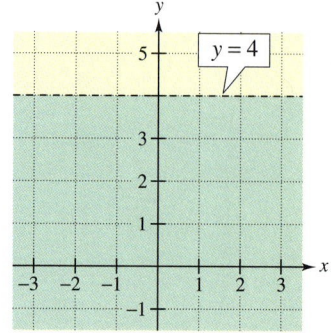

FIGURE 11.11

We summarize the guidelines for sketching the graph of a linear inequality in two variables as follows.

Guidelines for Graphing a Linear Inequality

1. Replace the inequality sign by an equal sign, and sketch the graph of the resulting equation. (Use a dashed line for $<$ or $>$, and a solid line for $\leq$ or $\geq$.)

2. Test one point in each of the half-planes formed by the graph in Step 1. If the point satisfies the inequality, shade the entire half-plane to denote that every point in the region satisfies the inequality.

EXAMPLE 3   *Sketching the Graph of a Linear Inequality*

Sketch the graph of the linear inequality

$$x + y > 3. \qquad \text{Original inequality}$$

Solution

The graph of the corresponding equation

$$x + y = 3 \qquad \text{Corresponding equation}$$

is a line, as shown in Figure 11.12. Because the origin $(0, 0)$ does not satisfy the inequality, the graph consists of the half-plane lying above the line. (Try checking a point above the line. Regardless of which point you choose, you will see that it is a solution.)

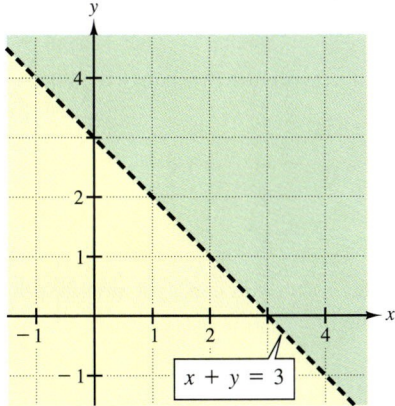

FIGURE 11.12

For a linear inequality in two variables, you can sometimes simplify the graphing procedure by writing the inequality in *slope-intercept form.* For instance, by writing $x + y > 1$ in the form $y > -x + 1$, you can see that the solution points lie *above* the line $y = -x + 1$, as shown in Figure 11.13. Similarly, by writing the inequality $4x - 3y > 12$ in the form

$$y < \frac{4}{3}x - 4$$

you can see that the solutions lie *below* the line $y = \frac{4}{3}x - 4$, as shown in Figure 11.14.

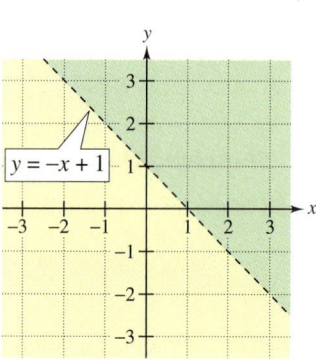

FIGURE 11.13   **FIGURE 11.14**

EXAMPLE 4   *Sketching the Graph of a Linear Inequality*

Use the slope-intercept form of a linear equation as an aid in sketching the graph of the inequality $2x - 3y \le 15$.

Solution

To begin, rewrite the inequality in slope-intercept form.

$$2x - 3y \le 15 \qquad \text{Original inequality}$$

$$-3y \le -2x + 15 \qquad \text{Subtract } 2x \text{ from both sides.}$$

$$y \ge \frac{2}{3}x - 5 \qquad \text{Slope-intercept form}$$

From this form, you can conclude that the solution is the half-plane lying *on or above* the line

$$y = \frac{2}{3}x - 5.$$

The graph is shown in Figure 11.15.

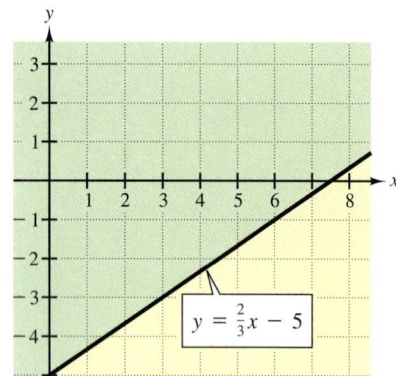

FIGURE 11.15

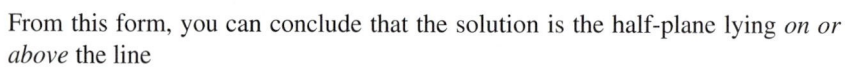

Technology

Graphing Utilities and Inequalities

Most graphing utilities can graph inequalities in two variables. For instance, you can sketch the graph of the inequality

$$y \leq \tfrac{1}{2}x - 3$$

on a *TI-82* with the following steps.

1. Set the viewing window. For this graph, you can use a standard setting by using the ZOOM feature and the STANDARD option.

2. Graph the inequality by entering the following steps.

   Press DRAW

   Cursor to "Shade("
   Enter the following: Shade(-10,0.5X−3)
   ENTER

   The graph of $y \leq \tfrac{1}{2}x - 3$ is shown below.

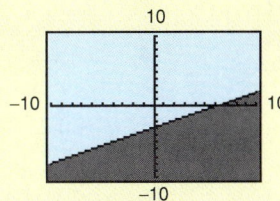

Group Activities   **Extending the Concept**

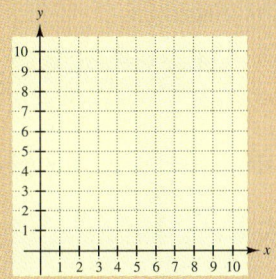

Using Inequalities   Try the following activity. One person picks a point with whole number coordinates on a grid like the one at left without revealing the coordinates. A second person writes the equation of a line passing through the grid region. The first person graphs the line on the grid and indicates whether the secret point lies above, below, or on the line. Continue writing and graphing lines until the second person is able to guess the coordinates of the secret point. Switch roles and try again. What is the fewest number of turns your team required to guess the point?

11.2 Exercises

Discussing the Concepts

1. List the four forms of a linear inequality in variables x and y.

2. What is meant by saying (x_1, y_1) is a solution of a linear inequality in x and y?

3. Explain the meaning of the term *half-plane*. Give an example of an inequality whose graph is a half-plane.

4. How does the solution set of $x - y > 1$ differ from the solution set of $x - y \geq 1$?

5. After graphing the boundary, how do you decide which half-plane is the solution set of a linear inequality?

6. Explain the difference between graphing the solution of the inequality $x \leq 3$ on the real number line and graphing it on a rectangular coordinate system.

Problem Solving

In Exercises 7–12, match the inequality with its graph. [The graphs are labeled (a), (b), (c), (d), (e), and (f).]

(a)
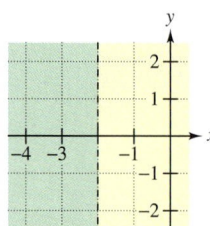

(b)
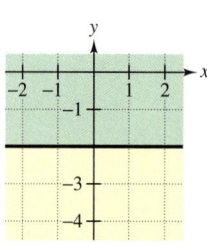

(c)
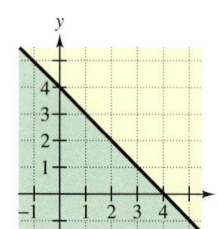

(d)
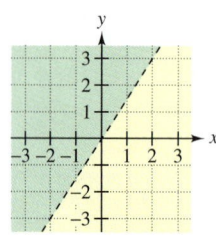

(e)
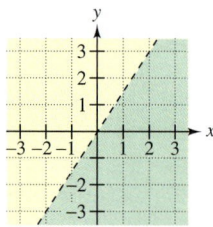

(f)
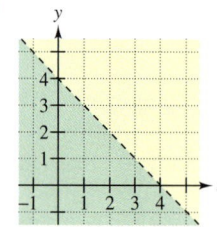

7. $y \geq -2$

8. $x < -2$

9. $3x - 2y < 0$

10. $3x - 2y > 0$

11. $x + y < 4$

12. $x + y \leq 4$

In Exercises 13–16, determine whether the points are solutions of the inequality.

Inequality	*Points*	
13. $x - 2y < 4$	(a) $(0, 0)$	(b) $(2, -1)$
	(c) $(3, 4)$	(d) $(5, 1)$
14. $-3x + 5y \geq 6$	(a) $(2, 8)$	(b) $(-10, -3)$
	(c) $(0, 0)$	(d) $(3, 3)$
15. $y > 0.2x - 1$	(a) $(0, 2)$	(b) $(6, 0)$
	(c) $(4, -1)$	(d) $(-2, 7)$
16. $y \geq |x - 3|$	(a) $(0, 0)$	(b) $(1, 2)$
	(c) $(4, 10)$	(d) $(5, -1)$

In Exercises 17–22, sketch the graph of the solution of the linear inequality.

17. $x \geq 2$

18. $y > 2$

19. $y \leq x + 1$

20. $y > 4 - x$

21. $x - 2y \geq 6$

22. $3x + 5y \leq 15$

In Exercises 23–26, use a graphing utility to graph (shade) the solution of the inequality.

23. $y \geq \frac{3}{4}x - 1$

24. $y \leq 9 - 1.5x$

25. $2x + 3y - 12 \leq 0$

26. $x - 3y + 9 \geq 0$

27. Represent all points above the line $y = x$ with an inequality.

28. Represent all points below the line $y = x$ with an inequality.

In Exercises 29–34, write an inequality for the shaded region shown in the figure.

29.

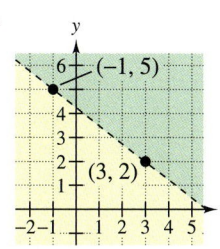

30.

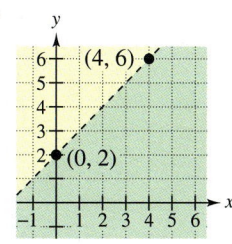

31.

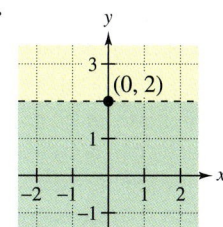

32.

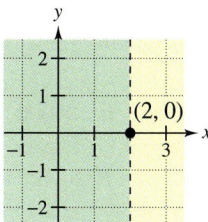

33.

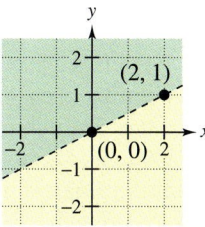

34.

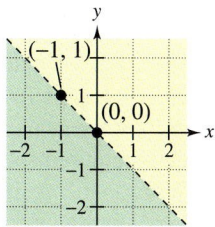

In Exercises 35–38, sketch the region containing the solution points that satisfy both inequalities. Explain your reasoning.

35. $x \geq 2$, $y \leq 4$

36. $y \leq x + 2$, $x \geq 1$

37. $3x + 2y \geq 2$, $y \geq -5$

38. $x + y \leq 5$, $x \geq 2$

In Exercises 39–42, use a graphing utility to graph (shade) the region containing the solution points that satisfy both inequalities.

39. $2x - 3y \leq 6$, $y \leq 4$

40. $6x + 3y \geq 12$, $y \leq 4$

41. $2x + y \leq 2$, $y \geq -4$

42. $x - 2y \geq -6$, $y \leq 6$

43. *Roasting a Turkey*   The time t (in minutes) that it takes to roast a turkey weighing p pounds at 350°F is given by the following inequalities.

For a turkey up to 6 pounds: $t \geq 20p$

For a turkey over 6 pounds: $t \geq 15p + 30$

Sketch the graphs of these inequalities. What are the coordinates for a 12-pound turkey that has been roasting for 3 hours and 40 minutes? Is this turkey fully cooked?

44. *Pizza and Soda Pop*   You and some friends go out for pizza. Together you have $26. You want to order two large pizzas with cheese at $8 each. Each additional topping costs $0.40, and each small soft drink costs $0.80. Write an inequality that represents the numbers of toppings x and drinks y that your group can afford. Sketch the graph of the inequality. What are the coordinates for an order of six soft drinks and two large pizzas with cheese, each with three additional toppings? Is this a solution of the inequality? (Assume there is no sales tax.)

Reviewing the Major Concepts

In Exercises 45–48, simplify the expression.

45. $(x^3 \cdot x^{-2})^{-3}$

46. $(5x^{-4}y^5)(-3x^2y^{-1})$

47. $\left(\dfrac{2x}{3y}\right)^{-2}$

48. $\left(\dfrac{7u^{-4}}{3v^{-2}}\right)\left(\dfrac{14u}{6v^2}\right)^{-1}$

49. *Geometry*   If a television set is advertised as having a 19-inch screen, it means that the diagonal measurement of the screen is 19 inches. If the screen is square, what are its height and width?

50. *Geometry*   A tennis court is 36 feet wide and 78 feet long. Find the length of the diagonal of the court.

Additional Problem Solving

In Exercises 51–54, determine whether the points are solutions of the inequality.

Inequality		Points
51. $3x + y \geq 10$	(a) $(1, 3)$	(b) $(-3, 1)$
	(c) $(3, 1)$	(d) $(2, 15)$
52. $x + y < 3$	(a) $(0, 6)$	(b) $(4, 0)$
	(c) $(0, -2)$	(d) $(1, 1)$
53. $y \leq 3 - |x|$	(a) $(-1, 4)$	(b) $(2, -2)$
	(c) $(6, 0)$	(d) $(5, -2)$
54. $y < -3.5x + 7$	(a) $(1, 5)$	(b) $(5, -1)$
	(c) $(-1, 4)$	(d) $(0, \frac{4}{3})$

In Exercises 55–66, sketch the graph of the solution of the linear inequality.

55. $y < 5$

56. $x < -3$

57. $y > \frac{1}{2}x$

58. $y \leq 2x$

59. $y - 1 > -\frac{1}{2}(x - 2)$

60. $y - 2 < -\frac{2}{3}(x - 1)$

61. $\dfrac{x}{3} + \dfrac{y}{4} \leq 1$

62. $\dfrac{x}{2} + \dfrac{y}{6} \geq 1$

63. $3x - 2y \geq 4$

64. $x + 3y \leq 5$

65. $0.2x + 0.3y < 2$

66. $x - 0.75y > 6$

In Exercises 67–70, use a graphing utility to graph (shade) the solution of the inequality.

67. $y \leq -\frac{2}{3}x + 6$

68. $y \geq \frac{1}{4}x + 3$

69. $x - 2y - 4 \geq 0$

70. $2x + 4y - 3 \leq 0$

In Exercises 71–74, sketch the region showing all points that satisfy both inequalities.

71. $x \leq 5, \ y \geq 2$

72. $y \geq 5 - x, \ x \geq 0$

73. $x + y \geq 4, \ y \leq 4$

74. $3x + y \leq 9, \ x \geq -1$

In Exercises 75–78, use a graphing utility to graph (shade) the region containing the solution points satisfying both inequalities.

75. $2x - 2y \leq 5, \ y \leq 6$

76. $y \geq -2, \ y \leq 8$

77. $2x + 3y \geq 12, \ y \geq 2$

78. $x - y \geq -4, \ y \leq 1$

79. *Geometry* The perimeter of a rectangle of length x and width y cannot exceed 500 feet. Write a linear inequality for this constraint and sketch its graph.

80. *Storage Space* A warehouse for storing chairs and tables has 1000 square feet of floor space. The amounts of space required for each chair and each table are 10 square feet and 15 square feet, respectively. Write a linear inequality for these space constraints if x is the number of chairs and y is the number of tables stored. Sketch a graph of the inequality.

81. *Diet Supplement* A dietitian is asked to design a special diet supplement using two foods. Each ounce of food X contains 30 units of calcium and each ounce of food Y contains 20 units of calcium. The minimum daily requirement in the diet is 300 units of calcium. Write an inequality that represents the different numbers of units of food X and food Y required. Sketch the graph of the inequality. From the graph, find several ordered pairs with positive integer coordinates that are solutions of the inequality.

82. *Weekly Pay* You have two part-time jobs. One is at a grocery store, which pays $9 per hour, and the other is mowing lawns, which pays $6 per hour. Between the two jobs, you want to earn at least $150 a week. Write an inequality that shows the different numbers of hours you can work at each job, and sketch the graph of the inequality. From the graph, find several ordered pairs with positive integer coordinates that are solutions of the inequality.

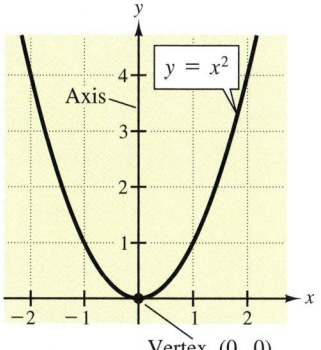

FIGURE 11.16

11.3   Graphs of Quadratic Functions

Graphs of Quadratic Functions ■ Sketching a Parabola ■
Writing the Equation of a Parabola ■ Application

Graphs of Quadratic Functions

In this section, you will study graphs of quadratic functions.

$$f(x) = ax^2 + bx + c$$ Quadratic function

Figure 11.16 shows the graph of a simple quadratic function, $y = x^2$.

> **Graphs of Quadratic Functions**
>
> The graph of $f(x) = ax^2 + bx + c$, $a \neq 0$, is a **parabola**. The completed-square form
>
> $$f(x) = a(x - h)^2 + k$$ Standard form
>
> is the **standard form** of the function. The **vertex** of the parabola occurs at the point (h, k), and the vertical line passing through the vertex is the **axis** of the parabola.

Every parabola is *symmetric* about its axis, which means that if it were folded along its axis, the two parts would match.

If a is positive, the graph of $f(x) = ax^2 + bx + c$ opens up, and if a is negative, the graph opens down, as shown in Figure 11.17.

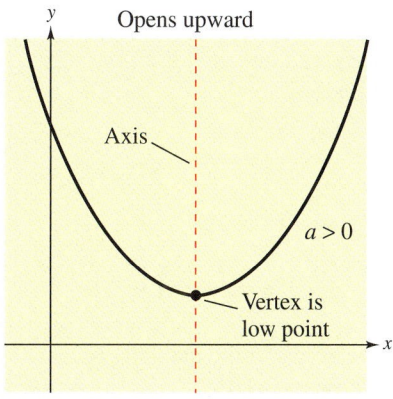
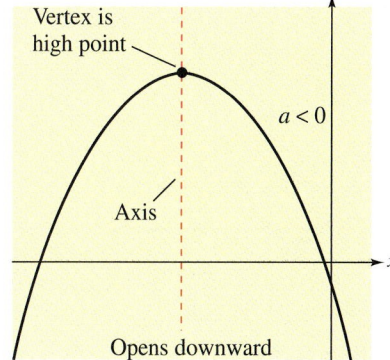

FIGURE 11.17

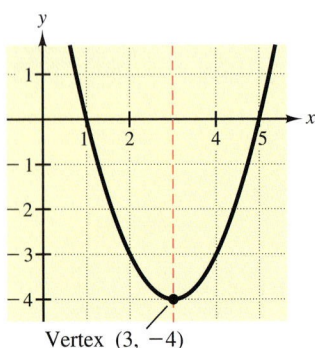

Vertex $(3, -4)$

FIGURE 11.18

EXAMPLE 1   *Finding a Vertex by Completing the Square*

Find the vertex of the graph of $f(x) = x^2 - 6x + 5$.

Solution

Begin by writing the function in standard form.

$$f(x) = x^2 - 6x + 5 \qquad \text{Original function}$$
$$f(x) = x^2 - 6x + (-3)^2 - (-3)^2 + 5 \qquad \text{Add and subtract } (-3)^2.$$
$$f(x) = (x^2 - 6x + 9) - 9 + 5 \qquad \text{Regroup terms.}$$
$$f(x) = (x - 3)^2 - 4 \qquad \text{Standard form}$$

From the standard form, you can see that the vertex of the parabola occurs at the point $(3, -4)$, as shown in Figure 11.18.

In Example 1, the vertex of the graph was found by *completing the square.* Another approach to finding the vertex is to complete the square once for a general function and then use the resulting formula for the vertex.

$$f(x) = ax^2 + bx + c \quad \Longrightarrow \quad f(x) = a\left(x + \frac{b}{2a}\right)^2 + c - \frac{b^2}{4a}$$

From this form you can see that the vertex occurs when $x = -b/2a$.

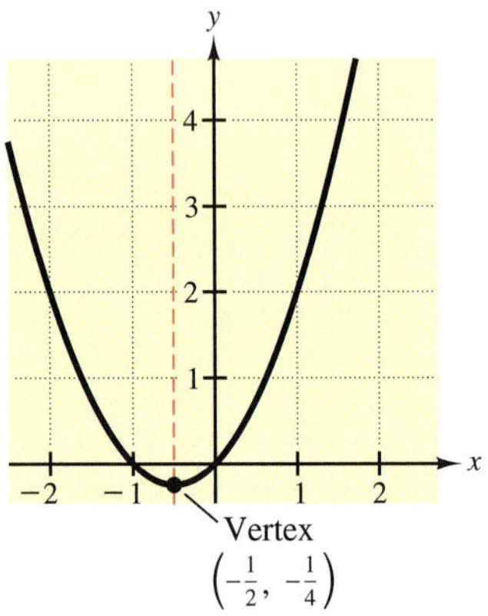

EXAMPLE 2   *Finding a Vertex with a Formula*

Find the vertex of the graph of $f(x) = x^2 + x$.

Solution

From the given function, it follows that $a = 1$ and $b = 1$. Thus, the x-coordinate of the vertex is

$$x = \frac{-b}{2a} = \frac{-1}{2(1)} = -\frac{1}{2}$$

and the y-coordinate is

$$f\left(-\frac{b}{2a}\right) = f\left(-\frac{1}{2}\right) = \left(-\frac{1}{2}\right)^2 + \left(-\frac{1}{2}\right) = \frac{1}{4} - \frac{1}{2} = -\frac{1}{4}.$$

Thus, the vertex of the parabola is $\left(-\frac{1}{2}, -\frac{1}{4}\right)$, and the parabola opens upward, as shown in Figure 11.19.

Vertex
$\left(-\frac{1}{2}, -\frac{1}{4}\right)$

FIGURE 11.19

Sketching a Parabola

NOTE The x- and y-intercepts are useful points to plot. Keep this in mind as you study the examples and do the exercises in this section.

Sketching a Parabola

1. Determine the vertex and axis of the parabola by completing the square or by formula.

2. Plot the vertex, axis, and a few additional points on the parabola. (Using the symmetry about the axis can reduce the number of points you need to plot.)

3. Use the fact that the parabola opens upward if $a > 0$ and opens downward if $a < 0$ to complete the sketch.

EXAMPLE 3 Sketching a Parabola

Sketch the graph of $x^2 - y + 6x + 8 = 0$.

Solution

Begin by writing the equation in standard form.

$$x^2 - y + 6x + 8 = 0 \qquad \text{Original equation}$$
$$-y = -x^2 - 6x - 8 \qquad \text{Subtract } x^2 + 6x + 8 \text{ from both sides.}$$
$$y = x^2 + 6x + 8 \qquad \text{Multiply both sides by } -1.$$
$$y = (x^2 + 6x + 3^2 - 3^2) + 8 \qquad \text{Add and subtract } 3^2.$$
$$y = (x^2 + 6x + 9) - 9 + 8 \qquad \text{Regroup terms.}$$
$$y = (x + 3)^2 - 1 \qquad \text{Standard form}$$

The vertex occurs at the point $(-3, -1)$ and the axis is given by the line $x = -3$. After plotting this information, calculate a few additional points on the parabola, as shown in the table. Note that the y-intercept is $(0, 8)$ and the x-intercepts are solutions to the equation $x^2 + 6x + 8 = (x + 4)(x + 2) = 0$. The graph of the parabola is shown in Figure 11.20. Note that it opens upward because the leading coefficient (in standard form) is positive.

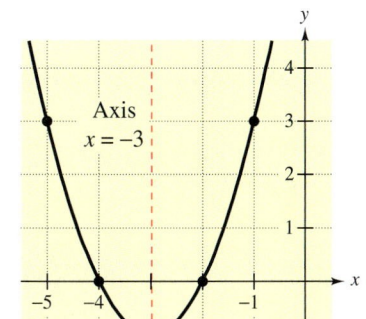

Vertex $(-3, -1)$

FIGURE 11.20

x-Value	-5	-4	-3	-2	-1
y-Value	3	0	-1	0	3
Solution Point	$(-5, 3)$	$(-4, 0)$	$(-3, -1)$	$(-2, 0)$	$(-1, 3)$

Writing the Equation of a Parabola

To write the equation of a parabola with a vertical axis, use the fact that its standard equation has the form $y = a(x - h)^2 + k$, where (h, k) is the vertex.

EXAMPLE 4 Writing the Equation of a Parabola

Write the equation of the parabola whose vertex is $(-2, 1)$ and whose y-intercept is $(0, -3)$, as shown in Figure 11.21.

Solution

Because the vertex occurs at $(h, k) = (-2, 1)$, you can write the following.

$$y = a(x - h)^2 + k \qquad \text{Standard form}$$

$$y = a[x - (-2)]^2 + 1 \qquad \text{Substitute } -2 \text{ for } h \text{ and } 1 \text{ for } k.$$

$$y = a(x + 2)^2 + 1 \qquad \text{Simplify.}$$

To find the value of a, use the fact that the y-intercept is $(0, -3)$.

$$y = a(x + 2)^2 + 1 \qquad \text{Standard form}$$

$$-3 = a(0 + 2)^2 + 1 \qquad \text{Substitute } 0 \text{ for } x \text{ and } -3 \text{ for } y.$$

$$-3 = 4a + 1 \qquad \text{Simplify.}$$

$$-4 = 4a \qquad \text{Subtract 1 from both sides.}$$

$$-1 = a \qquad \text{Divide both sides by 4.}$$

This implies that the standard form of the equation of the parabola is

$$y = -(x + 2)^2 + 1.$$

FIGURE 11.21

Technology

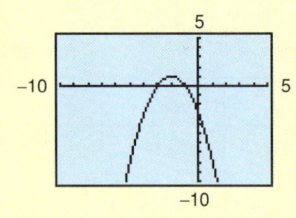

Once you have written an equation for a parabola, you can use a graphing utility to check your equation. For instance, the graph of

$$y = -(x + 2)^2 + 1$$

is shown at the left. By using the TRACE feature of the graphing utility, you can see that the vertex occurs at approximately $(-2, 1)$ and the y-intercept occurs at approximately $(0, -3)$.

Write the equation of the parabola whose vertex is $(3, -2)$ and whose y-intercept is $(0, 7)$. Graph the equation, and use the TRACE feature to check your equation.

Application

EXAMPLE 5   *An Application Involving a Minimum Point*

A suspension bridge is 100 feet long, as shown in Figure 11.22(a). The bridge is supported by cables attached at the tops of towers at each end of the bridge. Each cable hangs in the shape of a parabola (see Figure 11.22(b)) given by

$$y = 0.01x^2 - x + 35$$

where x and y are both measured in feet. (a) Find the distance between the lowest point of the cable and the roadbed of the bridge. (b) How tall are the towers?

Solution

a. Because $a = 0.01$ and $b = -1$, it follows that the vertex of the parabola occurs when $x = -b/2a = 1/(0.02) = 50$. At this x-value, the value of y is

$$y = 0.01(50)^2 - 50 + 35$$
$$= 10.$$

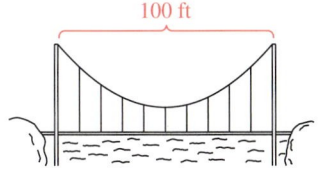

(a)

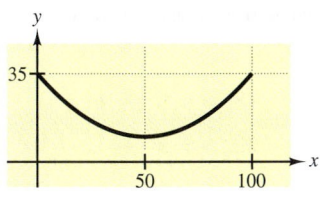

(b)

FIGURE 11.22

Thus, the minimum distance between the cable and the roadbed is 10 feet.

b. Because the vertex of the parabola occurs at the midpoint of the bridge, the two towers are located at the points where $x = 0$ and $x = 100$. Substituting an x-value of 0, you can find that the corresponding y-value is

$$y = 0.01(0)^2 - 0 + 35$$
$$= 35.$$

Therefore, the towers are each 35 feet high. (Try substituting $x = 100$ in the equation to see that you obtain the same y-value.)

Group Activities   P r o b l e m   S o l v i n g

Quadratic Modeling   The data in the table represents the average monthly temperature y in degrees Fahrenheit in Savannah, Georgia for the month x, with $x = 1$ corresponding to November   (Source: National Climate Data Center). Plot the data. Find a quadratic model for the data and use it to find the average temperatures for December and February. The actual average temperature for both December and February is 52°F. How well do you think the model fits the data? Use the model to predict the average temperature for June. How useful do you think the model would be for the whole year?

x	1	3	5
y	59	49	59

11.3 Exercises

Discussing the Concepts

1. In your own words, describe the graph of a quadratic function $f(x) = ax^2 + bx + c$.

2. Explain how to find the vertex of the graph of a quadratic function.

3. Explain how to find the x- and y-intercepts of the graph of a quadratic function.

4. Explain how to determine whether the graph of a quadratic function opens up or down.

5. How is the discriminant related to the graph of a quadratic function?

6. Is it possible for the graph of a quadratic function to have two y-intercepts? Explain.

Problem Solving

In Exercises 7–12, match the equation with its graph. [The graphs are labeled (a), (b), (c), (d), (e), and (f).]

(a)
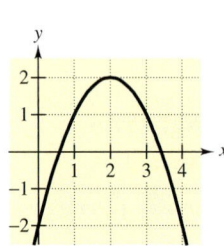

(b)
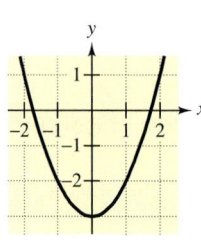

(c)
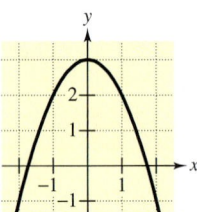

(d)
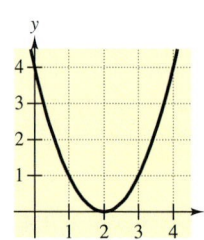

(e)
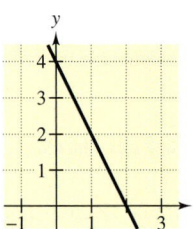

(f)
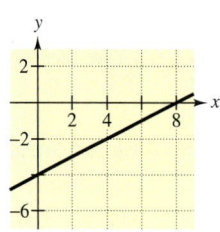

7. $f(x) = 4 - 2x$

8. $f(x) = \frac{1}{2}x - 4$

9. $f(x) = x^2 - 3$

10. $f(x) = -x^2 + 3$

11. $f(x) = (x - 2)^2$

12. $f(x) = 2 - (x - 2)^2$

In Exercises 13–16, state whether the graph opens up or down, and find the vertex.

13. $y = 2(x - 0)^2 + 2$

14. $y = -3(x + 5)^2 - 3$

15. $y = x^2 - 6$

16. $y = -(x + 1)^2$

In Exercises 17–20, find the x- and y-intercepts of the graph.

17. $y = 25 - x^2$

18. $y = x^2 + 4x$

19. $y = x^2 - 3x + 3$

20. $y = x^2 - 3x - 10$

In Exercises 21–24, write the equation in standard form and find the vertex of its graph.

21. $y = x^2 - 4x + 7$

22. $y = x^2 + 6x - 5$

23. $y = -x^2 + 2x - 7$

24. $y = -x^2 - 10x + 10$

In Exercises 25–32, sketch the graph of the equation. Identify the vertex and any x-intercepts. Use a graphing utility to verify your graph.

25. $f(x) = x^2 - 4$

26. $f(x) = -x^2 + 9$

27. $f(x) = -x^2 + 3x$

28. $f(x) = x^2 - 4x$

29. $f(x) = x^2 - 8x + 15$

30. $f(x) = -x^2 + 2x + 8$

31. $f(x) = \frac{1}{5}(3x^2 - 24x + 38)$

32. $f(x) = \frac{1}{5}(2x^2 - 4x + 7)$

In Exercises 33–38, write an equation of the parabola.

33.

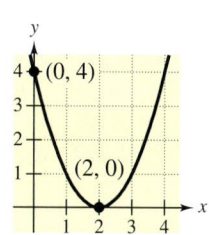

34.

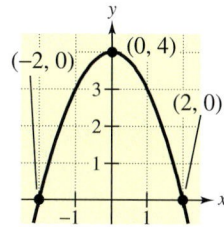

35.

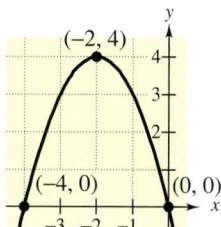

36.

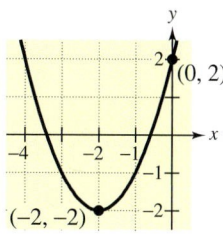

37.

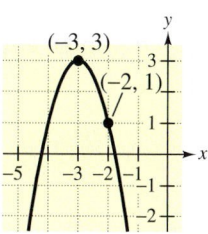

38.

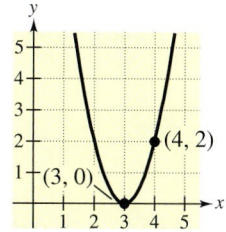

39. *Graphical Interpretation*   The height y (in feet) of a ball thrown by a child is given by

$$y = -\frac{1}{12}x^2 + 2x + 4$$

where x is the horizontal distance (in feet) from where the ball is thrown.

(a) Use a graphing utility to sketch the path of the ball.

(b) How high is the ball when it leaves the child's hand?

(c) How high is the ball when it is at its maximum height?

(d) How far from the child does the ball strike the ground?

40. *Maximum Height of a Diver*   The path of a diver is given by

$$y = -\frac{4}{9}x^2 + \frac{24}{9}x + 10$$

where y is the height in feet and x is the horizontal distance from the end of the diving board in feet (see figure). What is the maximum height of the diver?

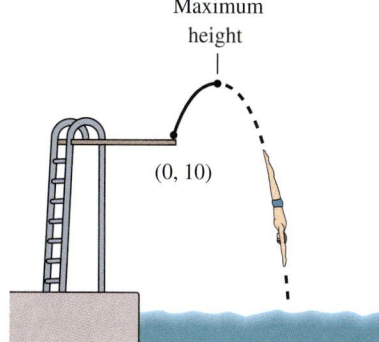

Maximum height

(0, 10)

41. *Graphical Interpretation*   A company manufactures radios that cost (the company) $60 each. For buyers who purchase 100 or fewer radios, the purchase price p is $90 per radio. To encourage large orders, the company will reduce the price *per radio* for orders over 100, as follows. If 101 radios are purchased, the price is $89.85 per unit. If 102 radios are purchased, the price is $89.70 per unit. If $(100 + x)$ radios are purchased, the price per unit is

$$p = 90 - x(0.15)$$

where x is the amount over 100 in the order.

(a) Show that the profit for orders over 100 is

$$P = (100 + x)[90 - x(0.15)] - (100 + x)60$$

$$= 3000 + 15x - \frac{3}{20}x^2.$$

(b) Use a graphing utility to graph the profit function.

(c) Find the vertex of the profit curve and determine the order size for maximum profit.

(d) Would you recommend this pricing scheme? Explain your reasoning.

42. *Graphical Interpretation* The advertising revenue for newspapers in the United States for the years 1985 through 1991 is approximated by the model

$$R = -1.03 + 7.11t - 0.38t^2$$

where R is revenue in billions of dollars and t represents the year, with $t = 5$ corresponding to 1985. (Source: McCann-Erickson, Inc.)

(a) Use a graphing utility to graph the revenue function.

(b) Determine the year when advertising revenue was greatest. Approximate the revenue that year.

Think About It In Exercises 43 and 44, write the equation of the parabola.

43.

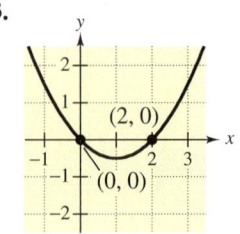

44.

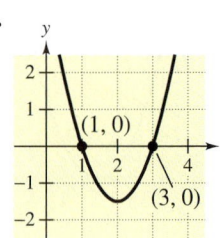

Reviewing the Major Concepts

In Exercises 45–48, solve the inequality and sketch its graph on the real number line.

45. $7 - 3x > 4 - x$ **46.** $2(x + 6) - 20 < 2$

47. $|x - 3| < 2$ **48.** $|x - 5| > 3$

In Exercises 49 and 50, solve the equation.

49. $x^2 - 4x + 3 = 0$ **50.** $2x^2 + 10x + 9 = 0$

Free-Falling Object In Exercises 51 and 52, find the time for an object to reach the ground when it is dropped from a height of s_0 feet. The height h (in feet) is given by

$$h = -16t^2 + s_0$$

where t is the time (in seconds).

51. $s_0 = 80$ **52.** $s_0 = 150$

Additional Problem Solving

In Exercises 53–56, state whether the graph opens up or down, and find the vertex.

53. $y = 4 - (x - 10)^2$ **54.** $y = 2(x - 12)^2 + 3$

55. $y = x^2 - 6x$ **56.** $y = -(x - 3)^2$

In Exercises 57–60, find the x- and y-intercepts of the graph.

57. $y = x^2 - 9x$ **58.** $y = x^2 - 49$

59. $y = 4x^2 - 12x + 9$ **60.** $y = 10 - x - 2x^2$

In Exercises 61–66, write the equation in standard form and find the vertex of its graph.

61. $y = x^2 + 6x + 5$ **62.** $y = x^2 - 4x + 5$

63. $y = -x^2 + 6x - 10$ **64.** $y = 4 - 8x - x^2$

65. $y = 2x^2 + 6x + 2$ **66.** $y = 3x^2 - 3x - 9$

In Exercises 67–82, sketch the graph of the equation. Identify the vertex and any x-intercepts. Use a graphing utility to verify your graph.

67. $f(x) = -x^2 + 4$ **68.** $f(x) = x^2 - 9$

69. $f(x) = x^2 - 3x$ **70.** $f(x) = -x^2 + 4x$

71. $f(x) = (x - 4)^2$ **72.** $f(x) = -(x + 4)^2$

73. $f(x) = 5 - \dfrac{x^2}{3}$ **74.** $f(x) = \dfrac{x^2}{3} - 2$

75. $f(x) = x^2 + 4x + 7$ **76.** $f(x) = x^2 + 4x + 2$

77. $f(x) = 2(x^2 + 6x + 8)$

78. $f(x) = -x^2 + 6x - 7$

79. $f(x) = -(x^2 + 6x + 5)$

80. $f(x) = 3x^2 - 6x + 4$

81. $f(x) = \frac{1}{2}(x^2 - 2x - 3)$

82. $f(x) = -\frac{1}{2}(x^2 - 6x + 7)$

Graphical and Algebraic Reasoning In Exercises 83–86, use a graphing utility to approximate the vertex of the graph. Then check your result algebraically.

83. $y = \frac{1}{6}(2x^2 - 8x + 11)$

84. $y = -\frac{1}{4}(4x^2 - 20x + 13)$

85. $y = -0.7x^2 - 2.7x + 2.3$

86. $y = 0.75x^2 - 7.50x + 23.00$

In Exercises 87–90, use a graphing utility to graph the two functions on the same screen. Do the graphs intersect? If so, approximate the point or points of intersection.

87. $y_1 = -x^2 + 6$, $y_2 = 2$

88. $y_1 = x^2 - 6x + 8$, $y_2 = 3$

89. $y_1 = \frac{1}{2}x^2 - 3x + \frac{13}{2}$, $y_2 = 3$

90. $y_1 = -2x^2 - 4x$, $y_2 = 1$

In Exercises 91–98, write an equation of the parabola $y = ax^2 + bx + c$ that satisfies the conditions.

91. Vertex: $(2, 1)$; $a = 1$

92. Vertex: $(-3, -3)$; $a = 1$

93. Vertex: $(-3, 4)$; $a = -1$

94. Vertex: $(3, -2)$; $a = -1$

95. Vertex: $(2, -4)$; point on graph: $(0, 0)$

96. Vertex: $(-2, -4)$; point on graph: $(0, 0)$

97. Vertex: $(3, 2)$; point on graph: $(1, 4)$

98. Vertex: $(-1, -1)$; point on graph: $(0, 4)$

99. *Graphical Estimation* The profit (in thousands of dollars) for a company is given by

$$P = 230 + 20s - \frac{1}{2}s^2$$

where s is the amount (in hundreds of dollars) spent on advertising. Use a graphing utility to graph the profit function and approximate the amount of advertising that yields a maximum profit. Verify the maximum algebraically.

100. *Bridge Design* A bridge is to be constructed over a gorge with the main supporting arch being a parabola (see figure). If the equation of the parabola is

$$y = 4\left(100 - \frac{x^2}{2500}\right)$$

where x and y are measured in feet,

(a) find the length of the road across the gorge.

(b) find the height of the parabolic arch at the center of the span.

(c) find the lengths of the vertical girders at intervals of 100 feet from the center of the bridge.

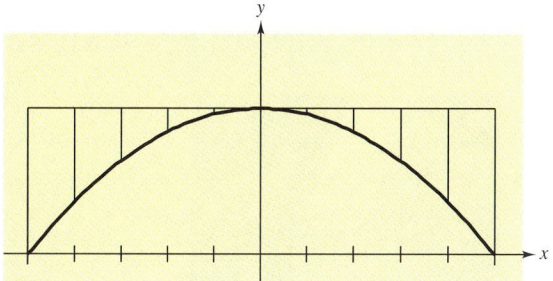

101. *Highway Design* A highway department engineer must design a parabolic arc to create a turn in a freeway around a city. The vertex of the parabola is placed at the origin, and the parabola must connect with roads represented by the equations

$$y = -0.4x - 100, \quad x < -500$$

$$y = 0.4x - 100, \quad x > 500$$

(see figure). Find an equation of the parabolic arc.

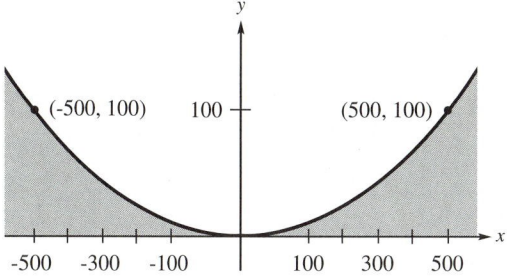

102. *Geometry* The area of a rectangle is given by the equation

$$A = \frac{2}{\pi}(100x - x^2), \quad 0 < x < 100$$

where x is the length of its base in feet. Use a graphing utility to sketch the graph of this equation and use the TRACE feature to approximate the value of x when A is maximum.

103. *Graphical Estimation* The cost of producing x units of a product is given by

$$C = 800 - 10x + \frac{1}{4}x^2, \quad 0 < x < 40.$$

Use a graphing utility to sketch the graph of this equation and use the TRACE feature to approximate the value of x when C is minimum.

Math Matters

Annual Salary Versus Hourly Wage

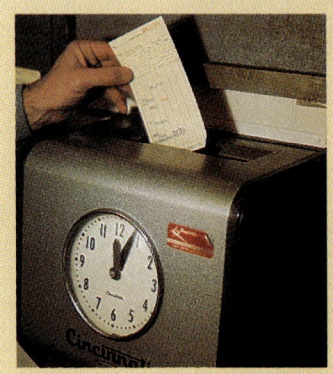

Here is a quick method that you can use to approximate the annual salary that corresponds to a given hourly wage.

$$\text{Hourly wage} \times 2 \times 1000 \approx \text{annual salary}$$

For example, $5 an hour is approximately $10,000 a year, and $12 hour is approximately $24,000. This approximation assumes that there are only fifty 40-hour weeks in a year. To obtain the exact conversion from hourly wage to annual salary, you should multiply by 40 (hours per week) and then multiply by 52 (weeks per year).

To approximate the hourly wage that corresponds to a given annual salary, use the following rule.

$$\text{Annual salary} \div 1000 \div 2 \approx \text{hourly wage}$$

For example, $12,000 a year is approximately $6 an hour, and $20,000 a year is approximately $10 an hour. To obtain the exact conversion from annual salary to hourly wage, you should divide by 52 (weeks per year) and then divide the result by 40 (hours per week).

Use the quick approximation technique to determine which of the following wages is greater.

(a) $7.50 an hour   or $17,500 a year

(b) $12.50 an hour   or $22,500 a year

MID-CHAPTER QUIZ

Take this quiz as you would take a quiz in class. After you are done, check your work against the answers given in the back of the book.

In Exercises 1–4, write an equation of the line passing through (x_1, y_1) with slope m.

1. $\left(0, -\frac{3}{2}\right)$, $m = 2$ **2.** $(0, 6)$, $m = -\frac{3}{2}$ **3.** $\left(\frac{5}{2}, 6\right)$, $m = -\frac{3}{4}$ **4.** $(-3.5, -1.8)$, $m = 3$

In Exercises 5–8, write an equation of the line passing through the points.

5. $\left(\frac{1}{3}, 1\right)$, $(4, 5)$ **6.** $(0, 0.8)$, $(3, -2.3)$ **7.** $(3, -1)$, $(10, -1)$ **8.** $\left(4, \frac{5}{3}\right)$, $(4, 8)$

9. Write an equation of the line that passes through the point $(3, 5)$ and is (a) parallel to and (b) perpendicular to the line $2x - 3y = 1$.

10. Decide whether the points are solutions of the inequality $2x - 3y \leq 4$. Explain your reasoning.

  (a) $(5, 2)$ (b) $(-2, 4)$ (c) $(2, -4)$ (d) $(3, 0)$

In Exercises 11 and 12, write an inequality for the shaded region.

11. See figure at right.   **12.** See figure at right.

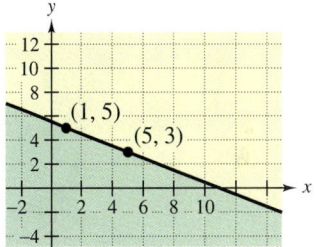

Figure for 11

In Exercises 13 and 14, sketch the region that is determined by both inequalities.

13. $2x + 3y \leq 9$, $\quad y \geq 1$ **14.** $2x - y \leq 4$, $\quad y \leq 4$

In Exercises 15 and 16, write an equation for the indicated parabola.

15. Vertex: $(3, -1)$; passes through the point $(5, 3)$

16. Vertex: $(5, 4)$; passes through the point $(3, 3)$

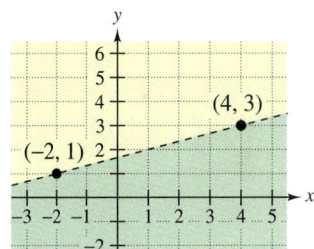

Figure for 12

In Exercises 17 and 18, sketch the graph of the quadratic function. Identify the vertex and the x-intercepts.

17. $y = -\frac{1}{4}(x^2 + 6x + 1)$ **18.** $y = 2x^2 - 4x - 7$

19. You purchase a used car for \$12,400. It is estimated that after 4 years its depreciated value will be \$5000. Assuming straight-line depreciation, write a linear function giving the value V of the car in terms of time t. State the domain of the function.

20. The path of a ball is given by $y = -0.005x^2 + x + 5$. Determine the maximum height of the ball.

11.4   Graphs of Second-Degree Equations

Circles ▪ Ellipses ▪ Hyperbolas ▪ Application

Circles

One of the first recognized female mathematicians, Hypatia (370–415 A.D.), wrote a textbook entitled *On the Conics of Apollonius*. Her death marked the end of major mathematical discoveries in Europe for several hundred years.

In Section 11.3, you learned that the graph of a second-degree equation of the form

$$y = ax^2 + bx + c$$

is a parabola. A parabola is one of four types of **conics** or **conic sections.** The other three types are circles, ellipses, and hyperbolas. All four types have equations that are of second degree. As indicated in Figure 11.23, the name "conic" relates to the fact that each of these figures can be obtained by intersecting a plane with a double-napped cone.

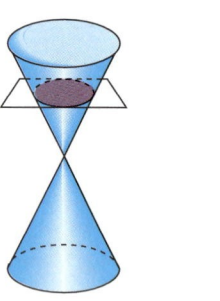 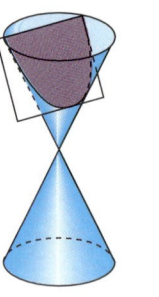 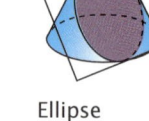 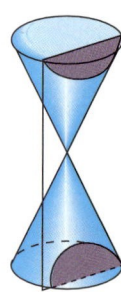

Circle   Parabola   Ellipse   Hyperbola
FIGURE 11.23

Conic sections occur in many practical applications. Reflective surfaces in satellite dishes, flashlights, and telescopes often are of parabolic shape. The orbits of planets are elliptical, and the orbits of comets are usually elliptical or hyperbolic. Ellipses and parabolas are also used in building archways and bridges.

A **circle** in the rectangular coordinate plane consists of all points (x, y) that are a given positive distance r from a fixed point, called the **center** of the circle. The positive distance r is the **radius** of the circle. If the center of the circle is the origin, as shown in Figure 11.24, the relationship between the coordinates of any point (x, y) on the circle and the radius r is given by

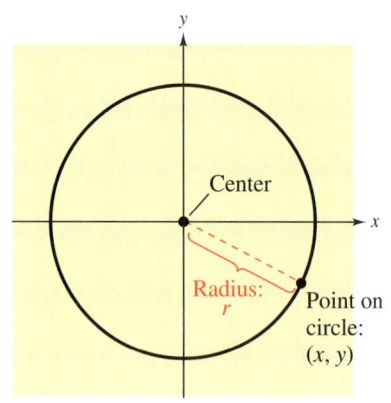

FIGURE 11.24

$$\text{Radius } = r = \sqrt{(x - 0)^2 + (y - 0)^2} \qquad \text{Distance Formula, center at } (0, 0)$$
$$= \sqrt{x^2 + y^2}$$
$$r = \sqrt{(x - h)^2 + (y - k)^2}. \qquad \text{Distance Formula, center at } (h, k)$$

By squaring both sides of this equation, you obtain the **standard form of the equation of a circle.**

> ### Standard Form of the Equation of a Circle
>
> The **standard form of the equation of a circle** is
>
> $$x^2 + y^2 = r^2 \qquad \text{Circle with center at } (0, 0)$$
> $$(x - h)^2 + (y - k)^2 = r^2. \qquad \text{Circle with center at } (h, k)$$
>
> The positive number r is the **radius** of the circle.

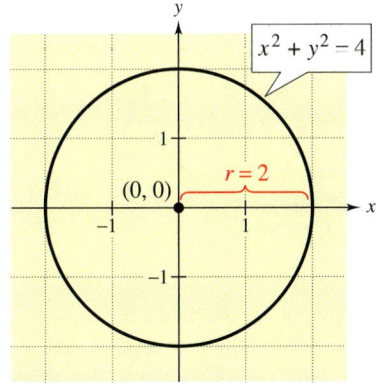

FIGURE 11.25

EXAMPLE 1   Finding an Equation of a Circle

Find an equation of the circle whose center is at $(0, 0)$ and whose radius is 2.

Solution

Use the standard form of the equation of a circle with center at the origin.

$$x^2 + y^2 = r^2 \qquad \text{Standard form with center at } (0, 0)$$
$$x^2 + y^2 = 2^2 \qquad \text{Substitute 2 for } r.$$
$$x^2 + y^2 = 4 \qquad \text{Equation of circle}$$

The circle given by this equation is shown in Figure 11.25.

To sketch the circle for a given equation, write the equation in standard form. From the standard form, you can identify the center and radius.

EXAMPLE 2   Finding the Center and Radius of a Circle

Identify the center and radius of the circle given by the equation, and sketch the circle.

$$x^2 + y^2 + 2x - 6y + 1 = 0$$

Solution

$$x^2 + y^2 + 2x - 6y + 1 = 0 \qquad \text{Original equation}$$
$$(x^2 + 2x) + (y^2 - 6y) = -1 \qquad \text{Group terms.}$$
$$(x^2 + 2x + 1) + (y^2 - 6y + 9) = -1 + 1 + 9 \qquad \text{Complete each square.}$$
$$(x + 1)^2 + (y - 3)^2 = 9 \qquad \text{Standard form}$$

From this standard form, you can see that the graph of the equation is a circle whose center is at $(-1, 3)$ and whose radius is 3, as shown in Figure 11.26.

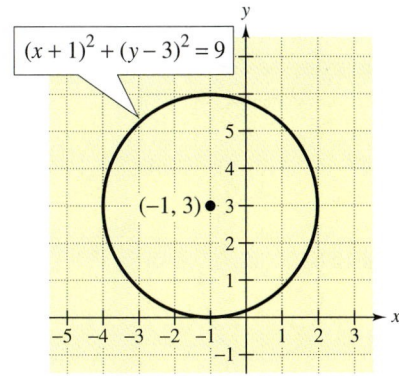

FIGURE 11.26

NOTE To trace an ellipse, place two thumbtacks at the foci, as shown in Figure 11.28. If the ends of a fixed length of string are fastened to the thumbtacks and the string is drawn taut with a pencil, the path traced by the pencil will be an ellipse.

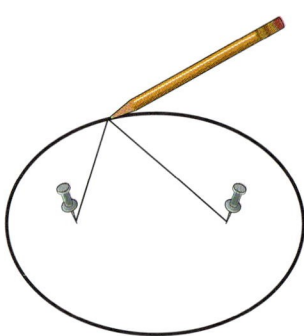

FIGURE 11.28

Ellipses

An **ellipse** is the set of all points (x, y) such that the sum of the distances between (x, y) and two distinct fixed points is a constant. As shown in Figure 11.27(a), each of the two fixed points is a **focus** of the ellipse. (The plural of focus is *foci*.) In this text, we restrict the study of ellipses to those whose centers are at the origin.

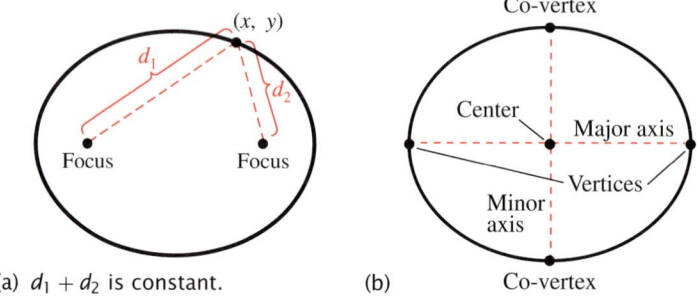

(a) $d_1 + d_2$ is constant.　　　(b)

FIGURE 11.27

The line through the foci intersects the ellipse at the **vertices,** as shown in Figure 11.27(b). The line segment joining the vertices is the **major axis,** and its midpoint is the **center** of the ellipse. The line segment perpendicular to the major axis at the center is the **minor axis** of the ellipse, and the points at which the minor axis intersects the ellipse are **co-vertices.**

Standard Form of the Equation of an Ellipse

The **standard form of the equation of an ellipse** with center at the origin and major and minor axes of lengths $2a$ and $2b$, respectively, is

$$\frac{x^2}{a^2} + \frac{y^2}{b^2} = 1 \quad \text{or} \quad \frac{x^2}{b^2} + \frac{y^2}{a^2} = 1, \qquad 0 < b < a.$$

The vertices lie on the major axis, a units from the center, and the co-vertices lie on the minor axis, b units from the center.

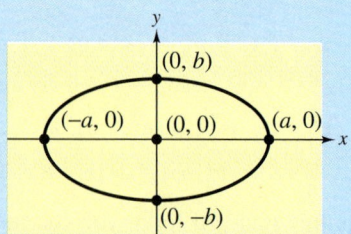

Major axis is horizontal.
Minor axis is vertical.

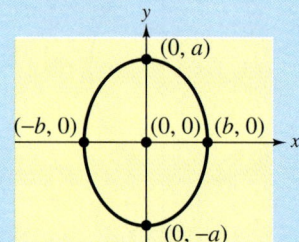

Major axis is vertical.
Minor axis is horizontal.

EXAMPLE 3 *Finding an Equation of an Ellipse*

Find an equation of the ellipse whose vertices are $(-3, 0)$ and $(3, 0)$ and whose co-vertices are $(0, -2)$ and $(0, 2)$.

Solution

Begin by plotting the vertices and co-vertices, as shown in Figure 11.29. The center of the ellipse is $(0, 0)$, because it is the point that lies halfway between the vertices (and halfway between the co-vertices). Thus, the equation of the ellipse has the form

$$\frac{x^2}{a^2} + \frac{y^2}{b^2} = 1.$$

For this ellipse, the major axis is horizontal. Thus, a is the distance between the center and either vertex, which implies that $a = 3$. Similarly, b is the distance between the center and either co-vertex, which implies that $b = 2$. Thus, the standard equation of the ellipse is

$$\frac{x^2}{3^2} + \frac{y^2}{2^2} = 1.$$

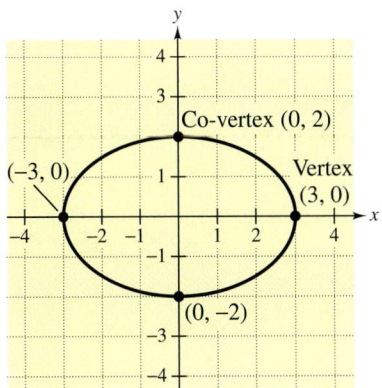

FIGURE 11.29

To sketch an ellipse, it helps to first write its equation in standard form.

EXAMPLE 4 *Sketching an Ellipse*

Sketch the ellipse given by $4x^2 + y^2 = 36$, and identify the vertices and co-vertices.

Solution

Begin by writing the equation in standard form.

$4x^2 + y^2 = 36$	Given equation
$\dfrac{4x^2}{36} + \dfrac{y^2}{36} = \dfrac{36}{36}$	Divide both sides by 36.
$\dfrac{x^2}{9} + \dfrac{y^2}{36} = 1$	Simplify.
$\dfrac{x^2}{3^2} + \dfrac{y^2}{6^2} = 1$	Standard form

Because the denominator of the y^2 term is larger than the denominator of the x^2 term, you can conclude that the major axis is vertical. Moreover, because $a = 6$, the vertices are $(0, -6)$ and $(0, 6)$. Finally, because $b = 3$, the co-vertices are $(-3, 0)$ and $(3, 0)$, as shown in Figure 11.30.

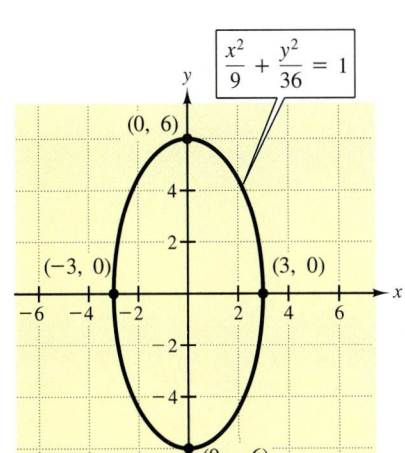

FIGURE 11.30

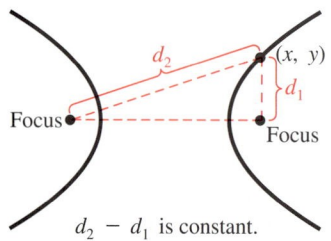

$d_2 - d_1$ is constant.

FIGURE 11.31

Hyperbolas

A **hyperbola** on the rectangular coordinate system consists of all points (x, y) such that the *difference* between (x, y) and two fixed points is a constant, as shown in Figure 11.31. The two fixed points are called the **foci** of the hyperbola. As with ellipses, we will consider only equations of hyperbolas whose foci lie on the x-axis or on the y-axis. The line on which the foci lie is called the **transverse axis** of the hyperbola.

Standard Form of the Equation of a Hyperbola

The **standard form of the equation of a hyperbola** whose center is at the origin is given by

$$\frac{x^2}{a^2} - \frac{y^2}{b^2} = 1 \qquad \text{Transverse axis is horizontal.}$$

$$\frac{y^2}{a^2} - \frac{x^2}{b^2} = 1 \qquad \text{Transverse axis is vertical.}$$

where a and b are positive real numbers. The **vertices** of the hyperbola lie on the transverse axis, a units from the center.

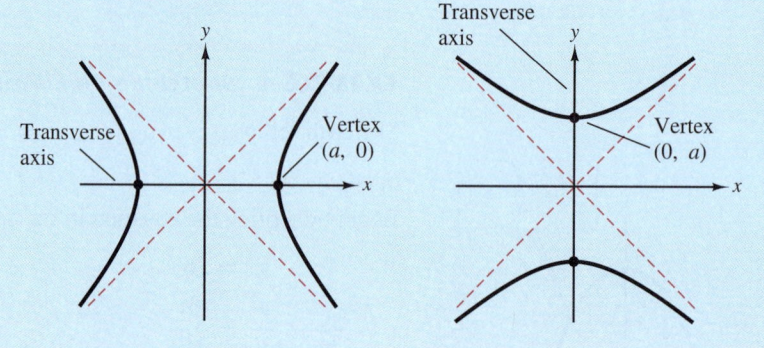

A hyperbola has two disconnected parts, each of which is a **branch** of the hyperbola. The two branches approach a pair of intersecting straight lines called **asymptotes** of the hyperbola. The two asymptotes intersect at the center of the hyperbola.

To sketch a hyperbola, form a **central rectangle** whose center is the origin and whose width and height are $2a$ and $2b$. Note in Figure 11.32 (on page 663) that the asymptotes pass through the corners of the central rectangle and the vertices of the hyperbola lie at the centers of opposite sides of the central rectangle.

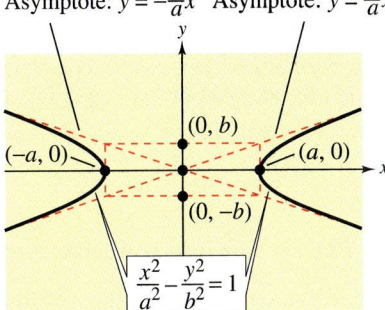

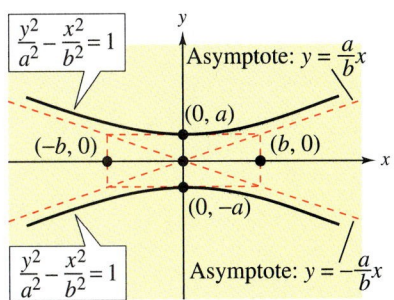

Transverse axis is horizontal.

Transverse axis is vertical.

FIGURE 11.32

EXAMPLE 5 Sketching a Hyperbola

Sketch the hyperbola given by $\dfrac{x^2}{36} - \dfrac{y^2}{16} = 1$.

Solution

From the standard form of the equation

$$\frac{x^2}{6^2} - \frac{y^2}{4^2} = 1$$

you can see that the center of the hyperbola is the origin and the transverse axis is horizontal. Therefore, the vertices lie six units to the left and right of the center at the points $(-6, 0)$ and $(6, 0)$. Because $a = 6$ and $b = 4$, you can sketch the hyperbola by first drawing a central rectangle whose width is $2a = 12$ and whose height is $2b = 8$, as shown in Figure 11.33(a). Next, draw the asymptotes of the hyperbola through the corners of the central rectangle and plot the vertices. Finally, draw the hyperbola, as shown in Figure 11.33(b).

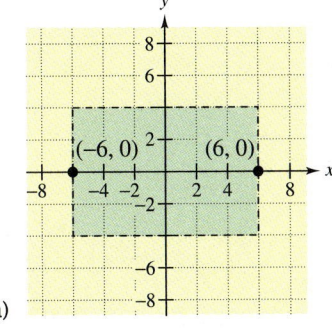

(a)

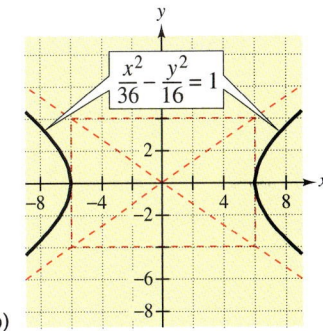

(b)

FIGURE 11.33

Finding an equation of a hyperbola is a little more difficult than finding equations of the other three types of conics. However, if you know the vertices and the asymptotes, you can find the values of a and b, which enable you to write the equation. Notice in Example 6, that the key to this procedure is knowing that the central rectangle has a width of $2b$ and a height of $2a$.

EXAMPLE 6   Finding the Equation of a Hyperbola

Find an equation of the hyperbola with a vertical transverse axis whose vertices are $(0, 3)$ and $(0, -3)$ and whose asymptotes are given by $y = \frac{3}{5}x$ and $y = -\frac{3}{5}x$.

Solution

To begin, sketch the lines that represent the asymptotes, as shown in Figure 11.34(a). Note that these two lines intersect at the origin, which implies that the center of the hyperbola is $(0, 0)$. Next, plot the two vertices at the points $(0, 3)$ and $(0, -3)$. Because you know where the vertices are located, you can sketch the central rectangle of the hyperbola, as shown in Figure 11.34(a). Note that the corners of the central rectangle occur at the points

$$(-5, 3), \quad (5, 3), \quad (-5, -3), \quad \text{and} \quad (5, -3).$$

Because the width of the central rectangle is $2b = 10$, it follows that $b = 5$. Similarly, because the height of the central rectangle is $2a = 6$, it follows that $a = 3$. Now that you know the values of a and b, you can conclude that the standard form of the equation of the hyperbola is

$$\frac{y^2}{3^2} - \frac{x^2}{5^2} = 1.$$

The graph is shown in Figure 11.34(b).

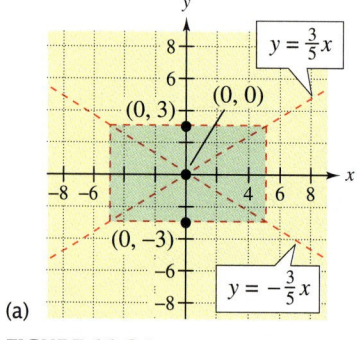

(a)

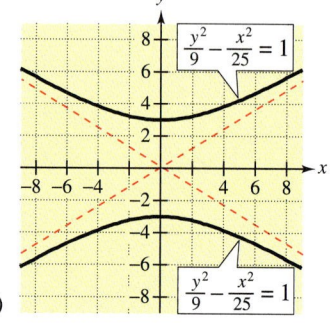

(b)

FIGURE 11.34

Application

EXAMPLE 7   *An Application Involving an Ellipse*

You are responsible for designing a semielliptical archway, as shown in Figure 11.35. The height of the archway is 10 feet and its width is 30 feet. Find an equation of the ellipse and use the equation to sketch an accurate diagram of the archway.

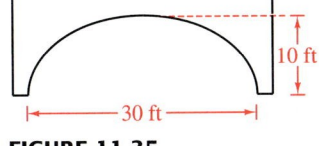

FIGURE 11.35

Solution

To make the equation simple, place the origin at the center of the ellipse. This means that the standard form of the equation is

$$\frac{x^2}{a^2} + \frac{y^2}{b^2} = 1.$$

Because the major axis is horizontal, it follows that $a = 15$ and $b = 10$, which implies that the equation is

$$\frac{x^2}{15^2} + \frac{y^2}{10^2} = 1.$$

In order to make an accurate sketch of the semiellipse, it is helpful to solve this equation for y as follows.

$$\frac{x^2}{15^2} + \frac{y^2}{10^2} = 1$$

$$\frac{x^2}{225} + \frac{y^2}{100} = 1$$

$$\frac{y^2}{100} = 1 - \frac{x^2}{225}$$

$$y^2 = 100\left(1 - \frac{x^2}{225}\right)$$

$$y = 10\sqrt{1 - \frac{x^2}{225}}$$

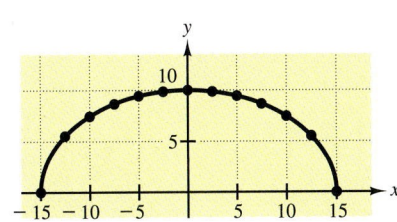

FIGURE 11.36

Next, calculate several y-values for the archway, as shown in the table. Then use the values in the table to sketch the archway, as shown in Figure 11.36.

x-Value	±15	±12.5	±10	±7.5	±5	±2.5	0
y-Value	0	5.53	7.45	8.66	9.43	9.86	10

Technology

Graphing Conic Sections

Most graphing utilities are designed to sketch the graphs of equations in which y is isolated on the left side of the equation. Thus, if you want to sketch the graph of the circle given by $x^2 + y^2 = 36$, you should solve the equation for y and obtain the following *two* equations.

$$y = \sqrt{36 - x^2} \qquad \text{and} \qquad y = -\sqrt{36 - x^2}$$

The first of these two equations represents the upper half of the circle and the second represents the lower half of the circle. Try graphing both the upper and lower halves of the circle on the same screen.

With a *standard* setting, the graph should appear as shown in the figure on the left. The reason the graph does not look like a circle is that with the standard setting the tick marks on the x-axis are farther apart than the tick marks on the y-axis. To correct this, choose a *square* setting, as shown in the figure on the right.

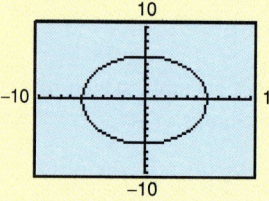 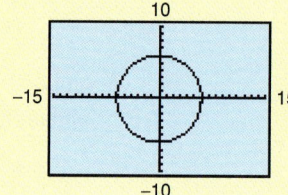

Group Activities   Extending the Concept

Identifying Conic Sections   Cut cone-shaped pieces of styrofoam to demonstrate how to obtain each type of conic section: circle, parabola, ellipse, and hyperbola. Discuss how you could write directions for someone else to form each conic section. Compile a list of real-life situations and/or everyday objects in which conic sections may be seen.

11.4 Exercises

Discussing the Concepts

1. Name the four types of conics.

2. Define a circle and give the standard form of the equation of a circle centered at the origin.

3. Define an ellipse and give the standard form of the equation of an ellipse centered at the origin.

4. Define a hyperbola and give the standard form of the equation of a hyperbola centered at the origin.

5. From its equation, how can you determine the lengths of the axes of an ellipse?

6. Explain how the central rectangle of a hyperbola can be used to sketch its asymptotes.

Problem Solving

In Exercises 7–12, match the equation with its graph. [The graphs are labeled (a), (b), (c), (d), (e), and (f).]

(a)
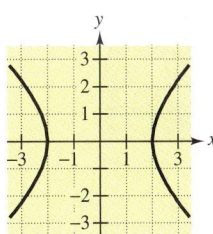

(b)
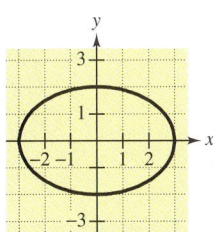

(c)
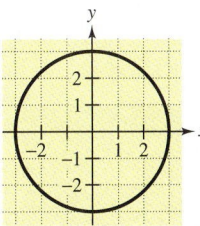

(d)
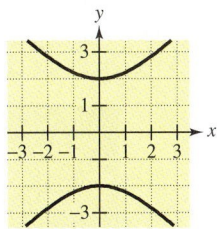

(e)
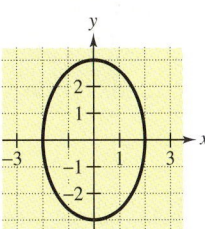

(f)
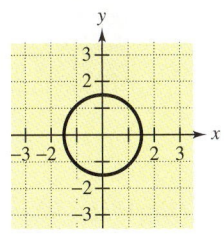

7. $x^2 + y^2 = 9$

8. $4x^2 + 4y^2 = 9$

9. $\dfrac{x^2}{4} + \dfrac{y^2}{9} = 1$

10. $\dfrac{x^2}{9} + \dfrac{y^2}{4} = 1$

11. $x^2 - y^2 = 4$

12. $x^2 - y^2 = -4$

In Exercises 13–16, find an equation of the circle with center at $(0, 0)$ that satisfies the given criteria.

13. Radius: 5

14. Radius: $\frac{5}{2}$

15. Passes through the point $(5, 2)$

16. Passes through the point $(-1, -4)$

In Exercises 17–20, identify the center and radius of the circle and sketch its graph.

17. $x^2 + y^2 = 16$

18. $x^2 + y^2 = 25$

19. $25x^2 + 25y^2 - 144 = 0$

20. $\dfrac{x^2}{4} + \dfrac{y^2}{4} - 1 = 0$

In Exercises 21–24, write the standard form of the equation of the ellipse, centered at the origin.

	Vertices	Co-vertices
21.	$(-4, 0), (4, 0)$	$(0, -3), (0, 3)$
22.	$(-4, 0), (4, 0)$	$(0, -1), (0, 1)$
23.	$(0, -4), (0, 4)$	$(-3, 0), (3, 0)$
24.	$(0, -5), (0, 5)$	$(-1, 0), (1, 0)$

25. Write the standard form of the equation of the ellipse, centered at the origin, with a horizontal major axis of length 20 and a vertical minor axis of length 12.

26. Write the standard form of the equation of the ellipse, centered at the origin, with a horizontal minor axis of length 30 and a vertical major axis of length 50.

In Exercises 27–30, sketch the ellipse. Identify its vertices and co-vertices.

27. $\dfrac{x^2}{16} + \dfrac{y^2}{4} = 1$ **28.** $\dfrac{x^2}{9} + \dfrac{y^2}{25} = 1$

29. $4x^2 + y^2 - 4 = 0$ **30.** $4x^2 + 9y^2 - 36 = 0$

In Exercises 31–34, sketch the hyperbola. Identify its vertices and asymptotes.

31. $\dfrac{x^2}{9} - \dfrac{y^2}{25} = 1$ **32.** $\dfrac{x^2}{4} - \dfrac{y^2}{9} = 1$

33. $\dfrac{y^2}{9} - \dfrac{x^2}{25} = 1$ **34.** $\dfrac{y^2}{4} - \dfrac{x^2}{9} = 1$

In Exercises 35–38, find an equation of the hyperbola centered at the origin.

	Vertices	*Asymptotes*	
35.	$(-4, 0), (4, 0)$	$y = 2x$	$y = -2x$
36.	$(-2, 0), (2, 0)$	$y = \frac{1}{3}x$	$y = -\frac{1}{3}x$
37.	$(0, -4), (0, 4)$	$y = \frac{1}{2}x$	$y = -\frac{1}{2}x$
38.	$(0, -2), (0, 2)$	$y = 3x$	$y = -3x$

In Exercises 39 and 40, sketch the circle and identify its center and radius.

39. $(x - 2)^2 + (y - 3)^2 = 4$

40. $(x + 4)^2 + (y - 3)^2 = 25$

In Exercises 41 and 42, write the equation in completed-square form. Then sketch its graph.

41. $x^2 + y^2 - 4x - 2y + 1 = 0$

42. $x^2 + y^2 + 6x - 4y - 3 = 0$

In Exercises 43–46, use a graphing utility to graph the equation.

43. $x^2 + y^2 = 30$ **44.** $4x^2 + 4y^2 = 45$

45. $6x^2 - y^2 = 40$ **46.** $x^2 + 4y^2 = 18$

47. *Satellite Orbit*   Find an equation of the circular orbit of a satellite 500 miles above the surface of the earth. Place the origin of the rectangular coordinate system at the center of the earth and assume the radius of the earth to be 4000 miles.

48. *Architecture*   The top portion of a stained-glass window is in the form of a pointed Gothic arch (see figure). Each side of the arch is an arc of a circle that has a radius of 12 feet and a center at the base of the opposite arch. Find an equation of one of the circles and use it to determine the height of the point of the arch above the horizontal base of the window.

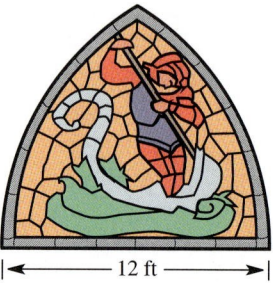

|←——— 12 ft ———→|

49. *Height of an Arch*   A *semielliptical* arch for a tunnel under a river has a width of 100 feet and a height of 40 feet (see figure). Determine the height of the arch 5 feet from the edge of the tunnel.

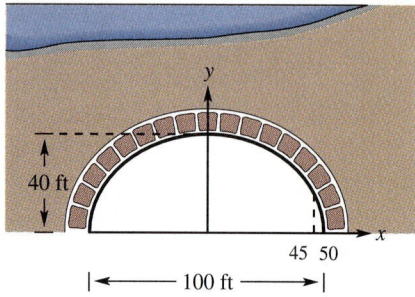

50. *Hyperbolic Mirror*   A hyperbolic mirror (used in some telescopes) has the property that a light ray directed at the first focus will be reflected to the second focus. The foci of a hyperbolic mirror are $(\pm 12, 0)$. Find the vertex of the mirror if its mount has coordinates $(12, 12)$.

Reviewing the Major Concepts

In Exercises 51–54, sketch the graph of the equation.

51. $y = 2x - 3$

52. $y = -\frac{3}{4}x + 2$

53. $y = x^2 - 4x + 4$

54. $y = 9 - x^2$

55. *Geometry*   Approximate the radius of a circle whose area is 3 square feet.

56. *Geometry*   The perimeter of a rectangle is 72 inches and the length of the diagonal is $12\sqrt{5}$ inches. Find the dimensions of the rectangle.

Additional Problem Solving

In Exercises 57–60, find an equation of the circle with center at $(0, 0)$ that satisfies the given criteria.

57. Radius: $\frac{2}{3}$

58. Radius: 7

59. Passes through the point $(0, 8)$

60. Passes through the point $(-2, 0)$

In Exercises 61–64, identify the center and radius of the circle and sketch its graph.

61. $x^2 + y^2 = 36$

62. $x^2 + y^2 = 10$

63. $4x^2 + 4y^2 = 1$

64. $9x^2 + 9y^2 = 64$

In Exercises 65–70, write the standard form of the equation of the ellipse, centered at the origin.

	Vertices	Co-vertices
65.	$(-2, 0), (2, 0)$	$(0, -1), (0, 1)$
66.	$(-10, 0), (10, 0)$	$(0, -4), (0, 4)$
67.	$(0, -2), (0, 2)$	$(-1, 0), (1, 0)$
68.	$(0, -8), (0, 8)$	$(-4, 0), (4, 0)$

69. The major axis is vertical with length 10, and the minor axis has length 6.

70. The major axis is horizontal with length 24, and the minor axis has length 10.

In Exercises 71–74, sketch the ellipse. Identify its vertices and co-vertices.

71. $\dfrac{x^2}{4} + \dfrac{y^2}{16} = 1$

72. $\dfrac{x^2}{25} + \dfrac{y^2}{9} = 1$

73. $\dfrac{x^2}{25/9} + \dfrac{y^2}{16/9} = 1$

74. $\dfrac{x^2}{1} + \dfrac{y^2}{1/4} = 1$

In Exercises 75–80, sketch the hyperbola. Identify its vertices and asymptotes.

75. $x^2 - y^2 = 9$

76. $x^2 - y^2 = 1$

77. $y^2 - x^2 = 9$

78. $y^2 - x^2 = 1$

79. $4y^2 - x^2 + 16 = 0$

80. $4y^2 - 9x^2 - 36 = 0$

In Exercises 81–84, write an equation of the hyperbola centered at the origin.

	Vertices	Asymptotes	
81.	$(-9, 0), (9, 0)$	$y = \frac{3}{2}x$	$y = -\frac{3}{2}x$
82.	$(-1, 0), (1, 0)$	$y = \frac{1}{2}x$	$y = -\frac{1}{2}x$
83.	$(0, -1), (0, 1)$	$y = 2x$	$y = -2x$
84.	$(0, -5), (0, 5)$	$y = x$	$y = -x$

In Exercises 85–94, identify the graph of the equation as a line, circle, parabola, ellipse, or hyperbola.

85. $y = 2x^2 - 8x + 2$

86. $y = 10 - \frac{3}{2}x$

87. $4x^2 + 9y^2 = 36$

88. $4x^2 + 4y^2 = 36$

89. $4x^2 - 9y^2 = 36$

90. $x^2 - 4y + 2x = 0$

91. $x^2 + y^2 - 1 = 0$

92. $2x^2 + 2y^2 = 9$

93. $3x + 2 = 0$

94. $y^2 = x^2 + 2$

In Exercises 95–98, use a graphing utility to graph the equation.

95. $x^2 - 2y^2 = 4$

96. $9x^2 + 9y^2 = 64$

97. $3x^2 + y^2 - 12 = 0$

98. $5x^2 - 2y^2 + 10 = 0$

99. A rectangle centered at the origin with sides parallel to the coordinate axes is placed in a circle of radius 25 inches centered at the origin (see figure). The length of the rectangle is $2x$ inches.

(a) Show that the width and area of the rectangle are given by $2\sqrt{625 - x^2}$ and $4x\sqrt{625 - x^2}$, respectively.

(b) Use a graphing utility to graph the area function. Approximate the value of x for which the area is maximum.

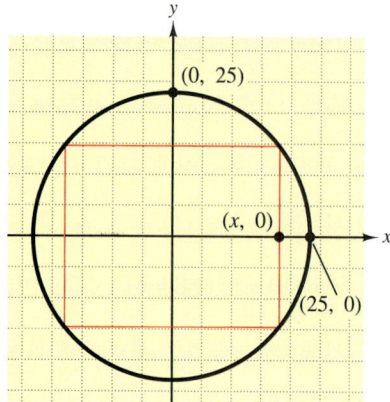

100. *Height of an Arch*   A *semicircular* arch for a tunnel under a river has a diameter of 100 feet (see figure). Determine the height of the arch 5 feet from the edge of the tunnel.

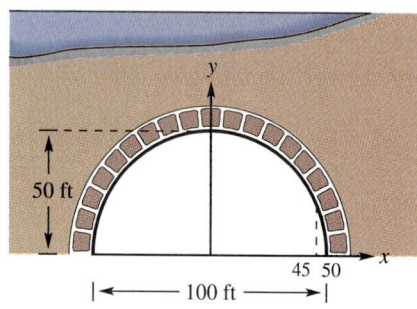

101. *Area*   The area A of the ellipse

$$\frac{x^2}{a^2} + \frac{y^2}{b^2} = 1$$

is given by $A = \pi ab$. Find the equation of an ellipse with area 301.59 square units and $a + b = 20$.

102. Sketch a graph of the ellipse that consists of all points (x, y) such that the sum of the distances between (x, y) and two fixed points is 15 units and for which the foci are located at the centers of the two sets of concentric circles in the figure.

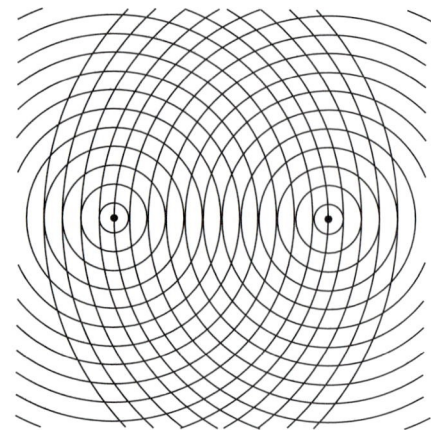

11.5	**Variation**
	Direct Variation ▪ Inverse Variation ▪ Joint Variation

Direct Variation

In the mathematical model for direct variation, y is a *linear* function of x. Specifically, $y = kx$.

STUDY TIP

To use this mathematical model in applications involving direct variation, you are usually given specific values of x and y, which then enable you to find the value for the constant k.

Direct Variation

The following statements are equivalent. The number k is the **constant of proportionality.**

1. y **varies directly** as x.

2. y is **directly proportional** to x.

3. $y = kx$ for some constant k.

EXAMPLE 1 *Direct Variation*

Assume that the total revenue R (in dollars) obtained from selling x units of a product is directly proportional to the number of units sold. When 10,000 units are sold, the total revenue is $142,500.

a. Find a model that relates the total revenue R to the number of units sold x.

b. Find the total revenue obtained from selling 12,000 units.

Solution

a. Because the total revenue is directly proportional to the number of units sold, the linear model is $R = kx$. To find the value of the constant k, substitute 142,500 for R and 10,000 for x

$$142,500 = k(10,000) \qquad \text{Substitute for } R \text{ and } x.$$

which implies that $k = 142,500/10,000 = 14.25$. Thus, the equation relating the total revenue to the total number of units sold is

$$R = 14.25x. \qquad \text{Direct variation model}$$

b. When $x = 12,000$, the total revenue is

$$R = 14.25(12,000) = \$171,000.$$

EXAMPLE 2   Direct Variation

Hooke's Law for springs states that the distance a spring is stretched (or compressed) is proportional to the force on the spring. A force of 20 pounds stretches a particular spring 5 inches.

a. Find a mathematical model that relates the distance the spring is stretched to the force applied to the spring.

b. How far will a force of 30 pounds stretch the spring?

Solution

a. For this problem, let d represent the distance (in inches) that the spring is stretched and let F represent the force (in pounds) that is applied to the spring. Because the distance d is proportional to the force F, the model is

$$d = kF.$$

To find the value of the constant k, use the fact that $d = 5$ when $F = 20$. Substituting these values into the given model produces

$$5 = k(20) \qquad \text{Substitute 5 for } d \text{ and 20 for } F.$$

implying that $k = \frac{5}{20} = \frac{1}{4}$. Thus, the equation relating distance and force is

$$d = \frac{1}{4}F. \qquad \text{Direct variation model}$$

b. When $F = 30$, the distance is

$$d = \frac{1}{4}(30) = 7.5 \text{ inches.} \qquad \text{See Figure 11.37.}$$

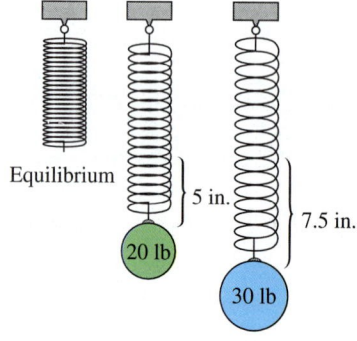

Equilibrium

5 in.

20 lb

7.5 in.

30 lb

FIGURE 11.37

In Example 2, you can get a clearer understanding of Hooke's Law by using the model $d = \frac{1}{4}F$ to create a table. From the table, you can see what it means for the distance to be "proportional to the force."

Force, F	10 lb	20 lb	30 lb	40 lb	50 lb	60 lb
Distance, d	2.5 in.	5.0 in.	7.5 in.	10.0 in.	12.5 in.	15.0 in.

In Examples 1 and 2, the direct variations were such that an *increase* in one variable corresponded to an *increase* in the other variable. There are, however, other applications of direct variation in which an increase in one variable corresponds to a *decrease* in the other variable. For instance, in the model $y = -2x$, an increase in x will yield a decrease in y.

A second type of direct variation relates one variable to a *power* of another.

Direct Variation as *n*th Power

The following statements are equivalent.

1. *y* **varies directly as the *n*th power** of *x*.

2. *y* is **directly proportional to the *n*th power** of *x*.

3. $y = kx^n$ for some constant *k*.

EXAMPLE 3   *Direct Variation as a Power*

The distance a ball rolls down an inclined plane is directly proportional to the square of the time it rolls. Assume that, during the first second, a ball rolls down a particular plane a distance of 6 feet.

a. Find a mathematical model that relates the distance traveled to the time.

b. How far will the ball roll during the first 2 seconds?

Solution

a. Letting *d* be the distance (in feet) that the ball rolls and letting *t* be the time (in seconds), we obtain the model

$$d = kt^2.$$

Because $d = 6$ when $t = 1$, it follows that $k = 6$. Therefore, the equation relating distance to time is

$$d = 6t^2.$$ *Direct variation as 2nd power model*

b. When $t = 2$, the distance traveled is

$$d = 6(2)^2 = 6(4) = 24 \text{ feet.}$$ *See Figure 11.38.*

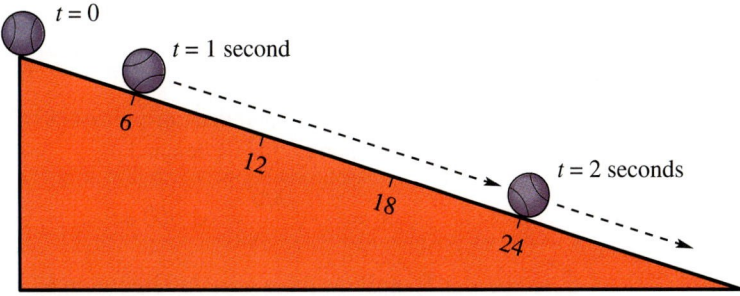

FIGURE 11.38

Inverse Variation

A third type of variation is called **inverse variation.** For this type of variation, we say that one of the variables is inversely proportional to the other variable.

NOTE If x and y are related by an equation of the form $y = k/x^n$, we say that y varies inversely as the nth power of x (or y is inversely proportional to the nth power of x).

Inverse Variation

The following statements are equivalent.

1. y **varies inversely** as x.

2. y is **inversely proportional** to x.

3. $y = \dfrac{k}{x}$ for some constant k.

EXAMPLE 4   Inverse Variation

The marketing department of a large company has found that the demand for one of its products varies inversely as the price of the product. (When the price is low, more people are willing to buy the product than when the price is high.) When the price of the product is \$7.50, the monthly demand is 50,000 units. Approximate the monthly demand if the price is reduced to \$6.00.

Solution

Let x represent the number of units that are sold each month (the demand), and let p represent the price per unit (in dollars). Because the demand is inversely proportional to the price, the model is

$$x = \frac{k}{p}.$$

By substituting $x = 50{,}000$ when $p = 7.50$, you obtain $k = (7.5)(50{,}000) = 375{,}000$. Thus, the model is

$$x = \frac{375{,}000}{p}. \qquad \text{Inverse variation model}$$

To find the demand that corresponds to a price of \$6.00, substitute $p = 6$ into the equation and obtain a demand of

$$x = \frac{375{,}000}{6} = 62{,}500 \text{ units.}$$

Thus, if the price were lowered from \$7.50 per unit to \$6.00 per unit, the monthly demand could be expected to increase from 50,000 units to 62,500 units.

Some applications of variation involve problems with *both* direct and inverse variation in the same model.

EXAMPLE 5   *Direct and Inverse Variation*

A company determines that the demand for one of its products is directly proportional to the amount spent on advertising and inversely proportional to the price of the product. When $40,000 is spent on advertising and the price per unit is $20, the monthly demand is 10,000 units.

a. If the amount of advertising were increased to $50,000, how much could the price be increased to maintain a monthly demand of 10,000 units?

b. If you were in charge of the advertising department, would you recommend this increased expense in advertising?

Solution

a. Let x represent the number of units that are sold each month (the demand), let a represent the amount spent on advertising (in dollars), and let p represent the price per unit (in dollars). Because the demand is directly proportional to the advertising and inversely proportional to the price, the model is

$$x = \frac{ka}{p}.$$

By substituting $x = 10,000$ when $a = 40,000$ and $p = 20$, you obtain $k = (10,000)(20)/(40,000) = 5$. Thus, the model is

$$x = \frac{5a}{p}. \qquad \text{Direct and inverse variation model}$$

To find the price that corresponds to a demand of 10,000 and an advertising expense of $50,000, substitute $x = 10,000$ and $a = 50,000$ into the model and solve for p.

$$10,000 = \frac{5(50,000)}{p} \qquad \Longrightarrow \qquad p = \frac{5(50,000)}{10,000} = \$25$$

b. The total revenue from selling 10,000 units at $20 is $200,000, and the revenue from selling 10,000 units at $25 is $250,000. Thus, increasing the advertising expense from $40,000 to $50,000 would increase the revenue by $50,000. This implies that you should recommend the increased expense in advertising.

Amount of Advertising	Price	Revenue
$40,000	$20.00	$10,000 \times 20 = \$200,000$
$50,000	$25.00	$10,000 \times 25 = \$250,000$

Joint Variation

NOTE If x, y, and z are related by the equation $z = kx^n y^m$, we say that z varies jointly as the nth power of x and the mth power of y.

Joint Variation
The following statements are equivalent.
1. z **varies jointly** as x and y.
2. z is **jointly proportional** to x and y.
3. $z = kxy$ for some constant k.

EXAMPLE 6   Joint Variation

The *simple interest* for a certain savings account is jointly proportional to the time and the principal. After one quarter (3 months), the interest for a principal of $6000 is $120. How much interest would a principal of $7500 earn in 5 months?

Solution

To begin, let I represent the interest earned (in dollars), let P represent the principal (in dollars), and let t represent the time (in years). Because the interest is jointly proportional to the time and the principal, the model is

$$I = ktP.$$

Because $I = 120$ when $P = 6000$ and $t = \frac{1}{4}$, it follows that $k = 120/\left(6000 \cdot \frac{1}{4}\right) = 0.08$. Thus, the model that relates interest to time and principal is

$$I = 0.08tP. \qquad \text{\color{red}Joint variation model}$$

To find the interest earned on a principal of $7500 over a 5-month period of time, substitute $P = 7500$ and $t = \frac{5}{12}$ into the model and obtain an interest of $I = 0.08(\frac{5}{12})(7500) = \$250.00$.

Group Activities   Exploring with Technology

Investigating Variation Models   Use a graphing utility to sketch the graphs of the direct variation, direct variation as nth power, and inverse variation models for different values of k (and n as appropriate). Summarize the basic characteristics of the graph for each type of model. Does understanding the basic graphs of these models shed light on the models themselves?

11.5 Exercises

Discussing the Concepts

1. Suppose the constant of proportionality is positive and y varies directly as x. If one of the variables increases, how will the other change? Explain.

2. Suppose the constant of proportionality is positive and y varies inversely as x. If one of the variables increases, how will the other change? Explain.

3. If y varies directly as the square of x and x is doubled, how will y change? Use the properties of exponents to explain your answer.

4. If y varies inversely as the square of x and x is doubled, how will y change? Use the properties of exponents to explain your answer.

Problem Solving

In Exercises 5–10, write a model for the statement.

5. I varies directly as V.

6. V varies directly as the cube of x.

7. p varies inversely as d.

8. S is inversely proportional to the square of v.

9. *Boyle's Law* If the temperature of a gas is constant, its absolute pressure P is inversely proportional to its volume V.

10. *Newton's Law of Universal Gravitation* The gravitational attraction F between two particles of masses m_1 and m_2 is proportional to the product of the masses and inversely proportional to the square of the distance r between the particles.

In Exercises 11–14, write a sentence using variation terminology to describe the formula.

11. *Area of a Triangle:* $A = \frac{1}{2}bh$

12. *Area of a Circle:* $A = \pi r^2$

13. *Volume of a Right Circular Cylinder:* $V = \pi r^2 h$

14. *Average Speed:* $r = d/t$

In Exercises 15–22, find the constant of proportionality and give the equation relating the variables.

15. s varies directly as t, and $s = 20$ when $t = 4$.

16. h is directly proportional to r, and $h = 28$ when $r = 12$.

17. F is directly proportional to the square of x, and $F = 500$ when $x = 40$.

18. v varies directly as the square root of s, and $v = 24$ when $s = 16$.

19. n varies inversely as m, and $n = 32$ when $m = 1.5$.

20. q is inversely proportional to p, and $q = \frac{3}{2}$ when $p = 50$.

21. d varies directly as the square of x and inversely with r, and $d = 3000$ when $x = 10$ and $r = 4$.

22. z is directly proportional to x and inversely proportional to the square root of y, and $z = 720$ when $x = 48$ and $y = 81$.

23. *Hooke's Law* A baby weighing $10\frac{1}{2}$ pounds compresses the spring of a baby scale 7 millimeters (see figure). What weight will compress the spring 12 millimeters?

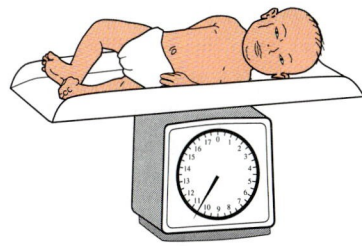

24. *Hooke's Law* A force of 50 pounds compresses the spring of a scale 1.5 inches. Find the distance the spring will be compressed when a 20-pound object is placed on the scale.

25. *Stopping Distance* The stopping distance d of an automobile is directly proportional to the square of its speed s. On a certain road surface, a car requires 75 feet to stop when its speed is 30 miles per hour. Estimate the stopping distance if the brakes are applied when the car is traveling at 50 miles per hour under similar road conditions.

26. *Distance* Neglecting air resistance, the distance d that an object falls varies directly as the square of the time t it has been falling. If an object falls 64 feet in 2 seconds, determine the distance it will fall in 6 seconds.

27. *Power Generation* The power P generated by a wind turbine varies directly as the cube of the wind speed w. The turbine generates 750 watts of power in a 25-mile-per-hour wind. Find the power it generates in a 40-mile-per-hour wind.

28. *Velocity of a Stream* The diameter d of the largest particle that can be moved by a stream is directly proportional to the square of the velocity v of the stream. A stream with a velocity of $\frac{1}{4}$ mile per hour can move coarse sand particles about 0.02 inch in diameter. What must the velocity be to carry particles with a diameter of 0.12 inch?

29. *Demand Function* A company has found that the daily demand x for its product is inversely proportional to the price p. When the price is $5, the demand is 800 units. Approximate the demand if the price is increased to $6.

30. *Weight of an Astronaut* The gravitational force F with which an object is attracted to the earth is inversely proportional to the square of its distance r from the center of the earth. If an astronaut weighs 190 pounds on the surface of the earth ($r \approx 4000$ miles), what will the astronaut weigh 1000 miles above the earth's surface?

31. *Amount of Illumination* The illumination I from a light source varies inversely as the square of the distance d from the light source. If you raise a lamp from 18 inches to 36 inches over your desk (see figure), the illumination will change by what factor?

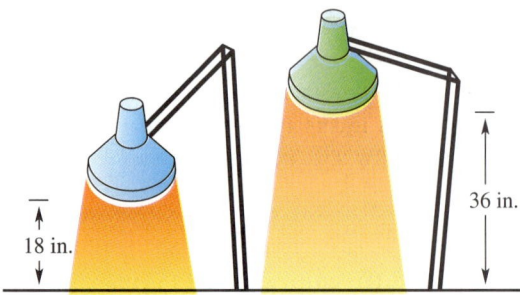

18 in.   36 in.

32. *Frictional Force* The frictional force F between the tires and the road required to keep a car on a curved section of a highway is directly proportional to the square of the speed s of the car. If the speed of the car is doubled, the required frictional force will change by what factor?

33. *Load of a Beam* The load that can be safely supported by a horizontal beam varies jointly as the width of the beam and the square of its depth and inversely as the length of the beam. A beam with width 3 inches, depth 8 inches, and length 10 feet can safely support 2000 pounds. Determine the safe load of a beam made from the same material if its depth is increased to 10 inches.

34. *Best Buy* The prices of 9-inch, 12-inch, and 15-inch diameter pizzas at a certain pizza shop are $6.78, $9.78, and $12.18, respectively. You might expect that the price of a certain size of pizza is directly proportional to its area. Is this the case for this pizza shop? If not, which size of pizza is the best buy?

Reviewing the Major Concepts

In Exercises 35–38, solve for y in terms of x.

35. $3x + 4y - 5 = 0$

36. $-2x - 3y + 6 = 0$

37. $-2x^2 + 3y + 2 = 0$

38. $x^2 - 2y = 9$

39. Find two positive consecutive odd integers whose product is 255.

40. Find two positive consecutive even integers, the sum of whose squares is 452.

Additional Problem Solving

In Exercises 41–48, write a model for the statement.

41. V is directly proportional to t.

42. C varies directly as r.

43. u is directly proportional to the square of v.

44. s varies directly as the cube of t.

45. P is inversely proportional to the square root of $1 + r$.

46. A varies inversely as the fourth power of t.

47. A varies jointly as l and w.

48. V varies jointly as h and the square of r.

In Exercises 49–52, write a sentence using variation terminology to describe the formula.

49. *Volume of a Sphere:* $V = \frac{4}{3}\pi r^3$

50. *Surface Area of a Sphere:* $A = 4\pi r^2$

51. *Area of an Ellipse:* $A = \pi ab$

52. *Height of a Cylinder:* $h = V/(\pi r^2)$

In Exercises 53–58, find the constant of proportionality and give the equation relating the variables.

53. H is directly proportional to u, and $H = 100$ when $u = 40$.

54. M varies directly as the cube of n, and $M = 0.012$ when $n = 0.2$.

55. g varies inversely as the square root of z, and $g = \frac{4}{5}$ when $z = 25$.

56. u varies inversely as the square of v, and $u = 40$ when $v = \frac{1}{2}$.

57. F varies jointly as x and y, and $F = 500$ when $x = 15$ and $y = 8$.

58. V varies jointly as h and the square of b, and $V = 288$ when $h = 6$ and $b = 12$.

59. *Revenue*   The total revenue R is directly proportional to the number of units sold x. When 500 units are sold, the revenue is \$3875.

  (a) Find the revenue when 635 units are sold.

  (b) Interpret the constant of proportionality.

60. *Hooke's Law*   A force of 50 pounds stretches a spring 3 inches.

  (a) How far will a 20-pound force stretch the spring?

  (b) What force will stretch the spring 1.5 inches?

61. *Free-Falling Object*   The velocity v of a free-falling object is proportional to the time that it has fallen. The constant of proportionality is the acceleration due to gravity. Find the acceleration due to gravity if the velocity of a falling object is 96 feet per second after the object has fallen 3 seconds.

62. *Free-Falling Object*   The distance d that a free-falling object has fallen is proportional to the square of the time that it has fallen. An object falls 144 feet in 3 seconds. Find the constant of proportionality.

63. *Travel Time*   The travel time between two cities is inversely proportional to the average speed. If a train travels between two cities in 3 hours at an average speed of 65 miles per hour, how long would it take at an average speed of 80 miles per hour? What does the constant of proportionality measure in this problem?

64. *Predator-Prey*   The number N of prey t months after a natural predator is introduced into the test area is inversely proportional to $t + 1$. If $N = 500$ when $t = 0$, find N when $t = 4$.

65. *Weight*   A person's weight on the moon varies directly with his or her weight on earth. Neil Armstrong, the first man on the moon, weighed 360 pounds on earth, including his equipment. On the moon he weighed only 60 pounds, with equipment. If the first woman in space, Valentina V. Tereshkova, had landed on the moon and weighed 54 pounds, with equipment, how much would she have weighed on earth?

66. *Snowshoes*   When a person walks, the pressure P on each sole varies inversely with the area A of the sole. Denise is trudging through deep snow, wearing boots that have a sole area of 29 square inches each. The sole pressure is 4 pounds per square inch. If Denise were wearing snowshoes, each with an area 11 times that of her boot soles, what would be the pressure on each snowshoe? The constant of variation is Denise's weight. What does she weigh?

67. *Reading a Graph*   The graph shows the percent p of oil that remained in Chedabucto Bay, Nova Scotia after an oil spill. The cleaning of the spill was left primarily to natural actions such as wave motion, evaporation, photochemical decomposition, and bacterial decomposition. After about a year, the percent that remained varied inversely with time. Find a model that relates p and t, where t is the number of years since the spill. Then use the model to find the amount of oil that remained $6\frac{1}{2}$ years after the spill.

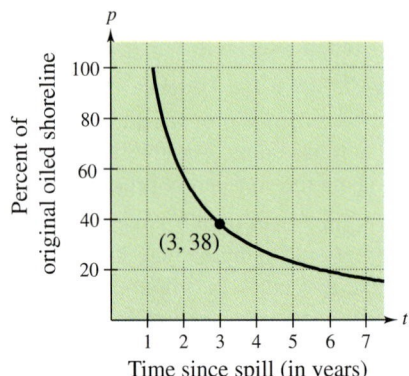

Percent of original oiled shoreline

Time since spill (in years)

68. *Reading a Graph*   The graph shows the temperature of the water in the north central Pacific Ocean. At depths greater than 900 meters, the water temperature varies inversely with the water depth. Find a model that relates the temperature T and the depth d. What is the temperature at a depth of 4385 meters?

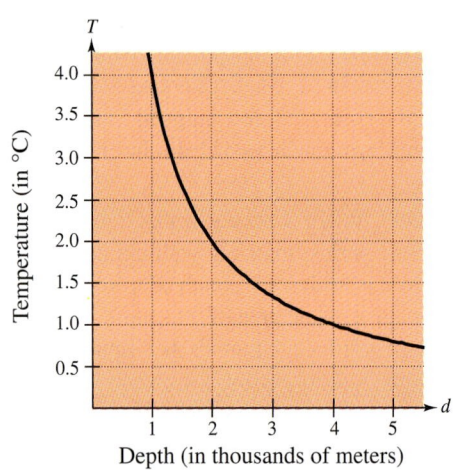

Temperature (in °C)

Depth (in thousands of meters)

In Exercises 69–72, complete the table and plot the resulting points.

x	2	4	6	8	10
$y = kx^2$					

69. $k = 1$ **70.** $k = 2$

71. $k = \frac{1}{2}$ **72.** $k = \frac{1}{4}$

In Exercises 73–76, complete the table and plot the resulting points.

x	2	4	6	8	10
$y = \dfrac{k}{x^2}$					

73. $k = 2$ **74.** $k = 5$

75. $k = 10$ **76.** $k = 20$

In Exercises 77–80, determine whether the variation model is of the form $y = kx$ or $y = k/x$, and find k.

77.

x	10	20	30	40	50
y	$\frac{2}{5}$	$\frac{1}{5}$	$\frac{2}{15}$	$\frac{1}{10}$	$\frac{2}{25}$

78.

x	10	20	30	40	50
y	2	4	6	8	10

79.

x	10	20	30	40	50
y	-3	-6	-9	-12	-15

80.

x	10	20	30	40	50
y	60	30	20	15	12

CHAPTER PROJECT: Mass Transportation

Mass transportation includes taxis, vans, buses, trains, subways, and airplanes. In 1991, approximately $800 billion was spent on mass transportation in the U.S.

The table below shows the amounts (in billions of dollars) spent on mass transportation for taxis and buses from 1978 through 1991. In this table, t represents the year, with $t = 0$ representing 1970. (Source: Eno Foundation for Transportation)

Year, t	8	9	10	11	12	13	14
Amount Spent on Taxis	4.4	4.5	5.2	5.2	5.6	5.3	5.5
Amount Spent on Buses	6.7	7.5	9.3	10.3	10.6	11.6	13.2

Year, t	15	16	17	18	19	20	21
Amount Spent on Taxis	5.6	6.0	6.4	6.9	7.1	7.5	7.9
Amount Spent on Buses	13.5	14.5	15.1	15.7	15.9	16.7	17.5

Use the information in the table to investigate the following questions.

1. *Linear Modeling* Plot the points in the table to form scatter plots for the amounts spent on taxis and buses. Draw the line that you think best approximates each scatter plot. Then find the equation of each line.

2. *Linear Modeling* Use a graphing utility or computer program to find the equation of the line that best fits each of the data sets. To do this with a *TI-82*, use the following steps.

   • Activate the STAT mode and clear previously entered data.

   • Enter the data for taxis or buses.

   • Use the linear regression program to find the best-fitting line.

   How do the equations for the regression lines compare with the ones you found in Question 1?

3. *Graphical Interpretation* What are the slopes of the linear models found in Question 1 or 2? Interpret these slopes in the context of the data. Which slope is greater? What real-life implications does this have?

4. *Prediction* Use the linear models found in Question 1 or 2 to predict the amounts of money that was spent on taxis and buses in 1994.

5. *Correlation* Which of the linear models that you found fits its set of data better? How can you tell?

6. *Research Project* Use your school's library or some other reference source to find the revenues for two other forms of mass transportation. Repeat Questions 1 and 2 for the data. Write a paragraph that interprets the data.

CHAPTER SUMMARY

After studying this chapter, you should have acquired the following skills. These skills are keyed to the Review Exercises that begin on page 683. Answers to odd-numbered Review Exercises are given in the back of the book.

- Write the equation of a line given its slope and a point on the line. *(Section 11.1)*

  Review Exercises 1–8

- Write the equation of a line passing through two points. *(Section 11.1)*

  Review Exercises 9–14

- Write the equation of a line through a given point that is parallel or perpendicular to a given line. *(Section 11.1)*

  Review Exercises 15–18

- Sketch pairs of lines using a graphing utility and determine if the lines are parallel, perpendicular, or neither. *(Section 11.1)*

  Review Exercises 19–22

- Translate real-life situations into algebraic equations or inequalities. *(Sections 11.1, 11.2, 11.5)*

  Review Exercises 23, 25, 26, 39, 71–74

- Solve real-life problems modeled by equations. *(Sections 11.1, 11.3)*

  Review Exercises 24, 57–60

- Sketch the graphs of inequalities. *(Section 11.2)*

  Review Exercises 27–32

- Sketch, with and without a graphing utility, the graphs of solution points that satisfy pairs of inequalities. *(Section 11.2)*

  Review Exercises 33–38

- Solve problems involving geometry. *(Section 11.2)*

  Review Exercise 40

- Identify and sketch conics represented by equations. *(Sections 11.3, 11.4)*

  Review Exercises 41–50

- Determine the equations of conics given points, dimensions, and/or other characteristics of the conics. *(Sections 11.3, 11.4)*

  Review Exercises 51–56

- Match equations with their graphs. *(Sections 11.1, 11.2, 11.3, 11.4)*

  Review Exercises 61–66

- Find the constant of proportionality and write an equation that relates the variables. *(Section 11.5)*

  Review Exercises 67–70

REVIEW EXERCISES

In Exercises 1–8, write an equation of the line passing through the point with the specified slope.

1. $(1, -4)$, $m = 2$

2. $(-5, -5)$, $m = 3$

3. $(-1, 4)$, $m = -4$

4. $(5, -2)$, $m = -2$

5. $\left(\frac{5}{2}, 4\right)$, $m = -\frac{2}{3}$

6. $\left(-2, -\frac{4}{3}\right)$, $m = \frac{3}{2}$

7. $(7, 8)$, m is undefined.

8. $(-6, 5)$, $m = 0$

In Exercises 9–14, write an equation of the line passing through the points.

9. $(-6, 0)$, $(0, -3)$

10. $(-2, -3)$, $(4, 6)$

11. $(0, 10)$, $(6, 10)$

12. $(-10, 2)$, $(4, -7)$

13. $\left(\frac{4}{3}, \frac{1}{6}\right)$, $\left(4, \frac{7}{6}\right)$

14. $\left(\frac{5}{2}, 0\right)$, $\left(\frac{5}{2}, 5\right)$

In Exercises 15–18, find equations of the line passing through the point that are (a) parallel and (b) perpendicular to the given line.

15. $\left(\frac{3}{5}, -\frac{4}{5}\right)$, $3x + y = 2$

16. $(-1, 5)$, $2x + 4y = 1$

17. $(12, 1)$, $5x = 3$

18. $\left(\frac{3}{8}, 3\right)$, $4x - 3y = 12$

In Exercises 19–22, use a graphing utility on a square setting to sketch the lines. Decide whether the lines are parallel, perpendicular, or neither.

19. $y = 2x - 3$
$y = \frac{1}{2}(4 - x)$

20. $y = \frac{2}{3}x + 3$
$y = \frac{2}{9}(3x - 4)$

21. $2x - 3y - 5 = 0$
$x + 2y - 6 = 0$

22. $4x + 3y - 6 = 0$
$3x - 4y - 8 = 0$

23. *Cost and Profit*   A company produces a product for which the variable cost is $8.55 per unit and the fixed costs are $25,000. The product is sold for $12.60, and the company can sell all it produces.

(a) Write the cost C as a linear function of x, the number of units produced.

(b) Write the profit P as a linear function of x, the number of units sold.

24. *Velocity*   The velocity of a ball thrown upward from ground level is given by $v = -32t + 80$, where t is time in seconds and v is velocity in feet per second.

(a) Find the velocity when $t = 2$.

(b) Find the time when the ball reaches its maximum height. (*Hint:* Find the time when $v = 0$.)

(c) Find the velocity when $t = 3$.

Rocker Arm Construction   In Exercises 25 and 26, consider the rocker arm shown in the figure.

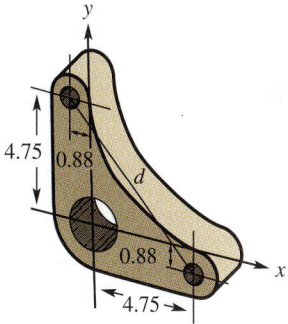

25. Find an equation of the line through the centers of the two small bolt holes in the rocker arm.

26. Find the distance d between the centers of the two small bolt holes in the rocker arm.

In Exercises 27–32, sketch the graph of the inequality.

27. $x - 2 \geq 0$

28. $y + 3 < 0$

29. $2x + y < 1$

30. $3x - 4y > 2$

31. $x \leq 4y - 2$

32. $(y - 3) \geq \frac{2}{3}(x - 5)$

In Exercises 33–36, sketch a graph of the solution points that satisfy both inequalities.

33. $x \geq 5$, $y \leq 2$

34. $y \leq \frac{1}{2}x + 1$, $x \geq 2$

35. $x - 2y \geq 2$, $y \geq -1$

36. $x + y \geq 2$, $x \leq 2$

In Exercises 37 and 38, use a graphing utility to graph the region containing the solution points that satisfy both inequalities.

37. $x + y \geq 0$, $y \leq 5$ **38.** $4x - 3y \geq 2$, $y \geq 1$

39. *Weekly Pay* You have two part-time jobs. One is at a grocery store, which pays $8 per hour, and the other is mowing lawns, which pays $10 per hour. Between the two jobs, you want to earn at least $200 a week. Write an inequality that shows the different numbers of hours you can work at each job, and sketch the graph of the inequality. From the graph, find several ordered pairs with positive integer coordinates that are solutions of the inequality.

40. *Geometry* The perimeter of a rectangle of length x and width y cannot exceed 800 feet. Write a linear inequality for this constraint and sketch its graph.

In Exercises 41–50, identify and sketch the conic.

41. $x^2 - 2y = 0$ **42.** $x^2 + y^2 = 64$

43. $x^2 - y^2 = 64$ **44.** $x^2 + 4y^2 = 64$

45. $y = x(x - 6)$ **46.** $y = 9 - (x - 3)^2$

47. $\dfrac{x^2}{25} + \dfrac{y^2}{4} = 1$ **48.** $\dfrac{x^2}{25} - \dfrac{y^2}{4} = -1$

49. $4x^2 + 4y^2 - 9 = 0$ **50.** $x^2 + 9y^2 - 9 = 0$

In Exercises 51–56, find the general form of the equation for the conic meeting the given criteria.

51. *Parabola:* vertex: $(5, 0)$;
passes through the point $(1, 1)$

52. *Parabola:* vertex: $(-2, 5)$;
passes through the point $(0, 1)$

53. *Ellipse:* vertices: $(0, -5)$, $(0, 5)$;
co-vertices: $(-2, 0)$, $(2, 0)$

54. *Ellipse:* vertices: $(-10, 0)$, $(10, 0)$;
co-vertices: $(0, -6)$, $(0, 6)$

55. *Circle:* center: $(0, 0)$;
radius: 20

56. *Hyperbola:* vertices: $(0, -4)$, $(0, 4)$;
asymptotes: $y = 2x$, $y = -2x$

57. *Graphical Estimation* The profit (in thousands of dollars) for a certain product is given by

$$P = 320 + 10s - \frac{1}{2}s^2$$

where s is the amount (in hundreds of dollars) spent on advertising. Use a graphing utility to graph the profit function and approximate the amount of advertising that yields a maximum profit. Verify the maximum algebraically.

58. *Graphical Interpretation* The height y (in feet) of a ball thrown by a child is given by

$$y = -\frac{1}{10}x^2 + 3x + 6$$

where x is the horizontal distance (in feet) from where the ball is thrown.

(a) Use a graphing utility to graph the path of the ball.

(b) How high is the ball when it leaves the child's hand?

(c) What is the maximum height of the ball?

(d) How far from the child does the ball hit the ground?

59. *Graphical Estimation* The enrollment E (in millions) in public elementary schools in the United States for the years 1970 through 1990 is approximated by the model

$$E = 27.611 - 0.586t + 0.026t^2$$

where t represents the calendar year, with $t = 0$ corresponding to 1970. (Source: U.S. National Center for Education Statistics)

(a) Use a graphing utility to graph the enrollment function.

(b) Use the graph of part (a) to approximate the year when the enrollment was minimum.

(c) Predict the enrollment in 1999 if this trend continues.

60. *Graphical Estimation* The number N of operable nuclear power units generating electricity in the United States for the years 1985 through 1992 is approximated by the model

$$N = 47.601 + 12.613t - 0.625t^2$$

where t represents the calendar year, with $t = 5$ corresponding to 1985. (Source: U.S. Energy Information Administration)

(a) Use a graphing utility to graph the function.

(b) Use the graph of part (a) to approximate the year when the number of operable generating units was maximum.

(c) What was the number of operable generating units in 1995?

In Exercises 61–66, match the equation with its graph. [The graphs are labeled (a), (b), (c), (d), (e), and (f).]

(a)

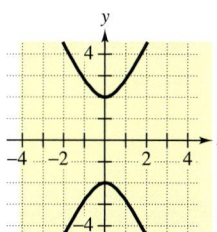

(b)

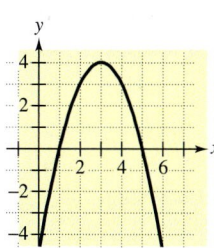

(c)

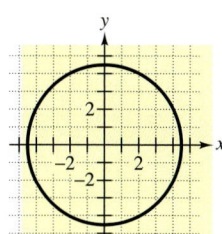

(d)

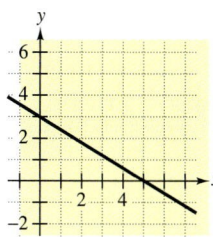

(e)

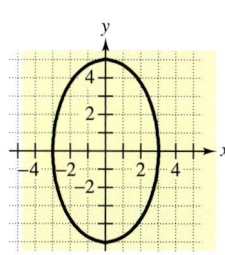

(f)

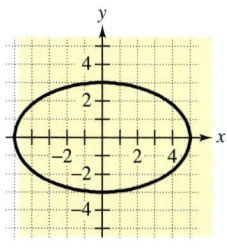

61. $4x^2 + 4y^2 = 81$

62. $3x + 5y = 15$

63. $\dfrac{y^2}{4} - x^2 = 1$

64. $\dfrac{x^2}{25} + \dfrac{y^2}{9} = 1$

65. $y = -x^2 + 6x - 5$

66. $\dfrac{x^2}{9} + \dfrac{y^2}{25} = 1$

In Exercises 67–70, find the constant of proportionality and write an equation that relates the variables.

67. y varies directly as the cube root of x, and $y = 12$ when $x = 9$.

68. r varies inversely as s, and $r = 45$ when $s = \frac{3}{5}$.

69. T varies jointly as r and the square of s, and $T = 5000$ when $r = 0.09$ and $s = 1000$.

70. D is directly proportional to the cube of x and inversely proportional to y, and $D = 810$ when $x = 3$ and $y = 25$.

71. *Hooke's Law* A force of 100 pounds stretches a spring 4 inches. Find the force required to stretch the same spring 6 inches.

72. *Stopping Distance* The stopping distance d of an automobile is directly proportional to the square of its speed s. How will the stopping distance be changed by doubling the speed of the car?

73. *Demand Function* A company has found that the daily demand x for its product varies inversely as the square root of the price p. When the price is $25, the demand is approximately 1000 units. Approximate the demand if the price is increased to $28.

74. *Weight of an Astronaut* The gravitational force F with which an object is attracted to the earth is inversely proportional to the square of its distance r from the center of the earth. If an astronaut weighs 200 pounds on the surface of the earth ($r \approx 4000$ miles), what will the astronaut weigh 500 miles above the earth's surface?

CHAPTER TEST

Take this test as you would take a test in class. After you are done, check your work against the answers given in the back of the book.

In Exercises 1–4, write an equation of the line.

1. The line has a slope of -2 and passes through $(2, -4)$.

2. The line passes through $(25, -15)$ and $(75, 10)$.

3. The line is horizontal and passes through $(5, -1)$.

4. The line is vertical and passes through $(-2, 4)$.

5. Find the slope of a line perpendicular to the line given by $5x + 3y - 9 = 0$.

6. After 4 years, a \$26,000 car will have depreciated to a value of \$10,000. Write a linear equation that gives the value V in terms of t, the number of years. When will the car be worth \$16,000? Explain your reasoning.

7. Sketch the graph of the inequality $x + 2y \le 4$.

8. Sketch the graph of the inequality $3x - y \ge 6$.

9. Sketch the graph of $y = -2(x - 2)^2 + 8$. Label its vertex and intercepts.

10. Write an equation of the parabola shown at the right.

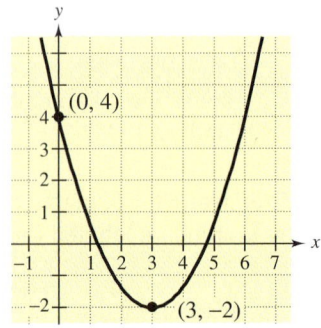

Figure for 10

11. The revenue R for a chartered bus trip is given by $R = -\frac{1}{20}(n^2 - 240n)$, where $80 \le n \le 160$ and n is the number of passengers. How many passengers will produce a maximum revenue? Explain your reasoning.

12. Write an equation of the circle shown at the right.

13. Write the standard form of the equation of the ellipse centered at the origin with vertices $(0, -10)$ and $(0, 10)$ and co-vertices $(-3, 0)$ and $(3, 0)$.

14. Write the standard form of the equation of the hyperbola centered at the origin with vertices $(-3, 0)$ and $(3, 0)$ and asymptotes $y = \frac{1}{2}x$ and $y = -\frac{1}{2}x$.

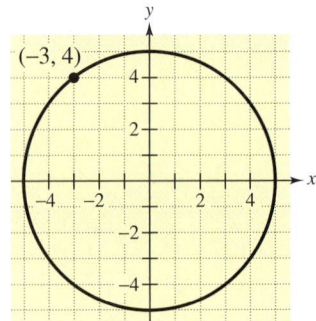

Figure for 12

In Exercises 15–18, sketch the graph of the equation.

15. $x^2 + y^2 = 9$

16. $\dfrac{x^2}{9} + \dfrac{y^2}{16} = 1$

17. $\dfrac{x^2}{9} - \dfrac{y^2}{16} = 1$

18. $\dfrac{x}{3} - \dfrac{y}{4} = 1$

19. Write a mathematical model for the statement, "S varies directly as the square of x and inversely as y."

20. Find the constant of proportionality if v varies directly as the square root of u and $v = \frac{3}{2}$ when $u = 36$.

Systems of Equations

12

- Introduction to Systems of Equations
- Systems of Linear Equations in Two Variables
- Systems of Linear Equations in Three Variables
- Matrices and Systems of Linear Equations
- Linear Systems and Determinants

Diagnostic, Inc. and Invacare Corporation are two American healthcare companies. The annual revenue R (in millions of dollars) for Diagnostic, Inc. from 1988 through 1993 can be modeled by

$$R = -564.34 + 79.67t$$

where $t = 0$ represents 1980.

The annual revenue R (in millions of dollars) for Invacare Corporation during that same period can be modeled by

$$R = -179.49 + 40.41t$$

where $t = 0$ represents 1980.

From the graph at the right, notice that the revenue for Diagnostic, Inc. was increasing at a faster rate than the revenue for Invacare Corporation. In 1990, both companies earned about $230 million. The table at the right confirms this estimate.

t	8	9	10	11	12	13
Diagnostic	71.42	150.89	230.36	309.83	389.30	468.77
Invacare	150.79	191.20	231.61	272.02	312.43	352.84

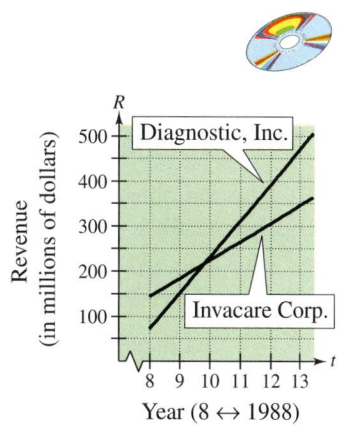

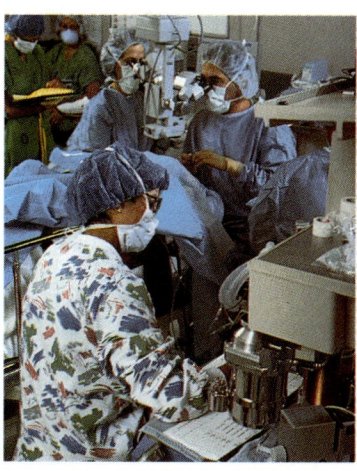

The chapter project related to this information is on page 748.

687

12.1 | Introduction to Systems of Equations

Systems of Equations ▪ Solving Systems of Equations by Substitution ▪ Solving Systems of Equations by Graphing ▪ Applications

Systems of Equations

Many problems in business and science involve **systems of equations** that consist of two or more equations, each involving two or more variables.

$ax + by = c$	Equation 1
$dx + ey = f$	Equation 2

A **solution** of such a system is an ordered pair (x, y) of real numbers that satisfies *each* equation in the system. When you find the set of all solutions of the system of equations, you are **solving the system of equations.**

EXAMPLE 1 *Checking Solutions of a System of Equations*

Which of the ordered pairs is a solution of the system: (a) (3, 3) or (b) (4, 2)?

$x + y = 6$	Equation 1
$2x - 5y = -2$	Equation 2

Solution

a. To determine whether the ordered pair (3, 3) is a solution of the system of equations, you should substitute $x = 3$ and $y = 3$ into *each* of the equations. Substituting into Equation 1 produces

$$3 + 3 = 6. \checkmark \qquad \text{Substitute 3 for } x \text{ and 3 for } y.$$

Similarly, substituting into Equation 2 produces

$$2(3) - 5(3) \neq -2. \times \qquad \text{Substitute 3 for } x \text{ and 3 for } y.$$

Because the ordered pair (3, 3) fails to check in *both* equations, you can conclude that it *is not* a solution of the original system of equations.

b. By substituting the coordinates of the ordered pair (4, 2) into the original equations, you can determine that it is a solution of the first equation

$$4 + 2 = 6 \checkmark \qquad \text{Substitute 4 for } x \text{ and 2 for } y.$$

and is also a solution of the second equation

$$2(4) - 5(2) = -2. \checkmark \qquad \text{Substitute 4 for } x \text{ and 2 for } y.$$

Thus, (4, 2) *is* a solution of the original system of equations.

Solving Systems of Equations by Substitution

One way to solve a system of two equations in two variables is to convert the system to *one* equation in *one* variable by an appropriate substitution.

EXAMPLE 2   *The Method of Substitution: One-Solution Case*

Solve the system of equations.

$$-x + y = 3 \qquad\qquad \text{Equation 1}$$
$$3x + y = -1 \qquad\qquad \text{Equation 2}$$

Solution

Begin by solving for y in Equation 1.

$$y = x + 3 \qquad\qquad \text{Solve for } y \text{ in Equation 1.}$$

Next, substitute this expression for y in Equation 2.

$$3x + y = -1 \qquad\qquad \text{Equation 2}$$
$$3x + (x + 3) = -1 \qquad\qquad \text{Substitute } x + 3 \text{ for } y.$$
$$4x = -4 \qquad\qquad \text{Simplify.}$$
$$x = -1 \qquad\qquad \text{Divide both sides by 4.}$$

NOTE The term **back-substitute** implies that you work backwards. After finding a value for one of the variables, substitute that value back into one of the equations in the original (or revised) system to find the value of the other variable.

At this point, you know that the x-coordinate of the solution is -1. To find the y-coordinate, *back-substitute* the x-value into the revised Equation 1.

$$y = x + 3 \qquad\qquad \text{Revised Equation 1}$$
$$y = -1 + 3 \qquad\qquad \text{Substitute } -1 \text{ for } x.$$
$$y = 2 \qquad\qquad \text{Simplify.}$$

The solution is $(-1, 2)$. Check this in the original system of equations.

When you use substitution, it does not matter which variable you solve for first. Whether you solve for y first or x first, you will obtain the same solution. When making your choice, you should choose the variable that is easier to work with. For instance, in the system

$$3x - 2y = 1 \qquad\qquad \text{Equation 1}$$
$$x + 4y = 3 \qquad\qquad \text{Equation 2}$$

it is easier to solve for x first (in the second equation). But in the system

$$2x + y = 5 \qquad\qquad \text{Equation 1}$$
$$3x - 2y = 11 \qquad\qquad \text{Equation 2}$$

it is easier to solve for y first (in the first equation).

The steps for using substitution are summarized as follows.

The Method of Substitution

1. Solve one of the equations for one variable in terms of the other.

2. Substitute the expression found in Step 1 into the other equation to obtain an equation in one variable.

3. Solve the equation obtained in Step 2.

4. Back-substitute the solution from Step 3 into the expression obtained in Step 1 to find the value of the other variable.

5. Check the solution in the original system.

Both of the equations in Example 2 are linear. That is, the variables x and y appear in the first power only. The method of substitution can also be used to solve a system of equations in which one or both of the equations is nonlinear.

EXAMPLE 3   The Method of Substitution: Two-Solution Case

Solve the system of equations.

$$x^2 + y^2 = 25 \qquad \text{Equation 1}$$
$$-x + y = -1 \qquad \text{Equation 2}$$

Solution

$$y = x - 1 \qquad \text{Solve for } y \text{ in Equation 2.}$$
$$x^2 + y^2 = 25 \qquad \text{Equation 1}$$
$$x^2 + (x - 1)^2 = 25 \qquad \text{Substitute } x - 1 \text{ for } y.$$
$$2x^2 - 2x - 24 = 0 \qquad \text{Simplify.}$$
$$2(x - 4)(x + 3) = 0 \qquad \text{Factor.}$$
$$x = 4, \ -3 \qquad \text{Solve for } x.$$

Finally, back-substitute these values of x into the revised second equation to solve for y. For $x = 4$, you have

$$y = 4 - 1 = 3$$

which implies that $(4, 3)$ is a solution. For $x = -3$, you have

$$y = -3 - 1 = -4$$

which implies that $(-3, -4)$ is a solution. Check these in the original system.

Solving Systems of Equations by Graphing

A system of two equations in two variables can have exactly one solution, more than one solution, or no solution. In practice, you can gain insight about the location and number of solutions of a system of equations by sketching the graph of each equation in the same coordinate plane. The solutions of the system correspond to the **points of intersection** of the graphs. For instance, Figure 12.1 shows the graphs of three pairs (systems) of equations.

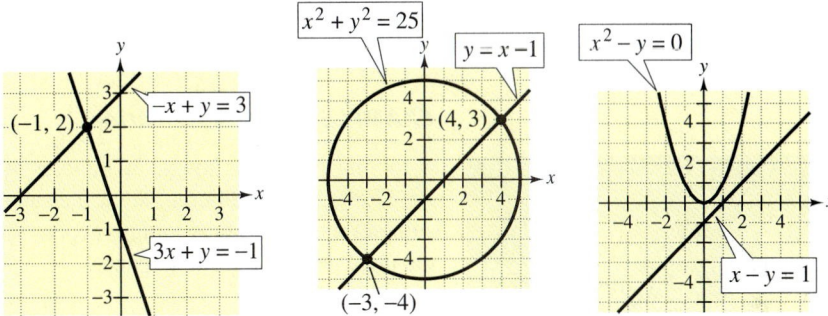

FIGURE 12.1

EXAMPLE 4   *The Graphical Method of Solving a System*

Use the graphical method to solve the system of equations.

$$2x + 3y = 7 \qquad \text{Equation 1}$$
$$2x - 5y = -1 \qquad \text{Equation 2}$$

Solution

Because both equations in the system are linear, you know that they have graphs that are straight lines. To sketch these lines, write each equation in slope-intercept form, as follows.

$$y = -\frac{2}{3}x + \frac{7}{3} \qquad \text{Equation 1}$$

$$y = \frac{2}{5}x + \frac{1}{5} \qquad \text{Equation 2}$$

The lines corresponding to these two equations are shown in Figure 12.2. From this figure, it appears that the two lines intersect in a single point, and that the coordinates of the point are approximately $(2, 1)$. You can check these as follows.

$$2(2) + 3(1) = 7 \qquad \text{Solution checks in Equation 1.} \checkmark$$
$$2(2) - 5(1) = -1 \qquad \text{Solution checks in Equation 2.} \checkmark$$

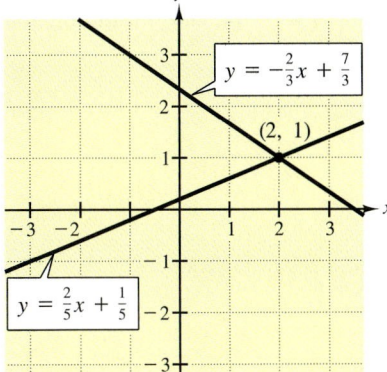

FIGURE 12.2

Applications

To model a real-life situation with a system of equations, you can use the same basic problem-solving strategy that has been used throughout the text.

| Write a verbal model. | → | Assign labels. | → | Write an algebraic model. | → | Solve the algebraic model. | → | Answer the question. |

After answering the question, remember to check the answer in the original statement of the problem.

EXAMPLE 5   An Application

A roofing contractor bought 30 bundles of shingles and four rolls of roofing paper for $528. A second purchase (at the same prices) cost $140 for eight bundles of shingles and one roll of roofing paper. Find the price per bundle of shingles and the price per roll of roofing paper.

Solution

Verbal Model:

$$\boxed{\text{Cost of } 30 \text{ bundles}} + \boxed{\text{Cost of } 4 \text{ rolls}} = \boxed{528}$$

$$\boxed{\text{Cost of } 8 \text{ bundles}} + \boxed{\text{Cost of } 1 \text{ roll}} = \boxed{140}$$

Labels: Price of bundle of shingles $= x$ (dollars)
Price of roll of roofing paper $= y$ (dollars)

System: $30x + 4y = 528$ Equation 1
$8x + y = 140$ Equation 2

Solving the second equation for y produces $y = 140 - 8x$, and substituting this expression into the first equation produces the following.

$$30x + 4(140 - 8x) = 528$$
$$30x + 560 - 32x = 528$$
$$-2x = -32$$
$$x = 16$$

Back-substituting $x = 16$ into the revised second equation produces

$$y = 140 - 8(16)$$
$$= 12.$$

Thus, you can conclude that the price of shingles is $16 per bundle and the price of roofing paper is $12 per roll. Check this in the original statement of the problem.

In 1990, there were about 90 thousand construction companies that specialized in single-family construction. These companies had an average of 4 employees each.

The total cost C of producing x units of a product usually has two components—the initial cost and the cost per unit. When enough units have been sold so that the total revenue R equals the total cost, the sales are said to have reached the **break-even point.** You can find this break-even point by setting C equal to R and solving for x. In other words, the break-even point corresponds to the point of intersection of the cost and revenue graphs.

EXAMPLE 6   An Application: Break-Even Analysis

A small business invests \$14,000 in equipment to produce a product. Each unit of the product costs \$0.80 to produce and is sold for \$1.50. How many items must be sold before the business breaks even?

Solution

Verbal Model:

| Total cost | = | Cost per unit | · | Number of units | + | Initial cost |

| Total revenue | = | Price per unit | · | Number of units |

Labels: Total cost $= C$ (dollars)
             Cost per unit $= 0.80$ (dollars per unit)
             Number of units $= x$ (units)
             Initial cost $= 14{,}000$ (dollars)
             Total revenue $= R$ (dollars)
             Price per unit $= 1.50$ (dollars per unit)

System: $C = 0.80x + 14{,}000$ Equation 1
            $R = 1.50x$ Equation 2

Because the break-even point occurs when $R = C$, you have

$$1.5x = 0.8x + 14{,}000$$
$$0.7x = 14{,}000$$
$$x = 20{,}000.$$

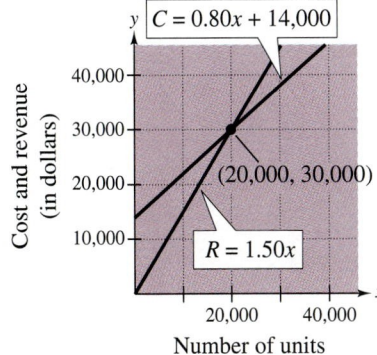

y $\boxed{C = 0.80x + 14{,}000}$

Cost and revenue (in dollars)

40,000 —
30,000 —
20,000 —   (20,000, 30,000)
10,000 — $\boxed{R = 1.50x}$

20,000   40,000   *x*

Number of units

FIGURE 12.3

Thus, it follows that the business must sell 20,000 units before it breaks even. Note in Figure 12.3 that sales less than the break-even point correspond to a loss for the business, whereas sales greater than the break-even point correspond to a profit for the business. The following table helps confirm this conclusion.

Units, x	0	5000	10,000	15,000	20,000	25,000
Profit, P	−\$14,000	−\$10,500	−\$7000	−\$3500	\$0	\$3500

EXAMPLE 7 An Interest Rate Problem

A total of $12,000 is invested in two funds paying 6% and 8% simple interest. If the interest for 1 year is $880, how much of the $12,000 was invested in each fund?

Solution

Verbal Model:

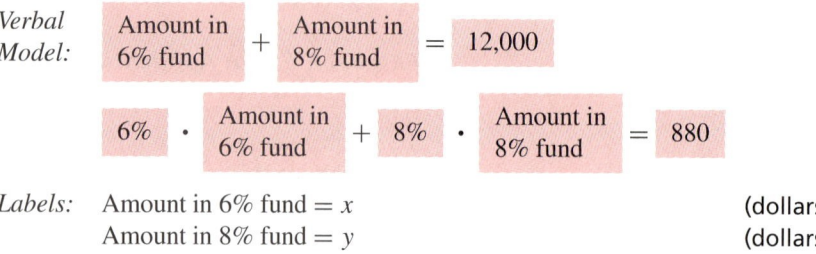

Labels: Amount in 6% fund $= x$ (dollars)
          Amount in 8% fund $= y$ (dollars)

System: $x + \quad y = 12{,}000$ Equation 1
             $0.06x + 0.08y = \quad 880$ Equation 2

To solve this system with a graphing utility, solve each equation for y to get

$$y = 12{,}000 - x \qquad \text{Equation 1}$$
$$y = 11{,}000 - 0.75x. \qquad \text{Equation 2}$$

Graph both equations on the same screen, as shown in Figure 12.4. You can estimate that the two graphs intersect when

$$x = 4000 \quad \text{and} \quad y = 8000.$$

Check this in the original statement of the problem.

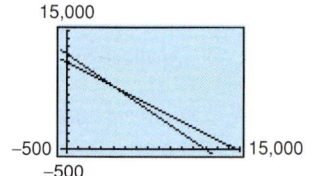

FIGURE 12.4

Group Activities   You Be the Instructor

Problem Posing   You want to create several systems of equations with relatively simple solutions that students can use for practice. Discuss how to create a system of equations that has a given solution. Illustrate your method by creating a system of linear equations that has one of the following solutions: $(1, 4)$, $(-2, 5)$, $(-3, 1)$, or $(4, 2)$. Trade your system for that of another group member, and verify that each other's systems have the desired solutions.

12.1 Exercises

Discussing the Concepts

1. What is meant by a solution of a system of equations in two variables?

2. List and explain the basic steps in solving a system of equations by substitution.

3. When solving a system of equations by substitution, how do you recognize that the system has no solution?

4. What does it mean to *back-substitute* when solving a system of equations?

5. Give a geometric description of the solution of a system of equations in two variables.

6. Describe any advantages of substitution over the graphical method of solving a system of equations.

Problem Solving

In Exercises 7–10, determine whether each ordered pair is a solution of the system of equations.

7. $x + 2y = 9$ (a) $(1, 4)$
   $-2x + 3y = 10$ (b) $(3, -1)$

8. $5x - 4y = 34$ (a) $(0, 3)$
   $x - 2y = 8$ (b) $(6, -1)$

9. $-2x + 7y = 46$ (a) $(-3, 2)$
   $3x + y = 0$ (b) $(-2, 6)$

10. $-5x - 2y = 23$ (a) $(-3, -4)$
    $x + 4y = -19$ (b) $(3, 7)$

In Exercises 11–20, solve the system by substitution.

11. $x - 2y = 0$
    $3x + 2y = 8$

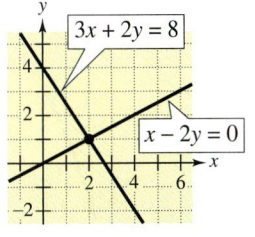

12. $x - 3y = -2$
    $5x + 3y = 17$

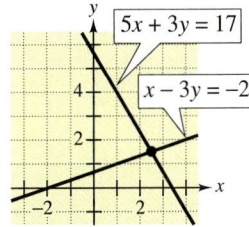

13. $x + y = 2$
    $x^2 - y = 0$

14. $2x + y = 10$
    $x^2 + y^2 = 25$

15. $x + y = 3$
    $2x - y = 0$

16. $-x + y = 5$
    $x - 4y = 0$

17. $4x - 14y = -15$
    $18x - 12y = 9$

18. $5x - 24y = -12$
    $17x - 24y = 36$

19. $x^2 + y^2 = 25$
    $2x - y = -5$

20. $x^2 - y^2 = 16$
    $3x - y = 12$

In Exercises 21–26, use the graphs of the equations to determine whether the system has any solutions. Find any solutions that exist.

21. $x + y = 4$
    $x + y = -1$

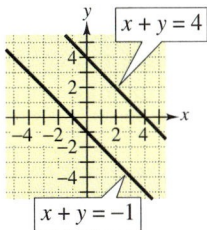

22. $-x + y = 5$
    $x + 2y = 4$

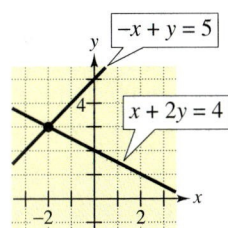

23. $5x - 3y = 4$
    $2x + 3y = 3$

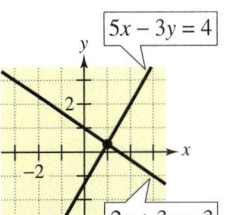

24. $2x - y = 4$
    $-4x + 2y = -12$

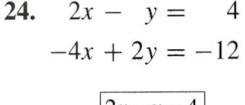
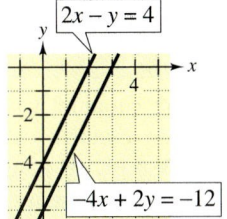

25. $x - 2y = 4$
$x^2 - y = 0$

26. $y = \sqrt{x - 2}$
$x - 2y = 1$

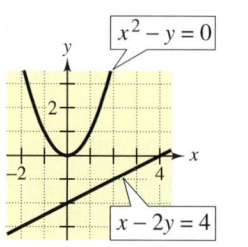

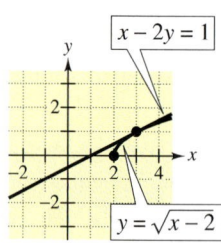

In Exercises 27–32, use a graphing utility to graph the equations and approximate any solutions of the system of equations.

27. $-2x + y = 1$
$x - 3y = 2$

28. $5x - 6y = -30$
$5x + 4y = 20$

29. $x^2 - y^2 = 12$
$x - 2y = 0$

30. $x^2 + y^2 = 20$
$x + 3y = 10$

31. $y = x^3$
$y = x^3 - 3x^2 + 3x$

32. $y = \frac{1}{5}(24 - x)$
$y = \sqrt{64 - x^2}$

In Exercises 33 and 34, use a graphing utility to graph each equation in the system. The graphs appear parallel. Yet, from the slope-intercept forms of the lines, you find that the slopes are not equal and thus the graphs intersect. Find the point of intersection of the two lines.

33. $x - 100y = -200$
$3x - 275y = 198$

34. $35x - 33y = 0$
$12x - 11y = 92$

35. *Break-Even Analysis* A small business invests $8000 in equipment to produce a product. Each unit of the product costs $1.20 to produce and is sold for $2.00. How many items must be sold before the business breaks even?

36. *Break-Even Analysis* A business invests $50,000 in equipment to produce a product. Each unit of the product costs $19.25 to produce and is sold for $35.95. How many items must be sold before the business breaks even?

37. *Hyperbolic Mirror* In a hyperbolic mirror, light rays directed to one focus are reflected to the other focus. The mirror illustrated below has the equation

$$\frac{x^2}{36} - \frac{y^2}{64} = 1.$$

At which point on the mirror will light from the point $(0, 10)$ reflect to the focus?

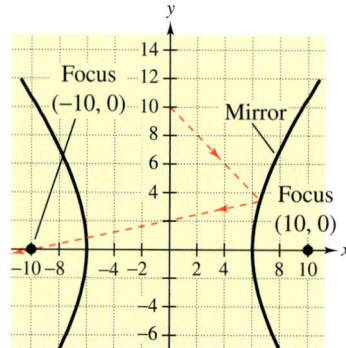

38. *Miniature Golf* You are playing miniature golf. Your golf ball is at the point $(-15, 25)$ (see figure). The wall at the end of the enclosed area is part of a hyperbola whose equation is

$$\frac{x^2}{19} - \frac{y^2}{81} = 1.$$

Using the reflective property of hyperbolas given in Exercise 37, at which point on the wall must your ball hit for it to go into the hole? (The ball bounces off a wall only once.)

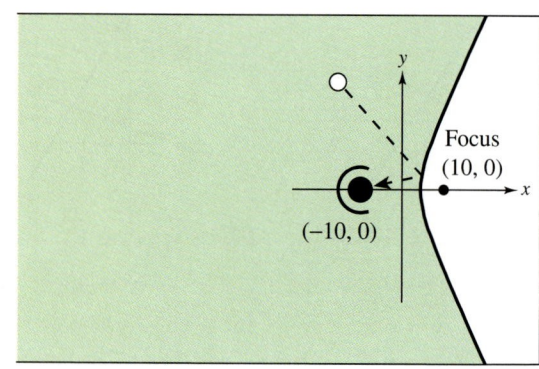

Reviewing the Major Concepts

In Exercises 39–42, find the general form of the equation of the line through the two points.

39. $(-1, -2)$, $(3, 6)$ **40.** $(1, 5)$, $(6, 0)$

41. $\left(\frac{3}{2}, 8\right)$, $\left(\frac{11}{2}, \frac{5}{2}\right)$ **42.** $(0, 2)$, $(7.3, 15.4)$

43. *Work Rate*   Machine A can complete a job in 4 hours and Machine B can complete it in 6 hours. How long will it take both machines working together to complete the job?

44. *Geometry*   Draw a rectangle whose diagonal has a length of 5 units. Do all such rectangles have the same dimensions? Explain.

Additional Problem Solving

In Exercises 45–66, solve the system by substitution.

45. $x \qquad = 4$
$x - 2y = -2$

46. $\qquad y = 2$
$x - 6y = -6$

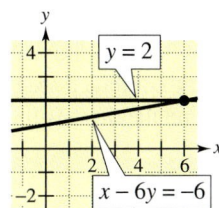

47. $7x + 8y = 24$
$x - 8y = 8$

48. $x - y = 0$
$5x - 2y = 6$

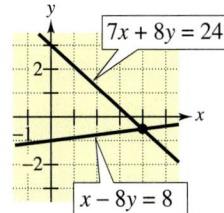

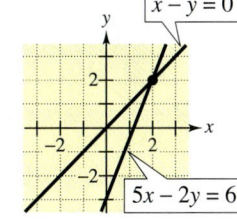

49. $y = \sqrt{8 - x}$
$y = -\frac{1}{5}(x - 14)$

50. $9x^2 + 4y^2 = 36$
$3x - 2y = -6$

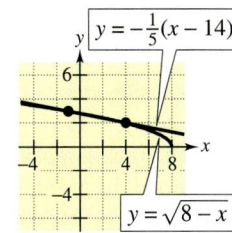

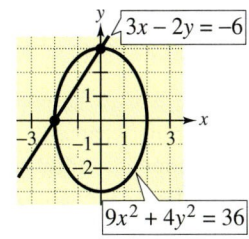

51. $x + y = 2$
$x - 4y = 12$

52. $x - 2y = -1$
$x - 5y = 2$

53. $x + 6y = 19$
$x - 7y = -7$

54. $x - 5y = -6$
$4x - 3y = 10$

55. $2x + 5y = 29$
$5x + 2y = 13$

56. $-13x + 16y = 10$
$5x + 16y = -26$

57. $y = 2x^2$
$y = -2x + 12$

58. $y = 5x^2$
$y = -15x - 10$

59. $x^2 + y = 9$
$x - y = -3$

60. $x - y^2 = 0$
$x - y = 2$

61. $3x + 2y = 90$
$xy = 300$

62. $x + 2y = 40$
$xy = 150$

63. $x^2 + y^2 = 100$
$x = 12$

64. $x^2 + y^2 = 169$
$x + y = 7$

65. $y = \sqrt{4 - x}$
$x + 3y = 6$

66. $x^2 + 2y = 6$
$x - y = -4$

In Exercises 67–78, use a graphing utility to graph the equations and approximate any solutions of the system.

67. $2x - 5y = 20$
$4x - 5y = 40$

68. $5x + 3y = 24$
$x - 2y = 10$

69. $x - y = -3$
$2x - y = 6$

70. $x = 4$
$y = 3$

71. $-5x + 3y = 15$
$x + y = 1$

72. $9x + 4y = 8$
$7x - 4y = 8$

73. $y = x^2$
$y = 4x - x^2$

74. $y = 8 - x^2$
$y = 6 - x$

75. $\sqrt{x} - y = 0$
$x - 5y = -6$

76. $x^2 + y = 4$
$x + y = 6$

77. $16x^2 + 9y^2 = 144$
$4x + 3y = 12$

78. $x^2 - y^2 = 1$
$\frac{1}{2}x^2 + y^2 = 1$

In Exercises 79–82, find two positive integers that satisfy the given requirements.

79. The sum of two numbers is 80 and their difference is 18.

80. The sum of a larger number and twice a smaller number is 61 and the difference between the numbers is 7.

81. The sum of two numbers is 52 and the larger number is 8 less than twice the smaller number.

82. The sum of two numbers is 160 and the larger number is three times the smaller number.

Geometry In Exercises 83–86, find the dimensions of the rectangle meeting the specified conditions.

	Perimeter	Condition
83.	50 feet	The length is 5 feet greater than the width.
84.	320 inches	The width is 20 inches less than the length.
85.	68 yards	The width is $\frac{7}{10}$ of the length.
86.	90 meters	The length is $1\frac{1}{2}$ times the width.

87. *Simple Interest* A combined total of $20,000 is invested in two bonds that pay 8% and 9.5% simple interest. The annual interest is $1675. How much is invested in each bond?

88. *Simple Interest* A combined total of $12,000 is invested in two bonds that pay 8.5% and 10% simple interest. The annual interest is $1140. How much is invested in each bond?

89. *Busing Boundary* To be eligible to ride the school bus to East High School, a student must live at least 1 mile from the school (see figure). Describe the portion of Clarke Street from which the residents are *not* eligible to ride the school bus. Use a coordinate system in which the school is at $(0, 0)$ and each unit represents 1 mile.

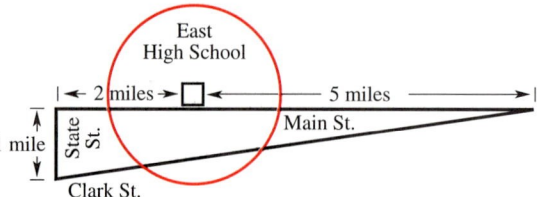

90. *Dimensions of a Corral* You have 250 feet of fencing to enclose two corrals of equal size (see figure). The combined area of the corrals is 2400 square feet. Find the dimensions of each corral.

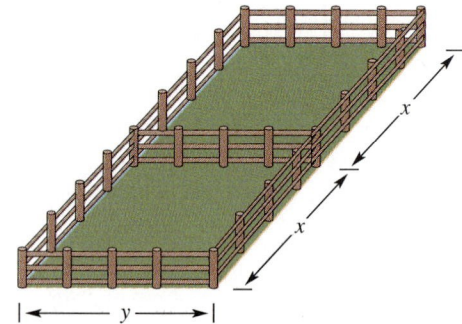

91. *Comparing Populations* From 1982 to 1988, the northeastern part of the United States grew at a slower rate than the western part. Two models that represent the populations of the two regions are

$$P = 49{,}094.5 + 106.9t + 9.7t^2 \qquad \text{Northeast}$$
$$P = 43{,}331.9 + 907.8t \qquad \text{West}$$

where P is the population in thousands and t is the calendar year with $t = 3$ corresponding to 1983. Use a graphing utility to determine when the population of the West overtook the population of the Northeast. (Source: U.S. Bureau of Census)

92. *Geometry* A theorem from geometry states that if a triangle is inscribed in a circle so that one side of the triangle is a diameter of the circle, the triangle is a right triangle (see figure). Show that this theorem is true for the circle

$$x^2 + y^2 = 100$$

and the triangle formed by the lines

$$y = 0, \quad y = \frac{1}{2}x + 5, \quad \text{and} \quad y = -2x + 20.$$

(Find the vertices of the triangle and verify that it is a right triangle.)

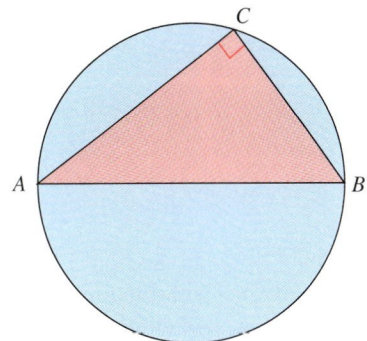

Math Matters   Circle Problem of Apollonius

Apollonius and Euclid are the two most famous mathematicians of the classical Greek period. Euclid lived about 300 B.C. and Apollonius lived about 100 years later. In one of the writings of Apollonius, he poses the following problem.

Suppose you are given three circles that have no points in common and no circle is inside either of the other two circles, as shown in the figure at the left. Is it always possible to find a fourth circle that is tangent to each of the given circles? The answer is yes. In fact, it can be shown that the problem always has eight distinct solutions. Four are shown here. Can you find the other four? (The answer is given in the back of the book.)

12.2 Systems of Linear Equations in Two Variables

The Method of Elimination ▪ Graphical Interpretation of Solutions ▪ Applications

The Method of Elimination

In Section 12.1, you studied two ways of solving a system of equations—substitution and graphing. In this section, you will study a third way—the **method of elimination.**

The key step in the method of elimination is to obtain, for one of the variables, coefficients that differ only in sign so that when the two equations are *added* this variable is eliminated. Notice how this is accomplished in Example 1—when the two equations are added, the y-terms are eliminated.

EXAMPLE 1 *The Method of Elimination*

Solve the system of linear equations.

$$3x + 2y = 4 \qquad \text{Equation 1}$$
$$5x - 2y = 8 \qquad \text{Equation 2}$$

Solution

Begin by noting that the coefficients of y differ only in sign. By adding the two equations, you can eliminate y.

$$
\begin{array}{llr}
3x + 2y = & 4 & \text{Equation 1} \\
5x - 2y = & 8 & \text{Equation 2} \\
\hline
8x \quad\;\; = & 12 & \text{Add equations}
\end{array}
$$

Thus, $x = \frac{3}{2}$. By back-substituting this value into the first equation, you can solve for y as follows.

$$3x + 2y = 4 \qquad \text{Equation 1}$$
$$3\left(\frac{3}{2}\right) + 2y = 4 \qquad \text{Substitute } \tfrac{3}{2} \text{ for } x.$$
$$2y = -\frac{1}{2} \qquad \text{Simplify.}$$
$$y = -\frac{1}{4} \qquad \text{Divide both sides by 2.}$$

The solution is $\left(\frac{3}{2}, -\frac{1}{4}\right)$. Check this solution in the original system of linear equations.

Try using substitution to solve the system given in Example 1. Which method do you think is easier: substitution or elimination? Many people find that the method of elimination is more efficient.

The Method of Elimination

1. Obtain coefficients for x (or y) that differ only in sign by multiplying all terms of one or both equations by suitably chosen constants.

2. Add the equations to eliminate one variable and solve the resulting equation.

3. Back-substitute the value obtained in Step 2 into either of the original equations and solve for the other variable.

4. Check your solution in both of the original equations.

To obtain coefficients (for one of the variables) that differ only in sign, you often need to multiply one or both of the equations by a suitable constant.

EXAMPLE 2   The Method of Elimination

Solve the system of linear equations.

$$4x - 5y = 13 \qquad \text{Equation 1}$$
$$3x - y = 7 \qquad \text{Equation 2}$$

Solution

To obtain coefficients of y that differ only in sign, multiply Equation 2 by -5.

$$
\begin{array}{ll}
4x - 5y = 13 & \text{Equation 1} \\
-15x + 5y = -35 & \text{Multiply Equation 2 by } -5. \\
\hline
-11x = -22 & \text{Add equations.}
\end{array}
$$

Thus, $x = 2$. Back-substitute this value into Equation 2 and solve for y.

$$3x - y = 7 \qquad \text{Equation 2}$$
$$3(2) - y = 7 \qquad \text{Substitute 2 for } x.$$
$$y = -1 \qquad \text{Solve for } y.$$

The solution is $(2, -1)$. Check this in the original system of equations.

$$4(2) - 5(-1) = 13 \qquad \text{Solution checks in Equation 1.} \checkmark$$
$$3(2) - (-1) = 7 \qquad \text{Solution checks in Equation 2.} \checkmark$$

Graphical Interpretation of Solutions

As you observed in Section 12.1, it is possible for a *general* system of equations to have exactly one solution, two or more solutions, or no solution. For a system of *linear* equations, the second of these three possibilities is different.

Specifically, if a system of linear equations has two different solutions, it must have an *infinite* number of solutions. To see why this is true, consider the following graphical interpretations of a system of two linear equations in two variables. (Remember that the graph of a linear equation in two variables is a straight line.)

Graphical Interpretation of Solutions

For a system of two linear equations in two variables, the number of solutions is given by one of the following.

Number of Solutions	*Graphical Interpretation*
1. Exactly one solution	The two lines intersect at one point.
2. Infinitely many solutions	The two lines are identical.
3. No solution	The two lines are parallel.

These three possibilities are shown in Figure 12.5.

Consistent   Inconsistent

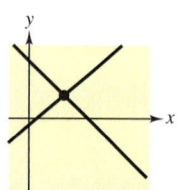 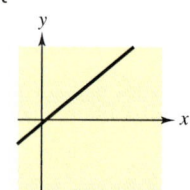 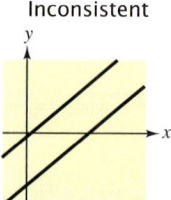

Two lines that intersect: single point of intersection

Two lines that coincide: infinitely many points of intersection

Two parallel lines: no point of intersection

FIGURE 12.5

A system of linear equations is **consistent** if it has at least one solution, and it is **inconsistent** if it has no solution. Here are some examples.

Consistent System	*Consistent System*	*Inconsistent System*
$x + y = 2$	$x + y = 2$	$x + y = 2$
$x + 2y = 4$	$2x + 2y = 4$	$x + y = 4$

Try using a graphing utility to determine how many solutions each of these systems has.

Example 3 shows how the method of elimination can be used to determine that a system of linear equations has no solution.

EXAMPLE 3 The Method of Elimination: No-Solution Case

Solve the system of linear equations.

$$3x + 9y = 8 \qquad \text{Equation 1}$$
$$2x + 6y = 7 \qquad \text{Equation 2}$$

Solution

To obtain coefficients of x that differ only in sign, multiply the first equation by 2 and multiply the second equation by -3.

$$\begin{aligned} 3x + 9y &= 8 \\ 2x + 6y &= 7 \end{aligned} \quad \Rightarrow \quad \begin{aligned} 6x + 18y &= 16 && \text{Multiply Equation 1 by 2.} \\ -6x - 18y &= -21 && \text{Multiply Equation 2 by } -3. \\ 0 &= -5 && \text{False statement} \end{aligned}$$

Because there are no values of x and y for which $0 = -5$, you can conclude that the system is inconsistent and has no solution. The lines corresponding to the two equations given in this system are shown in Figure 12.6. Note that the two lines are parallel, and therefore have no point of intersection.

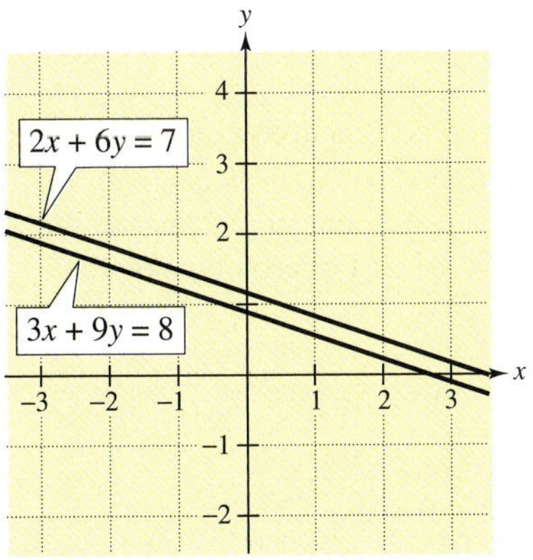

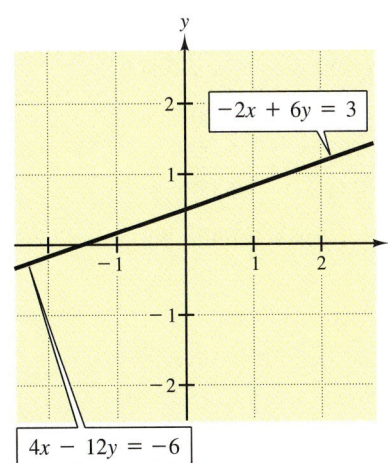

FIGURE 12.6

Example 4 shows how the method of elimination works with a system that has infinitely many solutions.

EXAMPLE 4 The Method of Elimination: Many-Solutions Case

Solve the system of linear equations.

$$-2x + 6y = 3 \qquad \text{Equation 1}$$
$$4x - 12y = -6 \qquad \text{Equation 2}$$

Solution

To obtain coefficients of x that differ only in sign, multiply the first equation by 2.

$$\begin{aligned} -2x + 6y &= 3 \\ 4x - 12y &= -6 \end{aligned} \quad \Rightarrow \quad \begin{aligned} -4x + 12y &= 6 && \text{Multiply Equation 1 by 2.} \\ 4x - 12y &= -6 && \text{Equation 2} \\ 0 &= 0 && \text{Add equations.} \end{aligned}$$

Because the two equations turn out to be equivalent, the system has infinitely many solutions. The solution set consists of all points (x, y) lying on the line $-2x + 6y = 3$, as shown in Figure 12.7.

FIGURE 12.7

Applications

To determine which real-life problems can be solved using a system of linear equations, consider the following. (1) Does the problem involve more than one unknown quantity? (2) Are there two (or more) equations or conditions to be satisfied? If one or both of these conditions occur, the appropriate mathematical model for the problem may be a system of linear equations.

EXAMPLE 5 A Mixture Problem

A company with two stores buys six large delivery vans and five small delivery vans. The first store receives four of the large vans and two of the small vans for a total cost of $160,000. The second store receives two of the large vans and three of the small vans for a total cost of $128,000. What is the cost of each type of van?

Solution

The two unknowns in this problem are the costs of the two types of vans.

Verbal Model:

$$4\left(\text{Cost of large van}\right) + 2\left(\text{Cost of small van}\right) = \$160{,}000$$

$$2\left(\text{Cost of large van}\right) + 3\left(\text{Cost of small van}\right) = \$128{,}000$$

Labels: Cost of large van $= x$ (dollars)
Cost of small van $= y$ (dollars)

System: $4x + 2y = 160{,}000$ Equation 1
$2x + 3y = 128{,}000$ Equation 2

To solve this system of linear equations, use the method of elimination. To obtain coefficients of x that differ only in sign, multiply the second equation by -2.

$$\begin{array}{rl} 4x + 2y = 160{,}000 \\ 2x + 3y = 128{,}000 \end{array} \quad\Longrightarrow\quad \begin{array}{rl} 4x + 2y = 160{,}000 \\ -4x - 6y = -256{,}000 \\ \hline -4y = -96{,}000 \end{array}$$

The cost of each small van is $y = \$24{,}000$. Back-substitute this value into Equation 1 to find the cost of each large van.

$$\begin{aligned} 4x + 2y &= 160{,}000 &&\text{Equation 1} \\ 4x + 2(24{,}000) &= 160{,}000 &&\text{Substitute 24,000 for } y. \\ 4x &= 112{,}000 &&\text{Simplify.} \\ x &= 28{,}000 &&\text{Divide both sides by 4.} \end{aligned}$$

The cost of each large van is $x = \$28{,}000$. Check this solution in the original statement of the problem.

EXAMPLE 6 *An Application Involving Two Speeds*

You take a motorboat trip on a river (18 miles upstream and 18 miles back downstream). You run the motor at the same speed going up and down the river, but because of the current of the river, the trip takes longer going upstream than it does downstream. You do not know the speed of the river's current, but you know that the trip upstream takes $1\frac{1}{2}$ hours and the trip downstream takes only 1 hour. From this information, can you determine the speed of the current?

Solution

Verbal Model:

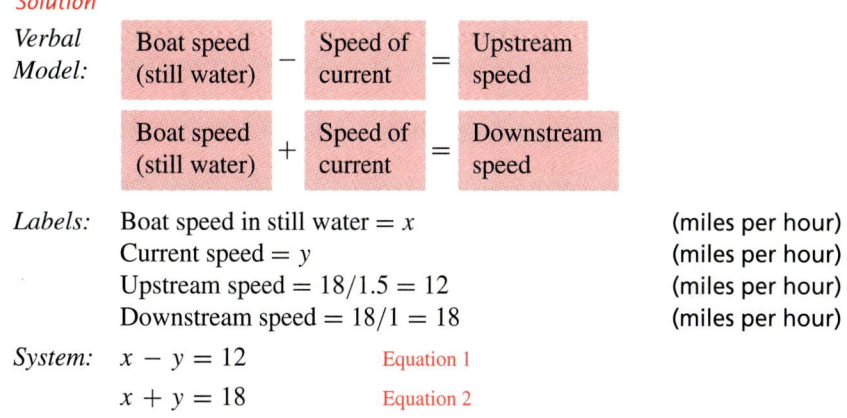

Labels: Boat speed in still water $= x$ (miles per hour)
        Current speed $= y$ (miles per hour)
        Upstream speed $= 18/1.5 = 12$ (miles per hour)
        Downstream speed $= 18/1 = 18$ (miles per hour)

System: $x - y = 12$ Equation 1
            $x + y = 18$ Equation 2

To solve this system of linear equations, use the method of elimination to obtain $x = 15$ miles per hour and $y = 3$ miles per hour. Check this in the original statement of the problem.

Group Activities   Problem Solving

Fitting a Line to Data   The median existing home price in the United States in 1988 was $90,600, and in 1994 it was $107,600. Suppose you wish to fit a linear equation to the points representing this data: $(8, 90,600)$ and $(14, 107,600)$. You remember that one form of an equation for a line is $y = mx + b$, but you can't remember how to use the methods for finding the equation of a line that you used in previous chapters. Write a system of equations that could be solved to find the values of m and b. Solve the system and write the linear equation that fits the data. Interpret the slope in the context of the data. By your model, what was the median existing home price in 1991? The actual median price in 1991 was $99,700. How does the value obtained from your model compare?   (Source: National Association of Realtors)

12.2 Exercises

Discussing the Concepts

1. Give a geometric description of the solution of a system of linear equations in two variables.

2. Explain what is meant by an *inconsistent* system of linear equations.

3. Is it possible for a consistent system of linear equations to have exactly two solutions? Explain.

4. In your own words, explain how to solve a system of linear equations by elimination.

5. How can you recognize that a system of linear equations has no solution? Give an example.

6. When might substitution be better than elimination for solving a system of linear equations?

Problem Solving

In Exercises 7–10, use a graphing utility to graph the equations in the system. Use the graphs to determine whether the system is consistent or inconsistent. If it is consistent, determine the number of solutions.

7. $\frac{1}{3}x - \frac{1}{2}y = 1$
$-2x + 3y = 6$

8. $x + y = 5$
$x - y = 5$

9. $-2x + 3y = 6$
$x - y = -1$

10. $2x - 4y = 9$
$x - 2y = 4.5$

In Exercises 11 and 12, determine whether the system is consistent or inconsistent.

11. $x + 2y = 6$
$x + 2y = 3$

12. $x - 2y = 3$
$2x - 4y = 6$

In Exercises 13–22, solve the system of linear equations by elimination. Identify and label each line with its equation and label the point of intersection (if any).

13. $-x + 2y = 1$
$x - y = 2$

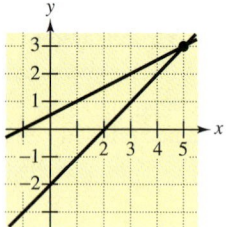

14. $3x + y = 3$
$2x - y = 7$

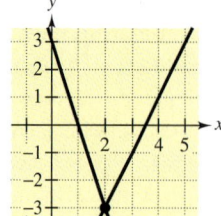

15. $x + y = 0$
$3x - 2y = 10$

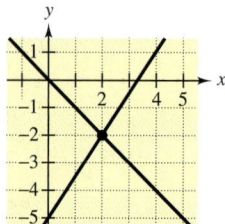

16. $-x + 2y = 2$
$3x + y = 15$

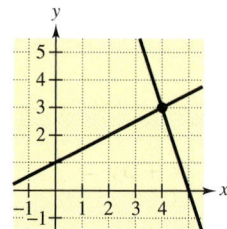

17. $x - y = 1$
$-3x + 3y = 8$

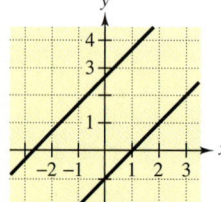

18. $3x + 4y = 2$
$0.6x + 0.8y = 1.6$

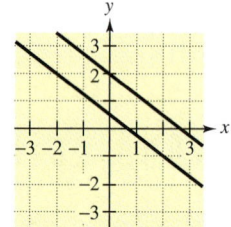

19. $x - 3y = 5$
$-2x + 6y = -10$

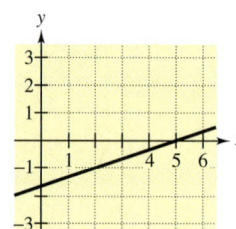

20. $x - 4y = 5$
$5x + 4y = 7$

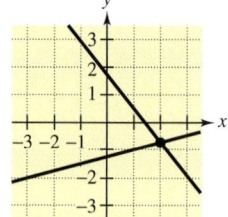

21. $2x - 8y = -11$
$\quad 5x + 3y = \quad 7$

22. $3x + 4y = \quad 0$
$\quad 9x - 5y = 17$

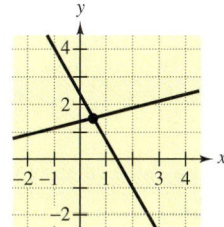
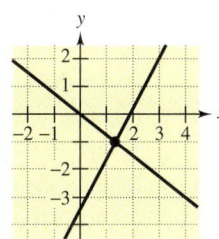

In Exercises 23–28, solve the system of linear equations by the method of elimination.

23. $3x - 2y = 5$
$\quad x + 2y = 7$

24. $7r - \quad s = -25$
$\quad 2r + 5s = \quad 14$

25. $\quad 12x - \quad 5y = 2$
$\quad -24x + 10y = 6$

26. $-2x + 3y = \quad 9$
$\quad 6x - 9y = -27$

27. $2x + 3y = \quad 0$
$\quad 3x + 5y = -1000$

28. $0.4u + 1v = \quad 800$
$\quad 0.7u + 2v = 1850$

In Exercises 29–32, solve the system by any convenient method.

29. $\frac{3}{2}x + 2y = 12$
$\quad \frac{1}{4}x + \quad y = \quad 4$

30. $\quad 4x + y = -2$
$\quad -6x + y = \quad 18$

31. $y = \quad 5x - \quad 3$
$\quad y = -2x + 11$

32. $3x + 2y = 5$
$\quad y = 2x + 13$

In Exercises 33 and 34, find a system of linear equations that has the given solution. (There are many correct answers.)

33. $\left(3, -\frac{3}{2}\right)$

34. $(-8, 12)$

35. *Break-Even Analysis*   You are planning to open a restaurant. You need an initial investment of $85,000. Each week your costs will be about $7400. If your weekly revenue is $8100, how many weeks will it take to break even?

36. *Investing in Two Funds*   A total of $4500 is invested in two funds paying 4% and 5% annual interest. The combined annual interest is $210. How much of the $4500 is invested in each fund?

37. *Average Speed*   A truck travels for 4 hours at an average speed of 42 miles per hour. How much longer must the truck travel at an average speed of 55 miles per hour so that the average speed for the entire trip will be 50 miles per hour?

38. *Air Speed*   An airplane flying into a headwind travels the 3000-mile flying distance between two cities in 6 hours and 15 minutes. On the return flight, the distance is traveled in 5 hours. Find the speed of the plane in still air and the speed of the wind, assuming that both remain constant throughout the round trip.

39. *Rope Length*   You must cut a rope that is 160 inches long into two pieces so that one piece is four times as long as the other. Find the length of each piece.

40. *Driving Distance*   Two people share the driving on a trip of 300 miles. One person drives three times as far as the other. Find the distance that each person drives.

41. *Vietnam Veterans Memorial*   "The Wall" in Washington, D.C., designed by Maya Ling Lin when she was a student at Yale University, has two vertical, triangular sections of black granite with a common side (see figure). The top of each section is level with the ground. The bottoms of the two sections can be modeled by the equations $y = \frac{2}{25}x - 10$ and $y = -\frac{5}{61}x - 10$ when the x-axis is superimposed on the top of the wall. Each unit on the coordinate system represents 1 foot. How deep is the memorial at the point where the two sections meet? How long is each section?

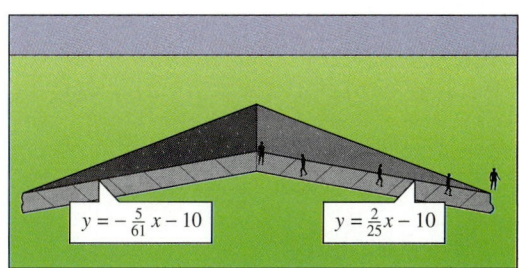

42. *Mathematical Modeling*   The total employment (in thousands) in aircraft industries in the United States from 1990 through 1992 is given in the table. (Source: U.S. Bureau of Labor Statistics)

Year	1990	1991	1992
Employment	898	838	758

(a) Plot the data given in the table, where $x = 0$ corresponds to 1990.

(b) The line $y = mx + b$ that best fits the data is given by the solution of the following system.

$$3b + 3m = 2494$$
$$3b + 5m = 2354$$

Solve the system and find the equation of the required line. Graph the line on the coordinate system of part (a).

(c) Explain the meaning of the slope of the line in the context of this problem.

Reviewing the Major Concepts

In Exercises 43–46, find an equation of the line through the two points.

43. $(-2, 7), (5, 5)$

44. $(3.5, 4), (10, 6)$

45. $(6, 3), (10, 3)$

46. $(4, -2), (4, 5)$

In Exercises 47 and 48, find the number of units x that produces a maximum revenue R. Confirm your result graphically with a graphing utility.

47. $R = 900x - 0.1x^2$

48. $R = 400x - x^2$

Additional Problem Solving

In Exercises 49–66, solve the system of linear equations by the method of elimination.

49. $2x + y = 9$
$3x - y = 16$

50. $-x + 2y = 9$
$x + 3y = 16$

51. $4x + y = -3$
$-4x + 3y = 23$

52. $-3x + 5y = -23$
$2x - 5y = 22$

53. $x - 3y = 2$
$3x - 7y = 4$

54. $2s - t = 9$
$3s + 4t = -14$

55. $2u + 3v = 8$
$3u + 4v = 13$

56. $4x - 3y = 25$
$-3x + 8y = 10$

57. $\frac{2}{3}r - s = 0$
$10r + 4s = 19$

58. $x - y = -\frac{1}{2}$
$4x - 48y = -35$

59. $0.7u - v = -0.4$
$0.3u - 0.8v = 0.2$

60. $0.15x - 0.35y = -0.5$
$-0.12x + 0.25y = 0.1$

61. $5x + 7y = 25$
$x + 1.4y = 5$

62. $12b - 13m = 2$
$-6b + 6.5m = -2$

63. $\frac{3}{2}x - y = 4$
$-x + \frac{2}{3}y = -1$

64. $12x - 3y = 6$
$4x - y = 2$

65. $2x = 25$
$4x - 10y = 0.52$

66. $6x - 6y = 25$
$3y = 11$

In Exercises 67–70, solve the system by any convenient method.

67. $2x - y = 20$
$-x + y = -5$

68. $3x - 2y = -20$
$5x + 6y = 32$

69. $3y = 2x + 21$
$x = 50 - 4y$

70. $x + 2y = 4$
$\frac{1}{2}x + \frac{1}{3}y = 1$

In Exercises 71 and 72, determine the value of k such that the system of linear equations is inconsistent.

71. $5x - 10y = 40$
$-2x + ky = 30$

72. $12x - 18y = 5$
$-18x + ky = 10$

In Exercises 73–78, decide whether the system is consistent or inconsistent.

73. $4x - 5y = 3$
         $-8x + 10y = -6$

74. $4x - 5y = 3$
         $-8x + 10y = 14$

75. $-2x + 5y = 3$
         $5x + 2y = 8$

76. $x + 10y = 12$
         $-2x + 5y = 2$

77. $-10x + 15y = 25$
         $2x - 3y = -24$

78. $4x - 5y = 28$
         $-2x + 2.5y = -14$

In Exercises 79–82, use a graphing utility to sketch the graphs of the two equations. Use the graphs to approximate the solution of the system.

79. $5x + 4y = 35$
         $-x + 3y = 12$

80. $5x - 4y = 0$
         $-3x + 8y = 14$

81. $4x - y = 3$
         $6x + 2y = 1$

82. $x - 6y = 2$
         $2x + 3y = 9$

Geometry   In Exercises 83 and 84, find the dimensions of the rectangle meeting the specified conditions.

Perimeter	Condition
83. 220 meters	The length is 120% of the width.
84. 280 feet	The width is 75% of the length.

In Exercises 85–88, determine how many coins of each type will yield the given value.

Coins	Value
85. 15 dimes and quarters	$3.00
86. 32 dimes and quarters	$6.95
87. 25 nickels and dimes	$1.95
88. 50 nickels and quarters	$7.50

89. *Gasoline Mixture*   Twelve gallons of regular unleaded gasoline plus 8 gallons of premium unleaded gasoline cost $23.08. The price of premium unleaded is 11 cents more per gallon than the price of regular unleaded. Find the price per gallon for each grade of gasoline.

90. *Ticket Sales*   Eight hundred tickets were sold for a theater production and the receipts for the performance were $8600. The tickets for adults and students sold for $12.50 and $7.50, respectively. How many of each kind of ticket were sold?

91. *Nut Mixture*   Ten pounds of mixed nuts sell for $6.95 per pound. The mixture is obtained from two kinds of nuts, with one variety priced at $5.65 per pound and the other priced at $8.95 per pound. How many pounds of each variety of nuts were used in the mixture?

92. *Hay Mixture*   How many tons of hay at $125 per ton must be mixed with hay at $75 per ton to have 100 tons of hay with a value of $90 per ton?

93. *Alcohol Mixture*   How many liters of a 40% alcohol solution must be mixed with a 65% solution to obtain 20 liters of a 50% solution?

94. *Acid Mixture*   Fifty gallons of a 70% acid solution is obtained by mixing an 80% solution with a 50% solution. How many gallons of each solution must be used to obtain the desired mixture?

95. *Best-Fitting Line*   The slope and y-intercept of the line $y = mx + b$ that "best fits" the three noncollinear points $(0, 0)$, $(1, 1)$, and $(2, 3)$ are given by the following system of linear equations.

$$3b + 3m = 4$$
$$3b + 6m = 7$$

(a) Solve the system and find the equation of the "best-fitting" line.

(b) Plot the three points and sketch the graph of the "best-fitting" line.

96. *Best-Fitting Line*   The slope and y-intercept of the line $y = mx + b$ that "best fits" the three points $(0, 3)$, $(1, 2)$, and $(2, 1)$ are given by the following system of linear equations.

$$3b + 3m = 6$$
$$3b + 5m = 4$$

(a) Solve the system and find the equation of the "best-fitting" line.

(b) Plot the three points and sketch the graph of the "best-fitting" line.

12.3   Systems of Linear Equations in Three Variables

Row-Echelon Form ▪ The Method of Elimination ▪ Applications

Row-Echelon Form

The method of elimination can be applied to a system of linear equations in more than two variables. In fact, this method easily adapts to computer use for solving systems of linear equations with dozens of variables.

When the method of elimination is used to solve a system of linear equations, the goal is to rewrite the system in a form to which back-substitution can be applied. For instance, consider the following two systems of linear equations.

$$\begin{array}{rcl} x - 2y + 2z &=& 9 \\ -x + 3y &=& -4 \\ 2x - 5y + z &=& 10 \end{array} \qquad \begin{array}{rcl} x - 2y + 2z &=& 9 \\ y + 2z &=& 5 \\ z &=& 3 \end{array}$$

The system on the right is said to be in **row-echelon form,** which means that it has a "stair-step" pattern with leading coefficients of 1. Which of these two systems do you think is easier to solve? After comparing the two systems, it should be clear that it is easier to solve the system on the right.

EXAMPLE 1   Using Back-Substitution

Solve the system of linear equations.

$$\begin{array}{rcl} x - 2y + 2z &=& 9 \qquad \text{Equation 1} \\ y + 2z &=& 5 \qquad \text{Equation 2} \\ z &=& 3 \qquad \text{Equation 3} \end{array}$$

Solution

From Equation 3, you know the value of z. To solve for y, substitute $z = 3$ into Equation 2 to obtain

$$\begin{array}{rcl} y + 2(3) &=& 5 \qquad \text{Substitute 3 for } z. \\ y &=& -1. \qquad \text{Solve for } y. \end{array}$$

Finally, substitute $y = -1$ and $z = 3$ into Equation 1 to obtain

$$\begin{array}{rcl} x - 2(-1) + 2(3) &=& 9 \qquad \text{Substitute } -1 \text{ for } y \text{ and 3 for } z. \\ x &=& 1. \qquad \text{Solve for } x. \end{array}$$

The solution is $x = 1$, $y = -1$, and $z = 3$, which can also be written as the **ordered triple** $(1, -1, 3)$. Check this in the original system of equations.

The Method of Elimination

Two systems of equations are **equivalent** if they have the same solution set. To solve a system that is not in row-echelon form, first convert it to an *equivalent* system that is in row-echelon form. To see how this is done, let's take another look at the method of elimination, as applied to a system of two linear equations.

EXAMPLE 2   *The Method of Elimination*

NOTE   As shown in Example 2, rewriting a system of linear equations in row-echelon form usually involves a *chain* of equivalent systems, each of which is obtained by using one of the three basic row operations. This process is called **Gaussian elimination,** after the German mathematician Carl Friedrich Gauss (1777–1855).

Solve the system of linear equations.

$$3x - 2y = -1 \qquad \text{Equation 1}$$
$$x - y = 0 \qquad \text{Equation 2}$$

Solution

$$x - y = 0$$
$$3x - 2y = -1$$

You can interchange two equations in the system.

$$-3x + 3y = 0$$

Multiply the first equation by -3.

$$-3x + 3y = 0$$
$$3x - 2y = -1$$
$$\overline{ y = -1}$$

You can add the multiple of the first equation to the second equation to obtain a new second equation.

$$x - y = 0$$
$$y = -1$$

New system in row-echelon form

Now, using back-substitution, you can determine that the solution is $y = -1$ and $x = -1$, which can be written as the ordered pair $(-1, -1)$. Check this in the original system of equations.

Operations That Produce Equivalent Systems

Each of the following **row operations** on a system of linear equations produces an *equivalent* system of linear equations.

1. Interchange two equations.

2. Multiply one of the equations by a nonzero constant.

3. Add a multiple of one of the equations to another equation to replace the latter equation.

EXAMPLE 3 Using Elimination to Solve a System

Solve the system of linear equations.

$$x - 2y + 2z = 9 \qquad \text{Equation 1}$$
$$-x + 3y = -4 \qquad \text{Equation 2}$$
$$2x - 5y + z = 10 \qquad \text{Equation 3}$$

Solution

There are many ways to begin, but we suggest working from the upper left corner, saving the x in the upper left position and eliminating the other x's from the first column.

$$x - 2y + 2z = 9$$
$$y + 2z = 5$$
$$2x - 5y + z = 10$$

> Adding the first equation to the second equation produces a new second equation.

$$x - 2y + 2z = 9$$
$$y + 2z = 5$$
$$-y - 3z = -8$$

> Adding -2 times the first equation to the third equation produces a new third equation.

Now that all but the first x have been eliminated from the first column, go to work on the second column. (You need to eliminate y from the third equation.)

$$x - 2y + 2z = 9$$
$$y + 2z = 5$$
$$-z = -3$$

> Adding the second equation to the third equation produces a new third equation.

Finally, you need a coefficient of 1 for z in the third equation.

$$x - 2y + 2z = 9$$
$$y + 2z = 5$$
$$z = 3$$

> Multiplying the third equation by -1 produces a new third equation.

This is the same system that was solved in Example 1, and, as in that example, you can conclude that the solution is

$$x = 1, \quad y = -1, \quad \text{and} \quad z = 3.$$

In Example 3, you can check the solution by substituting $x = 1$, $y = -1$, and $z = 3$ into each equation, as follows.

Equation 1: $(1) - 2(-1) + 2(3) = 9$ ✓
Equation 2: $-(1) + 3(-1) = -4$ ✓
Equation 3: $2(1) - 5(-1) + (3) = 10$ ✓

EXAMPLE 4   *Using Elimination to Solve a System*

Solve the following system of linear equations.

$$4x + y - 3z = 11 \qquad \text{Equation 1}$$
$$2x - 3y + 2z = 9 \qquad \text{Equation 2}$$
$$x + y + z = -3 \qquad \text{Equation 3}$$

Solution

$$x + y + z = -3$$
$$2x - 3y + 2z = 9$$
$$4x + y - 3z = 11$$

Interchange the first and third equations.

$$x + y + z = -3$$
$$- 5y = 15$$
$$4x + y - 3z = 11$$

Adding −2 times the first equation to the second equation produces a new second equation.

$$x + y + z = -3$$
$$- 5y = 15$$
$$- 3y - 7z = 23$$

Adding −4 times the first equation to the third equation produces a new third equation.

$$x + y + z = -3$$
$$y = -3$$
$$- 3y - 7z = 23$$

Multiplying the second equation by $-\frac{1}{5}$ produces a new second equation.

$$x + y + z = -3$$
$$y = -3$$
$$- 7z = 14$$

Adding 3 times the second equation to the third equation produces a new third equation.

$$x + y + z = -3$$
$$y = -3$$
$$z = -2$$

Multiplying the third equation by $-\frac{1}{7}$ produces a new third equation.

Now that the system of equations is in row-echelon form, you can see that $z = -2$ and $y = -3$. Moreover, by back-substituting these values into Equation 1, you can determine that $x = 2$. Therefore, the solution is

$$x = 2, \qquad y = -3, \qquad \text{and} \qquad z = -2.$$

You can check this solution as follows.

Equation 1: $4(2) + (-3) - 3(-2) = 11$ ✔

Equation 2: $2(2) - 3(-3) + 2(-2) = 9$ ✔

Equation 3: $(2) + (-3) + (-2) = -3$ ✔

The next example involves an inconsistent system—one that has no solution. The key to recognizing an inconsistent system is that at some stage in the elimination process, you obtain an absurdity such as $0 = -2$.

EXAMPLE 5   *An Inconsistent System*

Solve the system of linear equations.

$$\begin{array}{rl}
x - 3y + z = & 1 \qquad \text{Equation 1} \\
2x - y - 2z = & 2 \qquad \text{Equation 2} \\
x + 2y - 3z = & -1 \qquad \text{Equation 3}
\end{array}$$

Solution

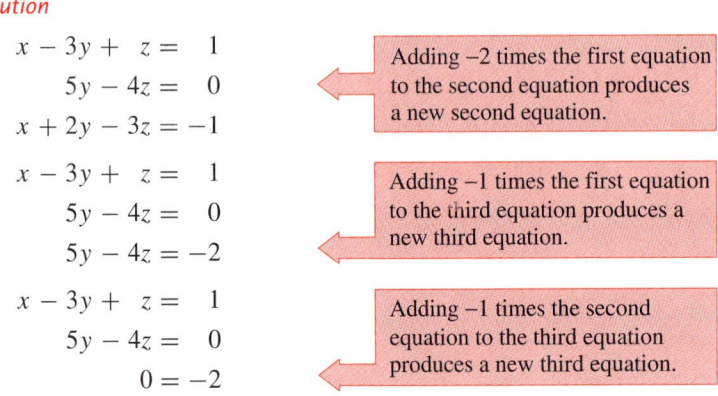

$$\begin{array}{rl}
x - 3y + z = & 1 \\
5y - 4z = & 0 \\
x + 2y - 3z = & -1
\end{array}$$

Adding -2 times the first equation to the second equation produces a new second equation.

$$\begin{array}{rl}
x - 3y + z = & 1 \\
5y - 4z = & 0 \\
5y - 4z = & -2
\end{array}$$

Adding -1 times the first equation to the third equation produces a new third equation.

$$\begin{array}{rl}
x - 3y + z = & 1 \\
5y - 4z = & 0 \\
0 = & -2
\end{array}$$

Adding -1 times the second equation to the third equation produces a new third equation.

Because the third "equation" is impossible, you can conclude that this system is inconsistent and therefore has no solution. Moreover, because this system is equivalent to the original system, you can conclude that the original system also has no solution.

As with a system of linear equations in two variables, the solution(s) of a system of linear equations in more than two variables must fall into one of three categories.

The Number of Solutions of a Linear System

For a system of linear equations, exactly one of the following is true.

1. There is exactly one solution.

2. There are infinitely many solutions.

3. There is no solution.

EXAMPLE 6   A System with Infinitely Many Solutions

Solve the system of linear equations.

$$x + y - 3z = -1 \qquad \text{Equation 1}$$
$$y - z = 0 \qquad \text{Equation 2}$$
$$-x + 2y = 1 \qquad \text{Equation 3}$$

Solution

Begin by rewriting the system in row-echelon form.

$$x + y - 3z = -1$$
$$y - z = 0$$
$$3y - 3z = 0$$

> Adding the first equation to the third equation produces a new third equation.

$$x + y - 3z = -1$$
$$y - z = 0$$
$$0 = 0$$

> Adding -3 times the second equation to the third equation produces a new third equation.

This means that Equation 3 depends on Equations 1 and 2 in the sense that it gives us no additional information about the variables. Thus, the original system is equivalent to the system

$$x + y - 3z = -1$$
$$y - z = 0.$$

In this last equation, solve for y in terms of z to obtain $y = z$. Back-substituting for y in the previous equation produces $x = 2z - 1$. Finally, letting $z = a$, the solutions to the given system are all of the form

$$x = 2a - 1, \quad y = a, \quad \text{and} \quad z = a$$

where a is a real number. Thus, every ordered triple of the form

$$(2a - 1, a, a), \qquad a \text{ is a real number}$$

is a solution of the system.

In Example 6, there are other ways to write the same infinite set of solutions. For instance, the solutions could have been written as

$$\left(b, \tfrac{1}{2}(b + 1), \tfrac{1}{2}(b + 1)\right), \qquad b \text{ is a real number.}$$

Try convincing yourself of this by substituting $a = 0$, $a = 1$, $a = 2$, and $a = 3$ into the solution listed in Example 6. Then substitute $b = -1$, $b = 1$, $b = 3$, and $b = 5$ into the solution listed above. In both cases, you should obtain the same ordered triples. Thus, when comparing descriptions of an infinite solution set, keep in mind that there is more than one way to describe the set.

Applications

EXAMPLE 7   An Application: Moving Object

The height at time t of an object that is moving in a (vertical) line with constant acceleration a is given by the **position equation**

$$s = \tfrac{1}{2}at^2 + v_0 t + s_0.$$

The height s is measured in feet, the acceleration a is measured in feet per second squared, the time t is measured in seconds, v_0 is the initial velocity (at time $t = 0$), and s_0 is the initial height. Find the values of a, v_0, and s_0, if $s = 164$ feet at 1 second, $s = 180$ feet at 2 seconds, and $s = 164$ feet at 3 seconds.

Solution

By substituting the three values of t and s into the position equation, you obtain three linear equations in a, v_0, and s_0.

When $t = 1$, $s = 164$: $\tfrac{1}{2}a(1)^2 + v_0(1) + s_0 = 164$

When $t = 2$, $s = 180$: $\tfrac{1}{2}a(2)^2 + v_0(2) + s_0 = 180$

When $t = 3$, $s = 164$: $\tfrac{1}{2}a(3)^2 + v_0(3) + s_0 = 164$

By multiplying the first and third equations by 2, this system can be rewritten

$$\begin{aligned} a + 2v_0 + 2s_0 &= 328 & &\text{Equation 1}\\ 2a + 2v_0 + s_0 &= 180 & &\text{Equation 2}\\ 9a + 6v_0 + 2s_0 &= 328 & &\text{Equation 3}\end{aligned}$$

and you can apply elimination to obtain

$$\begin{aligned} a + 2v_0 + 2s_0 &= 328\\ -2v_0 - 3s_0 &= -476\\ 2s_0 &= 232.\end{aligned}$$

From the third equation, $s_0 = 116$, so that back-substitution into the second equation yields

$$\begin{aligned} -2v_0 - 3(116) &= -476\\ -2v_0 &= -128\\ v_0 &= 64.\end{aligned}$$

Finally, back-substituting $s_0 = 116$ and $v_0 = 64$ into the first equation yields

$$\begin{aligned} a + 2(64) + 2(116) &= 328\\ a &= -32.\end{aligned}$$

Thus, the position equation for this object is $s = -16t^2 + 64t + 116$.

EXAMPLE 8   Data Analysis: Curve-Fitting

Find a quadratic equation $y = ax^2 + bx + c$ whose graph passes through the points $(-1, 3)$, $(1, 1)$, and $(2, 6)$.

Solution

Because the graph of $y = ax^2 + bx + c$ passes through the points $(-1, 3)$, $(1, 1)$, and $(2, 6)$, you can write the following.

$$\text{When } x = -1,\ y = 3:\quad a(-1)^2 + b(-1) + c = 3$$
$$\text{When } x = 1,\ y = 1:\quad a(1)^2 + b(1) + c = 1$$
$$\text{When } x = 2,\ y = 6:\quad a(2)^2 + b(2) + c = 6$$

This produces the following system of linear equations.

$$a - b + c = 3 \qquad \text{Equation 1}$$
$$a + b + c = 1 \qquad \text{Equation 2}$$
$$4a + 2b + c = 6 \qquad \text{Equation 3}$$

The solution of this system is $a = 2$, $b = -1$, and $c = 0$. Thus, the equation of the parabola is $y = 2x^2 - x$, as shown in Figure 12.8.

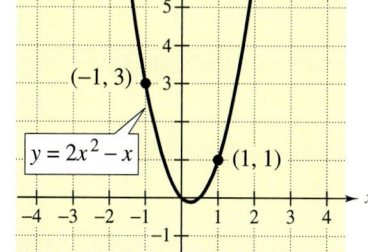

$y = 2x^2 - x$

$(-1, 3)$

$(2, 6)$

$(1, 1)$

FIGURE 12.8

Group Activities   Problem Solving

Fitting a Quadratic Model   The data in the table represents the United States government's annual net receipts y (in billions of dollars) from individual income taxes for the year x from 1990 through 1992, where $x = 0$ corresponds to 1990.   (Source: U.S. Department of the Treasury)

x	0	1	2
y	467	468	476

Use a system of three linear equations to find a quadratic model that fits the data. According to your model, what were the annual net receipts from individual income taxes in 1993? The actual annual net receipts for 1993 were $510 billion. How does the value obtained from your quadratic model compare? Suppose you had been involved in planning the 1993 federal budget and had used this model to estimate how much federal income could be expected from 1993 individual income taxes. When you review the actual 1993 tax receipts and see that the model wasn't completely accurate, how do you evaluate the model's prediction performance? Are you satisfied with it? Why or why not?

12.3 Exercises

Discussing the Concepts

1. Give an example of a system of linear equations that is in row-echelon form.

2. Show how to use back-substitution to solve the system you found for Exercise 1.

3. Describe the row operations that are performed on a system of linear equations to produce an equivalent system of equations.

4. Are the following two systems of equations equivalent? Give reasons for your answer.

$$
\begin{aligned}
x + 3y - z &= 6 \\
2x - y + 2z &= 1 \\
3x + 2y - z &= 2
\end{aligned}
\qquad
\begin{aligned}
x + 3y - z &= 6 \\
-7y + 4z &= 1 \\
-7y - 4z &= -16
\end{aligned}
$$

Problem Solving

In Exercises 5–8, use back-substitution to solve the system of linear equations.

5.
$$
\begin{aligned}
x - 2y + 4z &= 4 \\
3y - z &= 2 \\
z &= -5
\end{aligned}
$$

6.
$$
\begin{aligned}
5x + 4y - z &= 0 \\
10y - 3z &= 11 \\
z &= 3
\end{aligned}
$$

7.
$$
\begin{aligned}
x - 2y + 4z &= 4 \\
y &= 3 \\
y + z &= 2
\end{aligned}
$$

8.
$$
\begin{aligned}
x &= 10 \\
3x + 2y &= 2 \\
x + y + 2z &= 0
\end{aligned}
$$

In Exercises 9–18, use Gaussian elimination to solve the system of linear equations.

9.
$$
\begin{aligned}
x + z &= 4 \\
y &= 2 \\
4x + z &= 7
\end{aligned}
$$

10.
$$
\begin{aligned}
x - y + 2z &= -4 \\
3x + y - 4z &= -6 \\
2x + 3y - 4z &= 4
\end{aligned}
$$

11.
$$
\begin{aligned}
x &= 3 \\
-x + 3y &= 3 \\
y + 2z &= 4
\end{aligned}
$$

12.
$$
\begin{aligned}
x + y + z &= -3 \\
4x + y - 3z &= 11 \\
2x - 3y + 2z &= 9
\end{aligned}
$$

13.
$$
\begin{aligned}
x + 2y + 6z &= 5 \\
-x + y - 2z &= 3 \\
x - 4y - 2z &= 1
\end{aligned}
$$

14.
$$
\begin{aligned}
2x + y + 3z &= 1 \\
2x + 6y + 8z &= 3 \\
6x + 8y + 18z &= 5
\end{aligned}
$$

15.
$$
\begin{aligned}
0.2x + 1.3y + 0.6z &= 0.1 \\
0.1x + 0.3z &= 0.7 \\
2x + 10y + 8z &= 8
\end{aligned}
$$

16.
$$
\begin{aligned}
0.3x - 0.1y + 0.2z &= 0.35 \\
2x + y - 2z &= -1 \\
2x + 4y + 3z &= 10.5
\end{aligned}
$$

17.
$$
\begin{aligned}
x + 4y - 2z &= 2 \\
-3x + y + z &= -2 \\
5x + 7y - 5z &= 6
\end{aligned}
$$

18.
$$
\begin{aligned}
2x + z &= 1 \\
5y - 3z &= 2 \\
6x + 20y - 9z &= 11
\end{aligned}
$$

Curve-Fitting In Exercises 19 and 20, find the equation of the parabola

$$y = ax^2 + bx + c$$

that passes through the given points.

19. $(0, 5), (1, 6), (2, 5)$

20. $(0, -4), (1, 1), (2, 10)$

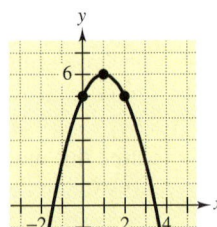

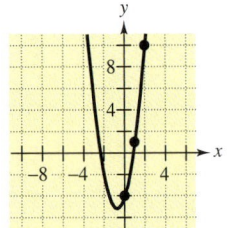

Curve-Fitting   In Exercises 21 and 22, find the equation of the circle

$$x^2 + y^2 + Dx + Ey + F = 0$$

that passes through the given points.

21. $(0, 0), (2, -2), (4, 0)$ **22.** $(0, 0), (0, 6), (-3, 3)$

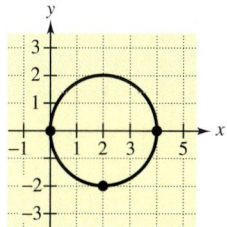   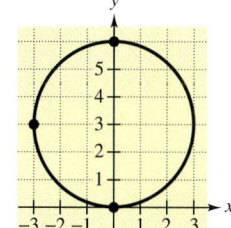

Vertical Motion   In Exercises 23–26, find the position equation, $s = \frac{1}{2}at^2 + v_0t + s_0$, for an object that has the indicated heights at the specified times.

23. $s = 128$ feet at $t = 1$ second

  $s = 80$ feet at $t = 2$ seconds

  $s = 0$ feet at $t = 3$ seconds

24. $s = 48$ feet at $t = 1$ second

  $s = 64$ feet at $t = 2$ seconds

  $s = 48$ feet at $t = 3$ seconds

25. $s = 32$ feet at $t = 1$ second

  $s = 32$ feet at $t = 2$ seconds

  $s = 0$ feet at $t = 3$ seconds

26. $s = 10$ feet at $t = 0$ seconds

  $s = 54$ feet at $t = 1$ second

  $s = 46$ feet at $t = 3$ seconds

27. *Crop Spraying*   A mixture of 12 gallons of chemical A, 16 gallons of chemical B, and 26 gallons of chemical C is required to kill a certain destructive crop insect. Commercial spray X contains 1, 2, and 2 parts, respectively, of these chemicals. Commercial spray Y contains only chemical C. Commercial spray Z contains only chemicals A and B in equal amounts. How much of each type of commercial spray is needed to get the desired mixture?

28. *Chemistry*   A chemist needs 10 liters of a 25% acid solution. The solution is to be mixed from three solutions whose concentrations are 10%, 20%, and 50%. How many liters of each solution should the chemist use to satisfy the following?

(a) Use as little as possible of the 50% solution.

(b) Use as much as possible of the 50% solution.

(c) Use 2 liters of the 50% solution.

29. *School Orchestra*   The table shows the percents of each section of the North High School orchestra that were chosen to participate in the city orchestra, the county orchestra, and the state orchestra. Thirty members of the city orchestra, 17 members of the county orchestra, and 10 members of the state orchestra are from North. How many members are in each section of North High's orchestra?

Orchestra	String	Wind	Percussion
City Orchestra	40%	30%	50%
County Orchestra	20%	25%	25%
State Orchestra	10%	15%	25%

30. *Diagonals of a Polygon*   The total numbers of sides and diagonals of regular polygons with three, four, and five sides are three, six, and ten, as shown in the figure. Find a quadratic function $y = ax^2 + bx + c$ that fits this data. Then check to see if it gives the correct answer for a polygon with six sides.

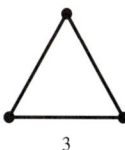   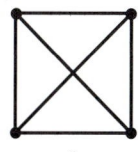

3   6

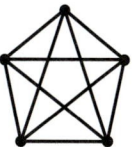   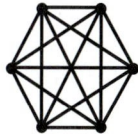

10   15

Reviewing the Major Concepts

In Exercises 31–34, find the domain of the function.

31. $f(x) = x^3 - 2x$

32. $g(x) = \sqrt[3]{x}$

33. $h(x) = \sqrt{16 - x^2}$

34. $A(x) = \dfrac{3}{(36 - x^2)}$

35. *Predator-Prey* The number N of prey animals t months after a predator is introduced into a test area is inversely proportional to $t + 1$. When $t = 0$, $N = 300$. Find N when $t = 5$.

36. *Geometry* Find the area A of the rectangle in the figure as a function of x.

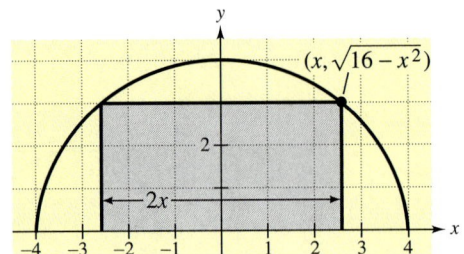

Additional Problem Solving

In Exercises 37 and 38, perform the row operation and write the equivalent system.

37. Add Equation 1 to Equation 2.

$x - 2y + 3z = 5$ Equation 1
$-x + y + 5z = 4$ Equation 2
$2x \qquad - 3z = 0$ Equation 3

38. Add -2 times Equation 1 to Equation 3.

$x - 2y + 3z = 5$ Equation 1
$-x + y + 5z = 4$ Equation 2
$2x \qquad - 3z = 0$ Equation 3

In Exercises 39 and 40, decide whether each ordered triple is a solution of the system.

39.
$x + 3y + 2z = 1$
$5x - y + 3z = 16$
$-3x + 7y + z = -14$

(a) $(0, 3, -2)$ (b) $(12, 5, -13)$
(c) $(1, -2, 3)$ (d) $(-2, 5, -3)$

40.
$3x - y + 4z = -10$
$-x + y + 2z = 6$
$2x - y + z = -8$

(a) $(-2, 4, 0)$ (b) $(0, -3, 10)$
(c) $(1, -1, 5)$ (d) $(7, 19, -3)$

In Exercises 41–56, use Gaussian elimination to solve the system of linear equations.

41.
$x + y + z = 6$
$2x - y + z = 3$
$3x \qquad - z = 0$

42.
$x + y + z = 2$
$-x + 3y + 2z = 8$
$4x + y \qquad = 4$

43.
$x + y + 8z = 3$
$2x + y + 11z = 4$
$x \qquad + 3z = 0$

44.
$x + 6y + 2z = 9$
$3x - 2y + 3z = -1$
$5x - 5y + 2z = 7$

45.
$2x \qquad + 2z = 2$
$5x + 3y \qquad = 4$
$3y - 4z = 4$

46.
$6y + 4z = -12$
$3x + 3y \qquad = 9$
$2x \qquad - 3z = 10$

47.
$3x - y - 2z = 5$
$2x + y + 3z = 6$
$6x - y - 4z = 9$

48.
$2x - 4y + z = 0$
$3x \qquad + 2z = -1$
$-6x + 3y + 2z = -10$

49.
$5x + 2y \qquad = -8$
$z = 5$
$3x - y + z = 9$

50.
$y + z = 5$
$2x \qquad + 4z = 4$
$2x - 3y \qquad = -14$

51.
$x + 2y - 2z = 4$
$2x + 5y - 7z = 5$
$3x + 7y - 9z = 10$

52.
$2x + 6y - 4z = 8$
$3x + 10y - 7z = 12$
$-2x - 6y + 5z = -3$

53. $2x + y - z = 4$
$\quad\quad\quad y + 3z = 2$
$\quad 3x + 2y \quad\quad = 4$

54. $x - 2y - z = 3$
$\quad 2x + y - 3z = 1$
$\quad x + 8y - 3z = -7$

55. $3x + y + z = 2$
$\quad 4x \quad\quad + 2z = 1$
$\quad 5x - y + 3z = 0$

56. $2x \quad\quad + 3z = 4$
$\quad 5x + y + z = 2$
$\quad 11x + 3y - 3z = 0$

Curve-Fitting   In Exercises 57–60, find the equation of the parabola $y = ax^2 + bx + c$ that passes through the given points.

57. $(1, 0), (2, -1), (3, 0)$ **58.** $(1, 2), (2, 1), (3, -4)$

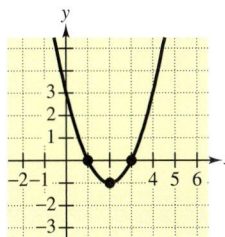   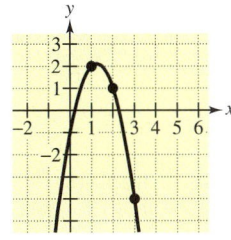

59. $(-1, -3), (1, 1), (2, 0)$

60. $(-1, -1), (1, 1), (2, -4)$

Curve-Fitting   In Exercises 61–64, find the equation of the circle $x^2 + y^2 + Dx + Ey + F = 0$ that passes through the given points.

61. $(0, 0), (0, 2), (3, 0)$ **62.** $(3, -1), (-2, 4), (6, 8)$

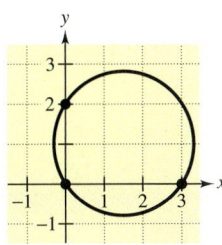   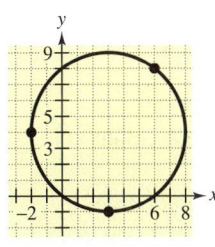

63. $(-3, 5), (4, 6), (5, 5)$

64. $(5, 13), (17, 5), (10, 12)$

65. *Graphical Estimation*   The table gives the amounts y, in millions of short tons, of newsprint produced in the years 1990 through 1992 in the United States. (Source: American Paper Institute)

Year	1990	1991	1992
y	6.6	6.8	7.1

(a) Find the equation of the parabola $y = at^2 + bt + c$ that passes through the three points, letting $t = 0$ correspond to 1990.

(b) Use a graphing utility to graph the model found in part (a).

(c) Use the model of part (a) to predict newsprint production in 1999 if the trend continues.

66. *Rewriting a Fraction*   The fraction $1/(x^3 - x)$ can be written as a sum of three fractions as follows.

$$\frac{1}{x^3 - x} = \frac{A}{x} + \frac{B}{x + 1} + \frac{C}{x - 1}$$

The numbers A, B, and C are the solutions of the following system.

$$A + B + C = 0$$
$$-B + C = 0$$
$$-A \quad\quad = 1$$

Solve the system and verify that the sum of the three resulting fractions is the original fraction.

In Exercises 67 and 68, find a system of linear equations in three variables with integer coefficients that has the given point as a solution. (*Note:* There are many correct answers.)

67. $(4, -3, 2)$ **68.** $(5, 7, -10)$

MID-CHAPTER QUIZ

Take this quiz as you would take a quiz in class. After you are done, check your work against the answers given in the back of the book.

1. Which is the solution of the system $5x - 12y = 2$ and $2x + 1.5y = 26$: $(1, -2)$ or $(10, 4)$? Explain your reasoning.

In Exercises 2–4, graph the equations in the system. Use the graphs to determine the number of solutions of the system.

2. $-6x + 9y = 9$
$2x - 3y = 6$

3. $x - 2y = -4$
$3x - 2y = 4$

4. $y = x - 1$
$y = 1 + 2x - x^2$

In Exercises 5–8, solve the system of equations graphically.

5. $x = 4$
$2x - y = 6$

6. $y = \frac{1}{3}(1 - 2x)$
$y = \frac{1}{3}(5x - 13)$

7. $2x + 7y = 16$
$3x + 2y = 24$

8. $7x - 17y = -169$
$x^2 + y^2 = 169$

In Exercises 9–12, use substitution to solve the system. Use a graphing utility to check the solution.

9. $2x - 3y = 4$
$y = 2$

10. $y = 5 - x^2$
$y = 2(x + 1)$

11. $5x - y = 32$
$6x - 9y = 18$

12. $0.2x + 0.7y = 8$
$-x + 2y = 15$

In Exercises 13–16, use Gaussian elimination to solve the linear system.

13. $x + 10y = 18$
$5x + 2y = 42$

14. $3x + 11y = 38$
$7x - 5y = -34$

15. $a + b + c = 1$
$4a + 2b + c = 2$
$9a + 3b + c = 4$

16. $x + 4z = 17$
$-3x + 2y - z = -20$
$x - 5y + 3z = 19$

In Exercises 17 and 18, find a system of linear equations that has the unique solution. (There are many correct answers.)

17. $(10, -12)$

18. $(1, 3, -7)$

19. Twenty gallons of a 30% brine solution is obtained by mixing a 20% solution with a 50% solution. Let x represent the number of gallons of the 20% solution and let y represent the number of gallons of the 50% solution. Write a system of equations that models this problem and solve the system.

20. Find the equation of the parabola $y = ax^2 + bx + c$ that passes through the points $(1, 2)$, $(-1, -4)$, and $(2, 8)$.

12.4 Matrices and Systems of Linear Equations

Matrices ▪ Elementary Row Operations ▪
Solving a System of Linear Equations

Matrices

In this section, you will study a streamlined technique for solving systems of linear equations. This technique involves the use of a rectangular array of real numbers called a **matrix.** (The plural of matrix is *matrices*.) Here is an example of a matrix.

$$\begin{bmatrix} 3 & -2 & 4 & 1 \\ 0 & 1 & -1 & 2 \\ 2 & 0 & -3 & 0 \end{bmatrix}$$

This matrix has three rows and four columns, which means that its **order** is 3×4, which is read as "3 by 4." Each number in the matrix is an **entry** of the matrix.

EXAMPLE 1   Examples of Matrices

The following matrices have the indicated orders.

a. Order: 2×3 **b.** Order: 2×2 **c.** Order: 3×2

$$\begin{bmatrix} 1 & -2 & 4 \\ 0 & 1 & -2 \end{bmatrix} \qquad \begin{bmatrix} 0 & 0 \\ 0 & 0 \end{bmatrix} \qquad \begin{bmatrix} 1 & -3 \\ -2 & 0 \\ 4 & -2 \end{bmatrix}$$

A matrix with the same number of rows as columns is called a **square matrix.** For instance, the 2×2 matrix in part (b) is square.

A matrix derived from a system of linear equations (each written in standard form) is the **augmented matrix** of the system. Moreover, the matrix derived from the coefficients of the system (but that does not include the constant terms) is the **coefficient matrix** of the system. Here is an example.

System	*Coefficient Matrix*	*Augmented Matrix*

$$\begin{array}{r} x - 4y + 3z = 5 \\ -x + 3y - z = -3 \\ 2x \quad - 4z = 6 \end{array} \qquad \begin{bmatrix} 1 & -4 & 3 \\ -1 & 3 & -1 \\ 2 & 0 & -4 \end{bmatrix} \qquad \left[\begin{array}{ccc:c} 1 & -4 & 3 & 5 \\ -1 & 3 & -1 & -3 \\ 2 & 0 & -4 & 6 \end{array} \right]$$

Note the use of 0 for the missing y-variable in the third equation, and also note the fourth column of constant terms in the augmented matrix.

When forming either the coefficient matrix or the augmented matrix of a system, you should begin by vertically aligning the variables in the equations.

Given System	Align Variables	Form Augmented Matrix
$x + 3y = 9$	$x + 3y \quad\quad = 9$	
$-y + 4z = -2$	$-y + 4z = -2$	$\begin{bmatrix} 1 & 3 & 0 & \vdots & 9 \\ 0 & -1 & 4 & \vdots & -2 \\ 1 & 0 & -5 & \vdots & 0 \end{bmatrix}$
$x - 5z = 0$	$x \quad\quad - 5z = 0$	

EXAMPLE 2 Forming Coefficient and Augmented Matrices

Form the coefficient matrix and the augmented matrix for each system of linear equations.

a. $-x + 5y = 2$ **b.** $3x + 2y - z = 1$ **c.** $x = 3y - 1$
      $7x - 2y = -6$ $x + 2z = -3$ $2y - 5 = 9x$
                         $-2x - y = 4$

Solution

	System	Coefficient Matrix	Augmented Matrix
a.	$-x + 5y = 2$ $7x - 2y = -6$	$\begin{bmatrix} -1 & 5 \\ 7 & -2 \end{bmatrix}$	$\begin{bmatrix} -1 & 5 & \vdots & 2 \\ 7 & -2 & \vdots & -6 \end{bmatrix}$
b.	$3x + 2y - z = 1$ $x + \quad 2z = -3$ $-2x - y \quad = 4$	$\begin{bmatrix} 3 & 2 & -1 \\ 1 & 0 & 2 \\ -2 & -1 & 0 \end{bmatrix}$	$\begin{bmatrix} 3 & 2 & -1 & \vdots & 1 \\ 1 & 0 & 2 & \vdots & -3 \\ -2 & -1 & 0 & \vdots & 4 \end{bmatrix}$
c.	$x - 3y = -1$ $-9x + 2y = 5$	$\begin{bmatrix} 1 & -3 \\ -9 & 2 \end{bmatrix}$	$\begin{bmatrix} 1 & -3 & \vdots & -1 \\ -9 & 2 & \vdots & 5 \end{bmatrix}$

EXAMPLE 3 Forming Linear Systems from Their Matrices

Write systems of linear equations that are represented by the following matrices.

a. $\begin{bmatrix} 3 & -5 & \vdots & 4 \\ -1 & 2 & \vdots & 0 \end{bmatrix}$ **b.** $\begin{bmatrix} 1 & 3 & \vdots & 2 \\ 0 & 1 & \vdots & -3 \end{bmatrix}$ **c.** $\begin{bmatrix} 2 & 0 & -8 & \vdots & 1 \\ -1 & 1 & 1 & \vdots & 2 \\ 5 & -1 & 7 & \vdots & 3 \end{bmatrix}$

Solution

a. $3x - 5y = 4$ **b.** $x + 3y = 2$ **c.** $2x \quad\quad - 8z = 1$
     $-x + 2y = 0$ $y = -3$ $-x + y + z = 2$
                                      $5x - y + 7z = 3$

Elementary Row Operations

In Section 12.3, you studied three operations that can be used on a system of linear equations to produce an equivalent system: (1) interchange two rows, (2) multiply a row by a nonzero constant, and (3) add a multiple of a row to another row. In matrix terminology, these three operations correspond to **elementary row operations.**

Elementary Row Operations

Any of the following **elementary row operations** performed on an augmented matrix will produce a matrix that is row-equivalent to the original matrix. Two matrices are **row-equivalent** if one can be obtained from the other by a sequence of elementary row operations.

1. Interchange two rows.

2. Multiply a row by a nonzero constant.

3. Add a multiple of a row to another row.

EXAMPLE 4   Elementary Row Operations

a. Interchange the first and second rows.

   Original Matrix

$$\begin{bmatrix} 0 & 1 & 3 & 4 \\ -1 & 2 & 0 & 3 \\ 2 & -3 & 4 & 1 \end{bmatrix}$$

   New Row-Equivalent Matrix

$$\begin{matrix} R_2 \\ R_1 \end{matrix} \begin{bmatrix} -1 & 2 & 0 & 3 \\ 0 & 1 & 3 & 4 \\ 2 & -3 & 4 & 1 \end{bmatrix}$$

b. Multiply the first row by $\frac{1}{2}$.

   Original Matrix

$$\begin{bmatrix} 2 & -4 & 6 & -2 \\ 1 & 3 & -3 & 0 \\ 5 & -2 & 1 & 2 \end{bmatrix}$$

   New Row-Equivalent Matrix

$$\tfrac{1}{2}R_1 \rightarrow \begin{bmatrix} 1 & -2 & 3 & -1 \\ 1 & 3 & -3 & 0 \\ 5 & -2 & 1 & 2 \end{bmatrix}$$

c. Add -2 times the first row to the third row.

   Original Matrix

$$\begin{bmatrix} 1 & 2 & -4 & 3 \\ 0 & 3 & -2 & -1 \\ 2 & 1 & 5 & -2 \end{bmatrix}$$

   New Row-Equivalent Matrix

$$-2R_1 + R_3 \rightarrow \begin{bmatrix} 1 & 2 & -4 & 3 \\ 0 & 3 & -2 & -1 \\ 0 & -3 & 13 & -8 \end{bmatrix}$$

In Section 12.3, Gaussian elimination was used with back-substitution to solve a system of linear equations. Example 5 demonstrates the matrix version of Gaussian elimination. The two methods are essentially the same. The basic difference is that with matrices you do not need to keep writing the variables.

EXAMPLE 5 Solving a System of Linear Equations

Linear System   *Associated Augmented Matrix*

$$\begin{aligned} x - 2y + 2z &= 9 \\ -x + 3y &= -4 \\ 2x - 5y + z &= 10 \end{aligned} \qquad \begin{bmatrix} 1 & -2 & 2 & \vdots & 9 \\ -1 & 3 & 0 & \vdots & -4 \\ 2 & -5 & 1 & \vdots & 10 \end{bmatrix}$$

Add the first equation to the second equation.

Add the first row to the second row $(R_1 + R_2)$.

$$\begin{aligned} x - 2y + 2z &= 9 \\ y + 2z &= 5 \\ 2x - 5y + z &= 10 \end{aligned} \qquad R_1 + R_2 \rightarrow \begin{bmatrix} 1 & -2 & 2 & \vdots & 9 \\ 0 & 1 & 2 & \vdots & 5 \\ 2 & -5 & 1 & \vdots & 10 \end{bmatrix}$$

Add -2 times the first equation to the third equation.

Add -2 times the first row to the third row $(-2R_1 + R_3)$.

$$\begin{aligned} x - 2y + 2z &= 9 \\ y + 2z &= 5 \\ -y - 3z &= -8 \end{aligned} \qquad -2R_1 + R_3 \rightarrow \begin{bmatrix} 1 & -2 & 2 & \vdots & 9 \\ 0 & 1 & 2 & \vdots & 5 \\ 0 & -1 & -3 & \vdots & -8 \end{bmatrix}$$

Add the second equation to the third equation.

Add the second row to the third row $(R_2 + R_3)$.

$$\begin{aligned} x - 2y + 2z &= 9 \\ y + 2z &= 5 \\ -z &= -3 \end{aligned} \qquad R_2 + R_3 \rightarrow \begin{bmatrix} 1 & -2 & 2 & \vdots & 9 \\ 0 & 1 & 2 & \vdots & 5 \\ 0 & 0 & -1 & \vdots & -3 \end{bmatrix}$$

Multiply the third equation by -1.

Multiply the third row by -1.

$$\begin{aligned} x - 2y + 2z &= 9 \\ y + 2z &= 5 \\ z &= 3 \end{aligned} \qquad -R_3 \rightarrow \begin{bmatrix} 1 & -2 & 2 & \vdots & 9 \\ 0 & 1 & 2 & \vdots & 5 \\ 0 & 0 & 1 & \vdots & 3 \end{bmatrix}$$

At this point, you can use back-substitution to find that the solution is $x = 1$, $y = -1$, and $z = 3$.

The last matrix in Example 5 is in **row-echelon form.** The term *echelon* refers to the stair-step pattern formed by the nonzero elements of the matrix.

Solving a System of Linear Equations

> **Gaussian Elimination with Back-Substitution**
>
> To use matrices and Gaussian elimination to solve a system of linear equations, use the following steps.
>
> 1. Write the augmented matrix of the system of linear equations.
>
> 2. Use elementary row operations to rewrite the augmented matrix in row-echelon form.
>
> 3. Write the system of linear equations corresponding to the matrix in row-echelon form, and use back-substitution to find the solution.

When you perform Gaussian elimination with back-substitution, we suggest that you operate from *left to right by columns,* using elementary row operations to obtain zeros in all entries directly below the leading 1's.

EXAMPLE 6   Gaussian Elimination with Back-Substitution

Solve the system of linear equations.

$$2x - 3y = -2$$
$$x + 2y = 13$$

Solution

$$\begin{bmatrix} 2 & -3 & \vdots & -2 \\ 1 & 2 & \vdots & 13 \end{bmatrix}$$ 　　Augmented matrix for system of linear equations.

$$\begin{matrix} R_2 \\ R_1 \end{matrix} \begin{bmatrix} 1 & 2 & \vdots & 13 \\ 2 & -3 & \vdots & -2 \end{bmatrix}$$ 　　First column has leading 1 in upper left corner.

$$-2R_1 + R_2 \rightarrow \begin{bmatrix} 1 & 2 & \vdots & 13 \\ 0 & -7 & \vdots & -28 \end{bmatrix}$$ 　　First column has a zero under its leading 1.

$$-\tfrac{1}{7}R_2 \rightarrow \begin{bmatrix} 1 & 2 & \vdots & 13 \\ 0 & 1 & \vdots & 4 \end{bmatrix}$$ 　　Second column has leading 1 in second row.

The system of linear equations that corresponds to the (row-echelon) matrix is

$$x + 2y = 13$$
$$y = 4.$$

Using back-substitution, you can find that the solution of the system is $x = 5$ and $y = 4$. Check this solution in the original system of linear equations.

EXAMPLE 7   Gaussian Elimination with Back-Substitution

Solve the system of linear equations.

$$\begin{aligned} 3x + 3y \phantom{{}- 3z} &= 9 \\ 2x \phantom{{}+ 3y} - 3z &= 10 \\ 6y + 4z &= -12 \end{aligned}$$

Solution

$$\begin{bmatrix} 3 & 3 & 0 & \vdots & 9 \\ 2 & 0 & -3 & \vdots & 10 \\ 0 & 6 & 4 & \vdots & -12 \end{bmatrix}$$

Augmented matrix for system of linear equations.

$$\tfrac{1}{3}R_1 \rightarrow \begin{bmatrix} 1 & 1 & 0 & \vdots & 3 \\ 2 & 0 & -3 & \vdots & 10 \\ 0 & 6 & 4 & \vdots & -12 \end{bmatrix}$$

First column has leading 1 in upper left corner.

$$-2R_1 + R_2 \rightarrow \begin{bmatrix} 1 & 1 & 0 & \vdots & 3 \\ 0 & -2 & -3 & \vdots & 4 \\ 0 & 6 & 4 & \vdots & -12 \end{bmatrix}$$

First column has zeros under its leading 1.

$$-\tfrac{1}{2}R_2 \rightarrow \begin{bmatrix} 1 & 1 & 0 & \vdots & 3 \\ 0 & 1 & \tfrac{3}{2} & \vdots & -2 \\ 0 & 6 & 4 & \vdots & -12 \end{bmatrix}$$

Second column has leading 1 in second row.

$$-6R_2 + R_3 \rightarrow \begin{bmatrix} 1 & 1 & 0 & \vdots & 3 \\ 0 & 1 & \tfrac{3}{2} & \vdots & -2 \\ 0 & 0 & -5 & \vdots & 0 \end{bmatrix}$$

Second column has zero under its leading 1.

$$-\tfrac{1}{5}R_3 \rightarrow \begin{bmatrix} 1 & 1 & 0 & \vdots & 3 \\ 0 & 1 & \tfrac{3}{2} & \vdots & -2 \\ 0 & 0 & 1 & \vdots & 0 \end{bmatrix}$$

Third column has leading 1 in third row.

The system of linear equations that corresponds to this (row-echelon) matrix is

$$\begin{aligned} x + y \phantom{{}+ \tfrac{3}{2}z} &= 3 \\ y + \tfrac{3}{2}z &= -2 \\ z &= 0. \end{aligned}$$

Using back-substitution, you can find that the solution is

$$x = 5, \quad y = -2, \quad \text{and} \quad z = 0.$$

Check this in the original system, as follows.

$$3(5) + 3(-2) \phantom{{}- 3(0)} = 9 \qquad \text{Substitute in Equation 1.} \ \checkmark$$
$$2(5) \phantom{{}+ 3(-2)} - 3(0) = 10 \qquad \text{Substitute in Equation 2.} \ \checkmark$$
$$6(-2) + 4(0) = -12 \qquad \text{Substitute in Equation 3.} \ \checkmark$$

EXAMPLE 8   A System with No Solution

Solve the system of linear equations.

$$6x - 10y = -4$$
$$9x - 15y = 5$$

Solution

$$\begin{bmatrix} 6 & -10 & \vdots & -4 \\ 9 & -15 & \vdots & 5 \end{bmatrix}$$ Augmented matrix for system of linear equations.

$$\frac{1}{6}R_1 \rightarrow \begin{bmatrix} 1 & -\frac{5}{3} & \vdots & -\frac{2}{3} \\ 9 & -15 & \vdots & 5 \end{bmatrix}$$ First column has leading 1 in upper left corner.

$$-9R_1 + R_2 \rightarrow \begin{bmatrix} 1 & -\frac{5}{3} & \vdots & -\frac{2}{3} \\ 0 & 0 & \vdots & 11 \end{bmatrix}$$ First column has a zero under its leading 1.

The "equation" that corresponds to the second row of this matrix is $0 = 11$. Because this does not make sense, the system of equations has no solution.

EXAMPLE 9   A System with Infinitely Many Solutions

Solve the system of linear equations.

$$12x - 6y = -3$$
$$-8x + 4y = 2$$

Solution

$$\begin{bmatrix} 12 & -6 & \vdots & -3 \\ -8 & 4 & \vdots & 2 \end{bmatrix}$$ Augmented matrix for system of linear equations.

$$\frac{1}{12}R_1 \rightarrow \begin{bmatrix} 1 & -\frac{1}{2} & \vdots & -\frac{1}{4} \\ -8 & 4 & \vdots & 2 \end{bmatrix}$$ First column has leading 1 in upper left corner.

$$8R_1 + R_2 \rightarrow \begin{bmatrix} 1 & -\frac{1}{2} & \vdots & -\frac{1}{4} \\ 0 & 0 & \vdots & 0 \end{bmatrix}$$ First column has a zero under its leading 1.

Because the second row of the matrix is all zeros, you can conclude that the system of equations has an infinite number of solutions, represented by all points (x, y) on the line $x - \frac{1}{2}y = -\frac{1}{4}$. Because this line can be written as

$$x = -\frac{1}{4} + \frac{1}{2}y$$

you can write the solution set as

$$\left(-\frac{1}{4} + \frac{1}{2}a, \ a \right), \qquad \text{where } a \text{ is any real number.}$$

Technology

Graphing Utilities and Matrices

Most graphing utilities are capable of working with matrices. For example, to write the matrix shown at the left in row-echelon form, use the following steps.

1. Use the MATRIX feature to enter the matrix.

2. To use a *TI-82* to write the matrix in row-echelon form, use the elementary row operations that appear in the MATRIX MATH menu, as follows.

$$\begin{bmatrix} 1 & -2 & 2 & 9 \\ -1 & 3 & 0 & -4 \\ 2 & -5 & 1 & 10 \end{bmatrix}$$

row+([A],1,2) ENTER   Add row 1 to row 2.

STO ▷ [A] ENTER   Store the result as matrix A.

*row+(−2,[A],1,3) ENTER   Multiply row 1 by −2 and add to row 3.

STO ▷ [A] ENTER   Store the result as matrix A.

row+([A],2,3) ENTER   Add row 2 to row 3.

STO ▷ [A] ENTER   Store the result as matrix A.

*row(−1,[A],3) ENTER   Multiply row 3 by −1.

STO ▷ [A] ENTER   Store the result as matrix A.

3. The result that is displayed on the screen should be as follows.

```
[ [ 1  −2  2  9 ]
  [ 0   1  2  5 ]
  [ 0   0  1  3 ] ]
```

At this point, you can use back-substitution to solve the corresponding system of equations.

Group Activities

Exploring with Technology

Analyzing Solutions to Systems of Equations   Use a graphing utility to graph each system of equations given in Example 6, Example 8, and Example 9. Verify the solution given in each example and explain how you may reach the same conclusion by using the graph. Summarize how you may conclude that a system has a unique solution, no solution, or infinitely many solutions when you use Gaussian elimination.

12.4 Exercises

Discussing the Concepts

1. Describe the three elementary row operations that can be performed on an augmented matrix.

2. What is the relationship between the three elementary row operations on an augmented matrix and the row operations on a system of linear equations?

3. What is meant by saying that two augmented matrices are *row-equivalent*?

4. Give an example of a matrix in *row-echelon form*.

5. Describe the row-echelon form of an augmented matrix that corresponds to a system of linear equations that is inconsistent.

6. Describe the row-echelon form of an augmented matrix that corresponds to a system of linear equations that has an infinite number of solutions.

Problem Solving

In Exercises 7–10, state the order of the matrix.

7. $\begin{bmatrix} 3 & -2 \\ -4 & 0 \\ 2 & -7 \end{bmatrix}$

8. $\begin{bmatrix} 4 & 0 & -5 \\ -1 & 8 & 9 \\ 0 & -3 & 4 \end{bmatrix}$

9. $\begin{bmatrix} 5 & -8 \\ 7 & 15 \end{bmatrix}$

10. $\begin{bmatrix} -2 & 5 \\ 0 & -1 \end{bmatrix}$

In Exercises 11–14, write the augmented matrix for the system of linear equations.

11. $4x - 5y = -2$
    $-x + 8y = 10$

12. $8x + 3y = 25$
    $3x - 9y = 12$

13. $x + 10y - 3z = 2$
    $5x - 3y + 4z = 0$
    $2x + 4y = 6$

14. $9x - 3y + z = 13$
    $12x - 8z = 5$

In Exercises 15–18, write the system of linear equations represented by the augmented matrix. (Use variables x, y, z, and w.)

15. $\left[\begin{array}{cc:c} 4 & 3 & 8 \\ 1 & -2 & 3 \end{array}\right]$

16. $\left[\begin{array}{cc:c} 9 & -4 & 0 \\ 6 & 1 & -4 \end{array}\right]$

17. $\left[\begin{array}{ccc:c} 1 & 0 & 2 & -10 \\ 0 & 3 & -1 & 5 \\ 4 & 2 & 0 & 3 \end{array}\right]$

18. $\left[\begin{array}{cccc:c} 5 & 8 & 2 & 0 & -1 \\ -2 & 15 & 5 & 1 & 9 \\ 1 & 6 & -7 & 0 & -3 \end{array}\right]$

In Exercises 19 and 20, fill in the blank(s) by using elementary row operations to form a row-equivalent matrix.

19. $\begin{bmatrix} 1 & 4 & 3 \\ 2 & 10 & 5 \end{bmatrix}$

    $\begin{bmatrix} 1 & 4 & 3 \\ 0 & \blacksquare & -1 \end{bmatrix}$

20. $\begin{bmatrix} 2 & 4 & 8 & 6 \\ 1 & -1 & -3 & 2 \\ 2 & 6 & 4 & 9 \end{bmatrix}$

    $\begin{bmatrix} 1 & \blacksquare & \blacksquare & \blacksquare \\ 1 & -1 & -3 & 2 \\ 2 & 6 & 4 & 9 \end{bmatrix}$

    $\begin{bmatrix} 1 & 2 & 4 & 3 \\ 0 & \blacksquare & -7 & -1 \\ 0 & 2 & \blacksquare & \blacksquare \end{bmatrix}$

In Exercises 21–24, convert the matrix to row-echelon form. (*Note:* There is more than one correct answer.)

21. $\begin{bmatrix} 1 & 2 & 3 \\ 2 & -1 & -4 \end{bmatrix}$

22. $\begin{bmatrix} 10 & 30 & 60 & 10 \\ -4 & -9 & 3 & 5 \end{bmatrix}$

23. $\begin{bmatrix} 1 & 1 & -1 & 3 \\ 2 & 1 & 2 & 5 \\ 3 & 2 & 1 & 8 \end{bmatrix}$

24. $\begin{bmatrix} 1 & -3 & -2 & -8 \\ 1 & 3 & -2 & 17 \\ 1 & 2 & -2 & -5 \end{bmatrix}$

In Exercises 25 and 26, write the system of linear equations represented by the augmented matrix. Then use back-substitution to find the solution. (Use variables x, y, and z.)

25. $\begin{bmatrix} 1 & 5 & \vdots & 3 \\ 0 & 1 & \vdots & -2 \end{bmatrix}$

26. $\begin{bmatrix} 1 & 5 & -3 & \vdots & 0 \\ 0 & 1 & 0 & \vdots & 6 \\ 0 & 0 & 1 & \vdots & -5 \end{bmatrix}$

In Exercises 27–34, use matrices to solve the system of linear equations.

27. $6x - 4y = 2$
$\quad\;\; 5x + 2y = 7$

28. $2x - y = -0.1$
$\quad\;\; 3x + 2y = 1.6$

29. $x - 2y - z = 6$
$\qquad\quad y + 4z = 5$
$\quad 4x + 2y + 3z = 8$

30. $x \qquad\quad - 3z = -2$
$\quad 3x + y - 2z = 5$
$\quad 2x + 2y + z = 4$

31. $x + y - 5z = 3$
$\quad x \qquad - 2z = 1$
$\quad 2x - y - z = 0$

32. $2x \qquad + 3z = 3$
$\quad 4x - 3y + 7z = 5$
$\quad 8x - 9y + 15z = 9$

33. $2x \qquad + 4z = 1$
$\quad x + y + 3z = 0$
$\quad x + 3y + 5z = 0$

34. $3x + y - 2z = 2$
$\quad 6x + 2y - 4z = 1$
$\quad -3x - y + 2z = 1$

35. *Simple Interest* A corporation borrowed $1,500,000 to expand its product line. Some of the money was borrowed at 8%, some at 9%, and the remainder at 12%. The total annual interest payment to the lenders was $133,000. If the amount borrowed at 8% was four times the amount borrowed at 12%, how much was borrowed at each rate?

36. *Nut Mixture* A grocer wishes to mix three kinds of nuts priced at $3.50, $4.50, and $6.00 per pound to obtain 50 pounds of a mixture priced at $4.95 per pound. How many pounds of each variety should the grocer use if half the mixture is to be composed of the two cheapest varieties?

Curve-Fitting In Exercises 37 and 38, find the equation of the parabola $y = ax^2 + bx + c$ that passes through the given points.

37.

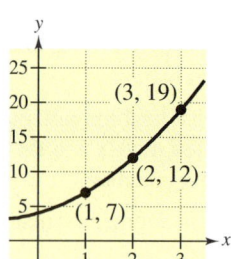

38.

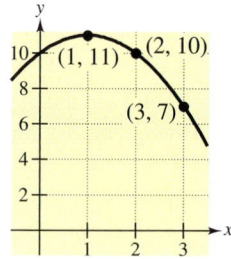

39. *Mathematical Modeling* A videotape of the path of a ball thrown by a baseball player was analyzed on a television set with a grid on the screen (see figure). The tape was paused three times and the position of the coordinates of the ball were measured each time. The coordinates were approximately $(0, 6)$, $(25, 18.5)$, and $(50, 26)$. (The x-coordinate measures the horizontal distance from the player and the y-coordinate is the height of the ball above the ground. Both are measured in feet.)

(a) Find the equation of the parabola $y = ax^2 + bx + c$ that passes through the three points.

(b) Use a graphing utility to graph the parabola. Approximate the maximum height of the ball and the point at which the ball struck the ground.

(c) Find algebraically the maximum height of the ball and the point at which it struck the ground.

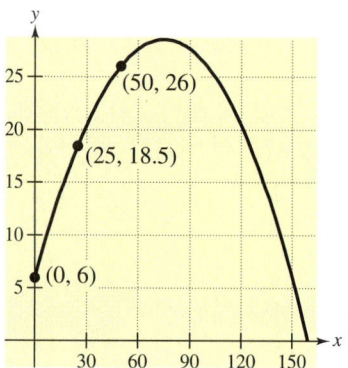

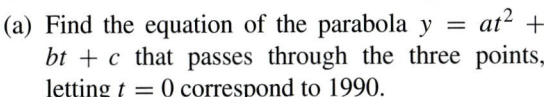

40. *Reading a Graph* The bar graph at the right gives the gross private savings y (in billions of dollars) in the years 1990 through 1992 in the United States. (Source: U. S. Bureau of Economic Analysis)

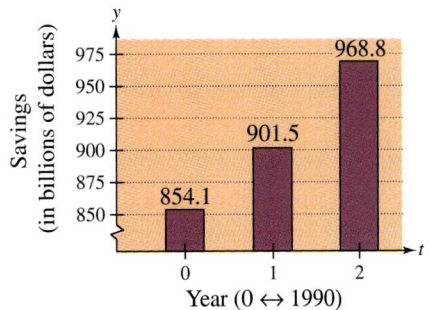

(a) Find the equation of the parabola $y = at^2 + bt + c$ that passes through the three points, letting $t = 0$ correspond to 1990.

(b) Use a graphing utility to graph the model found in part (a).

(c) Use the model of part (a) to predict gross private savings in 1999 if the trend continues.

Reviewing the Major Concepts

In Exercises 41–44, evaluate the function.

41. $f(x) = \frac{1}{3}x^2$

(a) $f(6)$

(b) $f\left(\frac{3}{4}\right)$

42. $f(x) = 3 - 2x$

(a) $f(5)$

(b) $f(x+3) - f(3)$

43. $g(x) = \dfrac{x}{x+10}$

(a) $g(5)$

(b) $g(c-6)$

44. $h(x) = \sqrt{x-4}$

(a) $h(16)$

(b) $h(t+3)$

45. *Cost* The inventor of a new game believes that the variable cost for producing the game is $5.75 per unit and the fixed costs are $12,000. Let x represent the number of games produced. Express the total cost C as a function of x.

46. *Geometry* The length of a rectangle is $1\frac{1}{2}$ times its width. Express the length L of the diagonal of the rectangle as a function of its width w.

Additional Problem Solving

In Exercises 47–50, fill in the blanks by using elementary row operations to form a row-equivalent matrix.

47.
$$\begin{bmatrix} 9 & -18 & 6 \\ 2 & 8 & 15 \end{bmatrix}$$
$$\begin{bmatrix} 1 & & \\ 2 & 8 & 15 \end{bmatrix}$$

48.
$$\begin{bmatrix} 3 & 6 & 8 \\ 4 & -3 & 6 \end{bmatrix}$$
$$\begin{bmatrix} 3 & 6 & 8 \\ 1 & -9 & \end{bmatrix}$$

49.
$$\begin{bmatrix} 1 & 1 & 4 & -1 \\ 3 & 8 & 10 & 3 \\ -2 & 1 & 12 & 6 \end{bmatrix}$$
$$\begin{bmatrix} 1 & 1 & 4 & -1 \\ 0 & 5 & & \\ 0 & 3 & & \end{bmatrix}$$
$$\begin{bmatrix} 1 & 1 & 4 & -1 \\ 0 & 1 & -\frac{2}{5} & \frac{6}{5} \\ 0 & 3 & & \end{bmatrix}$$

50.
$$\begin{bmatrix} 2 & 3 & -5 & 6 \\ 5 & -7 & 12 & 9 \\ -4 & 6 & 9 & 5 \end{bmatrix}$$
$$\begin{bmatrix} 2 & 3 & -5 & 6 \\ 5 & -7 & 12 & 9 \\ 0 & 12 & & \end{bmatrix}$$

In Exercises 51–56, convert the matrix to row-echelon form. (*Note:* There is more than one correct answer.)

51.
$$\begin{bmatrix} 4 & 6 & 1 \\ -2 & 2 & 5 \end{bmatrix}$$

52.
$$\begin{bmatrix} 3 & 2 & 6 \\ 2 & 3 & -3 \end{bmatrix}$$

53.
$$\begin{bmatrix} 1 & 1 & 0 & 5 \\ -2 & -1 & 2 & -10 \\ 3 & 6 & 7 & 14 \end{bmatrix}$$

54.
$$\begin{bmatrix} 1 & 2 & -1 & 3 \\ 3 & 7 & -5 & 14 \\ -2 & -1 & -3 & 8 \end{bmatrix}$$

55. $\begin{bmatrix} 1 & -1 & -1 & 1 \\ 4 & -4 & 1 & 8 \\ -6 & 8 & 18 & 0 \end{bmatrix}$

56. $\begin{bmatrix} 1 & -3 & 0 & -7 \\ -3 & 10 & 1 & 23 \\ 4 & -10 & 2 & -24 \end{bmatrix}$

In Exercises 57–60, write the system of linear equations represented by the augmented matrix. Then use back-substitution to find the solution. (Use variables x, y, and z.)

57. $\begin{bmatrix} 1 & -2 & \vdots & 4 \\ 0 & 1 & \vdots & -3 \end{bmatrix}$

58. $\begin{bmatrix} 1 & 5 & \vdots & 0 \\ 0 & 1 & \vdots & -1 \end{bmatrix}$

59. $\begin{bmatrix} 1 & -1 & 2 & \vdots & 4 \\ 0 & 1 & -1 & \vdots & 2 \\ 0 & 0 & 1 & \vdots & -2 \end{bmatrix}$

60. $\begin{bmatrix} 1 & 2 & -2 & \vdots & -1 \\ 0 & 1 & 1 & \vdots & 9 \\ 0 & 0 & 1 & \vdots & -3 \end{bmatrix}$

In Exercises 61–74, use matrices to solve the system of linear equations.

61. $x + 2y = 7$
$3x + y = 8$

62. $2x + 6y = 16$
$2x + 3y = 7$

63. $-x + 2y = 1.5$
$2x - 4y = 3$

64. $x - 3y = 5$
$-2x + 6y = -10$

65. $x - 3y + 2z = 8$
$2y - z = -4$
$x \quad\quad + z = 3$

66. $2y + z = 3$
$-4y - 2z = 0$
$x + y + z = 2$

67. $2x - y + 3z = 24$
$2y - z = 14$
$7x - 5y = 6$

68. $2x + 4y = 10$
$2x + 2y + 3z = 3$
$-3x + y + 2z = -3$

69. $x + 3y = 2$
$2x + 6y = 4$
$2x + 5y + 4z = 3$

70. $2x + 2y + z = 8$
$2x + 3y + z = 7$
$6x + 8y + 3z = 22$

71. $2x + 4y + 5z = 5$
$x + 3y + 3z = 2$
$2x + 4y + 4z = 2$

72. $-2x - 2y - 15z = 0$
$x + 2y + 2z = 18$
$3x + 3y + 22z = 2$

73. $3x + 3y + z = 4$
$2x + 6y + z = 5$
$-x - 3y + 2z = -5$

74. $2x + y - 2z = 4$
$3x - 2y + 4z = 6$
$-4x + y + 6z = 12$

Investment Portfolio In Exercises 75 and 76, consider an investor with a portfolio totaling \$500,000 that is to be allocated among the following types of investments: certificates of deposit, municipal bonds, blue-chip stocks, and growth or speculative stocks. How much should be allocated to each type of investment?

75. The certificates of deposit pay 10% annually, and the municipal bonds pay 8% annually. Over a 5-year period, the investor expects the blue-chip stocks to return 12% annually, and expects the growth stocks to return 13% annually. The investor wants a combined annual return of 10% and also wants to have only one-fourth of the portfolio invested in stocks.

76. The certificates of deposit pay 9% annually, and the municipal bonds pay 5% annually. Over a 5-year period, the investor expects the blue-chip stocks to return 12% annually, and expects the growth stocks to return 14% annually. The investor wants a combined annual return of 10% and also wants to have only one-fourth of the portfolio invested in stocks.

77. *Curve-Fitting* Find the equation of the parabola

$$y = ax^2 + bx + c$$

that passes through the points $(1, 1)$, $(-3, 17)$, and $\left(2, -\frac{1}{2}\right)$.

78. *Curve-Fitting* Find the equation of the circle

$$x^2 + y^2 + Dx + Ey + F = 0$$

that passes through the points $(1, 1)$, $(3, 3)$, and $(4, 2)$.

79. The sum of three positive numbers is 33. The second number is 3 greater than the first, and the third is four times the first. Find the three numbers.

80. *Investments*   An inheritance of $16,000 was divided among three investments yielding a total of $990 in simple interest per year. The interest rates for the three investments were 5%, 6%, and 7%. Find the amount placed in each investment if the 5% and 6% investments were $3000 and $2000 less than the 7% investment, respectively.

81. *Rewriting a Fraction*   The fraction

$$\frac{2x^2 - 9x}{(x - 2)^3}$$

can be written as a sum of three fractions as follows.

$$\frac{2x^2 - 9x}{(x - 2)^3} = \frac{A}{x - 2} + \frac{B}{(x - 2)^2} + \frac{C}{(x - 2)^3}$$

The numbers A, B, and C are the solutions of the system

$$
\begin{aligned}
4A - 2B + C &= 0 \\
-4A + B \phantom{{}+ C} &= -9 \\
A \phantom{{}- 2B + C} &= 2.
\end{aligned}
$$

Solve the system and verify that the sum of the three resulting fractions is the original fraction.

82. *Rewriting a Fraction*   The fraction

$$\frac{x + 1}{x(x^2 + 1)}$$

can be written as a sum of two fractions as follows.

$$\frac{x + 1}{x(x^2 + 1)} = \frac{A}{x} + \frac{Bx + C}{x^2 + 1}.$$

The numbers A, B, and C are the solutions of the system

$$
\begin{aligned}
2A + B + C &= 2 \\
2A + B - C &= 0 \\
5A + 4B + 2C &= 3.
\end{aligned}
$$

Solve the system and verify that the sum of the two resulting fractions is the original fraction.

C A R E E R   I N T E R V I E W

Richard S. Schroeder

Electrical Engineer

Lorain Products

Lorain, OH 44502

Lorain Products designs and manufactures power systems for the telecommunications industry. These power supplies convert the power delivered by an electric utility into a form that a telecommunication system can use. I work in Lorain Products' Custom Power unit, and it is my responsibility to design power supplies that meet customers' specifications, can be manufactured easily, and meet the quoted cost. A typical power supply design takes about one year to go from an idea to production, and a typical supply will have a production run of five years.

Because power supplies are vital to business communications throughout the world, it is my job as a designer to ensure that the power systems that Lorain Products supplies are reliable. To ensure reliability, I often use a circuit analysis technique that results in a system of linear equations. In a recent design I needed to find the currents in a certain circuit to check that none of the circuit's components would operate beyond their limits. My analysis of this circuit resulted in the following system of two equations in two unknowns.

$$
\begin{aligned}
5.99 I_1 + I_2 &= 2.5 \\
-I_1 + 101 I_2 &= 14.3
\end{aligned}
$$

This system can easily be solved to show that $I_1 = 0.393$ amps and $I_2 = 0.146$ amps. Knowing the currents I_1 and I_2 and the current ratings for each part of the circuit, I can tell if the circuit will last a long time. Systems of equations are quite useful in my line of work.

12.5 | Linear Systems and Determinants

The Determinant of a Matrix ▪ Cramer's Rule ▪
Applications of Determinants

The Determinant of a Matrix

Associated with each square matrix is a real number called its **determinant.** The use of determinants arose from special number patterns that occur during the solution of systems of linear equations. For instance, the system

$$a_1x + b_1y = c_1$$
$$a_2x + b_2y = c_2$$

has a solution given by

$$x = \frac{c_1b_2 - c_2b_1}{a_1b_2 - a_2b_1} \qquad \text{and} \qquad y = \frac{a_1c_2 - a_2c_1}{a_1b_2 - a_2b_1}$$

provided that $a_1b_2 - a_2b_1 \neq 0$. Note that the denominator of each fraction is the same. We call this denominator the **determinant** of the coefficient matrix of the system.

Arthur Cayley (1821–1895) is credited with creating the theory of matrices. Determinants had been studied as rectangular arrays of numbers since the middle of the 18th century. Thus, the use and basic properties of matrices were well established when Cayley first published articles introducing matrices as distinct entities.

Coefficient Matrix   *Determinant*

$$A = \begin{bmatrix} a_1 & b_1 \\ a_2 & b_2 \end{bmatrix} \qquad \det(A) = a_1b_2 - a_2b_1$$

The determinant of the matrix A can also be denoted by vertical bars on both sides of the matrix, as indicated in the following definition.

Definition of the Determinant of a 2 x 2 Matrix

$$\det(A) = \begin{vmatrix} a_1 & b_1 \\ a_2 & b_2 \end{vmatrix} = a_1b_2 - a_2b_1$$

A convenient method for remembering the formula for the determinant of a 2×2 matrix is shown in the following diagram.

$$\det(A) = \begin{vmatrix} a_1 & b_1 \\ a_2 & b_2 \end{vmatrix} = a_1b_2 - a_2b_1$$

Note that the determinant is given by the difference of the products of the two diagonals of the matrix.

EXAMPLE 1   The Determinant of a 2 x 2 Matrix

Find the determinant of each matrix.

a. $A = \begin{bmatrix} 2 & -3 \\ 1 & 4 \end{bmatrix}$ **b.** $B = \begin{bmatrix} -1 & 2 \\ 2 & -4 \end{bmatrix}$ **c.** $C = \begin{bmatrix} 1 & 3 \\ 2 & 5 \end{bmatrix}$

Solution

NOTE Notice in Example 1 that the determinant of a matrix can be positive, zero, or negative.

a. $\det(A) = \begin{vmatrix} 2 & -3 \\ 1 & 4 \end{vmatrix} = 2(4) - 1(-3) = 8 + 3 = 11$

b. $\det(B) = \begin{vmatrix} -1 & 2 \\ 2 & -4 \end{vmatrix} = (-1)(-4) - 2(2) = 4 - 4 = 0$

c. $\det(C) = \begin{vmatrix} 1 & 3 \\ 2 & 5 \end{vmatrix} = 1(5) - 2(3) = 5 - 6 = -1$

One way to evaluate the determinant of a 3×3 matrix, called **expanding by minors,** allows you to write the determinant of a 3×3 matrix in terms of three 2×2 determinants. The **minor** of an entry in a 3×3 matrix is the determinant of the 2×2 matrix that remains after deletion of the row and column in which the entry occurs. Here are two examples.

Given Determinant	*Entry*	*Minor of Entry*	*Value of Minor*
$\begin{vmatrix} 1 & -1 & 3 \\ 0 & 2 & 5 \\ -2 & 4 & -7 \end{vmatrix}$	1	$\begin{vmatrix} 2 & 5 \\ 4 & -7 \end{vmatrix}$	$2(-7) - 4(5) = -34$
$\begin{vmatrix} 1 & -1 & 3 \\ 0 & 2 & 5 \\ -2 & 4 & -7 \end{vmatrix}$	-1	$\begin{vmatrix} 0 & 5 \\ -2 & -7 \end{vmatrix}$	$0(-7) - (-2)(5) = 10$

Technology

A graphing utility can be used to evaluate the determinant of a square matrix. To find the determinant of

$\begin{bmatrix} 2 & -3 \\ 1 & 4 \end{bmatrix}$

first use the MATRIX feature to enter the matrix, and then evaluate the determinant, as follows.

det [A] ENTER

Check the result against Example 1(a). Evaluate the determinant of the 3×3 matrix at the right using a graphing utility. Then finish the evaluation of the determinant by expanding by minors to check the result.

Expanding by Minors

$$\det(A) = \begin{vmatrix} a_1 & b_1 & c_1 \\ a_2 & b_2 & c_2 \\ a_3 & b_3 & c_3 \end{vmatrix}$$

$$= a_1(\text{minor of } a_1) - b_1(\text{minor of } b_1) + c_1(\text{minor of } c_1)$$

$$= a_1 \begin{vmatrix} b_2 & c_2 \\ b_3 & c_3 \end{vmatrix} - b_1 \begin{vmatrix} a_2 & c_2 \\ a_3 & c_3 \end{vmatrix} + c_1 \begin{vmatrix} a_2 & b_2 \\ a_3 & b_3 \end{vmatrix}$$

This pattern is called **expanding by minors** along the first row. A similar pattern can be used to expand by minors along any row or column.

$$\begin{bmatrix} + & - & + \\ - & + & - \\ + & - & + \end{bmatrix}$$

FIGURE 12.9 Sign Pattern for 3 × 3 Matrix

The *signs* of the terms used in expanding by minors follows the alternating pattern shown in Figure 12.9. For instance, the signs used to expand by minors along the second row are $-, +, -$, as follows.

$$\det(A) = \begin{vmatrix} a_1 & b_1 & c_1 \\ a_2 & b_2 & c_2 \\ a_3 & b_3 & c_3 \end{vmatrix}$$

$$= -a_2(\text{minor of } a_2) + b_2(\text{minor of } b_2) - c_2(\text{minor of } c_2)$$

EXAMPLE 2   Finding the Determinant of a 3 x 3 Matrix

Find the determinant of $A = \begin{bmatrix} -1 & 1 & 2 \\ 0 & 2 & 3 \\ 3 & 4 & 2 \end{bmatrix}$.

Solution

By expanding by minors along the *first column*, you obtain the following.

$$\det(A) = \begin{vmatrix} -1 & 1 & 2 \\ 0 & 2 & 3 \\ 3 & 4 & 2 \end{vmatrix}$$

$$= (-1)\begin{vmatrix} 2 & 3 \\ 4 & 2 \end{vmatrix} - (0)\begin{vmatrix} 1 & 2 \\ 4 & 2 \end{vmatrix} + (3)\begin{vmatrix} 1 & 2 \\ 2 & 3 \end{vmatrix}$$

$$= (-1)(4 - 12) - (0)(2 - 8) + (3)(3 - 4)$$

$$= 8 - 0 - 3$$

$$= 5$$

STUDY TIP

Note in the expansion in Example 2 that a zero entry will always yield a zero term when expanding by minors. Thus, when you are evaluating the determinant of a matrix, you should choose to expand along the row or column that has the most zero entries.

EXAMPLE 3   Finding the Determinant of a 3 x 3 Matrix

Find the determinant of $A = \begin{bmatrix} 1 & 2 & 1 \\ 3 & 0 & 2 \\ 4 & 0 & -1 \end{bmatrix}$.

Solution

By expanding by minors along the *second column*, you obtain the following.

$$\det(A) = \begin{vmatrix} 1 & 2 & 1 \\ 3 & 0 & 2 \\ 4 & 0 & -1 \end{vmatrix}$$

$$= -(2)\begin{vmatrix} 3 & 2 \\ 4 & -1 \end{vmatrix} + (0)\begin{vmatrix} 1 & 1 \\ 4 & -1 \end{vmatrix} - (0)\begin{vmatrix} 1 & 1 \\ 3 & 2 \end{vmatrix}$$

$$= -(2)(-3 - 8) + 0 - 0$$

$$= 22$$

Cramer's Rule

So far in this chapter, you have studied three methods for solving a system of linear equations: substitution, elimination (with equations), and elimination (with matrices). We now look at one more method, called **Cramer's Rule,** which is named after Gabriel Cramer (1704–1752). This rule uses determinants to write the solution of a system of linear equations.

Cramer's Rule

1. For the system of linear equations

$$a_1 x + b_1 y = c_1$$
$$a_2 x + b_2 y = c_2$$

the solution is given by

$$x = \frac{D_x}{D} = \frac{\begin{vmatrix} c_1 & b_1 \\ c_2 & b_2 \end{vmatrix}}{\begin{vmatrix} a_1 & b_1 \\ a_2 & b_2 \end{vmatrix}}, \qquad y = \frac{D_y}{D} = \frac{\begin{vmatrix} a_1 & c_1 \\ a_2 & c_2 \end{vmatrix}}{\begin{vmatrix} a_1 & b_1 \\ a_2 & b_2 \end{vmatrix}}$$

provided that $D \neq 0$.

2. For the system of linear equations

$$a_1 x + b_1 y + c_1 z = d_1$$
$$a_2 x + b_2 y + c_2 z = d_2$$
$$a_3 x + b_3 y + c_3 z = d_3$$

the solution is given by

$$x = \frac{D_x}{D} = \frac{\begin{vmatrix} d_1 & b_1 & c_1 \\ d_2 & b_2 & c_2 \\ d_3 & b_3 & c_3 \end{vmatrix}}{\begin{vmatrix} a_1 & b_1 & c_1 \\ a_2 & b_2 & c_2 \\ a_3 & b_3 & c_3 \end{vmatrix}}, \qquad y = \frac{D_y}{D} = \frac{\begin{vmatrix} a_1 & d_1 & c_1 \\ a_2 & d_2 & c_2 \\ a_3 & d_3 & c_3 \end{vmatrix}}{\begin{vmatrix} a_1 & b_1 & c_1 \\ a_2 & b_2 & c_2 \\ a_3 & b_3 & c_3 \end{vmatrix}},$$

$$z = \frac{D_z}{D} = \frac{\begin{vmatrix} a_1 & b_1 & d_1 \\ a_2 & b_2 & d_2 \\ a_3 & b_3 & d_3 \end{vmatrix}}{\begin{vmatrix} a_1 & b_1 & c_1 \\ a_2 & b_2 & c_2 \\ a_3 & b_3 & c_3 \end{vmatrix}}$$

provided that $D \neq 0$.

EXAMPLE 4 Using Cramer's Rule for a 2 x 2 System

Use Cramer's Rule to solve the system of linear equations.

$$4x - 2y = 10$$
$$3x - 5y = 11$$

Solution

Begin by finding the determinant of the coefficient matrix, $D = -14$.

$$x = \frac{D_x}{D} = \frac{\begin{vmatrix} 10 & -2 \\ 11 & -5 \end{vmatrix}}{-14} = \frac{(-50) - (-22)}{-14} = \frac{-28}{-14} = 2$$

$$y = \frac{D_y}{D} = \frac{\begin{vmatrix} 4 & 10 \\ 3 & 11 \end{vmatrix}}{-14} = \frac{44 - 30}{-14} = \frac{14}{-14} = -1$$

The solution is $(2, -1)$. Check this in the original system of equations.

EXAMPLE 5 Using Cramer's Rule for a 3 x 3 System

Use Cramer's Rule to solve the system of linear equations.

$$-x + 2y - 3z = 1$$
$$2x + \quad\quad z = 0$$
$$3x - 4y + 4z = 2$$

Solution

The determinant of the coefficient matrix is $D = 10$.

NOTE When using Cramer's Rule, remember that the method *does not* apply if the determinant of the coefficient matrix is zero. For instance, the system

$$6x - 10y = -4$$
$$9x - 15y = \quad 5$$

has no solution (see Example 8 in Section 8.4), and the determinant of the coefficient matrix of this system is zero.

$$x = \frac{D_x}{D} = \frac{\begin{vmatrix} 1 & 2 & -3 \\ 0 & 0 & 1 \\ 2 & -4 & 4 \end{vmatrix}}{10} = \frac{8}{10} = \frac{4}{5}$$

$$y = \frac{D_y}{D} = \frac{\begin{vmatrix} -1 & 1 & -3 \\ 2 & 0 & 1 \\ 3 & 2 & 4 \end{vmatrix}}{10} = \frac{-15}{10} = -\frac{3}{2}$$

$$z = \frac{D_z}{D} = \frac{\begin{vmatrix} -1 & 2 & 1 \\ 2 & 0 & 0 \\ 3 & -4 & 2 \end{vmatrix}}{10} = \frac{-16}{10} = -\frac{8}{5}$$

The solution is $\left(\frac{4}{5}, -\frac{3}{2}, -\frac{8}{5}\right)$. Check this in the original system of equations.

Applications of Determinants

In addition to Cramer's Rule, determinants have many other practical applications. For instance, you can use a determinant to find the area of a triangle whose vertices are given by three points on a rectangular coordinate system.

Area of a Triangle

The area of a triangle with vertices (x_1, y_1), (x_2, y_2), and (x_3, y_3) is

$$\text{Area} = \pm \frac{1}{2} \begin{vmatrix} x_1 & y_1 & 1 \\ x_2 & y_2 & 1 \\ x_3 & y_3 & 1 \end{vmatrix}$$

where the symbol $(\pm)$ indicates that the appropriate sign should be chosen to yield a positive area.

EXAMPLE 6   *Finding the Area of a Triangle*

Find the area of the triangle whose vertices are $(2, 0)$, $(1, 3)$, and $(3, 2)$, as shown in Figure 12.10.

Solution

Choose $(x_1, y_1) = (2, 0)$, $(x_2, y_2) = (1, 3)$, and $(x_3, y_3) = (3, 2)$. To find the area of the triangle, evaluate the determinant

$$\begin{vmatrix} x_1 & y_1 & 1 \\ x_2 & y_2 & 1 \\ x_3 & y_3 & 1 \end{vmatrix} = \begin{vmatrix} 2 & 0 & 1 \\ 1 & 3 & 1 \\ 3 & 2 & 1 \end{vmatrix}$$

$$= 2 \begin{vmatrix} 3 & 1 \\ 2 & 1 \end{vmatrix} - 0 \begin{vmatrix} 1 & 1 \\ 3 & 1 \end{vmatrix} + 1 \begin{vmatrix} 1 & 3 \\ 3 & 2 \end{vmatrix}$$

$$= 2(1) - 0 + 1(-7)$$

$$= -5.$$

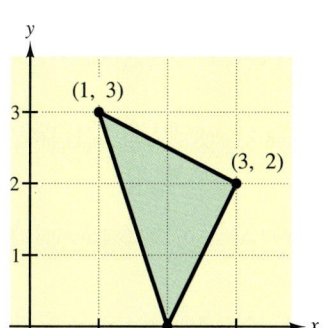

FIGURE 12.10

Using this value, you can conclude that the area of the triangle is

$$\text{Area} = -\frac{1}{2} \begin{vmatrix} 2 & 0 & 1 \\ 1 & 3 & 1 \\ 3 & 2 & 1 \end{vmatrix} = -\frac{1}{2}(-5) = \frac{5}{2}.$$

NOTE   To see the benefit of the "determinant formula," try finding the area of the triangle in Example 6 using the standard formula: Area $= \frac{1}{2}$(base)(height).

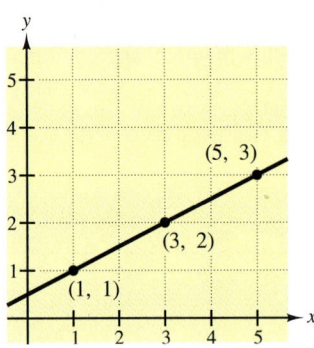

FIGURE 12.11

Suppose the three points in Example 6 had been on the same line. What would have happened had we applied the area formula to three such points? The answer is that the determinant would have been zero. Consider for instance, the three collinear points $(1, 1)$, $(3, 2)$, and $(5, 3)$, as shown in Figure 12.11. The area of the "triangle" that has these three points as vertices is

$$\frac{1}{2}\begin{vmatrix} 1 & 1 & 1 \\ 3 & 2 & 1 \\ 5 & 3 & 1 \end{vmatrix} = \frac{1}{2}\left(1\begin{vmatrix} 2 & 1 \\ 3 & 1 \end{vmatrix} - 1\begin{vmatrix} 3 & 1 \\ 5 & 1 \end{vmatrix} + 1\begin{vmatrix} 3 & 2 \\ 5 & 3 \end{vmatrix}\right)$$

$$= \frac{1}{2}[-1 - (-2) + (-1)]$$

$$= 0.$$

This result is generalized as follows.

Test for Collinear Points

Three points (x_1, y_1), (x_2, y_2), and (x_3, y_3) are collinear (lie on the same line) if and only if

$$\begin{vmatrix} x_1 & y_1 & 1 \\ x_2 & y_2 & 1 \\ x_3 & y_3 & 1 \end{vmatrix} = 0.$$

EXAMPLE 7   Testing for Collinear Points

Determine whether the points $(-2, -2)$, $(1, 1)$, and $(7, 5)$ lie on the same line. (See Figure 12.12.)

Solution

Letting $(x_1, y_1) = (-2, -2)$, $(x_2, y_2) = (1, 1)$, and $(x_3, y_3) = (7, 5)$, you have

$$\begin{vmatrix} x_1 & y_1 & 1 \\ x_2 & y_2 & 1 \\ x_3 & y_3 & 1 \end{vmatrix} = \begin{vmatrix} -2 & -2 & 1 \\ 1 & 1 & 1 \\ 7 & 5 & 1 \end{vmatrix}$$

$$= -2\begin{vmatrix} 1 & 1 \\ 5 & 1 \end{vmatrix} - (-2)\begin{vmatrix} 1 & 1 \\ 7 & 1 \end{vmatrix} + 1\begin{vmatrix} 1 & 1 \\ 7 & 5 \end{vmatrix}$$

$$= -2(-4) - (-2)(-6) + 1(-2)$$

$$= -6.$$

Because the value of this determinant *is not* zero, you can conclude that the three points *do not* lie on the same line.

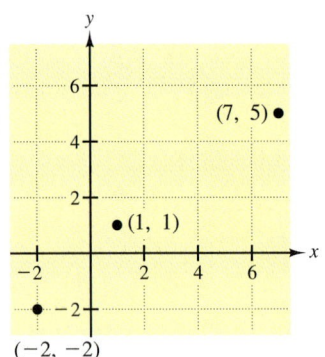

FIGURE 12.12

Two-Point Form of the Equation of a Line

An equation of the line passing through the distinct points (x_1, y_1) and (x_2, y_2) is given by

$$\begin{vmatrix} x & y & 1 \\ x_1 & y_1 & 1 \\ x_2 & y_2 & 1 \end{vmatrix} = 0.$$

EXAMPLE 8 Finding an Equation of a Line

Find an equation of the line passing through $(-2, 1)$ and $(3, -2)$.

Solution

$$\begin{vmatrix} x & y & 1 \\ -2 & 1 & 1 \\ 3 & -2 & 1 \end{vmatrix} = 0$$

$$x\begin{vmatrix} 1 & 1 \\ -2 & 1 \end{vmatrix} - y\begin{vmatrix} -2 & 1 \\ 3 & 1 \end{vmatrix} + 1\begin{vmatrix} -2 & 1 \\ 3 & -2 \end{vmatrix} = 0$$

$$3x + 5y + 1 = 0$$

Therefore, an equation of the line is $3x + 5y + 1 = 0$.

Group Activities   Extending the Concept

Determinant of a 3 x 3 Matrix   There is an alternative method for evaluating the determinant of a 3×3 matrix A. (This method works *only* for 3×3 matrices.) To apply this method, copy the first and second columns of A to form fourth and fifth columns. The determinant of A is then obtained by adding the products of three diagonals and subtracting the products of three diagonals.

$$|A| = \begin{vmatrix} 0 & 2 & 1 \\ 3 & -1 & 2 \\ 4 & -4 & 1 \end{vmatrix} \begin{matrix} 0 & 2 \\ 3 & -1 \\ 4 & -4 \end{matrix} = 0 + 16 - 12 - (-4) - 0 - 6 = 2$$

Try using this technique to find the determinants of the matrices in Examples 2 and 3. Do you think this method is easier than expanding by minors?

12.5 Exercises

Discussing the Concepts

1. Explain the difference between a square matrix and its determinant.

2. Is it possible to find the determinant of a 2×3 matrix? Explain your answer.

3. What is meant by the *minor* of an entry of a square matrix?

4. What conditions must be met in order to use Cramer's Rule to solve a system of linear equations?

Problem Solving

In Exercises 5–8, find the determinant of the matrix.

5. $\begin{bmatrix} 2 & 1 \\ 3 & 4 \end{bmatrix}$

6. $\begin{bmatrix} -3 & 1 \\ 5 & 2 \end{bmatrix}$

7. $\begin{bmatrix} -7 & 6 \\ \frac{1}{2} & 3 \end{bmatrix}$

8. $\begin{bmatrix} -1.2 & 4.5 \\ 0.4 & -0.9 \end{bmatrix}$

In Exercises 9 and 10, evaluate the determinant of the matrix six different ways by expanding by minors along each row and column.

9. $\begin{bmatrix} 2 & 3 & -1 \\ 6 & 0 & 0 \\ 4 & 1 & 1 \end{bmatrix}$

10. $\begin{bmatrix} 10 & 2 & -4 \\ 8 & 0 & -2 \\ 4 & 0 & 2 \end{bmatrix}$

In Exercises 11–14, evaluate the determinant of the matrix. Expand by minors on the row or column that appears to make the computation easiest.

11. $\begin{bmatrix} 2 & 4 & 6 \\ 0 & 3 & 1 \\ 0 & 0 & -5 \end{bmatrix}$

12. $\begin{bmatrix} 3 & 2 & 2 \\ 2 & 2 & 2 \\ -4 & 4 & 3 \end{bmatrix}$

13. $\begin{bmatrix} 6 & 8 & -7 \\ 0 & 0 & 0 \\ 4 & -6 & 22 \end{bmatrix}$

14. $\begin{bmatrix} 0.1 & 0.2 & 0.3 \\ -0.3 & 0.2 & 0.2 \\ 5 & 4 & 4 \end{bmatrix}$

In Exercises 15–18, use a graphing utility to evaluate the determinant of the matrix.

15. $\begin{bmatrix} 5 & -3 & 2 \\ 7 & 5 & -7 \\ 0 & 6 & -1 \end{bmatrix}$

16. $\begin{bmatrix} -\frac{1}{2} & -1 & 6 \\ 8 & -\frac{1}{4} & -4 \\ 1 & 2 & 1 \end{bmatrix}$

17. $\begin{bmatrix} 0.2 & 0.8 \\ -8 & -5 \end{bmatrix}$

18. $\begin{bmatrix} 250 & -125 \\ 60 & -50 \end{bmatrix}$

In Exercises 19–26, use Cramer's Rule to solve the system of equations. (If it is not possible, state the reason.)

19. $3x + 4y = -2$
$5x + 3y = 4$

20. $18x + 12y = 13$
$30x + 24y = 23$

21. $-0.4x + 0.8y = 1.6$
$2x - 4y = 5$

22. $-0.4x + 0.8y = 1.6$
$0.2x + 0.3y = 2.2$

23. $4x - y + z = -5$
$2x + 2y + 3z = 10$
$5x - 2y + 6z = 1$

24. $4x - 2y + 3z = -2$
$2x + 2y + 5z = 16$
$8x - 5y - 2z = 4$

25. $3x + 4y + 4z = 11$
$4x - 4y + 6z = 11$
$6x - 6y = 3$

26. $2x + 3y + 5z = 4$
$3x + 5y + 9z = 7$
$5x + 9y + 17z = 13$

In Exercises 27–30, use a graphing utility and Cramer's Rule to solve the system of equations.

27. $-3x + 10y = 22$
$9x - 3y = 0$

28. $3x + 7y = 3$
$7x + 25y = 11$

29. $x + y - z = 2$
$6x + 4y + 3z = 4$
$3x + 6z = -3$

30. $3x + 2y - 5z = -10$

$\quad 6x \qquad - z = \quad 8$

$\qquad -y + 3z = \quad -2$

Geometry In Exercises 31 and 32, use a determinant to find the area of the triangle with the given vertices.

31. $(0, 3), (4, 0), (8, 5)$

32. $(-4, 2), (1, 5), (4, -4)$

33. *Geometry* A large region of forest has been infested with gypsy moths. The region is roughly triangular, as shown in the figure. Approximate the number of square miles in this region.

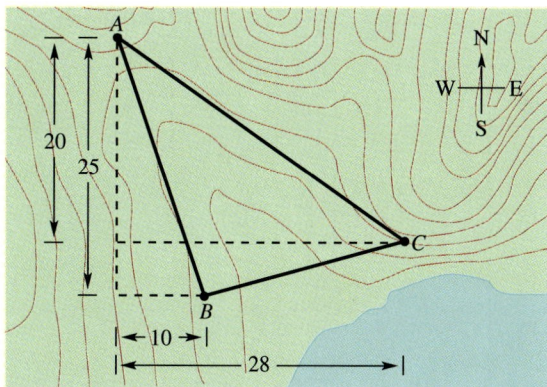

34. *Geometry* You have purchased a triangular tract of land, as shown in the figure. How many square feet are there in the tract of land?

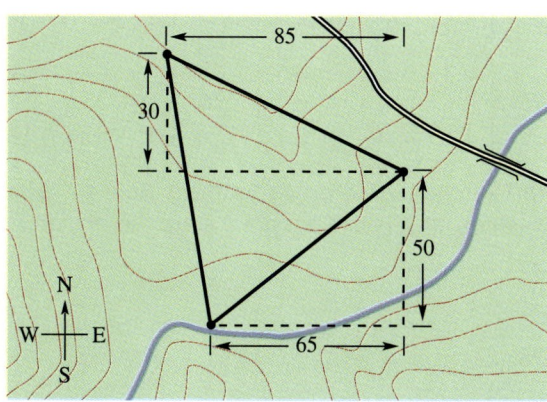

Collinear Points In Exercises 35 and 36, decide whether the points are collinear.

35. $(-1, -5), (1, -1), (4, 5)$

36. $(-1, 8), (1, 2), (2, 0)$

Equation of a Line In Exercises 37 and 38, find an equation of the line through the points.

37. $\left(-2, \frac{3}{2}\right), (3, -3)$ **38.** $(2, 3.6), (8, 10)$

Curve-Fitting In Exercises 39 and 40, use Cramer's Rule to find the equation of the parabola that passes through the points.

39. $(-2, 6), (2, -2), (4, 0)$

40. $(-1, 0), (1, 4), (4, -5)$

41. *Mathematical Modeling* The table gives the merchandise exports y_1 and the merchandise imports y_2 (in billions of dollars) for the years 1990 through 1992 in the United States. (Source: U.S. Bureau of Census)

Year	1990	1991	1992
y_1	393.6	421.7	448.2
y_2	495.3	487.1	532.5

(a) Find a quadratic model for exports. Let $t = 0$ represent 1990.

(b) Find a quadratic model for imports. Let $t = 0$ represent 1990.

(c) Use a graphing utility to graph the models found in parts (a) and (b).

(d) Find a model for the merchandise trade balance (merchandise exports − merchandise imports).

42. (a) Use Cramer's Rule to solve the system of linear equations.

$$kx + (1 - k)y = 1$$
$$(1 - k)x + \quad ky = 3$$

(b) For what value(s) of k will the system be inconsistent?

Reviewing the Major Concepts

In Exercises 43–46, sketch a graph of the equation.

43. $y = 3 - \frac{1}{2}x$

44. $4x^2 + 4y^2 = 25$

45. $\dfrac{y^2}{25} + x^2 = 1$

46. $\dfrac{x^2}{9} - \dfrac{y^2}{25} = 1$

47. *Defective Units*   A quality control engineer for a certain buyer found two defective units in a sample of 75. At that rate, what is the expected number of defective units in a shipment of 10,000 units?

48. *Geometry*   Find the perimeter and area of a right triangle whose legs are 4 inches and 5 inches.

Additional Problem Solving

In Exercises 49–56, find the determinant.

49. $\begin{bmatrix} 5 & 2 \\ -6 & 3 \end{bmatrix}$

50. $\begin{bmatrix} 2 & -2 \\ 4 & 3 \end{bmatrix}$

51. $\begin{bmatrix} 5 & -4 \\ -10 & 8 \end{bmatrix}$

52. $\begin{bmatrix} 4 & -3 \\ 0 & 0 \end{bmatrix}$

53. $\begin{bmatrix} 2 & 6 \\ 0 & 3 \end{bmatrix}$

54. $\begin{bmatrix} -2 & 3 \\ 6 & -9 \end{bmatrix}$

55. $\begin{bmatrix} 0.3 & 0.5 \\ 0.5 & 0.3 \end{bmatrix}$

56. $\begin{bmatrix} \frac{2}{3} & \frac{5}{6} \\ 14 & -2 \end{bmatrix}$

In Exercises 57–60, evaluate the determinant of the matrix six different ways by expanding by minors along each row and column.

57. $\begin{bmatrix} 2 & -5 & 0 \\ 4 & 7 & 0 \\ -7 & 25 & 3 \end{bmatrix}$

58. $\begin{bmatrix} 8 & 7 & 6 \\ -4 & 0 & 0 \\ 5 & 1 & 4 \end{bmatrix}$

59. $\begin{bmatrix} 1 & 1 & 2 \\ 3 & 1 & 0 \\ -2 & 0 & 3 \end{bmatrix}$

60. $\begin{bmatrix} 2 & 1 & 3 \\ 1 & 4 & 4 \\ 1 & 0 & 2 \end{bmatrix}$

In Exercises 61–72, evaluate the determinant of the matrix. Expand by minors on the row or column that appears to make the computation easiest.

61. $\begin{bmatrix} -2 & 2 & 3 \\ 1 & -1 & 0 \\ 0 & 1 & 4 \end{bmatrix}$

62. $\begin{bmatrix} 2 & 3 & 1 \\ 0 & 5 & -2 \\ 0 & 0 & -2 \end{bmatrix}$

63. $\begin{bmatrix} 1 & 4 & -2 \\ 3 & 6 & -6 \\ -2 & 1 & 4 \end{bmatrix}$

64. $\begin{bmatrix} 2 & -1 & 0 \\ 4 & 2 & 1 \\ 4 & 2 & 1 \end{bmatrix}$

65. $\begin{bmatrix} -3 & 2 & 1 \\ 4 & 5 & 6 \\ 2 & -3 & 1 \end{bmatrix}$

66. $\begin{bmatrix} -3 & 4 & 2 \\ 6 & 3 & 1 \\ 4 & -7 & -8 \end{bmatrix}$

67. $\begin{bmatrix} 1 & 4 & -2 \\ 3 & 2 & 0 \\ -1 & 4 & 3 \end{bmatrix}$

68. $\begin{bmatrix} -0.4 & 0.4 & 0.3 \\ 0.2 & 0.2 & 0.2 \\ 0.3 & 0.2 & 0.2 \end{bmatrix}$

69. $\begin{bmatrix} 0.4 & 0.3 & 0.3 \\ -0.2 & 0.6 & 0.6 \\ 3 & 1 & 1 \end{bmatrix}$

70. $\begin{bmatrix} \frac{1}{2} & \frac{3}{2} & \frac{1}{2} \\ 4 & 8 & 10 \\ -2 & -6 & 12 \end{bmatrix}$

71. $\begin{bmatrix} x & y & 1 \\ 3 & 1 & 1 \\ -2 & 0 & 1 \end{bmatrix}$

72. $\begin{bmatrix} x & y & 1 \\ -2 & -2 & 1 \\ 1 & 5 & 1 \end{bmatrix}$

In Exercises 73–76, use a graphing utility to evaluate the determinant of the matrix.

73. $\begin{bmatrix} 35 & 15 & 70 \\ -8 & 20 & 3 \\ -5 & 6 & 20 \end{bmatrix}$

74. $\begin{bmatrix} 3 & -1 & 2 \\ 1 & -1 & 2 \\ -2 & 3 & 10 \end{bmatrix}$

75. $\begin{bmatrix} 0.3 & -0.2 & 0.5 \\ 0.6 & 0.4 & -0.3 \\ 1.2 & 0 & 0.7 \end{bmatrix}$

76. $\begin{bmatrix} \frac{3}{2} & -\frac{3}{4} & 1 \\ 10 & 8 & 7 \\ 12 & -4 & 12 \end{bmatrix}$

In Exercises 77–86, use Cramer's Rule to solve the system. (If it is not possible, state the reason.)

77. $x + 2y = 5$
$-x + \ y = 1$

78. $2x - \ y = -10$
$3x + 2y = \ -1$

79. $20x + 8y = 11$
$12x - 24y = 21$

80. $13x - 6y = 17$
$26x - 12y = 8$

81. $3u + 6v = 5$
$6u + 14v = 11$

82. $3x_1 + 2x_2 = 1$
$2x_1 + 10x_2 = 6$

83. $3a + 3b + 4c = 1$
$3a + 5b + 9c = 2$
$5a + 9b + 17c = 4$

84. $14x_1 - 21x_2 - 7x_3 = 10$
$-4x_1 + 2x_2 - 2x_3 = 4$
$56x_1 - 21x_2 + 7x_3 = 5$

85. $5x - 3y + 2z = 2$
$2x + 2y - 3z = 3$
$x - 7y + 8z = -4$

86. $3x + 2y + 5z = 4$
$4x - 3y - 4z = 1$
$-8x + 2y + 3z = 0$

In Exercises 87–90, use a graphing utility and Cramer's Rule to solve the system of equations.

87. $4x - y = -2$
$-2x + y = 3$

88. $4x + 8y = 6$
$8x + 26y = 19$

89. $3x + y + z = 6$
$x - 4y + 2z = -1$
$x - 3y + z = 0$

90. $3x - 2y + 3z = 8$
$x + 3y + 6z = -3$
$x + 2y + 9z = -5$

Geometry In Exercises 91–94, use a determinant to find the area of the triangle with the given vertices.

91. $(0, 0), (3, 1), (1, 5)$

92. $(-2, -3), (2, -3), (0, 4)$

93. $(-2, 1), (3, -1), (1, 6)$

94. $\left(0, \frac{1}{2}\right), \left(\frac{5}{2}, 0\right), (4, 3)$

Geometry In Exercises 95 and 96, find the area of the shaded region of the figure.

95.

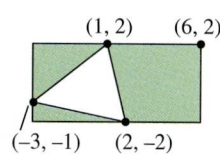

96.

Collinear Points In Exercises 97–100, decide whether the points are collinear.

97. $(-1, 11), (0, 8), (2, 2)$

98. $(-1, -1), (1, 9), (2, 13)$

99. $\left(-2, \frac{1}{3}\right), (2, 1), \left(3, \frac{1}{5}\right)$

100. $\left(0, \frac{1}{2}\right), \left(1, \frac{7}{6}\right), \left(9, \frac{13}{2}\right)$

Equation of a Line In Exercises 101–104, find an equation of the line through the points.

101. $(0, 0), (5, 3)$

102. $(-4, 3), (2, 1)$

103. $(10, 7), (-2, -7)$

104. $\left(-\frac{1}{2}, 3\right), \left(\frac{5}{2}, 1\right)$

Curve-Fitting In Exercises 105–108, use Cramer's Rule to find the equation of the parabola that passes through the points.

105. $(0, 1), (1, -3), (-2, 21)$

106. $(2, 3), \left(-1, \frac{9}{2}\right), (-2, 9)$

107. $(1, -1), (-1, -5), \left(\frac{1}{2}, \frac{1}{4}\right)$

108. $(-2, 6), (1, 9), (3, 1)$

In Exercises 109 and 110, solve the equation.

109. $\begin{vmatrix} 5 - x & 4 \\ 1 & 2 - x \end{vmatrix} = 0$

110. $\begin{vmatrix} 4 - x & -2 \\ 1 & 1 - x \end{vmatrix} = 0$

CHAPTER PROJECT: Healthcare

The table below gives models for the annual revenues of four companies in the medical supply industry from 1988 through 1993. The revenue R is listed in millions of dollars and t represents the year, with $t = 8$ corresponding to 1988.   (Source: SciMed Life Systems, Nellcor, St. Jude Medical, and Diagnostic Products)

Company	Model Revenue Equation	Years
SciMed Life Systems, Inc.	$R = -371.43 + 49.31t$	$8 \leq t \leq 13$
Nellcor, Inc.	$R = -103.67 + 24.64t$	$8 \leq t \leq 13$
St. Jude Medical, Inc.	$R = -110.69 + 28.63t$	$8 \leq t \leq 13$
Diagnostic Products Corp.	$R = -52.13 + 12.64t$	$8 \leq t \leq 13$

Each of these models was created using a linear regression program on a graphing utility. Use the information in the table to answer the following questions.

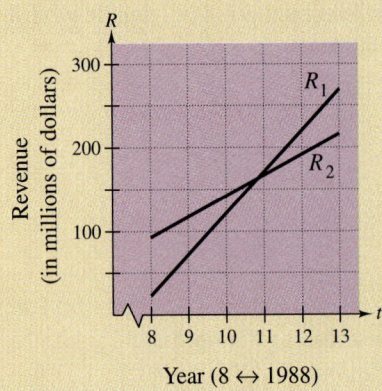

Revenue (in millions of dollars)

Year (8 ↔ 1988)

1. *Graphical Interpretation*   Which of the equations in the table are represented in the graph at the left? What is the point of intersection of the lines? What does the point of intersection represent?

2. *Graphical Interpretation*   Use a graphing utility to graph the equations for Nellcor, Inc. and St. Jude Medical, Inc. for the given years. Do the equations appear parallel? What does this tell you about the two companies from 1988 through 1993?

3. *Equal Revenues*   During which year did Diagnostic Products Corp. and St. Jude Medical, Inc. have about the same revenue? Does the answer make sense? Explain your reasoning.

4. *A System with Three Equations*   Sketch the graph of the following system of equations. Is there a single point of intersection?

$$\begin{aligned} 2x + 2y &= 6 \\ x + 2y &= 5 \\ -2x + y &= -5 \end{aligned}$$

5. *Numerical and Graphical Reasoning*   Create a table showing the revenues for each year from 1988 to 1993 for SciMed Life Systems, Inc., Nellcor, Inc., and St. Jude Medical, Inc. Graph the three equations in the same coordinate plane. Write a paragraph relating the information in the table to the positions and intersections of the lines of the graph.

6. *Research Project*   Use your school's library or some other reference source to find the 1994 revenues for the four companies listed above. How closely did the models predict the 1994 revenues?

CHAPTER SUMMARY

After studying this chapter, you should have acquired the following skills. These skills are keyed to the Review Exercises that begin on page 750. Answers to odd-numbered Review Exercises are given in the back of the book.

- Solve systems of equations using the method of substitution. *(Sections 12.1, 12.2)* — **Review Exercises 1–10**

- Solve systems of equations graphically. *(Section 12.1)* — **Review Exercises 11–16**

- Solve systems of equations using a graphing utility. *(Sections 12.1, 12.2)* — **Review Exercises 17–20**

- Solve systems of linear equations using the method of elimination. *(Sections 12.2, 12.3)* — **Review Exercises 21–24**

- Solve systems of linear equations using matrices and elementary row operations. *(Section 12.4)* — **Review Exercises 25–28**

- Evaluate the determinant of a 2 × 2 matrix. *(Section 12.5)* — **Review Exercises 29, 30**

- Evaluate the determinant of a 3 × 3 matrix by expanding by minors. *(Section 12.5)* — **Review Exercises 31, 32**

- Evaluate the determinant of a 3 × 3 matrix using any appropriate method. *(Section 12.5)* — **Review Exercises 33, 34**

- Solve systems of linear equations using Cramer's Rule. *(Section 12.5)* — **Review Exercises 35–40**

- Evaluate determinants to find equations of lines and areas of triangles. *(Section 12.5)* — **Review Exercises 41–48**

- Create systems of equations having given solutions. *(Sections 12.1, 12.2)* — **Review Exercises 49, 50**

- Model real-life situations with equations and solve. *(Sections 12.1, 12.2, 12.3, 12.4)* — **Review Exercises 51–56, 61**

- Find equations of parabolas and circles passing through points. *(Sections 12.3, 12.4, 12.5)* — **Review Exercises 57–60**

REVIEW EXERCISES

In Exercises 1–10, use substitution to solve the system.

1. $2x + 3y = 1$
$x + 4y = -2$

2. $3x - 7y = 10$
$-2x + y = -14$

3. $-5x + 2y = 4$
$10x - 4y = 7$

4. $5x + 2y = 3$
$2x + 3y = 10$

5. $3x - 7y = 5$
$5x - 9y = -5$

6. $24x - 4y = 20$
$6x - y = 5$

7. $y = 5x^2$
$y = -15x - 10$

8. $y^2 = 16x$
$4x - y = -24$

9. $x^2 + y^2 = 1$
$x + y = -1$

10. $x^2 + y^2 = 100$
$x + y = 0$

In Exercises 11–16, solve the system graphically.

11. $2x - y = 0$
$-x + y = 4$

12. $x = y + 3$
$x = y + 1$

13. $\dfrac{x^2}{16} + \dfrac{y^2}{4} = 1$
$y = x + 2$

14. $\dfrac{x^2}{100} + \dfrac{y^2}{25} = 1$
$y = -x - 5$

15. $\dfrac{x^2}{25} + \dfrac{y^2}{9} = 1$
$\dfrac{x^2}{25} - \dfrac{y^2}{9} = 1$

16. $x^2 + y^2 = 16$
$-x^2 + \dfrac{y^2}{16} = 1$

In Exercises 17–20, use a graphing utility to solve the system.

17. $5x - 3y = 3$
$2x + 2y = 14$

18. $8x + 5y = 1$
$3x - 4y = 18$

19. $x^2 + y^2 = 25$
$y^2 - x^2 = 7$

20. $x^2 + y = 9$
$x^2 - y^2 = 7$

In Exercises 21–24, use elimination to solve the system.

21. $x + y = 0$
$2x + y = 0$

22. $4x + y = 1$
$x - y = 4$

23. $-x + y + 2z = 1$
$2x + 3y + z = -2$
$5x + 4y + 2z = 4$

24. $2x + 3y + z = 10$
$2x - 3y - 3z = 22$
$4x - 2y + 3z = -2$

In Exercises 25–28, use matrices and elementary row operations to solve the system.

25. $5x + 4y = 2$
$-x + y = -22$

26. $2x - 5y = 2$
$3x - 7y = 1$

27. $x + 2y + 6z = 4$
$-3x + 2y - z = -4$
$4x + 2z = 16$

28. $2x_1 + 3x_2 + 3x_3 = 3$
$6x_1 + 6x_2 + 12x_3 = 13$
$12x_1 + 9x_2 - x_3 = 2$

In Exercises 29 and 30, find the determinant of the matrix.

29. $\begin{bmatrix} 7 & 10 \\ 10 & 15 \end{bmatrix}$

30. $\begin{bmatrix} -3.4 & 1.2 \\ -5 & 2.5 \end{bmatrix}$

In Exercises 31 and 32, evaluate the determinant of the matrix six different ways by expanding by minors along each row and column.

31. $\begin{bmatrix} 8 & 6 & 3 \\ 6 & 3 & 0 \\ 3 & 0 & 2 \end{bmatrix}$

32. $\begin{bmatrix} 7 & -1 & 10 \\ -3 & 0 & -2 \\ 12 & 1 & 1 \end{bmatrix}$

In Exercises 33 and 34, find the determinant of the matrix.

33. $\begin{bmatrix} 8 & 3 & 2 \\ 1 & -2 & 4 \\ 6 & 0 & 5 \end{bmatrix}$

34. $\begin{bmatrix} 4 & 0 & 10 \\ 0 & 10 & 0 \\ 10 & 0 & 34 \end{bmatrix}$

In Exercises 35–40, solve the system of linear equations by using Cramer's Rule. (If it is not possible, state the reason.)

35. $7x + 12y = 63$
$2x + 3y = 15$

36. $12x + 42y = -17$
$30x - 18y = 19$

37. $3x - 2y = 16$
$12x - 8y = -5$

38. $4x + 24y = 20$
$-3x + 12y = -5$

39. $-x + y + 2z = 1$
$2x + 3y + z = -2$
$5x + 4y + 2z = 4$

40. $2x_1 + x_2 + 2x_3 = 4$
$2x_1 + 2x_2 = 5$
$2x_1 - x_2 + 6x_3 = 2$

In Exercises 41–44, use a determinant to find the equation of the line through the points.

41. $(-4, 0), (4, 4)$

42. $(2, 5), (6, -1)$

43. $\left(-\frac{5}{2}, 3\right), \left(\frac{7}{2}, 1\right)$

44. $(-0.8, 0.2), (0.7, 3.2)$

In Exercises 45–48, use a determinant to find the area of the triangle with the given vertices.

45. $(1, 0), (5, 0), (5, 8)$

46. $(-4, 0), (4, 0), (0, 6)$

47. $(1, 2), (4, -5), (3, 2)$

48. $\left(\frac{3}{2}, 1\right), \left(4, -\frac{1}{2}\right), (4, 2)$

In Exercises 49 and 50, create a system of equations having the given solution.

49. $\left(\frac{2}{3}, -4\right)$

50. $(-10, 12)$

51. *Break-Even Analysis*   A small business invests $25,000 in equipment to produce a product. Each unit of the product costs $3.75 to produce and is sold for $5.25. How many items must be sold before the business breaks even?

52. *Geometry*   The perimeter of a rectangle is 480 meters and its length is 150% of its width. Find the dimensions of the rectangle.

53. *Acid Mixture*   One hundred gallons of a 60% acid solution is obtained by mixing a 75% solution with a 50% solution. How many gallons of each solution must be used to obtain the desired mixture?

54. *Rope Length*   You must cut a rope that is 128 inches long into two pieces such that one piece is three times as long as the other. Find the length of each piece.

55. *Flying Speeds*   Two planes leave Pittsburgh and Philadelphia at the same time, each going to the other city. Because of the wind, one plane flies 25 miles per hour faster than the other. Find the ground speed of each plane if the cities are 275 miles apart and the planes pass one another after 40 minutes.

56. *Investments*   An inheritance of $20,000 was divided among three investments yielding $1780 in interest per year. The interest rates for the three investments were 7%, 9%, and 11%. Find the amount placed in each investment if the second and third were $3000 and $1000 less than the first, respectively.

Curve-Fitting   In Exercises 57 and 58, find an equation of the parabola passing through the points.

57. $(0, -6), (1, -3), (2, 4)$

58. $(-5, 0), (1, -6), (2, 14)$

Curve-Fitting   In Exercises 59 and 60, find an equation of the circle $x^2 + y^2 + Dx + Ey + F = 0$ passing through the points.

59. $(2, 2), (5, -1), (-1, -1)$

60. $(4, 2), (1, 3), (-2, -6)$

61. *Mathematical Modeling*   A child throws a softball over a garage. The location of the eaves and the peak of the roof are given by $(0, 10), (15, 15)$, and $(30, 10)$.

(a) Find the equation of the parabola for the path of the ball if the ball follows a path 1 foot over the eaves and the peak of the roof.

(b) Use a graphing utility to graph the path.

(c) How far from the edge of the garage is the child standing if the ball is at a height of 5 feet when it leaves the child's hand?

CHAPTER TEST

Take this test as you would take a test in class. After you are done, check your work against the answers given in the back of the book.

1. Which ordered pair is the solution of the system at the right: $(3, -4)$ or $\left(1, \frac{1}{2}\right)$?

$$2x - 2y = 1$$
$$-x + 2y = 0$$

System for 1

In Exercises 2–13, use the indicated method to solve the system.

2. *Substitution:*
$$5x - y = 6$$
$$4x - 3y = -4$$

3. *Substitution:*
$$x + y = 8$$
$$xy = 12$$

4. *Graphical:*
$$x - 2y = -1$$
$$2x + 3y = 12$$

5. *Elimination:*
$$3x - 4y = -14$$
$$-3x + y = 8$$

6. *Elimination:*
$$8x + 3y = 3$$
$$4x - 6y = -1$$

7. *Elimination:*
$$x + 2y - 4z = 0$$
$$3x + y - 2z = 5$$
$$3x - y + 2z = 7$$

8. *Matrices:*
$$x - 3z = -10$$
$$-2y + 2z = 0$$
$$x - 2y = -7$$

9. *Matrices:*
$$x - 3y + z = -3$$
$$3x + 2y - 5z = 18$$
$$y + z = -1$$

10. *Cramer's Rule:*
$$2x - 7y = 7$$
$$3x + 7y = 13$$

11. *Graphical:*
$$x - 2y = -3$$
$$2x + 3y = 22$$

12. *Any Method:*
$$3x - 2y + z = 12$$
$$x - 3y = 2$$
$$-3x - 9z = -6$$

13. *Any Method:*
$$4x + y + 2z = -4$$
$$3y + z = 8$$
$$-3x + y - 3z = 5$$

14. Describe the number of possible solutions of a system of linear equations.

15. Evaluate the determinant of A, as shown at the right.

16. Find the value of a such that the system at the right is inconsistent.

$$A = \begin{bmatrix} 3 & -2 & 0 \\ -1 & 5 & 3 \\ 2 & 7 & 1 \end{bmatrix}$$

Matrix for 15

17. Find a system of linear equations with integer coefficients that has the solution $(5, -3)$. (The problem has many correct answers.)

18. Two people share the driving on a 200-mile trip. One person drives four times as far as the other. Write a system of linear equations that models the problem. Find the distance each person drives.

$$5x - 8y = 3$$
$$3x + ay = 0$$

System for 16

19. Find the equation of the parabola $y = ax^2 + bx + c$ that passes through the points $(0, 4)$, $(1, 3)$, and $(2, 6)$.

20. Find the area of the triangle with vertices $(0, 0)$, $(5, 4)$, and $(6, 0)$.

Exponential and Logarithmic Functions

13

- Exponential Functions
- Composite and Inverse Functions
- Logarithmic Functions
- Properties of Logarithms
- Solving Exponential and Logarithmic Equations
- Applications

Radiocarbon, also known as carbon 14, is a radioactive isotope of carbon. Plants and animals absorb radiocarbon into their systems while they are alive. Once the organism dies, it no longer absorbs the isotope. From that time on, the radiocarbon decays exponentially.

The *half-life* of radiocarbon, which is the amount of time it takes for half of the radiocarbon to decay, is 5700 years. By measuring the amounts of radiocarbon remaining in prehistoric plants and animals that lived up to 50,000 years ago, scientists can determine their ages.

The table and graph at the right show the time it takes for 16 grams of radiocarbon to decay to 1 gram. Notice that after 11,400 years, one-fourth of the radiocarbon is still present.

Grams	16	8	4	2	1
Years	0	5700	11,400	17,100	22,800

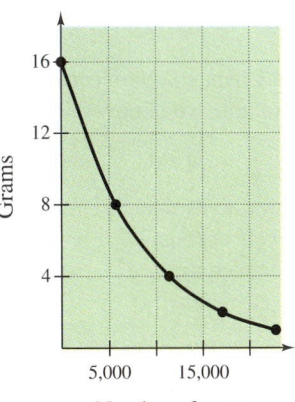

The chapter project related to this information is on page 818.

753

13.1	**Exponential Functions**
	Exponential Functions ▪ Graphs of Exponential Functions ▪ The Natural Exponential Function ▪ Compound Interest

Exponential Functions

In this section, you will study a new type of function called an **exponential function.** Whereas polynomial and rational functions have terms with variable bases and constant exponents, exponential functions have terms with *constant bases* and *variable exponents.* Here are some examples.

Polynomial or Rational Function   *Exponential Function*

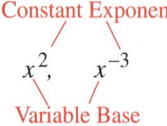   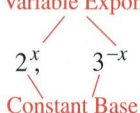

NOTE The base $a = 1$ is excluded because $f(x) = 1^x = 1$ is a constant function, *not* an exponential function.

Definition of Exponential Function

The **exponential function** f with base a is denoted by

$$f(x) = a^x$$

where $a > 0$, $a \neq 1$, and x is any real number.

In Chapter 9, you learned to evaluate a^x for integer and rational values of x. For example, you know that

$$a^3 = a \cdot a \cdot a \qquad \text{and} \qquad a^{2/3} = \left(\sqrt[3]{a}\right)^2.$$

However, to evaluate a^x for any real number x, you need to interpret forms with *irrational* exponents. For the purpose of this text, it is sufficient to think of a number such as

$$a^{\sqrt{2}}$$

where $\sqrt{2} \approx 1.414214$, as the number that has the successively closer approximations

$$a^{1.4}, \; a^{1.41}, \; a^{1.414}, \; a^{1.4142}, \; a^{1.41421}, \; a^{1.414214}, \ldots.$$

The properties of exponents that were discussed in Section 9.1 can be extended to cover exponential functions, as described on page 755.

Properties of Exponential Functions

1. $a^x \cdot a^y = a^{x+y}$ **2.** $(a^x)^y = a^{xy}$

3. $\dfrac{a^x}{a^y} = a^{x-y}$ **4.** $a^{-x} = \dfrac{1}{a^x} = \left(\dfrac{1}{a}\right)^x$

To evaluate exponential functions with a calculator, you can use the exponential key $\boxed{y^x}$ (where y is the base and x is the exponent) or $\boxed{\wedge}$. For example, to evaluate $3^{-1.3}$, you can use the following keystrokes.

Keystrokes	*Display*	
3 $\boxed{y^x}$ 1.3 $\boxed{+/-}$ $\boxed{=}$	0.239741	Scientific
3 $\boxed{\wedge}$ $\boxed{(}$ $\boxed{(-)}$ 1.3 $\boxed{)}$ $\boxed{\text{ENTER}}$	0.239741	Graphing

EXAMPLE 1   *Evaluating Exponential Functions*

Evaluate each function at the indicated values of x. Use a calculator only if it is necessary or more efficient.

Function	*Values*
a. $f(x) = 2^x$	$x = 3, \ x = -4, \ x = \pi$
b. $g(x) = 12^x$	$x = 3, \ x = -0.1, \ x = \frac{5}{7}$
c. $h(x) = (1.085)^x$	$x = 0, \ x = -3$

Solution

	Evaluation	*Comment*
a.	$f(3) = 2^3 = 8$	Calculator is not necessary.
	$f(-4) = 2^{-4} = \dfrac{1}{2^4} = \dfrac{1}{16}$	Calculator is not necessary.
	$f(\pi) = 2^\pi \approx 8.825$	Calculator is necessary.
b.	$g(3) = 12^3 = 1728$	Calculator is more efficient.
	$g(-0.1) = 12^{-0.1} \approx 0.7800$	Calculator is necessary.
	$g\left(\dfrac{5}{7}\right) = 12^{5/7} \approx 5.900$	Calculator is necessary.
c.	$h(0) = (1.085)^0 = 1$	Calculator is not necessary.
	$h(-3) = (1.085)^{-3} \approx 0.7829$	Calculator is more efficient.

Graphs of Exponential Functions

The basic nature of the graph of an exponential function can be determined by the point-plotting method or by using a graphing utility.

EXAMPLE 2 The Graphs of Exponential Functions

In the same coordinate plane, sketch the graphs of the following functions. Determine the domains and ranges.

a. $f(x) = 2^x$ **b.** $g(x) = 4^x$

Solution

The table lists some values of each function, and Figure 13.1 shows their graphs. From the graphs, you can see that the domain of each function is the set of all real numbers and that the range of each function is the set of all positive real numbers.

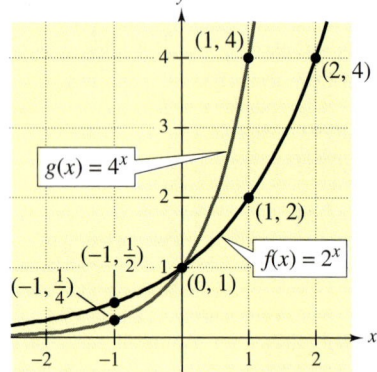

FIGURE 13.1

x	-2	-1	0	1	2	3
2^x	$\frac{1}{4}$	$\frac{1}{2}$	1	2	4	8
4^x	$\frac{1}{16}$	$\frac{1}{4}$	1	4	16	64

You know from your study of functions in Chapter 7 that the graph of $h(x) = f(-x) = 2^{-x}$ is a reflection of the graph of $f(x) = 2^x$ in the y-axis. This is reinforced in the following example.

EXAMPLE 3 The Graphs of Exponential Functions

In the same coordinate plane, sketch the graph of each function.

a. $f(x) = 2^{-x}$ **b.** $g(x) = 4^{-x}$

Solution

The table lists some values of each function, and Figure 13.2 shows their graphs.

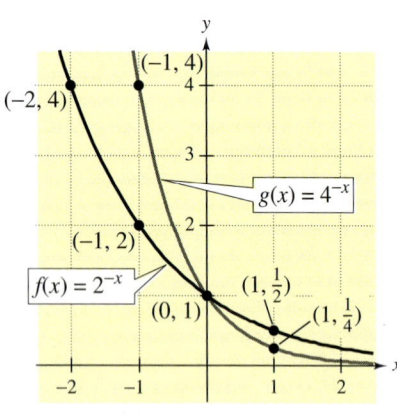

FIGURE 13.2

x	-3	-2	-1	0	1	2
2^{-x}	8	4	2	1	$\frac{1}{2}$	$\frac{1}{4}$
4^{-x}	64	16	4	1	$\frac{1}{4}$	$\frac{1}{16}$

Examples 2 and 3 suggest that for $a > 1$, the graph of $y = a^x$ increases and the graph of $y = a^{-x}$ decreases. The graphs shown in Figure 13.3 are typical of the graphs of exponential functions. Note that each has a y-intercept at $(0, 1)$.

Graph of $y = a^x$

- Domain: $(-\infty, \infty)$
- Range: $(0, \infty)$
- Intercept: $(0, 1)$
- Increasing

Graph of $y = a^{-x}$

- Domain: $(-\infty, \infty)$
- Range: $(0, \infty)$
- Intercept: $(0, 1)$
- Decreasing

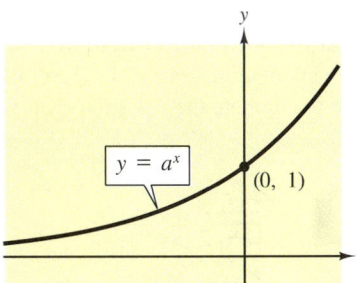

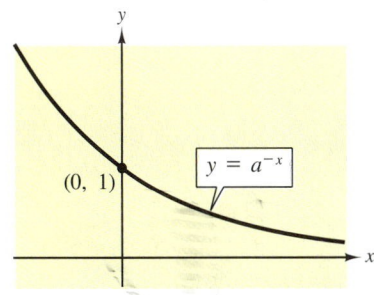

FIGURE 13.3 Characteristics of the Exponential Functions a^x and $a^{-x}(a > 1)$

EXAMPLE 4   An Application: Radioactive Decay

Let y represent the mass of a particular radioactive element whose half-life is 25 years. The initial mass is 10 grams. After t years, the mass (in grams) is given by

$$y = 10\left(\tfrac{1}{2}\right)^{t/25}, \quad t \ge 0.$$

How much of the initial mass remains after 120 years?

Solution

When $t = 120$, the mass is given by

$$y = 10\left(\tfrac{1}{2}\right)^{120/25} \qquad \text{Substitute 120 for } t.$$

$$= 10\left(\tfrac{1}{2}\right)^{4.8} \qquad \text{Simplify.}$$

$$\approx 0.359 \text{ gram.} \qquad \text{Use a calculator.}$$

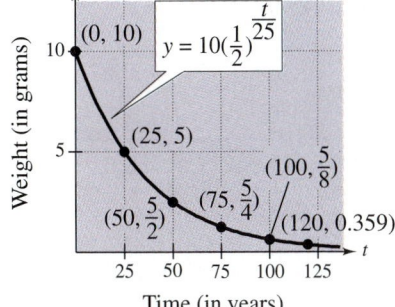

FIGURE 13.4

Thus, after 120 years, the mass has decayed from an initial amount of 10 grams to only 0.359 gram. Note in Figure 13.4 that the graph of the function shows the 25-year half-life. That is, after 25 years the mass is 5 grams (half of the original), after another 25 years the mass is 2.5 grams, and so on.

The Natural Exponential Function

So far, we have used integers or rational numbers as bases of exponential functions. In many applications of exponential functions, the convenient choice for a base is the following irrational number, denoted by the letter "*e*."

$$e \approx 2.71828\ldots\ldots \qquad \text{Natural base}$$

This number is called the **natural base.** The function

$$f(x) = e^x \qquad \text{Natural exponential function}$$

is called the **natural exponential function.** Be sure you understand that for this function, *e* is the constant number 2.71828 . . . , and *x* is a variable. To evaluate the natural exponential function, you need a calculator, preferably one having a natural exponential key $\boxed{e^x}$. Here are some examples of how to use such a calculator to evaluate the natural exponential function.

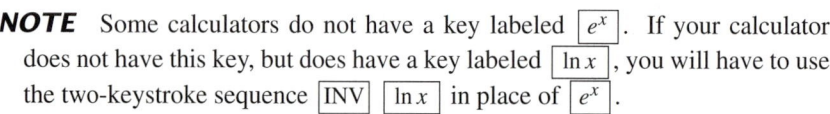

Value	Keystrokes	Display	
e^2	2 $\boxed{e^x}$	7.3890561	Scientific
e^2	$\boxed{e^x}$ 2 $\boxed{\text{ENTER}}$	7.3890561	Graphing
e^{-3}	3 $\boxed{+/-}$ $\boxed{e^x}$	0.049787	Scientific
e^{-3}	$\boxed{e^x}$ $\boxed{(}$ $\boxed{(-)}$ 3 $\boxed{)}$ $\boxed{\text{ENTER}}$	0.049787	Graphing
$e^{0.32}$	.32 $\boxed{e^x}$	1.3771278	Scientific
$e^{0.32}$	$\boxed{e^x}$ .32 $\boxed{\text{ENTER}}$	1.3771278	Graphing

NOTE Some calculators do not have a key labeled $\boxed{e^x}$. If your calculator does not have this key, but does have a key labeled $\boxed{\ln x}$, you will have to use the two-keystroke sequence $\boxed{\text{INV}}$ $\boxed{\ln x}$ in place of $\boxed{e^x}$.

After evaluating the natural exponential function at several values, as shown in the table, you can sketch its graph, as shown in Figure 13.5.

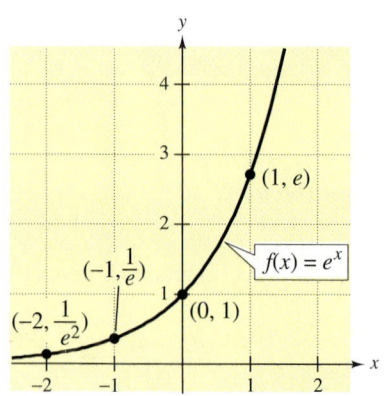

FIGURE 13.5

x	-1.5	-1.0	-0.5	0.0	0.5	1.0	1.5
$f(x) = e^x$	0.223	0.368	0.607	1.000	1.649	2.718	4.482

From the graph, notice the following properties of the natural exponential function.

- The domain is the set of all real numbers.
- The range is the set of positive real numbers.
- The *y*-intercept is (0, 1).

Compound Interest

One of the most familiar uses of exponential functions involves **compound interest.** Suppose a principal P is invested at an annual interest rate r (in decimal form), compounded once a year. If the interest is added to the principal at the end of the year, the balance is

$$A = P + Pr = P(1 + r).$$

This pattern of multiplying the previous principal by $(1 + r)$ is then repeated each successive year, as shown below.

Time in Years	Balance at Given Time
0	$A = P$
1	$A = P(1 + r)$
2	$A = P(1 + r)(1 + r) = P(1 + r)^2$
3	$A = P(1 + r)^2(1 + r) = P(1 + r)^3$
$\vdots$	$\vdots$
t	$A = P(1 + r)^t$

To account for more frequent compounding of interest (such as quarterly or monthly compounding), let n be the number of compoundings per year and let t be the number of years. Then the rate per compounding is r/n and the account balance after t years is

$$A = P\left(1 + \frac{r}{n}\right)^{nt}.$$

EXAMPLE 5   *Finding the Balance for Compound Interest*

A sum of $10,000 is invested at an annual interest rate of 7.5%, compounded monthly. Find the balance in the account after 10 years.

Solution

Using the formula for compound interest, with $P = 10{,}000$, $r = 0.075$, $n = 12$ (for monthly compounding), and $t = 10$, you obtain the following balance.

$$A = 10{,}000\left(1 + \frac{0.075}{12}\right)^{12(10)} \approx \$21{,}120.65$$

A second method that banks use to compute interest is called **continuous compounding.** The formula for the balance for this type of compounding is

$$A = Pe^{rt}.$$

The formulas for both types of compounding are summarized on page 760.

Formulas for Compound Interest

After t years, the balance A in an account with principal P and annual interest rate r (in decimal form) is given by the following formulas.

1. For n compoundings per year: $A = P\left(1 + \dfrac{r}{n}\right)^{nt}$

2. For continuous compounding: $A = Pe^{rt}$

EXAMPLE 6   Comparing Two Types of Compounding

A total of $15,000 is invested at an annual interest rate of 8%. Find the balance after 6 years if the interest is compounded

a. quarterly   and   **b.** continuously.

Solution

a. Letting $P = 15{,}000$, $r = 0.08$, $n = 4$, and $t = 6$, the balance after 6 years at quarterly compounding is

$$A = 15{,}000\left(1 + \frac{0.08}{4}\right)^{4(6)}$$

$$= \$24{,}126.56.$$

b. Letting $P = 15{,}000$, $r = 0.08$, and $t = 6$, the balance after 6 years at continuous compounding is

$$A = 15{,}000e^{0.08(6)}$$

$$= \$24{,}241.12.$$

Note that the balance is greater with continuous compounding than with quarterly compounding.

NOTE   Example 6 illustrates the following general rule. For a given principal, interest rate, and time, the more often the interest is compounded per year, the greater the balance will be. Moreover, the balance obtained by continuous compounding is larger than the balance obtained by compounding n times per year.

Group Activities   E x p l o r i n g   w i t h   T e c h n o l o g y

Finding a Pattern   Use a graphing utility to investigate the function $f(x) = k^x$ for different values of k. Discuss the effect that k has on the shape of the graph.

13.1 Exercises

Discussing the Concepts

1. Describe some differences between exponential functions and polynomial or rational functions.

2. Explain why 1^x is not an exponential function.

3. Compare the graphs of $f(x) = 3^x$ and $g(x) = \left(\frac{1}{3}\right)^x$.

4. Describe some applications of the exponential functions $f(x) = a^x$ and $g(x) = a^{-x}$.

5. *True or False?* $e = 271{,}801/99{,}990$. Explain.

6. Without using a calculator, how do you know that $2^{\sqrt{2}}$ is greater than 2, but less than 4?

Problem Solving

In Exercises 7–10, simplify the expression.

7. $2^x \cdot 2^{x-1}$

8. $\dfrac{3^{2x+3}}{3^{x+1}}$

9. $(2e^x)^3$

10. $\sqrt{4e^{6x}}$

In Exercises 11–14, evaluate the function as indicated. Use a calculator only if it is necessary or more efficient. (Round to three decimal places.)

11. $f(x) = 3^x$
   (a) $x = -2$
   (b) $x = 0$
   (c) $x = 1$

12. $F(x) = 3^{-x}$
   (a) $x = -2$
   (b) $x = 0$
   (c) $x = 1$

13. $g(x) = 10e^{-0.5x}$
   (a) $x = -4$
   (b) $x = 4$
   (c) $x = 8$

14. $f(z) = \dfrac{100}{1 + e^{-0.05z}}$
   (a) $z = 0$
   (b) $z = 10$
   (c) $z = 20$

In Exercises 15–20, match the function with its graph. [The graphs are labeled (a), (b), (c), (d), (e), and (f).]

(a)

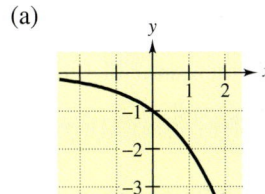

(b)

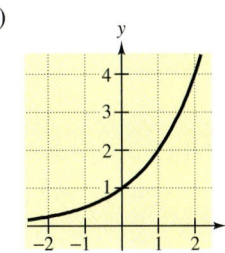

(c)

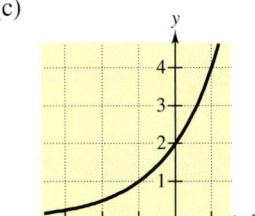

(d)

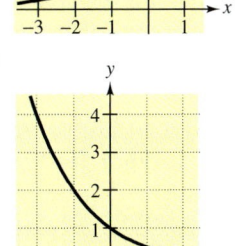

(e)

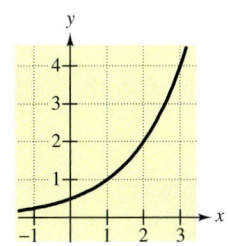

(f)

15. $f(x) = 2^x$

16. $f(x) = -2^x$

17. $f(x) = 2^{-x}$

18. $f(x) = 2^x - 1$

19. $f(x) = 2^{x-1}$

20. $f(x) = 2^{x+1}$

In Exercises 21–28, sketch the graph of the function.

21. $f(x) = 3^x$

22. $f(x) = 3^{-x} = \left(\frac{1}{3}\right)^x$

23. $g(x) = 3^x - 2$

24. $g(x) = 3^x + 1$

25. $f(t) = 2^{-t^2}$

26. $f(t) = 2^{t^2}$

27. $f(x) = -2^{0.5x}$

28. $h(t) = -2^{-0.5t}$

In Exercises 29–34, use a graphing utility to graph the function.

29. $y = 5^{x/3}$

30. $y = 5^{(x-2)/3}$

31. $f(x) = e^{0.2x}$

32. $f(x) = e^{-0.2x}$

33. $P(t) = 100e^{-0.1t}$

34. $A(t) = 1000e^{0.08t}$

35. *Depreciation* After t years, the value of a car that cost $16,000 is given by

$$V(t) = 16,000 \left(\tfrac{3}{4}\right)^t.$$

Sketch a graph of the function and determine the value of the car 2 years after it was purchased.

36. *Depreciation* Suppose straight-line depreciation is used to determine the value of the car in Exercise 35. If the car depreciates $3000 per year, the model for its value after t years is given by

$$V(t) = 16,000 - 3000t.$$

(a) Sketch the graph of this line on the same coordinate axes used for the graph in Exercise 35.

(b) If you were selling the car after owning it 2 years, which depreciation model would you prefer?

(c) If you sell the car after 4 years, which model would be to your advantage?

Creating a Table In Exercises 37 and 38, complete the table for P dollars invested at rate r for t years and compounded n times per year.

n	1	4	12	365	Continuous
A					

	Principal	Rate	Time
37.	$P = \$100$	$r = 8\%$	$t = 20$ years
38.	$P = \$400$	$r = 8\%$	$t = 50$ years

Creating a Table In Exercises 39 and 40, complete the table to find the principal P that yields a balance of A dollars when invested at rate r for t years, compounded n times per year.

n	1	4	12	365	Continuous
P					

	Balance	Rate	Time
39.	$A = \$5000$	$r = 7\%$	$t = 10$ years
40.	$A = \$100,000$	$r = 9\%$	$t = 20$ years

41. *Graphical Interpretation* An investment of $500 in two different accounts with respective interest rates of 6% and 8% is compounded continuously. The balances in the accounts after t years are modeled by

$$A_1 = 500e^{0.06t} \quad \text{and} \quad A_2 = 500e^{0.08t}.$$

(a) Use a graphing utility to graph each of the models.

(b) Use a graphing utility to graph the function $A_2 - A_1$ in the same window as the graphs of part (a).

(c) Use the graphs to discuss the rates of increase of the balances in the two accounts.

42. *Graphical Estimation* From 1970 through 1991, the number of recreational boats B (in millions) in the United States can be modeled by

$$B = 6.10(1.04)^t + 2.62$$

where $t = 0$ represents 1970. (Source: National Marine Manufacturers Association)

(a) Use a graphing utility to graph the model.

(b) Approximate the number of recreational boats owned in the United States in 1990.

Reviewing the Major Concepts

In Exercises 43–46, simplify the expression.

43. $(4x + 3y) - 3(5x + y)$

44. $(-15u + 4v) + 5(3u - 9v)$

45. $2x^2 + (2x - 3)^2 + 12x$

46. $y^2 - (y + 2)^2 + 4y$

47. *Geometry* A circle has a circumference of 1 inch.

(a) Find the radius of the circle.

(b) Find the area of the circle.

48. *Geometry* A square has a perimeter of 1 inch. What is its area?

Additional Problem Solving

In Exercises 49–52, evaluate the expression. (Round to three decimal places.)

49. $4^{\sqrt{3}}$

50. $6^{-\pi}$

51. $e^{1/3}$

52. $e^{-1/3}$

In Exercises 53–60, evaluate the function as indicated. Use a calculator only if it is necessary or more efficient. (Round to three decimal places.)

53. $g(x) = 5^x$
  (a) $x = -1$
  (b) $x = 1$
  (c) $x = 3$

54. $G(x) = 5^{-x}$
  (a) $x = -1$
  (b) $x = 1$
  (c) $x = \sqrt{3}$

55. $f(t) = 500 \left(\frac{1}{2}\right)^t$
  (a) $t = 0$
  (b) $t = 1$
  (c) $t = \pi$

56. $g(s) = 1200 \left(\frac{2}{3}\right)^s$
  (a) $s = 0$
  (b) $s = 2$
  (c) $s = 4$

57. $f(x) = 1000(1.05)^{2x}$
  (a) $x = 0$
  (b) $x = 5$
  (c) $x = 10$

58. $P(t) = \dfrac{10{,}000}{(1.01)^{12t}}$
  (a) $t = 2$
  (b) $t = 10$
  (c) $t = 20$

59. $f(x) = e^x$
  (a) $x = -1$
  (b) $x = 0$
  (c) $x = \frac{1}{2}$

60. $A(t) = 200e^{0.1t}$
  (a) $t = 10$
  (b) $t = 20$
  (c) $t = 40$

In Exercises 61–66, match the function with its graph. [The graphs are labeled (a), (b), (c), (d), (e), and (f).]

(a)

(b)

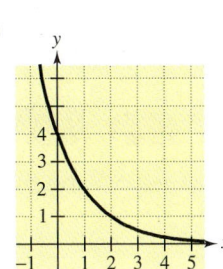

(c)

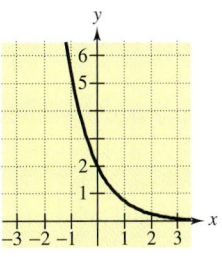

(d)

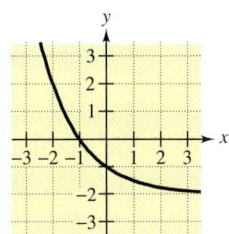

(e)

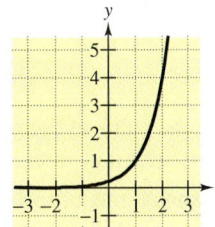

(f)

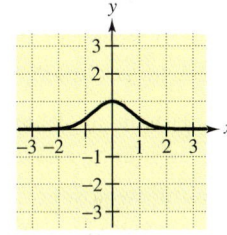

61. $f(x) = \left(\frac{1}{2}\right)^x - 2$

62. $f(x) = \left(\frac{1}{2}\right)^{x-2}$

63. $f(x) = 4^x - 1$

64. $f(x) = 4^{x-1}$

65. $f(x) = e^{-x^2}$

66. $f(x) = 2e^{-x}$

In Exercises 67–78, sketch the graph of the function.

67. $h(x) = \frac{1}{2}(3^x)$

68. $h(x) = \frac{1}{2}(3^{-x})$

69. $h(x) = 2^{0.5x}$

70. $g(t) = 2^{-0.5t}$

71. $f(x) = 4^{x-5}$

72. $f(x) = 4^{x+1}$

73. $g(x) = 4^x - 5$

74. $g(x) = 4^x + 1$

75. $f(x) = -\left(\frac{1}{3}\right)^x$

76. $f(x) = \left(\frac{3}{4}\right)^x + 1$

77. $g(t) = 200 \left(\frac{1}{2}\right)^t$

78. $h(y) = 27 \left(\frac{2}{3}\right)^y$

In Exercises 79–88, use a graphing utility to graph the function.

79. $y = 3^{-x/2}$

80. $y = 3^{-x/2} + 2$

81. $y = 5^{-x/3}$

82. $y = 5^{-x/3} + 2$

83. $y = 500(1.06)^t$

84. $y = 100(1.06)^{-t}$

85. $y = 3e^{0.2x}$

86. $y = 50e^{-0.05x}$

87. $y = 6e^{-x^2/3}$

88. $g(t) = \dfrac{10}{1 + e^{-0.5t}}$

89. *Population Growth*   The population of the United States (in recent years) can be approximated by the exponential function

$$P(t) = 203(1.0118)^{t-1970}$$

where t is the year and P is the population in millions. Use the model to approximate the population in the years (a) 1995 and (b) 2000.

90. *Property Value*   Suppose the value of a piece of property doubles every 15 years. If you buy the property for $64,000, its value t years after the date of purchase should be

$$V(t) = 64{,}000(2)^{t/15}.$$

Use the model to approximate the value of the property (a) 5 years and (b) 20 years after it is purchased.

91. *Inflation Rate*   Suppose the annual rate of inflation averages 5% over the next 10 years. With this rate of inflation, the approximate cost C of goods or services during any year in that decade will be given by

$$C(t) = P(1.05)^t, \quad 0 \le t \le 10$$

where t is time in years and P is the present cost. If the price of an oil change for your car is presently $19.95, estimate the price 10 years from now.

92. *Price and Demand*   The daily demand x and price p for a certain product are related by

$$p = 25 - 0.4e^{0.02x}.$$

Find the prices for demands of (a) $x = 100$ units and (b) $x = 125$ units.

Creating a Table   In Exercises 93–96, complete the table for P dollars invested at rate r for t years, compounded n times per year.

n	1	4	12	365	Continuous
A					

	Principal	Rate	Time
93.	$P = \$2000$	$r = 9\%$	$t = 10$ years
94.	$P = \$1500$	$r = 7\%$	$t = 2$ years

	Principal	Rate	Time
95.	$P = \$5000$	$r = 10\%$	$t = 40$ years
96.	$P = \$10{,}000$	$r = 9.5\%$	$t = 30$ years

Creating a Table   In Exercises 97 and 98, complete the table to determine the principal P that will yield a balance of A dollars when invested at rate r for t years, compounded n times per year.

n	1	4	12	365	Continuous
P					

	Balance	Rate	Time
97.	$A = \$1{,}000{,}000$	$r = 10.5\%$	$t = 40$ years
98.	$A = \$2500$	$r = 7.5\%$	$t = 2$ years

99. On the same set of coordinate axes, sketch the graph of the following functions. Which functions are exponential?

(a) $f(x) = 2x$ (b) $f(x) = 2x^2$
(c) $f(x) = 2^x$ (d) $f(x) = 2^{-x}$

100. *Savings Plan*   Suppose you decide to start saving pennies according to the following pattern. You save 1 penny the first day, 2 pennies the second day, 4 the third day, 8 the fourth day, and so on. Each day you save twice the number of pennies as the previous day. Which function in Exercise 99 models this problem, and how many pennies do you save on the thirtieth day? (In the next chapter you will learn how to find the total number saved.)

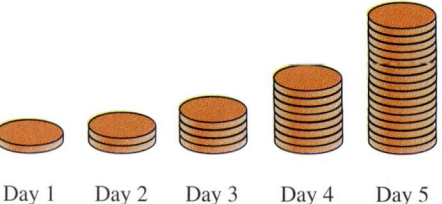

Day 1   Day 2   Day 3   Day 4   Day 5

101. *Graphical Estimation*   A parachutist jumps from a plane and opens the parachute at a height of 2000 feet (see figure). The height of the parachutist is

$$h = 1950 + 50e^{-1.6t} - 20t$$

where h is the height in feet and t is the time in seconds. (The time $t = 0$ corresponds to the time when the parachute is opened.)

(a) Use a graphing utility to graph the function.

(b) Find the height of the parachutist when $t = 0$, 25, 50, and 75.

(c) Approximate the time when the parachutist will reach the ground.

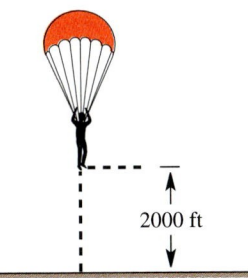

2000 ft

102. *Reading a Graph*   The popularity of 8-track cartridges increased until about 1980. From 1980 through 1985, the number n (in millions) of 8-track cartridges sold each year can be modeled by

$$n = 88 \left(\frac{10}{21} \right)^{t}$$

where $t = 0$ corresponds to 1980 (see figure). How many were sold in 1985?   (Source: Recording Industry Association of America)

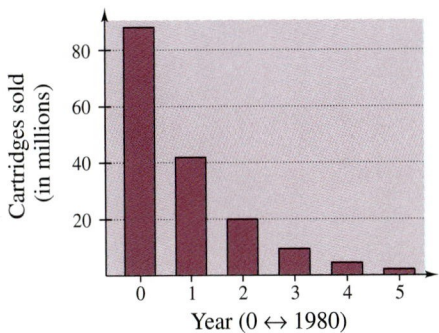

Cartridges sold (in millions)

Year (0 ↔ 1980)

103. *Creating a Table*   The median price of a home in the United States for the years 1987 through 1992 is given in the following table.   (Source: Chicago Title Insurance Company)

Year	1987	1988	1989
Price	$99,260	$121,920	$129,800

Year	1990	1991	1992
Price	$131,200	$134,300	$141,000

A model for this data is given by

$$y = 131{,}368e^{0.0102t^3}$$

where t is time in years, with $t = 0$ representing 1990.

(a) Use the model to complete the table and compare the results with the actual data.

Year	1987	1988	1989	1990	1991	1992
Price						

(b) Use a graphing utility to graph the model.

(c) If the model were used to predict home prices in the years ahead, would the predictions be increasing at a higher rate or a lower rate? Do you think the model would be reliable for predicting the future prices of homes? Explain.

104. *Identifying Graphs*   Identify the graphs of $y_1 = e^{0.2x}$, $y_2 = e^{0.5x}$, and $y_3 = e^x$ in the accompanying figure. Describe the effect on the graph of $y = e^{kx}$ when $k > 0$ is changed.

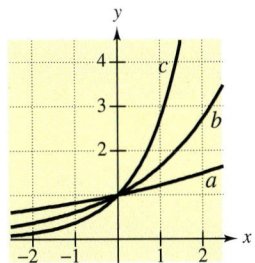

13.2 Composite and Inverse Functions

Composition of Functions ■ Inverse and One-to-One Functions ■
Finding Inverse Functions ■ Graphs of Inverse Functions

Composition of Functions

Two functions can be combined to form another function called the **composition** of
the two functions. For instance, if $f(x) = 2x^2$ and $g(x) = x - 1$, the composition
of f with g is given by

$$f(g(x)) = f(x - 1) = 2(x - 1)^2.$$

This composition is denoted by $f \circ g$.

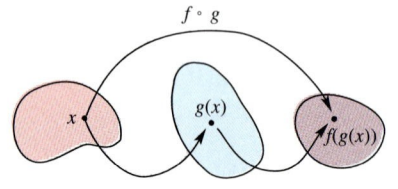

Domain of g Domain of f

FIGURE 13.6

Definition of Composition of Two Functions

The **composition** of the functions f and g is given by

$$(f \circ g)(x) = f(g(x)).$$

The domain of the **composite function** $(f \circ g)$ is the set of all x in the
domain of g such that $g(x)$ is in the domain of f. (See Figure 13.6.)

EXAMPLE 1   Forming the Composition of Two Functions

Find the composition of f with g. Evaluate the composite function when $x = 1$.

$$f(x) = 2x + 4 \quad \text{and} \quad g(x) = 3x - 1$$

Solution

The composition of f with g is given by

$$\begin{aligned}
(f \circ g)(x) &= f(g(x)) \\
&= f(3x - 1) \\
&= 2(3x - 1) + 4 \\
&= 6x - 2 + 4 \\
&= 6x + 2.
\end{aligned}$$

When $x = 1$, the value of this function is

$$(f \circ g)(1) = 6(1) + 2 = 8.$$

The composition of f with g is generally *not* the same as the composition of g with f. This is illustrated in Example 2.

EXAMPLE 2   *Comparing the Compositions of Functions*

Given $f(x) = 2x - 3$ and $g(x) = x^2 + 1$, find each of the following.

a. $(f \circ g)(x)$ **b.** $(g \circ f)(x)$

Solution

a. The composition of f with g is as follows.

$$(f \circ g)(x) = f(g(x)) \qquad \text{Definition of } f \circ g$$
$$= f(x^2 + 1) \qquad \text{Definition of } g(x)$$
$$= 2(x^2 + 1) - 3 \qquad \text{Definition of } f(x)$$
$$= 2x^2 + 2 - 3 \qquad \text{Simplify.}$$
$$= 2x^2 - 1 \qquad \text{Simplify.}$$

b. The composition of g with f is as follows.

$$(g \circ f)(x) = g(f(x)) \qquad \text{Definition of } g \circ f$$
$$= g(2x - 3) \qquad \text{Definition of } f(x)$$
$$= (2x - 3)^2 + 1 \qquad \text{Definition of } g(x)$$
$$= 4x^2 - 12x + 9 + 1 \qquad \text{Simplify.}$$
$$= 4x^2 - 12x + 10 \qquad \text{Simplify.}$$

Note that $(f \circ g)(x) \neq (g \circ f)(x)$.

EXAMPLE 3   *Finding the Domain of a Composite Function*

Find the domain of the composition of $f(x) = x^2$ with $g(x) = \sqrt{x}$.

Solution

The composition of f with g is given by

$$(f \circ g)(x) = f(g(x))$$
$$= f(\sqrt{x})$$
$$= (\sqrt{x})^2$$
$$= x, \qquad x \geq 0.$$

The domain of g consists of all nonnegative real numbers, $[0, \infty)$. This implies that the domain of the composition of f with g is this same set. That is, the domain of $f \circ g$ is $[0, \infty)$.

Inverse and One-to-One Functions

In Section 7.5, you learned that a function can be represented by a set of ordered pairs. For instance, the function $f(x) = x + 2$ from the set $A = \{1, 2, 3, 4\}$ to the set $B = \{3, 4, 5, 6\}$ can be written as follows.

$$f(x) = x + 2: \qquad \{(1, 3), (2, 4), (3, 5), (4, 6)\}$$

By interchanging the first and second coordinates of each of these ordered pairs, you can form another function that is called the **inverse function** of f. The inverse function is denoted by f^{-1}. It is a function from the set B to the set A, and can be written as follows.

$$f^{-1}(x) = x - 2: \qquad \{(3, 1), (4, 2), (5, 3), (6, 4)\}$$

Interchanging the ordered pairs for a function f will only produce another function when f is **one-to-one.** A function f is **one-to-one** if each value of the dependent variable corresponds to exactly one value of the independent variable.

Horizontal Line Test for Inverse Functions

A function f has an inverse function if and only if f is one-to-one. Graphically, a function f has an inverse function if and only if no *horizontal* line intersects the graph of f at more than one point.

EXAMPLE 4　*Applying the Horizontal Line Test*

a. The function $f(x) = x^3 - 1$ has an inverse because no horizontal line intersects its graph at more than one point, as shown in Figure 13.7.

b. The function $f(x) = x^2 - 1$ does not have an inverse because it is possible to find a horizontal line that intersects the graph of f at more than one point, as shown in Figure 13.8.

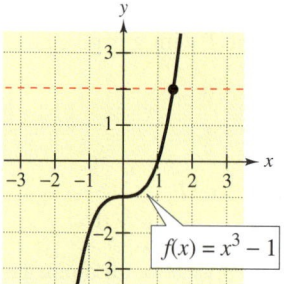

FIGURE 13.7

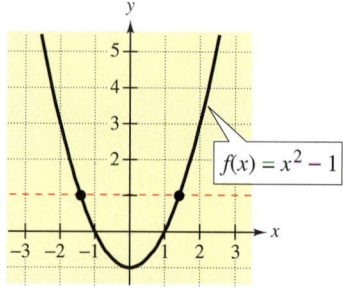

FIGURE 13.8

Definition of the Inverse of a Function

Let f and g be two functions such that

$$f(g(x)) = x \qquad \text{for every } x \text{ in the domain of } g$$

and

$$g(f(x)) = x \qquad \text{for every } x \text{ in the domain of } f.$$

The function g is the **inverse** of the function f, and is denoted by f^{-1} (read "f-inverse"). Thus, $f(f^{-1}(x)) = x$ and $f^{-1}(f(x)) = x$. The domain of f is equal to the range of f^{-1}, and vice versa.

If the function g is the inverse of the function f, it must also be true that the function f is the inverse of the function g. For this reason, you can refer to the functions f and g as being *inverses of each other.*

EXAMPLE 5   *Verifying Inverse Functions*

Show that each function is an inverse of the other.

$$f(x) = x^3 + 1 \qquad \text{and} \qquad g(x) = \sqrt[3]{x - 1}$$

Solution

Begin by noting that the domain and range of both functions is the entire set of real numbers. To show that f and g are inverses of each other, you need to show that $f(g(x)) = x$ and $g(f(x)) = x$, as follows.

$$
\begin{aligned}
f(g(x)) &= f\left(\sqrt[3]{x-1}\right) \\
&= \left(\sqrt[3]{x-1}\right)^3 + 1 \\
&= (x - 1) + 1 \\
&= x \\
g(f(x)) &= g(x^3 + 1) \\
&= \sqrt[3]{(x^3 + 1) - 1} \\
&= \sqrt[3]{x^3} \\
&= x
\end{aligned}
$$

Note that the two functions f and g "undo" each other in the following verbal sense. The function f first cubes the input x and then adds 1, whereas the function g first subtracts 1, and then takes the cube root of the result.

Finding Inverse Functions

You can find the inverse of a simple function by inspection. For instance, the inverse of $f(x) = 10x$ is $f^{-1}(x) = x/10$. For more complicated functions, however, it is best to use the following steps for finding the inverse of a function. The key step in these guidelines is switching the roles of x and y. This step corresponds to the fact that inverse functions have ordered pairs with the coordinates reversed.

Finding the Inverse of a Function

1. In the equation for $f(x)$, replace $f(x)$ by y.

2. Interchange the roles of x and y.

3. If the new equation does not represent y as a function of x, the function f does not have an inverse function. If the new equation does represent y as a function of x, solve the new equation for y.

4. Replace y by $f^{-1}(x)$.

EXAMPLE 6   Finding the Inverse of a Function

Find the inverse of $f(x) = 2x + 3$.

Solution

$$f(x) = 2x + 3 \qquad \text{Original function}$$
$$y = 2x + 3 \qquad \text{Replace } f(x) \text{ by } y.$$
$$x = 2y + 3 \qquad \text{Interchange } x \text{ and } y.$$
$$y = \frac{x - 3}{2} \qquad \text{Solve for } y.$$
$$f^{-1}(x) = \frac{x - 3}{2} \qquad \text{Replace } y \text{ by } f^{-1}(x).$$

Thus, the inverse of $f(x) = 2x + 3$ is

$$f^{-1}(x) = \frac{x - 3}{2}.$$

NOTE   Note in Step 3 of the guidelines for finding the inverse of a function that it is possible that a function has no inverse. For instance, the function $f(x) = x^2$ has no inverse.

Graphs of Inverse Functions

The graphs of a function f and its inverse f^{-1} are related to each other in the following way. If the point (a, b) lies on the graph of f, the point (b, a) must lie on the graph of f^{-1}, and vice versa. This means that the graph of f^{-1} is a reflection of the graph of f in the line $y = x$, as shown in Figure 13.9.

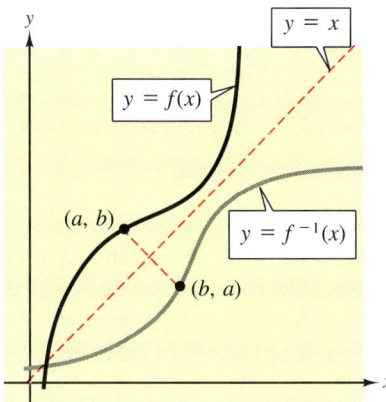

FIGURE 13.9

EXAMPLE 7   The Graphs of f and f^{-1}

Sketch the graphs of the inverse functions

$$f(x) = 2x - 3 \qquad \text{and} \qquad f^{-1}(x) = \frac{1}{2}(x + 3)$$

on the same rectangular coordinate system and show that the graphs are reflections of each other in the line $y = x$.

Solution

The graphs of f and f^{-1} are shown in Figure 13.10. Visually, it appears that the graphs are reflections of each other. You can further verify this by testing a few points on each graph. Note in the following list that if the point (a, b) is on the graph of f, the point (b, a) is on the graph of f^{-1}.

$f(x) = 2x - 3$	$f^{-1}(x) = \frac{1}{2}(x + 3)$
$(-1, -5)$	$(-5, -1)$
$(0, -3)$	$(-3, 0)$
$(1, -1)$	$(-1, 1)$
$(2, 1)$	$(1, 2)$
$(3, 3)$	$(3, 3)$

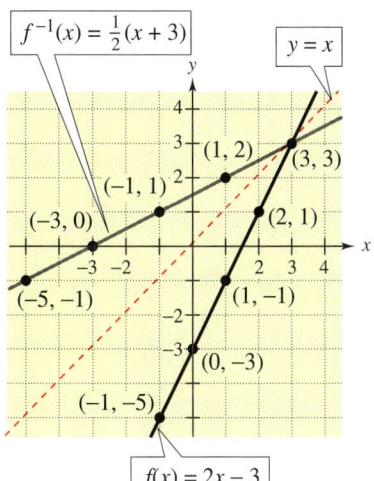

FIGURE 13.10

At the bottom of page 770, we mentioned that the function $f(x) = x^2$ has no inverse. What we mean is that *assuming the domain of f is the entire real line,* the function $f(x) = x^2$ has no inverse. If, however, you restrict the domain of f to the nonnegative real numbers, f does have an inverse.

EXAMPLE 8   The Graphs of f and f^{-1}

Graphically show that each function is an inverse of the other.

$$f(x) = x^2, \quad x \geq 0, \quad \text{and} \quad f^{-1}(x) = \sqrt{x}$$

Solution

The graphs of f and f^{-1} are shown in Figure 13.11. Visually, it appears that the graphs are reflections of each other in the line $y = x$. You can further verify this by testing a few points on each graph. Note in the following list that if the point (a, b) is on the graph of f, the point (b, a) is on the graph of f^{-1}.

$f(x) = x^2, \quad x \geq 0$	$f^{-1}(x) = \sqrt{x}$
$(0, 0)$	$(0, 0)$
$(1, 1)$	$(1, 1)$
$(2, 4)$	$(4, 2)$
$(3, 9)$	$(9, 3)$

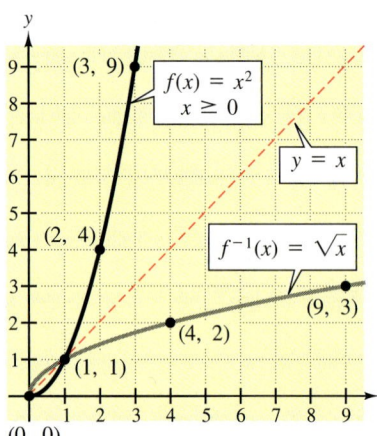

FIGURE 13.11

Group Activities   You Be the Instructor

Error Analysis   Suppose you are an algebra instructor and one of your students hands in the following solutions. Find and correct the errors and discuss how you can help your student avoid such errors in the future.

a. If $f(x) = 2x - 1$ and $g(x) = x^3 + 1$, find $(f \circ g)(2)$.

$$(f \circ g)(2) = (2 \cdot 2 - 1)(2^3 + 1)$$
$$= (4 - 1)(8 + 1)$$
$$= (3)(9)$$
$$= 27$$

b. If $f(x) = 3x^2 + x$ and $g(x) = x - 2$, find $(f \circ g)(1)$.

$$(f \circ g)(1) = f(1) - 2$$
$$= [3(1)^2 + 1] - 2$$
$$= (3 + 1) - 2$$
$$= 2$$

13.2 Exercises

Discussing the Concepts

1. Give an example showing that the composite functions $(f \circ g)(x)$ and $(g \circ f)(x)$ are not necessarily the same.

2. Describe how to find the inverse function of a function given by a set of ordered pairs. Give an example.

3. Describe how to find the inverse function of a function given by an equation in x and y. Give an example.

4. Give an example of a function that does not have an inverse function.

5. Explain the Horizontal Line Test. What is the relationship between this test and a function being one-to-one?

6. Describe the relationship between the graph of a function and its inverse.

Problem Solving

In Exercises 7–10, find the indicated composites.

7. $f(x) = x - 3$, $\quad g(x) = x^2$
   (a) $(f \circ g)(4)$ (b) $(g \circ f)(7)$
   (c) $(f \circ g)(x)$ (d) $(g \circ f)(x)$

8. $f(x) = |x|$, $\quad g(x) = 2x + 5$
   (a) $(f \circ g)(-2)$ (b) $(g \circ f)(-4)$
   (c) $(f \circ g)(x)$ (d) $(g \circ f)(x)$

9. $f(x) = \sqrt{x}$, $\quad g(x) = x + 5$
   (a) $(f \circ g)(4)$ (b) $(g \circ f)(9)$
   (c) $(f \circ g)(x)$ (d) $(g \circ f)(x)$

10. $f(x) = \dfrac{4}{x^2 - 4}$, $\quad g(x) = \dfrac{1}{x}$
    (a) $(f \circ g)(-2)$ (b) $(g \circ f)(1)$
    (c) $(f \circ g)(x)$ (d) $(g \circ f)(x)$

In Exercises 11–14, use the functions f and g to find the indicated values.

$f = \{(-2, 3), \ (-1, 1), \ (0, 0), \ (1, -1), \ (2, -3)\}$,

$g = \{(-3, 1), \ (-1, -2), \ (0, 2), \ (2, 2), \ (3, 1)\}$

11. (a) $f(1)$
    (b) $g(-1)$
    (c) $(g \circ f)(1)$

12. (a) $g(0)$
    (b) $f(2)$
    (c) $(f \circ g)(0)$

13. (a) $(f \circ g)(-3)$
    (b) $(g \circ f)(-2)$

14. (a) $(f \circ g)(2)$
    (b) $(g \circ f)(2)$

In Exercises 15 and 16, find the domain of the compositions (a) $f \circ g$ and (b) $g \circ f$.

15. $f(x) = \sqrt{x}$
    $g(x) = x - 2$

16. $f(x) = \dfrac{x}{x - 4}$
    $g(x) = \sqrt{x}$

17. *Sales Bonus*   You are a sales representative for a clothing manufacturer. You are paid an annual salary plus a bonus of 2% of your sales over \$200,000. Consider the two functions

   $f(x) = x - 200{,}000 \quad \text{and} \quad g(x) = 0.02x$.

   If x is greater than \$200,000, which of the following represents your bonus? Explain.

   (a) $f(g(x))$ (b) $g(f(x))$

18. *Geometry*   You are standing on a bridge over a calm pond and drop a pebble, causing ripples of concentric circles in the water. The radius (in feet) of the outer ripple is given by

   $r(t) = 0.6t$

   where t is time in seconds after the pebble hits the water. The area of the circle is given by the function

   $A(r) = \pi r^2$.

   Find an equation for the composite function $A(r(t))$. Interpret this composite function in the context of the problem.

In Exercises 19–22, find the inverse of the function f informally.

19. $f(x) = 5x$ **20.** $f(x) = x - 5$

21. $f(x) = x^7$ **22.** $f(x) = x^{1/5}$

In Exercises 23–26, verify algebraically that the functions f and g are inverses of each other.

23. $f(x) = x + 15$
    $g(x) = x - 15$

24. $f(x) = 2x - 1$
    $g(x) = \frac{1}{2}(x + 1)$

25. $f(x) = \sqrt[3]{x + 1}$
    $g(x) = x^3 - 1$

26. $f(x) = \dfrac{1}{x - 3}$
    $g(x) = 3 + \dfrac{1}{x}$

In Exercises 27–30, match the graph with the graph of its inverse. [The graphs of the inverse functions are labeled (a), (b), (c), and (d).]

(a)
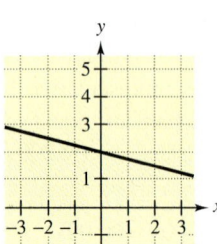

(b)
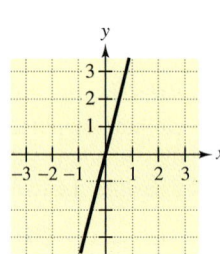

(c)
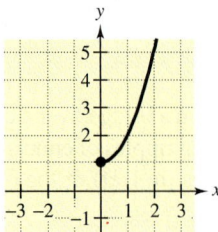

(d)
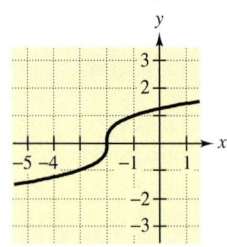

27.
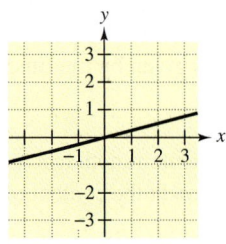

28.
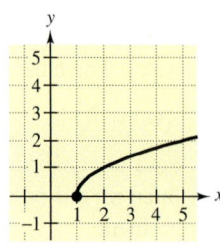

29.
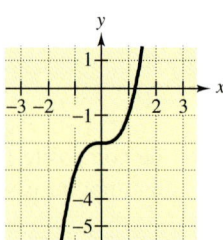

30.
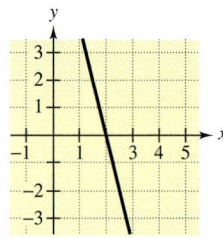

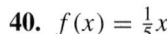 In Exercises 31–34, use a graphing utility to verify that the functions are inverses of each other.

31. $f(x) = \frac{1}{3}x$
    $g(x) = 3x$

32. $f(x) = \frac{1}{5}x - 1$
    $g(x) = 5x + 5$

33. $f(x) = \sqrt[3]{x + 2}$
    $g(x) = x^3 - 2$

34. $f(x) = \sqrt{x + 1}$
    $g(x) = x^2 - 1, \ x \geq 0$

In Exercises 35–38, find the inverse of the function.

35. $g(x) = 3 - 4x$ **36.** $h(x) = \sqrt{x + 5}$

37. $f(t) = t^3 - 1$ **38.** $f(s) = \dfrac{2}{3 - s}$

In Exercises 39–42, decide whether the function has an inverse.

39. $f(x) = x^2 - 2$ **40.** $f(x) = \frac{1}{5}x$

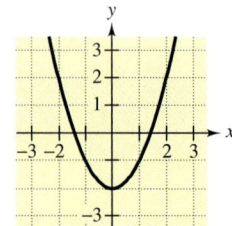

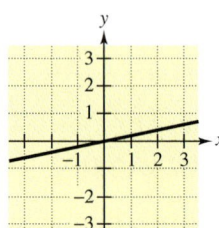

41. $g(x) = \sqrt{25 - x^2}$ **42.** $g(x) = |x - 4|$

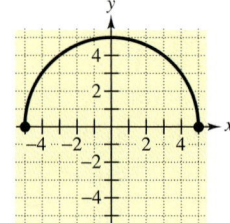

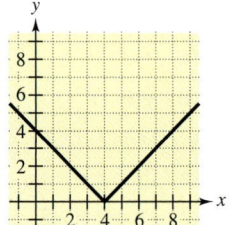

In Exercises 43–46, place a restriction on the domain of the function so that the graph of the restricted function satisfies the Horizontal Line Test. Then find the inverse of the restricted function. (*Note:* There is more than one correct answer.)

43. $f(x) = x^4$

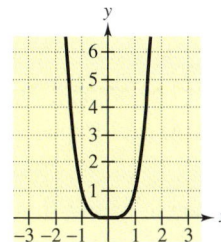

44. $f(x) = 9 - x^2$

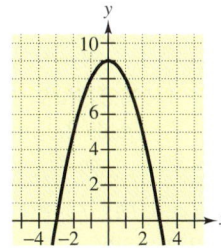

45. $f(x) = (x - 2)^2$

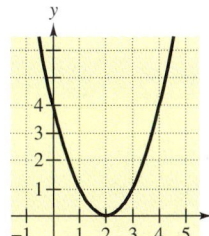

46. $f(x) = |x - 2|$

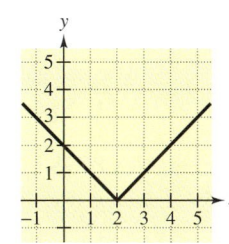

Graphical Reasoning In Exercises 47 and 48, use the graph of f to sketch the graph of f^{-1}. Explain the relationship between the graph of f and the graph of f^{-1}.

47.

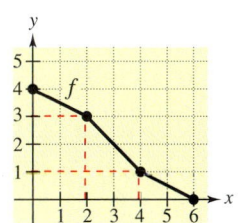

48.

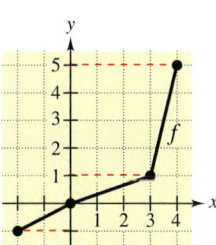

True or False? In Exercises 49 and 50, is the statement true or false? If true, explain your reasoning. If false, give an example to show why.

49. If the inverse of f exists, the y-intercept of f is an x-intercept of f^{-1}.

50. There exists no function f such that $f = f^{-1}$.

Reviewing the Major Concepts

In Exercises 51–54, solve the equation by the specified method.

51. $3x^2 + 9x - 12 = 0$ *Factoring*

52. $x^2 + 3x - 2 = 0$ *Completing the square*

53. $4x^2 - 7 = -6x$ *Quadratic Formula*

54. $x^2 - 10 = 0$ *Extracting square roots*

In Exercises 55 and 56, use $y = -x^2 + 4x$.

55. (a) Does the graph open up or down? Explain.

(b) Find the x-intercepts algebraically.

(c) Find the coordinates of the vertex of the parabola.

56. Use a graphing utility to graph the equation and verify the results of Exercise 55 graphically.

Additional Problem Solving

In Exercises 57–62, find the indicated composites.

57. $f(x) = x + 1$, $g(x) = 2x$

(a) $(f \circ g)(3)$ (b) $(g \circ f)(3)$

(c) $(f \circ g)(x)$ (d) $(g \circ f)(x)$

58. $f(x) = x^2 - 3x$, $g(x) = 5x + 3$

(a) $(f \circ g)(-1)$ (b) $(g \circ f)(3)$

(c) $(f \circ g)(x)$ (d) $(g \circ f)(x)$

59. $f(x) = |x - 3|$, $g(x) = 3x$

   (a) $(f \circ g)(1)$ (b) $(g \circ f)(2)$

   (c) $(f \circ g)(x)$ (d) $(g \circ f)(x)$

60. $f(x) = x + 5$, $g(x) = x^3$

   (a) $(f \circ g)(2)$ (b) $(g \circ f)(-3)$

   (c) $(f \circ g)(x)$ (d) $(g \circ f)(x)$

61. $f(x) = \dfrac{1}{x - 3}$, $g(x) = \sqrt{x}$

   (a) $(f \circ g)(49)$ (b) $(g \circ f)(12)$

   (c) $(f \circ g)(x)$ (d) $(g \circ f)(x)$

62. $f(x) = \sqrt{x + 6}$, $g(x) = 2x - 3$

   (a) $(f \circ g)(3)$ (b) $(g \circ f)(-2)$

   (c) $(f \circ g)(x)$ (d) $(g \circ f)(x)$

In Exercises 63–66, use the functions f and g to find the indicated values.

$f = \{(0, 1), (1, 2), (2, 5), (3, 10), (4, 17)\}$,

$g = \{(5, 4), (10, 1), (2, 3), (17, 0), (1, 2)\}$

63. (a) $f(3)$ **64.** (a) $g(2)$

   (b) $(g \circ f)(3)$ (b) $(f \circ g)(10)$

65. (a) $(g \circ f)(4)$ **66.** (a) $(f \circ g)(1)$

   (b) $(f \circ g)(2)$ (b) $(g \circ f)(0)$

In Exercises 67–70, find the domain of the compositions (a) $f \circ g$ and (b) $g \circ f$.

67. $f(x) = x^2 + 1$ **68.** $f(x) = 2 - 3x$

   $g(x) = 2x$ $g(x) = 5x + 3$

69. $f(x) = \dfrac{9}{x + 9}$ **70.** $f(x) = \sqrt{x - 5}$

   $g(x) = x^2$ $g(x) = x^2$

71. *Production Cost* The daily cost of producing x units of a product is $C(x) = 8.5x + 300$. The number of units produced in t hours during a day is given by $x(t) = 12t$, $0 \le t \le 8$. Find, simplify, and interpret $(C \circ x)(t)$.

72. *Rebate and Discount* The suggested retail price of a new car is p dollars. The dealership is advertising a factory rebate of \$2000 and a 5% discount.

   (a) Write a function R in terms of p, giving the cost of the car after receiving the rebate from the factory.

   (b) Write a function S in terms of p, giving the cost of the car after receiving the dealership discount.

   (c) Form the composite functions $(R \circ S)(p)$ and $(S \circ R)(p)$ and interpret each.

   (d) Find $(R \circ S)(26{,}000)$ and $(S \circ R)(26{,}000)$. Which yields the smaller cost for the car? Explain.

In Exercises 73–80, find the inverse of the function f informally.

73. $f(x) = 6x$ **74.** $f(x) = \frac{1}{3}x$

75. $f(x) = x + 10$ **76.** $f(x) = 10 - x$

77. $f(x) = \sqrt[3]{x}$ **78.** $f(x) = x^5$

79. $f(x) = 2x - 1$ **80.** $f(x) = \frac{1}{2}x + 3$

In Exercises 81–88, verify algebraically that the functions f and g are inverses of each other.

81. $f(x) = 10x$ **82.** $f(x) = \frac{2}{3}x$

   $g(x) = \frac{1}{10}x$ $g(x) = \frac{3}{2}x$

83. $f(x) = 1 - 2x$ **84.** $f(x) = 3 - x$

   $g(x) = \frac{1}{2}(1 - x)$ $g(x) = 3 - x$

85. $f(x) = 2 - 3x$ **86.** $f(x) = -\frac{1}{4}x + 3$

   $g(x) = \frac{1}{3}(2 - x)$ $g(x) = -4(x - 3)$

87. $f(x) = 1/x$ **88.** $f(x) = x^7$

   $g(x) = 1/x$ $g(x) = \sqrt[7]{x}$

In Exercises 89–92, use a graphing utility to verify that the functions are inverses of each other.

89. $f(x) = 3x + 4$ **90.** $f(x) = |x - 2|$, $x \ge 2$

   $g(x) = \frac{1}{3}(x - 4)$ $g(x) = x + 2$, $x \ge 0$

91. $f(x) = \frac{1}{8}x^3$

$g(x) = 2\sqrt[3]{x}$

92. $f(x) = \sqrt{4-x}$

$g(x) = 4 - x^2, \quad x \geq 0$

In Exercises 93–102, find the inverse of the function.

93. $f(x) = 8x$

94. $f(x) = \dfrac{x}{10}$

95. $g(x) = x + 25$

96. $f(x) = 7 - x$

97. $g(t) = -\frac{1}{4}t + 2$

98. $g(t) = 6t + 1$

99. $h(x) = \sqrt{x}$

100. $h(t) = t^5$

101. $g(s) = \dfrac{5}{s}$

102. $f(x) = \dfrac{4}{x-2}$

In Exercises 103 and 104, find the inverse of the function f. Use a graphing utility to sketch the graph of f and f^{-1}.

103. $f(x) = x^3 + 1$

104. $f(x) = \sqrt{x^2 - 4}, \quad x \geq 2$

In Exercises 105–114, use a graphing utility to graph the function and determine whether the function is one-to-one.

105. $f(x) = \frac{1}{4}x^3$

106. $f(x) = x^2 - 2$

107. $f(t) = \sqrt[3]{5 - t}$

108. $h(t) = 4 - \sqrt[3]{t}$

109. $g(x) = x^4$

110. $f(x) = (x+2)^5$

111. $h(t) = \dfrac{5}{t}$

112. $g(t) = \dfrac{5}{t^2}$

113. $f(s) = \dfrac{4}{s^2 + 1}$

114. $f(x) = \dfrac{1}{x-2}$

In Exercises 115 and 116, use the graph of f to sketch the graph of f^{-1}.

115.

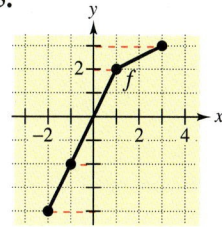

116.

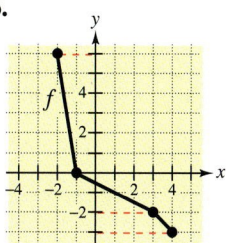

117. Consider the function $f(x) = 3 - 2x$.

(a) Find $f^{-1}(x)$.

(b) Find $(f^{-1})^{-1}(x)$.

118. Consider the functions $f(x) = 4x$ and $g(x) = x + 6$.

(a) Find $(f \circ g)(x)$.

(b) Find $(f \circ g)^{-1}(x)$.

(c) Find $f^{-1}(x)$ and $g^{-1}(x)$.

(d) Find $(g^{-1} \circ f^{-1})(x)$ and compare the result with that of part (b).

119. *Hourly Wage* Your wage is \$9.00 per hour plus \$0.65 for each unit produced per hour. Thus, your hourly wage y in terms of the number of units produced is given by $y = 9 + 0.65x$.

(a) Determine the inverse of the function.

(b) What does each variable represent in the inverse function?

(c) Determine the number of units produced when your hourly wage averages \$14.20.

120. *Cost* Suppose you need 100 pounds of two commodities costing \$0.50 and \$0.75 per pound.

(a) Verify that your total cost y is given by $y = 0.50x + 0.75(100 - x)$, where x is the number of pounds of the less expensive commodity.

(b) Find the inverse of the cost function. What does each variable represent in the inverse function?

(c) Use the context of the problem to determine the domain of the inverse function.

(d) Determine the number of pounds of the less expensive commodity purchased if the total cost is \$60.

True or False? In Exercises 121 and 122, decide whether the statement is true or false. If true, explain your reasoning. If false, give an example to show why.

121. If the inverse of f exists, the domains of f and f^{-1} are the same.

122. If the inverse of f exists and its graph passes through the point $(2, 2)$, the graph of f^{-1} also passes through the point $(2, 2)$.

13.3 Logarithmic Functions

Logarithmic Functions ▪ Graphs of Logarithmic Functions ▪ The Natural Logarithmic Function ▪ Change of Base

Logarithmic Functions

In Section 13.2, you were introduced to the concept of the inverse of a function. Moreover, you saw that if a function has the property that no horizontal line intersects the graph of the function more than once, the function must have an inverse. By looking back at the graphs of the exponential functions introduced in Section 13.1, you will see that every function of the form

$$f(x) = a^x$$

Exponential functions have inverses.

passes the horizontal line test, and therefore must have an inverse. This inverse function is called the **logarithmic function with base** *a*.

Definition of Logarithmic Function

Let *a* and *x* be positive real numbers such that $a \neq 1$. The **logarithm of** *x* **with base** *a* is denoted by $\log_a x$ and is defined as follows.

$$y = \log_a x \qquad \text{if and only if} \qquad x = a^y$$

The function $f(x) = \log_a x$ is the **logarithmic function with base** *a*.

From the definition of a logarithmic function, you can see that the equations $y = \log_a x$ and $x = a^y$ are equivalent.

Logarithmic Equation	*Exponential Equation*
$y = \log_a x$	$x = a^y$

The first equation is in *logarithmic* form and the second is in *exponential* form. From these equivalent equations it should also be clear that *a logarithm is an exponent.* For instance, because the exponent in the expression

$$2^3 = 8$$

The exponent of 2^3 is 3.

is 3, the value of the logarithm $\log_2 8$ is 3. That is,

$$\log_2 8 = 3.$$

A logarithm is an exponent.

Therefore, to evaluate the logarithmic expression $\log_a x$, you need to ask the question, "To what power must *a* be raised to obtain *x*?"

EXAMPLE 1   *Evaluating Logarithms*

Evaluate each logarithm.

a. $\log_2 16$ **b.** $\log_3 9$ **c.** $\log_4 2$

Solution

In each case you should answer the question, "To what power must the base be raised to obtain the given number?"

a. The power to which 2 must be raised to obtain 16 is 4. That is,

$$2^4 = 16 \qquad \Longrightarrow \qquad \log_2 16 = 4.$$

b. The power to which 3 must be raised to obtain 9 is 2. That is,

$$3^2 = 9 \qquad \Longrightarrow \qquad \log_3 9 = 2.$$

c. The power to which 4 must be raised to obtain 2 is $\frac{1}{2}$. That is,

$$4^{1/2} = 2 \qquad \Longrightarrow \qquad \log_4 2 = \frac{1}{2}.$$

Each of the logarithms in Example 2 involves a special important case. Study this example carefully.

EXAMPLE 2   *Evaluating Logarithms*

Evaluate each logarithm.

a. $\log_5 1$ **b.** $\log_{10} \dfrac{1}{10}$ **c.** $\log_3(-1)$ **d.** $\log_4 0$

Solution

a. The power to which 5 must be raised to obtain 1 is 0. That is,

$$5^0 = 1 \qquad \Longrightarrow \qquad \log_5 1 = 0.$$

b. The power to which 10 must be raised to obtain $\frac{1}{10}$ is -1. That is,

$$10^{-1} = \frac{1}{10} \qquad \Longrightarrow \qquad \log_{10} \frac{1}{10} = -1.$$

c. There is no power to which 3 can be raised to obtain -1. The reason for this is that for any value of x, 3^x is a positive number. Thus, $\log_3(-1)$ is undefined.

d. There is no power to which 4 can be raised to obtain 0. Thus, $\log_4 0$ is undefined.

The following properties of logarithms follow directly from the definition of the logarithmic function with base a.

Properties of Logarithms

Let a and x be positive real numbers such that $a \neq 1$. Then the following properties are true.

1. $\log_a 1 = 0$ because $a^0 = 1$.

2. $\log_a a = 1$ because $a^1 = a$.

3. $\log_a a^x = x$ because $a^x = a^x$.

The logarithmic function with base 10 is called the **common logarithmic function.** On most calculators, this function can be evaluated with the common logarithmic key $\boxed{\log}$, as illustrated in the next example.

EXAMPLE 3   *Evaluating Common Logarithms*

Evaluate each logarithm. Use a calculator only if necessary.

a. $\log_{10} 100$ **b.** $\log_{10} 0.01$ **c.** $\log_{10} 5$

Solution

a. Because $10^2 = 100$, it follows that

$$\log_{10} 100 = 2.$$

b. Because $10^{-2} = \frac{1}{100} = 0.01$, it follows that

$$\log_{10} 0.01 = -2.$$

c. There is no simple power to which 10 can be raised to obtain 5, so you should use a calculator to evaluate $\log_{10} 5$.

Keystrokes	*Display*	
5 $\boxed{\log}$	0.69897	Scientific
$\boxed{\log}$ 5 $\boxed{\text{ENTER}}$	0.69897	Graphing

Thus, rounded to three decimal places, $\log_{10} 5 \approx 0.699$.

NOTE Be sure you see that a logarithm can be zero or negative, *but* you cannot take the logarithm of zero or a negative number.

Graphs of Logarithmic Functions

To sketch the graph of $y = \log_a x$, we can use the fact that the graphs of inverse functions are reflections of each other in the line $y = x$.

EXAMPLE 4   *Graphs of Exponential and Logarithmic Functions*

On the same rectangular coordinate system, sketch the graphs of the following.

a. $f(x) = 2^x$ **b.** $g(x) = \log_2 x$

Solution

a. Begin by making a table of values for $f(x) = 2^x$.

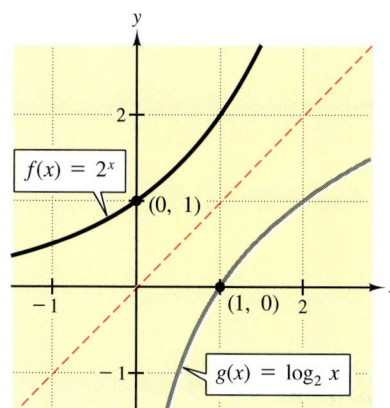

$f(x) = 2^x$

(0, 1)

-1 (1, 0)   2

$g(x) = \log_2 x$

FIGURE 13.12   Inverse Functions

x	-2	-1	0	1	2	3
$f(x) = 2^x$	$\frac{1}{4}$	$\frac{1}{2}$	1	2	4	8

By plotting these points and connecting them with a smooth curve, you obtain the graph shown in Figure 13.12.

b. Because $g(x) = \log_2 x$ is the inverse function of $f(x) = 2^x$, the graph of g is obtained by reflecting the graph of f in the line $y = x$, as shown in Figure 13.12.

STUDY TIP

In Example 4, the inverse property of logarithmic functions was used to sketch the graph of $g(x) = \log_2 x$. You could also use a standard point-plotting approach or a graphing utility.

Notice from the graph of $g(x) = \log_2 x$, shown in Figure 13.12, that the domain of the function is the set of positive numbers and the range is the set of all real numbers. The basic characteristics of the graph of a logarithmic function are summarized in Figure 13.13. In this figure, note that the graph has one x-intercept, at $(1, 0)$. Also note that the y-axis is a vertical asymptote of the graph.

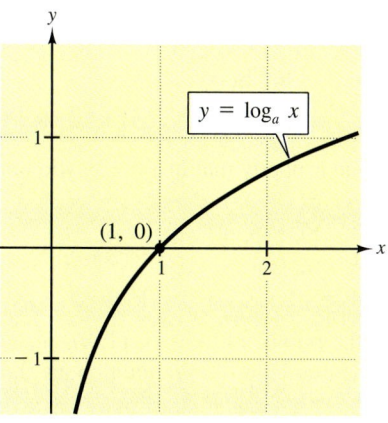

$y = \log_a x$

(1, 0)

Graph of $y = \log_a x$, $a > 1$
- Domain: $(0, \infty)$
- Range: $(-\infty, \infty)$
- Intercept: $(1, 0)$
- Increasing

FIGURE 13.13

The Natural Logarithmic Function

As with exponential functions, the most widely used base for logarithmic functions is the number e. The logarithmic function with base e is the **natural logarithmic function** and is denoted by the special symbol $\ln x$, which is read as "el en of x."

The Natural Logarithmic Function

The function defined by

$$f(x) = \log_e x = \ln x$$

where $x > 0$, is called the **natural logarithmic function.**

NOTE On most calculators, the natural logarithm key is denoted by $\boxed{\ln}$. For instance, on a scientific calculator, you can evaluate $\ln 2$ as

$2 \boxed{\ln}$

and on a graphing calculator, you can evaluate it as

$\boxed{\ln}\ \ 2\ \boxed{\text{ENTER}}$.

In either case, you should obtain a display of 0.6931472.

The three properties of logarithms listed earlier in this section are also valid for natural logarithms.

Properties of Natural Logarithms

Let x be a positive real number. Then the following properties are true.

1. $\ln 1 = 0$ because $e^0 = 1$.

2. $\ln e = 1$ because $e^1 = e$.

3. $\ln e^x = x$ because $e^x = e^x$.

EXAMPLE 5 *Evaluating the Natural Logarithmic Function*

Evaluate each of the following. Then incorporate the results into a graph of the natural logarithmic function.

a. $\ln e^2$ **b.** $\ln \dfrac{1}{e}$

Solution

Using the property that $\ln e^x = x$, you obtain the following.

a. $\ln e^2 = 2$

b. $\ln \dfrac{1}{e} = \ln e^{-1} = -1$

Using the points, $(1/e, -1)$, $(1, 0)$, $(e, 1)$, and $(e^2, 2)$, you can sketch the graph of the natural logarithmic function, as shown in Figure 13.14.

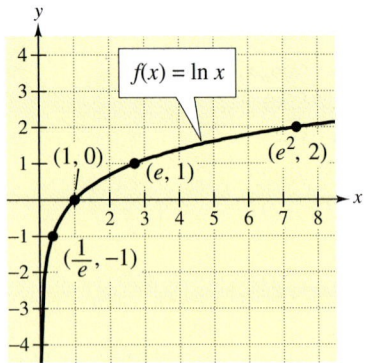

FIGURE 13.14

Change of Base

Although 10 and e are the most frequently used bases, you occasionally need to evaluate logarithms with other bases. In such cases the following **change-of-base formula** is useful.

Change-of-Base Formula

Let a, b, and x be positive real numbers such that $a \neq 1$ and $b \neq 1$. Then $\log_a x$ is given as follows.

$$\log_a x = \frac{\log_b x}{\log_b a} \quad \text{or} \quad \log_a x = \frac{\ln x}{\ln a}$$

The usefulness of this change-of-base formula is that you can use a calculator that has only the common logarithm key $\boxed{\log}$ and the natural logarithm key $\boxed{\ln}$ to evaluate logarithms to any base.

EXAMPLE 6 *Changing the Base to Evaluate Logarithms*

a. Use *common* logarithms to evaluate $\log_3 5$.

b. Use *natural* logarithms to evaluate $\log_6 2$.

Solution

Using the change-of-base formula, you can convert to common and natural logarithms by writing

$$\log_3 5 = \frac{\log_{10} 5}{\log_{10} 3} \quad \text{and} \quad \log_6 2 = \frac{\ln 2}{\ln 6}.$$

Now, use the following keystrokes.

a.

Keystrokes	Display	
5 $\boxed{\log}$ $\boxed{\div}$ 3 $\boxed{\log}$ $\boxed{=}$	1.4649735	Scientific
$\boxed{\log}$ 5 $\boxed{\div}$ $\boxed{\log}$ 3 $\boxed{\text{ENTER}}$	1.4649735	Graphing

Thus, $\log_3 5 \approx 1.465$.

b.

Keystrokes	Display	
2 $\boxed{\ln}$ $\boxed{\div}$ 6 $\boxed{\ln}$ $\boxed{=}$	0.3868528	Scientific
$\boxed{\ln}$ 2 $\boxed{\div}$ $\boxed{\ln}$ 6 $\boxed{\text{ENTER}}$	0.3868528	Graphing

Thus, $\log_6 2 \approx 0.387$.

At this point, you have been introduced to all the basic types of functions that are covered in this course: polynomial functions, radical functions, rational functions, exponential functions, and logarithmic functions. The only other common types of functions are *trigonometric functions*, which you will study if you go on to take a course in trigonometry or precalculus.

Group Activities   Reviewing the Major Concepts

Comparing Models   Suppose you work for a research and development firm that deals with a wide variety of disciplines. Your supervisor has asked your group to give a presentation to your department on four basic kinds of mathematical models. Identify each of the models shown below. Develop a presentation describing the types of data sets that each model would best represent. Include distinctions in domain, range, and intercepts and a discussion of the types of applications to which each model is suited.

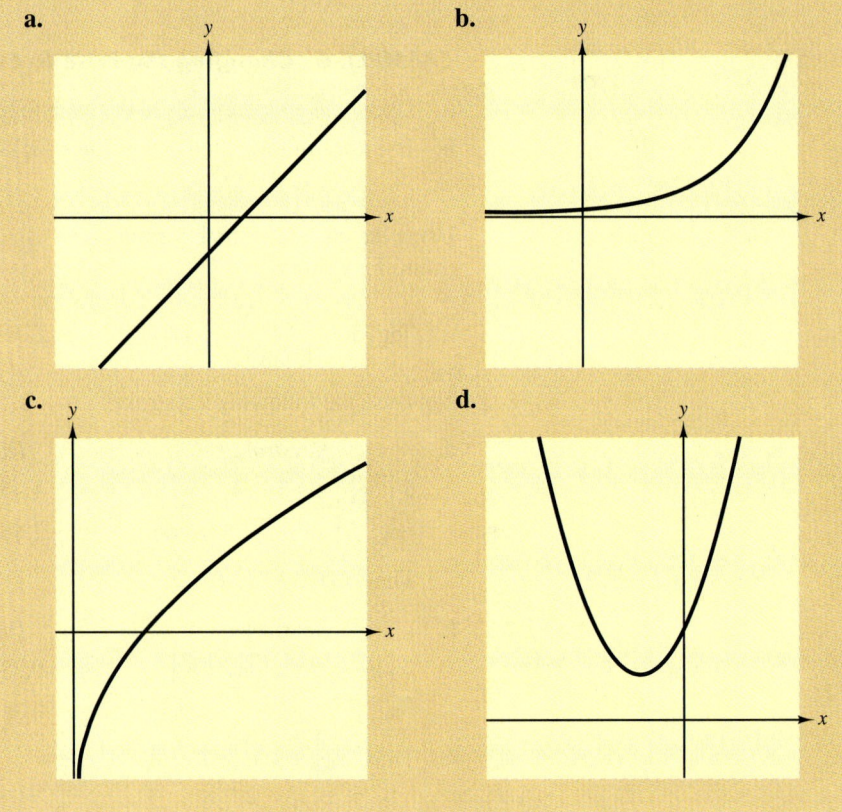

a.

b.

c.

d.

13.3 Exercises

Discussing the Concepts

1. Write "logarithm of x with base 5" symbolically.

2. Explain the relationship between $f(x) = 2^x$ and $g(x) = \log_2 x$.

3. Explain why $\log_a a = 1$.

4. Explain why $\log_a a^x = x$.

5. What are common and natural logarithms?

6. Describe how to use a calculator to find the logarithm of a number if the base is not 10 or e.

Problem Solving

In Exercises 7–10, write the logarithmic equation in exponential form.

7. $\log_5 25 = 2$

8. $\log_3 \frac{1}{243} = -5$

9. $\log_{36} 6 = \frac{1}{2}$

10. $\log_{32} 4 = \frac{2}{5}$

In Exercises 11–14, write the exponential equation in logarithmic form.

11. $3^{-2} = \frac{1}{9}$

12. $5^4 = 625$

13. $8^{2/3} = 4$

14. $81^{3/4} = 27$

In Exercises 15–22, evaluate the expression without a calculator. (If it is not possible, state the reason.)

15. $\log_2 8$

16. $\log_3 27$

17. $\log_2 \frac{1}{4}$

18. $\log_3 \frac{1}{9}$

19. $\log_2 -3$

20. $\log_3 1$

21. $\log_9 3$

22. $\log_{25} 125$

In Exercises 23–26, use the properties of natural logarithms to evaluate the expression.

23. $\ln e^2$

24. $\ln e^{-4}$

25. $\ln e$

26. $\ln(e^2 \cdot e^4)$

In Exercises 27–30, use a calculator to evaluate the logarithm. Round to four decimal places.

27. $\log_{10} 31$

28. $\log_{10} \dfrac{\sqrt{3}}{2}$

29. $\ln 0.75$

30. $\ln \left(\sqrt{3} + 1\right)$

In Exercises 31–36, match the function with its graph. [The graphs are labeled (a), (b), (c), (d), (e), and (f).]

(a)
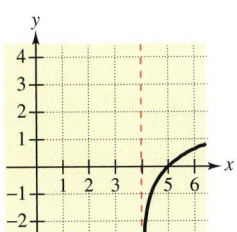

(b)
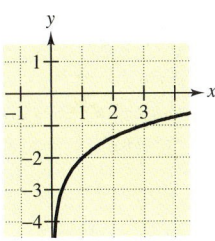

(c)
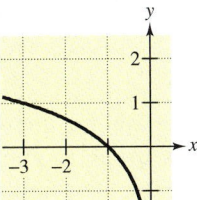

(d)
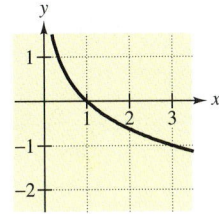

(e)
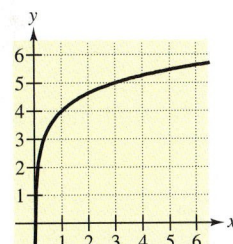

(f)
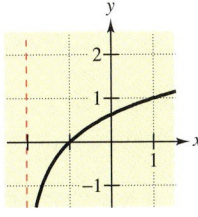

31. $f(x) = 4 + \log_3 x$

32. $f(x) = -2 + \log_3 x$

33. $f(x) = -\log_3 x$

34. $f(x) = \log_3(-x)$

35. $f(x) = \log_3(x - 4)$

36. $f(x) = \log_3(x + 2)$

In Exercises 37 and 38, describe the relationship between the graphs of f and g.

37. $f(x) = \log_3 x$

$g(x) = 3^x$

38. $f(x) = \log_4 x$

$g(x) = 4^x$

In Exercises 39–42, sketch the graph of the function.

39. $f(x) = \log_5 x$

40. $g(x) = \log_8 x$

41. $f(x) = 3 + \log_2 x$

42. $f(x) = -2 + \log_3 x$

In Exercises 43–46, use a graphing utility to sketch the graph of the function.

43. $y = 5 \log_{10} x$

44. $y = 5 \log_{10} (x - 3)$

45. $f(x) = 3 \ln x$

46. $h(x) = 2 + \ln x$

In Exercises 47–50, use a calculator to evaluate the logarithm by means of the change-of-base formula.

47. $\log_8 132$

48. $\log_5 510$

49. $\log_2 0.72$

50. $\log_3(1 + e^2)$

51. *Creating a Table*   The time t in years for an investment to double in value when compounded continuously at annual interest rate r is given by

$$t = \frac{\ln 2}{r}.$$

Complete the table, which shows the "doubling time" for several annual interest rates.

r	0.07	0.08	0.09	0.10	0.11	0.12
t						

52. *Intensity of Sound*   The relationship between the number of decibels B and the intensity of a sound I in watts per meter squared is given by

$$B = 10 \log_{10}\left(\frac{I}{10^{-16}}\right).$$

Determine the number of decibels of a sound with an intensity of 10^{-4} watts per centimeter squared.

Reviewing the Major Concepts

In Exercises 53 and 54, use a graphing utility to graph each equation and use the graph to approximate any points of intersection. Find the solution of the system of equations algebraically.

53. $x + 3y = 11$

$x^2 - 3y = -5$

54. $x + 2y = -7$

$-x - y^2 = -17$

Geometry   In Exercises 55 and 56, find the area of the shaded region.

55.

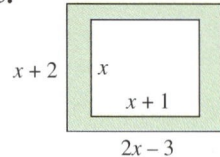

56.

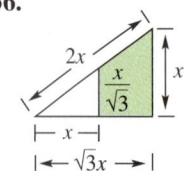

In Exercises 57 and 58, use the following information. From 1986 through 1990, the number of compact discs C (in millions) and the number of record albums R (in millions) sold in the United States can be approximated by the models

$C = 4.3t^2 - 9.8t - 42.7$ and $R = -28.8t + 301$

where $t = 6$ represents 1986.   (Source: Recording Industry Association of America)

57. How many compact discs and how many records were sold in 1986?

58. During which year was the number of compact discs sold approximately equal to the number of record albums sold?

Additional Problem Solving

In Exercises 59–64, write the logarithmic equation in exponential form.

59. $\log_6 36 = 2$

60. $\log_{10} 10,000 = 4$

61. $\log_4 \frac{1}{16} = -2$

62. $\log_8 \frac{1}{8} = -1$

63. $\log_8 4 = \frac{2}{3}$

64. $\log_{16} 8 = \frac{3}{4}$

In Exercises 65–70, write the exponential equation in logarithmic form.

65. $7^2 = 49$

66. $6^4 = 1296$

67. $25^{-1/2} = \frac{1}{5}$

68. $6^{-3} = \frac{1}{216}$

69. $4^0 = 1$

70. $10^{0.12} \approx 1.318$

In Exercises 71–82, evaluate the expression without a calculator. (If it is not possible, state the reason.)

71. $\log_4 1$

72. $\log_5 (-6)$

73. $\log_{10} 10$

74. $\log_8 8$

75. $\log_{10} 1000$

76. $\log_{10} \frac{1}{100}$

77. $\log_4(-4)$

78. $\log_2 0$

79. $\log_{16} 8$

80. $\log_{144} 12$

81. $\log_7 7^4$

82. $\log_5 5^3$

In Exercises 83–86, use the properties of natural logarithms to evaluate the expression.

83. $\ln 1$

84. $\ln 8^0$

85. $\ln \dfrac{e^3}{e^2}$

86. $\ln \dfrac{1}{e^3}$

In Exercises 87–92, use a calculator to evaluate the logarithm. Round to four decimal places.

87. $\log_{10} \left(\sqrt{2} + 4 \right)$

88. $\log_{10} 5310$

89. $\log_{10} 0.85$

90. $\log_{10} 0.345$

91. $\ln 25$

92. $\ln 6.57$

In Exercises 93 and 94, describe the relationship between the graphs of f and g.

93. $f(x) = \log_6 x$
 $g(x) = 6^x$

94. $f(x) = \log_{(1/2)} x$
 $g(x) = \left(\frac{1}{2} \right)^x$

In Exercises 95–100, sketch the graph of the function.

95. $g(t) = -\log_2 t$

96. $h(s) = -2\log_3 s$

97. $g(x) = \log_2(x - 3)$

98. $h(x) = \log_3(x + 1)$

99. $f(x) = \log_{10}(10x)$

100. $g(x) = \log_4(4x)$

In Exercises 101–108, use a graphing utility to sketch the graph of the function.

101. $y = -3 + 5\log_{10} x$

102. $y = 5\log_{10}(3x)$

103. $h(t) = -2\ln t$

104. $f(x) = 4\ln x$

105. $g(x) = \ln(x + 6)$

106. $h(x) = -\ln(x - 2)$

107. $f(x) = \ln(-x)$

108. $g(x) = -\frac{3}{2}\ln x$

In Exercises 109–116, use a calculator to evaluate the logarithm by means of the change-of-base formula. Use (a) the common logarithm key and (b) the natural logarithm key.

109. $\log_3 7$

110. $\log_7 4$

111. $\log_{15} 1250$

112. $\log_{20} 125$

113. $\log_4 \sqrt{42}$

114. $\log_{12} 0.6$

115. $\log_{(1/2)} 4$

116. $\log_{(1/3)}(0.015)$

117. *American Elk* The antler spread a (in inches) and shoulder height h (in inches) of an adult male American elk are related by the model

$$h = 116\log_{10}(a + 40) - 176.$$

Approximate the shoulder height of a male American elk with an antler spread of 55 inches.

118. *Tornadoes* Most tornadoes last less than 1 hour and travel less than 20 miles. The speed of the wind S (in miles per hour) near the center of the tornado is related to the distance the tornado travels d (in miles) by the model

$$S = 93 \log_{10} d + 65.$$

On March 18, 1925, a large tornado struck portions of Missouri, Illinois, and Indiana, covering a distance of 220 miles. Approximate the speed of the wind near the center of this tornado.

In Exercises 119–124, answer the question for the function $f(x) = \log_{10} x$. (Do not use a calculator.)

119. What is the domain of f?

120. Find the inverse function of f.

121. Describe the values of $f(x)$ for $1000 \leq x \leq 10{,}000$.

122. Describe the values of x, given that $f(x)$ is negative.

123. By what amount will x increase, given that $f(x)$ is increased by one unit?

124. Find the ratio of a to b, given that $f(a) = 3f(b)$.

Math Matters

Animal Species

There are approximately 6 million different species of living organisms on earth—animals, plants, single-cell organisms, fungi, and so on. Of those that are animals, by far most are insects. What percent of the different species on earth are insects? (The answer is given in the back of the book.)

Type of Animal	Examples	Number of Species
Insect	Ant, Butterfly, Honey Bee	950,000
Arachnid	Spider, Scorpion, Mite	110,000
Mollusk	Snail, Clam, Squid	45,000
Fish	Tuna, Salmon, Trout	30,000
Crustacean	Shrimp, Lobster, Crab	26,000
Bird	Robin, Ostrich, Eagle	8,650
Worm	Earthworm, Flatworm, Roundworm	6,700
Echinoderm	Starfish, Sea Urchin, Sea Cucumber	6,000
Reptile	Snake, Lizard, Alligator	5,180
Mammal	Human, Mouse, Whale	4,230
Amphibian	Frog, Toad, Salamander	3,000

MID-CHAPTER QUIZ

Take this quiz as you would take a quiz in class. After you are done, check your work against the answers given in the back of the book.

1. Given $f(x) = \left(\frac{4}{3}\right)^x$, find (a) $f(2)$, (b) $f(0)$, (c) $f(-1)$, and (d) $f(1.5)$.

2. Find the domain and range of $g(x) = 2^{-0.5x}$.

In Exercises 3–6, sketch the graph of the function.

3. $y = \frac{1}{2}(4^x)$ **4.** $y = 5(2^{-x})$ **5.** $f(t) = 12e^{-0.4t}$ **6.** $g(x) = 100(1.08)^x$

7. You deposit \$750 at $7\frac{1}{2}\%$ interest, compounded n times per year or continuously. Find the balance A after 20 years.

n	1	4	12	365	Continuous compounding
A					

8. A gallon of milk costs \$2.23 now. If the price increases by 4% each year, what will the price be after 5 years?

9. Given $f(x) = 2x - 3$ and $g(x) = x^3$, find the indicated composition.

    (a) $(f \circ g)(-2)$ (b) $(g \circ f)(4)$ (c) $(f \circ g)(x)$ (d) $(g \circ f)(x)$

10. Verify algebraically and graphically that $f(x) = 3 - 5x$ and $g(x) = \frac{1}{5}(3-x)$ are inverses of each other.

In Exercises 11 and 12, find the inverse of the function.

11. $h(x) = 10x + 3$ **12.** $g(t) = \frac{1}{2}t^3 + 2$

13. Write the logarithmic equation $\log_4\left(\frac{1}{16}\right) = -2$ in exponential form.

14. Write the exponential equation $3^4 = 81$ in logarithmic form.

15. Evaluate $\log_5 125$ without the aid of a calculator.

16. Write a paragraph comparing the graphs of $f(x) = \log_5 x$ and $g(x) = 5^x$.

In Exercises 17 and 18, use a graphing utility to sketch the graph of the function.

17. $f(t) = \frac{1}{2}\ln t$ **18.** $h(x) = 3 - \ln x$

19. Use the graph of f at the right to determine h and k if $f(x) = \log_5(x-h)+k$.

20. Use a calculator and the change-of-base formula to evaluate $\log_6 450$.

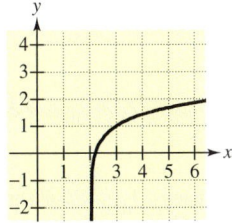

Figure for 19

13.4 Properties of Logarithms

Properties of Logarithms ▪ Rewriting Logarithmic Expressions ▪ Application

Properties of Logarithms

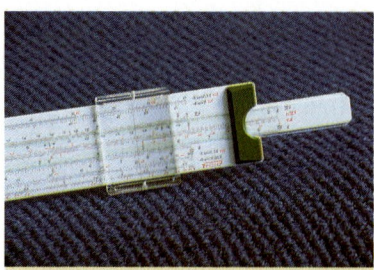

You know from the previous section that the logarithmic function with base a is the *inverse* of the exponential function with base a. Thus, it makes sense that each property of exponents should have a corresponding property of logarithms. For instance, the exponential property

$$a^0 = 1 \qquad \text{Exponential property}$$

has the corresponding logarithmic property

$$\log_a 1 = 0. \qquad \text{Corresponding logarithmic property}$$

In this section you will study the logarithmic properties that correspond to the following three exponential properties:

Base a	*Natural Base*
1. $a^n a^m = a^{n+m}$	$e^n e^m = e^{n+m}$
2. $\dfrac{a^n}{a^m} = a^{n-m}$	$\dfrac{e^n}{e^m} = e^{n-m}$
3. $(a^n)^m = a^{nm}$	$(e^n)^m = e^{nm}$

Before electronic hand-held calculators became available in the 1970s, mathematicians, engineers, and scientists relied on a tool called the slide rule. Created by Edmund Gunter (1581–1626), the slide rule uses logarithms to quickly multiply and divide numbers.

Properties of Logarithms

Let a be a positive real number such that $a \neq 1$, and let n be a real number. If u and v are real numbers, variables, or algebraic expressions such that $u > 0$ and $v > 0$, the following properties are true.

Logarithm with Base a	*Natural Logarithm*
1. $\log_a(uv) = \log_a u + \log_a v$	$\ln(uv) = \ln u + \ln v$
2. $\log_a \dfrac{u}{v} = \log_a u - \log_a v$	$\ln \dfrac{u}{v} = \ln u - \ln v$
3. $\log_a u^n = n \log_a u$	$\ln u^n = n \ln u$

NOTE   There is no general property of logarithms that can be used to simplify $\log_a(u + v)$. Specifically,

$$\log_a(u + v) \ \text{does not equal} \ \log_a u + \log_a v.$$

EXAMPLE 1   *Using Properties of Logarithms*

Use the fact that $\ln 2 \approx 0.693$, $\ln 3 \approx 1.099$, and $\ln 5 \approx 1.609$ to approximate each of the following.

a. $\ln \dfrac{2}{3}$ **b.** $\ln 10$ **c.** $\ln 30$

Solution

a.

$$\ln \frac{2}{3} = \ln 2 - \ln 3 \qquad\qquad \text{Property 2}$$

$$\approx 0.693 - 1.099 \qquad\qquad \text{Substitute for } \ln 2 \text{ and } \ln 3.$$

$$= -0.406 \qquad\qquad\qquad \text{Simplify.}$$

b.

$$\ln 10 = \ln(2 \cdot 5) \qquad\qquad \text{Factor.}$$

$$= \ln 2 + \ln 5 \qquad\qquad \text{Property 1}$$

$$\approx 0.693 + 1.609 \qquad\qquad \text{Substitute for } \ln 2 \text{ and } \ln 5.$$

$$= 2.302 \qquad\qquad\qquad \text{Simplify.}$$

c.

$$\ln 30 = \ln(2 \cdot 3 \cdot 5) \qquad\qquad \text{Factor.}$$

$$= \ln 2 + \ln 3 + \ln 5 \qquad\qquad \text{Property 1}$$

$$\approx 0.693 + 1.099 + 1.609 \qquad\qquad \text{Substitute for } \ln 2, \ln 3, \text{ and } \ln 5.$$

$$= 3.401 \qquad\qquad\qquad \text{Simplify.}$$

NOTE   When using the properties of logarithms, it helps to state the properties *verbally*. For instance, the verbal form of the property $\ln(uv) = \ln u + \ln v$ is: *The log of a product is the sum of the logs of the factors.* Similarly, the verbal form of the property $\ln(u/v) = \ln u - \ln v$ is: *The log of a quotient is the difference of the logs of the numerator and denominator.*

STUDY TIP

Remember that you can verify results such as those given in Example 2 with a calculator.

EXAMPLE 2   *Using Properties of Logarithms*

Use the properties of logarithms to verify that $-\ln 2 = \ln \frac{1}{2}$.

Solution

Using Property 3, you can write the following.

$$-\ln 2 = (-1)\ln 2 \qquad\qquad \text{Rewrite coefficient as } -1.$$

$$= \ln 2^{-1} \qquad\qquad\qquad \text{Property 3}$$

$$= \ln \frac{1}{2} \qquad\qquad\qquad \text{Rewrite } 2^{-1} \text{ as } \tfrac{1}{2}.$$

Rewriting Logarithmic Expressions

In Examples 1 and 2, the properties of logarithms were used to rewrite logarithmic expressions involving the log of a *constant*. A more common use of the properties is to rewrite the log of a *variable expression*.

EXAMPLE 3   Rewriting the Logarithm of a Product

Use the properties of logarithms to rewrite $\log_{10} 7x^3$.

Solution

$$\log_{10} 7x^3 = \log_{10} 7 + \log_{10} x^3 \qquad \text{Property 1}$$
$$= \log_{10} 7 + 3\log_{10} x \qquad \text{Property 3}$$

When you rewrite a logarithmic expression as in Example 3, you are **expanding** the expression. The reverse procedure is demonstrated in Example 4, and is called **condensing** a logarithmic expression.

EXAMPLE 4   Condensing a Logarithmic Expression

Use the properties of logarithms to condense $\ln x - \ln 3$.

Solution

Using Property 2, you can write

$$\ln x - \ln 3 = \ln \frac{x}{3}. \qquad \text{Property 2}$$

NOTE   In Example 4, try confirming the result by substituting values of x.

EXAMPLE 5   Expanding a Logarithmic Expression

Expand the logarithmic expression.

$$\log_2 3xy^2, \qquad x > 0, \quad y > 0$$

Solution

$$\log_2 3xy^2 = \log_2 3 + \log_2 x + \log_2 y^2 \qquad \text{Property 1}$$
$$= \log_2 3 + \log_2 x + 2\log_2 y \qquad \text{Property 3}$$

Sometimes expanding or condensing logarithmic expressions involves several steps. In the next example, be sure that you can justify each step in the solution. Also, notice how different the expanded expression is from the original.

EXAMPLE 6 *Expanding a Logarithmic Expression*

Expand the logarithmic expression.

$$\ln \sqrt{x^2 - 1}, \qquad x > 1$$

Solution

$$
\begin{aligned}
\ln \sqrt{x^2 - 1} &= \ln(x^2 - 1)^{1/2} && \text{Rewrite using fractional exponent.} \\
&= \tfrac{1}{2} \ln(x^2 - 1) && \text{Property 3} \\
&= \tfrac{1}{2} \ln(x - 1)(x + 1) && \text{Factor.} \\
&= \tfrac{1}{2} [\ln(x - 1) + \ln(x + 1)] && \text{Property 1} \\
&= \tfrac{1}{2} \ln(x - 1) + \tfrac{1}{2} \ln(x + 1) && \text{Distributive Property}
\end{aligned}
$$

EXAMPLE 7 *Condensing a Logarithmic Expression*

Use the properties of logarithms to condense the expression.

$$\ln 2 - 2 \ln x$$

Solution

$$
\begin{aligned}
\ln 2 - 2 \ln x &= \ln 2 - \ln x^2, & x > 0 && \text{Property 3} \\
&= \ln \frac{2}{x^2}, & x > 0 && \text{Property 2}
\end{aligned}
$$

When you expand or condense a logarithmic expression, it is possible to change the domain of the expression. For instance, the domain of the function

$$f(x) = 2 \ln x \qquad \text{Domain is the set of positive real numbers.}$$

is the set of positive real numbers, whereas the domain of

$$g(x) = \ln x^2 \qquad \text{Domain is the set of nonzero real numbers.}$$

is the set of nonzero real numbers. Thus, when you expand or condense a logarithmic expression, you should check to see whether the rewriting has changed the domain of the expression. In such cases, you should restrict the domain appropriately. For instance, you can write

$$f(x) = 2 \ln x = \ln x^2, \qquad x > 0.$$

Application

EXAMPLE 8   An Application: Human Memory Model

Students participating in a psychological experiment attended several lectures on a subject. Every month for a year after that, the students were tested to see how much of the material they remembered. The average scores for the group were given by the **human memory model**

$$f(t) = 80 - \ln(t + 1)^9, \qquad 0 \le t \le 12$$

where t is the time in months. Find the average scores for the group after 2 months and 8 months.

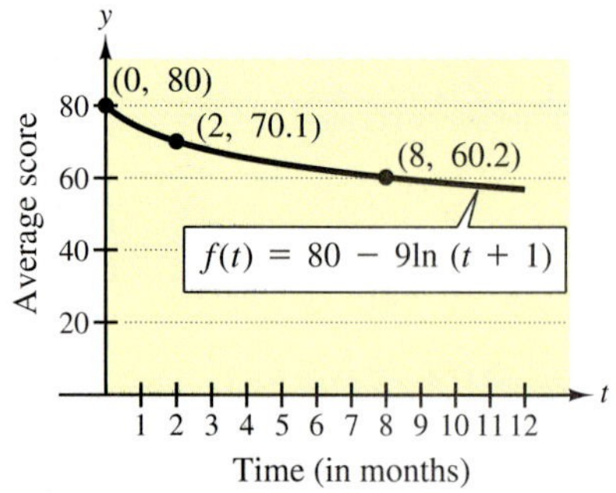

Solution

To make the calculations easier, rewrite the model as

$$f(t) = 80 - 9\ln(t + 1), \qquad 0 \le t \le 12.$$

After 2 months, the average score will be

$$f(2) = 80 - 9\ln 3 \approx 70.1 \qquad \text{Average score after 2 months}$$

and after 8 months, the average score will be

$$f(8) = 80 - 9\ln 9 \approx 60.2. \qquad \text{Average score after 8 months}$$

The graph of the function is shown in Figure 13.15.

FIGURE 13.15   Human Memory Model

The function graph shows y = Average score vs t = Time (in months), with points (0, 80), (2, 70.1), (8, 60.2), and the equation $f(t) = 80 - 9\ln(t + 1)$.

Group Activities   Problem Solving

x	5	7	9	11	13
y	522	1682	3370	5375	12,789

Mathematical Modeling   The data in the table represents the annual number y of new AIDS cases in females reported in the United States for the year x from 1985 through 1993, with $x = 5$ corresponding to 1998.   (Source: National Center for Health Statistics)

a.   Plot the data. Would a linear model fit the points well?

b.   Add a third row to the table giving the values of $\ln y$.

c.   Plot the coordinate pairs $(x, \ln y)$. Would a linear model fit these points well? If so, draw the best-fitting line and find its equation.

d.   Describe the shapes of the two scatter plots. Using your knowledge of logarithms, explain why the second scatter plot is so different from the first.

13.4 Exercises

Discussing the Concepts

1. State the properties of logarithms verbally.

2. Is it true that $\log_4(x^2 + 9) = \log_4 x^2 + \log_4 9$? Explain.

3. If $f(x) = \log_a x$, does $f(ax) = 1 + f(x)$? Explain.

4. If $f(x) = \log_a x$, does $f(a^n) = n$?

Problem Solving

In Exercises 5–8, evaluate the expression without using a calculator. (If it is not possible, state the reason.)

5. $\log_5 5^2$

6. $\log_6 2 + \log_6 3$

7. $\log_2 \frac{1}{8}$

8. $\log_3(-3)$

In Exercises 9–14, use $\log_{10} 3 \approx 0.477$ and $\log_{10} 12 \approx 1.079$ to approximate the value of the expression. Use a calculator to verify your results.

9. $\log_{10} 9$

10. $\log_{10} \frac{1}{4}$

11. $\log_{10} 36$

12. $\log_{10} 144$

13. $\log_{10} \sqrt{36}$

14. $\log_{10} 5^0$

In Exercises 15–22, use the properties of logarithms to expand the expression as a sum, difference, or multiple of logarithms.

15. $\log_2 3x$

16. $\log_3 x^3$

17. $\log_5 x^{-2}$

18. $\log_2 \sqrt{s}$

19. $\ln \frac{5}{x-2}$

20. $\ln \sqrt[3]{\frac{x^2}{x+1}}$

21. $\ln \sqrt{x(x+2)}$

22. $\ln \left(\frac{x+1}{x-1}\right)^2$

In Exercises 23–30, use the properties of logarithms to condense the expression.

23. $\log_{12} x - \log_{12} 3$

24. $\log_6 12 + \log_6 y$

25. $-2 \log_5 2x$

26. $10 \log_4 z$

27. $5 \ln 2 - \ln x + 3 \ln y$

28. $4 \ln 2 + 2 \ln x - \frac{1}{2} \ln y$

29. $\frac{1}{2}(\ln 8 + \ln 2x)$

30. $5 \left[\ln x - \frac{1}{2} \ln(x+4)\right]$

In Exercises 31–36, simplify the expression.

31. $\log_4 \frac{4}{x}$

32. $\log_3(3^2 \cdot 4)$

33. $\log_5 \sqrt{50}$

34. $\log_2 \sqrt{22}$

35. $\ln 3e^2$

36. $\ln \frac{6}{e^5}$

In Exercises 37 and 38, use a graphing utility to verify that the expressions are equivalent.

37. $\ln\left[10/(x^2 + 1)\right]^2$ and $2\left[\ln 10 - \ln(x^2 + 1)\right]$

38. $\ln \sqrt{x(x + 1)}$ and $\frac{1}{2}[\ln x + \ln(x + 1)]$

39. *Think About It* Explain how you could show that $(\ln x)/(\ln y) \neq \ln(x/y)$.

40. *Think About It* Approximate the natural logarithms of as many integers as possible between 1 and 20 using $\ln 2 \approx 0.6931$, $\ln 3 \approx 1.0986$, and $\ln 5 \approx 1.6094$. (Do not use a calculator.)

Biology In Exercises 41 and 42, use the following information. The energy E (in kilocalories per gram molecule) required to transport a substance from the outside to the inside of a living cell is

$$E = 1.4(\log_{10} C_2 - \log_{10} C_1)$$

where C_1 is the concentration of the substance outside the cell and C_2 is the concentration inside.

41. Condense the expression.

42. The concentration of a particular substance inside a cell is twice the concentration outside the cell. How much energy is required to transport the substance from outside to inside the cell?

Reviewing the Major Concepts

In Exercises 43–48, use the properties of exponents to simplify the expression.

43. $(x^2 \cdot x^3)^4$

44. $4^{-2} \cdot x^2$

45. $\dfrac{15y^{-3}}{10y^2}$

46. $\left(\dfrac{3x^2}{2y}\right)^{-2}$

47. $\dfrac{3x^2y^3}{18x^{-1}y^2}$

48. $(x^2 + 1)^0$

49. *Ticket Prices* A service organization paid $288 for a block of tickets to a game. The block contained three more tickets than the organization needed for its members. By inviting three more people to share in the cost, the organization lowered the price per ticket by $8. How many people will attend?

50. *Geometry* Sketch a diagram of a right triangle whose perimeter is 12 inches.

Additional Problem Solving

In Exercises 51–58, evaluate the expression without using a calculator. (If it is not possible, state the reason.)

51. $\log_4 2 + \log_4 8$

52. $\log_5 50 - \log_5 2$

53. $\log_3 9$

54. $\log_6 \sqrt{6}$

55. $\ln e^5 - \ln e^2$

56. $\ln e^4$

57. $\ln 1$

58. $\log_{10} 1$

In Exercises 59–70, approximate the logarithm given that $\log_4 2 = 0.5000$ and $\log_4 3 \approx 0.7925$. Do not use a calculator.

59. $\log_4 4$

60. $\log_4 8$

61. $\log_4 6$

62. $\log_4 24$

63. $\log_4 \frac{3}{2}$

64. $\log_4 \frac{9}{2}$

65. $\log_4 \sqrt{2}$

66. $\log_4 \sqrt[3]{9}$

67. $\log_4 3 \cdot 2^4$

68. $\log_4 \sqrt{3 \cdot 2^5}$

69. $\log_4 3^0$

70. $\log_4 4^3$

In Exercises 71–84, use the properties of logarithms to expand the expression as a sum, difference, or multiple of logarithms.

71. $\log_3 11x$

72. $\ln 5x$

73. $\ln y^3$

74. $\log_7 x^2$

75. $\log_2 \dfrac{z}{17}$

76. $\log_{10} \dfrac{7}{y}$

77. $\log_3 \sqrt[3]{x+1}$

78. $\log_4 \dfrac{1}{\sqrt{t}}$

79. $\ln 3x^2 y$

80. $\ln y(y-1)^2$

81. $\log_2 \dfrac{x^2}{x-3}$

82. $\log_5 \sqrt{\dfrac{x}{y}}$

83. $\ln \sqrt[3]{x(x+5)}$

84. $\ln \left[3x(x-5)\right]^2$

In Exercises 85–98, use the properties of logarithms to condense the expression.

85. $\log_2 3 + \log_2 x$

86. $\log_5 2x + \log_5 3y$

87. $\log_{10} 4 - \log_{10} x$

88. $\ln 10x - \ln z$

89. $4 \ln b$

90. $-5 \ln (x+3)$

91. $\frac{1}{3} \ln (2x+1)$

92. $-\frac{1}{2} \log_3 5y$

93. $\log_3 2 + \frac{1}{2} \log_3 y$

94. $\ln 6 - 3 \ln z$

95. $2 \ln x + 3 \ln y - \ln z$

96. $4 \ln 3 - 2 \ln x - \ln y$

97. $4(\ln x + \ln y)$

98. $2[\ln x - \ln (x+1)]$

True or False? In Exercises 99–104, determine whether the equation is true or false. Explain.

99. $\ln e^{2-x} = 2 - x$

100. $\log_2 8x = 3 + \log_2 x$

101. $\log_8 4 + \log_8 16 = 2$

102. $\log_3 (u + v) = \log_3 u + \log_3 v$

103. $\log_3 (u + v) = \log_3 u \cdot \log_3 v$

104. $\dfrac{\log_6 10}{\log_6 3} = \log_6 10 - \log_6 3$

In Exercises 105 and 106, use a graphing utility to graph the two equations in the same viewing rectangle. Use the graphs to verify that the expressions are equivalent.

105. $y_1 = \ln[x^2(x+2)], \quad x > 0$

$y_2 = 2 \ln x + \ln(x+2)$

106. $y_1 = \ln\left(\dfrac{\sqrt{x}}{x-3}\right)$

$y_2 = \dfrac{1}{2} \ln x - \ln(x-3)$

107. *Intensity of Sound*   The relationship between the number of decibels B and the intensity of a sound I in watts per meter squared is given by

$$B = 10 \log_{10}\left(\dfrac{I}{10^{-16}}\right).$$

Use the properties of logarithms to write the formula in simpler form, and determine the number of decibels of a sound with an intensity of 10^{-10} watts per centimeter squared.

108. *Human Memory Model*   Students participating in a psychological experiment attended several lectures on a subject. Every month for a year after that, the students were tested to see how much of the material they remembered. The average scores for the group were given by the human memory model

$$f(t) = 80 - \log_{10}(t+1)^{12}, \quad 0 \le t \le 12$$

where t is the time in months.

(a) Find the average scores for the group after 2 months and 8 months.

(b) Use a graphing utility to graph the function.

True or False?   In Exercises 109–114, determine if the statement is true or false given that $f(x) = \ln x$. If false, state why or give an example to show that it is false.

109. $f(0) = 0$

110. $f(2x) = \ln 2 + \ln x, \quad x > 0$

111. $f(x-3) = \ln x - \ln 3, \quad x > 3$

112. $\sqrt{f(x)} = \frac{1}{2} \ln x$

113. If $f(u) = 2f(v)$, then $v = u^2$.

114. If $f(x) > 0$, then $x > 1$.

C A R E E R   I N T E R V I E W

Mary Kay Brown

Research Assistant

Dartmouth Medical School

Hanover, NH 03755

A question that is often asked in the lab is "How much drug does it take to kill 50% of a cell population?" To find out, we set up an assay using different drug concentrations to treat the same number of cells. After a period of several days, the percent of cells that survive at each drug concentration is calculated and graphed. However, there is such a wide range of concentrations (from 0.005 to 100 units in this example) that a linear-scale graph is difficult to read. So we graph this data using a *logarithmic scale*, which gives the same results as graphing $\log_{10}$ of each concentration. The logarithmic scale spreads the points out in a nice curve, and it is easy to see that a drug concentration between one and two units killed 50% of the cell population.

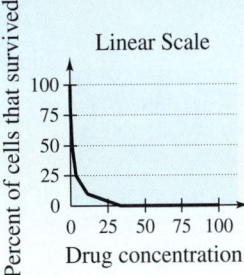

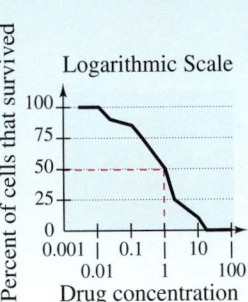

13.5	**Solving Exponential and Logarithmic Equations**

Exponential and Logarithmic Equations ▪ Solving Exponential Equations ▪ Solving Logarithmic Equations ▪ Application

Exponential and Logarithmic Equations

So far in this chapter, we have focused on the definitions, graphs, and properties of exponential and logarithmic functions. In this section, you will study procedures for *solving equations* that involve exponential or logarithmic expressions. As a simple example, consider the exponential equation $2^x = 16$. By rewriting this equation in the form $2^x = 2^4$, you can see that the solution is $x = 4$. To solve this equation, you can use one of the following properties.

Properties of Exponential and Logarithmic Equations

Let a be a positive real number such that $a \neq 1$, and let x and y be real numbers. Then the following properties are true.

1. $a^x = a^y$ if and only if $x = y$.

2. $\log_a x = \log_a y$ if and only if $x = y$ $(x > 0, \ y > 0)$.

EXAMPLE 1   *Solving Exponential and Logarithmic Equations*

Solve each equation.

a. $4^{x+2} = 4^5$ **b.** $\ln(2x - 3) = \ln 11$

Solution

a.

$4^{x+2} = 4^5$	Original equation
$x + 2 = 5$	Property 1
$x = 3$	Subtract 2 from both sides.

The solution is 3. Check this in the original equation.

b.

$\ln(2x - 3) = \ln 11$	Original equation
$2x - 3 = 11$	Property 2
$2x = 14$	Add 3 to both sides.
$x = 7$	Divide both sides by 2.

The solution is 7. Check this in the original equation.

Solving Exponential Equations

Technology

You can use logarithmic and exponential functions to derive other functions. Use the definitions of the logarithmic and natural logarithmic functions to show that $e^{(\ln x)} = x$.

In Example 1(a), you were able to solve the given equation because both sides of the equation were written in exponential form (with the same base). However, if only one side of the equation is written in exponential form, it is more difficult to solve the equation. For example, how would you solve the following equation?

$$2^x = 7$$

To solve this equation, you must find the power to which 2 can be raised to obtain 7. To do this, you can use one of the following inverse properties of exponents and logarithms.

Inverse Properties of Exponents and Logarithms	
Base a	*Natural Base e*
1. $\log_a(a^x) = x$	$\ln(e^x) = x$
2. $a^{(\log_a x)} = x$	$e^{(\ln x)} = x$

EXAMPLE 2 Solving an Exponential Equation

Solve $2^x = 7$.

Solution

To isolate the x, take the $\log_2$ of both sides of the equation, as follows.

$2^x = 7$	Original equation
$x = \log_2 7$	Inverse property

The solution is $x = \log_2 7 \approx 2.807$. Check this in the original equation.

EXAMPLE 3 Solving an Exponential Equation

Solve $2e^x = 10$.

Solution

$2e^x = 10$	Original equation
$e^x = 5$	Divide both sides by 2.
$x = \ln 5$	Inverse property

The solution is $x = \ln 5 \approx 1.609$. Check this in the original equation.

Technology

Graphical Check of Solutions

Remember that you can use a graphing utility to solve equations graphically or check solutions that are obtained algebraically. For instance, to check the solutions in Examples 2 and 3, graph both sides of the equations, as shown below.

Graph $y = 2^x$ and $y = 7$. Then approximate the intersection of the two graphs to be $x \approx 2.807$.

Graph $y = 2e^x$ and $y = 10$. Then approximate the intersection of the two graphs to be $x \approx 1.609$.

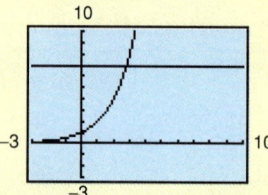

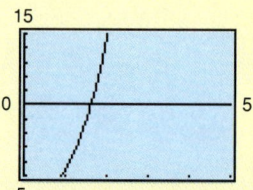

EXAMPLE 4 Solving an Exponential Equation

Solve $5 + e^{x+1} = 20$.

Solution

$5 + e^{x+1} = 20$	Original equation
$e^{x+1} = 15$	Subtract 5 from both sides.
$x + 1 = \ln 15$	Inverse property
$x = -1 + \ln 15$	Subtract 1 from both sides.

The solution is $x = -1 + \ln 15 \approx 1.708$. Check this in the original equation, as follows.

Check

$5 + e^{x+1} = 20$	Original equation
$5 + e^{1.708+1} \stackrel{?}{=} 20$	Substitute 1.708 for x.
$5 + e^{2.708} \stackrel{?}{=} 20$	Simplify.
$5 + 14.999 \approx 20$	Solution checks.

Solving Logarithmic Equations

You know how to solve an exponential equation by *taking the logarithms of both sides*. To solve a logarithmic equation, you need to **exponentiate** both sides. For instance, to solve a logarithmic equation such as

$$\ln x = 2$$

you can exponentiate both sides of the equation as follows.

$\ln x = 2$	Original equation
$x = e^2$	Inverse property

This procedure is demonstrated in the next three examples. We suggest the following guidelines for solving exponential and logarithmic equations.

Solving Exponential and Logarithmic Equations

1. To solve an exponential equation, first isolate the exponential expression, then **take the logarithms of both sides of the equation** and solve for the variable.

2. To solve a logarithmic equation, first isolate the logarithmic expression, then **exponentiate both sides of the equation** and solve for the variable.

EXAMPLE 5   *Solving a Logarithmic Equation*

Solve $2 \ln x = 5$.

Solution

$2 \ln x = 5$	Original equation
$\ln x = \dfrac{5}{2}$	Divide both sides by 2.
$x = e^{5/2}$	Inverse property

The solution is $x = e^{5/2} \approx 12.182$. Check this in the original equation, as follows.

Check

$2 \ln x = 5$	Original equation
$2 \ln(12.182) \stackrel{?}{=} 5$	Substitute 12.182 for x.
$2(2.49996) \stackrel{?}{=} 5$	Use a calculator.
$4.99992 \approx 5$	Solution checks. ✔

EXAMPLE 6   Solving a Logarithmic Equation

Solve $3 \log_{10} x = 6$.

Solution

$3 \log_{10} x = 6$	Original equation
$\log_{10} x = 2$	Divide both sides by 3.
$x = 100$	Inverse property

The solution is $x = 100$. Check this in the original equation.

EXAMPLE 7   Solving a Logarithmic Equation

Solve $20 \ln 0.2x = 30$.

Solution

$20 \ln 0.2x = 30$	Original equation
$\ln 0.2x = 1.5$	Divide both sides by 20.
$0.2x = e^{1.5}$	Inverse property
$x = 5e^{1.5}$	Divide both sides by 0.2.

The solution is $x = 5e^{1.5} \approx 22.408$. Check this in the original equation.

The next example uses logarithmic properties as part of the solution.

EXAMPLE 8   Solving a Logarithmic Equation

Solve $\log_{10} 2x - \log_{10}(x - 3) = 1$.

Solution

$\log_{10} 2x - \log_{10}(x - 3) = 1$	Original equation
$\log_{10} \dfrac{2x}{x - 3} = 1$	Condense the left side.
$\dfrac{2x}{x - 3} = 10$	Inverse property
$2x = 10x - 30$	Multiply both sides by $x - 3$.
$-8x = -30$	Subtract $10x$ from both sides.
$x = \dfrac{15}{4}$	Divide both sides by -8.

The solution is $x = \frac{15}{4}$. Check this in the original equation.

Application

EXAMPLE 9 An Application: Compound Interest

A deposit of $5000 is placed in a savings account for 2 years. The interest for the account is compounded continuously. At the end of 2 years, the balance in the account is $5867.56. What is the annual interest rate for this account?

Solution

Using the formula for continuously compounded interest, $A - Pe^{rt}$, you have the following solution.

Formula: $A = Pe^{rt}$

Labels: Principal $= P = 5000$ (dollars)
Amount $= A = 5867.56$ (dollars)
Time $= t = 2$ (years)
Annual interest rate $= r$ (percent in decimal form)

Equation: $5867.56 = 5000e^{2r}$

$$\frac{5867.56}{5000} = e^{2r}$$

$$1.1735 \approx e^{2r}$$

$$\ln 1.1735 \approx \ln(e^{2r})$$

$$0.16 \approx 2r$$

$$0.08 \approx r$$

The rate is 8%. Check this in the original statement of the problem.

Group Activities   Reviewing the Major Concepts

Solving Equations   Solve each equation.

a. $x^2 - 3x - 4 = 0$

b. $e^{2x} - 3e^x - 4 = 0$

c. $(\ln x)^2 - 3 \ln x - 4 = 0$

Explain your strategy. What are the similarities among the three equations? One of the equations has only one solution. Explain why.

13.5 Exercises

Discussing the Concepts

1. State the three basic properties of logarithms.

2. Which equation requires logarithms for its solution:

$$2^{x-1} = 32 \quad \text{or} \quad 2^{x-1} = 30?$$

3. Explain how to solve $10^{2x-1} = 5316$.

4. In your own words, state the guidelines for solving exponential and logarithmic equations.

Problem Solving

In Exercises 5–10, determine whether the x-values are solutions of the equation.

5. $3^{2x-5} = 27$

    (a) $x = 1$

    (b) $x = 4$

6. $4^{x+3} = 16$

    (a) $x = -1$

    (b) $x = 0$

7. $e^{x+5} = 45$

    (a) $x = -5 + \ln 45$

    (b) $x = -5 + e^{45}$

8. $2^{3x-1} = 324$

    (a) $x \approx 3.1133$

    (b) $x \approx 2.4327$

9. $\log_9(6x) = \frac{3}{2}$

    (a) $x = 27$

    (b) $x = \frac{9}{2}$

10. $\ln(x + 3) = 2.5$

    (a) $x = -3 + e^{2.5}$

    (b) $x \approx 9.1825$

In Exercises 11–18, solve the equation.

11. $3^{x-1} = 3^7$

12. $4^{2x} = 64$

13. $2^{x+2} = \frac{1}{16}$

14. $3^{2-x} = 9$

15. $\log_3(2 - x) = 2$

16. $\log_4(x-4) = \log_4 12$

17. $\ln(2x - 3) = \ln 15$

18. $\ln 3x = \ln 24$

In Exercises 19–22, simplify the expression.

19. $\ln e^{2x-1}$

20. $\log_3 3^{x^2}$

21. $10^{\log_{10} 2x}$

22. $e^{\ln(x+1)}$

In Exercises 23–32, solve the exponential equation. (Round the solution to two decimal places.)

23. $4^x = 8$

24. $2^x = 1.5$

25. $\frac{1}{2} e^{3x} = 20$

26. $6e^{-x} = 3$

27. $5(2)^{3x} - 4 = 13$

28. $-16 + 0.2(10)^x = 35$

29. $8 - 12e^{-x} = 7$

30. $4 - 2e^x = -23$

31. $\dfrac{8000}{(1.03)^t} = 6000$

32. $\dfrac{500}{1 + e^{-0.1t}} = 400$

In Exercises 33–42, solve the logarithmic equation. (Round the solution to two decimal places.)

33. $\log_2 x = 4.5$

34. $\log_4(25x) = 7$

35. $4\log_3 x = 28$

36. $5\log_{10}(x + 2) = 18$

37. $16 \ln x = 30$

38. $\ln\left(\frac{1}{2}t\right) = \frac{1}{4}$

39. $1 - 2\ln x = -4$

40. $-5 + 2\ln 3x = 5$

41. $\log_2(x - 1) + \log_2(x + 3) = 3$

42. $\log_{10}(25x) - \log_{10}(x - 1) = 2$

In Exercises 43–46, use a graphing utility to approximate the x-intercept of the graph.

43. $y = 10^{x/2} - 5$

44. $y = 2e^x - 21$

45. $y = 6\ln(0.4x) - 13$

46. $y = 5\log_{10}(x + 1) - 3$

47. *Doubling Time* At 9% interest, compounded continuously, how long does it take an investment to double?

48. *Human Memory Model* The average score A for students who took a test t months after the completion of a course is given by the memory model

$$A = 80 - \log_{10}(t + 1)^{12}.$$

How long after completing the course will the average score fall to $A = 72$?

49. *Friction*   To restrain an untrained horse, a person partially wraps the rope around a cylindrical post in the corral (see figure). If the horse is pulling on the rope with a force of 200 pounds, the force F in pounds required by the person is

$$F = 200e^{-0.2\pi\theta/180}$$

where θ is the angle of wrap in degrees. Find the smallest value of θ if F cannot exceed 80 pounds.

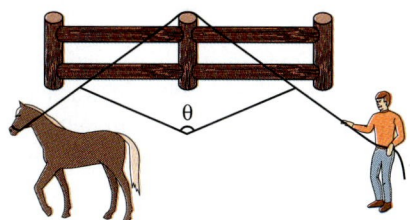

50. *Military Personnel*   The number N (in thousands) of United States military personnel on active duty for the years 1988 through 1992 is modeled by the equation

$$N = 2123.53 - 44.15e^t, \quad -2 \le t \le 2$$

where t is time in years, with $t = 0$ corresponding to 1990. (Source: U.S. Department of Defense)

(a) Use a graphing utility to graph the equation over the specified domain.

(b) Use the graph of part (a) to estimate the value of t when $N = 2000$.

(c) Answer the question in part (b) algebraically. Compare your result with the result obtained graphically.

Reviewing the Major Concepts

In Exercises 51–54, simplify the expression.

51. $\sqrt{24x^2y^3}$

52. $\sqrt[3]{9} \cdot \sqrt[3]{15}$

53. $(12a^{-4}b^6)^{1/2}$

54. $(16^{1/3})^{3/4}$

55. *Balance* $4000 is deposited at 6%, compounded monthly. Find the account balance after 5 years.

56. *Demand for a Product*   The demand per day x and price per unit p for a certain product are related by

$$p = 30 - \sqrt{0.5(x - 1)}.$$

Find the demand if the price is $26.76.

Additional Problem Solving

In Exercises 57–72, solve the equation. (Do not use a calculator.)

57. $2^x = 2^5$

58. $5^x = 5^3$

59. $3^{x+4} = 3^{12}$

60. $10^{1-x} = 10^4$

61. $4^{x-1} = 16$

62. $3^{2x} = 81$

63. $5^x = \frac{1}{125}$

64. $4^{x+1} = \frac{1}{64}$

65. $\log_5 2x = \log_5 36$

66. $\log_2(x + 3) = \log_2 7$

67. $\ln 5x = \ln 22$

68. $\ln(2x - 3) = \ln 17$

69. $\log_3 x = 4$

70. $\log_5 x = 3$

71. $\log_{10} 2x = 6$

72. $\log_2(3x - 1) = 5$

In Exercises 73–92, solve the exponential equation. (Round the solution to two decimal places.)

73. $2^x = 45$

74. $5^x = 212$

75. $10^{2y} = 52$

76. $12^{x-1} = 1500$

77. $\frac{1}{5}4^{x+2} = 300$

78. $3(2^{t+4}) = 350$

79. $4 + e^{2x} = 150$

80. $500 - e^{x/2} = 35$

81. $23 - 5e^{x+1} = 3$

82. $2e^x + 5 = 115$

83. $300e^{x/2} = 9000$

84. $1000^{0.12x} = 25,000$

85. $6000e^{-2t} = 1200$

86. $10,000e^{-0.1t} = 4000$

87. $32(1.5)^x = 640$

88. $250(1.04)^x = 1000$

89. $\dfrac{1600}{(1.1)^x} = 200$

90. $\dfrac{5000}{(1.05)^x} = 250$

91. $4(1 + e^{x/3}) = 84$

92. $50(3 - e^{2x}) = 125$

In Exercises 93–108, solve the logarithmic equation. (Round the solution to two decimal places.)

93. $\log_{10} x = 0$

94. $\ln x = 1$

95. $\log_{10} x = 3$

96. $\log_{10} x = -2$

97. $\log_{10} 4x = \frac{3}{2}$

98. $\log_{10}(x + 3) = \frac{5}{3}$

99. $\ln x = 2.1$

100. $\ln 2x = 3$

101. $\frac{2}{3} \ln(x + 1) = -1$

102. $8 \ln(3x - 2) = 1.5$

103. $2 \log_{10}(x + 5) = 15$

104. $-1 + 3 \log_{10} \dfrac{x}{2} = 8$

105. $\ln x^2 = 6$

106. $\ln \sqrt{x} = 6.5$

107. $\log_{10} x + \log_{10}(x - 3) = 1$

108. $\log_{10} x + \log_{10}(x + 1) = 0$

In Exercises 109–112, use a graphing utility to approximate the point of intersection of the graphs.

109. $y_1 = 2$
$y_2 = e^x$

110. $y_1 = 2$
$y_2 = \ln x$

111. $y_1 = 3$
$y_2 = 2 \ln(x + 3)$

112. $y_1 = 200$
$y_2 = 1000e^{-x/2}$

113. *Intensity of Sound* The relationship between the number of decibels B and the intensity of a sound I in watts per meter squared is given by

$$B = 10 \log_{10} \left(\dfrac{I}{10^{-16}} \right).$$

Determine the intensity of a sound I if it registers 75 decibels on an intensity meter.

114. *Doubling Time* Solve the exponential equation

$$10{,}000 = 5000e^{10r}$$

for r to determine the interest rate required for an investment of $5000 to double in value when compounded continuously for 10 years.

115. *Muon Decay* A muon is an elementary particle that is similar to an electron, but much heavier. Muons are unstable—they quickly decay to form electrons and other particles. In an experiment conducted in 1943, the number of muon decays m (of an original 5000 muons) was related to the time T (in microseconds) by the model

$$T = 15.7 - 2.48 \ln m.$$

How many decays were recorded when $T = 2.5$?

116. *Oceanography* Oceanographers use the density d (in grams per cubic centimeter) of seawater to obtain information about the circulation of water masses and the rates at which waters of different densities mix. For water with a salinity of 30%, the water temperature T (in degrees Celsius) is related to the density by

$$T = 7.9 \ln(1.0245 - d) + 61.84.$$

Find the densities of the subantarctic water and the antarctic bottom water shown in the figure.

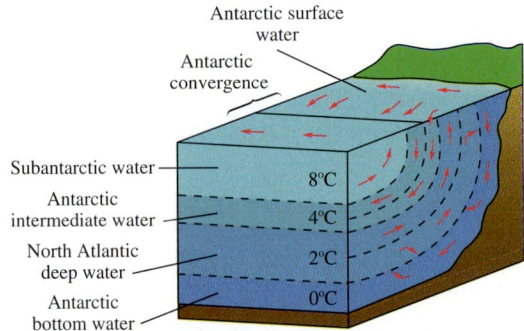

This cross section shows complex currents at various depths in the South Atlantic Ocean off Antarctica.

13.6	**Applications**

Compound Interest ▪ Growth and Decay ▪ Intensity Models

Compound Interest

In Section 13.1, you are introduced to the following two formulas for compound interest. In these formulas, A is the balance, P is the principal, r is the annual interest rate (in decimal form), and t is the time in years.

n Compoundings per Year

$$A = P\left(1 + \frac{r}{n}\right)^{nt}$$

Continuous Compounding

$$A = Pe^{rt}$$

EXAMPLE 1   *Finding the Annual Interest Rate*

An investment of $50,000 is made in an account that compounds interest quarterly. After 4 years, the balance in the account is $71,381.07. What is the annual interest rate for this account?

Solution

Formula: $A = P\left(1 + \dfrac{r}{n}\right)^{nt}$

Labels: Principal $= P = 50{,}000$ (dollars)
Amount $= A = 71{,}381.07$ (dollars)
Time $= t = 4$ (years)
Number of compoundings per year $= n = 4$
Annual interest rate $= r$ (percent in decimal form)

Equation: $71{,}381.07 = 50{,}000\left(1 + \dfrac{r}{4}\right)^{(4)(4)}$

$$\frac{71{,}381.07}{50{,}000} = \left(1 + \frac{r}{4}\right)^{16}$$

$$1.42762 \approx \left(1 + \frac{r}{4}\right)^{16}$$

$(1.42762)^{1/16} \approx 1 + \dfrac{r}{4}$ Raise both sides to $\frac{1}{16}$ power.

$$1.0225 \approx 1 + \frac{r}{4}$$

$$0.09 \approx r$$

NOTE Notice in Example 1 that you can "get rid of" the 16th power on the right side of the equation by raising both sides of the equation to the $\frac{1}{16}$ power.

The annual interest rate is approximately 9%. Check this in the original problem.

EXAMPLE 2   Doubling Time for Continuous Compounding

An investment is made in a trust fund at an annual interest rate of 8.75%, compounded continuously. How long will it take for the investment to double?

Solution

$A = Pe^{rt}$	Formula for continuous compounding
$2P = Pe^{0.0875t}$	Substitute known values.
$2 = e^{0.0875t}$	Divide both sides by P.
$\ln 2 = 0.0875t$	Inverse property
$\dfrac{\ln 2}{0.0875} = t$	Divide both sides by 0.0875.
$7.92 \approx t$	

NOTE   In Example 2, note that you do not need to know the principal to find the doubling time. In other words, the doubling time is the same for *any* principal.

It will take approximately 7.92 years for the investment to double. Check this in the original problem.

EXAMPLE 3   Finding the Type of Compounding

You deposit $1000 in an account. At the end of 1 year your balance is $1077.63. If the bank tells you that the annual interest rate for the account is 7.5%, how was the interest compounded?

Solution

If the interest had been compounded continuously at 7.5%, the balance would have been

$$A = 1000e^{(0.075)(1)} = \$1077.88.$$

Because the actual balance is slightly less than this, you should use the formula for interest that is compounded n times per year.

$$A = 1000\left(1 + \frac{0.075}{n}\right)^n = 1077.63$$

At this point, it is not clear what you should do to solve the equation for n. However, by completing a table, you can see that $n = 12$. Thus, the interest was compounded monthly.

n	1	4	12	365
$\left(1 + \dfrac{0.075}{n}\right)^n$	1.075	1.07714	1.07763	1.07788

In Example 3, notice that an investment of $1000 compounded monthly produced a balance of $1077.63 at the end of 1 year. Because $77.63 of this amount is interest, the **effective yield** for the investment is

$$\text{Effective yield} = \frac{\text{year's interest}}{\text{amount invested}} = \frac{77.63}{1000} = 0.07763 = 7.763\%.$$

In other words, the effective yield for an investment collecting compound interest is the *simple interest rate* that would yield the same balance at the end of 1 year.

EXAMPLE 4 *Finding the Effective Yield*

An investment is made in an account that pays 6.75% interest, compounded continuously. What is the effective yield for this investment?

Solution

Notice that you do not have to know the principal or the time that the money will be left in the account. Instead, you can choose an arbitrary principal, such as $1000. Then, because effective yield is based on the balance at the end of 1 year, you can use the following formula.

$$A = Pe^{rt} = 1000e^{0.0675(1)} = 1069.83$$

Now, because the account would earn $69.83 in interest after 1 year for a principal of $1000, you can conclude that the effective yield is

$$\text{Effective yield} = \frac{69.83}{1000} = 0.06983 = 6.983\%.$$

Growth and Decay

The balance in an account earning *continuously* compounded interest is one example of a quantity that increases over time according to the **exponential growth model** $y = Ce^{kt}$.

Exponential Growth and Decay

The mathematical model for exponential growth or decay is given by

$$y = Ce^{kt}.$$

For this model, t is the time, C is the original amount of the quantity, and y is the amount after the time t. The number k is a constant that is determined by the rate of growth. If $k > 0$, the model represents **exponential growth,** and if $k < 0$, it represents **exponential decay.**

One common application of exponential growth is in modeling the growth of a population, as shown in Example 5.

EXAMPLE 5   Population Growth

A country's population was 2 million in 1980 and 3 million in 1990. What would you predict the population of the country to be in the year 2000?

Solution

If you assumed a *linear growth model*, you would simply predict the population in the year 2000 to be 4 million. If, however, you assumed an *exponential growth model*, the model would have the form

$$y = Ce^{kt}.$$

In this model, let $t = 0$ represent the year 1980. The given information about the population can be described by the following table.

t (years)	0	10	20
Ce^{kt} (million)	$Ce^{k(0)} = 2$	$Ce^{k(10)} = 3$	$Ce^{k(20)} = ?$

To find the population when $t = 20$, you must first find the values of C and k. From the table, you can use the fact that $Ce^{k(0)} = Ce^0 = 2$ to conclude that $C = 2$. Then, using this value of C, you solve for k as follows.

$$Ce^{k(10)} = 3 \qquad \text{From table}$$

$$2e^{10k} = 3 \qquad \text{Substitute value of } C.$$

$$e^{10k} = \frac{3}{2} \qquad \text{Divide both sides by 2.}$$

$$10k = \ln \frac{3}{2} \qquad \text{Inverse property}$$

$$k = \frac{1}{10} \ln \frac{3}{2} \qquad \text{Divide both sides by 10.}$$

$$k \approx 0.0405 \qquad \text{Simplify.}$$

Finally, you can use this value of k to conclude that the population in the year 2000 is given by

$$2e^{0.0405(20)} \approx 2(2.25) = 4.5 \text{ million.}$$

Figure 13.16 graphically compares the exponential growth model with a linear growth model.

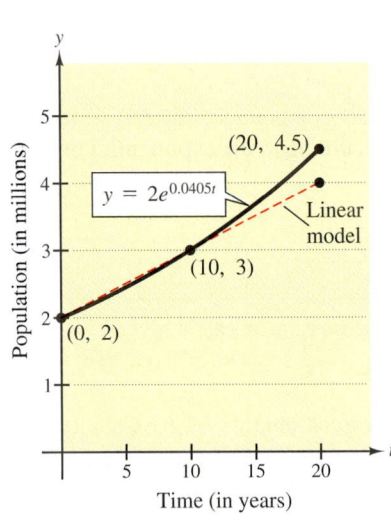

FIGURE 13.16   Population Models

EXAMPLE 6 Radioactive Decay

Radioactive iodine is a by-product of some types of nuclear reactors. Its **half-life** is 60 days. That is, after 60 days, a given amount of radioactive iodine will have decayed to half the original amount. Suppose a nuclear accident occurs and releases 20 grams of radioactive iodine. How long will it take for the radioactive iodine to decay to a level of 1 gram?

Solution

To solve this problem, use the model for exponential decay.

$$y = Ce^{kt}$$

Next, use the information given in the problem to set up the following table.

t (days)	0	60	?
Ce^{kt} (grams)	$Ce^{k(0)} = 20$	$Ce^{k(60)} = 10$	$Ce^{k(t)} = 1$

Because $Ce^{k(0)} = Ce^0 = 20$, you can conclude that $C = 20$. Then, using this value of C, you can solve for k, as follows.

$Ce^{k(60)} = 10$	From table
$20e^{60k} = 10$	Substitute value of C.
$e^{60k} = \dfrac{1}{2}$	Divide both sides by 20.
$60k = \ln \dfrac{1}{2}$	Inverse property
$k = \dfrac{1}{60} \ln \dfrac{1}{2} \approx -0.01155$	Divide both sides by 60 and simplify.

Finally, you can use this value of k to find the time when the amount is 1 gram, as follows.

$Ce^{kt} = 1$	From table
$20e^{-0.01155t} = 1$	Substitute values of C and k.
$e^{-0.01155t} = \dfrac{1}{20}$	Divide both sides by 20.
$-0.01155t = \ln \dfrac{1}{20}$	Inverse property
$t = \dfrac{1}{-0.01155} \ln \dfrac{1}{20} \approx 259.4$ days	Divide both sides by -0.01155 and simplify.

Thus, 20 grams of radioactive iodine will have decayed to 1 gram after about 259.4 days. This solution is shown graphically in Figure 13.17.

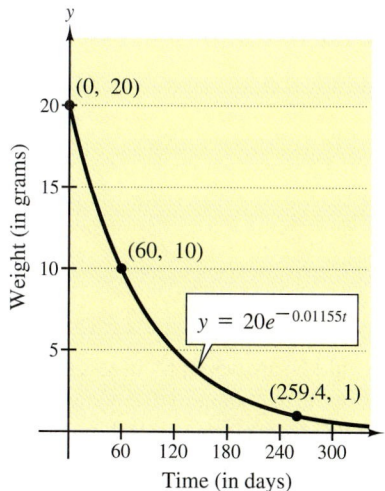

FIGURE 13.17 Radioactive Decay

Intensity Models

On the **Richter scale,** the magnitude R of an earthquake can be measured by the **intensity model**

$$R = \log_{10} I$$

where I is the intensity of the shock wave.

Earthquakes take place along faults in the earth's crust. The 1989 earthquake in California took place along the San Andreas Fault.

EXAMPLE 7   Earthquake Intensity

In 1906, San Francisco experienced an earthquake that measured 8.6 on the Richter scale. In 1989, another earthquake, which measured 7.7 on the Richter scale, struck the same area. Compare the intensities of these two earthquakes.

Solution

The intensity of the 1906 earthquake is given as follows.

$$8.6 = \log_{10} I \qquad \text{Given}$$
$$10^{8.6} = I \qquad \text{Inverse property}$$

The intensity of the 1989 earthquake can be found in a similar way.

$$7.7 = \log_{10} I \qquad \text{Given}$$
$$10^{7.7} = I \qquad \text{Inverse property}$$

The ratio of these two intensities is

$$\frac{I \text{ for } 1906}{I \text{ for } 1989} = \frac{10^{8.6}}{10^{7.7}} = 10^{8.6-7.7} = 10^{0.9} \approx 7.94.$$

Thus, the 1906 earthquake had an intensity that was about eight times greater than the 1989 earthquake.

Group Activities   You Be the Instructor

Problem Posing   Write a problem that could be answered by investigating the exponential growth model $y = 10e^{0.08t}$ or the exponential decay model $y = 5e^{-0.25t}$. Exchange your problem for that of another group member, and solve one another's problems.

13.6 Exercises

Discussing the Concepts

1. If the equation $y = Ce^{kt}$ models exponential growth, what must be true about k?

2. If the equation $y = Ce^{kt}$ models exponential decay, what must be true about k?

3. The formulas for periodic and continuous compounding have the four variables A, P, r, and t in common. Explain what each variable measures.

4. What is meant by the effective yield of an investment? Explain how it is computed.

5. In your own words, explain what is meant by the half-life of a radioactive isotope.

6. If the reading on the Richter scale is increased by 1, the intensity of the earthquake is increased by what factor?

Problem Solving

Annual Interest Rate   In Exercises 7–10, find the annual interest rate.

	Principal	Balance	Time	Compounding
7.	$1000	$36,581.00	40 years	Daily
8.	$200	$314.85	5 years	Yearly
9.	$750	$8267.38	30 years	Continuous
10.	$2000	$4234.00	10 years	Continuous

Doubling Time   In Exercises 11–14, find the time for an investment to double. Explain how to use a graphing utility to graphically check your result.

	Principal	Rate	Compounding
11.	$6000	8%	Quarterly
12.	$500	$5\frac{1}{4}\%$	Monthly
13.	$2000	10.5%	Daily
14.	$100	6%	Continuous

Type of Compounding   In Exercises 15–18, determine the type of compounding.

	Principal	Balance	Time	Rate
15.	$750	$1587.75	10 years	7.5%
16.	$10,000	$73,890.56	20 years	10%
17.	$100	$141.48	5 years	7%
18.	$4000	$4788.76	2 years	9%

In Exercises 19–22, find the effective yield.

	Rate	Compounding
19.	8%	Continuous
20.	9.5%	Daily
21.	7%	Monthly
22.	8%	Yearly

In Exercises 23–26, find the principal that must be deposited in an account to obtain the given balance.

	Balance	Rate	Time	Compounding
23.	$10,000	9%	20	Continuous
24.	$5000	8%	5	Continuous
25.	$750	6%	3	Daily
26.	$3000	7%	10	Monthly

Exponential Growth and Decay   In Exercises 27–30, find the constant k such that the graph of $y = Ce^{kt}$ passes through the given points.

27.

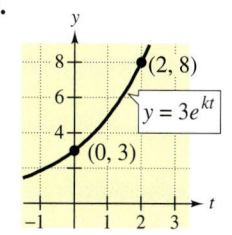

28.

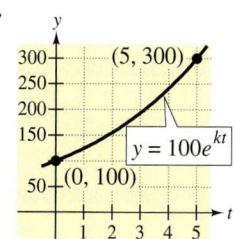

29.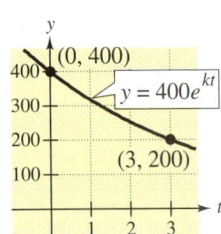

30.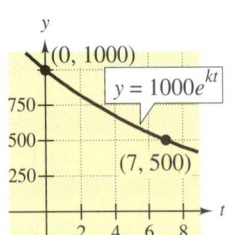

Population of a City In Exercises 31–34, the population of a city for the year 1992 and the predicted population for the year 2000 are given. Find the constants C and k to obtain the exponential growth model $y = Ce^{kt}$ for the population growth. (Let $t = 0$ correspond to the year 1992.) Use the model to predict the population of the city in the year 2005. (Source: U.S. Bureau of the Census)

City	1992	2000
31. Los Angeles	10.1 million	10.7 million
32. Houston	2.4 million	2.7 million
33. Dhaka, Bangladesh	4.6 million	6.5 million
34. Lagos, Nigeria	8.5 million	12.5 million

35. ***Rate of Growth***

(a) Compare the values of k in Exercises 31 and 33. Which is larger? Explain.

(b) What variable in the continuous compounding interest formula is equivalent to k in the model for population growth? Use your answer to give an interpretation of k.

36. ***Graphical Estimation*** The figure shows the population P (in billions) of the world as projected by the Population Reference Bureau. The Bureau's projection can be modeled by

$$P = \frac{11.14}{1 + 1.101e^{-0.051t}}$$

where $t = 0$ represents 1990.

(a) Estimate the world population in 2020.

(b) Use a graphing utility to estimate the year when the population will be twice the 1990 population.

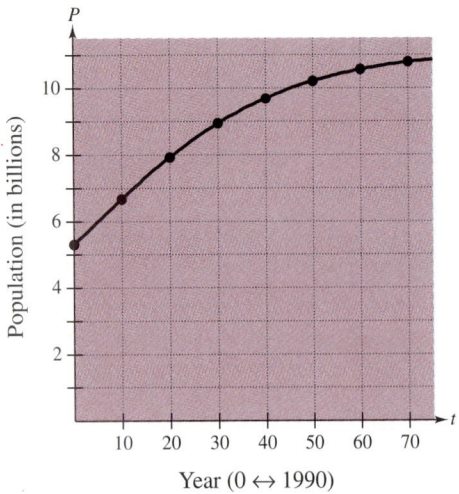

Figure for 36

37. ***Radioactive Decay*** Radioactive radium (Ra^{226}) has a half-life of 1620 years. If you start with 5 grams of the isotope, how much remains after 1000 years?

38. ***Radioactive Decay*** The isotope Pu^{230} has a half-life of 24,360 years. If you start with 10 grams of the isotope, how much remains after 10,000 years?

39. ***Earthquake Intensity*** On March 27, 1964, Alaska had an earthquake that measured 8.4 on the Richter scale. On February 9, 1971, an earthquake in the San Fernando Valley measured 6.6 on the Richter scale. Compare the intensities of these two earthquakes.

40. ***Earthquake Intensity*** On March 10, 1933, Long Beach, California had an earthquake that measured 6.2 on the Richter scale. On December 3, 1988, an earthquake in Pasadena, California measured 5.0 on the Richter scale. Compare the intensities of these two earthquakes.

Acidity Model In Exercises 41 and 42, use the acidity model $pH = -\log_{10}[H^+]$, where acidity (pH) is a measure of the hydrogen ion concentration $[H^+]$ (measured in moles of hydrogen per liter) of solution.

41. Find the pH of a solution that has a hydrogen ion concentration of 9.2×10^{-8}.

42. Compute the hydrogen ion concentration if the pH of a solution is 4.7.

Reviewing the Major Concepts

In Exercises 43–46, simplify the compound fraction.

43. $\dfrac{\left(\dfrac{9}{x}\right)}{\left(\dfrac{6}{x}+2\right)}$

44. $\dfrac{\left(1+\dfrac{2}{x}\right)}{\left(x-\dfrac{4}{x}\right)}$

45. $\dfrac{\left(\dfrac{4}{x^2-9}+\dfrac{2}{x-2}\right)}{\left(\dfrac{1}{x+3}+\dfrac{1}{x-3}\right)}$

46. $\dfrac{\left(\dfrac{1}{x+1}+\dfrac{1}{2}\right)}{\left(\dfrac{3}{2x^2+4x+2}\right)}$

47. *Equal Parts*   Find two real numbers that divide the real number line between $x/2$ and $4x/3$ into three equal parts.

48. *Capacitance*   When two capacitors, with capacitance C_1 and C_2, respectively, are connected in series, the equivalent capacitance is given by

$$\frac{1}{\left(\dfrac{1}{C_1}+\dfrac{1}{C_2}\right)}.$$

Simplify this compound fraction.

Additional Problem Solving

Annual Interest Rate   In Exercises 49–54, find the annual interest rate.

	Principal	Balance	Time	Compounding
49.	$500	$1004.83	10 years	Monthly
50.	$3000	$21,628.70	20 years	Quarterly
51.	$5000	$22,405.68	25 years	Daily
52.	$10,000	$110,202.78	30 years	Daily
53.	$1500	$24,666.97	40 years	Continuous
54.	$7500	$15,877.50	15 years	Continuous

Doubling Time   In Exercises 55–60, find the time for an investment to double.

	Principal	Rate	Compounding
55.	$300	5%	Yearly
56.	$10,000	9.5%	Yearly
57.	$6000	7%	Quarterly
58.	$500	9%	Daily
59.	$1500	7.5%	Continuous
60.	$12,000	4%	Continuous

In Exercises 61–66, find the effective yield.

	Rate	Compounding
61.	6%	Quarterly
62.	9%	Quarterly
63.	8%	Monthly
64.	$5\frac{1}{4}\%$	Daily
65.	7.5%	Continuous
66.	4%	Continuous

67. *Doubling Time*   Is it necessary to know the principal P to find the doubling time in Exercises 11–14 and Exercises 55–60? Explain.

68. *Effective Yield*

(a) Is it necessary to know the principal P to find the effective yield in Exercises 61–66? Explain.

(b) When the interest is compounded more frequently, what inference can you make about the difference between the effective yield and the annual interest rate?

In Exercises 69–74, find the principal that must be deposited in an account to obtain the given balance.

	Balance	Rate	Time	Compounding
69.	$25,000	7%	30 years	Monthly
70.	$8000	6%	2 years	Monthly
71.	$1000	5%	1 year	Daily
72.	$100,000	9%	40 years	Daily
73.	$500,000	8%	25 years	Continuous
74.	$1,000,000	10%	50 years	Continuous

Balance After Monthly Deposits   In Exercises 75–78, you make monthly deposits of P dollars in a savings account at an annual interest rate r, compounded continuously. Find the balance A after t years given that

$$A = \frac{P(e^{rt} - 1)}{e^{r/12} - 1}.$$

	Principal	Interest Rate	Time
75.	$P = 30$	$r = 8\%$	$t = 10$ years
76.	$P = 100$	$r = 9\%$	$t = 30$ years
77.	$P = 50$	$r = 10\%$	$t = 40$ years
78.	$P = 20$	$r = 7\%$	$t = 20$ years

Reading a Graph   In Exercises 79 and 80, you make monthly deposits of $30 in a savings account at an annual interest rate of 8%, compounded continuously (see figure).

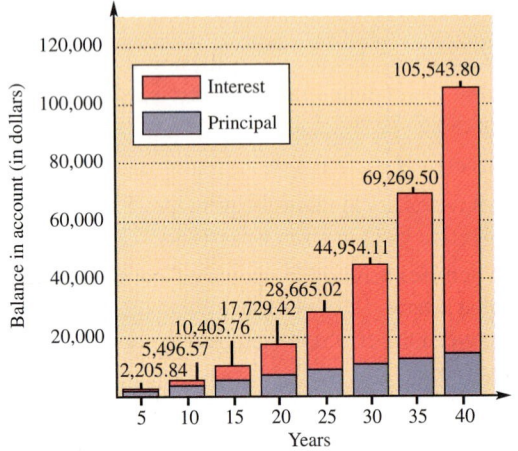

79. Find the total amount that you deposited in the account in 20 years and the total interest earned.

80. Find the total amount that you deposited in the account in 40 years and the total interest earned.

81. *Radioactive Decay*   The isotope Pu^{230} has a half-life of 24,360 years. If you start with 2 grams of this isotope, how much remains after 30,000 years?

82. *Radioactive Decay*   The isotope C^{14} has a half-life of 5730 years. If you start with 5 grams of this isotope, how much remains after 1000 years?

Population of a City   In Exercises 83–86, the population of a city for the year 1992 and the predicted population for the year 2000 are given. Find the constants C and k to obtain the exponential growth model $y = Ce^{kt}$ for the population growth. (Let $t = 0$ correspond to the year 1992.) Use the model to predict the population of the city in the year 2005. (Source: U.S. Bureau of the Census)

	City	1992	2000
83.	Chicago	6.5 million	6.6 million
84.	San Francisco	4.0 million	4.2 million
85.	Mexico City, Mexico	21.6 million	27.9 million
86.	Sao Paulo, Brazil	19.3 million	25.4 million

87. *Graphical Interpretation*   The population p of a certain species t years after it is introduced into a new habitat is given by

$$p(t) = \frac{5000}{1 + 4e^{-t/6}}.$$

(a) Use a graphing utility to graph the population function.

(b) Determine the population size that was introduced into the habitat.

(c) Determine the population size after 9 years.

(d) After how many years will the population be 2000?

88. *Carbon 14 Dating* C^{14} dating assumes that the carbon dioxide on earth today has the same radioactive content as it did centuries ago. If this is true, the amount of C^{14} absorbed by a tree that grew several centuries ago should be the same as the amount of C^{14} absorbed by a tree growing today. A piece of ancient charcoal contains only 15% as much of the radioactive carbon as a piece of modern charcoal. How long ago did the tree burn to make the ancient charcoal if the half-life of C^{14} is 5730 years? (Round your answer to the nearest 100 years.)

89. *Depreciation*   A car that cost $22,000 new has a depreciated value of $16,500 after 1 year. Find the value of the car when it is 3 years old by using the exponential model $y = Ce^{kt}$.

90. *Graphical Estimation*   After x years, the value y of a truck that cost $32,000 is given by $y = 32,000(0.8)^x$.

(a) Use a graphing utility to graph the equation.

(b) Graphically approximate the value after 1 year.

(c) Graphically approximate the time when the truck's value will be $16,000.

91. *Graphical Estimation*   Annual sales y of a product x years after it is introduced are approximated by

$$y = \frac{2000}{1 + 4e^{-x/2}}.$$

(a) Use a graphing utility to graph the equation.

(b) Graphically approximate sales when $x = 4$.

(c) Graphically approximate the time when annual sales will be 1100 units.

(d) Graphically estimate the maximum level annual sales will approach.

92. *Advertising Effect*   The sales S (in thousands of units) of a product after spending x hundred dollars in advertising are given by

$$S = 10(1 - e^{kx}).$$

(a) Find S as a function of x if 2500 units are sold when $500 is spent on advertising.

(b) How many units will be sold if advertising expenditures are raised to $700?

Earthquake Intensity   In Exercises 93 and 94, compare the earthquake intensities.

93. On August 16, 1906, Chile had an earthquake that measured 8.6 on the Richter scale. On December 7, 1988, an earthquake in Armenia, USSR measured 6.8 on the Richter scale.

94. On September 19, 1985, Mexico City, Mexico had an earthquake that measured 8.1 on the Richter scale. On August 20, 1988, an earthquake in Nepal measured 6.5 on the Richter scale.

Acidity Model   In Exercises 95 and 96, use the acidity model pH $= -\log_{10}[\text{H}^+]$, where acidity (pH) is a measure of the hydrogen ion concentration $[\text{H}^+]$ (measured in moles of hydrogen per liter) of solution.

95. A certain fruit has a pH of 2.5 and an antacid tablet has a pH of 9.5. The hydrogen ion concentration of the fruit is how many times the concentration of the tablet?

96. If the pH of a solution is decreased by one unit, the hydrogen ion concentration is increased by what factor?

97. *Comparing Models*   The figure gives the earnings per share of common stock for the years 1981 through 1993 for Automatic Data Processing, Inc. A list of models ($t = 1$ represents 1981) for the data is also given. For each of the models, (a) use a graphing utility to obtain its graph, and (b) find the sum of the squares of the differences between the actual data and the approximations given by the model. Use this sum to determine which model "best fits" the data. (Source: Automatic Data Processing, Inc.)

Linear:	$E = 0.146t + 0.007$
Quadratic:	$E = 0.009t^2 + 0.018t + 0.325$
Exponential:	$E = 0.301(1.165)^t$
Exponential:	$E = 0.301e^{0.153t}$
Logarithmic:	$E = 0.201 + 0.228t - 0.444 \ln t$

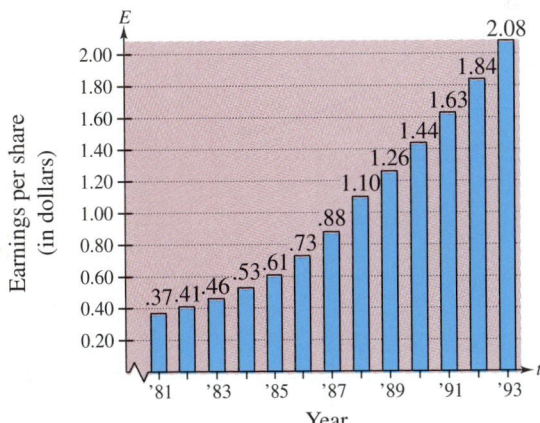

CHAPTER PROJECT: Radioactivity

Radioactive elements occur naturally. Some of these elements have proven to be useful. For instance, carbon 14 dating enables people to determine the ages of prehistoric animals and plants. Modern medicine uses iodine 131 to diagnose thyroid conditions, iron 59 to study red blood cells, technetium 99 to assess heart damage, and sodium 24 to study the circulatory system.

In large quantities, however, radioactive elements can cause serious illness. For instance, radon 222 is a radioactive gas that is produced by the decay of uranium 238. Overexposure to radon 222 can lead to lung cancer.

The nucleus of a radioactive element decays into other elements until the nucleus of the new element becomes stable. The uranium 238 decay series includes 14 radioactive elements. When the last radioactive element in this series, polonium 210, decays, it forms lead 206, a stable nonradioactive element.

Investigate the following questions.

1. Consider the exponential decay model $y = Ce^{kt}$, where y is the amount of radioactive substance left after time t, C is the beginning amount of the substance, and k is the rate constant of the radioactive element. The half-life of iodine 131 is 8.1 days. Determine its rate constant k.

2. Radon 222 is a radioactive gas that can be found in the basements of homes. It is one of the radioactive elements produced by the decay of uranium 238. Use the model for exponential decay $y = Ce^{kt}$ and the information in the table to determine the half-life of radon 222.

t (days)	0	2	?
Ce^{kt} (grams)	$Ce^{kt} = 10$	$Ce^{kt} = 6.9565$	$Ce^{kt} = 5$

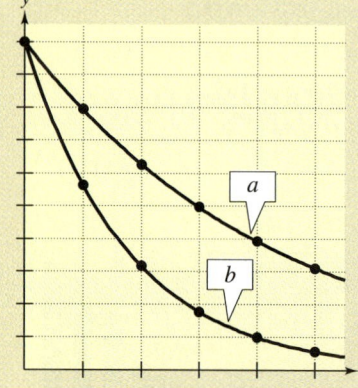

3. Technetium 99 has a half-life of 6.0 hours and sodium 24 has a half-life of 14.8 hours. Decay models for both are graphed on the same figure at the left. Determine the scales used for the x- and y-axes and write a paragraph describing how to determine which graph is the model for technetium 99 and which graph is the model for sodium 24.

4. The radioactive element iron 59 has a half-life of 45.1 days. Determine an algebraic decay model for 10 grams of the element. Use the model to complete the table below and draw a graph of the model.

t (days)	0	45.1	90.2	?	?
y (grams)	10	?	?	2.150	0.462

CHAPTER SUMMARY

After studying this chapter, you should have acquired the following skills. These skills are keyed to the Review Exercises that begin on page 820. Answers to odd-numbered Review Exercises are given in the back of the book.

- Evaluate exponential and logarithmic functions for given values of the variable. *(Sections 13.1, 13.3)* — **Review Exercises 1–10**

- Match exponential and logarithmic functions with their graphs. *(Sections 13.1, 13.3)* — **Review Exercises 11–16**

- Sketch the graphs of exponential and logarithmic functions. *(Sections 13.1, 13.3)* — **Review Exercises 17–26**

- Graph exponential and logarithmic functions using a graphing utility. *(Sections 13.1, 13.3)* — **Review Exercises 27–34**

- Find composite functions. *(Sections 13.2, 13.5)* — **Review Exercises 35–38**

- Find composite functions and determine their domains. *(Section 13.2)* — **Review Exercises 39, 40**

- Determine if functions have inverses using a graphing utility. *(Section 13.2)* — **Review Exercises 41–44**

- Find the inverses of functions. *(Section 13.2)* — **Review Exercises 45–50**

- Write exponential equations in logarithmic form and logarithmic equations in exponential form. *(Section 13.3)* — **Review Exercises 51–54**

- Evaluate logarithmic expressions. *(Sections 13.3, 13.4)* — **Review Exercises 55–62**

- Rewrite logarithmic expressions in expanded form using the properties of logarithms. *(Section 13.4)* — **Review Exercises 63–68**

- Rewrite logarithmic expressions in condensed form using the properties of logarithms. *(Section 13.4)* — **Review Exercises 69–74**

- Graphically verify that two logarithmic expressions are equivalent using a graphing utility. *(Section 13.4)* — **Review Exercises 75, 76**

- Decide whether exponential and logarithmic equations are true or false. *(Section 13.4)* — **Review Exercises 77–82**

- Approximate values of logarithmic expressions given the values of specific logarithms. *(Section 13.4)* — **Review Exercises 83–88**

- Evaluate logarithmic expressions using the change-of-base formula. *(Section 13.3)* — **Review Exercises 89–92**

- Solve exponential and logarithmic equations. *(Section 13.5)* — **Review Exercises 93–110**

- Solve real-life problems involving exponential and logarithmic functions. *(Sections 13.1, 13.3, 13.4, 13.5, 13.6)* — **Review Exercises 111–122**

REVIEW EXERCISES

In Exercises 1–10, evaluate the function as indicated.

1. $f(x) = 2^x$
  (a) $x = -3$ (b) $x = 1$ (c) $x = 2$

2. $g(x) = 2^{-x}$
  (a) $x = -2$ (b) $x = 0$ (c) $x = 2$

3. $g(t) = e^{-t/3}$
  (a) $t = -3$ (b) $t = \pi$ (c) $t = 6$

4. $h(s) = 1 - e^{0.2s}$
  (a) $s = 0$ (b) $s = 2$ (c) $s = \sqrt{10}$

5. $f(x) = \log_3 x$
  (a) $x = 1$ (b) $x = 27$ (c) $x = 0.5$

6. $g(x) = \log_{10} x$
  (a) $x = 0.01$ (b) $x = 0.1$ (c) $x = 30$

7. $f(x) = \ln x$
  (a) $x = e$ (b) $x = \frac{1}{3}$ (c) $x = 10$

8. $h(x) = \ln x$
  (a) $x = e^2$ (b) $x = \frac{5}{4}$ (c) $x = 1200$

9. $g(x) = \ln e^{3x}$
  (a) $x = -2$ (b) $x = 0$ (c) $x = 7.5$

10. $f(x) = \log_2 \sqrt{x}$
  (a) $x = 4$ (b) $x = 64$ (c) $x = 5.2$

In Exercises 11–16, match the function with its graph. [The graphs are labeled (a), (b), (c), (d), (e), and (f).]

11. $f(x) = 2^x$ **12.** $f(x) = 2^{-x}$

13. $f(x) = -2^x$ **14.** $f(x) = 2^x + 1$

15. $f(x) = \log_2 x$ **16.** $f(x) = \log_2(x - 1)$

(a)
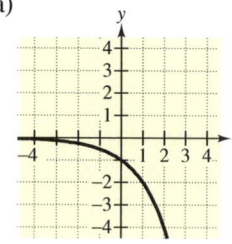

(b)
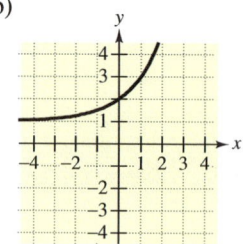

(c)
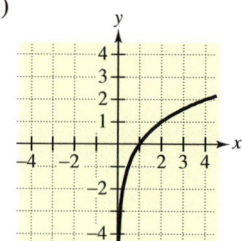

(d)
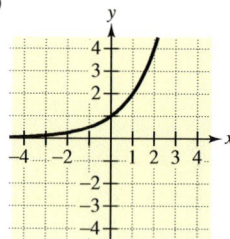

(e)
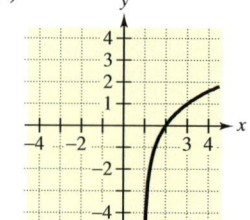

(f)
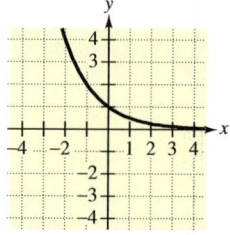

In Exercises 17–26, sketch the graph of the function.

17. $y = 3^{x/2}$ **18.** $y = 3^{x/2} - 2$

19. $f(x) = 3^{-x/2}$ **20.** $f(x) = -3^{-x/2}$

21. $f(x) = 3^{-x^2}$ **22.** $g(t) = 3^{|t|}$

23. $f(x) = -2 + \log_3 x$ **24.** $f(x) = 2 + \log_3 x$

25. $y = \log_2(x - 4)$ **26.** $y = 2 \log_4(x + 1)$

In Exercises 27–34, use a graphing utility to graph the function.

27. $y = 5e^{-x/4}$ **28.** $y = 6 - e^{x/2}$

29. $f(x) = e^{x+2}$ **30.** $h(t) = \dfrac{8}{1 + e^{-t/5}}$

31. $g(x) = \ln 2x$ **32.** $f(x) = 3 + \ln x$

33. $g(t) = 2 - \ln(t - 1)$ **34.** $g(x) = \ln(x - 5)$

In Exercises 35–38, find $(f \circ g)(x)$ and $(g \circ f)(x)$.

35. $f(x) = x + 2$, $g(x) = x^2$

36. $f(x) = \sqrt[3]{x}$, $g(x) = x + 2$

37. $f(x) = \sqrt{x + 1}$, $g(x) = x^2 - 1$

38. $f(x) = e^x$, $g(x) = \ln x$

In Exercises 39 and 40, find the domains of the compositions (a) $(f \circ g)$ and (b) $(g \circ f)$.

39. $f(x) = \sqrt{x-4}, \quad g(x) = 2x$

40. $f(x) = \dfrac{2}{x-4}, \quad g(x) = x^2$

In Exercises 41–44, use a graphing utility to decide if the function has an inverse. Explain your reasoning.

41. $f(x) = x^2 - 25$ **42.** $f(x) = \frac{1}{4}x^3$

43. $h(x) = 4\sqrt[3]{x}$ **44.** $g(x) = \sqrt{9 - x^2}$

In Exercises 45–50, find the inverse of the function. (If it is not possible, state the reason.)

45. $f(x) = \frac{1}{4}x$ **46.** $f(x) = 2x - 3$

47. $h(x) = \sqrt{x}$ **48.** $g(x) = x^2 + 2, \quad x \geq 0$

49. $f(t) = |t + 3|$ **50.** $h(t) = t$

In Exercises 51 and 52, write the exponential equation in logarithmic form.

51. $4^3 = 64$ **52.** $25^{3/2} = 125$

In Exercises 53 and 54, write the logarithmic equation in exponential form.

53. $\ln e = 1$ **54.** $\log_3 \frac{1}{9} = -2$

In Exercises 55–62, evaluate the expression.

55. $\log_{10} 1000$ **56.** $\log_9 3$

57. $\log_3 \frac{1}{9}$ **58.** $\log_4 \frac{1}{16}$

59. $\ln e^7$ **60.** $\log_a \dfrac{1}{a}$

61. $\ln 1$ **62.** $\ln e^{-3}$

In Exercises 63–68, use the properties of logarithms to expand the expression.

63. $\log_4 6x^4$ **64.** $\log_{10} 2x^{-3}$

65. $\log_5 \sqrt{x+2}$ **66.** $\ln \sqrt[3]{\frac{1}{5}x}$

67. $\ln \dfrac{x+2}{x-2}$ **68.** $\ln x(x-3)^2$

In Exercises 69–74, use properties of logarithms to condense the expression.

69. $\log_4 x - \log_4 10$ **70.** $5 \log_2 y$

71. $4(1 + \ln x + \ln x)$ **72.** $\log_8 16x + \log_8 2x^2$

73. $-2(\ln 2x - \ln 3)$ **74.** $-\frac{2}{3} \ln 3y$

In Exercises 75 and 76, use a graphing utility to graphically verify that the expressions are equivalent.

75. $y_1 = \ln \left(\dfrac{x+1}{x-1} \right)^2, \quad x > 1$

     $y_2 = 2[\ln(x+1) - \ln(x-1)]$

76. $y_1 = \ln \sqrt{x^2 - 4}, \quad x > 2$

     $y_2 = \frac{1}{2}[\ln(x+2) + \ln(x-2)]$

True or False? In Exercises 77–82, decide whether the equation is true or false. Explain your reasoning.

77. $\log_2 4x = 2 \log_2 x$ **78.** $\dfrac{\ln 5x}{\ln 10x} = \ln \dfrac{1}{2}$

79. $\log_{10} 10^{2x} = 2x$ **80.** $e^{\ln t} = t$

81. $\log_4 \dfrac{16}{x} = 2 - \log_4 x$

82. $e^{2x} - 1 = (e^x + 1)(e^x - 1)$

In Exercises 83–88, approximate the logarithm given that $\log_5 2 \approx 0.43068$ and $\log_5 3 \approx 0.6826$.

83. $\log_5 18$ **84.** $\log_5 \sqrt{6}$

85. $\log_5 \frac{1}{2}$ **86.** $\log_5 \frac{2}{3}$

87. $\log_5(12)^{2/3}$ **88.** $\log_5(5^2 \cdot 6)$

In Exercises 89–92, evaluate the logarithm using the change-of-base formula. Round each logarithm to three decimal places.

89. $\log_4 9$ **90.** $\log_{1/2} 5$

91. $\log_{12} 200$ **92.** $\log_3 0.28$

In Exercises 93–98, solve the equation.

93. $2^x = 64$ **94.** $3^{x-2} = 81$

95. $4^{x-3} = \frac{1}{16}$

96. $\log_2 2x = \log_2 100$

97. $\log_3 x = 5$

98. $\log_5(x - 10) = 2$

In Exercises 99–110, solve the equation. (Round your answer to two decimal places.)

99. $3^x = 500$

100. $8^x = 1000$

101. $2e^{x/2} = 45$

102. $100e^{-0.6x} = 20$

103. $\dfrac{500}{(1.05)^x} = 100$

104. $25(1 - e^t) = 12$

105. $\log_{10} 2x = 1.5$

106. $\frac{1}{3} \log_2 x + 5 = 7$

107. $\ln x = 7.25$

108. $\ln x = -0.5$

109. $\log_2 2x = -0.65$

110. $\log_5(x + 1) = 4.8$

Creating a Table In Exercises 111–114, complete the table to determine the balance A for P dollars invested at rate r for t years, and compounded n times per year.

n	1	4	12	365	Continuous
A					

	Principal	Interest Rate	Time
111.	$P = \$500$	$r = 7\%$	$t = 30$ years
112.	$P = \$100$	$r = 5\frac{1}{4}\%$	$t = 60$ years
113.	$P = \$10{,}000$	$r = 10\%$	$t = 20$ years
114.	$P = \$2500$	$r = 8\%$	$t = 1$ year

Creating a Table In Exercises 115 and 116, complete the table to determine the principal P that will yield a balance of A dollars when invested at rate r for t years and compounded n times per year.

n	1	4	12	365	Continuous
P					

	Balance	Interest Rate	Time
115.	$A = \$50{,}000$	$r = 8\%$	$t = 40$ years
116.	$A = \$1000$	$r = 6\%$	$t = 1$ year

117. *Inflation Rate* If the annual rate of inflation averages 5% over the next 10 years, the approximate cost C of goods or services during any year in that decade will be given by

$$C(t) = P(1.05)^t, \quad 0 \le t \le 10$$

where t is the time in years and P is the present cost. If the price of an oil change is presently $19.95, when will it cost $25.00?

118. *Doubling Time* Find the time for an investment of $1000 to double in value when invested at 8% compounded monthly.

119. *Product Demand* The demand x and price p for a product are related by

$$p = 25 - 0.4e^{0.02x}.$$

Approximate the demand when the price is $16.97.

120. *Sound Intensity* The relationship between the number of decibels B and the intensity of a sound I in watts per meter squared is given by

$$B = 10\log_{10}\left(\frac{I}{10^{-16}}\right).$$

Determine the intensity of a sound in watts per meter squared if the decibel level is 125.

121. *Graphical Interpretation* The population p of a certain species t years after it is introduced into a new habitat is given by

$$p(t) = \frac{600}{1 + 2e^{-0.2t}}.$$

Use a graphing utility to graph the function. Use the graph to determine the limiting size of the population in this habitat.

122. *Deer Herd* The state Parks and Wildlife Department releases 100 deer into a wilderness area. The population P of the herd can be modeled by

$$P = \frac{500}{1 + 4e^{-0.36t}}$$

where t is measured in years.

(a) Find the population after 5 years.

(b) After how many years will the population be 250?

CHAPTER TEST

Take this test as you would take a test in class. After you are done, check your work against the answers given in the back of the book.

1. Evaluate $f(t) = 54 \left(\frac{2}{3}\right)^t$ when $t = -1, 0, \frac{1}{2}$, and 2.

2. Sketch a graph of the function $f(x) = 2^{x/3}$.

3. Write the logarithmic equation $\log_5 125 = 3$ in exponential form.

4. Write the exponential equation $4^{-2} = \frac{1}{16}$ in logarithmic form.

5. Evaluate $\log_8 2$ without the aid of a calculator.

6. Describe the relationship between the graphs of $f(x) = \log_5 x$ and $g(x) = 5^x$.

7. Use the properties of logarithms to expand $\log_4 \left(5x^2/\sqrt{y}\right)$.

8. Use the properties of logarithms to condense $\ln x - 4 \ln y$.

9. Simplify $\log_5 5^3 \cdot 6$.

In Exercises 10–13, solve the equation.

10. $\log_4 x = 64$

11. $10^{3y} = 832$

12. $400e^{0.08t} = 1200$

13. $3 \ln(2x - 3) = 10$

14. Determine the balance after 20 years if $2000 is invested at 7% compounded (a) quarterly and (b) continuously.

15. Determine the principal that will yield $100,000 when invested at 9% compounded quarterly for 25 years.

16. A principal of $500 yields a balance of $1006.88 in 10 years when the interest is compounded continuously. What is the annual interest rate?

17. A car that cost $18,000 new has a depreciated value of $14,000 after 1 year. Find the value of the car when it is 3 years old by using the exponential model $y = Ce^{kt}$.

In Exercises 18–20, the population of a certain species t years after it is introduced into a new habitat is given by

$$p(t) = \frac{2400}{1 + 3e^{-t/4}}.$$

18. Determine the population size that was introduced into the habitat.

19. Determine the population after 4 years.

20. After how many years will the population be 1200?

CUMULATIVE TEST: CHAPTERS 11–13

Take this test as you would take a test in class. After you are done, check your work against the answers given in the back of the book.

1. Find an equation of the line through $(-4, 0)$ and $(4, 6)$.

In Exercises 2–4, sketch the graph of the equation.

2. $x^2 + y^2 = 8$ **3.** $\dfrac{x^2}{1} + \dfrac{y^2}{4} = 1$ **4.** $\dfrac{x^2}{1} - \dfrac{y^2}{4} = 1$

5. Graph the inequality $5x + 2y > 10$.

6. Find an equation of the parabola shown at the right.

7. A semicircular arch is positioned over the roadway onto the grounds of an estate. The roadway is 10 feet wide and the arch is sitting on pillars that are 8 feet tall. Find the maximum height of a truck that can be driven onto the estate if its width is 8 feet.

8. The stopping distance d of a car is directly proportional to the square of its speed s. On a certain type of pavement, a car requires 50 feet to stop when its speed is 25 miles per hour. Estimate the stopping distance when the speed of the car is 40 miles per hour. Explain your reasoning.

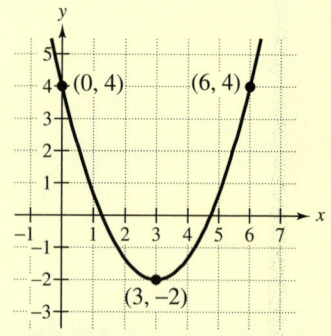

Figure for 6

In Exercises 9–12, solve the system of equations by the specified method.

9. *Graphical:* $x - y = 1$
                       $2x + y = 5$

10. *Substitution:* $4x + 2y = 8$
                           $x - 5y = 13$

11. *Elimination:* $4x - 3y = 8$
                          $-2x + y = -6$

12. *Cramer's Rule:* $2x - y = 4$
                            $3x + y = -5$

13. The sum of three positive numbers is 44. The second number is 4 greater than the first, and the third number is three times the first. Find the numbers.

14. Graph $y = 4e^{-x^2/4}$.

15. Evaluate $\log_4 \frac{1}{16}$ without using a calculator.

16. Describe the relationship between the graphs of $f(x) = e^x$ and $g(x) = \ln x$.

17. Use the properties of logarithms to condense $3(\log_2 x + \log_2 y) - \log_2 z$.

18. Solve each equation.

(a) $\log_x \left(\frac{1}{9}\right) = -2$ (b) $4 \ln x = 10$

(c) $500(1.08)^t = 2000$ (d) $3(1 + e^{2x}) = 20$

19. Determine the effective yield of an 8% interest rate compounded continuously.

20. Determine the length of time for an investment of $1000 to quadruple in value if the investment earns 9% compounded continuously.

Additional Topics in Algebra

14

- Sequences
- Arithmetic Sequences
- Geometric Sequences
- The Binomial Theorem
- Counting Principles
- Probability

In 1991, 73.2% of the households in the United States held interest-earning accounts such as savings accounts, money market deposit accounts, certificates of deposit, and interest-earning checking accounts at financial institutions. Other interest-earning assets include money market funds, government securities, corporate and municipal bonds, stocks and mutual fund shares, U.S. savings bonds, and IRA and Keogh accounts.

The balance A of a $1000 savings account with an interest rate of 3%, compounded quarterly, increases in value, as long as there are no withdrawals, according to the formula

$$A = 1000 \left(1 + \frac{0.03}{4}\right)^{4t}.$$

The increase in the balance of the account is shown in the table and graph at the right.

Year	1	5	10	15
Balance	$1030.34	$1161.18	$1348.35	$1565.70

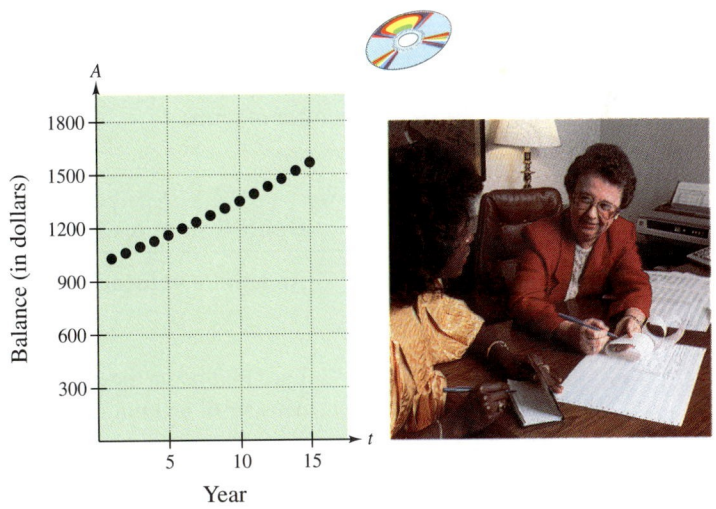

The chapter project related to this information is on page 881.

825

14.1	**Sequences**
	Sequences ▪ Factorial Notation ▪ Sigma Notation

Sequences

Suppose you were given the following choices of a contract offer for the next 5 years of employment.

Contract A $20,000 the first year and a $2200 raise each year

Contract B $20,000 the first year and a 10% raise each year

Year	Contract A	Contract B
1	$20,000	$20,000
2	$22,200	$22,000
3	$24,400	$24,200
4	$26,600	$26,620
5	$28,800	$29,282
Total	$122,000	$122,102

Which contract offers the largest salary over the 5-year period? The salaries for each contract are shown at the left. The salaries for contract A represent the first five terms of an **arithmetic sequence,** and the salaries for contract B represent the first five terms of a **geometric sequence.** Notice that after 5 years the geometric sequence represents a better contract offer than the arithmetic sequence.

A mathematical **sequence** is simply an ordered list of numbers. Each number in the list is a **term** of the sequence. A sequence can have a finite number of terms or an infinite number of terms. For instance, the sequence of positive odd integers that are less than 15 is a *finite* sequence

$$1, \ 3, \ 5, \ 7, \ 9, \ 11, \ 13 \qquad \text{Finite sequence}$$

whereas the sequence of positive odd integers is an *infinite* sequence.

$$1, \ 3, \ 5, \ 7, \ 9, \ 11, \ 13, \ . \ . \ . \qquad \text{Infinite sequence}$$

Note that the three dots indicate that the sequence continues and has an infinite number of terms.

Definition of an Infinite Sequence

An **infinite sequence** is an ordered list of real numbers.

$$a_1, \ a_2, \ a_3, \ a_4, \ a_5, \ . \ . \ . \ , \ a_n, \ . \ . \ .$$

Each number in the sequence is a **term** of the sequence, and the sequence consists of an infinite number of terms.

NOTE Sometimes it is convenient to begin subscripting an infinite sequence with 0 instead of 1. In such cases, the terms of the sequence are denoted by

$$a_0, \ a_1, \ a_2, \ a_3, \ a_4, \ a_5, \ . \ . \ . \ , \ a_n, \ . \ . \ . \ .$$

Technology

Most graphing utilities have a "sequence graphing mode" that allows you to plot the terms of a sequence as points on a rectangular coordinate system. For instance, the graph of the first six terms of the sequence given by

$$a_n = n^2 - 1$$

is shown below.

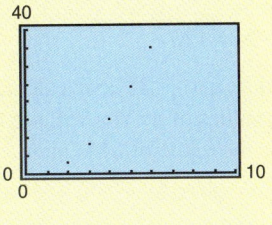

EXAMPLE 1   *Finding the Terms of a Sequence*

Write the first six terms of the sequence whose nth term is

$$a_n = n^2 - 1.$$ Begin sequence with $n = 1$.

Solution

$$a_1 = (1)^2 - 1 = 0 \qquad a_2 = (2)^2 - 1 = 3 \qquad a_3 = (3)^2 - 1 = 8$$
$$a_4 = (4)^2 - 1 = 15 \qquad a_5 = (5)^2 - 1 = 24 \qquad a_6 = (6)^2 - 1 = 35$$

To represent the entire sequence, you can write the following.

$$0, \ 3, \ 8, \ 15, \ 24, \ 35, \ \ldots, \ n^2 - 1, \ \ldots$$

EXAMPLE 2   *Finding the Terms of a Sequence*

Write the first six terms of the sequence whose nth term is

$$a_n = 3(2^n).$$ Begin sequence with $n = 0$.

Solution

$$a_0 = 3(2^0) = 3 \cdot 1 = 3 \qquad a_1 = 3(2^1) = 3 \cdot 2 = 6$$
$$a_2 = 3(2^2) = 3 \cdot 4 = 12 \qquad a_3 = 3(2^3) = 3 \cdot 8 = 24$$
$$a_4 = 3(2^4) = 3 \cdot 16 = 48 \qquad a_5 = 3(2^5) = 3 \cdot 32 = 96$$

The entire sequence can be written as follows.

$$3, \ 6, \ 12, \ 24, \ 48, \ 96, \ \ldots, \ 3(2^n), \ \ldots$$

EXAMPLE 3   *A Sequence Whose Terms Alternate in Sign*

Write the first six terms of the sequence whose nth term is

$$a_n = \frac{(-1)^n}{2n - 1}.$$ Begin sequence with $n = 1$.

Solution

$$a_1 = \frac{(-1)^1}{2(1) - 1} = -\frac{1}{1} \qquad a_2 = \frac{(-1)^2}{2(2) - 1} = \frac{1}{3} \qquad a_3 = \frac{(-1)^3}{2(3) - 1} = -\frac{1}{5}$$
$$a_4 = \frac{(-1)^4}{2(4) - 1} = \frac{1}{7} \qquad a_5 = \frac{(-1)^5}{2(5) - 1} = -\frac{1}{9} \qquad a_6 = \frac{(-1)^6}{2(6) - 1} = \frac{1}{11}$$

The entire sequence can be written as follows.

$$-1, \ \frac{1}{3}, \ -\frac{1}{5}, \ \frac{1}{7}, \ -\frac{1}{9}, \ \frac{1}{11}, \ \ldots, \ \frac{(-1)^n}{2n - 1}, \ \ldots$$

Factorial Notation

Some very important sequences in mathematics involve terms that are defined with special types of products called **factorials.**

Definition of Factorial

If n is a positive integer, n **factorial** is defined as

$$n! = 1 \cdot 2 \cdot 3 \cdot 4 \cdot \cdots \cdot (n-1) \cdot n.$$

As a special case, zero factorial is defined as $0! = 1$.

The first several factorial values are as follows.

$0! = 1$ $1! = 1$
$2! = 1 \cdot 2 = 2$ $3! = 1 \cdot 2 \cdot 3 = 6$
$4! = 1 \cdot 2 \cdot 3 \cdot 4 = 24$ $5! = 1 \cdot 2 \cdot 3 \cdot 4 \cdot 5 = 120$

Many calculators have a factorial key, denoted by $\boxed{n!}$. If your calculator has such a key, try using it to evaluate $n!$ for several values of n. You will see that the value of n does not have to be very large before the value of $n!$ is huge. For instance,

$$10! = 3,628,800.$$

EXAMPLE 4   *A Sequence Involving Factorials*

Write the first six terms of the sequence whose nth term is

$$a_n = \frac{1}{n!}.$$ Begin sequence with $n = 0$.

Solution

$$a_0 = \frac{1}{0!} = \frac{1}{1} = 1 \qquad\qquad a_1 = \frac{1}{1!} = \frac{1}{1} = 1$$

$$a_2 = \frac{1}{2!} = \frac{1}{2} \qquad\qquad a_3 = \frac{1}{3!} = \frac{1}{1 \cdot 2 \cdot 3} = \frac{1}{6}$$

$$a_4 = \frac{1}{4!} = \frac{1}{1 \cdot 2 \cdot 3 \cdot 4} = \frac{1}{24} \qquad a_5 = \frac{1}{5!} = \frac{1}{1 \cdot 2 \cdot 3 \cdot 4 \cdot 5} = \frac{1}{120}$$

The entire sequence can be written as follows.

$$1, \ 1, \ \frac{1}{2}, \ \frac{1}{6}, \ \frac{1}{24}, \ \frac{1}{120}, \ \cdots, \ \frac{1}{n!}, \ \cdots$$

Factorials follow the same conventions for order of operation as do exponents. For instance, $2n!$ means $2(n!)$, not $(2n)!$. Notice how these conventions are used in the next example.

EXAMPLE 5   *A Sequence Involving Factorials*

Write the first six terms of the sequence whose nth term is

$$a_n = \frac{2n!}{(2n)!}.$$

Begin sequence with $n = 0$.

Solution

$$a_0 = \frac{2(0\,!)}{(2 \cdot 0\,)!} = \frac{2(0!)}{0!} = \frac{2}{1} = 2$$

$$a_1 = \frac{2(1\,!)}{(2 \cdot 1\,)!} = \frac{2(1!)}{2!} = \frac{2}{2} = 1$$

$$a_2 = \frac{2(2\,!)}{(2 \cdot 2\,)!} = \frac{2(2!)}{4!} = \frac{4}{24} = \frac{1}{6}$$

$$a_3 = \frac{2(3\,!)}{(2 \cdot 3\,)!} = \frac{2(3!)}{6!} = \frac{12}{720} = \frac{1}{60}$$

$$a_4 = \frac{2(4\,!)}{(2 \cdot 4\,)!} = \frac{2(4!)}{8!} = \frac{48}{40,320} = \frac{1}{840}$$

$$a_5 = \frac{2(5\,!)}{(2 \cdot 5\,)!} = \frac{2(5!)}{10!} = \frac{240}{3,628,800} = \frac{1}{15,120}$$

The entire sequence can be written as follows.

$$2,\ 1,\ \frac{1}{6},\ \frac{1}{60},\ \frac{1}{840},\ \frac{1}{15,120},\ \cdots,\ \frac{2n!}{(2n)!},\ \cdots$$

In Example 5, the numerators and denominators were multiplied before reducing the fractions. When you are finding the terms of a sequence, reducing is often easier if you leave the numerator and denominator in factored form. For instance, notice the cancellation in the following fraction.

$$a_5 = \frac{2(5!)}{10!}$$

$$= \frac{2 \cdot 1 \cdot 2 \cdot 3 \cdot 4 \cdot 5}{1 \cdot 2 \cdot 3 \cdot 4 \cdot 5 \cdot 6_3 \cdot 7 \cdot 8 \cdot 9 \cdot 10}$$

$$= \frac{1}{3 \cdot 7 \cdot 8 \cdot 9 \cdot 10}$$

$$= \frac{1}{15,120}$$

Sigma Notation

Many applications involve finding the sum of the first n terms of a sequence. A convenient shorthand notation for such a sum is **sigma notation.** This name comes from the use of the uppercase Greek letter sigma, written as Σ.

Definition of Sigma Notation

The sum of the first n terms of the sequence whose nth term is a_n is

$$\sum_{i=1}^{n} a_i = a_1 + a_2 + a_3 + a_4 + \cdots + a_n$$

where i is the **index of summation,** n is the **upper limit of summation,** and 1 is the **lower limit of summation.**

NOTE In Example 6, the index of summation is i and the summation begins with $i = 1$. Any letter can be used as the index of summation, and the summation can begin with any integer. For instance, in Example 7, the index of summation is k and the summation begins with $k = 0$.

EXAMPLE 6 Sigma Notation for Sums

Find the sum $\displaystyle\sum_{i=1}^{6} 2i$.

Solution

$$\sum_{i=1}^{6} 2i = 2(1) + 2(2) + 2(3) + 2(4) + 2(5) + 2(6)$$

$$= 2 + 4 + 6 + 8 + 10 + 12$$

$$= 42$$

EXAMPLE 7 Sigma Notation for Sums

Find the sum $\displaystyle\sum_{k=0}^{8} \frac{1}{k!}$.

Solution

$$\sum_{k=0}^{8} \frac{1}{k!} = \frac{1}{0!} + \frac{1}{1!} + \frac{1}{2!} + \frac{1}{3!} + \frac{1}{4!} + \frac{1}{5!} + \frac{1}{6!} + \frac{1}{7!} + \frac{1}{8!}$$

$$= 1 + 1 + \frac{1}{2} + \frac{1}{6} + \frac{1}{24} + \frac{1}{120} + \frac{1}{720} + \frac{1}{5040} + \frac{1}{40,320}$$

$$\approx 2.71828$$

Note that this sum is approximately $e = 2.71828\ldots\ldots$

EXAMPLE 8   *Sigma Notation for Sums*

a. $\displaystyle\sum_{i=1}^{4} 5 = 5 + 5 + 5 + 5 = 20$

b. $\displaystyle\sum_{i=0}^{6} \frac{1}{2^i} = \frac{1}{2^0} + \frac{1}{2^1} + \frac{1}{2^2} + \frac{1}{2^3} + \frac{1}{2^4} + \frac{1}{2^5} + \frac{1}{2^6}$

$$= \frac{1}{1} + \frac{1}{2} + \frac{1}{4} + \frac{1}{8} + \frac{1}{16} + \frac{1}{32} + \frac{1}{64}$$

$$= \frac{127}{64}$$

EXAMPLE 9   *Writing a Sum in Sigma Notation*

Write the sum in sigma notation.

$$\frac{2}{2} + \frac{2}{3} + \frac{2}{4} + \frac{2}{5} + \frac{2}{6}$$

Solution

To write this sum in sigma notation, you must find a pattern for the terms. After examining the terms, you can see that they have numerators of 2 and denominators that range over the integers from 2 to 6. Thus, one possible sigma notation is

$$\sum_{i=1}^{5} \frac{2}{i+1} = \frac{2}{2} + \frac{2}{3} + \frac{2}{4} + \frac{2}{5} + \frac{2}{6}.$$

Group Activities

Communicating Mathematically

Finding a Pattern   You learned in this section that a sequence is an ordered list of numbers. Study the following sequence and see if you can guess what its next term should be.

Z, O, T, T, F, F, S, S, E, N, T, E, T, . . .

(*Hint:* you might try to figure out what numbers the letters represent.) Construct another sequence with letters. Can the other members of your group guess the next term?

14.1 Exercises

Discussing the Concepts

1. Give an example of an infinite sequence.

2. State the definition of n factorial.

3. The nth term of a sequence is $a_n = (-1)^n n$. Which terms of the sequence are negative?

In Exercises 4–6, decide whether the statement is true. Explain your reasoning.

4. $\sum_{i=1}^{4} (i^2 + 2i) = \sum_{i=1}^{4} i^2 + \sum_{i=1}^{4} 2i$

5. $\sum_{k=1}^{4} 3k = 3 \sum_{k=1}^{4} k$

6. $\sum_{j=1}^{4} 2^j = \sum_{j=3}^{6} 2^{j-2}$

Problem Solving

In Exercises 7–14, write the first five terms of the sequence. (Begin with $n = 1$.)

7. $a_n = 2n$

8. $a_n = (-1)^{n+1} 3n$

9. $a_n = \left(-\dfrac{1}{2}\right)^n$

10. $a_n = \left(\dfrac{2}{3}\right)^{n-1}$

11. $a_n = \dfrac{2n}{3n+2}$

12. $a_n = \dfrac{5n}{4n+3}$

13. $a_n = \dfrac{2^n}{n!}$

14. $a_n = \dfrac{n!}{(n-1)!}$

In Exercises 15–18, match the sequence with the graph of its first 10 terms. [The graphs are labeled (a), (b), (c), and (d).]

15. $a_n = \dfrac{6}{n+1}$

16. $a_n = \dfrac{6n}{n+1}$

17. $a_n = (0.6)^{n-1}$

18. $a_n = \dfrac{3^n}{n!}$

(a)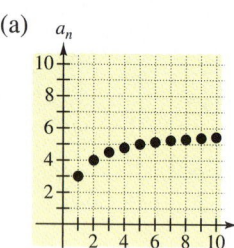

(b)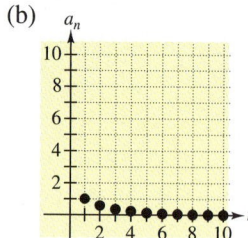

(c)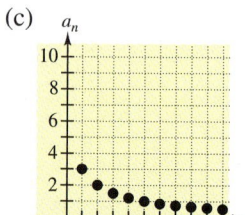

(d)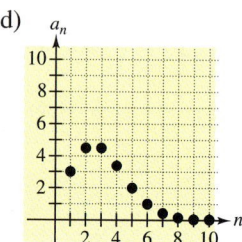

In Exercises 19 and 20, use a graphing utility to graph the first 10 terms of the sequence.

19. $a_n = (-0.8)^{n-1}$

20. $a_n = \dfrac{2n^2}{n^2+1}$

In Exercises 21 and 22, find the indicated term of the sequence.

21. $a_n = (-1)^n (5n - 3)$

$a_{15} = $

22. $a_n = \dfrac{n^2}{n!}$

$a_{12} = $

In Exercises 23–26, simplify the expression.

23. $\dfrac{25!}{27!}$

24. $\dfrac{20!}{15! \cdot 5!}$

25. $\dfrac{(n+1)!}{(n-1)!}$

26. $\dfrac{(3n)!}{(3n+2)!}$

In Exercises 27–34, find the sum.

27. $\displaystyle\sum_{k=1}^{6} 3k$

28. $\displaystyle\sum_{k=1}^{4} 5k$

29. $\displaystyle\sum_{i=0}^{6} (2i + 5)$

30. $\displaystyle\sum_{j=3}^{7} (6j - 10)$

31. $\displaystyle\sum_{m=2}^{6} \frac{2m}{2(m - 1)}$

32. $\displaystyle\sum_{k=1}^{5} \frac{10k}{k + 2}$

33. $\displaystyle\sum_{i=0}^{4} (i! + 4)$

34. $\displaystyle\sum_{k=1}^{6} \left(\frac{1}{2k} - \frac{1}{2k - 1} \right)$

In Exercises 35–40, write the sum using sigma notation. (Begin with $k = 0$ or $k = 1$.)

35. $2 + 4 + 6 + 8 + 10$

36. $24 + 30 + 36 + 42$

37. $\dfrac{4}{1 + 3} + \dfrac{4}{2 + 3} + \dfrac{4}{3 + 3} + \cdots + \dfrac{4}{20 + 3}$

38. $\dfrac{1}{2^3} - \dfrac{1}{4^3} + \dfrac{1}{6^3} - \dfrac{1}{8^3} + \cdots + \dfrac{1}{14^3}$

39. $\dfrac{2}{4} + \dfrac{4}{5} + \dfrac{6}{6} + \dfrac{8}{7} + \cdots + \dfrac{40}{23}$

40. $\left(2 + \dfrac{1}{1}\right) + \left(2 + \dfrac{1}{2}\right) + \left(2 + \dfrac{1}{3}\right) + \cdots + \left(2 + \dfrac{1}{25}\right)$

41. *Compound Interest*   A deposit of \$500 is made in an account that earns 7% interest compounded yearly. The balance in the account after N years is given by

$$A_N = 500(1 + 0.07)^N, \quad N = 1, 2, 3, \ldots.$$

(a) Compute the first eight terms of this sequence.

(b) Find the balance in this account after 40 years by computing A_{40}.

(c) Use a graphing utility to graph the first 40 terms of the sequence.

(d) The terms of the sequence are increasing. Is the rate of growth of the terms increasing? Explain.

42. *Depreciation*   At the end of each year, the value of a car with an initial cost of \$16,000 is three-fourths what it was at the beginning of the year. Thus, after n years its value is given by

$$a_n = 16,000 \left(\frac{3}{4}\right)^n, \quad n = 1, 2, 3, \ldots.$$

(a) Find the value of the car 3 years after it was purchased by computing a_3.

(b) Find the value of the car 6 years after it was purchased by computing a_6. Is this value half of what it was after 3 years?

Reviewing the Major Concepts

In Exercises 43–46, simplify the expression.

43. $(x + 10)^2, \ x \neq -10$

44. $\dfrac{18(x - 3)^5}{(x - 3)^2}$

45. $(a^2)^{-4}, \ a \neq 0$

46. $(8x^3)^{1/3}$

47. Find an equation of the line through $(2, 3)$ and $(5, 6)$.

48. *Volleyball Court*   A volleyball court is 60 feet long and 30 feet wide. To be assured that the court is rectangular, you check the diagonals of the court. How long should each be?

Additional Problem Solving

In Exercises 49–60, write the first five terms of the sequence. (Begin with $n = 1$.)

49. $a_n = (-1)^n 2n$

50. $a_n = 3n$

51. $a_n = \left(\dfrac{1}{2}\right)^n$

52. $a_n = \left(\dfrac{1}{3}\right)^n$

53. $a_n = \dfrac{1}{n + 1}$

54. $a_n = \dfrac{3}{2n + 1}$

55. $a_n = \dfrac{(-1)^n}{n^2}$

56. $a_n = \dfrac{1}{\sqrt{n}}$

57. $a_n = 5 - \dfrac{1}{2^n}$

58. $a_n = 7 + \dfrac{1}{3^n}$

59. $a_n = 2 + (-2)^n$

60. $a_n = \dfrac{1 + (-1)^n}{n^2}$

In Exercises 61–64, use a graphing utility to graph the first 10 terms of the sequence.

61. $a_n = \frac{1}{2}n$

62. $a_n = 10\left(\frac{3}{4}\right)^{n-1}$

63. $a_n = 3 - \frac{4}{n}$

64. $a_n = \frac{n+2}{n}$

In Exercises 65–72, simplify the expression.

65. $\dfrac{5!}{4!}$

66. $\dfrac{18!}{17!}$

67. $\dfrac{10!}{12!}$

68. $\dfrac{5!}{8!}$

69. $\dfrac{n!}{(n+1)!}$

70. $\dfrac{(n+2)!}{n!}$

71. $\dfrac{(2n)!}{(2n-1)!}$

72. $\dfrac{(2n+2)!}{(2n)!}$

In Exercises 73–88, find the sum.

73. $\displaystyle\sum_{i=0}^{4} (2i+3)$

74. $\displaystyle\sum_{i=2}^{7} (4i-1)$

75. $\displaystyle\sum_{j=1}^{5} \frac{(-1)^{j+1}}{j}$

76. $\displaystyle\sum_{j=0}^{3} \frac{1}{j^2+1}$

77. $\displaystyle\sum_{k=1}^{6} (-8)$

78. $\displaystyle\sum_{n=3}^{12} 10$

79. $\displaystyle\sum_{i=1}^{8} \left(\frac{1}{i} - \frac{1}{i+1}\right)$

80. $\displaystyle\sum_{k=1}^{5} \left(\frac{2}{k} - \frac{2}{k+2}\right)$

81. $\displaystyle\sum_{n=0}^{5} \left(-\frac{1}{3}\right)^n$

82. $\displaystyle\sum_{n=0}^{6} \left(\frac{3}{2}\right)^n$

83. $\displaystyle\sum_{n=1}^{6} n(n+1)$

84. $\displaystyle\sum_{n=0}^{5} 2n^2$

85. $\displaystyle\sum_{j=2}^{6} (j!-j)$

86. $\displaystyle\sum_{j=0}^{4} \frac{6}{j!}$

87. $\displaystyle\sum_{k=1}^{6} \ln k$

88. $\displaystyle\sum_{k=2}^{4} \frac{k}{\ln k}$

In Exercises 89–100, write the sum using sigma notation. (Begin with $k=0$ or $k=1$.)

89. $1+2+3+4+5$

90. $8+9+10+11+12+13$

91. $\dfrac{1}{2(1)} + \dfrac{1}{2(2)} + \dfrac{1}{2(3)} + \dfrac{1}{2(4)} + \cdots + \dfrac{1}{2(10)}$

92. $\dfrac{3}{1+1} + \dfrac{3}{1+2} + \dfrac{3}{1+3} + \dfrac{3}{1+4} + \cdots + \dfrac{3}{1+50}$

93. $\dfrac{1}{1^2} + \dfrac{1}{2^2} + \dfrac{1}{3^2} + \dfrac{1}{4^2} + \cdots + \dfrac{1}{20^2}$

94. $\dfrac{1}{2^0} + \dfrac{1}{2^1} + \dfrac{1}{2^2} + \dfrac{1}{2^3} + \cdots + \dfrac{1}{2^{12}}$

95. $\dfrac{1}{3^0} - \dfrac{1}{3^1} + \dfrac{1}{3^2} - \dfrac{1}{3^3} + \cdots - \dfrac{1}{3^9}$

96. $\left(-\frac{2}{3}\right)^0 + \left(-\frac{2}{3}\right)^1 + \left(-\frac{2}{3}\right)^2 + \cdots + \left(-\frac{2}{3}\right)^{20}$

97. $\frac{1}{2} + \frac{2}{3} + \frac{3}{4} + \frac{4}{5} + \frac{5}{6} + \cdots + \frac{11}{12}$

98. $\frac{2}{4} + \frac{4}{7} + \frac{6}{10} + \frac{8}{13} + \frac{10}{16} + \cdots + \frac{20}{31}$

99. $1+1+2+6+24+120+720$

100. $1+1+\frac{1}{2}+\frac{1}{6}+\frac{1}{24}+\frac{1}{120}+\frac{1}{720}$

Arithmetic Mean In Exercises 101–104, find the arithmetic mean of the set. The *arithmetic mean* $\bar{x}$ of a set of n measurements $x_1, x_2, x_3, \ldots, x_n$ is

$$\bar{x} = \frac{1}{n}\sum_{i=1}^{n} x_i.$$

101. 1, 2, 3, 4, 5

102. 66, 70, 74, 78

103. 5, 8, 11, 14, 17, 20

104. 17, 22, 27, 32, 37

105. *Soccer Ball* The number of degrees a_n in each angle of a regular n-sided polygon is

$$a_n = \frac{180(n-2)}{n}, \quad n \geq 3.$$

The surface of a soccer ball is made of regular hexagons and pentagons. If a soccer ball is taken apart and flattened, as shown in the figure at the top of the next page, the sides of the hexagons do not meet each other. Use the terms a_5 and a_6 to explain why there are gaps between adjacent hexagons.

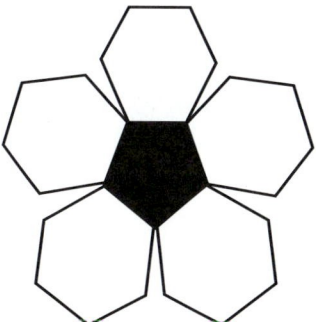

Figure for 105

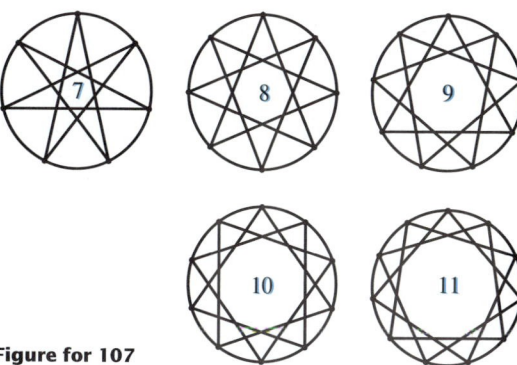

Figure for 107

106. *Stars*   The number of degrees d_n in each tip of the n-pointed stars in the figure is given by

$$d_n = \frac{180(n - 4)}{n}, \quad n \geq 5.$$

Write the first six terms of this sequence.

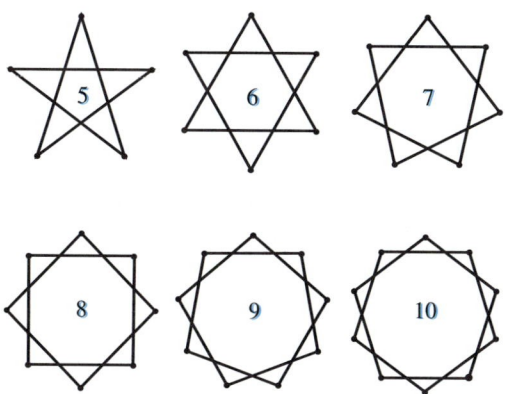

107. *Stars*   The stars in Exercise 106 are formed by placing n equally spaced points on a circle and connecting each point with the second point from it on the circle. The stars in the figure at the top of the next column are formed in a similar way except each point is connected with the third point from it. For these stars, the number of degrees in a tip is

$$d_n = \frac{180(n - 6)}{n}, \quad n \geq 7.$$

Write the first five terms of this sequence.

14.2	**Arithmetic Sequences**

Arithmetic Sequences ▪ The Sum of an Arithmetic Sequence ▪ Application

Arithmetic Sequences

A sequence whose consecutive terms have a common difference is called an **arithmetic sequence**.

Definition of an Arithmetic Sequence

A sequence is called **arithmetic** if the differences between consecutive terms are the same. Thus, the sequence

$$a_1, \ a_2, \ a_3, \ a_4, \ \ldots, \ a_n, \ \ldots$$

is arithmetic if there is a number d such that

$$a_2 - a_1 = d, \quad a_3 - a_2 = d, \quad a_4 - a_3 = d$$

and so on. The number d is the **common difference** of the sequence.

EXAMPLE 1   *Examples of Arithmetic Sequences*

a. The sequence whose nth term is $3n + 2$ is arithmetic. For this sequence, the common difference between consecutive terms is 3.

$$5, \ 8, \ 11, \ 14, \ \ldots, \ 3n + 2, \ \ldots$$

$$8 - 5 = 3$$

b. The sequence whose nth term is $7 - 5n$ is arithmetic. For this sequence, the common difference between consecutive terms is -5.

$$2, \ -3, \ -8, \ -13, \ \ldots, \ 7 - 5n, \ \ldots$$

$$-3 - 2 = -5$$

c. The sequence whose nth term is $\frac{1}{4}(n + 3)$ is arithmetic. For this sequence, the common difference between consecutive terms is $\frac{1}{4}$.

$$1, \ \frac{5}{4}, \ \frac{3}{2}, \ \frac{7}{4}, \ \ldots, \ \frac{n+3}{4}, \ \ldots$$

$$\tfrac{5}{4} - 1 = \tfrac{1}{4}$$

The nth Term of an Arithmetic Sequence

The nth term of an arithmetic sequence has the form

$$a_n = a_1 + (n - 1)d$$

where d is the common difference between the terms of the sequence, and a_1 is the first term.

EXAMPLE 2 *Finding the nth Term of an Arithmetic Sequence*

Find a formula for the nth term of the arithmetic sequence whose common difference is 2 and whose first term is 5.

Solution

You know that the formula for the nth term is of the form $a_n = a_1 + (n - 1)d$. Moreover, because the common difference is $d = 2$, and the first term is $a_1 = 5$, the formula must have the form

$$a_n = 5 + 2(n - 1).$$

Thus, the formula for the nth term is

$$a_n = 2n + 3.$$

The sequence therefore has the following form.

$$5, \ 7, \ 9, \ 11, \ 13, \ \ldots, \ 2n + 3, \ \ldots$$

If you know the nth term and the common difference of an arithmetic sequence, you can find the $(n+1)$th term by using the following **recursion formula.**

$$a_{n+1} = a_n + d$$

EXAMPLE 3 *Using a Recursion Formula*

The 12th term of an arithmetic sequence is 52 and the common difference is 3. What is the 13th term of the sequence?

Solution

$$a_{13} = a_{12} + 3 = 52 + 3 = 55$$

The Sum of an Arithmetic Sequence

The sum of the first n terms of an arithmetic sequence is called the **nth partial sum** of the sequence. For instance, the 5th partial sum of the arithmetic sequence whose nth term is $3n + 4$ is

$$\sum_{i=1}^{5} (3i + 4) = 7 + 10 + 13 + 16 + 19 = 65.$$

A formula for the nth partial sum of an arithmetic sequence is given below.

STUDY TIP

You can use the formula for the nth partial sum of an arithmetic sequence to find the sum of consecutive numbers. For instance, the sum of the integers from 1 to 100 is

$$\sum_{i=1}^{100} i = \frac{100}{2}(1 + 100)$$
$$= 50(101)$$
$$= 5050.$$

The nth Partial Sum of an Arithmetic Sequence

The nth partial sum of the arithmetic sequence whose nth term is a_n is

$$\sum_{i=1}^{n} a_i = a_1 + a_2 + a_3 + a_4 + \cdots + a_n$$
$$= n\left(\frac{a_1 + a_n}{2}\right).$$

In other words, to find the sum of the first n terms of an arithmetic sequence, find the average of the first and nth terms, and multiply by n.

NOTE This summation form of an arithmetic sequence is also called a **series**.

EXAMPLE 4   Finding the nth Partial Sum

Find the sum of the first 20 terms of the arithmetic sequence whose nth term is $4n + 1$.

Solution

The first term of this sequence is $a_1 = 4(1) + 1 = 5$ and the 20th term is $a_{20} = 4(20) + 1 = 81$. Therefore, the sum of the first 20 terms is given by

$$\sum_{i=1}^{n} a_i = \frac{n}{2}(a_1 + a_n)$$
$$\sum_{i=1}^{20} (4i + 1) = \frac{20}{2}(a_1 + a_{20})$$
$$= 10(5 + 81)$$
$$= 10(86)$$
$$= 860.$$

Application

EXAMPLE 5 An Application: Total Sales

Your business sells $100,000 worth of products during its first year. You have a goal of increasing annual sales by $25,000 each year for 9 years. If you meet this goal, how much will you sell during your first 10 years of business?

Solution

The annual sales during the first 10 years form the following arithmetic sequence.

$100,000, $125,000, $150,000, $175,000, $200,000,

$225,000, $250,000, $275,000, $300,000, $325,000

Using the formula for the *n*th partial sum of an arithmetic sequence, you find the total sales during the first 10 years as follows.

$$\text{Total sales} = \frac{10}{2}(100,000 + 325,000) = 5(425,000) = \$2,125,000$$

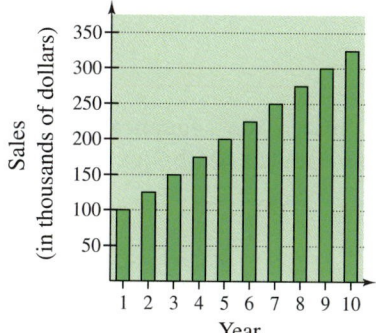

FIGURE 14.1

From the bar graph shown in Figure 14.1, notice that the annual sales for this company follows a *linear growth* pattern. In other words, saying that a quantity increases arithmetically is the same as saying that it increases linearly.

Group Activities   Extending the Concept

6	1	8
7	5	3
2	9	4

Using Arithmetic Sequences   A magic square is a square table of positive integers in which each row, column, and diagonal adds up to the same number. One example is shown at left. In addition, the values in the middle row, in the middle column, and along both diagonals form arithmetic sequences. See if you can complete the following magic squares.

a.

	11	14
	10	
		15

b.

8		
	9	
	13	

c.

		20
	13	
6		

14.2 Exercises

Discussing the Concepts

1. In your own words, explain what makes a sequence arithmetic.

2. The second and third terms of an arithmetic sequence are 12 and 15, respectively. What is the first term?

3. Explain how the first two terms of an arithmetic sequence can be used to find the nth term.

4. Explain what is meant by the recursion formula.

5. Explain what is meant by the nth partial sum of a sequence.

6. Explain how to find the sum of the integers from 100 to 200.

Problem Solving

In Exercises 7–10, find the common difference of the arithmetic sequence.

7. 10, 22, 34, 46, 58, . . .

8. 4, $\frac{9}{2}$, 5, $\frac{11}{2}$, 6, . . .

9. $\frac{7}{2}$, $\frac{9}{4}$, 1, $-\frac{1}{4}$, $-\frac{3}{2}$, . . .

10. $\frac{5}{2}$, $\frac{11}{6}$, $\frac{7}{6}$, $\frac{1}{2}$, $-\frac{1}{6}$, . . .

In Exercises 11–18, determine whether the sequence is arithmetic. If it is, find the common difference.

11. 10, 8, 6, 4, 2, . . .

12. 1, 2, 4, 8, 16, . . .

13. 3, $\frac{5}{2}$, 2, $\frac{3}{2}$, 1, . . .

14. $\frac{1}{3}$, $\frac{2}{3}$, $\frac{4}{3}$, $\frac{8}{3}$, $\frac{16}{3}$, . . .

15. -12, -8, -4, 0, 4, . . .

16. $\frac{9}{4}$, 2, $\frac{7}{4}$, $\frac{3}{2}$, $\frac{5}{4}$, . . .

17. $\ln 4$, $\ln 8$, $\ln 12$, $\ln 16$, . . .

18. e, e^2, e^3, e^4, . . .

In Exercises 19–22, write the first five terms of the arithmetic sequence. (Begin with $n = 1$.)

19. $a_n = -5n + 45$

20. $a_n = 3n + 1$

21. $a_n = \frac{3}{5}n + 1$

22. $a_n = \frac{3}{4}(n + 1) - 2$

In Exercises 23 and 24, write the first five terms of the arithmetic sequence defined recursively.

23. $a_1 = 25$

$a_{k+1} = a_k + 3$

24. $a_1 = 12$

$a_{k+1} = a_k - 6$

In Exercises 25–28, match the sequence with its graph. [The graphs are labeled (a), (b), (c), and (d).]

(a)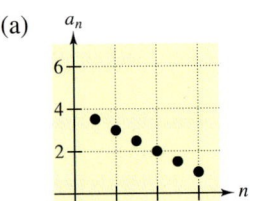

(b)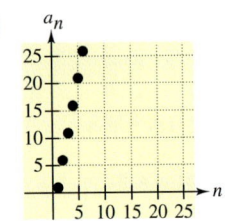

(c)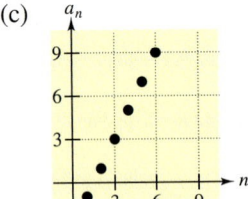

(d)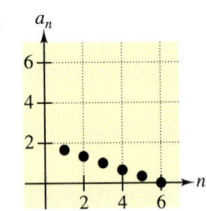

25. $a_n = -\frac{1}{3}n + 2$

26. $a_n = 5n - 4$

27. $a_n = 2n - 3$

28. $a_n = -\frac{1}{2}n + 4$

In Exercises 29–34, find a formula for the nth term of the arithmetic sequence.

29. $a_1 = 3, d = \frac{3}{2}$

30. $a_1 = 12, d = -3$

31. $a_3 = 20, d = -4$

32. $a_6 = 5, d = \frac{3}{2}$

33. $a_1 = 5, a_5 = 13$

34. $a_2 = 93, a_6 = 65$

In Exercises 35 and 36, find the sum.

35. $\sum_{k=1}^{10} 5k$

36. $\sum_{n=1}^{30} \left(\frac{1}{2}n + 2\right)$

In Exercises 37 and 38, use a graphing utility to find the sum.

37. $\displaystyle\sum_{j=1}^{25} (750 - 30j)$

38. $\displaystyle\sum_{i=1}^{60} \left(300 - \tfrac{8}{3}i\right)$

In Exercises 39–42, find the nth partial sum of the arithmetic sequence.

39. $2, 8, 14, 20, \ldots ,\quad n = 25$

40. $500, 480, 460, 440, \ldots ,\quad n = 20$

41. $0.5, 0.9, 1.3, 1.7, \ldots ,\quad n = 10$

42. $a_1 = 15,\ a_{100} = 312,\quad n = 100$

43. Find the sum of the multiples of 6 from 12 to 240.

44. *Clock Chimes*   A clock chimes once at 1:00, twice at 2:00, three times at 3:00, and so on. How many times does the clock chime in a 12-hour period?

45. *Free-Falling Object*   A free-falling object will fall 16 feet during the first second, 48 more feet during the second, 80 more feet during the third, and so on. What is the total distance the object will fall in 8 seconds if this pattern continues?

46. *Pile of Logs*   Logs are stacked in a pile as shown in the figure. The top row has 15 logs and the bottom row has 21 logs. How many logs are in the stack?

Reviewing the Major Concepts

In Exercises 47–50, solve the system of equations.

47.
$$y = x^2$$
$$-3x + 2y = 2$$

48.
$$x - y^3 = 0$$
$$x - 2y^2 = 0$$

49.
$$-x + y = 1$$
$$x + 2y - 2z = 3$$
$$3x - y + 2z = 3$$

50.
$$2x + y - 2z = 1$$
$$x \qquad\ - z = 1$$
$$3x + 3y + z = 12$$

51. *Ticket Sales*   Twelve hundred tickets are sold for a total of \$21,120. Adult tickets cost \$20 and children's tickets cost \$12.50. How many of each type of ticket were sold?

52. *Best-Fitting Line*   The slope and y-intercept of the line $y = mx + b$ that "best fits" the four points $(-1, 5)$, $(0, 3)$, $(2, 3)$, and $(4, 0)$ are given by the solution of the system of linear equations.

$$4b + 5m = 11$$
$$5b + 21m = 1$$

(a) Solve the system and find the equation of the best-fitting line.

(b) Plot the three points and sketch the graph of the best-fitting line.

Additional Problem Solving

In Exercises 53–58, find the common difference of the arithmetic sequence.

53. $2, 5, 8, 11, \ldots$

54. $-8, 0, 8, 16, \ldots$

55. $100, 94, 88, 82, \ldots$

56. $3200, 2800, 2400, 2000, \ldots$

57. $1, \tfrac{5}{3}, \tfrac{7}{3}, 3, \ldots$

58. $\tfrac{1}{2}, \tfrac{5}{4}, 2, \tfrac{11}{4}, \ldots$

In Exercises 59–70, determine whether the sequence is arithmetic. If it is, find the common difference.

59. $2, 4, 6, 8, \ldots$

60. $2, 6, 10, 14, \ldots$

61. $2, \tfrac{7}{2}, 5, \tfrac{13}{2}, \ldots$

62. $5, 13, 21, 29, 37, \ldots$

63. $32, 16, 8, 4, \ldots$

64. $32, 16, 0, -16, \ldots$

65. $\frac{1}{3}$, $\frac{1}{2}$, $\frac{2}{3}$, $\frac{5}{6}$, 1, . . .

66. $\frac{1}{3}$, $\frac{2}{3}$, $\frac{4}{3}$, $\frac{8}{3}$, $\frac{16}{3}$, . . .

67. 3.2, 4, 4.8, 5.6, . . .

68. 8, 4, 2, 1, 0.5, 0.25, . . .

69. 1, $\sqrt{2}$, $\sqrt{3}$, 2, $\sqrt{5}$, . . .

70. 1, 4, 9, 16, 25, . . .

In Exercises 71–78, write the first five terms of the arithmetic sequence. (Begin with $n = 1$.)

71. $a_n = 3n + 4$

72. $a_n = 5n - 4$

73. $a_n = -2n + 8$

74. $a_n = -10n + 100$

75. $a_n = \frac{5}{2}n - 1$

76. $a_n = \frac{2}{3}n + 2$

77. $a_n = -\frac{1}{4}(n - 1) + 4$

78. $a_n = 4(n + 2) + 24$

⊞ In Exercises 79–82, use a graphing utility to graph the first 10 terms of the sequence.

79. $a_n = -2n + 21$

80. $a_n = \frac{3}{2}n + 1$

81. $a_n = \frac{3}{5}n + \frac{3}{2}$

82. $a_n = -25n + 500$

In Exercises 83–90, find a formula for the nth term of the arithmetic sequence.

83. $a_1 = 3, d = \frac{1}{2}$

84. $a_1 = -1, d = 1.2$

85. $a_1 = 64, d = -8$

86. $a_1 = 1000, d = -25$

87. $a_3 = 16, a_4 = 20$

88. $a_5 = 30, a_4 = 25$

89. $a_1 = 50, a_3 = 30$

90. $a_{10} = 32, a_{12} = 48$

In Exercises 91–94, write the first five terms of the arithmetic sequence defined recursively.

91. $a_1 = 9$

$a_{k+1} = a_k - 3$

92. $a_1 = 8$

$a_{k+1} = a_k + 7$

93. *First term:* -10

(k+1)st term: $a_k + 6$

94. *First term:* -20

(k+1)st term: $a_k + 7$

In Exercises 95–100, match the sequence with its graph. [The graphs are labeled (a), (b), (c), (d), (e), and (f).]

(a)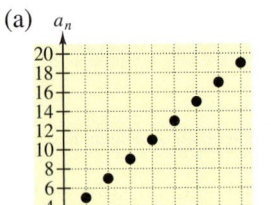

(b)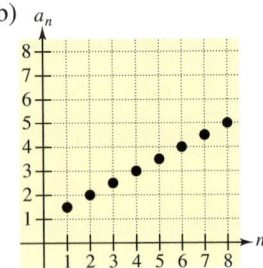

(c)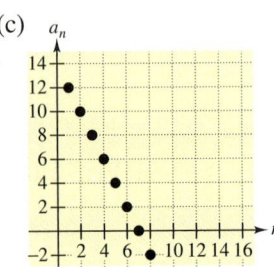

(d)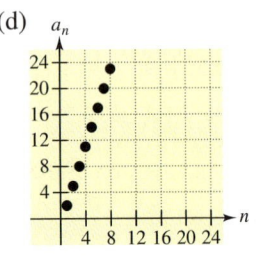

(e)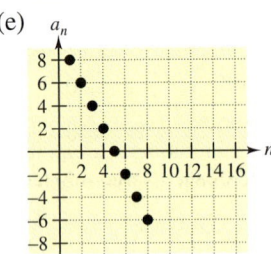

(f)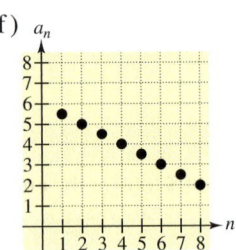

95. $a_n = \frac{1}{2}n + 1$

96. $a_n = -\frac{1}{2}n + 6$

97. $a_n = -2n + 10$

98. $a_n = 2n + 3$

99. $a_1 = 12$

$a_{k+1} = a_k - 2$

100. $a_1 = 2$

$a_{k+1} = a_k + 3$

In Exercises 101–106, find the sum.

101. $\displaystyle\sum_{n=1}^{20} n$

102. $\displaystyle\sum_{n=1}^{30} 4n$

103. $\displaystyle\sum_{n=1}^{50} (2n + 3)$

104. $\displaystyle\sum_{n=1}^{100} (4n - 1)$

105. $\displaystyle\sum_{n=1}^{500} \frac{n}{2}$

106. $\displaystyle\sum_{n=1}^{600} \frac{2n}{3}$

⊞ In Exercises 107 and 108, use a graphing utility to find the sum.

107. $\displaystyle\sum_{n=1}^{40} (1000 - 25n)$ **108.** $\displaystyle\sum_{n=1}^{20} (500 - 10n)$

In Exercises 109–116, find the nth partial sum of the arithmetic sequence.

109. 5, 12, 19, 26, 33, . . . , $n = 12$

110. 2, 12, 22, 32, 42, . . . , $n = 20$

111. 200, 175, 150, 125, 100, . . . , $n = 8$

112. 800, 785, 770, 755, 740, . . . , $n = 25$

113. -50, -38, -26, -14, -2, . . . , $n = 50$

114. -16, -8, 0, 8, 16, . . . , $n = 30$

115. 1, 4.5, 8, 11.5, 15, . . . , $n = 12$

116. 2.2, 2.8, 3.4, 4.0, 4.6, . . . , $n = 12$

117. Find the sum of the first 75 positive integers.

118. Find the sum of the integers from 35 to 100.

119. Find the sum of the first 50 even positive integers.

120. Find the sum of the first 100 positive odd integers.

121. *Salary Increases*   In your new job you are told that your starting salary will be $36,000 with an increase of $2000 at the end of each 5 years. How much will you be paid through the end of your first 6 years of employment with the company?

122. *Daily Wages*   Suppose that you receive 25 cents on the first day of the month, 50 cents the second day, 75 cents the third day, and so on. Determine the total amount that you will receive during a month with 30 days.

123. *Ticket Prices*   There are 20 rows of seats on the main floor of a concert hall—20 seats in the first row, 21 seats in the second row, 22 seats in the third row, and so on. How much should you charge per ticket in order to obtain $15,000 for the sale of all of the seats on the main floor?

124. *Baling Hay*   In the first two trips baling hay around a large field (see figure), a farmer obtains 93 bales and 89 bales, and the farmer estimates that the same pattern will continue. Estimate the total number of bales made if there are another 6 trips around the field.

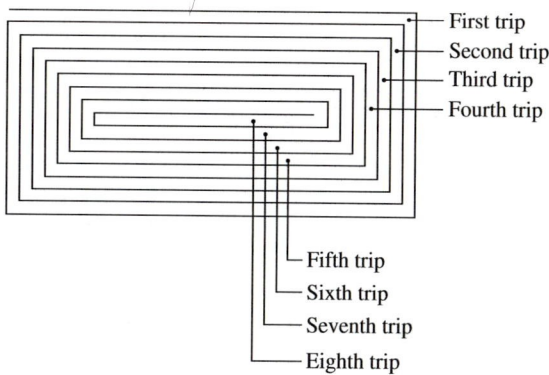

125. *Pattern Recognition*

(a) Compute the sums of positive odd integers.

$1 + 3 = $

$1 + 3 + 5 = $

$1 + 3 + 5 + 7 = $

$1 + 3 + 5 + 7 + 9 = $

$1 + 3 + 5 + 7 + 9 + 11 = $

(b) Use the sums of part (a) to make a conjecture about the sums of positive odd integers. Check your conjecture for the sum

$1 + 3 + 5 + 7 + 9 + 11 + 13 = $ 　.

(c) Verify your conjecture algebraically.

126. *Think About It*   Each term of an arithmetic sequence is multiplied by a constant C. Is the resulting sequence arithmetic? If so, how does the common difference compare with the common difference of the original sequence?

14.3	**Geometric Sequences**

Geometric Sequences ▪ The Sum of a Geometric Sequence ▪ Applications

Geometric Sequences

In Section 14.2, you studied sequences whose consecutive terms have a common *difference*. In this section, you will study sequences whose consecutive terms have a common *ratio*.

Definition of a Geometric Sequence

A sequence is called **geometric** if the ratios of consecutive terms are the same. Thus, the sequence a_1, a_2, a_3, a_4, . . . , a_n, . . . is geometric if there is a number r, $r \neq 0$, such that

$$\frac{a_2}{a_1} = r, \quad \frac{a_3}{a_2} = r, \quad \frac{a_4}{a_3} = r$$

and so on. The number r is the **common ratio** of the sequence.

EXAMPLE 1   *Examples of Geometric Sequences*

a. The sequence whose nth term is 2^n is geometric. For this sequence, the common ratio between consecutive terms is 2.

$$2, \ 4, \ 8, \ 16, \ . . . , \ 2^n, \ . . .$$

$$\frac{4}{2} = 2$$

b. The sequence whose nth term is $4(3^n)$ is geometric. For this sequence, the common ratio between consecutive terms is 3.

$$12, \ 36, \ 108, \ 324, \ . . . , \ 4(3^n), \ . . .$$

$$\frac{36}{12} = 3$$

c. The sequence whose nth term is $\left(-\frac{1}{3}\right)^n$ is geometric. For this sequence, the common ratio between consecutive terms is $-\frac{1}{3}$.

$$-\frac{1}{3}, \ \frac{1}{9}, \ -\frac{1}{27}, \ \frac{1}{81}, \ . . . , \ \left(-\frac{1}{3}\right)^n, \ . . .$$

$$\frac{1/9}{-1/3} = -\frac{1}{3}$$

NOTE If you know the nth term of a geometric sequence, the $(n + 1)$th term can be found by multiplying by r. That is,

$$a_{n+1} = ra_n.$$

The nth Term of a Geometric Sequence

The nth term of a geometric sequence has the form

$$a_n = a_1 r^{n-1}$$

where r is the common ratio of consecutive terms of the sequence. Thus, every geometric sequence can be written in the following form.

$$a_1, \ a_1 r, \ a_1 r^2, \ a_1 r^3, \ a_1 r^4, \ \ldots, \ a_1 r^{n-1}, \ \ldots$$

EXAMPLE 2 Finding the nth Term of a Geometric Sequence

Find a formula for the nth term of the geometric sequence whose common ratio is 3 and whose first term is 1. What is the eighth term of this sequence?

Solution

The formula for the nth term is of the form $a_1 r^{n-1}$. Moreover, because the common ratio is $r = 3$, and the first term is $a_1 = 1$, the formula must have the form

$$a_n = a_1 r^{n-1} = (1)(3)^{n-1} = 3^{n-1}.$$

The sequence therefore has the following form.

$$1, \ 3, \ 9, \ 27, \ 81, \ \ldots, \ 3^{n-1}, \ \ldots$$

The eighth term of the sequence is $a_8 = 3^{8-1} = 3^7 = 2187$.

EXAMPLE 3 Finding the nth Term of a Geometric Sequence

Find a formula for the nth term of the geometric sequence whose first two terms are 4 and 2.

Solution

Because the common ratio is

$$\frac{a_2}{a_1} = \frac{2}{4} = \frac{1}{2}$$

the formula for the nth term must be

$$a_n = a_1 r^{n-1} = 4\left(\frac{1}{2}\right)^{n-1}.$$

The sequence therefore has the following form.

$$4, \ 2, \ 1, \ \frac{1}{2}, \ \frac{1}{4}, \ \ldots, \ 4\left(\frac{1}{2}\right)^{n-1}, \ \ldots$$

The Sum of a Geometric Sequence

Some of the early work in representing functions by series was done by the French mathematician Joseph Fourier (1768–1830). Fourier's work is important in the history of calculus, partly because it forced 18th-century mathematicians to question the then-prevailing narrow concept of a function. Both Cauchy and Dirichlet were motivated by Fourier's work in series, and in 1837 Dirichlet published the general definition of a function that is used today.

The nth Partial Sum of a Geometric Sequence

The nth partial sum of the geometric sequence whose nth term is $a_n = a_1 r^{n-1}$ is given by

$$\sum_{i=1}^{n} a_1 r^{i-1} = a_1 + a_1 r + a_1 r^2 + a_1 r^3 + \cdots + a_1 r^{n-1} = a_1 \left(\frac{r^n - 1}{r - 1} \right).$$

EXAMPLE 4 *Finding the nth Partial Sum*

Find the sum.

$$1 + 2 + 4 + 8 + 16 + 32 + 64 + 128$$

Solution

This is a geometric sequence whose common ratio is $r = 2$. Because the first term of the sequence is $a_1 = 1$, it follows that the sum is

$$\sum_{i=1}^{8} 2^{i-1} = (1)\left(\frac{2^8 - 1}{2 - 1} \right) = \frac{256 - 1}{2 - 1} = 255.$$

EXAMPLE 5 *Finding the nth Partial Sum*

Find the sum of the first five terms of the geometric sequence whose nth term is $a_n = \left(\frac{2}{3} \right)^n$.

Solution

$$\begin{aligned}
\sum_{i=1}^{5} \left(\frac{2}{3} \right)^i &= \frac{2}{3} \left[\frac{(2/3)^5 - 1}{(2/3) - 1} \right] \qquad \text{Substitute } \tfrac{2}{3} \text{ for } a_1 \text{ and } \tfrac{2}{3} \text{ for } r. \\
&= \frac{2}{3} \left[\frac{(32/243) - 1}{-1/3} \right] \\
&= \frac{2}{3} \left(-\frac{211}{243} \right) (-3) \\
&= \frac{422}{243} \\
&\approx 1.737
\end{aligned}$$

Applications

EXAMPLE 6   An Application: A Lifetime Salary

You have accepted a job that pays a salary of $28,000 the first year. During the next 39 years, suppose you receive a 6% raise each year. What will your total salary be over the 40-year period?

Solution

Using a geometric sequence, your salary during the first year will be

$$a_1 = 28{,}000.$$

Then, with a 6% raise, your salary during the next 2 years will be as follows.

$$a_2 = 28{,}000 + 28{,}000(0.06) = 28{,}000(1.06)^1$$

$$a_3 = 28{,}000(1.06) + 28{,}000(1.06)(0.06) = 28{,}000(1.06)^2$$

From this pattern, you can see that the common ratio of the geometric sequence is $r = 1.06$. Using the formula for the nth partial sum of a geometric sequence, you will find that the total salary over the 40-year period is given by

$$\text{Total salary} = a_1 \left(\frac{r^n - 1}{r - 1} \right)$$

$$= 28{,}000 \left[\frac{(1.06)^{40} - 1}{1.06 - 1} \right]$$

$$= 28{,}000 \left[\frac{(1.06)^{40} - 1}{0.06} \right]$$

$$\approx \$4{,}333{,}335.$$

The bar graph in Figure 14.2 illustrates your salary during the 40-year period.

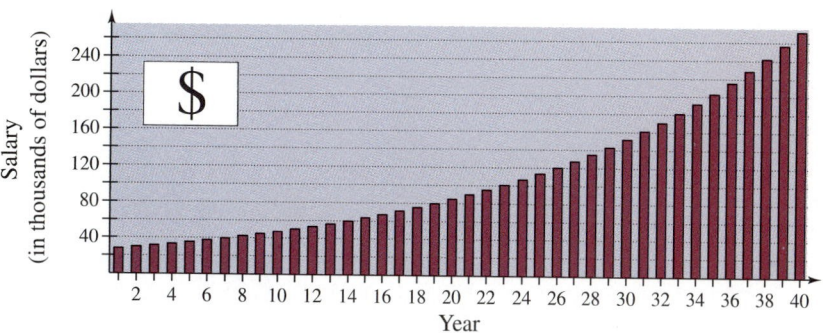

FIGURE 14.2

EXAMPLE 7   An Application: Increasing Annuity

You deposit $100 in an account each month for 2 years. The account pays an annual interest rate of 9%, compounded monthly. What is your balance at the end of 2 years? (This type of savings plan is called an **increasing annuity.**)

Solution

The first deposit would earn interest for the full 24 months, the second deposit would earn interest for 23 months, the third deposit would earn interest for 22 months, and so on. Using the formula for compound interest, you can see that the total of the 24 deposits would be

$$\text{Total} = a_1 + a_2 + \cdots + a_{24}$$
$$= 100(1.0075)^1 + 100(1.0075)^2 + \cdots + 100(1.0075)^{24}$$
$$= 100(1.0075)\left(\frac{1.0075^{24} - 1}{1.0075 - 1}\right) \quad a_1\left(\frac{r^n - 1}{r - 1}\right)$$
$$= \$2638.49.$$

Group Activities   Extending the Concept

Annual Revenue   The two bar graphs below show the annual revenues for two companies. One company's revenue grew at an arithmetic rate, whereas the other grew at a geometric rate. Which company had the greatest revenue during the 10-year period? Which company would you rather own? Explain.

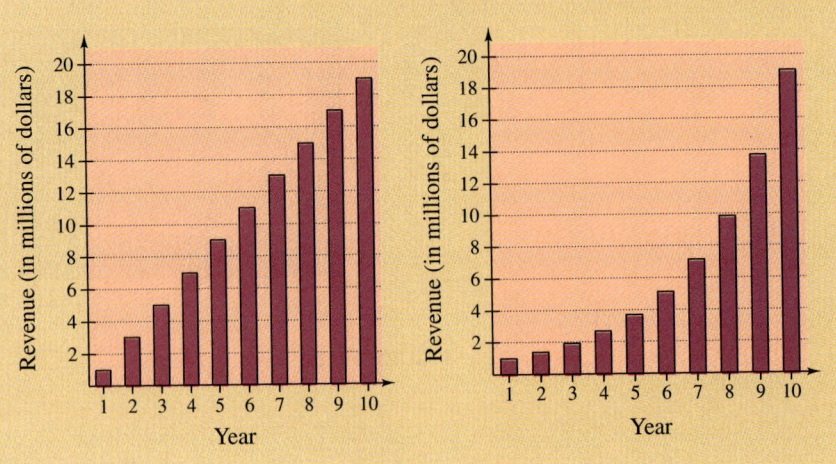

14.3 Exercises

Discussing the Concepts

1. In your own words, explain what makes a sequence geometric.

2. What is the general formula for the nth term of a geometric sequence?

3. The second and third terms of a geometric sequence are 6 and 3, respectively. What is the first term?

4. Give an example of a geometric sequence whose terms alternate in sign.

5. Explain why the terms of a geometric sequence decrease when $0 < r < 1$.

6. In your own words, describe an increasing annuity.

Problem Solving

In Exercises 7–10, find the common ratio of the geometric sequence.

7. $1,\ -3,\ 9,\ -27,\ \ldots$

8. $5,\ -\frac{5}{2},\ \frac{5}{4},\ -\frac{5}{8},\ \ldots$

9. $1,\ \pi,\ \pi^2,\ \pi^3,\ \ldots$

10. $50(1.06),\ 50(1.06)^2,\ 50(1.06)^3,\ 50(1.06)^4,\ \ldots$

In Exercises 11–14, determine whether the sequence is geometric. If it is, find the common ratio.

11. $5,\ 10,\ 20,\ 40,\ \ldots$

12. $54,\ -18,\ 6,\ -2,\ \ldots$

13. $1,\ 8,\ 27,\ 64,\ 125,\ \ldots$

14. $12,\ 7,\ 2,\ -3,\ -8,\ \ldots$

In Exercises 15–18, write the first five terms of the geometric sequence.

15. $a_1 = 4,\ r = -\frac{1}{2}$

16. $a_1 = 4,\ r = \frac{3}{2}$

17. $a_1 = 20,\ r = 1.07$

18. $a_1 = 100,\ r = \dfrac{1}{1.05}$

In Exercises 19–22, find the nth term of the geometric sequence.

19. $a_1 = 120,\ r = -\frac{1}{3},\ a_{10} =$

20. $a_1 = 5,\ r = \sqrt{3},\ a_9 =$

21. $a_1 = 200,\ r = 1.2,\ a_{12} =$

22. $a_1 = 1200,\ a_2 = 1000,\ a_6 =$

In Exercises 23–26, find the formula for the nth term of the geometric sequence. (Begin with $n = 1$.)

23. $a_1 = 25,\ r = 4$

24. $a_1 = 12,\ r = -\frac{4}{3}$

25. $1,\ \frac{3}{2},\ \frac{9}{4},\ \frac{27}{8},\ \ldots$

26. $4,\ -6,\ 9,\ -\frac{27}{2},\ \ldots$

In Exercises 27–30, match the sequence with its graph. [The graphs are labeled (a), (b), (c), and (d).]

(a)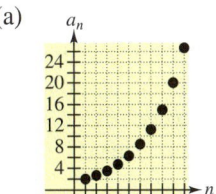

(b)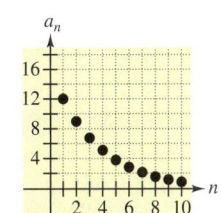

(c)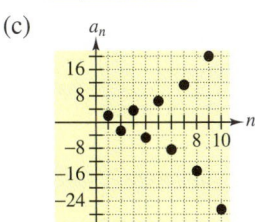

(d)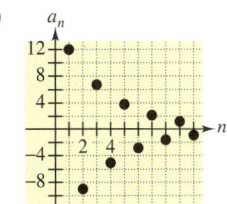

27. $a_n = 12\left(\frac{3}{4}\right)^{n-1}$

28. $a_n = 12\left(-\frac{3}{4}\right)^{n-1}$

29. $a_n = 2\left(\frac{4}{3}\right)^{n-1}$

30. $a_n = 2\left(-\frac{4}{3}\right)^{n-1}$

In Exercises 31 and 32, use a graphing utility to graph the first 10 terms of the sequence.

31. $a_n = 20(-0.6)^{n-1}$

32. $a_n = 4(1.4)^{n-1}$

In Exercises 33–36, find the sum.

33. $\displaystyle\sum_{i=1}^{12} 4(-2)^{i-1}$ **34.** $\displaystyle\sum_{i=1}^{20} 16\left(\frac{1}{2}\right)^{i-1}$

35. $\displaystyle\sum_{i=1}^{8} 6(0.1)^{i-1}$ **36.** $\displaystyle\sum_{i=1}^{24} 1000(1.06)^{i-1}$

In Exercises 37 and 38, use a graphing utility to find the sum.

37. $\displaystyle\sum_{i=1}^{30} 100(0.75)^{i-1}$ **38.** $\displaystyle\sum_{i=1}^{24} 5000(1.08)^{-(i-1)}$

In Exercises 39–42, find the nth partial sum of the geometric sequence.

39. 4, 12, 36, 108, . . . , $n = 8$

40. $\frac{1}{36}$, $-\frac{1}{12}$, $\frac{1}{4}$, $-\frac{3}{4}$, . . . , $n = 20$

41. 60, -15, $\frac{15}{4}$, $-\frac{15}{16}$, . . . , $n = 12$

42. 50, 50(1.04), 50(1.04)^2, 50(1.04)^3, . . . , $n = 18$

43. *Depreciation*   Your company buys a machine for $250,000. During the next 5 years, the machine depreciates at the rate of 25% per year. (That is, at the end of each year, the depreciated value is 75% of what it was at the beginning of the year.)

(a) Find a formula for the nth term of the geometric sequence that gives the value of the machine n full years after it was purchased.

(b) Find the depreciated value of the machine at the end of 5 full years.

(c) During which year did the machine depreciate the most?

44. *Population Increase*   A city of 500,000 people is growing at a rate of 1% per year. (That is, at the end of each year, the population is 1.01 times the population at the beginning of the year.)

(a) Find a formula for the nth term of the geometric sequence that gives the population t years from now.

(b) Estimate the population 20 years from now.

Increasing Annuity   In Exercises 45 and 46, find the balance in an increasing annuity when P dollars is invested each month for t years, compounded monthly at rate r.

45. $P = \$75$, $t = 30$ years, $r = 6\%$

46. $P = \$100$, $t = 25$ years, $r = 8\%$

47. *Geometry*   A square has 12-inch sides. A new square is formed by connecting the midpoints of the sides of the square, and two of the triangles are shaded (see figure). This process is repeated five more times. What is the total area of the shaded region?

48. *Bungee Jumping*   A bungee jumper stretches a cord 100 feet. Successive bounces stretch the cord 75% of each previous length (see figure). Find the total distance traveled by the bungee jumper during 10 bounces.

$$100 + 2(100)(0.75) + \cdots + 2(100)(0.75)^{10}$$

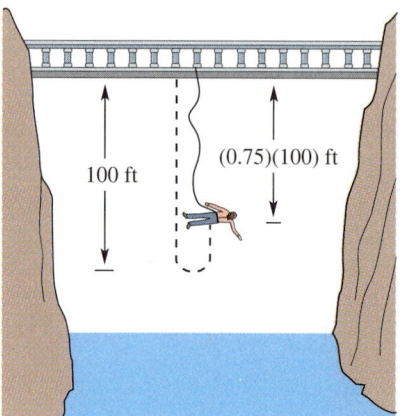

Reviewing the Major Concepts

In Exercises 49–52, evaluate the determinant.

49. $\begin{bmatrix} 10 & 25 \\ 6 & -5 \end{bmatrix}$ **50.** $\begin{bmatrix} 3 & 7 \\ -2 & 6 \end{bmatrix}$

51. $\begin{bmatrix} 3 & -2 & 1 \\ 0 & 5 & 3 \\ 6 & 1 & 1 \end{bmatrix}$ **52.** $\begin{bmatrix} 4 & 3 & 5 \\ 3 & 2 & -2 \\ 5 & -2 & 0 \end{bmatrix}$

53. *Geometry* Use a determinant to find the area of the triangle with vertices $(-5, 8)$, $(10, 0)$, and $(3, -4)$.

54. Solve the system of linear equations.

$$\begin{aligned} 3x - 2y + z &= 1 \\ x + 5y - 6z &= 4 \\ 4x - 3y + 2z &= 2 \end{aligned}$$

Additional Problem Solving

In Exercises 55–62, find the common ratio of the geometric sequence.

55. 2, 6, 18, 54, . . . **56.** 5, −10, 20, −40, . . .

57. 54, 18, 6, 2, . . . **58.** 12, −6, 3, $-\frac{3}{2}$, . . .

59. 1, $-\frac{3}{2}$, $\frac{9}{4}$, $-\frac{27}{8}$, . . .

60. 9, 6, 4, $\frac{8}{3}$, . . .

61. e, e^2, e^3, e^4, . . .

62. 1.1, $(1.1)^2$, $(1.1)^3$, $(1.1)^4$, . . .

In Exercises 63–70, determine whether the sequence is geometric. If it is, find the common ratio.

63. 10, 15, 20, 25, . . .

64. 10, 20, 40, 80, . . .

65. 64, 32, 16, 8, . . .

66. 64, 32, 0, −32, . . .

67. 1, $-\frac{2}{3}$, $\frac{4}{9}$, $-\frac{8}{27}$, . . .

68. $\frac{1}{3}$, $-\frac{2}{3}$, $\frac{4}{3}$, $-\frac{8}{3}$, . . .

69. $10(1 + 0.02)$, $10(1 + 0.02)^2$, $10(1 + 0.02)^3$, . . .

70. 1, 0.2, 0.04, 0.008, . . .

In Exercises 71–78, write the first five terms of the geometric sequence.

71. $a_1 = 4, r = 2$ **72.** $a_1 = 3, r = 4$

73. $a_1 = 6, r = \frac{1}{3}$ **74.** $a_1 = 4, r = \frac{1}{2}$

75. $a_1 = 1, r = -\frac{1}{2}$ **76.** $a_1 = 32, r = -\frac{3}{4}$

77. $a_1 = 1000, r = 1.01$ **78.** $a_1 = 4000, r = \dfrac{1}{1.01}$

In Exercises 79–86, find the nth term of the geometric sequence.

79. $a_1 = 6, r = \frac{1}{2}, a_{10} =$

80. $a_1 = 8, r = \frac{3}{4}, a_8 =$

81. $a_1 = 3, r = \sqrt{2}, a_{10} =$

82. $a_1 = 500, r = 1.06, a_{40} =$

83. $a_1 = 4, a_2 = 3, a_5 =$

84. $a_1 = 1, a_2 = 9, a_7 =$

85. $a_1 = 1, a_3 = \frac{9}{4}, a_6 =$

86. $a_2 = 12, a_3 = 16, a_4 =$

In Exercises 87–94, find the formula for the nth term of the geometric sequence. (Begin with $n = 1$.)

87. $a_1 = 2, r = 3$ **88.** $a_1 = 5, r = 4$

89. $a_1 = 1, r = 2$ **90.** $a_1 = 1, r = -5$

91. $a_1 = 4, r = -\frac{1}{2}$ **92.** $a_1 = 9, r = \frac{2}{3}$

93. $a_1 = 8, a_2 = 2$ **94.** $a_1 = 18, a_2 = 8$

In Exercises 95–100, find the sum.

95. $\displaystyle\sum_{i=1}^{10} 2^{i-1}$ **96.** $\displaystyle\sum_{i=1}^{6} 3^{i-1}$

97. $\displaystyle\sum_{i=1}^{12} 3\left(\frac{3}{2}\right)^{i-1}$ **98.** $\displaystyle\sum_{i=1}^{20} 12\left(\frac{2}{3}\right)^{i-1}$

99. $\displaystyle\sum_{i=1}^{15} 3\left(-\frac{1}{3}\right)^{i-1}$ **100.** $\displaystyle\sum_{i=1}^{8} 8\left(-\frac{1}{4}\right)^{i-1}$

In Exercises 101 and 102, use a graphing utility to find the sum.

101. $\displaystyle\sum_{i=1}^{20} 100(1.1)^i$ **102.** $\displaystyle\sum_{i=1}^{40} 50(1.07)^i$

In Exercises 103–110, find the nth partial sum of the geometric sequence.

103. 1, −3, 9, −27, 81, . . . , $n = 10$

104. 3, −6, 12, −24, 48, . . . , $n = 12$

105. 8, 4, 2, 1, $\frac{1}{2}$, . . . , $n = 15$

106. 9, 6, 4, $\frac{8}{3}$, $\frac{16}{9}$, . . . , $n = 10$

107. 1, $\sqrt{2}$, 2, $2\sqrt{2}$, 4, . . . , $n = 12$

108. 40, −10, $\frac{5}{2}$, $-\frac{5}{8}$, $\frac{5}{32}$, . . . , $n = 10$

109. 3, 3(1.06), 3(1.06)^2, 3(1.06)^3, . . . , $n = 20$

110. 10, 10(1.08), 10(1.08)^2, 10(1.08)^3, . . . , $n = 40$

111. *Power Supply*   The electrical power for an implanted medical device decreases by 0.1% each day.

(a) Find a formula for the nth term of the geometric sequence that gives the power n days after the device is implanted.

(b) What percent of the initial power is still available 1 year after the device is implanted?

(c) The power supply needs to be changed when half the power is depleted. Use a graphing utility to graph the first 750 terms and estimate when the power source should be changed.

112. *Cooling*   The temperature of an item is 70° when it is placed in a freezer. Its temperature n hours after being placed in the freezer is 20% less than 1 hour earlier.

(a) Find a formula for the nth term of the geometric sequence that gives the temperature of the item n hours after being placed in the freezer.

(b) Find the temperature of the item 6 hours after being placed in the freezer.

(c) Use a graphing utility to estimate the time when the item freezes. Explain your reasoning.

113. *Salary Increases*   You accept a job that pays a salary of $30,000 the first year. During the next 39 years, you receive a 5% raise each year. What is your *total* salary over the 40-year period?

114. *Salary Increases*   You accept a job that pays a salary of $30,000 the first year. During the next 39 years, you receive a 5.5% raise each year. What is your *total* salary over the 40-year period?

Increasing Annuity   In Exercises 115–118, find the balance in an increasing annuity when P dollars is invested each month for t years, compounded monthly at rate r.

115. $P = \$100$, $t = 10$ years, $r = 9\%$

116. $P = \$50$, $t = 5$ years, $r = 7\%$

117. $P = \$30$, $t = 40$ years, $r = 8\%$

118. $P = \$200$, $t = 30$ years, $r = 10\%$

119. *Salary*   You start work at a company that pays $0.01 for the first day, $0.02 for the second day, $0.04 for the third day, and so on. If the daily wage keeps doubling, what would your total income be for working (a) 29 days and (b) 30 days?

120. *Salary*   You start work at a company that pays $0.01 for the first day, $0.03 for the second day, $0.09 for the third day, and so on. If the daily wage keeps tripling, what would your total income be for working (a) 25 days and (b) 26 days?

121. *Geometry*   A square has 12-inch sides. The square is divided into nine smaller squares and the center square is shaded (see figure). Each of the eight unshaded squares is then divided into nine smaller squares and each center square is shaded. This process is repeated four more times. What is the total area of the shaded region?

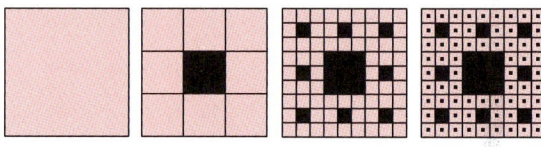

122. *Number of Ancestors*   The number of direct ancestors a person has had is as follows.

$$2 + 2^2 + 2^3 + 2^4 + \cdots + 2^n + \cdots.$$

This formula is valid provided the person has no common ancestors. (A common ancestor is one to whom you are related in more than one way.) During the past 2000 years, suppose your ancestry can be traced through 66 generations. During that time, your total number of ancestors would be

$$2 + 2^2 + 2^3 + 2^4 + \cdots + 2^{66}.$$

Considering the total, do you think that you have had no common ancestors in the past 2000 years?

Math Matters   **D u o – D i c e***

Ordinary six-sided dice

Twelve-sided Duo-Dice

Ordinary playing dice are made in the shape of cubes. On the sides of each cube, the dice are imprinted with one, two, three, four, five, or six dots, as shown in the figure. On normal playing dice, these dots are usually recessed to prevent the paint from wearing off. These recessions, however, cause the dice to be unbalanced. The heaviest side is the side with only one recession, and the lightest side is the side with six recessions. In practice, this unbalancing is enough to create an increased probability that the lightest side (the side with six recessions) will turn up.

To prevent the unbalancing caused by recessed dots, you can make the dice in the shape of dodecahedrons, as shown in the figure. On two opposing sides, one dot is recessed. On two other opposing sides, two dots are recessed, and so on. Dice made with this pattern are perfectly balanced (because each pair of opposing sides has the same number of recessed dots). When two of these dice are used in any game that requires two six-sided dice, the probability of rolling a given total is the same as if you used two perfectly balanced, six-sided dice.

*Duo-Dice are patented by Roland E. Larson (one of the authors of this text). (Patent Number: 4,465,279, August 14, 1984) If you would like a free pair of these dice, write to Roland E. Larson, Department of Mathematics, Pennsylvania State University, Erie, Pennsylvania 16563.

MID-CHAPTER QUIZ

Take this quiz as you would take a quiz in class. After you are done, check your work against the answers given in the back of the book.

In Exercises 1 and 2, write the first five terms of the sequence.

1. $a_n = 32\left(\dfrac{1}{4}\right)^{n-1}$ (Begin with $n = 1$.)   **2.** $a_n = \dfrac{(-3)^n n}{n+4}$ (Begin with $n = 1$.)

In Exercises 3–6, find the sum.

3. $\displaystyle\sum_{k=1}^{4} 10k$ **4.** $\displaystyle\sum_{i=1}^{10} 4$ **5.** $\displaystyle\sum_{j=1}^{5} \dfrac{60}{j+1}$ **6.** $\displaystyle\sum_{n=1}^{8} 8\left(-\dfrac{1}{2}\right)$

In Exercises 7 and 8, write the sum using sigma notation.

7. $\dfrac{2}{3(1)} + \dfrac{2}{3(2)} + \dfrac{2}{3(3)} + \cdots + \dfrac{2}{3(20)}$ **8.** $\dfrac{1}{1^3} - \dfrac{1}{2^3} + \dfrac{1}{3^3} - \cdots + \dfrac{1}{25^3}$

In Exercises 9 and 10, find the common difference of the arithmetic sequence.

9. $1, \frac{3}{2}, 2, \frac{5}{2}, 3, \ldots$ **10.** $100, 94, 88, 82, 76, \ldots$

In Exercises 11 and 12, find a formula for a_n.

11. *Arithmetic, $a_1 = 20$, $a_4 = 11$*   **12.** *Geometric, $a_1 = 32$, $r = -\frac{1}{4}$*

In Exercises 13–16, find the sum.

13. $\displaystyle\sum_{n=1}^{50} (3n + 5)$ **14.** $\displaystyle\sum_{n=1}^{300} \dfrac{n}{5}$

15. $\displaystyle\sum_{i=1}^{8} 9\left(\dfrac{2}{3}\right)^{i-1}$ **16.** $\displaystyle\sum_{j=1}^{20} 500(1.06)^{j-1}$

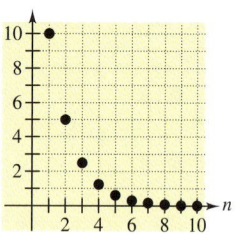

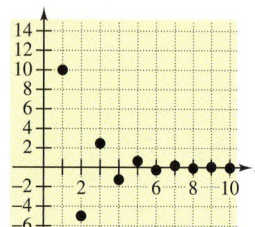

Figure for 18

17. Find the 12th term of $625, -250, 100, -40, \ldots$.

18. Match $a_n = 10\left(\frac{1}{2}\right)^{n-1}$ and $b_n = 10\left(-\frac{1}{2}\right)^{n-1}$ with the graphs at the right.

19. The temperature of a coolant decreases by $25.75°F$ the first hour. For each subsequent hour, the temperature decreases by $2.25°F$ less than it decreased the previous hour. How much does the temperature decrease during the 10th hour?

20. The sequence given by $a_n = 2^{n-1}$ is geometric. Describe the sequence given by $b_n = \ln a_n$.

14.4	**The Binomial Theorem**
	Binomial Coefficients ▪ Pascal's Triangle ▪ Binomial Expansions

Binomial Coefficients

Recall that a **binomial** is a polynomial that has two terms. In this section, you will study a formula that provides a quick method of raising a binomial to a power. To begin, let's look at the expansion of $(x + y)^n$ for several values of n.

$$(x + y)^0 = 1$$
$$(x + y)^1 = x + y$$
$$(x + y)^2 = x^2 + 2xy + y^2$$
$$(x + y)^3 = x^3 + 3x^2y + 3xy^2 + y^3$$
$$(x + y)^4 = x^4 + 4x^3y + 6x^2y^2 + 4xy^3 + y^4$$
$$(x + y)^5 = x^5 + 5x^4y + 10x^3y^2 + 10x^2y^3 + 5xy^4 + y^5$$

There are several observations you can make about these expansions.

1. In each expansion, there are $n + 1$ terms.

2. In each expansion, x and y have symmetrical roles. The powers of x decrease by 1 in successive terms, whereas the powers of y increase by 1.

3. The sum of the powers of each term is n. For instance, in the expansion of $(x + y)^5$, the sum of the powers of each term is 5.

$$4 + 1 = 5 \quad\quad 3 + 2 = 5$$
$$(x + y)^5 = x^5 + 5x^4y^1 + 10x^3y^2 + 10x^2y^3 + 5xy^4 + y^5$$

4. The coefficients increase and then decrease in a symmetric pattern.

The coefficients of a binomial expansion are called **binomial coefficients.** To find them, you can use the following theorem.

The Binomial Theorem

In the expansion of $(x + y)^n$

$$(x + y)^n = x^n + nx^{n-1}y + \cdots + {}_nC_r x^{n-r}y^r + \cdots + nxy^{n-1} + y^n$$

the coefficient of $x^{n-r}y^r$ is given by

$$_nC_r = \frac{n!}{(n-r)!\,r!}.$$

EXAMPLE 1 Finding Binomial Coefficients

Find each binomial coefficient.

a. $_8C_2$ **b.** $_{10}C_3$ **c.** $_7C_0$ **d.** $_8C_8$

Solution

a. $_8C_2 = \dfrac{8!}{6! \cdot 2!} = \dfrac{(8 \cdot 7) \cdot \cancel{6!}}{\cancel{6!} \cdot 2!} = \dfrac{8 \cdot 7}{2 \cdot 1} = 28$

b. $_{10}C_3 = \dfrac{10!}{7! \cdot 3!} = \dfrac{(10 \cdot 9 \cdot 8) \cdot \cancel{7!}}{\cancel{7!} \cdot 3!} = \dfrac{10 \cdot 9 \cdot 8}{3 \cdot 2 \cdot 1} = 120$

c. $_7C_0 = \dfrac{7!}{7! \cdot 0!} = 1$

d. $_8C_8 = \dfrac{8!}{0! \cdot 8!} = 1$

NOTE When $r \neq 0$ and $r \neq n$, as in parts (a) and (b) above, there is a simple pattern for evaluating binomial coefficients.

$$_8C_2 = \overbrace{\dfrac{8 \cdot 7}{\underbrace{2 \cdot 1}_{\text{2 factorial}}}}^{\text{2 factors}} \quad \text{and} \quad _{10}C_3 = \overbrace{\dfrac{10 \cdot 9 \cdot 8}{\underbrace{3 \cdot 2 \cdot 1}_{\text{3 factorial}}}}^{\text{3 factors}}$$

EXAMPLE 2 Finding Binomial Coefficients

Find each binomial coefficient.

a. $_7C_3$ **b.** $_7C_4$ **c.** $_{12}C_1$ **d.** $_{12}C_{11}$

Solution

a. $_7C_3 = \dfrac{7 \cdot 6 \cdot 5}{3 \cdot 2 \cdot 1} = 35$

b. $_7C_4 = \dfrac{7 \cdot 6 \cdot 5 \cdot 4}{4 \cdot 3 \cdot 2 \cdot 1} = 35$

c. $_{12}C_1 = \dfrac{12!}{11!1!} = \dfrac{(12) \cdot \cancel{11!}}{\cancel{11!} \cdot 1!} = \dfrac{12}{1} = 12$

d. $_{12}C_{11} = \dfrac{12!}{1! \cdot 11!} = \dfrac{(12) \cdot \cancel{11!}}{1! \cdot \cancel{11!}} = \dfrac{12}{1} = 12$

NOTE In Example 2, it is not a coincidence that the results in parts (a) and (b) are the same, *and* the results in parts (c) and (d) are the same. In general it is true that

$$_nC_r = {_nC_{n-r}}.$$

This shows the symmetric property of binomial coefficients.

Pascal's Triangle

There is a convenient way to remember a pattern for binomial coefficients. By arranging the coefficients in a triangular pattern, you obtain the following array, which is called **Pascal's Triangle.** This triangle is named after the famous French mathematician Blaise Pascal (1623–1662).

$$
\begin{array}{ccccccccccccc}
 & & & & & & 1 & & & & & & \\
 & & & & & 1 & & 1 & & & & & \\
 & & & & 1 & & 2 & & 1 & & & & \\
 & & & 1 & & 3 & & 3 & & 1 & & & \\
 & & 1 & & 4 & & 6 & & 4 & & 1 & & \\
 & 1 & & 5 & & 10 & & 10 & & 5 & & 1 & \\
1 & & 6 & & 15 & & 20 & & 15 & & 6 & & 1 \\
\end{array}
$$

1   7   21   35   35   21   7   1

NOTE   The top row in Pascal's Triangle is called the *zero row* because it corresponds to the binomial expansion

$$(x + y)^0 = 1.$$

Similarly, the next row is called the *first row* because it corresponds to the binomial expansion

$$(x + y)^1 = 1(x) + 1(y).$$

In general, the *nth row* in Pascal's Triangle gives the coefficients of $(x + y)^n$.

The first and last numbers in each row of Pascal's Triangle are 1. Every other number in each row is formed by adding the two numbers immediately above the number. Pascal noticed that numbers in this triangle are precisely the same numbers that are the coefficients of binomial expansions, as follows.

$$(x + y)^0 = 1$$
$$(x + y)^1 = 1x + 1y$$
$$(x + y)^2 = 1x^2 + 2xy + 1y^2$$
$$(x + y)^3 = 1x^3 + 3x^2y + 3xy^2 + 1y^3$$
$$(x + y)^4 = 1x^4 + 4x^3y + 6x^2y^2 + 4xy^3 + 1y^4$$
$$(x + y)^5 = 1x^5 + 5x^4y + 10x^3y^2 + 10x^2y^3 + 5xy^4 + 1y^5$$
$$(x + y)^6 = 1x^6 + 6x^5y + 15x^4y^2 + 20x^3y^3 + 15x^2y^4 + 6xy^5 + 1y^6$$
$$(x + y)^7 = 1x^7 + 7x^6y + 21x^5y^2 + 35x^4y^3 + 35x^3y^4 + 21x^2y^5 + 7xy^6 + 1y^7$$

EXAMPLE 3   *Using Pascal's Triangle*

You can use the seventh row of Pascal's Triangle to find the eighth row.

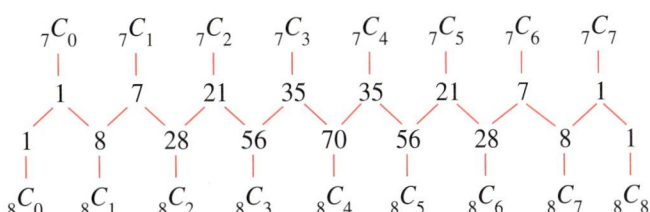

Binomial Expansions

As mentioned at the beginning of this section, when you write out the coefficients for a binomial that is raised to a power, you are **expanding a binomial.** The formulas for binomial coefficients give you an easy way to expand binomials, as demonstrated in the next three examples.

EXAMPLE 4   *Expanding a Binomial*

Write the expansion for the expression.

$$(x + 1)^3$$

Solution

The binomial coefficients from the third row of Pascal's Triangle are

$$1, 3, 3, 1.$$

Therefore, the expansion is as follows.

$$(x + 1)^3 = (1)x^3 + (3)x^2(1) + (3)x(1^2) + (1)(1^3)$$
$$= x^3 + 3x^2 + 3x + 1$$

To expand binomials representing *differences*, rather than sums, you alternate signs. Here are two examples.

$$(x - 1)^3 = x^3 - 3x^2 + 3x - 1$$
$$(x - 1)^4 = x^4 - 4x^3 + 6x^2 - 4x + 1$$

EXAMPLE 5   *Expanding a Binomial*

Write the expansion for the expression.

$$(x - 3)^4$$

Solution

The binomial coefficients from the fourth row of Pascal's Triangle are

$$1, 4, 6, 4, 1.$$

Therefore, the expansion is as follows.

$$(x - 3)^4 = (1)x^4 - (4)x^3(3) + (6)x^2(3^2) - (4)x(3^3) + (1)(3^4)$$
$$= x^4 - 12x^3 + 54x^2 - 108x + 81$$

EXAMPLE 6 Expanding a Binomial

Write the expansion for $(x - 2y)^4$.

Solution

Use the fourth row of Pascal's Triangle, as follows.

$$(x - 2y)^4 = (1)x^4 - (4)x^3(2y) + (6)x^2(2y)^2 - (4)x(2y)^3 + (1)(2y)^4$$
$$= x^4 - 8x^3y + 24x^2y^2 - 32xy^3 + 16y^4$$

EXAMPLE 7 Finding a Term in the Binomial Expansion

Find the sixth term of $(a + 2b)^8$.

Solution

From the Binomial Theorem, you can see that the $(r + 1)$th term is $_nC_r x^{n-r} y^r$. So in this case, $6 = r + 1$ means that $r = 5$. Because $n = 8$, $x = a$, and $y = 2b$, the sixth term in the binomial expansion is

$$_8C_5 a^{8-5}(2b)^5 = 56 \cdot a^3 \cdot (2b)^5$$
$$= 56(2^5)a^3b^5$$
$$= 1792a^3b^5$$

Group Activities   Extending the Concept

Finding a Pattern   By adding the terms in each of the rows of Pascal's Triangle, you obtain the following.

Row 0: $1 = 1$
Row 1: $1 + 1 = 2$
Row 2: $1 + 2 + 1 = 4$
Row 3: $1 + 3 + 3 + 1 = 8$
Row 4: $1 + 4 + 6 + 4 + 1 = 16$

Find a pattern for this sequence. Then use th
terms in the 10th row of Pascal's Triang
actually adding the terms of the 10th r

um of the
answer by

14.4 Exercises

Discussing the Concepts

1. How many terms are in the expansion of $(x + y)^n$?

2. Describe the pattern for the exponents with base x in the expansion of $(x + y)^n$.

3. How do the expansions of $(x + y)^n$ and $(x - y)^n$ differ?

4. What is the relationship between $_nC_r$ and $_nC_{n-r}$?

5. Which of the following is equal to $_{11}C_5$? Explain.

(a) $\dfrac{11 \cdot 10 \cdot 9 \cdot 8 \cdot 7}{5 \cdot 4 \cdot 3 \cdot 2 \cdot 1}$ (b) $\dfrac{11 \cdot 10 \cdot 9 \cdot 8 \cdot 7}{6 \cdot 5 \cdot 4 \cdot 3 \cdot 2 \cdot 1}$

6. In your own words, explain how to form the rows of Pascal's Triangle.

Problem Solving

In Exercises 7–10, evaluate $_nC_r$.

7. $_6C_4$ 8. $_7C_3$

9. $_{20}C_{20}$ 10. $_{200}C_1$

In Exercises 11–14, use a graphing utility to evaluate $_nC_r$.

11. $_{30}C_6$ 12. $_{25}C_{10}$

13. $_{52}C_5$ 14. $_{100}C_8$

In Exercises 15–18, evaluate the binomial coefficient $_nC_r$. Also, evaluate its symmetric coefficient $_nC_{n-r}$.

15. $_{15}C_3$ 16. $_9C_4$

17. $_{25}C_5$ 18. $_{30}C_3$

19. Find the ninth row of Pascal's Triangle.

9. Use Pascal's Triangle to evaluate $_nC_r$.

(a) $_6C_2$ (b) $_9C_3$

ises 21–24, use the Binomial Theorem to ex- xpression.

23.

22. $(x - 5)^4$

24. $(2x + y)^5$

In Exe the exp

25. $(x +$ e Pascal's Triangle to expand

$(r - s)^7$

27. $(x - 2)^6$ 28. $(2x + 3)^5$

In Exercises 29–32, find the coefficient of the given term of the expression.

Expression	Term
29. $(x + 1)^{10}$	x^7
30. $(x + 3)^{12}$	x^9
31. $(x - y)^{15}$	x^4y^{11}
32. $(x - 3y)^{14}$	x^3y^{11}

Probability In Exercises 33–36, use the Binomial Theorem to expand the expression. In the study of probability, it is sometimes necessary to use the expansion $(p + q)^n$, where $p + q = 1$.

33. $\left(\dfrac{1}{2} + \dfrac{1}{2}\right)^5$ 34. $\left(\dfrac{2}{3} + \dfrac{1}{3}\right)^4$

35. $\left(\dfrac{1}{4} + \dfrac{3}{4}\right)^4$ 36. $(0.4 + 0.6)^6$

In Exercises 37–40, use the Binomial Theorem to approximate the quantity accurate to two decimal places. For example,

$(1.02)^{10} \approx 1 + 10(0.02) + 45(0.02)^2 \approx 1.22$.

37. $(1.02)^8$ 38. $(2.005)^{10}$

39. $(2.99)^{12}$ 40. $(1.98)^9$

Reviewing the Major Concepts

In Exercises 41–44, rewrite the logarithmic equation in exponential form.

41. $\log_4 64 = 3$ **42.** $\log_3 \frac{1}{81} = -4$

43. $\ln 1 = 0$ **44.** $\ln 5 \approx 1.6094$

45. *Graphical Estimation* After t years, the value of a car is given by $V(t) = 22{,}000(0.8)^t$. Graphically determine when the value of the car will be $15,000.

46. *Interest* Find the balance when $10,000 is invested at $7\frac{1}{2}\%$ compounded monthly for 15 years.

Additional Problem Solving

In Exercises 47–54, evaluate $_nC_r$.

47. $_{10}C_5$ **48.** $_{12}C_9$

49. $_{18}C_{18}$ **50.** $_{15}C_0$

51. $_{50}C_{48}$ **52.** $_{75}C_1$

53. $_{25}C_4$ **54.** $_{18}C_5$

In Exercises 55–60, use a graphing utility to evaluate $_nC_r$.

55. $_{12}C_7$ **56.** $_{40}C_5$

57. $_{200}C_{10}$ **58.** $_{500}C_6$

59. $_{25}C_{12}$ **60.** $_{1000}C_2$

In Exercises 61–66, evaluate the binomial coefficient $_nC_r$. Also, evaluate its symmetric coefficient $_nC_{n-r}$.

61. $_5C_2$ **62.** $_8C_6$

63. $_{12}C_5$ **64.** $_{14}C_8$

65. $_{10}C_0$ **66.** $_{25}C_{25}$

In Exercises 67–70, use Pascal's Triangle to evaluate the binomial coefficient.

67. $_7C_3$ **68.** $_9C_5$

69. $_8C_4$ **70.** $_{10}C_6$

In Exercises 71–78, use the Binomial Theorem to expand the expression.

71. $(x + 1)^5$ **72.** $(x + 2)^5$

73. $(x - 4)^6$ **74.** $(x - 8)^4$

75. $(x + y)^4$ **76.** $(u + v)^6$

77. $(x - y)^5$ **78.** $(4t - 1)^4$

In Exercises 79–84, use Pascal's Triangle to expand the expression.

79. $(a + 2)^3$ **80.** $(x + 3)^5$

81. $(2x - 1)^4$ **82.** $(4 - 3y)^3$

83. $(2y + z)^6$ **84.** $(2t - s)^5$

In Exercises 85–90, find the coefficient of the given term of the expression.

Expression	Term
85. $(x - 1)^{10}$	x^7
86. $(x - 2)^8$	x^4
87. $(2x + y)^{12}$	$x^3 y^9$
88. $(x + y)^{10}$	$x^7 y^3$
89. $(x^2 - 3)^4$	x^4
90. $(3 - y^3)^5$	y^9

91. *Patterns in Pascal's Triangle* Use each encircled group of numbers to form a 2×2 matrix. Find the determinant of each matrix. Describe the pattern.

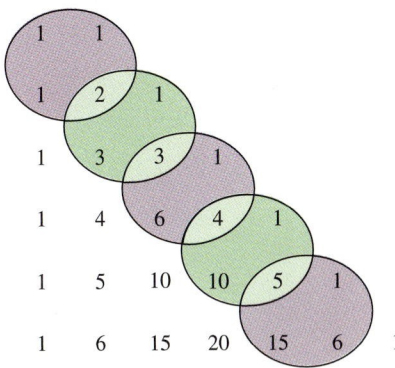

14.5	**Counting Principles**

Simple Counting Problems ▪ Counting Principles ▪
Permutations ▪ Combinations

Simple Counting Problems

The last two sections of this chapter contain a brief introduction to some of the basic counting principles and their application to probability. In the next section, you will see that much of probability has to do with counting the number of ways an event can occur. Examples 1, 2, and 3 describe some simple cases.

EXAMPLE 1   A Random Number Generator

A random number generator (on a computer) selects an integer from 1 to 30. Find the number of ways each event can occur.

a.   An even integer is selected.

b.   A number that is less than 12 is selected.

c.   A prime number is selected.

d.   A perfect square is selected.

Solution

a.   Because half of the numbers from 1 to 30 are even, this event can occur in 15 different ways.

b.   The positive integers that are less than 12 are as follows.

$$\{1,\ 2,\ 3,\ 4,\ 5,\ 6,\ 7,\ 8,\ 9,\ 10,\ 11\}$$

Because this set has 11 members, you can conclude that there are 11 different ways this event can happen.

c.   The prime numbers between 1 and 30 are as follows.

$$\{2,\ 3,\ 5,\ 7,\ 11,\ 13,\ 17,\ 19,\ 23,\ 29\}$$

Because this set has 10 members, you can conclude that there are 10 different ways this event can happen.

d.   The perfect square numbers between 1 and 30 are as follows.

$$\{1,\ 4,\ 9,\ 16,\ 25\}$$

Because this set has five members, you can conclude that there are five different ways this event can happen.

EXAMPLE 2   *Selecting Pairs of Numbers at Random*

Eight pieces of paper are numbered from 1 to 8 and placed in a box. One piece of paper is drawn from the box, its number is written down, and the piece of paper is replaced in the box. Then, a second piece of paper is drawn from the box, and its number is written down. Finally, the two numbers are added together. How many different ways can a total of 12 be obtained?

Solution

To solve this problem, count the different ways that a total of 12 can be obtained using two numbers between 1 and 8.

$$\boxed{\text{First number}} \ + \ \boxed{\text{Second number}} \ = \ \boxed{12}$$

After considering the various possibilities, you can see that this equation can be solved in the following five ways.

First Number: 4   5   6   7   8
Second Number: 8   7   6   5   4

Thus, a total of 12 can be obtained in five different ways.

Solving counting problems can be tricky. Often, seemingly minor changes in the statement of a problem can affect the answer. For instance, compare the counting problem in the next example with that given in Example 2.

EXAMPLE 3   *Selecting Pairs of Numbers at Random*

Eight pieces of paper are numbered from 1 to 8 and placed in a box. Two pieces of paper are drawn from the box, and the numbers on them are written down and totaled. How many different ways can a total of 12 be obtained?

Solution

To solve this problem, count the different ways that a total of 12 can be obtained *using two different numbers* between 1 and 8.

$$\boxed{\text{First number}} \ + \ \boxed{\text{Second number}} \ = \ \boxed{12}$$

After considering the various possibilities, you can see that this equation can be solved in the following four ways.

First Number: 4   5   7   8
Second Number: 8   7   5   4

Thus, a total of 12 can be obtained in four different ways.

NOTE   The difference between the counting problems in Examples 2 and 3 can be distinguished by saying that the random selection in Example 2 occurs *with replacement,* whereas the random selection in Example 3 occurs *without replacement,* which eliminates the possibility of choosing two 6's.

Counting Principles

The first three examples in this section are considered simple counting problems in which you can *list* each possible way that an event can occur. When it is possible, this is always the best way to solve a counting problem. However, some events can occur in so many different ways that it is not feasible to write out the entire list. In such cases, you must rely on formulas and counting principles. The most important of these is called the **Fundamental Counting Principle.**

NOTE The Fundamental Counting Principle can be extended to three or more events. For instance, the number of ways that three events E_1, E_2, and E_3 can occur is $m_1 \cdot m_2 \cdot m_3$.

Fundamental Counting Principle

Let E_1 and E_2 be two events. The first event E_1 can occur in m_1 different ways. After E_1 has occurred, E_2 can occur in m_2 different ways. The number of ways that the two events can occur is

$$m_1 \cdot m_2.$$

EXAMPLE 4   *Applying the Fundamental Counting Principle*

The English alphabet contains 26 letters. Thus, the number of possible "two-letter words" is $26 \cdot 26 = 676$.

EXAMPLE 5   *Applying the Fundamental Counting Principle*

Telephone numbers in the United States have ten digits. The first three are the *area code* and the next seven are the *local telephone number*. How many different telephone numbers are possible within each area code? (A telephone number cannot have 0 or 1 as its first or second digit.)

Solution

There are only eight choices for the first and second digits because neither can be 0 or 1. For each of the other digits, there are 10 choices.

In 1991, there were 131 million active telephone numbers in use in the United States.

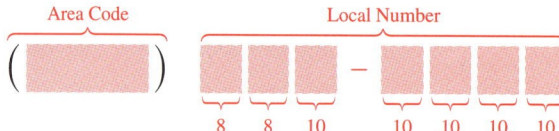

Thus, by the Fundamental Counting Principle, the number of local telephone numbers that are possible within each area code is

$$8 \cdot 8 \cdot 10 \cdot 10 \cdot 10 \cdot 10 \cdot 10 = 6,400,000.$$

Permutations

One important application of the Fundamental Counting Principle is in determining the number of ways that n elements can be arranged (in order). An ordering of n elements is called a **permutation** of the elements.

Definition of Permutation

A **permutation** of n different elements is an ordering of the elements such that one element is first, one is second, one is third, and so on.

EXAMPLE 6 Listing Permutations

The six possible permutations of the letters A, B, and C are as follows.

A, B, C	B, A, C	C, A, B
A, C, B	B, C, A	C, B, A

EXAMPLE 7 Finding the Number of Permutations of n Elements

How many permutations are possible for the letters A, B, C, D, E, and F?

Solution

1st position:	Any of the *six* letters.
2nd position:	Any of the remaining *five* letters.
3rd position:	Any of the remaining *four* letters.
4th position:	Any of the remaining *three* letters.
5th position:	Any of the remaining *two* letters.
6th position:	The *one* remaining letter.

Thus, the number of choices for the six positions are as follows.

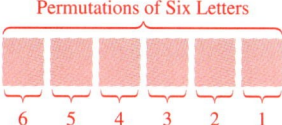

Permutations of Six Letters

By the Fundamental Counting Principle, the total number of permutations of the six letters is

$$6 \cdot 5 \cdot 4 \cdot 3 \cdot 2 \cdot 1 = 6! = 720.$$

The result obtained in Example 7 is generalized below.

Number of Permutations of n Elements

The number of permutations of n elements is given by

$$n \cdot (n-1) \cdot \cdots \cdot 4 \cdot 3 \cdot 2 \cdot 1 = n!.$$

In other words, there are $n!$ different ways that n elements can be ordered.

EXAMPLE 8 Finding the Number of Permutations

How many ways can you form a four-digit number using each of the digits 1, 3, 5, and 7 exactly once?

Solution

One way to solve this problem is simply to list the number of ways.

    1357, 1375, 1537, 1573, 1735, 1753

    3157, 3175, 3517, 3571, 3715, 3751

    5137, 5173, 5317, 5371, 5713, 5731

    7135, 7153, 7315, 7351, 7513, 7531

Another way to solve the problem is to use the formula for the number of permutations of four elements. By that formula, there are $4! = 24$ permutations.

EXAMPLE 9 Finding the Number of Permutations

You are a supervisor for eleven different employees. One of your responsibilities is to perform an annual evaluation for each employee, and then rank the eleven different performances. How many different rankings are possible?

Solution

Because there are eleven different employees, you have eleven choices for first ranking. After choosing the first ranking, you can choose any of the remaining ten for second ranking, and so on.

Rankings of Eleven Employees

  11 10 9 8 7 6 5 4 3 2 1

Thus, the number of different rankings is $11! = 39{,}916{,}800$.

Combinations

When one counts the number of possible permutations of a set of elements, order is important. The final topic in this section is a method of selecting subsets of a larger set in which order is *not important*. Such subsets are called **combinations of *n* elements taken *r* at a time.** For instance, the combinations

$\qquad$ {A, B, C} $\qquad$ and $\qquad$ {B, A, C}

are equivalent because both sets contain the same three elements, and the order in which the elements are listed is *not important*. Hence, you would count only one of the two sets. A common example of how a combination occurs is a card game in which the player is free to reorder the cards after they have been dealt.

EXAMPLE 10   Combination of n Elements Taken r at a Time

In how many different ways can three letters be chosen from the letters A, B, C, D, and E? (The order of the three letters is not important.)

Solution

The following subsets represent the different combinations of three letters that can be chosen from five letters.

$\qquad$ {A, B, C} $\qquad$ {A, B, D}

$\qquad$ {A, B, E} $\qquad$ {A, C, D}

$\qquad$ {A, C, E} $\qquad$ {A, D, E}

$\qquad$ {B, C, D} $\qquad$ {B, C, E}

$\qquad$ {B, D, E} $\qquad$ {C, D, E}

From this list, you can conclude that there are 10 different ways that three letters can be chosen from five letters. Because order is not important, the set {B, C, A} is not chosen. It is represented by the set {A, B, C}.

The formula for the number of combinations of *n* elements taken *r* at a time is as follows.

Number of Combinations of *n* Elements Taken *r* at a Time

The number of combinations of *n* elements taken *r* at a time is

$$_nC_r = \frac{n!}{(n-r)!\,r!}.$$

Note that the formula for $_nC_r$ is the same one given for binomial coefficients. To see how this formula is used, consider the counting problem given in Example 10. In that problem, you need to find the number of combinations of five elements taken three at a time. Thus, $n = 5$, $r = 3$, and the number of combinations is

$$_5C_3 = \frac{5!}{2!3!}$$
$$= \frac{5 \cdot 4 \cdot 3}{3 \cdot 2 \cdot 1}$$
$$= 10$$

which is the same as the answer obtained in Example 10.

EXAMPLE 11   Combinations of n Elements Taken r at a Time

A standard poker hand consists of five cards dealt from a deck of 52. How many different poker hands are possible? (After the cards are dealt, the player may reorder them, and therefore order is not important.)

Solution

Use the formula for the number of combinations of 52 elements taken five at a time, as follows.

$$_{52}C_5 = \frac{52!}{47!5!}$$
$$= \frac{52 \cdot 51 \cdot 50 \cdot 49 \cdot 48}{5 \cdot 4 \cdot 3 \cdot 2 \cdot 1}$$
$$= 2,598,960$$

Thus, there are almost 2.6 million different hands.

Group Activities   P r o b l e m   S o l v i n g

Applying Counting Methods   The Boston Market restaurant chain offers individual rotisserie chicken meals with two side items. Customers can choose either dark or white meat and can select side items from a list of 15. How many different meals are available if two different side items are to be ordered? Of the 15 side items, nine are hot and six are cold. How many different meals are available if a customer wishes to order one hot and one cold side item? (Source: Boston Market, Inc.)

14.5 Exercises

Discussing the Concepts

1. State the Fundamental Counting Principle.

2. When you use the Fundamental Counting Principle, what are you counting?

3. Give examples of a permutation and a combination.

4. Without calculating the numbers, determine which of the following is greater. Explain.

    (a) The combination of 10 elements taken 6 at a time

    (b) The permutation of 10 elements taken 6 at a time

Problem Solving

Random Selection   In Exercises 5–10, find the number of ways the specified event can occur when one or more marbles are selected from a bowl containing 10 marbles numbered 0 through 9.

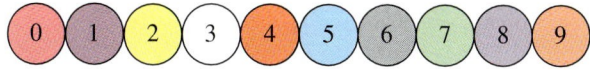

5. One number is drawn and it is even.

6. One number is drawn and it is prime.

7. Two marbles are drawn one after the other. The first is replaced before the second is drawn. The sum of the numbers is 10.

8. Two marbles are drawn one after the other. The first is replaced before the second is drawn. The sum of the numbers is 7.

9. Two marbles are drawn without replacement. The sum of the numbers is 10.

10. Two marbles are drawn without replacement. The sum of the numbers is 7.

11. *Staffing Choices*   A small grocery store needs to open another checkout line. Three people who can run the cash register are available and two people are available to bag groceries. How many different ways can the additional checkout line be staffed?

12. *Computer System*   You are in the process of purchasing a new computer system. You must choose one of three monitors, one of two computers, and one of two keyboards. How many different configurations of the system are available to you?

13. *Taking a Trip*   Five people are taking a long trip in a car. Two sit in the front seat and three in the back seat. Three of the people agree to share the driving. In how many different arrangements can the five people sit?

14. *Aircraft Boarding*   Eight people are boarding an aircraft. Three have tickets for first class and board before those in the economy class. In how many different ways can the eight people board the aircraft?

15. *Permutations*   List all the permutations of the letters X, Y, and Z.

16. *Permutations*   List all the permutations of the letters A, B, C, and D.

17. *Seating Arrangement*   In how many ways can five children be seated in a single row of chairs?

18. *Combination Lock*   A combination lock will open when the right choice of three numbers (from 1 to 40, inclusive) is selected. How many different lock combinations are possible?

19. *Work Assignments*   Eight workers are assigned to eight different tasks. In how many ways can this be done assuming there are no restrictions in making the assignments?

20. *Work Assignments*   Eight workers are available for five different tasks. In how many ways can five workers be selected from the eight and assigned to the tasks if there are no restrictions in making the assignments?

21. *Number of Subsets*   List all the subsets with two elements that can be formed from the set of letters {A, B, C, D, E, F}.

22. *Number of Subsets*   List all the subsets with three elements that can be formed from the set of letters {A, B, C, D, E, F}.

23. *Relationships*   As the size of a group increases, the number of relationships increases dramatically (see figure). Determine the number of two-person relationships in a group that has the following numbers.

(a) 3   (b) 4

(c) 6   (d) 8

(e) 10   (f) 12

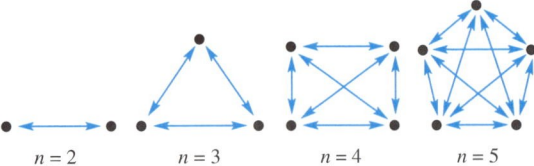

$n = 2$ $n = 3$ $n = 4$ $n = 5$

24. *Committee Selection*   Three students are selected from a class of 20 to form a fundraising committee. In how many ways can the committee be formed?

25. *Number of Triangles*   Eight points are located in the coordinate plane such that no three are collinear. How many different triangles can be formed having their vertices as three of the eight points?

26. *Test Questions*   A student is required to answer any nine questions from 12 questions on an exam. In how many ways can the student select the questions?

27. *Defective Units*   A shipment of 10 microwave ovens contains two defective units. In how many ways can a vending company purchase three of these units and receive (a) all good units, (b) two good units, and (c) one good unit?

28. *Job Applicants*   An employer interviews six people for four openings in the company. Four of the six people are women. If all six are qualified, in how many ways can the employer fill the four positions if (a) the selection is random and (b) exactly two are women?

Reviewing the Major Concepts

In Exercises 29–32, solve the equation. (Round your answer to two decimal places.)

29. $\sqrt{x-5} = 6$ **30.** $\dfrac{4}{t} + \dfrac{3}{2t} = 1$

31. $\log_2(x-5) = 6$ **32.** $e^{x/2} = 8$

33. Write the equation of the line that passes through $(3, 5)$ and $(6, 7)$.

34. *Radioactive Decay*   Carbon 14 has a half-life of 5730 years. If you start with 10 grams of this isotope, how much remains after 3000 years?

Additional Problem Solving

Random Selection   In Exercises 35–44, determine the number of ways the specified event can occur when one or more marbles are selected from a bowl containing 20 marbles numbered 1 through 20.

35. One number is drawn, and it is odd.

36. One number is drawn, and it is even.

37. One number is drawn, and it is prime.

38. One number is drawn, and it is greater than 12.

39. One number is drawn, and it is divisible by 3.

40. One number is drawn, and it is divisible by 6.

41. Two marbles are drawn one after the other. The first is replaced before the second is drawn. The sum of the numbers is 8.

42. Two marbles are drawn one after the other. The first is replaced before the second is drawn. The sum of the numbers is 15.

43. Two marbles are drawn without replacement. The sum of the numbers is 8.

44. Three marbles are drawn, one after another. The first and second are replaced before drawing the second and third. The sum of the numbers is 15.

45. License Plates   How many distinct automobile license plates can be formed by using a four-digit number followed by two letters?

46. Task Assignment   Four people are assigned to four different tasks. In how many ways can the assignments be made if one of the four is not qualified for the first task?

47. Permutations   List all the permutations of two letters selected from the letters A, B, C, and D.

48. Permutations   List all the permutations of two letters selected from the letters A, B, and C.

49. Posing for a Photograph   In how many ways can four children line up in one row to have their picture taken?

50. Seating Arrangement   In how many ways can six people be seated in a six-passenger car?

51. Choosing Officers   From a pool of 10 candidates, the offices of president, vice-president, secretary, and treasurer will be filled. In how many ways can the offices be filled if each of the 10 can hold any one of the offices?

52. Time Management Study   There are eight steps in accomplishing a certain task and these steps can be performed in any order. Management wants to test each possible order to determine which is least time consuming.

(a) How many different orders will have to be tested?

(b) How many different orders will have to be tested if one step in accomplishing the task must be done first? (The other seven steps can be performed in any order.)

53. Number of Subsets   List all the subsets with three elements that can be formed from the set of letters {A, B, C, D, E}.

54. Number of Subsets   List all the subsets with two elements that can be formed from the set of letters {A, B, C, D, E, F}.

55. Identification Numbers   In a statistical study, each participant is given an identification label consisting of a letter of the alphabet followed by a single digit. How many distinct labels are possible?

56. Test Questions   A student may answer any seven questions from a total of 10 questions on an exam. In how many different ways can the student select the questions?

57. Committee Selection   In how many ways can a committee of five be formed from a group of 30 people?

58. Menu Selection   A group of four people go out to dinner at a restaurant. There are nine entrees on the menu and the four people decide that no two will order the same thing. How many ways can the four order from the nine entrees?

59. Basketball Lineup   A high school basketball team has 15 players. In how many different ways can the coach choose the starting lineup of five? (Assume each player can play each position.)

60. Softball League   Six churches form a softball league. If each team must play every other team twice during the season, what is the total number of league games played?

61. Group Selection   Four people are to be selected from four couples. In how many ways can this be done if

(a) there are no restrictions?

(b) one person from each couple must be selected?

62. Geometry   Three points that are not on a line determine three lines. How many lines are determined by seven points, no three of which are on a line?

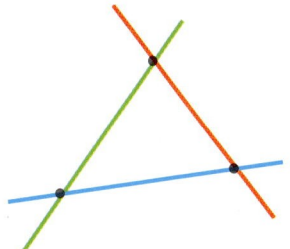

63. Diagonals of a Polygon   Find the number of diagonals of each polygon. (A line segment connecting any two nonadjacent vertices of a polygon is called a *diagonal* of the polygon.)

(a) Pentagon   (b) Hexagon   (c) Octagon

14.6 Probability

The Probability of an Event ▪
Using Counting Methods to Find Probabilities

The Probability of an Event

The **probability of an event** is a number from 0 to 1 that indicates the likelihood that the event will occur. An event that is certain to occur has a probability of 1. An event that cannot occur has a probability of 0. An event that is equally likely to occur or not occur has a probability of $\frac{1}{2}$ or 0.5.

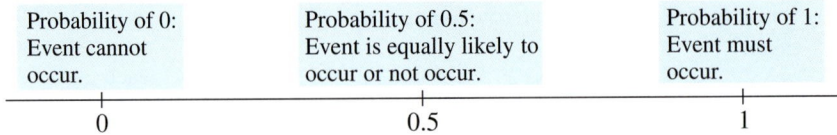

Probability of 0:
Event cannot
occur.

Probability of 0.5:
Event is equally likely to
occur or not occur.

Probability of 1:
Event must
occur.

0   0.5   1

The Probability of an Event

Consider a **sample space** S that is composed of a finite number of outcomes, each of which is equally likely to occur. A subset E of the sample space is an **event.** The probability that an outcome E will occur is

$$P = \frac{\text{number of outcomes in event}}{\text{number of outcomes in sample space}}.$$

EXAMPLE 1 Finding the Probability of an Event

a. You are dialing a friend's phone number but cannot remember the last digit. If you choose a digit at random, the probability that it is correct is

$$P = \frac{\text{number of correct digits}}{\text{number of possible digits}} = \frac{1}{10}.$$

b. On a multiple-choice test, you know that the answer to a question is not (a) or (d), but you are not sure about (b), (c), and (e). If you guess, the probability that you are wrong is

$$P = \frac{\text{number of wrong answers}}{\text{number of possible answers}} = \frac{2}{3}.$$

EXAMPLE 2 *Conducting a Poll*

In 1990, the Center for Disease Control took a survey of 11,631 high school students. The students were asked whether they considered themselves to be a good weight, underweight, or overweight. The results are shown in Figure 14.3.

Female

Good weight
3082

Overweight
1776

Underweight
366

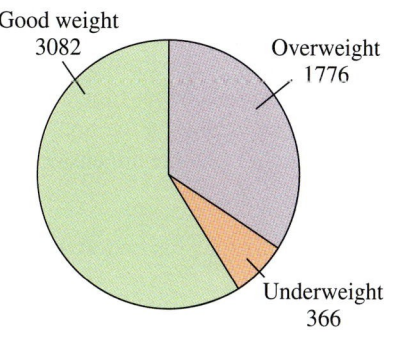

Male

Good weight
4389

Underweight
1057

Overweight
961

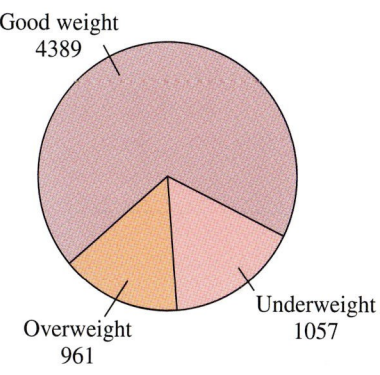

FIGURE 14.3

a. If you choose a female at random from those surveyed, the probability that she said she was underweight is

$$P = \frac{\text{number of females who answered "underweight"}}{\text{number of females in survey}}$$

$$= \frac{366}{3082 + 366 + 1776}$$

$$= \frac{366}{5224}$$

$$\approx 0.07.$$

b. If you choose a person who answered "underweight" from those surveyed, the probability that the person is female is

$$P = \frac{\text{number of females who answered "underweight"}}{\text{number in survey who answered "underweight"}}$$

$$= \frac{366}{366 + 1057}$$

$$= \frac{366}{1423}$$

$$\approx 0.26.$$

Polls such as the one described in Example 2 are often used to make inferences about a population that is larger than the sample. For instance, from Example 2, you might infer that 7% of *all* high school girls consider themselves to be underweight. When you make such an inference, it is important that those surveyed are representative of the entire population.

EXAMPLE 3 *Using Area to Find Probability*

You have just stepped into the tub to take a shower when one of your contact lenses falls out. (You have not yet turned on the water.) Assuming the lens is equally likely to land anywhere on the bottom of the tub, what is the probability that it lands in the drain? Use the dimensions in Figure 14.4 to answer the question.

Solution

Because the area of the tub bottom is $(26)(50) = 1300$ square inches and the area of the drain is

$$\pi(1^2) = \pi \qquad \text{Area of drain}$$

square inches, the probability that the lens lands in the drain is about

$$P = \frac{\pi}{1300} \approx 0.0024.$$

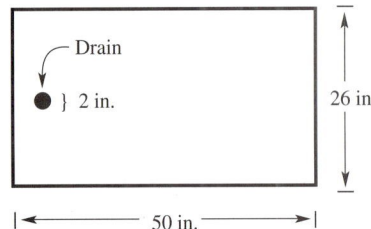

FIGURE 14.4

EXAMPLE 4 *The Probability of Inheriting Certain Genes*

Common parakeets have genes that can produce any one of four feather colors:

Green:	BBCC, BBCc, BbCC, BbCc
Blue:	BBcc, Bbcc
Yellow:	bbCC, bbCc
White:	bbcc

Use the *Punnett square* in Figure 14.5 to find the probability that an offspring to two green parents (both with BbCc feather genes) will be yellow. Note that each parent passes along a B or b gene and a C or c gene.

Solution

The probability that an offspring will be yellow is

$$P = \frac{\text{number of yellow possibilities}}{\text{number of possibilities}}$$

$$= \frac{3}{16}.$$

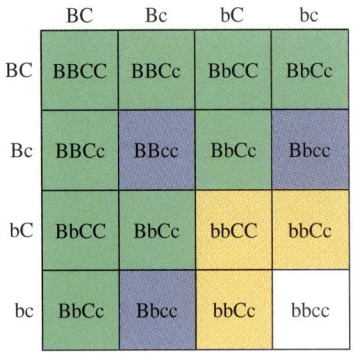

FIGURE 14.5

Standard 52-Card Deck

A♠	A♥	A♦	A♣
K♠	K♥	K♦	K♣
Q♠	Q♥	Q♦	Q♣
J♠	J♥	J♦	J♣
10♠	10♥	10♦	10♣
9♠	9♥	9♦	9♣
8♠	8♥	8♦	8♣
7♠	7♥	7♦	7♣
6♠	6♥	6♦	6♣
5♠	5♥	5♦	5♣
4♠	4♥	4♦	4♣
3♠	3♥	3♦	3♣
2♠	2♥	2♦	2♣

FIGURE 14.6

Using Counting Methods to Find Probabilities

EXAMPLE 5   *The Probability of a Royal Flush*

Five cards are dealt at random from a standard deck of 52 playing cards (see Figure 14.6). What is the probability that the cards are 10-J-Q-K-A of the same suit?

Solution

On page 868, you saw that the number of possible five-card hands from a deck of 52 cards is $_{52}C_5 = 2,598,960$. Because only four of these five-card hands are 10-J-Q-K-A of the same suit, the probability that the hand contains these cards is

$$P = \frac{4}{2,598,960} = \frac{1}{649,740}.$$

EXAMPLE 6   *Conducting a Survey*

In 1990, a survey was conducted of 500 adults who had worn Halloween costumes. Each person was asked how he or she acquired a Halloween costume: created it, rented it, bought it, or borrowed it. The results are shown in Figure 14.7. What is the probability that the first four people who were polled all created their costumes?

Solution

To answer this question, you need to use the formula for the number of combinations *twice*. First, find the number of ways to choose four people from 360 who created their own costumes.

$$_{360}C_4 = \frac{360 \cdot 359 \cdot 358 \cdot 357}{4 \cdot 3 \cdot 2 \cdot 1} = 688,235,310$$

Next, find the number of ways to choose four people from the 500 who were surveyed.

$$_{500}C_4 = \frac{500 \cdot 499 \cdot 498 \cdot 497}{4 \cdot 3 \cdot 2 \cdot 1} = 2,573,031,125$$

The probability that all of the first four people surveyed created their own costumes is the ratio of these two numbers.

$$P = \frac{\text{number of ways to choose 4 from 360}}{\text{number of ways to choose 4 from 500}}$$

$$= \frac{688,235,310}{2,573,031,125}$$

$$\approx 0.267$$

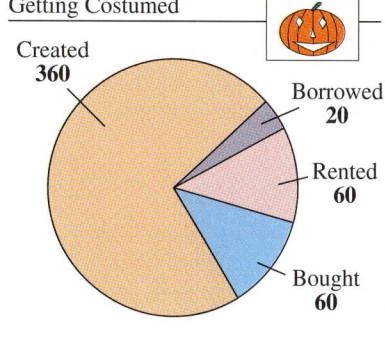

Getting Costumed

Created **360**

Borrowed **20**

Rented **60**

Bought **60**

FIGURE 14.7

EXAMPLE 7 Forming a Committee

To obtain input from 200 company employees, the management of a company selected a committee of five. Of the 200 employees, 56 were from minority groups. None of the 56, however, was selected to be on the committee. Does this indicate that the management's selection was biased?

Solution

Part of the solution is similar to that of Example 6. If the five committee members were selected at random, the probability that all five would be nonminority is

$$P = \frac{\text{number of ways to choose 5 from 144 nonminority employees}}{\text{number of ways to choose 5 from 200 employees}}$$

$$= \frac{_{144}C_5}{_{200}C_5}$$

$$= \frac{481{,}008{,}528}{2{,}535{,}650{,}040}$$

$$\approx 0.19.$$

Thus, if the committee were chosen at random (that is, without bias), the likelihood that it would have no minority members is about 0.19. Although this does not *prove* that there was bias, it does suggest it.

Group Activities   Extending the Concept

Probability of Guessing Correctly   You are taking a chemistry test and are asked to arrange the first 10 elements in the order in which they appear on the periodic table of elements. Suppose that you have no idea of the correct order and simply guess. Does the following computation represent the probability that you guess correctly?

Solution

You have 10 choices for the first element, nine choices for the second, eight choices for the third, and so on. The number of different orders is $10! = 3{,}628{,}800$, which means that your probability of guessing correctly is

$$P = \frac{1}{3{,}628{,}800}.$$

14.6 Exercises

Discussing the Concepts

1. The probability of an event must be a real number in what interval? Is the interval open or closed?

2. The probability of an event is $\frac{3}{4}$. What is the probability that the event *does not* occur? Explain.

3. What is the sum of the probabilities of all elements in a sample space? Explain.

4. The weather forecast indicates that the probability of rain is 40%. Explain what this means.

Problem Solving

In Exercises 5–8, determine the sample space for the experiment.

5. One letter from the alphabet is chosen.

6. A six-sided die is tossed twice and the sum is recorded.

7. Two county supervisors are selected from five supervisors, A, B, C, D, and E.

8. A salesperson makes a presentation in three homes. In each home there may be a sale (denote by Y) or there may be no sale (denote by N).

In Exercises 9 and 10, you are given the probability that an event will occur. Find the probability that the event will not occur.

9. $p = 0.35$

10. $p = 0.8$

Coin Tossing In Exercises 11–14, a coin is tossed three times. Find the probability of the specified event. Use the following sample space.

HHH, HHT, HTH, THH
HTT, THT, TTH, TTT

11. The event of getting two heads

12. The event of getting a tail on the second toss

13. The event of getting at least one head

14. The event of getting no more than two heads

Reading a Graph In Exercises 15 and 16, use the circle graph, which shows the number of people in the United States in 1990 with each blood type. (Source: American Association of Blood Banks)

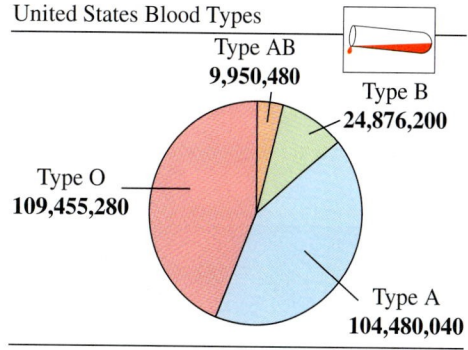

United States Blood Types

Type AB
9,950,480

Type B
24,876,200

Type O
109,455,280

Type A
104,480,040

Figure for 15 and 16

15. A person is selected at random from the United States population. What is the probability that the person does not have blood type B?

16. What is the probability that a person selected at random from the United States population does have blood type B? How is this probability related to the probability found in Exercise 15?

17. *Random Selection* Twenty marbles numbered 1 through 20 are placed in a bag, and one is selected. Find the probability of the specified event.

    (a) The number is 12.

    (b) The number is prime.

    (c) The number is odd.

    (d) The number is less than 6.

18. *Class Election* Three people are running for class president. It is estimated that the probability that candidate A will win is 0.5 and the probability that candidate B will win is 0.3. What is the probability that the third candidate will win?

19. *Meteorites* The largest meteorite that ever landed in the United States was found in the Willamette Valley of Oregon in 1902. Earth contains 57,510,000 square miles of land and 139,440,000 square miles of water. What is the probability that a meteorite that hits the earth will fall onto land? What is the probability that a meteorite that hits the earth will fall into water?

20. *Estimating Pi* A coin of diameter d is dropped onto a paper that contains a grid of squares d units on a side (see figure).

(a) Find the probability that the coin covers a vertex of one of the squares in the grid.

(b) Repeat the experiment 100 times and use the results to approximate π.

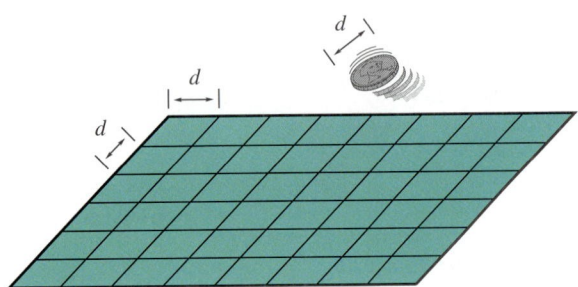

In Exercises 21–26, the sample spaces are large, and therefore you should use the counting principles discussed in Section 10.5.

21. *Game Show* On a game show, you are given five digits to arrange in the proper order to form the price of a car. If you arrange them correctly, you win the car. Find the probability of winning if you know the correct position of only one digit and must guess the positions of the other digits.

22. *Shelving Books* A parent instructs a young child to place a five-volume set of books on a bookshelf. Find the probability that the books are in correct order if the child places them on the shelf at random.

23. *Defective Units* A shipment of 10 food processors to a certain store contained two defective units. If you purchase two of these food processors as birthday gifts for friends, determine the probability that you get both defective units.

24. *Book Selection* Four books are selected at random from a shelf containing six novels and four autobiographies. Find the probability that the four autobiographies are selected.

25. *Card Selection* Five cards are selected from a standard deck of 52 cards. Find the probability that the four aces are selected.

26. *Card Selection* Five cards are selected from a standard deck of 52 cards. Find the probability of getting all hearts.

Reviewing the Major Concepts

In Exercises 27–30, describe the relationship between the graphs of f and g.

27. $g(x) = f(x) - 4$

28. $g(x) = f(x - 4)$

29. $g(x) = -f(x)$

30. $g(x) = f(-x)$

In Exercises 31 and 32, use a graphing utility to solve the system of equations.

31. $\quad 5x - 2y = -25$
$\quad -3x + 7y = \;\; 44$

32. $\quad 6x + 2y = 20$
$\quad 3x - \;\; y = 14$

Additional Problem Solving

In Exercises 33–36, determine the sample space for the experiment.

33. A taste tester must taste and rank three brands of yogurt, A, B, and C, according to preference.

34. A coin and a die are tossed.

35. A basketball tournament between two teams consists of three games. For each game, your team may win (denote by W) or lose (denote by L).

36. Two students are randomly selected from four students, A, B, C, and D.

In Exercises 37 and 38, you are given the probability that an event will not occur. Find the probability that the event will occur.

37. $p = 0.82$ **38.** $p = 0.13$

Playing Cards   In Exercises 39–42, a card is drawn from a standard deck of playing cards. Find the probability of drawing the indicated card.

39. A red card   **40.** A queen

41. A face card   **42.** A black face card

Tossing a Die   In Exercises 43–46, a six-sided die is tossed. Find the probability of the specified event.

43. The number is a 5.   **44.** The number is a 7.

45. The number is no more than 5.

46. The number is at least 1.

Reading a Graph   In Exercises 47–50, use the circle graphs, which show for a certain college the numbers of incoming freshmen in each average high school grade category for the years 1970 and 1992.

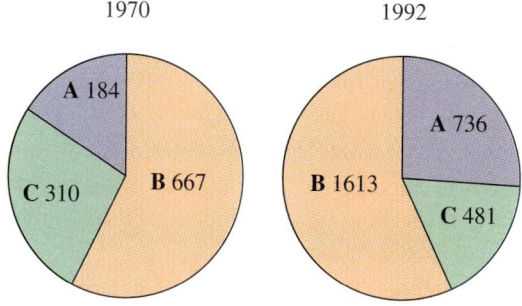

1970   1992

A 184
B 667
C 310

A 736
B 1613
C 481

47. A person is selected at random from the 1970 freshman class. What is the probability that the person's high school average was an A?

48. A person is selected at random from the 1992 freshman class. What is the probability that the person's high school average was an A?

49. What is the probability that a person selected from the 1970 freshman class did not have a high school average grade of C?

50. What is the probability that a person selected from the 1992 freshman class did not have a high school average grade of B?

Geometry   A child uses a spring-loaded device to shoot a marble into the square box shown in the figure. The base of the square is horizontal and the marble has an equal likelihood of coming to rest at any point on the base. In Exercises 51–54, find the probability that the marble comes to rest in the specified region.

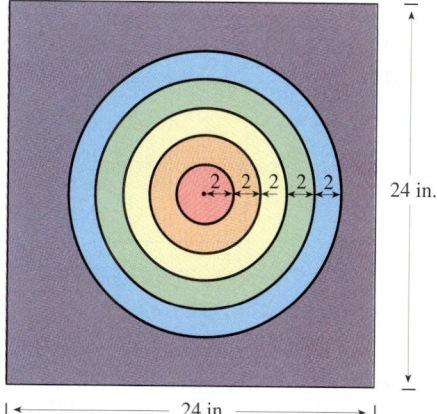

2 2 2 2 2   24 in.

|← 24 in. →|

51. The red center

52. The blue ring

53. The purple border

54. Not in the yellow ring

55. *Multiple-Choice Test*   A student takes a multiple-choice test in which there are five choices for each question. Find the probability that the first question is answered correctly given the following conditions.

(a) The student has no idea of the answer and guesses at random.

(b) The student can eliminate two of the choices and guesses from the remaining choices.

(c) The student knows the answer.

56. *Multiple-Choice Test*   A student takes a multiple-choice test in which there are four choices for each question. Find the probability that the first question is answered correctly given the following conditions.

(a) The student has no idea of the answer and guesses at random.

(b) The student can eliminate two of the choices and guesses from the remaining choices.

(c) The student knows the answer.

57. *Girl or Boy?*   The genes that determine the sex of a human baby are denoted by XX (female) and XY (male). Complete the Punnett square below. Then use the result to explain why it is equally likely that a newborn baby will be a boy or a girl.

Female

	X	X
X	?	?
Y	?	?

Male

58. *Blood Types*   There are four basic human blood types: A (AA or Ao), B (BB or Bo), AB (AB), and O (oo). Complete the Punnett square below. What is the blood type of each parent? What is the probability that their offspring will have blood type A? B? AB? O?

	A	o
B	?	?
o	?	?

59. *Continuing Education*   In a high school graduating class of 325 students, 255 are going to continue their education. What is the probability that a student selected at random from the class will not be furthering his or her education?

60. *Study Questions*   An instructor gives his class a list of four study questions for the next exam. Two of the four study questions will be on the exam. Find the probability that a student who knows the material relating to three of the four questions will be able to answer both questions selected for the exam.

In Exercises 61–66, the sample spaces are large, and therefore you should use the counting principles discussed in Section 10.5.

61. *Lottery*   You buy a lottery ticket inscribed with a five-digit number. On the designated day, five digits are randomly selected from the digits 0 through 9, inclusive. What is the probability that you have a winning ticket?

62. *Game Show*   On a game show, you are given four digits to arrange in proper order to form the price of a grandfather clock. What is the probability of winning given the following conditions?

(a) You guess the position of each digit.

(b) You know the first digit, but must guess the remaining three.

63. *Preparing for a Test*   An instructor gives her class a list of 10 study problems, from which she will select eight to be answered on an exam. If you know how to solve eight of the problems, find the probability you will be able to answer all eight test questions.

64. *Committee Selection*   A committee of three students is to be selected from a group of three girls and five boys. Find the probability that the committee is composed entirely of girls.

65. *Defective Units*   A shipment of 12 compact disc players contains two defective units. A husband and wife buy three of these players to give to their children as gifts.

(a) What is the probability that all three are good players?

(b) What is the probability that they buy at least one defective player?

66. *Card Selection*   Five cards are selected from a standard deck of 52 cards. Find the probability that two aces and three queens are selected.

CHAPTER PROJECT: Savings Account Interest

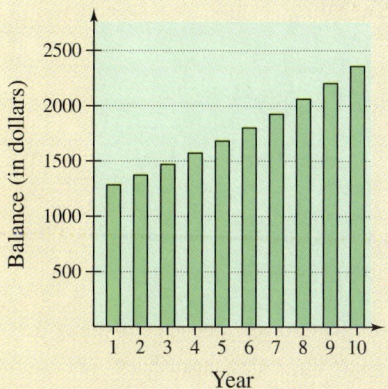

Figure for 1

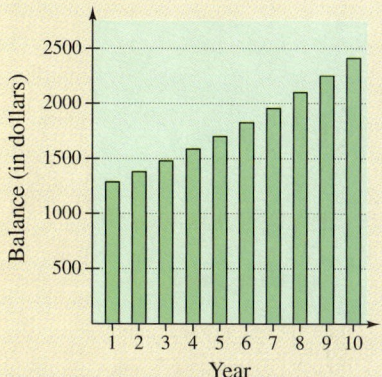

Figure for 2

Saving money requires careful planning and determination. Six factors that should be considered when devising a savings plan are: present assets, present debts, present income, present expenses, future income, and future goals. The type of savings account depends on the amount of deposit and the length of time the money will be in the account.

A compound interest formula to determine a savings account balance is

$$A = P\left(1 + \frac{r}{n}\right)^{nt}$$

where A is the balance in the account after t years, P is the principal, and r is the annual interest rate compounded n times per year. Use this information to investigate the following questions.

1. Use the formula for compound interest to find the balance for each of the first 10 years for a savings account with a principal of $1200 and 7% interest, compounded annually.

2. Use the formula for compound interest to find the balance for each of the first 10 years for a savings account with a principal of $1200 and 7% interest, compounded monthly.

3. In Questions 1 and 2, each savings account had a principal of $1200. Write a paragraph that answers the following questions.

   • What was the difference in the balances of the two accounts at the end of 10 years?

   • How do you account for this difference?

   • Which would you prefer: an account that compounds yearly, an account that compounds monthly, or an account that compounds daily? What other factors should you consider when choosing an account?

4. Use the information from Questions 1 and 2 to create a double bar graph of the annual account balances for the two savings accounts. For each account, in which year was the earned interest the least? For each account, in which year was the earned interest the greatest? How can you tell this from the graph?

5. Use the formula $R = I/P$, where R is the *annual interest rate,* I is the earned interest, and P is the principal, to find the annual interest rate that takes into consideration the effect of compounding for the third year for each account in Questions 1 and 2. Which account has a higher rate?

6. Use the formula for compound interest to find the balances in a savings account for each of the first five years with a principal of $10,000 and 8.5% interest compounded yearly, quarterly, and monthly. Create a triple bar graph to display the information.

CHAPTER SUMMARY

After studying this chapter, you should have acquired the following skills. These skills are keyed to the Review Exercises that begin on page 883. Answers to odd-numbered Review Exercises are given in the back of the book.

- Write sums in sigma notation. *(Section 14.1)* **Review Exercises 1–4**

- Simplify expressions involving factorials. *(Section 14.1)* **Review Exercises 5–8**

- Write the first several terms of an arithmetic sequence and of a geometric sequence. *(Sections 14.2, 14.3)* **Review Exercises 9–12, 15–18**

- Find the common difference of an arithmetic sequence and the common ratio of a geometric sequence. *(Sections 14.2, 14.3)* **Review Exercises 13, 14, 19, 20**

- Find formulas for the *n*th terms of an arithmetic sequence and a geometric sequence. *(Sections 14.2, 14.3)* **Review Exercises 21–30**

- Match arithmetic and geometric sequences with their graphs. *(Sections 14.2, 14.3)* **Review Exercises 31–36**

- Graph the first several terms of sequences using a graphing utility. *(Sections 14.1, 14.2, 14.3)* **Review Exercises 37–40**

- Evaluate sums expressed in sigma notation. *(Sections 14.1, 14.2, 14.3)* **Review Exercises 41–56**

- Find and evaluate the sums of arithmetic and geometric sequences from verbal statements. *(Sections 14.2, 14.3)* **Review Exercises 57–62**

- Evaluate binomial coefficients using the Binomial Theorem or Pascal's Triangle. *(Section 14.4)* **Review Exercises 63–66**

- Evaluate binomial coefficients using a graphing utility. *(Section 14.4)* **Review Exercises 67, 68**

- Expand binomial expressions using the Binomial Theorem. *(Section 14.4)* **Review Exercises 69–74**

- Find the coefficients of specified terms of binomial expressions. *(Section 14.4)* **Review Exercises 75, 76**

- Calculate the numbers of ways events can occur using the Fundamental Counting Principle, permutations, or combinations. *(Section 14.5)* **Review Exercises 77–80**

- Find the probabilities of the occurrences of specified events. *(Section 14.6)* **Review Exercises 81–86**

REVIEW EXERCISES

In Exercises 1–4, use sigma notation to write the sum.

1. $[5(1) - 3] + [5(2) - 3] + [5(3) - 3] + [5(4) - 3]$

2. $[9 - 2(1)] + [9 - 2(2)] + [9 - 2(3)] + [9 - 2(4)]$

3. $\dfrac{1}{3(1)} + \dfrac{1}{3(2)} + \dfrac{1}{3(3)} + \dfrac{1}{3(4)} + \dfrac{1}{3(5)} + \dfrac{1}{3(6)}$

4. $\left(-\frac{1}{3}\right)^0 + \left(-\frac{1}{3}\right)^1 + \left(-\frac{1}{3}\right)^2 + \left(-\frac{1}{3}\right)^3 + \left(-\frac{1}{3}\right)^4$

In Exercises 5–8, simplify the expression.

5. $\dfrac{20!}{18!}$

6. $\dfrac{50!}{53!}$

7. $\dfrac{n!}{(n-3)!}$

8. $\dfrac{(n-1)!}{(n+1)!}$

In Exercises 9–12, write the first five terms of the arithmetic sequence. (Begin with $n = 1$.)

9. $a_n = 132 - 5n$

10. $a_n = 2n + 3$

11. $a_n = \frac{3}{4}n + \frac{1}{2}$

12. $a_n = -\frac{3}{5}n + 1$

In Exercises 13 and 14, find the common difference of the arithmetic sequence.

13. 30, 27.5, 25, 22.5, 20, . . .

14. 9, 12, 15, 18, 21, . . .

In Exercises 15–18, write the first five terms of the geometric sequence.

15. $a_1 = 10$, $r = 3$

16. $a_1 = 2$, $r = -5$

17. $a_1 = 100$, $r = -\frac{1}{2}$

18. $a_1 = 12$, $r = \frac{1}{6}$

In Exercises 19 and 20, find the common ratio of the geometric sequence.

19. 8, 12, 18, 27, $\frac{81}{2}$, . . .

20. 81, −54, 36, −24, 16, . . .

In Exercises 21–30, find a formula for the nth term of the specified sequence.

21. Arithmetic sequence: $a_1 = 10$, $d = 4$

22. Arithmetic sequence: $a_1 = 32$, $d = -2$

23. Arithmetic sequence: $a_1 = 1000$, $a_2 = 950$

24. Arithmetic sequence: $a_1 = 12$, $a_2 = 20$

25. Geometric sequence: $a_1 = 1$, $r = -\frac{2}{3}$

26. Geometric sequence: $a_1 = 100$, $r = 1.07$

27. Geometric sequence: $a_1 = 24$, $a_2 = 48$

28. Geometric sequence: $a_1 = 16$, $a_2 = -4$

29. Geometric sequence: $a_1 = 12$, $a_4 = -\frac{3}{2}$

30. Geometric sequence: $a_2 = 1$, $a_3 = \frac{1}{3}$

In Exercises 31–36, match the sequence with its graph. [The graphs are labeled (a), (b), (c), (d), (e), and (f).]

(a)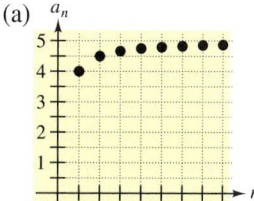

(b)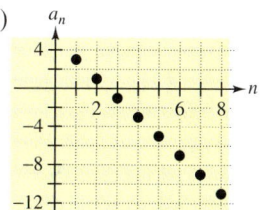

(c)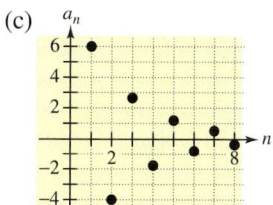

(d)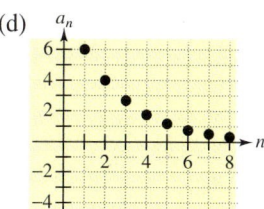

(e)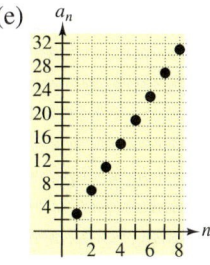

(f)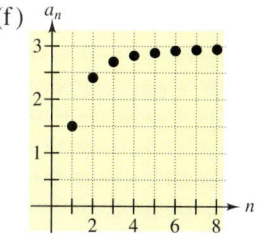

31. $a_n = 5 - \dfrac{1}{n}$

32. $a_n = \dfrac{3n^2}{n^2 + 1}$

33. $a_n = 5 - 2n$

34. $a_n = 4n - 1$

35. $a_n = 6\left(\dfrac{2}{3}\right)^{n-1}$

36. $a_n = 6\left(-\dfrac{2}{3}\right)^{n-1}$

In Exercises 37–40, use a graphing utility to graph the first 10 terms of the sequence.

37. $a_n = \dfrac{3n}{n + 1}$

38. $a_n = \dfrac{3}{n + 1}$

39. $a_n = 5\left(\dfrac{3}{4}\right)^{n-1}$

40. $a_n = 5\left(-\dfrac{3}{4}\right)^{n-1}$

In Exercises 41–56, evaluate the sum.

41. $\displaystyle\sum_{k=1}^{4} 7$

42. $\displaystyle\sum_{k=1}^{4} \dfrac{(-1)^k}{k}$

43. $\displaystyle\sum_{n=1}^{4} \left(\dfrac{1}{n} - \dfrac{1}{n + 1}\right)$

44. $\displaystyle\sum_{n=1}^{4} \left(\dfrac{1}{n} - \dfrac{1}{n + 2}\right)$

45. $\displaystyle\sum_{k=1}^{12} (7k - 5)$

46. $\displaystyle\sum_{k=1}^{10} (100 - 10k)$

47. $\displaystyle\sum_{j=1}^{100} \dfrac{j}{4}$

48. $\displaystyle\sum_{j=1}^{50} \dfrac{3j}{2}$

49. $\displaystyle\sum_{n=1}^{12} 2^n$

50. $\displaystyle\sum_{n=1}^{12} (-2)^n$

51. $\displaystyle\sum_{k=1}^{8} 5\left(-\dfrac{3}{4}\right)^k$

52. $\displaystyle\sum_{k=1}^{10} 4\left(\dfrac{3}{2}\right)^k$

53. $\displaystyle\sum_{i=1}^{8} (1.25)^{i-1}$

54. $\displaystyle\sum_{i=1}^{8} (-1.25)^{i-1}$

55. $\displaystyle\sum_{n=1}^{120} 500(1.01)^n$

56. $\displaystyle\sum_{n=1}^{40} 1000(1.1)^n$

57. Find the sum of the first 50 positive integers that are multiples of 4.

58. Find the sum of the integers from 225 to 300.

59. *Auditorium Seating*   Each row in a small auditorium has three more seats than the preceding row. Find the seating capacity of the auditorium if the front row seats 22 people and there are 12 rows of seats.

60. *Depreciation*   A company pays $120,000 for a machine. During the next 5 years, the machine depreciates at a rate of 30% per year. (That is, at the end of each year, the depreciated value is 70% of what it was at the beginning of the year.)

(a) Find a formula for the nth term of the geometric sequence that gives the value of the machine n full years after it was purchased.

(b) Find the depreciated value of the machine at the end of 5 full years.

61. *Population Increase*   A city of 85,000 people is growing at the rate of 1.2% per year. (That is, at the end of each year, the population is 1.012 times the population at the beginning of the year.)

(a) Find a formula for the nth term of the geometric sequence that gives the population n years from now.

(b) Estimate the population 50 years from now.

62. *Salary Increase*   You accept a job that pays a salary of $32,000 the first year. During the next 39 years, you receive a 5.5% raise each year. What is your total salary over the 40-year period?

In Exercises 63–66, evaluate $_nC_r$.

63. $_8C_3$

64. $_{12}C_2$

65. $_{12}C_0$

66. $_{100}C_1$

In Exercises 67 and 68, use a graphing utility to evaluate $_nC_r$.

67. $_{40}C_4$

68. $_{15}C_9$

In Exercises 69–74, use the Binomial Theorem to expand the expression. Simplify your answer.

69. $(x + 1)^{10}$

70. $(u - v)^9$

71. $(y - 2)^6$

72. $(x + 3)^5$

73. $\left(\dfrac{1}{2} - x\right)^8$

74. $(3x - 2y)^4$

In Exercises 75 and 76, find the coefficient of the given term of the expression.

Expression	Term
75. $(x-3)^{10}$	x^5
76. $(2x-3y)^5$	x^2y^3

77. *Morse Code* In Morse code, all characters are transmitted using a sequence of *dots* and *dashes*. How many different characters can be formed by using a sequence of three dots and dashes? (These can be repeated. For example, dash-dot-dot represents the letter *d*.)

78. *Forming Line Segments* How many straight line segments can be formed by five points of which no three are collinear?

79. *Committee Selection* Determine the number of ways a committee of five people can be formed from a group of 15 people.

80. *Program Listing* There are seven participants in a piano recital. In how many orders can their names be listed in the program?

81. *Rolling a Die* Find the probability of obtaining a number greater than 4 when a single six-sided die is rolled.

82. *Coin Tossing* Find the probability of obtaining at least one head when a coin is tossed four times.

83. *Book Selection* A child who does not know how to read carries a four-volume set of books to a bookshelf. Find the probability that the child will put the books on the shelf in the correct order.

84. *Rolling a Die* Are the chances of rolling a 3 with one six-sided die the same as the chances of rolling a total of 6 with two six-sided dice? If not, which has the greater probability of occurring?

85. *Hospital Inspection* As part of a monthly inspection at a hospital, the inspection team randomly selects reports from eight of the 84 nurses who are on duty. What is the probability that none of the reports selected will be from the 10 most experienced nurses on duty?

86. *Target Shooting* An archer shoots an arrow at the target shown in the figure. Suppose that the arrow is equally likely to hit any point on the target. What is the probability that the arrow hits the bull's-eye? What is the probability that the arrow hits the blue ring?

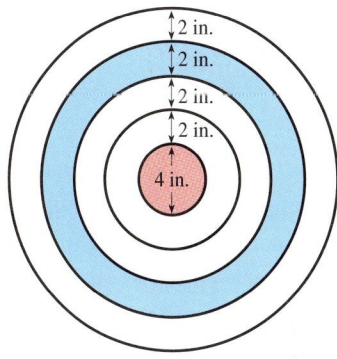

CHAPTER TEST

Take this test as you would take a test in class. After you are done, check your work against the answers given in the back of the book.

1. Write the first five terms of the sequence $a_n = \left(-\frac{2}{3}\right)^{n-1}$. (Begin with $n = 1$.)

2. Evaluate: $\displaystyle\sum_{j=0}^{4} (3j + 1)$

3. Evaluate: $\displaystyle\sum_{n=1}^{5} (3 - 4n)$

4. Use sigma notation to write $\dfrac{2}{3(1) + 1} + \dfrac{2}{3(2) + 1} + \cdots + \dfrac{2}{3(12) + 1}$.

5. Write the first five terms of the arithmetic sequence whose first term is $a_1 = 12$ and whose common difference is $d = 4$.

6. Find a formula for the nth term of the arithmetic sequence whose first term is $a_1 = 5000$ and whose common difference is $d = -100$.

7. Find the sum of the first 50 positive integers that are multiples of 3.

8. Find the common ratio of the geometric sequence: $2, \ -3, \ \frac{9}{2}, \ -\frac{27}{4}, \ \ldots$.

9. Find a formula for the nth term of the geometric sequence whose first term is $a_1 = 4$ and whose common ratio is $r = \frac{1}{2}$.

10. Evaluate: $\displaystyle\sum_{n=1}^{8} 2(2^n)$

11. Evaluate: $\displaystyle\sum_{n=1}^{10} 3\left(\frac{1}{2}\right)^n$

12. Fifty dollars is deposited each month in an increasing annuity that pays 8%, compounded monthly. What is the balance after 25 years?

13. Evaluate: $_{20}C_3$

14. Explain how to use Pascal's Triangle to expand $(x - 2)^5$.

15. Find the coefficient of the term $x^3 y^5$ in the expansion of $(x + y)^8$.

16. How many license plates can consist of one letter followed by three digits?

17. Four students are randomly selected from a class of 25 to answer questions from a reading assignment. In how many ways can the four be selected?

18. The weather report indicates that the probability of snow tomorrow is 0.75. What is the probability that it will not snow?

19. A card is drawn from a standard deck of playing cards. Find the probability that it is a red face card.

20. Suppose two spark plugs require replacement in a four-cylinder engine. If the mechanic randomly removes two plugs, find the probability that they are the two defective plugs.

Appendices

Introduction to Logic
Statements • *Truth Tables*

Statements

In everyday speech and in mathematics we make inferences that adhere to common **laws of logic**. These laws (or methods of reasoning) allow us to build an algebra of statements by using logical operations to form compound statements from simpler ones. One of the primary goals of logic is to determine the truth value (true or false) of a compound statement knowing the truth value of its simpler component statements. For instance, we will learn that the compound statement "The temperature is below freezing and it is snowing" is true only if both component statements are true.

Definition of a Statement	1. A **statement** is a sentence to which only one truth value (either true or false) can be meaningfully assigned.
	2. An **open statement** is a sentence that contains one or more variables and becomes a statement when each variable is replaced by a specific item from a designated set.

NOTE: In the above definition, the word *statement* can be replaced by the word *proposition*.

EXAMPLE 1 ■ **Statements, Nonstatements, and Open Statements**

Statement	*Truth Value*
A square is a rectangle.	T
-3 is less than -5.	F

Nonstatement	*Truth Value*
Do your homework.	No truth value can be meaningfully assigned.
Did you call the police?	No truth value can be meaningfully assigned.

Open Statement	*Truth Value*
x is an irrational number.	We need a value of x.
She is a computer science major.	We need a specific person. ■

Symbolically, we represent statements by lowercase letters p, q, r, and so on. Statements can be changed or combined to form **compound statements** by means of the three logical operations **and**, **or**, and **not**, which we represent by $\wedge$ (and), $\vee$ (or), and $\sim$ (not). In logic we use the word *or* in the *inclusive* sense (meaning "and/or" in everyday language). That is, the statement "p or q" is true if p is true, q is true, or both p and q are true. The following list summarizes the terms and symbols used with these three operations of logic.

Operations of Logic	Operation	Verbal Statement	Symbolic Form	Name of Operation
	$\sim$	not p	$\sim p$	**Negation**
	$\wedge$	p and q	$p \wedge q$	**Conjunction**
	$\vee$	p or q	$p \vee q$	**Disjunction**

Compound statements can be formed using more than one logical operation, as demonstrated in Example 2.

EXAMPLE 2 ■ **Forming Negations and Compound Statements**

The statements p and q are as follows.

> p: The temperature is below freezing.
> q: It is snowing.

Write the verbal form for each of the following.

(a) $p \wedge q$ (b) $\sim p$

(c) $\sim(p \vee q)$ (d) $\sim p \wedge \sim q$

Solution

(a) The temperature is below freezing and it is snowing.

(b) The temperature is not below freezing.

(c) It is not true that the temperature is below freezing or it is snowing.

(d) The temperature is not below freezing and it is not snowing.   ■

EXAMPLE 3 ■ **Forming Compound Statements**

The statements p and q are as follows.

> p: The temperature is below freezing.
> q: It is snowing.

(a) Write the symbolic form for: *The temperature is not below freezing or it is not snowing*.

(b) Write the symbolic form for: *It is not true that the temperature is below freezing and it is snowing*.

Solution

(a) The symbolic form is: $\sim p \vee \sim q$

(b) The symbolic form is: $\sim(p \wedge q)$ ■

Truth Tables

To determine the truth value of a compound statement, we use charts called **truth tables**. The following tables represent the three basic logical operations.

TABLE 1 Negation

p	q	$\sim p$	$\sim q$
T	T	F	F
T	F	F	T
F	T	T	F
F	F	T	T

TABLE 2 Conjunction

p	q	$p \wedge q$
T	T	T
T	F	F
F	T	F
F	F	F

TABLE 3 Disjunction

p	q	$p \vee q$
T	T	T
T	F	T
F	T	T
F	F	F

For the sake of uniformity, all truth tables with two component statements will have T and F values for p and q assigned in the order shown in the first column of these three tables. Truth tables for several operations can be combined into one chart by using the same two first columns. For each operation a new column is added. Such an arrangement is especially useful with compound statements that involve more than one logical operation and for showing that two statements are logically equivalent.

Logical Equivalence   Two compound statements are **logically equivalent** if they have identical truth tables. Symbolically, we denote the equivalence of the statements p and q by writing $p \equiv q$.

EXAMPLE 4 ■ Logical Equivalence

Use a truth table to show the logical equivalence of the statements $\sim p \wedge \sim q$ and $\sim(p \vee q)$.

Solution

TABLE 4

p	q	$\sim p$	$\sim q$	$\sim p \wedge \sim q$	$p \vee q$	$\sim(p \vee q)$
T	T	F	F	F	T	F
T	F	F	T	F	T	F
F	T	T	F	F	T	F
F	F	T	T	T	F	T

Identical

Since the fifth and seventh columns in Table 4 are identical, the two given statements are logically equivalent.   ■

The equivalence established in Example 4 is one of two well-known rules in logic called **DeMorgan's Laws**. Verification of the second of DeMorgan's Laws is left as an exercise.

DeMorgan's Laws	1. $\sim(p \vee q) \equiv \sim p \wedge \sim q$
	2. $\sim(p \wedge q) \equiv \sim p \vee \sim q$

Compound statements that are true, no matter what the truth values of component statements, are called **tautologies**. One simple example is the statement "p or not p," as shown in Table 5.

TABLE 5 $p \vee \sim p$ **is a tautology**

p	$\sim p$	$p \vee \sim p$
T	F	T
F	T	T

A.1	**EXERCISES**

In Exercises 1–12, classify the sentence as a statement, a nonstatement, or an open statement.

1. All dogs are brown.

2. Can I help you?

3. That figure is a circle.

4. Substitute 4 for x.

5. x is larger than 4.

6. 8 is larger than 4.

7. $x + y = 10$

8. $12 + 3 = 14$

9. Hockey is fun to watch.

10. One mile is greater than one kilometer.

11. It is more than one mile to the school.

12. Come to the party.

In Exercises 13–20, determine whether the open statement is true for the given values of x.

Open Statement	Values of x	
13. $x^2 - 5x + 6 = 0$	(a) $x = 2$	(b) $x = -2$
14. $x^2 - x - 6 = 0$	(a) $x = 2$	(b) $x = -2$
15. $x^2 \le 4$	(a) $x = -2$	(b) $x = 0$
16. $|x - 3| = 4$	(a) $x = -1$	(b) $x = 7$
17. $4 - |x| = 2$	(a) $x = 0$	(b) $x = 1$
18. $\sqrt{x^2} = x$	(a) $x = 3$	(b) $x = -3$
19. $\dfrac{x}{x} = 1$	(a) $x = -4$	(b) $x = 0$
20. $\sqrt[3]{x} = -2$	(a) $x = 8$	(b) $x = -8$

In Exercises 21–24, write the verbal form for each of the following.

(a) $\sim p$ (b) $\sim q$ (c) $p \wedge q$ (d) $p \vee q$

21. p: The sun is shining.
q: It is hot.

22. p: The car has a radio.
q: The car is red.

23. p: Lions are mammals.
q: Lions are carnivorous.

24. p: Twelve is less than fifteen.
q: Seven is a prime number.

In Exercises 25–28, write the verbal form for each of the following.

(a) $\sim p \wedge q$ (b) $\sim p \vee q$ (c) $p \wedge \sim q$ (d) $p \vee \sim q$

25. p: The sun is shining.
q: It is hot.

26. p: The car has a radio.
q: The car is red.

27. p: Lions are mammals.
q: Lions are carnivorous.

28. p: Twelve is less than fifteen.
q: Seven is a prime number.

In Exercises 29–32, write the symbolic form of the given compound statement. In each case let p represent the statement "It is four o'clock," and let q represent the statement "It is time to go home."

29. It is four o'clock and it is not time to go home.

30. It is not four o'clock or it is not time to go home.

31. It is not four o'clock or it is time to go home.

32. It is four o'clock and it is time to go home.

In Exercises 33–36, write the symbolic form of the given compound statement. In each case let p represent the statement "The dog has fleas," and let q represent the statement "The dog is scratching."

33. The dog does not have fleas or the dog is not scratching.

34. The dog has fleas and the dog is scratching.

35. The dog does not have fleas and the dog is scratching.

36. The dog has fleas or the dog is not scratching.

In Exercises 37–42, write the negation of the given statement.

37. The bus is not blue. **38.** Frank is not six feet tall.

39. x is equal to 4. **40.** x is not equal to 4.

41. The Earth is not flat. **42.** The Earth is flat.

In Exercises 43–48, construct a truth table for the given compound statement.

43. $\sim p \wedge q$ **44.** $\sim p \vee q$

45. $\sim p \vee \sim q$ **46.** $\sim p \wedge \sim q$

47. $p \vee \sim q$ **48.** $p \wedge \sim q$

In Exercises 49–54, use a truth table to determine whether the given statements are logically equivalent.

49. $\sim p \wedge q, \quad p \vee \sim q$ **50.** $\sim(p \wedge \sim q), \quad \sim p \vee q$

51. $\sim(p \vee \sim q), \quad \sim p \wedge q$ **52.** $\sim(p \vee q), \quad \sim p \vee \sim q$

53. $p \wedge \sim q, \quad \sim(\sim p \vee q)$ **54.** $p \wedge \sim q, \quad \sim(\sim p \wedge q)$

In Exercises 55–58, determine whether the statements are logically equivalent.

55. (a) The house is red and it is not made of wood.
(b) The house is red or it is not made of wood.

56. (a) It is not true that the tree is not green.
(b) The tree is green.

57. (a) The statement that the house is white or blue is not true.
(b) The house is not white and it is not blue.

58. (a) I am not twenty-five years old and I am not applying for this job.
(b) The statement that I am twenty-five years old and applying for this job is not true.

In Exercises 59–62, use a truth table to determine whether the given statement is a tautology.

59. $\sim p \wedge p$ **60.** $\sim p \vee p$

61. $\sim(\sim p) \vee \sim p$ **62.** $\sim(\sim p) \wedge \sim p$

63. Use a truth table to verify the second of DeMorgan's Laws:

$$\sim(p \wedge q) \equiv \sim p \vee \sim q$$

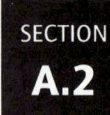

SECTION
A.2

Implications, Quantifiers, and Venn Diagrams
Implications • Logical Quantifiers • Venn Diagrams

Implications

A statement of the form "If p, then q," is called an **implication** (or a conditional statement) and is denoted by

$$p \to q.$$

We call p the **hypothesis** and q the **conclusion**. There are many different ways to express the implication $p \to q$, as shown in the following list.

Different Ways of Stating Implications	The implication $p \to q$ has the following equivalent verbal forms.
	1. If p, then q. 4. q follows from p.
	2. p implies q. 5. q is necessary for p.
	3. p, only if q. 6. p is sufficient for q.

Normally, we think of the implication $p \to q$ as having a cause-and-effect relationship between the hypothesis p and the conclusion q. However, you should be careful not to confuse the truth value of the component statements with the truth value of the implication. The following truth table should help you keep this distinction in mind.

TABLE 6 Implication

p	q	$p \to q$
T	T	T
T	F	F
F	T	T
F	F	T

Note in Table 6 that the implication $p \to q$ is false only when p is true and q is false. This is like a promise. Suppose you promise a friend that "If the sun shines, I will take you fishing." The only way you can break your promise is for the sun to shine (p is true) and you do not take your friend fishing (q is false). If the sun doesn't shine (p is false), you have no obligation to go fishing, and hence, the promise cannot be broken.

EXAMPLE 1   ■   **Finding Truth Values of Implications**

Give the truth value of each implication.

(a) If 3 is odd, then 9 is odd.

(b) If 3 is odd, then 9 is even.

(c) If 3 is even, then 9 is odd.

(d) If 3 is even, then 9 is even.

Solution

	Hypothesis	*Conclusion*	*Implication*
(a)	T	T	T
(b)	T	F	F
(c)	F	T	T
(d)	F	F	T

The next example shows how to write an implication as a disjunction.

EXAMPLE 2   ■   **Identifying Equivalent Statements**

Use a truth table to show the logical equivalence of the following statements.

(a) If I get a raise, I will take my family on a vacation.

(b) I will not get a raise *or* I will take my family on a vacation.

Solution

We let p represent the statement "I will get a raise," and let q represent the statement "I will take my family on a vacation." Then, we can represent the statement in part (a) as $p \rightarrow q$ and the statement in part (b) as $\sim p \vee q$. The logical equivalence of these two statements is shown in the following truth table.

TABLE 7 $p \rightarrow q \equiv \sim p \vee q$

p	q	$\sim p$	$\sim p \vee q$	$p \rightarrow q$
T	T	F	T	T
T	F	F	F	F
F	T	T	T	T
F	F	T	T	T

└─ Identical ─┘

Because the fourth and fifth columns of the truth table are identical, we can conclude that the two statements $p \rightarrow q$ and $\sim p \vee q$ are equivalent.   ■

From Table 7 and the fact that $\sim(\sim p) \equiv p$, we can write the **negation of an implication**. That is, since $p \rightarrow q$ is equivalent to $\sim p \vee q$, it follows that the negation of $p \rightarrow q$ must be $\sim(\sim p \vee q)$, which by DeMorgan's Laws can be written as follows.

$$\sim(p \rightarrow q) \equiv p \wedge \sim q$$

For the implication $p \rightarrow q$, there are three important associated implications.

1. The **converse** of $p \rightarrow q$: $q \rightarrow p$

2. The **inverse** of $p \rightarrow q$: $\sim p \rightarrow \sim q$

3. The **contrapositive** of $p \rightarrow q$: $\sim q \rightarrow \sim p$

From Table 8 you can see that these four statements yield two pairs of logically equivalent implications.

TABLE 8

p	q	$\sim p$	$\sim q$	$p \rightarrow q$	$\sim q \rightarrow \sim p$	$q \rightarrow p$	$\sim p \rightarrow \sim q$
T	T	F	F	T	T	T	T
T	F	F	T	F	F	T	T
F	T	T	F	T	T	F	F
F	F	T	T	T	T	T	T

Identical   Identical

NOTE: The connective "$\rightarrow$" is used to determine the truth values in the last three columns of Table 8.

EXAMPLE 3 ■ Writing the Converse, Inverse, and Contrapositive

Write the converse, inverse, and contrapositive for the implication "If I get a B on my test, then I will pass the course."

Solution

(a) *Converse:* If I pass the course, then I got a B on my test.

(b) *Inverse:* If I do not get a B on my test, then I will not pass the course.

(c) *Contrapositive:* If I do not pass the course, then I did not get a B on my test. ■

NOTE: In Example 3 be sure you see that neither the converse nor the inverse are logically equivalent to the original implication. To see this, consider that the original implication simply states that if you get a B on your test, then you will pass the course. The converse is not true because knowing that you passed the course does not imply that you got a B on the test. After all, you might have gotten an A on the test!

A **biconditional statement**, denoted by $p \leftrightarrow q$, is the conjunction of the implications $p \rightarrow q$ and $q \rightarrow p$. We often write a biconditional statement as "p if and only if q," or in shorter form as "p iff q." A biconditional statement is true when both components are true and when both components are false, as shown in the following truth table.

TABLE 9   Biconditional Statement: p if and only if q

p	q	$p \rightarrow q$	$q \rightarrow p$	$p \leftrightarrow q$	$(p \rightarrow q) \wedge (q \rightarrow p)$
T	T	T	T	T	T
T	F	F	T	F	F
F	T	T	F	F	F
F	F	T	T	T	T

The following list summarizes some of the laws of logic that we have discussed up to this point.

Laws of Logic

1. For every statement p, either p is true or p is false.   Law of Excluded Middle
2. $\sim(\sim p) \equiv p$ Law of Double Negation
3. $\sim(p \vee q) \equiv \sim p \wedge \sim q$ DeMorgan's Law
4. $\sim(p \wedge q) \equiv \sim p \vee \sim q$ DeMorgan's Law
5. $p \rightarrow q \equiv \sim p \vee q$ Law of Implication
6. $p \rightarrow q \equiv \sim q \rightarrow \sim p$ Law of Contraposition

Logical Quantifiers

Logical quantifiers are words such as *some*, *all*, *every*, *each*, *one*, and *none*. Here are some examples of statements with quantifiers.

Some isosceles triangles are right triangles.

Every painting on display is for sale.

Not all corporations have male chief executive officers.

All squares are parallelograms.

Being able to recognize the negation of a statement involving a quantifier is one of the most important skills in logic. For instance, consider the statement "All dogs are brown." In order for this statement to be false, we do not have to show that *all* dogs are not brown, we must simply find at least one dog that is not brown. Thus, the negation of the statement is "Some dogs are brown."

Next we list some of the more common negations involving quantifiers.

Negating Statements with Quantifiers

Statement	Negation
1. All p are q.	Some p are not q.
2. Some p are q.	No p is q.
3. Some p are not q.	All p are q.
4. No p is q.	Some p are q.

When using logical quantifiers, the word *all* can be replaced by the words *each* or *every*. For instance, the following are equivalent.

All p are q.   Each p is q.   Every p is q.

Similarly, the word *some* can be replaced by the words *at least one*. For instance, the following are equivalent.

Some p are q.   At least one p is q.

EXAMPLE 4 ■ **Negating Quantifying Statements**

Write the negation of each of the following.

(a) All students study.

(b) Not all prime numbers are odd.

(c) At least one mammal can fly.

(d) Some bananas are not yellow.

Solution

(a) Some students do not study.

(b) All prime numbers are odd.

(c) No mammals can fly.

(d) All bananas are yellow.

Venn Diagrams

Venn diagrams are figures that are used to show relationships between two or more sets of objects. They can help us interpret quantifying statements. Study the following Venn diagrams in which the circle marked A represents people over six feet tall and the circle marked B represents the basketball players.

1. All basketball players are over six feet tall.

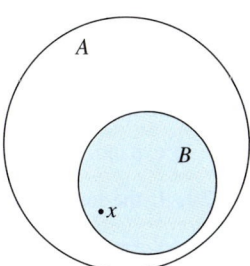

2. Some basketball players are over six feet tall.

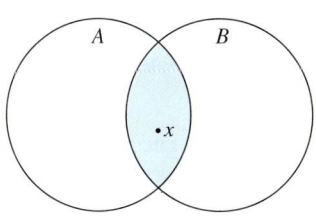

3. Some basketball players are not over six feet tall.

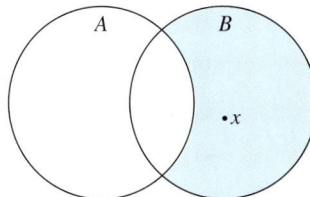

4. No basketball player is over six feet tall.

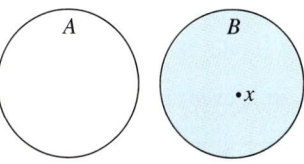

A.2   EXERCISES

In Exercises 1–4, write the verbal form for each of the following.

(a) $p \to q$ (b) $q \to p$ (c) $\sim q \to \sim p$ (d) $p \to \sim q$

1. p: The engine is running.
   q: The engine is wasting gasoline.

2. p: The student is at school.
   q: It is nine o'clock.

3. p: The integer is even.
   q: It is divisible by two.

4. p: The person is generous.
   q: The person is rich.

In Exercises 5–10, write the symbolic form of the compound statement. Let p represent the statement "The economy is expanding," and let q represent the statement "Interest rates are low."

5. If interest rates are low, then the economy is expanding.

6. If interest rates are not low, then the economy is not expanding.

7. An expanding economy implies low interest rates.

8. Low interest rates are sufficient for an expanding economy.

9. Low interest rates are necessary for an expanding economy.

10. The economy will expand only if interest rates are low.

In Exercises 11–20, give the truth value of the implication.

11. If 4 is even, then 12 is even.

12. If 4 is even, then 2 is odd.

13. If 4 is odd, then 3 is odd.

14. If 4 is odd, then 2 is odd.

15. If $2n$ is even, then $2n + 2$ is odd.

16. If $2n + 1$ is even, then $2n + 2$ is odd.

17. $3 + 11 > 16$ only if $2 + 3 = 5$.

18. $\frac{1}{6} < \frac{2}{3}$ is necessary for $\frac{1}{2} > 0$.

19. $x = -2$ follows from $2x + 3 = x + 1$.

20. If $2x = 224$, then $x = 10$.

In Exercises 21–26, write the converse, inverse, and contrapositive of the statement.

21. If the sky is clear, then you can see the eclipse.

22. If the person is nearsighted, then he is ineligible for the job.

23. If taxes are raised, then the deficit will increase.

24. If wages are raised, then the company's profits will decrease.

25. It is necessary to have a birth certificate to apply for the visa.

26. The number is divisible by three only if the sum of its digits is divisible by three.

In Exercises 27–40, write the negation of the statement.

27. Paul is a junior or senior.

28. Jack is a senior and he plays varsity basketball.

29. If the temperature increases, then the metal rod will expand.

30. If the test fails, then the project will be halted.

31. We will go to the ocean only if the weather forecast is good.

32. Completing the pass on this play is necessary if we are going to win the game.

33. Some students are in extracurricular activities.

34. Some odd integers are not prime numbers.

35. All contact sports are dangerous.

36. All members must pay their dues prior to June 1.

37. No child is allowed at the concert.

38. No contestant is over the age of twelve.

39. At least one of the $20 bills is counterfeit.

40. At least one unit is defective.

In Exercises 41–48, construct a truth table for the compound statement.

41. $\sim(p \rightarrow \sim q)$

42. $\sim q \rightarrow (p \rightarrow q)$

43. $\sim(q \rightarrow p) \wedge q$

44. $p \rightarrow (\sim p \vee q)$

45. $[(p \vee q) \wedge (\sim p)] \rightarrow q$

46. $[(p \rightarrow q) \wedge (\sim q)] \rightarrow p$

47. $(p \leftrightarrow \sim q) \rightarrow \sim p$

48. $(p \vee \sim q) \leftrightarrow (q \rightarrow \sim p)$

In Exercises 49–56, use a truth table to show the logical equivalence of the two statements.

49. $q \to p$ $\qquad\qquad$ $\sim p \to \sim q$

50. $\sim p \to q$ $\qquad\qquad$ $p \vee q$

51. $\sim(p \to q)$ $\qquad\qquad$ $p \wedge \sim q$

52. $(p \vee q) \to q$ $\qquad\qquad$ $p \to q$

53. $(p \to q) \vee \sim q$ $\qquad\qquad$ $p \vee \sim p$

54. $q \to (\sim p \vee q)$ $\qquad\qquad$ $q \vee \sim q$

55. $p \to (\sim p \wedge q)$ $\qquad\qquad$ $\sim p$

56. $\sim(p \wedge q) \to \sim q$ $\qquad\qquad$ $p \vee \sim q$

57. Select the statement that is logically equivalent to the statement "If a number is divisible by six, then it is divisible by two."
  (a) If a number is divisible by two, then it is divisible by six.
  (b) If a number is not divisible by six, then it is not divisible by two.
  (c) If a number is not divisible by two, then it is not divisible by six.
  (d) Some numbers are divisible by six and not divisible by two.

58. Select the statement that is logically equivalent to the statement "It is not true that Pam is a conservative and a Democrat."
  (a) Pam is a conservative and a Democrat.
  (b) Pam is not a conservative and not a Democrat.
  (c) Pam is not a conservative or she is not a Democrat.
  (d) If Pam is not a conservative, then she is a Democrat.

59. Select the statement that is *not* logically equivalent to the statement "Every citizen over the age of 18 has the right to vote."
  (a) Some citizens over the age of 18 have the right to vote.
  (b) Each citizen over the age of 18 has the right to vote.
  (c) All citizens over the age of 18 have the right to vote.
  (d) No citizen over the age of 18 can be restricted from voting.

60. Select the statement that is *not* logically equivalent to the statement "It is necessary to pay the registration fee to take the course."
  (a) If you take the course, then you must pay the registration fee.
  (b) If you do not pay the registration fee, then you cannot take the course.
  (c) If you pay the registration fee, then you may take the course.
  (d) You may take the course only if you pay the registration fee.

In Exercises 61–70, sketch a Venn diagram and shade the region that illustrates the given statement. Let A be a circle that represents people who are happy, and let B be a circle that represents college students.

61. All college students are happy.

62. All happy people are college students.

63. No college students are happy.

64. No happy people are college students.

65. Some college students are not happy.

66. Some happy people are not college students.

67. At least one college student is happy.

68. At least one happy person is not a college student.

69. Each college student is sad.

70. Each sad person is not a college student.

In Exercises 71–74, state whether the statement follows from the given Venn diagram. Assume that each area shown in the Venn diagram is non-empty. (*Note:* Use only the information given in the diagram. Do not be concerned with whether the statement is actually true or false.)

71. (a) All toads are green.
  (b) Some toads are green.

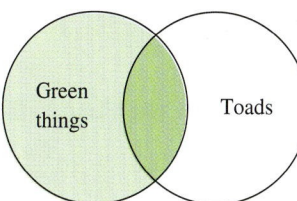

72. (a) All men are company presidents.

(b) Some company presidents are women.

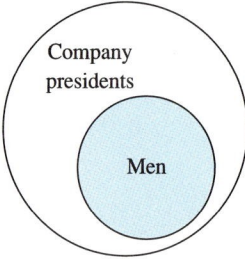

73. (a) All blue cars are old.

(b) Some blue cars are not old.

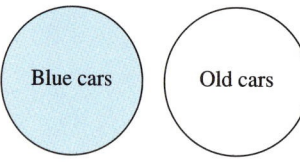

74. (a) No football players are over six feet tall.

(b) Every football player is over six feet tall.

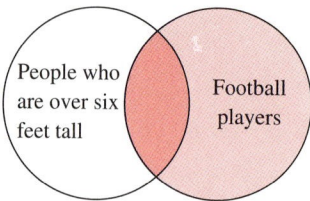

SECTION
A.3

Logical Arguments

Arguments • *Venn Diagrams and Arguments* • *Proofs*

Arguments

An **argument** is a collection of statements, listed in order. The last statement is called the **conclusion** and the other statements are called the **premises**. An argument is **valid** if the conjunction of all the premises implies the conclusion. The most common type of argument takes the following form.

Premise #1: $p \rightarrow q$
Premise #2: p
Conclusion: q

This form of argument is called the **Law of Detachment** or *Modus Ponens*. It is illustrated in the following example.

EXAMPLE 1   ■   A Valid Argument

Show that the following argument is valid.

Premise #1:   If Sean is a freshman, then he is taking algebra.
Premise #2:   Sean is a freshman.
Conclusion: Therefore, Sean is taking algebra.

Solution

We let p represent the statement "Sean is a freshman," and q represent the statement "Sean is taking algebra." Then the argument fits the Law of Detachment, which can be written as follows.

$$[(p \to q) \land p] \to q$$

The validity of this argument is shown in the following truth table.

TABLE 10   Law of Detachment

p	q	$p \to q$	$(p \to q) \land p$	$[(p \to q) \land p] \to q$
T	T	T	T	T
T	F	F	F	T
F	T	T	F	T
F	F	T	F	T

Keep in mind that the validity of an argument has nothing to do with the truthfulness of the premises or conclusion. For instance, the following argument is valid—the fact that it is fanciful does not alter its validity.

Premise #1:   If I snap my fingers, elephants will stay out of my house.
Premise #2:   I am snapping my fingers.
Conclusion: Therefore, elephants will stay out of my house.

We have discussed the most common form of logical argument. This and three other commonly used forms of valid arguments are summarized in the following list.

Four Types of Valid Arguments

Name	Pattern	
1. **Law of Detachment** or *Modus Ponens*	Premise #1:	$p \to q$
	Premise #2:	p
	Conclusion:	q
2. **Law of Contraposition** or *Modus Tollens*	Premise #1:	$p \to q$
	Premise #2:	$\sim q$
	Conclusion:	$\sim p$
3. **Law of Transitivity** or *Syllogism*	Premise #1:	$p \to q$
	Premise #2:	$q \to r$
	Conclusion:	$p \to r$
4. **Law of Disjunctive Syllogism**	Premise #1:	$p \lor q$
	Premise #2:	$\sim p$
	Conclusion:	q

EXAMPLE 2 ■ **An Invalid Argument**

Determine whether the following argument is valid.

Premise #1: If John is elected, the income tax will be increased.
Premise #2: The income tax was increased.
Conclusion: Therefore, John was elected.

Solution

This argument has the following form.

Pattern		*Implication*
Premise #1:	$p \rightarrow q$	$[(p \rightarrow q) \wedge q] \rightarrow p$
Premise #2:	q	
Conclusion:	p	

This is not one of the four valid forms of arguments that we listed. We can construct a truth table to verify that the argument is invalid, as follows.

TABLE 11 An Invalid Argument

p	q	$p \rightarrow q$	$(p \rightarrow q) \wedge q$	$[(p \rightarrow q) \wedge q] \rightarrow p$
T	T	T	T	T
T	F	F	F	T
F	T	T	T	F
F	F	T	F	T

■

An invalid argument, like the one in Example 2, is called a **fallacy**. Other common fallacies are given in the following example.

EXAMPLE 3 ■ **Common Fallacies**

Each of the following arguments is invalid.

(a) *Arguing from the Converse:* If the football team wins the championship, then students will skip classes. The students skipped classes. Therefore, the football team won the championship.

(b) *Arguing from the Inverse:* If the football team wins the championship, then students will skip classes. The football team did not win the championship. Therefore, the students did not skip classes.

(c) *Arguing from False Authority:* Wheaties are best for you because Joe Montana eats them.

(d) *Arguing from an Example:* Beta Brand products are not reliable because my Beta Brand snowblower does not start in cold weather.

(e) *Arguing from Ambiguity:* If automobile carburetors are modified, the automobile will pollute. Brand X automobiles have modified carburetors. Therefore, Brand X automobiles pollute.

(f) *Arguing by False Association:* Joe was running through the alley when the fire alarm went off. Therefore, Joe started the fire. ■

EXAMPLE 4 ■ A Valid Argument

Determine whether the following argument is valid.

Premise #1:	You like strawberry pie or you like chocolate pie.
Premise #2:	You do not like strawberry pie.
Conclusion:	Therefore, you like chocolate pie.

Solution

This argument has the following form.

Premise #1:	$p \lor q$
Premise #2:	$\sim p$
Conclusion:	q

This argument is a disjunctive syllogism, which is one of the four common types of valid arguments. ■

In a valid argument, the conclusion drawn from the premise is called a **valid conclusion**.

EXAMPLE 5 ■ Making Valid Conclusions

Given the following two premises, which of the conclusions are valid?

| Premise #1: | If you like boating, then you like swimming. |
| Premise #2: | If you like swimming, then you are a scholar. |

(a) Conclusion: If you like boating, then you are a scholar.

(b) Conclusion: If you do not like boating, then you are not a scholar.

(c) Conclusion: If you are not a scholar, then you do not like boating.

Solution

(a) This conclusion is valid. It follows from the Law of Transitivity (or syllogism).

(b) This conclusion is invalid. The fallacy stems from arguing from the inverse.

(c) This conclusion is valid. It follows from the Law of Contraposition. ■

FIGURE A.1

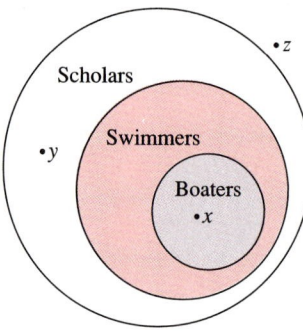

Venn Diagrams and Arguments

Venn diagrams can be used to informally test the validity of an argument. For instance, a Venn diagram for the premises in Example 5 is shown in Figure A.1. In this figure the validity of Conclusion (a) is seen by choosing a boater x in all three sets. Conclusion (b) is seen to be invalid by choosing a person y who is a scholar but does not like boating. Finally, person z indicates the validity of Conclusion (c).

Venn diagrams work well for testing arguments that involve quantifiers, as shown in the next two examples.

EXAMPLE 6 ■ **Using a Venn Diagram to Show that an Argument Is Not Valid**

Use a Venn diagram to test the validity of the following argument.

Premise #1:	Some plants are green.
Premise #2:	All lettuce is green.
Conclusion:	Therefore, lettuce is a plant.

Solution

FIGURE A.2

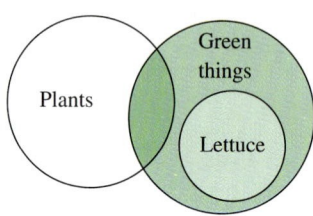

From the Venn diagram shown in Figure A.2, we can see that this is not a valid argument. Remember that even though the conclusion is true (lettuce is a plant), this does not imply that the argument is true. ■

NOTE: When you are using Venn diagrams, you must remember to draw the most general case. For example, in Figure A.2 the circle representing plants is not drawn entirely within the circle representing green things because we are told that only *some* plants are green.

EXAMPLE 7 ■ **Using a Venn Diagram to Show that an Argument Is Valid**

Use a Venn diagram to test the validity of the following argument.

Premise #1:	All good tennis players are physically fit.
Premise #2:	Some golfers are good tennis players.
Conclusion:	Therefore, some golfers are physically fit.

FIGURE A.3

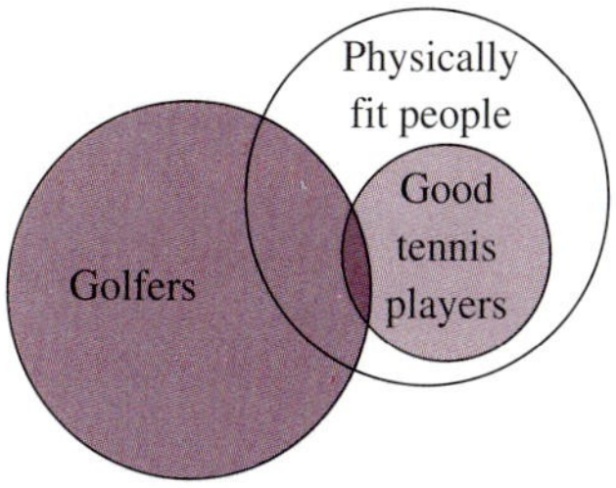

Solution

Since the set of golfers intersects the set of good tennis players, we see from Figure A.3 that the set of golfers must also intersect the set of physically fit people. Therefore, the argument is valid. ■

Proofs

What does the word *proof* mean to you? In mathematics we use the word *proof* to simply mean a valid argument. Many proofs involve more than two premises and a conclusion. For instance, the proof in Example 8 involves three premises and a conclusion.

EXAMPLE 8 ■ **A Proof by Contraposition**

Use the following three premises to prove that "It is not snowing today."

> Premise #1:   If it is snowing today, Greg will go skiing.
> Premise #2:   If Greg is skiing today, then he is not studying.
> Premise #3:   Greg is studying today.

Solution

We let p represent the statement "It is snowing today," let q represent "Greg is skiing," and let r represent "Greg is studying today." Thus, the given premises have the following form.

> Premise #1: $p \rightarrow q$
> Premise #2: $q \rightarrow \sim r$
> Premise #3: r

By noting that $r \equiv \sim(\sim r)$, reordering the premises, and writing the contrapositives of the first and second premises, we can obtain the following valid argument.

> Premise #3: r
> Contrapositive of Premise #2: $r \rightarrow \sim q$
> Contrapositive of Premise #1: $\sim q \rightarrow \sim p$
> Conclusion: $\sim p$

Thus, we can conclude $\sim p$. That is, "It is not snowing today." ■

A.3 EXERCISES

In Exercises 1–4, use a truth table to show that the given argument is valid.

1. Premise #1: $p \rightarrow \sim q$
Premise #2: q
Conclusion: $\sim p$

2. Premise #1: $p \leftrightarrow q$
Premise #2: p
Conclusion: q

3. Premise #1: $p \vee q$
Premise #2: $\sim p$
Conclusion: q

4. Premise #1: $p \wedge q$
Premise #2: $\sim p$
Conclusion: q

In Exercises 5–8, use a truth table to show that the given argument is invalid.

5. Premise #1: $\sim p \rightarrow q$
Premise #2: p
Conclusion: $\sim q$

6. Premise #1: $p \rightarrow q$
Premise #2: $\sim p$
Conclusion: $\sim q$

7. Premise #1: $p \vee q$
Premise #2: q
Conclusion: p

8. Premise #1: $\sim(p \wedge q)$
Premise #2: q
Conclusion: p

In Exercises 9–22, determine whether the argument is valid or invalid.

9. Premise #1: If taxes are increased, then businesses will leave the state.
Premise #2: Taxes are increased.
Conclusion: Therefore, businesses will leave the state.

10. Premise #1: If a student does the homework, then a good grade is certain.
Premise #2: Liza does the homework.
Conclusion: Therefore, Liza will receive a good grade for the course.

11. Premise #1: If taxes are increased, then businesses will leave the state.
Premise #2: Businesses are leaving the state.
Conclusion: Therefore, taxes were increased.

12. Premise #1: If a student does the homework, then a good grade is certain.
Premise #2: Liza received a good grade for the course.
Conclusion: Therefore, Liza did her homework.

13. Premise #1: If the doors are kept locked, then the car will not be stolen.
Premise #2: The car was stolen.
Conclusion: Therefore, the car doors were unlocked.

14. Premise #1: If Jan passes the exam, she is eligible for the position.
Premise #2: Jan is not eligible for the position.
Conclusion: Therefore, Jan did not pass the exam.

15. Premise #1:   All cars manufactured by the Ford Motor Company are reliable.
    Premise #2:   Lincolns are manufactured by Ford.
    Conclusion: Therefore, Lincolns are reliable cars.

16. Premise #1:   Some cars manufactured by the Ford Motor Company are reliable.
    Premise #2:   Lincolns are manufactured by Ford.
    Conclusion: Therefore, Lincolns are reliable.

17. Premise #1:   All federal income tax forms are subject to the Paperwork Reduction Act of 1980.
    Premise #2:   The 1040 Schedule A form is subject to the Paperwork Reduction Act of 1980.
    Conclusion: Therefore, the 1040 Schedule A form is a federal income tax form.

18. Premise #1:   All integers divisible by six are divisible by three.
    Premise #2:   Eighteen is divisible by six.
    Conclusion: Therefore, eighteen is divisible by three.

19. Premise #1:   Eric is at the store or the handball court.
    Premise #2:   He is not at the store.
    Conclusion: Therefore, he must be at the handball court.

20. Premise #1:   The book must be returned within two weeks or you pay a fine.
    Premise #2:   The book was not returned within two weeks.
    Conclusion: Therefore, you must pay a fine.

21. Premise #1:   It is not true that it is a diamond and it sparkles in the sunlight.
    Premise #2:   It does sparkle in the sunlight.
    Conclusion: Therefore, it is a diamond.

22. Premise #1:   Either I work tonight or I pass the mathematics test.
    Premise #2:   I'm going to work tonight.
    Conclusion: Therefore, I will fail the mathematics test.

In Exercises 23–30, determine which conclusion is valid from the given premises.

23. Premise #1:   If seven is a prime number, then seven does not divide evenly into twenty-one.
    Premise #2:   Seven divides evenly into twenty-one.
    (a) Conclusion: Therefore, seven is a prime number.
    (b) Conclusion: Therefore, seven is not a prime number.
    (c) Conclusion: Therefore, twenty-one divided by seven is three.

24. Premise #1:   If the fuel is shut off, then the fire will be extinguished.
    Premise #2:   The fire continues to burn.
    (a) Conclusion: Therefore, the fuel was not shut off.
    (b) Conclusion: Therefore, the fuel was shut off.
    (c) Conclusion: Therefore, the fire becomes hotter.

25. Premise #1: It is necessary that interest rates be lowered for the economy to improve.

Premise #2: Interest rates were not lowered.

(a) Conclusion: Therefore, the economy will improve.

(b) Conclusion: Therefore, interest rates are irrelevant to the performance of the economy.

(c) Conclusion: Therefore, the economy will not improve.

26. Premise #1: It will snow only if the temperature is below 32° at some level of the atmosphere.

Premise #2: It is snowing.

(a) Conclusion: Therefore, the temperature is below 32° at ground level.

(b) Conclusion: Therefore, the temperature is above 32° at some level of the atmosphere.

(c) Conclusion: Therefore, the temperature is below 32° at some level of the atmosphere.

27. Premise #1: Smokestack emissions must be reduced or acid rain will continue as an environmental problem.

Premise #2: Smokestack emissions have not decreased.

(a) Conclusion: Therefore, the ozone layer will continue to be depleted.

(b) Conclusion: Therefore, acid rain will continue as an environmental problem.

(c) Conclusion: Therefore, stricter automobile emission standards must be enacted.

28. Premise #1: The library must upgrade its computer system or service will not improve.

Premise #2: Service at the library has improved.

(a) Conclusion: Therefore, the computer system was upgraded.

(b) Conclusion: Therefore, more personnel were hired for the library.

(c) Conclusion: Therefore, the computer system was not upgraded.

29. Premise #1: If Rodney studies, then he will make good grades.

Premise #2: If he makes good grades, then he will get a good job.

(a) Conclusion: Therefore, Rodney will get a good job.

(b) Conclusion: Therefore, if Rodney doesn't study, then he won't get a good job.

(c) Conclusion: Therefore, if Rodney doesn't get a good job, then he didn't study.

30. Premise #1: It is necessary to have a ticket and an ID card to get into the arena.

Premise #2: Janice entered the arena.

(a) Conclusion: Therefore, Janice does not have a ticket.

(b) Conclusion: Therefore, Janice has a ticket and an ID card.

(c) Conclusion: Therefore, Janice has an ID card.

In Exercises 31–34, use a Venn diagram to test the validity of the argument.

31. Premise #1: All numbers divisible by ten are divisible by five.

Premise #2: Fifty is divisible by ten.

Conclusion: Therefore, fifty is divisible by five.

32. Premise #1: All human beings require adequate rest.

Premise #2: All infants are human beings.

Conclusion: Therefore, all infants require adequate rest.

33. Premise #1:   No person under the age of eighteen is eligible to vote.
     Premise #2:   Some college students are eligible to vote.
     Conclusion: Therefore, some college students are under the age of eighteen.

34. Premise #1:   Every amateur radio operator has a radio license.
     Premise #2:   Jackie has a radio license.
     Conclusion: Therefore, Jackie is an amateur radio operator.

In Exercises 35–38, use the premises to prove the given conclusion.

35. Premise #1:   If Sue drives to work, then she will stop at the grocery store.
     Premise #2:   If she stops at the grocery store, then she'll buy milk.
     Premise #3:   Sue drove to work today.
     Conclusion: Therefore, Sue will get milk.

36. Premise #1:   If Bill is patient, then he will succeed.
     Premise #2:   Bill will get bonus pay if he succeeds.
     Premise #3:   Bill did not get bonus pay.
     Conclusion: Therefore, Bill is not patient.

37. Premise #1:   If this is a good product, then we should buy it.
     Premise #2:   Either it was made by XYZ Corporation, or we will not buy it.
     Premise #3:   It is not made by XYZ Corporation.
     Conclusion: Therefore, it is not a good product.

38. Premise #1:   If the book is returned within two weeks, then there is no fine.
     Premise #2:   You pay a fine or you may not check out another book.
     Premise #3:   You are allowed to check out another book.
     Conclusion: Therefore, the book was not returned within two weeks.

Stem-and-Leaf Plots

Statistics is the branch of mathematics that studies techniques for collecting, organizing, and interpreting data. In this section, you will study several ways to organize and interpret data.

One type of plot that can be used to organize sets of numbers by hand is a **stem-and-leaf plot.** A set of test scores and the corresponding stem-and-leaf plot are shown below.

Test Scores	*Stems*	*Leaves*
93, 70, 76, 58, 86, 93, 82, 78, 83, 86,	5	8
64, 78, 76, 66, 83, 83, 96, 74, 69, 76,	6	4 4 6 9
64, 74, 79, 76, 88, 76, 81, 82, 74, 70	7	0 0 4 4 4 6 6 6 6 6 8 8 9
	8	1 2 2 3 3 3 6 6 8
	9	3 3 6

Note that the *leaves* represent the units digits of the numbers and the *stems* represent the tens digits. Stem-and-leaf plots can also be used to compare two sets of data, as shown in the following example.

EXAMPLE 1 ■ Comparing Two Sets of Data

Use a stem-and-leaf plot to compare the test scores given above with the following test scores. Which set of test scores is better?

90, 81, 70, 62, 64, 73, 81, 92, 73, 81, 92, 93, 83, 75, 76, 83, 94, 96, 86, 77, 77, 86, 96, 86, 77, 86, 87, 87, 79, 88

Solution

Begin by ordering the second set of scores.

62, 64, 70, 73, 73, 75, 76, 77, 77, 77, 79, 81, 81, 81, 83, 83, 86, 86, 86, 86, 87, 87, 88, 90, 92, 92, 93, 94, 96, 96

Now that the data has been ordered, you can construct a *double* stem-and-leaf plot by letting the leaves to the right of the stems represent the units digits for the first group of test scores and letting the leaves to the left of the stems represent the units digits for the second group of test scores.

Leaves (2nd Group)	Stems	Leaves (1st Group)
	5	8
4 2	6	4 4 6 9
9 7 7 7 6 5 3 3 0	7	0 0 4 4 4 6 6 6 6 6 8 8 9
8 7 7 6 6 6 6 3 3 1 1 1	8	1 2 2 3 3 3 6 6 8
6 6 4 3 2 2 0	9	3 3 6

By comparing the two sets of leaves, you can see that the second group of test scores is better than the first group. ■

EXAMPLE 2 ■   Using a Stem-and-Leaf Plot

Table B.1 shows the percent of the population of each state and the District of Columbia that was at least 65 years old in 1989. Use a stem-and-leaf plot to organize the data. (*Source:* U.S. Bureau of Census)

TABLE B.1

AK	4.1	AL	12.7	AR	14.8	AZ	13.1	CA	10.6
CO	9.8	CT	13.6	DC	12.5	DE	11.8	FL	18.0
GA	10.1	HI	10.7	IA	15.1	ID	11.9	IL	12.3
IN	12.4	KS	13.7	KY	12.7	LA	11.1	MA	13.8
MD	10.8	ME	13.4	MI	11.9	MN	12.6	MO	13.9
MS	12.4	MT	13.2	NC	12.1	ND	13.9	NE	13.9
NH	11.4	NJ	13.2	NM	10.5	NV	10.9	NY	13.0
OH	12.8	OK	13.3	OR	13.9	PA	15.1	RI	14.8
SC	11.1	SD	14.4	TN	12.6	TX	10.1	UT	8.6
VA	10.8	VT	11.9	WA	11.9	WI	13.4	WV	14.6
WY	9.8								

Solution

Begin by ordering the numbers, as shown below.

4.1, 8.6, 9.8, 9.8, 10.1, 10.1, 10.5, 10.6, 10.7, 10.8, 10.8,
10.9, 11.1, 11.1, 11.4, 11.8, 11.9, 11.9, 11.9, 11.9, 12.1, 12.3,
12.4, 12.4, 12.5, 12.6, 12.6, 12.7, 12.7, 12.8, 13.0, 13.1, 13.2,
13.2, 13.3, 13.4, 13.4, 13.6, 13.7, 13.8, 13.9, 13.9, 13.9, 13.9,
14.4, 14.6, 14.8, 14.8, 15.1, 15.1, 18.0

Next construct the stem-and-leaf plot using the leaves to represent the digits to the right of the decimal points.

Stems	Leaves
4.	1 Alaska has the lowest percent.
5.	
6.	
7.	
8.	6
9.	8 8
10.	1 1 5 6 7 8 8 9
11.	1 1 4 8 9 9 9 9
12.	1 3 4 4 5 6 6 7 7 8
13.	0 1 2 2 3 4 4 6 7 8 9 9 9 9
14.	4 6 8 8
15.	1 1
16.	
17.	
18.	0 Florida has the highest percent.

■

Histograms and Frequency Distributions

With data such as that given in Example 2, it is useful to group the numbers into intervals and plot the frequency of the data in each interval. For instance, the **frequency distribution** and **histogram** shown in Figure B.1 represent the data given in Example 2.

Frequency Distribution

Interval	Tally
[4, 6)		
[6, 8)	
[8, 10)	|||
[10, 12)	JHT JHT JHT I
[12, 14)	JHT JHT JHT JHT IIII
[14, 16)	JHT I
[16, 18)	
[18, 20)		

FIGURE B.1

Histogram

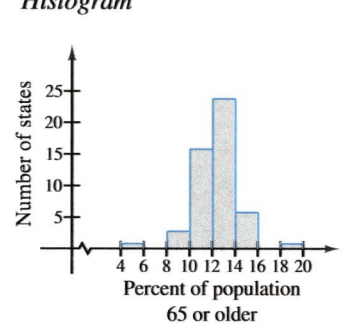

Percent of population 65 or older

Technology Note

Try using a computer or graphing calculator to create a histogram for the data at the right. How does the histogram change when the intervals change?

A histogram has a portion of a real number line as its horizontal axis. A **bar graph** is similar to a histogram, except that the rectangles (bars) can be either horizontal or vertical and the labels of the bars are not necessarily numbers.

Another difference between a bar graph and a histogram is that the bars in a bar graph are usually separated by spaces, whereas the bars in a histogram are not separated by spaces.

EXAMPLE 3 ■ Constructing a Bar Graph

The data below shows the average monthly precipitation (in inches) in Houston, Texas. Construct a bar graph for this data. What can you conclude? (*Source:* PC USA)

January	3.2	February	3.3	March	2.7
April	4.2	May	4.7	June	4.1
July	3.3	August	3.7	September	4.9
October	3.7	November	3.4	December	3.7

Solution

To create a bar graph, begin by drawing a vertical axis to represent the precipitation and a horizontal axis to represent the months. The bar graph is shown in Figure B.2. From the graph, you can see that Houston receives a fairly consistent amount of rain throughout the year—the driest month tends to be March and the wettest month tends to be September.

FIGURE B.2

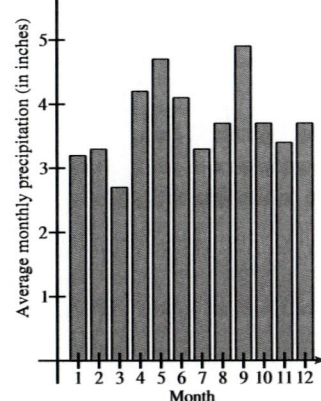

NOTE Bar graphs are used in exercises and examples throughout this book. For example, see Exercises 93, 94, and 95 on page 25 and Exercises 35–38 on page 222. The material presented here allows for a more in-depth discussion.

Line Graphs

A **line graph** is similar to a standard coordinate graph. Line graphs are usually used to show trends over periods of time.

EXAMPLE 4 ■   Constructing a Line Graph

The following data shows the number of immigrants (in thousands) to the United States per decade. Construct a line graph of the data. What can you conclude?

Decade	Number	Decade	Number
1851–1860	2598	1861–1870	2315
1871–1880	2812	1881–1890	5247
1891–1900	3688	1901–1910	8795
1911–1920	5736	1921–1930	4107
1931–1940	528	1941–1950	1035
1951–1960	2515	1961–1970	3322
1971–1980	4493	1981–1990	6447

NOTE   Line graphs are used in exercises throughout this book. For example, see Exercise 9 on page 270, Exercises 99 and 101 on page 417, and Exercises 55 and 56 on page 428. The material presented here allows for a more in-depth discussion.

Solution

Begin by drawing a vertical axis to represent the number of immigrants in thousands. Then label the horizontal axis with decades and plot the points shown in the table. Finally, connect the points with line segments, as shown in Figure B.3. From the line graph, you can see that the number of immigrants hit a low point during the depression of the 1930's. Since then the numbers have steadily increased.

FIGURE B.3

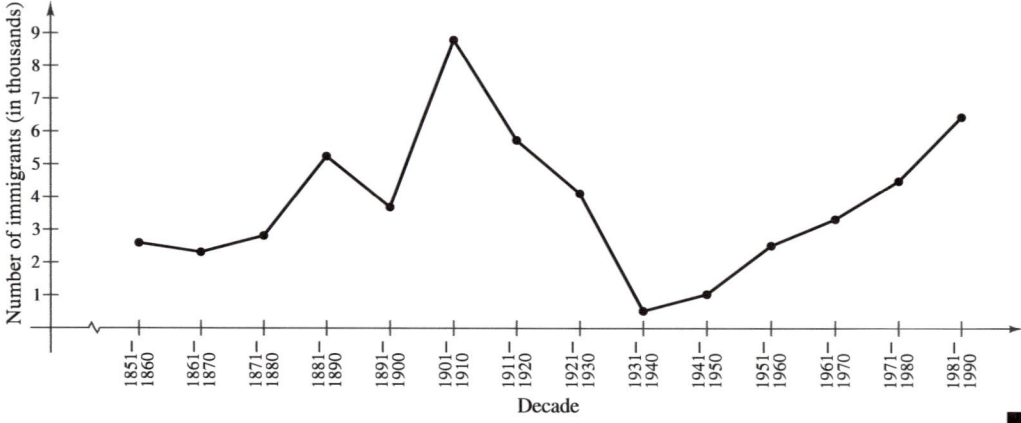

Choosing an Appropriate Graph

Line graphs and bar graphs are commonly used for displaying data. When you are using a graph to organize and present data, you must first decide which type of graph to use.

EXAMPLE 5   ■   Organizing Data with a Graph

Listed below are the daily average numbers of miles walked by people while working at their jobs. Organize the data graphically. (*Source:* American Podiatry Association)

Occupation	Miles Walked per Day
Mail Carrier	4.4
Medical Doctor	3.5
Nurse	3.9
Police Officer	6.8
Television Reporter	4.2

STUDY TIP

Here are some guidelines to use when you must decide which type of graph to use.

1. Use a bar graph when the data falls into distinct categories and you want to compare totals.

2. Use a line graph when you want to show the relationship between consecutive amounts or data over time.

Solution

You can use a bar graph because the data falls into distinct categories, and it would be useful to compare totals. The bar graph shown in Figure B.4 is horizontal. This makes it easier to label each bar. Also notice that the occupations are listed in order of the number of miles walked.

FIGURE B.4

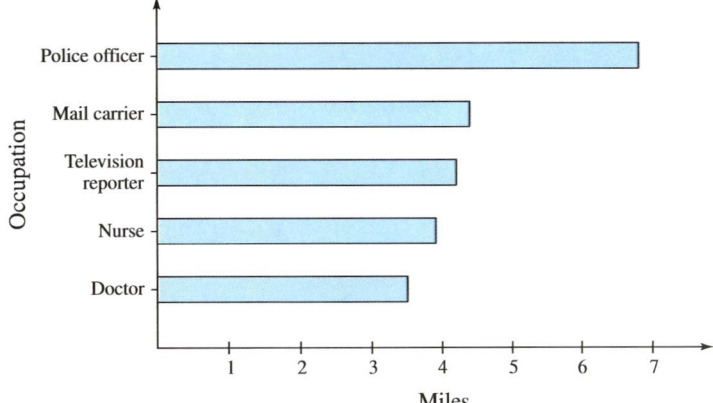

Technology Note

Most graphing utilities have built-in statistical programs that can create scatter plots. Use your graphing utility to plot the points given in the table below. The data shows the number of people P (in millions) in the United States who were part of the labor force from 1980 through 1990. In the table, t represents the year, with $t = 0$ corresponding to 1980. (*Source:* U.S. Bureau of Labor Statistics)

t	P
0	109
1	110
2	112
3	113
4	115
5	117

t	P
6	120
7	122
8	123
9	126
10	126

Scatter Plots

Many real-life situations involve finding relationships between two variables, such as the year and the number of people in the labor force. In a typical situation, data is collected and written as a set of ordered pairs. The graph of such a set is called a **scatter plot.**

FIGURE B.5

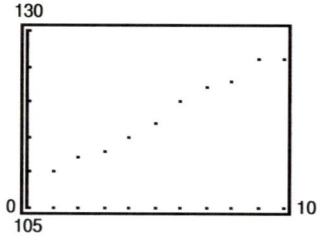

From the scatter plot in Figure B.5, it appears that the points describe a relationship that is nearly linear. (The relationship is not *exactly* linear because the labor force did not increase by precisely the same amount each year.) A mathematical equation that approximates the relationship between t and P is called a *mathematical model*. When developing a mathematical model, you strive for two (often conflicting) goals—accuracy and simplicity.

Consider a collection of ordered pairs of the form (x, y). If y tends to increase as x increases, the collection is said to have a **positive correlation.** If y tends to decrease as x increases, the collection is said to have a **negative correlation.** Figure B.6 shows three examples: one with a positive correlation, one with a negative correlation, and one with no (discernible) correlation.

FIGURE B.6

NOTE Scatter plots are discussed on page 268 and used in chapter projects throughout this book. The idea of correlation is included in several chapter projects. For example, see the Chapter 7 project on page 441 and the Chapter 11 project on page 681. The material presented here allows for a more in-depth discussion.

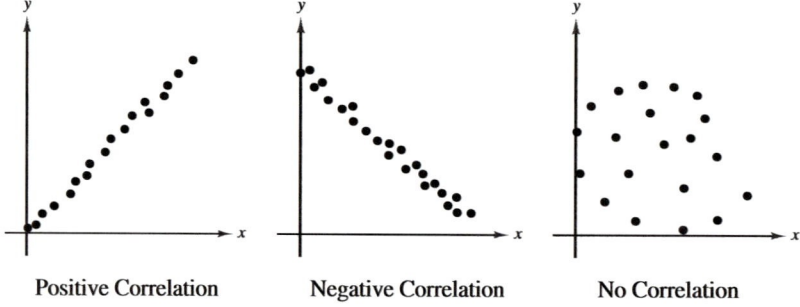

Positive Correlation   Negative Correlation   No Correlation

FIGURE B.7

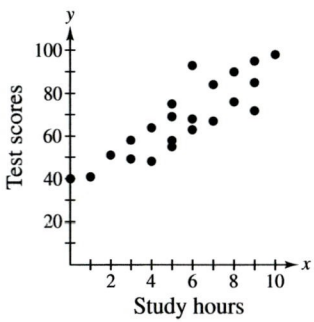

Study hours

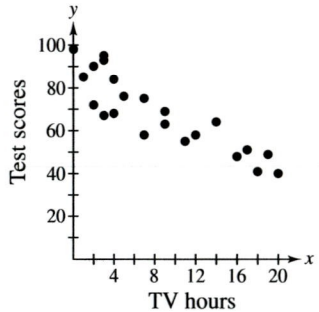

TV hours

EXAMPLE 6 ■ Interpreting Scatter Plots

On a Friday, 22 students in a class were asked to keep track of the numbers of hours they spent studying for a test on Monday and the numbers of hours they spent watching television. The numbers are shown below. Construct a scatter plot for each set of data. Then determine whether the points are positively correlated, negatively correlated, or have no discernible correlation. What can you conclude? (The first coordinate is the number of hours and the second coordinate is the score obtained on Monday's test.)

Study Hours: (0, 40), (1, 41), (2, 51), (3, 58), (3, 49), (4, 48), (4, 64), (5, 55), (5, 69), (5, 58), (5, 75), (6, 68), (6, 63), (6, 93), (7, 84), (7, 67), (8, 90), (8, 76), (9, 95), (9, 72), (9, 85), (10, 98)

TV Hours: (0, 98), (1, 85), (2, 72), (2, 90), (3, 67), (3, 93), (3, 95), (4, 68), (4, 84), (5, 76), (7, 75), (7, 58), (9, 63), (9, 69), (11, 55), (12, 58), (14, 64), (16, 48), (17, 51), (18, 41), (19, 49), (20, 40)

Solution

Scatter plots for the two sets of data are shown in Figure B.7. The scatter plot relating study hours and test scores has a positive correlation. This means that the more a student studied, the higher his or her score tended to be. The scatter plot relating television hours and test scores has a negative correlation. This means that the more time a student spent watching television, the lower his or her score tended to be.   ■

Fitting a Line to Data

Finding a linear model that represents the relationship described by a scatter plot is called **fitting a line to data.** You can do this graphically by simply sketching the line that appears to fit the points, finding two points on the line, and then finding the equation of the line that passes through the two points.

EXAMPLE 7 ■ Fitting a Line to Data

Find a linear model that relates the year with the number of people P (in millions) who were part of the United States labor force from 1980 through 1990. In Table B.2, t represents the year, with $t = 0$ corresponding to 1980. (*Source:* U.S. Bureau of Labor Statistics)

TABLE B.2

t	0	1	2	3	4	5	6	7	8	9	10
P	109	110	112	113	115	117	120	122	123	126	126

Solution

After plotting the data from Table B.2, draw the line that you think best represents the data, as shown in Figure B.8. Two points that lie on this line are (0, 109) and (9, 126). Using the point-slope form, you can find the equation of the line to be

$$P = \frac{17}{9}t + 109. \qquad \text{Linear model}$$

FIGURE B.8

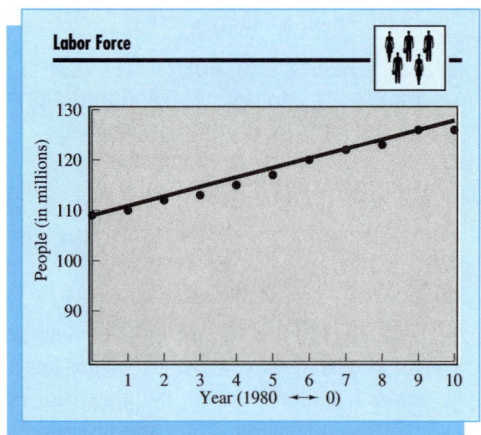

NOTE Fitting a line to data is incorporated throughout this book. For example, see the group activities on pages 634 and 705, and Exercises 95–97 on page 638. The material presented here allows for a more in-depth discussion.

Once you have found a model, you can measure how well the model fits the data by comparing the actual values with the values given by the model, as shown in Table B.3.

TABLE B.3

t	0	1	2	3	4	5	6	7	8	9	10
Actual → P	109	110	112	113	115	117	120	122	123	126	126
Model → P	109	110.9	112.8	114.7	116.6	118.4	120.3	122.2	124.1	126	127.9

The sum of the squares of the differences between the actual values and the model's values is the **sum of the squared differences.** The model that has the least sum is called the **least squares regression line** for the data. For the model in Example 7, the sum of the squared differences is 13.81. The least squares regression line for the data is

NOTE Built-in regression features of calculators are mentioned in the chapter projects throughout this book. For example, see the Chapter 7 project on page 441.

$$P = 1.864t + 108.2. \qquad \text{Best-fitting linear model}$$

Its sum of squared differences is 4.7.

Many calculators have "built-in" least squares regression programs. If your calculator has such a program, enter the data in Table B.2 and use it to find the least squares regression line.

B EXERCISES

1. Construct a stem-and-leaf plot for the following exam for a class of 30 students. The scores are for a 100-point exam.

77, 100, 77, 70, 83, 89, 87, 85, 81, 84, 81, 78, 89, 78, 88, 85, 90, 92, 75, 81, 85, 100, 98, 81, 78, 75, 85, 89, 82, 75

2. *Education Expenses* The table shows the per capita expenditures for public elementary and secondary education in the 50 states and the District of Columbia in 1991. Use a stem-and-leaf plot to organize the data. (*Source:* National Education Association)

AK	1626	AL	694	AR	668	AZ	892	CA	918
CO	841	CT	1151	DC	1010	DE	891	FL	862
GA	859	HI	784	IA	846	ID	725	IL	788
IN	925	KS	906	KY	725	LA	758	MA	866
MD	944	ME	1062	MI	926	MN	990	MO	742
MS	671	MT	983	NC	813	ND	719	NE	757
NH	881	NJ	1223	NM	915	NV	1004	NY	1186
OH	861	OK	776	OR	925	PA	889	RI	892
SC	835	SD	716	TN	618	TX	905	UT	828
VA	941	VT	992	WA	1095	WI	928	WV	883
WY	1178								

In Exercises 3 and 4, use the following set of data, which lists students' scores on a 100-point exam.

93, 84, 100, 92, 66, 89, 78, 52, 71, 85, 83, 95, 98, 99, 93, 81, 80, 79, 67, 59, 90, 55, 77, 62, 90, 78, 66, 63, 93, 87, 74, 96, 72, 100, 70 ,73

3. Use a stem-and-leaf plot to organize the data.

4. Draw a histogram to represent the data.

5. Complete the following frequency distribution table and draw a histogram to represent the data.

44, 33, 17, 23, 16, 18, 44, 47, 18, 20, 25, 27, 18, 29, 29, 28, 27, 18, 36, 22, 32, 38, 33, 41, 49, 48, 45, 38, 49, 15

Interval	Tally
[15, 22)	
[22, 29)	
[29, 36)	
[36, 43)	
[43, 50)	

6. *Snowfall* The data below shows the seasonal snowfall (in inches) at Erie, Pennsylvania, for the years 1960 through 1989 (the amounts are listed in order by year). How would you organize this data? Explain your reasoning. (*Source:* National Oceanic and Atmospheric Administration)

69.6, 42.5, 75.9, 115.9, 92.9, 84.8, 68.6, 107.9, 79.7, 85.6, 120.0, 92.3, 53.7, 68.6, 66.7, 66.0, 111.5, 142.8, 76.5, 55.2, 89.4, 71.3, 41.2, 110.0, 106.3, 124.9, 68.2, 103.5, 76.5, 114.9

7. *Travel to the United States* The data below gives the places of origin and the numbers of travelers (in millions) to the United States in 1991. Construct a horizontal bar graph for this data. (*Source:* U.S. Travel and Tourism Administration)

Canada 18.9 Mexico 7.0 Europe 7.4
Latin America 2.0 Other 6.8

8. *Fruit Crops* The data below shows the cash receipts (in millions of dollars) from fruit crops for farmers in 1990. Construct a bar graph for this data. (*Source:* U.S. Department of Agriculture)

Apples	1159	Peaches	365
Grapefruit	317	Pears	266
Grapes	1668	Plums and Prunes	293
Lemons	278	Strawberries	560
Oranges	1707		

Handling Garbage In Exercises 9–14, use the line graph given below. (*Source:* Franklin Associates)

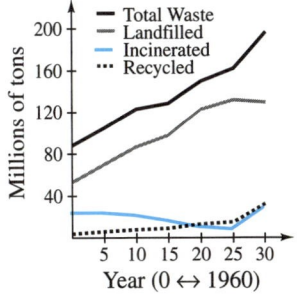

9. Estimate the total waste in 1980 and 1990.

10. Estimate the amount of incinerated garbage in 1970.

11. Which quantities increased every year?

12. During which time period did the amount of incinerated garbage decrease?

13. *Reasoning*   What is the relationship between the four quantities in the line graph?

14. *Think About It*   Why do you think landfill garbage is decreasing?

15. *College Attendance*   The following table shows the enrollment in a liberal arts college. Construct a line graph for the data.

Year	1985	1986	1987	1988
Enrollment	1675	1704	1710	1768

Year	1989	1990	1991	1992
Enrollment	1833	1918	1967	1972

16. *Oil Imports*   The table shows the crude oil imports into the United States in millions of barrels for the years 1982 through 1990. Construct a line graph for the data and state what information it reveals. (*Source:* U.S. Energy Information Administration)

Year	1982	1983	1984	1985	1986
Oil Imports	1273	1215	1254	1168	1525

Year	1987	1988	1989	1990
Oil Imports	1706	1869	2133	2145

17. *Stock Market*   The list below shows stock prices for selected companies in March of 1994. Draw a graph that best represents the data. Explain why you chose that type of graph.

Company	Stock Price
Sears, Roebuck	$48
Wal-Mart Stores	$28
JC Penney	$55
K Mart Corp.	$19
The Gap, Inc.	$45

18. *Net Profit*   The table shows the net profits (in millions of dollars) of Blockbuster Entertainment for the years 1988 through 1993. Draw a graph that best represents the net profit and explain why you chose that type of graph.

Year	1988	1989	1990	1991	1992	1993
Net Profit	15.5	44.2	68.7	93.7	142.0	243.6

19. *Videocassette Recorders*   The average numbers (out of 100) of people who owned videocassette recorders in selected years from 1980 to 1992 are given in the table. Organize the data graphically. Explain your reasoning. (*Source:* The Roper Organization)

Year	1980	1983	1985	1987	1988	1990	1992
Number	3	10	19	50	64	65	68

20. *Owning Cats*   The average numbers (out of 100) of cat owners who state various reasons for owning a cat are listed below. Organize the data graphically. Explain your reasoning. (*Source:* Gallup Poll)

Reason for Owning a Cat	Number
Have a pet to play with	93
Companionship	84
Help children learn responsibility	78
Have a pet to communicate with	62
Security	51

Interpreting a Scatter Plot   In Exercises 21–24, use the scatter plot shown. The scatter plot compares the number of hits *x* made by 30 softball players during the first half of the season with the number of runs batted in *y*.

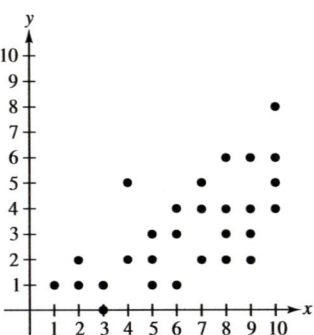

21. Do *x* and *y* have a positive correlation, a negative correlation, or no correlation?

22. Why does the scatter plot show only 28 points?

23. From the scatter plot, does it appear that players with more hits tend to have more runs batted in?

24. Can a player have more runs batted in than hits? Explain.

In Exercises 25–28, decide whether a scatter plot relating the two quantities would tend to have a positive, negative, or no correlation. Explain.

25. The age and value of a car

26. A student's study time and test scores

27. The height and age of a pine tree

28. A student's height and test scores

In Exercises 29–32, use the data in the table, which shows the relationship between the altitude *A* (in thousands of feet) and the air pressure *P* (in pounds per square foot).

A	0	5	10	15	20	25
P	14.7	12.3	10.2	8.4	6.8	5.4

A	30	35	40	45	50
P	4.5	3.5	2.8	2.1	1.8

29. Sketch a scatter plot of the data.

30. How are *A* and *P* related?

31. Estimate the air pressure at 42,500 feet.

32. Estimate the altitude at which the air pressure is 5.0 pounds per square foot.

Crop Yield   In Exercises 33–36, use the data in the table, where *x* is the number of units of fertilizer applied to sample plots and *y* is the yield (in bushels) of a crop.

x	0	1	2	3	4	5	6	7	8
y	58	60	59	61	63	66	65	67	70

33. Sketch a scatter plot of the data.

34. Determine whether the points are positively correlated, are negatively correlated, or have no discernible correlation.

35. Sketch a linear model that you think best represents the data. Find an equation of the line you sketched. Use the line to predict the yield if 10 units of fertilizer are used.

36. Can the model found in Exercise 35 be used to predict yields for arbitrarily large values of *x*? Explain.

Speed of Sound   In Exercises 37–40, use the data in the table, where *h* is altitude in thousands of feet and *v* is speed of sound in feet per second.

h	0	5	10	15	20	25	30	35
v	1116	1097	1077	1057	1036	1015	995	973

37. Sketch a scatter plot of the data.

38. Determine whether the points are positively correlated, are negatively correlated, or have no discernible correlation.

39. Sketch a linear model that you think best represents the data. Find an equation of the line you sketched. Use the line to predict the speed of sound at an altitude of 27,000 feet.

40. The speed of sound at an altitude of 70,000 feet is approximately 971 feet per second. What does this suggest about the validity of using the model in Exercise 39 to extrapolate beyond the data given in the table?

In Exercises 41–44, use a graphing utility to find the least squares regression line for the data. Sketch a scatter plot and the regression line.

41. (0, 23), (1, 20), (2, 19), (3, 17), (4, 15), (5, 11), (6, 10)

42. (4, 52.8), (5, 54.7), (6, 55.7), (7, 57.8), (8, 60.2), (9, 63.1), (10, 66.5)

43. (−10, 5.1), (−5, 9.8), (0, 17.5), (2, 25.4), (4, 32.8), (6, 38.7), (8, 44.2), (10, 50.5)

44. (−10, 213.5), (−5, 174.9), (0, 141.7), (5, 119.7), (8, 102.4), (10, 87.6)

45. *Advertising* The management of a department store ran an experiment to determine if a relationship existed between sales S (in thousands of dollars) and the amount spent on advertising x (in thousands of dollars). The following data was collected.

x	1	2	3	4	5	6	7	8
S	405	423	455	466	492	510	525	559

Use a graphing utility to find the least squares regression line. Use the equation to estimate sales if $4500 is spent on advertising. Make a scatter plot of the data and sketch the graph of the regression line.

Exploring Congruence and Similarity

Identifying Congruent Figures

NOTE   One way to compare real-life objects is to determine whether they are congruent or similar. For instance, in manufacturing, producing congruent parts allows them to be used interchangeably.

Two figures are *congruent* if they have the same shape and the same size. Each of the triangles below on the left are congruent to each of the other triangles. The triangles on the right are not congruent to each other.

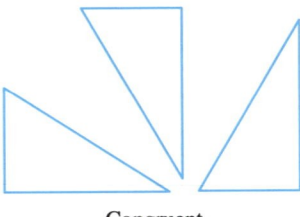

Congruent

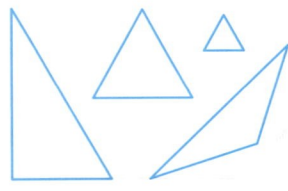

Not congruent

Notice that two figures can be congruent without having the same orientation. If two figures are congruent, then either one can be moved (and turned or flipped if neccessary) so that it coincides with the other figure.

EXAMPLE 1   ■   Dividing Regions into Congruent Parts

Divide the region into two congruent parts.

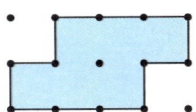

Solution

There are many solutions to this problem. Some of the solutions are shown below. Can you think of others?

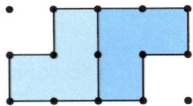   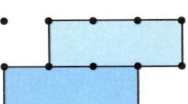   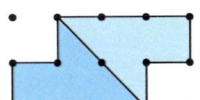   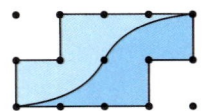   ■

Identifying Similar Figures

Two figures are *similar* if they have the same shape. (They may or may not have the same size.) Each of the quadrilaterals below on the left is similar to the other. The quadrilaterals on the right are not similar to each other.

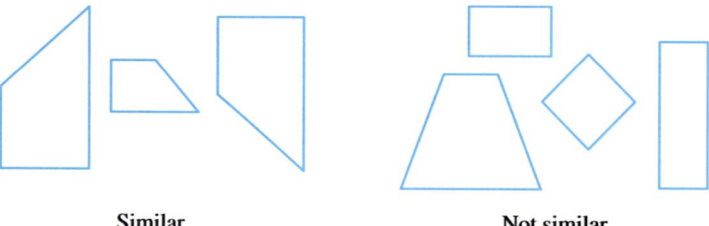

Similar   Not similar

EXAMPLE 2 ■   Determining Similarity

Two of the figures are similar. Which two are they?

a.   **b.**   **c.**

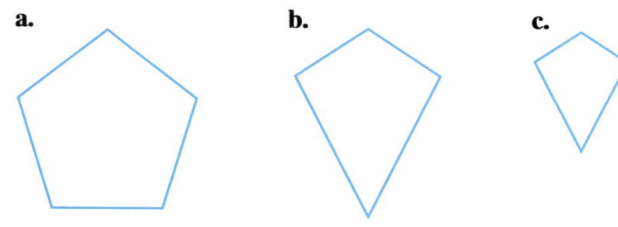

Solution

The first figure has 5 sides and the other two figures have 4 sides. Because similar figures must have the same shape, the first figure is not similar to either of the others. Because you are told that two figures are similar, it follows that the second and third figures are similar.   ■

EXAMPLE 3 ■   Determining Similarity

You wrote an essay on Euclid, the Greek mathematician who is famous for writing a geometry book titled *Elements of Geometry*. You are making a copy of the essay using a photocopier that is set at 75% reduction. Is each image on the copied pages similar to its original?

Solution

Every image on a copied page *is* similar to its original. The copied pages are smaller, but that doesn't matter because similar figures do not have to be the same.

■

EXAMPLE 4   ■   **Drawing an Object to Scale**

You are drawing a floor plan of a building. You choose a scale of $\frac{1}{8}$-inch to 1-foot. That is, $\frac{1}{8}$ inch of the floor plan represents 1 foot of the actual building. What dimensions should you draw for a room that is 12 feet wide and 18 feet long?

Solution

Because each foot is represented as $\frac{1}{8}$ inch, the width of the room should be

$$12\left(\tfrac{1}{8}\right) = \tfrac{12}{8} = \tfrac{3}{2} = 1\tfrac{1}{2}$$

and the length of the room should be

$$18\left(\tfrac{1}{8}\right) = \tfrac{18}{8} = \tfrac{9}{4} = 2\tfrac{1}{4}$$

The scale dimensions of the room are $1\frac{1}{2}$ inches by $2\frac{1}{4}$ inches.   ■

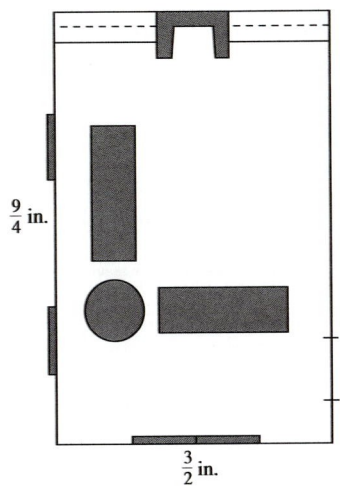

$\frac{9}{4}$ in.

$\frac{3}{2}$ in.

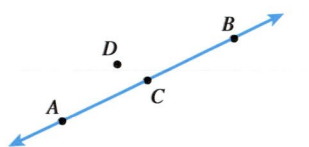

C is between A and B. D is not between A and B.

Reading and Using Definitions

A definition uses *known* words to describe a *new* word. If no words were known, then no new words could be defined. Hence, some words such as **point, line,** and **plane** must be commonly understood without being defined. Some statements such as "a point lies on a line" and "point C lies between points A and B" are also not defined.

■ **Segments and Rays**	Consider the line, $\overleftrightarrow{AB}$, that contains the points A and B. (In geometry, the word *line* means a *straight line*.)

The **line segment** (or simply **segment**) $\overline{AB}$ consists of the *endpoints* A and B and all points on the line $\overleftrightarrow{AB}$ that lie between A and B.

The **ray** $\overrightarrow{AB}$ consists of the *initial point* A and all points on the line $\overleftrightarrow{AB}$ that lie on the same side of A as B lies. If C is between A and B, then $\overrightarrow{CA}$ and $\overrightarrow{CB}$ are **opposite** rays.

Points, segments, or rays that lie on the same line are **collinear.**

Lines are drawn with two arrowheads, line segments are drawn with no arrowhead, and rays are drawn with a single arrowhead.

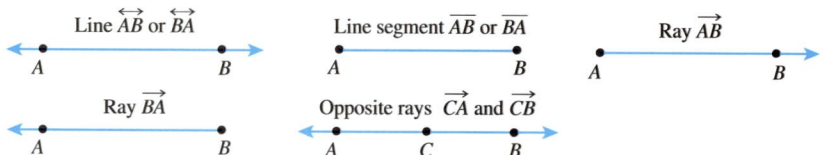

It follows that $\overline{AB}$ and $\overline{BA}$ denote the same segment, but $\overrightarrow{AB}$ and $\overrightarrow{BA}$ do not denote the same ray. No length is given to lines or rays because each is infinitely long. The *length* of the line segment $\overline{AB}$ is denoted by AB.

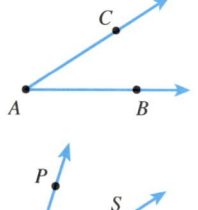

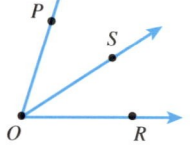

The top angle can be denoted by $\angle A$ or by $\angle BAC$. In the lower figure, the angle $\angle ROS$ should not be denoted by $\angle O$ because the figure contains three angles whose vertex is O.

Angles

An **angle** consists of two different rays that have the same initial point. The rays are the *sides* of the angle. The angle that consists of the rays $\overrightarrow{AB}$ and $\overrightarrow{AC}$ is denoted by $\angle BAC$, $\angle CAB$, or by $\angle A$. The point A is the *vertex* of the angle.

The measure of $\angle A$ is denoted by $m\angle A$. Angles are classified as **acute, right, obtuse,** and **straight.**

Acute	$0° < m\angle A < 90°$
Right	$m\angle A = 90°$
Obtuse	$90° < m\angle A < 180°$
Straight	$m\angle A = 180°$

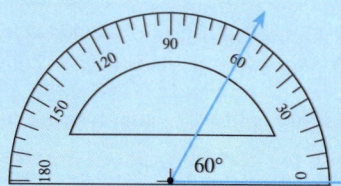

The measure of an angle can be approximated with a protractor.

In geometry, *unless specifically stated otherwise*, angles are assumed to have a measure that is greater than $0°$ and less than or equal to $180°$.

Every nonstraight angle has an **interior** and an **exterior.** A point D is in the interior of $\angle A$ if it is between points that lie on each side of the angle. Two angles (such as $\angle ROS$ and $\angle SOP$ shown above) are **adjacent** if they share a common vertex and side, but have no common interior points.

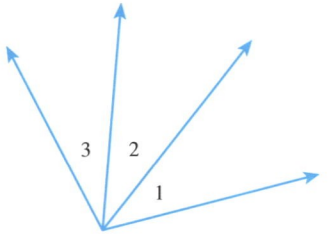

$\angle 1$ and $\angle 2$ are adjacent.
$\angle 1$ and $\angle 3$ are not adjacent.

Segment and Angle Congruence

Two segments are **congruent**, $\overline{AB} \cong \overline{CD}$, if they have the same length.
Two angles are **congruent**, $\angle P \cong \angle Q$, if they have the same measure.

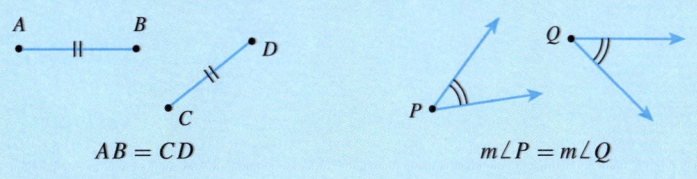

$AB = CD$ $m\angle P = m\angle Q$

NOTE Building a vocabulary in geometry helps you read and communicate ideas clearly.

Definitions can always be interpreted "forward" and "backward." For instance, the definition of congruent segments means (1) if two segments have the same measure, then they are congruent, *and* (2) if two segments are congruent, then they have the same measure. You learned that two figures are congruent if they have the same shape and size.

NOTE If two triangles are congruent, then you know that they share many properties.

Definition of Congruent Triangles

If $\triangle ABC$ is **congruent** to $\triangle PQR$, then there is a correspondence between their angles and sides such that corresponding angles are congruent and corresponding sides are congruent. The notation $\triangle ABC \cong \triangle PQR$ indicates the congruence *and* the correspondence, as shown below.

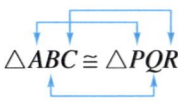

$$\triangle ABC \cong \triangle PQR$$

Corresponding angles are	*Corresponding sides are*
$\angle A \cong \angle P$	$\overline{AB} \cong \overline{PQ}$
$\angle B \cong \angle Q$	$\overline{BC} \cong \overline{QR}$
$\angle C \cong \angle R$	$\overline{CA} \cong \overline{RP}$

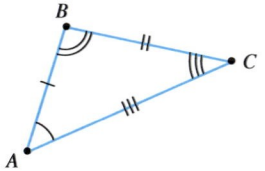

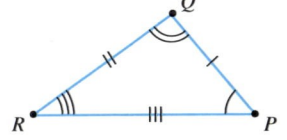

EXAMPLE 5 ■ Naming Congruent Parts

You and a friend have identical drafting triangles, as shown below. Name all congruent parts.

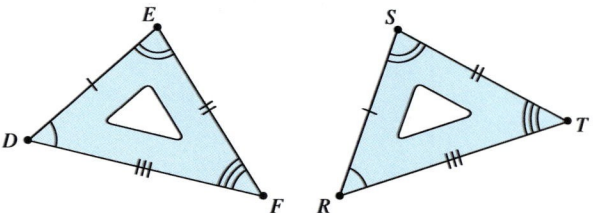

Solution

Given that $\triangle DEF \cong \triangle RST$, the congruent angles and sides are as follows.

Angles: $\angle D \cong \angle R,$ $\angle E \cong \angle S,$ $\angle F \cong \angle T$

Sides: $\overline{DE} \cong \overline{RS},$ $\overline{EF} \cong \overline{ST},$ $\overline{FD} \cong \overline{TR},$ ■

Identifying Types of Triangles

A triangle can be classified by relationships among its sides or among its angles, as shown in the following definitions.

Classification by Sides

1. An **equilateral triangle** has three congruent sides.
2. An **isosceles triangle** has at least two congruent sides.
3. A **scalene triangle** has no sides congruent.

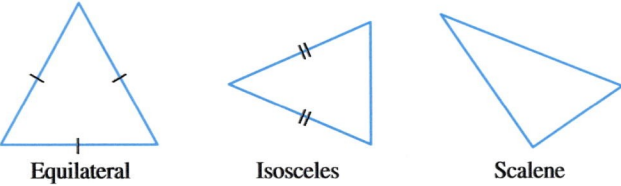

Equilateral   Isosceles   Scalene

Classification by Angles

1. An **acute triangle** has three acute angles. If these angles are all congruent, then the triangle is also **equiangular.**

2. A **right triangle** has exactly one right angle.

3. An **obtuse triangle** has exactly one obtuse angle.

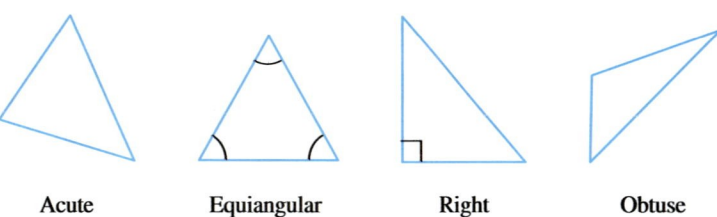

Acute   Equiangular   Right   Obtuse

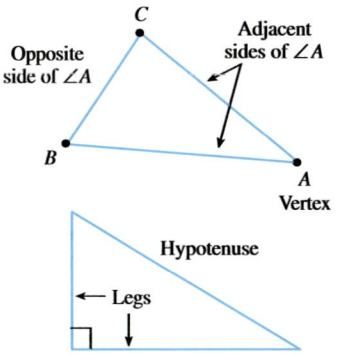

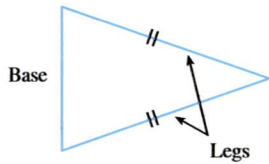

In △ABC, each of the points A, B, and C is a **vertex** of the triangle. (The plural of vertex is *vertices*.) The side $\overline{BC}$ is the side *opposite* $\angle A$. Two sides that share a common vertex are *adjacent sides*.

The sides of right triangles and isosceles triangles are given special names. In a right triangle, the sides adjacent to the right angle are the **legs** of the triangle. The side opposite the right angle is the **hypotenuse** of the triangle.

An isosceles triangle can have three congruent sides. If it has only two, then the two congruent sides are the **legs** of the triangle. The third side is the **base** of the triangle.

EXERCISES

1. Copy the region on a piece of dot paper. Then divide the region into two congruent parts. Can you find more than one way to do this?

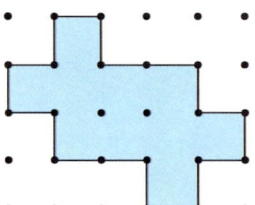

2. Two of the figures are congruent. Which are they?

a. **b.** **c.**

 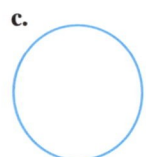

3. Two of the figures are similar. Which are they?

a. **b.** **c.**

 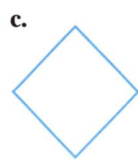

4. True or False? If two figures are congruent, then they are similar.

5. True or False? If two figures are similar, then they are congruent.

6. True or False? A triangle can be similar to a square.

7. True or False? Any two squares are similar.

8. *Landscape Design* You are designing a patio. Your plans use a scale of $\frac{1}{8}$ inch to 1 foot. The patio is 24 feet by 36 feet. What are its dimensions on the plans?

Independent Practice

In Exercises 9 and 10, copy the region on a piece of dot paper. Then divide the region into two congruent parts. How many different ways can you do this?

9.

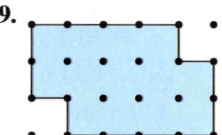

10.

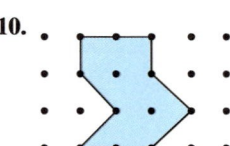

In Exercises 11 and 12, copy the region on a piece of paper. Then divide the region into four congruent parts.

11.

12.

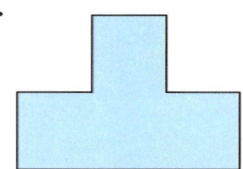

In Exercises 13–17, use the triangular grid below. In the grid, each small triangle has sides of one unit.

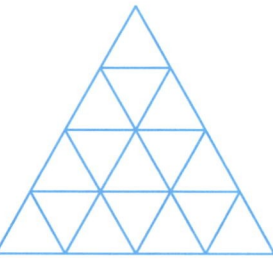

13. How many congruent triangles with 1-unit sides are in the grid?

14. How many congruent triangles with 2-unit sides are in the grid?

15. How many congruent triangles with 3-unit sides are in the grid?

16. Does the grid contain triangles that are not similar to each other?

17. *Architecture*   The Pentagon, near Washington D.C., covers a region that is about 1200 feet by 1200 feet. About how large would a $\frac{1}{8}$-inch to 1-foot scale drawing of the Pentagon be? Would such a scale be reasonable?

In Exercises 18–21, match the description with its correct notation.

a. $\overline{PQ}$ **b.** PQ **c.** $\overleftrightarrow{PQ}$ **d.** $\overrightarrow{PQ}$

18. The line through P and Q

19. The ray from P through Q

20. The segment between P and Q

21. The length of the segment between P and Q

22. The point R is between points S and T. Which of the following are true?

a. R, S, and T are collinear.

b. $\overrightarrow{SR}$ is the same as $\overrightarrow{ST}$.

c. $\overline{ST}$ is the same as $\overline{TS}$.

d. $\overrightarrow{ST}$ is the same as $\overrightarrow{TS}$.

In Exercises 23–25, use the figure below.

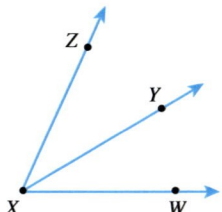

23. The figure shows three angles whose vertex is X. Write two names for each angle. Which two angles are adjacent?

24. Is Y in the interior or exterior of $\angle WXZ$?

25. Which is the best estimate for $m\angle WXY$?

a. $15°$ **b.** $30°$ **c.** $45°$

In Exercises 26–31, match the triangle with its name.

a. Equilateral   **b.** Scalene   **c.** Obtuse

d. Equiangular   **e.** Isosceles   **f.** Right

26. Side lengths: 2 cm, 3 cm, 4 cm

27. Angle measures: $60°$, $60°$, $60°$

28. Side lengths: 3 cm, 2 cm, 3 cm

29. Side lengths: $30°$, $60°$, $90°$

30. Side lengths: 4 cm, 4 cm, 4 cm

31. Side lengths: $20°$, $145°$, $15°$

Independent Practice

In Exercises 32–34, use the figure in which $\triangle LMP \cong \triangle ONQ$.

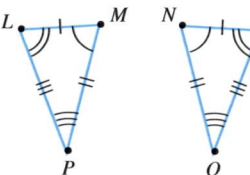

32. Name three pairs of congruent angles.

33. Name three pairs of congruent sides.

34. If $\triangle LMP$ is isosceles, explain why $\triangle ONQ$ must be isosceles?

35. If $\triangle ABC \cong \triangle TUV$, then $m\angle C = \boxed{?}$.

36. If $\triangle PQR \cong \triangle XYZ$, then $\angle P \cong \boxed{?}$.

37. If $\triangle LMN \cong \triangle TUV$, then $\overline{LN} \cong \boxed{?}$.

38. If $\triangle DEF \cong \triangle NOP$, then $DE = \boxed{?}$.

39. Copy and complete the table. Write *yes* if it is possible to sketch a triangle with both characteristics. Write *no* if it is not possible. Illustrate your results with sketches. (The first is done for you.)

	Scalene	Isosceles	Equilateral
Acute	Yes	?	?
Obtuse	?	?	?
Right	?	?	?

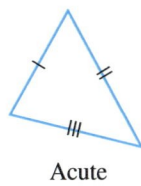

Acute and Scalene

Algebra   In Exercises 40–43, $\triangle ABC$ is isosceles with $\overline{AC} \cong \overline{BC}$. Solve for x. Then decide whether the triangle is equilateral. (The figures are not necessarily drawn to scale.)

40.

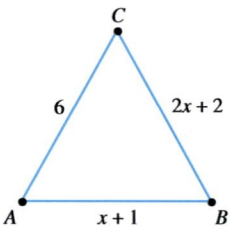

41.

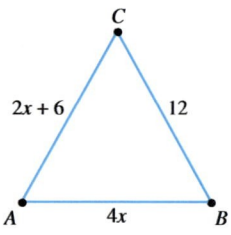

42.

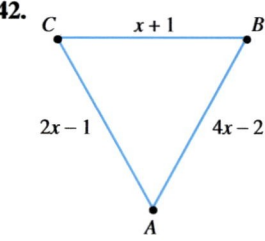

43.

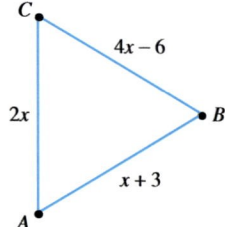

Coordinate Geometry   In Exercises 44 and 45, use the figure below.

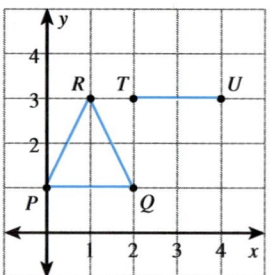

44. Find a location of S so that $\triangle PQR \cong \triangle PQS$.

45. Find two locations of V so that $\triangle PQR \cong \triangle TUV$.

46. *Logical Reasoning*   Arrange sixteen toothpicks as shown below. What is the least number of toothpicks you must remove to create four congruent triangles? (Each toothpick must be the side of at least one triangle.) Sketch your result.

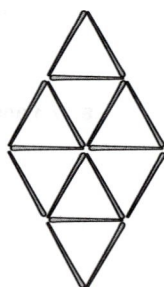

47. *Logical Reasoning*   Show how you could arrange six toothpicks to form four congruent triangles. Each triangle has one toothpick for each side, and you cannot bend, break, or overlap the toothpicks.

Angles

Identifying Special Pairs of Angles

You have been introduced to several definitions concerning angles. For instance, you know that two angles are *adjacent* if they share a common vertex and side but have no common interior points. Here are some other definitions for pairs of angles.

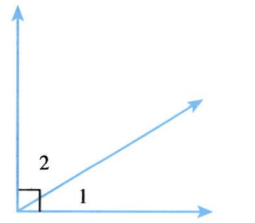

∠1 and ∠2 are complementary angles.

Two angles are **vertical angles** if their sides form two pairs of opposite rays.

Two adjacent angles are a **linear pair** if their noncommon sides are opposite rays.

Two angles are **complementary** if the sum of their measures is 90°. Each angle is the *complement* of the other.

Two angles are **supplementary** if the sum of their measures is 180°. Each angle is the *supplement* of the other.

EXAMPLE 1   ■   Identifying Special Pairs of Angles

Use the terms defined above to describe relationships between the labeled angles in the figure at the left.

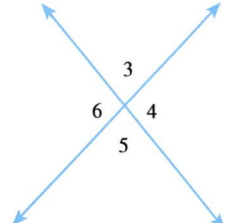

Solution

a. ∠3 and ∠5 are vertical angles. So are ∠4 and ∠6.

b. There are four sets of linear pairs:

∠3 and ∠4, ∠4 and ∠5, ∠5 and ∠6, and ∠3 and ∠6.

Each of these pairs are also supplementary angles.   ■

In Example 1b, note that the linear pairs are also supplementary. This result is stated in the following postulate.

Linear Pair Postulate If two angles form a linear pair, then they are supplementary, i.e., the sum of their measures is 180°.

Vertical Angles Theorem If two angles are vertical angles, then they are congruent.

Angles Formed by a Transversal

A **transversal** is a line that intersects two or more coplanar lines at different points. The angles that are formed when the transversal intersects the lines have the following names.

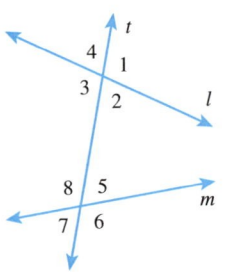

Angles Formed by a Transversal

In the figure at the left, the transversal t intersects the lines l and m.

Two angles are **corresponding angles** if they occupy corresponding positions, such as $\angle 1$ and $\angle 5$.

Two angles are **alternate interior angles** if they lie between l and m on opposite sides of t, such as $\angle 2$ and $\angle 8$.

Two angles are **alternate exterior angles** if they lie outside l and m on opposite sides of t, such as $\angle 1$ and $\angle 7$.

Two angles are **consecutive interior angles** if they lie between l and m on the same side of t, such as $\angle 2$ and $\angle 5$.

EXAMPLE 2 ■ Naming Pairs of Angles

In the figure at the left, how is $\angle 9$ related to the other angles?

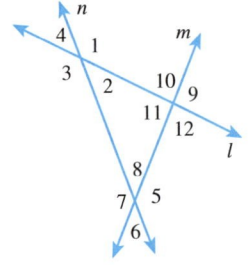

Solution

You can consider that $\angle 9$ is formed by the transversal l as it intersects m and n, or you can consider $\angle 9$ to be formed by the transversal m as it intersects l and n. Considering one or the other of these, you have the following.

a. $\angle 9$ and $\angle 10$ are a linear pair. So are $\angle 9$ and $\angle 12$.

b. $\angle 9$ and $\angle 11$ are vertical angles.

c. $\angle 9$ and $\angle 7$ are alternate exterior angles. So are $\angle 9$ and $\angle 3$.

d. $\angle 9$ and $\angle 5$ are corresponding angles. So are $\angle 9$ and $\angle 1$. ■

Algebraic Property

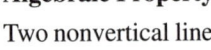

Two nonvertical lines are parallel if and only if they have the same slope.

To help build understanding involving angles formed by a transversal, consider relationships between two lines. **Parallel lines** are coplanar lines that do not intersect. **Intersecting lines** are coplanar and have exactly one point in common. If intersecting lines meet at right angles, they are perpendicular; otherwise they are **oblique.**

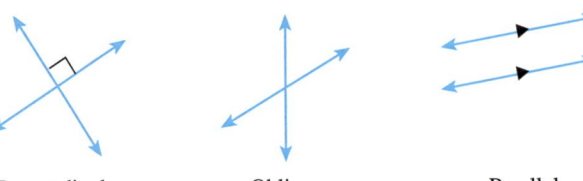

Perpendicular   Oblique   Parallel

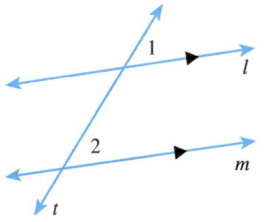

$l \parallel m, \angle 1 \cong \angle 2$

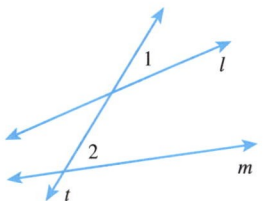

$l \nparallel m, \angle 1 \ncong \angle 2$

Many of the angles formed by a transversal that intersects *parallel* lines are congruent. The following postulate and theorems list useful results.

Corresponding Angles Postulate If two parallel lines are cut by a transversal, then the pairs of corresponding angles are congruent.

Note that the hypothesis of this postulate states that the lines must be parallel, as shown in the top figure at the left.

Alternate Interior Angles Theorem If two parallel lines are cut by a transversal, then the pairs of alternate interior angles are congruent.

Consecutive Interior Angles Theorem If two parallel lines are cut by a transversal, then the pairs of consecutive interior angles are supplementary.

Alternate Exterior Angles Theorem If two parallel lines are cut by a transversal, then the pairs of alternate exterior angles are congruent.

Perpendicular Transversal Theorem If a transversal is perpendicular to one of two parallel lines, then it is perpendicular to the second.

EXAMPLE 3 ■ **Using Properties of Parallel Lines**

In the figure, lines r and s are parallel lines cut by a transversal, l. Find the measure of each labeled angle.

Solution

$\angle 1$ and the given angle are alternate exterior angles and are congruent. So $m\angle 1 = 75°$. $\angle 5$ and the given angle are vertical angles. Since vertical angles are congruent, they have the same measure. $m\angle 5 = 75°$. Similarly, $\angle 1 \cong \angle 4$ and $m\angle 1 = m\angle 4 = 75°$. There are several sets of linear pairs including:

$\angle 1$ and $\angle 2$; $\angle 3$ and $\angle 4$; $\angle 5$ and $\angle 6$, $\angle 5$ and $\angle 7$

Each of these pairs are also supplementary angles; the sum of the measures of each pair of angles is $180°$. Since one angle of each pair measures $75°$, the supplements each measure $105°$. So $\angle 2$, $\angle 3$, $\angle 6$, and $\angle 7$ each measure $105°$. ■

Measures of Angles of a Triangle

The word "triangle" means "three angles." When the sides of a triangle are extended, however, other angles are formed. The original three angles of the triangle are the **interior angles.** The angles that are adjacent to interior angles are the **exterior angles** of the triangle. Each vertex has a pair of exterior angles.

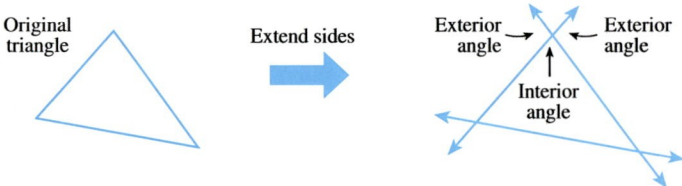

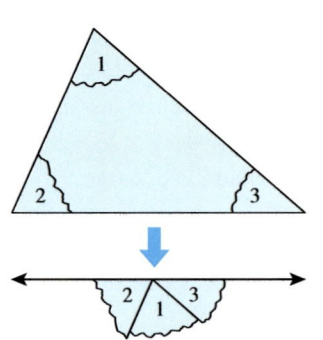

You could cut a triangle out of a piece of paper. Tear off the three angles and place them adjacent to each other, as shown at the left. What do you observe? (You could perform this investigation by measuring with a protractor or using a computer drawing program.) You should arrive at the conclusion given in the following theorem.

Triangle Sum Theorem

The sum of the measures of the interior angles of a triangle is 180°.

EXAMPLE 4 ■ Using the Triangle Sum Theorem

In the triangle at the left, find $m\angle 1$, $m\angle 2$, and $m\angle 3$.

Solution

To find the measure of $\angle 3$, use the Triangle Sum Theorem, as follows.

$$m\angle 3 = 180° - (51° + 42°) = 87°$$

Knowing the measure of $\angle 3$, you can use the Linear Pair Postulate to write $m\angle 2 = 180° - 87° = 93°$. Using the Triangle Sum Theorem, you have

$$m\angle 1 = 180° - (28° + 93°) = 59°. \qquad ■$$

The next theorem is one that you might have anticipated from the investigation on the previous page. As shown in the figure at the left, if you had torn only two of the angles from the paper triangle, you could put them together to exactly cover one of the exterior angles.

Exterior Angle Theorem The measure of an exterior angle of a triangle is equal to the sum of the measures of the two remote (nonadjacent) interior angles.

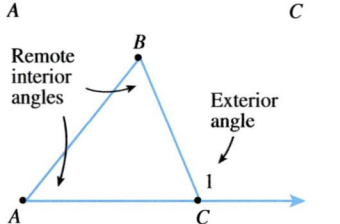

C   EXERCISES

In Exercises 1–6, sketch a pair of angles that fits the description. Label the angles as $\angle 1$ and $\angle 2$.

1. A linear pair of angles

2. Supplementary angles for which $\angle 1$ is acute

3. Acute vertical angles

4. Adjacent congruent complementary angles

5. Obtuse vertical angles

6. Adjacent congruent supplementary angles

In Exercises 7–12, use the figure below to determine relationships among the given angles.

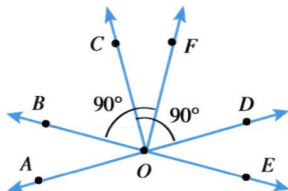

7. $\angle AOC$ and $\angle COD$
8. $\angle AOB$ and $\angle BOC$
9. $\angle BOC$ and $\angle COE$
10. $\angle AOB$ and $\angle EOD$
11. $\angle BOC$ and $\angle COF$
12. $\angle AOB$ and $\angle AOE$

Independent Practice

Logical Reasoning   In Exercises 13–18, use the following information to decide whether the statement is true or false. (*Hint:* Make a sketch.)

Vertical angles: $\angle 1$ and $\angle 2$; Linear pairs: $\angle 1$ and $\angle 3$, $\angle 1$ and $\angle 4$.

13. If $m\angle 3 = 30°$, then $m\angle 4 = 150°$.

14. If $m\angle 1 = 150°$, then $m\angle 4 = 30°$.

15. $\angle 2$ and $\angle 3$ are congruent.

16. $m\angle 3 + m\angle 1 = m\angle 4 + m\angle 2$

17. $\angle 3 \cong \angle 4$

18. $m\angle 3 = 180° - m\angle 2$

Integrated Review

Algebra   In Exercises 19–24, find the value of x.

19.

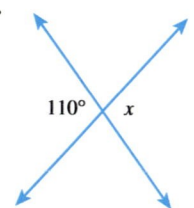

$110°$ x

20.

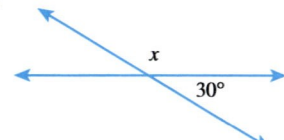

x $30°$

21.

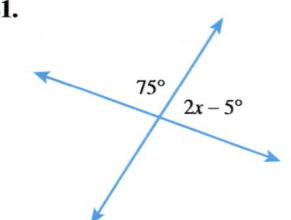

$75°$ $2x - 5°$

22.

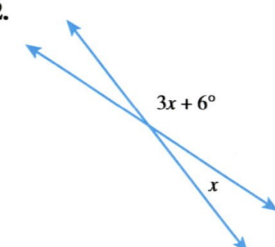

$3x + 6°$ x

23.

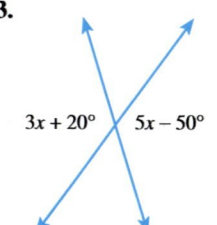

$3x + 20°$ $5x - 50°$

24.

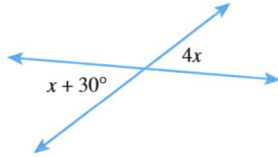

25. In the figure below, P, S, and T are collinear. If $m\angle P = 40°$ and $m\angle QST = 110°$, what is $m\angle Q$? (*Hint:* $m\angle P + m\angle Q + m\angle PSQ = 180°$)

   a. $40°$ **b.** $55°$ **c.** $70°$ **d.** $110°$ **e.** $140°$

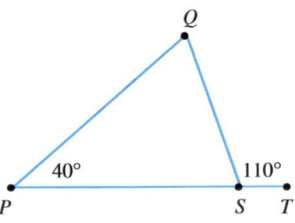

In Exercise 26–29, use the figure below.

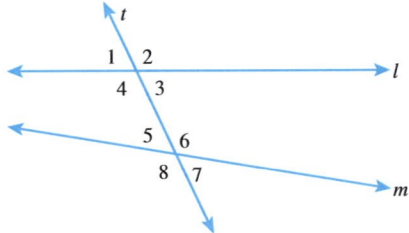

26. Name two corresponding angles.

27. Name two alternate interior angles.

28. Name two alternate exterior angles.

29. Name two consecutive interior angles.

Independent Practice

In Exercises 30–33, $l_1 \parallel l_2$. Find the measures of $\angle 1$ and $\angle 2$. Explain your reasoning.

30.

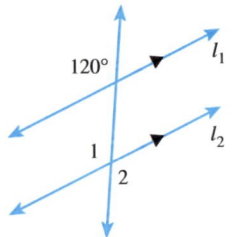

31.

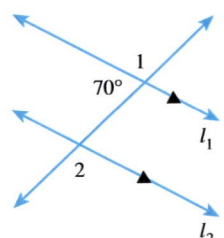

32.

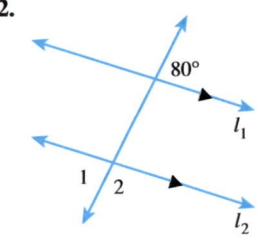

33.

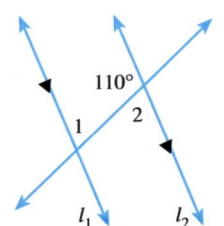

In Exercises 34–36, $m \parallel n$ and $k \parallel l$. Determine the values of a and b.

34.

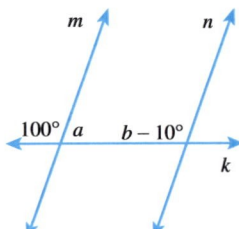

35.

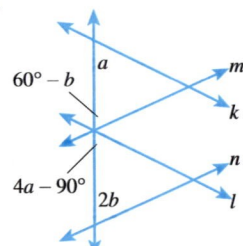

36.

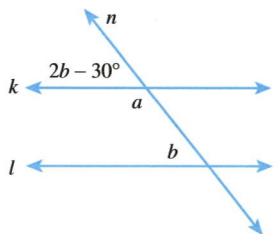

In Exercises 37–39, use the figure below.

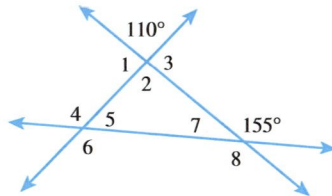

37. Name the interior angles of the triangle.

38. Name the exterior angles of the triangle.

39. Two angle measures are given in the figure. Find the measure of the eight labeled angles.

In Exercises 40–43, use the figure below in which $\triangle ABC \cong \triangle DEF$.

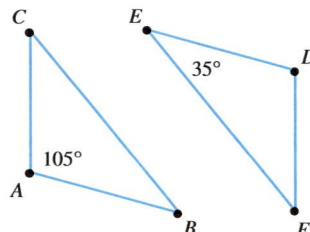

40. What is the measure of $\angle D$?

41. What is the measure of $\angle B$?

42. What is the measure of $\angle C$?

43. What is the measure of $\angle F$?

44. True or False?   A right triangle can have an obtuse angle.

45. True or False?   A triangle that has two $60°$ angles must be equiangular.

46. True or False?   If a right triangle has two congruent angles, then it must have two $45°$ angles.

In Exercises 47 and 48, find the measure of each labeled angle.

47.

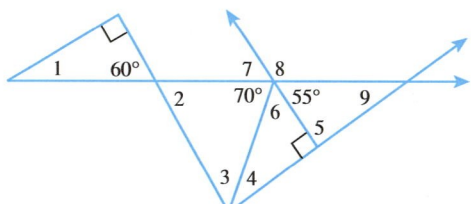

48.

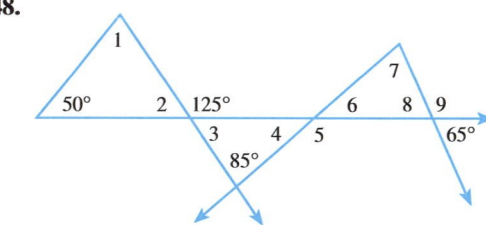

In Exercises 49 and 50, draw and label a right triangle, $\triangle ABC$, for which the right angle is $\angle C$. What is $m\angle B$?

49. $m\angle A = 13°$ **50.** $m\angle A = 47°$

In Exercises 51 and 52, draw two noncongruent, isosceles triangles that have an exterior angle with the given measure.

51. $130°$ **52.** $145°$

Integrated Review

Algebra   In Exercises 53–56, find the measures of the interior angles.

53.
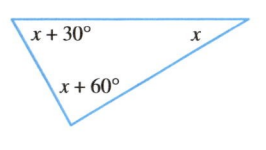

54.
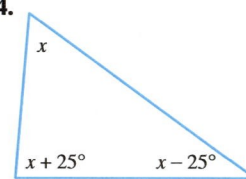

55.
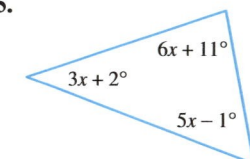

56.
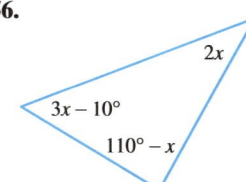

Programs

Programs for the *TI-82* are given in several sections of the text. This appendix contains additional programming hints for the *TI-82* and translations of the text programs for the following graphics calculators from Texas Instruments, Casio, and Sharp:

- *TI-80*
- *TI-81*
- *TI-82*
- *TI-85*
- *Sharp EL-9200*
- *Sharp EL-9300*
- *Casio fx-6300G*
- *Casio fx-7700G*
- *Casio fx-7700GE*
- *Casio fx-9700GE*
- *Casio CFX-9800G.*

Similar programs can be written for other brands and models of calculators.

Enter a program in your calculator, then refer to the text discussion and apply the program as appropriate. Section references are provided to help you locate the text discussion of the programs and their uses.

To illustrate the power and versatility of programmable calculators, a variety of types of programs is presented, including a simulation program (Graph Reflection Program) and a tutorial program (Reflections and Shifts Program).

Evaluating an Algebraic Expression (Section 4.1)

The program, shown in the marginal Programming note on page 218, can be used to evaluate an algebraic expression in one variable at several values of the variable.

Note: On the *TI-82,* the "Lbl" and "Goto" commands may be entered through the "CTL" menu accessed by pressing the PRGM key. The "Disp" and "Input" commands may be entered through the "I/O" menu accessed by pressing the PRGM key. The symbol "Y_1" may be entered through the "Function" menu accessed by pressing the Y-VARS key. Keystroke sequences required for similar commands on other calculators will vary. Consult the user's manual for your calculator.

TI-80
TI-82

PROGRAM:EVALUATE
:Lbl A
:Input "ENTER X ",X
:Disp Y_1
:Goto A

To use this program, enter an expression in Y_1. Expressions may also be evaluated directly on the home screen.

TI-81

Prgm1: EVALUATE
:Lbl 1
:Disp "ENTER X"
:Input X
:Disp Y_1
:Goto 1

To use this program, enter an expression in Y_1. Expressions may also be evaluated on the home screen.

TI-85

PROGRAM:Evaluate
:Lbl A
:Input "Enter x ",x
:Disp y1
:Goto A

To use this program, enter an expression in y1.

Sharp EL-9200
Sharp EL-9300

Evaluate
- - - - - - - - - - - - - - - - REAL
Goto top
Label equation
Y=f(X)
Return
Label top
Input X
Gosub equation
Print Y
Goto top
End

To use this program, replace f(X) with your expression in X.

Casio fx-6300G

EVALUATE:
Lbl 1:
"X="?→X:
"F(X)="◢Prog 0◢
Goto 1

To use this program, write the expression as Prog 0.

Casio fx-7700G

EVALUATE
Lbl 1
"X="?→X
"F(X)=":f₁◢
Goto 1

To use this program, enter an expression in f_1.

Casio fx-7700GE
Casio fx-9700GE
Casio CFX-9800G

EVALUATE↵
Lbl 1↵
"X="?→X↵
"F(X)=":Y1◢
Goto 1

To use this program, enter an expression in Y1. On the *fx-9700GE* and *CFX-9800G,* expressions may also be evaluated in a table from the TABLE & GRAPH MENU.

Simple Interest Program (Section 4.5)

The program, shown in the marginal Programming note on page 257, can be used to find the amount of simple interest earned by a given principal at a given annual interest rate for a certain amount of time.

Note: The program requires a line of code that restricts the displayed result to two decimal places. For instance, the *TI-82* program requires "Fix 2" as the first line. To do this, press the MODE key, cursor to "2" on the line beginning with "Float," and press ENTER. Similarly, the last line of the program may be entered by pressing the MODE key, cursoring to "Float," and pressing ENTER. On the *TI-82,* the → symbol represents pressing the STO▷ key. For additional keystroke instructions, see previous programs in this appendix. Keystroke sequences required for similar commands on other calculators will vary. Consult the user's manual for your calculator.

TI-80
TI-81
TI-82

```
PROGRAM:SIMPINT
:Fix 2
:Disp "PRINCIPAL"
:Input P
:Disp "INTEREST RATE"
:Disp "IN DECIMAL FORM"
:Input R
:Disp "NO. OF YEARS"
:Input T
:PRT→ I
:Disp "THE INTEREST IS"
:Disp I
:Float
```

TI-85

```
PROGRAM:SimpInt
:Fix 2
:Disp "Principal"
:Input P
:Disp "Interest rate"
:Disp "in decimal form"
:Input R
:Disp "No. of years"
:Input T
:P*R*T→ I
:Disp "The interest is"
:Disp I
:Float
```

Sharp EL-9200
Sharp EL-9300

```
SimpInt
- - - - - - - - - - - - - - - - REAL
Input principal
Print "Interest rate
Print "in decimal form
Input rate
Print "No. of years
Input time
interest=principal*rate*time
Print interest
```

Casio fx-6300G

```
SIMPINT:
Fix 2:"PRINCIPAL"?→P:
"INT"◢"RATE AS"◢ "DEC."?→R:
"NO. YEARS"?→T:PRT→I:
"INTEREST IS"◢ I:Norm
```

Casio fx-7700G

SIMPINT
Fix 2
"PRINCIPAL"?→P
"INTEREST RATE"
"IN DECIMAL FORM"?→R
"NO. OF YEARS"?→T
PRT→ I
"THE INTEREST IS":I
Norm

Casio fx-7700GE
Casio fx-9700GE
Casio CFX-9800G

SIMPINT↵
Fix 2↵
"PRINCIPAL"?→P↵
"INTEREST RATE"↵
"IN DECIMAL FORM"?→R↵
"NO. OF YEARS"?→T↵
PRT→ I↵
"THE INTEREST IS":I↵
Norm

Reflections and Shifts Program (Section 7.6)

The program, referenced in the marginal Programming note on page 436, will sketch a graph of the function $y = R(x + H)^2 + V$, where $R = \pm 1$, H is an integer between -6 and 6, and V is an integer between -3 and 3. This program gives you practice working with reflections, horizontal shifts, and vertical shifts.

Note: On the *TI-82,* the "int" and "rand" commands may be entered through the "NUM" and "PRB" menus, respectively, accessed by pressing the MATH key. The "=" and "<" symbols may be entered through the "TEST" menu accessed by pressing the TEST key. Other commands such as "If," "Then," "Else," and "End" may be entered through the "CTL" menu accessed by pressing the PRGM key. The commands "Xmin," "Xmax," "Xscl," "Ymin," "Ymax," and "Yscl" may be entered through the "Window" menu accessed by pressing the VARS key. The commands "DispGraph" and "Pause" may be entered through the "I/O" and "CTL" menus, respectively, accessed by pressing the PRGM key. For additional keystroke instructions, see previous programs in this appendix. Keystroke sequences for similar commands on other calculators will vary. Consult the user's manual for your calculator.

TI-80
TI-82

PROGRAM:PARABOLA
:-6+int (12rand)→H
:-3+int (6rand)→V
:rand→R
:If R<0.5
:Then
:-1→R
:Else
:1→R
:End
:"R(X+H)^2+V"→Y$_1$
:-9→Xmin
:9→Xmax
:1→Xscl
:-6→Ymin
:6→Ymax
:1→Yscl
:DispGraph
:Pause
:Disp "Y=R(X+H)^2+V"
:Disp "R=",R
:Disp "H=",H
:Disp "V=",V

Press ENTER after the graph to display the values of *R, H,* and *V.*

TI-81

Prgm2: PARABOLA
:Rand→H
:-6+Int (12H)→H
:Rand→V
:-3+Int (6V)→V
:Rand→R
:If R<0.5
:-1→R
:If R>0.49
:1→R
:"R(X+H)^2+V"→Y$_1$
:-9→Xmin
:9→Xmax
:1→Xscl
:-6→Ymin
:6→Ymax
:1→Yscl
:DispGraph
:Pause
:Disp "Y=R(X+H)^2+V"
:Disp "R="
:Disp R
:Disp "H="
:Disp H
:Disp "V="
:Disp V
:End

Press ENTER after the graph to display the values of *R, H,* and *V.*

TI-85

PROGRAM:Parabola
:rand→H
:-6+int(12H)→H
:rand→V
:-3+int(6V)→V
:rand→R
:If R<0.5
:-1→R
:If R>0.49
:1→R
:y1=R(x+H)^2+V
:-9→xMin
:9→xMax
:1→xScl
:-6→yMin
:6→yMax
:1→yScl
:DispG
:Pause
:Disp "Y=R(X+H)^2+V"
:Disp "R=",R
:Disp "H=",H
:Disp "V=",V

Press ENTER after the graph to display the values of *R, H,* and *V.*

Casio fx-6300G

PARABOLA:
-6+Int (12Ran#)→H:
-3+Int (6Ran#)→V:
-1→R:Ran#<0.5⇒1→R:
Range -9,9,1,-6,6,1:
Graph Y=R(X+H)^2+V ◢
"Y=R(X+H)^2+V" ◢
"R=" ◢R ◢
"H=" ◢H ◢
"V=" ◢V

Sharp EL-9200
Sharp EL-9300

Parabola
- - - - - - - - - - - - - - - - - - REAL
H=int (random*12)− 6
V=int (random*6)−3
S=(random*2)−1
R=S/abs S
Range -9,9,1,-6,6,1
Graph R(X+H)^2+V
Wait
Print "Y=R(X+H)^2+V
Print R
Print H
Print V
End

Press ENTER after the graph to display the values of *R, H,* and *V.*

Casio fx-7700G

PARABOLA
-6+Int (12Ran#)→H
-3+Int (6Ran#)→V
-1→R:Ran#<0.5⇒1→R
Range -9,9,1,-6,6,1
Graph Y=R(X+H)^2+V◢
"Y=R(X+H)^2+V"
"R=":R ◢
"H=":H ◢
"V=":V

Casio fx-7700GE
Casio fx-9700GE
Casio CFX-9800G

PARABOLA↵
-6+Int (12Ran#)→H↵
-3+Int (6Ran#)→V↵
Ran#→R↵
R<0.5⇒-1→R↵
R≥0.5⇒1→R↵
Range -9,9,1,-6,6,1↵
Graph Y=R(X+H)²+V▲
"Y=R(X+H)²+V"↵
"R=":R▲
"H=":H▲
"V=":V

Press ENTER after the graph to display the values of *R, H,* and *V.*

Quadratic Formula Program (Section 10.3)

The program, shown in the marginal Programming note on page 594, will display the solutions to quadratic equations or the words "No Real Solution" if
complex solutions are not available. To use the program, write the quadratic
equation in standard form and enter the values of *a, b,* and *c.*

Note: On the *TI-82,* the "Prompt" command may be entered through the "I/O"
menu accessed by pressing the PRGM key. For additional keystroke instructions,
see previous programs in this appendix. Keystroke sequences for similar
commands on other calculators will vary. Consult the user's manual for your
calculator.

TI-80

```
PROGRAM:QUADRAT
:Disp "AX²+BX+C=0"
:Input "ENTER A ",A
:Input "ENTER B ",B
:Input "ENTER C ",C
:B² – 4AC→D
:If D≥0
:Then
:(-B+√D)/(2A)→M
:Disp M
:(-B–√D)/(2A)→N
:Disp N
:Else
:Disp "NO REAL SOLUTION"
:End
```

TI-81

```
Prgm3: QUADRAT
:Disp "AX²+BX+C=0"
:Disp "ENTER A"
:Input A
:Disp "ENTER B"
:Input B
:Disp "ENTER C"
:Input C
:B² – 4AC→D
:If D<0
:Goto 1
:((-B+√D)/(2A))→M
:Disp M
:((-B–√D)/(2A))→N
:Disp N
:End
:Lbl 1
:Disp "NO REAL SOLUTION"
:End
```

TI-82

PROGRAM:QUADRAT
:Disp "AX^2+BX+C=0"
:Prompt A
:Prompt B
:Prompt C
:B^2 − 4AC→D
:If D≥0
:Then
:(-B+$\sqrt{D}$)/(2A)→M
:Disp M
:(-B−$\sqrt{D}$)/(2A)→N
:Disp N
:Else
:Disp "NO REAL SOLUTION"
:End

TI-85

PROGRAM:Quadrat
:Disp "AX^2+BX+C=0"
:Input "Enter A ",A
:Input "Enter B ",B
:Input "Enter C ",C
:B^2 − 4*A*C→D
:Disp ((-B+$\sqrt{D}$)/(2A))
:Disp ((-B−$\sqrt{D}$)/(2A))

This program gives both real and complex solutions. Solutions to quadratic equations are also available directly by using the POLY function.

Sharp EL-9200
Sharp EL-9300

Quadratic
- - - - - - - - - - - - - - - COMPLEX
Input A
Input B
Input C
D=B^2 − 4AC
x1=(-B+$\sqrt{D}$)/(2A)
x2=(-B−$\sqrt{D}$)/(2A)
Print x1
Print x2
X=x1
Y=x2
End

This program is written in complex mode, so both real and complex solutions are given. The solutions are also stored under variables X and Y so they can be used in the calculator mode.

Casio fx-6300G

QUADRATIC:
"AX^2+BX+C=0"◢
"A="?→A:
"B="?→B:
"C="?→C:
B^2 − 4AC→D:
D<0⇒Goto1:
"X="◢ (-B+$\sqrt{D}$)÷(2A) ◢
"OR X="◢ (-B−$\sqrt{D}$)÷(2A):
Goto 2:
Lbl 1:
"NO REAL SOLUTION"
Lbl 2

Casio fx-7700G

QUADRATIC
"AX^2+BX+C=0"
"A="?→A
"B="?→B
"C="?→C
B^2 − 4AC→D
D<0⇒Goto1
"X=":(-B+√D)÷(2A) ◢
"OR X=":(-B−√D)÷(2A)
Goto 2
Lbl 1
"NO REAL SOLUTION"
Lbl 2

Casio fx-7700GE
Casio fx-9700GE
Casio CFX-9800G

QUADRATIC↵
"AX^2+BX+C=0"↵
"A="?→A↵
"B="?→B↵
"C="?→C↵
B^2 − 4AC→D↵
(-B+√D)÷(2A)◢
(-B−√D)÷(2A)

Both real and complex solutions are given. Solutions to quadratic equations are also available directly from the EQUATION MENU.

Two-Point Form of a Line (Section 11.1)

The program, shown in the marginal Programming note on page 630, will display the slope and y-intercept for the line that passes through the two points, (x_1, y_1) and (x_2, y_2), entered by the user.

Note: For help with *TI-82* keystrokes, see previous programs in this appendix.

TI-80
TI-81
TI-82

PROGRAM:TWOPTFM
:Disp "ENTER X1,Y1"
:Input X
:Input Y
:Disp "ENTER X2,Y2"
:Input C
:Input D
:(D−Y)/(C−X)→M
:M*(-X)+Y→B
:Disp "SLOPE ="
:Disp M
:Disp "Y-INT ="
:Disp B

TI-85

PROGRAM:TwoPtFrm
:Disp "Enter X1,Y1"
:Input X
:Input Y
:Disp "Enter X2,Y2"
:Input C
:Input D
:(D−Y)/(C−X)→M
:M*(-X)+Y→B
:Disp "Slope ="
:Disp M
:Disp "Y−int ="
:Disp B

Sharp EL-9200
Sharp EL-9300

TwoPtForm
- - - - - - - - - - - - - - - - - REAL
Print "Enter X1,Y1
Input x
c=x
Input y
d=y
Print "Enter X2,Y2
Input x
Input y
m=(d−y)÷(c−x)
b=m∗(-x)+y
Print "Slope
Print m
Print "Y-int
Print b

Casio fx-7700G

TWOPTFORM
"ENTER X1,Y1"?→X:?→Y
"ENTER X2,Y2"?→C:?→D
(D−Y)÷(C−X)→M
M×(-X)+Y→B
"SLOPE =":M◢
"Y-INT =":B

Casio fx-6300G

TWOPTFORM:
"ENTER X1,Y1"?→X:?→Y:
"ENTER X2,Y2"?→C:?→D:
(D−Y)÷(C−X)→M:M×(-X)+Y→B:
"SLOPE ="◢ M ◢"Y-INT ="◢ B

Casio fx-7700GE
Casio fx-9700GE
Casio CFX-9800G

TWOPTFORM↵
"ENTER X1,Y1"?→X:?→Y↵
"ENTER X2,Y2"?→C:?→D↵
(D−Y)÷(C−X)→M↵
M×(-X)+Y→B↵
"SLOPE =":M◢
"Y-INT =":B

Systems of Linear Equations (Section 12.2)

The program, shown in the marginal Programming note on page 704, will display the solution of a system of two linear equations in two variables of the form

$$ax + by = c$$
$$dx + ey = f$$

if a unique solution exists.

Note: For help with *TI-82* keystrokes, see previous programs in this appendix.

TI-80

```
PROGRAM:SOLVE
:Disp "AX+BY=C"
:Input "ENTER A ",A
:Input "ENTER B ",B
:Input "ENTER C ",C
:Disp "DX+EY=F"
:Input "ENTER D ",D
:Input "ENTER E ",E
:Input "ENTER F ",F
:If AE−DB=0
:Then
:Disp "NO SOLUTION"
:Else
:(CE−BF)/(AE−DB)→X
:(AF−CD)/(AE−DB)→Y
:Disp X
:Disp Y
:End
```

TI-81

```
Prgm4: SOLVE
:Disp "AX+BY=C"
:Disp "DX+EY=F"
:Disp "Enter A,B,C,D,E,F""
:Input A
:Input B
:Input C
:Input D
:Input E
:Input F
:If AE−DB=0
:Goto 1
:(CE−BF)/(AE−DB)→X
:(AF−CD)/(AE−DB)→Y
:Disp X
:Disp Y
:End
:Lbl 1
:Disp "NO SOLUTION"
```

TI-82

```
PROGRAM:SOLVE
:Disp "AX+BY=C"
:Prompt A
:Prompt B
:Prompt C
:Disp "DX+EY=F"
:Prompt D
:Prompt E
:Prompt F
:If AE−DB=0
:Then
:Disp "NO SOLUTION"
:Else
:(CE−BF)/(AE−DB)→X
:(AF−CD)/(AE−DB)→Y
:Disp X
:Disp Y
:End
```

TI-85

```
PROGRAM: Solve
:Disp "ax+by=c"
:Input "Enter a ",A
:Input "Enter b ",B
:Input "Enter c ",C
:Disp "dx+ey=f"
:Input "Enter d ",D
:Input "Enter e ",E
:Input "Enter f ",F
:If A*E−D*B==0
:Goto A
:(C*E−B*F)/(A*E−D*B)→X
:(A*F−C*D)/(A*E−D*B)→Y
:Disp X
:Disp Y
:Stop
:Lbl A
:Disp "No Solution"
```

Sharp EL-9200
Sharp EL-9300

```
Solve
- - - - - - - - - - - - - - - - - - REAL
Print "AX+BY=C
Print "DX+EY=F
Input A
Input B
Input C
Input D
Input E
Input F
If A*E−D*B=0 Goto 1
X=(C*E−B*F)/(A*E−D*B)
Y=(A*F−C*D)/(A*E−D*B)
Print X
Print Y
End
Label 1
Print "No solution
End
```

Equations must be entered in the form: $Ax + By = C$; $Dx + Ey = F$. Uppercase letters are used so that the values can be accessed in the calculation mode of the calculator.

Casio fx-6300G

```
SOLVE:
"AX+BY=C":
"DX+EY=F":
"A="?→A:
"B="?→B:
"C="?→C:
"D="?→D:
"E="?→E:
"F="?→F:
AE−DB=0⇒Goto 1:
"X="◢(CE−BF)÷(AE−DB)◢
"Y="◢(AF−CD)÷(AE−DB):
Goto 2:
Lbl 1:
"NO UNIQUE SOLUTION":
Lbl 2
```

Casio fx-7700G

```
SOLVE
"AX+BY=C"
"DX+EY=F"
"ENTER A,B,C,D,E,F"
"A="?→A
"B="?→B
"C="?→C
"D="?→D
"E="?→E
"F="?→F
AE−DB=0⇒Goto 1
"X=":(CE−BF)÷(AE−DB)◢
"Y=":(AF−CD)÷(AE−DB)
Goto 2
Lbl 1
"NO UNIQUE SOLUTION"
Lbl 2
```

Casio fx-7700GE
Casio fx-9700GE
Casio CFX-9800G

```
SOLVE↵
"AX+BY=C"↵
"DX+EY=F"↵
"ENTER A,B,C,D,E,F"↵
"A="?→A↵
"B="?→B↵
"C="?→C↵
"D="?→D↵
"E="?→E↵
"F="?→F↵
AE−DB=0⇒Goto 1↵
"X=":(CE−BF)÷(AE−DB) ◢
"Y=":(AF−CD)÷(AE−DB)↵
Goto 2↵
Lbl 1↵
"NO UNIQUE SOLUTION"↵
Lbl 2
```

Solutions to systems of linear equations are also available directly from the EQUATION MENU.

Graph Reflection Program (Section 13.2)

The program, shown in the marginal Programming note on page 771, will graph a function f and its reflection in the line $y = x$.

   Note: On the *TI-82,* the "While" command may be entered through the "CTL" menu accessed by pressing the PRGM key. The "Pt-On(" command may be entered through the "POINTS" menu accessed by pressing the DRAW key. For additional keystroke instructions, see previous programs in this appendix. Keystroke sequences required for similar commands on other calculators will vary. Consult the user's manual for your calculator.

TI-80

```
PROGRAM:REFLECT
:47Xmin/63→Ymin
:47Xmax/63→Ymax
:Xscl→Yscl
:"X"→Y₂
:DispGraph
:(Xmax−Xmin)/62→N
:Xmin→X
:Lbl A
:Pt-On(Y₁,X)
:X+N→X
:If X>Xmax
:Stop
:Goto A
```

To use this program, enter the function in Y_1 and set a viewing rectangle.

TI-81

```
Prgm5:REFLECT
:2Xmin/3→Ymin
:2Xmax/3→Ymax
:Xscl→Yscl
:"X"→Y₂
:DispGraph
:(Xmax−Xmin)/95→N
:Xmin→X
:Lbl 1
:Pt-On(Y₁,X)
:X+N→X
:If X>Xmax
:End
:Goto 1
```

To use this program, enter the function in Y_1 and set a viewing rectangle.

TI-82

PROGRAM:REFLECT
:63Xmin/95→Ymin
:63Xmax/95→Ymax
:Xscl→Yscl
:"X"→Y_2
:DispGraph
:(Xmax−Xmin)/94→N
:Xmin→X
:While X≤Xmax
:Pt-On(Y_1,X)
:X+N→X
:End

To use this program, enter the function in Y_1 and set a viewing rectangle.

TI-85

PROGRAM:Reflect
:63xMin/127→yMin
:63xMax/127→yMax
:xScl→yScl
:y2=x
:DispG
:(xMax−xMin)/126→N
:xMin→x
:Lbl A
:PtOn(y1,x)
:x+N→X
:If x>xMax
:Stop
:Goto A

To use this program, enter the function in y1 and set a viewing rectangle.

Sharp EL-9200
Sharp EL-9300

Reflection
- - - - - - - - - - - - - - - - - - REAL
Goto top
Label eqtn
Y=X^3+X+1
Return
Label rng
xmin=-10
xmax=10
xstp=(xmax−xmin)/10
ymin=2xmin/3
ymax=2xmax/3
ystp=xstp
Range xmin,xmax,xstp,ymin,
    ymax,ystp
Return
Label top
Gosub rng
Graph X
step=(xmax−xmin)/(94*2)
X=xmin
Label 1
Gosub eqtn
Plot X,Y
Plot Y,X
X=X+step
If X<=xmax Goto 1
End

To use this program, enter a function in X in the third line of the program.

Casio fx-6300G

REFLECTION
"GRAPH -A to A"◢
A="?→A:
Range -A,A,1,-2A÷3,2A÷3,1:
-A→B:
Lbl 1:
B→X:
Prog 0:
Ans→Y:
Plot B,Y:
B+A÷24→B:
B≤A⇒Goto 1:
-A→B:
Lbl 2:
B→X:
Prog 0:
Ans→Y:
Plot Y,B:
B+A÷24→B:
B≤A⇒Goto 2:Graph Y=X

To use this program, write the function as Prog 0 and set a viewing rectangle.

Casio fx-7700G

REFLECTION
"GRAPH -A TO A"
"A="?→A
Range -A,A,1,-2A÷3,2A÷3,1
Graph Y=f_1
-A→B
Lbl 1
B→X
Plot f_1,B
B+A÷32→B
B≤A⇒Goto 1:Graph Y=X

To use this program, enter the function in f_1 and set a viewing rectangle.

Casio fx-7700GE
Casio fx-9700GE

REFLECTION↵
63Xmin÷127→A↵
63Xmax÷127→B↵
Xscl→C↵
Range , , ,A,B,C↵
(Xmax−Xmin)÷126→N↵
Xmax→M↵
Xmin→D↵
Graph Y=f_1↵
Lbl 1↵
D→X↵
Plot f_1,D↵
D+N→D↵
D≤M⇒Goto 1:Graph Y=X

To use the program, enter a function in f_1 and set a viewing rectangle.

Casio CFX-9800G

REFLECTION↵
63Xmin÷95→A↵
63Xmax÷95→B↵
Xscl→C↵
Range , , ,A,B,C↵
(Xmax−Xmin)÷94→N↵
Xmax→M↵
Xmin→D↵
Graph Y=f_1↵
Lbl 1↵
D→X↵
Plot f_1,D↵
D+N→D↵
D≤M⇒Goto 1:Graph Y=X

To use the program, enter a function in f_1 and set a viewing rectangle.

Appendix E	**Absolute Value Equations and Inequalities**
	Solving Equations Involving Absolute Value ▪ Solving Inequalities Involving Absolute Value

Solving Equations Involving Absolute Value

Consider the **absolute value equation**

$$|x| = 3.$$

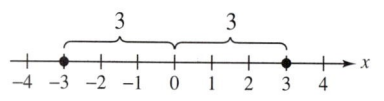

FIGURE E.1

The only solutions of this equation are -3 and 3, because these are the only two real numbers whose distance from zero is 3. (See Figure E.1.) In other words, the absolute value equation $|x| = 3$ has exactly two solutions: $x = -3$ and $x = 3$.

Solving an Absolute Value Equation

Let x be a variable or a variable expression and let a be a real number such that $a \geq 0$. The solutions of the equation $|x| = a$ are given by $x = -a$ and $x = a$. That is,

$$|x| = a \implies x = -a \quad \text{and} \quad x = a.$$

EXAMPLE 1 *Solving Absolute Value Equations*

Solve each absolute value equation.

a. $|x| = 10$ **b.** $|x| = 0$ **c.** $|y| = -1$

Solution

a. This equation is equivalent to the two linear equations

$$x = -10 \quad \text{and} \quad x = 10. \qquad \text{Equivalent linear equations}$$

Thus, the absolute value equation has two solutions: -10 and 10.

b. This equation is equivalent to the two linear equations

$$x = 0 \quad \text{and} \quad x = 0. \qquad \text{Equivalent linear equations}$$

Because both equations are the same, you can conclude that the absolute value equation has only one solution: 0.

c. This absolute value equation has *no solution* because it is not possible for the absolute value of a real number to be negative.

STUDY TIP

The strategy for solving absolute value equations is to *rewrite* the equation in *equivalent forms* that can be solved by previously learned methods. This is a common strategy in mathematics. That is, when you encounter a new type of problem, you try to rewrite the problem so that it can be solved by techniques you already know.

EXAMPLE 2   Solving Absolute Value Equations

Solve $|3x + 4| = 10$.

Solution

$$|3x + 4| = 10 \qquad \text{Original equation}$$

$$3x + 4 = -10 \quad \text{or} \quad 3x + 4 = 10 \qquad \text{Equivalent equations}$$

$$3x + 4 - 4 = -10 - 4 \qquad 3x + 4 - 4 = 10 - 4 \qquad \text{Subtract 4 from both sides.}$$

$$3x = -14 \qquad\qquad 3x = 6 \qquad \text{Combine like terms.}$$

$$x = -\frac{14}{3} \qquad\qquad x = 2 \qquad \text{Divide both sides by 3.}$$

The solutions are $-\frac{14}{3}$ and 2. Check these in the original equation.

NOTE   When you are solving absolute value equations, remember that it is possible that they have no solution. For instance, the equation

$$|3x + 4| = -10$$

has no solution because the absolute value of a real number cannot be negative. Do not make the mistake of trying to solve such an equation by writing the "equivalent" linear equations as $3x + 4 = -10$ and $3x + 4 = 10$. These equations have solutions, but they are both extraneous.

The equation in the next example is not given in the **standard form**

$$|ax + b| = c, \qquad c \geq 0.$$

Notice that the first step in solving such an equation is to write it in standard form.

EXAMPLE 3   An Absolute Value Equation in Nonstandard Form

Solve $|2x - 1| + 3 = 8$.

Solution

$$|2x - 1| + 3 = 8 \qquad \text{Original equation}$$

$$|2x - 1| = 5 \qquad \text{Standard form}$$

$$2x - 1 = -5 \quad \text{or} \quad 2x - 1 = 5 \qquad \text{Equivalent equations}$$

$$2x = -4 \qquad\qquad 2x = 6 \qquad \text{Add 1 to both sides.}$$

$$x = -2 \qquad\qquad x = 3 \qquad \text{Divide both sides by 2.}$$

The solutions are -2 and 3. Check these in the original equation.

If two algebraic expressions are equal in absolute value, they must either be *equal* to each other or be the *opposites* of each other. Thus, you can solve equations of the form

$$|ax + b| = |cx + d|$$

by forming the two linear equations

Expressions opposite

$$ax + b = -(cx + d) \quad \text{and} \quad ax + b = cx + d.$$

Expressions equal

EXAMPLE 4 Solving an Equation Involving Absolute Values

Solve $|3x - 4| = |7x - 16|$.

Solution

$$|3x - 4| = |7x - 16| \qquad \text{Original equation}$$

$$3x - 4 = -(7x - 16) \quad \text{or} \quad 3x - 4 = 7x - 16 \qquad \text{Equivalent equations}$$

$$3x - 4 = -7x + 16 \qquad\qquad 3x = 7x - 12$$

$$10x = 20 \qquad\qquad\qquad -4x = -12$$

$$x = 2 \qquad\qquad\qquad\quad x = 3 \qquad \text{Solutions}$$

The solutions are 2 and 3. Check these in the original equation.

EXAMPLE 5 Solving an Equation Involving Absolute Values

NOTE When solving equations of the form

$$|ax + b| = |cx + d|$$

it is possible that one of the resulting equations will not have a solution. Note this occurrence in Example 5.

Solve $|x + 5| = |x + 11|$.

Solution

By equating the expression $(x + 5)$ to the opposite of $(x + 11)$, you obtain

$$x + 5 = -(x + 11)$$

$$x + 5 = -x - 11$$

$$x = -x - 16$$

$$2x = -16$$

$$x = -8.$$

However, by setting the two expressions equal to each other, you obtain

$$x + 5 = x + 11$$

$$x = x + 6$$

$$0 = 6$$

which makes no sense. Therefore, the original equation has only one solution: -8. Check this solution in the original equation.

Solving Inequalities Involving Absolute Value

To see how to solve inequalities involving absolute value, consider the following comparisons.

$$|x| = 2 \qquad\qquad |x| < 2 \qquad\qquad |x| > 2$$

$$x = -2 \text{ and } x = 2 \qquad -2 < x < 2 \qquad x < -2 \text{ or } x > 2$$

These comparisons suggest the following rule for solving inequalities involving absolute value.

Solving an Absolute Value Inequality

Let x be a variable or an algebraic expression and let a be a real number such that $a > 0$.

1. The solutions of $|x| < a$ are all values of x that lie between $-a$ and a. That is,

$$|x| < a \quad \text{if and only if} \quad -a < x < a.$$

2. The solutions of $|x| > a$ are all values of x that are *less than* $-a$ or *greater than* a. That is,

$$|x| > a \quad \text{if and only if} \quad x < -a \text{ or } x > a.$$

These rules are also valid if $<$ is replaced by $\leq$ and $>$ is replaced by $\geq$.

Technology

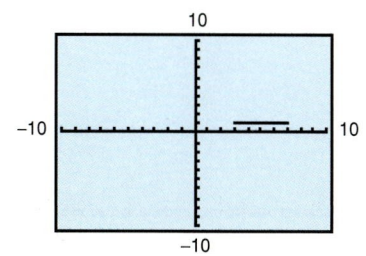

Most graphing utilities can sketch the graph of an absolute value inequality. For instance, the following steps show how to sketch the graph of

$$|x - 5| < 2$$

on a *TI-82*.

Y= Y$_1$= ABS (X−5) < 2

ZOOM 6

The graph produced by these steps is shown at the left. Notice that the graph occurs as an interval *above* the x-axis.

EXAMPLE 6   Solving an Absolute Value Inequality

Solve $|x - 5| < 2$.

Solution

$$|x - 5| < 2$$ Original inequality

$$-2 < x - 5 < 2$$ Equivalent double inequality

$$-2 + 5 < x - 5 + 5 < 2 + 5$$ Add 5 to all three parts.

$$3 < x < 7$$ Combine like terms.

The solution set consists of all real numbers that are greater than 3 and less than 7. The interval notation for this solution set is $(3, 7)$. The graph of this solution set is shown in Figure E.2.

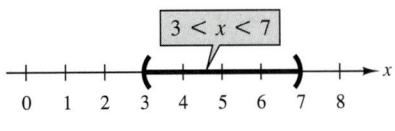

FIGURE E.2

STUDY TIP

In Example 6, note that absolute value inequalities of the form $|x| < a$ (or $|x| \le a$) can be solved with a double inequality. Inequalities of the form $|x| > a$ (or $|x| \ge a$) cannot be solved with a double inequality. Instead, you must solve two separate inequalities, as demonstrated in Example 7.

EXAMPLE 7   Solving an Absolute Value Inequality

Solve $|3x - 4| \ge 5$.

Solution

$$|3x - 4| \ge 5$$ Original inequality

$$3x - 4 \le -5 \quad \text{or} \quad 3x - 4 \ge 5$$ Equivalent inequalities

$$3x - 4 + 4 \le -5 + 4 \qquad 3x - 4 + 4 \ge 5 + 4$$ Add 4 to both sides.

$$3x \le -1 \qquad\qquad 3x \ge 9$$ Combine like terms.

$$\frac{3x}{3} \le \frac{-1}{3} \qquad\qquad \frac{3x}{3} \ge \frac{9}{3}$$ Divide both sides by 3.

$$x \le -\frac{1}{3} \qquad\qquad x \ge 3$$ Simplify.

The solution set consists of all real numbers that are less than or equal to $-\frac{1}{3}$ or greater than or equal to 3. The interval notation for the solution set is $\left(-\infty, -\frac{1}{3}\right] \cup [3, \infty)$. The symbol $\cup$ is called a **union** symbol, and it is used to denote the combining of two sets. The graph is shown in Figure E.3.

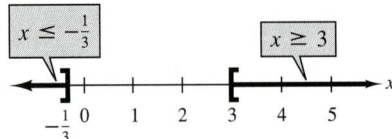

FIGURE E.3

EXAMPLE 8   Solving an Absolute Value Inequality

$\left\|2 - \dfrac{x}{3}\right\| \le 0.01$	Original inequality
$-0.01 \le 2 - \dfrac{x}{3} \le 0.01$	Equivalent double inequality
$-2.01 \le -\dfrac{x}{3} \le -1.99$	Subtract 2 from all three parts.
$6.03 \ge x \ge 5.97$	Multiply all three parts by -3 and reverse both inequality symbols.
$5.97 \le x \le 6.03$	Solution set in standard form

The solution set consists of all real numbers that are greater than or equal to 5.97 and less than or equal to 6.03. The interval notation for the solution set is [5.97, 6.03]. The graph is shown in Figure E.4.

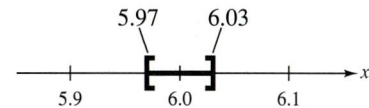

FIGURE E.4

Group Activities   You Be the Instructor

Error Analysis   Suppose you are teaching a class in algebra and one of your students hands in the following solution.

$$|3x - 4| \ge -5$$

$3x - 4 \le -5$	or	$3x - 4 \ge 5$
$3x - 4 + 4 \le -5 + 4$		$3x - 4 + 4 \ge 5 + 4$
$3x \le -1$		$3x \ge 9$
$\dfrac{3x}{3} \le \dfrac{-1}{3}$		$\dfrac{3x}{3} \ge \dfrac{9}{3}$
$x \le -\dfrac{1}{3}$		$x \ge 3$

What is wrong with this solution? What could you say to help your students avoid this type of error?

E Exercises

Discussing the Concepts

1. Give a graphical description of the absolute value of a real number.

2. Give an example of an absolute value equation that has only one solution.

3. In your own words, explain how to solve an absolute value equation. Illustrate your explanation with an example.

4. Describe the solution of the inequality $|x| > 3$.

5. The graph of the inequality $|x - 3| < 2$ can be described as *all real numbers that are within two units of 3*. Give a similar description of $|x - 4| < 1$.

6. The graph of the inequality $|y - 1| > 3$ can be described as *all real numbers greater than three units from 1*. Give a similar description of $|y + 2| > 4$.

Problem Solving

In Exercises 7–10, determine whether the value is a solution of the equation.

	Equation	*Value*
7.	$|4x + 5| = 10$	$x = -3$
8.	$|2x - 16| = 10$	$x = 3$
9.	$|6 - 2w| = 2$	$w = 4$
10.	$\left|\frac{1}{2}t + 4\right| = 8$	$t = 6$

In Exercises 11–14, transform the absolute value equation into two linear equations.

11. $|x - 10| = 17$ **12.** $|7 - 2t| = 5$

13. $|4x + 1| = \frac{1}{2}$ **14.** $|22k + 6| = 9$

In Exercises 15–24, solve the equation. (Some of the equations may have no solution.)

15. $|t| = 45$ **16.** $|s| = 16$

17. $|2s + 3| = 25$ **18.** $|7a + 6| = 8$

19. $|3x + 4| = -16$ **20.** $4|x + 5| = 9$

21. $\left|\frac{2}{3}x + 4\right| = 9$ **22.** $|3.2 - 1.054x| = 2$

23. $|x + 2| = |3x - 1|$ **24.** $|x - 2| = |2x - 15|$

Think About It In Exercises 25 and 26, write a single equation that is equivalent to the two equations.

25. $2x + 3 = 5, 2x + 3 = -5$

26. $4x - 6 = 7, 4x - 6 = -7$

In Exercises 27–30, determine whether the x-values are solutions of the inequality.

	Inequality	*Values*	
27.	$|x| < 3$	(a) $x = 2$	(b) $x = -4$
		(c) $x = 4$	(d) $x = -1$
28.	$|x| \le 5$	(a) $x = -7$	(b) $x = -4$
		(c) $x = 4$	(d) $x = 9$
29.	$|x - 7| \ge 3$	(a) $x = 9$	(b) $x = -4$
		(c) $x = 11$	(d) $x = 6$
30.	$|x - 3| > 5$	(a) $x = 16$	(b) $x = 3$
		(c) $x = -2$	(d) $x = -3$

In Exercises 31–34, transform the absolute value inequality into a double inequality or two separate inequalities.

31. $|y + 5| < 3$ **32.** $|6x + 7| \le 5$

33. $|7 - 2h| \ge 9$ **34.** $|8 - x| > 25$

In Exercises 35–44, solve the inequality and sketch the graph of the solution.

35. $|y| < 4$ **36.** $|x| < 6$

37. $|y + 2| < 4$ **38.** $|x + 3| < 6$

39. $|x| > 6$ **40.** $|y| \ge 8$

41. $|y + 2| \ge 4$ **42.** $|x + 3| \ge 6$

43. $|2x + 3| > 9$ **44.** $|7r - 3| > 11$

In Exercises 45–48, solve the inequality. (If it is not possible, state the reason.)

45. $|0.2x - 3| < 4$

46. $|1.5t - 8| \leq 16$

47. $\left| \dfrac{z}{10} - 3 \right| > 8$

48. $\left| \dfrac{x}{8} + 1 \right| < 0$

49. *Body Temperature* Physicians consider an adult's body temperature to be normal if it is between 97.6°F and 99.6°F. Write an absolute value inequality that describes this normal temperature range.

50. *Accuracy of Measurements* In woodshop class, you must cut several pieces of wood within $\frac{3}{16}$ inch of the instructor's specifications. Let $(s - x)$ represent the difference between the specification s and the measured length x of a cut piece.

(a) Write an absolute value inequality that describes the values of x that are within specifications.

(b) The length of one of the pieces of wood is specified to be $s = 5\frac{1}{8}$ inches. Describe the acceptable lengths for this piece.

Reviewing the Major Concepts

51. What is $7\frac{1}{2}\%$ of 25?

52. What is 150% of 6000?

53. 225 is what percent of 150?

54. What percent of 240 is 160?

55. 0.5% of what number is 400?

56. 48% of what number is 132?

Additional Problem Solving

In Exercises 57–72, solve the equation. (Some of the equations may have no solution.)

57. $|h| = 0$

58. $|x| = -82$

59. $|x - 16| = 5$

60. $|z - 100| = 100$

61. $|32 - 3y| = 16$

62. $|3 - 5x| = 13$

63. $|3x - 2| = -5$

64. $|4 - 3x| = 0$

65. $|5x - 3| + 8 = 22$

66. $|20 - 5t| = 50$

67. $|0.32x - 2| = 4$

68. $\left| \dfrac{3}{2} - \dfrac{4}{5}x \right| = 1$

69. $|x + 8| = |2x + 1|$

70. $|10 - 3x| = |x + 7|$

71. $|5x + 4| = |3x + 8|$

72. $|45 - 4x| = |32 - 3x|$

In Exercises 73–76, determine whether the x-values are solutions of the inequality.

Inequality	Values
73. $|x| \geq 3$	(a) $x = 2$ (b) $x = -4$
	(c) $x = 4$ (d) $x = -1$
74. $|x| > 5$	(a) $x = -7$ (b) $x = -4$
	(c) $x = 4$ (d) $x = 9$
75. $|x - 7| < 3$	(a) $x = 9$ (b) $x = -4$
	(c) $x = 11$ (d) $x = 6$
76. $|x - 3| \leq 5$	(a) $x = 16$ (b) $x = 3$
	(c) $x = -2$ (d) $x = -3$

In Exercises 77–80, sketch a graph that represents the statement.

77. All x greater than -2 *and* less than 5

78. All x greater than or equal to 3 *and* less than 10

79. All x less than or equal to 4 *or* greater than 7

80. All x less than -6 *or* greater than or equal to 6

In Exercises 81–94, solve the inequality. Use a graphing utility to verify the solution.

81. $|y - 2| \leq 4$

82. $|x - 3| \leq 6$

83. $|y| \geq 4$

84. $|x| \geq 6$

85. $|y - 2| > 4$

86. $|x - 3| > 6$

87. $|2x| < 14$

88. $|4z| \leq 9$

89. $\left| \dfrac{y}{3} \right| \leq 3$

90. $\left| \dfrac{t}{2} \right| < 4$

91. $|3x + 2| > 4$

92. $|2x - 1| \geq 3$

93. $|x - 5| + 3 > 5$

94. $|a + 1| - 4 < 0$

Think About It In Exercises 95 and 96, complete the statement so that the solution is $0 \leq x \leq 6$.

95. $|x - 3| \leq$ ▮

96. $|2x - 6| \leq$ ▮

In Exercises 97–102, solve the inequality. (If it is not possible, state the reason.)

97. $\dfrac{|x + 2|}{10} \le 8$ **98.** $\dfrac{|y - 16|}{4} < 30$

99. $|6t + 15| \ge 30$ **100.** $|3t + 1| > 5$

101. $\dfrac{|s - 3|}{5} > 4$ **102.** $\dfrac{|a + 6|}{2} \ge 16$

In Exercises 103–106, match the inequality with its graph. [The graphs are labeled (a), (b), (c), and (d).]

(a)
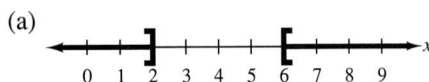

(b)

(c)

(d)

103. $|x - 4| \le 4$ **104.** $|x - 4| < 1$

105. $\frac{1}{2}|x - 4| > 4$ **106.** $|2(x - 4)| \ge 4$

In Exercises 107–112, write an absolute value inequality that represents the given interval.

107.
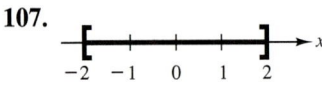

108.

109.

110.

111.

112.

113. *Temperature* The operating temperature for an electronic device must satisfy the inequality

$|t - 72| < 10$

where t is given in degrees Fahrenheit. Sketch the graph of the solution of the inequality.

114. *Time Study* A time study was conducted to determine the length of time required to perform a particular task in a manufacturing process. The time required by approximately two-thirds of the workers in the study satisfied the inequality

$\left| \dfrac{t - 15.6}{1.9} \right| < 1$

where t is time in minutes. Sketch the graph of the solution of the inequality.

In Exercises 115–118, write an absolute value inequality that represents the verbal statement.

115. The set of all real numbers x whose distance from 0 is less than 3

116. The set of all real numbers x whose distance from 0 is more than 2

117. The set of all real numbers x whose distance from 5 is more than 6

118. The set of all real numbers x whose distance from 16 is less than 5

PHOTO CREDITS

CHAPTER P

Section P.1 *(page 9)*

7. (a) $2, \frac{9}{3}$ (b) $-3, 2, \frac{9}{3}$ (c) $-3, 2, -\frac{3}{2}, \frac{9}{3}, 4.5$

9. $2 < 5$ **11.** $-\frac{9}{2} < -3$

13. $4 > -\frac{7}{2}$ **15.** $-4.6 < 1.5$

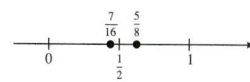

17. 4 **19.** 3 **21.** 3.4 **23.** -23.6

25. $|-4| > |3|$ **27.** $-|-48.5| < |-48.5|$

29.   **31.** $-4.5, \ 20.5$

33. (a) $\frac{8}{4}$ (b) $\frac{8}{4}$ (c) $-\frac{5}{2}, 6.5, -4.5, \frac{8}{4}, \frac{3}{4}$

35. $-4 < -1$ **37.** $-2 < \frac{3}{2}$

39. $\frac{1}{3} < 4$ **41.** $-2\pi > -10$

43. $\frac{7}{16} < \frac{5}{8}$ **45.** $0 > -\frac{7}{16}$

47. 3 **49.** -5 **51.** $\frac{5}{2}$ **53.** 7 **55.** $\frac{7}{2}$

57. -4.09 **59.** -3.2 **61.** $|-15| = |15|$

63. $|32| < |-50|$ **65.** $\left|\frac{3}{16}\right| < \left|\frac{3}{2}\right|$

67. $|-\pi| > -|-2\pi|$ **69.**

71. $-5.5, \ 1.5$ **73.** False. $|0| = 0$

75. True   **77.** True

Section P.2 *(page 16)*

7. 165 **9.** 100 **11.** Composite   **13.** Prime

15. $2 \cdot 3 \cdot 5 \cdot 7$ **17.** $3 \cdot 5 \cdot 5 \cdot 7$ **19.** 5

21. 6 **23.** $12, 24, 36, 48, 60$ **25.** 90 **27.** 126

29. $63 = 3 \cdot 3 \cdot 7$ **31.** 2. It is divisible only by
     $1375 = 5 \cdot 5 \cdot 5 \cdot 11$ 1 and itself.

33. 1955 **35.** 5525 **37.** 2744 **39.** Prime

41. Composite   **43.** Prime   **45.** $2 \cdot 2 \cdot 2 \cdot 3 \cdot 5$

47. $2 \cdot 2 \cdot 2 \cdot 2 \cdot 2 \cdot 2 \cdot 3$ **49.** $3 \cdot 5 \cdot 13 \cdot 13$

51. 4 **53.** 2 **55.** 1 **57.** $14, 28, 42, 56, 70$

59. 30 **61.** 300 **63.** 196 **65.** Relatively prime

67. Not relatively prime.   **69.** Not relatively prime.
     Common factor of 31 Common factor of 17

71. $114, 115, 116, 117, 118, 119, 120, 121, 122, 123$

73. (a)
```
 1  2  3  4  5  6  7  8  9  10
11 12 13 14 15 16 17 18 19 20
21 22 23 24 25 26 27 28 29 30
31 32 33 34 35 36 37 38 39 40
41 42 43 44 45 46 47 48 49 50
51 52 53 54 55 56 57 58 59 60
61 62 63 64 65 66 67 68 69 70
71 72 73 74 75 76 77 78 79 80
81 82 83 84 85 86 87 88 89 90
91 92 93 94 95 96 97 98 99 100
```
   (b) Prime

Mid-Chapter Quiz *(page 18)*

1. $-2.5 > -4$ **2.** $\frac{3}{16} < \frac{3}{8}$

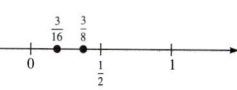

3. $-3.1 < 2.7$ **4.** $2\pi > 6$

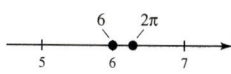

5. -0.75 **6.** 25.2

7. $\left|\frac{7}{2}\right| = |-3.5|$ **8.** $\left|\frac{3}{4}\right| > -|0.75|$

9. $\frac{3}{2}, -\frac{5}{2}$ **10.** $\frac{3}{4}, -\frac{9}{4}$

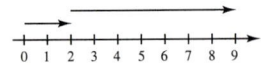

11. 11,011   **12.** 3335   **13.** Prime

14. Composite   **15.** $2 \cdot 3 \cdot 59$ **16.** $3 \cdot 3 \cdot 3 \cdot 5 \cdot 7$

17. 10   **18.** 69   **19.** 84   **20.** 90

Section P.3 *(page 23)*

7. 9   **9.** -2

11. -19 **13.** 30   **15.** -16 **17.** 32

19. $12 + (-9) = 3$ **21.** $-4 + 4 = 0$ **23.** 35

25. 1500   **27.** -610 **29.** 16   **31.** -50

33. 500   **35.** $12°F$ **37.** 7000 feet   **39.** 0

41. 8   **43.** -30 **45.** -16 **47.** -36

49. 2225   **51.** 4558   **53.** 898   **55.** 99

57. -1034 **59.** -6 **61.** 0   **63.** 8   **65.** 37

67. -17 **69.** 12   **71.** 0   **73.** -13 **75.** -11

77. 1997   **79.** -38 **81.** 35   **83.** 1000

85. 1550   **87.** 125   **89.** -15 **91.** $77

93.

Day	Tuesday	Wednesday	Thursday	Friday
Daily Gain/Loss	+21	+16	−17	−28

95. (a) 21,000   (b) 131,000   (c) 8000

Section P.4 *(page 34)*

7. $2 + 2 + 2 = 6$ **9.** 210   **11.** -32 **13.** 90

15. -36 **17.** $-47,400$ **19.** 3   **21.** -6

23. Division by zero is undefined.   **25.** 4   **27.** -32

29. 47   **31.** 3   **33.** -20 **35.** $-24°$

37. (a) 82 points

(b)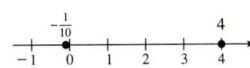

(c) $73 - 82 = -9$
$77 - 82 = -5$
$87 - 82 = 5$
$91 - 82 = 9$
The sum is 0.

39. $5 + 5 + 5 + 5 = 20$

41. $(-3) + (-3) + (-3) + (-3) + (-3) = -15$

43. 72   **45.** -48 **47.** -9920 **49.** -30

51. 90   **53.** 0   **55.** 3380   **57.** $-62,352$

59. 6   **61.** 27   **63.** -29 **65.** 0   **67.** -6

69. Division by zero is undefined.   **71.** -26

73. 47   **75.** 110   **77.** -3 **79.** -5 **81.** 4

83. 4540   **85.** 86   **87.** 1045   **89.** $-532,000$

91. $6000   **93.** $208   **95.** 1280 square feet

97. 594 cubic inches   **99.** $-7°$ per hour

101. Negative   **103.** Answers will vary.

Review Exercises *(page 39)*

1. $-\frac{1}{10} < 4$ **3.** $-3 > -7$

5. -152 **7.** $\frac{7}{3}$ **9.** 8.5   **11.** -8.5

13. $|-84| = |84|$ **15.** $\left|\frac{3}{10}\right| > -\left|\frac{4}{5}\right|$ **17.** Prime

19. Composite   **21.** $2 \cdot 3 \cdot 3 \cdot 3 \cdot 7$

23. $2 \cdot 2 \cdot 13 \cdot 31$ **25.** 18   **27.** 21   **29.** 144

31. 840   **33.** 100   **35.** 11   **37.** 240

39. -268 **41.** -38 **43.** 140   **45.** 45

47. 1500   **49.** -2358 **51.** -18

53. Division by zero is undefined.   **55.** 0

57. 7   **59.** -3 **61.** -469 **63.** $-15,869$

65. 789   **67.** 1162   **69.** -102

71. George Halas, Don Shula, Tom Landry, Earl Lambeau, Chuck Noll, Paul Brown, Bud Grant, Chuck Knox, Steve Owen, Hank Stram

73. (a) 6   (b) 2178

75. (a) $5030, $3090, $4510, $5700

(b) $14,850, $3480

(c) $18,330

77. 32,000 miles   **79.** 0.6

81. False. If the absolute value of the negative integer is less than the positive integer, the sum will be positive.

83. (a) 86,500,000

(b) 45,000,000

(c) 77,700,000, 81,500,000, 86,500,000, 92,500,000, 91,500,000, 104,500,000. Increased.

(d) The bar representing the winning candidate must be taller than the stacked bars of the other candidates. No.

Chapter Test (page 42)

1. (a) 8, $\frac{12}{4}$ (b) -10, 8, $\frac{12}{4}$ (c) -10, 8, $\frac{3}{4}$, $\frac{12}{4}$, 6.5

2. $-\frac{3}{5} > -|-2|$ **3.** -8, 28   **4.** $2 \cdot 3 \cdot 3 \cdot 13$

5. 30   **6.** 168   **7.** -4 **8.** 10

9. 10   **10.** 47   **11.** -160 **12.** 8

13. 3   **14.** 1   **15.** $450,450

16. 128 cubic feet   **17.** 15 feet   **18.** $46.7 million

19. False. The sum of two positive numbers is positive.

20. Each odd number can be written as an even number plus 1. (For example, $17 = 16 + 1$.) Therefore, the sum of two odd numbers can be written as the sum of two even numbers plus 2, which is even.

CHAPTER 1

Section 1.1 (page 52)

7. $\frac{1}{4}$ **9.** $\frac{5}{16}$ **11.** $\frac{3}{5}$ **13.** $\frac{3}{5}$ **15.** $\frac{3}{5}$ **17.** $-\frac{1}{2}$

19. 6   **21.** $\frac{5}{6}$ **23.** $-\frac{1}{24}$ **25.** $\frac{4}{3}$ **27.** $\frac{5}{6}$

29. $\frac{23}{5}$ **31.** $\frac{37}{10}$ **33.** $\frac{55}{6}$ **35.** $-\frac{53}{12}$ **37.** $\frac{3}{10}$

39. $\$1\frac{5}{8}$ **41.** 14   **43.** -23 **45.** $332,050

47. $\frac{2}{3}$ **49.** $\frac{2}{25}$ **51.** $\frac{14}{11}$ **53.** $\frac{3}{8}$ **55.** -1

57. $\frac{4}{3}$ **59.** $-\frac{1}{12}$ **61.** $\frac{9}{16}$ **63.** $-\frac{1}{12}$ **65.** $-\frac{41}{24}$

67. $\frac{17}{48}$ **69.** $-\frac{5}{4}$ **71.** $\frac{13}{12}$ **73.** $\frac{26}{3}$ **75.** $-\frac{115}{11}$

77. $-\frac{17}{16}$ **79.** $-\frac{121}{12}$ **81.** $\frac{1013}{40} = 25.325$ tons

83. $\frac{5}{8}$ **85.** ≈ 1 **87.** $\frac{3}{6} + \frac{4}{5} = \frac{13}{10}$

Section 1.2 (page 63)

7. $\frac{3}{8}$ **9.** $-\frac{3}{8}$ **11.** $\frac{27}{40}$ **13.** $\frac{121}{12}$ **15.** $\frac{1}{7}$

17. $\frac{7}{4}$ **19.** $\frac{2}{5}$ **21.** $\frac{1}{2}$ **23.** $-\frac{8}{27}$ **25.** $\frac{5}{6}$

27. $\frac{10}{7}$ **29.** 0.75   **31.** $0.\overline{45}$ **33.** 2.27

35. -57.02 **37.** 39.08   **39.** -9.47 **41.** 4.7 hours

43. ≈ 14 minutes   **45.** -20 **47.** $\frac{1}{80}$ **49.** $-\frac{111}{10}$

51. $\frac{1}{12}$ mile   **53.** $\frac{1}{3}$ **55.** $\frac{21}{20}$ **57.** $-\frac{3}{16}$ **59.** $\frac{12}{5}$

61. 1   **63.** $\frac{56}{3}$ **65.** Division by zero is undefined.

67. 5   **69.** $-\frac{16}{3}$ **71.** -90 **73.** $\frac{5}{2}$ **75.** 0.7

77. 0.5625   **79.** $0.\overline{6}$ **81.** $0.58\overline{3}$ **83.** 106.65

85. 0.04   **87.** -14.00 **89.** 0.01

91. A profit of $432.37   **93.** $11.85

95. 48 breadsticks   **97.** $623.68

99. The product is greater than 20, because the factors are greater than factors that yield a product of 20.

101. False

103. "If the product of two numbers is 15, then 15 divided by the first number yields the second."

Mid-Chapter Quiz (page 66)

1. $\frac{5}{6}$ **2.** 0.429   **3.** No. $\frac{1}{3} = 0.\overline{3}$ **4.** $-\frac{76}{9}$

5. $\frac{6}{11}$ **6.** $\frac{11}{8}$ **7.** $\frac{41}{12}$ **8.** $\frac{81}{20}$ **9.** $\frac{1}{4}$

10. $\frac{7}{24}$ **11.** $\frac{10}{7}$ **12.** $-\frac{25}{3}$ **13.** $-\frac{10}{3}$ **14.** $\frac{57}{215}$

15. $\frac{6}{35}$ **16.** $\frac{59}{49}$ **17.** 31.35   **18.** 9.49

19. $\approx \$0.14$ per ounce   **20.** $1212.50

Section 1.3 (page 73)

7. 2^5 **9.** $(-3)(-3)(-3)(-3)(-3)(-3)$

11. False. $-2^4 = (-1)2^4 = -16$ **13.** 9   **15.** $\frac{1}{64}$

17. -1.728 **19.** 4   **21.** 210   **23.** 0.0084

25. Division by zero is undefined.   **27.** 26   **29.** 28

31. $-\frac{11}{2}$ **33.** -32 **35.** $\frac{8}{3}$ **37.** -1 **39.** 1.19

41. $24^2 = (4 \cdot 6)^2 = 4^2 \cdot 6^2$ **43.** 57,500,000 square miles

45. 36 square units   **47.** $V = 7^3$ cubic inches

49. 0   **51.** $\frac{3}{4}$ **53.** $3\frac{19}{24}$ yards   **55.** $(-5)^4$

57. $\left(\frac{5}{8}\right)^5$ **59.** $(9.8)(9.8)(9.8)$

61. $\left(-\frac{1}{2}\right)\left(-\frac{1}{2}\right)\left(-\frac{1}{2}\right)\left(-\frac{1}{2}\right)\left(-\frac{1}{2}\right)$ **63.** Negative

65. Negative   **67.** 64   **69.** -125 **71.** -125

73. $\frac{16}{81}$ **75.** Division by zero is undefined.   **77.** 13

79. 5840   **81.** 7.32   **83.** 8   **85.** -9 **87.** 17

89. -64 **91.** 9   **93.** $\frac{7}{3}$ **95.** $\frac{5}{4}$ **97.** $\frac{5}{6}$

99. 366.12   **101.** True. $-9 \neq 9$
$$-3^2 = -(3)(3) = -9$$
$$(-3)(-3) = 9$$

103. 0.07, \$31,500   **105.** \$11,070

Section 1.4   *(page 83)*

7. Commutative Property of Multiplication

9. Additive Identity Property

11. Associative Property of Multiplication

13. Multiplicative Inverse Property

15. Associative Property of Addition

17. $5(v + u)$ or $(u + v)5$

19. $6x + 6 \cdot 2$ **21.** (a) -50 (b) $\frac{1}{50}$

23. (a) 1   (b) -1 **25.** $(10 + 8) + 2$ **27.** $2(3 \cdot 4)$

29. $3 \cdot 6 + 3 \cdot 10 = 48$ **31.** $\frac{2}{3} \cdot 9z + \frac{2}{3} \cdot 24 = 6z + 16$

33. $3 + 10(x + 1)$
$$= 3 + 10x + 10 \qquad \text{Distributive Property}$$
$$= 3 + 10 + 10x \qquad \text{Commutative Property of Addition}$$
$$= (3 + 10) + 10x \qquad \text{Associative Property of Addition}$$
$$= 13 + 10x \qquad \text{Addition of Real Numbers}$$

35. $5(x + 3) = 5x + 5 \cdot 3$ **37.** $\frac{8}{0}$ is undefined.

39. (a) $x(1 + 0.06) = 1.06x$ (b) \$27.51   **41.** 0

43. 10   **45.** $\frac{7}{30}$

47. Commutative Property of Addition

49. Additive Inverse Property

51. Associative Property of Multiplication

53. Distributive Property   **55.** Distributive Property

57. Associative Property of Addition

59. $7(3 + x)$ or $(x + 3)7$ **61.** $(3x + 2y) + 5$

63. $4 \cdot 25 + y \cdot 25$ **65.** (a) $-2x$ (b) $\frac{1}{2x}$

67. (a) $-ab$ (b) $\frac{1}{ab}$ **69.** $x + (3 + 2)$

71. $(2 \cdot 3)y$ **73.** $3(2x) - 3 \cdot 4 = 6x - 12$

75. $\frac{3}{5}(10y) - \frac{3}{5}(45) = 6y - 27$

77. $7x + 9 + 2x$
$$= 7x + 2x + 9 \qquad \text{Commutative Property of Addition}$$
$$= (7x + 2x) + 9 \qquad \text{Associative Property of Addition}$$
$$= (7 + 2)x + 9 \qquad \text{Distributive Property}$$
$$= 9x + 9 \qquad \text{Addition of Real Numbers}$$
$$= 9(x + 1) \qquad \text{Distributive Property}$$

79. $a + b + 2c + 12$ **81.** No

Review Exercises   *(page 88)*

1. 10   **3.** 15   **5.** $\frac{2}{5}$ **7.** $\frac{3}{4}$ **9.** $\frac{1}{9}$

11. $\frac{103}{96}$ **13.** $\frac{5}{4}$ **15.** $\frac{17}{8}$ **17.** \$36   **19.** $\frac{1}{9}$

21. $-\frac{3}{5}$ **23.** $-\frac{1}{12}$ **25.** 1   **27.** $\frac{2}{3}$ **29.** $\frac{6}{7}$

31. Division by zero is undefined.   **33.** 21   **35.** 1.875

37. $0.41\overline{6}$ **39.** \$3.52   **41.** \$0.065   **43.** 343

45. -343 **47.** $2^2 < 2^4$ **49.** $\frac{3}{4} > \left(\frac{3}{4}\right)^2$ **51.** 796.11

53. 1841.74   **55.** (a) \$6750   (b) \$9250   **57.** 0

59. 13.1   **61.** $\frac{81}{625}$ **63.** 160   **65.** 54   **67.** $\frac{37}{8}$

69. $\frac{1}{20}$ **71.** $\frac{10}{3}$ **73.** $\frac{2}{15}$ **75.** 4

77. Additive Inverse Property

79. Commutative Property of Multiplication

81. Multiplicative Identity Property

83. Associative Property of Addition

85. Distributive Property

87.

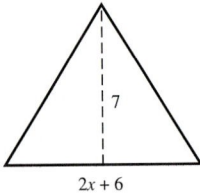

$$\tfrac{1}{2}(2x + 6)(7) = (x + 3)7 = 7x + 21$$

Chapter Test   *(page 90)*

1. $\frac{5}{12}$ **2.** 0.556   **3.** $-3^4 = (-1)(3^4)$

4. Exponentiation, multiplication, subtraction

5. $\frac{1}{2}$ **6.** $\frac{17}{24}$ **7.** $\frac{3}{10}$ **8.** $-\frac{45}{2}$ **9.** $\frac{7}{12}$

10. -27 **11.** -64 **12.** $-\frac{4}{9}$ **13.** 33   **14.** $\frac{1}{8}$

15. Distributive Property

16. Multiplicative Inverse Property

17. Associative Property of Addition

18. Commutative Property of Multiplication   **19.** \$10.45

20.

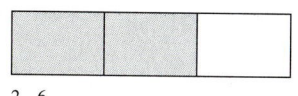

$\frac{2}{3}, \frac{6}{9}$

111.

t	1	5	10
$10\left(2^{t-1}\right)$	\$0.10	\$1.60	\$51.20

t	20	25
$10\left(2^{t-1}\right)$	\$52,428.80	\$1,677,721.60

113. $k\pi L\dfrac{a^4}{2}$ **115.** $3(4^3)$; 192

CHAPTER 2

Section 2.1   *(page 100)*

7. Variable: x **9.** Constant: π **11.** $3x^2, 5$
Constant: 3

13. -6 **15.** $2u^4$ **17.** $(2u)^4$

19. $3^5(a-b)^5$ **21.** $2 \cdot 2 \cdot x \cdot x \cdot x \cdot x$

23. $a^2 \cdot a^2 \cdot a^2 = a \cdot a \cdot a \cdot a \cdot a \cdot a$

25. $3(r+s)(r+s) \cdot 3(r+s) \cdot 3(r+s)$

27. u^6 **29.** $-5z^5$ **31.** t^8 **33.** $5u^5v^5$

35. $-8s^3$ **37.** $(x-2y)^4$ **39.** $-72x^7 + 5x^2$

41. u^6v^{11} **43.** $x^5 \cdot x^3 = x^{5+3} = x^8 \neq x^{15}$

45. $-3x^3 \neq -27x^3 = (-3x)^3$ **47.** 10^4

49. $P\left(1+\dfrac{r}{4}\right)^4$ **51.** $n = 1$ is the only solution.

53. 20   **55.** Distributive Property

57. (a) \$24,300   (b) \$4301   **59.** $\frac{5}{3}, -3y^3$

61. $2x, -3y, 1$ **63.** $3(x+5), 10$

65. $\dfrac{3}{x+2}, -3x, 4$ **67.** $-\frac{1}{3}$ **69.** $-\frac{3}{2}$ **71.** -120

73. -4.7 **75.** a^3b^2 **77.** $4x^3y^2$ **79.** $3^3(x-y)^2$

81. $\left(\dfrac{x^2}{2}\right)^3$ **83.** $4 \cdot y \cdot y \cdot z \cdot z \cdot z$

85. $(-2y)(-2y)(-2y)$ **87.** $5 \cdot x \cdot x \cdot x \cdot x \cdot x \cdot x \cdot x$

89. $(ab)(ab)(ab)$ **91.** $(x+y)(x+y)$

93. $\left(\dfrac{a}{3s}\right)\left(\dfrac{a}{3s}\right)\left(\dfrac{a}{3s}\right)\left(\dfrac{a}{3s}\right)$ **95.** $5x^7$ **97.** $-8x^3$

99. $-6ab^7$ **101.** $-54x^3 + 5x^2$ **103.** $a^{10}b^{11}$

105. $\dfrac{49x^4}{729y^5}$ **107.** 4   **109.** 8390

Section 2.2   *(page 108)*

7. Commutative Property of Addition

9. Associative Property of Addition

11. Multiplicative Inverse Property

13. Distributive Property

15. $(x+1) - (x+1) = 0$
Additive Inverse Property

17. $v(2) = 2v$
Commutative Property of Multiplication

19. $(t+5)(t-2) = t(t-2) + 5(t-2)$
Distributive Property

21. $-10x + 5y$ **23.** $3x + 6$ **25.** $x^2 + x^2y + xy^2$

27. $z^3 - 2z^2 + 2z$ **29.** $6x^2, 6; -3xy, -3; y^2, 1$

31. $16t^3, 3t^3; 4, -5$ **33.** $-2y$ **35.** $x^2 - xy + 4$

37. $2\left(\dfrac{1}{x}\right) + 8$ **39.** Variable factors are not alike.
$x^2y \neq xy^2$

41. 416   **43.** $4x + 12$ **45.** $4x^6$ **47.** $5z^7$

49. $P(1+r)^4$ **51.** Associative Property of Multiplication

53. Multiplicative Identity Property

55. Commutative Property of Multiplication

57. Additive Inverse and Additive Identity Properties

59. $5x\left(\dfrac{1}{5x}\right) = 1$
Multiplicative Inverse Property

61. $(2z-3) - (2z-3) = 0$
Additive Inverse Property

63. $12 + (8-x) = (12+8) - x$
Associative Property of Addition

65. $3x^2y + 3x$ **67.** $-u + v$ **69.** $8x + x^2$

71. $2x^2 + 18x$ **73.** $-12y^2 + 16y$ **75.** $a^3 + a^2b + ab^2$

77. $6x^2y, -4x^2y$ **79.** $-2x + 4y$ **81.** $11x^2 + 4$

83. $3x^2 + 2x^2y + 3xy^2 + y^2$ **85.** $5\left(\dfrac{1}{x}\right) - 4x$

87. False. $3(x - 4) = 3x - 12$ **89.** True

91. 39.9   **93.** $a(b + c) = ab + ac$

95. $a \odot b = 2 \cdot a + b \neq a + 2 \cdot b = b \odot a$
 Not commutative
$$a \odot (b \odot c) = a \odot (2b + c) = 2a + 2b + c$$
$$(a \odot b) \odot c = (2a + b) \odot c$$
$$= 2(2a + b) + c = 4a + 2b + c$$
 Not associative

Section 2.3 *(page 118)*

 7. $-12x$ **9.** $2x$ **11.** -4 **13.** $4n - 3$

15. $2x - 3y$ **17.** $2x - 17$ **19.** $4t^2 - 11t$ **21.** $\dfrac{5x}{6}$

23. $\dfrac{x}{3}$ **25.** x **27.** (a) 0   (b) 7

29. (a) 3   (b) -20

31. (a) 10   (b) Division by zero is undefined.

33. (a) $\frac{15}{2}$ (b) 10

35. (a)

x	-1	0	1	2	3	4
$3x - 2$	-5	-2	1	4	7	10

 (b) 3   (c) $\frac{2}{3}$

37. $21,589.25   **39.** $\frac{1}{2}n(10 + n)$, 3600 square units

41. $(x + y)^2$, 169 square units   **43.** $8x - 20$ **45.** $x - 4$

47. $2,362,000   **49.** $-35a$ **51.** $6x^2$ **53.** $9a$

55. $-6x^5$ **57.** $-3x - 5$ **59.** $8x + 26$ **61.** $-\frac{3}{8}y + 9$

63. $10x - 7x^2$ **65.** $3x^2 + 5x$ **67.** $-12r + 19s$

69. $\dfrac{7z}{5}$ **71.** $-\dfrac{11x}{12}$ **73.** (a) -5 (b) 25

75. (a) 6   (b) 0   **77.** (a) 17   (b) 4

79. (a) 0   (b) Division by zero is undefined.

81. (a)

x	-1	0	1	2	3	4
$3 - 2x$	5	3	1	-1	-3	-5

 (b) -2 (c) $-\frac{3}{2}$

83. (a) $5x^2$ (b) 2000 square feet   **85.** 9375 square feet

Mid-Chapter Quiz *(page 122)*

1. (a) -5 (b) $\frac{5}{16}$ **2.** (a) $(3y)^4$ (b) $2^3(x - 3)^2$

3. x^7 **4.** v^{10} **5.** $9y^5$

6. $8(x - 4)^6$ **7.** $\dfrac{10z^3}{21y^4}$ **8.** $\left(\dfrac{x}{y}\right)^7$

9. Associative Property of Multiplication

10. Distributive Property

11. Multiplicative Inverse Property

12. Commutative Property of Addition

13. $y^2 + 4xy + y$ **14.** $3\left(\dfrac{1}{u}\right) + 3u$ **15.** $8a - 7b$

16. $-8x - 66$ **17.** (a) 0   (b) 10   (c) 0

18. (a) 2   (b) 0   (c) Division by zero is undefined.

19. $\frac{1}{10}k\pi r^4 h$ **20.** 45,700

Section 2.4 *(page 131)*

 7. $x + 5$ **9.** $x - 6$ **11.** $2x$ **13.** $\dfrac{x}{50}$

15. $3x + 5$ **17.** The sum of three times a number and $\frac{1}{3}$

19. The sum of a number and 1 is divided by 2.

21. The product of $\frac{1}{2}$ and the sum of a number and 1

23. $(x + 3)x = x^2 + 3x$ **25.** $(9 - x)(3) = 27 - 3x$

27. $0.10n$ **29.** $8.25n$ **31.** $\dfrac{100}{r}$

33. s^2 square inches   **35.** $n + (n + 1) + (n + 2) = 3n + 3$

37. (a)

n	0	1	2	3	4	5
$2n - 1$	-1	1	3	5	7	9
Differences		2	2	2	2	2

 (b) a

39. $-y^4 + 2y^2$ **41.** (a) 7   (b) 16   **43.** 5   **45.** d

47. e   **49.** b   **51.** $25 + x$ **53.** $\frac{1}{4}x$ **55.** $\frac{1}{3}x$

57. $5x + 8$ **59.** $10(x + 4)$ **61.** $|x + 4|$ **63.** $x^2 + 1$

65. A number decreased by 10

67. Seven times a number increased by 4

69. Three times the difference of 2 and a number

71. The square of a number increased by 5

73. $(6 + n)(5) = 30 + 5n$

75. $\dfrac{8(x + 24)}{2} = 4x + 96$ **77.** $0.06L$ **79.** $15m + 2n$

81. $5w$ **83.** $3x(6x - 1) = 18x^2 - 3x$

85. $\frac{1}{2}(12)(5x^2 + 2) = 30x^2 + 12$

87. $(2n + 1) + (2n + 3) = 4n + 4$

89. (a)

n	0	1	2	3	4	5
$2n + 5$	5	7	9	11	13	15
Differences		2	2	2	2	2

(b) a

91. $a = 4, b = 1$

Section 2.5 *(page 141)*

7. (a) Solution (b) Not a solution

9. (a) Not a solution (b) Solution

11. (a) Solution (b) Solution

13. (a) Not a solution (b) Not a solution

15. (a) Solution (b) Not a solution

17. (a) Solution (b) Not a solution

19. $4x = -28$ Given equation

$\dfrac{4x}{4} = \dfrac{-28}{4}$ Divide both sides by 4.

$x = -7$ Solution

21. $2x - 2 = x + 3$ Given equation

$-x + 2x - 2 = -x + x + 3$ Subtract x from both sides.

$x - 2 = 3$ Combine like terms.

$x - 2 + 2 = 3 + 2$ Add 2 to both sides.

$x = 5$ Solution

23. 2 **25.** 10 **27.** $x + 6 = 94$ **29.** $4(x + 6) = 100$

31. $1.75n = 2000$ **33.** $3r + 25 = 160$

35. The sum of a number and 8 is 25.

37. Two times the difference of a number and 5 is 12.

39. $x + 10$ **41.** $4 + 2x$ **43.** 7.5

45. Perimeter: $6x$; Area: $\dfrac{9x^2}{4}$

47. (a) Not a solution (b) Solution

49. (a) Solution (b) Not a solution

51. (a) Solution (b) Not a solution

53. (a) Not a solution (b) Solution

55. (a) Solution (b) Not a solution

57. (a) Solution (b) Solution

59. $\frac{2}{3}x = 12$ Given equation

$\frac{3}{2}\left(\frac{2}{3}x\right) = \frac{3}{2}(12)$ Multiply both sides by $\frac{3}{2}$.

$x = 18$ Solution

61. $x = -2(x + 3)$ Given equation

$x = -2x - 6$ Distributive Property

$2x + x = 2x - 2x - 6$ Add $2x$ to both sides.

$3x = 0 - 6$ Additive Inverse Property

$3x = -6$ Combine like terms.

$\dfrac{3x}{3} = \dfrac{-6}{3}$ Divide both sides by 3.

$x = -2$ Solution

63. $3650 + x = 4532$ **65.** $0.25x = 105.75$

67. $x + 12 = 45$ **69.** $2n - 14 = \dfrac{n}{3}$

71. $2n + (2n + 2) + (2n + 4) = 18$

73. $2l + 2\left(\frac{1}{3}l\right) = 96$ **75.** $x + 45 = 375$

77. $750,000 - 3D = 75,000$ **79.** $4r + 24 = 200$

81. Nine minus the square of a number is zero.

83. The product of 10 and the difference of a number and 3 is 8 times the number.

85. 15 dollars **87.** 15 dollars **89.** 6000 feet

Review Exercises *(page 147)*

1. $4; -\frac{1}{2}x^3, -\frac{1}{2}$ **3.** $y^2, 1; -10yz, -10; \frac{2}{3}z^2, \frac{2}{3}$

5. $(5z)^3$ **7.** $a^2(b - c)^2$ **9.** x^6

11. $-2t^6$ **13.** $-5x^3y^4$ **15.** $-64y^7$ **17.** $P\left(\frac{9}{10}\right)^5$

19. Multiplicative Inverse Property

21. Commutative Property of Multiplication

23. Associative Property of Addition **25.** $4x + 12y$

27. $-10u + 15v$ **29.** $8x^2 + 5xy$ **31.** $a - 3b$

33. $-2a$ **35.** $11p - 3q$ **37.** $\frac{15}{4}s - 5t$

39. $x^2 + 2xy + 4$ **41.** $3x - 3y + 3xy$ **43.** $3\left(1 + \dfrac{r}{n}\right)^2$

45. $5u - 10$ **47.** $5s - r$ **49.** $8x - 32$ **51.** $-2x + 4y$

53. $10z - 1$ **55.** $2z - 2$ **57.** (a) 5   (b) 5

59. (a) 1   (b) -2 **61.** $\frac{2}{3}x + 5$ **63.** $2x - 10$

65. $50 + 7x$ **67.** $\dfrac{x + 10}{8}$ **69.** $x^2 + 64$

71. A number plus 3

73. A number decreased by 2 is divided by 3.

75. $0.28I$ **77.** $w^2 + 3w$

79. $(2n - 1) + (2n + 1) + (2n + 3) = 6n + 3$ **81.** $9x^2$

83. (a)

n	0	1	2	3	4	5
$n^2 + 3n + 2$	2	6	12	20	30	42
Differences		4	6	8	10	12
Differences			2	2	2	2

(b) Third row: entries increase by 2
Fourth row: constant 2

85. (a) Not a solution   (b) Solution

87. (a) Not a solution   (b) Solution

89. (a) Solution   (b) Not a solution

91. (a) Not a solution   (b) Solution

93. (a) Solution   (b) Solution

95. $x + \dfrac{1}{x} = \dfrac{37}{6}$ **97.** $6x - \frac{1}{2}(6x) = \frac{1}{2}(6x) = 24$

Chapter Test *(page 150)*

1. $2x^2, 2; -7xy, -7; 3y^3, 3$ **2.** $x^3(x + y)^2$

3. Associative Property of Multiplication

4. Commutative Property of Addition

5. Additive Inverse Property

6. Multiplicative Identity Property

7. $3x + 24$ **8.** $-3y + 2y^2$ **9.** c^8 **10.** $-10u^4v$

11. $-a - 7b$ **12.** $8u - 8v$ **13.** $4z - 4$

14. $18 - 2t$ **15.** (a) 25   (b) -31

16. Division by zero is undefined.   **17.** $\frac{1}{5}n + 2$

18. (a) Perimeter: $2w + 2(2w - 4)$; Area: $w(2w - 4)$

(b) Perimeter: $6w - 8$; Area: $2w^2 - 4w$

(c) Perimeter: unit of length; Area: square units

(d) Perimeter: 64 feet; Area: 240 square feet

19. $3n + 2m$ **20.** (a) Not a solution   (b) Solution

CHAPTER 3

Section 3.1 *(page 160)*

7. 8   **9.** 3

11.

$5x + 15 = 0$	Given equation
$5x + 15 - 15 = 0 - 15$	Subtract 15 from both sides.
$5x = -15$	Combine like terms.
$\dfrac{5x}{5} = \dfrac{-15}{5}$	Divide both sides by 5.
$x = -3$	Solution

13. $\frac{11}{4}$ **15.** 4   **17.** -6 **19.** $\frac{2}{3}$ **21.** $\frac{1}{3}$ **23.** 1

25. 0   **27.** $\frac{5}{6}$ **29.** No solution   **31.** 2   **33.** 7.71

35. 8.99   **37.** 36 feet   **39.** 30   **41.** 17

43. $-9x + 11y$ **45.** $2(x - 10)$ **47.** 13   **49.** 4

51. $-2x = 8$ Given equation

$\dfrac{-2x}{-2} = \dfrac{8}{-2}$ Divide both sides by -2.

$x = -4$ Solution

53. -3 **55.** 2   **57.** 2   **59.** -2 **61.** 0

63. No solution   **65.** $\frac{2}{5}$ **67.** 0   **69.** $\frac{2}{3}$

71. 30   **73.** $\frac{5}{3}$ **75.** $\frac{1}{6}$ **77.** 123.00

79. 3.51   **81.** 35, 37   **83.** 80 inches $\times$ 40 inches

85. 3 feet, 3 feet, 6 feet   **87.** 6 hours

89. (a)

t	1	1.5	2
Width	300	240	200
Length	300	360	400
Area	90,000	86,400	80,000

t	3	4	5
Width	150	120	100
Length	450	480	500
Area	67,500	57,600	50,000

(b) The area decreases.

Section 3.2 *(page 170)*

7.

Percent	Parts out of 100	Decimal	Fraction
40%	40	0.40	$\frac{2}{5}$

9.

Percent	Parts out of 100	Decimal	Fraction
15.5%	15.5	0.155	$\frac{31}{200}$

11. 0.125   **13.** 2.50   **15.** 7.5%   **17.** 62%

19. 80%   **21.** 35%   **23.** $37\frac{1}{2}\%$ **25.** 45

27. 0.42   **29.** 2200   **31.** 500%   **33.** $425

35. $71\frac{2}{3}\%$ **37.** 80   **39.** 6252

41. Vehicular accidents: 3600
Falls: 1760
Acts of violence: 1280
Sports injuries: 1040
Other: 320

43.

$2x - 5 = x + 9$	Given equation
$2x - 5 + 5 = x + 9 + 5$	Add 5 to both sides.
$2x = x + 14$	Combine like terms.
$-x + 2x = -x + x + 14$	Subtract x from both sides.
$x = 14$	Combine like terms.

45.

$2x + \frac{3}{2} = \frac{3}{2}$	Given equation
$2x + \frac{3}{2} - \frac{3}{2} = \frac{3}{2} - \frac{3}{2}$	Subtract $\frac{3}{2}$ from both sides.
$2x = 0$	Combine like terms.
$\frac{2x}{2} = \frac{0}{2}$	Divide both sides by 2.
$x = 0$	Solution

47. $14.67

49.

Percent	Parts out of 100	Decimal	Fraction
63%	63	0.63	$\frac{63}{100}$

51.

Percent	Parts out of 100	Decimal	Fraction
60%	60	0.60	$\frac{3}{5}$

53. 0.0075   **55.** 1.25   **57.** 20%   **59.** 250%

61. 25%   **63.** $83\frac{1}{3}\%$ **65.** 77.52   **67.** 176

69. 2100   **71.** 132   **73.** 360   **75.** 72%

77. 12.5%   **79.** $41\frac{2}{3}\%$ **81.** $9750   **83.** $312.50

85. 10,210 eligible voters

87. (a) 934,925
(b) 97,854
(c) The number of male biologists increased faster.

89. 75% reduction

Math Matters *(page 174)*

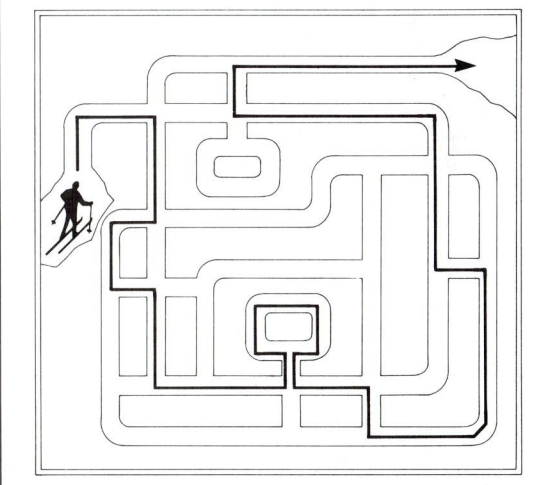

Abbot, R. *Mad Mazes*.
Holbrook, MA: Bob Adams, Inc.
Copyright ©1990. Used by permission.

Section 3.3 *(page 182)*

7. 2   **9.** 10   **11.** 5   **13.** 2   **15.** 3

17. No solution   **19.** -4 **21.** 1   **23.** 3   **25.** $\frac{5}{2}$

27. $-\frac{2}{5}$ **29.** 10   **31.** $\frac{16}{3}$ **33.** $\frac{4}{11}$ **35.** 4.8 hours

37. $x = 4$ feet   **39.** 14   **41.** -3 **43.** 30

45. 6   **47.** 10   **49.** 2   **51.** -10 **53.** -5

55. -7 **57.** $\frac{8}{5}$ **59.** 1   **61.** $\frac{1}{2}$ **63.** 3   **65.** $-\frac{3}{2}$

67. 0   **69.** $-\frac{10}{3}$ **71.** $-\frac{16}{7}$ **73.** $\frac{4}{3}$ **75.** 6

77. (a) Each of the n bricks is 8 inches. Each of the $(n-1)$ mortar joints is $\frac{1}{2}$ inch. The total length is 93 inches.

(b) 11

79. 25 quarts   **81.** $1\frac{1}{3}$ quarts

83. Concentration (or price) of each part of the mixture.

Mid-Chapter Quiz *(page 185)*

1. 40   **2.** 8   **3.** $\frac{19}{2}$ **4.** 0   **5.** $-\frac{1}{3}$

6. $\frac{40}{13}$ **7.** 36   **8.** 11   **9.** 2.06   **10.** 51.23

11. 15.5   **12.** 30   **13.** 200%   **14.** 455

15. 12 meters × 18 meters   **16.** 10 hours

17. 6 square meters, 12 square meters, 24 square meters

18. 93   **19.** \$495.37   **20.** 44.6%

Section 3.4 *(page 193)*

5. $\frac{4}{1}$ **7.** $\frac{2}{3}$ **9.** $\frac{3}{2}$ **11.** $\frac{1}{4}$ **13.** $\frac{3}{8}$ **15.** $\frac{3}{4}$

17. $\frac{2}{1}$ **19.** $\frac{12}{1}$ **21.** \$0.0395   **23.** \$0.0645

25. $27\frac{3}{4}$-ounce jar   **27.** 12   **29.** 16   **31.** 27

33. $22\frac{2}{9}$ gallons   **35.** 20 pints   **37.** 250 miles

39. $\frac{5}{2}$ **41.** 35   **43.** 2

45. 48 inches, 48 inches, 24 inches   **47.** $\frac{1}{2}$ **49.** $\frac{9}{1}$

51. $\frac{5}{4}$ **53.** $\frac{7}{15}$ **55.** $\frac{2}{1}$ **57.** $\frac{8}{3}$ **59.** $\frac{11}{8}$ **61.** $\frac{23}{1}$

63. $\frac{3}{2}$ **65.** $\frac{100}{49}$ **67.** 16-ounce package

69. 2-liter bottle   **71.** 50   **73.** $\frac{10}{3}$ **75.** $\frac{1}{2}$ **77.** $\frac{14}{5}$

79. 250 blocks   **81.** \$1142.31   **83.** 64 pounds

85. $6\frac{2}{3}$ feet   **87.** \$183   **89.** \$41

Section 3.5 *(page 204)*

7. All real numbers greater than or equal to 3.

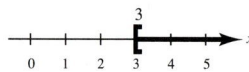

9. $-2 \le x < 1$ **11.** $0 < x \le 6$

13. (a) Yes   (b) No   (c) Yes   (d) No

15. (a) Yes   (b) No   (c) No   (d) Yes

17. b   **19.** c   **21.** d   **23.** $x \ge 0$ **25.** $z \ge 3$

27. $t \ge 5$

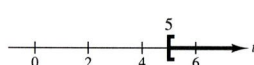

29. $x < 3$

31. $x < 4$

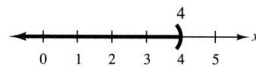

33. $x > \frac{1}{2}$

35. $y > 1$

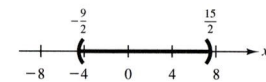

37. $x \ge -\frac{10}{3}$

39. $-\frac{9}{2} < x < \frac{15}{2}$

41. $m < 24{,}062.50$

43. Mars is farther from the sun than Mercury.

45. (a) $-\frac{1}{2} > -7$ (b) $-\frac{1}{3} < -\frac{1}{6}$

47. 36   **49.** 2   **51.** Perimeter: $2x^2 + 7x - 3$
Area: $\frac{1}{2}x^2(5x - 3)$

53. (a) No   (b) Yes   (c) Yes   (d) Yes

55. (a) Yes   (b) No   (c) Yes   (d) No

57. $x \le 2$

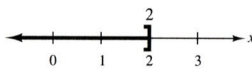

59. $x > -4$

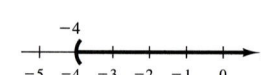

61. $n < 3$

63. $x \le 18$

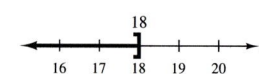

65. $x > 6$

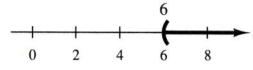

67. $x \ge 4$

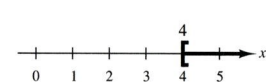

69. $x > \frac{11}{3}$

71. $y > \frac{7}{8}$

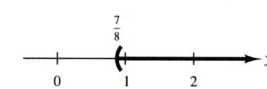

73. $z \leq \frac{3}{11}$

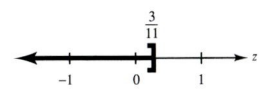

75. $x > -\frac{2}{5}$

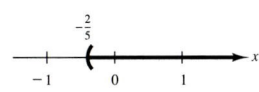

59. $x \geq 2$

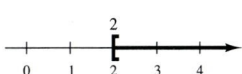

61. $x > 10$

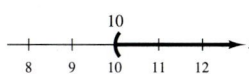

77. $-1 < x < 3$

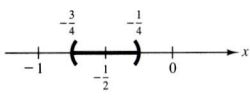

79. $-16 < x < 8$

63. $t > 4$

65. $y \leq -1$

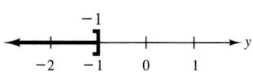

81. $-\frac{3}{4} < x < -\frac{1}{4}$

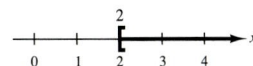

83. $x \geq 4$ **85.** $y \leq 25$

67. $-6 < x \leq 6$

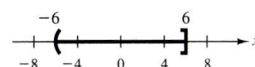

69. $-7 < x < -1$

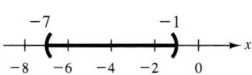

87. $A < C$, Transitive Property

89. 0 minutes $\leq t \leq 12.4$ minutes

91. 180 miles $\leq x \leq 260$ miles

71. $z \geq 10$ **73.** $8 < y < 12$ **75.** $V < 12$

Chapter Test *(page 211)*

1. $\frac{21}{4}$ **2.** 7   **3.** 10   **4.** 10   **5.** 11.03

6. 50   **7.** 4 hours   **8.** $37\frac{1}{2}\%$, 0.375

9. 1200   **10.** 36%   **11.** $200   **12.** $\frac{5}{9}$

13. $\frac{15}{2}$ **14.** 6   **15.** 110 miles

16. $x \leq 4$

17. $x < -6$

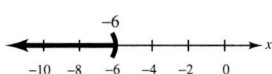

Review Exercises *(page 209)*

1. 35   **3.** 28

5.

$10x - 12 = 18$	Given equation
$10x - 12 + 12 = 18 + 12$	Add 12 to both sides.
$10x = 30$	Combine like terms.
$\dfrac{10x}{10} = \dfrac{30}{10}$	Divide both sides by 10.
$x = 3$	Solution

7. 5   **9.** 4   **11.** 3   **13.** $\frac{4}{3}$ **15.** 20

17. No solution   **19.** 20   **21.** 7   **23.** $\frac{1}{3}$ **25.** 20

27. 7.99   **29.** 224.31   **31.** 480 miles, 720 miles

33.

Percent	Parts out of 100	Decimal	Fraction
35%	35	0.35	$\frac{7}{20}$

35. 20   **37.** 400   **39.** 60%   **41.** 3.5%   **43.** $\frac{1}{8}$

45. $\frac{4}{3}$ **47.** $\frac{7}{2}$ **49.** 10

51. $1687   **53.** 200 miles

55. $x \geq 2$

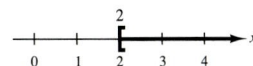

57. $x < 3$

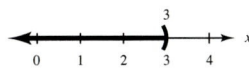

18. $-1 < x \leq 2$

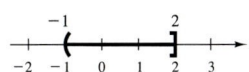

19. $-1 \leq x < 5$

20. (a) $y \leq 10$ (b) $t \geq 10$

Cumulative Test: Chapters P–3
(page 212)

1. $-\frac{3}{4} < \left| -\frac{7}{8} \right|$ **2.** 1200   **3.** $-\frac{11}{24}$ **4.** $-\frac{25}{12}$

5. 8   **6.** $3^3(x + y)^2$ **7.** $-2x^2 + 6x$

8. Associative Property of Addition   **9.** $15x^7$

10. a^8b^7 **11.** $7x^2 - 6x - 2$ **12.** 6   **13.** $\frac{52}{3}$

14. 5   **15.** $-5 \le x < 1$ **16.** $624.91

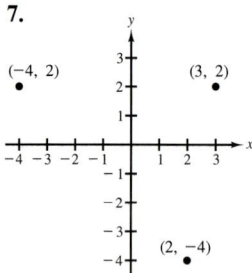

17. $\frac{3}{4}$ **18.** 246,248   **19.** $920   **20.** $57,000

CHAPTER 4

Section 4.1   *(page 221)*

7.

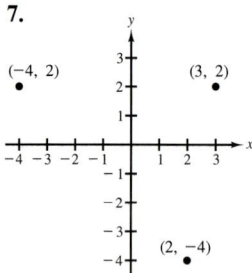

9.

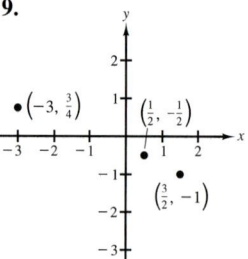

11. A: $(5, 2)$
    B: $(-3, 4)$
    C: $(2, -5)$
    D: $(-2, -2)$

13.

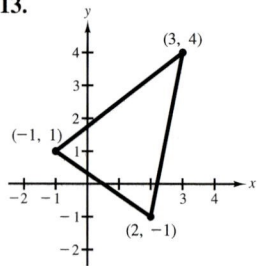

15.

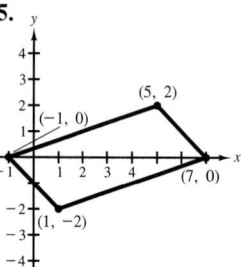

17. Quadrant II   **19.** Quadrant III

21. Quadrants II or III

23. (a)   (b) No
                      (c) June, July, August

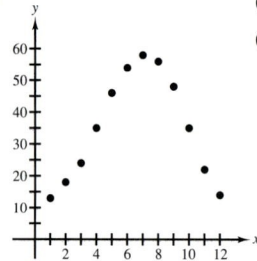

25. (a) Solution point
     (b) Not a solution point
     (c) Not a solution point
     (d) Solution point

27. (a) Not a solution point
     (b) Solution point
     (c) Solution point
     (d) Not a solution point

29.

x	-2	0	2	4	6
$y = 3x - 4$	-10	-4	2	8	14

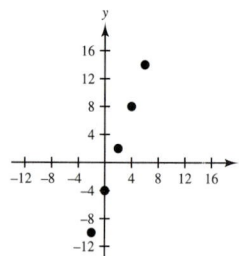

31.

x	-5	$\frac{3}{2}$	5	10	20
$y = -\frac{3}{2}x + 5$	$\frac{25}{2}$	$\frac{11}{4}$	$-\frac{5}{2}$	-10	-25

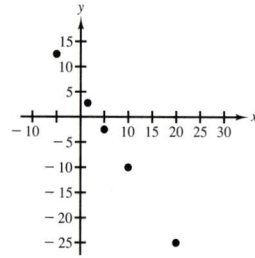

33. Coefficient of x **35.** $1,800,000$

37. Between 600,000 and 650,000; 60–65%   **39.** 6

41. 144   **43.** $19,250

45.

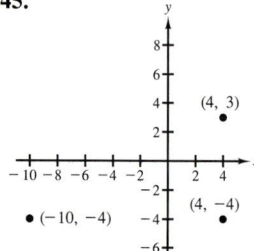

47.

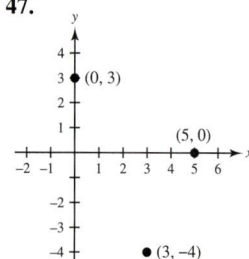

49. A: $(-1, 3)$
B: $(5, -3)$
C: $(2, 1)$
D: $(-1, -2)$

51.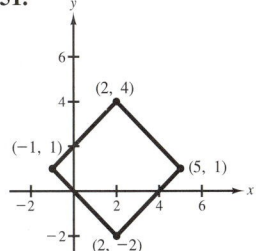

53.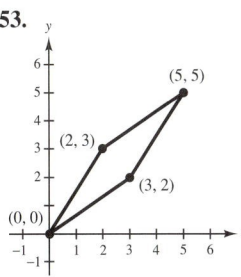

55. (a)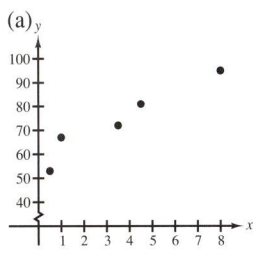

(b) Score increases with increased study time.

57. (a) Solution point
(b) Not a solution point
(c) Not a solution point
(d) Solution point

59. (a) Solution point
(b) Solution point
(c) Not a solution point
(d) Solution point

61.

x	-2	-1	0	1	2
$y = 2x - 1$	-5	-3	-1	1	3

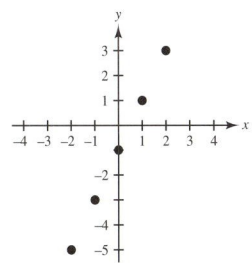

63.

x	-2	-1	0	1	2
$y = 4 - x^2$	0	3	4	3	0

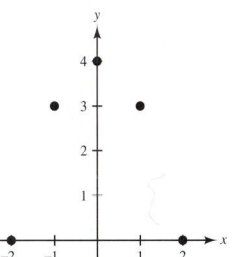

65.

x	100	150	200	250	300
$y = 35x + 5000$	8500	$10{,}250$	$12{,}000$	$13{,}750$	$15{,}500$

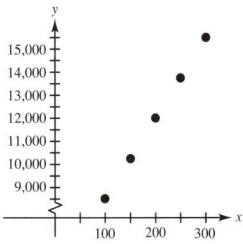

67. Declining balances   **69.** $1000

71. Between $4000 and $5000   **73.** $8\frac{1}{2}\%$

75. (a) and (b)

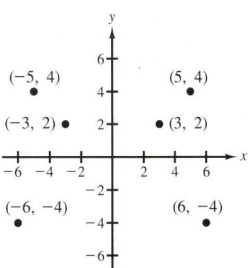

(c) Reflection in the y-axis

Section 4.2 *(page 232)*

7. d   **9.** a   **11.** $y = \frac{1}{4}(12 - 3x)$

13. $y = \frac{1}{2}(x-8)$ **15.**

x	-2	-1	0	1	2
y	11	10	9	8	7

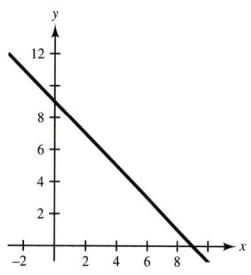

39. Answers will vary.   **41.** $C = 500 + 5x$

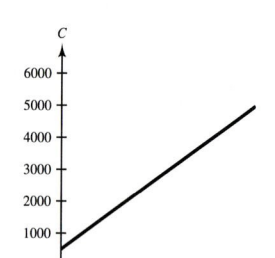

17. $(-2, 0), (0, 4)$ **19.** $(-3, 0), (3, 0), (0, -3)$

21. $(1, 0), (0, -1)$ **23.** $(2, 0), (0, 4)$

25. $(2, 0), (0, -1)$

43. $\frac{42}{5}$ **45.** 40   **47.** 15 inches $\times$ 25 inches

49.

x	-2	-1	0	1	2
y	-10	-6	-2	2	6

27.

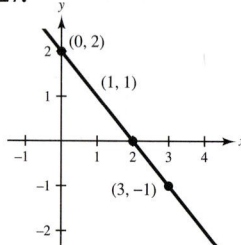

29.

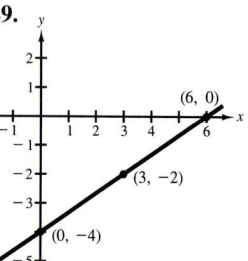

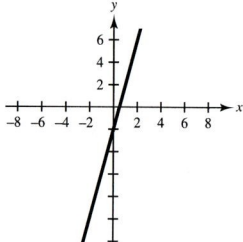

31.

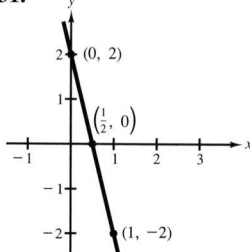

33.

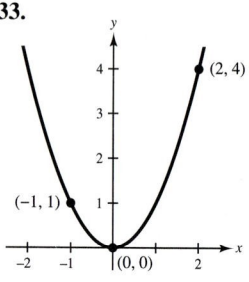

51.

x	-3	-2	-1	0	1
y	2	1	0	1	2

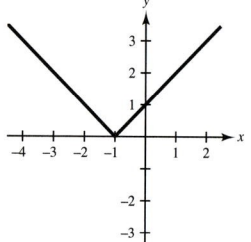

35.

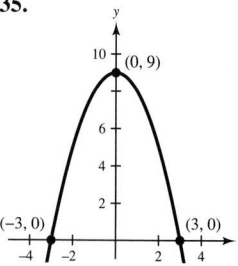

37.

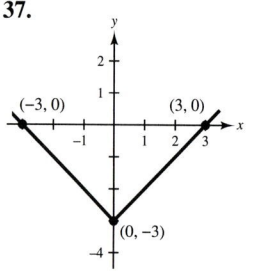

53. $(-4, 0), (4, 0), (0, 4)$ **55.** $\left(-\frac{1}{3}, 0\right), (0, 2)$

57. $\left(\frac{9}{2}, 0\right), \left(0, \frac{3}{2}\right)$ **59.** $(4, 0), (0, -6)$

61.

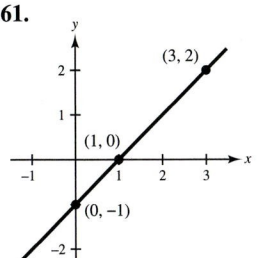

63.

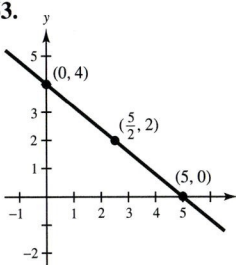

Section 4.3 *(page 241)*

7.

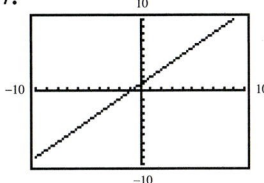

9.

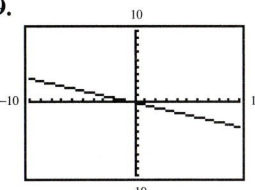

65.

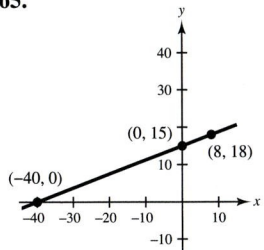

67.

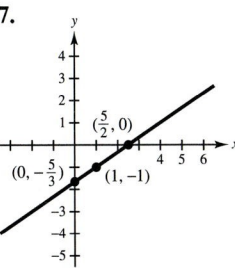

11.

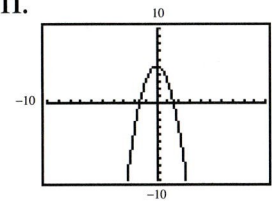

13.

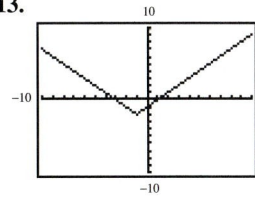

69.

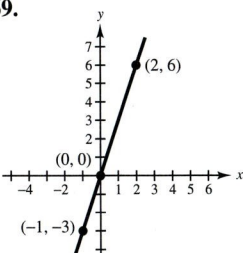

71.

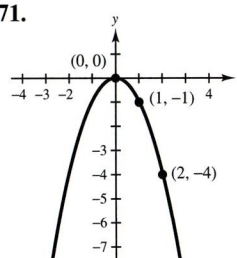

15.

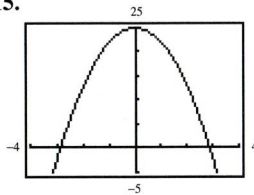

17.

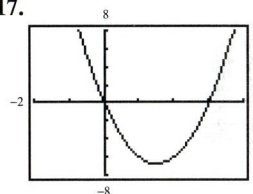

19.

$$\text{Xmin} = -15$$
$$\text{Xmax} = 15$$
$$\text{Xscl} = 1$$
$$\text{Ymin} = -10$$
$$\text{Ymax} = 10$$
$$\text{Yscl} = 1$$

21.

$$\text{Xmin} = -5$$
$$\text{Xmax} = 20$$
$$\text{Xscl} = 5$$
$$\text{Ymin} = -5$$
$$\text{Ymax} = 20$$
$$\text{Yscl} = 5$$

73.

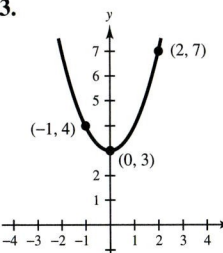

75.

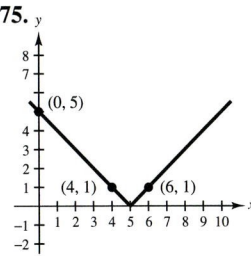

23.

$$\text{Xmin} = -8$$
$$\text{Xmax} = 8$$
$$\text{Xscl} = 4$$
$$\text{Ymin} = -10$$
$$\text{Ymax} = 10$$
$$\text{Yscl} = 1$$

25. b **27.** d

77. (a)

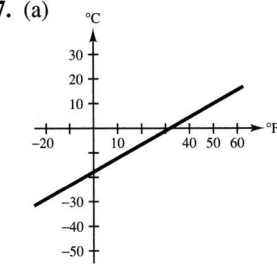

(b) Horizontal axis:
Fahrenheit temperature when
$C = 0$.
Vertical axis:
Celsius temperature when
$F = 0$.

29.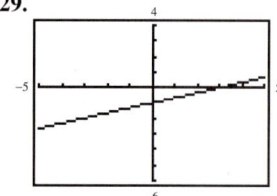

31.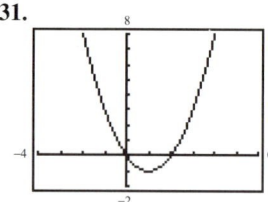

Commutative Property
of Addition

Distributive Property

33. $\left(\frac{5}{2}, 0\right)$, $(0, -5)$ **35.** $(-2, 0)$, $\left(\frac{1}{2}, 0\right)$, $(0, -1)$

37.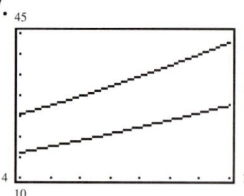

39.

Xmin = 4
Xmax = 20
Xscl = 1
Ymin = 14
Ymax = 22
Yscl = 1

41.

Xmin = −20
Xmax = −4
Xscl = 1
Ymin = −16
Ymax = −8
Yscl = 1

43. $(0, 2)$, $(5, 6)$, $(-5, -2)$

45. $(0, 8)$, $(4, 5)$, $(-4, 11)$ **47.** v^5 **49.** $-9x$

51.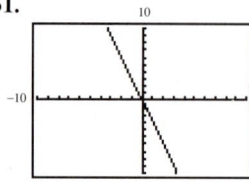

53.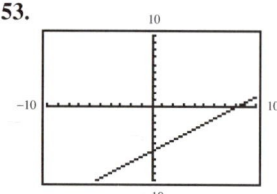

55.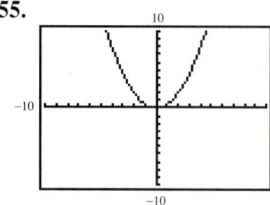

57.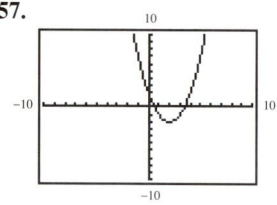

59.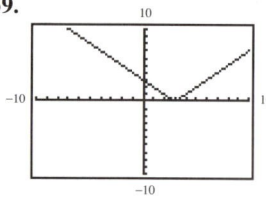

61.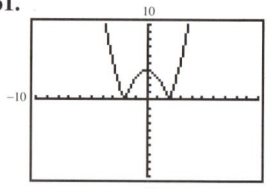

63.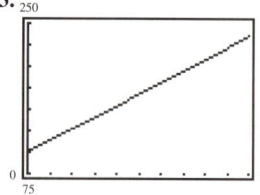

65.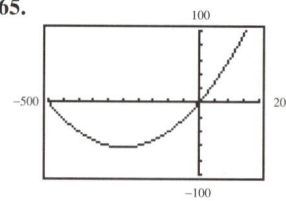

67.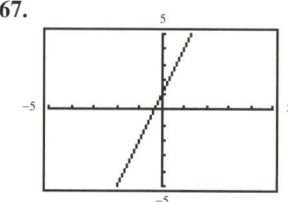

69.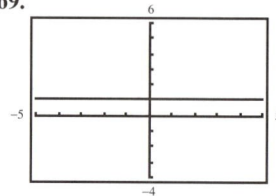

Associative Property
of Addition

Multiplicative Inverse
Property

71. $(-3, 0)$, $(3, 0)$, $(0, 9)$ **73.** $(-8, 0)$, $(4, 0)$, $(0, 4)$

75. Triangle

77. Square

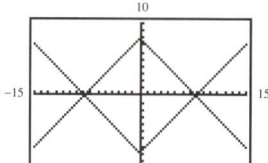

79.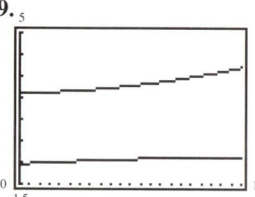

Mid-Chapter Quiz *(page 244)*

1.

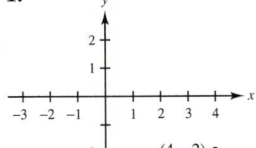

2. I, II

3. (a) Solution

(b) Not a solution

(c) Solution

(d) Not a solution

4.

x	-2	0	2	4	6
$y = -x + 3$	5	3	1	-1	-3

5. 30%

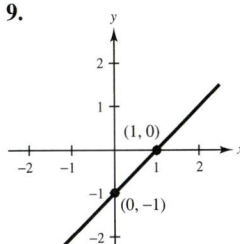

6. 0   **7.** $(12, 0), (0, -4)$ **8.** $\left(\frac{3}{2}, 0\right), (0, 6)$

9.

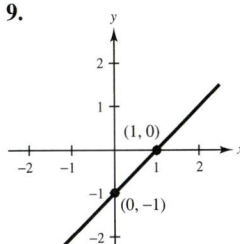

10.

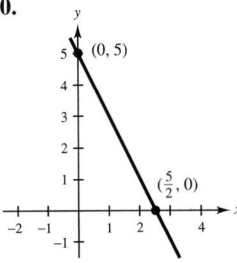

11.

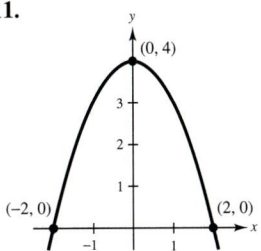

12.

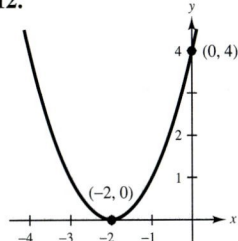

13.

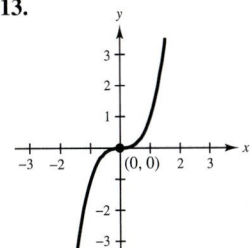

14.

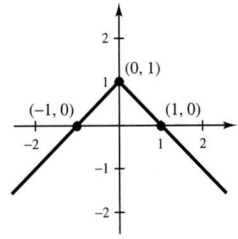

15.

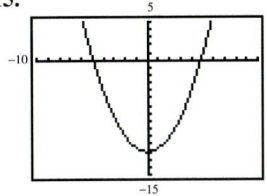

16.

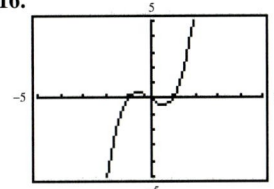

17. $(7, 0), (0, -1)$

18.

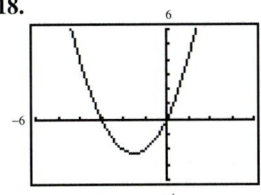

19.

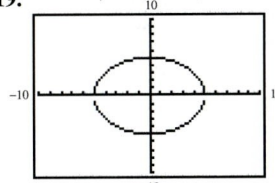

Distributive Property   SQUARE setting

20. $y = 3000 - 500x$

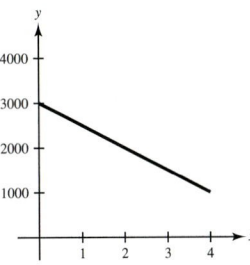

Section 4.4 *(page 252)*

	Cost	Selling Price	Markup	Markup Rate
7.	$26.97	$49.95	$22.98	85.2%
9.	$40.98	$74.38	$33.40	81.5%

	List Price	Sale Price	Discount	Discount Rate
11.	$39.95	$29.95	$10.00	25%
13.	$23.69	$18.95	$4.74	20%

15. 2 hours   **17.** 7 minutes; $1.18

19. (a) $87, $1537   (b) $1037

21. 12.5%   **23.** Local store at $47.99

25. (a) $y = 300 + 0.03x$

(b)

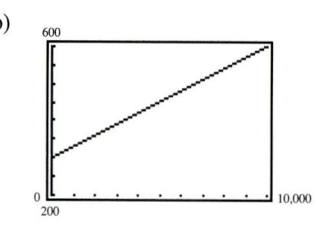

(c) $465

27. $\frac{4}{3}$ **29.** 54   **31.** $65t$

	Cost	Selling Price	Markup	Markup Rate
33.	$69.29	$125.98	$56.69	81.8%
35.	$13,250.00	$15,900.00	$2650.00	20%
37.	$107.97	$199.96	$91.99	85.2%

	List Price	Sale Price	Discount	Discount Rate
39.	$189.99	$159.99	$30.00	15.8%
41.	$119.96	$59.98	$59.98	50%
43.	$995.00	$695.00	$300.00	30.2%

45. 3 hours   **47.** 18%   **49.** 23 minutes, $6.50

51. $960.70   **53.** $35,714.29   **55.** 10.75 hours

57. $60.00

Section 4.5 *(page 263)*

7. 784 square feet   **9.** 2.39 meters   **11.** $540

13. $15,975   **15.** $12\pi \approx 37.70$ cubic meters   **17.** $\frac{2A}{b}$

19. $\frac{S}{1+R}$ **21.** 165 miles   **23.** 5.6 hours

25. 2112 feet per second   **27.** 0.176 hour $\approx$ 10.6 minutes

29. 360 meters per minute

31.

Corn x	Soybeans $100 - x$	Price per Ton of the Mixture
0	100	$200
20	80	$185
40	60	$170
60	40	$155
80	20	$140
100	0	$125

(a) Decreases; (b) Decreases; (c) Average of the two prices

33. 32 dimes, 18 quarters

35. 16 dozen roses, 8 dozen carnations

37. -30 **39.** -4 **41.** 13   **43.** $2915

45. 8 meters   **47.** 144 square feet

49. 24 square inches   **51.** 96 cubic inches   **53.** 11%

55. $1200   **57.** $\frac{E}{I}$ **59.** $\frac{V}{wh}$ **61.** $\frac{A - P}{Pt}$

63. $\frac{3V + \pi h^3}{3\pi h^2}$ **65.** 28 miles   **67.** 1154 miles per hour

69. Solution 1: 25 gallons   **71.** Solution 1: 5 quarts
Solution 2: 75 gallons   Solution 2: 5 quarts

73. $\frac{8}{7}$ gallons   **75.** 30 pounds at $2.49 per pound
70 pounds at $3.89 per pound

77. 8 nickels, 12 dimes

79. Candidates A and B: 250 votes each
Candidate C: 500 votes

81. $3\frac{3}{7}$ hours   **83.** 15

Review Exercises *(page 270)*

1.

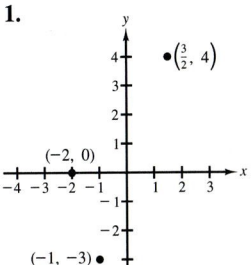

3. II **5.** II

7. (a) Solution
  (b) Not a solution

9. (a) $125 billion
  (b) 1989
  (c) 12%

11.

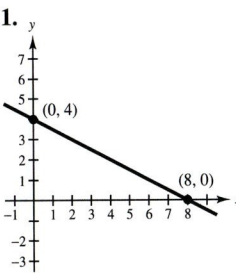

13.

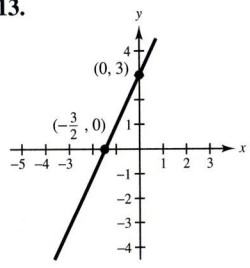

15.

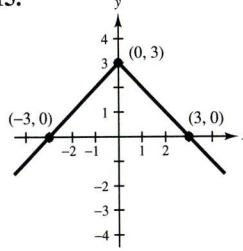

17.

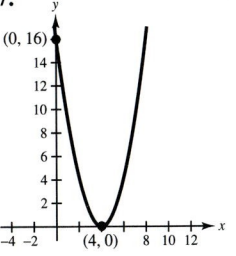

19.

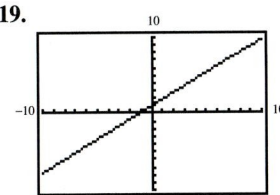

21.

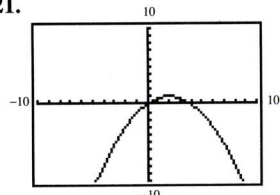

23.

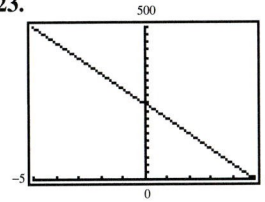

25. $(-5.2, 0)$, $(5.2, 0)$, $(0, 5.2)$

27. $y = 125 + 0.27x$ **29.** $114.75

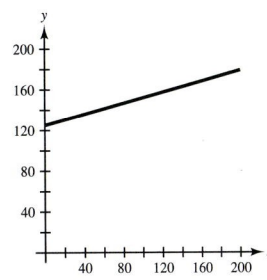

31. $181.35   **33.** Local store at $90.38   **35.** 9.37 hours

37. 30 feet $\times$ 26 feet   **39.** $285,714

41. $\dfrac{2A}{r^2}$ **43.** $\dfrac{30}{11} \approx 2.7$ hours

Chapter Test *(page 272)*

1.

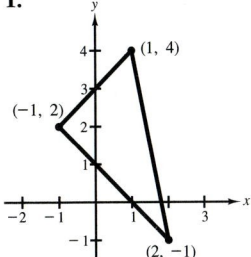

2.

x	2	4	6	8	10	12
$y = 0.75x + 4$	5.5	7	8.5	10	11.5	13

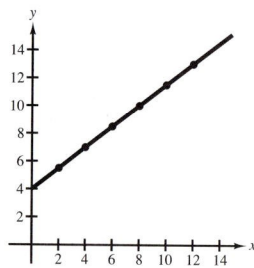

3. (a) Not a solution
(b) Solution
(c) Solution
(d) Not a solution

4. $(-4, 0)$, $(0, 3)$

5. 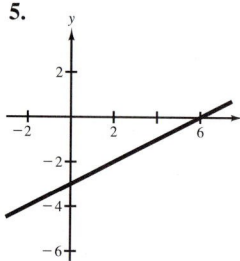　**6.**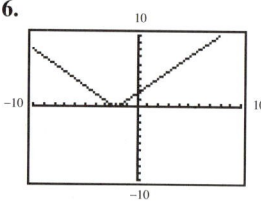

7. 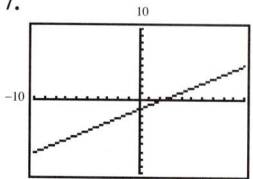　**8.**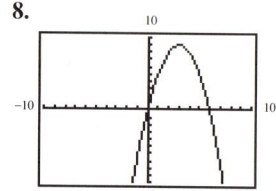

9. 40%　**10.** 8 dimes and 12 quarters

11. 4　**12.** 48 miles per hour　**13.** $\frac{36}{7} \approx 5.1$ hours

14. 30, 31, 32　**15.** $\frac{S - C}{C}$　**16.** \$40　**17.** \$6250

18. Colors not available　**19.** Colors not available
20. Colors not available

CHAPTER 5

Section 5.1 (page 280)

7. Polynomial: $5 - 32x$
Standard Form: $-32x + 5$
Degree: 1
Leading Coefficient: -32

9. Polynomial: $8x + 2x^5 - x^2 - 1$　**11.** Binomial
Standard Form: $2x^5 - x^2 + 8x - 1$
Degree: 5
Leading Coefficient: 2

13. Monomial　**15.** No. Term is not of the form ax^k (k must be nonnegative).

17. Yes　**19.** $5x^3 - 10$　**21.** $6x^2$　**23.** $2x^3 + 2x^2 + 8$
25. $y^4 + 4$　**27.** $4z^2 - z - 2$　**29.** 1　**31.** $4x^2 + 8$
33. $-x^2 - 2x + 3$　**35.** $4t^3 - 3t^2 + 15$
37. $3x^3 + 4x + 10$　**39.** $-3x^3 + 1$
41. $2x^4 + 9x + 2$　**43.** $12z + 8$
45. (a) $T = 0.02t^2 + 0.35t + 115.9$
(b)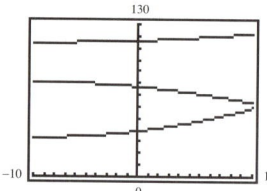

(c) B: Decreasing
P: Increasing
T: Increasing

47. $10z + 4$　**49.** $21x^2 - 8x$　**51.** $(3z)^4$; Degree: 4
53. 4^2m^4; Degree: 4　**55.** $3 \cdot 3 \cdot x \cdot x \cdot x \cdot x \cdot y \cdot y \cdot y$
57. $-6t + 8$　**59.** Polynomial: $x^3 - 4x^2 + 9$
Standard Form: $x^3 - 4x^2 + 9$
Degree: 3
Leading Coefficient: 1

61. Polynomial: 10
Standard Form: 10
Degree: 0
Leading Coefficient: 10

63. Polynomial: $3r + \pi r^2$ **65.** Yes
Standard Form: $\pi r^2 + 3r$
Degree: 2
Leading Coefficient: π

67. No. The first term is not of the form ax^k (k must be nonnegative).

69. $3^4 - 2x^3 - 3x^2 - 5x$ **71.** $3n^2 - 2$ **73.** $3x^2 + 2$

75. $4x^2 + 2x + 2$ **77.** $2b^3 - b^2$ **79.** $\frac{3}{2}y^2 + \frac{5}{4}$

81. $-2x^4 - 5x^3 - 4x^2 + 6x - 10$ **83.** $-2x^3$

85. $5x^3 - 6x^2$ **87.** $x^2 - 2x + 2$ **89.** $-u^2 + 5$

91. $-2x - 20$ **93.** $3x^3 - 2x + 2$

95. $8x^3 + 29x^2 + 11$ **97.** $4t^2 + 20$

99. $6v^2 + 90v + 30$ **101.** $2x^2 - 2x$ **103.** $6x$

105. (a) $-x^2 + 60x - 100$
(b)
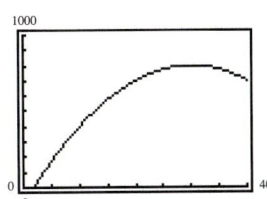
(c) $800

Section 5.2   *(page 291)*

7. $-2x^2$ **9.** $6b^3$ **11.** $18x^3$ **13.** $x^2 + 7x + 12$

15. $10x^2 - x - 21$ **17.** $-12x - 12x^3 + 24x^4$

19. $2st - 4t^2$ **21.** $x^4 - 5x^3 - 2x^2 + 11x - 5$

23. $x^3 - 8$ **25.** $x^3 + 27$ **27.** $x^4 - x^2 + 4x - 4$

29. $x^5 + 5x^4 - 3x^3 + 8x^2 + 11x - 12$ **31.** $x^2 - 4$

33. $x^2 + 12x + 36$ **35.** $4x^2 - 20xy + 25y^2$

37. $u^2 + v^2 - 2uv + 6u - 6v + 9$ **39.** $x^2 + 10x$

41. (a)
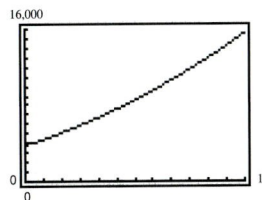
(b) $72.9076t^3 + 8680.11166t^2 + 143,782.8026t + 863,782.0956$
(c) $3,242,529.38 million

43. $2x(x + 2) = 2x^2 + 4x$ **45.** $7x - 8$ **47.** $-4z + 12$

49. $29,090.91   **51.** $\frac{5}{2}x^2$ **53.** $4t^3$ **55.** $-27t^3$

57. $3y - y^2$ **59.** $-x^3 + 4x$ **61.** $6t^2 - 15t$

63. $3x^3 - 6x^2 + 3x$ **65.** $2x^3 - 4x^2 + 16x$

67. $30x^3 + 12x^2$ **69.** $x^2 - 16x + 63$

71. $6x^2 - 7x - 5$ **73.** $x^2 + 3xy + 2y^2$

75. $12x^5 - 6x^4$ **77.** $-x^2 + 17x$

79. $4u^3 + 4u^2 - 5u - 3$ **81.** $x^2 - 25$

83. $y^2 - 81$ **85.** $4x^2 - 9y^2$ **87.** $a^2 - 4a + 4$

89. $9x^2 + 12x + 4$ **91.** $64 - 48z + 9z^2$

93. $x^2 + y^2 + 2xy + 2x + 2y + 1$ **95.** $8x$ **97.** Yes

99. $x^3 + 6x^2 + 12x + 8$ **101.** (a) $6w$ (b) $2w^2$

103. $2x[2(x + 1)] = 4x^2 + 4x$

105. $4x = (x + 4)(x + 3) - x^2 - 3x - 12$

107. (a)
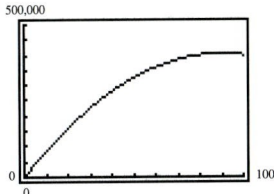
(b) $900x - 0.5x^2$
(c) $325,000; Increase

Mid-Chapter Quiz   *(page 295)*

1. No. The term $-3/x$ is not of the form ax^k (k must be nonnegative).

2. Degree: 4
Leading coefficient: -3

3. $3x^5 - 3x + 1$ **4.** False.
$(x - 1)(x + 5) = x^2 + 4x - 5$

5. $4y + 4$ **6.** $-v^3 + v^2 + 6v - 5$ **7.** $3s - 11$

8. $3x^2 + 5x - 4$ **9.** $50r^3$ **10.** $-8m^4$ **11.** $\dfrac{16a^4}{9}$

12. $2y^2 + 7y - 15$ **13.** $16 - 24x + 9x^2$ **14.** $4u^2 - 9$

15. $5x^4 + 3x^3 - 2x + 2$ **16.** $2x^3 - 4x^2 + 3x + 1$

17. $6x^3 - x^2 - 33x - 5$

18. $5x^5 - 21x^4 + 18x^3 + 3x^2 - 9x$

19. $10x + 36$ **20.** $x^2 + 4x$

Section 5.3 *(page 302)*

7. x^3 **9.** $\dfrac{4^4}{x^2}$ **11.** $-\dfrac{x}{2}$ **13.** $\dfrac{3s^3}{2r^2}$ **15.** $\dfrac{1}{2z}$

17. $\dfrac{4u^6}{3v^2}$ **19.** $z + 1$ **21.** $8z^2 + 3z - 2$ **23.** $x - 2$

25. $y + 2$ **27.** $4x - 1 + \dfrac{2}{x + 1}$

29. $x^2 - 2x + 5 + \dfrac{3}{x - 2}$ **31.** $2x$

33. Error; you can only cancel common factors of the numerator and denominator.

35. Valid   **37.** 15.5   **39.** 455   **41.** $4n^2 + 8n + 3$

43. $2y^2$ **45.** $\dfrac{1}{z^3}$ **47.** 1   **49.** $\dfrac{27}{ab}$ **51.** $4z^2$

53. $\dfrac{8b}{3}$ **55.** $-\dfrac{11y}{2}$ **57.** $\dfrac{3y}{x^3}$ **59.** $z - 3$

61. $5x - 2$ **63.** $-5z^2 - 2z$ **65.** $-4x + 3$

67. $m^2 + 3 - \dfrac{4}{m}$ **69.** $x + 5$ **71.** $3x - 1$

73. $2z - 1$ **75.** $6t + 1$ **77.** $x^2 + 2x + 4$

79. $x^2 + 2x - 3$ **81.** $7 - \dfrac{11}{x + 2}$ **83.** $x - 3 + \dfrac{18}{x + 3}$

85. $x + 2 + \dfrac{2}{2x + 3}$ **87.** $2z + \dfrac{5}{2} + \dfrac{5}{2(2z - 1)}$

89. $3t^2 + t + 1 - \dfrac{4}{t + 2}$ **91.** $x^3 + x^2 + x + 1$

93. $5uv$

95. (a) $1 + \dfrac{10}{t + 8}$

(b)

t	0	10	20	30	40	50	60
$\dfrac{t + 18}{t + 8}$	2.25	1.56	1.36	1.26	1.21	1.17	1.15

(c) Approaches 1

97.

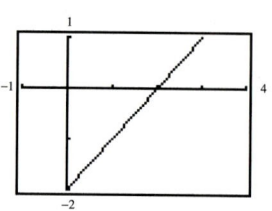

The first expression is not defined at $x = 1$.

99. b, e

Section 5.4 *(page 310)*

7. $\dfrac{5}{x^4}$ **9.** $\dfrac{z^4}{2}$ **11.** $\dfrac{1}{9}$ **13.** $\dfrac{9}{16}$ **15.** 0.0048

17. 239.66   **19.** x^5 **21.** $\dfrac{1}{4x^4}$ **23.** $\dfrac{a^6}{64b^9}$

25. $-2x^3$ **27.** $\dfrac{10}{x}$ **29.** $-\dfrac{81}{16}$ **31.** 1,090,000

33. 0.00852   **35.** 1.637×10^9 **37.** 4.35×10^{-4}

39. 4.984×10^{12} **41.** 1.2885×10^{10}

43. 8.45 minutes   **45.** 7   **47.**

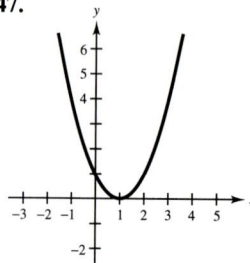

49. (a) -4 (b) 10   **51.** $-\dfrac{1}{64}$ **53.** 16   **55.** $\dfrac{1}{128}$

57. $\dfrac{9}{4}$ **59.** $\dfrac{1}{y^6}$ **61.** $\dfrac{1}{x^2}$ **63.** $\dfrac{1}{y^6}$ **65.** $\dfrac{1}{s^2}$

67. $\dfrac{1}{b^5}$ **69.** $\dfrac{1}{9x^4y^2}$ **71.** 1   **73.** $\dfrac{x^2}{9z^4}$ **75.** 1

77. $-\dfrac{375}{64}$ **79.** 6.21   **81.** 0.0867   **83.** 8003.05

85. 9.3×10^7 **87.** 4.392×10^{-3} **89.** 3.0981×10^6
91. 3.35544×10^{32} **93.** 1.15743×10^{-22}
95. 1.38×10^{-16}
97. (a)

t	0	2	4
$24{,}000(1.2)^{-t}$	\$24,000	\$16,667	\$11,574

t	6	8
$24{,}000(1.2)^{-t}$	\$8053	\$5582

(b) (c) 31 years

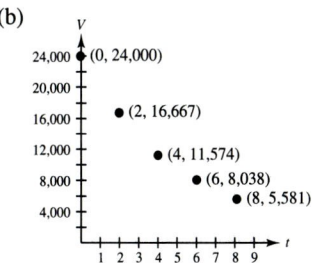

99. 1

Review Exercises   *(page 315)*

1. Polynomial: $10x - 4 - 5x^3$
Standard Form: $-5x^3 + 10x - 4$
Degree: 3
Leading Coefficient: -5
3. Polynomial: $2(x - 5) + 10$
Standard Form: $2x$
Degree: 1
Leading Coefficient: 2
5. $x^4 + x^2 + 2$ **7.** $\frac{3}{8} + \frac{1}{8} = \frac{4}{8}$ **9.** $(2x)^4 = 2^4 x^4$
11. $3^{-2} = \frac{1}{3^2} = \frac{1}{9}$ **13.** $\frac{7 - x}{7} = 1 - \frac{x}{7}$
15. $-3(x - 2) = -3x + 6$ **17.** $(x + 2)^2 = x^2 + 4x + 4$
19. $3x - 1$ **21.** $-t^2 + 2t$ **23.** $-x^2 + 2x$
25. $-5x^3 - 5x - 2$ **27.** $7y^2 - y + 6$ **29.** $2x^2 + 8x$
31. $2x^2 + 2x - 12$ **33.** $2x^3 + 13x^2 + 19x + 6$
35. $u^2 - 6u + 5$ **37.** $x + 3$ **39.** $\frac{7}{3}(x + 4)$
41. $x + 2$ **43.** $8x + 5 + \dfrac{2}{3x - 2}$

45. $2x^2 + 4x + 3 + \dfrac{5}{x - 1}$ **47.** $x^2 - 2$
49. $x^2 + 6x + 9$ **51.** $16x^2 - 56x + 49$ **53.** $u^2 - 36$
55. $9t^2 - 1$ **57.** $a^2 - b^2 - 2a + 1$ **59.** $\frac{1}{16}$ **61.** $\frac{1}{36}$
63. $\frac{125}{27}$ **65.** $-\frac{2}{5}$ **67.** 9×10^6 **69.** 3.7×10^4
71. $\dfrac{1}{9a^4}$ **73.** $\dfrac{x^4}{y^6}$ **75.** $\dfrac{1}{t^3}$ **77.** $\dfrac{25}{y^2}$
79. $80 - 2x^2$ **81.** $2x^2 + 8x + 8$ **83.** (a) $4x - 6$
(b) $x^2 - 3x$
85. $10p^3 - 20p^4 + 10p^5$ **87.** $x^2 - y^2 = (x + y)(x - y)$

Chapter Test   *(page 318)*

1. Degree: 4   **2.** $z^4 + 2z^2 - 3$
Leading coefficient: -3
3. $2z^2 - 3z + 15$ **4.** $7u^3 - 1$ **5.** $-y^2 + 8y + 3$
6. $-6x^2 + 12x$ **7.** $10b^2 + b - 3$ **8.** $9x^3$
9. $2z^3 + z^2 - z + 10$ **10.** $x^2 - 10x + 25$
11. $4x^2 - 9$ **12.** $3x + 5$ **13.** $x^2 + 2x + 3$
14. $2x^2 + 4x - 3 - \dfrac{2}{2x + 1}$ **15.** $\dfrac{2a}{3}$ **16.** $\dfrac{x^4}{9y^6}$
17. (a) $\frac{3}{8}$ (b) $22{,}500{,}000{,}000$ **18.** $2x^2 + 11x - 6$
19. $384{,}000{,}000$ **20.** 1.013×10^5

CHAPTER 6

Section 6.1   *(page 325)*

7. 6   **9.** $2x$ **11.** $3yz^2$ **13.** 1   **15.** $3(x + 1)$
17. $8(t - 2)$ **19.** $-5(5x + 2)$ **21.** $2x(6x - 1)$
23. $10ab(1 + a)$ **25.** $25(4 + 3z - 2z^2)$
27. $(x - 3)(x + 5)$ **29.** $(z + 5)(z^2 + 1)$
31. $(x - 5)(x + y)$ **33.** $(t - 3)(t^2 + 2)$
35. $-5(2x - 1)$ **37.** $-(x^2 - 2x - 4)$
39. $\frac{1}{4}(2x + 3)$ **41.** $44 - h$
43.   **45.**

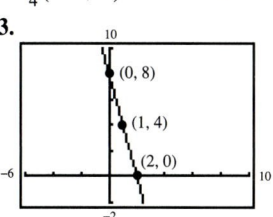

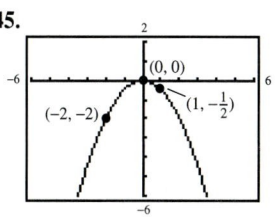

47. 3%   **49.** 2   **51.** z^2 **53.** $u^2 v$ **55.** $14a^2 b^2$

57. $6(z-1)$ **59.** $6(4y^2-3)$ **61.** $x(x+1)$

63. $u(25u-14)$ **65.** $2x^3(x+3)$

67. No common factor

69. $-5r(2r^2+7)$ **71.** $8a^3 b^3(2+3a)$

73. $4(3x^2+4x-2)$ **75.** $3x^2(3x^2+2x+6)$

77. $5u(2u+1)$ **79.** $(s+10)(t-8)$

81. $a(b+2)(a-1)$ **83.** $(a+b)(c+7)(c+8)$

85. $(a-4)(a+b)$ **87.** $(y-4)(x+2)$

89. $(x+2)(x^2+1)$ **91.** $(z+3)(z^2-2)$

93. $-3(x-1000)$ **95.** $-2(x^2-6x-2)$

97. $\frac{1}{5}(10y-1)$ **99.** $\frac{1}{16}(14x+5y)$

101. 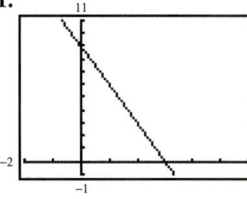   **103.**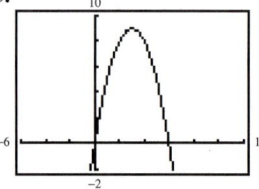

105. $x+1$ **107.** $2\pi r(r+h)$

109. $R = x(900-0.1x)$; $p = 900-0.1x$

Section 6.2   *(page 333)*

7. $x^2+4x+3 = (x+3)(x+1)$

9. $y^2-2y-15 = (y+3)(y-5)$

11. $(x+1)(x+11)$ **13.** $(x+12)(x+1)$
$(x-1)(x-11)$ $\quad\quad\quad (x-12)(x-1)$
$\quad\quad\quad\quad\quad\quad\quad\quad (x+6)(x+2)$
$\quad\quad\quad\quad\quad\quad\quad\quad (x-6)(x-2)$
$\quad\quad\quad\quad\quad\quad\quad\quad (x+4)(x+3)$
$\quad\quad\quad\quad\quad\quad\quad\quad (x-4)(x-3)$

15. $(x+4)(x+2)$ **17.** $(x-3)(x+2)$ **19.** $(x-9)(x-8)$

21. Prime   **23.** $(u+2)(u-24)$ **25.** $(a+5b)(a-3b)$

27. $\pm 8, \pm 16$ **29.** $5, -7$ **31.** $3(x+5)(x+2)$

33. Prime   **35.** $2xy(x+3y)(x-y)$

37. $(x+3)(x+1)$ **39.** $(x+3)(x+2)$

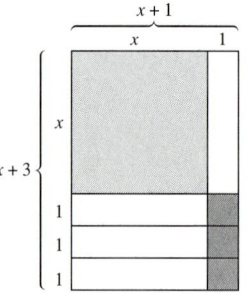   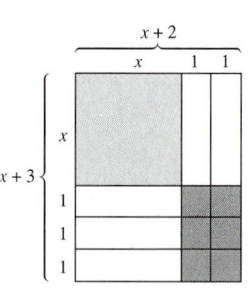

41. $-a^3+a^2$ **43.** $u^2-5u-24$

45.

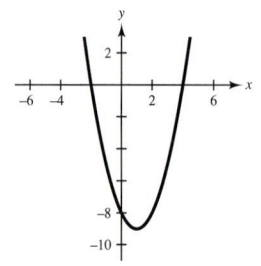

47. \$3,975,000   **49.** $a^2+a-6 = (a+3)(a-2)$

51. $z^2-5z+6 = (z-3)(z-2)$

53. $(x-8)(x-5)$ **55.** $(z-4)(z-3)$

57. $(x+5)(x-3)$ **59.** $(x+10)(x-7)$

61. Prime   **63.** $(x+15)(x+4)$

65. $(x+2y)(x-y)$ **67.** $(x+5y)(x+3y)$

69. $(x-9z)(x+2z)$ **71.** $x(x-10)(x-3)$

73. $4(y-3)(y+1)$ **75.** $10x(x+2y)(x+3y)$

77.   **79.**

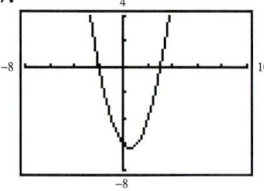   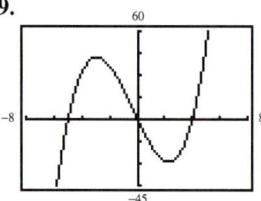

81. (a) $4x(x-2)(x-3)$

(b)

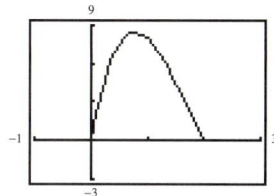

0.785 foot

Math Matters *(page 335)*

The first three prime numbers that are greater than or equal to 1000 are 1009, 1013, and 1019.

Section 6.3 *(page 341)*

7. $5x^2 + 18x + 9 = (x+3)(5x+3)$

9. $4z^2 - 13z + 3 = (z-3)(4z-1)$

11. $(5x+3)(x+1)$ $\quad (5x-3)(x-1)$
$(5x+1)(x+3)$ $\quad (5x-1)(x-3)$

13. $(5x+12)(x+1)$ $\quad (5x-12)(x-1)$
$(5x+6)(x+2)$ $\quad (5x-6)(x-2)$
$(5x+4)(x+3)$ $\quad (5x-4)(x-3)$
$(5x+1)(x+12)$ $\quad (5x-1)(x-12)$
$(5x+2)(x+6)$ $\quad (5x-2)(x-6)$
$(5x+3)(x+4)$ $\quad (5x-3)(x-4)$

15. $(2x+3)(x+1)$ $\quad$ **17.** $(2y-1)(y-1)$

19. Prime $\quad$ **21.** $(8z-5)(2z-3)$

23. $-(2x-3)(x+1)$ $\quad$ **25.** $(1+6x)(1-10x)$

27. $x(x-3)$ $\quad$ **29.** $(v+7)(v-6)$ $\quad$ **31.** Prime

33. $\pm 11, \pm 13, \pm 17, \pm 31$ $\quad$ **35.** $\pm 1, \pm 4, \pm 11$

37. $(3x+1)(x+2)$ $\quad$ **39.** $(5x-2)(3x-1)$

---

41. (a) $y_1 = y_2$ $\qquad$ **43.** $(2x+1)(x+2)$

(b)

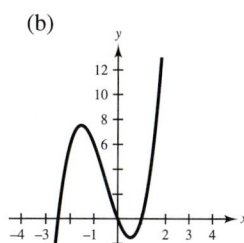

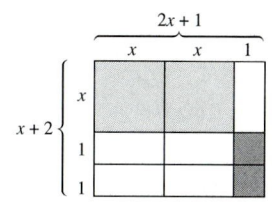

(c) $\left(-\frac{5}{2}, 0\right), (0,0), (1,0)$

45. $-3t^2 - 6t$ $\quad$ **47.** $x^2 + 8x + 16$

49. 120 miles $\leq x \leq$ 165 miles

51. $5a^2 + 12a - 9 = (a+3)(5a-3)$

53. $5x^2 + 19x + 12 = (x+3)(5x+4)$

55. $(4y+1)(y+1)$ $\quad$ **57.** $(2x-3)(x+1)$ $\quad$ **59.** Prime

61. $(5a-2)(3a+4)$ $\quad$ **63.** $(3u-2)(6u+1)$

65. $(5t+6)(2t-3)$ $\quad$ **67.** $(5m-3)(3m+5)$

69. Prime $\quad$ **71.** $(2-3x)(2+x)$

73. $-(6x+5)(x-2)$ $\quad$ **75.** $3y(5y+6)$

77. $(u-3)(u+9)$ $\quad$ **79.** $(x+10)(x-4)$

81. $2(3x-2)(x+2)$ $\quad$ **83.** $y^2(3-2y)(5+y)$

85. $9(u+3)(u-1)$ $\quad$ **87.** $(2x+3)(x-1)$

89. $(3x+4)(2x-1)$ $\quad$ **91.** $(3a+5)(a+2)$

93. $(8x-3)(2x+1)$ $\quad$ **95.** $-1, -7, -10, -22$

97. $y_1 = y_2$ $\qquad\qquad$ **99.** $y_1 = y_2$

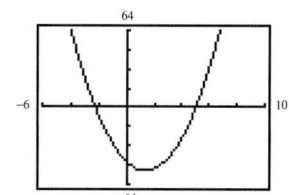

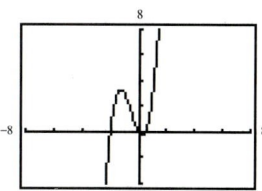

101. $l = 3x + 1$

Mid-Chapter Quiz *(page 344)*

1. $\frac{1}{3}(2x-3)$ $\quad$ **2.** $xy(x-y)$ $\quad$ **3.** $(y+7)(y-6)$

4. $(x - 1)(2x + 1)$ **5.** $10(x^2 + 7)$

6. $2a^2 b(a - 2b)$ **7.** $(x + 2)(x - 3)$

8. $(t - 3)(t^2 + 1)$ **9.** $(y + 6)(y + 5)$

10. $(u + 6)(u - 5)$ **11.** $x(x - 6)(x + 5)$

12. $2y(x + 8)(x - 4)$ **13.** Prime

14. $(3 + z)(2 - 5z)$ **15.** $(3x - 2)(2x + 1)$

16. $2s^2(5s^2 - 7s + 1)$ **17.** $7, 8, 13$ **18.** $16, 21$

19. $(3x + 1)(x + 6)$ $(3x - 1)(x - 6)$ **20.** $x + 5$

$(3x + 6)(x + 1)$ $(3x - 6)(x - 1)$

$(3x + 2)(x + 3)$ $(3x - 2)(x - 3)$

$(3x + 3)(x + 2)$ $(3x - 3)(x - 2)$

Section 6.4   *(page 351)*

7. $(x + 6)(x - 6)$ **9.** $\left(u + \frac{1}{2}\right)\left(u - \frac{1}{2}\right)$

11. $2(x + 6)(x - 6)$ **13.** $(x^2 + 1)(x + 1)(x - 1)$

15. $(x - 2)^2$ **17.** $(5y - 1)^2$ **19.** $\left(b + \frac{1}{2}\right)^2$

21. $(x - 3y)^2$ **23.** ± 20 **25.** ± 2

27. $(x - 2)(x^2 + 2x + 4)$ **29.** $(1 + 2t)(1 - 2t + 4t^2)$

31. $x^2(x - 4)$ **33.** $(1 - 2x)^2$ **35.** $(x - 1)(2x - 1)$

37. $(3t + 4)(3t - 4)$ **39.** $(2y - 5)(y + 1)$

41. $2(t - 2)(t^2 + 2t + 4)$ **43.** $(x + 1)^2 = x^2 + 2x + 1$

45. $\pi(R - r)(R + r)$ **47.** -1 **49.** $\frac{16}{9}$ **51.** 6955

53. $(9 + x)(9 - x)$ **55.** $\left(t + \frac{1}{4}\right)\left(t - \frac{1}{4}\right)$

57. $(4y + 3)(4y - 3)$ **59.** $-z(10 + z)$

61. $(y^2 + 9)(y + 3)(y - 3)$ **63.** $2(2 + 5x)(2 - 5x)$

65. $(z + 3)^2$ **67.** $(2t + 1)^2$ **69.** $\left(2x - \frac{1}{4}\right)^2$

71. $(2y + 5z)^2$ **73.** $(3a - 2b)^2$ **75.** ± 2 **77.** ± 12

79. 9 **81.** 4 **83.** $(y + 4)(y^2 - 4y + 16)$

85. $(3u + 2)(9u^2 - 6u + 4)$ **87.** $y^2(y + 5)(y - 5)$

89. $z^2\left(z + \frac{2}{3}\right)\left(z - \frac{2}{3}\right)$ **91.** $(x - 1)^2$ **93.** $(2v + 1)^2$

95. $2x(2 - x)(1 + x)$ **97.** $(9x + 1)(x + 1)$

99. $(x - 1)(x - 3)$ **101.** $(1 + x^2)(5 - x)$

103. $x(x^2 + 1)(x - 4)$ **105.** $(y + 1)(y + 3)$

107. $-z(z + 16)$ **109.** $3u(u + 3)(u - 3)$

111. $2(4x^2 + 1)$ **113.** $\left(y + \frac{1}{2}\right)\left(y^2 - \frac{1}{2}y + \frac{1}{4}\right)$

115. $2(2x - 1)(4x^2 + 2x + 1)$

117. $(x - 4)(x^2 - 5x + 7)$

119. $u(u^2 + 2u + 3)$ **121.** $(x^2 + 9)(x + 3)(x - 3)$

123. $(1 + x^2)(1 + x)(1 - x)$

125. $y_1 = y_2$ **127.** $y_1 = y_2$

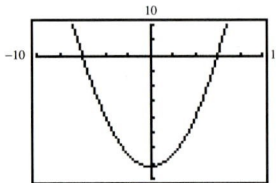

129. 441 **131.** 3599 **133.** $(x + 3)^2 + (1)^2$

135. $x(x + 4) = x^2 + 4x$

Section 6.5   *(page 361)*

7. $0, 5$ **9.** $-3, 0, 1$ **11.** $-4, 4$ **13.** $-3, 3$

15. $-\frac{1}{2}, 0$ **17.** $-2, 4$ **19.** $-\frac{1}{2}, 3$ **21.** $-2, 7$

23. $-2, 3$ **25.** $-3, 2, 3$ **27.** $t = 10$

29. $(-3, 0), (1, 0)$ **31.** $(0, 0), (3, 0)$

33.   **35.**

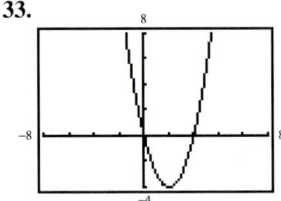   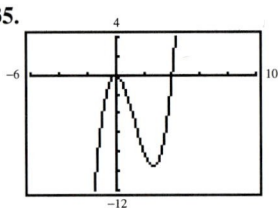

$(0, 0), (4, 0)$ $(0, 0), (4, 0)$

37. $8, 9$ **39.** 3 **41.** $\frac{3}{2}$ **43.** $\$750$ **45.** $0, 3$

47. $-25, 0, 3$ **49.** $-1, 2$ **51.** $-10, 10$ **53.** $-\frac{3}{2}, \frac{3}{2}$

55. $-2, 8$ **57.** $0, 2$ **59.** $-2, 8$ **61.** 1 **63.** -7

65. $\frac{3}{2}$ **67.** $-\frac{5}{3}, 1$ **69.** $-5, 3$ **71.** $-\frac{3}{2}, 1$

73. $0, -5$ **75.** $-4, 0, \frac{3}{2}$ **77.** $-3, -1, 1$

79. $(-3, 0), (4, 0)$ **81.** $\left(-\frac{5}{2}, 0\right), (0, 0), (1, 0)$

83.   **85.**

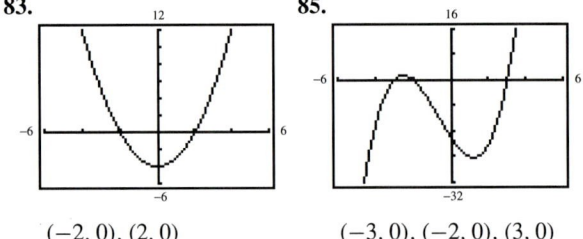

$(-2, 0), (2, 0)$ $(-3, 0), (-2, 0), (3, 0)$

87. $15, 16$ **89.** 9 inches $\times$ 12 inches

91. (a)

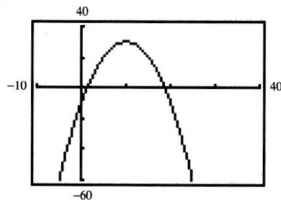

(b) 5, 15

(c) 5, 15

93. (a) $V = lwh$

$V = (x)(x)(2)$

$V = 2x^2$

(b)

x	2	4	6	8
V	8	32	72	128

(c) 14 inches $\times$ 14 inches

95. 0, 1

Review Exercises *(page 367)*

1. 10 **3.** $9ab$ **5.** $5x^2(1 + 2x)$ **7.** $4a(2 - 3a^2)$

9. $-6(x + 1)(3x - 1)$ **11.** $(y + 3)(y^2 + 2)$

13. $(a + 10)(a - 10)$ **15.** $(u + 3)(u - 1)$

17. $(x - 4)^2$ **19.** $(3s + 2)^2$ **21.** $(2x + 1)(2x + 3)$

23. $(10 + x)(5 - x)$ **25.** $\frac{1}{6}(2x + 5)$

27. $(x + 4)(3x + 2)$ **29.** $(x + 1)^2(x - 1)^2$

31. $(x - 7)(x + 4)$ **33.** $(3x + 2)(2x + 1)$

35. $3u(2u + 5)(u - 2)$ **37.** $(5x + 2y)(2x + y)$

39. $st(s + t)(s - t)$ **41.** $(2x - 1)(x - 1)$

43. $(3 - 2t)(9 + 6t + 4t^2)$ **45.** $-4a(2a + 1)^2$

47. $(x^2 + 1)(x + 2)$ **49.** $(x + 2)(x - 2)(x - 4)$

51. $\pm 6, \pm 10$ **53.** ± 12 **55.** $\pm 2, \pm 5, \pm 10, \pm 23$

57. $\pm 4, \pm 7, \pm 11, \pm 17, \pm 28, \pm 59$

59. $5, -7$ **61.** $2, -6$

63. $y_1 = y_2$ **65.** $y_1 = y_2$

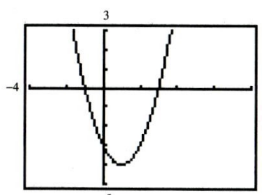 

67. (a)

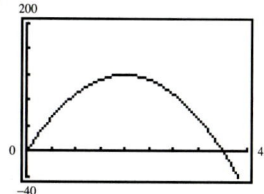

(b) $2x(x - 1)(2x - 3)$

(c) $x \approx 0.4$

69. $0, \frac{3}{2}$ **71.** $-9, 9$ **73.** 6 **75.** $-1, \frac{3}{4}$

77. $3, 4$ **79.** $-5, -1, 1$

81. (a)

(b) 20 (c) 20

83. 40 inches $\times$ 60 inches

85. 10 inches $\times$ 10 inches $\times$ 5 inches

Chapter Test *(page 369)*

1. $7x^2(1 - 2x)$ **2.** $(z + 7)(z - 3)$ **3.** $(t - 5)(t + 1)$

4. $(3x - 4)(2x - 1)$ **5.** $3y(2y + 5)(y + 5)$

6. $(2 + 5v)(2 - 5v)$ **7.** $(2x - 5)^2$

8. $(-z - 5)(z + 13)$ **9.** $(x + 2)(x + 3)(x - 3)$

10. $(4 + z^2)(2 + z)(2 - z)$ **11.** $\frac{1}{12}(8x - 9)$ **12.** ± 6

13. 36 **14.** $3x^2 - 3x - 6 = 3(x + 1)(x - 2)$

15. $-4, \frac{3}{2}$ **16.** $0, 2$ **17.** $-3, \frac{2}{3}$ **18.** $-\frac{3}{2}, 2$

19. 7 inches $\times$ 12 inches **20.** $\frac{5}{4}$ seconds

Cumulative Test: Chapters 4–6
(page 370)

1. Because $x = -2$, the point must lie in Quadrant II or Quadrant III.

2. (a) Not a solution

(b) Solution

(c) Solution

(d) Not a solution

3.

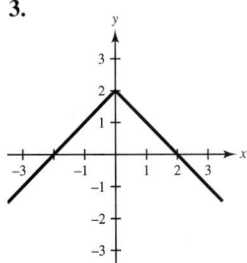

$(-2, 0), (2, 0), (0, 2)$

4.

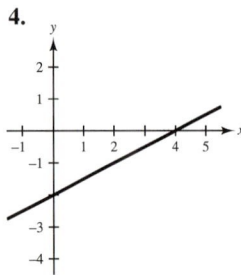

$(4, 0), (0, -2)$

5.

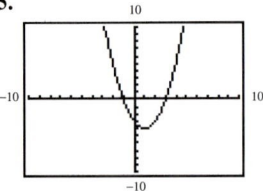

$(-1, 0), (3, 0)$

6.

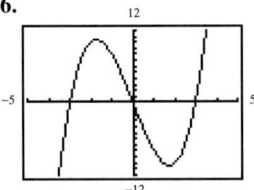

$(-3, 0), (0, 0), (3, 0)$

7. $-5x^2 + 5$ **8.** $-42z^4$ **9.** $3x^2 - 7x - 20$

10. $25x^2 - 9$ **11.** $25x^2 + 60x + 36$ **12.** $x + 12$

13. $x + 1 + \dfrac{2}{x - 4}$ **14.** $\dfrac{81}{16}$ **15.** $2u(u - 3)$

16. $(x + 2)(x - 6)$ **17.** $x(x + 4)^2$

18. $(x + 2)^2(x - 2)$ **19.** $0, 12$ **20.** $-\frac{3}{5}, 3$ **21.** $\dfrac{4}{x^2}$

22. $8000 at 7.5%; $4000 at 9%

23. $C = 125 + 0.35x$; $149.50

CHAPTER 7

Section 7.1   *(page 381)*

7.

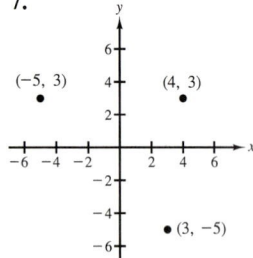

9.

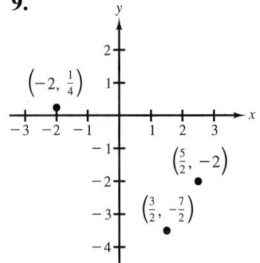

11. A: $(4, -2)$
B: $(-3, -2.5)$
C: $(3, 0.75)$

13.

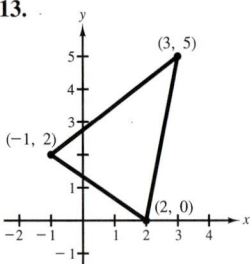

15.

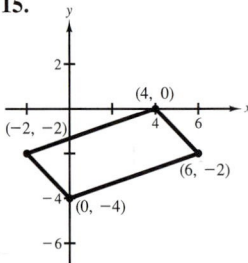

17. Quadrant III   **19.** Quadrants I and II

21. Quadrants II and IV   **23.** $(-5, 2)$ **25.** $(10, 0)$

27.

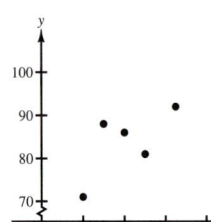

29. (a) Solution
(b) Not a solution
(c) Not a solution
(d) Solution

31.

x	-2	0	2	4	6
$y = 5x - 1$	-11	-1	9	19	29

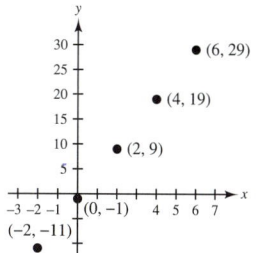

33.

x	-2	0	2	4	6
$y = 4x^2 + x - 2$	12	-2	16	66	148

35.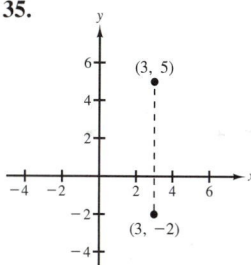

Distance: 7
Vertical line

37.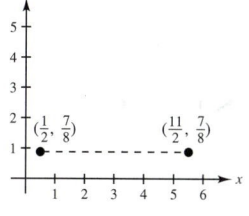

Distance: 5
Horizontal line

39. 5   **41.** 15   **43.** $(x, y) = (10, 2)$

6, 8, 10

45.

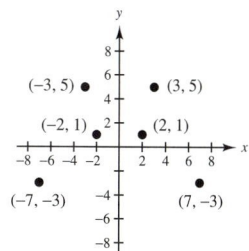

Reflection through the y-axis

47. $-\frac{1}{2} < x < \frac{1}{2}$ **49.** $-6 \le x \le 6$ **51.** \$29,018

53.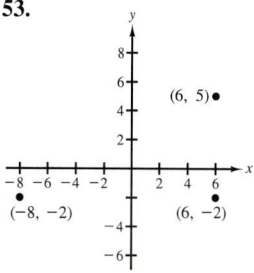

55.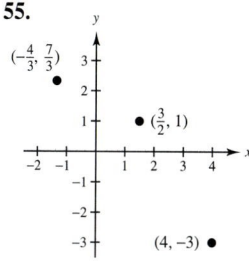

57. A: $(-2, 4)$
B: $(0, -2)$
C: $(4, -2)$

59.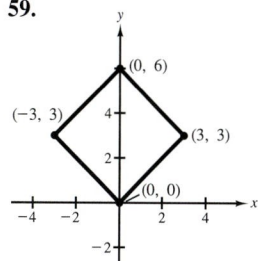

61.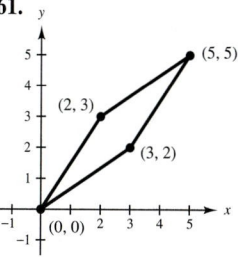

63. (a) Quadrant IV   **65.** Quadrants II and III
(b) Quadrant II

67. Quadrants I and III   **69.** $(3, -4)$ **71.** $(-10, -10)$

73.

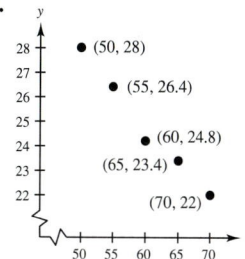

75. (a) Solution   **77.** (a) Not a solution
(b) Solution   (b) Solution
(c) Solution   (c) Solution
(d) Not a solution   (d) Not a solution

79.

x	-4	$\frac{2}{5}$	4	8	12
$y = -\frac{5}{2}x + 4$	14	3	-6	-16	-26

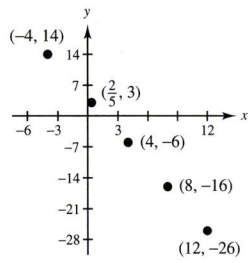

81.

x	100	150	200	250	300
$y = 28x + 3000$	5800	7200	8600	10,000	11,400

83.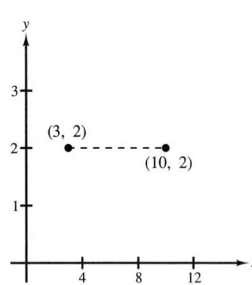

Distance: 7
Horizontal line

85.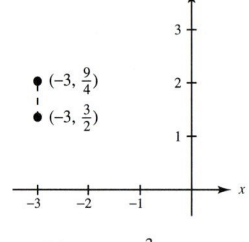

Distance: $\frac{3}{4}$
Vertical line

87. $(x, y) = (4, -4)$ **89.** $\sqrt{61}$ **91.** $\sqrt{29}$
8, 7, $\sqrt{113}$

93. Not collinear   **95.** Collinear

97. $3 + \sqrt{26} + \sqrt{29} \approx 13.48$

99.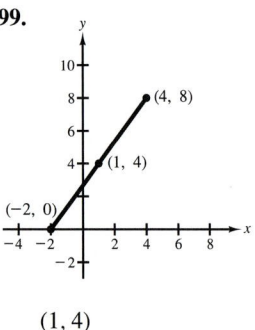

$(1, 4)$

101.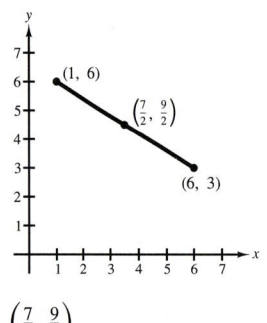

$\left(\frac{7}{2}, \frac{9}{2}\right)$

103. $(-3, -4) \implies (-1, 1)$ **105.** 18.55 feet
$(1, -3) \implies (3, 2)$
$(-2, -1) \implies (0, 4)$

Section 7.2 *(page 392)*

5. e   **7.** f   **9.** d

11.

x	-4	-2	0	2	4
y	11	7	3	-1	-5

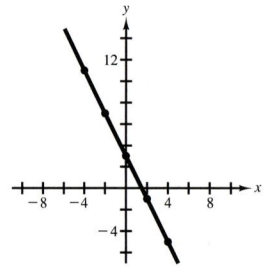

13.

x	± 2	-1	0	2	± 3
y	0	3	4	0	-5

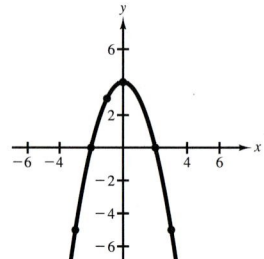

15.

x	-4	-2	0	2	4
y	0	2	4	2	0

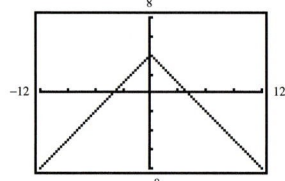

17.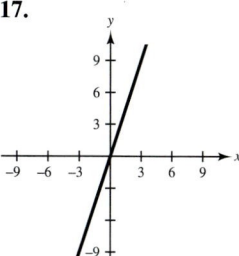

19.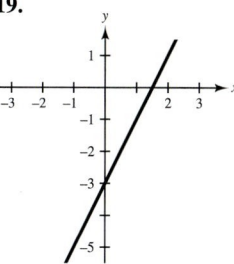

21.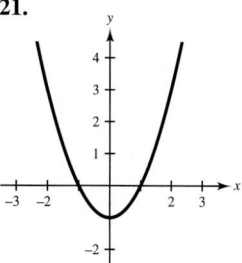

23.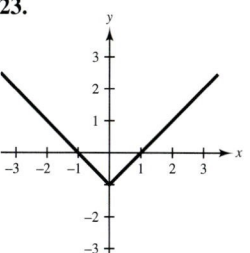

25. $(10, 0), (0, 5)$ **27.** $(\pm 5, 0), (0, -25)$

29.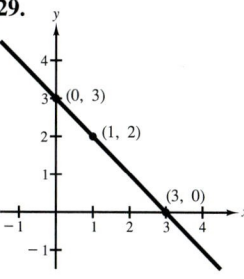

31.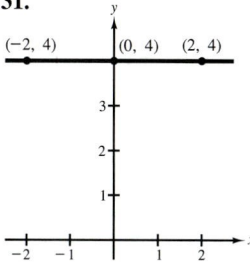

33.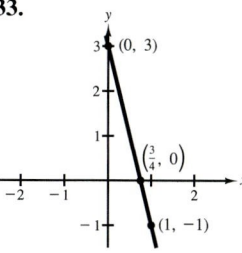

35.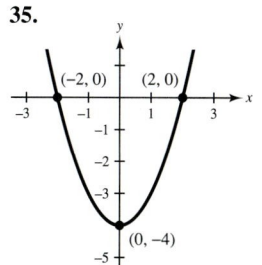

37.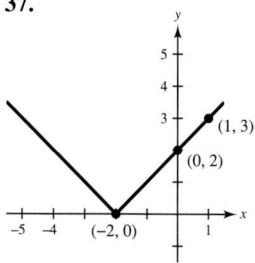

39. (a)

x	0	3	6	9	12
F	0	4	8	12	16

(b)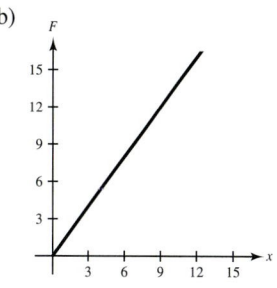

(c) Length doubles

41. a. It is difficult to assess the increase in sales.

43. Multiplicative Inverse Property

45. Distributive Property   **47.** 72 pounds

49.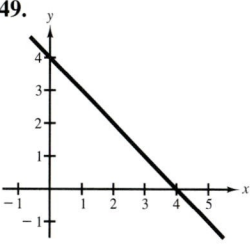

51.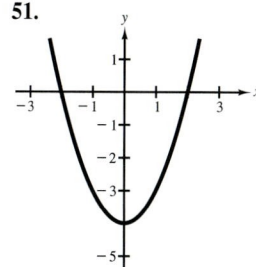

53.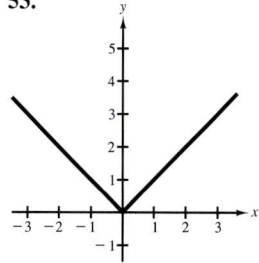

55.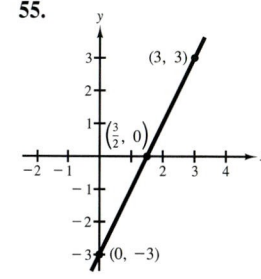

57.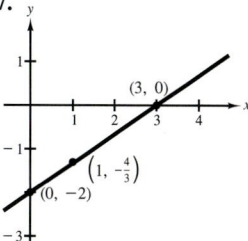

59.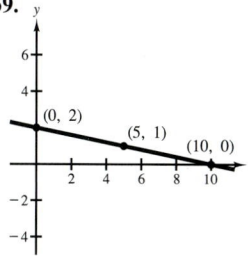

61.

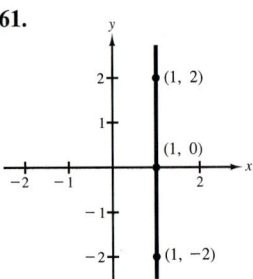

63.

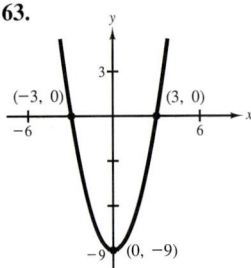

65.

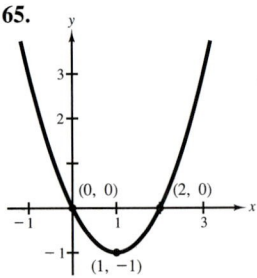

67.

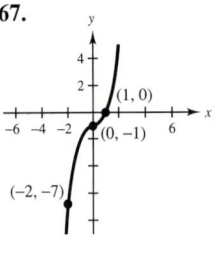

69.

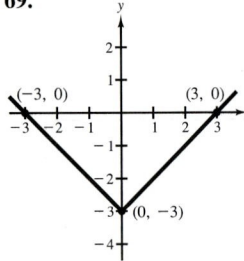

71.

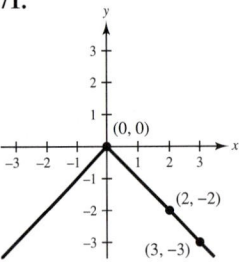

73. (0, 3)   **75.** (2, 0), (0, 2)

77.

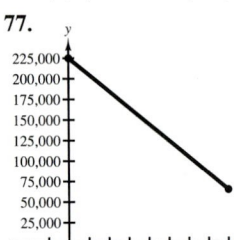

79.

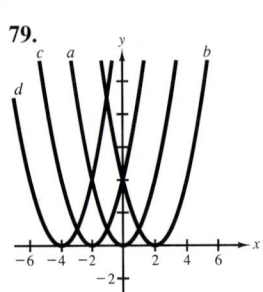

Horizontal translations

Section 7.3 *(page 401)*

5. c   **7.** d   **9.** a

11.

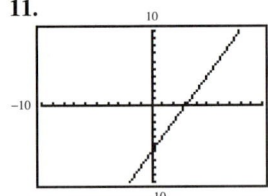

13.

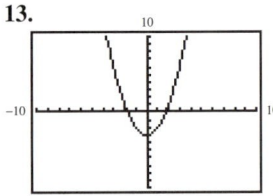

15.

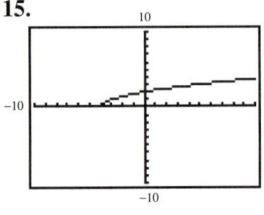

17.

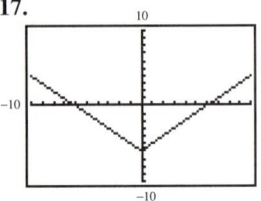

19.

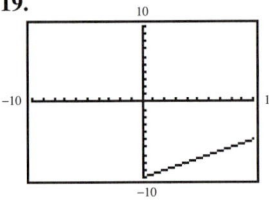

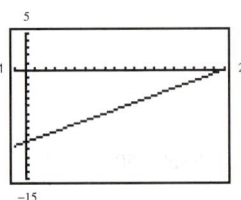

21.

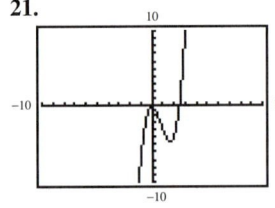

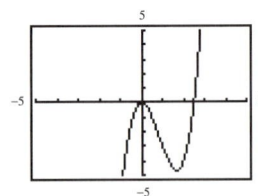

23.

Xmin = −2
Xmax = 8
Xscl = 1
Ymin = −1
Ymax = 17
Yscl = 1

25.

Xmin = −5
Xmax = 5
Xscl = 1
Ymin = −3
Ymax = 6
Yscl = 1

27.

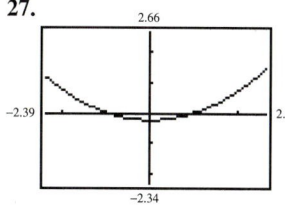

$(0, -0.25)$

29.

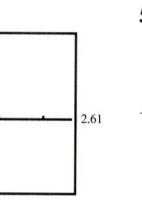

$(-1, -1)$

31. Evaluate the expression for values of x near the lowest point on the graph.

33. $(-2.5, 0), (3, 0)$ **35.** $(0.68, 0)$

37. Identical graphs
Distributive Property

39. Identical graphs
Associative Property of Addition

41.

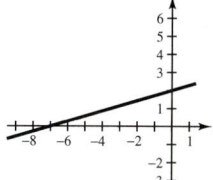

$$\begin{array}{l} \text{Xmin} = 20 \\ \text{Xmax} = 80 \\ \text{Xscl} = 5 \\ \text{Ymin} = 20 \\ \text{Ymax} = 500 \\ \text{Yscl} = 50 \end{array}$$

43. $y = 4 - 3x$ **45.** $y = \frac{1}{3}(4 - x^2)$

47.

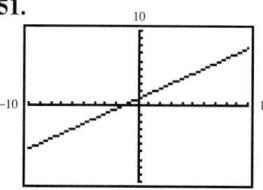

49. $9.35 + 0.75q$

51.

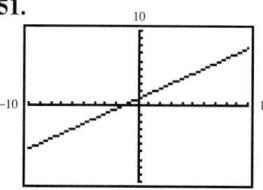

53.

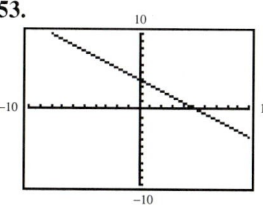

55.

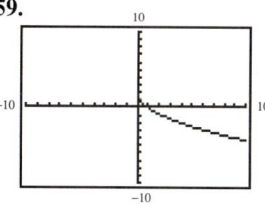

57.

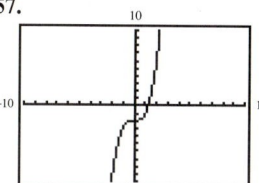

59.

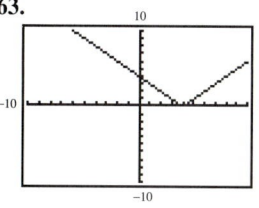

61.

63.

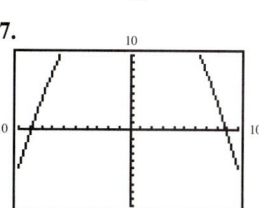

65.

 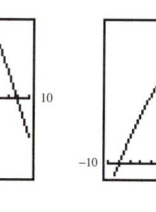

67.

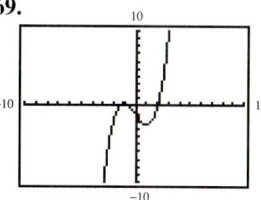 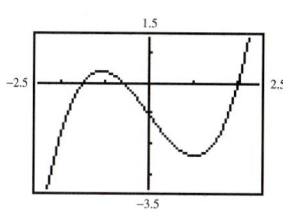

69.

71.

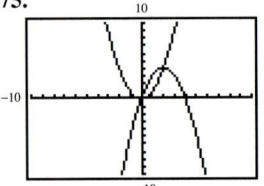

Xmin = −10
Xmax = 40
Xscl = 5
Ymin = −10
Ymax = 100
Yscl = 10

73.

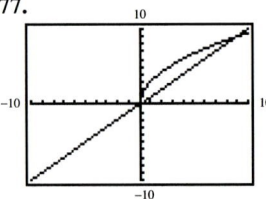

Xmin = −2
Xmax = 10
Xscl = 1
Ymin = −5
Ymax = 5
Yscl = 1

75.

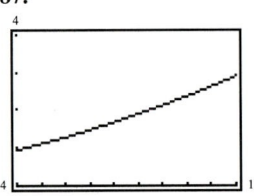

Intersect twice

77.

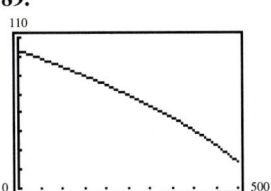

Intersect twice

79. $(-0.65, 0), (4.65, 0)$ **81.** No intercepts

83. $y = 2 - x$

85.

Xmin = 0
Xmax = 50
Xscl = 5
Ymin = −10
Ymax = 100
Yscl = 20

Utilizes the entire calculator display

87.

Model: $2.89

89.

Mid-Chapter Quiz *(page 405)*

1.

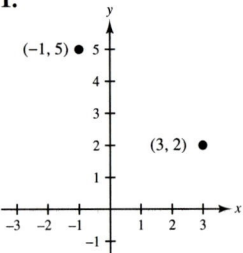

Distance: 5

2.

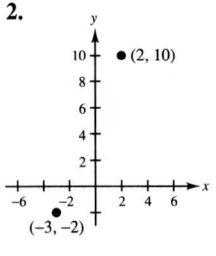

Distance: 13

3. Quadrants I and II   **4.** $(10, -3)$

5. (a) Not a solution   **6.** $(-8, 0), (0, 6)$
   (b) Solution
   (c) Solution
   (d) Solution

7.

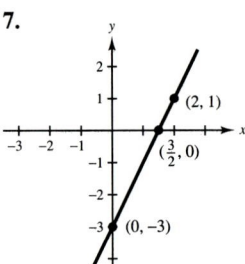

8.

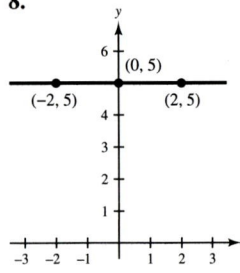

9.

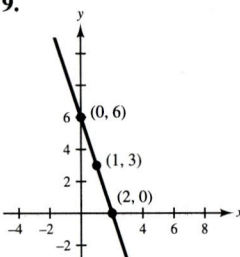

10.

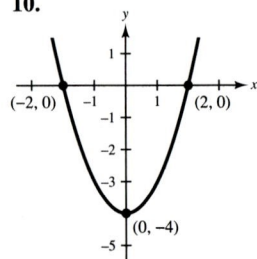

11.

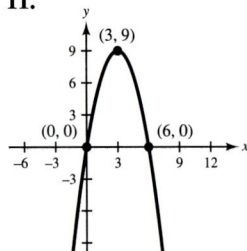

12.

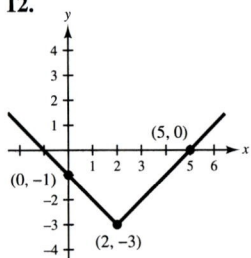

13.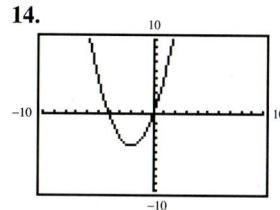

14.

15.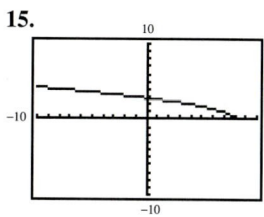

16.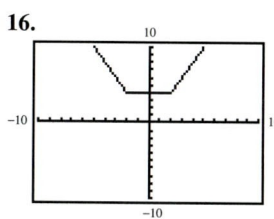

17.

| Xmin = -2 |
| Xmax = 8 |
| Xscl = 1 |
| Ymin = -4 |
| Ymax = 24 |
| Yscl = 4 |

18.

| Xmin = -4 |
| Xmax = 4 |
| Xscl = 1 |
| Ymin = -2 |
| Ymax = 12 |
| Yscl = 2 |

19. $(-1.23, 0)$, $(2.23, 0)$

20. Associative Property of Addition

Section 7.4 *(page 414)*

7. $\frac{2}{3}$ **9.** -2 **11.** Undefined

13.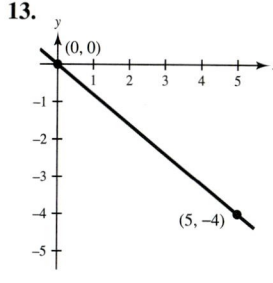

$m = -\frac{4}{5}$; falls

15.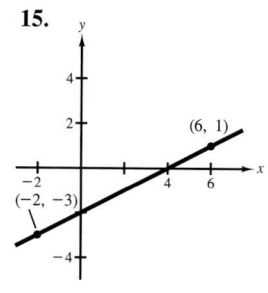

$m = \frac{1}{2}$; rises

17.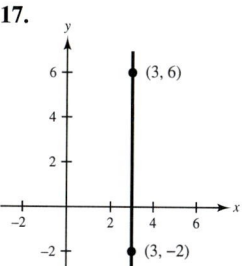

m is undefined; vertical

19.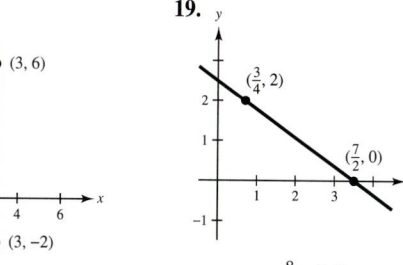

$m = -\frac{8}{11}$; falls

21. $(0, 2)$, $(1, 2)$ **23.** $(4, -1)$, $(5, 2)$

25.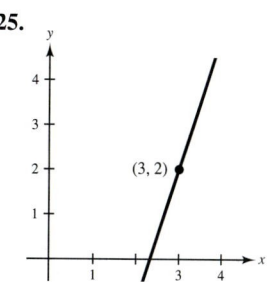

27.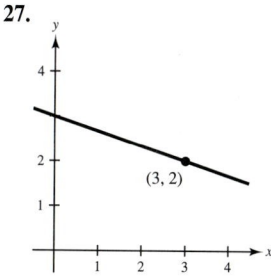

29. Line with slope m_2

31.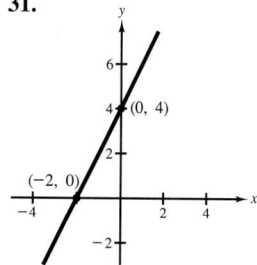

33.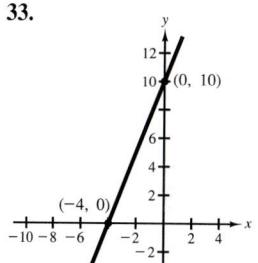

35. $y = 3x - 2$

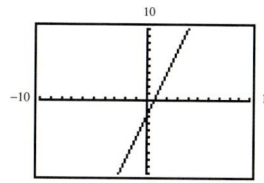

37. $y = -\frac{3}{2}x + 1$

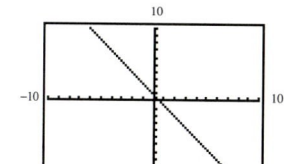

39. Parallel

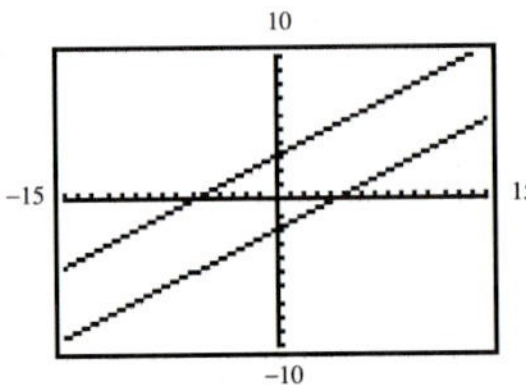

41. Perpendicular

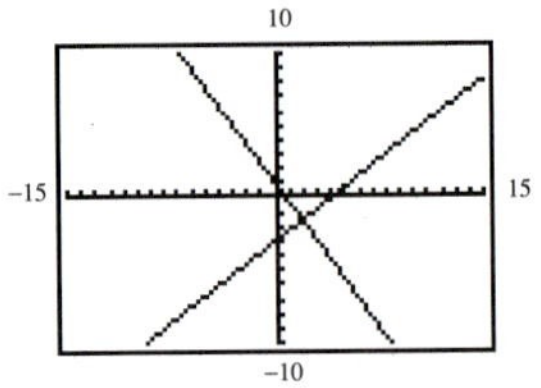

43. (a) $y = 8x + 140$

(b)

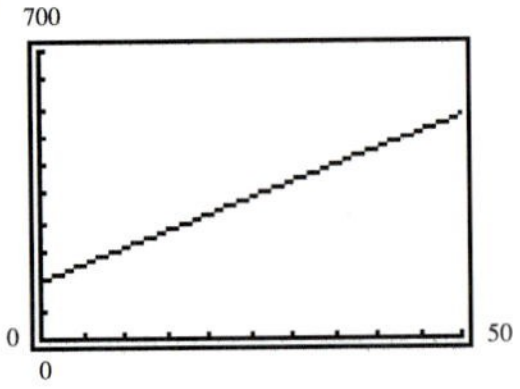

(c) $m = 8$, 8 feet

45. 15 **47.** −6 **49.** 2.4 hours **51.** (a) L_3 (b) L_2 (c) L_1

53.

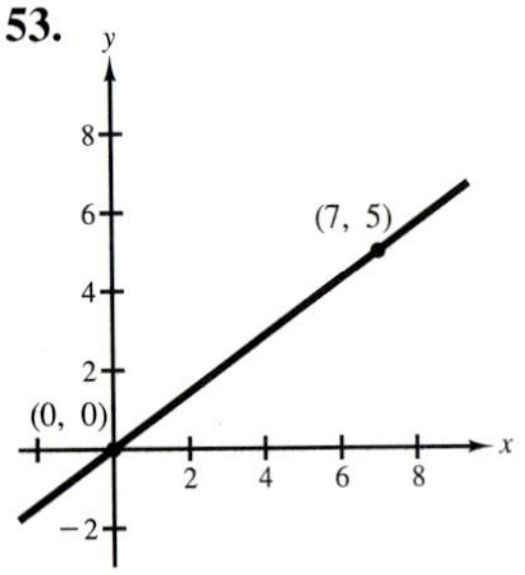

$m = \frac{5}{7}$; rises

55.

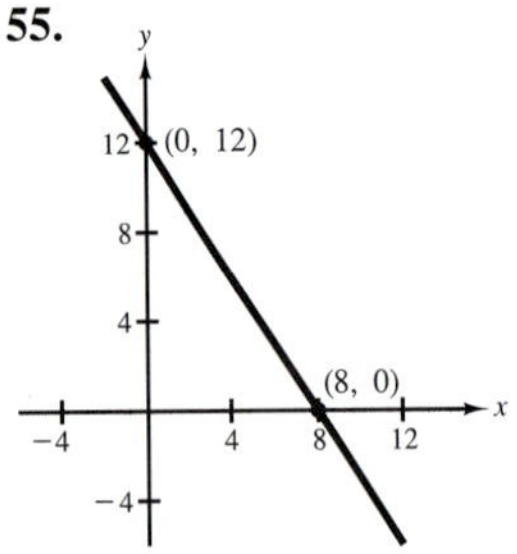

$m = -\frac{3}{2}$; falls

57.

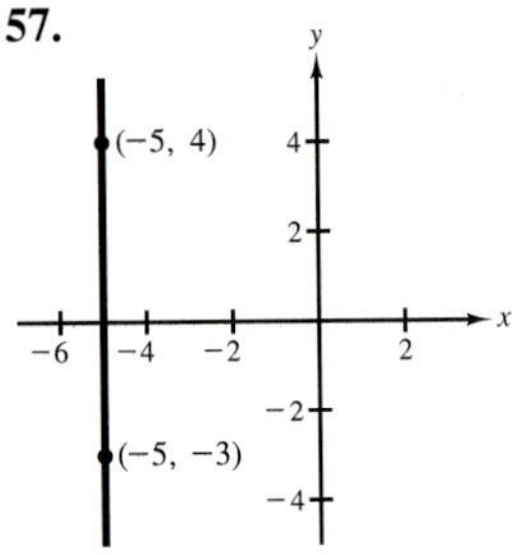

m is undefined; vertical

59.

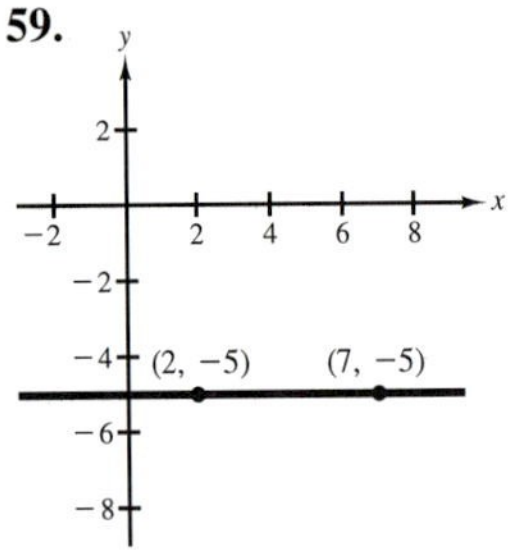

$m = 0$; horizontal

61.

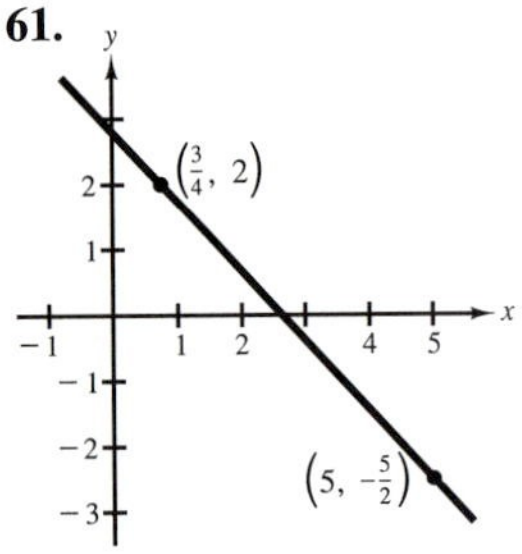

$m = -\frac{18}{17}$; falls

63.

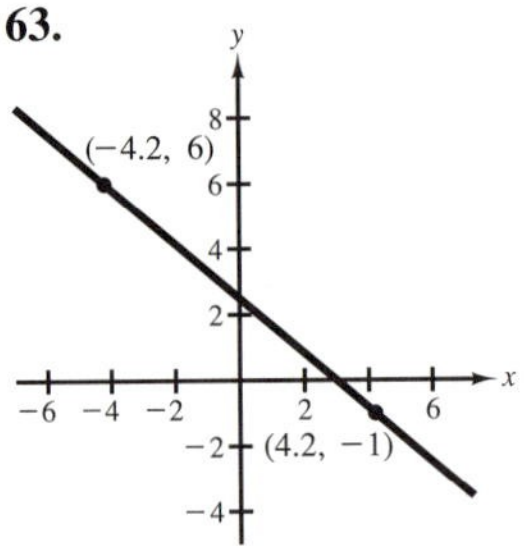

$m = -\frac{5}{6}$; falls

65.

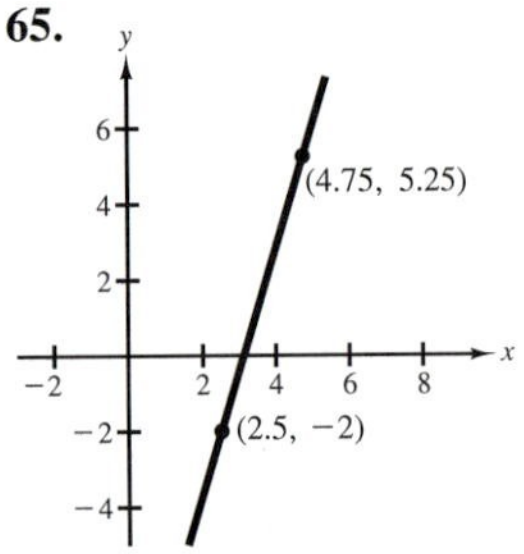

$m = \frac{29}{9}$; rises

67. $x = 1$ **69.** $y = -15$

71. (1, 2), (2, 1) **73.** (−2, 4), (1, 8) **75.** (4, 0), (4, 1)

77.

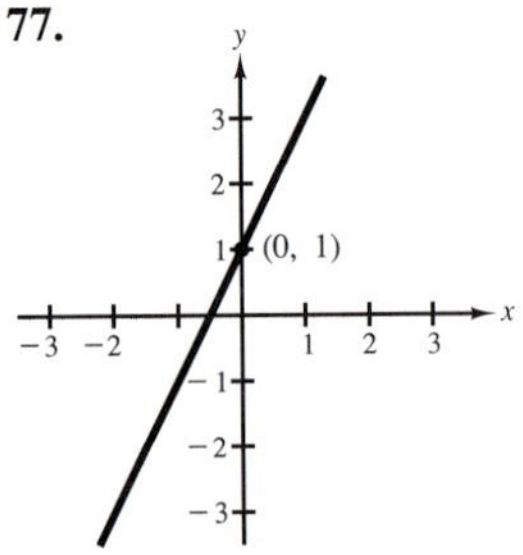

79.

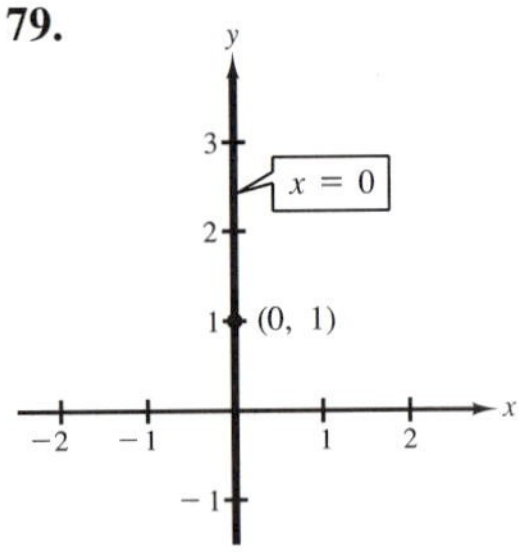

81.

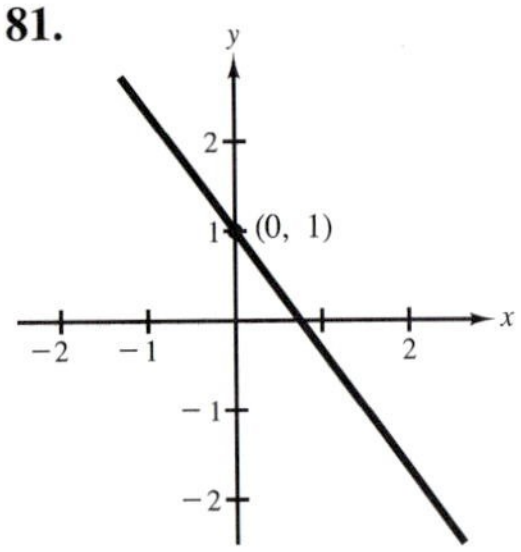

83. $y = -x$

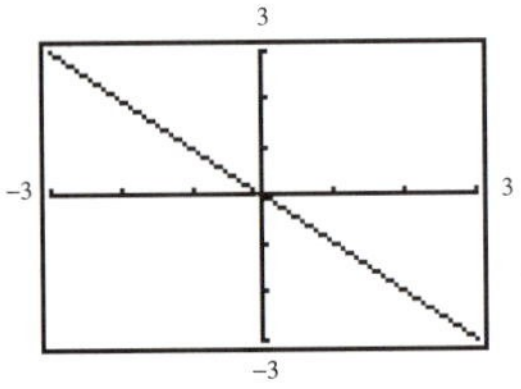

85. $y = \frac{1}{4}x + \frac{1}{2}$

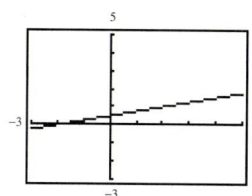

87. $y = -\frac{2}{3}x + 2$

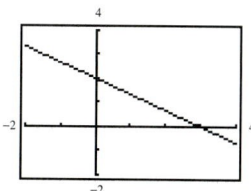

89. $y = 2$

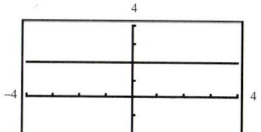

91. Perpendicular

93. Parallel   **95.** y_2 and y_3 are perpendicular.

97. The lines are parallel.   **99.** (a) 1991
                                          (b) 1992

101. 1989

Section 7.5 *(page 426)*

7. Domain: $\{-2, 0, 1\}$
      Range: $\{-1, 0, 1, 4\}$

9. Domain: $\{0, 2, 4, 5, 6\}$
      Range: $\{-3, 0, 5, 8\}$

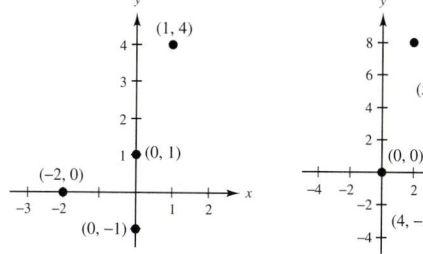

11. $(3, 150), (2, 100), (8, 400), (6, 300), \left(\frac{1}{2}, 25\right)$

13. (1990, Cincinnati), (1991, Minnesota), (1992, Toronto), (1993, Toronto)

15. Not a function   **17.** Not a function

19. (a) Function   **21.** Function
     (b) Not a function
     (c) Function
     (d) Not a function

23. Not a function   **25.** (a) $3(2) + 5 = 11$
                                (b) $3(-2) + 5 = -1$
                                (c) $3(k) + 5$
                                (d) $3(k + 1) + 5 = 3k + 8$

27. (a) 29   **29.** (a) 2   **31.** (a) 2
     (b) 11   (b) -2 (b) $\dfrac{2(x - 6)}{x}$
     (c) $12a - 2$ (c) 10
     (d) $12a + 5$ (d) -8

33. All real values of x such that $x \neq 3$

35. All real values of x such that $x \geq \frac{1}{2}$

37. $P(x) = 4x$ **39.** $V(x) = x(24 - 2x)^2$
                              $= 4x(12 - x)^2$

41. $1 - 6x$ **43.** $22 - 3x$ **45.** $\frac{1}{4}, \frac{1}{5}, \frac{20}{9} \approx 2.2$ hours

47. Function   **49.** Not a function   **51.** Function

53. Not a function   **55.** Function   **57.** Function

59. Function   **61.** Not a function

63. (a) 8   **65.** (a) 2
     (b) $\frac{2}{9}$ (b) 3
     (c) $2y^2$ (c) $\sqrt{\frac{31}{3}}$
     (d) 26   (d) $\sqrt{5(z + 1)}$

67. (a) 0   **69.** (a) -3.84 **71.** (a) 0
     (b) $-\frac{3}{2}$ (b) -4.2 (b) $\frac{7}{4}$
     (c) $-\frac{5}{2}$ (c) 3
     (d) $\dfrac{3(x + 4)}{x - 1}$ (d) 0

73. Domain: $\{0, 2, 4, 6\}$
      Range: $\{0, 1, 8, 27\}$

75. Domain: all real numbers r such that $r \geq 0$
      Range: all real numbers C such that $C \geq 0$

77. All real values of x

79. All real values of t such that $t \neq 0$ and $t \neq -2$

81. All real values of x such that $x \geq -4$

83. All real values of t **85.** $V(x) = x^3$

87. $C(x) = 1.95x + 8000$

89. (a) 10,680 pounds (b) 8010 pounds

91. (a) Correct (b) Not correct

Math Matters *(page 430)*

At first, it might seem that the answer is for the spider to walk straight down the wall, across the floor, and up the other wall. The total length of this path is 42 feet. There is, however, a shorter path. To see this, imagine that the sides of the room are unfolded as shown in the figure. By imposing a rectangular coordinate system on the unfolded walls, we see that the spider starts at the point $(0, -1)$ and wants to travel to the point $(24, 31)$. The distance between these two points is $\sqrt{24^2 + 32^2} = 40$. Thus, the shortest distance is not the 42-foot path, but the diagonal path, running across five of the six surfaces of the room, as shown in the figure.

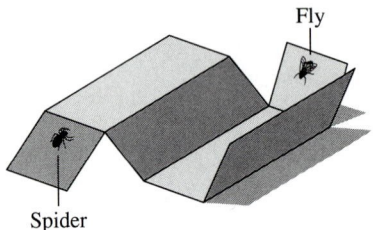

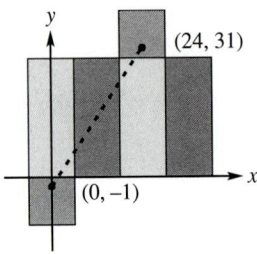

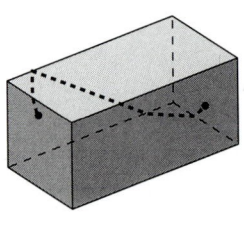

Section 7.6 *(page 426)*

7.

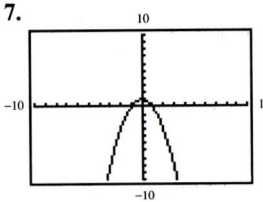

Domain: $-\infty < x < \infty$
Range: $-\infty < y \leq 1$

9.

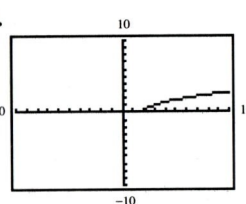

Domain: $2 \leq x < \infty$
Range: $0 \leq y < \infty$

11. Function **13.** Not a function

15. Function. For each value of y there corresponds one value of x.

17. b **19.** a

21.

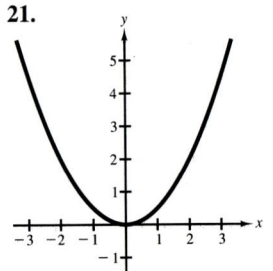

Domain: $-\infty < x < \infty$
Range: $0 \leq y < \infty$

23.

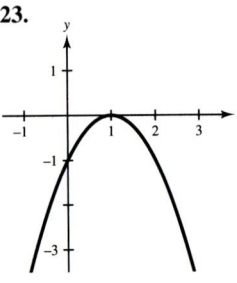

Domain: $-\infty < x < \infty$
Range: $-\infty < y \leq 0$

25.

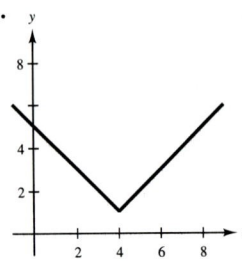

Domain: $-\infty < s < \infty$
Range: $1 \leq y < \infty$

27.

Domain: $-\infty < x < \infty$
Range: $-\infty < y \leq 3$

29. b   **31.** Vertical shift two units upward

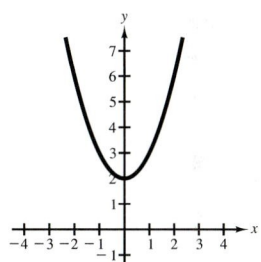

33. Horizontal shift
two units to the left

35. Reflection in
the x-axis

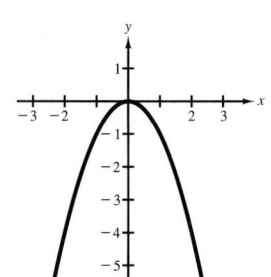

37. $y = -\sqrt{x}$ **39.** $y = \sqrt{x + 2}$

41. $y = \sqrt{-x}$ **43.** (a)

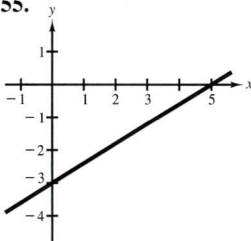

(b) $x = 50$. The figure is a square.

45. -4 **47.** 6   **49.** $0 < t < 23$

51. Function   **53.** Not a function

55.

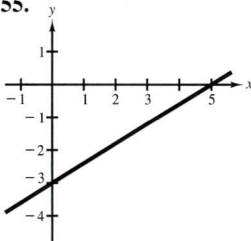

Function

57.

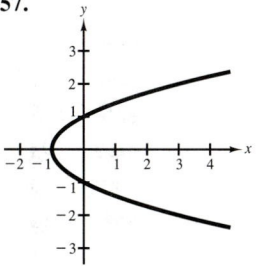

Not a function

59.

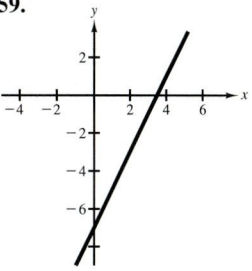

Domain: $-\infty < x < \infty$
Range: $-\infty < y < \infty$

61.

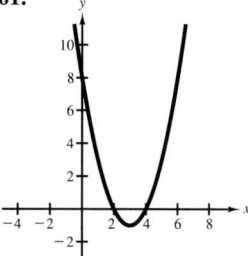

Domain: $-\infty < x < \infty$
Range: $-1 \le y < \infty$

63.

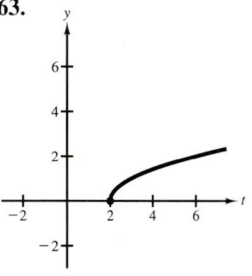

Domain: $2 \le t < \infty$
Range: $0 \le y < \infty$

65.

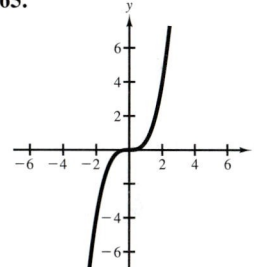

Domain: $-\infty < s < \infty$
Range: $-\infty < y < \infty$

67.

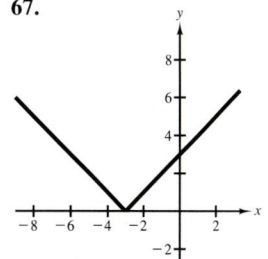

Domain: $-\infty < x < \infty$
Range: $0 \le y < \infty$

69.

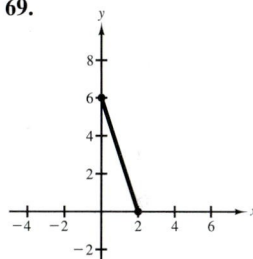

Domain: $0 \le x \le 2$
Range: $0 \le y \le 6$

71.

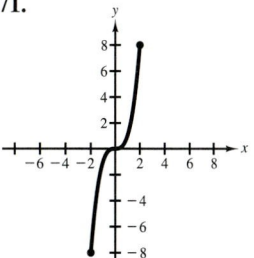

Domain: $-2 \le x \le 2$
Range: $-8 \le y \le 8$

73.

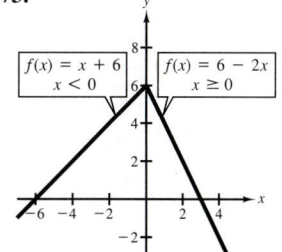

$f(x) = x + 6$
$x < 0$

$f(x) = 6 - 2x$
$x \ge 0$

Domain: $-\infty < x < \infty$
Range: $-\infty < y \le 6$

75.

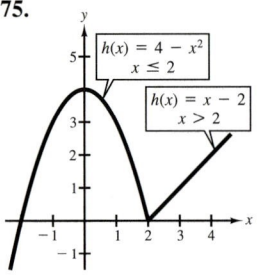

$h(x) = 4 - x^2$
$x \le 2$

$h(x) = x - 2$
$x > 2$

Domain: $-\infty < x < \infty$
Range: $-\infty \le y < \infty$

77. Vertical shift three units upward

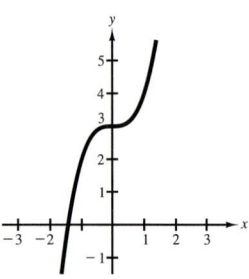

79. Horizontal shift three units to the right

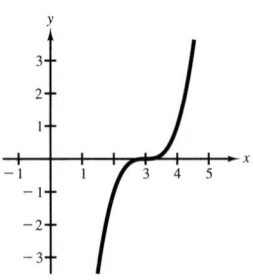

81. Reflection in the x-axis followed by a horizontal shift one unit to the right followed by a vertical shift two units upward

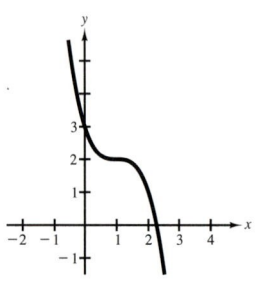

83. Horizontal shift five units to the right

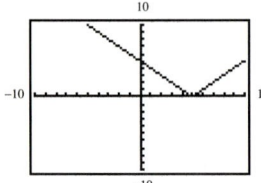

85. Vertical shift five units downward

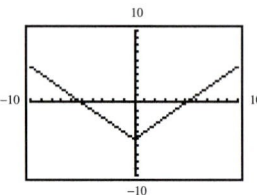

87. Reflection in the x-axis

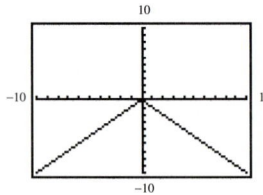

89. $h(x) = (x + 3)^2$

91. $h(x) = -x^2$ **93.** $h(x) = -(x + 3)^2$

95. $h(x) = -x^2 + 2$

97. (a)

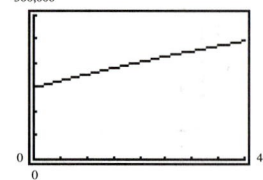

(b) 1970

(c)

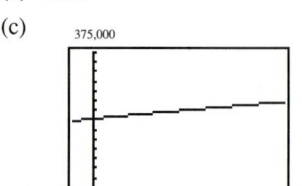

Review Exercises *(page 443)*

1.

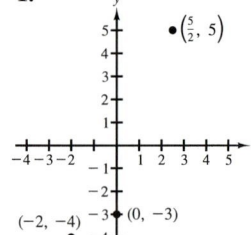

$\left(\frac{5}{2}, 5\right)$

$(-2, -4)$ $(0, -3)$

3.

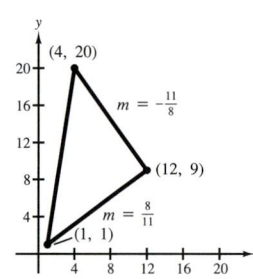

$(4, 20)$

$m = -\frac{11}{8}$

$(12, 9)$

$m = \frac{8}{11}$

$(1, 1)$

5. Quadrant IV **7.** Quadrants I and IV

9.

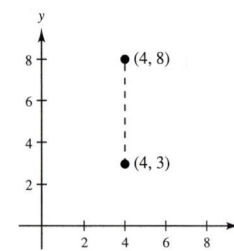

$(4, 8)$

$(4, 3)$

Distance: 5

11.

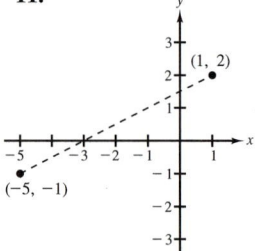

$(1, 2)$

$(-5, -1)$

Distance: $3\sqrt{5}$

13. (a) Solution

(b) Not a solution

(c) Not a solution

(d) Solution

15.

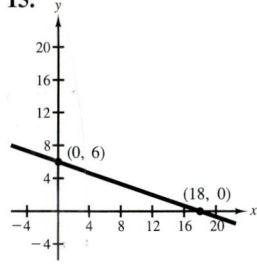

17.

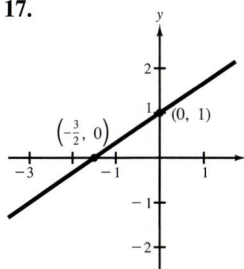

19.

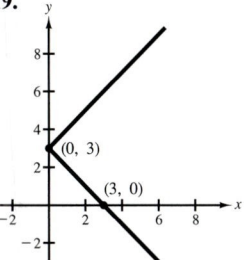

21. $\frac{2}{7}$ **23.** 0   **25.** $-\frac{3}{4}$ **27.** $\frac{3}{2}$ **29.** $(1, -1), (0, 2)$

31. $(7, 6), (11, 11)$ **33.** $(3, 0), (3, 5)$

35. $y = \frac{5}{2}x - 2$ **37.** $y = -\frac{1}{2}x + 1$

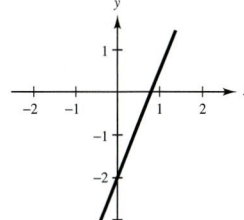

39.

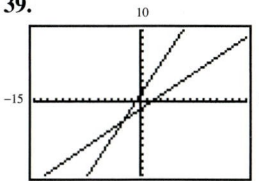

Neither

41.

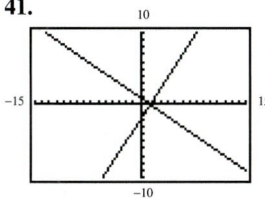

Perpendicular

43. Not a function   **45.** Function

47. Intercept: $(0, 0)$ **49.** Intercept: $(0, 0), (3, 0)$

Not a function   Function

51. (a) 29

(b) 3

(c) $\dfrac{36 - 5t}{2}$

(d) $4 - \frac{5}{2}(x + h)$

53. (a) $2\sqrt{2}$

(b) 0

(c) $\dfrac{2\sqrt{3}}{3}$

(d) $\sqrt{5 - 5z}$

55. (a) -3

(b) 2

(c) 0

(d) -7

57. (a) -2

(b) $\dfrac{2(6 - x)}{x}$

59. All real values of x

61. All real values of x such that $x \leq \frac{5}{2}$ **63.** c

65.

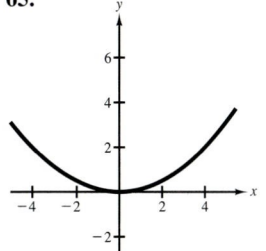

67.

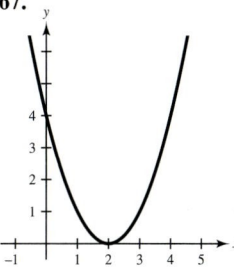

69.

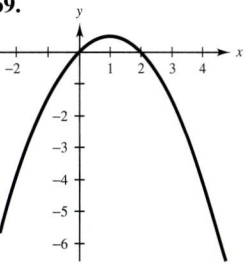

71.

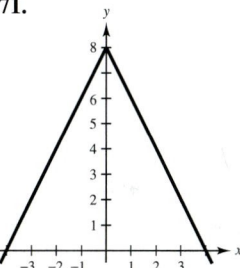

73.

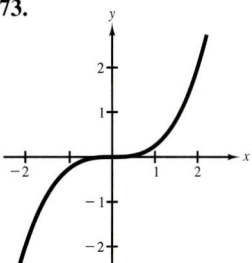

75.

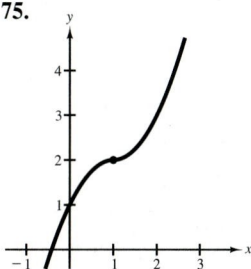

77. Reflection in the
x-axis

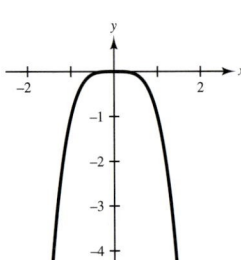

79. Horizontal shift one
unit to the right

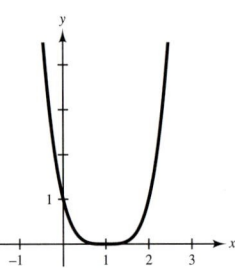

81. (a)

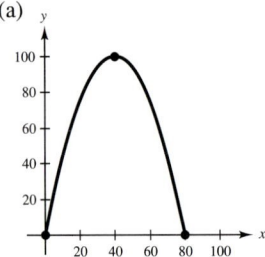

(b) 100 feet
(c) 80 feet

83. (a) $k = \frac{1}{8}$

(b) 1953.1 kw

Chapter Test *(page 446)*

1. Quadrant IV

2.

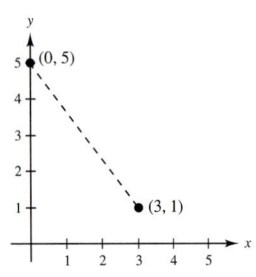

Distance: 5

3. $(-1, 0)$, $(0, -3)$

4.

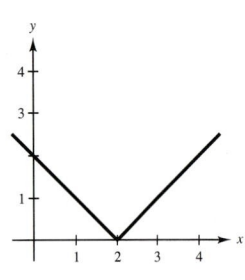

5. $-\frac{2}{3}$ **6.** Undefined

7.

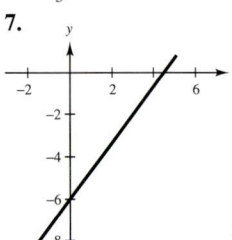

8.

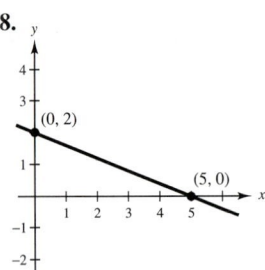

9. $y = -\frac{5}{3}x + 3$ **10.** Not a function

$\frac{3}{5}$

11. Not a function   **12.** Function

13. -2 **14.** 7   **15.** $\dfrac{x + 2}{x - 1}$

16. All real values of t such that $t \leq 9$

17. All real values of x such that $x \neq 4$

18.

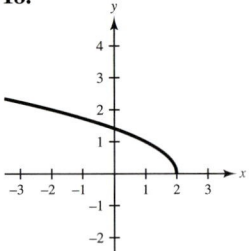

19.

$(0, 5)$, $(-1.94, 0)$, $(8.60, 0)$

20. (a) $y = |x - 2|$

(b) $y = |x| - 2$

(c) $y = 2 - |x|$

CHAPTER 8

Section 8.1 *(page 454)*

7. $(-\infty, 8) \cup (8, \infty)$ **9.** $(-\infty, \infty)$

11. $(-\infty, 0) \cup (0, 3) \cup (3, \infty)$

13. (a) 1   **15.** (a) $\frac{25}{22}$

    (b) -8 (b) 0

    (c) Undefined   (c) Undefined

    (d) 0   (d) Undefined

17. $\{1, 2, 3, 4, \ldots\}$ **19.** $x + 3$ **21.** $x + 2$

23. $\dfrac{x}{5}$ **25.** $\dfrac{6x}{5y^3}$ **27.** $\dfrac{3y^2}{y^2 + 1}$ **29.** $\dfrac{y - 8}{15}$

31. $\dfrac{-1}{2x + 3}$ **33.** $\dfrac{3(m - 2n)}{m + 2n}$

35.

x	-2	-1	0	1	2	3	4
$\dfrac{x^2 - x - 2}{x - 2}$	-1	0	1	2	Undefined	4	5
$x + 1$	-1	0	1	2	3	4	5

37. $\dfrac{x}{3(x + 3)}$ **39.** $\dfrac{107.1 + 12.64t + 0.54t^2}{31.6 + 0.51t - 0.14t^2}$

41. $42x^2 - 60x$ **43.** $121 - x^2$ **45.** $\left(\dfrac{y}{3}\right)\left(\dfrac{y}{3}\right)\left(\dfrac{y}{3}\right)\left(\dfrac{y}{3}\right)$

47. $(-\infty, -4) \cup (-4, \infty)$ **49.** $(-\infty, \infty)$

51. $(-\infty, -4) \cup (-4, 4) \cup (4, \infty)$

53. $(-\infty, \infty)$ **55.** $(0, \infty)$

57. $(3)(x + 16)^2$ **59.** $(x)(x - 2)$ **61.** $6y$

63. x **65.** $\dfrac{1}{2}$ **67.** $-\dfrac{1}{3}$ **69.** $\dfrac{1}{a + 3}$

71. $\dfrac{x}{x - 7}$ **73.** $\dfrac{y(y + 2)}{y + 6}$ **75.** $\dfrac{5x + 4}{5x + 2}$

77. $\dfrac{3xy + 5}{y^2}$ **79.** $\dfrac{u - 2v}{u - v}$

81. Evaluating both sides when $x = 10$ yields $\frac{3}{2} \neq 9$.

83. $x^n - 2$ **85.** (a) $\dfrac{2500 + 9.25x}{x}$ **87.** π

           (b) $x > 0$

           (c) \$34.25

Section 8.2 *(page 463)*

5. (a) 0   **7.** x^2 **9.** $(-1)(2 + x)$

    (b) Undefined

    (c) $\frac{3}{2}$

    (d) $\frac{1}{24}$

11. $\dfrac{3x}{2}$ **13.** 24   **15.** $-\dfrac{x + 8}{x^2}$ **17.** $\dfrac{(u + v)^2}{(u - v)^2}$

19. $\dfrac{3}{2x}$ **21.** $\dfrac{3}{2(a + b)}$ **23.** $\dfrac{(x + 3)(4x + 1)}{(3x - 1)(x - 1)}$

25. $\dfrac{3x}{10}$ **27.** $\dfrac{x + 4}{3}$ **29.** $\dfrac{1}{4}$

31.

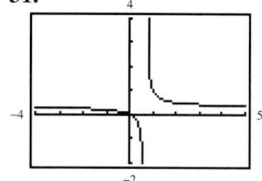

33.

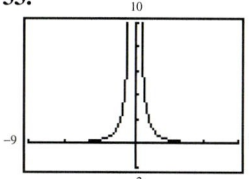

35. $\dfrac{2w^2 + 3w}{6}$ **37.** $3x(x - 7)$ **39.** $(2t + 13)(2t - 13)$

41. \$720   **43.** $(x + 2)^2$ **45.** $u + 1$ **47.** $\dfrac{s^3}{6}$

49. $24u^2$ **51.** $\dfrac{2uv(u + v)}{3(3u + v)}$ **53.** -1

55. $\dfrac{2(r + 2)}{11r}$ **57.** $(u - 2v)(u + 2v)$ **59.** $2t + 5$

61. $\dfrac{(x - 1)(2x + 1)}{(3x - 2)(x + 2)}$ **63.** $\dfrac{3y^2}{2ux^2}$ **65.** $x^4 y(x + 2y)$

67. $-\dfrac{5x}{2}$ **69.** $\dfrac{(x + 2)(x + 3)}{x}$ **71.** $\dfrac{(x + 1)(2x - 5)}{x}$

73. (a) 1/20 minute   **75.** $x/[4(3x - 2)]$

    (b) $x/20$ minutes

    (c) 7/4 minutes

Section 8.3 (page 472)

5. $\dfrac{3}{2}$ **7.** $-\dfrac{2}{9}$ **9.** $1, x \neq -4$ **11.** $20x^3$

13. $15x^2(x + 5)$ **15.** $6x(x + 2)(x - 2)$

17. $\dfrac{2n^2(n + 8)}{6n^2(n - 4)}$ **19.** $\dfrac{(x - 8)(x - 5)}{(x + 5)(x - 5)^2}$

$\dfrac{10(n - 4)}{6n^2(n - 4)}$ $\dfrac{9x(x + 5)}{(x + 5)(x - 5)^2}$

21. $\dfrac{25 - 12x}{20x}$ **23.** 0 **25.** $\dfrac{5(5x + 22)}{x + 4}$

27. $\dfrac{x^2 + 3x + 9}{x(x^2 - 9)}$ **29.** $\dfrac{5u - 2v}{(u - v)^2}$ **31.** $\dfrac{x}{x - 1}$

33. 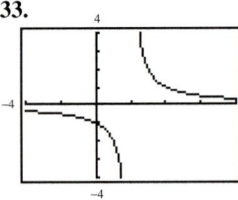   **35.** $\dfrac{x}{2(3x + 1)}$

37. $\dfrac{3}{4}$ **39.** $y - x$ **41.** $\dfrac{R_1 R_2}{R_1 + R_2}$

43.

x	-3	-2	-1	0	1	2	3
$\dfrac{\left(1 - \dfrac{1}{x}\right)}{\left(1 - \dfrac{1}{x^2}\right)}$	$\dfrac{3}{2}$	2	Undef.	Undef.	Undef.	$\dfrac{2}{3}$	$\dfrac{3}{4}$
$\dfrac{x}{x + 1}$	$\dfrac{3}{2}$	2	Undef.	0	$\dfrac{1}{2}$	$\dfrac{2}{3}$	$\dfrac{3}{4}$

45. $54x^{10}$ **47.** 1 **49.** $\dfrac{25}{x^4}$

51. $2500 < 1500 + 0.04x$ **53.** $\dfrac{2 + y}{2}$
$x > \$25,000$

55. $-\dfrac{3}{a}$ **57.** $\dfrac{x + 6}{3}$ **59.** $-\dfrac{4}{3}$ **61.** $\dfrac{-25}{9}$

63. $36y^3$ **65.** $30x^2(x - 1)$

67. $\dfrac{3v^2}{6v^2(v + 1)}$ **69.** $\dfrac{4x(x - 5)}{(x + 5)^2(x - 5)}$

$\dfrac{8(v + 1)}{6v^2(v + 1)}$ $\dfrac{(x - 2)(x + 5)}{(x + 5)^2(x - 5)}$

71. $\dfrac{7(a + 2)}{a^2}$ **73.** $\dfrac{3(x + 2)}{x - 8}$ **75.** 1

77. $\dfrac{1}{2x(x - 3)}$ **79.** $\dfrac{x^2 - 7x - 15}{(x + 3)(x - 2)}$ **81.** $\dfrac{x - 2}{x(x + 1)}$

83. $\dfrac{5(x + 1)}{(x + 5)(x - 5)}$ **85.** $\dfrac{4}{x^2(x^2 + 1)}$ **87.** $\dfrac{4x}{(x - 4)^2}$

89. $\dfrac{y - x}{xy}$ **91.** $\dfrac{2(4x^2 + 5x - 3)}{x^2(x + 3)}$ **93.** $-4x - 1$

95. $\dfrac{5(x + 3)}{2x(5x - 2)}$ **97.** $\dfrac{y + 1}{y - 3}$ **99.** $\dfrac{x(x + 6)}{3x^3 + 10x - 30}$

101. $-\dfrac{1}{2(h + 2)}$ **103.** $\dfrac{5t}{12}$ **105.** $\dfrac{5x}{24}$

107. $\dfrac{x}{4}, \dfrac{x}{3}, \dfrac{5x}{12}$ **109.** (a) 19.6%
(b) $\dfrac{288(MN - P)}{N(MN + 12P)}$

Mid-Chapter Quiz (page 476)

1. $(-\infty, 0) \cup (0, 4) \cup (4, \infty)$ **2.** (a) 0
(b) $\dfrac{9}{2}$
(c) Undefined
(d) $\dfrac{8}{9}$

3. $\dfrac{3}{2}y$ **4.** $\dfrac{2u^2}{9v}$ **5.** $-\dfrac{2x + 1}{x}$ **6.** $\dfrac{z + 3}{2z - 1}$

7. $\dfrac{7 + 3ab}{a}$ **8.** $\dfrac{n^2}{m + n}$ **9.** $\dfrac{t}{2}$ **10.** $\dfrac{5x}{x - 2}$

11. $\dfrac{8x}{3(x - 1)(x^2 + 2x - 3)}$ **12.** $\dfrac{4(u - v)^2}{5uv}$

13. $-\dfrac{3t}{2}$ **14.** $\dfrac{2(x + 1)}{3x}$ **15.** $\dfrac{x(1 - 3x)}{4(x + 5)}$

16. $\dfrac{4x^4 + x^3 - 18x^2 + 8}{x^2(x^2 - 4)}$ **17.** $\dfrac{5(2 - x)}{4x - 15}$

18. $\dfrac{2(x^2 + 9)}{x + 3}$ **19.** (a) $\dfrac{6000 + 10.50x}{x}$ **20.** $\dfrac{8(x + 2)}{(x + 4)^2}$
(b) $\$22.50$

Section 8.4 (page 484)

7. $-10z^2 - 6$ **9.** $\dfrac{5}{2}x - 4 + \dfrac{7}{2}y$ **11.** $x - 5$

13. $x + 7$ **15.** $x^2 + 4$ **17.** $x - 4 + \dfrac{32}{x + 4}$

19. $\dfrac{6}{5}z + \dfrac{41}{25} + \dfrac{41}{25(5z-1)}$ **21.** $x^3 - x + \dfrac{x}{x^2+1}$

23. $x + 2$ **25.** $x^2 - x + 4 - \dfrac{17}{x+4}$

27. $\dfrac{1}{10}x + \dfrac{41}{50} + \dfrac{291}{250(x-0.2)}$ **29.** $(x-7)(x-8)$

31. $(x^3 - 3x + 1)(x - 1)$ **33.**

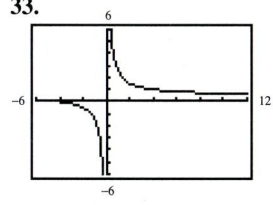

35.

x-Values	Polynomial Values	Divisors	Remainders
-2	-8	$x+2$	-8
-1	-1	$x+1$	-1
0	0	x	0
$\frac{1}{2}$	$-\frac{9}{8}$	$x-\frac{1}{2}$	$-\frac{9}{8}$
1	-2	$x-1$	-2
2	0	$x-2$	0

37. $2(x+4)$ **39.** $x^2 + 2x + 1$ **41.** $16 - 25z^2$

43. $\frac{5}{2}x(2x+9)$ **45.** $3z + 5$ **47.** $\frac{5}{2}z^2 + z - 3$

49. $7x^2 - 2x$ **51.** $4z^2 + \dfrac{3}{2}z - 1$ **53.** $m^3 + 2m - \dfrac{7}{m}$

55. $4(x+7)$ **57.** $x + 10$ **59.** $y + 3$ **61.** $6t - 5$

63. $4x - 1$ **65.** $x^2 - 5x + 25$ **67.** $2 + \dfrac{5}{x+2}$

69. $5x - 8 + \dfrac{19}{x+2}$ **71.** $4x - \dfrac{25}{3} + \dfrac{35}{3(3x+2)}$

73. $2x^2 + x + 4 + \dfrac{6}{x-3}$

75. $x^5 + x^4 + x^3 + x^2 + x + 1$ **77.** $x^{2n} + x^n + 4$

79. $x^3 - 3x + 1$ **81.** $5x^2 - 25x + 125 - \dfrac{613}{x+5}$

83. $(2a-5)(a+9)$ **85.** $(15x+10)\left(x-\dfrac{4}{5}\right)$

87. $(2t^2 + 5t - 6)(t+5)$ **89.** -8 **91.** $x^2 - 3$

Section 8.5   *(page 493)*

5.

x	0	0.5	0.9	0.99	0.999
y	-4	-8	-40	-400	-4000

x	2	1.5	1.1	1.01	1.001
y	4	8	40	400	4000

x	2	5	10	100	1000
y	4	1	0.4444	0.0404	0.0040

7. Domain: $(-\infty, 0) \cup (0, \infty)$
Horizontal asymptote: $y = 0$
Vertical asymptote: $x = 0$

9. Domain: $(-\infty, \infty)$
Horizontal asymptote: $y = 0$
Vertical asymptotes: None

11. Domain: $(-\infty, -1) \cup (-1, 1) \cup (1, \infty)$
Horizontal asymptote: $y = 5$
Vertical asymptotes: $x = -1, x = 1$

13. d   **15.** a

17.   **19.**

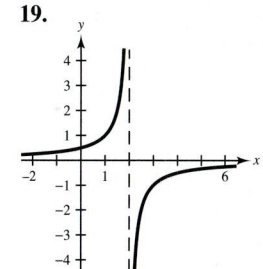

21.   **23.**

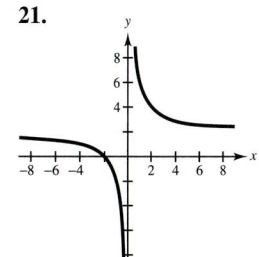
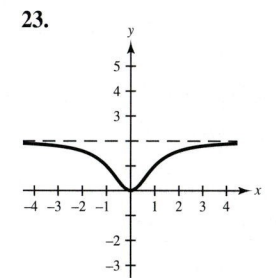

25.

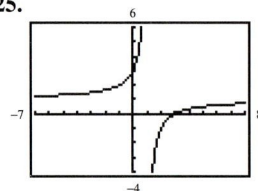

Domain: $(-\infty, 1) \cup (1, \infty)$
Horizontal asymptote: $y = 1$
Vertical asymptote: $x = 1$

27.

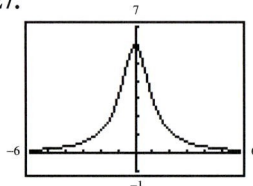

Domain: $(-\infty, \infty)$
Horizontal asymptote: $y = 0$
Vertical asymptotes: None

29.

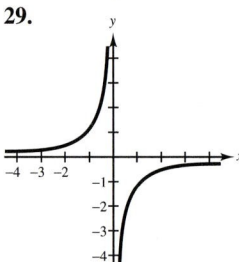

31.

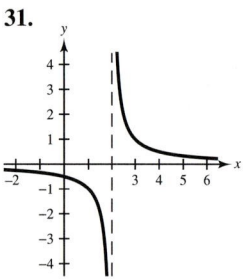

33. (a) $A = \dfrac{2500 + 0.50x}{x}$

(b) \$3, \$0.75

(c)

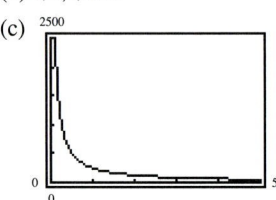

$A = \$0.50$

35. $x \geq 6$ **37.** $1 < x < 5$ **39.** $30 \leq n \leq 150$

41. Domain: $(-\infty, 3) \cup (3, \infty)$
Horizontal asymptote: $y = 2$
Vertical asymptote: $x = 3$

43. Domain: $(-\infty, -3) \cup (-3, 3) \cup (3, \infty)$
Horizontal asymptote: $y = 0$
Vertical asymptotes: $x = -3, x = 3$

45. Domain: $(-\infty, -8) \cup (8, \infty)$
Horizontal asymptote: $y = 1$
Vertical asymptote: $x = -8$

47. Domain: $(-\infty, 0) \cup (0, 1) \cup (1, \infty)$
Horizontal asymptote: $y = 0$
Vertical asymptotes: $t = 0, t = 1$

49. Domain: $(-\infty, \infty)$
Horizontal asymptote: $y = 2$
Vertical asymptotes: None

51.

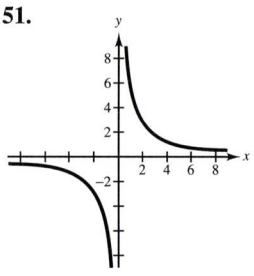

53.

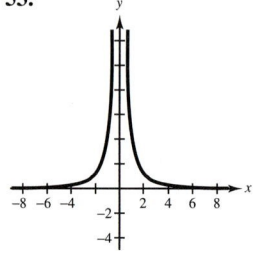

55.

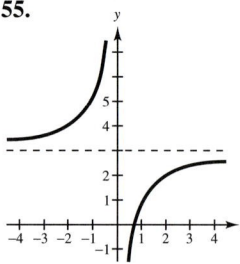

57.

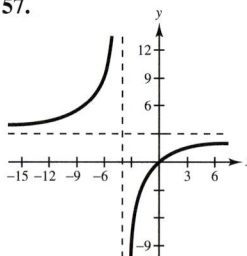

59.

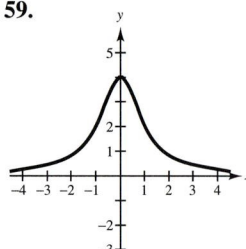

61.

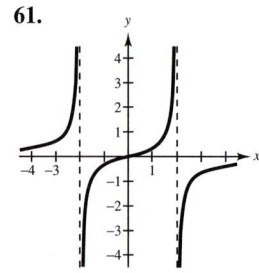

63.

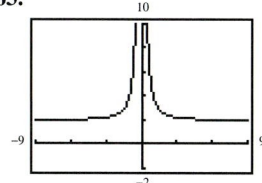

Domain: $(-\infty, 0) \cup (0, \infty)$

65.

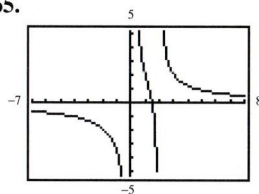

Domain: $(-\infty, 0) \cup (0, 2) \cup (2, \infty)$

67.

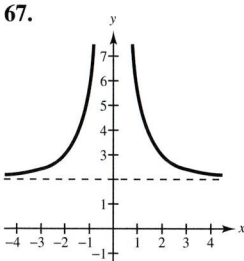

69.

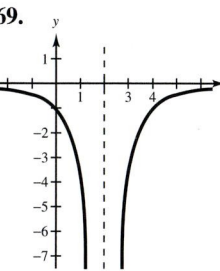

71.

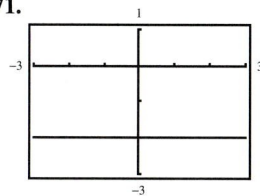

The fraction is not reduced to lowest terms.

73. (a) $C = 0$. The chemical is eliminated from the body.

(b)

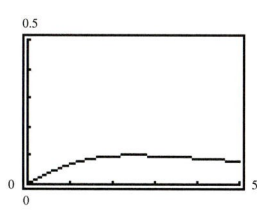

$t \approx 2.5$

75. (a) area = (length)(width)

$$400 = x \cdot y$$
$$y = \frac{400}{x}$$
$$P = 2(x + y)$$
$$P = 2\left(x + \frac{400}{x}\right)$$

(b) $0 < x$

(c)

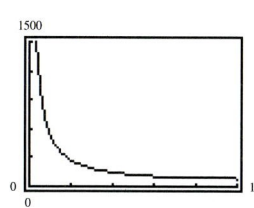

20 units × 20 units

Section 8.6 *(page 503)*

7. (a) Not a solution

(b) Not a solution

(c) Not a solution

(d) Solution

9. (a) Not a solution

(b) Solution

(c) Solution

(d) Not a solution

11. $\frac{3}{2}$ **13.** 10   **15.** $\frac{1}{3}$ **17.** 3   **19.** $2, -\frac{11}{10}$

21. No solution   **23.** ± 4 **25.** $0, 2$ **27.** $2, 3$

29. $(-2, 0)$ **31.** $(-1, 0), (1, 0)$

33.

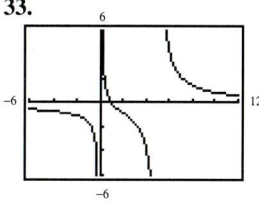

$(1, 0)$

35.

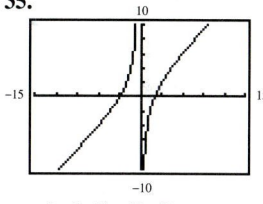

$(-3, 0), (2, 0)$

37. 40 miles per hour   **39.** 12   **41.** $11\frac{1}{4}$ hours

43. (a) 1989

(b) $y = 43.31 - \dfrac{275.25}{9} + \dfrac{654.53}{9^2} \approx 20.8$ billion

(c) $28.66 billion

45. $\frac{5}{2}$ **47.** $-7, 6$ **49.** 24, 26   **51.** $\frac{1}{4}$ **53.** $-\frac{8}{3}$

55. 10   **57.** 61   **59.** $\frac{18}{5}$ **61.** 3   **63.** $-\frac{11}{5}$ **65.** $\frac{4}{3}$

67. No solution   **69.** $\frac{3}{4}$ **71.** 3   **73.** ± 6 **75.** $8, -9$

77. 3, 13   **79.** -5 **81.** No solution   **83.** $8, -3$

85.

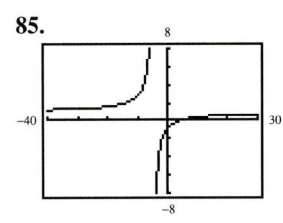

(4, 0)

87.

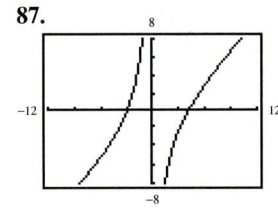

(−3, 0), (4, 0)

89. 8   **91.** (a)

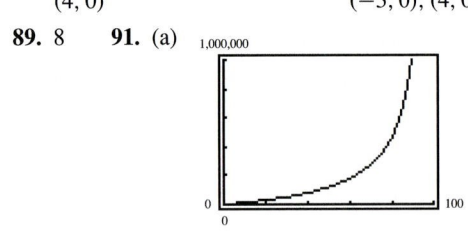

(b) 85%

93. 8 miles per hour; 10 miles per hour

95. 4 miles per hour   **97.** 10

99. 3 hours; $\dfrac{15}{8}$ minutes; $\dfrac{5}{3}$ hours

101. 15 hours; $22\frac{1}{2}$ hours

Review Exercises *(page 509)*

1. $(-\infty, 8) \cup (8, \infty)$

3. $(-\infty, 1) \cup (1, 6) \cup (6, \infty)$

5. $\dfrac{2x^3}{5}$ **7.** $\dfrac{b-3}{6(b-4)}$ **9.** -9 **11.** $\dfrac{x}{2(x+5)}$

13. b   **15.** a   **17.** $\dfrac{y}{8x}$ **19.** $12z(z-6)$ **21.** $-\dfrac{1}{4}$

23. $3x^2$ **25.** $\dfrac{125y}{x}$ **27.** $\dfrac{x(x-1)}{x-7}$ **29.** $-\dfrac{7}{9}$

31. $-\dfrac{13}{48}$ **33.** $\dfrac{4x+3}{(x+5)(x-12)}$

35. $\dfrac{5x^3 - 5x^2 - 31x + 13}{(x+2)(x-3)}$ **37.** $\dfrac{x+24}{x(x^2+4)}$

39. $\dfrac{6(x-9)}{(x+3)^2(x-3)}$ **41.** $\dfrac{6(x+5)}{x(x+7)}$ **43.** $\dfrac{3t^2}{5t-2}$

45. $-\dfrac{a^2 - a - 16}{(4a^2 + 16a + 1)(a-4)}$ **47.** $2x^2 - \dfrac{1}{2}$

49. $2x^2 + \dfrac{4}{3}x - \dfrac{8}{9} + \dfrac{10}{9(3x-1)}$ **51.** $x^2 - 2$

53. $x^2 + 5x - 7$ **55.** $x^3 + 3x^2 + 6x + 18 + \dfrac{29}{x-3}$

57.

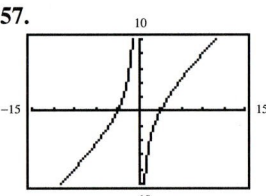

59.

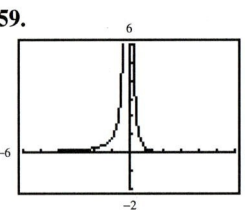

61.

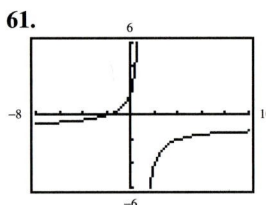

63.

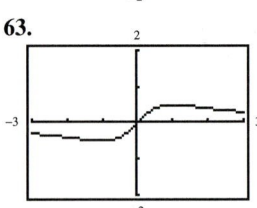

65.

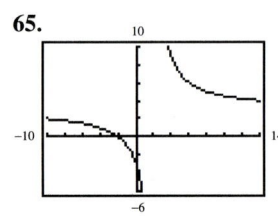

67.

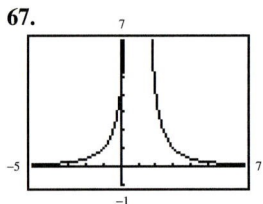

69.

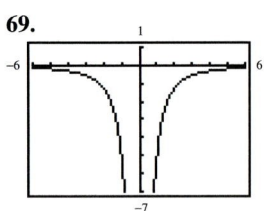

71.

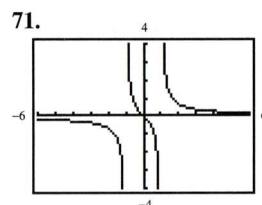

73.

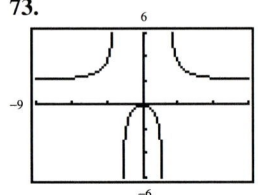

75. -40 **77.** $\frac{3}{2}$

79. 5   **81.** 6, −4,   **83.** 2, −6,   **85.** −2, 2

87. $-\frac{9}{5}, 3$ **89.**

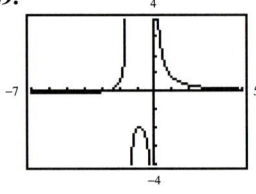

(−3, 0)

91. 56 miles per hour   **93.** 4

Chapter Test *(page 512)*

1. $(-\infty, -5) \cup (-5, 5) \cup (5, \infty)$ **2.** $x^3(x+3)^2(x-3)$

3. (a) $-\dfrac{1}{3}$ **4.** $\dfrac{5z}{3}$ **5.** $\dfrac{4}{y+4}$ **6.** $\dfrac{(2x+3)^2}{x+1}$

  (b) $\dfrac{2a+3}{5}$

7. $\dfrac{14y^6}{15}$ **8.** $\dfrac{x^3}{4}$ **9.** $-(3x+1)$

10. $\dfrac{-2x^2+2x+1}{x+1}$ **11.** $\dfrac{5x^2-15x-2}{(x-3)(x+2)}$

12. $\dfrac{5x^3+x^2-7x-5}{x^2(x+1)^2}$ **13.** 4 **14.** $t^2+3-\dfrac{6t-6}{t^2-2}$

15. $2x^3+6x^2+3x+9+\dfrac{20}{x-3}$

16. (a) (b)

17. 22 **18.** $-1, -\dfrac{15}{2}$ **19.** No solution

20. $6\dfrac{2}{3}$ hours, 10 hours

CHAPTER 9

Section 9.1 *(page 520)*

7. $\dfrac{1}{25}$ **9.** -32 **11.** $\dfrac{3}{2}$ **13.** 1 **15.** 100,000

17. $\dfrac{1}{16}$ **19.** $\dfrac{16}{15}$ **21.** y^2 **23.** t^2 **25.** $-\dfrac{12}{xy^3}$

27. $\dfrac{y^4}{9x^4}$ **29.** $\dfrac{3x^5}{y^4}$ **31.** $\dfrac{81v^8}{u^6}$ **33.** $\dfrac{ab}{b-a}$

35. 5.75×10^7 **37.** 9.461×10^{15} **39.** 8.99×10^{-5}

41. 28,200,000,000 **43.** 13,000,000

45. 0.00000000048 **47.** 1.3×10^{11} **49.** 6×10^6

51. 3.30×10^8 **53.** 3.33×10^5 **55.** $(x-1)(x-2)$

57. $(11x-5)(x+1)$ **59.** 7650 **61.** $-\dfrac{1}{1000}$

63. $-\dfrac{1}{243}$ **65.** 64 **67.** 1 **69.** 729 **71.** $\dfrac{1}{64}$

73. $\dfrac{3}{16}$ **75.** $\dfrac{64}{121}$ **77.** z^2 **79.** x^6 **81.** a

83. $\dfrac{1}{4x^4}$ **85.** $\dfrac{10}{x}$ **87.** $\dfrac{b^5}{a^5}$ **89.** x^8y^{12}

91. $\dfrac{v^2}{uv^2+1}$ **93.** $\dfrac{x^{12}y^8}{16}$ **95.** 3.6×10^6

97. 3.81×10^{-8} **99.** 60,000,000

101. 0.0000001359 **103.** 6.8×10^5 **105.** 4×10^3

107. 9×10^{15} **109.** 4.70×10^{11} **111.** 3.46×10^{10}

113. 9.3×10^7 **115.** 1.6×10^{-5} years ≈ 8.4 minutes

117. \$17,159.53

Section 9.2 *(page 530)*

7. 7 **9.** Cube root **11.** Square root **13.** 9

15. $\dfrac{3}{4}$ **17.** Not a real number **19.** $-\dfrac{2}{3}$ **21.** 5

23. 3 **25.** -0.3 **27.** 5 **29.** $\dfrac{1}{4}$ **31.** $16^{1/2}=4$

33. $27^{2/3}=9$ **35.** $\sqrt[4]{256^3}=64$ **37.** 1.0420

39. 4.3004 **41.** t^2 **43.** 3 **45.** $\dfrac{1}{2}$ **47.** x^3

49. $x^{1/4}$ **51.** $r=0.128$ **53.** $[0, \infty)$ **55.** $\dfrac{5}{56}$

57. 5 **59.** $h+4$ **61.** 8 **63.** Not a real number

65. 0.4 **67.** 13 **69.** 10 **71.** $-\dfrac{1}{4}$

73. Not a real number **75.** Irrational **77.** Rational

79. 7 **81.** 8 **83.** $\dfrac{4}{9}$ **85.** $\dfrac{3}{11}$ **87.** 8.5440

89. -0.1248 **91.** 0.0038 **93.** 66.7213 **95.** $|t|$

97. y^3 **99.** $\dfrac{4}{9}$ **101.** $\dfrac{9y^{3/2}}{x^{2/3}}$ **103.** $\dfrac{3y^2}{4z^{4/3}}$

105. $c^{1/2}$ **107.** $\sqrt[8]{y}$ **109.** $2x^{3/2}-3x^{1/2}$

111. $1+5y$ **113.** $(0, \infty)$

115. **117.**

Domain: $(0, \infty)$ Domain: $(-\infty, \infty)$

119. 1, 4, 5, 6, 9, 0 **121.** 0.026 inch
No

Section 9.3 *(page 538)*

7. $\sqrt[3]{110}$ **9.** $\sqrt{\frac{15}{31}}$ **11.** $3\sqrt{35}$ **13.** $\frac{10\sqrt[3]{121}}{11}$

15. $2\sqrt{5}$ **17.** $2\sqrt[3]{3}$ **19.** $\frac{1}{2}\sqrt{15}$ **21.** $\frac{1}{3}\sqrt[5]{15}$

23. $3x^2\sqrt{x}$ **25.** $|x|\sqrt[4]{3y^2}$ **27.** $\frac{2}{y}\sqrt[5]{x^2}$

29. $\frac{4a^2}{|b|}\sqrt{2}$ **31.** $\frac{\sqrt{3}}{3}$ **33.** $\frac{\sqrt[4]{20}}{2}$ **35.** $\frac{\sqrt{y}}{y}$

37. $\frac{\sqrt[3]{18xy^2}}{3y}$ **39.** $2\sqrt{2}$ **41.** $30\sqrt[3]{2}$ **43.** $13\sqrt{y}$

45.

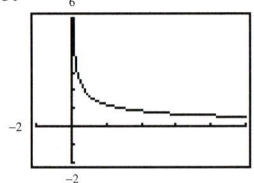

47. $3\sqrt{5}$

49. 89.44 cycles per second   **51.** 1   **53.** $4rs^2$

55. Base: 22 inches   **57.** $\sqrt{30}$
Height: 10 inches

59. $\sqrt[5]{\frac{152}{3}}$ **61.** $3\sqrt[4]{11}$ **63.** $\frac{1}{3}\sqrt{35}$ **65.** $3\sqrt{3}$

67. $10\sqrt[4]{3}$ **69.** $\frac{1}{7}\sqrt{15}$ **71.** $\frac{1}{4}\sqrt[3]{35}$ **73.** 0.02

75. $20\sqrt[3]{300}$ **77.** $4y^2\sqrt{3}$ **79.** $xy\sqrt[3]{x}$ **81.** $2xy\sqrt[5]{y}$

83. $\frac{1}{5}\sqrt{13}$ **85.** $\frac{3a\sqrt[3]{2a}}{b^3}$ **87.** $4\sqrt{3}$ **89.** $\frac{3}{2}\sqrt[3]{2}$

91. $\frac{2\sqrt{x}}{x}$ **93.** $\frac{2\sqrt{3b}}{b^2}$ **95.** $\frac{a^2\sqrt[3]{a^2b}}{b}$

97. $-5\sqrt[4]{7}-11\sqrt[4]{3}$ **99.** $24\sqrt{2}-6$ **101.** $12\sqrt{x}$

103. $(10-z)\sqrt[3]{z}$ **105.** $\frac{2\sqrt{5}}{5}$ **107.** $\frac{9\sqrt{5}}{5}$ **109.** $>$

111. $>$ **113.** $3\sqrt{13}$ **115.** 2.22 seconds   **117.** 1

Math Matters *(page 541)*

1,000,000,000 nanoseconds in a second

1000 millimeters in a meter

1000 watts in a kilowatt

Mid-Chapter Quiz *(page 542)*

1. $-\frac{1}{144}$ **2.** $\frac{64}{27}$ **3.** $\frac{5}{3}$ **4.** -4 **5.** $3t^{3/2}$

6. $\frac{3x^6}{16y^3}$ **7.** $\frac{2}{3u^3}$ **8.** x^2+4 **9.** (a) 1.34×10^7
(b) 7.5×10^{-4}

10. (a) 8.1×10^{13} **11.** (a) $5\sqrt{6}$ **12.** (a) $3|x|\sqrt{3}$
(b) 2×10^{-4} (b) $3\sqrt[3]{2}$ (b) $3x\sqrt{x}$

13. (a) $\frac{\sqrt[4]{5}}{2}$ **14.** (a) $\frac{2u\sqrt{10u}}{3}$ **15.** (a) $\frac{\sqrt{6}}{3}$
(b) $\frac{2\sqrt{6}}{7}$ (b) $\frac{16}{u^2}$ (b) $4\sqrt{3}$

16. (a) $\frac{2\sqrt{5x}}{x}$ **17.** $4\sqrt{2y}$ **18.** $6x\sqrt[3]{5x^2}+4x\sqrt[3]{5x}$
(b) $\frac{\sqrt[3]{12a^2}}{2a}$

19. $\sqrt{5^2+12^2}=\sqrt{169}=13$ **20.** $23+8\sqrt{2}$

Section 9.4 *(page 547)*

5. $3\sqrt{2}$ **7.** $2\sqrt{5}-\sqrt{15}$ **9.** -1 **11.** $8\sqrt{5}+24$

13. $y+4\sqrt{y}$ **15.** $45x-17\sqrt{x}-6$ **17.** $2-7\sqrt[3]{4}$

19. $\sqrt{3}+3\sqrt{3}$ **21.** $4-3x$ **23.** $2u+\sqrt{2u}$

25. $\sqrt{11}+\sqrt{3},8$ **27.** $\sqrt{x}+3,x-9$ **29.** $\frac{\sqrt{22}+2}{3}$

31. $4\left(3-\sqrt{7}\right)$ **33.** $\frac{\left(\sqrt{5}+1\right)t^{3/2}}{2},\ t>0$

35. $-\frac{\sqrt{u+v}\left(\sqrt{u-v}+\sqrt{u}\right)}{v}$ **37.** (a) 0   (b) -1

39.

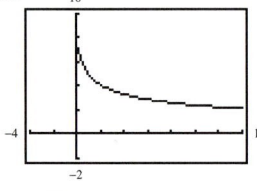

41. $192\sqrt{2}$ **43.** 6

45. $-\frac{2}{3},5$ **47.** Vertical shift four units upward

49. Reflection in the x-axis   **51.** 4   **53.** $2\sqrt{10}+8\sqrt{2}$

55. 4   **57.** 4   **59.** $\sqrt{15}+3\sqrt{3}-5\sqrt{5}-15$

61. $2x+20\sqrt{2x}+100$ **63.** $9x-25$ **65.** $x-y$

67. $\sqrt[3]{4x^2}+10\sqrt[3]{2x}+25$ **69.** $\frac{1-2\sqrt{x}}{3}$

71. $\dfrac{-1+\sqrt{3y}}{4}$　　**73.** $2-\sqrt{5}, -1$

75. $\sqrt{2u}+\sqrt{3}, 2u-3$　　**77.** $\dfrac{6-\sqrt{2}}{17}$　　**79.** $\dfrac{4\sqrt{7}+11}{3}$

81. $\dfrac{(\sqrt{15}+\sqrt{3})\,x}{4}$　　**83.** $\dfrac{(\sqrt{x}-5)(2\sqrt{x}+1)}{4x-1}$

85. (a) $2\sqrt{3}-4$　　**87.**

(b) 0

89. $\dfrac{\sqrt{3}}{\sqrt{4}}$

Section 9.5 *(page 556)*

5. (a) Not a solution　　**7.** (a) Not a solution

(b) Not a solution　　(b) Solution

(c) Not a solution　　(c) Not a solution

(d) Solution　　(d) Not a solution

9. 400　　**11.** No solution　　**13.** $\dfrac{44}{3}$　　**15.** $-\dfrac{2}{3}$

17. No solution　　**19.** 7　　**21.** $1, 3$　　**23.** 1.407

25. 4.840　　**27.** 10　　**29.** $\dfrac{5\sqrt{3}}{2}\approx 4.33$

31.

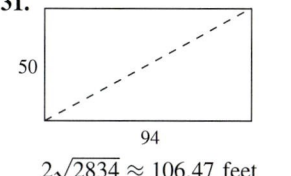

$2\sqrt{2834}\approx 106.47$ feet

33.

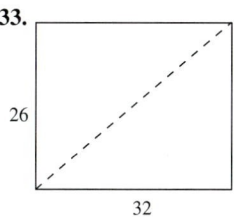

$10\sqrt{17}\approx 41.23$ feet

35. 64 feet　　**37.** 1.82 feet　　**39.** $-\dfrac{4x}{3}$　　**41.** $\dfrac{2u}{9v^6}$

43. $\dfrac{4}{9x^4y^2}$　　**45.** The store, at $191.96　　**47.** 49

49. 525　　**51.** 90　　**53.** $\dfrac{14}{25}$　　**55.** No solution

57. 4　　**59.** -15　　**61.** $4, -12$　　**63.** 8　　**65.** 1.569

67. 1.978　　**69.** $4\sqrt{2}\approx 5.66$　　**71.** $5\sqrt{15}\approx 19.36$

73. c　　**75.** d　　**77.** f　　**79.** $8\sqrt{6}\approx 19.596$

81. $40\sqrt{2}\approx 56.57$ feet per second

83. 56.25 feet　　**85.** 500 units

87. $r=\sqrt{\dfrac{A}{\pi}}$　　**89.** $26{,}250$

Section 9.6 *(page 566)*

7. $2i$　　**9.** $3\sqrt{3}\,i$　　**11.** $\sqrt{7}\,i$　　**13.** $10i$

15. $3\sqrt{2}\,i$　　**17.** -4　　**19.** $-\sqrt{15}+\sqrt{10}$

21. $10+4i$　　**23.** $-14+20i$　　**25.** $-3+49i$

27. -36　　**29.** $-40-5i$　　**31.** $-24-2\sqrt{3}\,i$

33. $-2+8i, 68$　　**35.** $1-\sqrt{3}\,i, 4$　　**37.** $-\dfrac{24}{53}+\dfrac{84}{53}i$

39. $-10i$　　**41.** $23\sqrt{2}$　　**43.** $\dfrac{4}{5}\sqrt{10}$　　**45.** $1489.65

47. $-12i$　　**49.** $\dfrac{2}{5}i$　　**51.** $0.3i$　　**53.** $2\sqrt{2}\,i$

55. $-3\sqrt{6}$　　**57.** $-2\sqrt{3}-3$　　**59.** $a=2, b=-3$

61. $a=-4, b=-2\sqrt{2}$　　**63.** $-14-40i$　　**65.** 13

67. $9-7i$　　**69.** -20　　**71.** $-36i$　　**73.** $27i$

75. $-65-10i$　　**77.** $20-12i$　　**79.** $-14+42i$

81. $-7-24i$　　**83.** 9　　**85.** $2-i, 5$

87. $5+\sqrt{6}\,i, 31$　　**89.** $-10i, 100$　　**91.** $2+2i$

93. $-\dfrac{6}{5}+\dfrac{2}{5}i$　　**95.** $\dfrac{8}{5}-\dfrac{1}{5}i$　　**97.** $1-\dfrac{6}{5}i$

99. $-\dfrac{53}{25}+\dfrac{29}{25}i$　　**101.** (a) Solution

(b) Solution

103. (a) Solution　　**105.** $2a$　　**107.** $2bi$

(b) Solution

Review Exercises *(page 570)*

1. $\dfrac{1}{72}$　　**3.** $\dfrac{125}{8}$　　**5.** 3.6×10^7　　**7.** 500　　**9.** $2x$

11. $\dfrac{x^6}{y^8}$　　**13.** $\dfrac{1}{t^3}$　　**15.** $\dfrac{27}{y^3}$　　**17.** 1.2　　**19.** $\dfrac{5}{6}$

21. 12　　**23.** $11{,}414.13$　　**25.** 10.63　　**27.** $49^{1/2}=7$

29. $\sqrt[3]{216}=6$　　**31.** 81　　**33.** 125　　**35.** 0.04

37. $x^{7/12}$　　**39.** $\dfrac{3}{x^{1/4}y^{2/5}}$　　**41.** $6\sqrt{10}$

43. $0.5x^2\sqrt{y}$　　**45.** $2ab\sqrt[3]{6b}$　　**47.** $\dfrac{\sqrt{30}}{6}$

49. $\dfrac{\sqrt{3x}}{2x}$ **51.** $\dfrac{\sqrt[3]{4x^2}}{x}$ **53.** $\dfrac{7+\sqrt{7}}{7}$

55. $-24\sqrt{10}$ **57.** $11\sqrt{x}$ **59.** $9-x$

61.

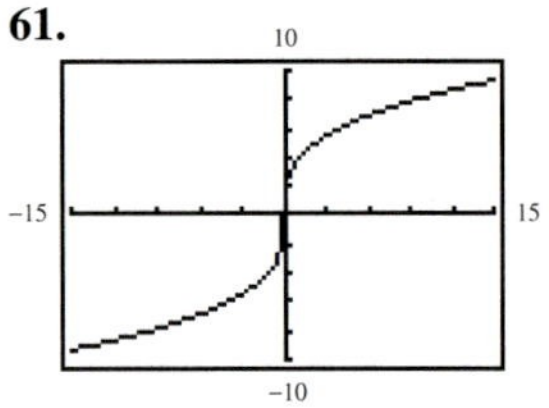

63.

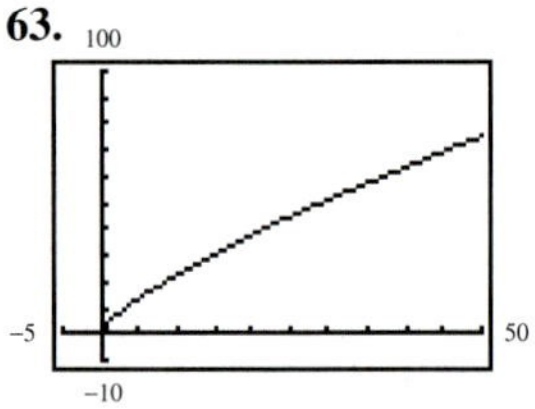

65. 225 **67.** 105 **69.** $-3, -5$ **71.** 5

73. $\frac{3}{32}$ **75.** 1.978 **77.** $4\sqrt{3}\,i$ **79.** $10-9\sqrt{3}\,i$

81. $\frac{3}{4}-\sqrt{3}\,i$ **83.** $8-3i$ **85.** -90 **87.** 25

89. $59+74i$ **91.** $\frac{9}{17}+\frac{2}{17}i$ **93.** $23+8\sqrt{2}$

95. 1.37 feet

Chapter Test *(page 572)*

1. (a) $\frac{3}{8}$ (b) 3.0×10^{-5} **2.** (a) $\frac{1}{9}$ (b) 6 **3.** 3.2×10^{-5}

4. 30,400,000 **5.** (a) $\dfrac{3}{5t}$ (b) $\dfrac{y^2}{xy^2+1}$ **6.** (a) $x^{1/3}$ (b) 25

7. (a) $\dfrac{4}{3}\sqrt{2}$ (b) $2\sqrt[3]{3}$ **8.** $\dfrac{\sqrt{6}}{2}$ **9.** $-10\sqrt{3x}$

10. $5\sqrt{3x}+3\sqrt{5}$ **11.** $16-8\sqrt{2x}+2x$

12. $7\sqrt{3}\,(4y+3)$ **13.** No solution **14.** 9

15. $2-2i$ **16.** $-5-12i$ **17.** $-8+4i$

18. $13+13i$ **19.** $-2-5i$ **20.** 100 feet

CHAPTER 10

Section 10.1 *(page 579)*

7. 0, 3 **9.** 6 **11.** 1, 6 **13.** $\frac{5}{3}, 6$ **15.** ±3

17. ±8 **19.** $9, -17$ **21.** $\dfrac{-1\pm5\sqrt{2}}{2}$ **23.** $\pm2i$

25. $-6\pm\frac{11}{3}i$ **27.** $1\pm3\sqrt{3}\,i$ **29.** $0, \frac{5}{2}$ **31.** ±10

33. $5\pm10i$ **35.** $\pm1, \pm2\sqrt{2}i$ **37.** $4, \frac{25}{4}$

39. $\pm2\sqrt{2}, \pm2i$ **41.** $\pm\sqrt{15}, \pm\sqrt{5}\,i$

43.

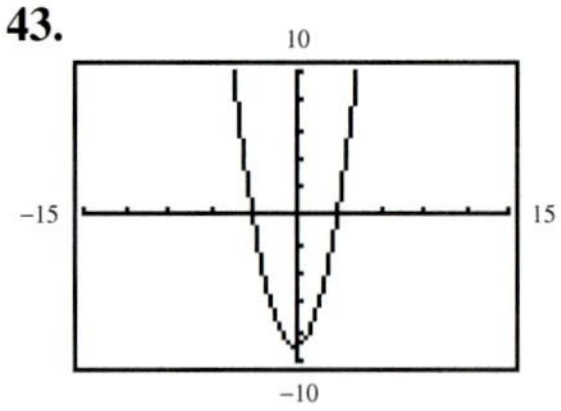

$(-3, 0), (3, 0)$

45.

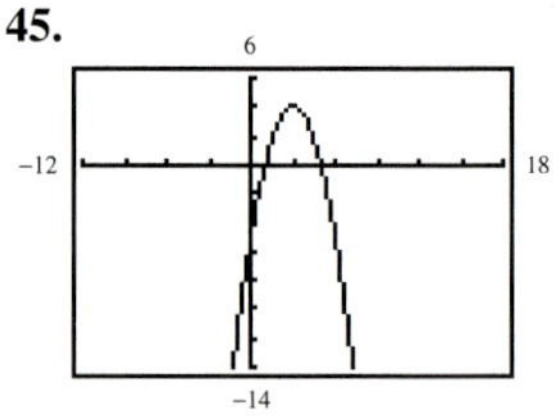

$(1, 0), (5, 0)$

47.

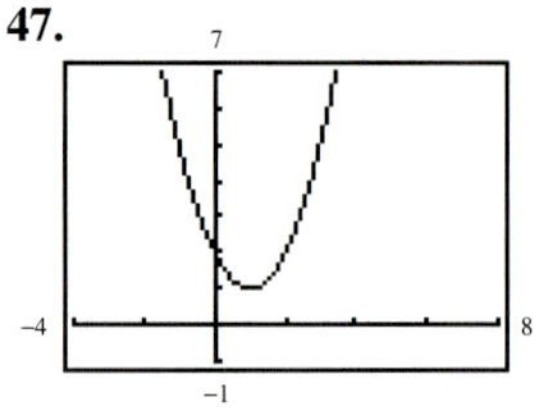

$1\pm i$

49.

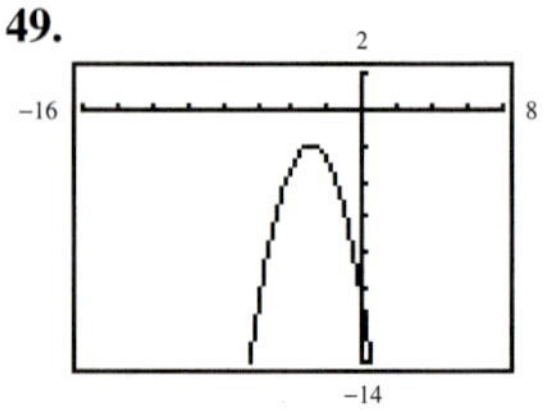

$-3\pm\sqrt{2}\,i$

51. 1978 **53.** $(4x+11)(4x-11)$

55. $(x-10)(x-4)$ **57.** 3 miles per hour **59.** 9, 12

61. $\pm\frac{5}{2}$ **63.** Prime **65.** $\frac{1}{2}, -\frac{5}{6}$ **67.** ±8

69. $\pm\frac{4}{5}$ **71.** $\pm\frac{15}{2}$ **73.** 2.5, 3.5 **75.** $2\pm\sqrt{7}$

77. $\dfrac{3\pm7\sqrt{2}}{4}$ **79.** $\pm\sqrt{17}\,i$ **81.** $3\pm5i$

83. $\frac{3}{2}\pm\frac{5}{2}i$ **85.** $\frac{2}{3}\pm\frac{1}{3}i$ **87.** $-\dfrac{7}{3}\pm\dfrac{\sqrt{38}}{3}i$

89. ±30 **91.** $\pm30i$ **93.** ±3 **95.** $-5, 15$

97. $5\pm10i$ **99.** $\pm1, \pm2$ **101.** $\pm\sqrt{2}, \pm\sqrt{3}$

103. $\pm2, \pm i$ **105.** $\pm1, \pm\sqrt{5}$ **107.** 4 **109.** $\frac{1}{2}, 1$

111. $-\frac{3}{4}, 2$ **113.** $-8, 27$

115.

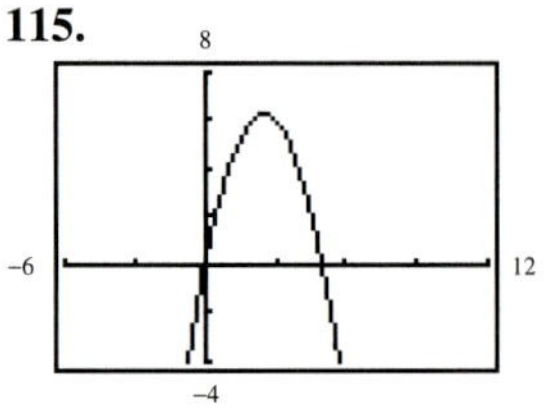

$(0, 0), (5, 0)$

117.

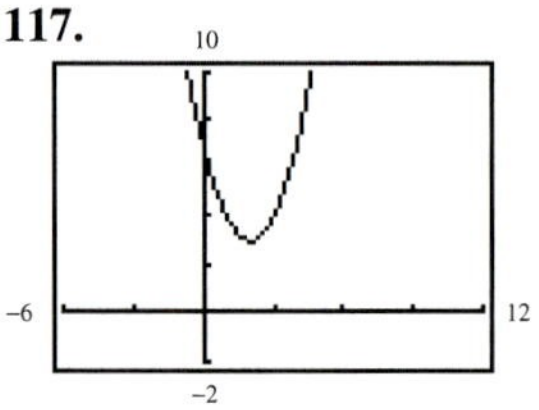

$2\pm\sqrt{3}\,i$

119. $(x-5)(x+2)=x^2-3x-10=0$

121. $\left[x-\left(1+\sqrt{2}\right)\right]\left[x-\left(1-\sqrt{2}\right)\right]=x^2-2x-1=0$
$=x^2-2x+3=0$

123. 4 seconds **125.** 9 seconds **127.** 1987

129. $0.06=6\%$ **131.** $\frac{5}{2}\sqrt{5}\approx5.590$ seconds

Section 10.2 *(page 587)*

7. 16 **9.** $\frac{25}{4}$ **11.** 0, 6 **13.** $-3, -4$

15. $2 + \sqrt{7} \approx 4.65$ **17.** $-1 + \sqrt{2}\,i \approx -1 + 1.41i$

$2 - \sqrt{7} \approx -0.65$ $\qquad -1 - \sqrt{2}\,i \approx -1 - 1.41i$

19. $\dfrac{1 + 2\sqrt{7}}{3} \approx 2.10$ **21.** $\dfrac{-5 + \sqrt{13}}{2} \approx -0.70$

$\dfrac{1 - 2\sqrt{7}}{3} \approx -1.43$ $\qquad \dfrac{-5 - \sqrt{13}}{2} \approx -4.30$

23. $\dfrac{-4 + \sqrt{10}}{2} \approx -0.42$ **25.** $\dfrac{-5 + \sqrt{17}}{2} \approx -0.44$

$\dfrac{-4 - \sqrt{10}}{2} \approx -3.58$ $\qquad \dfrac{-5 - \sqrt{17}}{2} \approx -4.56$

27. $1 \pm \sqrt{3}$ **29.** $4 \pm 2\sqrt{2}$

31. 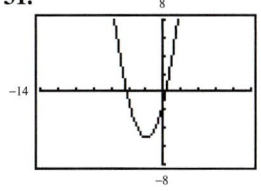 **33.**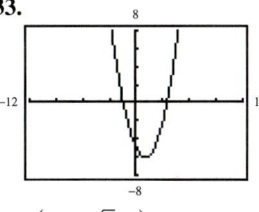

$(-2 \pm \sqrt{5}, 0)$ $\qquad\qquad (1 \pm \sqrt{6}, 0)$

35. $30\sqrt{6} \approx 73.48$ feet **37.** $2xy\sqrt{2x}$

39. $(|x| - 4)\sqrt{3}$ **41.** 15 minutes **43.** 100

45. $\frac{9}{25}$ **47.** $\frac{9}{100}$ **49.** $\frac{1}{25}$ **51.** 0, 25

53. 1, 7 **55.** $-6, 4$ **57.** $-3, 6$ **59.** $\frac{3}{2}, 4$

61. $-2 + \sqrt{7} \approx 0.65$ **63.** $2 + \sqrt{3} \approx 3.73$

$-2 - \sqrt{7} \approx -4.65$ $\qquad 2 - \sqrt{3} \approx 0.27$

65. $5 + 3\sqrt{3} \approx 10.20$ **67.** $-10 + 3\sqrt{10} \approx -0.51$

$5 - 3\sqrt{3} \approx -0.20$ $\qquad -10 - 3\sqrt{10} \approx -19.49$

69. $\dfrac{-3 + \sqrt{17}}{2} \approx 0.56$ **71.** $\dfrac{1}{2} + \dfrac{\sqrt{3}}{2}\,i \approx 0.5 + 0.87i$

$\dfrac{-3 - \sqrt{17}}{2} \approx -3.56$ $\qquad \dfrac{1}{2} - \dfrac{\sqrt{3}}{2}\,i \approx 0.5 - 0.87i$

73. $\dfrac{-9 + \sqrt{21}}{6} \approx -0.74$ **75.** $\dfrac{-1 + \sqrt{10}}{2} \approx 1.08$

$\dfrac{-9 - \sqrt{21}}{6} \approx -2.26$ $\qquad \dfrac{-1 - \sqrt{10}}{2} \approx -2.08$

77. $-1 \pm 2i$ **79.** $\dfrac{7 + \sqrt{57}}{2} \approx 7.27$

$\dfrac{7 - \sqrt{57}}{2} \approx -0.27$

81. $-8.20, 2.20$ **83.** $-2.30, 1.30$

85. (a) $x^2 + 8x$ **87.** 8 feet × 20 feet

(b) $x^2 + 8x + 16$

(c) $(x + 4)^2$

89. 42 units or 58 units

91. 20 feet × 35 feet or 15 feet × $46\frac{2}{3}$ feet

93. 6.53 feet × 6.53 feet × 3 feet

Section 10.3 *(page 595)*

5. $2x^2 + 2x - 7 = 0$ **7.** $-x^2 + 10x - 5 = 0$

9. 4, 7 **11.** $-\frac{3}{2}$ **13.** Two distinct irrational solutions

15. Two complex solutions **17.** $1 \pm \sqrt{5}$

19. $-2 \pm \sqrt{3}$ **21.** $-\dfrac{3}{2} \pm \dfrac{\sqrt{3}}{2}\,i$ **23.** $\dfrac{-1 \pm \sqrt{5}}{3}$

25. $\dfrac{-1 \pm \sqrt{10}}{5}$ **27.** $-15, 0$ **29.** $-4 \pm 3i$

31. 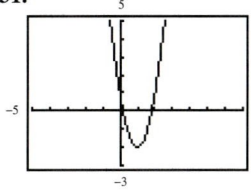 **33.**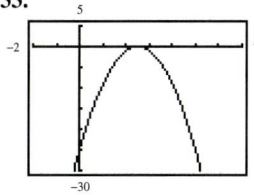

$(0.18, 0), (1.82, 0)$ $\qquad\qquad (2.50, 0)$

35. $\dfrac{5 \pm \sqrt{185}}{8}$ **37.** $\dfrac{3 + \sqrt{17}}{2}$ **39.** (a) $c < 9$

(b) $c = 9$

(c) $c > 9$

41. (a) $c < 16$ **43.** 5.1 inches × 11.4 inches

(b) $c = 16$

(c) $c > 16$

45. $x - 9$ **47.** $4t + 12\sqrt{t} + 9$

49. 30% solution: $13\frac{1}{3}$ gallons **51.** $-2, -4$ **53.** $\frac{2}{3}$

60% solution: $6\frac{2}{3}$ gallons

55. $-\frac{1}{2}, \frac{2}{3}$ **57.** Two complex solutions

59. One rational solution **61.** $-3 \pm 2\sqrt{3}$

63. $5 \pm \sqrt{2}$ **65.** $\dfrac{1 \pm \sqrt{3}}{2}$ **67.** $\dfrac{-2 \pm \sqrt{10}}{2}$

69. $\dfrac{1 \pm \sqrt{5}}{5}$ **71.** $\dfrac{3}{4} \pm \dfrac{\sqrt{3}}{4}\,i$ **73.** $\dfrac{3 \pm \sqrt{13}}{6}$

75. ± 13 **77.** $\frac{9}{5}, \frac{21}{5}$ **79.** 8, 16 **81.** 0.372, 3.228

83. 0.200, 99.800   **85.** (a) 2.5 seconds

(b) $\dfrac{5 + 5\sqrt{3}}{4} \approx 3.415$ seconds

87.

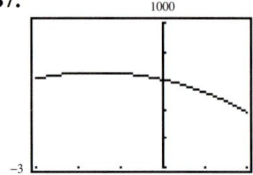

89. 606,870

Mid-Chapter Quiz *(page 598)*

1. ± 6 **2.** $-4, \frac{5}{2}$ **3.** $\pm 2\sqrt{3}$ **4.** $-1, 7$

5. $-5 \pm \sqrt{6}$ **6.** $\dfrac{-3 \pm \sqrt{19}}{2}$ **7.** $-2 \pm \sqrt{10}$

8. $\dfrac{3 \pm \sqrt{105}}{12}$ **9.** $-\dfrac{5}{2} \pm \dfrac{\sqrt{3}}{2} i$ **10.** $-2, 10$

11. $-3, 10$ **12.** $-2, 5$ **13.** $\dfrac{3}{2}$

14. $\dfrac{-5 \pm \sqrt{10}}{3}$ **15.** $-4, 1$ **16.** 2

17.

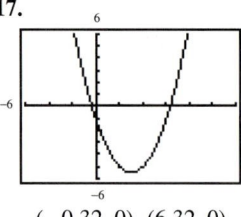

$(-0.32, 0), (6.32, 0)$

18.

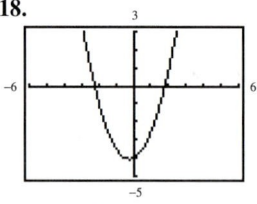

$(-2.24, 0), (1.79, 0)$

19. 50 units   **20.** 35 meters $\times$ 65 meters

Section 10.4 *(page 605)*

7. 15, 16   **9.** 14, 16   **11.** 108 square inches

13. 440 meters   **15.** 8%   **17.** 2.59%

19. (a) 89 miles   **21.** 100 feet $\times$ 125 feet

(b) 178 miles

23. 18 dozen, $1.20 per dozen   **25.** 48

27. (a) $d = \sqrt{(3 + x)^2 + (4 + x)^2}$

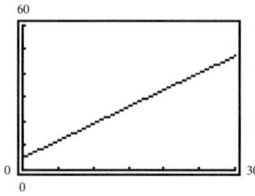

(b) $\dfrac{-7 + \sqrt{199}}{2} \approx 3.55$ meters

29. 400 miles per hour   **31.** 9.1 hours, 11.1 hours

33. 3 seconds   **35.** 9.5 seconds   **37.** 30 units

39. $\sqrt{\dfrac{A}{P}} - 1$ **41.** $-10x^2$ **43.** $4x^2 - 60x + 225$

45. $2300   **47.** 21, 23   **49.** 12, 13

51. 180 square kilometers   **53.** 210 inches

55. 12 inches $\times$ 24 inches   **57.** 6%   **59.** 7.5%

61. 5   **63.** 15.86 miles   **65.** 6.8 days, 9.8 days

67. 4.7 seconds   **69.** $\sqrt{\dfrac{S}{4\pi}}$ **71.** $\sqrt{\dfrac{12I - a^2 M}{M}}$

73. 46 miles per hour

75. (a) $b = 20 - a$

$A = \pi a b$

$A = \pi a (20 - a)$

(b)

a	4	7	10	13	16
A	201.1	285.9	314.2	285.9	201.1

(c) 7.9, 12.1

(d)

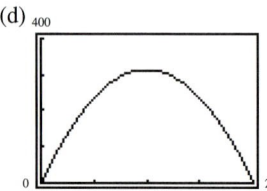

Section 10.5 *(page 617)*

7. $0, \frac{5}{2}$ **9.** 1, 3

11. (a) Not a solution   **13.** (a) Not a solution

(b) Solution   (b) Solution

(c) Not a solution   (c) Not a solution

(d) Solution   (d) Not a solution

15. Negative: $(-\infty, 4)$
Positive: $(4, \infty)$

17. Negative: $(-1, 5)$
Positive: $(-\infty, -1)$ or $(5, \infty)$

19. $x \geq -3$

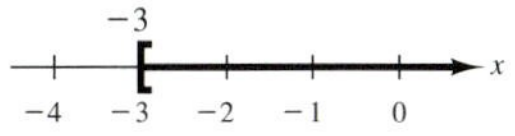

21. $x < 0$ or $x > 2$

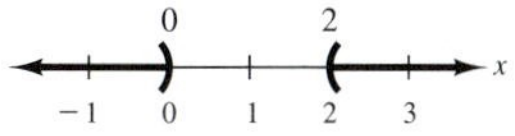

23. $-5 \leq x \leq 2$

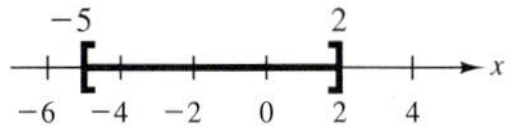

25. No solution

27. $-\infty < x < \infty$

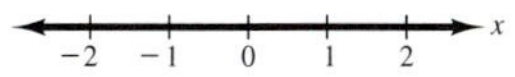

29. $-2 < x < 0$ or $x > 2$

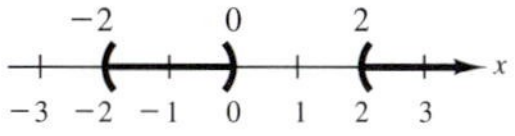

31. 3

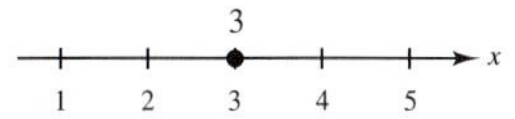

33. 0, −5

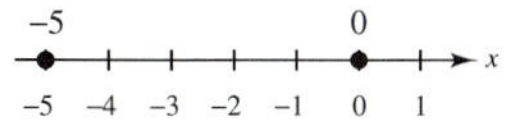

35. $x > 3$

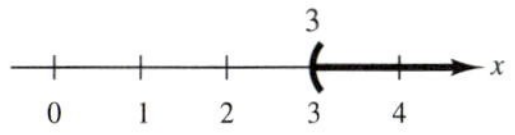

37. $x < 3$ or $x \geq 6$

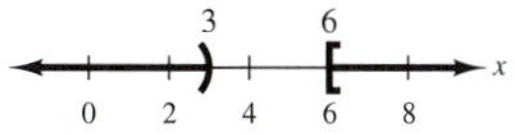

39. $x < -4$ or $x > \frac{3}{2}$ **41.** $-5 < x < 2.5$

43. $3 < t < 5$ **45.** $x < \frac{3}{2}$ **47.** $1 < x < 5$

49. 10 inches × 15 inches **51.** $\pm\frac{9}{2}$ **53.** $\frac{5}{2}$

55. Negative: $(-\infty, 12)$
Positive: $(12, \infty)$

57. Negative: $(0, 4)$
Positive: $(-\infty, 0) \cup (4, \infty)$

59. $x > 8$

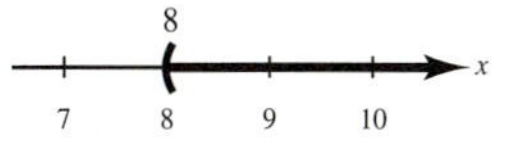

61. $0 < x < 2$

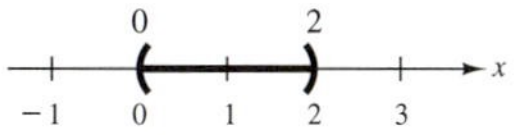

63. $x \leq -2$ or $x \geq 2$

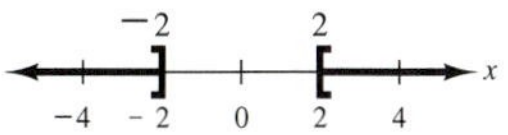

65. $u < -\frac{1}{2}$ or $u > 4$

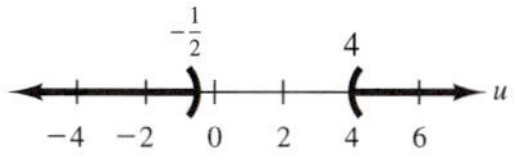

67. $x < 2 - \sqrt{2}$ or $x > 2 + \sqrt{2}$

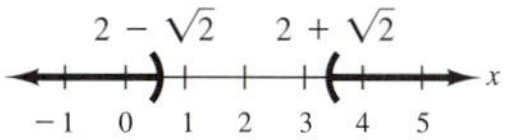

69. No solution

71. $x < 5 - \sqrt{6}$ or $x > 5 + \sqrt{6}$

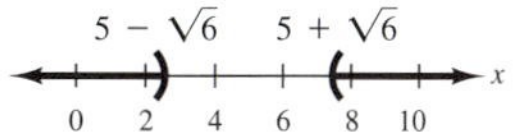

73. All real numbers

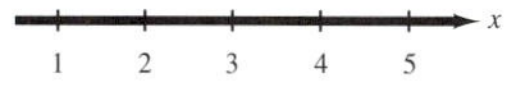

75. $x < 3$

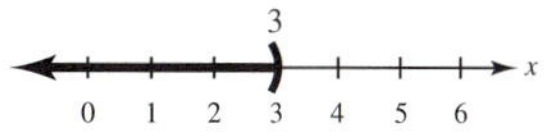

77. $x < 3$

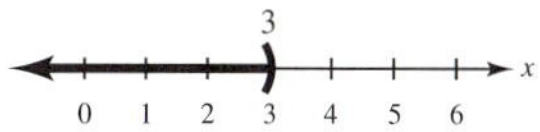

79. $0 \leq x < 4$

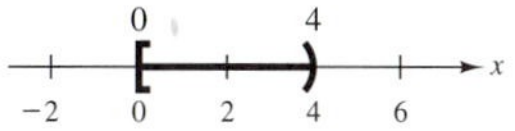

81. $y \leq 3$ or $y > \frac{11}{2}$

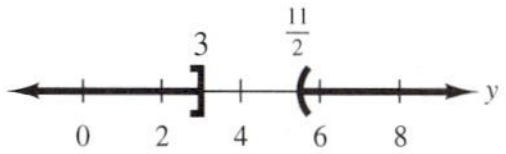

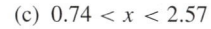

83. $4 < x < 7$

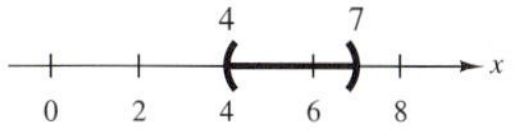

85. $-3 < x < 7$

87. $3 < x < 4$ **89.** $0.64 < t < 4.86$ **91.** $12 < l < 20$

93. (a) $A = \pi x^2 + \pi(12 - x)^2, 0 < x < 12$

(b)

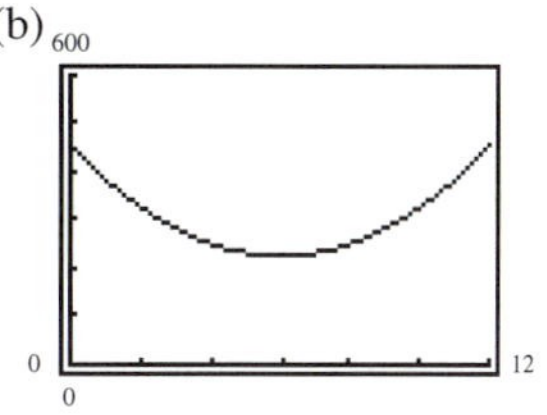

(c) $0.74 < x < 2.57$

Review Exercises *(page 622)*

1. 0, −12 **3.** $-10, \frac{8}{3}$ **5.** $\pm\frac{1}{2}$ **7.** $-\frac{4}{3}, -\frac{4}{3}$

9. −9, 10 **11.** ±100 **13.** $\pm\frac{3}{2}$ **15.** −4, 36

17. $3 \pm 2\sqrt{3}$ **19.** $\frac{3}{2} \pm \frac{\sqrt{3}}{2}i$ **21.** $\frac{-5 \pm \sqrt{19}}{2}$

23. 5, −6 **25.** $3, -\frac{7}{2}$ **27.** $\frac{10}{3} \pm \frac{5\sqrt{2}}{3}i$

29. $3 \pm 5\sqrt{10}$ **31.** −7, 12 **33.** −2, 20 **35.** $3 \pm i$

37. $-\dfrac{1}{2}, -1$ **39.** $\dfrac{3 \pm \sqrt{17}}{2}$ **41.** $1, 1 \pm \sqrt{6}$

43. $-\dfrac{2}{3}, 3$ **45.**

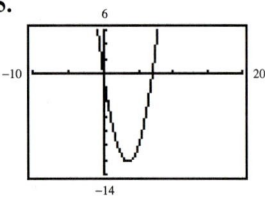

$(0, 0), (7, 0)$

47.

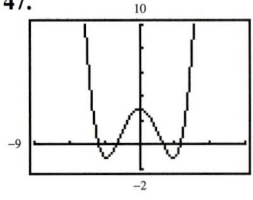

$(\pm 2, 0), \left(\pm 2\sqrt{3}, 0\right)$

49.

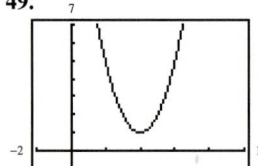

$4 \pm i$

51. 11, 12 **53.** (a) 200 feet

(b) Dropped

(c) $\dfrac{5\sqrt{2}}{2} \approx 3.54$ seconds

55. (a) $P = 2(l + w)$ (b) Substitute $24 - l$ for w.

$48 = 2(l + w)$

$24 = l + w$

$w = 24 - l$

(c)

l	2	4	6	8	10	12	14	16	18
A	44	80	108	128	140	144	140	128	108

57. 15 **59.** 10.7 hours, 13.7 hours **61.** 48

63. 19 hours, 21 hours **65.** $(x + 3)(x - 3) = 0$

$x^2 - 9 = 0$

67. $(x + 5)(x + 1) = 0$ **69.** $y = x^2 - 4x + 3$

$x^2 + 6x + 5 = 0$

71. $y = x^2 + 5x + 4$ **73.** $y = \frac{1}{9}x^2 - 1$

75. $x < 3$ **77.** $0 < x < 7$

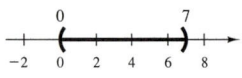

79. $x \le -2$ or $x \ge 6$ **81.** $-4 < x < \frac{5}{2}$

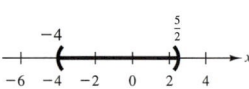

83. $x \le 0$ or $x > \frac{7}{2}$ **85.** 1995

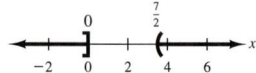

Chapter Test *(page 625)*

1. $-5, 10$ **2.** $-\frac{3}{8}, 3$ **3.** 1.7, 2.3

4. $-3 \pm 9i$ **5.** $C = \dfrac{9}{4}$ **6.** $\dfrac{3 \pm \sqrt{3}}{2}$

7. -56; two complex solutions **8.** $\dfrac{4 \pm \sqrt{7}}{3}$

9. $\dfrac{2 \pm 3\sqrt{2}}{2}$ **10.** $\dfrac{7 + \sqrt{61}}{2}$

11. $(x + 4)(x - 5) = x^2 - x - 20 = 0$

12. $[x - (i - 1)][x - (-i - 1)] = x^2 + 2x + 2 = 0$

13. $0 < x < 3$ **14.** $x \le -2$ or $x \ge 6$

15. $2 < x < \frac{11}{4}$ **16.** $-8 \le u < 3$

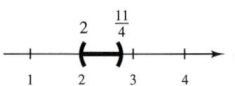 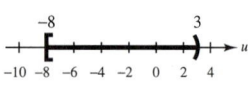

17. 14, 15 **18.** 12×20 feet **19.** 50 miles per hour

20. $\dfrac{\sqrt{10}}{2} \approx 1.58$ seconds

Cumulative Test: Chapters 7–10 *(page 626)*

1. $\dfrac{x(x + 2)(x + 4)}{9(x - 4)}$ **2.** $\dfrac{3x + 5}{x(x + 3)}$ **3.** $x + y$

4. $-4 + 3\sqrt{2}\, i$ **5.** $-\dfrac{2y^6}{3x^4}$ **6.** $t^{1/2}$

7. $x^2 - 3x + 9$ **8.** $\sqrt{10} + 2$

9.
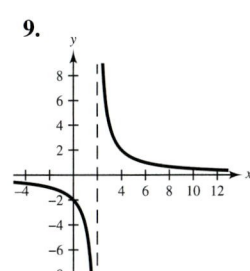

10.
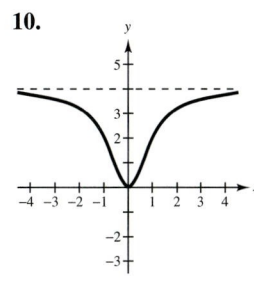

11. 2 **12.** 6 **13.** $5 \pm 5\sqrt{2}i$ **14.** $\dfrac{-3 \pm \sqrt{33}}{3}$

15. (a) $x \le 0$ or $x \ge 3$ **16.** $m = \frac{3}{4}$, distance $= 10$

(b) 2

(c) $\sqrt{c^2 - 3c}$

17.
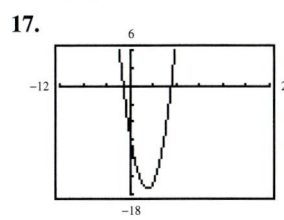

$(-1.12, 0), (7.12, 0)$

18. $y = x^2 - 4x - 12$

19. Function **20.** $r_2 = \dfrac{\sqrt{15}\,r_1}{5}$

CHAPTER 11

Section 11.1 *(page 635)*

7. b **9.** a **11.** $m = 6, (-4, 3)$ **13.** $m = \frac{3}{2}, (0, 0)$

15. $y = -\frac{1}{2}x$ **17.** $y + 4 = 3x$

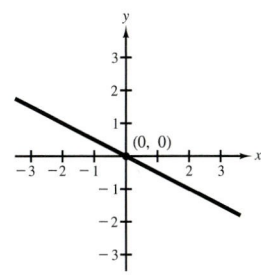

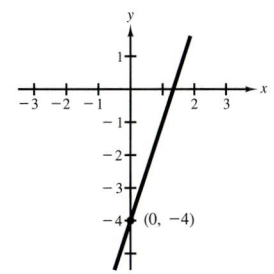

19. $y - \frac{7}{2} = -4(x + 2)$ **21.** $3x - 2y = 0$

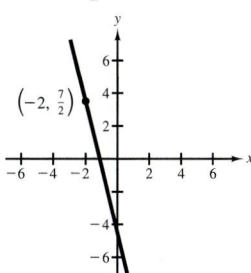

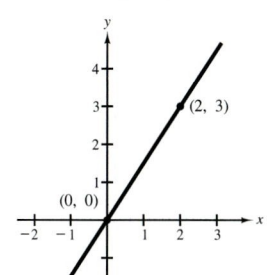

23. $3x + 7y - 15 = 0$ **25.** $14x + 6y - 39 = 0$

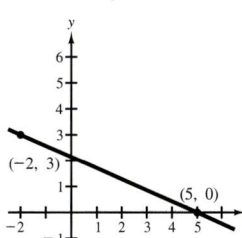

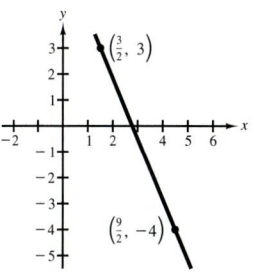

27. $f(x) = \frac{1}{2}x + 3$ **29.** $f(x) = 3$

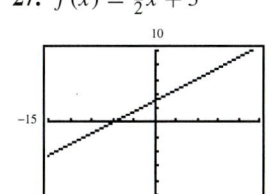

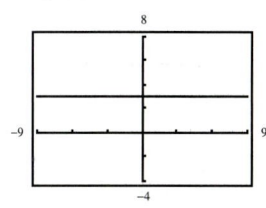

31. (a) $3x - y - 5 = 0$ **33.** (a) $x + y + 1 = 0$

(b) $x + 3y - 5 = 0$ (b) $x - y + 9 = 0$

35. (a) $y - 2 = 0$ **37.**

(b) $x + 1 = 0$

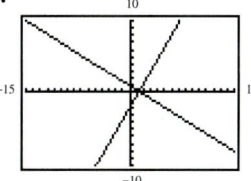

Perpendicular

39.
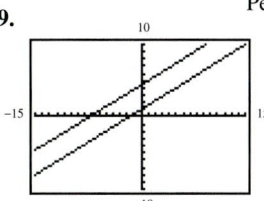

Parallel

41.

x	0	50	100	500	1,000
C	5000	6000	7000	15,000	25,000

43. (a) $P = 1200 - 15x$ **45.** $5x(1 - 4x)$

(b) 42

(c) 48

47. $(3x - 5)(5x + 3)$

49. (a)

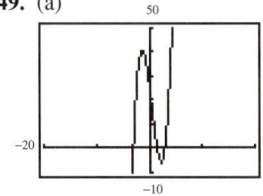

(b) $(2x - 3)(x + 3)(x - 3)$

$= (2x - 3)(x^2 - 9)$

$= 2x^3 - 3x^2 - 18x + 27$

51. $m = \frac{3}{4}, \left(-2, -\frac{5}{8}\right)$ **53.** $m = \frac{2}{3}, (0, -2)$

55. $m = \frac{5}{2}, (0, 12)$ **57.** $y + 6 = \frac{1}{2}(x)$

59. $y - 8 = -2(x + 2)$ **61.** $y = -\frac{2}{3}(x - 5)$

63. $y - 5 = 0(x + 8)$ or $y - 5 = 0$ **65.** $y - \frac{5}{2} = \frac{4}{3}\left(x - \frac{3}{4}\right)$

67. $2x + 5y = 0$ **69.** $5x + 34y - 67 = 0$

71. $2x - 15 = 0$ **73.** $4x + 5y - 11 = 0$

75. (a) $4x - y - 5 = 0$ **77.** (a) $3x + 10y + 22 = 0$

(b) $x + 4y - 31 = 0$ (b) $10x - 3y - 72 = 0$

79. Parallel   **81.** Perpendicular   **83.** Parallel

85. $\frac{x}{3} + \frac{y}{2} = 1$ **87.** $-\frac{6x}{5} - \frac{3y}{7} = 1$

89. $S = 1500 + 0.03M$ **91.** (a) $S = 0.70L$

(b) \$94.50

93. (a) $N = 1500 + 60t$ **95.** $C = 0.32t + 2.3$

(b) 2700   \$7.1 billion

(c) 1800

97. (a) and (b)   (c) $y = 4x + 19$ (d) 87

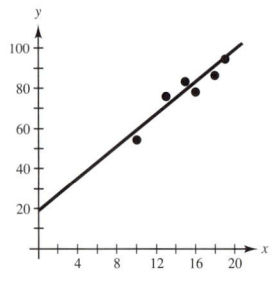

Section 11.2   *(page 644)*

7. b   **9.** d   **11.** f   **13.** (a) Solution

(b) Not a solution

(c) Solution

(d) Solution

15. (a) Solution   **17.**

(b) Not a solution

(c) Not a solution

(d) Solution

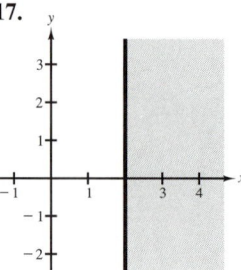

19.

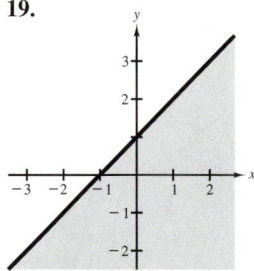

21.

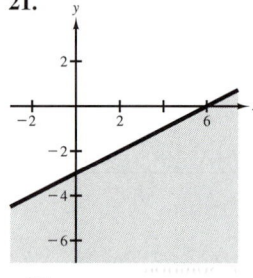

23.

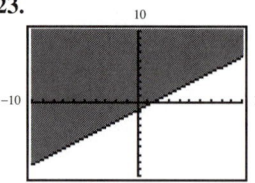

25.

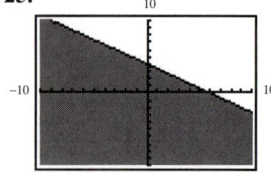

27. $y > x$ **29.** $3x + 4y - 17 > 0$

31. $y < 2$ **33.** $x - 2y < 0$

35.

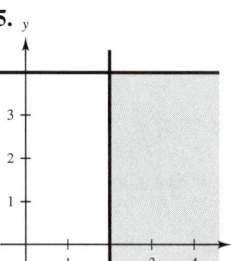

37.

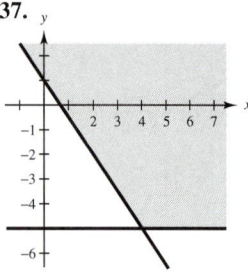

39.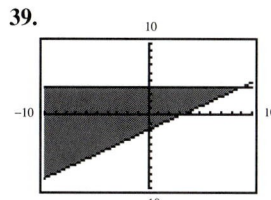

41.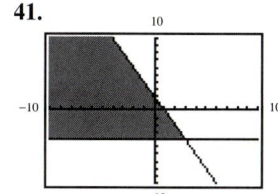

45. $\dfrac{1}{x^3}$ **47.** $\dfrac{9y^2}{4x^2}$

43.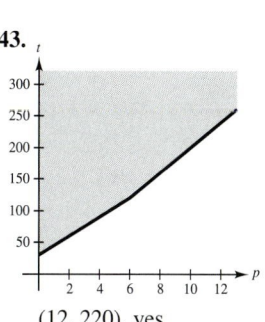

(12, 220), yes

49. $\dfrac{19\sqrt{2}}{2} \approx 13.435$ inches

51. (a) Not a solution
(b) Not a solution
(c) Solution
(d) Solution

53. (a) Not a solution
(b) Solution
(c) Not a solution
(d) Solution

55.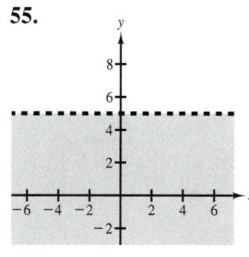

57.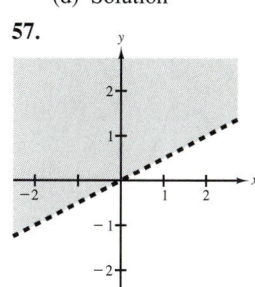

59.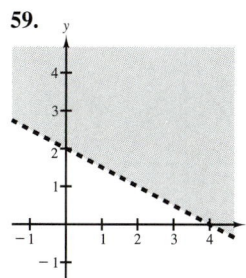

61.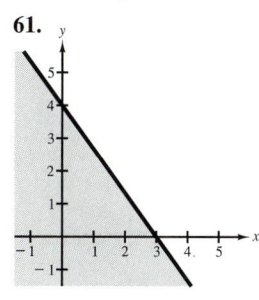

63.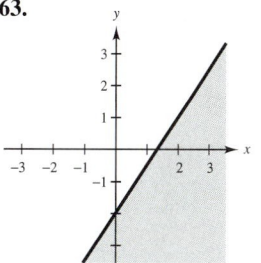

65.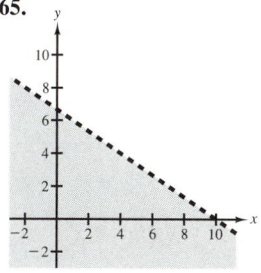

67.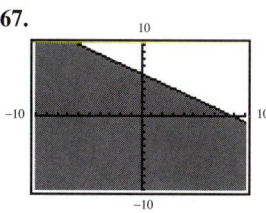

69.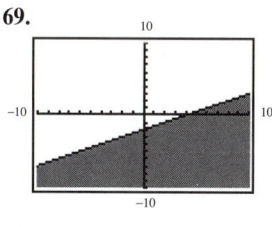

71.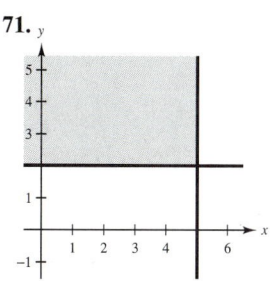

73.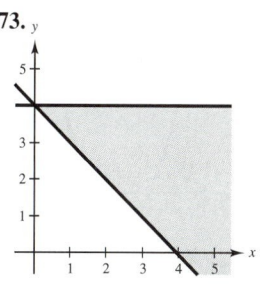

75.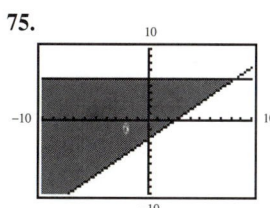

77.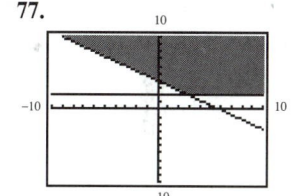

79. $2x + 2y \leq 500$
(Note: x and y cannot be negative.)

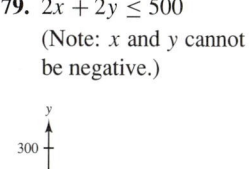

81. $30x + 20y \geq 300$
(Note: x and y cannot be negative.)

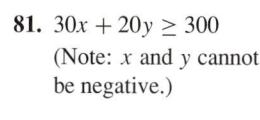

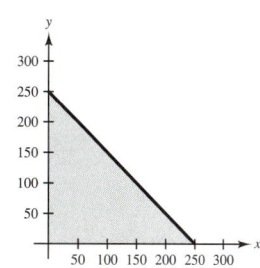

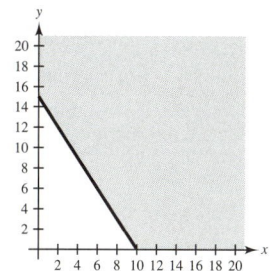

Section 11.3 *(page 652)*

7. e **9.** b **11.** d **13.** Up, $(0, 2)$

15. Up, $(0, -6)$ **17.** $(\pm 5, 0)$, $(0, 25)$ **19.** $(0, 3)$

21. $y = (x - 2)^2 + 3$, $(2, 3)$

23. $y = -(x - 1)^2 - 6$, $(1, -6)$

25. 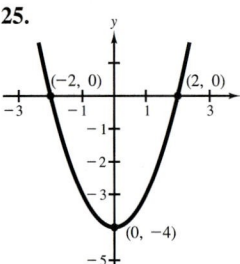 **27.**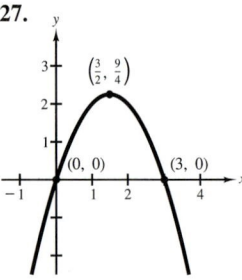

29. 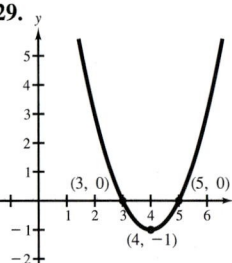 **31.**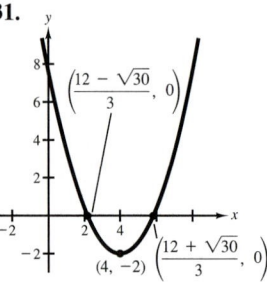

33. $y = (x - 2)^2$ **35.** $y = 4 - (x + 2)^2$

37. $y = -2(x + 3)^2 + 3$

39. (a) 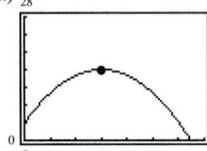 (b) 4 feet
(c) 16 feet
(d) $12 + 8\sqrt{3} \approx 25.9$ feet

41. (a) Revenue = (number sold) · (price per unit)
$$= (100 + x)[90 - x(0.15)]$$
$$= 9000 + 75x - 0.15x^2$$
Cost = (number sold) · (cost per unit)
$$= 60(100 + x) = 6000 + 60x$$
Profit = Revenue − Cost
$$= 3000 + 15x - 0.15x^2$$

(b)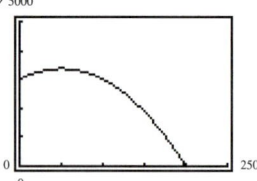

(c) Vertex; $(50, 3375)$; 150 units

(d) Yes for orders up to 150 units

43. $y = \frac{1}{2}x(x - 2)$ **45.** $x < \frac{3}{2}$

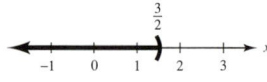

47. $1 < x < 5$ **49.** 1, 3

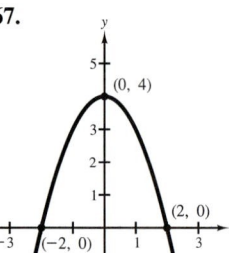 (for 47)

51. $\sqrt{5} \approx 2.236$ seconds **53.** Down, $(10, 4)$

55. Up, $(3, -9)$ **57.** $(0, 0)$, $(9, 0)$ **59.** $\left(\frac{3}{2}, 0\right)$, $(0, 9)$

61. $y = (x + 3)^2 - 4$, $(-3, -4)$

63. $y = -(x - 3)^2 - 1$, $(3, -1)$

65. $y = 2\left(x + \frac{3}{2}\right)^2 - \frac{5}{2}$, $\left(-\frac{3}{2}, -\frac{5}{2}\right)$

67.   **69.**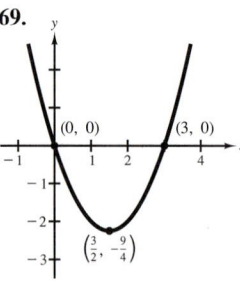

71. 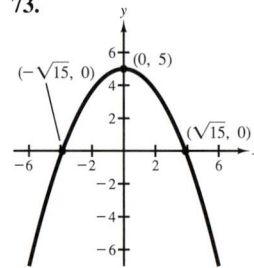 **73.**

75.

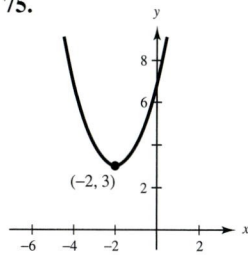

77.

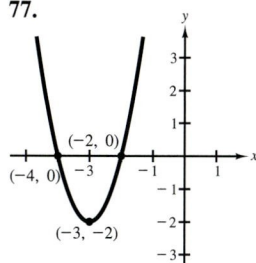

103.

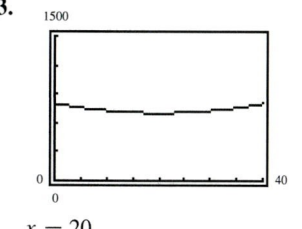

$x = 20$

Math Matters *(page 656)*

(a) $17,500 a year   (b) $12.50 an hour

79.

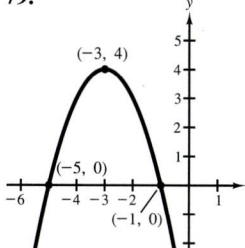

81.

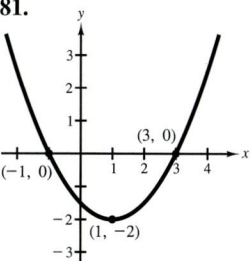

Mid-Chapter Quiz *(page 657)*

1. $4x - 2y - 3 = 0$ **2.** $3x + 2y - 12 = 0$

3. $6x + 8y - 63 = 0$ **4.** $30x - 10y + 87 = 0$

5. $12x - 11y + 7 = 0$ **6.** $31x + 30y - 24 = 0$

7. $y + 1 = 0$ **8.** $x - 4 = 0$

9. (a) $2x - 3y + 9 = 0$ **10.** (a) Solution

    (b) $3x + 2y - 19 = 0$ (b) Solution

                             (c) Not a solution

                             (d) Not a solution

83.

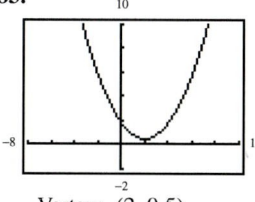

Vertex: $(2, 0.5)$

85.

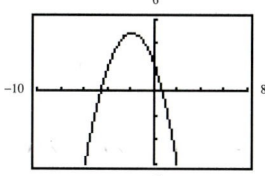

Vertex: $(-1.9, 4.9)$

11. $x + 2y \leq 11$ **12.** $x - 3y > -5$

13.

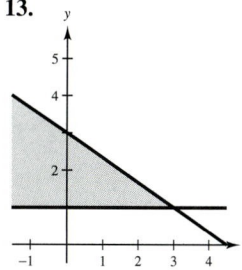

14.

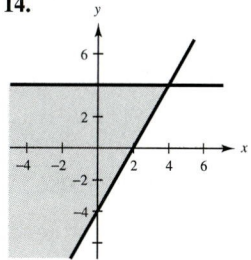

87.

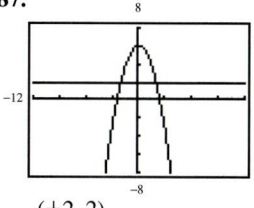

$(\pm 2, 2)$

89.

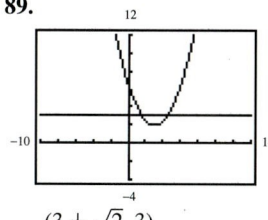

$(3 \pm \sqrt{2}, 3)$

91. $y = x^2 - 4x + 5$ **93.** $y = -x^2 - 6x - 5$

95. $y = x^2 - 4x$ **97.** $y = \frac{1}{2}x^2 - 3x + \frac{13}{2}$

99.

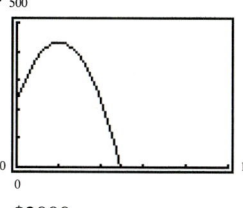

$2000

101. $y = \frac{1}{2500}x^2$

15. $y = (x - 3)^2 - 1$ **16.** $y = -\frac{1}{4}(x - 5)^2 + 4$

17.

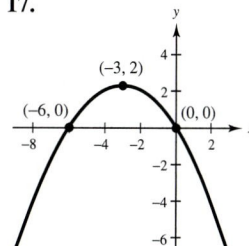

18.

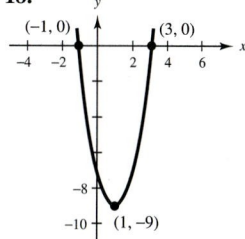

19. $V = -1850t + 12{,}400, 0 \le t \le 4$ **20.** 55 feet

Section 11.4 *(page 667)*

7. c   **9.** e   **11.** a

13. $x^2 + y^2 = 25$ **15.** $x^2 + y^2 = 29$

17. Center: $(0, 0)$ **19.** Center: $(0, 0)$
    Radius: $r = 4$ Radius: $r = \frac{12}{5}$

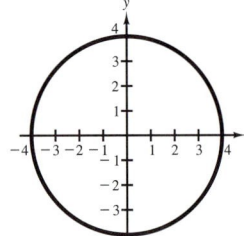

21. $\frac{x^2}{16} + \frac{y^2}{9} = 1$ **23.** $\frac{x^2}{9} + \frac{y^2}{16} = 1$ **25.** $\frac{x^2}{100} + \frac{y^2}{36} = 1$

27.
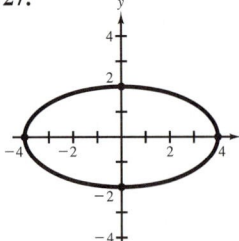
Vertices: $(\pm 4, 0)$
Co-vertices: $(0, \pm 2)$

29.
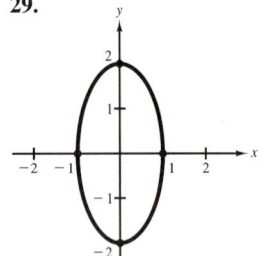
Vertices: $(0, \pm 2)$
Co-vertices: $(\pm 1, 0)$

31.
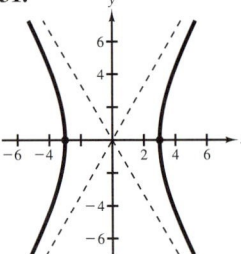
Vertices: $(\pm 3, 0)$
Asymptotes: $y = \pm \frac{5}{3}x$

33.
Vertices: $(0, \pm 3)$
Asymptotes: $y = \pm \frac{3}{5}x$

35. $\frac{x^2}{16} - \frac{y^2}{64} = 1$ **37.** $\frac{y^2}{16} - \frac{x^2}{64} = 1$

39. Center: $(2, 3)$
    radius: $r = 2$

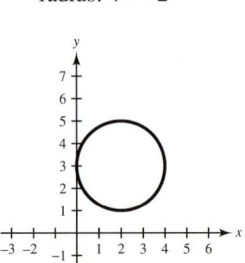

41. $(x - 2)^2 + (y - 1)^2 = 4$

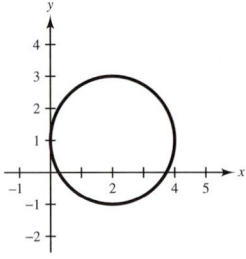

43.
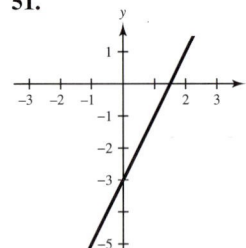

45.
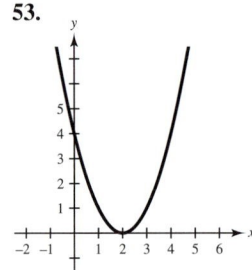

47. $x^2 + y^2 = 4500^2$ **49.** $\sqrt{304} \approx 17.4$ feet

51.

53.

55. $\sqrt{\dfrac{3}{\pi}} \approx 0.9722$ **57.** $x^2 + y^2 = \frac{4}{9}$ **59.** $x^2 + y^2 = 64$

61. Center: $(0, 0)$ **63.** Center: $(0, 0)$
    radius: $r = 6$ radius: $r = \frac{1}{2}$

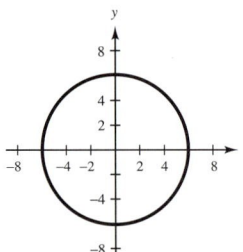
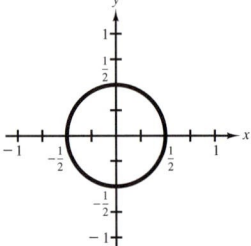

65. $\frac{x^2}{4} + \frac{y^2}{1} = 1$ **67.** $\frac{x^2}{1} + \frac{y^2}{4} = 1$ **69.** $\frac{x^2}{9} + \frac{y^2}{25} = 1$

71.

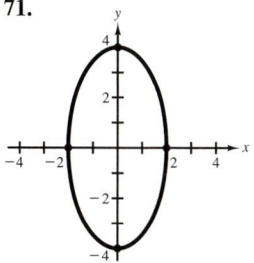

Vertices: $(0, +4)$
Co-vertices: $(\pm 2, 0)$

73.

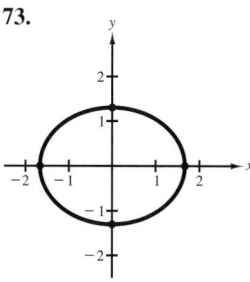

Vertices: $\left(\pm \frac{5}{3}, 0\right)$
Co-vertices: $\left(0, \pm \frac{4}{3}\right)$

75.

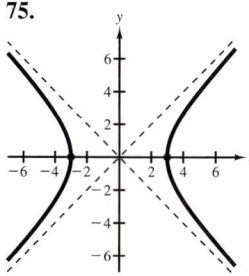

Vertices: $(\pm 3, 0)$
Asymptotes: $y = \pm x$

77.

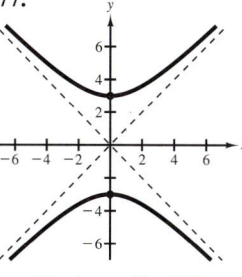

Vertices: $(0, \pm 3)$
Asymptotes: $y = \pm x$

79.

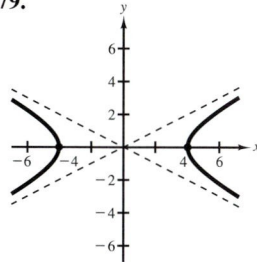

Vertices: $(\pm 4, 0)$
Asymptotes: $y = \pm \frac{1}{2}x$

81. $\dfrac{x^2}{81} - \dfrac{y^2}{36} = 1$

83. $\dfrac{y^2}{1} - \dfrac{x^2}{\frac{1}{4}} = 1$ **85.** Parabola **87.** Ellipse

89. Hyperbola **91.** Circle **93.** Line

95.

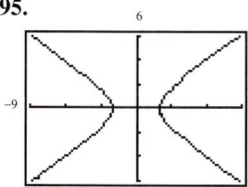

97.

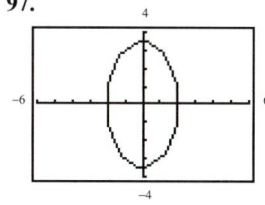

99. (a)
$$x^2 + y^2 = 25^2$$
$$y^2 = 625 - x^2$$
$$y = \sqrt{625 - x^2}$$
$$\text{width} = 2y = 2\sqrt{625 - x^2}$$
$$\text{Area} = (2x)(2y)$$
$$= 4x\sqrt{625 - x^2}$$

(b)

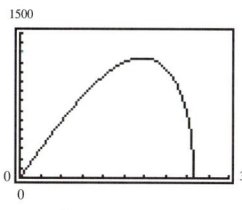

$x \approx 17.68$

101. $\dfrac{x^2}{144} + \dfrac{y^2}{64} = 1$

Section 11.5 *(page 677)*

5. $I = kV$ **7.** $p = \dfrac{k}{d}$ **9.** $P = \dfrac{k}{V}$

11. The area of a triangle is proportional to the product of the base and height.

13. The volume of a right circular cylinder is proportional to the product of the square of the radius and the height.

15. $s = 5t$ **17.** $F = \dfrac{5}{16}x^2$ **19.** $n = \dfrac{48}{m}$

21. $d = \dfrac{120x^2}{r}$ **23.** 18 pounds **25.** $208\frac{1}{3}$ feet

27. 3072 watts **29.** 667 units **31.** $\frac{1}{4}$

33. 3125 pounds **35.** $y = \frac{1}{4}(5 - 3x)$

37. $y = \frac{2}{3}(x^2 - 1)$ **39.** 15, 17 **41.** $V = kt$

43. $u = kv^2$ **45.** $P = \dfrac{k}{\sqrt{1 + r}}$ **47.** $A = klw$

49. The volume of a sphere varies directly as the cube of the radius.

51. The area of an ellipse varies jointly as the semimajor axis and the semiminor axis.

53. $H = \frac{5}{2}u$ **55.** $g = \dfrac{4}{\sqrt{z}}$ **57.** $F = \dfrac{25}{6}xy$

59. (a) $4921.25 **61.** 32 feet per second squared
(b) Price per unit

63. 2.44 hours, distance   **65.** 324 pounds

67. $p = \dfrac{114}{t}$, 17.5%

69.

x	2	4	6	8	10
$y = kx^2$	4	16	36	64	100

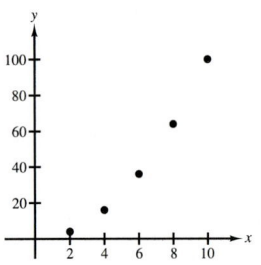

71.

x	2	4	6	8	10
$y = kx^2$	8	32	72	128	200

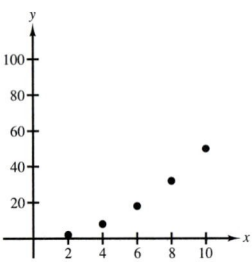

73.

x	2	4	6	8	10
$y = \dfrac{k}{x^2}$	$\dfrac{1}{2}$	$\dfrac{1}{8}$	$\dfrac{1}{18}$	$\dfrac{1}{32}$	$\dfrac{1}{50}$

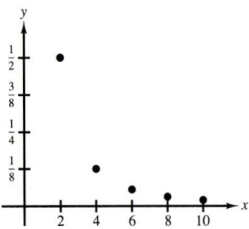

75.

x	2	4	6	8	10
$y = \dfrac{k}{x^2}$	$\dfrac{5}{2}$	$\dfrac{5}{8}$	$\dfrac{5}{18}$	$\dfrac{5}{32}$	$\dfrac{1}{10}$

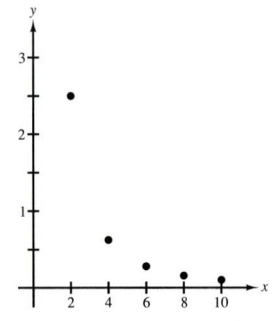

77. $y = \dfrac{4}{x}$ **79.** $y = -\dfrac{3}{10}x$

Review Exercises *(page 683)*

1. $2x - y - 6 = 0$ **3.** $4x + y = 0$

5. $2x + 3y - 17 = 0$ **7.** $x - 7 = 0$

9. $x + 2y + 6 = 0$ **11.** $y - 10 = 0$

13. $9x - 24y - 8 = 0$

15. (a) $3x + y - 1 = 0$ **17.** (a) $x - 12 = 0$

(b) $x - 3y - 3 = 0$ (b) $y - 1 = 0$

19. 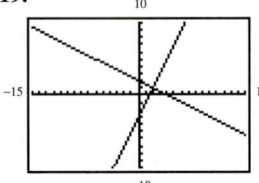   **21.**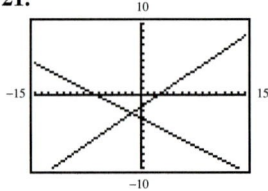

Perpendicular   Neither

23. $C = 8.55x + 25{,}000$ **25.** $y = -x + 3.87$

$P = 4.05x - 25{,}000$

27. 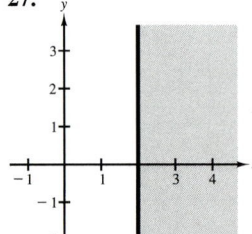   **29.**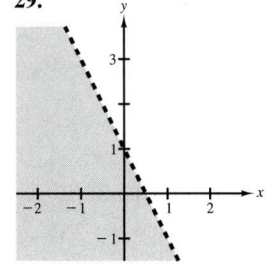

31.
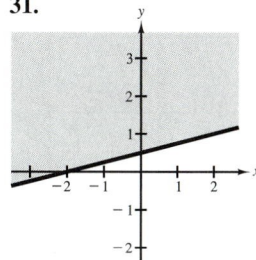

33.
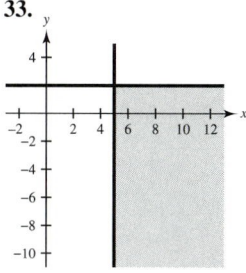

35.
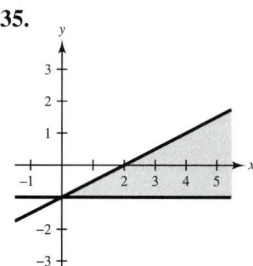

37.
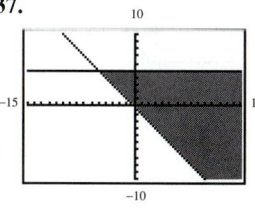

39. x hours at the grocery store
y hours mowing the lawn
$8x + 10y \geq 200$

41. Parabola

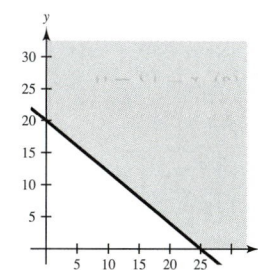

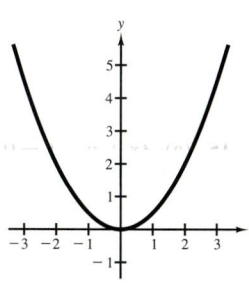

43. Hyperbola

45. Parabola

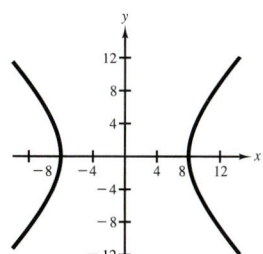

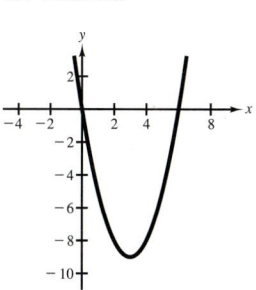

47. Ellipse

49. Circle

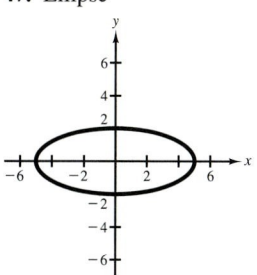

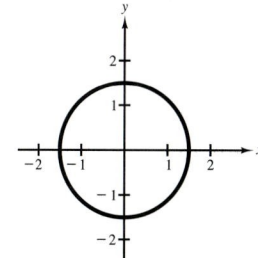

51. $y = \frac{1}{16}x^2 - \frac{5}{8}x + \frac{25}{16}$ **53.** $\frac{x^2}{4} + \frac{y^2}{25} = 1$

55. $x^2 + y^2 = 400$ **57.**
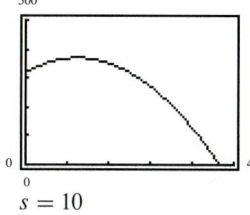
$s = 10$

59. (a)
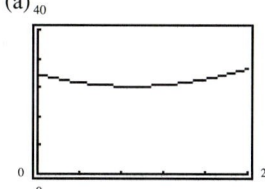
(b) 1981

(c) 32.5 million

61. c   **63.** a   **65.** b   **67.** $y = 4\sqrt[3]{3}x$

69. $T = \frac{1}{18}rs^2$ **71.** 150 pounds   **73.** 945 units

Chapter Test *(page 686)*

1. $2x + y = 0$ **2.** $x - 2y - 55 = 0$ **3.** $y + 1 = 0$

4. $x + 2 = 0$ **5.** $\frac{3}{5}$ **6.** $V = -4000t + 26{,}000$

$t = 2.5$

7.

8.

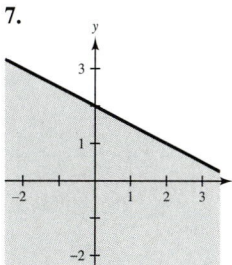

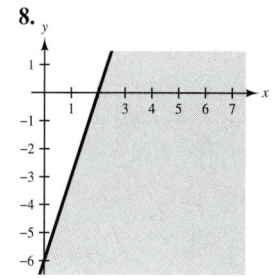

9.

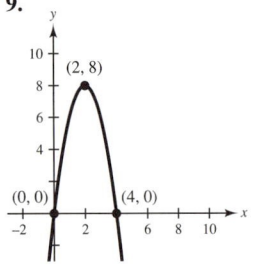

10. $y = \frac{2}{3}(x - 3) - 2$

11. 120   **12.** $x^2 + y^2 = 25$ **13.** $\dfrac{x^2}{9} + \dfrac{y^2}{100} = 1$

14. $\dfrac{x^2}{9} - \dfrac{y^2}{\frac{9}{4}} = 1$

15.

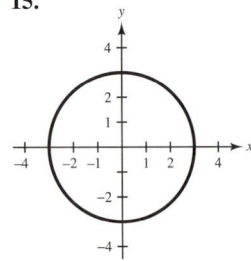

16.

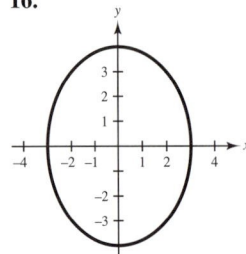

17.

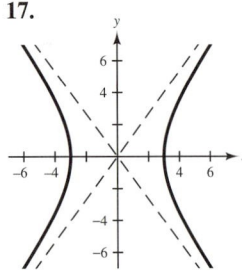

18.

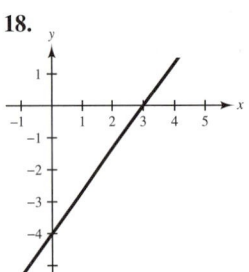

19. $S = \dfrac{kx^2}{y}$ **20.** $v = \dfrac{1}{4}\sqrt{u}$

CHAPTER 12

Section 12.1 *(page 695)*

7. (a) Solution   **9.** (a) Not a solution

  (b) Not a solution   (b) Solution

11. $(2, 1)$ **13.** $(-2, 4), (1, 1)$ **15.** $(1, 2)$

17. $\left(\frac{3}{2}, \frac{3}{2}\right)$ **19.** $(0, 5), (-4, -3)$ **21.** No solution

23. $\left(1, \frac{1}{3}\right)$ **25.** No solution

27.

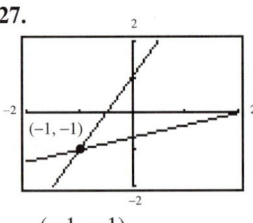

$(-1, -1)$

29.

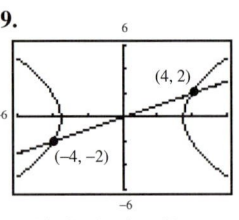

$(4, 2), (-4, -2)$

31.

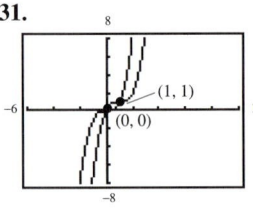

$(0, 0), (1, 1)$

33.

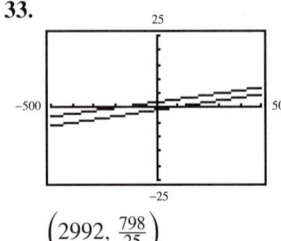

$\left(2992, \frac{798}{25}\right)$

35. 10,000 units   **37.** $\left(\dfrac{-90 + 96\sqrt{2}}{7}, \dfrac{160 - 96\sqrt{2}}{7}\right)$

39. $2x - y = 0$ **41.** $22x + 16y - 161 = 0$

43. 2.4 hours   **45.** $(4, 3)$ **47.** $\left(4, -\frac{1}{2}\right)$

49. $(-1, 3), (4, 2)$ **51.** $(4, -2)$ **53.** $(7, 2)$

55. $\left(\frac{1}{3}, \frac{17}{3}\right)$ **57.** $(2, 8), (-3, 18)$ **59.** $(2, 5), (-3, 0)$

61. $(10, 30), (20, 15)$ **63.** No solution   **65.** $(0, 2), (3, 1)$

67.

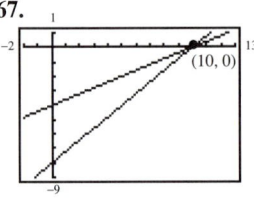

$(10, 0)$

69.

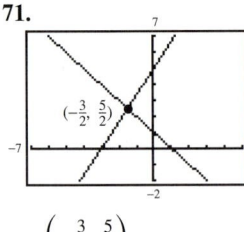

$(9, 12)$

71.

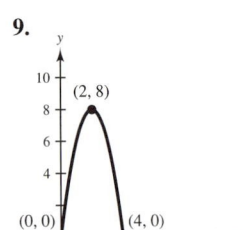

$\left(-\frac{3}{2}, \frac{5}{2}\right)$

73.

$(0, 0), (2, 4)$

75.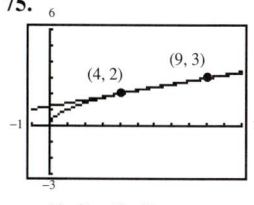

(4, 2), (9, 3)

77.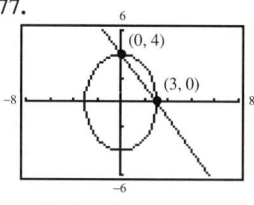

(0, 4), (3, 0)

79. 31, 49 **81.** 20, 32 **83.** 10 feet × 15 feet

85. 14 yards × 20 yards **87.** $15,000 at 8%

$5000 at 9.5%

89. Between the points **91.** 1987

$\left(-\frac{3}{5}, -\frac{4}{5}\right)$ and $\left(\frac{4}{5}, -\frac{3}{5}\right)$

Math Matters *(page 699)*

1. 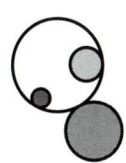 **2.**

3.  **4.**

Section 12.2 *(page 706)*

7.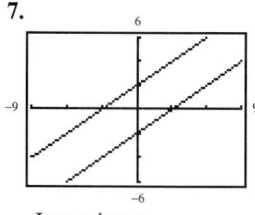

Inconsistent

9.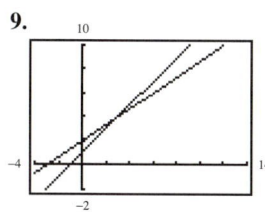

One solution

11. Inconsistent **13.** (5, 3)

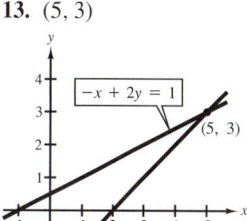

15. (2, −2) **17.** Inconsistent

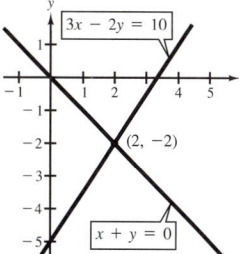

19. Infinitely many solutions **21.** $\left(\frac{1}{2}, \frac{3}{2}\right)$

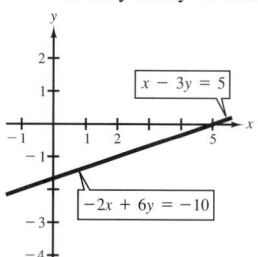

23. (3, 2) **25.** Inconsistent **27.** (3000, −2000)

29. (4, 3) **31.** (2, 7) **33.** $x + 2y = 0$
$4x + 2y = 9$

35. 121 weeks **37.** 6.4 hours **39.** 32 inches, 128 inches

41. Depth: 10 feet; Length of sections: 122 feet, 125 feet

43. $2x + 7y - 45 = 0$ **45.** $y - 3 = 0$

47. 4500 units **49.** (5, −1)

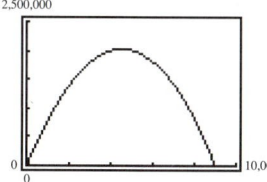

51. $(-2, 5)$ **53.** $(-1, -1)$ **55.** $(7, -2)$ **57.** $\left(\frac{3}{2}, 1\right)$

59. $(-2, -1)$ **61.** Infinitely many solutions

63. Inconsistent   **65.** $\left(\frac{25}{2}, \frac{1237}{250}\right)$ **67.** $(15, 10)$

69. $(6, 11)$ **71.** $k = 4$ **73.** Consistent

75. Consistent   **77.** Inconsistent

79.

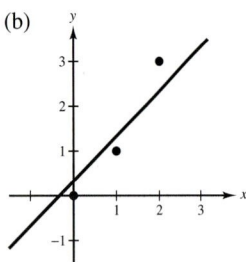

$(3, 5)$

81.

$(1/2, -1)$

83. 50 meters × 60 meters   **85.** 5 dimes, 10 quarters

87. 11 nickels, 14 dimes   **89.** Regular unleaded: \$1.11
Premium unleaded: \$1.22

91. \$5.65 variety: 6.1 pounds   **93.** 40% solution: 12 liters
\$8.95 variety: 3.9 pounds   65% solution: 8 liters

95. (a) $y = x + \frac{1}{3}$ (b)

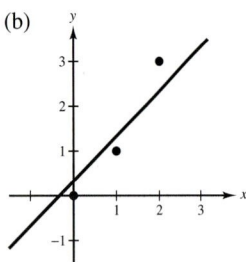

Section 12.3 *(page 718)*

5. $(22, -1, -5)$ **7.** $(14, 3, -1)$ **9.** $(1, 2, 3)$

11. $(3, 2, 1)$ **13.** Inconsistent   **15.** $(1, -1, 2)$

17. $\left(\frac{6}{13}a + \frac{10}{13}, \frac{5}{13}a + \frac{4}{13}, a\right)$ **19.** $y = -x^2 + 2x + 5$

21. $x^2 + y^2 - 4x = 0$ **23.** $s = -16t^2 + 144$

25. $s = -16t^2 + 48t$ **27.** Spray X: 20 gallons
Spray Y: 18 gallons
Spray Z: 16 gallons

29. String: 50   **31.** $(-\infty, \infty)$ **33.** $[-4, 4]$
Wind: 20
Percussion: 8

35. 50   **37.** $x - 2y + 3z = 5$ **39.** (a) Not a solution
$\quad\quad\quad\; - y + 8z = 9$ (b) Solution
$\quad\; 2x \quad\quad - 3z = 0$ (c) Solution
(d) Not a solution

41. $(1, 2, 3)$ **43.** Inconsistent   **45.** $(-4, 8, 5)$

47. $(2, -1, 1)$ **49.** $(0, -4, 5)$ **51.** Inconsistent

53. Inconsistent   **55.** $\left(\frac{1}{4}, \frac{5}{4}, 0\right)$

57. $y = x^2 - 4x + 3$ **59.** $y = 2x - x^2$

61. $x^2 + y^2 - 3x - 2y = 0$

63. $x^2 + y^2 - 2x - 4y - 20 = 0$

65. (a) $y = 0.05t^2 + 0.15t + 6.6$

(b)

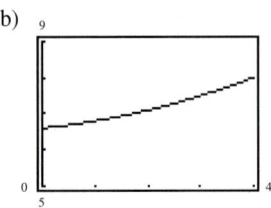

(c) 12 million short tons

67. $x + 2y - z = -4$
$\quad\quad\; y + 2z = 1$
$3x + y + 3z = 15$

Mid-Chapter Quiz *(page 722)*

1. $(10, 4)$ **2.**

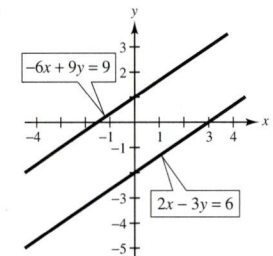

No solution

3.

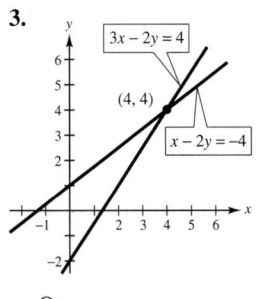

One

4.

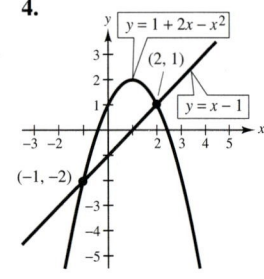

Two solutions

75.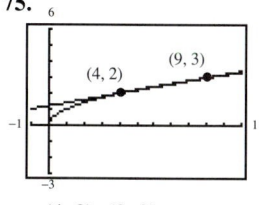

(4, 2), (9, 3)

77.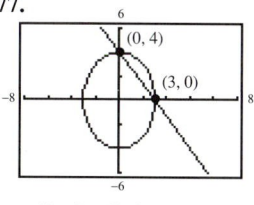

(0, 4), (3, 0)

79. 31, 49   **81.** 20, 32   **83.** 10 feet × 15 feet

85. 14 yards × 20 yards   **87.** $15,000 at 8%

$5000 at 9.5%

89. Between the points   **91.** 1987

$\left(-\frac{3}{5}, -\frac{4}{5}\right)$ and $\left(\frac{4}{5}, -\frac{3}{5}\right)$

Math Matters *(page 699)*

1.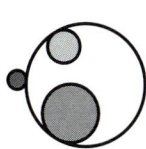

2.

3.

4.

Section 12.2 *(page 706)*

7.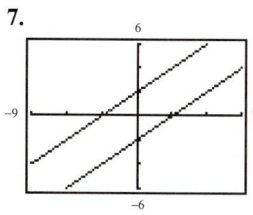

Inconsistent

9.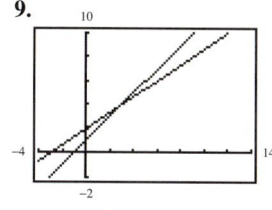

One solution

11. Inconsistent   **13.** (5, 3)

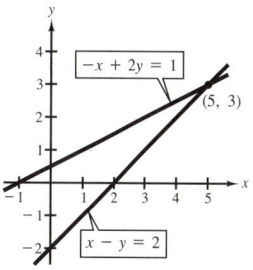

15. (2, −2)   **17.** Inconsistent

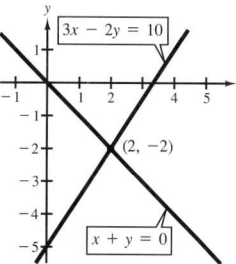

19. Infinitely many solutions   **21.** $\left(\frac{1}{2}, \frac{3}{2}\right)$

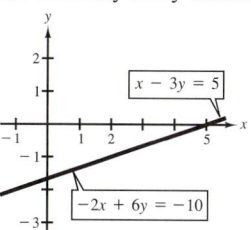

23. (3, 2)   **25.** Inconsistent   **27.** (3000, −2000)

29. (4, 3)   **31.** (2, 7)   **33.** $x + 2y = 0$

$4x + 2y = 9$

35. 121 weeks   **37.** 6.4 hours   **39.** 32 inches, 128 inches

41. Depth: 10 feet; Length of sections: 122 feet, 125 feet

43. $2x + 7y − 45 = 0$ **45.** $y − 3 = 0$

47. 4500 units   **49.** (5, −1)

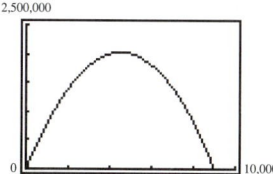

51. $(-2, 5)$ **53.** $(-1, -1)$ **55.** $(7, -2)$ **57.** $\left(\frac{3}{2}, 1\right)$

59. $(-2, -1)$ **61.** Infinitely many solutions

63. Inconsistent **65.** $\left(\frac{25}{2}, \frac{1237}{250}\right)$ **67.** $(15, 10)$

69. $(6, 11)$ **71.** $k = 4$ **73.** Consistent

75. Consistent **77.** Inconsistent

79.

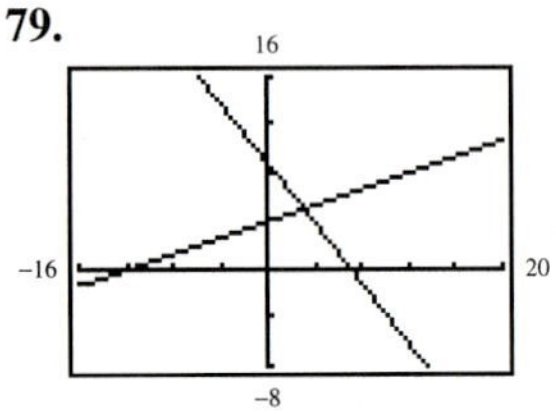

$(3, 5)$

81.

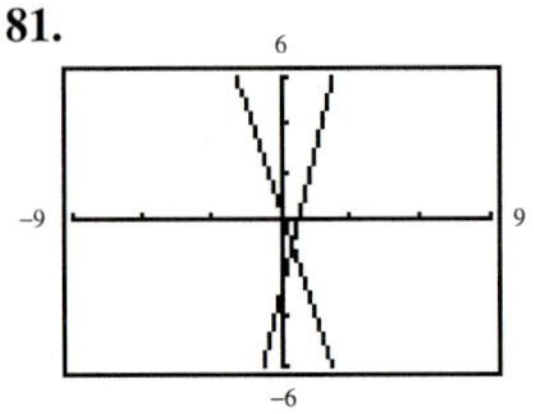

$(1/2, -1)$

83. 50 meters × 60 meters **85.** 5 dimes, 10 quarters

87. 11 nickels, 14 dimes **89.** Regular unleaded: $1.11
Premium unleaded: $1.22

91. $5.65 variety: 6.1 pounds
$8.95 variety: 3.9 pounds

93. 40% solution: 12 liters
65% solution: 8 liters

95. (a) $y = x + \frac{1}{3}$ (b)

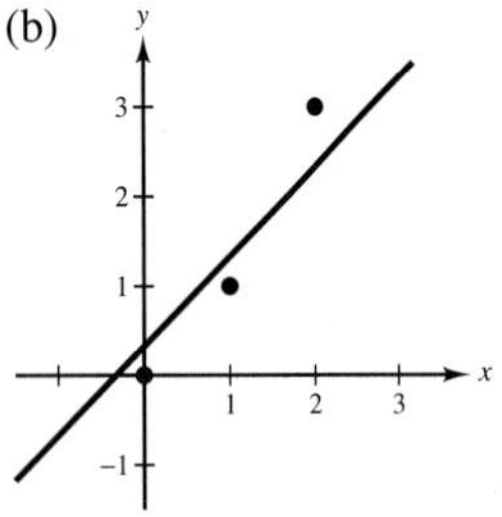

Section 12.3 *(page 718)*

5. $(22, -1, -5)$ **7.** $(14, 3, -1)$ **9.** $(1, 2, 3)$

11. $(3, 2, 1)$ **13.** Inconsistent **15.** $(1, -1, 2)$

17. $\left(\frac{6}{13}a + \frac{10}{13}, \frac{5}{13}a + \frac{4}{13}, a\right)$ **19.** $y = -x^2 + 2x + 5$

21. $x^2 + y^2 - 4x = 0$ **23.** $s = -16t^2 + 144$

25. $s = -16t^2 + 48t$ **27.** Spray X: 20 gallons
Spray Y: 18 gallons
Spray Z: 16 gallons

29. String: 50
Wind: 20
Percussion: 8

31. $(-\infty, \infty)$ **33.** $[-4, 4]$

35. 50 **37.**
$$\begin{aligned} x - 2y + 3z &= 5 \\ -y + 8z &= 9 \\ 2x \qquad - 3z &= 0 \end{aligned}$$

39. (a) Not a solution
(b) Solution
(c) Solution
(d) Not a solution

41. $(1, 2, 3)$ **43.** Inconsistent **45.** $(-4, 8, 5)$

47. $(2, -1, 1)$ **49.** $(0, -4, 5)$ **51.** Inconsistent

53. Inconsistent **55.** $\left(\frac{1}{4}, \frac{5}{4}, 0\right)$

57. $y = x^2 - 4x + 3$ **59.** $y = 2x - x^2$

61. $x^2 + y^2 - 3x - 2y = 0$

63. $x^2 + y^2 - 2x - 4y - 20 = 0$

65. (a) $y = 0.05t^2 + 0.15t + 6.6$

(b)

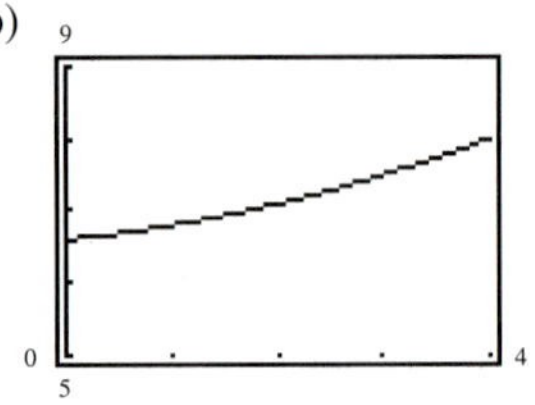

(c) 12 million short tons

67.
$$\begin{aligned} x + 2y - z &= -4 \\ y + 2z &= 1 \\ 3x + y + 3z &= 15 \end{aligned}$$

Mid-Chapter Quiz *(page 722)*

1. $(10, 4)$ **2.**

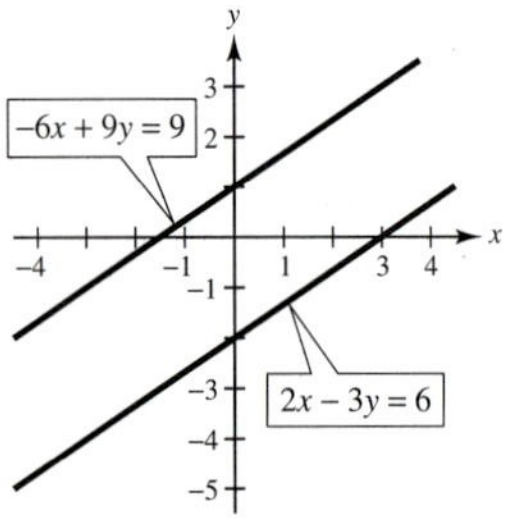

No solution

3.

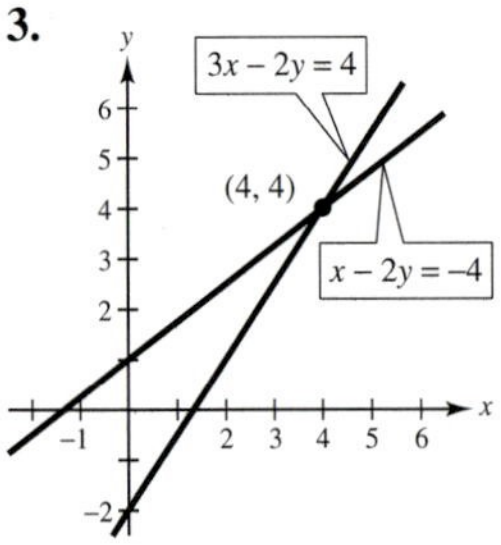

One

4.

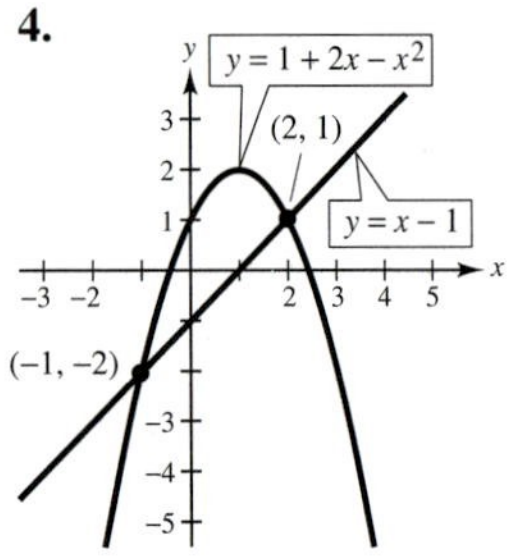

Two solutions

5.

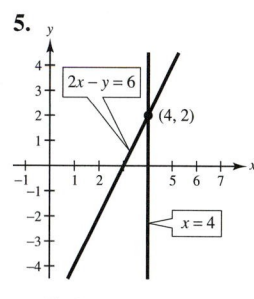

$(4, 2)$

6.

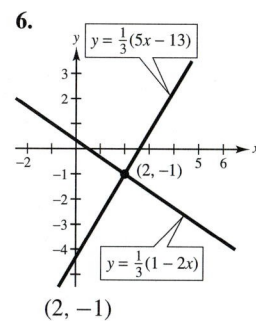

$(2, -1)$

7.

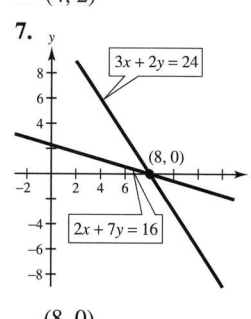

$(8, 0)$

8.

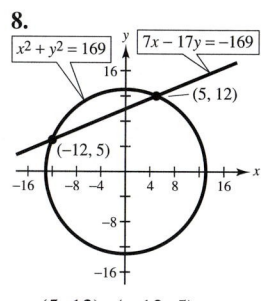

$(5, 12), (-12, 5)$

9. $(5, 2)$ **10.** $(1, 4), (-3, -4)$ **11.** $\left(\frac{90}{13}, \frac{34}{13}\right)$

12. $(5, 10)$ **13.** $(8, 1)$ **14.** $(-2, 4)$

15. $\left(\frac{1}{2}, -\frac{1}{2}, 1\right)$ **16.** $(5, -1, 3)$

17. $\quad x + y = -2$
$\quad 2x - y = 32$

18. $\quad x + y - z = 11$
$\quad x + 2y - z = 14$
$\quad -2x + y + z = -6$

19. $\quad x + \quad y = 20$
$\quad 0.2x + 0.5y = 6$

20% solution: $13\frac{1}{3}$ gallons
50% solution: $6\frac{2}{3}$ gallons

20. $y = x^2 + 3x - 2$

Section 12.4 *(page 731)*

7. 3×2 **9.** 2×2 **11.** $\begin{bmatrix} 4 & -5 & -2 \\ -1 & 8 & 10 \end{bmatrix}$

13. $\begin{bmatrix} 1 & 10 & -3 & 2 \\ 5 & -3 & 4 & 0 \\ 2 & 4 & 0 & 6 \end{bmatrix}$ **15.** $4x + 3y = 8$
$\quad x - 2y = 3$

17. $\quad x \quad + 2z = -10$
$\quad 3y - z = 5$
$\quad 4x + 2y \quad = 3$

19. $\begin{bmatrix} 1 & 4 & 3 \\ 0 & 2 & -1 \end{bmatrix}$

21. $\begin{bmatrix} 1 & 2 & 3 \\ 0 & 1 & 2 \end{bmatrix}$ **23.** $\begin{bmatrix} 1 & 1 & -1 & 3 \\ 0 & 1 & -4 & 1 \\ 0 & 0 & 0 & 0 \end{bmatrix}$

25. $x + 5y = 3$
$\quad y = -2$
$(13, -2)$

27. $(1, 1)$ **29.** $(2, -3, 2)$

31. $(2a + 1, 3a + 2, a)$ **33.** Inconsistent

35. 8%: \$800,000
9%: \$500,000
12%: \$200,000

37. $y = x^2 + 2x + 4$

39. (a) $y = -\frac{1}{250}x^2 + \frac{3}{5}x + 6$

(b)

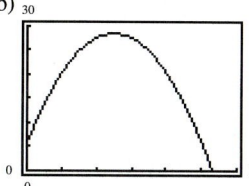

(c) Maximum height: 28.5 feet
Range: 159.4 feet

41. (a) 12
(b) $\frac{3}{16}$

43. (a) $\frac{1}{3}$
(b) $\dfrac{c - 6}{c + 4}$

45. $C = 5.75x + 12,000$

47. $\begin{bmatrix} 1 & -2 & \frac{2}{3} \\ 2 & 8 & 15 \end{bmatrix}$ **49.** $\begin{bmatrix} 1 & 1 & 4 & -1 \\ 0 & 5 & -2 & 6 \\ 0 & 3 & 20 & 4 \end{bmatrix}$

$\begin{bmatrix} 1 & 1 & 4 & -1 \\ 0 & 1 & -\frac{2}{5} & \frac{6}{5} \\ 0 & 3 & 20 & 4 \end{bmatrix}$

51. $\begin{bmatrix} 1 & \frac{3}{2} & \frac{1}{4} \\ 0 & 1 & \frac{11}{10} \end{bmatrix}$ **53.** $\begin{bmatrix} 1 & 1 & 0 & 5 \\ 0 & 1 & 2 & 0 \\ 0 & 0 & 1 & -1 \end{bmatrix}$

55. $\begin{bmatrix} 1 & -1 & -1 & 1 \\ 0 & 1 & 6 & 3 \\ 0 & 0 & 1 & \frac{4}{5} \end{bmatrix}$ **57.** $x - 2y = 4$
$\quad y = -3$
$(-2, -3)$

59. $x - y + 2z = 4$
$\quad y - z = 2$
$\quad z = -2$
$(8, 0, -2)$

61. $\left(\frac{9}{5}, \frac{13}{5}\right)$

63. Inconsistent   **65.** $(1, -1, 2)$ **67.** $(8, 10, 6)$

69. $(-12a - 1, 4a + 1, a)$ **71.** $(-1, -2, 3)$

73. $\left(1, \frac{2}{3}, -1\right)$ **75.** Certificates of deposit: $250{,}000 - \frac{1}{2}s$
Municipal bonds: $125{,}000 + \frac{1}{2}s$
Blue-chip stocks: $125{,}000 - s$
Growth stocks: s

77. $y = \frac{1}{2}x^2 - 3x + \frac{7}{2}$ **79.** 5, 8, 20

81. $\dfrac{2x^2 - 9x}{(x-2)^3} = \dfrac{2}{x-2} - \dfrac{1}{(x-2)^2} - \dfrac{10}{(x-2)^3}$

Section 12.5 *(page 744)*

5. 5 **7.** −24 **9.** −24 **11.** −30 **13.** 0
15. 248 **17.** 5.4 **19.** (2, −2)
21. Not possible; $D = 0$ **23.** (−1, 3, 2) **25.** $\left(1, \frac{1}{2}, \frac{3}{2}\right)$
27. $\left(\frac{22}{27}, \frac{22}{9}\right)$ **29.** $\left(\frac{1}{3}, 1, -\frac{2}{3}\right)$ **31.** 16
33. 250 square miles **35.** Collinear
37. $9x + 10y + 3 = 0$ **39.** $y = \frac{1}{2}x^2 - 2x$
41. (a) $y_1 = -0.8t^2 + 28.9t + 393.6$
(b) $y_2 = 26.8t^2 - 35t + 495.3$
(c)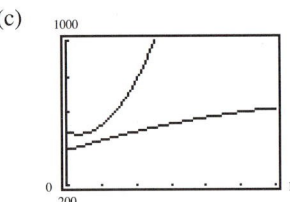
(d) $y = -27.6t^2 + 63.9t - 101.7$

43. 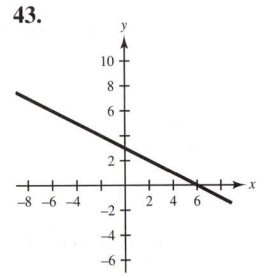 **45.**

47. 267 **49.** 27 **51.** 0 **53.** 6 **55.** −0.16
57. 102 **59.** −2 **61.** 3 **63.** 0 **65.** −75
67. −58 **69.** 0 **71.** $x - 5y + 2$ **73.** 19,185
75. 0 **77.** (1, 2) **79.** $\left(\frac{3}{4}, -\frac{1}{2}\right)$ **81.** $\left(\frac{2}{3}, \frac{1}{2}\right)$
83. (1, −2, 1) **85.** $\left(\frac{13}{16}, \frac{11}{16}, 0\right)$ **87.** $\left(\frac{1}{2}, 4\right)$

89. $\left(2, \frac{1}{2}, -\frac{1}{2}\right)$ **91.** 7 **93.** $\frac{31}{2}$ **95.** 31 square units
97. Collinear **99.** Not collinear **101.** $3x - 5y = 0$
103. $7x - 6y - 28 = 0$ **105.** $y = 2x^2 - 6x + 1$
107. $y = -3x^2 + 2x$ **109.** 1, 6

Review Exercises *(page 750)*

1. (2, −1) **3.** Inconsistent **5.** (−10, −5)
7. (−1, 5), (−2, 20) **9.** (0, −1), (−1, 0)

11.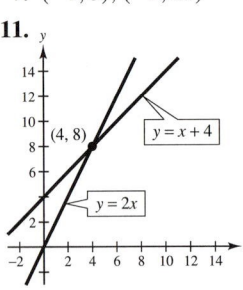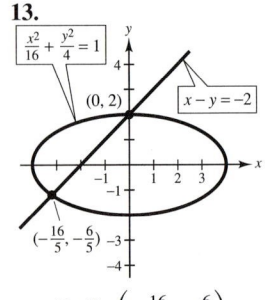
(4, 8)

13.
$(0, 2), \left(-\frac{16}{5}, -\frac{6}{5}\right)$

15.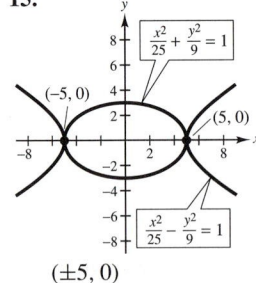
(±5, 0)

17. (3, 4)

19. (±3, ±4) **21.** (0, 0) **23.** (2, −3, 3)
25. (10, −12) **27.** $\left(\frac{24}{5}, \frac{22}{5}, -\frac{8}{5}\right)$ **29.** 5
31. −51 **33.** 1 **35.** (−3, 7) **37.** Inconsistent
39. (2, −3, 3) **41.** $x - 2y + 4 = 0$
43. $2x + 6y - 13 = 0$ **45.** 16 **47.** 7
49. $3x + y = -2$ **51.** 16,667 units
$\quad\;\; 6x + y = 0$
53. 75% solution: 40 gallons
50% solution: 60 gallons
55. 193.75 miles per hour, 218.75 miles per hour
57. $y = 2x^2 + x - 6$
59. $x^2 + y^2 - 4x + 2y - 4 = 0$

61. (a) $y = -\frac{1}{45}x^2 + \frac{2}{3}x + 11$

(b)

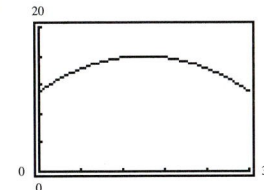

(c) $7\frac{1}{4}$ feet

Chapter Test *(page 752)*

1. $\left(1, \frac{1}{2}\right)$ **2.** $(2, 4)$ **3.** $(2, 6), (6, 2)$ **4.** $(3, 2)$

5. $(-2, 2)$ **6.** $\left(\frac{1}{4}, \frac{1}{3}\right)$ **7.** $(2, -1, 0)$ **8.** $(-1, 3, 3)$

9. $(2, 1, -2)$ **10.** $\left(4, \frac{1}{7}\right)$ **11.** $(5, 4)$

12. $(5, 1, -1)$ **13.** $\left(-\frac{11}{5}, \frac{56}{25}, \frac{32}{25}\right)$

14. Inconsistent   **15.** -62 **16.** $-\frac{24}{5}$
One solution
Infinitely many solutions

17. $x + 2y = -1$ **18.** $x + y = 200$
$\quad\ x +\ y =\ \ 2$ $\qquad 4x - y =\ \ 0$
$\qquad\qquad\qquad\qquad$ 40 miles, 60 miles

19. $y = 2x^2 - 3x + 4$ **20.** 12

CHAPTER 13

Section 13.1 *(page 761)*

7. 2^{2x-1} **9.** $8e^{3x}$ **11.** (a) $\frac{1}{9}$ **13.** (a) 73.89
$\qquad\qquad\qquad\qquad\qquad$ (b) 1 $\qquad\quad$ (b) 1.35
$\qquad\qquad\qquad\qquad\qquad$ (c) 3 $\qquad\quad$ (c) 0.18

15. b   **17.** e   **19.** f

21.

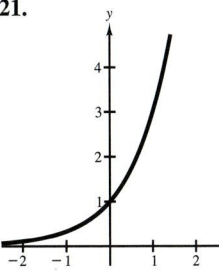

23.

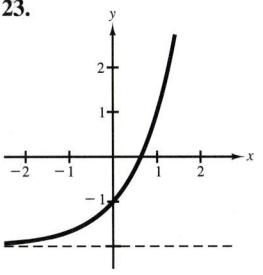

25.

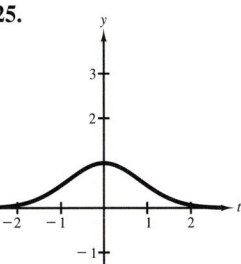

27.

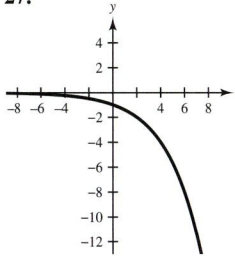

29.

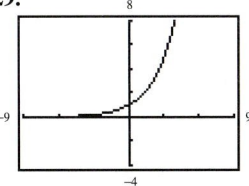

31.

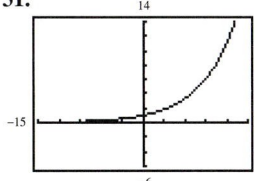

33.

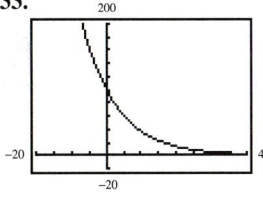

35.

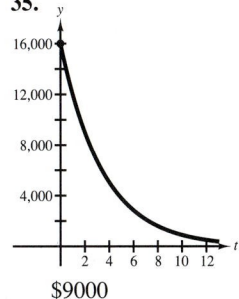

$9000

37.

n	1	4	12	365	Continuous
A	$466.10	$487.54	$492.68	$495.23	$495.30

39.

n	1	4	12	365	Continuous
A	$2541.75	$2498.00	$2487.98	$2483.09	$2482.93

41. (a)

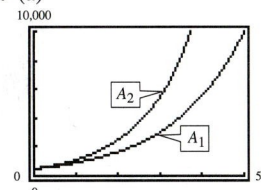

(b)

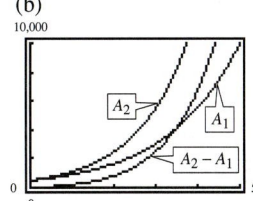

(c) The difference between the functions increases at an increasing rate.

43. $-11x$ **45.** $6x^2 + 9$ **47.** (a) $\dfrac{1}{2\pi}$ **49.** 11.036

(b) $\dfrac{1}{4\pi}$

51. 1.396   **53.** (a) $\frac{1}{5}$ **55.** (a) 500

(b) 5   (b) 250

(c) 125   (c) 56.657

57. (a) 1000   **59.** (a) 0.368   **61.** d   **63.** a   **65.** f

(b) 1628.895   (b) 1

(c) 2653.298   (c) 1.649

67.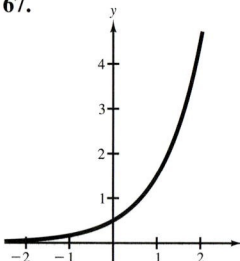

69.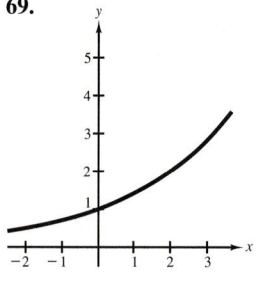

71.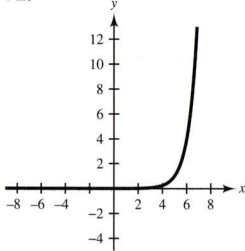

73.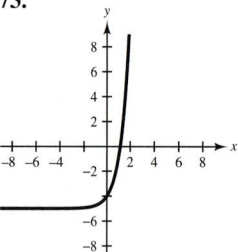

75.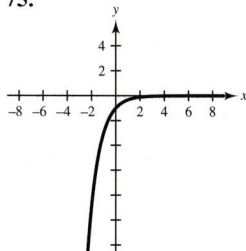

77.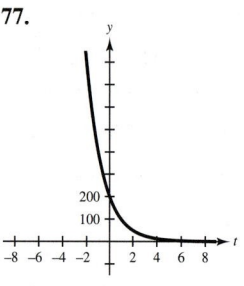

79.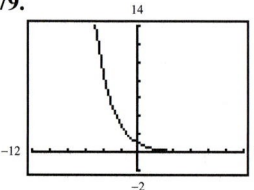

81.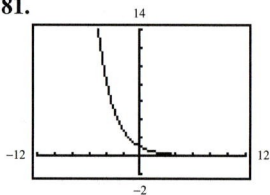

83.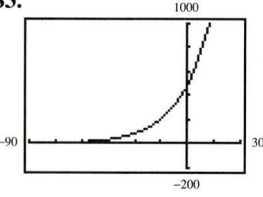

85.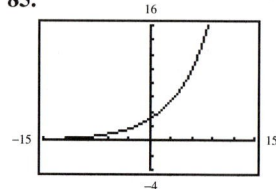

87.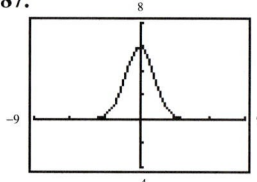

89. (a) 272.184 million

(b) 288.627 million

91. \$32.50

93.

n	1	4	12
A	\$4734.73	\$4870.38	\$4902.71

n	365	Continuous
A	\$4918.66	\$4919.21

95.

n	1	4	12
A	\$226,296.28	\$259,889.34	\$268,503.32

n	365	Continuous
A	\$272,841.23	\$272,990.75

97.

n	1	4	12
A	\$18,429.30	\$15,830.43	\$15,272.04

n	365	Continuous
A	\$15,004.64	\$14,995.58

99.

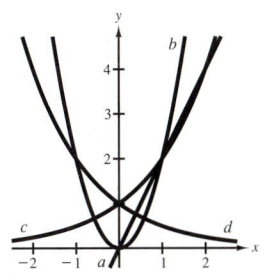

$f(x) = 2^x$ and $f(x) = 2^{-x}$ are exponential.

101. (a)

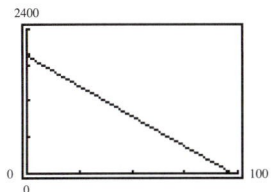

(b)

t	0	25	50	75
h	2000 ft	1450 ft	950 ft	450 ft

Ground level: 97.5 seconds

103. (a)

Year	1987	1988	1989	1990
Price	$99,744	$121,074	$130,035	$131,368

Year	1991	1992
Price	$132,715	$142,537

(b)

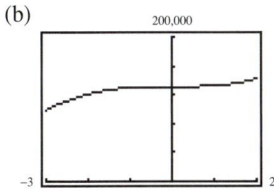

(c) Increasing at a higher rate. Home prices probably will not increase at a higher rate indefinitely.

Section 13.2 *(page 773)*

7. (a) 13
   (b) 16
   (c) $x^2 - 3$
   (d) $(x - 3)^2$

9. (a) 3
   (b) 8
   (c) $\sqrt{x + 5}$
   (d) $\sqrt{x + 5}$

11. (a) -1
   (b) -2
   (c) -2

13. (a) -1
   (b) 1

15. (a) $[2, \infty)$
   (b) $[0, \infty)$

17. $g(f(x)) = 0.02(x - 200,000)$
   $x > 200,000$

19. $f^{-1}(x) = \frac{1}{5}x$ **21.** $f^{-1}(x) = \sqrt[7]{x}$

23. $f(g(x)) = (x - 15) + 15 = x$
   $g(f(x)) = (x + 15) - 15 = x$

25. $f(g(x)) = \sqrt[3]{(x^3 - 1) + 1}$
   $= \sqrt[3]{x^3} = x$
   $g(f(x)) = \left(\sqrt[3]{x + 1}\right)^3 - 1$
   $= x + 1 - 1 = x$

27. b   **29.** d   **31.**

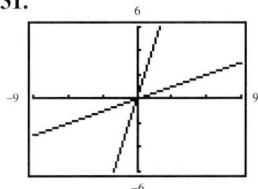

33.

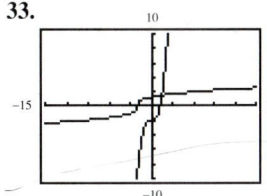

35. $g^{-1}(x) = \frac{1}{4}(3 - x)$

37. $f^{-1}(t) = \sqrt[3]{t + 1}$ **39.** No   **41.** No

43. $x \geq 0$, $f^{-1}(x) = \sqrt[4]{x}$ **45.** $x \geq 2$, $f^{-1}(x) = \sqrt{x} + 2$

47.

x	0	1	3	4
f^{-1}	6	4	2	0

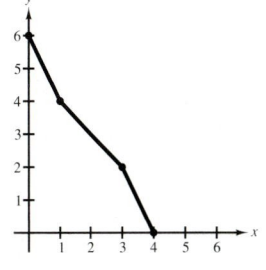

49. True   **51.** $-4, 1$ **53.** $\dfrac{-3 \pm \sqrt{37}}{4}$

55. (a) Down
(b) $(0, 0), (4, 0)$
(c) $(2, 4)$

57. (a) 7
(b) 8
(c) $2x + 1$
(d) $2(x + 1)$

59. (a) 0
(b) 3
(c) $3|x - 1|$
(d) $3|x - 3|$

61. (a) $\frac{1}{4}$
(b) $\frac{1}{3}$
(c) $\dfrac{1}{\sqrt{x} - 3}$
(d) $\dfrac{1}{\sqrt{x - 3}}$

63. (a) 10
(b) 1

65. (a) 0
(b) 10

67. (a) $(-\infty, \infty)$
(b) $(-\infty, \infty)$

69. (a) $(-\infty, \infty)$
(b) $(-\infty, -9) \cup (-9, \infty)$

71. $(C \circ x)(t) = 102t + 300$
Total cost after t hours of production

73. $f^{-1}(x) = \frac{1}{6}x$ **75.** $f^{-1}(x) = x - 10$

77. $f^{-1}(x) = x^3$ **79.** $f^{-1}(x) = \frac{1}{2}(x + 1)$

81. $f(g(x)) = 10\left(\frac{1}{10}x\right) = x$

$g(f(x)) = \frac{1}{10}(10x) = x$

83. $f(g(x)) = 1 - 2\left[\frac{1}{2}(1 - x)\right]$

$\qquad = 1 - (1 - x) = x$

$g(f(x)) = \frac{1}{2}[1 - (1 - 2x)]$

$\qquad = \frac{1}{2}(2x) = x$

85. $f(g(x)) = 2 - 3\left[\frac{1}{3}(2 - x)\right]$

$\qquad = 2 - (2 - x) = x$

$g(f(x)) = \frac{1}{3}[2 - (2 - 3x)]$

$\qquad = \frac{1}{3}(3x) = x$

87. $f(g(x)) = \dfrac{1}{\left(\frac{1}{x}\right)} = x$

$g(f(x)) = \dfrac{1}{\left(\frac{1}{x}\right)} = x$

89.

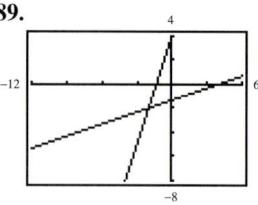

91.

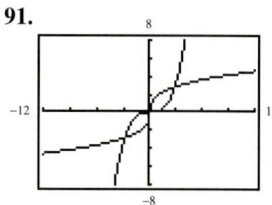

93. $f^{-1}(x) = \frac{1}{8}x$ **95.** $g^{-1}(x) = x - 25$

97. $g^{-1}(t) = -4(t - 2)$ **99.** $h^{-1}(x) = x^2,\ x \geq 0$

101. $g^{-1}(s) = \dfrac{5}{s}$

103. $f^{-1}(x) = \sqrt[3]{x - 1}$ **105.**

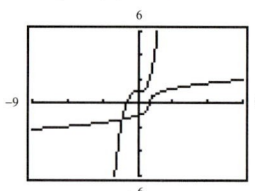

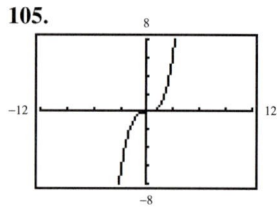

Yes

107.

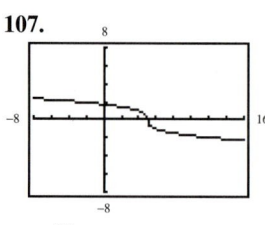

Yes

109.

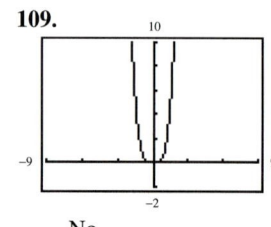

No

111.

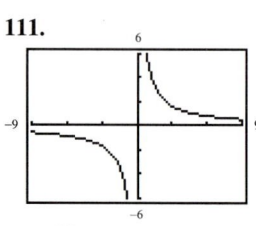

Yes

113.

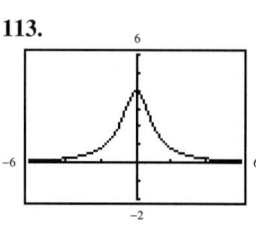

No

115.

x	-4	-2	2	3
f^{-1}	-2	-1	1	3

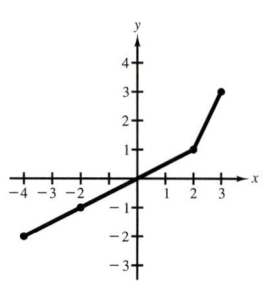

117. (a) $f^{-1}(x) = \frac{1}{2}(3-x)$

(b) $(f^{-1})^{-1}(x) = 3 - 2x$

119. (a) $y = \frac{20}{13}(x-9)$ (b) x: hourly wage (c) 8

y: number of units produced

121. False. $f(x) = \sqrt{x-1}$; domain: $[1, \infty)$

$f^{-1}(x) = x^2 + 1$; domain: $[0, \infty)$

Section 13.3 *(page 785)*

7. $5^2 = 25$ **9.** $36^{1/2} = 6$ **11.** $\log_3 \frac{1}{9} = -2$

13. $\log_8 4 = \frac{2}{3}$ **15.** 3 **17.** -2

19. There is no power to which 2 can be raised to obtain -3.

21. $\frac{1}{2}$ **23.** 2 **25.** 1 **27.** 1.4914 **29.** -0.2877

31. e **33.** d **35.** a **37.** $f^{-1} = g$

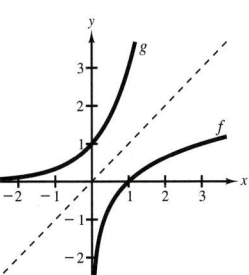

39.

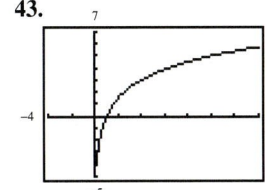

41.

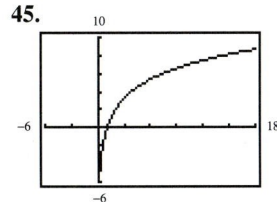

43.

45.

47. 2.3481 **49.** -0.4739

51.

r	0.07	0.08	0.09	0.10	0.11	0.12
t	9.90	8.66	7.70	6.93	6.30	5.78

53.

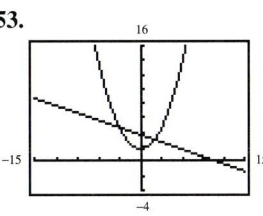

$\left(-3, \frac{14}{3}\right), (2, 3)$

55. $x^2 - 6$

57. Compact discs: 53.3 million **59.** $6^2 = 36$

Records: 128.2 million

61. $4^{-2} = \frac{1}{16}$ **63.** $8^{2/3} = 4$ **65.** $\log_7 49 = 2$

67. $\log_{25} \frac{1}{5} = -\frac{1}{2}$ **69.** $\log_4 1 = 0$ **71.** 0

73. 1 **75.** 3

77. There is no power to which 4 can be raised to obtain -4.

79. $\frac{3}{4}$ **81.** 4 **83.** 0 **85.** 1 **87.** 0.7335

89. -0.0706 **91.** 3.2189

93. $f^{-1} = g$

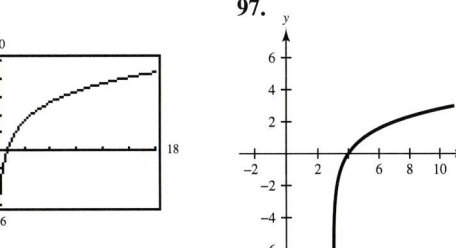

95.

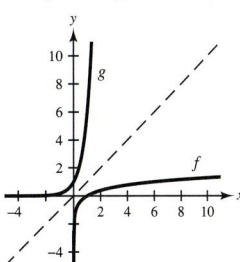

97.

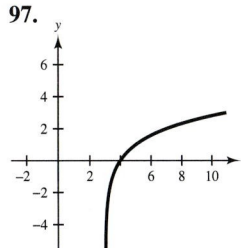

99.

101.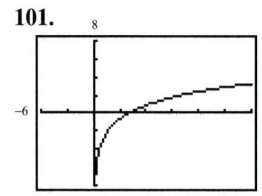

103.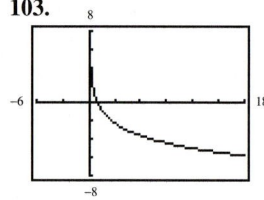

105.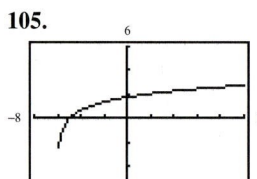

107.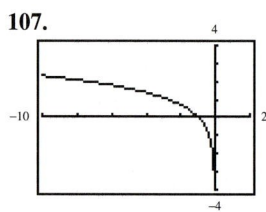

109. 1.7712 **111.** 2.6332 **113.** 1.3481 **115.** -2

117. 53.4 inches **119.** $0 < x < \infty$

121. $3 \le f(x) \le 4$ **123.** A factor of 10

Math Matters *(page 788)*

79.5%

Mid-Chapter Quiz *(page 789)*

1. (a) $\frac{16}{9}$

   (b) 1

   (c) $\frac{3}{4}$

   (d) $\frac{8\sqrt{3}}{9}$

2. Domain: $(-\infty, \infty)$
   Range: $(0, \infty)$

3.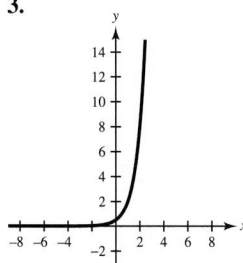

4.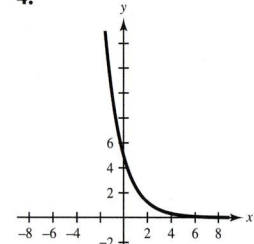

5.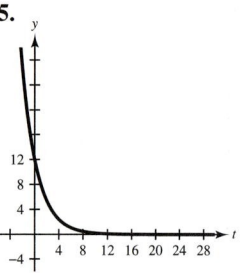

6.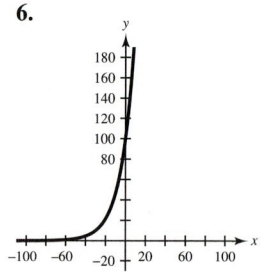

7.

n	1	4	12	365	Continuous
A	\$3185.89	\$3314.90	\$3345.61	\$3360.75	\$3361.27

8. \$2.71 **9.** (a) -19

   (b) 125

   (c) $2x^3 - 3$

   (d) $(2x - 3)^3$

10. $f(g(x)) = 3 - 5\left[\frac{1}{5}(3 - x)\right] = 3 - 3 + x = x$

   $g(f(x)) = \frac{1}{5}[3 - (3 - 5x)] = \frac{1}{5}(5x) = x$

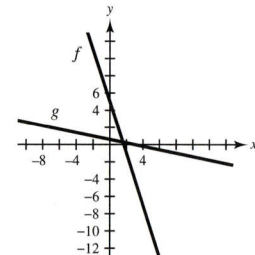

11. $h^{-1}(x) = \frac{1}{10}(x - 3)$ **12.** $g^{-1}(t) = \sqrt[3]{2(t - 2)}$

13. $4^{-2} = \frac{1}{16}$ **14.** $\log_3 81 = 4$ **15.** 3

16. $f^{-1}(x) = g(x)$

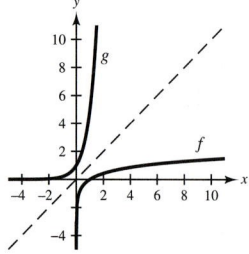

17.

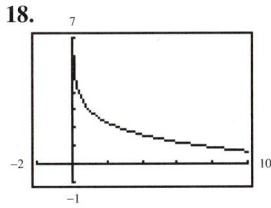

18.

19. $f(x) = \log_5(x - 3) + 1$ **20.** 3.4096

Section 13.4 (page 795)

5. 2   **7.** -3 **9.** 0.954   **11.** 1.556

13. 0.778   **15.** $\log_2 3 + \log_2 x$ **17.** $-2\log_5 x$

19. $\ln 5 - \ln(x - 2)$ **21.** $\frac{1}{2}[\ln x + \ln(x + 2)]$

23. $\log_{12} \dfrac{x}{3}$ **25.** $\log_5(2x)^{-2}$ **27.** $\ln \dfrac{32y^3}{x}$

29. $\ln 4\sqrt{x}$ **31.** $1 - \log_4 x$ **33.** $1 + \frac{1}{2}\log_5 2$

35. $2 + \ln 3$ **37.**

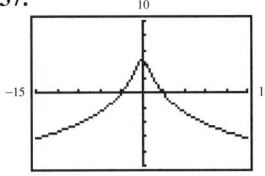

39. Evaluate when $x = 3$ and $y = 2$.

41. $E = 1.4\log_{10}\dfrac{C_2}{C_1}$ **43.** x^{20} **45.** $\dfrac{3}{2y^5}$

47. $\frac{1}{6}x^3 y$ **49.** 12   **51.** 2   **53.** 2   **55.** 3

57. 0   **59.** 1   **61.** 1.2925   **63.** 0.2925

65. 0.2500   **67.** 2.7925   **69.** 0

71. $\log_3 11 + \log_3 x$ **73.** $3\ln y$ **75.** $\log_2 z - \log_2 17$

77. $\frac{1}{3}\log_3(x + 1)$ **79.** $\ln 3 + 2\ln x + \ln y$

81. $2\log_2 x - \log_2(x - 3)$ **83.** $\frac{1}{3}[\ln x + \ln(x + 5)]$

85. $\log_2 3x$ **87.** $\log_{10}\dfrac{4}{x}$ **89.** $\ln b^4$ **91.** $\ln\sqrt[3]{2x + 1}$

93. $\log_3 2\sqrt{y}$ **95.** $\ln\dfrac{x^2 y^3}{z}$ **97.** $\ln(xy)^4$ **99.** True

101. True   **103.** False   **105.**

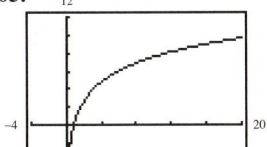

107. $10(\log_{10} I + 16)$, 60 decibels

109. False. 0 is not in the domain of f.

111. False. $f(x - 3) = \ln(x - 3)$

113. False. If $v = u^2$, then $f(v) = \ln u^2 = 2\ln u = 2f(u)$.

Section 13.5 (page 804)

5. (a) Not a solution   **7.** (a) Solution
(b) Solution   (b) Not a solution

9. (a) Not a solution   **11.** 8   **13.** -6 **15.** -7
(b) Solution

17. 9   **19.** $2x - 1$ **21.** $2x$ **23.** $\frac{3}{2}$ **25.** 1.23

27. 0.59   **29.** 2.48   **31.** 9.73   **33.** 22.63

35. 2187   **37.** 6.52   **39.** 12.18   **41.** 2.46

43.

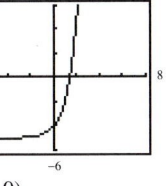

$(1.40, 0)$

45.

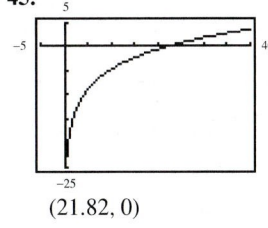

$(21.82, 0)$

47. 7.70 years   **49.** 262.5°   **51.** $2|x|y\sqrt{6y}$

53. $\dfrac{2\sqrt{3}b^3}{a^2}$ **55.** \$5395.40   **57.** 5   **59.** 8

61. 3   **63.** -3 **65.** 18   **67.** $\frac{22}{5}$ **69.** 81

71. 500,000   **73.** 5.49   **75.** 0.86   **77.** 3.28

79. 2.49   **81.** 0.39   **83.** 6.80   **85.** 0.80

87. 7.39   **89.** 21.82   **91.** 8.99   **93.** 1

95. 1000   **97.** 7.91   **99.** 8.17   **101.** -0.78

103. 31,622,771.60   **105.** ± 20.09 **107.** 5

109.

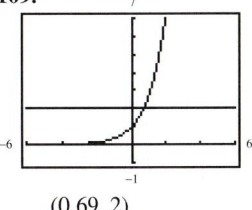

$(0.69, 2)$

111.

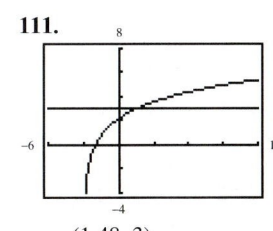

$(1.48, 3)$

113. 3.16×10^{-9} **115.** 205

Section 13.6 *(page 813)*

7. 9% **9.** 8% **11.** 8.75 years **13.** 6.60 years

15. Continuous **17.** Quarterly **19.** 8.33%

21. 7.23% **23.** $1652.99 **25.** $626.46

27. $k = \frac{1}{2} \ln \frac{8}{3} \approx 0.4904$ **29.** $k = -\frac{\ln 2}{3} \approx -0.2310$

31. $y = 10.1e^{0.0072t}$ **33.** $y = 4.6e^{0.0432t}$
11.1 million   8.1 million

35. (a) k is larger in Exercise 33, because the population of Dhaka is increasing faster than the population of Los Angeles.

(b) k corresponds to r; k gives the annual percentage rate of growth.

37. 3.2595 grams **39.** 63 times as great **41.** 7.04

43. $\dfrac{9}{2(x+3)}$ **45.** $\dfrac{x^2 + 2x - 13}{x(x-2)}$ **47.** $\dfrac{7x}{9}, \dfrac{19x}{18}$

49. 7% **51.** 6% **53.** 7% **55.** 14.21 years

57. 9.99 years **59.** 9.24 years **61.** 6.136%

63. 8.300% **65.** 7.788% **67.** No **69.** $3080.15

71. $951.23 **73.** $67,667.64 **75.** $5,496.57

77. $320,250.81 **79.** Total deposits: $7200.00
Total interest: $10,529.42

81. 0.8517 grams **83.** $y = 6.5e^{0.0019t}$
6.7 million

85. $y = 21.6e^{0.0320t}$
32.7 million

87. (a) 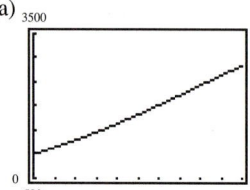   (b) 1000
(c) 2642
(d) 5.88 years

89. $9281.25

91. (a)   (b) 1300 units
(c) 3 years
(d) 2000 units

93. The one in Chile is 63 times as great. **95.** 10^7 times

97. (a)

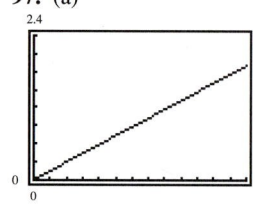   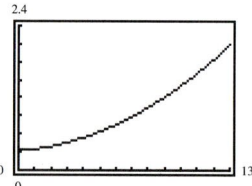

$E = 0.146t + 0.007$ $E = 0.009t^2 + 0.018t + 0.325$

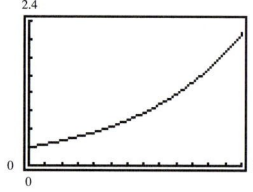   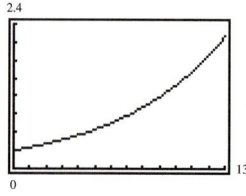

$E = 0.301(1.165)^t$ $E = 0.301e^{0.153t}$

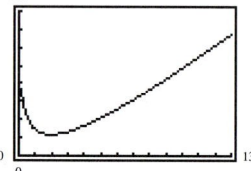

$E = 0.201 + 0.228t - 0.444 \ln t$

(b) $E = 0.146t + 0.007$.
Sum of Squares: 0.1725

$E = 0.009t^2 + 0.018t + 0.325$.
Sum of Squares: 0.0076

$E = 0.301(1.165)^t$.
Sum of Squares: 0.0315

$E = 0.301e^{0.153t}$.
Sum of Squares: 0.0313

$E = 0.201 + 0.228t - 0.444 \ln t$.
Sum of Squares: 0.0209

Quadratic

Review Exercises *(page 820)*

1. (a) $\frac{1}{8}$ **3.** (a) 2.718 **5.** (a) 0
(b) 2   (b) 0.351   (b) 3
(c) 4   (c) 0.135   (c) −0.631

7. (a) 1
(b) −1.099
(c) 2.303

9. (a) −6
(b) 0
(c) 22.5

11. d   **13.** a   **15.** c

17.

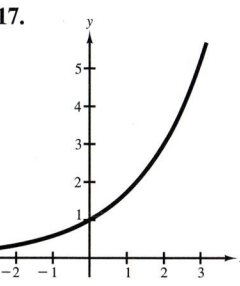

19.

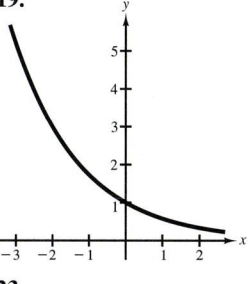

21.

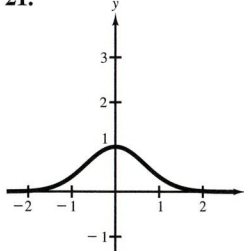

23.

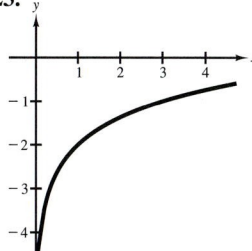

25.

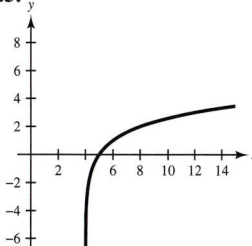

27.

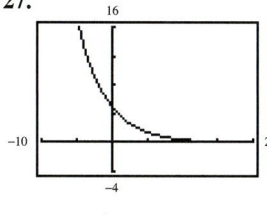

29.

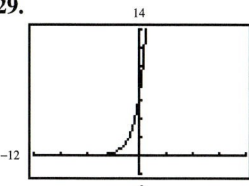

31.

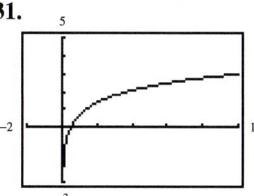

33.

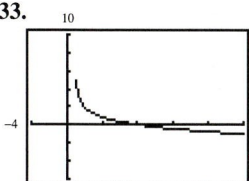

35. $(f \circ g)(x) = x^2 + 2$
$(g \circ f)(x) = (x + 2)^2$

37. $(f \circ g)(x) = x$
$(g \circ f)(x) = x$

39. (a) $[2, \infty)$
(b) $[4, \infty)$

41.

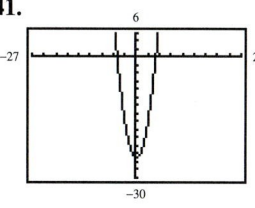

Not one-to-one

43.

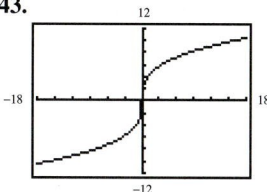

One-to-one

45. $f^{-1}(x) = 4x$ **47.** $h^{-1}(x) = x^2, x \geq 0$

49. Not one-to-one   **51.** $\log_4 64 = 3$ **53.** $e^1 = e$

55. 3   **57.** −2   **59.** 7   **61.** 0

63. $\log_4 6 + 4 \log_4 x$ **65.** $\frac{1}{5} \log_5(x + 2)$

67. $\ln(x + 2) - \ln(x - 2)$ **69.** $\log_4 \dfrac{x}{10}$

71. $4 + \ln x^8$ **73.** $\ln\left(\dfrac{3}{2x}\right)^2$

75.

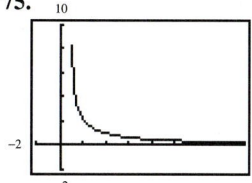

77. False.
$\log_2 4x = 2 + \log_2 x$

79. True   **81.** True   **83.** 1.79588   **85.** −0.43068

87. 1.02931   **89.** 1.585   **91.** 2.132   **93.** 6

95. 1   **97.** 125   **99.** 5.66   **101.** 6.23

103. 32.99   **105.** 15.81   **107.** 1408.10   **109.** 0.32

111.

n	1	4	12
A	\$3806.13	\$4009.59	\$4058.25

n	365	Continuous
A	\$4082.26	\$4083.08

113.

n	1	4	12
A	\$67,275.00	\$72,095.68	\$73,280.74

n	365	Continuous
A	\$73,870.32	\$73,890.56

115.

n	1	4	12
P	\$2301.55	\$2103.50	\$2059.87

n	365	Continuous
P	\$2038.82	\$2038.11

117. 4.6 years **119.** 150 units

121.

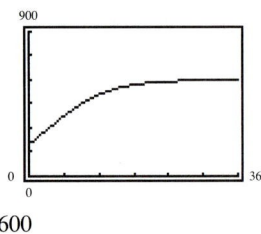

600

Chapter Test *(page 823)*

1. $f(-1) = 81$
$f(0) = 54$
$f\left(\frac{1}{2}\right) = 18\sqrt{6} \approx 44.09$
$f(2) = 24$

2.

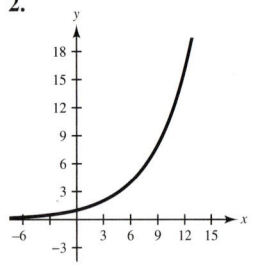

3. $5^3 = 125$ **4.** $\log_4 \frac{1}{16} = -2$ **5.** $\frac{1}{3}$

6. $g = f^{-1}$ **7.** $\log_4 5 + 2\log_4 x - \frac{1}{2}\log_4 y$

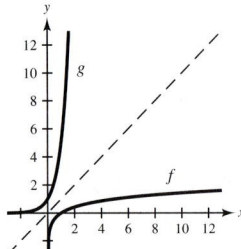

8. $\ln \dfrac{x}{y^4}$ **9.** $3 + \log_5 6$ **10.** 3 **11.** 0.973

12. 13.733 **13.** 15.516 **14.** (a) \$8012.78
(b) \$8110.40

15. \$10,806.08 **16.** 7% **17.** \$8469.14
18. 600 **19.** 1141 **20.** 4.4 years

Cumulative Test: Chapters 11–13 *(page 824)*

1. $3x - 4y + 12 = 0$ **2.**

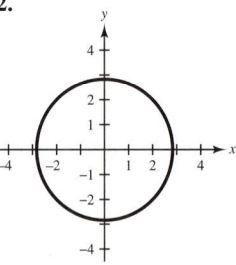

3.

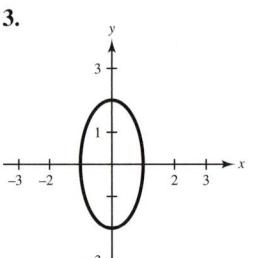

4.

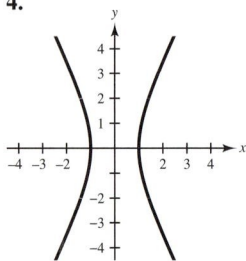

5.

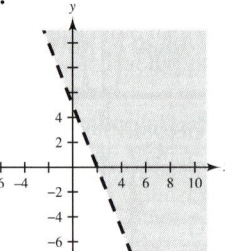

6. $y = \frac{2}{3}(x-3)^2 - 2$

7. 11 feet **8.** 128 feet **9.** $(2, 1)$ **10.** $(3, -2)$
11. $(5, 4)$ **12.** $\left(-\frac{1}{5}, -\frac{22}{5}\right)$ **13.** 8, 12, 24
14. **15.** -2

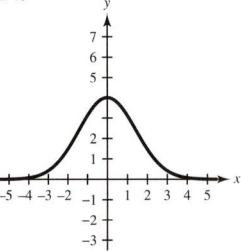

16. Reflections in the line $y = x$ because
$f^{-1}(x) = g(x)$.

17. $\log_2\left(\dfrac{x^3 y^3}{z}\right)$ **18.** (a) 3   **19.** 8.329%

                      (b) 12.18

                      (c) 18.01

                      (d) 0.87

20. 15.40 years

CHAPTER 14

Section 14.1 *(page 832)*

7. 2, 4, 6, 8, 10   **9.** $-\dfrac{1}{2}, \dfrac{1}{4}, -\dfrac{1}{8}, \dfrac{1}{16}, -\dfrac{1}{32}$

11. $\dfrac{2}{5}, \dfrac{4}{8}, \dfrac{6}{11}, \dfrac{8}{14}, \dfrac{10}{17}$ **13.** $2, 2, \dfrac{4}{3}, \dfrac{2}{3}, \dfrac{4}{15}$

15. c   **17.** b   **19.**

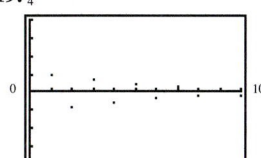

21. -72 **23.** $\dfrac{1}{702}$ **25.** $n(n+1)$

27. 63   **29.** 77   **31.** 7.283   **33.** 54

35. $\displaystyle\sum_{k=1}^{5} 2k$ **37.** $\displaystyle\sum_{k=1}^{20} \dfrac{4}{k+3}$ **39.** $\displaystyle\sum_{k=1}^{20} \dfrac{2k}{k+3}$

41. (a) \$502.92, \$505.85, \$508.80, \$511.77, \$514.75,
      \$517.76, \$520.78, \$523.82

    (b) \$2019.37   (c)

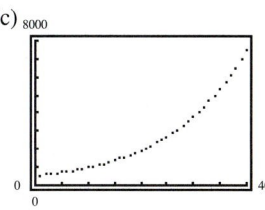

    (d) Yes. Investment earning compound interest increases
       at an increasing rate.

43. $x^2 + 20x + 100$ **45.** $\dfrac{1}{a^8}$ **47.** $x - y + 1 = 0$

49. $-2, 4, -6, 8, -10$ **51.** $\dfrac{1}{2}, \dfrac{1}{4}, \dfrac{1}{8}, \dfrac{1}{16}, \dfrac{1}{32}$

53. $\dfrac{1}{2}, \dfrac{1}{3}, \dfrac{1}{4}, \dfrac{1}{5}, \dfrac{1}{6}$ **55.** $-1, \dfrac{1}{4}, -\dfrac{1}{9}, \dfrac{1}{16}, -\dfrac{1}{25}$

57. $\dfrac{9}{2}, \dfrac{19}{4}, \dfrac{39}{8}, \dfrac{79}{16}, \dfrac{159}{32}$ **59.** $0, 4, 0, 4, 0$

61. 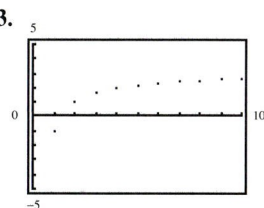   **63.**

65. 5   **67.** $\dfrac{1}{132}$ **69.** $\dfrac{1}{n+1}$ **71.** $2n$ **73.** 35

75. $\dfrac{47}{60}$ **77.** -48 **79.** $\dfrac{8}{9}$ **81.** $\dfrac{182}{243}$ **83.** 112

85. 852   **87.** 6.579   **89.** $\displaystyle\sum_{k=1}^{5} k$ **91.** $\displaystyle\sum_{k=1}^{10} \dfrac{1}{2k}$

93. $\displaystyle\sum_{k=1}^{20} \dfrac{1}{k^2}$ **95.** $\displaystyle\sum_{k=0}^{9} \dfrac{(-1)^k}{3^k}$ **97.** $\displaystyle\sum_{k=1}^{11} \dfrac{k}{k+1}$ **99.** $\displaystyle\sum_{k=0}^{6} k!$

101. 3   **103.** 12.5   **105.** $a_5 = 108°, a_6 = 120°$
                                 $a_5 + 2a_6 = 348° < 360°$

107. $25.7°, 45°, 60°, 72°, 81.8°$

Section 14.2 *(page 840)*

7. 12   **9.** $-\dfrac{5}{4}$ **11.** Arithmetic, $d = -2$

13. Arithmetic, $d = -\dfrac{1}{2}$ **15.** Arithmetic, $d = 4$

17. Not arithmetic   **19.** 40, 35, 30, 25, 20

21. $\dfrac{8}{5}, \dfrac{11}{5}, \dfrac{14}{5}, \dfrac{17}{5}, 4$ **23.** 25, 28, 31, 34, 37

25. d   **27.** c   **29.** $3 + \dfrac{3}{2}(n-1)$

31. $28 - 4(n-1)$ **33.** $5 + 2(n-1)$ **35.** 275

37. 9000   **39.** 1850   **41.** 23   **43.** 4914

45. 1024 feet   **47.** $\left(-\dfrac{1}{2}, \dfrac{1}{4}\right), (2, 4)$ **49.** $(1, 2, 1)$

51. Adults: 816   **53.** 3   **55.** -6 **57.** $\dfrac{2}{3}$
    Children: 384

59. Arithmetic, $d = 2$ **61.** Arithmetic, $d = \dfrac{3}{2}$

63. Not arithmetic   **65.** Arithmetic, $d = \dfrac{1}{6}$

67. Arithmetic, $d = 0.8$ **69.** Not arithmetic

71. 7, 10, 13, 16, 19   **73.** 6, 4, 2, 0, -2

75. $\dfrac{3}{2}, 4, \dfrac{13}{2}, 9, \dfrac{23}{2}$ **77.** $4, \dfrac{15}{4}, \dfrac{7}{2}, \dfrac{13}{4}, 3$

79. 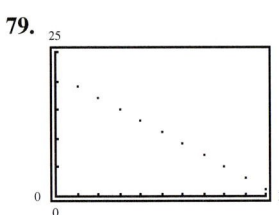 **81.**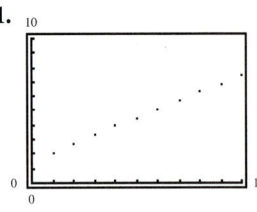

83. $3 + \frac{1}{2}(n - 1)$ **85.** $64 - 8(n - 1)$

87. $8 + 4(n - 1)$ **89.** $50 - 10(n - 1)$

91. $9, 6, 3, 0, -3$ **93.** $-10, -4, 2, 8, 14$

95. b **97.** e **99.** c **101.** 210 **103.** 2700

105. 62,625 **107.** 19,500 **109.** 522 **111.** 900

113. 12,200 **115.** 243 **117.** 2850 **119.** 2550

121. \$246,000 **123.** \$25.42

125. (a) $4, 9, 16, 25, 36$ (b) 49

(c) $\displaystyle\sum_{k=1}^{n}[1 + 2(k - 1)] = n^2$

Section 14.3 *(page 849)*

7. -3 **9.** π **11.** Geometric, $r = 2$

13. Not geometric **15.** $4, -2, 1, -\frac{1}{2}, \frac{1}{4}$

17. $20, 21.40, 22.90, 24.50, 26.22$ **19.** -0.0061

21. 1486.02 **23.** $25(4)^{n-1}$ **25.** $\left(\frac{3}{2}\right)^{n-1}$ **27.** b

29. a **31.** 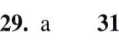 **33.** -5460

35. 6.67 **37.** 399.93 **39.** 13,120 **41.** 48.00

43. (a) $250,000(0.75)^n$ **45.** \$75,715.32

(b) \$59,326.17

(c) The first year

47. 70.875 square inches **49.** -200 **51.** -60

53. 58 **55.** 3 **57.** $\frac{1}{3}$ **59.** $-\frac{3}{2}$ **61.** e

63. Not geometric **65.** Geometric, $r = \frac{1}{2}$

67. Geometric, $r = -\frac{2}{3}$ **69.** Geometric, $r = 1.02$

71. $4, 8, 16, 32, 64$ **73.** $6, 2, \frac{2}{3}, \frac{2}{9}, \frac{2}{27}$

75. $1, -\frac{1}{2}, \frac{1}{4}, -\frac{1}{8}, \frac{1}{16}$

77. 1000, 1010, 1020.10, 1030.30, 1040.60

79. $\frac{3}{256}$ **81.** $48\sqrt{2}$ **83.** $\frac{81}{256}$ **85.** $\frac{243}{32}$

87. $2(3)^{n-1}$ **89.** 2^{n-1} **91.** $4\left(-\frac{1}{2}\right)^{n-1}$

93. $8\left(\frac{1}{4}\right)^{n-1}$ **95.** 1023 **97.** 772.478

99. 2.250 **101.** 6300.250 **103.** $-14,762$

105. 16.000 **107.** 152.095 **109.** 110.357

111. (a) $P(0.999)^n$ (b) 69.4%

(c)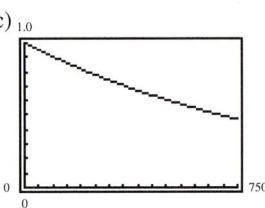

700 days

113. \$3,623,993.23 **115.** \$19,496.56 **117.** \$105,428.44

119. (a) \$53,687.09 (b) \$107,374.18

121. 72.969 square inches

Mid-Chapter Quiz *(page 854)*

1. $32, 8, 2, \frac{1}{2}, \frac{1}{8}$ **2.** $-\frac{3}{5}, 3, -\frac{81}{7}, \frac{81}{2}, -135$

3. 100 **4.** 40 **5.** 87 **6.** -32

7. $\displaystyle\sum_{k=1}^{20}\frac{2}{3k}$ **8.** $\displaystyle\sum_{k=1}^{25}\frac{(-1)^{k-1}}{k^3}$ **9.** $\frac{1}{2}$ **10.** -6

11. $20 - 3(n - 1)$ **12.** $32\left(-\frac{1}{4}\right)^{n-1}$ **13.** 4075

14. 9030 **15.** 25.947 **16.** 18,392.796 **17.** -0.026

18. a_n: upper graph **19.** 5.5° **20.** Arithmetic

b_n: lower graph

Section 14.4 *(page 860)*

7. 15 **9.** 1 **11.** 593,775

13. 2,598,960 **15.** 455 **17.** 53,130

19. $1, 9, 36, 84, 126, 126, 84, 36, 9, 1$

21. $x^6 + 18x^5 + 135x^4 + 540x^3 + 1215x^2 + 1458x + 729$

23. $u^3 - 6u^2v + 12uv^2 - 8v^3$

25. $x^8 + 8x^7y + 28x^6y^2 + 56x^5y^3 + 70x^4y^4 + 56x^3y^5 + 28x^2y^6 + 8xy^7 + y^8$

27. $x^6 - 12x^5 + 60x^4 - 160x^3 + 240x^2 - 192x + 64$

29. 120   **31.** -1365

33. $\frac{1}{32} + \frac{5}{32} + \frac{10}{32} + \frac{10}{32} + \frac{5}{32} + \frac{1}{32}$

35. $\frac{1}{256} + \frac{12}{256} + \frac{54}{256} + \frac{108}{256} + \frac{81}{256}$

37. 1.17   **39.** 510,568.79   **41.** $4^3 = 64$ **43.** $e^0 = 1$

45.

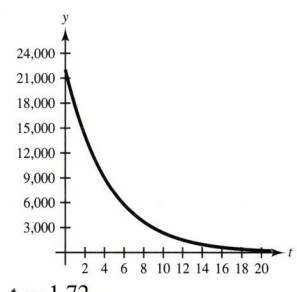

$t = 1.72$

47. 252   **49.** 1

51. 1225   **53.** 12,650   **55.** 792

57. 22,451,004,309,013,280   **59.** 5,200,300

61. 10   **63.** 792   **65.** 1   **67.** 35   **69.** 70

71. $x^5 + 5x^4 + 10x^3 + 10x^2 + 5x + 1$

73. $x^6 - 24x^5 + 240x^4 - 1280x^3 + 3840x^2 - 6144x + 4096$

75. $x^4 + 4x^3y + 6x^2y^2 + 4xy^3 + y^4$

77. $x^5 - 5x^4y + 10x^3y^2 - 10x^2y^3 + 5xy^4 - y^5$

79. $a^3 + 6a^2 + 12a + 8$

81. $16x^4 - 32x^3 + 24x^2 - 8x + 1$

83. $64y^6 + 192y^5z + 240y^4z^2 + 160y^3z^3 + 60y^2z^4 + 12yz^5 + z^6$

85. -120 **87.** 1760   **89.** 54

91. 1, 3, 6, 10, 15
The difference between consecutive determinants increases by 1.

Section 14.5 *(page 869)*

5. 5   **7.** 9   **9.** 8   **11.** 6   **13.** 72

15. $xyz, xzy, yxz, yzx, zxy, zyx$ **17.** 120   **19.** 40,320

21. $\{A, B\}, \{A, C\}, \{A, D\}, \{A, E\}, \{A, F\}, \{B, C\}, \{B, D\},$
$\{B, E\}, \{B, F\}, \{C, D\}, \{C, E\}, \{C, F\}, \{D, E\}, \{D, F\},$
$\{E, F\}$

23. (a) 3   (b) 6   **25.** 56   **27.** (a) 56
   (c) 15   (d) 28   (b) 56
   (e) 45   (f) 66   (c) 8

29. 41   **31.** 69   **33.** $2x - 3y + 9 = 0$

35. 10   **37.** 8   **39.** 6   **41.** 7   **43.** 6

45. 6,760,000   **47.** AB   BA   **49.** 24   **51.** 5040
   AC   CA
   AD   DA
   BC   CB
   BD   DB
   CD   DC

53. $\{A, B, C\}, \{A, B, D\}, \{A, B, E\}, \{A, C, D\}, \{A, C, E\},$
   $\{A, D, E\}, \{B, C, D\}, \{B, C, E\}, \{B, D, E\}, \{C, D, E\}$

55. 260   **57.** 142,506   **59.** 3003

61. (a) 70   (b) 16   **63.** (a) 5   (b) 9   (c) 20

Section 14.6 *(page 877)*

5. $\{A, B, C, D, E, \ldots, X, Y, Z\}$

7. $\{AB, AC, AD, AE, BC, BD, BE, CD, CE, DE\}$

9. 0.65   **11.** $\frac{3}{8}$ **13.** $\frac{7}{8}$ **15.** $\frac{9}{10}$

17. (a) $\frac{1}{20}$ (b) $\frac{2}{5}$ (c) $\frac{1}{2}$ (d) $\frac{1}{4}$

19. $\frac{1917}{6565}, \frac{4648}{6565}$ **21.** $\frac{1}{24}$ **23.** $\frac{1}{45}$ **25.** $\frac{1}{54,145}$

27. g is a vertical shift of f four units downward.

29. g is a reflection of f in the x-axis.

31.

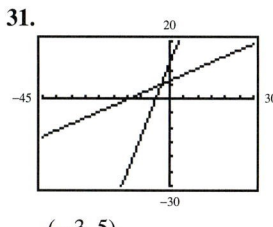

$(-3, 5)$

33. $\{ABC, ACB, BAC, BCA, CAB, CBA\}$

35. $\{WWW, WWL, WLW, WLL, LWW, LWL, LLW, LLL\}$

37. 0.18   **39.** $\frac{1}{2}$ **41.** $\frac{3}{13}$ **43.** $\frac{1}{6}$ **45.** $\frac{5}{6}$

47. 0.158   **49.** 0.733   **51.** 0.022   **53.** 0.455

55. (a) $\frac{1}{5}$ (b) $\frac{1}{3}$ (c) 1

57.

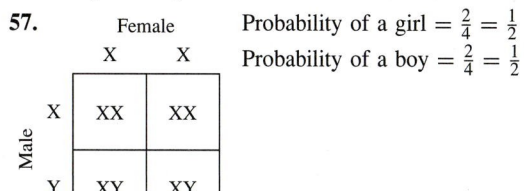

Probability of a girl $= \frac{2}{4} = \frac{1}{2}$
Probability of a boy $= \frac{2}{4} = \frac{1}{2}$

59. $\dfrac{14}{65}$ **61.** $\dfrac{1}{100,000}$ **63.** $\dfrac{1}{45}$ **65.** (a) $\dfrac{6}{11}$ (b) $\dfrac{5}{11}$

Review Exercises *(page 883)*

1. $\displaystyle\sum_{k=1}^{4}(5k-3)$ **3.** $\displaystyle\sum_{n=1}^{6}\dfrac{1}{3n}$ **5.** 380

7. $n(n-1)(n-2)$ **9.** 127, 122, 117, 112, 107

11. $\dfrac{5}{4}, 2, \dfrac{11}{4}, \dfrac{7}{2}, \dfrac{17}{4}$ **13.** -2.5 **15.** 10, 30, 90, 270, 810

17. $100, -50, 25, -\dfrac{25}{2}, \dfrac{25}{4}$ **19.** $\dfrac{3}{2}$ **21.** $10+4(n-1)$

23. $1000-50(n-1)$ **25.** $\left(-\dfrac{2}{3}\right)^{n-1}$ **27.** $24(2)^{n-1}$

29. $12\left(-\dfrac{1}{2}\right)^{n-1}$ **31.** a **33.** b **35.** d

37. 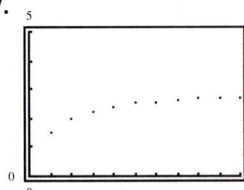 **39.**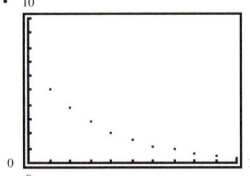

41. 28 **43.** $\dfrac{4}{5}$ **45.** 486 **47.** 1262.5 **49.** 8190

51. -1.928 **53.** 19.842 **55.** 116,169.538

57. 5100 **59.** 462 **61.** (a) $85,000(1.012)^{t}$

(b) 154,328

63. 56 **65.** 1 **67.** 91,390

69. $x^{10}+10x^{9}+45x^{8}+120x^{7}+210x^{6}+252x^{5}+210x^{4}+$ $120x^{3}+45x^{2}+10x+1$

71. $y^{6}-12y^{5}+60y^{4}-160y^{3}+240y^{2}-192y+64$

73. $x^{8}-4x^{7}+7x^{6}-7x^{5}+\dfrac{35}{8}x^{4}-\dfrac{7}{4}x^{3}+\dfrac{7}{16}x^{2}-\dfrac{1}{16}x+\dfrac{1}{256}$

75. $-61,236$ **77.** 8 **79.** 3003 **81.** $\dfrac{1}{3}$

83. $\dfrac{1}{24}$ **85.** 0.346

Chapter Test *(page 886)*

1. $1, -\dfrac{2}{3}, \dfrac{4}{9}, -\dfrac{8}{27}, \dfrac{16}{81}$ **2.** 35 **3.** -45 **4.** $\displaystyle\sum_{k=1}^{12}\dfrac{2}{3k+1}$

5. 12, 16, 20, 24, 28 **6.** $5000-100(n-1)$

7. 3825 **8.** $-\dfrac{3}{2}$ **9.** $4\left(\dfrac{1}{2}\right)^{n-1}$ **10.** 1020

11. $\dfrac{3069}{1024}$ **12.** \$47,868.33 **13.** 1140

14. $x^{5}-10x^{4}+40x^{3}-80x^{2}+80x-32$ **15.** 56

16. 26,000 **17.** 12,650 **18.** 0.25 **19.** $\dfrac{3}{26}$ **20.** $\dfrac{1}{6}$

APPENDIX A

Section A.1 *(page A5)*

1. Statement **3.** Open statement **5.** Open statement

7. Open statement **9.** Nonstatement

11. Open statement **13.** (a) True **15.** (a) True

(b) False (b) True

17. (a) False **19.** (a) True

(b) False (b) False

21. (a) The sun is not shining.

(b) It is not hot.

(c) The sun is shining and it is hot.

(d) The sun is shining or it is hot.

23. (a) Lions are not mammals.

(b) Lions are not carnivorous.

(c) Lions are mammals and lions are carnivorous.

(d) Lions are mammals or lions are carnivorous.

25. (a) The sun is not shining and it is hot.

(b) The sun is not shining or it is hot.

(c) The sun is shining and it is not hot.

(d) The sun is shining or it is not hot.

27. (a) Lions are not mammals and lions are carnivorous.

(b) Lions are not mammals or lions are carnivorous.

(c) Lions are mammals and lions are not carnivorous.

(d) Lions are mammals or lions are not carnivorous.

29. $p \wedge \sim q$ **31.** $\sim p \vee q$ **33.** $\sim p \vee \sim q$

35. $\sim p \wedge q$ **37.** The bus is blue.

39. x is not equal to 4. **41.** The earth is flat.

43.

p	q	$\sim p$	$\sim p \wedge q$
T	T	F	F
T	F	F	F
F	T	T	T
F	F	T	F

45.

p	q	$\sim p$	$\sim q$	$\sim p \vee \sim q$
T	T	F	F	F
T	F	F	T	T
F	T	T	F	T
F	F	T	T	T

47.

p	q	$\sim q$	$p \vee \sim q$
T	T	F	T
T	F	T	T
F	T	F	F
F	F	T	T

49. Not logically equivalent   **51.** Logically equivalent

53. Logically equivalent   **55.** Not logically equivalent

57. Logically equivalent   **59.** Not a tautology

61. A tautology

63.

| | | | | | Identical | |
p	q	$\sim p$	$\sim q$	$p \wedge q$	$\sim(p \wedge q)$	$\sim p \vee \sim q$
T	T	F	F	T	F	F
T	F	F	T	F	T	T
F	T	T	F	F	T	T
F	F	T	T	F	T	T

Section A.2   *(page A13)*

1. (a) If the engine is running, then the engine is wasting gasoline.
   (b) If the engine is wasting gasoline, then the engine is running.
   (c) If the engine is not wasting gasoline, then the engine is not running.
   (d) If the engine is running, then the engine is not wasting gasoline.

3. (a) If the integer is even, then it is divisible by 2.
   (b) If it is divisible by 2, then the integer is even.
   (c) If it is not divisible by 2, then the integer is not even.
   (d) If the integer is even, then it is not divisible by 2.

5. $q \to p$ **7.** $p \to q$ **9.** $p \to q$ **11.** True

13. True   **15.** False   **17.** True   **19.** True

21. Converse:
   If you can see the eclipse, then the sky is clear.

   Inverse:
   If the sky is not clear, then you cannot see the eclipse.

   Contrapositive:
   If you cannot see the eclipse, then the sky is not clear.

23. Converse:
   If the deficit increases, then taxes were raised.

   Inverse:
   If taxes are not raised, then the deficit will not increase.

   Contrapositive:
   If the deficit does not increase, then taxes were not raised.

25. Converse:
   It is necessary to apply for the visa to have a birth certificate.

   Inverse:
   It is not necessary to have a birth certificate to not apply for the visa.

   Contrapositive:
   It is not necessary to apply for the visa to not have a birth certificate.

27. Paul is not a junior and not a senior.

29. The temperature will increase and the metal rod will not expand.

31. We will go to the ocean and the weather forecast is not good.

33. No students are in extracurricular activities.

35. Some contact sports are not dangerous.

37. Some children are allowed at the concert.

39. None of the $20 bills is counterfeit.

41.

p	q	$\sim q$	$p \to \sim q$	$\sim(p \to \sim q)$
T	T	F	F	T
T	F	T	T	F
F	T	F	T	F
F	F	T	T	T

43.

p	q	$q \to p$	$\sim(q \to p)$	$\sim(q \to p) \bigwedge q$
T	T	T	F	F
T	F	T	F	F
F	T	F	T	T
F	F	T	F	F

45.

p	q	$\sim p$	$p \bigvee q$	$(p \bigvee q) \bigwedge (\sim p)$
T	T	F	T	F
T	F	F	T	F
F	T	T	T	T
F	F	T	F	F

$[(p \bigvee q) \bigwedge (\sim p)] \to q$
T
T
T
T

47.

p	q	$\sim p$	$\sim q$	$p \leftrightarrow (\sim q)$	$(p \leftrightarrow \sim q) \to \sim p$
T	T	F	F	F	T
T	F	F	T	T	F
F	T	T	F	T	T
F	F	T	T	F	T

49.

p	q	$\sim p$	$\sim q$	$q \to p$	$\sim p \to \sim q$
T	T	F	F	T	T
T	F	F	T	T	T
F	T	T	F	F	F
F	F	T	T	T	T

Identical

51.

p	q	$\sim q$	$p \to q$	$\sim(p \to q)$	$p \bigwedge \sim q$
T	T	F	T	F	F
T	F	T	F	T	T
F	T	F	T	F	F
F	F	T	T	F	F

Identical

53.

p	q	$\sim p$	$\sim q$	$p \to q$	$(p \to q) \bigvee \sim q$
T	T	F	F	T	T
T	F	F	T	F	T
F	T	T	F	T	T
F	F	T	T	T	T

$p \bigvee \sim p$
T
T
T
T

Identical

55.

p	q	$\sim p$	$\sim p \bigwedge q$	$p \to (\sim p \bigwedge q)$
T	T	F	F	F
T	F	F	F	F
F	T	T	T	T
F	F	T	F	T

Identical

57. (c)   **59.** (a)   **61.**

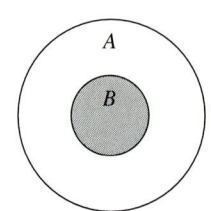

63.   **65.**

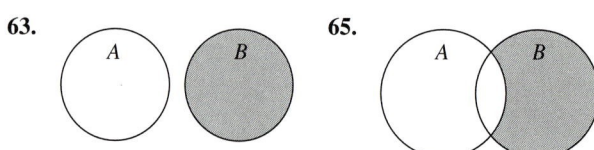

67. 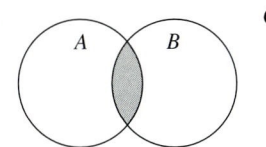   **69.**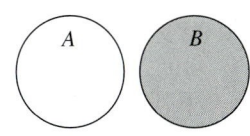

71. (a) Statement does not follow
(b) Statement follows

73. (a) Statement does not follow
(b) Statement does not follow

Section A.3 *(page A22)*

1.

p	q	~p	~q	p → ~q	(p → ~q) ⋀ q
T	T	F	F	F	F
T	F	F	T	T	F
F	T	T	F	T	T
F	F	T	T	T	F

[(p → ~q) ⋀ q] → ~p
T
T
T
T

3.

p	q	~p	p ⋁ q	(p ⋁ q) ⋀ ~p
T	T	F	T	F
T	F	F	T	F
F	T	T	T	T
F	F	T	F	F

[(p ⋁ q) ⋀ ~p] → q
T
T
T
T

5.

p	q	~p	~q	~p → q	(~p → q) ⋀ p
T	T	F	F	T	T
T	F	F	T	T	T
F	T	T	F	T	F
F	F	T	T	F	F

[(~p → q) ⋀ p] → ~q
F
T
T
T

7.

p	q	p ⋁ q	(p ⋁ q) ⋀ q	[(p ⋁ q) ⋀ q] → p
T	T	T	T	T
T	F	T	F	T
F	T	T	T	F
F	F	F	F	T

9. Valid   **11.** Invalid   **13.** Valid   **15.** Valid
17. Invalid   **19.** Valid   **21.** Invalid
23. (b)   **25.** (c)   **27.** (b)   **29.** (c)
31. Valid

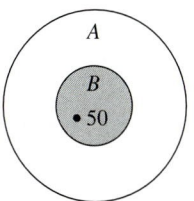

A: All numbers divisible by five
B: All numbers divisible by ten

33. Invalid

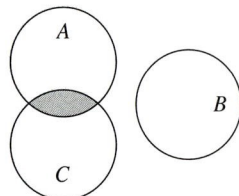

A: People eligible to vote
B: People under the age of 18
C: College students

35. Let p represent the statement "Sue drives to work," let q represent "Sue will stop at the grocery store," and let r represent "She'll buy milk."

First write:

> Premise #1: $p \rightarrow q$
> Premise #2: $q \rightarrow r$
> Premise #3: p

Reorder the premises:

> Premise #3: p
> Premise #1: $p \rightarrow q$
> Premise #2: $q \rightarrow r$
> Conclusion: r

Then we can conclude r. That is, "Sue will get milk."

37. Let p represent "This is a good product," let q represent "We will buy it," and let r represent "The product was made by XYZ Corporation."

First write:

> Premise #1: $p \rightarrow q$
> Premise #2: $r \lor \sim q$
> Premise #3: $\sim r$

Note that $p \rightarrow q \equiv \sim q \rightarrow \sim p$, and reorder the premises:

> Premise #2: $r \lor \sim q$
> Premise #3: $\sim r$
> (Conclusion from Premise #2, Premise #3 : $\sim q$)
> Premise #1: $\sim q \rightarrow \sim p$
> Conclusion: $\sim p$

Then we can conclude $\sim p$. That is, "It is not a good product."

APPENDIX B *(page A35)*

1.

Stems	Leaves
7	0 5 5 5 7 7 8 8 8
8	1 1 1 1 2 3 4 5 5 5 5 5 7 8 9 9 9
9	0 2 8
10	0 0

3.

Stems	Leaves
5	2 5 9
6	2 3 6 6 7
7	0 1 2 3 4 7 8 8 9
8	0 1 3 4 5 7 9
9	0 0 2 3 3 3 5 6 8 9
10	0 0

5. Frequency Distribution

Interval	Tally
[15, 22)	ⅲⅱ III
[22, 29)	ⅲⅱ I
[29, 36)	ⅲⅱ
[36, 43)	IIII
[43, 50)	ⅲⅱ II

Histogram

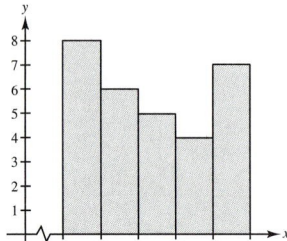

7.

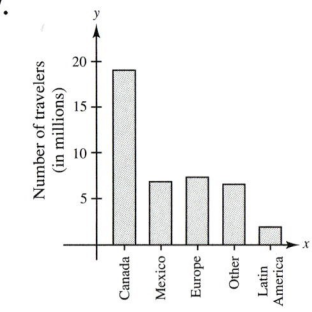

9. 150 million tons, 195 million tons

11. Total waste, recycled waste

13. Total waste equals the sum of the other three quantities.

15.

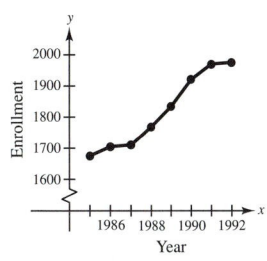

17.

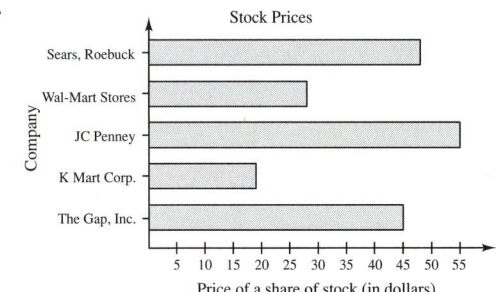

19.

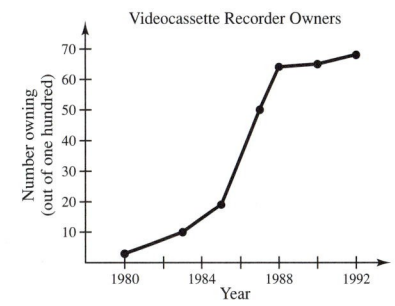

21. Positive correlation   **23.** Yes

25. Negative correlation   **27.** Positive correlation

29.

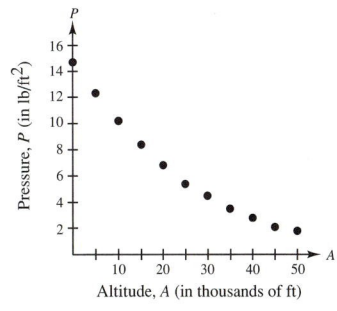

31. 2.45 pounds per square inch

33.

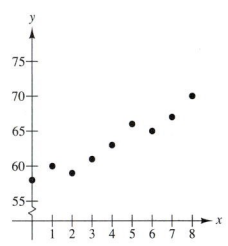

35. $y = 57.49 + 1.43x$; 71.8

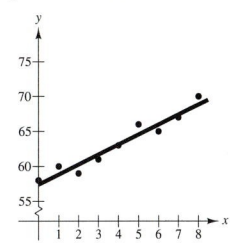

37.

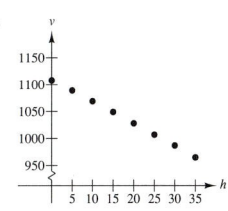

39. $v = 1117.3 - 4.1h$; 1006.6

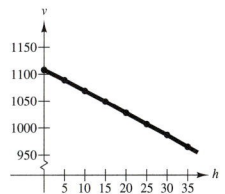

41. $y = -2.179x + 22.964$

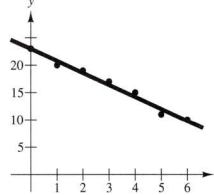

43. $y = 2.378x + 23.546$

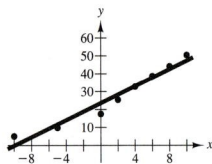

45. $S = 384.1 + 21.2x$; $479,500

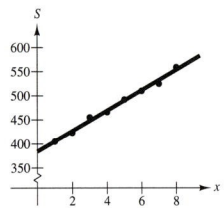

APPENDIX C (*page A39*)

1. Answers vary.

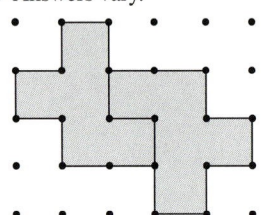

3. (a), (b)   **5.** False   **7.** True

9. Answers vary.

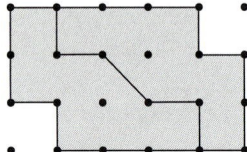

11.

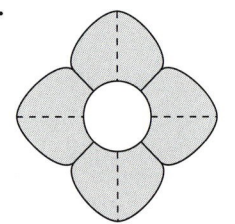

13. 16   **15.** 3   **17.** 12.5 ft by 12.5 ft, no

19. (d)   **21.** (b)

23. $\angle ZXW$ or $\angle WXZ$, $\angle ZXY$ or $\angle YXZ$,
$\angle YXW$ or $\angle WXY$, $\angle ZXY$ and $\angle YXW$

25. (b)   **27.** (d)   **29.** (f)   **31.** (c)

33. $\overline{LM} \cong \overline{NO}$, $\overline{MP} \cong \overline{NQ}$, $\overline{LP} \cong \overline{OQ}$

35. $m\angle V$ **37.** $\overline{TV}$

39.

	Scalene	Isosceles	Equilateral
Acute	Yes	Yes	Yes
Obtuse	Yes	Yes	No
Right	Yes	Yes	No

41. 3; yes   **43.** 3; yes   **45.** (3, 5)(3, 1)

47. Form a tetrahedron.

Section C.2 (*page A49*)

1.   **3.**

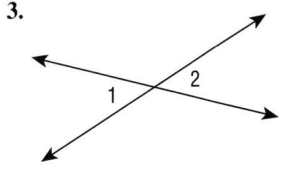

5.

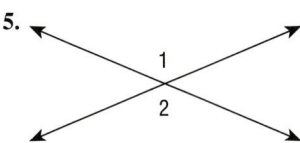

7. Adjacent $\cong$ suppl. $\angle$ **9.** Adjacent suppl. $\angle$

10. Adjacent compl. $\angle$ **13.** False

15. True if $m\angle 2 = m\angle 3 = 90°$; false otherwise

17. True   **19.** 110°   **21.** 55°   **23.** 35°

25. (c)   **27.** $\angle 3$ and $\angle 5$ *or* $\angle 4$ and $\angle 6$

29. $\angle 4$ and $\angle 5$ *or* $\angle 3$ and $\angle 6$

31. $m\angle 1 = 110°$ because it forms a linear pair with the given
angle; $m\angle 2 = 110°$ by Alternate Exterior Angles Theorem

33. $m\angle 1 = 70°$ by Consecutive Interior Angles Theorem;
$m\angle 2 = 70°$ because it forms a linear pair with the given
angle, or by Alternate Interior Angles Theorem

35. $a = 30°; b = 20°$ **37.** $\angle 2, \angle 5, \angle 7$

39. $m\angle 1 = m\angle 3 = 70°$; $m\angle 4 = m\angle 6 = 135°$, $m\angle 2 = 110°$
     $m\angle 5 = 45°$, $m\angle 7 = 25°$, $m\angle 8 = 155°$

41. $35°$ **43.** $40°$

45. True. The third angle must be $180° - 2(60°) = 60°$.

47. $m\angle 1 = 30°$, $m\angle 2 = 60°$, $m\angle 3 = 50°$, $m\angle 4 = 35°$,
     $m\angle 5 = 90°$, $m\angle 6 = 55°$, $m\angle 7 = 55°$, $m\angle 8 = 125°$,
     $m\angle 9 = 35°$

49. $77°$ **51.**

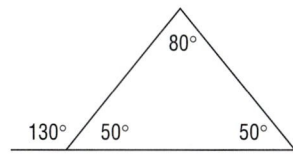

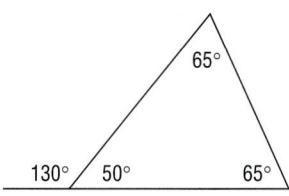

53. $30°, 60°, 90°$ **55.** $38°, 59°, 83°$

APPENDIX E   *(page A80)*

7. Not a solution   **9.** Solution

11. $x - 10 = 17$ **13.** $4x + 1 = \frac{1}{2}$
     $x - 10 = -17$ $4x + 1 = -\frac{1}{2}$

15. $45, -45$ **17.** $11, -14$ **19.** No solution

21. $-\frac{39}{2}, \frac{15}{2}$ **23.** $\frac{3}{2}, -\frac{1}{4}$ **25.** $|2x + 3| = 5$

27. (a) Solution   **29.** (a) Not a solution
     (b) Not a solution   (b) Solution
     (c) Not a solution   (c) Solution
     (d) Solution   (d) Not a solution

31. $-3 < y + 5 < 3$ **33.** $7 - 2h \geq 9$
                           $7 - 2h \leq -9$

35. $-4 < y < 4$ **37.** $-6 < y < 2$

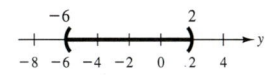

39. $x > 6$ or $x < -6$ **41.** $y \geq 2$ or $y \leq -6$

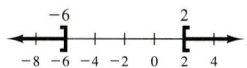

43. $x > 3$ or $x < -6$ **45.** $-5 < x < 35$

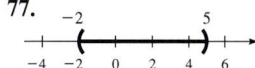

47. $z > 110$ or $z < -50$ **49.** $|t - 98.6| < 1$

51. 1.875 **53.** 150% **55.** $80{,}000$ **57.** 0

59. $21, 11$ **61.** $\frac{16}{3}, 16$ **63.** No solution   **65.** $-\frac{11}{5}, \frac{17}{5}$

67. $18.75, -6.25$ **69.** $-3, 7$ **71.** $2, -\frac{3}{2}$

73. (a) Not a solution   **75.** (a) Solution
     (b) Solution   (b) Not a solution
     (c) Solution   (c) Not a solution
     (d) Not a solution   (d) Solution

77. 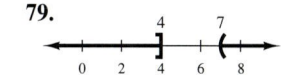   **79.**

81. $-2 < y < 6$ **83.** $y \geq 4$ or $y \leq -4$

85. $y > 6$ or $y < -2$ **87.** $-7 < x < 7$

89. $-9 \leq y \leq 9$ **91.** $x > \frac{2}{3}$ or $x < -2$

93. $x > 7$ or $x < 3$ **95.** 3 **97.** $-82 \leq x \leq 78$

99. $t \geq \frac{5}{2}$ or $t \leq -\frac{15}{2}$ **101.** $s > 23$ or $s < -17$

103. d   **105.** b   **107.** $|x| \leq 2$ **109.** $|x - 10| < 3$

111. $|x - 19| < 3$ **113.**

115. $|x| < 3$ **117.** $|x - 5| > 6$

INDEX OF APPLICATIONS

Construction Problems

Consumer Applications

Time and Distance Applications

U.S. Demographics Applications

Graphing Linear Equations

A **linear equation in two variables** is an equation of first degree in both variables.

The **graph of an equation** is the set of all points in the rectangular coordinate system whose coordinates are solutions of the equation.

The **graph of a linear equation in two variables** is a straight line.

To find the **x-intercepts** (where the graph intersects the x-axis), let y be zero and solve the equation for x.

To find the **y-intercepts** (where the graph intersects the y-axis), let x be zero and solve the equation for y.

Slope, $m = \dfrac{y_2 - y_1}{x_2 - x_1} = \dfrac{\text{Change in } y}{\text{Change in } x}$.

Slope-intercept form of the equation of a line:
$y = mx + b$, m is the slope, and $(0, b)$ is the y-intercept.

Point-slope form of the equation of a line:
$y - y_1 = m(x - x_1)$, m is the slope, and (x_1, y_1) is a point on the line.

Graphs of Systems of Linear Equations

The **solution** of a system of linear equations is an ordered pair (a, b) that satisfies each of the equations. A system of linear equations may have no solution, exactly one solution, or infinitely many solutions.

Graphs of Quadratic Equations

Standard form of a quadratic equation:
$ax^2 + bx + c = 0$, a, b, and c are real numbers with $a \neq 0$. A quadratic equation can be solved by factoring, extracting square roots, completing the square, or using the Quadratic Formula.

Extracting square roots: If $u^2 = d$, where $d > 0$, then $u = \pm\sqrt{d}$.

Quadratic Formula: $x = \dfrac{-b \pm \sqrt{b^2 - 4ac}}{2a}$

Discriminant: $b^2 - 4ac$
If $b^2 - 4ac > 0$, then the equation has two real number solutions.
If $b^2 - 4ac = 0$, then the equation has one (repeated) real number solution.
If $b^2 - 4ac < 0$, then the equation has no real number solutions.

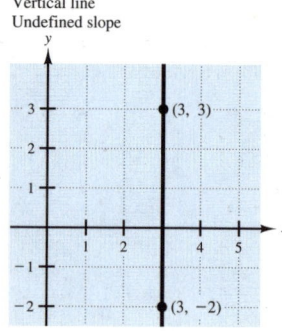

Vertical line
Undefined slope

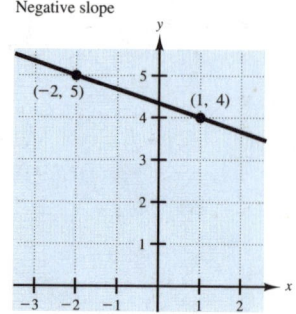

Line falls
Negative slope

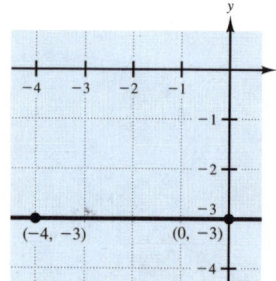

Horizontal line
Zero slope

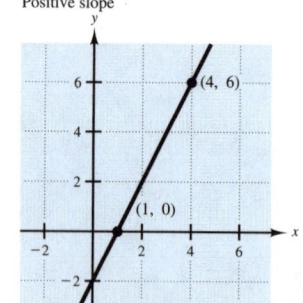

Line rises
Positive slope

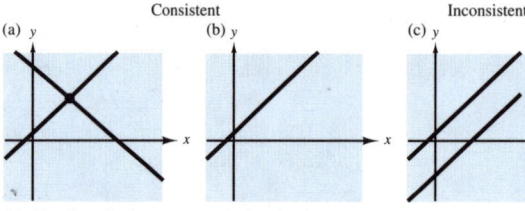

(a) Two lines that intersect at a single point
(b) Two lines that coincide with infinitely many points of intersection
(c) Two parallel lines with no point of intersection

The graph of $y = ax^2 + bx + c$, $a \neq 0$, is called a **parabola** which opens up if $a > 0$ and opens down if $a < 0$.

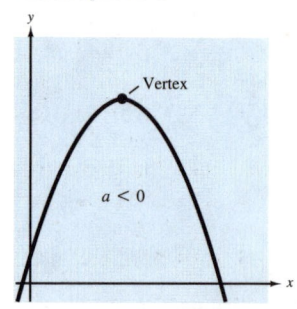

Parabola Opens Up

Parabola Opens Down